FEATURES AND BENEFITS
Mathematics: Applications and Connections, Course 1

S0-AIV-983

		See page(s):

Integration

... of mathematical topics helps students to see mathematics as a whole.
Algebra lessons prepare students for first-year algebra. — 22
Other integrated topics:
- geometry — 145
- measurement — 164
- statistics — 163
- probability — 316
- proportional reasoning — 317

Applications

... show students how mathematics relates to the real world around them.
- Lessons open with an application or connection relevant to teenagers — 232
- *"When am I ever going to use this?"* answers the age-old question. — 152
- *School to Career* demonstrates careers in which mathematics is used. — 185
- *Math in the Media* helps students interpret mathematics in print. — 279

Connections

... to other subject areas help students appreciate the role of mathematics in the other courses they are taking, such as science, social studies, and literature.
- *Chapter Projects* — 177
- Connection examples and exercises — 168, 219
- *Interdisciplinary Investigations* — 392-393

Problem Solving

... activities and applications are integrated into every chapter.
- problem-solving strategies through *Thinking Labs* — 150-151
- 4-step problem-solving plan used throughout each chapter — 4
- Study Hints — 207

Labs and Investigations

... provide students with an opportunity to explore, create mathematical models, and work cooperatively with a partner or group.
- Optional Hands-On Labs and Technology Labs provide a preview of the next lesson or can extend the concepts in the preceding lesson. — 181, 197
- Mini-labs introduce or reinforce concepts within a lesson. — 202
- Activities from *Interactive Mathematics: Activities and Investigations* provide additional options for extension activities in real-world settings. — 2d

Ample Practice and Review

... reinforces new skills and concepts.
- Extra Practice for each lesson in the back of the text. — 548-594
- Mixed Review exercises in each exercise set — 205
- *Let the Games Begin* helps students maintain skills in previous lessons — 121

Test Preparation and Assessment

... provides practice for local, state, and national tests.
- Test Practice questions in each exercise set — 139
- Standardized Test Practice at the end of each chapter includes multiple-choice and free-response questions. — 174-175
- Alternative Assessment options in the Study Guide and Assessment — 173
- A free-response test for each chapter in the back of the text. — 595-607

Technology

... strand prepares students to function in a technological society through a variety of instruction and activities, including Technology Labs.
- scientific calculators — 340
- graphing calculators — 26
- spreadsheets — 144
- Internet Connections — 135

Teacher Support

... in the *Teacher's Wraparound Edition* makes it easy for you to organize, present, and enhance the content. The extensive set of resource materials denoted in the Teacher's Edition helps you increase each student's chance for success.

Glencoe
Mathematics
Applications and Connections

Course 1

 Glencoe
McGraw-Hill

New York, New York   Columbus, Ohio   Woodland Hills, California   Peoria, Illinois

Mathematics: Applications and Connections
Course 1

Student Edition
Teacher's Wraparound Edition
Spanish Student Edition

Applications

Classroom Games
Diversity Masters
Family Letters and Activities
Investigations and Projects Masters
School to Career Masters
Science and Mathematics Lab Manual

Meeting Individual Needs

Enrichment Masters
Investigations for the Special Education Student
Practice Masters
Spanish Study Guide and Assessment
Study Guide and Practice Workbook
Study Guide Masters
Transition Booklet

Technology/Multimedia

 CD-ROM Program

 Interactive Mathematics
Tools Software

Technology Masters

Assessment/Evaluation

Assessment and Evaluation Masters

 MindJogger Videoquizzes

 Test and Review Software

Manipulatives/Modeling

Hands-On Lab Masters
Glencoe Mathematics Classroom Manipulative Kit
Overhead Manipulative Resources
Glencoe Mathematics Student Manipulative Kit

Teaching Aids

Answer Key Masters
Block Scheduling Booklet

 5-Minute Check Transparencies
Teaching Transparencies

Lesson Planning Guide
Solutions Manual

Electronic Teacher's Classroom Resources

 All blackline masters on a single CD-ROM

Glencoe/McGraw-Hill

A Division of The **McGraw·Hill** *Companies*

Send all inquiries to:
Glencoe/McGraw-Hill
936 Eastwind Drive
Westerville, OH 43081-3374

ISBN: 0-02-833050-1 (Student Edition)
         0-02-833053-6 (Teacher's Wraparound Edition)

3 4 5 6 7 8 9 10 027/043 07 06 05 04 03 02 01 00 99

Dear Students, Teachers, and Parents,

Middle School math students are very special to us! That's why we wrote **Mathematics: Applications and Connections,** the first middle school math program designed specifically for you. The exciting, relevant content and up-to-date design will hold your interest and answer the question "When am I ever going to use this?"

As you page through your text, you'll notice the variety of ways math is presented for you. You'll see real-world applications as well as connections to other subjects like science, history, language arts, and music. You'll have opportunities to use technology tools such as the Internet, CD-ROM, graphing calculators, and computer applications like spreadsheets.

You'll appreciate the easy-to-follow lesson format. Each new concept is introduced with an interesting application or connection followed by clear explanations and examples. As you complete the exercises and solve interesting problems, you'll learn a great deal of useful math. You'll also have the opportunity to complete relevant Chapter Projects, Hands-On Labs, and Interdisciplinary Investigations. Test Practice, Test-Taking Tips, and Reading Math Study Hints will help you improve your test-taking skills.

Each day, as you use **Mathematics: Applications and Connections,** you'll see the practical value of math. You'll quickly grow to appreciate how often math is used in ways that relate directly to your life. If you don't already realize the importance of math in your life, you soon will!

Sincerely, The Authors

Kay McClain

Patricia S. Wilson

Patricia Frey

Linda Dritsas

Barbara Smith

Jack M. Ott

Ron Pelfrey

Beatrice Moore-Harris

David Molina

Meet Our Authors

William Collins

*Director of The Sisyphus
    Mathematics Learning Center
W.C. Overfelt High School
San Jose, CA*

William Collins provides mathematics support services to the students of Overfelt and the East Side community. He also has many years of experience as a mathematics instructor and mathematics department chairperson. Mr. Collins received his B.A. in Mathematics and his B.A. in Philosophy from Herbert H. Lehman College, CUNY, and his M.S. in Mathematics Education from California State University, Hayward. He is active in several professional mathematics organizations.

Linda Dritsas

*District Coordinator
Fresno Unified School District
Fresno, CA*

Linda Dritsas is highly involved with the Fresno Urban Systemic Initiative. She received her B.A. and M.A. from California State University, Fresno, where she also taught. Ms. Dritsas has published numerous articles, workbooks, and other supplementary materials. She received the Edward Begle Award from the California Mathematics Council. Ms. Dritsas has is an active member of NCTM and the Association for Supervision and Curriculum Development.

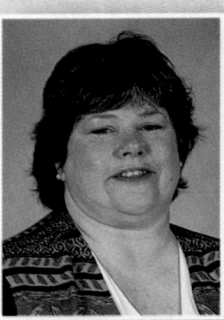

Patricia Frey

*Mathematics Department
    Chairperson
Buffalo Academy for Visual and
    Performing Arts
Buffalo, NY*

Patricia Frey received her B.A. from D'Youville College in Buffalo, and her M.Ed. from the State University of New York at Buffalo. Her publications include articles in mathematics education journals as well as mathematics, computer, and calculator curriculum journals. Ms. Frey is a Woodrow Wilson fellow and a New York State Math Mentor. She is an active member of NCTM and other professional mathematics organizations.

Arthur C. Howard

*Program Director for Secondary
    Mathematics
Aldine Independent School District
Houston, TX*

Arthur Howard is also an adjunct mathematics teacher at North Harris College and a member of the Executive Board of the Rice University School Mathematics Project. He received his B.S. in Mathematics and M.Ed. in Mathematics Education from the University of Houston. Mr. Howard is active in NCTM and has made presentations at local, state, regional, and national mathematics conferences.

Kay McClain

*Lecturer
George Peabody College
Vanderbilt University
Nashville, TN*

Kay McClain is currently working on a Ph.D. at Vanderbilt University. While a teacher at Mountain Brook Middle School in Birmingham, Alabama. she received the Presidential Award for Excellence in the Teaching of Mathematics and was a Woodrow Wilson Fellow. Ms. McClain received her B.A. for Auburn University and her Educational specialist degree from the University of Montevallo. She is an active member of NCTM.

David Molina

*Adjunct Professor of Mathematics
    Education
The University of Texas at Austin
Austin, TX*

Dr. David Molina is the Associate Director of the Charles A. Dana Center. He earned his B.S. in Mathematics from the University of Notre Dame and his M.A. and Ph.D. in Mathematics Education from The University of Texas at Austin. Dr. Molina is active in every area of mathematics education. He has published numerous articles about professional development and educational technology.

Beatrice Moore-Harris

Staff Development Specialist
Bureau of Education and Research
Houston, TX

Beatrice Moore-Harris is a former middle school mathematics teacher and Mathematics Supervisor for the Houston and Fort Worth Independent School Districts. She received her B.A. from Prairie View A&M University. Ms. Moore-Harris is also a consultant and one of four official spokespersons for NCTM. She is also an active member of several professional mathematics organizations.

Jack M. Ott

Distinguished Professor of
Mathematics Education
University of South Carolina
Columbia, SC

Jack Ott has taught grades 5-12 and college. He recently received the South Carolina Council of Teachers of Mathematics Award for Outstanding Contributions in Mathematics Education. Dr. Ott received his A.B. from Indiana Wesleyan University, his M.A. from Ball State University, and Ph.D. from The Ohio State University. He has written articles for The Mathematics Teacher and The Arithmetic Teacher and has been a speaker at national and state mathematics conferences.

Ronald Pelfrey

Mathematics Consultant
Lexington, KY

Ronald S. Pelfry is a former middle school and high school mathematics teacher and district-level mathematics supervisor. He received his B.S., M.A., and Ed.D. from the University of Kentucky. Dr. Pelfrey has been active with NCTM and its local and state affiliates in Kentucky. He has been an author of several articles on curriculum and assessment, a speaker, and both a regional and state conference chair.

Jack Price

Professor, Mathematics
Education
California State Polytechnic
University
Pomona, CA

Past-president of the National Council of Teacher of Mathematics, Dr. Price teaches mathematics and mathematics methods classes to prospective teachers. He received his B.A. from Eastern Michigan University and his M.Ed. and Ed. D. from Wayne State University in Detroit. Dr. Price also serves on the Expert Panel on Mathematics and Science Education of the U.S. Department of Education and the standing mathematics committee for NAEP.

Barbara Smith

Mathematics Consultant
Unionville-Chadds Ford School
District
Kennett Square, PA

Barbara Smith is a former mathematics teacher with 13 years experience at the middle school level and 3 years at the high school level. She received her B.S. from Grove City College and her M.Ed. from the University of Pittsburgh. Ms. Smith is an active member of NCTM and has held offices in several state and local mathematics organizations.

Patricia S. Wilson

Associate Professor of Mathematics
Education
University of Georgia
Athens, GA

Patricia Wilson, a former middle school mathematics teacher, is currently working with preservice and inservice teachers. She received the Excellence in Teaching Award from the College of Education at the University of Georgia. Dr. Wilson received her B.S. from Ohio University and her M.A. and Ph.D. from The Ohio State University. She is an active member of NCTM and other professional mathematics organizations.

Academic Consultants and Teacher Reviewers

Each of the Academic Consultants read all 39 chapters in Courses 1, 2, and 3, while each Teacher Reviewer read two chapters. The Consultants and Reviewers gave suggestions for improving the Student Editions and the Teacher's Wraparound Editions.

ACADEMIC CONSULTANTS

Richie Berman, Ph.D.
Mathematics Lecturer and Supervisor
University of California, Santa Barbara
Santa Barbara, California

Robbie Bonneville
Mathematics Coordinator
La Joya Unified School District
Alamo, Texas

Cindy J. Boyd
Mathematics Teacher
Abilene High School
Abilene, Texas

Gail Burrill
Mathematics Teacher
Whitnall High School
Hales Corners, Wisconsin

Georgia Cobbs
Assistant Professor
The University of Montana
Missoula, Montana

Gilbert Cuevas
Professor of Mathematics Education
University of Miami
Coral Gables, Florida

David Foster
Mathematics Director
Robert Noyce Foundation
Palo Alto, California

Eva Gates
Independent Mathematics
    Consultant
Pearland, Texas

Berchie Gordon-Holliday
Mathematics/Science Coordinator
Northwest Local School District
Cincinnati, Ohio

Deborah Grabosky
Mathematics Teacher
Hillview Middle School
Whittier, California

Deborah Ann Haver
Principal
Great Bridge Middle School
Virginia Beach, Virginia

Carol E. Malloy
Assistant Professor, Math Education
The University of North Carolina,
    Chapel Hill
Chapel Hill, North Carolina

Daniel Marks, Ed.D.
Associate Professor of Mathematics
Auburn University at Montgomery
Montgomery, Alabama

Melissa McClure
Mathematics Consultant
Teaching for Tomorrow
Fort Worth, Texas

TEACHER REVIEWERS

Course 1

Carleen Alford
Math Department Head
Onslow W. Minnis, Sr. Middle School
Richmond, Virginia

Margaret L. Bangerter
Mathematics Coordinator K-6
St. Joseph School District
St. Joseph, Missouri

Diana F. Brock
Sixth and Seventh Grade Math Teacher
Memorial Parkway Junior High
Katy, Texas

Mary Burkholder
Mathematics Department Chair
Chambersburg Area Senior High
Chambersburg, Pennsylvania

Eileen M. Egan
Sixth Grade Teacher
Howard M. Phifer Middle School
Pennsauken, New Jersey

Melisa R. Grove
Sixth Grade Math Teacher
King Philip Middle School
West Hartford, Connecticut

David J. Hall
Teacher
Ben Franklin Middle School
Baltimore, Maryland

Ms. Karen T. Jamieson, B.A., M.Ed.
Teacher
Thurman White Middle School
Henderson, Nevada

David Lancaster
Teacher/Mathematics Coordinator
North Cumberland Middle School
Cumberland, Rhode Island

Jane A. Mahan
Sixth Grade Math Teacher
Helfrich Park Middle School
Evansville, Indiana

Margaret E. Martin
Mathematics Teacher
Powell Middle School
Powell, Tennessee

Diane Duggento Sawyer
Mathematics Department Chair
Exeter Area Junior High
Exeter, New Hampshire

Susan Uhrig
Teacher
Monroe Middle School
Columbus, Ohio

Cindy Webb
Title 1 Math Demonstration Teacher
Federal Programs LISD
Lubbock, Texas

Katherine A. Yule
Teacher
Los Alisos Intermediate School
Mission Viejo, California

Course 2

Sybil Y. Brown
Math Teacher Support Team-USI
Columbus Public Schools
Columbus, Ohio

Ruth Ann Bruny
Mathematics Teacher
Preston Junior High School
Fort Collins, Colorado

BonnieLee Gin
Junior High Teacher
St. Mary of the Woods
Chicago, Illinois

Larry J. Gonzales
Math Department Chair
Desert Ridge Middle School
Albuquerque, New Mexico

Susan Hertz
Mathematics Teacher
Revere Middle School
Houston, Texas

Rosalin McMullan
Mathematics Teacher
Honea Path Middle School
Honea Path, South Carolina

Mrs. Susan W. Palmer
Teacher
Fort Mill Middle School
Fort Mill, South Carolina

Donna J. Parish
Teacher
Zia Middle School
Las Cruces, New Mexico

Ronald J. Pischke
Mathematics Coordinator
St. Mary of the Woods
Chicago, Illinois

Sister Edward William Quinn I.H.M.
Chairperson Elementary Mathematics
  Curriculum
Archdiocese of Philadelphia
Philadelphia, Pennsylvania

Marlyn G. Slater
Title I Math Specialist
Paradise Valley USD
Paradise Valley, Arizona

Sister Margaret Smith O.S.F.
Seventh and Eighth Grade Math Teacher
St. Mary's Elementary School
Lancaster, New York

Pamela Ann Summers
Coordinator, Secondary Math/Science
Lubbock ISD
Lubbock, Texas

Dora Swart
Teacher/Math Department Chair
W. F. West High School
Chehalis, Washington

Rosemary O'Brien Wisniewski
Middle School Math Chairperson
Arthur Slade Regional School
Glen Burnie, Maryland

Laura J. Young, Ed. D
Eighth Grade Mathematics Teacher
Edwards Middle School
Conyers, Georgia

Susan Luckie Youngblood
Teacher/Math Department Chair
Weaver Middle School
Macon, Georgia

Course 3

Beth Murphy Anderson
Mathematics Department Chair
Brownell Talbot School
Omaha, Nebraska

David S. Bradley
Mathematics Teacher
Thomas Jefferson Junior High School
Salt Lake City, Utah

Sandy Brownell
Math Teacher/Team Leader
Los Alamos Middle School
Los Alamos, New Mexico

Eduardo Cancino
Mathematics Specialist
Education Service Center, Region One
Edinburg, Texas

Sharon Cichocki
Secondary Math Coordinator
Hamburg High School
Hamburg, New York

Nancy W. Crowther
Teacher, retired
Sandy Springs Middle School
Atlanta, Georgia

Charlene Mitchell DeRidder, Ph.D.
Mathematics Supervisor K-12
Knox County Schools
Knoxville, Tennessee

Ruth S. Garrard
Mathematics Teacher
Davidson Fine Arts School
Augusta, Georgia

Lolita Gerardo
Secondary Math Teacher
Pharr San Juan Alamo High School
San Juan, Texas

Donna Jorgensen
Teacher of Mathematics/Science
Toms River Intermediate East
Toms River, New Jersey

Statha Kline-Cherry, Ed.D.
Director of Elementary Education
University of Houston – Downtown
Houston, Texas

Charlotte Laverne Sykes Marvel
Mathematics Instructor
Bryant Junior High School
Bryant, Arkansas

Albert H. Mauthe, Ed.D.
Supervisor of Mathematics
Norristown Area School District
Norristown, Pennsylvania

Barbara Gluskin McCune
Teacher
East Middle School
Farmington, Michigan

Laurie D. Newton
Teacher
Crossler Middle School
Salem, Oregon

Indercio Abel Reyes
Mathematics Teacher
PSJA Memorial High School
Alamo, Texas

Fernando Rosa
Mathematics Department Chair
Edinburg High School
Edinburg, Texas

Mary Ambriz Soto
Mathematics Coordinator
PSJA I.S.D.
Pharr, Texas

Judy L. Thompson
Eighth Grade Mathematics Teacher
Adams Middle School
North Platte, Nebraska

Karen A. Universal
Eighth Grade Mathematics Teacher
Cassadaga Valley Central School
Sinclairville, New York

Tommie L. Walsh
Teacher
S. Wilson Junior High School
Lubbock, Texas

Marcia K. Ziegler
Mathematics Teacher
Pharr-San Juan-Alamo North High School
Pharr, Texas

Student Advisory Board

The Student Advisory Board gave the editorial staff and design team feedback on the design, content, and covers of the Student Editions. We thank these students from Crestview Middle School in Columbus, Ohio, and McCord Middle School in Worthington, Ohio, for their hard work and creative suggestions in making *Mathematics: Applications and Connections* more student friendly.

Mara Flood

Anna Hoza

Neil Gardner

Lareva Summerall

Mike Leonelli

Chris Franklin

Kelly McLoughlin

Jennifer Shoemaker

Maghee Disch

CHAPTER 1

Problem Solving, Numbers, and Algebra

Let the GameS Begin

- EXPO Bingo, **31**

SCHOOL to CAREER

- Business, **27**

interNET CONNECTION

Test Practice

Applications, Connections, and
Integration Index, pages xxii–1.

CHAPTER 2

Statistics: Graphing Data

Table of Contents

Adding and Subtracting Decimals

Applications, Connections, and
Integration Index, pages xxii–1.

Interdisciplinary Investigation

"O" is for Olympics, **128**

Let the Games Begin

• The 1-Meter Dash, **121**

SCHOOL to CAREER

• Aviation, **99**

interNET CONNECTION

• Chapter Project, **93**
• School to Career, **99**
• Data Update, **107**
• Let the Games Begin, **121**
• Interdisciplinary Investigation, **129**

Test Practice

98, 104, 108, 111, 115,
117, 121, 126–127

MATH IN THE MEDIA

• Tiger, **98**

Table of Contents

Let the GameS Begin

- Scavenger Hunt, **166**

SCHOOL to CAREER

- Business, **160**

interNET CONNECTION

Test Practice

Using Number Patterns, Fractions, and Ratios

Applications, Connections, and
Integration Index, pages xxii–1.

Let the **GameS** Begin

- LCM Spin-Off, **209**

SCHOOL to CAREER

- Graphic Design, **185**

inter NET
CONNECTION

- Chapter Project, **177**
- School to Career, **185**
- Data Update, **196**
- Let the Games Begin, **209**

Test Practice

MATH **IN THE MEDIA**

- Peanuts, **205**

CHAPTER 6

Adding and Subtracting Fractions

Table of Contents

Interdisciplinary Investigation

Just "State" the Facts!, **264**

Let the Games Begin

• What Difference Does it Make?, **253**

School to Career

• Cartography, **235**

interNET CONNECTION

Test Practice

CHAPTER 7
Multiplying and Dividing Fractions

Applications, Connections, and
Integration Index, pages xxii–1.

Let the Games Begin

interNET CONNECTION

Test Practice

MATH IN THE MEDIA

Table of Contents

CHAPTER 8

Exploring Ratio, Proportion, and Percent

Let the Games Begin

- Fishin' for Matches, **343**

SCHOOL to CAREER

- Arts and Crafts, **328**

interNET CONNECTION

Test Practice

CHAPTER 9

Geometry: Investigating Patterns

Applications, Connections, and
Integration Index, pages xxii–1.

Table of Contents

Interdisciplinary Investigation

That Is One Humongous Pie!, **392**

Let the Games Begin

• Pento, **383**

SCHOOL to CAREER

• Architecture, **367**

interNET CONNECTION

• Chapter Project, **351**
• School to Career, **367**
• Let the Games Begin, **383**
• Interdisciplinary Investigation, **393**

Test Practice

355, 357, 361, 366, 373,
378, 382, 390–391

MATH IN THE MEDIA

• Peanuts, **382**

CHAPTER 10

Geometry: Understanding Area and Volume

Let the Games Begin

- Tiddlywink Target, **409**

interNET CONNECTION

Test Practice

Algebra: Investigating Integers

Let the **Games** Begin

SCHOOL to CAREER

inter NET CONNECTION

Test Practice

Applications, Connections, and
Integration Index, pages xxii–1.

Algebra: Exploring Equations

Table of Contents

Going in Circles, **510**

• Four in a Row, **491**

interNET
CONNECTION

Test Practice

MATH IN THE MEDIA

CHAPTER 13 — Using Probability

Let the **Games** Begin

- Mix & Match, **539**

SCHOOL to CAREER

- Biological Science, **519**

*inter*NET CONNECTION

- Chapter Project, **513**
- School to Career, **519**
- Data Update, **525**
- Let the Games Begin, **539**

Test Practice

518, 521, 525, 529,
534, 539, 544–545

Applications, Connections, and
Integration Index, pages xxii–1.

Table of Contents

Applications, Connections, and Integration Index

Applications, Connections, and Integration Index

Applications, Connections, and Integration Index

Applications, Connections, and Integration Index

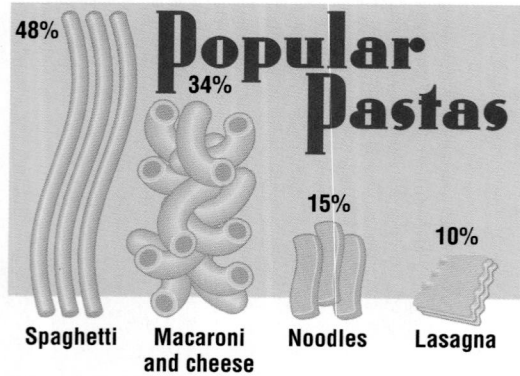

48%
Popular Pastas
34%
15%
10%

Spaghetti   Macaroni and cheese   Noodles   Lasagna

Source: National Pasta Association

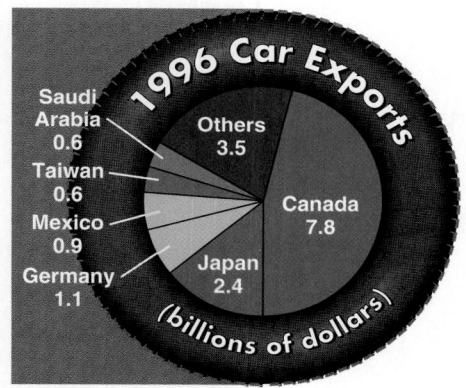

1996 Car Exports

Saudi Arabia 0.6
Taiwan 0.6
Mexico 0.9
Germany 1.1
Others 3.5
Canada 7.8
Japan 2.4
(billions of dollars)

Source: U.S. Census Bureau

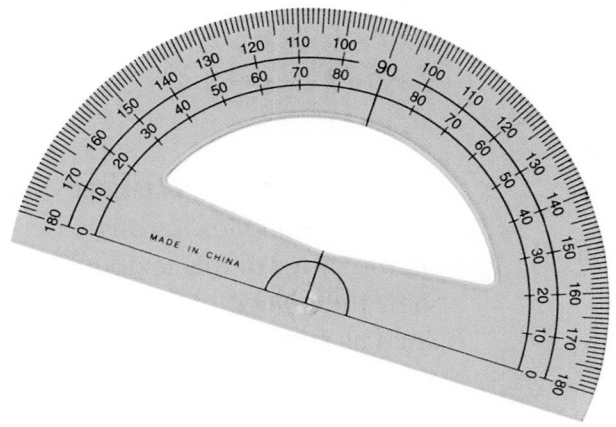

GLENCOE Online
MATHEMATICS

Vist the Glencoe Mathematics internet site for
Mathematics: Applications and Connections at
www.glencoe.com/sec/math/mac/mathnet

You'll find:

Group Activities

Games

interNET CONNECTION links to websites relevant to:
- Chapter Projects
- Interdisciplinary Investigations
- exercises

and much more!

About the Roller Coaster Hologram
The optimum viewing angle for the roller coaster hologram
on the cover of this textbook is a 45° angle. For best results,
view the hologram at this angle under a direct light source,
such as sunlight or incandescent lighting.

Table of Contents

Mathematics: Applications and Connections, Course 1

Pages T2-T31 provide a brief overview of the latest trends and research in mathematics education as well as to illustrate ways in which Glencoe has responded to these.

Chapter Overviews

Student Handbook

PATHWAYS TO SUCCESS

For the last several years, it's been difficult to find a newspaper or magazine that does not have an article about how poorly U.S. students score on international math exams compared to students from other countries such as Japan or Germany. For example, in the **Third International Mathematics and Science Study (TIMSS),** U.S. eighth grade students scored *below* the average for industrialized countries.

This result should not be surprising given the well-known study by James Flanders. He found that only 30-40% of the content in grades 6-8 of the most widely used K-8 mathematics series was new while algebra textbooks contained about 88% new content. (*Arithmetic Teacher*, "How Much of the Content in Mathematics Textbooks is New?", September 1987). This lack of new content in middle school combined with almost all new content in algebra 1 often leads to student frustration and failure in algebra 1. Flanders' research and the TIMSS data were both obtained before the first edition of *Mathematics: Applications and Connections* was published.

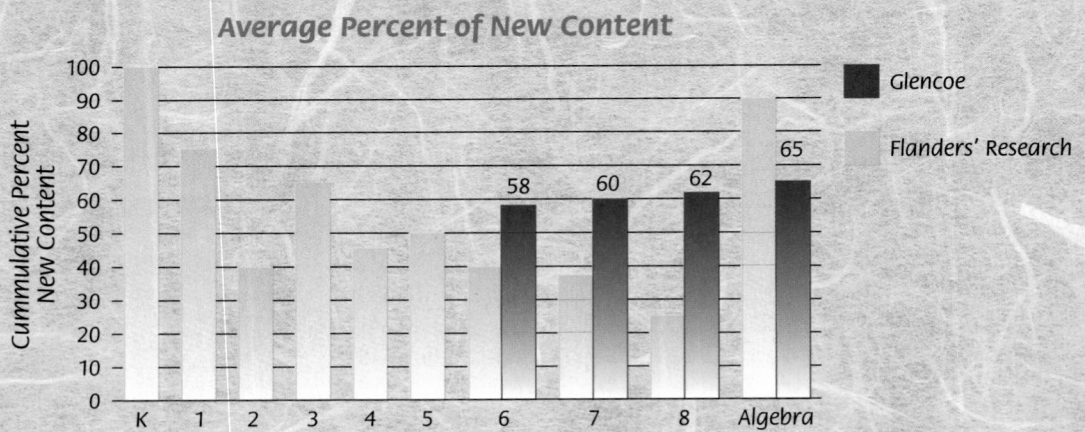

Note that the TIMSS data was collected when the programs described in the Flanders study were **very** widely used.

Glencoe changed this pattern with the publication of *Mathematics: Applications and Connections.*

Glencoe understands that the middle school experience is critical in preparing students for success in algebra 1 and geometry. ***Mathematics: Applications and Connections*** is designed to smooth the path to algebra and geometry by creating a program that has about the same amount of new material in Grade 6, Grade 7, and Grade 8 as well as in Glencoe's *Algebra 1.*

Here are some of the key findings from TIMSS and Glencoe's response.

TIMSS Finding	Glencoe's Response
Achievement U.S. eighth-grade students test at about the international average in algebra, fractions, statistics, and probability. U.S. students do not do as well in geometry, measurement, and proportionality.	All mathematics topics are integrated throughout Course 1-3 of ***Mathematics: Applications and Connections.***
Curriculum The content taught in U.S. eighth-grade classrooms is at a seventh-grade level in comparison to other countries.	***Mathematics: Applications and Connections*** meets or exceeds international standards in grades 6-8. It emphasizes geometry, measurement, and proportionality, as well as algebra, fractions, and statistics.
Curriculum Topic coverage in U.S. eighth-grade mathematics classes is not as focused as in Germany and Japan. (The U.S. curriculum is "a mile wide and an inch deep.)	***Mathematics: Applications and Connections*** follows a structured scope and sequence that introduces, reinforces, and extends topics needed for success in algebra 1 and geometry.

In addition to Glencoe's progressive response to the Flanders' research and the TIMSS, here are some other key features and benefits that middle school mathematics educators asked us to include in ***Mathematics: Applications and Connections.***

Feature	Benefit
Integrated Content There is an emphasis on integrating algebra, geometry, measurement, proportional reasoning, statistics, probability, technology, and problem solving.	Students will be prepared for success in algebra 1 and geometry, the gateway courses for success in college and careers.
Applications and Connections Relevant, real-life applications and interdisciplinary connections are a part of every lesson. Every application and connection is written in an engaging style and often accompanied by colorful, high-interest photos, graphs, tables, and charts.	Students are motivated to learn mathematics when it is interesting and related to their lives. Every page is designed to help you answer the question, "When am I ever going to use this?".
Test Preparation Every lesson has at least one multiple-choice *Test Practice* question. Every chapter has two pages of *Standardized Test Practice,* which includes multiple-choice test items, open-ended test items, and a *Test-Taking Tip.*	Test scores will increase because students will be prepared for success on state, national, and international standardized tests as well as classroom tests and end-of-course examinations.

As you examine ***Mathematics: Applications and Connections,*** you will see that this program will help you prepare your students for success in school and in their lives. As students use the program, they will repeatedly receive this message: "Math is for everyone...You can do it...You'll use it every day."

Mathematics: Applications and Connections— something good just got better!

BRIDGING THE GAP

From Elementary Mathematics to High School Mathematics

Students' experiences in middle school mathematics are crucial in preparing them for success in algebra and geometry.

Glencoe's *Mathematics: Applications and Connections* prepares all students for success in algebra and geometry. How? By introducing a variety of new concepts at the right time in the right way and by integrating them appropriately into all three courses. For example, integers are introduced in Chapter 11 of Course 1, in Chapter 5 of Course 2, and in Chapter 2 of Course 3. Because algebra and geometry are introduced in Chapter 1 in all three courses – and because both are reinforced throughout – students are much better prepared to take these courses in high school.

Manipulatives help students bridge the gap from the concrete to the abstract.

How does *Mathematics: Applications and Connections* help students bridge from the concrete, number-oriented mathematics in elementary school to abstract, symbol-centered algebra and geometry in high school? Many middle school students' learning styles lend themselves to concrete operations. They may not be able to grasp abstract concepts easily, but they can understand abstract concepts on a concrete level.

Shoe □ Jeff MacNelly

Hands-On Labs and Mini-Labs help students discover concepts on their own.

"Hands-on experiential learning in which students take an active role and assume responsibility for their own learning is an integral part of the instructional practices of this program."

Beatrice Moore-Harris, Author

Glencoe's *Mathematics: Applications and Connections* encourages students to **do** mathematics. **Hands-On Labs** give students hands-on experiences, with a partner or group, in discovering mathematical concepts for themselves. The **Hands-On Lab Masters** provide students with a way to record what they observe and discover in the Hands-On Labs. Students may also participate in shorter **Hands-On Mini-Labs** in which they investigate mathematical concepts within a lesson. **Hands-On Math** exercises provide even more opportunities for concrete learning.

The **Overhead Manipulative Resources** include a wealth of transparent manipulatives for use with overhead projectors such as a compass, spinner, counters, geoboard, and coordinate planes. A complete Teacher's Guide, correlated to the Hands-On Labs and Hands-On Mini-Labs, provides suggestions for demonstrations by the teacher or students.

The Glencoe **Mathematics Manipulative Kit** offers the tools students need to work through the Hands-On Labs and Hands-On Mini-Labs. A Teacher's Guide provides a correlation to all three courses.

Cooperative Learning in the Mathematics Classroom, part of the Glencoe Mathematics Professional Series, provides suggestions for implementing cooperative learning techniques in your classroom.

Cooperative learning helps students with academic achievement and interpersonal skills.

In cooperative learning, small groups of students work as teams to share ideas, solve problems, and justify conclusions. At the same time, they develop vital skills in cooperating with others toward a common goal. The increased communication that takes place in cooperative learning experiences clarifies and strengthens individual students' understanding.

Glencoe's *Mathematics: Applications and Connections* offers an abundance of ways for students to learn cooperatively. **Chapter Projects** provide open-ended activities that often include collecting and organizing data. Students work together to explore, analyze, and report their results. **Interdisciplinary Investigations** relate mathematics to other content areas students are studying. These long-term projects are ideal for cooperative learning groups. The **Investigations and Projects Masters** provide students with ways to organize their explorations in the Chapter Projects and Interdisciplinary Investigations.

Hands-On Labs, Technology Labs, Thinking Labs, and **Mini-Labs** throughout all three courses of Mathematics: Applications and Connections also provide excellent opportunities for cooperative learning experiences. **Let the Games Begin** features interesting and fun games to help students reinforce and extend concepts.

Mathematics: Applications and Connections
Exemplifies the NCTM Standards

Mathematics: Applications and Connections thoroughly integrates all thirteen curriculum standards for grades 5-8 as outlined in the NCTM Standards. For more information on the Standards, refer to pages 64-119 of the *Curriculum and Evaluation Standards for School Mathematics* © 1989 by the National Council of Teachers of Mathematics.

STANDARD 1: Mathematics as Problem Solving

Course 1   (Chapters 1-13)
4-37, 46-85, 94-121, 132-155, 157-169, 178-219, 228-257, 268-283, 285-301, 310-343, 352-385, 396-425, 434-467, 476-503, 514-539

Course 2   (Chapters 1-13)
4-35, 44-79, 88-121, 132-175, 184-217, 226-257, 268-307, 316-351, 360-397, 408-441, 450-481, 490-517, 528-553

Course 3   (Chapters 1-13)
4-47, 56-95, 104-129, 140-177, 186-223, 232-267, 278-321, 330-373, 382-417, 428-467, 476-507, 516-548, 560-591

STANDARD 2: Mathematics as Communication

Course 1   (Chapters 1-13)
4-37, 46-85, 94-121, 132-169, 178-219, 228-257 , 28-301, 310-343, 352-385, 396-425, 434-467, 476-503, 514-539

Course 2   (Chapters 1-13)
4-35, 44-79, 88-121, 132-175, 184-217, 226-257, 268-307, 316-351, 360-397, 408-441, 450-481, 490-517, 528-553

Course 3   (Chapters 1-13)
4-47, 56-95, 104-129, 140-177, 186-223, 232-267, 278-321, 330-373, 382-417, 428-467, 476-507, 516-548, 560-591

STANDARD 3: Mathematics as Reasoning

Course 1   (Chapters 1-13)
4-37, 46-85, 94-121, 132-169, 178-219, 228-257, 268-283, 285-301, 310-343, 352-385. 396-425, 434-467, 476-503, 514-539

Course 2   (Chapters 1-13)
4-35, 44-79, 88-121, 132-175, 184-217, 226-257, 268-307, 316-351, 360-397, 408-441,450-481, 490-517, 528-553

Course 3   (Chapters 1-13)
4-15, 17-47, 56-95, 104-129, 140-177, 186-223, 232-267, 278-321, 330-373, 382-417, 428-467, 476-507, 516-548, 560-591

STANDARD 4: Mathematical Connections

Course 1   (Chapters 1-13)
4-37, 46-85, 94-121, 133-139, 141-169, 178-180, 182-190, 193-219, 228-241, 243-257 , 268-270, 273-283, 285-301, 312-343, 352-361, 364-366, 370-373, 375-382, 396-414, 416-425, 434-439, 441-461, 464-467, 476-487, 492-493, 496-503, 514-539

Course 2   (Chapters 1-13)
4-35, 44-79, 88-121, 132-175, 184-217, 226-257, 268-307, 316-351, 360-397, 408-441, 450-481, 490-517, 528-553

Course 3   (Chapters 1-13)
4-15, 17-47, 56-95, 104-129, 140-177, 186-192, 194-212, 214-223, 232-267, 278-321, 330-373, 382-417, 428-467, 476-507, 516-548, 560-591

STANDARD 5: Number and Number Relationships

Course 1   (Chapters 1-13)
8-11, 20-25, 28-37, 46-49, 54-85, 94-121, 132-148, 152-169, 178-180, 182-184, 188-219, 228-241, 243-257, 268-301, 310-321, 324-343, 352-366, 370-373, 375-382, 396-414, 416-424, 434-461, 464-467, 476-503, 515-534, 536-539

Course 2   (Chapters 1-11, 13)
4-20, 24-35, 44-79, 88-121, 132-175, 184-217, 226-257, 268-307, 316-351,366-367, 370-373, 382-385, 408-411, 450-459, 464-468, 474-481, 528-533, 538-553

Course 3   (Chapters 1-13)
4-10, 17-25, 30-36, 56-65, 69-95, 104-129, 140-155, 158-177, 232-252, 257-267, 278-280, 286-300, 307-314, 318-321, 330-333, 335-350, 352-369, 382-389, 396-417, 433-444, 456-467,476-479, 504-507, 532-533, 586-587

STANDARD 6: Number Systems and Number Theory

Course 1   (Chapters 1-13)
28-31, 34-37, 50-53, 60-63, 94-121, 132-148, 150-163, 167-169, 178-180, 182-219, 228-241, 243-257, 268-301, 312-321, 324-343, 352-366, 370-373, 375-382, 396-414, 416-424, 434-461, 464-467, 476-503, 515-529, 531-534, 536-539

Course 2 (Chapters 1-11)

4-7, 44-53, 56-73, 77-79, 132-164, 169-175, 184-209, 212-214, 226-231, 234-238, 249-252, 268-287, 292-307, 316-329, 332-335, 339-341,376-379,410-422, 432-435, 450-455, 474-480

Course 3 (Chapters 1, 2, 6, 7, 9)

11-15, 17-25, 30-36, 44-47, 66-72, 78-83, 86-89, 232-238, 240-260, 281-285, 307-308, 382-384, 390-394

STANDARD 7: Computation and Estimation

Course 1 (Chapters 1-13)

4-37, 46-67, 71-77, 112-121, 132-143, 145-148, 152-163, 167-169, 178-180, 182-190, 193-196, 217-219, 228-241, 243-257, 268-301, 312-343, 352-366, 370-373, 375-382, 396-414, 416-425, 434-461, 464-467, 476-503, 514-534, 536-539

Course 2 (Chapters 1-13)

4-16, 21-23, 28-35, 50-63, 66-79,88-121, 132-175, 197-200, 202-205, 207-217, 228-237, 239-241, 246-248, 253-257, 268-279, 284-307, 317-351, 362-365, 369-373, 376-379, 382-385, 408-417, 419-426, 428-441, 450-458, 460-467, 469-481, 503-506, 510-517, 528-533, 538-545, 547-549, 551-553

Course 3 (Chapters 1-13)

4-16, 21-31, 37-47, 56-95, 104-129, 140-146, 148-167, 188-192, 194-199, 201-204, 206-212, 214-218, 232-238, 240-264, 278-299, 301-306, 309-321, 330-373, 382-417, 428-440, 42-444, 446-467, 476-481, 486-493, 495-507, 518-543, 546-548, 561-591

STANDARD 8: Patterns and Functions

Course 1 (Chapters 1, 2, 4-5, 7-13)

8-11, 28-31, 46-49, 54-57, 64-85, 133-139, 145-155, 178-184, 217-219, 280-283, 285-288, 296-301, 322-327, 334-336, 356-357, 384-385, 402-409, 416-417, 425, 440, 456-467, 476-486, 514-534, 536-539

Course 2 (Chapters 1, 3-7, 9, 11)

24-27, 94-97, 132-149, 207-211, 215-217, 249-257, 288, 360-361, 388-397, 459-467, 478-480

Course 3 (Chapters 1-10, 12, 13)

11-15, 37, 73-76, 107-110, 118-119, 140-147, 153-157, 162-177, 194-195, 200, 205-209, 220-223, 232-234, 239-244, 257-260, 278-280, 294-300, 305-308, 318-321, 334, 342-343, 352, 365, 370-373, 385-389, 396-407, 414-417, 428-467,528-533, 586-587

STANDARD 9: Algebra

Course 1 (Chapters 1, 2, 4-13)

16-31, 34-37, 46-57, 64-85, 133-139, 141-143, 145-148, 150-155, 161-163, 167-169, 178-180, 182-184, 197-201, 214-216, 243-245, 250-253 , 273-283, 285-294, 296-301, 317-321, 324-327, 330-333, 337-343, 370-373, 398-409, 416-420, 425, 434-461, 464-467, 476-503, 515-518

Course 2 (Chapters 1, 2, 4-9, 11-13)

8-23, 28-33, 56-63, 66-69, 138-141, 197-200, 202-205, 207-209, 212-214, 226-231, 234-245, 253-257, 284-287, 297-304, 317-320, 325-328, 332-335, 346-348, 362-365, 369-373, 376-379, 382-385, 450-453, 456-458, 469-481, 514-517, 538-541

Course 3 (Chapters 1-11, 13)

11-29, 32-47, 62-72, 78-89, 92-95, 107-117, 142-147, 153-155, 158-162, 168-177, 188-192, 196-199, 201-204, 210-212, 215-218, 242-252, 257-

259, 278-298, 301-321, 330-341, 344-373, 390-394, 398-401, 414-417, 428-440, 4422-449, 452-467, 476-479, 490-493, 560-585, 588-591

STANDARD 10: Statistics

Course 1 (Chapters 1-13)

4-7, 12-15, 28-31, 34-37, 46-85, 10-5-108, 112-115, 133-139, 141-148, 150-151, 157-159, 182-184, 188-190, 193-196, 206-219,242-245, 250-253, 268-270, 289-291, 298-301, 316, 330-343, 358-361, 410-411,434-439, 445-448, 456-461, 484-487, 500-503, 514-525, 530, 535-539

Course 2 (Chapters 2-6, 8, 11)

44-63, 66-73, 88-121, 172-175, 212-214, 253, 336-338, 342-345, 459-467, 469-472

Course 3 (Chapters 2-6, 8, 12)

56-65, 73-76, 104-113, 120-123, 126-129, 140-177, 194-195, 260-264, , 330-333, 441, 445, 450-451, 546-548

STANDARD 11: Probability

Course 1 (Chapters 1-4, 8, 11, 13)

34-37, 54-57, 64-67, 105-108, 132, 316, 434-436, 514-518, 522-539

Course 2 (Chapters 4, 8, 10, 11, 13)

165-168, 317-320, 436-441, 464-467, 528-533, 542-545, 547-549, 551-553

Course 3 (Chapters 4-6, 11, 12)

142-146, 200, 253-256, 476-479, 516-545

STANDARD 12: Geometry

Course 1 (Chapters 1, 3-8, 10-13)

34-37, 102-111, 141-143, 145-155, 161-163, 178-181, 188-190, 198-201, 228-234, 238-241, 246-249 , 268-272, 277-283, 296-301, 310-315, 317-320, 322-327, 334-336, 340-343, 398-425, 434-436, 445-448, 456-467, 484-487, 500-503, 522-534, 536-539

Course 2 (Chapters 1, 2, 5-10, 12)

17-20, 28-33, 56-59, 66-69, 191-195, 215-217, 254-257, 292-300, 317-320, 362-365, 369-385, 388-397, 410-441, 490-517

Course 3 (Chapters 1-11, 13)

8-10, 17-25, 38-42, 81-83, 92-95, 107-110, 118-123, 126-129, 148-151, 186-193, 196-223, 239, 245-248, 253-256, 296-299, 301-314, 342-343, 357-373, 382-384, 386-389, 396-401, 404-417, 433-435, 442-444, 446-449, 452-467, 476-503, 506

STANDARD 13: Measurement

Course 1 (Chapters 1, 3-13)

4-7, 12-15, 28-31, 34-37, 95-104, 109-112, 118-121, 133-136, 145-155, 161-166, 188-190, 197-213, 228-234, 238-241, 246-249, 254-257, 268-283, 292-301, 312-315, 317-327, 334-336, 340-343, 352-355, 358-366, 368-373, 379-382,434-436, 449-452, 459-461, 464-467, 484-487, 522-529, 531-534

Course 2 (Chapters 1-4, 7-9, 11-12)

4-7, 74-76, 118, 148-149, 272-279, 284-287, 289-300, 305-307, 325-328, 360-385, 388-397, 469-472, 490-517

Course 3 (Chapters 1-9, 11)

38-47, 56-58, 62-72, 92-95, 118-119, 126-129, 148-151, 186-193, 196-199, 201-205, 213-215, 249-252, 265-267, 278-280, 307-311, 331-373, 396-401, 404-417, 482-485, 499-507

Motivating Middle School Students

"When are we ever going to use it?"

"Why do I have to learn this?"

Although these may be just complaints from our middle school students, they are legitimate questions that deserve answers. We at Glencoe realize that effective mathematics programs that really motivate middle school students must have more than strong content. They must demonstrate the usefulness of mathematics in a way that relates to student interests.

Mathematics: Applications and Connections makes math a part of students' daily school lives.

- **Applications** begin lessons with a real-life situation that points out how mathematics is used in students' everyday lives as well as in the world about them.
- **When am I ever going to use this?** gives an example of when the math concepts of the lesson will be useful in students' lives.
- **Connection** examples and exercises show how mathematics is used in other courses they may be taking such as Language Arts, Life Science, Geography, Music, Health, Physical Education, and Art.
- **Integration** examples and exercises illustrate how the math they are learning relates to mathematics courses they may take later such as Algebra, Geometry, and Probability and Statistics.

"The curriculum must go beyond the basics—to be relevant, it must be of interest to students and emphasize the usefulness of mathematics."

Ron Pelfrey, Author

Mathematics: Applications and Connections
invites students to look beyond their textbook to
see mathematics in the world.

Let the Games Begin

Let the Games Begin gives students a fun
way to practice their math skills and learn to
interact positively with fellow students.

Family Activity

The **Family Activity** provides
an opportunity for students to
include their family members
in the math they are studying.

Cultural Kaleidoscope

Cultural Kaleidoscope introduces students
to a variety of world cultures.

School to Career

School to Career
features
demonstrate how
people use
mathematics in the
workplace and give
students
opportunities to
explore possible
avenues for future
careers.

Did You Know?

Did You Know? features little
known facts to capture students'
interest.

Math in the Media

Math in the Media includes cartoons,
graphs, advertisements, and other aspects
of the media that use mathematics to get a
point across.

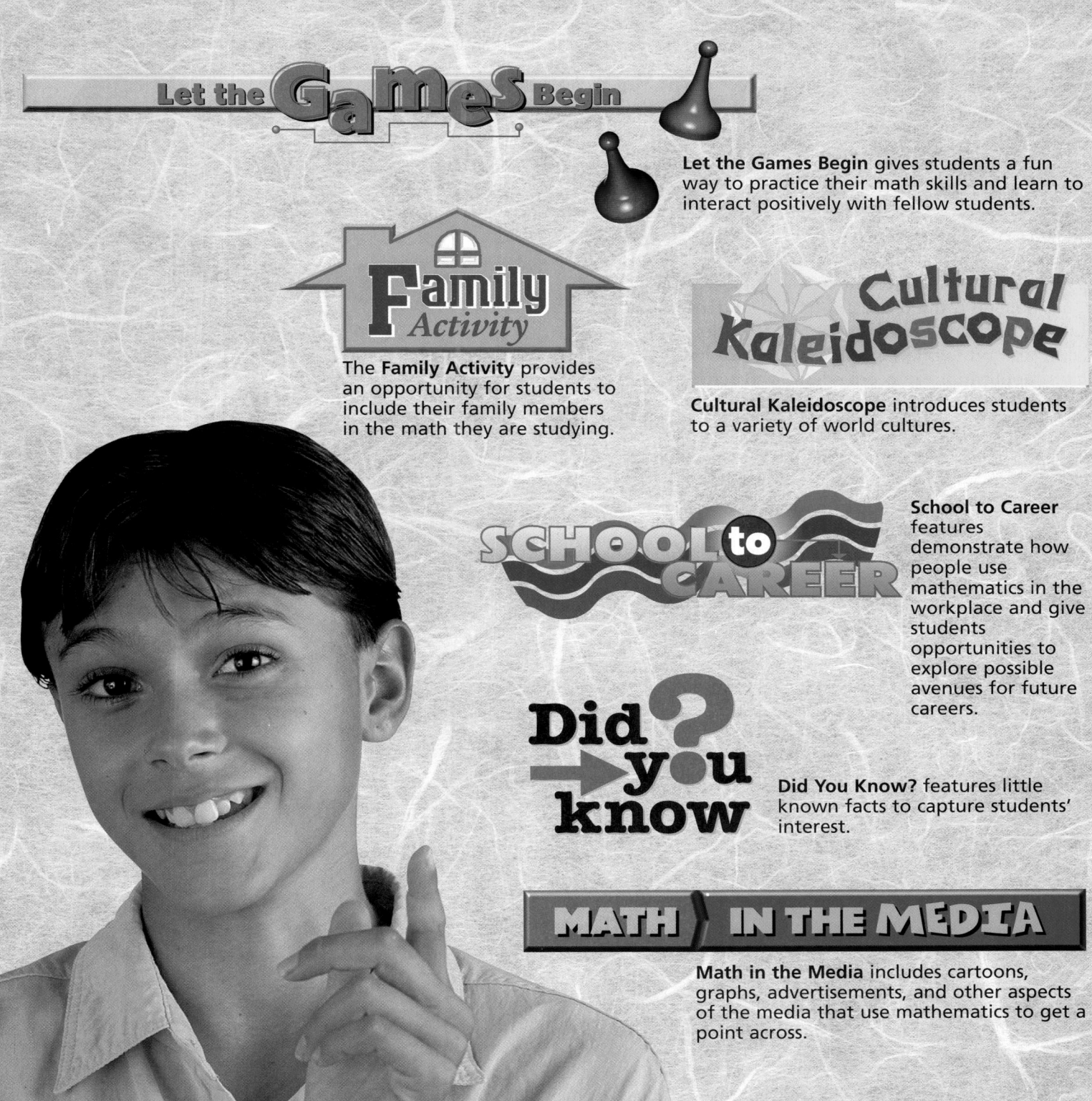

Developing Problem Solving

According to the NCTM Standards, "Problem solving is the process by which students experience the power and usefulness of mathematics in the world around them. It is also a method of inquiry and application ... to provide a consistent context for learning and applying mathematics. Problem situations can establish a 'need to know' and foster the motivation for the development of concepts."

Problem solving is an integral part of every lesson in every course of Glencoe's *Mathematics: Applications and Connections.* How is this accomplished?

The **first chapter** of each course focuses on problem solving.
- Course 1:
   Problem Solving, Numbers, and Algebra
- Course 2:
   Problem Solving, Algebra, and Geometry
- Course 3: **Problem Solving and Algebra**

Thinking Labs in every chapter focus on problem-solving strategies, such as solving a simpler problem and working backward, and present opportunities to solve nonroutine problems.

Frequent **Problem-Solving Study Hints** suggest using various problem-solving strategies to investigate and understand mathematical content and apply strategies to new problem situations.

Reading Math Study Hints provide tips on how to read and interpret the language or symbolism of mathematics.

> **Study Hint**
> Reading Math The symbols < and > always point to the lesser of the two numbers.

Critical Thinking exercises in every lesson give students practice in developing and applying higher-order thinking skills.

Glencoe's *Mathematics: Applications and Connections* links practical problem solving to students' real-life interests. Mathematics becomes a vital force in the lives of middle school students as their eyes are opened to the relationship between mathematics and sports, shopping, and other teen interests. They "take ownership" of their skills by writing their own problems and presenting class projects connected to real life.

"Problem solving is an integral component of this program. It requires students to think critically, examine new concepts, and then extend or generalize what they already know."

Linda Dritsas, Author

Most lessons begin with either a real-world **Application,** an interdisciplinary **Connection,** or content **Integration** that provides students with a reason to learn mathematics. Application, connection, and integration examples also occur throughout the texts to give students the opportunity to study completely worked-out problems.

Applications and Problem Solving exercises in every lesson directly link mathematics to real-world topics like entertainment, and to art, history, science, and other subjects.

The **Chapter Projects** and **Interdisciplinary Investigations** enable students to become more deeply engaged in a problem situation.

USING TECHNOLOGY

Is there a day that goes by when you don't encounter a computer-operated machine? The future world of our students will be even more involved in high technology. Clearly, technology is changing the workplace and the home at an increasingly rapid pace. Without technical mathematical skills, today's students will have little or no chance of finding good jobs.

Mathematics: Applications and Connections provides many opportunities for you to introduce your students to the world of technology as they learn mathematics.

Internet Connections throughout the Student Edition refer students to the Glencoe *Mathematics: Applications and Connections* site **www.glencoe.com/sec/math/mac/mathnet** that provides links to other websites on the Internet. These sites provide more information on the topics that students are studying.

Technology Labs and **Mini-Labs** give students hands-on opportunities to use a computer or a graphing calculator to solve problems. Students learn how to use a spreadsheet to organize data and make predictions. Graphing calculator programs provide an introduction to logic and a way to perform calculations in a timely manner.

Technology Study Hints provide helpful suggestions on how students can use scientific calculators, graphing calculators, or software in studying the topics of the lesson.

In the **Chapter Projects** and **Interdisciplinary Investigations,** students are encouraged to use the Internet and application software like word processing, publishing software, and spreadsheets.

The **CD-ROM Program** for *Mathematics: Applications and Connections* contains a variety of activities correlated to each chapter. These include a Chapter Introduction and a Resource Lesson for each lesson in the Student Edition. The activities also include Interactive Lessons and Extended Activities for some of the lessons in the chapter. Every chapter includes an Assessment Game that is a cumulative review of skills they have learned.

MindJogger Videoquizzes feature a game show format that provides an alternative assessment and an entertaining way for your students to review chapter concepts and problem-solving skills.

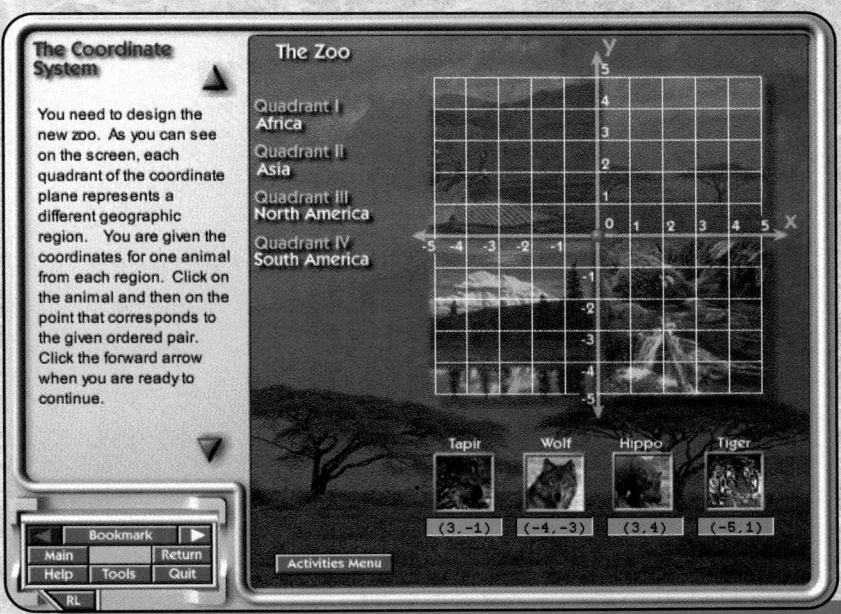

CD-ROM Program, Course 2, Chapter 7

Interactive Mathematics Tools Software helps students gain mathematical power through interactive activities that combine video, sound, animation, graphics, and text.

All of the blackline masters for *Mathematics: Applications and Connections* are available on the **Electronic Teacher's Classroom Resources** CD-ROM. There's no need to carry a large array of booklets from room to room—just print out the masters you wish to use from a single CD-ROM that works on both Macintosh and Windows formats.

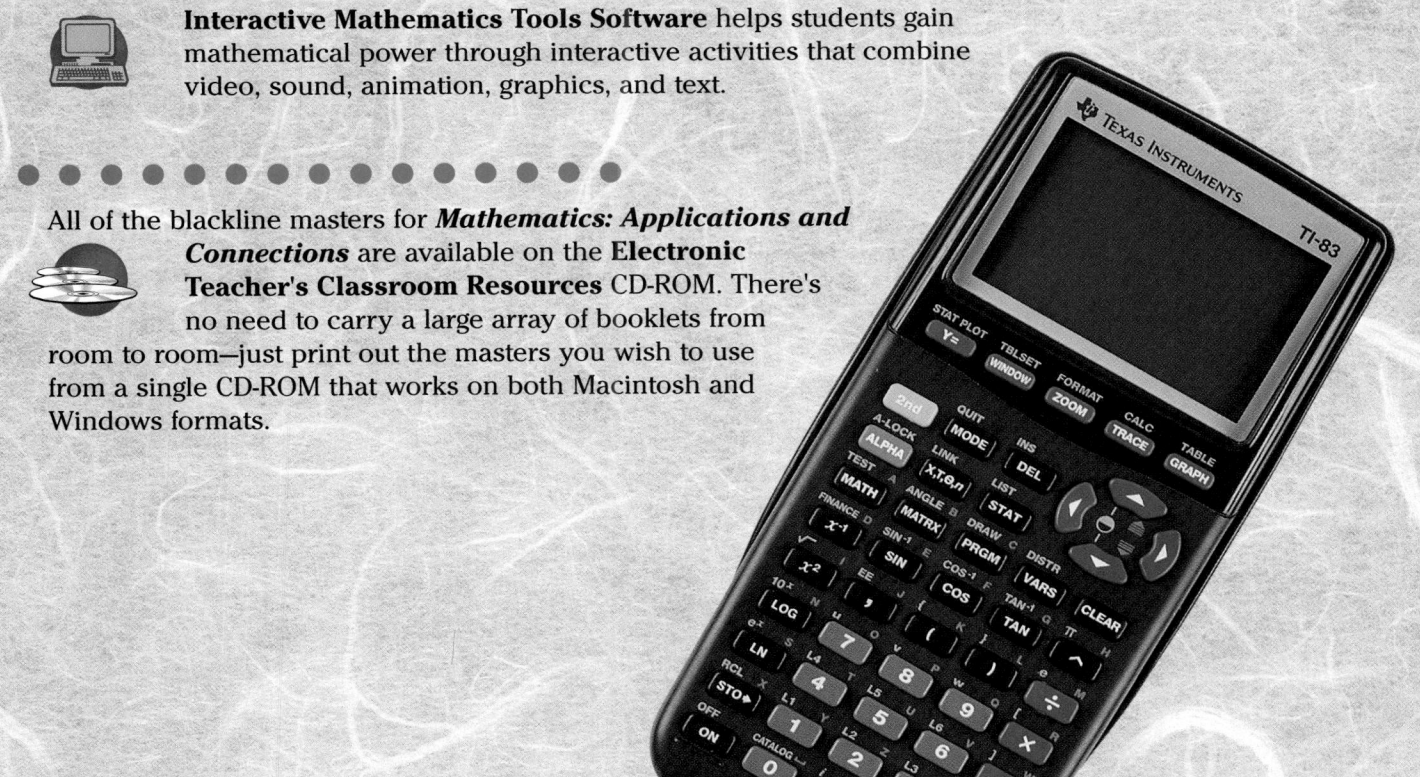

Assessment

The assessment tools built into *Mathematics: Applications and Connections* are designed to assess traditional basic skills as well as those skills that will be required for success in the 21st century. In addition to basic skills, *Mathematics: Applications and Connections* also helps you assess students' ability to organize information, apply previously-learned information, and make conjectures based on gathered data. The curriculum alignment in *Mathematics: Applications and Connections* also enables students to perform well on standardized testing at both state and national levels.

The following features and components help you accurately assess each student's achievement.

In the Student Editions...

- Every lesson has a **Mixed Review** that includes a **Test Practice** item.

- Every chapter has **Math Journals**, a **Chapter Project**, a **Mid-Chapter Self Test**, **Standardized Test Practice**, and a **Chapter Test**.

- Every chapter has a **Study Guide and Assessment** that includes Vocabulary, Understanding and Using Vocabulary, Examples and Review Exercises for each Objective, Applications and Problem Solving, a **Performance Task**, and a **Portfolio** suggestion.

- Since practically every mathematics test is also a reading test, several **Reading Math** study hints are included in every chapter.

In the Teacher's Wraparound Editions...

- Every lesson is correlated to the major national standardized tests: CAT, CTBS, ITBS, MAT, SAT, and Terra Nova.

- Every lesson includes a **5-Minute Check** that covers the previous lesson or chapter.

- Every lesson has a **Closing Activity** that involves writing, speaking, or modeling.

In the supplementary materials...

Assessment and Evaluation Masters include:

1. For each chapter: three Multiple Choice tests (Basic, Average, Honors), three Free-Response Tests (Basic, Average, Honors), Performance Assessment (includes a Scoring Guide), Standardized Test Practice, a Cumulative Review, four quizzes, and a Mid-Chapter Test.
2. Also Included: a Placement Test, two Semester Tests and a Final Test

MindJogger Videoquizzes (VHS) review each chapter by using a game show format. As students compete on teams, they hear and see each review problem as it is presented and then completely solved.

Test and Review Software (Windows & Macintosh) combines a test generator and test bank. You can easily create your own free-response, multiple choice, and open-ended tests for honors, average, and basic students.

5-Minute Check Transparencies provide a quick review of the previous lesson or chapter. There is one full-color transparency for every lesson. The 5-Minute Check is also printed in the Teacher's Wraparound Edition to make your lesson plans easier to prepare.

The CD-ROM Program includes guided practice and an Assessment Game that is similar to Trivial Pursuit.

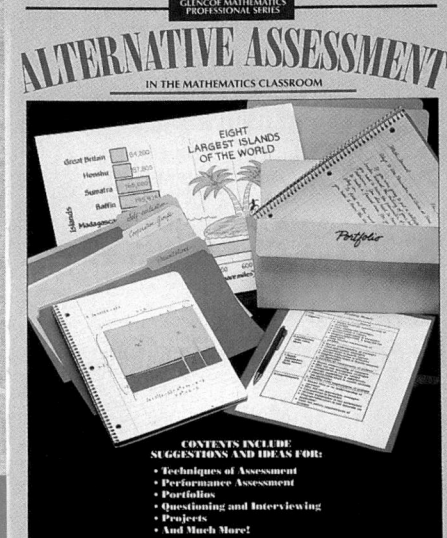

Alternative Assessment in the Mathematics Classroom, part of the Glencoe Mathematics Professional Series, gives an overview of the latest trends in assessment.

Using Projects

Doing mathematics is so much more effective than memorizing mathematics!

Mathematics: Applications and Connections gives you and your students several opportunities to engage in projects of varying lengths that put mathematics into motion.

The **Chapter Project** is introduced at the beginning of the chapter and sets in motion several activities that culminate

with **Completing the Chapter Project** in the Chapter Study Guide and Assessment. Students apply the mathematics they learn in the chapter to various real-life situations. They are often asked to represent their results as a working model or in statistical graph form. Throughout the chapter, **Working on the Chapter Project** exercises remind students of the next steps in their project. These activities not only implement the mathematics student are learning, but give students the opportunity to work together as a team.

Interdisciplinary Investigation

The **Interdisciplinary Investigations** found at the ends of Chapters 3, 6, 9, and 12 show how the mathematics students have learned can be applied to other courses they may be taking. Each investigation uses the skills from the previous three chapters. In addition to the problem they are assigned to investigate, they are given writing assignments that explore other areas such as Language Arts, Foreign Language, Science, Health, Physical Education, and Social Studies.

Interactive Mathematics: Activities and Investigations

offers an innovative approach to teaching and learning middle school mathematics. Each of the 18 units that make up this comprehensive, activity-based program may be used to enhance chapters in *Mathematics: Applications and Connections.*

The interleaf pages of each chapter in the Teacher's Wraparound Edition highlight two activities from *Interactive Mathematics: Activities and Investigations* that are appropriate for that lesson. However, there are many other activities that may be useful.

The chart below summarizes the mathematical focus of each unit of *Interactive Mathematics: Activities and Investigations.*

INTER·Active Mathematics
Activities and Investigations

	Unit	Title	Mathematical Focus
Course 1	1	From the Beginning	Building Math Power
	2	A Million to One	Number Sense
	3	Just the Right Size	Scale Drawings and Proportional Reasoning
	4	Through the Looking Glass	Spatial Visualization
	5	Get a Clue	Logical Reasoning
	6	The Road Not Taken	Graph Theory and Networks
Course 2	7	Take It From the Top	Building Math Power
	8	Data Sense	Statistics and Data Analysis
	9	Don't Fence Me In	Area and Perimeter
	10	Against the Odds	Probability
	11	Cycles	Algebra Patterns
	12	Treasure Island	Geometry and Measurement
Course 3	13	Start Your Engines	Building Math Power
	14	Run for Cover	Surface Area and Volume
	15	On the Move	Graphing and Functions
	16	Growing Pains	Linear and Exponential Growth
	17	Infinite Windows	Fractals and Chaos Theory
	18	Quality Control	Applied Data Analysis

*G*lencoe has correlated *Mathematics: Applications and Connections (MAC)* with the *Interactive Mathematics: Activities and Investigations (IMAI)* units to provide a unique opportunity for you to tailor a program to your middle school students. Integrating the two programs allows you to emphasize areas and mathematical ideas that interest and fit the unique needs of the students in your class.

MAC Course 1 Chapter	IMAI Units 1-6	Other IMAI Units	MAC Course 2 Chapter	IMAI Units 7-12	Other IMAI Units	MAC Course 3 Chapter	IMAI Units 13-18	Other IMAI Units
1	1, 5		1	7		1	15	8
2	6	8	2	7	2	2	15	
3	2		3	8		3	16	3
4	2	7, 9	4	11	18	4	18	8
5	2, 3		5	11	16	5	14	4
6	3	11	6	11	15	6	16, 17	
7	3	10	7	9, 10		7	16, 17	
8	3	10	8	10	3	8	16, 17	3, 12
9	4	12	9	12		9	17, 18	10
10	4	14	10	9		10	15, 16	
11		11	11	10		11	14	
12		11	12	9	4, 14	12	17	6, 10
13	6	10	13	10	6	13	16	5

Interactive Mathematics: Activities and Investigations is available as individual units or as a three book series that groups Units 1-6, 7-12, and 13-18. Each unit also has a Classroom Instructional Resources binder that includes teaching instructions, blackline masters, transparencies, and guidance on using projects in the classroom.

For more information on *Interactive Mathematics: Activities and Investigations,* contact your nearest Glencoe Regional Office or call 1-800-334-7344.

During the past few years, the use of games in the mathematics classroom has increased dramatically. These activities tend to be very motivational, and they often help improve students' attitudes toward mathematics in general. Because games are enjoyable, they allow students to learn and review mathematical skills and concepts while they unknowingly increase their problem-solving, logic, and computational skills.

Why Use Games? While mathematical games are not the only method you can use to provide individualized instruction or enrichment, they may be a very effective strategy. According to NCTM, most mathematical games serve one or more of the following functions.

- To develop concepts.
- To provide drill and reinforcement.
- To develop perceptual abilities.
- To provide opportunities for logical thinking and/or problem solving.

Therefore, mathematical games can be used in a number of different ways. You may want to use a game to introduce a new topic, thus encouraging students to employ discovery learning. You may want to use a game within the framework of your lesson instruction, for example, in the last few moments of a class period. Finally, you may want to use a game to review a concept previously taught.

The games presented in **Let the Games Begin** are designed to be useful and enjoyable to all students. Feel free to develop new and fun variations that will make each game your own.

Let the **Games** Begin

More information on games is available from Glencoe on the Internet at **www.glencoe.com/sec/math/mac/mathnet.**

In addition to the games presented in the Student Edition, more games are available in the **Classrooms Games** booklet.

Cuisenaire® Company of America, Inc.

Cuisenaire® Company of America, Inc.

Multiple Learning Styles

People learn in many different ways. There are several different learning styles that help us approach and solve problems. Everyone possesses varying degrees of each of these learning styles, but the ways in which they combine and blend are as varied as the personalities of the individuals. Glencoe's *Mathematics: Applications and Connections* provides you with ways to accommodate students with these diverse learning styles.

Learning Style	Characteristics of Students	Activities in Student Edition
verbal/ linguistic	read regularly, write clearly, and easily understand the written word	**Communicating Mathematics** exercises ask students to tell, write, and explain mathematical concepts. Students also express what they have learned in their **Math Journals.**
logical	use numbers, logic, and critical thinking skills	Clearly-written **Examples** present important concepts, and **Critical Thinking** exercises extend those concepts. **Thinking Labs** encourage students to practice their logical thinking skills by using various strategies.
visual/ spatial	think in terms of pictures and images	**Integration** and **Hands-On Lab** exercises ask students to draw or show mathematical concepts through modeling, coordinate grids, charts, and graphs.
auditory/ musical	have "good ears" and can produce rhythms and melodies	Multimedia software, such as the **CD-ROM Program, MindJogger Videoquizzes,** and **Interactive Mathematics Tools Software,** can be easily incorporated into lessons.
kinesthetic	learn from touch, movement, and manipulating objects	**Hands-On Labs** and **Mini-Labs** provide for physical involvement in learning.
interpersonal	understand and work well with other people	**Chapter Projects, Interdisciplinary Investigations,** as well as **Hands-On Labs, Technology Labs,** and **Mini-Labs,** allow students to collaborate with others.
intrapersonal	have a realistic understanding of their strengths and weaknesses	**Math Journals, Portfolios,** and **Family Activities** help students personalize mathematics.
naturalist	can distinguish among, classify, and use features of the environment	**Application** and **Connection** examples and exercises show students how mathematics relates to the world around them.

As mathematics teacher, you may want to assign activities to students that accommodate their strongest learning styles, but frequently ask them to use their weakest learning styles. Additional activities are provided in the bottom margins of the lesson notes in the Teacher's Wraparound Edition.

The resources available in *Mathematics: Applications and Connections* guarantee that your classroom will be a multisensory environment, providing multiple paths for student learning.

Involving Parents and the Community

When children enter 6th grade, for a variety of reasons, it may become difficult for parents or guardians to remain on top of what's going on at school. The curricula grow more specialized, the typical middle school student becomes more independent of his or her parents, and multiple teachers replace the primary teacher of the elementary school years.

When parents do have the opportunity to meet with teachers, they often ask what they can do to motivate their children. Here are some ideas to share with parents. Developed by Reginald Clark of the Academy for Educational Development in Washington, D.C., they are designed to foster positive attitudes and boost learning.

- Share the fact that there is an inverse correlation between excessive TV viewing and high achievement in school.
- Stress the importance of seeing that their children complete homework assignments.
- Urge them to provide time, space, and materials needed for homework and reading.
- Remind them how important it is to listen to their children read and/or to read to their children.

Each chapter in Glencoe's *Mathematics: Applications and Connections* contains a suggested **Family Activity** for your students to complete with a family member. The **Family Letters and Activities** provide a letter and activity for each chapter to encourage students' families to become active participants in their students' learning.

Fostering Community Involvement

There are many ways to involve the community in your mathematics classroom and to share your learning with them.

- Send photos and press releases to your local newspaper to inform the community about the exciting things your students are learning.
- Have a "math career day" and invite local people to describe to students how they use math in their jobs.
- Set up a "shadow day" in which students spend a half-day "shadowing" people in the workplace who use math in their careers.
- Find out about – or implement your own – math fairs and competitions that give your students a chance to shine in the "outside world."

Involving Parents and the Community in the Mathematics Classroom, one of the booklets in Glencoe's Mathematics Professional Series, presents additional suggestions on how parents and the community can be active participants in supporting mathematics instruction.

Classroom Management

Meet Susan Uhrig. Mrs. Uhrig has taught in the Columbus Public Schools in Columbus, Ohio, for more than 25 years. She currently uses *Mathematics: Applications and Connections* in her classroom at Monroe Middle School.

Just one look at her classroom tells you that Mrs. Uhrig enjoys teaching middle school students. Their work is displayed throughout her bright classroom. Positive sayings on the bulletin boards encourage students to do their best. Manipulatives are kept out so they are easy to use—evidence that Mrs. Uhrig encourages her students to do mathematics.

Mrs. Uhrig says that she enjoys the diverse personalities of her students. You never know what challenges lie ahead each day! She organizes her classroom in ways that really get her students involved and keep their parents informed. Here are some of her ideas.

- Write the **assignment** and **objective** for the day on the chalkboard for students to copy when they come into class.
- Keep a **master notebook** where assignments are written so students can refer to the notebook if they miss a class.
- Have students keep assignment **logs** throughout the year. This helps students get and stay organized.

- Keep **manipulatives** on a table so they are easy to find and use. Mrs. Uhrig assigns several students to help distribute and collect manipulatives and calculators.
- Use "homework coupons" as a **reward** for special behavior.
- Have students write in their **journals** every day. Mrs. Uhrig has her students do the Check for Understanding exercises and 5-Minute Checks in their journals and allows them to refer to their journals during tests.
- Communicate with **parents and guardians** early and often. At the beginning of the year, Mrs. Uhrig sends each parent a letter that indicates her classroom policies. Throughout the year, she communicates with parents though student logs, interim reports, and grade cards.

There are many teachers like Mrs. Uhrig who, through years of experience, have acquired classroom management techniques that help their classrooms run smoothly. They also possess a wealth of knowledge about how to teach mathematics. Glencoe allows you to network with these teachers through the **Classroom Vignettes** in the Teacher's Wraparound Edition. Teachers from across the country share their ides with you so that you may add to your own list of classroom management techniques.

Classroom Vignettes

Glencoe's unique **Classroom Vignettes** are classroom-tested teaching suggestions made by experienced mathematics educators. Each vignette was written by a teacher who uses Glencoe's *Mathematics: Applications and Connections* or by a Glencoe reviewer, consultant, or author. Glencoe thanks these outstanding mathematics educators for their unique contributions.

Research Activities

Glencoe's **Mathematics: Applications and Connections**, as well as the entire Glencoe Mathematics Series, is the product of ongoing, classroom-oriented research that involves students, teachers, curriculum supervisors, administrators, parents, and college-level mathematics educators.

The programs that make up the Glencoe Mathematics Series are currently being used by millions of students and tens of thousands of teachers. The key reason that Glencoe Mathematics programs are so successful in the classroom is the fact that each Glencoe author team is a mix of practicing classroom teachers, curriculum supervisors, and college-level mathematics educators. Glencoe's balanced author teams help ensure that Glencoe Mathematics programs are both practical and progressive.

Prior to publication of a Glencoe program, typical research activities include:
- a review of educational research and recommendations made by groups such as NCTM
- mail surveys of mathematics educators
- discussion groups involving mathematics teachers, department heads, and supervisors
- focus groups involving mathematics educators
- face-to-face interviews with mathematics educators
- telephone surveys of mathematics educators
- in-depth analyses of manuscript by a wide range of reviewers and consultants
- field tests in which students and teachers use pre-publication manuscript in the classroom

Feedback from teachers, curriculum supervisors, and even students who currently use Glencoe Mathematics programs is also incorporated as Glencoe plans and publishes new and revised programs. For example, Classroom Vignettes, which are printed in the *Teacher's Wraparound Editions*, are one result of this feedback.

All of this research and information is used by Glencoe's authors and editors to publish the best instructional resources possible.

Scope and Sequence

Mathematics: Applications and Connections is a comprehensive, well-balanced, three-course program that prepares middle school students for success in algebra and geometry. Through a carefully planned scope and sequence of mathematical topics, students encounter, practice, and extend their knowledge of mathematics to promote confidence and mastery.

The chart below shows the chapter titles for the three courses in ***Mathematics: Applications and Connections.*** The following pages present a detailed chart describing the depth to which each topic is covered.

Chapter	Course 1	Course 2	Course 3
1	Problem Solving, Numbers, and Algebra	Problem Solving, Algebra, and Geometry	Problem Solving and Algebra
2	**Statistics:** Graphing Data	Applying Decimals	**Algebra:** Using Integers
3	Adding and Subtracting Decimals	**Statistics:** Analyzing Data	Using Proportion and Percent
4	Multiplying and Dividing Decimals	Using Number Patterns, Fractions, and Percents	**Statistics:** Analyzing Data
5	Using Number Patterns, Fractions, and Ratios	**Algebra:** Using Integers	**Geometry:** Investigating Patterns
6	Adding and Subtracting Fractions	**Algebra:** Exploring Equations and Functions	Exploring Number Patterns
7	Multiplying and Dividing Fractions	Applying Fractions	**Algebra:** Using Rational Numbers
8	Exploring Ratio, Proportion, and Percent	Using Proportional Reasoning	Applying Proportional Reasoning
9	**Geometry:** Investigating Patterns	**Geometry:** Investigating Patterns	**Algebra:** Exploring Real Numbers
10	**Geometry:** Understanding Area and Volume	**Geometry:** Exploring Area	**Algebra:** Graphing Functions
11	**Algebra:** Investigating Integers	Applying Percents	**Geometry:** Using Area and Volume
12	**Algebra:** Exploring Equations	**Geometry:** Finding Volume and Surface Area	Investigating Discrete Math and Probability
13	Using Probability	Exploring Discrete Math and Probability	**Algebra:** Exploring Polynomials

Mathematics: Applications and Connections

Scope and Sequence

Course	1	2	3
PROBLEM SOLVING			
Develop a plan	●	●	●
Strategies			
Look for a pattern	●	●	●
Solve a simpler problem	●	●	●
Act it out		●	●
Guess and check	●	●	●
Draw a diagram	●	●	●
Make a table	●	●	●
Work backward	●	●	●
Choose the method of computation	●	●	●
Make a list	●	●	●
Eliminate the possibilities	●	●	●
Determine reasonable answers	●	●	●
Make a model	●	●	●
Use a graph	●	●	●
Use an equation	●	●	●
Use logical reasoning	●	●	●
Use the Pythagorean Theorem	●	●	●
Use a Venn diagram	●	●	●
Use a frequency table	●	●	●
Use a spreadsheet	●	●	●
Use proportional reasoning			●
NUMBER AND OPERATIONS			
Number Relationships			
Decimals			
Decimal concepts	●	●	●
Reading and writing	●	●	●
Decimal place value	●	●	●
Comparing and ordering	●	●	●
Rounding	●	●	●
Relating decimals and fractions	●	●	●
Relating decimals, ratios, and percents	●	●	●
Terminating and repeating decimals	●	●	●
Scientific notation		●	●
Powers of ten		●	●
Fractions			
Fraction concepts	●	●	●
Writing mixed numbers as fractions	●	●	●
Mixed numbers and improper fractions	●	●	●

Course	1	2	3
Equivalent fractions	●	●	●
Comparing and ordering fractions	●	●	●
Simplifying fractions	●	●	●
Least common denominator (LCD)	●	●	●
Rounding and estimating fractions	●	●	●
Relating fractions and decimals	●	●	●
Relating fractions and percents		●	●
Proportional Reasoning			
Ratio			
Concept of ratio	●	●	●
Reading and writing ratios	●	●	●
Simplifying ratios	●	●	●
Relating ratios and fractions	●	●	●
Relating ratios and rates	●	●	●
Ratio and probability	●	●	●
The Golden Ratio			●
Proportion			
Concept of proportion	●	●	●
Solving proportions	●	●	●
Property of proportion (cross product)	●	●	●
Scale drawings	●	●	●
Similar figures	●	●	●
Dilations		●	●
Indirect measurement		●	●
Percent			
Concept of percent	●	●	●
Writing fractions and decimals as percent	●	●	●
Percents greater than 100% or less than 1%	●	●	●
Find percent of a number	●	●	●
Percent one number is of another		●	●
Finding number when percent is known		●	●
Percent proportion		●	●
Relating percent and ratio		●	●
Percent equation		●	●
Capture/recapture		●	
Non-proportional relationships			●
Computations and Estimation			
Order of operations	●	●	●
Decimals			
Adding and subtracting	●	●	●
Multiplying by a whole number	●	●	●

Course	1	2	3
Multiplying two decimals	●	●	●
Dividing by a whole number	●	●	●
Dividing by decimals	●	●	●
Dividing with zeros in the quotient	●		
Fractions			
Adding and subtracting	●	●	●
Subtracting with renaming	●	●	●
Multiplying and dividing	●	●	●
Add and subtract mixed numbers	●	●	●
Multiply and divide mixed numbers	●	●	●
Percents			
Discount	●	●	●
Sales tax		●	
Simple interest		●	●
Percent of change		●	●
Integers			
Adding and subtracting		●	●
Multiplying and dividing		●	●
Estimation			
Whole numbers			
Rounding	●	●	●
Sums and differences	●	●	●
Products and quotients	●	●	●
Decimals			
Rounding	●	●	●
Sums and differences	●	●	●
Products and quotients	●	●	●
Fractions			
Sums and differences	●	●	●
Products and quotients	●	●	●
Percents	●	●	●
Use equivalent fractions, decimals, and percents			●
Strategies for estimating			
Rounding	●	●	●
Compatible numbers	●		
Capture-recapture		●	
Clustering	●		
Square roots		●	●
Area or volume	●	●	●
Mental math			
Divisibility patterns	●	●	●
Compatible numbers	●		
Solving equations mentally		●	
Finding percents			●
Powers of ten		●	
Using formulas	●	●	●

Number Systems and Number Theory

Course	1	2	3
Reading and writing whole numbers	●	●	●

Course	1	2	3
Place value of whole numbers	●	●	●
Place value of decimals	●	●	●
Comparing and ordering			
Whole numbers	●	●	●
Decimals	●	●	●
Fractions	●	●	●
Integers	●	●	●
Rationals			●
Positive exponents	●		
Negative exponents			●
Divisibility patterns	●		●
Prime and composite numbers	●	●	●
Relative primes			●
Prime factorization	●	●	●
Greatest common factor (GCF)	●	●	●
Least common multiple (LCM)	●	●	●
Scientific notation		●	●
Square roots		●	●
Factorials		●	●
Properties			
Properties of numbers		●	●
Distributive property	●	●	●
Property of proportions (cross products)	●	●	●
Properties of equality		●	●
Density property			●

PATTERNS AND FUNCTIONS

Course	1	2	3
Numeric patterns			
Sequences	●	●	●
Fibonacci sequence			●
Pascal's triangle		●	●
Divisibility patterns	●	●	●
Geometric patterns			
Recognizing geometry patterns	●	●	●
Tessellations	●	●	●
Fractals	●		
Represent relationships			
Tables	●	●	●
Graphs		●	●
Function rules		●	●
Analyze functional relationships			●
Use patterns and functions to solve problems	●	●	●

ALGEBRA

Course	1	2	3
Integers			
Reading and writing integers	●	●	●
Graphing integers on a number line	●	●	●
Comparing and ordering integers	●	●	●
Adding and subtracting integers	●	●	●

Course

Course	1	2	3
Multiplying and dividing integers	●	●	●
Absolute value		●	●
Rational numbers			
Identify and simplify rational numbers			●
Properties of rational numbers			●
Density property			●
Rational numbers and decimals			●
Scientific notation			●
Comparing and ordering			●
Solving equations with rational number solutions			●
Real numbers			
Identify and classify real numbers			●
Square roots		●	●
Irrational numbers			●
Density property			●
Functions			
Function machine	●		●
Function tables	●	●	●
Linear functions			●
Analyze tables and graphs	●	●	●
Equations and expressions			
Concepts of variable, expression, equation	●	●	●
Order of operations		●	●
Evaluate algebraic expressions	●	●	●
Write algebraic expressions and equations		●	●
Solve addition and subtraction equations	●	●	●
Solve multiplication and division equations	●	●	●
Solve two-step equations	●	●	●
Solve equations with two variables		●	●
Solve inequalities		●	●
Solve equations with concrete methods	●	●	●
Solve equations with informal methods	●	●	●
Solve equations with formal methods		●	●
Graphing			
Integers on a number line	●	●	●
Irrational numbers on a number line			●
Inequalities on a number line		●	●
Points on a coordinate plane	●	●	●
Transformations on a coordinate plane	●	●	●
Functions	●	●	●
Linear functions			●
Quadratic functions			●
Equations		●	●
Systems of equations			●
Using graphing calculators		●	●
Polynomials			
Model with algebra tiles			●
Represent and simplify polynomials			●
Add, subtract, and multiply polynomials			●

Course	1	2	3
Factor polynomials			●
Multiply binomials			●
Apply algebra to real-world and math problems	●	●	●
Use spreadsheets and formulas	●	●	●

STATISTICS

Course	1	2	3
Taking a survey	●	●	●
Analyzing survey data	●	●	●
Organizing data			
Using a table to organize data	●	●	●
Frequency tables	●	●	●
Using tables to solve problems	●	●	●
Using matrices to organize data			●
Constructing and interpreting graphs			
Bar graphs	●	●	●
Circle graphs	●	●	●
Line graphs	●	●	●
Stem-and-leaf plots	●	●	●
Box-and-whisker plots	●	●	●
Line plots		●	●
Histograms			●
Scatter plots		●	●
Maps that show statistics			●
Choosing an appropriate display			●
Interpreting data			
Clusters		●	●
Mean, median, and mode	●	●	●
Range and quartiles	●	●	●
Misleading graphs and statistics	●	●	●
Making predictions from statistics	●	●	●
Making predictions from graphs	●	●	●
Making predictions from a sample	●	●	●

PROBABILITY

Course	1	2	3
Outcomes	●	●	●
Simple event	●	●	●
Independent events	●	●	●
Dependent events		●	●
Complementary events	●		
Experimental probability	●	●	●
Theoretical probability		●	●
Tree diagrams	●	●	●
Counting principle		●	●
Permutations and combinations	●	●	●
Probability and ratio	●	●	●
Fair and unfair games	●	●	●
Simulations or experiments	●	●	●
Area models	●	●	●
Capture-recapture		●	

Course	1	2	3
Punnett squares			●

GEOMETRY

Constructions

	1	2	3
Congruent segments	●		●
Perpendicular lines		●	●
Parallel lines		●	●
Segment bisectors	●		
Congruent angles	●		●
Angle bisectors	●		
Polygons, inscribed	●	●	●
Congruent triangles			●

Angles

	1	2	3
Classify and measure angles	●	●	●
Sum of angle measures		●	●
Parallel lines and transversal		●	●

Polygons

	1	2	3
Identify polygons	●	●	●
Classify triangles and quadrilaterals	●	●	●
Identify congruent figures	●	●	●
Using polygons as networks			●

Triangles

	1	2	3
Determine congruent triangles			●
Right triangle relationships (trigonometry)			●
Pythagorean Theorem		●	●
Special right triangles			●

Similarity

	1	2	3
Corresponding parts of similar figures	●	●	●
Identify similar figures	●	●	●
Scale drawings	●	●	●
Dilations	●	●	

Circles

	1	2	3
Circumference (radius, diameter)	●	●	●
Area	●	●	●

Perimeter	●	●	●

Area

	1	2	3
Rectangles	●	●	●
Parallelograms (base, height)	●	●	●
Trapezoids		●	●
Triangles	●	●	●
Circles	●	●	●
Square roots and area of squares		●	●
Pick's Theorem			●
Area and probability	●	●	

Transformations

	1	2	3
Translations, reflections, and rotations	●	●	●
Dilations		●	●
On the coordinate plane	●	●	●
Tessellations	●	●	●

Course	1	2	3
Symmetry	●	●	●

Solids

	1	2	3
Identify, draw three-dimensional figures	●	●	●
Nets	●	●	●
Surface area		●	●
Volume	●	●	●

Coordinate Geometry

	1	2	3
Graphing ordered pairs	●	●	●
Distance in the coordinate plane			●
Transformations on the coordinate plane	●	●	●

Patterns

	1	2	3
Recognizing geometric patterns	●	●	●
Symmetry	●	●	●
Tessellations	●	●	●
Fractals		●	

Trigonometry			●
Inductive and deductive thinking			●

MEASUREMENT

Metric System

	1	2	3
Units of length, capacity, and mass	●	●	●
Changing units within the metric system	●	●	●

Customary system

	1	2	3
Units of length, capacity, and weight	●	●	●
Change units within the customary system	●	●	●

Time	●		
Non-standard units	●		
Perimeter and circumference	●	●	●

Area

	1	2	3
Irregular figures	●	●	●
Rectangles	●	●	●
Parallelograms	●	●	●
Triangles	●	●	●
Circles	●	●	●
Trapezoids		●	●

Surface area

	1	2	3
Rectangular prisms		●	●
Triangular prisms		●	●
Cylinders		●	●

Volume

	1	2	3
Rectangular prisms	●	●	●
Cylinders		●	●
Pyramids and circular cones		●	●

Relating area and perimeter	●		
Relating surface area and volume			●
Precision and significant digits			●
Indirect measurement		●	●

Legend ● Introduce ● Develop ● Reinforce

Planning Your

The charts on these two pages contain course planning guides for three types of classes. A Pacing Chart included in the interleaf of each chapter shows in greater detail the number of days suggested for all three types of pacing and which lessons are appropriate. This same pacing is repeated in the margin notes for each lesson for your convenience.

40- TO 50-MINUTE CLASS PERIODS

The chart below gives suggested pacing guides for two options, Standard and Honors, for four 9-week grading periods. The Standard option covers Chapters 1-12 while the Honors option covers Chapter 1-13.

The total number of days suggested for each option is 165 days. This allows for teacher flexibility in planning due to school cancellation or shortened class periods.

Grading Period	Standard		Honors	
	Chapter	Days	Chapter	Days
1	1	12	1	12
	2	14	2	14
	3	12	3	10
	4-1 to 4-2	4	4-1 to 4-4	7
2	4-3 to end	12	4-5A to end	8
	5	16	5	15
	6	13	6	12
			7-1 to 7-5	6
3	7	15	7-6 to end	8
	8	14	8	13
	9	14	9	12
			10-1 to 10-4B	8
4	10	13	10-5 to end	5
	11	15	11	13
	12	11	12	11
			13	11

Course of Study

BLOCK SCHEDULE, 90-MINUTE CLASS PERIODS

The chart below gives a suggested pacing guide for teaching Course 1 using block scheduling in one semester (classes meet every day) or during the entire year (classes meet every other day). A total of 85 class periods is suggested to cover Chapters 1-12.

Chapter	Class Periods
1	6
2	7
3	6
4	8
5	8
6	7
7	8
8	7
9	7
10	7
11	8
12	6

For more detailed descriptions for lesson planning and pacing, please refer to the **Lesson Planning Guide** and the **Block Scheduling Booklet** for **Course 1** of *Mathematics: Applications and Connections.*

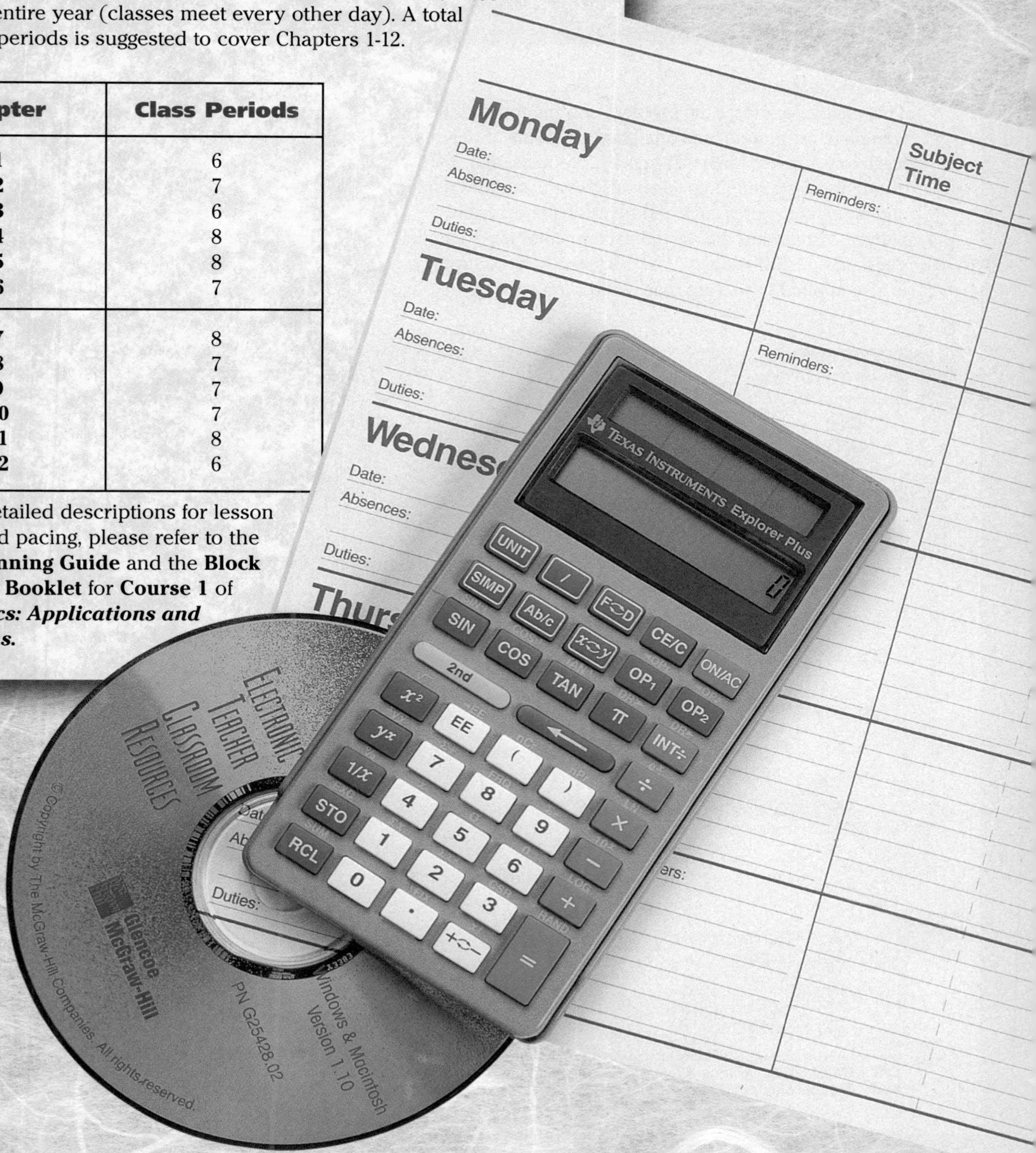

Problem Solving, Numbers, and Algebra

Previewing the Chapter

Overview

This chapter explores the problem-solving tools of a four-step plan, estimation, and algebraic equations. Students learn to estimate sums, differences, products, and quotients using rounding. They also evaluate numerical and algebraic expressions and investigate numerical patterns. Equations are solved using mental math and the guess-and-check strategy. The guess-and-check strategy is also applied to solving real-world problems.

Lesson (pages)	Lesson Objectives	NCTM Standards	Standardized Tests	State/Local Objectives
1-1 (4–7)	Solve problems using the four-step plan.	1–4, 7, 10, 13	MAT, SAT	
1-2 (8–11)	Solve problems using patterns.	1–5, 7, 8	MAT, SAT	
1-3 (12–15)	Estimate sums, differences, products, and quotients using rounding.	1–4, 7, 10, 13	CAT, CTBS, ITBS, SAT, TN	
1-4 (16–19)	Evaluate expressions using the order of operations.	1–4, 7, 9	CAT, CTBS, MAT, SAT, TN	
1-5A (20–21)	Model algebraic expressions.	1–5, 7, 9	CTBS, MAT, TN	
1-5 (22–25)	Evaluate numerical and simple algebraic expressions.	1–5, 7, 9	CAT, MAT, SAT	
1-5B (26)	Evaluate algebraic expressions by using a graphing calculator.	1–4, 7, 9	CAT, MAT, SAT	
1-6 (28–31)	Use powers and exponents in expressions.	1–10, 13		
1-7A (32–33)	Solve problems by using the guess-and-check strategy.	1–5, 7	MAT, SAT	
1-7 (34–37)	Solve equations by using mental math and guess and check.	1–7, 9–13		

CAT = California Achievement Tests, CTBS = Comprehensive Tests of Basic Skills, ITBS = Iowa Tests of Basic Skills,
MAT = Metropolitan Achievement Tests, SAT = Stanford Achievement Tests, TN = Terra Nova

CD-ROM

All of the blackline masters in the Teacher's Classroom Resources are available on the **Electronic Teacher's Classroom Resources** CD-ROM.

LESSON PLANNING GUIDE

BLACKLINE MASTERS (PAGE NUMBERS)

Lesson	Extra Practice (Student Edition)	Study Guide	Practice	Enrichment	Assessment & Evaluation	Classroom Games	Diversity	Hands-On Lab	School to Career	Science and Math Lab Manual	Technology	Transparencies A and B
1-1	p. 557	1	1	1								1-1
1-2	p. 557	2	2	2	15							1-2
1-3	p. 557	3	3	3								1-3
1-4	p. 558	4	4	4	14, 15						1	1-4
1-5A								39				
1-5	p. 558	5	5	5						1–4		1-5
1-5B												
1-6	p. 558	6	6	6	16	1–2					2	1-6
1-7A	p. 559											
1-7	p. 559	7	7	7	16		1	69	1			1-7
Study Guide/ Assessment					1–13, 17–19							

OTHER CHAPTER RESOURCES

Student Edition
Chapter Project, pp. 3, 15, 19, 41
School to Career, p. 27
Let the Games Begin, p. 31

Technology
CD-ROM Program
Interactive Mathematics Tools Software

Teacher's Classroom Resources

Applications
Family Letters and Activities, pp. 1–2
Investigations and Projects Masters, pp. 17–20
Meeting Individual Needs
Transition Booklet, pp. 11–22, 25–26
Investigations for the Special Education Student, pp. 1–2

Teaching Aids
Answer Key Masters
Block Scheduling Booklet
Lesson Planning Guide
Solutions Manual

Professional Publications
Glencoe Mathematics Professional Series

Planning the Chapter

MindJogger Videoquizzes
provide a unique format for reviewing concepts presented in the chapter.

ASSESSMENT RESOURCES

Student Edition
Mixed Review, pp. 11, 15, 19, 25, 30, 37
Mid-Chapter Self Test, p. 19
➡ Math Journal, pp. 6, 14
Study Guide and Assessment, pp. 38–41
Performance Task, p. 41
➥ Portfolio Suggestion, p. 41
Standardized Test Practice, pp. 42–43
Chapter Test, p. 595

Assessment and Evaluation Masters
Multiple-Choice Tests (Forms 1A, 1B, 1C), pp. 1–6
Free-Response Tests (Forms 2A, 2B, 2C), pp. 7–12
Performance Assessment, p. 13
Mid-Chapter Test, p. 14
Quizzes A–D, pp. 15–16
Standardized Test Practice, pp. 17–18
Cumulative Review, p. 19

Teacher's Wraparound Edition
5-Minute Check, pp. 4, 8, 12, 16, 22, 28, 34
➥ Building Portfolios, p. 2
➡ Math Journal, pp. 21, 26
Closing Activity, pp. 7, 11, 15, 19, 25, 31, 33, 37

Technology
Test and Review Software
MindJogger Videoquizzes
CD-ROM Program

MATERIALS AND MANIPULATIVES

Lesson 1-2
centimeter cubes*

Lesson 1-4
calculator

Lesson 1-5A
paper bags
popcorn

Lesson 1-5B
graphing calculator

Lesson 1-6
calculator

*Glencoe Manipulative Kit

PACING CHART

See pages T25–T27 for the Course Planning Calendar.

COURSE	DAY 1	DAY 2	DAY 3	DAY 4	DAY 5	DAY 6	DAY 7
Standard	Chapter Project	Lesson 1-1	Lesson 1-2	Lesson 1-3	Lesson 1-4	Lessons 1-5A & 1-5	
Honors	Chapter Project	Lesson 1-1	Lesson 1-2	Lesson 1-3	Lesson 1-4	Lessons 1-5 & 1-5B	
Block	Chapter Project & Lesson 1-1	Lessons 1-2 & 1-3	Lessons 1-4 & 1-5A	Lessons 1-5 & 1-6	Lessons 1-7A & 1-7	Study Guide and Assessment, Chapter Test	

The *Transition Booklet* (Skills 4–9, 11) can be used to practice rounding, adding, and subtracting whole numbers, and multiplying and dividing 1-digit numbers.

Interactive Mathematics:
Activities and Investigations

is an activity-based program that may be used as an enhancement for chapters in *Mathematics: Applications and Connections.*

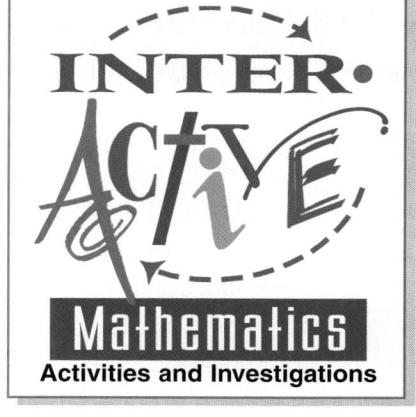

INTER·ACTIVE
Mathematics
Activities and Investigations

Unit 11, Activity One
Menu B
Use with Lesson 1-1.

Summary Students may work individually, in pairs, or in groups to convert a given amount of time into seconds, minutes, days, and weeks. After examination of the solutions by the class, students describe the methods they used to arrive at their solutions.

Math Connection Students examine time measurement units. After finding and interpreting information from the *Time Measurement Reference* in the Data Bank, they use logical reasoning and computation skills to solve conversion problems.

Unit 1, Activity Four
Make Up a Problem
Use with Lesson 1-4.

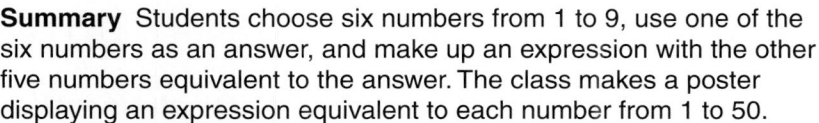

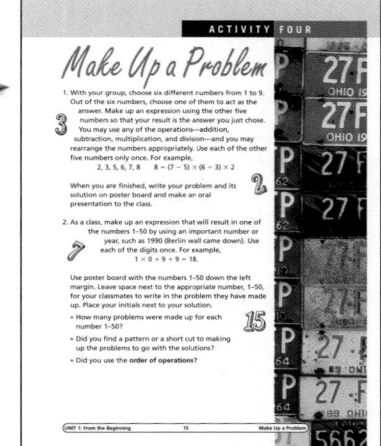

Summary Students choose six numbers from 1 to 9, use one of the six numbers as an answer, and make up an expression with the other five numbers equivalent to the answer. The class makes a poster displaying an expression equivalent to each number from 1 to 50.

Math Connection Students are asked to write an expression using addition, subtraction, multiplication, and division of whole numbers to represent a given solution. They use the order of operations during this activity.

DAY 8	DAY 9	DAY 10	DAY 11	DAY 12	DAY 13	DAY 14	DAY 15
Lesson 1-6	Lesson 1-7A	Lesson 1-7	Study Guide and Assessment	Chapter Test			
Lesson 1-6	Lesson 1-7A	Lesson 1-7	Study Guide and Assessment	Chapter Test			

Enhancing the Chapter

APPLICATIONS

Classroom Games, pp. 1–2

TEACHING SUGGESTIONS

Chapter 1 In-Class Game
(Lesson 1–6)

Double or Take Away

GET READY

Separate the students into pairs.
- Double or Take Away master, p. 2
- number cubes

GET SET

Make a copy of the Double or Take Away master on page 2 for each student in the class.

GO

- Player A rolls two number cubes and uses the numbers on the cubes to form a two-digit number. Then Player B can either double the number, subtract any square number, or subtract any cube number. Player A then either doubles the number, subtracts any square number, or subtracts any cube number. Play continues until someone reaches 0. A sample game is shown below if Player A rolls 6 and 3.

A	B
The number is 63.	Subtract 7^2 or 49. $63 - 49 = 14$
Double 14. 28	Subtract 5^2 or 25. $28 - 25 = 3$
Double 6. 6	Double 6. 12
Subtract 2^2 or 8. $12 - 8 = 4$	Subtract 2^2 or 4. $4 - 4 = 0$

- Player B is the winner of this round since he or she reached 0 first.
- Players alternate turns and receive 5 points when they win a round.

Variations:
- Have students find, for starting numbers 1 to 10, which player could win on the first move.
- Have students find, for starting numbers 1 to 100, which could give (a) a win on the first move, and (b) a win after two moves.

© Glencoe/McGraw-Hill   1   *Mathematics: Applications and Connections, Course 1*

Diversity Masters, p. 1

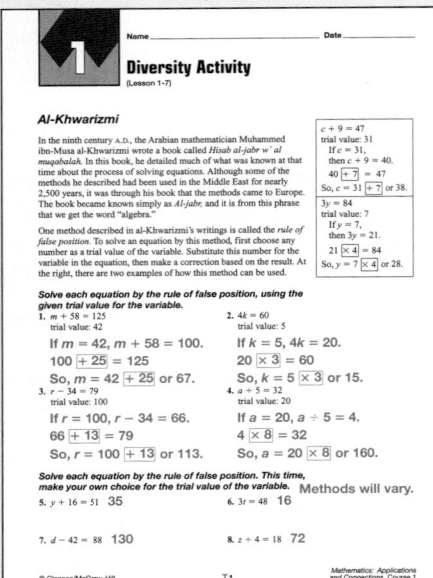

Name_____ Date_____

Diversity Activity
(Lesson 1–7)

Al-Khwarizmi

In the ninth century A.D., the Arabian mathematician Muhammed ibn-Musa al-Khwarizmi wrote a book called *Hisab al-jabr w' al muqabalah.* In this book, he detailed much of what was known at that time about the process of solving equations. Although some of the methods he described had been used in the Middle East for nearly 2,500 years, it was through his book that the methods came to Europe. The book became known simply as *Al-jabr,* and it is from this phrase that we get the word "algebra."

One method described in al-Khwarizmi's writings is called the *rule of false position.* To solve an equation by this method, first choose any number as a trial value of the variable. Substitute this number for the variable in the equation, then make a correction based on the result. At the right, there are two examples of how this method can be used.

$c + 9 = 47$
trial value: 31
If $c = 31$,
then $c + 9 = 40$.
$40 \boxed{+ 7} = 47$
So, $c = 31 \boxed{+ 7}$ or 38.

$3y = 84$
trial value: 7
If $y = 7$,
then $3y = 21$.
$21 \boxed{\times 4} = 84$
So, $y = 7 \boxed{\times 4}$ or 28.

Solve each equation by the rule of false position, using the given trial value for the variable.

1. $m + 58 = 125$
   trial value: 42
   If $m = 42$, $m + 58 = 100$.
   $100 \boxed{+ 25} = 125$
   So, $m = 42 \boxed{+ 25}$ or 67.

2. $4k = 60$
   trial value: 5
   If $k = 5$, $4k = 20$.
   $20 \boxed{\times 3} = 60$
   So, $k = 5 \boxed{\times 3}$ or 15.

3. $r - 34 = 79$
   trial value: 100
   If $r = 100$, $r - 34 = 66$.
   $66 \boxed{+ 13} = 79$
   So, $r = 100 \boxed{+ 13}$ or 113.

4. $a + 5 = 32$
   trial value: 20
   If $a = 20$, $a \div 5 = 4$.
   $4 \boxed{\times 8} = 32$
   So, $a = 20 \boxed{\times 8}$ or 160.

Solve each equation by the rule of false position. This time, make your own choice for the trial value of the variable. Methods will vary.

5. $y + 16 = 51$ 35

6. $3t = 48$ 16

7. $d - 42 = 88$ 130

8. $z + 4 = 18$ 72

© Glencoe/McGraw-Hill   T1   *Mathematics: Applications and Connections, Course 1*

School to Career Masters, p. 1

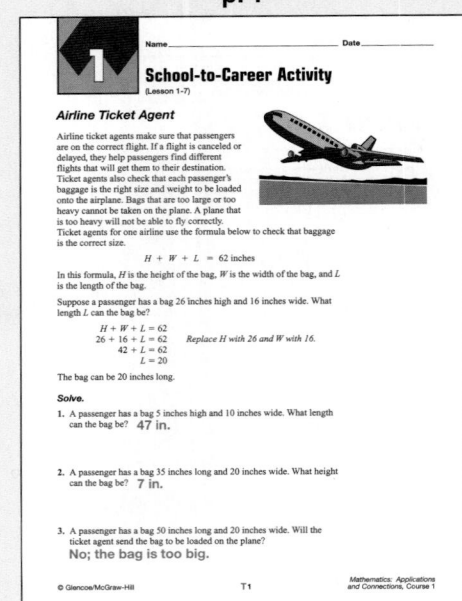

Name_____ Date_____

School-to-Career Activity
(Lesson 1–7)

Airline Ticket Agent

Airline ticket agents make sure that passengers are on the correct flight. If a flight is canceled or delayed, they help passengers find different flights that will get them to their destination. Ticket agents also check that each passenger's baggage is the right size and weight to be loaded onto the airplane. Bags that are too large or too heavy cannot be taken on the plane. A plane that is too heavy will not be able to fly correctly. Ticket agents for one airline use the formula below to check that baggage is the correct size.

$$H + W + L = 62 \text{ inches}$$

In this formula, H is the height of the bag, W is the width of the bag, and L is the length of the bag.

Suppose a passenger has a bag 26 inches high and 16 inches wide. What length L can the bag be?

$H + W + L = 62$
$26 + 16 + L = 62$ Replace H with 26 and W with 16.
$42 + L = 62$
$L = 20$

The bag can be 20 inches long.

Solve.

1. A passenger has a bag 5 inches high and 10 inches wide. What length can the bag be?   47 in.

2. A passenger has a bag 35 inches long and 20 inches wide. What height can the bag be?   7 in.

3. A passenger has a bag 50 inches long and 20 inches wide. Will the ticket agent send the bag to be loaded on the plane?
   No; the bag is too big.

© Glencoe/McGraw-Hill   T1   *Mathematics: Applications and Connections, Course 1*

Family Letters and Activities, pp. 1–2

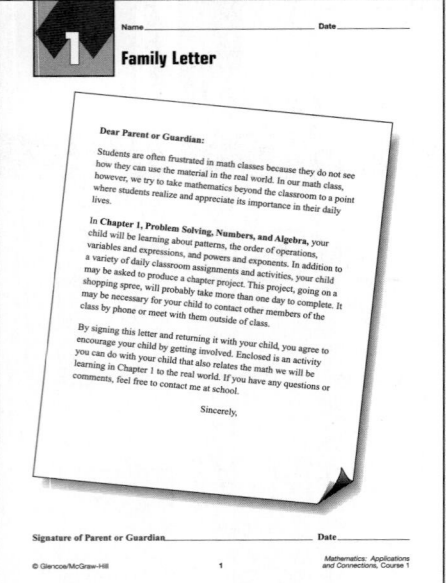

Name_____ Date_____

Family Letter

Dear Parent or Guardian:

Students are often frustrated in math classes because they do not see how they can use the material in the real world. In our math class, however, we try to take mathematics beyond the classroom to a point where students realize and appreciate its importance in their daily lives.

In **Chapter 1, Problem Solving, Numbers, and Algebra,** your child will be learning about patterns, the order of operations, variables and expressions, and powers and exponents. In addition to a variety of daily classroom assignments and activities, your child may be asked to produce a chapter project. This project, going on a shopping spree, will probably take more than one day to complete. It may be necessary for your child to contact other members of the class by phone or meet with them outside of class.

By signing this letter and returning it with your child, you agree to encourage your child by getting involved. Enclosed is an activity you can do with your child that also relates the math we will be learning in Chapter 1 to the real world. If you have any questions or comments, feel free to contact me at school.

Sincerely,

Signature of Parent or Guardian_____ Date_____

© Glencoe/McGraw-Hill   1   *Mathematics: Applications and Connections, Course 1*

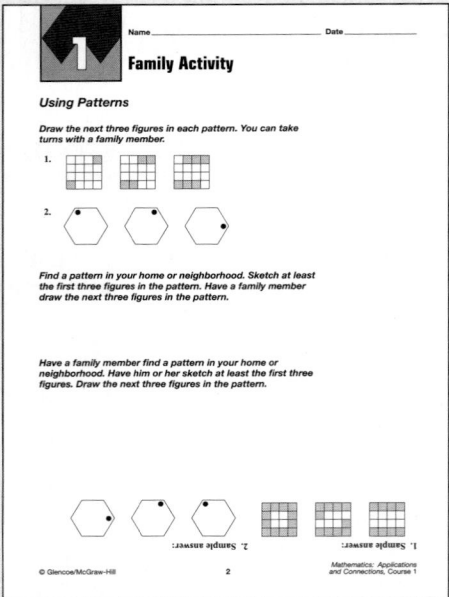

Name_____ Date_____

Family Activity

Using Patterns

Draw the next three figures in each pattern. You can take turns with a family member.

1.

2.

Find a pattern in your home or neighborhood. Sketch at least the first three figures in the pattern. Have a family member draw the next three figures in the pattern.

Have a family member find a pattern in your home or neighborhood. Have him or her sketch at least the first three figures. Draw the next three figures in the pattern.

1. Sample answer: 2. Sample answer:

© Glencoe/McGraw-Hill   2   *Mathematics: Applications and Connections, Course 1*

Science and Math Lab Manual, pp. 1–4

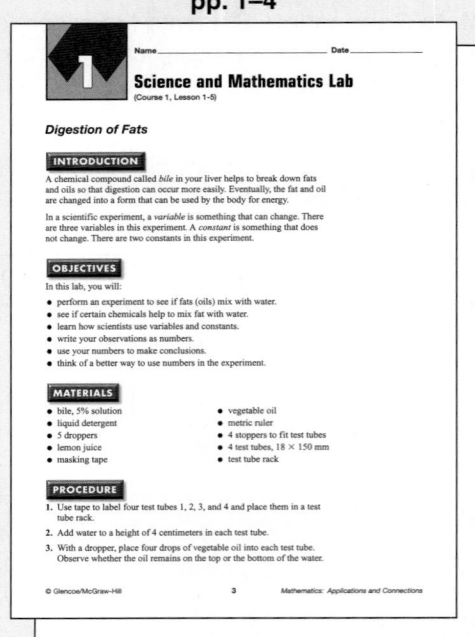

Name_____ Date_____

Science and Mathematics Lab
(Course 1, Lesson 1–5)

Digestion of Fats

INTRODUCTION

A chemical compound called *bile* in your liver helps to break down fats and oils so that digestion can occur more easily. Eventually, the fat and oil are changed into a form that can be used by the body for energy.

In a scientific experiment, a *variable* is something that can change. There are three variables in this experiment. A *constant* is something that does not change. There are two constants in this experiment.

OBJECTIVES

In this lab, you will:
- perform an experiment to see if fats (oils) mix with water.
- see if certain chemicals help to mix fat with water.
- learn how scientists use variables and constants.
- write your observations as numbers.
- use your numbers to make conclusions.
- think of a better way to use numbers in the experiment.

MATERIALS

- bile, 5% solution
- liquid detergent
- 5 droppers
- lemon juice
- masking tape
- vegetable oil
- metric ruler
- 4 stoppers to fit test tubes
- 4 test tubes, 18 × 150 mm
- test tube rack

PROCEDURE

1. Use tape to label four test tubes 1, 2, 3, and 4 and place them in a test tube rack.
2. Add water to a height of 4 centimeters in each test tube.
3. With a dropper, place four drops of vegetable oil into each test tube. Observe whether the oil remains on the top or the bottom of the water.

© Glencoe/McGraw-Hill   3   *Mathematics: Applications and Connections*

MANIPULATIVES/MODELING

Hands-On Lab Masters, p. 69

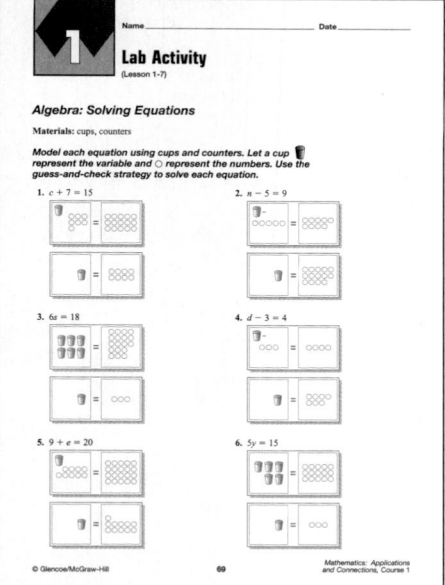

1

Name _____ Date _____

Lab Activity
(Lesson 1-7)

Algebra: Solving Equations

Materials: cups, counters

Model each equation using cups and counters. Let a cup represent the variable and ○ represent the numbers. Use the guess-and-check strategy to solve each equation.

1. $c + 7 = 15$
2. $n - 5 = 9$
3. $6a = 18$
4. $d - 3 = 4$
5. $9 + e = 20$
6. $5y = 15$

© Glencoe/McGraw-Hill — 69 — Mathematics: Applications and Connections, Course 1

ASSESSMENT/EVALUATION

Assessment and Evaluation Masters, pp. 14–16

1

Name _____ Date _____

Chapter 1 Mid-Chapter Test
(Lessons 1-1 through 1-4)

Use the four-step plan to solve each problem. Use a calculator if necessary.

1. Rayna studied 315 vocabulary words in three weeks for her French class. If Rayna studied the same number of words each day, how many vocabulary words did she study a day? **1. 15**

2. A computer diskette holds 1,440K of memory. If four programs are on the diskette and they use 32K, 54K, 68K, and 72K of memory, how much memory is left on the diskette? **2. 1,214K**

3. The Hartmans want to buy a 31-inch television that costs $750. They plan to make a down payment of $275 and pay the rest in five equal payments. What will be the amount of each payment? **3. $95**

Find the next three numbers in each pattern.

4. 6, 7, 9, 12, _?_, _?_, _?_ **4. 16, 21, 27**
5. 7, 8, 7, 8, _?_, _?_, _?_ **5. 7, 8, 7**
6. 8, 16, 32, _?_, _?_, _?_ **6. 64, 128, 256**
7. 35, 28, 21, _?_, _?_, _?_ **7. 14, 7, 0**

Estimate.

8. $721 - 53$ **8. $700 - 50 = 650$**
9. $905 + 273$ **9. $900 + 300 = 1,200$**
10. 27×9 **10. $30 \times 10 = 300$**
11. 95×24 **11. $100 \times 20 = 2000$**
12. $520 \div 5$ **12. $500 \div 5 = 100$**
13. $776 \div 42$ **13. $800 \div 40 = 20$**
14. $1,729 + 5,681$ **14. $2,000 + 6,000 = 8,000$**
15. $4,921 - 2,602$ **15. $5,000 - 3,000 = 2,000$**

Find the value of each expression.

16. $24 - 5 + 2 \times 8$ **16. 35**
17. $81 + 9 + 2 \times 4$ **17. 17**
18. $15 + 5 - 3$ **18. 0**
19. $6 \times 5 + 8 \times 7$ **19. 86**
20. $37 - 14 + 2 - 3 \times 7$ **20. 9**

© Glencoe/McGraw-Hill — 14 — Mathematics: Applications and Connections, Course 1

1

Name _____ Date _____

Chapter 1 Quiz A
(Lessons 1-1 and 1-2)

Use the four-step plan to solve each problem. Use a calculator if necessary.

1. On a trip to Florida, the Juarez family bought 4 adult plane tickets costing a total of $920. What was the cost of each ticket? **1. $230**

2. Jill saved $12 per week for 50 weeks. How much did she save in all? **2. $600**

Find the next three numbers in each pattern.

3. 222, 202, 182, _?_, _?_, _?_ **3. 162, 142, 122**
4. 3, 12, 48, _?_, _?_, _?_ **4. 192, 768, 3,072**

5. At the class play, 135 adult tickets were sold for $4 each, and 236 student tickets were sold for $2 each. How many more student tickets were sold than adult tickets? **5. 101**

Chapter 1 Quiz B
(Lessons 1-3 and 1-4)

Name _____ Date _____

Round each number to the underlined place-value position.

1. 7,499 **1. 7,000**
2. 495 **2. 500**

Estimate using rounding.

3. $865 + 276$ **3. $900 + 300 = 1,200$**
4. $78 - 56$ **4. $80 - 60 = 20$**
5. 994×27 **5. $1,000 \times 30 = 30,000$**
6. $4,106 \div 1,824$ **6. $4,000 \div 2,000 = 2$**

Find the value of each expression.

7. $5 + 7 \times 9$ **7. 68**
8. $36 - 27 \div 3 + 1$ **8. 28**
9. $6 \times 7 - 7 \times 3$ **9. 21**
10. $10 \div 2 + 15 \div 3$ **10. 10**

© Glencoe/McGraw-Hill — 15 — Mathematics: Applications and Connections, Course 1

TECHNOLOGY/MULTIMEDIA

Technology Masters, pp. 1–2

1

Name _____ Date _____

Calculator Activity
(Lesson 1-4)

Order of Operations

Scientific calculators follow the correct order of operations.

Examples 1 Evaluate $18 + 3 \times 2$.

Enter: 18 [+] 3 [×] 2 [=]

If the display is 24, then the calculator follows the correct order of operations. If the display is 42, use parentheses to show the operation to be performed first as in the following example.

2 Evaluate $18 + (3 \times 2)$.

Enter: 18 [+] [(] 3 [×] 2 [)] [=] 24

The correct answer is 24.

Use a calculator to find the value of each expression. Use parentheses if necessary.

1. $112 - 4 \times 23$ **20**
2. $15 + 12 \div 2$ **21**
3. $14 + 28 \div 7$ **18**
4. $25 + 10 - 5 \div 5$ **34**
5. $200 - 10 \times 10 \times 2$ **0**
6. $12 \div 4 \times 2 + 8$ **14**
7. $20 \div 2 \times 10$ **100**
8. $114 + 10 - 9 \times 9$ **43**
9. $28 + 42 \div 7 \div 2$ **31**
10. $125 - 100 - 25 \div 5$ **20**

CHALLENGE

11. $24 + 4 \times 2 + 8 + 4 + 4$ **18**
12. $75 + 5 \times 5 \div 10 \times 10$ **100**

© Glencoe/McGraw-Hill — T1 — Mathematics: Applications and Connections, Course 1

1

Name _____ Date _____

Graphing Calculator Activity
(Lesson 1-6)

Exponents

You can use a graphing calculator to evaluate expressions involving exponents.

Example 1 Evaluate $5^4 + 6^3$.

Enter: 5 [^] 4 [+] 6 [x²] [ENTER] 661

So, $5^4 + 6^3 = 661$.

You can also use a graphing calculator to evaluate algebraic expressions that involve exponents. You store the variables' values in the memory before evaluating all of the expressions.

Example 2 Evaluate $x^3 + y^4$ if $x = 2$ and $y = 5$.

Enter: 2 [STO▸] [ALPHA] X [ENTER]
5 [STO▸] [ALPHA] Y [ENTER]
[ALPHA] X [^] 3 [+] [ALPHA] Y [^] 4 [ENTER] 633

So, $x^3 + y^4 = 633$ when $x = 2$ and $y = 5$.

Evaluate each expression.

1. $6 + 3^4 + 1$ **88**
2. $4^7 + 10^7$ **10,001,024**
3. $7^3 - 1^9 - 2^8$ **278**
4. $4 + 3^5 - 2$ **240**
5. $27 - 5^2 + 7^3$ **51**
6. $6^5 - 5^3 + 4^4$ **43,787**
7. $9^4 + 16 + 2^5 - 24 + 3$ **6,557**
8. $10^3 - 0^7 + 8^3 - 7^4 + 6^2$ **5,816**

Evaluate each expression if $x = 6$ and $y = 3$.

9. x^4 **1,296**
10. $y^3 + y$ **245**
11. $x^3 + 5^3$ **341**
12. $x^3 + y^8$ **6,777**
13. $3x^2 + y^3$ **189**
14. $x^3 - 2y^4 - 126$ **1,259,586**
15. xy^3 **39,366**
16. $5y^3 + 5x^3$ **40,095**
17. $x^{10} - 10xy$ **60,465,996**
18. $4xy + 2y^3 + xy - x^4 + y^{10}$ **57,897**
19. $xy^3 + 5y + yx^2 + 5x$ **24,831**

© Glencoe/McGraw-Hill — T2 — Mathematics: Applications and Connections, Course 1

MEETING INDIVIDUAL NEEDS

Investigations for the Special Education Student, pp. 1–2

Use with:
Course 1-Chapter 1
Course 2-Chapter 1
Course 3-Chapter 2

Investigation 1 Teacher's Guide

Vacation Getaways

Overview

This investigation shows how mathematics is connected to history and social studies. Not only will students improve their skills with decimal operations and map reading, but will they learn about a country of their choice—its history, geography, culture, and interesting facts. In order to accomplish this task, students will research and devise a detailed itinerary for a trip to their country, including a daily schedule and costs for each activity planned. Once completed, trip itineraries will be shared with or displayed for the class, and each student will be asked to locate his or her country on a world map.

Activity Goals

Students will:
• research a country of their choice,
• organize and plan a trip to that country, and
• make a budget for the trip.

Planning the Instruction

Prerequisite Skills

Students should have a significant amount of practice computing with whole numbers and decimals, figuring distances on a map, understanding time zone differences, and possibly converting from one currency to another.

Materials

• investigation worksheet
• calculators
• access to research materials

Time Needed

six 45-minute periods

Procedure

1. Explain that students' task is to plan a trip to a country of their choice. The trip should be for a minimum of 5 days, and the cost should not exceed $5,000. While on the trip, they should visit at least one historical, one scenic, and one entertaining point of interest.

2. Requirements for this task may include: contacting travel agents, auto clubs, airlines, or hotels to obtain current rates; consulting the library for information on the geography and history of the area to be visited; or gathering information about the area from relatives who have been there.

3. Have students complete the Preliminary Plan worksheet by estimating answers based on prior knowledge of the country.

4. Once the Preliminary Plan is done, students will begin the research necessary to find accurate, current responses to the preliminary plan questions.

5. To complete their final itinerary, students will use the information they have gathered to write a detailed schedule for each individual day of the trip, including the cost of each event.

6. If desired, have students type their final itineraries and share them with the class. Then, display each itinerary in the classroom and have students mark their country on a large world map.

Adaptations and Variations	

The following are some ideas on how this investigation may be modified depending on student population.

LD • Allow more time.
• If there seems to be a difficulty with students completing their research, allow students to have "research assistants" such as an older sibling or friend outside class to help with research.
• Allow students to type their responses and itinerary.

PH • Help these students obtain research materials if access is difficult due to lack of mobility.
• If writing is a problem, allow students to type their work.

CD • Arrange for someone to help students with the verbal portion of their research.
• If students are to share a summary of their trip orally, have them instead write a summary for the teacher or another student to read aloud.

BD • Maintain close proximity during independent work time.

HI • Have a student sign each oral summary.

VI • Help students obtain enlarged print material. In addition, students may need an aide to help them with the visual portion of their research.

© Glencoe/McGraw-Hill — 1 — Mathematics: Applications and Connections

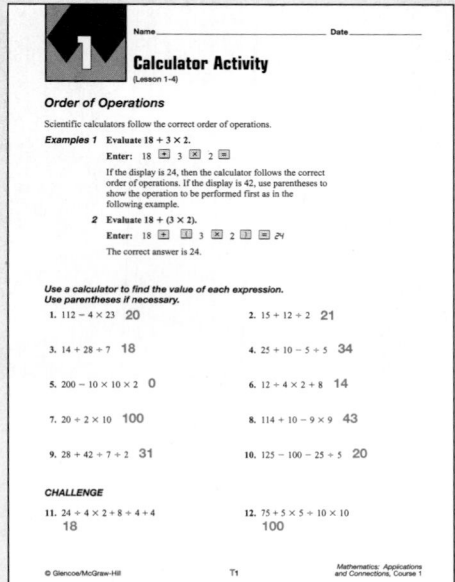

Theme: Shopping

Many people enjoy the convenience of shopping by catalog from the comfort of their homes. Shopping by mail-order catalog generates almost $50 billion in sales annually.

The Internet is the newest rage in catalog shopping. From zero sales in 1994 to about $560 million in 1996, shopping on the Internet continues to grow.

Question of the Day A retailer decides to base shipping and handling charges on the volume of an item. The volume of an item is r^3. The charge is $3 times the volume. Find the cost if $r = 2$. **$24**

Assess Prerequisite Skills

Ask students to read through the list of objectives presented in "What you'll learn in Chapter 1." You may wish to ask them what each of the objectives means or if they have experienced or used any of these math concepts before.

Building Portfolios

A portfolio is each student's collection of work that expresses how the student has grown as a mathematics student. It does not necessarily include the best work done by the student, but rather should demonstrate improvement. You may want to have each student attach an explanation to each paper in the portfolio to state why it was chosen to be in the collection. Students should update their portfolios with the completion of each chapter.

Math and the Family

In the *Family Letters and Activities* booklet (pp. 1–2), you will find a letter to the parents explaining what students will study in Chapter 1. An activity appropriate for the whole family is also available.

CHAPTER 1
Problem Solving, Numbers, and Algebra

What you'll learn in Chapter 1

- to solve real-world problems using the four-step plan,
- to estimate sums, differences, products, and quotients using rounding,
- to evaluate numerical and algebraic expressions, and
- to solve equations using mental math and the guess-and-check strategy.

OFFICIAL BASKETBALL FOR NCAA® CHAMPIONSHIPS

2 Chapter 1 Problem Solving, Numbers, and Algebra

 CD-ROM Program

Activities for Chapter 1
- Chapter 1 Introduction
- Interactive Lessons 1-3, 1-5
- Assessment Game
- Resource Lessons 1-1 through 1-7

CHAPTER Project

SHOP 'TIL YOU DROP!

In this project, you will use problem-solving strategies, estimation, and algebraic expressions to go on a shopping spree. You can use any catalog you wish for your shopping spree such as a sporting goods catalog or a novelty gift catalog.

Getting Started

- Look at the two shipping tables. Describe the method used by each catalog to determine the shipping charge.
- For each table, find the shipping charge for an item that costs $20 and weighs less than 5 pounds.

Table 1: Sporting Goods Catalog

Shipping Charges	
For the first item	$5.99
Each additional item	$1.99
Rush Orders	
Next day air	$21.99*
Express delivery (2–3 days)	$9.99*
*Plus standard shipping charges	

Table 2: Novelty Gift Catalog

Merchandise Total	Shipping Charges
up to $20.00	$4.95
$20.01 to $30.00	$5.95
$30.01 to $40.00	$6.95
$40.01 to $50.00	$7.95
$50.01 to $75.00	$8.95
$75.01 to $100.00	$9.95
$100.01 to $150.00	$12.95
$150.01 to $200.00	$15.95
1-Day Delivery	$9.99*
3-Day Delivery	$6.99*
*Plus standard shipping charges	

Technology Tips

- Use a **spreadsheet** to calculate the cost of the items and their shipping charges.
- Surf the **Internet** to see if any catalogs have websites.

interNET CONNECTION For up-to-date information on catalogs, visit:
www.glencoe.com/sec/math/mac/mathnet

Working on the Project

You can use what you'll learn in Chapter 1 to help you go on a shopping spree.

Page	Exercise
15	45
19	23
41	Alternative Assessment

Chapter 1 **3**

Instructional Resources ▶ ▶ ▶
A recording sheet to help students organize their data for the Chapter Project is shown at the right and is available in the *Investigations and Projects Masters*, p. 20.

CHAPTER Project
N O T E S

Objectives Students should
- learn to apply problem-solving strategies to real-life situations, such as catalog shopping.
- gain an understanding of estimation and algebraic expressions by applying them to catalog orders.

Project Pointer You may suggest that students begin a *Project Folder* to keep their work as they complete each stage of the Chapter Project. The completed project may also be added to their portfolios.

Using the Tables For Table 1, make sure students realize that the $1.99 is for *each* additional item. For example, if you buy six items, the shipping would be $5.99 + 5($1.99) = $15.94. For Table 2, emphasize that the shipping charges increase as the amount of money spent increases. It doesn't matter whether it's one item or twenty items.

Investigations and Projects Masters, p. 20

1 Name_____ Date_____
Chapter 1 Project

Shop 'Til You Drop!
Page 3, Getting Started

Page 15, Working on the Chapter Project, Exercise 45

Order Table	
Item Description	Price
Shipping Charges	
Total Charges	
Item Total	
Shipping Total	
Order Total	

Page 19, Working on the Chapter Project, Exercise 23
$(3 \times \$ \text{ Item } I) + (4 \times \$ \text{ Item } 2) + \$ \text{ Shipping}$
$(\times \$ \quad) + (\times \$ \quad) + \$ \quad = \$ \_\_\_\_$

© Glencoe/McGraw-Hill   20   Mathematics: Applications and Connections, Course 1

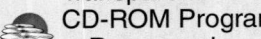

- *Study Guide Masters*, p. 1
- *Practice Masters*, p. 1
- *Enrichment Masters*, p. 1
- Transparencies 1-1, A and B

CD-ROM Program
- Resource Lesson 1-1

Recommended Pacing	
Standard	Day 2 of 12
Honors	Day 2 of 12
Block	Day 1 of 6

1 FOCUS

5-Minute Check

Find the value of each expression.
1. $17 + 5$ **22**
2. $65 + 12 + 37$ **114**
3. $86 - 49$ **37**
4. 6×52 **312**
5. $72 \div 4$ **18**

The 5-Minute Check is also available on **Transparency 1-1A** for this lesson.

Motivating the Lesson

Hands-On Activity Plan the solution to a story problem by writing each of the four steps on a separate sheet of paper. Mix up the sheets and ask students to put them in the correct order. Cover one of the sheets with a blank sheet and discuss how the exclusion of this step might affect other steps in the solution. Discuss each step in this way.

What you'll learn
You'll learn to solve problems using the four-step plan.

When am I ever going to use this?
The four-step plan gives you an organized method for solving problems.

The graph shows the number of years it took from the time a new technology was invented for it to be in one-half of U.S. homes. In what year was the telephone in one-half of U.S. homes?
This problem will be solved in Example 1.

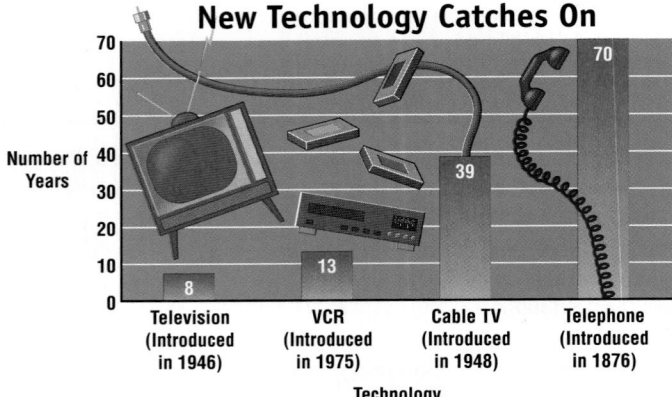

New Technology Catches On

Television (Introduced in 1946)	VCR (Introduced in 1975)	Cable TV (Introduced in 1948)	Telephone (Introduced in 1876)
8	13	39	70

Number of Years (vertical axis: 0 to 70)
Technology (horizontal axis)

You can use a four-step plan to help solve this problem and other problems.

1. *Explore*
   - Read the problem carefully.
   - Ask yourself questions like, "What facts do I know?" and "What do I need to find out?"

2. *Plan*
   - See how the facts relate to each other.
   - Make a plan for solving the problem.
   - Estimate the answer.

3. *Solve*
   - Use your plan to solve the problem.
   - If your plan does not work, revise it or make a new plan.

4. *Examine*
   - Reread the problem.
   - Ask, "Is my answer close to my estimate?"
   - Ask, "Does my answer make sense for the problem?"
   - If not, solve the problem another way.

4   **Chapter 1** Problem Solving, Numbers, and Algebra

✳ Cross-Curriculum Cue

Inform other teachers on your team that your students are studying problem-solving strategies. Suggestions for curriculum integration are:
Economics: consumer issues
Health: weekly caloric needs
Physical Science: formulas

Inventions Refer to the beginning of the lesson. In what year was the telephone in one-half of U.S. homes?

Explore The graph states that the telephone was invented in 1876 and that it took 70 years until one-half of U.S. homes had a telephone. You need to find the year when one-half of U.S. homes had a telephone.

Plan To solve the problem, add the number of years it took until one-half of U.S. homes had a telephone to the year the telephone was invented.

Solve

year telephone was invented	plus	number of years until one-half of U.S. homes had a telephone
1876	+	70

$$1876 + 70 = 1946$$

By 1946, one-half of U.S. homes had a telephone.

Examine $1946 - 70 = 1876$. So the answer is correct.

> **Did you know?** According to a survey, more than half of U.S. homes today have three or more telephones.

Example ② **APPLICATION**

Work Tyler delivers newspapers each weekday. If Tyler works every weekday for one year, how many days does he work?

Explore You know that Tyler delivers newspapers each weekday. You need to find out how many days Tyler works in one year.

Plan Multiply the number of weekdays in each week, 5, by the number of weeks in one year, 52, to find the number of days Tyler works in one year.

Solve

number of weekdays in each week	times	number of weeks in one year
5	×	52

In one year, Tyler works 5×52 or 260 days.

(continued on the next page)

Lesson 1-1 A Plan for Problem Solving **5**

2 TEACH

Transparency 1-1B contains a teaching aid for this lesson.

Thinking Algebraically When students are solving the exercises in this lesson, have them write out the problems as equations to prepare them for algebraic expressions.

In-Class Examples

For Example 1
Refer to the graph at the beginning of the lesson. How many years are there between the time when one-half of all U.S. homes had cable TV and the time when one-half owned VCRs? **one year**

For Example 2
To help out an injured news carrier, Tyler delivered newspapers on weekends for one-half of a year. Including his weekday deliveries, how many days did he work that year? **312 days**

Teaching Tip In Examples 1 and 2, remind students to work carefully through the four-step process, even if the answer seems obvious.

Check for Understanding

If students need additional practice or instruction after completing Exercises 1–5, one of these options may be helpful.
- Extra Practice, see p. 557
- Reteaching Activity
- *Study Guide Masters,* p. 1
- *Practice Masters,* p. 1

Assignment Guide

Core: 7–13 odd, 15
Enriched: 6–14 even, 15

Additional Answers

1. Explore—Decide what facts you know and what you need to find out. Plan—Make a plan for solving the problem and estimate the answer. Solve—Solve the problem. Make a new plan if necessary. Examine—Check to see if your answer makes sense.

2. If your answer does not make sense, you will need to solve the problem another way.

Study Guide Masters, p. 1

Name _____ Date _____

Study Guide

A Plan for Problem Solving

A one-year subscription to a popular magazine costs $24. Ted is thinking of sharing the cost of the subscription equally with two friends. How much will each have to pay?

Explore   What do you know?
          A one-year subscription costs $24.
          Ted will share the cost equally with two friends.
          What are you trying to find?
          How much each friend will have to pay.

Plan   Since Ted will share with two friends, there are 1 + 2 or 3
          people sharing the cost. Find the number of times 3 goes
          into $24.

Solve $24 ÷ 3 = $8   Each friend will pay $8 toward the subscription.

Examine   The answer makes sense. 3 × $8 = $24

Use the four-step plan to solve each problem.

1. A record weight for a red snapper was 46 pounds. A record weight for a sturgeon was 468 pounds. About how many times heavier than the red snapper was the sturgeon?

   about 10 times

2. The distance from Chicago to Cleveland is 344 miles. Lisa plans to drive this distance in two days. If she drives 180 miles today, how many miles will she have to drive tomorrow?

   164 miles

3. On a certain day, 525 people signed up to play softball. If 15 players are assigned to each team, how many teams can be formed?

   35 teams

4. The Farrells want to buy a VCR that costs $460. They plan to make a down payment of $100 and pay the rest in eight equal payments. What will be the amount of each payment?

   $45

5. Josita received $50 as a gift. She plans to buy two cassette tapes that cost $9 each and a deluxe headphone set that costs $25. How much money will she have left?

   $7

6. Tran wants to buy a bicycle that costs $192. He plans to save $6 each week. How many weeks will it take him to save enough money?

   32 weeks

© Glencoe/McGraw-Hill   T1   Mathematics: Applications and Connections, Course 1

Examine There are 2 weekend days each week, Saturday and Sunday. So, in one year, there are 2 × 52 or 104 weekend days. Subtract 104 from 365 to find the number of weekdays. The result is 261. The answer is reasonable. *Why don't the number of weekend days and weekdays add up to 365?*

CHECK FOR UNDERSTANDING

Communicating Mathematics

Read and study the lesson to answer each question.

1. *Tell* how each step of the four-step plan for problem solving helps you solve a problem. **See margin.**

2. *Explain* why you should check to see whether your answer makes sense for the problem. **See margin.**

3. *Write a Problem* with an answer of $500 that uses addition. Ask a classmate to solve your problem using the four-step plan. **See students' work.**

Guided Practice

Use the four-step plan to solve each problem.

4. *School* Mika studied 224 vocabulary words in two weeks for his French class. If Mika studied the same number of words each day, how many vocabulary words did he study per day? **16 words**

5. *Computers* In 1993, about 5 million computers had CD-ROM drives. In 1997, there were three times as many computers with CD-ROM drives than in 1993. How many computers had CD-ROM drives in 1997? **about 15 million**

EXERCISES

Applications and Problem Solving

Use the four-step plan to solve each problem.

6. *Computers* A computer disk holds 720K of memory. If three programs are on the disk and they use 27K, 34K, and 52K of memory, how much memory is left on the disk? **607K**

7. *Geography* On a map of Tennessee, each inch represents approximately 18 miles. Travis is planning to travel from Nashville to Knoxville. If the distance on the map from Nashville to Knoxville is about 10 inches, approximately how far will he travel? **180 miles**

8. *Earth Science* On the first day of summer, Barrow, Alaska, has 24 hours of daylight, and Honolulu, Hawaii, has 13 hours and 26 minutes of daylight. How much more daylight does Barrow have than Honolulu? **10 h 34 min**

9. *Money Matters* The Anstines want to buy a 27-inch television that costs $390. They plan to make a down payment of $150 and pay the rest in six equal payments. What will be the amount of each payment? **$40**

■ Reteaching the Lesson ■

Activity Separate the class into groups. Provide each group with a store flyer that lists prices for items the store sells. Tell each group to choose any combination of 3 items from the list and to get as close to $25 as possible.

10. **Postal Service** Refer to the graph.
   a. How many more cards are sent for Valentine's Day than Father's Day? **824 million**
   b. How many more cards are sent for Valentine's Day than Mother's Day? **770 million**

Holiday Mail
Average number of cards and letters mailed:

Valentine's Day 925 million
Mother's Day 155 million
Father's Day 101 million

Source: U.S Postal Service, Greeting Card Association

11. **Life Science** To estimate the temperature in degrees Fahrenheit, you can count the number of times a cricket chirps in 15 seconds and then add that number to 40. What is the temperature if a cricket chirps 25 times in 15 seconds? **65°F**

12. **Money Matters** The student council at West Boulevard Middle School is planning their annual winter dance. Tickets to the dance cost $3 each. There are 345 students in the school. If each student buys one ticket, how much money would the student council collect? **$1,035**

13. 84,480 pennies

13. **Measurement** Mrs. Hall challenged her class to determine the number of pennies it takes to reach one mile. If there are 16 pennies in one foot, how many pennies are in one mile? (*Hint*: 1 mile = 5,280 feet)

14. **Life Science** An adult male walrus weighs about 2,670 pounds. An adult female walrus weighs about 1,835 pounds. How many more pounds does an adult male walrus weigh than an adult female walrus? **835 pounds**

15. Yes; Sample answer: a team can score 7 field goals and 1 safety or 3 touchdowns, 2 two-point conversions, and 1 one-point conversion.

15. **Critical Thinking** Refer to the table. To score in either of the last two ways, a team must have scored a touchdown on the preceding play. Can a team score 23 points? Explain your answer.

Different Ways a Team Can Score in Football	
Touchdown	6 points
Field Goal	3 points
Safety	2 points
2-Point Conversion	2 points
1-Point Conversion	1 point

Lesson 1-1 A Plan for Problem Solving **7**

Extending the Lesson

Enrichment Masters, p. 1

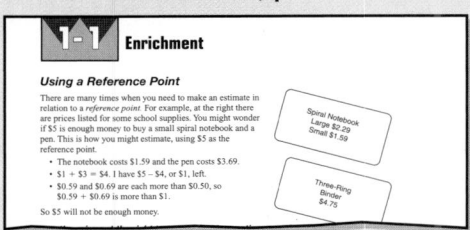

1-1 Enrichment

Using a Reference Point

There are many times when you need to make an estimate in relation to a *reference point*. For example, at the right there are prices listed for some school supplies. You might wonder if $5 is enough money to buy a small spiral notebook and a pen. This is how you might estimate, using $5 as the reference point.

Spiral Notebook Large $2.29 Small $1.59

Three-Ring Binder $4.75

• The notebook costs $1.59 and the pen costs $3.69.
• $1 + $3 = $4. I have $5 − $4, or $1, left.
• $0.59 and $0.69 are each more than $0.50, so $0.59 + $0.69 is more than $1.

So $5 will not be enough money.

Activity Have students ask their family members how they use mathematics in their jobs. As a class activity, use the responses to prepare a list of how mathematics is used on various jobs. Ask students to suggest other jobs they think require the use of mathematics.

Closing Activity
Speaking Separate students into groups of four. Within the groups, have each student explain one step of the problem-solving process.

Practice Masters, p. 1

Name_____ Date_____

1-1 Practice

A Plan for Problem Solving

Use the four-step plan to solve each problem.

1. Kara is a member of the school swim team. She swims 35 laps at each practice. The swim team practices 4 days per week. How many laps does Kara swim in one month?
560 laps

2. The school band is having a car wash to raise money to purchase new uniforms. If they charge $4 per car, how many cars must band members wash to raise $208?
52 cars

3. When the Breen family left for their vacation, the car's odometer read 5,289 miles. Upon their return, the odometer read 6,035 miles. How many miles did the Breen family travel during their vacation?
746 miles

4. Kyle is renewing his subscription to his favorite computer magazine. The cost is $24 for 12 issues. What is the cost of each issue?
$2 per issue

5. The sixth-graders at Danville Middle School collected aluminum cans for a recycling project. Room 6A collected 137 cans; Room 6B collected 98 cans; Room 6C collected 164 cans. How many aluminum cans were collected in all?
399 cans

6. Carla enjoys collecting baseball cards. Her album contains 115 pages. Each page holds 9 baseball cards. How many baseball cards will there be in her in all? album when it is
1,035 cards

7. Chi-Sun went shopping for hiking supplies. He bought a knapsack for $22, a canteen for $5 and a compass for $12. How much money did he spend in all?
$39

8. Eva has saved $64 from her babysitting money. She needs 4 times this amount to buy a new bike. How much does the new bike cost?
$256

Instructional Resources
- *Study Guide Masters*, p. 2
- *Practice Masters*, p. 2
- *Enrichment Masters*, p. 2
- Transparencies 1-2, A and B
- *Assessment and Evaluation Masters*, p. 15
- CD-ROM Program
  - Resource Lesson 1-2

Recommended Pacing	
Standard	Day 3 of 12
Honors	Day 3 of 12
Block	Day 2 of 6

1 FOCUS

5-Minute Check
(Lesson 1-1)

Use the four-step plan to solve each problem.

1. A warehouse manager agreed to supply 5 dozen widgets each to 100 different distributors. There are 6,000 widgets in stock. Can the orders be filled from stock? **yes**

2. The Ski Rack is having a half-price sale. What is the sale price of a cross-country package that regularly sells for $164? **$82**

The 5-Minute Check is also available on **Transparency 1-2A** for this lesson.

Motivating the Lesson

Communication Name a sport such as football and ask students to give at least one example of a pattern relating to the sport. For example, the markings on a football field are 10, 20, 30, 40, 50, 40, 30, 20, and 10.

1-2 Using Patterns

What you'll learn
You'll learn to solve problems using patterns.

When am I ever going to use this?
You can use patterns to make formations in marching band.

Patterns are all around us, even in a bee's honeycomb. Think about your day. What events follow a pattern? Do you always have the same class after lunch? Does your school day always end at the same time?

In this lesson, you will use patterns to solve problems.

Examples

1 Draw the next three figures in the pattern.

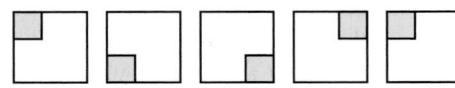

In the first figure, the top left corner is shaded. In the figures that follow, the shaded corner moves counterclockwise. The next three figures are drawn below.

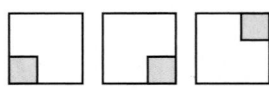

Cultural Kaleidoscope

Mancala is one of the most popular games throughout Africa and the Middle East. It is a board game of capture, using mathematical logic and patterns, that has been played for thousands of years.

2 Draw the next three figures in the pattern.

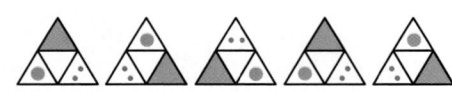

In the figures above, the triangle moves clockwise. The next three figures are drawn below.

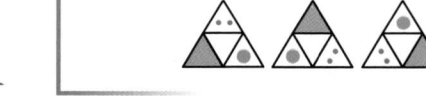

In Examples 1 and 2, geometric shapes were arranged in a pattern. Numbers can also be arranged in a pattern.

3 Find the next three numbers in the pattern
6, 11, 16, 21, _?_, _?_, _?_.

Study the pattern. How do you get each succeeding number?

Each number is 5 more than the number before it.

$21 + 5 = 26$ $26 + 5 = 31$ $31 + 5 = 36$

The next three numbers are 26, 31, and 36.

4 Find the next three numbers in the pattern 2, 4, 8, 16, _?_, _?_, _?_.

Study the pattern. How do you get each succeeding number?

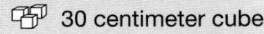

Each number is twice the number before it.

$16 \times 2 = 32$ $32 \times 2 = 64$ $64 \times 2 = 128$

The next three numbers are 32, 64, and 128.

When using patterns, making a table helps organize the information.

HANDS-ON MINI-LAB

Work with a partner.   30 centimeter cubes

A "staircase" that is 4 cubes high is shown at the right.
Notice that 10 cubes are needed to build the staircase.

Try This

1. Copy the chart below. Then use centimeter cubes to find the
number of cubes needed to build each staircase.

Height of Staircase	1	2	3	4	5	6	7
Number of Cubes Needed	1	?	?	10	?	?	?
		3	6		15	21	28

Talk About It

2. Describe the pattern you see in the number of cubes needed to
build each staircase.

3. Use the pattern to find the number of cubes needed to build
staircases that are 9 cubes and 10 cubes high without using
centimeter cubes. **45 cubes; 55 cubes**

2. Sample answer:
The number of
cubes needed to
build each
staircase are
increasing by 2, 3,
4, 5, and so on.

Classroom Vignette

"My students use blank transparencies and overhead pens to
write out solutions to problems. Then we display the solutions
with the overhead projector to evaluate the problem-solving
approach. It is a great way to show that there are often several
ways to solve a given problem."

Louise Chatterton

Louise Chatterton, Teacher
Penns Grove Middle School
Penns Grove, NJ

Transparency 1-2B contains a
teaching aid for this lesson.

Using the Mini-Lab Have
students use cubes to build other
shapes, such as pyramids or walls.
Keep track of the number of cubes
needed as they enlarge their
figures. Have them describe the
patterns.

Teaching Tip Point out to students
that there are many different types
of patterns. This lesson teaches
one-step sequential patterns.

In-Class Examples

For Example 1
Draw the next three figures in
the pattern.

For Example 2
Draw the next three figures in
the pattern.

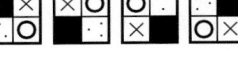

For Example 3
Find the next three numbers in
the pattern.
9, 16, 23, 30, _?_, _?_, _?_
37, 44, 51

For Example 4
Find the next three numbers in
the pattern.
2, 6, 18, _?_, _?_, _?_
54, 162, 486

Teaching Tip In Examples 2 and
3, point out that a pattern can be
formed by adding or multiplying.
Tell students that patterns can also
be formed by subtracting or
dividing.

In-Class Example

For Example 5

The band needs to make a triangular formation with 10 students at the base. How many students are needed? **55**

3 PRACTICE/APPLY

Check for Understanding

If students need additional practice or instruction after completing Exercises 1–6, one of these options may be helpful.
- Extra Practice, see p. 557
- Reteaching Activity
- *Study Guide Masters*, p. 2
- *Practice Masters*, p. 2

Additional Answer

3.

Number of Folds	1	2	3	4
Number of Thicknesses	2	4	8	16

Sample answer: The number of thicknesses doubles.

Study Guide Masters, p. 2

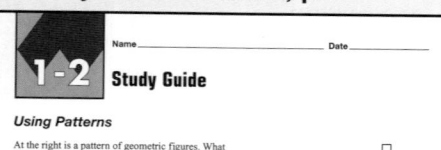

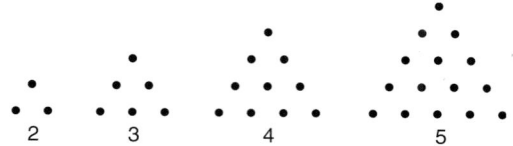

Geometry The marching band wants to make a triangular formation on the football field. Triangular formations with 2, 3, 4, and 5 students at the base look as follows:

How many students are needed to make a triangular formation with 8 students at the base?

Explore   You know the number of students needed to make a formation with 2, 3, 4, and 5 students at the base. You need to find the number of students needed to make a formation with 8 students at the base.

Plan   Make a table to organize the information. Then, study the table to find a pattern.

Solve

Number of Students at Base	2	3	4	5	6	7	8
Total Number of Students Needed	3	6	10	15	?	?	?

+3 +4 +5

The extra students needed increases by 1 each time. Use the pattern to complete the chart.

$15 + 6 = 21$ $21 + 7 = 28$ $28 + 8 = 36$

36 students will be needed to make a triangular formation with 8 students at the base.

Examine   Draw a triangular formation with 8 dots at the base. Check to see if there are 36 dots in the formation.

CHECK FOR UNDERSTANDING

Communicating Mathematics

Read and study the lesson to answer each question.

1. *Draw* the first five phases of a geometric pattern. Then write a rule for finding the next shape in the pattern. **See students' work.**

2. *Write* a number pattern with six numbers. Then write a rule for finding the next number in the pattern. **See students' work.**

HANDS-ON MATH

3. Fold a piece of paper 1 time. Count the number of thicknesses. How many thicknesses are there when the paper is folded 2 times? 3 times? 4 times? Describe the pattern in the number of thicknesses. **See margin.**

10   Chapter 1 Problem Solving, Numbers, and Algebra

Reteaching the Lesson

Activity Give students a number and have them use the constant key (OP₁ or OP₂ on the TI Explorer Plus calculator) to create a pattern of five numbers. Have them record the numbers. Then ask them to exchange papers with others to find the next two numbers in the pattern.

Error Analysis

Watch for students who think they find a pattern but who don't check that it works with all of the given numbers.

Prevent by encouraging students to write the operation for each succeeding number.

Find the next three numbers in each pattern. 5. 8, 4, 2

4. 0, 8, 16, 24, _?_, _?_, _?_ **32, 40, 48** 5. 128, 64, 32, 16, _?_, _?_, _?_

6. *Sports* Jamil and Mark are competing against each other in a tennis match. Their scores are shown below.

	Set 1	Set 2	Set 3	Set 4	Set 5
Jamil	6	2	6	4	6
Mark	3	6	3	6	?

If the pattern continues, what will Mark's score be in set 5? **3**

EXERCISES

Practice

8. 19, 22, 25

9. 512, 2,048, 8,192

10. 12, 6, 0

12. 243, 729, 2,187

Find the next three numbers in each pattern.

7. 3, 4, 3, 4, _?_, _?_, _?_ **3, 4, 3**

9. 2, 8, 32, 128, _?_, _?_, _?_

11. 7, 8, 10, 13, _?_, _?_, _?_ **17, 22, 28**

13. Find the next three numbers in the pattern 1, 3, 7, 13, ... **21, 31, 43**

8. 7, 10, 13, 16, _?_, _?_, _?_

10. 36, 30, 24, 18, _?_, _?_, _?_

12. 3, 9, 27, 81, _?_, _?_, _?_

Applications and Problem Solving

14. *Fitness* Marcie is training for the swim team. On the first day, she swims 5 laps. The second day she swims 6 laps. The third day she swims 8 laps, and on the fourth day she swims 11 laps. **a. 15 laps**
   a. If this pattern continues, how many laps will she swim on the fifth day?
   b. On which day will she reach her goal of 33 laps? **day 8**

15. *Transportation* Isabel needs to take the bus home from the mall. She forgot her bus schedule but she remembers that the bus comes every 20 minutes. She also remembers that there is a bus at 12:30 P.M. What time does the first bus after 5:00 P.M. arrive? **5:10 P.M.**

16. *Geometry* Find the next two shapes in the pattern.

17. *Critical Thinking* What is the next number in the pattern 1, 2, 2, 4, 8, 32, ... ? **256**

Mixed Review

18. *History* In July 1776, General George Washington had nearly 30,000 troops guarding New York City. After retreating through New Jersey to the Delaware River, only about 3,000 troops remained with him. How many troops did he lose? *(Lesson 1-1)* **27,000 troops**

19. [Test Practice] In 1994, Lucas Software had a profit of $2,763,000. In 1995, their profit increased by $184,000. How much profit did Lucas Software have in 1995? *(Lesson 1-1)* **C**
   A $2,579,000   B $2,763,000
   C $2,947,000   D $3,248,000

Extending the Lesson

Enrichment Masters, p. 2

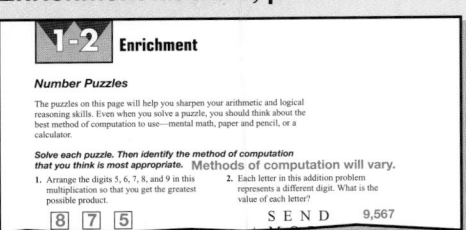

Activity The pattern 1, 1, 2, 3, 5, 8, ... is called the *Fibonacci sequence.* Each number is the sum of the previous two numbers. Find the first 20 numbers in the sequence. Look for other patterns in the sequence. **1, 1, 2, 3, 5, 8, 13, 21, 34, 55, 89, 144, 233, 377, 610, 987, 1,597, 2,584, 4,181, 6,765. Sample answer: Every fifth number is divisible by 5.**

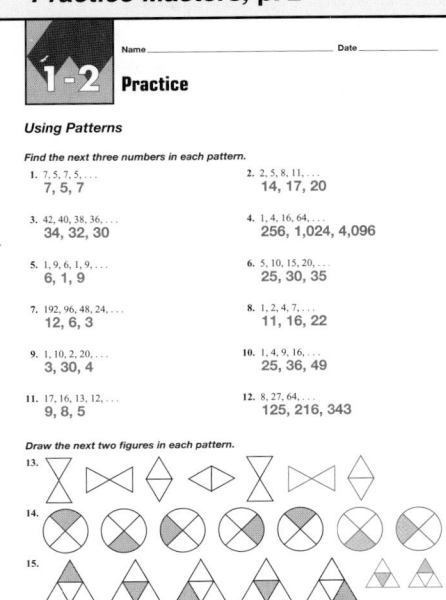

Instructional Resources
- *Study Guide Masters*, p. 3
- *Practice Masters*, p. 3
- *Enrichment Masters*, p. 3
- Transparencies 1-3, A and B

 CD-ROM Program
- Resource Lesson 1-3
- Interactive Lesson 1-3

Recommended Pacing	
Standard	Day 4 of 12
Honors	Day 4 of 12
Block	Day 2 of 6

1 FOCUS

5-Minute Check
(Lesson 1-2)

Find the next three numbers for each pattern.

1. 2, 5, 8, 11, _?_, _?_, _?_
   14, 17, 20

2. 1, 3, 9, 27, _?_, _?_, _?_
   81, 243, 729

3. 25, 21, 17, 13, _?_, _?_, _?_
   9, 5, 1

4. 192, 96, 48, 24, _?_, _?_, _?_
   12, 6, 3

5. How many students in a marching band are needed to make a triangle formation with 6 students at the base?
   21

 The 5-Minute Check is also available on **Transparency 1-3A** for this lesson.

Motivating the Lesson
Communication Have students read the opening paragraph of the lesson. Ask the following question. *Should you estimate or find an exact answer? Why?* Estimate; the question asks *about* how many wild horses are on public land in five states.

1-3 Estimation by Rounding

What you'll learn

You'll learn to estimate sums, differences, products, and quotients using rounding.

When am I ever going to use this?

Rounding can be used to estimate the total cost of items at a department store.

Wild horses can be found throughout the United States. The table shows the number of wild horses living on public lands in five western states. About how many wild horses are living on public lands in these five states?

State	Number of Wild Horses
Nevada	30,798
Wyoming	4,115
Oregon	1,891
Utah	1,884
California	1,745

To estimate the number of wild horses, you can use rounding. Round each number to the nearest thousand. Then add.

$$
\begin{array}{rcr}
30,798 & \rightarrow & 31,000 \\
4,115 & \rightarrow & 4,000 \\
1,891 & \rightarrow & 2,000 \\
1,884 & \rightarrow & 2,000 \\
+\ 1,745 & \rightarrow & +\ 2,000 \\
\hline
& & 41,000
\end{array}
$$

Did you know? Wild horses are descendants of horses that were brought to the United States by Spaniards 500 years ago.

About 41,000 horses are living on public lands in the five states.

Let's review the rules for rounding.

Rounding Whole Numbers	Look at the digit to the right of the place being rounded. • The digit remains the same if the digit to the right is 0, 1, 2, 3, or 4. • Round up if the digit to the right is 5, 6, 7, 8, or 9.

Example ① **Round 43 to the nearest ten.**

On the number line, 43 is closer to 40 than it is to 50.

Look at the digit to the right of the tens place. Since 3 is less than 5, the digit in the tens place remains the same. 43 rounded to the nearest ten is 40.

12 Chapter 1 Problem Solving Numbers, and Algebra

Multiple Learning Styles

Interpersonal Have students work in groups to practice estimation problems. Ask each student to think up a problem about going to a grocery with money to buy several items. Have them try out their problems on the other students in the group.

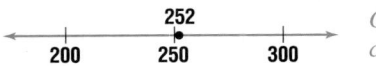

Examples

2 Round 252 to the nearest hundred.

```
        252
  |------●------|------|
 200    250    300
```

On the number line, 252 is closer to 300 than it is to 200.

Look at the digit to the right of the hundreds place. Since the digit is 5, round up. That is, increase the digit in the hundreds place by one. The number 252 rounded to the nearest hundred is 300.

3 Round 9,973 to the nearest thousand.

```
                  9,973
  |------|------|--●--|
 9,000  9,500  10,000
```

On the number line, 9,973 is closer to 10,000 than it is to 9,000.

Look at the digit to the right of the thousands place. Round up since 9 is greater than 5. Thus, 9,973 rounded to the nearest thousand is 10,000.

Estimation is a useful skill that provides a quick answer when an exact answer is not needed.

Examples

4 Estimate the product of 7 and 64.

$$
\begin{array}{ccc}
64 & \rightarrow & 60 \\
\times\ 7 & \rightarrow & \times\ 7 \\
\hline
 & & 420
\end{array}
$$

5 Estimate the quotient of 112 and 23.

$$
112 \div 23 \rightarrow 20)\overline{100}\ \ ^{5}
$$

Using estimation, you can also check to see whether an answer is reasonable.

Example

APPLICATION

6 **Money Matters** Eduardo is shopping at Calyan's department store. He purchases a bag of candy, a mechanical pencil, and a package of baseball cards. The total bill is $9.12. He gives the cashier a $20 bill and receives $9.46 in change. Is this amount reasonable?

Round to the nearest dollar amount. Then subtract mentally.

$$
\begin{array}{ccc}
\$20.00 & \rightarrow & \$20 \\
-\ 9.12 & \rightarrow & -\ 9 \\
\hline
 & & \$11
\end{array}
$$

Eduardo should have received about $11. So, the amount of $9.46 is *not* reasonable.

Lesson 1-3 Estimation by Rounding **13**

 Transparency 1-3B contains a teaching aid for this lesson.

Using Discussion Some students may be hesitant to make estimates using the methods presented, thinking that only exact answers are acceptable. Discuss situations in which estimates are more useful or descriptive than exact answers. **estimating the total cost of items to see if you have enough money**

In-Class Examples

For Example 1
a. Round 23 to the nearest ten. **20**

b. Round 86 to the nearest ten. **90**

For Example 2
a. Round 155 to the nearest hundred. **200**

b. Round 921 to the nearest hundred. **900**

For Example 3
a. Round 3,413 to the nearest thousand. **3,000**

b. Round 1,650 to the nearest thousand. **2,000**

For Example 4
a. Estimate the product of 3 and 18. **60**

b. Estimate the product of 13 and 92. **900**

For Example 5
a. Estimate the quotient of 157 and 42. **4**

b. Estimate the quotient of 96 and 12. **10**

For Example 6
Aaron went to a music store and bought 2 CDs at $11.59 each and a cassette for $8.39. He gave the cashier two $20 bills and got back $8.43 in change. Is this amount reasonable? **yes**

Teaching Tip Explain to students that estimates are useful when you don't need an exact answer and want to save time and work or when you don't have a pencil and paper handy.

Check for Understanding

If students need additional practice or instruction after completing Exercises 1–14, one of these options may be helpful.
- Extra Practice, see p. 557
- Reteaching Activity
- *Transition Booklet,* pp. 11–14, 17–18
- *Study Guide Masters,* p. 3
- *Practice Masters,* p. 3

Assignment Guide

Core: 15–43 odd, 46–52
Enriched: 16–42 even, 43, 44, 46–52

Additional Answers

1. On the number line, 957 is closer to 1,000 than to 900.

13. 500 ÷ 50 = 10; not reasonable

36. 120 ÷ 2 = 60; not reasonable

37. 5,000 + 5,000 = 10, 000; reasonable

38. 400 × 4 = 1,600; not reasonable

***Study Guide Masters,* p. 3**

Name_____ Date_____

1-3 Study Guide

Estimation by Rounding

You can use these rules to round whole numbers.

Rules for Rounding
Look at the digit to the right of the place being rounded.
• The digit remains the same if the digit to the right is 0, 1, 2, 3, or 4.
• Round up if the digit to the right is 5, 6, 7, 8, or 9.

Examples 1 Round 6,905 to the nearest ten, to the nearest hundred, and to the nearest thousand.

6,905   5 = 5, round 0 up to 1.   6,910
6,905   0 < 5, leave 9 unchanged.   6,900
6,905   9 > 5, round 6 up to 7.   7,000

2 Round $48.29 to the nearest dollar and to the nearest ten dollars.

$48.29   2 < 5, leave 8 unchanged. $48.00
$48.29   8 > 5, round 4 up to 5. $50.00

Round each number to the underlined place-value position.
1. 392 400   2. 47 50   3. 1,013 1,000   4. 674 700
5. 731 700   6. 8,204 8,200   7. 3,789 4,000   8. 606 600
9. 4,111 4,110   10. 7,619 8,000   11. 2,588 2,600   12. 5,045 5,050

Estimate. State whether the answer shown is reasonable.
13. $8.90 + $0.45 + $1.25 = $13.72 no   14. 812 − 193 = 619 yes
15. $7.89 × 3 = $23.67 yes   16. 1,164 + 582 = 3 no

© Glencoe/McGraw-Hill   T3   *Mathematics: Applications and Connections, Course 1*

Communicating Mathematics

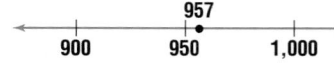

2. Sample answer: Round 517 to 500; round 192 to 200; 500 − 200 = 300 so 517 − 192 is about 300.

Read and study the lesson to answer each question.

1. *Tell* how you would round 957 to the nearest hundred by using the number line below. **See margin.**

```
        957
 <──┼────●──────┼──>
   900   950  1,000
```

2. *Explain* how you would use rounding to estimate 517 − 192.

3. *Write* about a situation when you would use estimation. **See students' work.**

Guided Practice

10. $7.00 + $9.00 = $16.00; not reasonable

11. 900 − 400 = 500; reasonable

12. 200 × 6 = 1,200; reasonable

Round each number to the underlined place-value position.

4. 3̲5 40   5. 4̲24 400   6. 5̲42 540
7. 2̲96 300   8. 2,0̲88 2,000   9. 9,̲922 10,000

Estimate. State whether the answer shown is reasonable.

10. $6.67 + $8.99 = $12.66   11. 898 − 413 = 485
12. 215 × 6 = 1,290   13. 468 ÷ 52 = 14 **See margin.**

14. *Food* The average U.S. citizen eats about 47 pints of ice cream per year. The average Italian citizen eats about 22 pints per year. About how many more pints of ice cream does the average U.S. citizen eat than the average Italian citizen? **about 30 pints**

Practice

Round each number to the underlined place-value position.

15. 9̲7 100   16. 5̲20 520   17. 5̲1 50   18. 1̲8 20
19. 4̲44 440   20. 8,1̲96 8,200   21. 5,̲500 5,500   22. 9̲27 930
23. 3,̲704 3,700   24. 3̲94 400   25. 5,6̲23 5,600   26. 3,6̲56 3,700
27. 9,̲716 10,000   28. 4,̲448 4,000   29. 1,̲888 2,000   30. 7,̲050 7,000

31. Round 9,438 to the nearest thousand. **9,000**

32. What is 6,612 rounded to the nearest hundred? **6,600**

33. 80 × 3 = 240; reasonable

34. $3.00 + $3.00 + $3.00 = $9.00; reasonable

35. 100 − 20 = 80; not reasonable

36–42. See margin.

Estimate. State whether the answer shown is reasonable.

33. 82 × 3 = 246   34. $3.32 + $3.34 + $3.21 = $9.87
35. 96 − 18 = 65   36. 121 ÷ 2 = 42
37. 4,901 + 5,002 = 9,903   38. 392 × 4 = 12,368
39. 765 − 234 = 531   40. 391 ÷ 23 = 17
41. $5.25 + $35.27 = $46.12   42. 2,914 × 3 = 8,742

Applications and Problem Solving

43. *Recycling* The average U.S. citizen uses about 190 pounds of plastic, 55 pounds of aluminum cans, 586 pounds of paper, and 325 pounds of glass per year. If these materials are recyclable, about how many pounds of these materials could be recycled by the average citizen in one year? **Sample answer: 200 + 60 + 600 + 300 = 1,160; about 1,160 lbs**

■ Reteaching the Lesson ■

Activity Write a list of prices on the chalkboard for various items from a department store. Prepare several gift coupons for amounts ranging from $100 to $2,000. Hand out these coupons to students and ask them to shop from the list, getting as close to the amount on the coupon as possible.

Additional Answers

39. 800 − 200 = 600; reasonable

40. 400 ÷ 20 = 20; reasonable

41. $5.00 + $35.00 = $40.00; not reasonable

42. 3,000 × 3 = 9,000; reasonable

44. Travel The chart shows the approximate distance, in miles, between some U.S. cities. Suppose you travel from Raleigh, North Carolina, to Nashville, Tennessee. If you average 52 miles per hour, about how many hours will it take you to get there?
about 10 hours

Mileage Chart

Approximate Mileages	Raleigh, NC	Albuquerque, NM	Nashville, TN	Minneapolis, MN	Portland, OR	Charleston, WV	Dallas, TX
Raleigh, NC		1,760	538	1,215	2,965	299	1,200
Albuquerque, NM	1,760		1,228	1,227	1,377	1,527	654
Nashville, TN	538	1,228		812	2,424	395	668
Minneapolis, MN	1,215	1,227	812		1,748	937	963
Portland, OR	2,965	1,377	2,424	1,748		2,572	2,041
Charleston, WV	299	1,527	395	937	2,572		1,141
Dallas, TX	1,200	654	668	963	2,041	1,141	

45. Working on the  **Refer to the tables on page 3.**

45a. $75 + $36 + $21 + $6 + $2 + $2 + $10 = $152

a. Suppose you place an order from the Sporting Goods Catalog. You order a pair of running shoes for $74.99, a hooded sweatshirt for $35.99, and a pair of shorts for $20.99. Estimate the total cost of the order including express delivery charges. **b. See students' work.**

b. Choose a catalog from which you would like to order. Imagine you have $200 to spend. List the items you would like to order and their prices. Estimate the total cost of the order including shipping charges.

46. See margin.

46. Critical Thinking You want to purchase three items at a store for $4.43, $5.42, and $6.45. You have $15. To see if you have enough money, you estimate the total cost by rounding the cost of each item to the nearest dollar and adding. Will you have a problem making this purchase? Explain.

Mixed Review

47. Patterns Find the next three numbers in the pattern 1, 2, 4, 7, ...
(Lesson 1-2) **11, 16, 22**

48. Geometry Find the next figure in the pattern. *(Lesson 1-2)*

49. School At Watson Middle School, the bells ring at 8:50 A.M., 8:54 A.M., 9:34 A.M., 9:38 A.M., and 10:18 A.M. each day. When do the next three bells ring? *(Lesson 1-2)* **10:22 A.M., 11:02 A.M., 11:06 A.M.**

50. Population In 1994, Denver had a population of 494,000 people, and Atlanta had a population of 396,000 people. How many more people were living in Denver than Atlanta? *(Lesson 1-1)* **98,000 people**

51. Test Practice The Changs' dot matrix printer prints four pages each minute. Hiro Chang has a history report that is 12 pages long. How long will it take the printer to print the report? *(Lesson 1-1)* **C**
A 6 min **B** 4 min **C** 3 min **D** 2 min

52. School Jill needs to hand out an equal number of counters to 23 students. If there are 522 counters, how many should she give each student? *(Lesson 1-1)* **22**

Lesson 1-3 Estimation by Rounding **15**

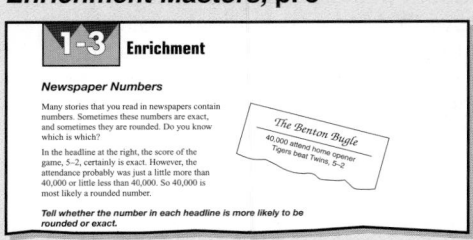

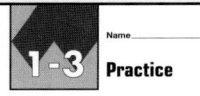
Lesson 1-3 **15**

* *Study Guide Masters*, p. 4
* *Practice Masters*, p. 4
* *Enrichment Masters*, p. 4
* Transparencies 1-4, A and B
* *Assessment and Evaluation Masters*, pp. 14, 15
* *Technology Masters*, p. 1
* CD-ROM Program
  * Resource Lesson 1-4

Recommended Pacing

Standard	Day 5 of 12
Honors	Day 5 of 12
Block	Day 3 of 6

1 FOCUS

5-Minute Check
(Lesson 1-3)

1. Round 535 to the nearest ten. **540**

Estimate. State whether the answer shown is reasonable.

2. $73 + 89 + 38 = 300$
   $70 + 90 + 40 = 200$; **not reasonable**

3. $879 - 585 = 294$
   $900 - 600 = 300$; **reasonable**

4. $1,679 \times 31 = 52,049$
   $2,000 \times 30 = 60,000$; **reasonable**

5. $399 \div 19 = 20$
   $400 \div 20 = 20$; **reasonable**

The 5-Minute Check is also available on Transparency **1-4A** for this lesson.

Motivating the Lesson

Hands-On Activity Have students use counters to represent $6 + 4 \times 2$. Discuss whether the correct answer is 20 or 14 and why. **14**

1-4 Order of Operations

What **you'll learn**

You'll learn to evaluate expressions using the order of operations.

When **am I ever going to use this?**

Knowing how to follow the order of operations can help you find the total cost of zoo tickets for your class.

Word Wise
order of operations

The San Diego Zoo has more than 3,200 animals. Each year, the animal food bill is more than $685,000. Many people help reduce this cost by adopting an animal. The table shows the costs for adopting various zoo animals. How much would it cost to adopt 1 pygmy hippopotamus, 4 fishing cats, and 3 tapirs? *This problem will be solved in Example 4.*

Animal	Cost ($)
Anteater	500
Spider Monkey	50
Pygmy Hippopotamus	100
Spectacled Owl	50
Tapir	100
Fishing Cat	50

When more than one operation is used, we need to know which one to perform first so that everyone gets the same result. Mathematicians have come up with rules called the **order of operations**.

Order of Operations	1. Multiply and divide in order from left to right.
	2. Add and subtract in order from left to right.

Examples

Find the value of each expression.

1 $15 + 7 - 3$

$15 + 7 - 3 = 22 - 3$ *Add 7 to 15.*
$\qquad\qquad\quad = 19$ *Subtract 3 from 22.*

2 $3 + 6 \times 4 - 2$

$3 + 6 \times 4 - 2 = 3 + 24 - 2$ *Multiply 6 and 4.*
$\qquad\qquad\qquad = 27 - 2$ *Add 3 and 24.*
$\qquad\qquad\qquad = 25$ *Subtract 2 from 27.*

3 $14 \div 7 + 12 \times 3 - 9$

$14 \div 7 + 12 \times 3 - 9 = 2 + 12 \times 3 - 9$ *Divide 14 by 7.*
$\qquad\qquad\qquad\qquad = 2 + 36 - 9$ *Multiply 12 and 3.*
$\qquad\qquad\qquad\qquad = 38 - 9$ *Add 2 and 36.*
$\qquad\qquad\qquad\qquad = 29$ *Subtract 9 from 38.*

Multiple Learning Styles

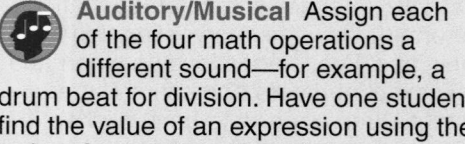

Auditory/Musical Assign each of the four math operations a different sound—for example, a drum beat for division. Have one student find the value of an expression using the order of operations. Then have another student make the sounds in the same order that the first student evaluated the expression. Have students discuss whether the sounds came in the correct order.

Example 4
CONNECTION

Life Science Refer to the beginning of the lesson.

a. Write an expression for the cost of adopting 1 pygmy hippopotamus, 4 fishing cats, and 3 tapirs.

b. Find the cost of adopting the animals.

a.

cost for 1 hippo	*plus*	*cost for 4 cats*	*plus*	*3 tapirs*
$1 \times \$100$	$+$	$4 \times \$50$	$+$	$3 \times \$100$

b. Find the value of the expression.

$$1 \times \$100 + 4 \times \$50 + 3 \times \$100 = \$100 + 4 \times \$50 + 3 \times \$100$$
$$= \$100 + \$200 + 3 \times \$100$$
$$= \$100 + \$200 + \$300$$
$$= \$600$$

The cost of adopting the animals is $600.

Does your calculator follow the order of operations?

TECHNOLOGY
MINI-LAB

Work with a partner.   calculator

Find the value of $5 + 8 \div 4 \times 3$.

Method 1
Use a calculator.

5 `+` 8 `÷` 4 `×` 3 `=` *11*

Method 2
Use paper and pencil.

Follow the order of operations.
$$5 + 8 \div 4 \times 3 = 5 + 2 \times 3$$
$$= 5 + 6$$
$$= 11$$

This calculator follows the order of operations.

Try This   1. See students' work.
1. Using your calculator, evaluate $3 + 2 \times 5 - 7$.
2. Evaluate this expression using paper and pencil and the order of operations. **6**

Talk About It
3. Compare your answers. Are they the same? If so, why are they the same? If not, why are they different? **See margin.**

Lesson 1-4 Order of Operations **17**

Transparency 1-4B contains a teaching aid for this lesson.

Using the Mini-Lab Scientific calculators follow the order of operations. Basic calculators may or may not. Find a calculator that does not follow the order of operations. Have students compare the answers found using this calculator with those found using a scientific calculator. Ask students what this shows about the importance of order of operations.

Teaching Tip To help students remember to multiply and divide before they add and subtract, suggest that they use the mnemonic **M**Y **D**EAR **A**UNT **S**ALLY. Remind students this does *not* necessarily mean to multiply before dividing or to add before subtracting.

In-Class Examples
For Example 1
Find the value of each expression.
a. $28 - 4 + 9$ **33**
b. $103 + 15 - 8$ **110**

For Example 2
Find the value of each expression.
a. $24 + 3 \times 4 - 15$ **21**
b. $15 - 8 + 6 \times 3$ **25**

For Example 3
Find the value of each expression.
a. $16 \div 4 + 6 \times 2 - 5$ **11**
b. $13 + 56 \div 7 - 2 \times 4$ **13**

For Example 4
Refer to the beginning of the lesson. Find the cost of adopting 3 spectacled owls, 2 tapirs, 4 spider monkeys, and 1 anteater. **$1,050**

Additional Answer for the Mini-Lab
3. Whereas one calculator follows the order of operations, the other one performs the operations as they are entered into the calculator.

Check for Understanding

If students need additional practice or instruction after completing Exercises 1–7, one of these options may be helpful.

- Extra Practice, see p. 558
- Reteaching Activity
- *Study Guide Masters,* p. 4
- *Practice Masters,* p. 4
- Interactive Mathematics Tools Software

Assignment Guide

Core: 9–21 odd, 24–27
Enriched: 8–20 even, 21, 22, 24–27
All: Self Test, 1–10

Have students share with the class the projects they chose and the steps involved. Ask students to brainstorm activities in everday life that require a specific order of operations.

Additional Answer

1. **Multiply and divide from left to right. Then add and subtract from left to right.**

Study Guide Masters, p. 4

1-4 Study Guide

Name_____ Date_____

Order of Operations

To evaluate expressions, use the order of operations.

Order of Operations
1. Multiply and divide in order from left to right.
2. Add and subtract in order from left to right.

Example Find the value of $2 \times 8 - 42 \div 7 + 4$.

$2 \times 8 - 42 \div 7 + 4 = 16 - 42 \div 7 + 4$ *Multiply 2 and 8.*
$= 16 - 6 + 4$ *Divide 42 by 7.*
$= 10 + 4$ *Subtract 6 from 16.*
$= 14$ *Add 10 and 4.*

Find the value of each expression.

1. $15 + 4 \times 5 - 6$ **29**
2. $40 \div 8 - 3$ **2**
3. $9 \times 3 + 3$ **30**
4. $20 - 15 \div 3$ **15**
5. $2 \times 4 \times 5$ **40**
6. $18 \div 2 + 9 \times 2$ **27**
7. $3 + 7 - 5 + 2$ **7**
8. $4 \times 9 - 9 \times 4$ **0**
9. $15 \div 3 + 3 \times 5$ **20**
10. $45 \div 9 + 5$ **1**
11. $18 \div 3 + 15 \div 3$ **26**
12. $50 - 5 \times 5 - 25$ **0**
13. $35 \div 7 - 5$ **0**
14. $12 \times 2 - 5 \times 4$ **4**
15. $4 + 18 \div 3$ **10**
16. $150 \div 10 - 3 \times 5$ **0**
17. $7 + 2 \times 5$ **17**
18. $3 \times 8 - 16$ **8**
19. $80 \times 2 \div 40 - 1$ **3**
20. $25 - 3 \times 4$ **13**
21. $2 + 4 \times 9 \div 12$ **5**
22. $45 - 40 + 1 \times 2$ **7**
23. $18 \div 2 \times 9$ **81**
24. $2 \times 5 \times 3$ **30**

© Glencoe/McGraw-Hill   T4   *Mathematics: Applications and Connections, Course 1*

Communicating Mathematics

Read and study the lesson to answer each question.

1. *List* each step of the order of operations. **See margin.**

2. *Identify* the first step when evaluating the expression $25 + 2 - 6 \div 3 \times 2$. **Divide 6 by 3.**

Guided Practice

Name the operation that should be done first. Then find the value of the expression.

3. $22 + 5 - 13$ **+; 14**
4. $13 - 4 + 7 \times 6$ **×; 51**
5. $9 \times 15 \div 5 + 6$ **×; 33**
6. $27 - 8 \div 4 \times 3$ **÷; 21**

7. *Money Matters* Mrs. Jamison's class is planning a trip to the San Diego Zoo. Tickets cost $6 for children and $15 for adults.
   a. Write an expression for the total cost of 12 adult tickets and 28 children's tickets. **12 × $15 + 28 × $6**
   b. Find the total cost of the tickets. **$348**

Practice

Find the value of each expression.

8. $16 + 5 - 7$ **14**
9. $15 \times 8 - 3$ **117**
10. $85 \div 17 \times 14$ **70**
11. $29 + 56 \div 4$ **43**
12. $17 - 3 \times 4 + 6$ **11**
13. $21 + 18 \div 6 - 9$ **15**
14. $68 + 19 - 7 \times 5$ **52**
15. $19 + 45 \div 3 - 11$ **23**
16. $26 \div 13 + 9 \times 6 - 21$ **35**
17. $9 + 12 - 6 \times 4 \div 2$ **9**

18. Evaluate the expression $67 - 21 \div 7 \times 8 + 10$. **53**

19. Find the value of $21 \div 3 + 10 \times 9 - 42$. **55**

20. What is the value of 144 divided by 12 plus 3 times 7 minus 6? **27**

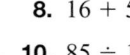

Ask a family member to help you make a list of the steps involved in a project around your home, such as painting a room. Discuss how the order of steps affects the result.

Applications and Problem Solving

21. *Nutrition* The table shows the number of fat grams in one scoop of various ice cream flavors at Baskin-Robbins.
    a. Yoko ordered 2 scoops of chocolate chip ice cream and 1 scoop of pralines 'n cream. Robert ordered 3 scoops of chocolate almond. Write an expression for the total number of fat grams in the two orders. **2 × 15 + 1 × 14 + 3 × 18**

Baskin-Robbins Flavor	Fat (g)
Chocolate Almond	18
Chocolate Chip	15
Pralines 'n Cream	14
Rainbow Sherbet	2

    b. Find the total number of fat grams in the two orders. **98 g**

18 Chapter 1 Problem Solving, Numbers, and Algebra

Reteaching the Lesson

Activity Play "First Operation." Write expressions on index cards. Then write the order of operations on the chalkboard. Each player, in turn, chooses a card, performs the first operation, rewrites the expression, and then passes the card on to the next player who checks it for accuracy. Play continues until the last operation is performed.

Error Analysis
Watch for students who forget to always multiply and divide before they add and subtract.
Prevent by placing parentheses around the parts to be solved first.

22. **Entertainment** The Music Club is planning a production of *The Wizard of Oz*. In order to determine the number of tickets they can sell, they need to know how many seats are in the auditorium. The auditorium has 3 sections, and each section has 24 rows. Sections A and C have 15 seats in each row and section B has 20 seats in each row.
   a. Write an expression for the total number of seats in the auditorium.
   b. How many seats are in the auditorium? **1,200 seats**

23. **Working on the CHAPTER Project**  Refer to the tables on page 3.
   a. You decide to order 3 *Star Trek* life-size cut-outs for $29.95 each from the Novelty Gift catalog. You also want 2 *Star Trek* T-shirts that cost $16.95 each. Write an expression for the total cost of the order including shipping. Find the total cost of the order.
   b. Find a catalog from which you would like to order. Select 3 of one item and 4 of a different item. Write an expression for the total cost of your order including shipping. Find the total cost of the order.

24. **Critical Thinking** Write an expression using four numbers and at least two of the operations $+$, $-$, $\times$, or $\div$ in which the value of the expression is 11. **Sample answer: $3 \times 2 + 6 - 1$**

22a. See Answer Appendix.
23a. See Answer Appendix.
23b. See students' work.

Mixed Review

25. Round 9,046 to the nearest thousand. *(Lesson 1-3)* **9,000**

26. **Patterns** Find the next three numbers in the pattern 5, 10, 15, 20, ... *(Lesson 1-2)* **25, 30, 35**

27. **Test Practice** Brandi can run a mile in 7 minutes. At this rate, how long will it take her to run 8 miles? *(Lesson 1-1)* **B**
   A 58 min **B** 56 min **C** 54 min **D** 15 min

CHAPTER 1

Mid-Chapter Self Test

Use the four-step plan to solve each problem. *(Lesson 1-1)*

1. Thomas purchased a CD game for his Sony PlayStation. The total amount, including tax, was $43. How much change did he receive from $50? **$7**

2. On average, an elephant's heart beats 30 times per minute. At this rate, how many times would an elephant's heart beat in 1 hour? **1,800 times**

Find the next three numbers in each pattern. *(Lesson 1-2)*

3. 5, 13, 21, 29, ?, ?, ? **37, 45, 53**   4. 3, 12, 48, 192, ?, ?, ? **768, 3,072, 12,288**

Round each number to the underlined place-value position. *(Lesson 1-3)*

5. 1\underline{5}4 **150**   6. \underline{9},511 **10,000**

7. Estimate 18,392 divided by 317. **$18,000 \div 300 = 60$**

Find the value of each expression. *(Lesson 1-4)*

8. $19 - 6 + 24$ **37**   9. $75 \div 3 - 8 \times 2$ **9**   10. $132 + 4 \times 9 \div 6 - 53$ **85**

Extending the Lesson

Enrichment Masters, p. 4

1-4 Enrichment

Operations Puzzles

Now that you have learned how to evaluate an expression using the order of operations, can you work backwards? In this activity, the value of the expression will be given to you. It is your job to decide what the operations or the numbers must be in order to arrive at that value.

Fill in each ☐ with +, −, ×, or ÷ to make a true statement.

1. 48 ☐ 3 × 12 = 12   2. 30 ☐ 15 × 3 = 6
3. 24 ÷ 12 ☐ 6 ÷ 3 = 4   4. 24 ☐ 12 ☐ 6 × 3 = 18
5. 4 × 16 ☐ 2 ☐ 8 = 24   6. 45 ☐ 3 ☐ 3 ☐ 9 = 3

Activity Have students work in groups to determine how they would evaluate expressions such as $3 + 5 \times n = 24$.

Exercise 23 asks students to advance to the next stage of work on the Chapter Project. You may wish to have students check their calculations by adding the item costs individually.

4 ASSESS

Closing Activity
Writing Have students work in pairs to write, exchange, and evaluate expressions.

Chapter 1, Quiz B (Lessons 1-3 and 1-4) is available in the *Assessment and Evaluation Masters*, p. 15.

Mid-Chapter Test (Lessons 1-1 through 1-4) is available in the *Assessment and Evaluation Masters*, p. 14.

Mid-Chapter Self Test

The Mid-Chapter Self Test reviews concepts in Lessons 1-1 through 1-4. Lesson references are given so students can review concepts not yet mastered.

Practice Masters, p. 4

Name_____ Date_____

1-4 Practice

Order of Operations

Name the operation that should be done first. Then find the value of the expression.
1. $17 - 2 \cdot 4$ **×; 9**   2. $24 \div 6 + 3$ **÷; 7**   3. $11 + 9 - 12$ **+; 8**
4. $30 + 6 \cdot 5$ **÷; 25**   5. $13 - 4 + 7$ **−; 16**   6. $8 \times 7 + 2$ **×; 58**
7. $6 \cdot 7 - 2$ **×; 40**   8. $36 \div 9 - 3$ **÷; 1**   9. $24 \times 2 \div 6$ **×; 8**

Find the value of each expression.
10. $2 + 4 \times 8$ **34**   11. $12 - 9 \div 3$ **9**   12. $8 \cdot 7 - 4$ **52**
13. $10 + 25 \div 5$ **15**   14. $6 + 24 \div 3 \times 2$ **22**   15. $15 - 9 \cdot 4 \div 12$ **12**
16. $8 \cdot 9 \div 6 - 4$ **8**   17. $54 \div 6 - 3 \times 2$ **3**   18. $2 \cdot 4 + 16 \div 8$ **10**
19. $6 + 21 \div 3 - 9$ **4**   20. $12 - 5 \times 6 \div 3$ **2**   21. $2 + 7 \cdot 8 \div 4$ **16**
22. $63 \div 9 + 12 - 2$ **17**   23. $3 \times 6 + 14 \div 2$ **25**   24. $7 \cdot 4 - 3 \cdot 8$ **4**
25. $9 + 18 \div 3 \times 5$ **39**   26. $7 + 8 - 3 + 5$ **17**   27. $7 \cdot 5 + 2 \cdot 3$ **41**
28. $20 - 5 \cdot 2$ **10**   29. $24 \div 8 + 2$ **5**   30. $18 + 24 \div 12 + 5$ **25**
31. $6 + 5 \cdot 2 + 4$ **20**   32. $75 \div 15 \cdot 4$ **20**   33. $45 + 22 \div 11$ **47**
34. $32 \cdot 4 \div 2$ **64**   35. $9 + 4 - 2 \times 3$ **7**   36. $39 - 9 \cdot 3 + 6$ **18**
37. $15 \times 2 + 5 \times 6$ **60**   38. $10 + 53 - 60$ **3**   39. $32 - 4 \times 2 + 10$ **34**

© Glencoe/McGraw-Hill   T4   Mathematics: Applications and Connections, Course 1

Objective Students model algebraic expressions.

Optional Resources
Hands-On Lab Masters
• worksheet, p. 39

MANAGEMENT TIPS

Recommended Time
30 minutes

Getting Started Demonstrate how the paper bag represents some number. Place random numbers of popcorn in the bag and then count them.

Activities 1 and 2 on page 20 demonstrate the sum of some number and a known number. Emphasize that this sum changes according to the value of the "some number."

Activities 3 and 4 on page 21 increase the unknown quantity. Ask students how they would represent *three times some number*.
three bags

Teaching Tip Make sure students realize that the popcorn is defining the unknown quantity. For example, placing four pieces of popcorn in each of two bags changes the equation from $2x$ to $2(4)$.

HANDS-ON
LAB

COOPERATIVE LEARNING

 paper bags

🍿 popcorn

1-5A Variables and Expressions

A Preview of Lesson 1-5

In algebra, we often use unknown values. Consider the expression *the sum of three and some number*. The "some number" is an unknown value. When you assign a value to "some number," then you can find the value of the expression.

TRY THIS

Work with a partner.

1 To model the expression *the sum of three and some number*, place three pieces of popcorn next to an empty paper bag. The three pieces of popcorn represent the known value, 3. The bag represents the unknown value. It can contain any number of popcorn pieces.

the sum of three and some number

2 To evaluate the expression *the sum of three and some number*, follow these steps.

• Have one partner assign a value to "some number" by placing any number of the remaining pieces of popcorn in the bag.

• Empty the bag and count the pieces of popcorn that are in the bag. Add this to the other 3 pieces of popcorn. This total number is the value of the expression.

ON YOUR OWN

1. What is the "some number?" **the unknown value**

2. In the expression *the sum of three and some number*, replace "some number" with its known value. What is the value of the expression? **See students' work.**

3. Find the value of *some number plus eight* if "some number" is 5. **13**

4. *Look Ahead* Find the value of $m + 7$ if $m = 12$. **19**

20 Chapter 1 Problem Solving, Numbers, and Algebra

Have students complete Exercises 1–3 and 5–7. Encourage students to draw pictures to help them visualize the problems.

Use Exercises 4 and 8 to determine whether students understand the use of variables.

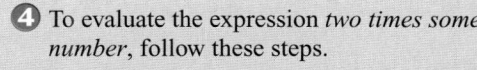

TRY THIS

Work with a partner.

3 To model the expression *two times some number,* place two bags in front of you. Each bag represents the same unknown value.

two times some number

4 To evaluate the expression *two times some number*, follow these steps.

- Have one partner assign a value to "some number" by placing the same number of pieces of popcorn in each bag.

- Empty the bags and count all of the pieces of popcorn. The number of pieces of popcorn is the value of the expression.

ON YOUR OWN

5. What is the "some number?"   **the unknown value**

6. In the expression *two times some number*, replace "some number" with its known value. What is the value of the expression?   **See students' work.**

7. How many bags would you need to model the expression *five times a number*?   **5 bags**

8. *Look Ahead* Find the value of $4 \times h$ if $h = 6$.   **24**

Lesson 1-5A   HANDS-ON LAB   **21**

Math Journal Have students write a paragraph explaining how the models show variables. Have them include a description of how they could use the models if they forget the pattern they discovered in this lab.

Hands-On Lab Masters, p. 39

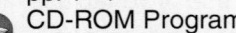

Integration: Algebra
Variables and Expressions

- *Study Guide Masters*, p. 5
- *Practice Masters*, p. 5
- *Enrichment Masters*, p. 5
- Transparencies 1-5, A and B
- *Science and Math Lab Manual*, pp. 1–4

- CD-ROM Program
  - Resource Lesson 1-5
  - Interactive Lesson 1-5

Recommended Pacing

Standard	Days 6 & 7 of 12
Honors	Days 6 & 7 of 12
Block	Day 4 of 6

1 FOCUS

5-Minute Check
(Lesson 1-4)

Find the value of each expression.
1. $12 - 4 + 6 - 7$ **7**
2. $63 \div 7 + 16 \times 2$ **41**
3. $81 \div 9 + 6 - 5 \times 3$ **0**
4. $18 + 56 \div 8 - 3$ **22**
5. $5 \cdot 8 + 6 - 81 \div 9$ **37**

The 5-Minute Check is also available on **Transparency 1-5A** for this lesson.

Motivating the Lesson
Problem Solving Ask students what further information they would need to solve the following problems.
- If a pound of grapes costs $0.79, what is the total cost? **How many pounds of grapes were purchased?**
- How many buses are needed for 365 students? **How many students fit on a bus?**

What you'll learn
You'll learn to evaluate numerical and simple algebraic expressions.

When am I ever going to use this?
Knowing how to evaluate algebraic expressions can help you find the miles per gallon achieved by a car.

Word Wise
algebra
variable
algebraic expression
evaluate

Western Park Mall offers a gift wrapping service. The cost for gift wrapping is found by adding the length and the width of the box and multiplying by 5 cents. The table below shows how to find the cost of wrapping the most common size boxes. Study the pattern.

Length	Width	Length + Width	Cost (cents)
10	4	10 + 4	(10 + 4) × 5
12	6	12 + 6	(12 + 6) × 5
16	8	16 + 8	(16 + 8) × 5
20	10	20 + 10	(20 + 10) × 5

In mathematics, **algebra** is a language of symbols. By extending the table, using the letters L and W, the following table results.

Length	Width	Length + Width	Cost (cents)
L	W	$L + W$	$(L + W) \times 5$

The letters L and W are called **variables**. A variable is a symbol, usually a letter, used to represent a number. Any letter may be used as a variable. In this problem, the letter L represents length and W represents width.

The expression $(L + W) \times 5$ is called an **algebraic expression**. Algebraic expressions are combinations of variables, numbers, and at least one operation. $(L + W) \times 5$ is an algebraic expression for the cost of gift wrapping

The variables can be replaced with any number. Once the variables have been replaced, you can **evaluate**, or find the value of, the algebraic expression.

Example

1 Evaluate $14 + c$ if $c = 32$.

$14 + c = 14 + 32$ *Replace c with 32.*
$= 46$ *Add 14 and 32.*

Multiple Learning Styles

Verbal/Linguistic Ask students to write word problems that can be written as algebraic expressions. Guide them in the selection of topics, such as problems dealing with money, weight, or distance. Have them review their problems to make sure all necessary information is there. Then ask them to write the algebraic expressions for their problems.

Example ② Evaluate $x - y$ if $x = 47$ and $y = 13$.

$$x - y = 47 - 13 \quad \textit{Replace x with 47 and y with 13.}$$
$$= 34 \quad \textit{Subtract 13 from 47.}$$

In algebra, there are several ways to show multiplication.

$3 \cdot m$ ⟦means⟧⟶ $3 \times m$

$3m$ ⟦means⟧⟶ $3 \times m$

mn ⟦means⟧⟶ $m \times n$

Be sure to use the order of operations when you evaluate algebraic expressions.

Examples ③ Evaluate $3m + 2 \cdot 4$ if $m = 12$.

$$3m + 2 \cdot 4 = 3 \times 12 + 2 \cdot 4 \quad \textit{Replace m with 12.}$$
$$= 36 + 2 \cdot 4 \quad \textit{Multiply 3 and 12.}$$
$$= 36 + 8 \quad \textit{Multiply 2 and 4.}$$
$$= 44 \quad \textit{Add 36 and 8.}$$

> **Study Hint**
> **Reading Math** The expression $3m + 2 \cdot 4$ is read as *three times m plus two times four.*

④ Evaluate $4 + mn$ if $m = 6$ and $n = 2$.

$$4 + mn = 4 + 6 \times 2 \quad \textit{Replace m with 6 and n with 2.}$$
$$= 4 + 12 \quad \textit{Multiply 6 and 2.}$$
$$= 16 \quad \textit{Add 4 and 12.}$$

APPLICATION ⑤ **Hobbies** To find distance traveled, you can use the expression $r \times t$, where r represents rate and t represents time. Find the distance a hot air balloon traveled if its rate was 13 miles per hour and it traveled for 4 hours.

Evaluate $r \times t$ if $r = 13$ and $t = 4$.

$$r \times t = 13 \times 4 \quad \textit{Replace r with 13 and t with 4.}$$
$$= 52 \quad \textit{Multiply 13 and 4.}$$

The hot air balloon traveled 52 miles.

Lesson 1–5 Integration: Algebra   Variables and Expressions   **23**

---

2 TEACH

 Transparency 1-5B contains a teaching aid for this lesson.

Reading Mathematics Algebra is used every day by scientists, engineers, and businesspeople. The variables and symbols used in algebra represent numbers and operations. Ask students what symbols they might find along a highway or at an airport and discuss what they mean.

In-Class Examples

For Example 1
Evaluate $c + 24$ if $c = 9$. **33**

For Example 2
Evaluate $x - y$ if $x = 18$ and $y = 6$. **12**

For Example 3
Evaluate $5m + 4 \cdot 5$ if $m = 8$. **60**

For Example 4
Evaluate $6 + mn$ if $m = 8$ and $n = 3$. **30**

For Example 5
How much would it cost to gift wrap a package that is 24 inches long and 18 inches wide if the algebraic expression for the cost of gift wrapping is $(L + W) \times \$0.06$? **\$2.52**

Teaching Tip Note that for a product of a number and a variable, the number is usually written first, followed by the variable. For example, the product of m and 3 is expressed as $3m$, not $m3$.

Teaching Tip Point out to students that if x is the variable, they should refrain from using $\times$ for multiplication to avoid confusion.

Lesson 1-5   **23**

Check for Understanding

If students need additional practice or instruction after completing Exercises 1–12, one of these options may be helpful.
- Extra Practice, see p. 558
- Reteaching Activity
- *Study Guide Masters,* p. 5
- *Practice Masters,* p. 5
- 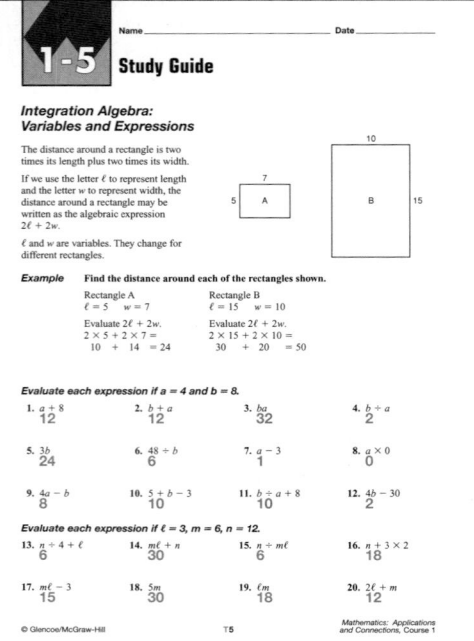 Interactive Mathematics Tools Software

Assignment Guide

Core: 13–39 odd, 41–46
Enriched: 14–38 even, 39–46

Study Guide Masters, p. 5

CHECK FOR UNDERSTANDING

Communicating Mathematics

Read and study the lesson to answer each question.

1. *Write* two algebraic expressions that mean the same as $5r$. $5 \times r$ or $5 \cdot r$

2. *Copy and complete* the chart.

Algebraic Expressions	Variables	Numbers	Operations
$4x - 2y$	x, y	4, 2	$-, \times$
$3r - 2s + 5t$	r, s, t	3, 2, 5	$-, +, \times$

3. *You Decide* Laura says that the value of $g + h \div 2$ when $g = 2$ and $h = 14$ is 8. Nicholas says the answer is 9. Who is correct? Explain your reasoning. **Nicholas;** $2 + 14 \div 2 = 9$

Guided Practice

Evaluate each expression if $a = 3$ and $b = 12$.

4. $a + 5$ **8**   5. $48 \div b$ **4**   6. $7ab$ **252**   7. $2a + b$ **18**

Evaluate each expression if $c = 5$, $d = 2$, and $f = 8$.

8. $f - d$ **6**   9. $32 \div d$ **16**   10. $6d + c$ **17**   11. $c + d + f$ **15**

12. *Geometry* The expression $\ell \times w$ gives the area of a rectangle. The letter ℓ represents the length of the rectangle, and w represents the width. Find the area of each rectangle.

a.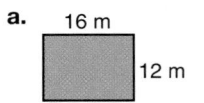
16 m
12 m
192 square meters

b.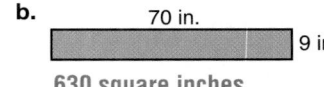
70 in.
9 in.
630 square inches

EXERCISES

Practice

Evaluate each expression if $n = 5$ and $p = 7$.

13. $n + 5$ **10**   14. $n \times 0$ **0**   15. $4p - 6$ **22**   16. $12 - n$ **7**

17. $3np$ **105**   18. $8 + p$ **15**   19. $2n \times 6$ **60**   20. $n \div 1$ **5**

21. $p \times n$ **35**   22. $2p - p$ **7**   23. $3n + p$ **22**   24. $21 \div 3 \times p$ **49**

Evaluate each expression if $r = 6$, $s = 12$, and $t = 3$.

25. $s - t$ **9**   26. $s \div 2$ **6**   27. rt **18**   28. $4 + 2r$ **16**

29. $t \times 18$ **54**   30. $48 \div s$ **4**   31. $s - r + t$ **9**   32. $rt - s$ **6**

33. $3s - 4t$ **24**   34. $4 + 2rt$ **40**   35. $3rt - 2s$ **30**   36. $s - 2r \div t$ **8**

37. *Algebra* Find the value of $3z \div y$ if $y = 3$ and $z = 9$. **9**

38. *Algebra* What is the value of $2xyz - 26$ if $x = 3$, $y = 6$, and $z = 10$? **334**

■ Reteaching the Lesson ■

Activity Ask students to write variables, numbers, and operation signs on individual paper squares. Have pairs of students take turns making up expressions by arranging the paper squares in different combinations, choosing numbers to replace the variables, and evaluating the expressions.

Applications and Problem Solving

39. Geometry To find the diameter of a circle, you can use the expression $2r$, where r is the length of the radius. Find the diameter of a circle whose radius is 6 inches. **12 inches**

6 in.

40. Travel To find the miles per gallon achieved by a car, you can use the expression $m = d \div g$ where m represents miles per gallon, d represents the distance traveled, and g represents the number of gallons of gasoline used.

   a. Find the miles per gallon for a car that travels 180 miles on 5 gallons of gasoline. **36 mpg**

   b. Find the miles per gallon for a car that travels 256 miles on 8 gallons of gasoline. **32 mpg**

41. Critical Thinking Christina and Shawon each have a calculator. Shawon starts at zero and adds three each time. Christina starts at 100 and subtracts seven each time. If they press their keys at the same time, will their displays ever show the same number at the same time? If so, what is the number? **yes; 30**

Mixed Review

42. Find the value of $36 \div 9 + 7 \times 2$. *(Lesson 1-4)* **18**

43. Evaluate $2 \times 18 \div 3 + 9$. *(Lesson 1-4)* **21**

44. Shopping The graph shows how much consumers spent shopping on-line in 1996. *(Lesson 1-3)*

Shopping On-Line

$140	Computer products
$126	Travel
$85	Entertainment
$46	Apparel
$45	Gifts/flowers
$39	Food/drink

(millions)

   a. Estimate the total amount spent shopping on-line.

   b. About how much more was spent on travel than on apparel? **about $80 million**

44a. Sample answer: $500 million

45. **Test Practice** What number is missing from this pattern?
   ..., ☐, 35, 41, 47, 53, ... *(Lesson 1-2)* **A**
   A 29
   B 31
   C 26
   D 24

46. Business In one month, Pam collected $51 from the customers on her paper route. If she owed the newspaper company $32, how much was her profit? *(Lesson 1-1)* **$19**

Extending the Lesson

Enrichment Masters, p. 5

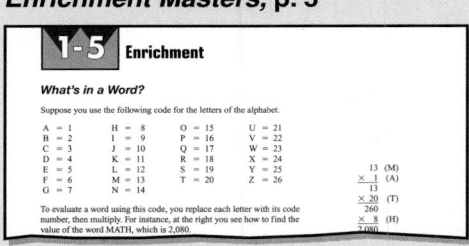

1-5 Enrichment

What's in a Word?

Suppose you use the following code for the letters of the alphabet.

A = 1	H = 8	O = 15	U = 21
B = 2	I = 9	P = 16	V = 22
C = 3	J = 10	Q = 17	W = 23
D = 4	K = 11	R = 18	X = 24
E = 5	L = 12	S = 19	Y = 25
F = 6	M = 13	T = 20	Z = 26
G = 7	N = 14		

To evaluate a word using this code, you replace each letter with its code number, then multiply. For instance, at the right you see how to find the value of the word MATH, which is 2,080.

13 (M)
× 1 (A)
13
× 20 (T)
260
× 8 (H)
2,080

Activity Write an algebraic expression for each problem.
• Mei Ling counted 24 robins at her feeder. She saw other birds that were not robins. **24 + x**
• Suppose Mei Ling counted a total of 36 birds. How would you change the expression? **24 + x = 36**

Closing Activity

Modeling Write an algebraic expression on the chalkboard. Have a student roll a number cube to get a number to replace each variable. Then have another student evaluate the expression.

Practice Masters, p. 5

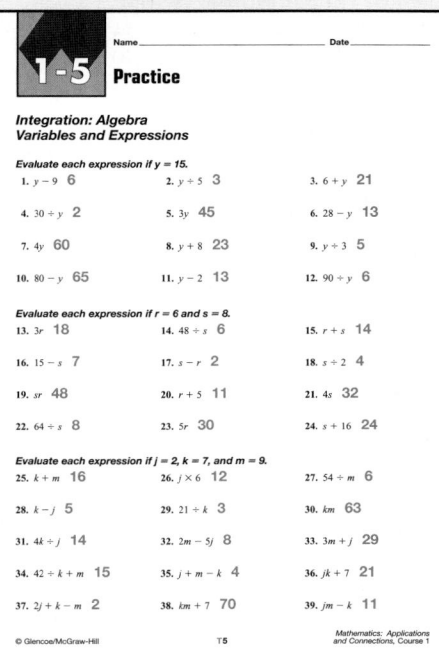

Name _____ Date _____

1-5 Practice

Integration: Algebra
Variables and Expressions

Evaluate each expression if y = 15.
1. $y - 9$ **6**   2. $y \div 5$ **3**   3. $6 + y$ **21**
4. $30 \div y$ **2**   5. $3y$ **45**   6. $28 - y$ **13**
7. $4y$ **60**   8. $y + 8$ **23**   9. $y + 3$ **5**
10. $80 - y$ **65**   11. $y - 2$ **13**   12. $90 \div y$ **6**

Evaluate each expression if r = 6 and s = 8.
13. $3r$ **18**   14. $48 \div s$ **6**   15. $r + s$ **14**
16. $15 - s$ **7**   17. $s - r$ **2**   18. $s \div 2$ **4**
19. sr **48**   20. $r + 5$ **11**   21. $4s$ **32**
22. $64 \div s$ **8**   23. $5r$ **30**   24. $s + 16$ **24**

Evaluate each expression if j = 2, k = 7, and m = 9.
25. $k + m$ **16**   26. $j \times 6$ **12**   27. $54 \div m$ **6**
28. $k - j$ **5**   29. $21 \div k$ **3**   30. km **63**
31. $4k \div j$ **14**   32. $2m - 5j$ **8**   33. $3m + j$ **29**
34. $42 \div k + m$ **15** 35. $j + m - k$ **4**   36. $jk + 7$ **21**
37. $2j + k - m$ **2**   38. $km + 7$ **70**   39. $jm - k$ **11**

© Glencoe/McGraw-Hill   T5   Mathematics: Applications and Connections, Course 1

GET READY

Objective Students evaluate algebraic expressions by using a graphing calculator.

Technology Resources
• TI-80, TI-81, TI-82, or TI-83 graphing calculator

MANAGEMENT TIPS

Recommended Time
40 minutes

Getting Started Allow students plenty of time to explore the various keys on their calculators. Have students practice using the alpha keys by entering their names. Also have them enter a few expressions to evaluate so they get used to how the different symbols look on the calculator screen.

Using a Graphing Calculator Discuss with students some of the advantages of using a graphing calculator for non-graphing calculations. Some advantages: larger screen, shows everything that is typed in and the answer, and the ability to store up to 26 numbers. Students can also re-evaluate the expression without re-keying every part of the expression by pressing ⎡2nd⎤ ⎡ENTER⎤ and using the arrow keys to find the appropriate numbers to replace.

ASSESS

After students answer Exercises 1–8, ask them to develop a problem that would be difficult to do mentally. Then have them work the problem on the graphing calculator.

Use Exercise 9 to determine whether students understand evaluating expressions on a calculator.

1-5B Evaluating Expressions

A Follow-Up of Lesson 1-5

✎ graphing calculator

Graphing calculators observe the order of operations. So, you can enter an expression just as it is written to evaluate it. You can also use parentheses keys on a graphing calculator to indicate multiplication.

The expression appears on the screen as you enter it. On TI calculators, the multiplication and division signs appear on the screen as * and /.

TRY THIS

Work with a partner.

Evaluate $3(x - 6) + 1$ if $x = 8$.

3 ⎡(⎤ 8 ⎡−⎤ 6 ⎡)⎤ ⎡+⎤ 1 ⎡ENTER⎤ *You can use arrow keys to go back and correct any error you make by typing over it, or by using the [INS] (insert) key, or by using the ⎡DEL⎤ (delete) key.*

If $x = 8$, the value of $3(x - 6) + 1$ is 7.

If you discover that you entered the expression incorrectly, you can use the REPLAY feature to correct your error and reevaluate without reentering your expression. Press ⎡2nd⎤ [ENTRY]. Then use the arrow keys to move to the location of the correction. After you make the correction, press ⎡ENTER⎤ to evaluate. You don't have to move the cursor to the end of the line.

ON YOUR OWN

Use a graphing calculator to evaluate each expression if $x = 4$, $y = 7$, and $z = 9$.

1. $12 - z$ **3**
2. $x + 9$ **13**
3. xy **28**
4. $2(18 - z)$ **18**
5. $x \div 2 + 12$ **14**
6. $\frac{2(z-x)}{(y-2)}$ **2**
7. $14 + \frac{3x}{2}$ **20**
8. $x(y + z) - x$ **60**

9. Evaluate the expression $5(3 + 8)$ with a scientific calculator and with a graphing calculator. Can you enter the expression the same way on both calculators? If not, explain the difference. **See Answer Appendix.**

26 Chapter 1 Problem Solving, Numbers, and Algebra

Math Journal Have students write a paragraph that compares a graphing calculator, a scientific calculator, and a pencil and paper for evaluating expressions. Have students explain how to use the alpha key on a graphing calculator.

BUSINESS

SCHOOL to CAREER

Gary C. Comer
MAIL-ORDER BUSINESS OWNER

Gary C. Comer is the founder and Chairman of the Board of Land's End, Inc., a mail-order business. When he first founded the company in 1963, there was one catalog in which sailboat hardware and equipment was sold. Today, there are eight different catalogs. Consumers can purchase unique clothing and accessories, and products for the home.

If you are interested in owning your own mail-order business, you should consider taking courses in business management, accounting, and mathematics. A bachelor's degree in marketing and advertising, communications, business management, or accounting is also helpful.

For more information:
The Direct Marketing Association
1120 Avenue of the Americas
New York, NY 10036-6700

interNET CONNECTION
www.glencoe.com/sec/
math/mac/mathnet

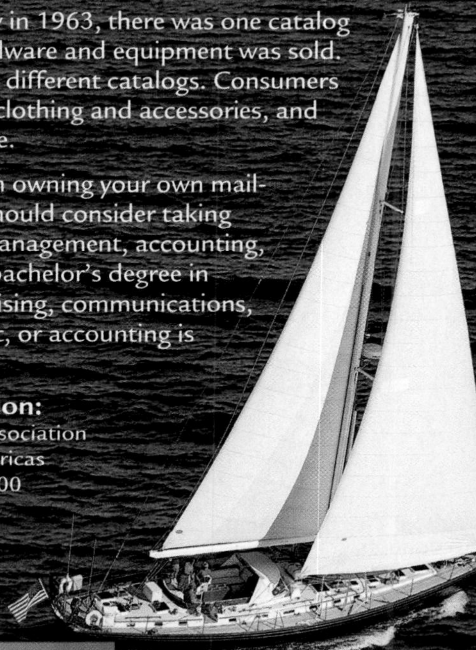

Someday, I'd like to start a mail-order business. I would travel around the world to find unusual products to sell in my catalog.

Your Turn
Design a catalog for a mail-order business you would like to start. Select a name for your company and describe and illustrate the products you want to sell.

School to Career: Business **27**

More About Gary C. Comer
- Before he started *Lands' End,* Gary Comer was an advertising copywriter and an avid sailor. He won the North American Star Boat Championship, finished second in the World Star Boat Championship, and finished third in the Pan American Games. Then he decided to start a business that involved sailboat racing.
- *Lands' End* originally sold sailboat equipment only. Now the company sells casual and tailored clothing, children's clothing, products for the home, and corporate gifts. On an average day, the catalog's phone lines handle between 40,000 and 50,000 calls. During the Christmas season, *Lands' End* receives 100,000 calls daily.

Motivating Students
Mail-order catalogs are a popular way of selling retail merchandise. Many companies have catalogs in addition to retail outlets, or stores. Other companies sell their merchandise exclusively through catalogs. To begin a discussion of mail-order businesses, ask students these questions.
- What mail-order catalogs have you read? **Sample answer:** *Spiegel, The Sharper Image, Cheryl & Co., L. L. Bean, Doubleday Book Club*
- Have you ever purchased anything from a catalog? If so, what? **Sample answer: sport cards, games, CDs, videotapes, clothing**
- Why might shopping by mail be preferable to shopping in a store? **Sample answer: convenience, many catalog companies take orders 24 hours a day, merchandise is delivered to your door, presents are delivered to the recipient**

Making the Math Connection
The owner of a mail-order business needs a thorough understanding of the mathematical aspects of retailing—keeping track of inventory, using sales figures to determine buying strategies, and computing profits and losses.

Working on *Your Turn*
Have students work in pairs. Make catalogs available to help students choose a design. Encourage them to use their creativity to illustrate and describe products so that customers would be eager to make purchases. Then have each pair give a brief oral presentation explaining their company and sharing their catalog with classmates.

*An additional School to Career activity is available on page 1 of the **School to Career Masters.***

Integration: Algebra
Powers and Exponents

Instructional Resources
- *Study Guide Masters*, p. 6
- *Practice Masters*, p. 6
- *Enrichment Masters*, p. 6
- Transparencies 1-6, A and B
- *Assessment and Evaluation Masters*, p. 16
- *Classroom Games*, pp. 1–2
- *Technology Masters*, p. 2

 CD-ROM Program
- Resource Lesson 1-6

Recommended Pacing	
Standard	Day 8 of 12
Honors	Day 8 of 12
Block	Day 4 of 6

FOCUS

5-Minute Check
(Lesson 1-5)

Evaluate each expression if r = 2 and s = 8.
1. $r \times 6$ **12**
2. $s \div 4$ **2**

Evaluate each expression if d = 3, f = 6, and g = 12.
3. $d + f + g$ **21**
4. $g \div d + f$ **10**
5. $6f - 4$ **32**

The 5-Minute Check is also available on **Transparency 1-6A** for this lesson.

Motivating the Lesson
Hands-On Activity After reading the beginning of the lesson, have students use their calculators to find the product of $10 \cdot 10 \cdot 10 \cdot 10$. **10,000**
Then have them choose any number to multiply by itself several times.

What you'll learn
You'll learn to use powers and exponents in expressions.

When am I ever going to use this?
Knowing how to use powers and exponents can help you determine the amount of water needed to fill an aquarium.

Word Wise
factor
exponent
base
power
squared
cubed

The state of Hawaii has an area of more than 10,000 square miles. You can express 10,000 as the product $10 \cdot 10 \cdot 10 \cdot 10$.

When two or more numbers are multiplied, each number is called a **factor** of the product. For $3 \times 12 = 36$, 3 and 12 are factors of 36.

When the same factor is repeated, you can use an **exponent** to simplify the notation. An exponent tells how many times a number, called the **base**, is used as a factor.

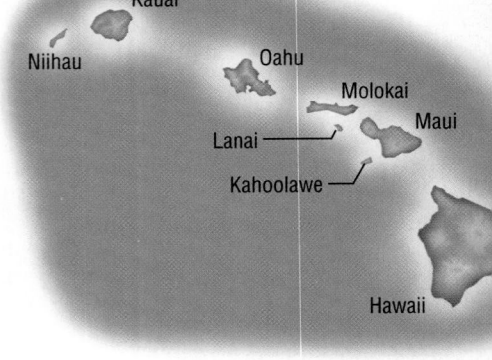

$$10,000 = \underbrace{10 \cdot 10 \cdot 10 \cdot 10}_{four\ factors} = 10^4 \leftarrow exponent$$
$$\uparrow base \quad 10^4 \text{ is a power of 10.}$$

A **power** is a number that is expressed using exponents. The powers 12^2, 5^3, and 7^4 are read as follows.

Symbols	Words
12^2	12 to the second power or 12 **squared.**
5^3	5 to the third power or 5 **cubed.**
7^4	7 to the fourth power.

 Examples

1 Write $8 \cdot 8 \cdot 8 \cdot 8$ using exponents.

The base is 8.
Since 8 is a factor four times, the exponent is 4.
$$8 \cdot 8 \cdot 8 \cdot 8 = 8^4$$

2 Write 4^3 as a product. Then evaluate.

The base is 4.
The exponent 3 means that 4 is a factor three times.
$$4^3 = 4 \cdot 4 \cdot 4$$
$$= 64$$

Since powers are a form of multiplication, the rules for the order of operations need to be expanded to include them.

Order of Operations	1. Do all powers before other operations. 2. Multiply and divide in order from left to right. 3. Add and subtract in order from left to right.

Multiple Learning Styles

Kinesthetic Dried beans and two number cubes will be needed. Roll the number cubes. The greatest number will be the base. The other number will be the exponent. As you roll the cubes, have students write the expression and model it with beans. Then have them count the beans to evaluate the expression. For example, $4^3 = 4$ groups of 4 groups of 4 = 64 beans.

Example ③ Evaluate $5 \cdot 3^2 - 8$.

$$5 \cdot 3^2 - 8 = 5 \cdot 9 - 8 \quad \textit{Evaluate } 3^2 \textit{ first. } 3^2 = 3 \cdot 3 \textit{ or } 9.$$
$$= 45 - 8 \quad \textit{Multiply 5 and 9.}$$
$$= 37 \quad \textit{Subtract 8 from 45.}$$

Powers and exponents are often used in algebra.

Examples

INTEGRATION

Algebra

④ Write $n \cdot n \cdot n \cdot n \cdot n$ using exponents.

The base is n.
Since n is a factor five times, the exponent is 5.
$n \cdot n \cdot n \cdot n \cdot n = n^5$

⑤ Write d^4 as a product.

The base is d.
The exponent 4 means that d is a factor four times.
$d^4 = d \cdot d \cdot d \cdot d$

⑥ Evaluate g^6 if $g = 3$.

$g^6 = 3^6 \quad \textit{Replace g with 3.}$
$= 3 \cdot 3 \cdot 3 \cdot 3 \cdot 3 \cdot 3$
$= 729$

You can also use a calculator to evaluate 3^6.

$3 \boxed{y^x} 6 \boxed{=} \; 729$

> **Study Hint**
>
> **Technology** Many calculators have a $\boxed{y^x}$ key. You can use this key to evaluate powers. To find 12^4, enter $12 \boxed{y^x} 4 \boxed{=}$.

CHECK FOR UNDERSTANDING

Communicating Mathematics

Read and study the lesson to answer each question. 1. See margin.

1. *Write* a paragraph explaining the terms *exponent, base,* and *power*.

2. *Demonstrate* how to evaluate 5^8 using a calculator. $5 \boxed{y^x} 8 \boxed{=} \; 390,625$

3. *You Decide* Esteban says that the value of $3^2 + 2^4 \cdot 6$ is 150. Tamika says the value is 105. Who is correct? Explain your reasoning.
Tamika; $3^2 + 2^4 \cdot 6 = 105$

Guided Practice

Write each product using exponents.

4. $2 \cdot 2 \cdot 2 \cdot 2 \cdot 2 \cdot 2$ 2^6 5. $m \cdot m \cdot m \cdot m \cdot m$ m^5 6. $3 \cdot 3 \cdot 5 \cdot 5 \cdot 5$ $3^2 \cdot 5^3$

Write each power as a product.

7. c^3 $c \cdot c \cdot c$ 8. 4^7 9. h^6 $h \cdot h \cdot h \cdot h \cdot h \cdot h$
$4 \cdot 4 \cdot 4 \cdot 4 \cdot 4 \cdot 4 \cdot 4$

Evaluate each expression.

10. 2^3 **8** 11. 10^5 **100,000** 12. $5^2 + 4 \cdot 2$ **33**

13. *Foreign Languages* Mandarin Chinese is spoken by more people in the world than any other language. An estimated 10^9 people speak this language. About how many people in the world speak Mandarin Chinese?
about 1,000,000,000 people

■ **Reteaching the Lesson** ■

Activity Have groups of students write word problems that use powers. Have them think of situations such as bacteria growth that doubles every hour or someone who tells two friends, who tell two friends, and so on. Trade problems between two groups and have the second group check the appropriateness of the problem and solve it.

Additional Answer

1. An exponent is a number that tells how many times a number, called the base, is used as a factor. A power is a number that is expressed using exponents.

 Transparency 1-6B contains a teaching aid for this lesson.

Reading Mathematics Write the words *factor, exponent, base, power, squared,* and *cubed* on an overhead transparency. Say the words. Then write an example of each as you say it again. Label the examples. Then give students a word and have them identify the corresponding example. Conclude by pointing out that the term *power* is sometimes used to refer to the exponent itself.

> **In-Class Examples**
>
> *For Example 1*
> Write $5 \cdot 5 \cdot 5$ using exponents.
> 5^3
>
> *For Example 2*
> Write 6^5 as a product. Then evaluate. $6 \cdot 6 \cdot 6 \cdot 6 \cdot 6 = 7,776$
>
> *For Example 3*
> Evaluate $2 \cdot 6^3 + 7$. **439**
>
> *For Example 4*
> Write $a \cdot a \cdot a \cdot a$ using exponents. a^4
>
> *For Example 5*
> Write k^6 as a product.
> $k \cdot k \cdot k \cdot k \cdot k \cdot k$
>
> *For Example 6*
> Evaluate r^3 if $r = 8$. **512**

Teaching Tip Encourage students to expand their order of operations' mnemonic to include powers. For example, they could say **P**lease, **M**y **D**ear **A**unt **S**ally.

Check for Understanding

If students need additional practice or instruction after completing Exercises 1–13, one of these options may be helpful.

• Extra Practice, see p. 558
• Reteaching Activity
• *Study Guide Masters,* p. 6
• *Practice Masters,* p. 6

EXERCISES

Practice

Write each product using exponents.

17. $3^2 \cdot 4^4$
18. $15^2 \cdot 8^2$
20. $y^3 \cdot z^2$
21. $6^3 \cdot 1^3$
24. $21 \cdot 21 \cdot 21 \cdot 21$
25. $7 \cdot 7 \cdot 7 \cdot 7 \cdot 7$
26. $9 \cdot 9 \cdot 9 \cdot 2 \cdot 2 \cdot 2 \cdot 2$

14. $11 \cdot 11 \cdot 11$ 11^3
15. $3 \cdot 3 \cdot 3 \cdot 3 \cdot 3 \cdot 3$ 3^6
16. $w \cdot w \cdot w$ w^3
17. $3 \cdot 3 \cdot 4 \cdot 4 \cdot 4 \cdot 4$
18. $15 \cdot 15 \cdot 8 \cdot 8$
19. $r \cdot r \cdot r \cdot r$ r^4
20. $y \cdot y \cdot y \cdot z \cdot z$
21. $6 \cdot 6 \cdot 6 \cdot 1 \cdot 1 \cdot 1$
22. $a \cdot a \cdot b \cdot b$ $a^2 \cdot b^2$

Write each power as a product. 27–30. See margin.

23. 14^2 $14 \cdot 14$
24. 21^4
25. 7^5
26. $9^3 \cdot 2^4$
27. 16^6
28. a^7
29. $x^3 \cdot y^4$
30. $b^2 \cdot c^4 \cdot d^3$

Evaluate each expression.

31. 10^3 $1,000$
32. 6^3 216
33. 2^5 32
34. 16^2 256
35. 3 squared 9
36. 4 cubed 64
37. $4^5 + 3$ $1,027$
38. $4^2 \cdot 3^3 \cdot 6$ $2,592$
39. $4 \cdot 5^2 + 3$ 103

40. *Algebra* What is the value of $x^3 + y^2$ if $x = 3$ and $y = 6$? 63

41. *Algebra* If $m = 4$ and $n = 5$, what is the value of $m^2 \times n^3$? $2,000$

Applications and Problem Solving

42. *Earth Science* A light year is the distance light travels in one year. The Milky Way galaxy is about 100,000 light years wide. Write 100,000 as a power with 10 as the base. 10^5

43. *Geometry* The volume of a cube can be found by using the expression $V = s^3$, where s is the length of a side. The aquarium shown at the right is in the shape of a cube. What is its volume? *The volume will be expressed in cubic units.* $5,832 \text{ in}^3$

18 in.
18 in.
18 in.

44. *Critical Thinking* Evaluate 10^3, 10^4, 10^5, and 10^6. Explain how you can evaluate 10^{20} without using a calculator. **See margin.**

Mixed Review

45. *Algebra* Evaluate the expression $ab - a$ if $a = 7$ and $b = 3$. *(Lesson 1-5)* 14

46. **Test Practice** The Marcos family is going to a high school football game. Tickets cost $5 for adults and $2 for students. Mrs. Marcos will need 2 adult tickets and 4 student's tickets. Which expression could be used to find the cost in dollars of the tickets? *(Lesson 1-4)*   **E**

A $2 + \$5 \times 4 + \$2$
B $\$5 + \$2 + 6$
C $\$2 + 4 \times \$5 + 2$
D $6 \times \$5 + \$2$
E $2 \times \$5 + 4 \times \$2$

30   **Chapter 1**   Problem Solving, Numbers, and Algebra

💡 **Investigations for the Special Education Student**

This blackline master booklet helps you plan for the needs of your special education students by providing long-term projects along with teacher notes. Investigation 1, *Vacation Getaways,* may be used with this chapter.

47. Use rounding to estimate 398 + 688 + 241. Explain whether 1,337 is a reasonable answer. *(Lesson 1-3)* **400 + 700 + 200 = 1,300; reasonable**

48. *Patterns* Find the next three numbers in the pattern 7, 12, 17, 22, ... *(Lesson 1-2)* **27, 32, 37**

49. *School* The number of students at Aurora Junior High is half the number it was twenty years ago. If there were 758 students twenty years ago, how many students are there now? *(Lesson 1-1)* **379 students**

Let the Games Begin

EXPO Bingo

Get Ready This game is for two to four players.

📋 calculator

Get Set Each player copies the EXPO bingo playing card onto a sheet of paper. Then each player selects 16 different powers and writes them in any of the upper right hand corner boxes. Choose a person to be the bingo caller.

EXPO BINGO CARD

5^2	12^2	3^3	2^4
7^2	2^3	8^2	3^2
1^3	4^3	9^2	11^2
6^2	14^2	5^3	15^2
10^3	3^4	13^2	2^2

Go ● The caller reads one power at a time. Each player marks the space by writing the equivalent number in the larger box.

● The first player with four powers listed in any row, column, or diagonal wins.

💻 interNET CONNECTION Visit www.glencoe.com/sec/math/mac/mathnet for more games.

Lesson 1–6 Integration: Algebra Powers and Exponents **31**

■ Extending the Lesson ■

Enrichment Masters, p. 6

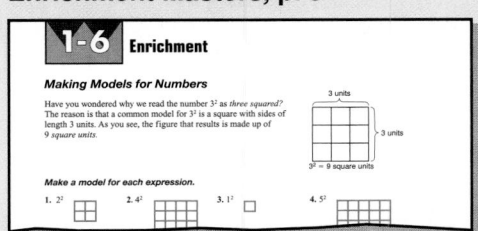

1-6 Enrichment

Making Models for Numbers

Have you wondered why we read the number 3² as *three squared*? The reason is that a common model for 3² is a square with sides of length 3 units. As you see, the figure that results is made up of 9 square units.

3 units

3 units

3² = 9 square units

Make a model for each expression.

1. 2²   2. 4²   3. 1²   4. 5²

Let the Games Begin

Students can use the game to practice their knowledge of powers and exponents. Make sure students know how to enter expressions with exponents on their calculators so that the game proceeds smoothly and fairly.
*Additional Resources for this game can be found on page 41 of the **Classroom Games**.*

Closing Activity
Speaking Write several expressions on the board that use the same factor multiple times. Ask student volunteers to describe how they would change these expressions to exponential expressions. Then have them evaluate the expressions using mental math or a calculator.

Chapter 1, Quiz C (Lessons 1-5 and 1-6) is available in the *Assessment and Evaluation Masters*, p. 16.

Practice Masters, p. 6

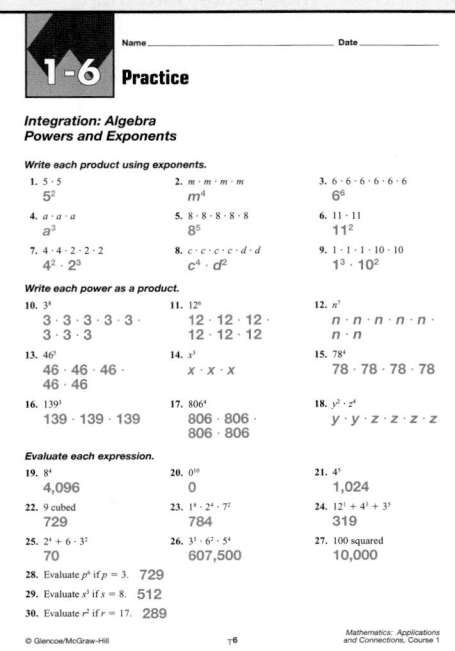

Name _____ Date _____

1-6 Practice

Integration: Algebra
Powers and Exponents

Write each product using exponents.

1. 5 · 5
   5^2
2. $m · m · m$
   m^4
3. 6 · 6 · 6 · 6 · 6 · 6
   6^6
4. $a · a · a$
   a^3
5. 8 · 8 · 8 · 8 · 8
   8^5
6. 11 · 11
   11^2
7. 4 · 4 · 2 · 2 · 2
   $4^2 · 2^3$
8. $c · c · c · c · d · d$
   $c^4 · d^2$
9. 1 · 1 · 1 · 10 · 10
   $1^3 · 10^2$

Write each power as a product.

10. 3^5
    3 · 3 · 3 · 3 · 3 · 3 · 3
11. 12^6
    12 · 12 · 12 · 12 · 12 · 12
12. n^7
    n · n · n · n · n · n · n
13. 46^3
    46 · 46 · 46
14. x^3
    x · x · x
15. 78^4
    78 · 78 · 78 · 78
16. 139^3
    139 · 139 · 139
17. 806^4
    806 · 806 · 806 · 806
18. $y^3 · z^4$
    y · y · y · z · z · z · z

Evaluate each expression.

19. 8^4
    4,096
20. 0^{10}
    0
21. 4^5
    1,024
22. 9 cubed
    729
23. $1^4 · 2^4 · 7^2$
    784
24. $12^1 + 4^3 + 3^7$
    319
25. $2^4 + 6 · 3^2$
    70
26. $3^3 · 6^2 · 5^4$
    607,500
27. 100 squared
    10,000
28. Evaluate p^6 if $p = 3$. 729
29. Evaluate s^3 if $s = 8$. 512
30. Evaluate r^2 if $r = 17$. 289

© Glencoe/McGraw-Hill   T6   Mathematics: Applications and Connections, Course 1

1 FOCUS

Objective Students solve problems by using the guess-and-check strategy.

Recommended Pacing	
Standard	Day 9 of 12
Honors	Day 9 of 12
Block	Day 5 of 6

1 FOCUS

Getting Started Have each student write *yes* or *no* on a scrap of paper. Ask them to guess how many students wrote *yes*. Then ask how they would check the total number of *yes* responses. **count the papers**

2 TEACH

Teaching Tip Encourage students to think of situations when they use guess-and-check. A common one is a multiple-choice math test. Instead of solving each problem and finding the answer among the choices, students often check each choice to see if it is an appropriate answer.

In-Class Example
Find a number such that the sum of that number and its double is 27. **9**

THINKING LAB

PROBLEM SOLVING

1-7A Guess and Check

A Preview of Lesson 1-7

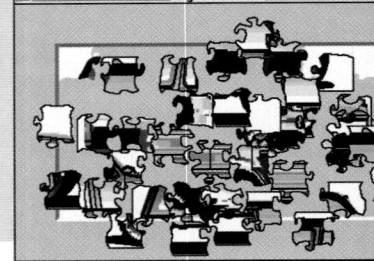
Jigsaw Puzzle

Yvonne and her friend Catalina are putting together a computer generated jigsaw puzzle. They are trying to decide the best way to start the puzzle. Let's listen in!

Yvonne

Do you know what the picture is supposed to look like?

I can't tell. Maybe we should start by finding the piece that will fit in the lower right-hand corner.

All right. Three of the four corner pieces are already in place. Let's look for the fourth corner piece.

It is the one that has a straight edge on the right side and on the bottom.

Yes! That piece looks like it will fit.

Catalina

THINK ABOUT IT

Work with a partner.

1. **Analyze** Catalina's and Yvonne's thinking. Do you agree with their thinking? Why or why not? **See students' work.**

2. **Think** of another way for Catalina and Yvonne to start the puzzle.

3. **Choose** the puzzle piece that will fit on the right side of the top left-hand puzzle piece. Explain your reasoning. **2–3. See margin.**

4. **Apply** what you have learned from Catalina's and Yvonne's situation to solve the following problem.

   Find an even number between 50 and 60 that is divisible by 2 and 3. **54**

32 Chapter 1 Problem Solving, Numbers, and Algebra

■ Reteaching the Lesson ■

Activity Have small groups of students use their calculators to create problems such as the following: *The product of a number and 14 is 210.* Have them exchange their problems with another group to solve using the guess-and-check strategy.

Additional Answers

2. Sample answer: Start by finding all the edge pieces.

3. The piece that is beside the top left-hand puzzle piece; it will connect to the corner piece.

ON YOUR OWN

5. The second step of the four-step plan for problem solving asks you to make a *plan* for solving the problem. *Tell* how you can use the **guess-and-check** strategy to help you make a plan for solving a problem. **See margin.**

6. *Write a Problem* that can be solved using the guess-and-check strategy. Then ask a classmate to solve the problem. **See students' work.**

7. *Explain* how you could use the guess-and-check strategy to answer Exercise 35 on page 36. **See margin.**

MIXED PROBLEM SOLVING

Solve. Use any strategy.

STRATEGIES
Look for a pattern.
Solve a simpler problem.
Act it out.
Guess and check.
Draw a diagram.
Make a chart.
Work backward.

8. *Money Matters* Kamaria earns $5 per hour baby-sitting. Last week, she worked 16 hours. How much money did Kamaria make? **$80**

9. *Life Science* Last week, the Mahoning Animal Shelter sent a total of 24 dogs and cats to new homes. There were 6 more dogs than cats. How many of each were adopted? **15 dogs, 9 cats**

10. *Entertainment* The chart below gives admission costs to Mount Carmel's health fair. Twelve people paid a total of $50 for admission. If 8 children attended the health fair, how many adults and senior citizens attended? **2 adults, 2 senior citizens**

11. There are 8 bones in the wrist, 14 in the fingers, and 5 in the palm.

Admission Costs	
Adults	$6
Children	$4
Senior Citizens	$3

11. *Life Science* Each hand in the human body has 27 bones. There are 6 more bones in your fingers than in your wrist. There are 3 fewer bones in your palm than in your wrist. How many bones are in each part of your hand?

12. *Money Matters* Refer to the chart below. Olivia and Imena each got a perm, a manicure, and a bottle of shampoo. About how much money did they spend? **about $180**

Alberto's Salon	
Cut/Style	$12
Perm	$56
Manicure	$17
Shampoo	$ 8

13. *Physical Science* Sound travels through air at a speed of 1,129 feet per second. At this rate, how far will sound travel in 1 minute? **67,740 feet per minute**

14. *Sports* Jamal's basketball team won twice as many games as they lost. They lost 8 games. How many games did they win? **16 games**

15. **Test Practice** Jorge purchased a new car. His loan, including interest, is $10,740. How much are his monthly payments if he has 60 payments to make? **B**

A $195
B $179
C $135
D $159
E Not Here

Lesson 1-7A THINKING **33**

Extending the Lesson

Activity Give pairs of students play money and coins. Have them create a word problem about the money that can be solved using the guess-and-check strategy. Then have them exchange their problems with others for solving.

Sample problem: *Kevin has seven $1 bills, three quarters, six dimes, eight nickels, and four pennies. He wants to buy a book for $5.95. How much will he have left after the purchase? (Assume tax is included in the price.)* **$2.84**

3 PRACTICE/APPLY

Check for Understanding
Use the results from Exercises 5–7 to determine if students comprehend the guess-and-check strategy.

Extra Practice If students need additional practice in problem solving, extra practice is available on the following pages.
• Guess and Check, see p. 559
• Mixed Problem Solving, see pp. 593–594

Assignment Guide
All: 5–15

4 ASSESS

Closing Activity
Writing Have students write word problems from everyday experiences that involve the guess-and-check strategy. Have them exchange problems and solve them.

Additional Answers
5. Sample answer: When making a plan to solve a problem, you can use the guess-and-check strategy to guess an answer and compare the answer to the problem. If your answer does not work, then you know to make another guess.

7. Sample answer: You could guess values for *p* to find what number divided by 5 is 45.

Integration: Algebra
Solving Equations

- *Study Guide Masters*, p. 7
- *Practice Masters*, p. 7
- *Enrichment Masters*, p. 7
- Transparencies 1-7, A and B
- *Assessment and Evaluation Masters*, p. 16
- *Diversity Masters*, p. 1
- *Hands-On Lab Masters*, p. 69
- *School to Career Masters*, p. 1

 CD-ROM Program
- Resource Lesson 1-7

Recommended Pacing

Standard	Day 10 of 12
Honors	Day 10 of 12
Block	Day 5 of 6

1 FOCUS

 5-Minute Check
(Lesson 1-6)

1. Write $6 \cdot 6 \cdot 6 \cdot 6$ using exponents. 6^4
2. Write 22^7 as a product.
   $22 \cdot 22 \cdot 22 \cdot 22 \cdot 22 \cdot 22 \cdot 22$
3. Evaluate $6^3 \cdot 8^2$. **13,824**
4. Evaluate r^3 if $r = 9$. **729**
5. In 1996, the population of Ecuador was estimated to be 117×10^5. About how many people lived there at that time? **about 11,700,000 people**

The 5-Minute Check is also available on **Transparency 1-7A** for this lesson.

Motivating the Lesson

Problem Solving Ask students what they need to know to determine whether the statement is true or false. *More red markers are sold each year than yellow markers.* **the number of red and yellow markers sold**

What you'll learn

You'll learn to solve equations by using mental math and guess and check.

When am I ever going to use this?

Knowing how to solve equations can help you determine the number of new students in your school.

Word Wise

equation
solve
solution

> **Study Hint**
> **Reading Math** The equation $x - 4 = 5$ is read as *x minus four is equal to five.*

On March 23, 1775, ___?___ said, "*I know not what course others may take; but as for me, give me liberty, or give me death!*"

This sentence is neither true nor false until you fill in the blank. If you answer Patrick Henry, the sentence is true. If you answer George Washington, Harriet Tubman, or any other historical figure, the sentence is false.

In mathematics, an **equation** is a sentence that contains an equal sign, $=$. Equations can be either true or false.

$$34 - 12 = 22 \quad \textit{This equation is true.}$$
$$45 \times 2 = 100 \quad \textit{This equation is false.}$$

An equation like $x - 4 = 5$ is neither true nor false until x is replaced with a number.

Replace x with 10.

$x - 4 = 5$
$10 - 4 \stackrel{?}{=} 5$
$6 = 5$

This sentence is false.

Replace x with 9.

$x - 4 = 5$
$9 - 4 \stackrel{?}{=} 5$
$5 = 5$

This sentence is true.

When you replace a variable with a value that makes the equation true, you **solve** the equation. The replacement that makes the equation true is called a **solution** of the equation. The solution of $x - 4 = 5$ is 9.

Example ① Which of the numbers 3, 4, or 5 is the solution of $y + 8 = 12$?

Replace y with 3.	*Replace y with 4.*	*Replace y with 5.*
$y + 8 = 12$	$y + 8 = 12$	$y + 8 = 12$
$3 + 8 \stackrel{?}{=} 12$	$4 + 8 \stackrel{?}{=} 12$	$5 + 8 \stackrel{?}{=} 12$
$11 = 12$	$12 = 12$ ✓	$13 = 12$
This sentence is false.	This sentence is true.	This sentence is false.

The solution is 4 because replacing y with 4 results in a true sentence.

In Example 1, $y + 8 = 12$ was solved by replacing y with a number until a true sentence resulted.

Some equations can be solved mentally by using basic facts or arithmetic skills you already know well.

Examples

Solve each equation using mental math.

2 $2m = 10$

$2 \times 5 = 10$ *You know that $2 \times 5 = 10$.*

$10 = 10$

The solution is 5.

3 $7 = 5 + d$

$7 = 5 + 2$ *You know that $5 + 2 = 7$.*

$7 = 7$

The solution is 2.

You can also solve equations by using guess and check.

Example

APPLICATION

4 **School** Last year, the enrollment in Pointview Middle School was 695 students. This year there are 748 students. If n represents the number of new students, the equation $695 + n = 748$ results. Find the number of new students.

Use guess and check.

Try 50.	*Try 52.*	*Try 53.*
$695 + n = 748$	$695 + n = 748$	$695 + n = 748$
$695 + 50 \stackrel{?}{=} 748$	$695 + 52 \stackrel{?}{=} 748$	$695 + 53 \stackrel{?}{=} 748$
$745 = 748$	$747 = 748$	$748 = 748$ ✓
This sentence is false.	This sentence is false.	This sentence is true.

Study Hint

Estimation You know that 695 is about 700 and 748 is about 750. Think: $700 + 50 = 750$. So the answer should be close to 50.

The solution is 53. So, there are 53 new students.

CHECK FOR UNDERSTANDING

Communicating Mathematics

Read and study the lesson to answer each question.

1. *Write* an example of an algebraic expression. Write an example of an equation. How are they alike? How are they different? **See margin.**

2. *State* an equation with a solution of 7. **Sample answer:** $x + 7 = 14$

Lesson 1-7 Integration: Algebra Solving Equations **35**

---

2 TEACH

Transparency 1-7B contains a teaching aid for this lesson.

Using Manipulatives Have students use a two-pan balance and small weights to determine which of the numbers 3, 5, or 7 is the solution of $x + 5 = 12$. **7**

In-Class Examples

For Example 1
Which of the numbers 7, 8, or 9 is the solution of $t + 21 = 28$? **7**

For Example 2
Solve $5s = 30$ using mental math. **6**

For Example 3
Solve $29 = 12 + a$ using mental math. **17**

For Example 4
Ellen purchased a basketball and a pair of basketball shoes for $93. The shoes cost $65. The equation $65 + b = 93$ represents the total cost of the two items where b represents the cost of the basketball. What is the cost of the basketball? **$28**

Teaching Tip Some students may need to review their multiplication skills. Stress the importance of being able to perform simple computations without a calculator.

3 PRACTICE/APPLY

Check for Understanding
If students need additional practice or instruction after completing Exercises 1–12, one of these options may be helpful.
- Extra Practice, see p. 559
- Reteaching Activity
- *Transition Booklet,* pp. 15–16, 19–22, 25–26
- *Study Guide Masters,* p. 7
- *Practice Masters,* p. 7

■ Reteaching the Lesson ■

Activity Play "Equation Concentration." Use index cards in two different colors— one for equations, the other for solutions. Place all cards facedown. Each player, in turn, turns over two cards to try to make a match. The player with the most matches at the end of the game wins.

Additional Answer

1. Sample answer: Both algebraic expressions and equations can contain variables, operations, and numbers. Whereas an equation contains an equals sign, an algebraic expression does not contain an equals sign.

Guided Practice

Tell whether the equation is *true* or *false* by replacing the variable with the given value.

3. $18 + j = 45; j = 29$ false   **4.** $y \div 7 = 35; y = 245$ true

Identify the solution to each equation from the list given.

5. $b + 5 = 12; 5, 6, 7$ **7**   **6.** $36 - h = 15; 21, 22, 23$ **21**

7. $72 \div d = 18; 3, 4, 5$ **4**   **8.** $16r = 96; 4, 5, 6$ **6**

Solve each equation mentally.

9. $4 + x = 13$ **9**   **10.** $5w = 20$ **4**   **11.** $z - 8 = 2$ **10**

12. *Shopping* Ana purchased a helmet and some reflectors for her bike. The cost of the helmet was \$25. She spent a total of \$32.   **a.** $\$25 + c = \$32$

a. Let c represent the cost of the reflectors. Write an equation that represents the total cost of the items.

b. Solve your equation to find the cost of the reflectors. **\$7**

EXERCISES

Practice

Tell whether the equation is *true* or *false* by replacing the variable with the given value.

13. $p + 11 = 27; p = 16$ **true**   **14.** $4w = 24; w = 6$ **true**

15. $t - 15 = 32; t = 47$ **true**   **16.** $57 \div d = 21; d = 3$ **false**

17. $93 = 3v; v = 32$ **false**   **18.** $y \div 39 = 6; y = 234$ **true**

Identify the solution to each equation from the list given.

19. $8 + h = 23; 14, 15, 16$ **15**   **20.** $t - 6 = 3; 5, 7, 9$ **9**

21. $42 = 6a; 7, 8, 9$ **7**   **22.** $23 + n = 56; 31, 32, 33$ **33**

23. $b = 65 \div 5; 11, 13, 15$ **13**   **24.** $31 - m = 19; 10, 11, 12$ **12**

25. $k + 38 = 45; 7, 9, 11$ **7**   **26.** $76 \div q = 19; 3, 4, 5$ **4**

27. $11 = d \div 4; 41, 43, 44$ **44**   **28.** $7 \div g = 1; 6, 7, 8$ **7**

Solve each equation mentally.

29. $m - 9 = 17$ **26**   **30.** $3 + j = 18$ **15**   **31.** $12 = d + 5$ **7**

32. $22 = 2y$ **11**   **33.** $24 \div w = 8$ **3**   **34.** $4n = 32$ **8**

35. $p \div 5 = 45$ **225**   **36.** $a = 4 + 13$ **17**   **37.** $18 \div s = 9$ **2**

38. *Algebra* Which of the numbers 3, 6, or 9 is the solution of $7h = 63$? **9**

39. *Algebra* Find the solution of $176 = 11s$. **16**

36 **Chapter 1** Problem Solving, Numbers, and Algebra

Study Guide Masters, p. 7

Name_____ Date_____

1-7 **Study Guide**

Integration: Algebra
Solving Equations

An **equation** is a mathematical sentence that contains an equals sign.
The **solution** is the number that makes the equation true.

Example

Warren spent \$120 for theater tickets. If each ticket cost \$20, how many tickets did Warren buy?

Let t equal the number of tickets.
The problem may be represented by this equation: $\$20 \times t = \$120$.

Guess and check.

Replace t with 7.   Replace t with 6.
$\$20 \times t = \$120$ $\$20 \times t = \$120$
$\$20 \times 7 = \$120$ $\$20 \times 6 = \$120$
$\$140 = \$120$ $\$120 = \$120$ ✔
The sentence is false.   The sentence is true.

The solution is 6. Warren bought 6 tickets.

Identify the solution to each equation from the list given.

1. $15 + r = 21$ 5, 6, 7 **6**   **2.** $y \div 6 = 7$ 42, 48, 56 **42**

3. $5m = 40$ 6, 7, 8 **8**   **4.** $n = 7 + 21$ 3, 14, 28 **28**

5. $9 = 63 \div b$ 7, 8, 9 **7**   **6.** $5w = 10$ 2, 25, 50 **2**

Solve each equation mentally.

7. $y = 10 + 6$ **16**   **8.** $p = 18 - 9$ **9**   **9.** $x \div 6 = 4$ **24**   **10.** $h \div 5 = 25$ **125**

11. $3r = 15$ **5**   **12.** $q - 7 = 16$ **23**   **13.** $18 = 9 + v$ **9**   **14.** $28 = 4p$ **7**

© Glencoe/McGraw-Hill   T7   *Mathematics: Applications and Connections, Course 1*

Classroom Vignette

"For practicing and reviewing for tests, I use small grease boards and markers. We review in a game format and students hold up their answers on the boards. They receive 1 point for each correct answer. Students bring in old tube socks to use as erasers."

Jan Wickboldt

Jan Wickboldt, Teacher
Clayton School
Clayton, WI

40. *Banking* Currency rates are calculated daily. If 140d represents the number of Spanish pesetas for each United States dollar, represented by d, how many pesetas will be received in exchange for $50? **7,000 pesetas**

41. *Food* Celina and Marcus went to Burger King for lunch. Celina ordered a Whopper with cheese, one order of onion rings, and water. There are 19 grams of fat in one order of onion rings. There is a total of 65 grams of fat in the entire meal.

    a. Let g represent the number of grams of fat in a Whopper with cheese. Write an equation to find how many grams of fat are in a Whopper with cheese. **19 + g = 65**

    b. Solve your equation to find the number of grams of fat in a Whopper with cheese. **46 grams**

42. *Critical Thinking* Translate each sentence into an equation. Then solve.

    a. *Eight less than m is equal to 5.* **m − 8 = 5; 13**

    b. *The sum of h and 7 is equal to 18.* **h + 7 = 18; 11**

    c. *The product of 4 and 16 is equal to n.* **4 × 16 = n; 64**

Mixed Review

43. [**Test Practice**] How is 2 × 3 × 3 × 11 expressed in exponential notation? *(Lesson 1-6)* **C**

    A $2 \times 2 \times 3 \times 11$

    B $2^3 \times 11$

    C $2 \times 3^2 \times 11$

    D $2 \times 2^3 \times 11$

44. *Algebra* Evaluate $3x - 2 \cdot 4 + y$ if x = 7 and y = 1. *(Lesson 1-5)* **14**

45. Find the value of $11 - 3 \div 3 + 2$. *(Lesson 1-4)* **12**

46. Sample answer: 40 miles per day

46. *Transportation* The steamer *Alaska* traveled 182 miles in 5 days. Use rounding to estimate how many miles it averaged each day. *(Lesson 1-3)*

47. *Sports* The graph shows the average cost of professional hockey, football, basketball, and baseball tickets. Camilia is planning to attend a hockey and a football game. About how much will she spend on the two tickets? *(Lesson 1-3)* **about $70**

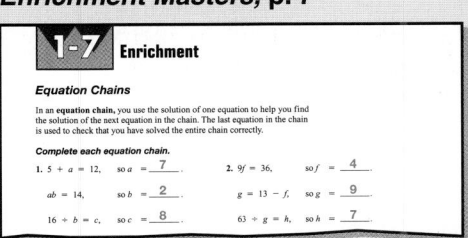

Source: Team Marketing Report

Sports Ticket Costs

NFL $35.74

NHL $34.79

NBA $31.80

MLB $11.13

Extending the Lesson

Enrichment Masters, p. 7

> **1-7 Enrichment**
>
> **Equation Chains**
>
> In an **equation chain**, you use the solution of one equation to help you find the solution of the next equation in the chain. The last equation in the chain is used to check that you have solved the entire chain correctly.
>
> **Complete each equation chain.**
>
> **1.** 5 + a = 12, so a = __7__
> ab = 14, so b = __2__
> 16 ÷ b = c, so c = __8__
>
> **2.** 9f = 36, so f = __4__
> g = 13 − f, so g = __9__
> 63 ÷ g = h, so h = __7__

Activity Ask students how they would solve this problem. Erik, Ana, and Gene collected 48 cans of food for the Hunger Drive. Ana brought in 23 cans. Erik collected 3 more cans than Gene. How many cans did they each collect?
48 = 23 + (3 + g) + g; Erik-14 cans; Gene-11 cans

Closing Activity

Modeling Have students use tiles or other small objects to model solutions to equations. For example, model 13 = 5 + r by laying out 13 tiles. Group 5 of these tiles together and count the rest. That is the answer, r.

Chapter 1, Quiz D (Lesson 1-7) is available in the *Assessment and Evaluation Masters,* p. 16.

Practice Masters, p. 7

> Name_____ Date_____
>
> **1-7 Practice**
>
> *Integration: Algebra*
> *Solving Equations*
>
> **Tell whether the equation is true or false by replacing the variable with the given value.**
>
> **1.** a − 53 = 32; a = 85 **true**
> **2.** f + 17 = 31; f = 14 **true**
> **3.** 108 = 6d; d = 13 **false**
> **4.** s + 15 = 9; s = 105 **false**
> **5.** k ÷ 21 = 8; k = 168 **true**
> **6.** 45 − p = 18; p = 27 **true**
> **7.** h × 17 = 119; h = 7 **true**
> **8.** 73 + m = 24; m = 49 **false**
>
> **Identify the solution to each equation from the list given.**
>
> **9.** b = 16 + 19; (35) 36, 37
> **10.** j = 64 − 46; 8, (18) 110
> **11.** 12n = 168; (14) 15, 16
> **12.** 56 + w = 81; 15, (25) 35
> **13.** m + 4 = 23; (92) 93, 94
> **14.** 114 = 6c; 17, 18, (19)
> **15.** 77 − g = 18; 49, (59) 69
> **16.** 18 = 126 ÷ y; 5, 6, (7)
>
> **Solve each equation mentally.**
>
> **17.** 8x = 56 **7**
> **18.** 12 − q = 7 **5**
> **19.** 17 = 8 + z **9**
> **20.** 7 = 21 + e **3**
> **21.** 12 ÷ 6 = u **2**
> **22.** 72 = 9t **8**
> **23.** r = 14 − 8 **6**
> **24.** y + 9 = 13 **4**
> **25.** 9 = n + 6 **54**
> **26.** 15 − h = 6 **9**
> **27.** 7s = 49 **7**
> **28.** p ÷ 2 = 12 **24**
>
> © Glencoe/McGraw-Hill   T7   Mathematics: Applications and Connections, Course 1

Vocabulary

This section provides a listing of the new terms, properties, and phrases that were introduced in this chapter. Have students define each term and provide an example or two of it, if appropriate.

Understanding and Using the Vocabulary

These exercises check students' understanding of the terms by using a variety of verbal formats including matching, completion, and true/false.

A complete glossary of terms appears on pages 642–648. The glossary also appears in Spanish on pages 649–656.

Vocabulary

After completing this chapter, you should be able to define each term, concept, or phrase and give an example or two of each.

Algebra
algebra (p. 22)
algebraic expression (p. 22)
equation (p. 34)
evaluate (p. 22)
solution (p. 34)
solve (p. 34)
variable (p. 22)

Number and Operations
base (p. 28)
cubed (p. 28)
exponent (p. 28)
factor (p. 28)
order of operations (p. 16)
power (p. 28)
squared (p. 28)

Problem Solving
guess and check (p. 32)

Understanding and Using the Vocabulary

Choose the correct term or number to complete each sentence.

1. When finding the value of an expression, first (<u>multiply and divide,</u> add and subtract) in order from left to right.
2. $(7, \underline{h + 2})$ is an algebraic expression.
3. The base of 4^5 is ($\underline{4}$, 5).
4. When the same factor is repeated, a(n) (variable, <u>exponent</u>) can simplify the notation.
5. The factors of 12 are (<u>3 and 4</u>, 6 and 6).
6. Four cubed is written as (4^2, $\underline{4^3}$).
7. $2 \cdot 2 \cdot 2 \cdot 2 \cdot 2$ can be written as ($\underline{2^5}$, 5^2).
8. A symbol, usually a letter, used to represent a number is called a (<u>variable</u>, power).
9. If $d = 4$, then the value of the expression $14 - d$ is ($\underline{10}$, 18).
10. ($\underline{3x = 6}$, $2x + y$) is an equation.
11. The solution of $16 - g = 5$ is (21, $\underline{11}$).

In Your Own Words

12. **Summarize** the rules for the order of operations.

12. Sample answer: Do powers first. Next, do all multiplication and division in order from left to right. Then, do all addition and subtraction in order from left to right.

38 Chapter 1 Problem Solving, Numbers, and Algebra

 MindJogger Videoquizzes

MindJogger Videoquizzes provide an alternative review of concepts presented in this chapter. Students work in teams to answer questions, gaining points for correct answers. The questions are presented in three rounds.
Round 1 Concepts–5 questions
Round 2 Skills–4 questions
Round 3 Problem Solving–4 questions

Objectives & Examples

Upon completing the chapter, you should be able to:

• solve problems using the four-step plan *(Lesson 1-1)*

Use the four-step plan to solve.
Migina studied 2 hours each day for 10 days. How many hours has she studied?

Explore	Migina studied 2 hours each day for 10 days. Find the total number of hours studied.
Plan	Multiply 2 by 10.
Solve	$2 \times 10 = 20$
	Migina studied 20 hours.
Examine	The answer makes sense and is reasonable.

• solve problems using patterns *(Lesson 1-2)*

Find the next three numbers for the pattern 1, 4, 7, 10, ?, ?, ?.

Each number is 3 more than the number before it.

$10 + 3 = 13 \quad 13 + 3 = 16 \quad 16 + 3 = 19$

The next three numbers are 13, 16, and 19.

• estimate sums, differences, products, and quotients using rounding *(Lesson 1-3)*

Estimate the product of 8 and 57.

$$\begin{array}{ccc} 57 & \rightarrow & 60 \\ \times\ 8 & \rightarrow & \times\ 8 \\ \hline & & 480 \end{array}$$

24. No; the total should be about $130.

Review Exercises

Use these exercises to review and prepare for the chapter test.

Use the four-step plan to solve each problem. 13. $39.92

13. *Money Matters* The Music Store sells compact discs for $9.98 each. If Lilla buys 4 CDs, how much will he spend?

14. *School* Tickets to the school dance cost $4 each. The dance committee collected a total of $352. How many students bought tickets? **88 students**

15. *History* In the 1932 presidential election, Franklin Roosevelt won with 472 electoral votes while Herbert Hoover had 59. How many electoral votes were cast? **531 votes**

Find the next three numbers or shapes in each pattern. 18. 225, 275, 325

16. 3, 6, 12, 24, ?, ?, ? **48, 96, 192**

17. 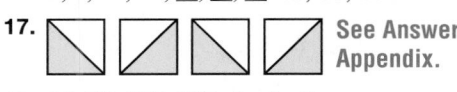 See Answer Appendix.

18. 25, 75, 125, 175, ?, ?, ?

19. 16, 24, 32, 40, ?, ?, ? **48, 56, 64**

Round each number to the underlined place-value position.

20. 4,650 **5,000**   21. 27 **30**

22. 5,321 **5,320**   23. 654 **650**

24. *Money Matters* Chapa purchases a coat for $91, a shirt for $17, and jeans for $24. The cashier asks for $163. Is this amount reasonable?

Chapter 1 Study Guide and Assessment **39**

Objectives & Examples

This section reviews the skills and concepts of the chapter and shows completely worked examples.

Review Exercises

These exercises provide practice for the corresponding objectives.

Assessment and Evaluation Masters, pp. 3–4

Name _____   Date _____

1   Chapter 1 Test, Form 1B

1. Find the next three numbers in the pattern 1, 5, 9, 13, ?, ?, ?.
   A. 3, 7, 9   B. 15, 17, 19   C. 17, 21, 25   D. 19, 25, 31.   1. __C__

2. Find the next three numbers in the pattern 3, 9, 27, ?, ?, ?.
   A. 54, 108, 216   B. 81, 243, 729
   C. 36, 45, 54   D. 30, 90, 270   2. __B__

3. Round 158 to the nearest ten.
   A. 100   B. 200   C. 150   D. 160   3. __D__

4. Round 785 to the nearest hundred.
   A. 700   B. 780   C. 790   D. 800   4. __D__

5. Round 3,450 to the nearest thousand.
   A. 3,000   B. 3,200   C. 4,000   D. 3,300   5. __A__

6. Estimate 3,912 − 1,122 using rounding.
   A. 3,000   B. 2,000   C. 1,000   D. 3,800   6. __A__

7. Estimate 17 × 391 using rounding.
   A. 4,000   B. 5,000   C. 6,000   D. 8,000   7. __D__

8. Find the value of 20 − 8 ÷ 4.
   A. 18   B. 3   C. 22   D. 48   8. __A__

9. Find the value of 28 + 6 × 4 − 2.
   A. 134   B. 50   C. 68   D. 40   9. __B__

10. Find the value of 51 − 2 × 13 + 14.
    A. 1,323   B. 3   C. 39   D. 651   10. __C__

11. Find the value of 42 × 3 − 10 ÷ 5.
    A. 23   B. 7   C. 124   D. 84   11. __C__

12. Evaluate mn if $m = 23$ and $n = 5$.
    A. 115   B. 235   C. 28   D. 523   12. __A__

13. Evaluate $3 + 2m$ if $m = 6$.
    A. 30   B. 29   C. 15   D. 11   13. __C__

© Glencoe/McGraw-Hill   3   Mathematics: Applications and Connections, Course 1

---

1   Chapter 1 Test, Form 1B (continued)

14. Evaluate $a + b − c$ if $a = 20$, $b = 10$, and $c = 5$.
    A. 5   B. 35   C. 25   D. 22   14. __C__

15. Write 7 · 7 · 7 · 7 · 7 · 7 using exponents.
    A. 6 × 7   B. 7⁶   C. 6⁷   D. 7 × 6   15. __B__

16. Evaluate 10⁵.
    A. 10,000   B. 50   C. 100,000   D. 1,000,000   16. __C__

17. Write 4³ as a product.
    A. 4 · 3   B. 3 · 3 · 3   C. 4 · 4 · 4   D. 4 · 4 · 4 · 4   17. __C__

18. Evaluate 5² · 3 − 2.
    A. 73   B. 28   C. 25   D. 60   18. __A__

19. The sum of a number and its double is 30. Use the guess-and-check strategy to find the number.
    A. 15   B. 10   C. 20   D. 5   19. __B__

20. Which number is the solution of $x − 7 = 42$?
    A. 29   B. 35   C. 49   D. 52   20. __C__

21. Which number is the solution of $42 = 7x$?
    A. 6   B. 35   C. 49   D. 294   21. __A__

22. Solve $5 + n = 14$ mentally.
    A. 10   B. 19   C. 8   D. 9   22. __D__

23. Solve $36 ÷ r = 9$ mentally.
    A. 5   B. 45   C. 27   D. 4   23. __D__

24. A television sells for $495 plus tax. The tax is $24. Use the four-step plan to find the total cost of the television.
    A. $471   B. $419   C. $529   D. $519   24. __D__

25. Wendie decided to start training for track. The first day, she jogged 6 laps. The second day, she jogged 12 laps. The third day, she jogged 18 laps. If this pattern continues, how many laps did she jog on the fourth day?
    A. 22   B. 24   C. 36   D. 30   25. __B__

© Glencoe/McGraw-Hill   4   Mathematics: Applications and Connections, Course 1

Assessment and Evaluation

Six forms of Chapter 1 Test are available in the *Assessment and Evaluation Masters*.

Chapter 1 Test, Form 1B, is shown at the right. Chapter 1 Test, Form 2B, is shown on the next page.

1A	Multiple Choice	Honors
1B	Multiple Choice	Average
1C	Multiple Choice	Basic
2A	Free Response	Honors
2B	Free Response	Average
2C	Free Response	Basic

Assessment and Evaluation Masters, pp. 9–10

Objectives & Examples

Review Exercises

● evaluate expressions using the order of operations *(Lesson 1-4)*

Find the value of $6 + 12 \div 3$.
$$6 + 12 \div 3 = 6 + 4$$
$$= 10$$

Find the value of each expression.
25. $9 - 3 \times 2 - 1$ **2**
26. $3 \times 7 + 5 \times 9 \div 3$ **36**
27. $1 + 8 \div 4 - 1$ **2**
28. $4 + 8 - 6 + 10$ **16**
29. $2 \times 8 \div 4 + 3$ **7**
30. $30 \div 3 + 16 \div 4$ **14**

● evaluate numerical and simple algebraic expressions *(Lesson 1-5)*

Evaluate $2a + b$ if $a = 7$ and $b = 6$.
$$2a + b = 2 \times 7 + 6$$
$$= 14 + 6$$
$$= 20$$

Evaluate each expression if $a = 4$, $b = 3$, and $c = 12$.
31. $c + a$ **16**   32. $8 - b$ **5**
33. $2a - 1$ **7**   34. $c - b - a$ **5**
35. $c + 3b$ **21**   36. $3a - 4b$ **0**
37. $5a + 2c$ **44**   38. $2ab - c$ **12**

● use powers and exponents in expressions *(Lesson 1-6)*

Evaluate each expression.
$$4^3 = 4 \cdot 4 \cdot 4$$
$$= 64$$
$$6 \cdot 3^2 = 6 \cdot 3 \cdot 3$$
$$= 6 \cdot 9$$
$$= 54$$

Evaluate each expression.
39. 4^6 **4,096**
40. $5^2 \cdot 3^2 \cdot 4^2$ **3,600**
41. x^3 if $x = 5$ **125**
42. ten squared **100**
43. Write $t \cdot t \cdot t \cdot t \cdot t$ using exponents. t^5
44. Write 3^6 as a product. $3 \cdot 3 \cdot 3 \cdot 3 \cdot 3 \cdot 3$

● solve equations by using mental math and guess and check *(Lesson 1-7)*

Solve $m - 2 = 7$ mentally.
$9 - 2 = 7$ *You know that $9 - 2 = 7$.*
$7 = 7$

The solution is 9.

Identify the solution to each equation from the list given.
45. $14a = 42$; 3, 4, 5 **3**
46. $b + 23 = 31$; 7, 8, 9 **8**
47. $9 = g \div 7$; 61, 62, 63 **63**
48. $y - 7 = 12$; 18, 19, 20 **19**

Solve each equation mentally.
49. $72 \div x = 8$ **9**
50. $m - 13 = 11$ **24**
51. $18 = 3w$ **6**
52. $12 + j = 21$ **9**

Test and Review Software

You may use this software, a combination of an item generator and item bank, to create your own tests or worksheets. Types of items include free response, multiple choice, short answer, and open ended.

CD-ROM Program

The CD-ROM Program contains an Assessment Game whose questions review the concepts in this chapter.

Applications & Problem Solving

53. *History* John F. Kennedy began his first term in the House of Representatives in January 1947. In November 1960, he was elected president of the United States. How long had he been in the legislature when he was elected president? *(Lesson 1-1)* **13 years 10 months**

54. *Fitness* Ricardo made a bar graph of the number of push-ups he did in 4 days. If the pattern continues, how many push-ups will he do on the seventh day? *(Lesson 1-2)* **32 push-ups**

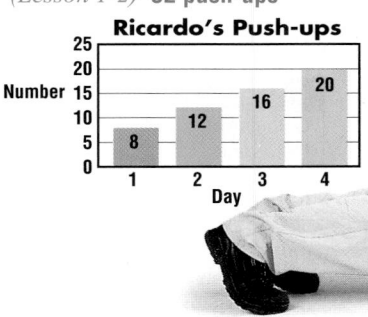

Ricardo's Push-ups

55. *Travel* Kanya and her family are planning to go to Hawaii with a tour group. The cost of the trip is $964 per person. There are 63 people in the tour group. About how much money is needed for the tour group to go to Hawaii? *(Lesson 1-3)* **about $63,000**

56. *Guess and Check* The combined weight of two brothers is 300 pounds. One brother weighs 50 pounds more than the other. How much does each brother weigh? *(Lesson 1-7A)* **125 pounds, 175 pounds**

Alternative Assessment

Performance Task

Suppose you are planning to buy a television that costs $410. The store has two payment plans. Plan 1 requires you to pay half of the purchase price up front and the remaining cost over 6 months, with a $5 fee added each month. Plan 2 requires you to make 12 equal payments of $50. Find the cost of each plan. **Plan 1: $440; Plan 2: $600**
Suppose you have $250 in savings and you can afford a $50 monthly payment. Which payment plan should you choose? Explain your choice. **See margin.**

A practice test for Chapter 1 is provided on page 595.

Completing the CHAPTER Project

Use the following checklist to complete your project

☑ The items and their costs are listed on the order form.

☑ The table of shipping charges is included.

☑ Pictures of the items are included.

☑ A list of advantages and disadvantages of mail-order shopping is included.

 PORTFOLIO Select one of the assignments from this chapter and place it in your portfolio. Attach a note to it explaining why you selected it.

Additional Answer for the Performance Task

Plan 1 is the better choice. See students' explanations.

Performance Assessment

Additional performance assessment tasks for this chapter are included in the *Assessment and Evaluation Masters* on page 13. A scoring guide is also provided on page 25.

Applications & Problem Solving

This section provides additional practice in solving real-world problems that involve the skills of this chapter.

Alternative Assessment

The **Performance Task** section provides students with a performance assessment opportunity to assess their work and understanding.

 **CHAPTER Project**

Students should complete the final stages of their project and prepare a class demonstration of their results. A scoring guide for the project is available in the *Investigations and Projects Masters*, p. 19.

PORTFOLIO Students should add to their portfolios at this time.

Assessment and Evaluation Masters, p. 13

Name_____ Date_____

Chapter 1 Performance Assessment

Instructions: Demonstrate your knowledge by giving a clear, concise solution to each problem. Be sure to include all relevant drawings and justify your answers. You may show your solution in more than one way or investigate beyond the requirements of the problem.

1. The number of Calories per serving of certain food items is given in the table below.

Food	Serving	Calories
Apple	1 large	117
Banana	1 large	176
Carrots	1 cup	42
Fried Chicken	½ breast	232
Egg	1 medium	77

a. Ayani had an apple, banana, and half chicken breast for lunch. Estimate the number of Calories he ate for lunch. Explain each step.
b. If Ayani had an apple and a banana on alternating days, how would you write that as a pattern?
c. Estimate the number of Calories in four eggs. Explain each step. Find the number of Calories in four eggs. How close was your estimate to the exact answer?
d. Kristin ate three eggs, a whole chicken breast, and a banana. How many Calories did she eat?
2. Write the order of operations, in your own words.
3. Explain one situation where you could use exponents in real life.

4. Let a bag, ☐, represent the phrase "some number" and a piece of popcorn, ☐, represent 1 for the following exercises.
a. Draw a picture to represent each phrase.
   (1) some number and five more
   (2) the sum of five and three times some number
b. Write a phrase representing the diagram below.

The Standardized Test Practice may be used to help students prepare for standardized tests. The test items are written in the same style as those in state proficiency tests and standardized tests like CAT, CTBS, ITBS, MAT, SAT, and Terra Nova. The test items cover skills and concepts covered up to this point in the text.

The pages can be used as an overnight assessment. After students have completed the pages, discuss how each problem can be solved, or provide copies of the solutions from the *Solutions Manual*.

Assessment and Evaluation Masters, p. 19

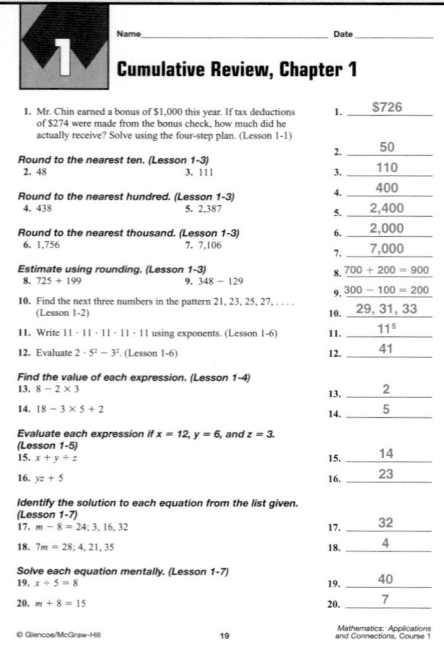

Section One: Multiple Choice

There are twelve multiple-choice questions in this section. Choose the best answer. If a correct answer is *not here*, mark the letter for Not Here.

1. What is the value of 2^5? **B**
   - A 10
   - B 32
   - C 25
   - D 64

2. There will be 120 people at the annual orchestra awards banquet. Each table seats 8 people. Which procedure can be used to find the number of tables needed? **H**
   - F Add 8 to 120.
   - G Multiply 8 and 120.
   - H Divide 120 by 8.
   - J Subtract 8 from 120.

3. Find the value of the expression $121 \div 11 + 4 \cdot 8$. **B**
   - A 120
   - B 43
   - C 12
   - D 60

4. By scoring 2,491 points during the 1995-1996 regular season, Michael Jordan scored more points than any other NBA player. What is this number rounded to the nearest hundred? **J**
   - F 2,490
   - G 2,400
   - H 2,590
   - J 2,500

5. Evaluate mn if $m = 120$ and $n = 50$. **D**
   - A 500
   - B 600
   - C 5,000
   - D 6,000

Please note that Questions 6–12 have five answer choices.

6. Kenyatta is planning to buy a car that costs $9,786. To have a sunroof installed, the cost is $548 more. What is the price of the car with a sunroof? **G**
   - F $9,334
   - G $10,334
   - H $10,024
   - J $11,334
   - K Not Here

7. A sign at the school bookstore read:

Item	Price
Calculator	$5
Notebook	$1
Combination Lock	$3
3-Ring Binder	$2

   Donavan purchased 3 notebooks, 2 binders, and a lock. Find the total cost of the items. **E**
   - A $8
   - B $7
   - C $12
   - D $9
   - E $10

8. The students at Grant Middle School are selling candy bars as a fundraiser. Each class that sells 400 candy bars will earn a field trip to a water park. Lori's class has sold 207 candy bars. How many more candy bars must her class sell to earn a field trip? **F**
   - F 193
   - G 184
   - H 173
   - J 194
   - K Not Here

◄◄◄ Instructional Resources

Another cumulative review is shown at the left and is available in the *Assessment and Evaluation Masters*, p. 19.

9. Cord Photography is having a sale on camera equipment.

Camera Package	
Camera	$178
Lens	$ 96
Camera Strap	$ 23

Which is the best estimate for the total amount of the camera package? **B**

A $100

B $300

C $400

D $200

E $600

10. The school cafeteria prepared 1,987 lunches last week. About how many lunches were prepared each day? **J**

F 600

G 700

H 500

J 400

K Not Here

11. Which is a reasonable estimate when 298 is subtracted from 702? **A**

A 400

B 500

C 900

D 1,000

E Not Here

12. Javiar drove his car for 4 hours at 45 miles per hour and for 3 hours at 50 miles per hour. Which expression could be used to find the total number of miles that Javier drove his car? **J**

F $45 \div 4 \times 50 \div 3$

G $45 + 50 \times 4$

H $45 + 4 \times 50 + 3$

J $45 \times 4 + 50 \times 3$

K Not Here

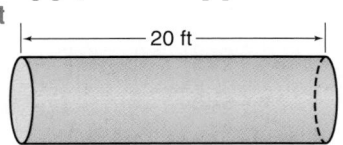

Test-Taking Tip

Many problems can be solved without much calculating if you understand the basic mathematical concepts. Always look carefully at what is asked, and think of possible shortcuts for solving the problem.

Section Two: Free Response

This section contains five questions for which you will provide short answers. Write your answers on your paper.

13. While installing gas pipes, Hector used pieces of pipe that measured 1 ft, 2 ft, 3 ft, and 4 ft. If he cut the pieces from a 20-ft pipe, how much pipe was left? **10 ft**

20 ft

14. Write $2 \times 3 \times 5 \times 5 \times 7$ in exponential notation. $2 \times 3 \times 5^2 \times 7$

15. Sue recycled 41 aluminum cans. Her friend Tammy gave her some more aluminum cans, and then she had recycled 62 cans. To find how many cans she was given, Sue wrote $c + 41 = 62$. What is the value of c? **21**

16. Find the value of $10 \times 10 + 10 \div 2$. **105**

17. Solve $t - 16 = 11$. **27**

Assessment and Evaluation Masters, pp. 17–18

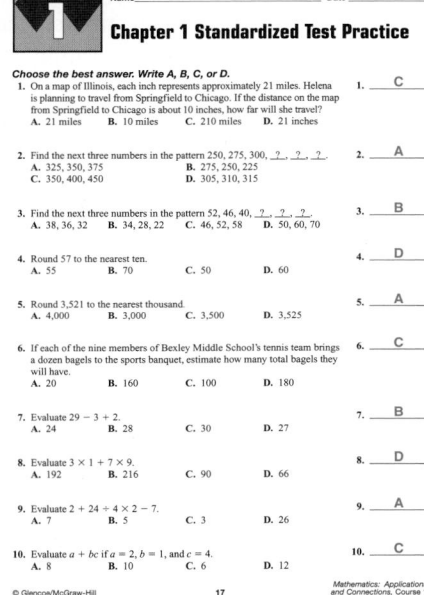

Name_____ Date_____

Chapter 1 Standardized Test Practice

Choose the best answer. Write A, B, C, or D.

1. On a map of Illinois, each inch represents approximately 21 miles. Helena is planning to travel from Springfield to Chicago. If the distance on the map from Springfield to Chicago is about 10 inches, how far will she travel?
   A. 21 miles   B. 10 miles   C. 210 miles   D. 21 inches   **1. C**

2. Find the next three numbers in the pattern 250, 275, 300, _?_, _?_, _?_.
   A. 325, 350, 375   B. 275, 250, 225   C. 350, 400, 450   D. 305, 310, 315   **2. A**

3. Find the next three numbers in the pattern 52, 46, 40, _?_, _?_, _?_.
   A. 38, 36, 32   B. 34, 28, 22   C. 46, 52, 58   D. 50, 60, 70   **3. B**

4. Round 57 to the nearest ten.
   A. 55   B. 70   C. 50   D. 60   **4. D**

5. Round 3,521 to the nearest thousand.
   A. 4,000   B. 3,000   C. 3,500   D. 3,525   **5. A**

6. If each of the nine members of Bexley Middle School's tennis team brings a dozen bagels to the sports banquet, estimate how many total bagels they will have.
   A. 20   B. 160   C. 100   D. 180   **6. C**

7. Evaluate $29 - 3 + 2$.
   A. 24   B. 28   C. 30   D. 27   **7. B**

8. Evaluate $3 \times 1 + 7 \times 9$.
   A. 192   B. 216   C. 90   D. 66   **8. D**

9. Evaluate $2 + 24 \div 4 \times 2 - 7$.
   A. 7   B. 5   C. 3   D. 26   **9. A**

10. Evaluate $a + bc$ if $a = 2$, $b = 1$, and $c = 4$.
    A. 8   B. 10   C. 6   D. 12   **10. C**

© Glencoe/McGraw-Hill   17   *Mathematics: Applications and Connections, Course 1*

Chapter 1 Standardized Test Practice (continued)

11. Evaluate $2r$ if $r = 37$.
    A. 39   B. 74   C. 18.5   D. 35   **11. B**

12. The area of a rectangle is found by multiplying length times width. Find the area of a rectangle 12 inches wide and 13 inches long.
    A. 25 sq. in.   B. 1 sq. in.   C. 144 sq. in.   D. 156 sq. in.   **12. D**

13. Write $3 \cdot 5 \cdot 5 \cdot 7$ using exponents.
    A. 525   B. $3 \cdot 5^2 \cdot 7$ C. $3 \cdot 25 \cdot 7$ D. $21 \cdot 25$ **13. B**

14. Write 3^4 as a product.
    A. 81   B. $3 \cdot 3 \cdot 3$ C. $3 \cdot 3 \cdot 3 \cdot 3$ D. $3 \cdot 3 \cdot 3 \cdot 3 \cdot 3$ **14. C**

15. Evaluate $4 \cdot 3^2 - 7^2$.
    A. 59   B. 5   C. 88   D. 0   **15. A**

16. Write $b \cdot b \cdot b \cdot b \cdot b \cdot b$ using exponents.
    A. b^6 B. $7b$ C. $6b$ D. b^7 **16. D**

17. Which number is the solution of $x + 12 = 19$?
    A. 6   B. 7   C. 8   D. 9   **17. B**

18. Solve $3x = 27$ using mental math.
    A. 24   B. 7   C. 8   D. 9   **18. D**

19. Solve $242 - c = 204$ using mental math.
    A. 36   B. 40   C. 38   D. 446   **19. C**

20. Cole purchased a new tennis racquet and tennis glove. The cost of the racquet was $120 and he spent $132 total. If g represents the cost of the glove, the equation $120 + g = 132$ results. Find the cost of the glove.
    A. $12   B. $15   C. $10   D. $22   **20. A**

© Glencoe/McGraw-Hill   18   *Mathematics: Applications and Connections, Course 1*

Instructional Resources ▶▶▶

Additional standardized test practice is shown at the right and is available in the *Assessment and Evaluation Masters,* pp. 17–18.

Statistics: Graphing Data

Previewing the Chapter

Overview
This chapter explores graphing and statistics. Students learn to interpret and make frequency tables, bar graphs, line graphs, circle graphs, and stem-and-leaf plots. They use line graphs to make predictions. They find the mean, median, mode, and range of a set of data and learn to recognize when statistics and graphs are misleading. Students learn to solve problems by interpreting pictographs and bar graphs and by organizing data into a table and into a box-and-whisker plot. Students learn to graph ordered pairs in the first quadrant.

Lesson (pages)	Lesson Objectives	NCTM Standards	Standardized Tests	State/Local Objectives
2-1 (46–49)	Make and interpret frequency tables.	1–5, 7–10		
2-2 (50–53)	Choose appropriate scales and intervals for frequency tables.	1–4, 6, 7, 9, 10		
2-3 (54–57)	Construct bar graphs and line graphs.	1–5, 7–11	MAT	
2-3B (58–59)	Solve problems by using a graph.	1–5, 7, 10	CAT, MAT	
2-4 (60–63)	Interpret circle graphs.	1–7, 10	CAT, MAT, SAT	
2-5 (64–67)	Make predictions from line graphs.	1–5, 7–11	MAT, SAT	
2-6 (68–70)	Construct stem-and-leaf plots.	1–5, 8–10		
2-7 (71–75)	Find the mean, median, mode, and range to describe a set of data.	1–5, 7–10	CTBS, ITBS, MAT, SAT, TN	
2-7B (76–77)	Display and summarize data in a box-and-whisker plot.	1–5, 7–10		
2-8 (78–81)	Recognize when statistics and graphs are misleading.	1–5, 8–10		
2-9 (82–85)	Use ordered pairs to locate points and organize data.	1–5, 8–10	MAT	

CAT = California Achievement Tests, CTBS = Comprehensive Tests of Basic Skills, ITBS = Iowa Tests of Basic Skills, MAT = Metropolitan Achievement Tests, SAT = Stanford Achievement Tests, TN = Terra Nova

Organizing the Chapter

LESSON PLANNING GUIDE

Lesson	Extra Practice (Student Edition)	BLACKLINE MASTERS (PAGE NUMBERS)										
		Study Guide	Practice	Enrichment	Assessment & Evaluation	Classroom Games	Diversity	Hands-On Lab	School to Career	Science and Math Lab Manual	Technology	Transparencies A and B
2-1	p. 559	8	8	8								2-1
2-2	p. 560	9	9	9								2-2
2-3	p. 560	10	10	10	43			70			4	2-3
2-3B	p. 560											
2-4	p. 561	11	11	11			2					2-4
2-5	p. 561	12	12	12	42, 43							2-5
2-6	p. 561	13	13	13								2-6
2-7	p. 562	14	14	14	44				2	5–8	3	2-7
2-7B								40				
2-8	p. 562	15	15	15		3–4						2-8
2-9	p. 562	16	16	16	44							2-9
Study Guide/ Assessment					29–41, 45–47							

OTHER CHAPTER RESOURCES

Student Edition
Chapter Project, pp. 45, 49, 57, 81, 89
Let the Games Begin, p. 75

Technology
 CD-ROM Program
Interactive Mathematics Tools Software

Teacher's Classroom Resources

Applications
Family Letters and Activities, pp. 3–4
Investigations and Projects Masters, pp. 21–24
Meeting Individual Needs
Investigations for the Special Education Student, pp. 3–5

Teaching Aids
Answer Key Masters
Block Scheduling Booklet
Lesson Planning Guide
Solutions Manual

Professional Publications
Glencoe Mathematics Professional Series

Planning the Chapter

MindJogger Videoquizzes provide a unique format for reviewing concepts presented in the chapter.

ASSESSMENT RESOURCES

Student Edition

Mixed Review, pp. 49, 53, 57, 63, 67, 70, 74, 81, 85
Mid-Chapter Self Test, p. 67
Math Journal, pp. 56, 69, 79
Study Guide and Assessment, pp. 86–89
Performance Task, p. 89
Portfolio Suggestion, p. 89
Standardized Test Practice, pp. 90–91
Chapter Test, p. 596

Assessment and Evaluation Masters

Multiple-Choice Tests (Forms 1A, 1B, 1C), pp. 29–34
Free-Response Tests (Forms 2A, 2B, 2C), pp. 35–40
Performance Assessment, p. 41
Mid-Chapter Test, p. 42
Quizzes A–D, pp. 43–44
Standardized Test Practice, pp. 45–46
Cumulative Review, p. 47

Teacher's Wraparound Edition

5-Minute Check, pp. 46, 50, 54, 60, 64, 68, 71, 78, 82
Building Portfolios, p. 44
Math Journal, p. 77
Closing Activity, pp. 49, 53, 57, 59, 63, 67, 70, 74, 81, 85

Technology

Test and Review Software
MindJogger Videoquizzes
CD-ROM Program

MATERIALS AND MANIPULATIVES

Lesson 2-1	**Lesson 2-4**	**Lesson 2-7**	**Lesson 2-7B**
newspaper	adding machine tape	bowls of beans	index cards
	cellophane tape	calculator	cardboard
	markers	index cards	markers
	tape measure*	spinners*†	push pins
	string		string

*Glencoe Manipulative Kit   †Glencoe Overhead Manipulative Resources

PACING CHART

See pages T25–T27 for the Course Planning Calendar.

COURSE	DAY 1	DAY 2	DAY 3	DAY 4	DAY 5	DAY 6	DAY 7
Standard	Chapter Project	Lesson 2-1	Lesson 2-2	Lesson 2-3	Lesson 2-3B	Lesson 2-4	Lesson 2-5
Honors	Chapter Project	Lesson 2-1	Lesson 2-2	Lesson 2-3	Lesson 2-3B	Lesson 2-4	Lesson 2-5
Block	Chapter Project & Lesson 2-1	Lessons 2-2 & 2-3	Lessons 2-3B & 2-4	Lessons 2-5 & 2-6	Lesson 2-7	Lessons 2-8 & 2-9	Study Guide and Assessment, Chapter Test

Interactive Mathematics:
Activities and Investigations

is an activity-based program that may be used as an enhancement for chapters in *Mathematics: Applications and Connections.*

Activities and Investigations

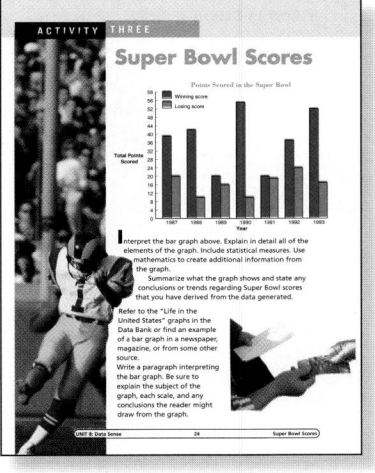

Unit 8, Activity Three
Use with Lesson 2-3.

Summary Students work in groups to generate a list of statements about a double bar graph of the winner's and loser's Super Bowl scores. Then students write individual reports about any conclusions or trends they have interpreted from the graph and group discussion.

Math Connection Students analyze a set of data that is in the form of a double bar graph. They select a measure of central tendency to determine whether there are any trends in Super Bowl scores.

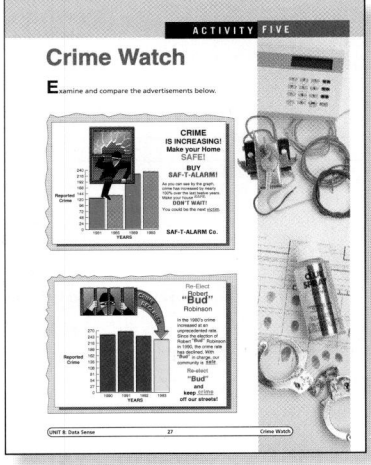

Unit 8, Activity Five
Use with Lesson 2-8.

Summary Students work in groups to discuss and analyze two advertisements showing different graphs in newspapers made using the same data. Each student then writes a letter to the editor of the newspaper describing the inconsistencies between the information presented in the two graphs.

Math Connection Students use two bar graphs that were constructed using the same data. A bar graph is used to compare quantities, with the length of each bar representing a number.

DAY 8	DAY 9	DAY 10	DAY 11	DAY 12	DAY 13	DAY 14	DAY 15
Lesson 2-6	Lesson 2-7		Lesson 2-8	Lesson 2-9	Study Guide and Assessment	Chapter Test	
Lesson 2-6	Lessons 2-7 & 2-7B		Lesson 2-8	Lesson 2-9	Study Guide and Assessment	Chapter Test	

Enhancing the Chapter

APPLICATIONS

Classroom Games, pp. 3–4

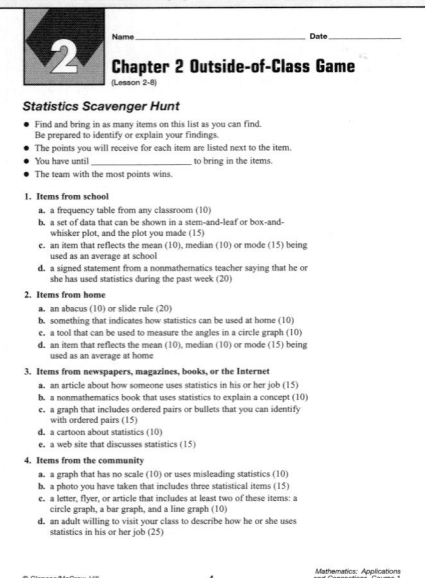

Diversity Masters, p. 2

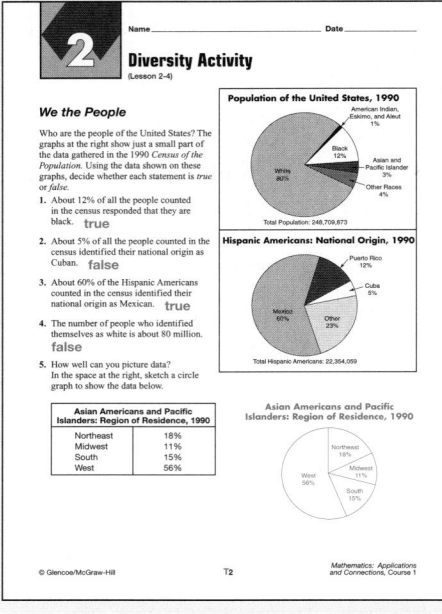

School to Career Masters, p. 2

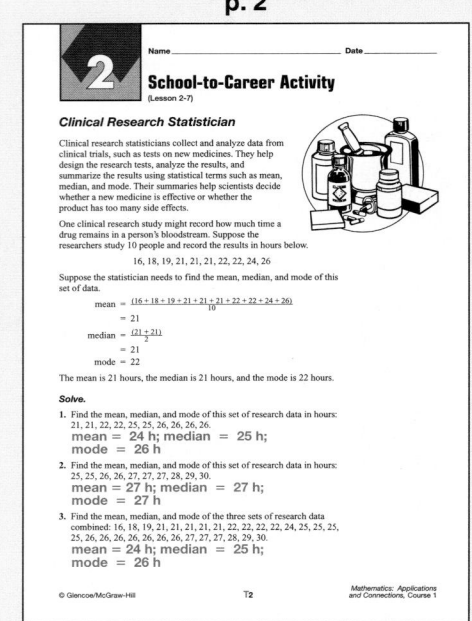

Family Letters and Activities, pp. 3–4

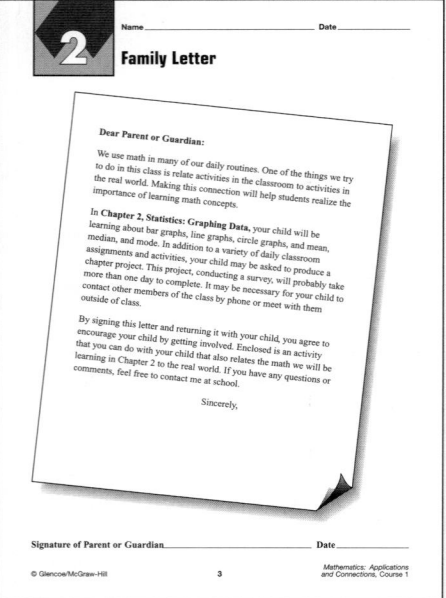

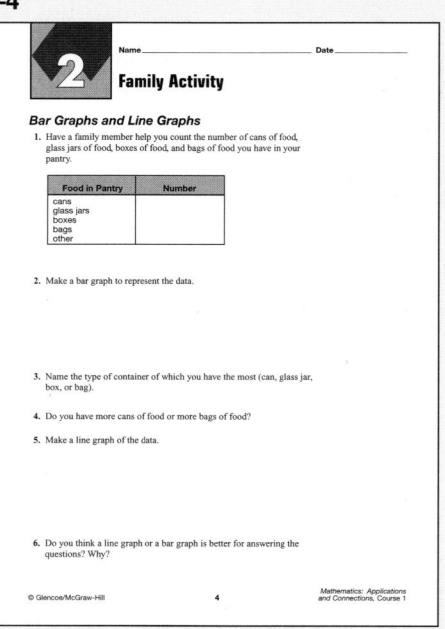

Science and Math Lab Manual, pp. 5–8

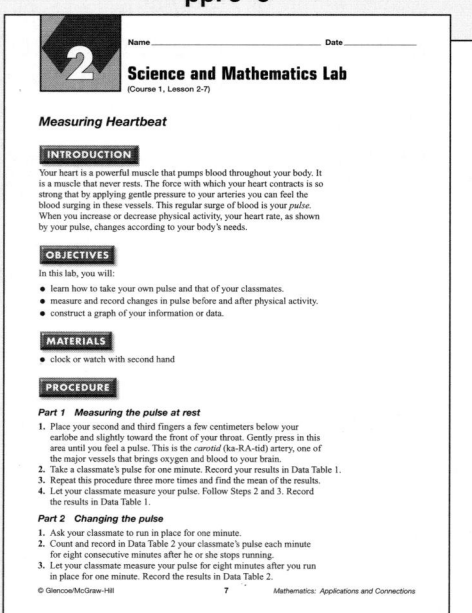

MANIPULATIVES/MODELING

Hands-On Lab Masters, p. 70

ASSESSMENT/EVALUATION

Assessment and Evaluation Masters, pp. 42–44

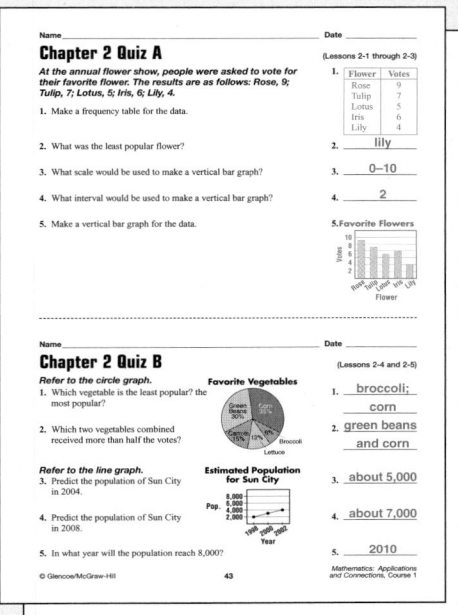

TECHNOLOGY/MULTIMEDIA

Technology Masters, pp. 3–4

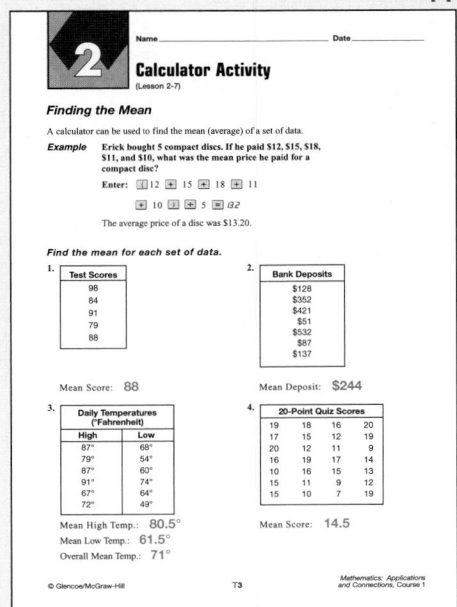

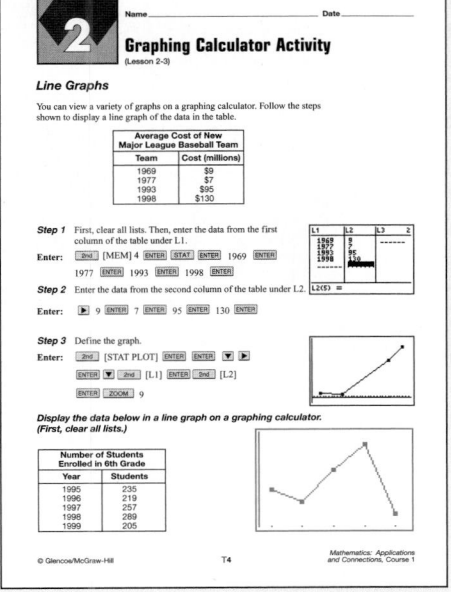

MEETING INDIVIDUAL NEEDS

Investigations for the Special Education Student, pp. 3–5

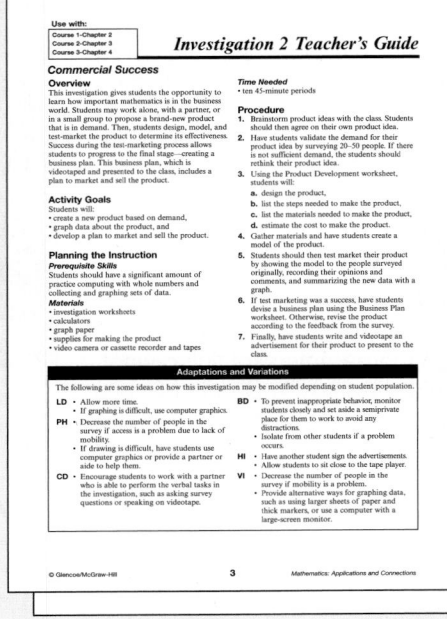

Statistics: Graphing Data

Theme: Pets

The Humane Society of the United States, founded in 1954, is an animal protection organization that also educates the public on the humane and proper care of pets.

In September 1997, the Society kicked off a new campaign called *First Strike* to educate the public on the connection between animal abuse and human violence.

Question of the Day
In 1996, there were approximately 66 million cats and 55 million dogs in households. If 32 million households had cats and 36 million households had dogs, what is the mean number of cats and dogs per household? **cats, 2.1; dogs, 1.5**

Assess Prerequisite Skills

Ask students to read through the list of objectives presented in "What you'll learn in Chapter 2." You may wish to ask them what each of the objectives means or if they have experienced or used any of these math concepts before.

 Building Portfolios

Encourage students to revise their portfolios as they study this chapter. They may wish to include copies of the various surveys and graphs used in this chapter. These are good for future reference as well as reminders of what students have learned.

 **Math and the Family**

In the *Family Letters and Activities* booklet (pp. 3–4), you will find a letter to the parents explaining what students will study in Chapter 2. An activity appropriate for the whole family is also available.

What you'll learn in Chapter 2

- to interpret and make frequency tables, bar graphs, line graphs, circle graphs, and stem-and-leaf plots,
- to use graphs to solve problems and make predictions,
- to find the mean, median, mode, and range of a set of data,
- to recognize when statistics and graphs are misleading, and
- to graph points in a coordinate system.

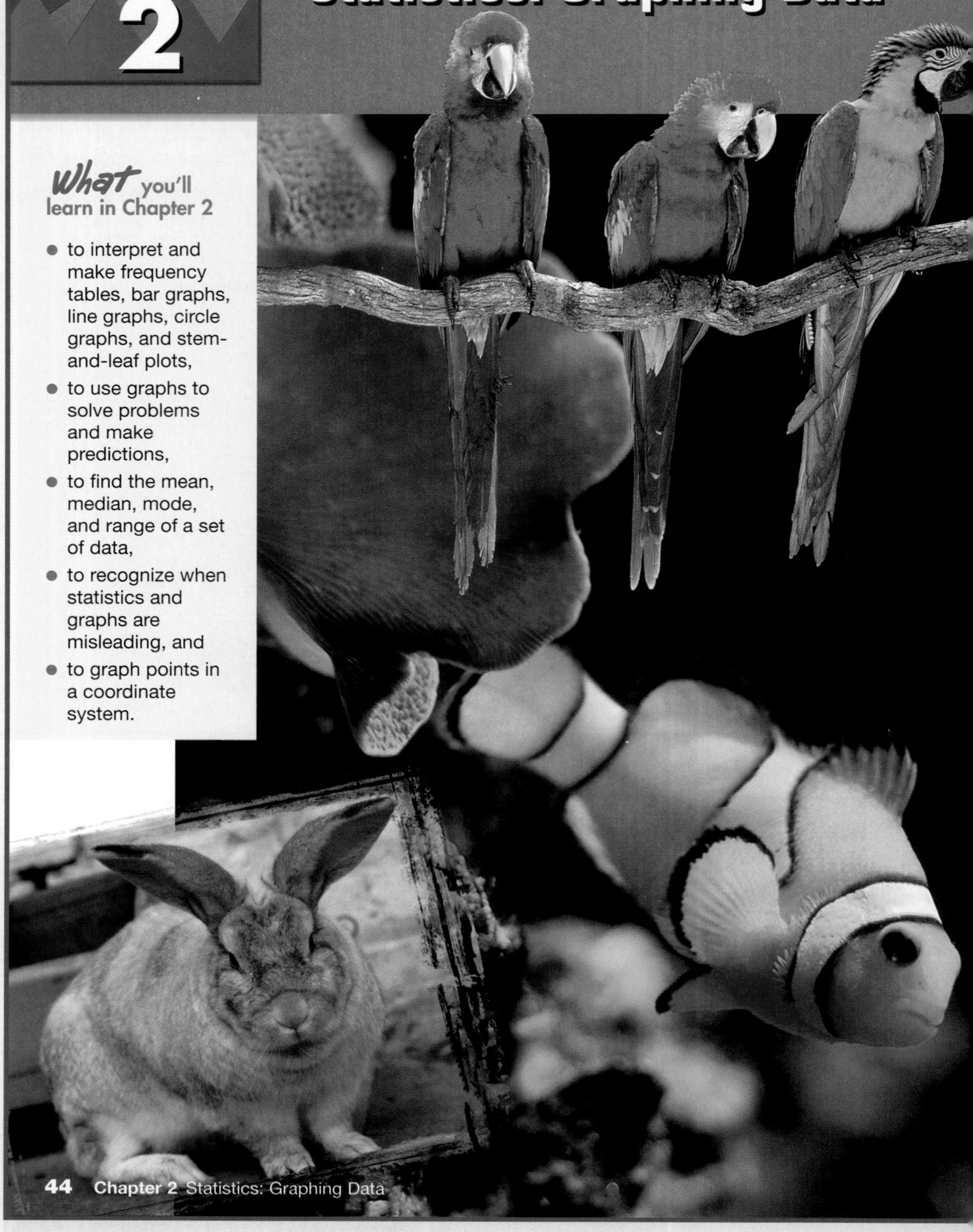

 **CD-ROM Program**

Activities for Chapter 2
- Interactive Lessons 2-1, 2-3
- Extended Activity 2-5
- Assessment Game
- Resource Lessons 2-1 through 2-9

CHAPTER Project

PETS ARE (HU)MANS' BEST FRIENDS

This project involves conducting a survey about pets. You will make tables and graphs from your survey and present them to the class. You should then compare the results of your survey to other students' surveys.

Getting Started

- Look at the class list below. How many students have no pets? How many pets are there altogether?
- Survey your classmates to determine the number and kind of pets they have.

Name	Number and Type of Pet	Total Pets
Kevin	2 dogs, 2 turtles	4
Carmen	2 dogs, 2 horses, 1 cat, 4 fish	9
Lawanda	1 dog, 3 cats	4
Heather	1 dog, 1 cat	2
Tiffany	2 dogs, 1 cat	3
Javier	1 cat, 1 hamster	2
Sonia	2 dogs, 2 cats, 20 fish	24
Matthew	1 dog	1
Denny	1 dog, 1 cat	2
Hisano	2 dogs, 1 turtle, 2 hamsters	5
Benito	1 dog, 1 cat, 6 fish, 1 hamster	9
Codi	4 cats, 3 dogs, 2 hamsters, 2 horses, 1 rabbit	12
Paquita	6 dogs, 3 cats, 3 birds, 2 snakes, 3 rats	17
Alvin	2 birds, 1 rabbit	3
Nikki	1 dog, 1 bird	2
Kyle	1 cat, 1 turtle	2
Malik	1 cat	1
Diego	1 dog, 1 snake, 1 fish	3
Cassi	5 rabbits, 2 dogs, 3 cats	10
Stephanie	1 dog, 2 cats, 1 horse	4
Lindsay	4 dogs, 5 horses, 5 cats	14
Tyrone	0	0
Michael	0	0
Judi	1 dog	1

Technology Tips

- Use **computer software** to make graphs of your data.
- Use an **electronic encyclopedia** to do your research.
- Use a **word processor.**

 interNET CONNECTION For up-to-date information on pets, visit:
www.glencoe.com/sec/math/mac/mathnet

Working on the Project

You can use what you'll learn in Chapter 2 to help you make your presentation.

Page	Exercise
49	10
57	17
81	15
89	Alternative Assessment

Chapter 2 · 45

Instructional Resources ▶▶▶

A recording sheet to help students organize their data for the Chapter Project is shown at the right and is available in the *Investigations and Projects Masters*, p. 24.

CHAPTER Project
NOTES

Objectives Students should
- make tables and graphs from a survey.
- compare different survey results.

Project Pointer You may suggest that students begin a *Project Folder* to keep their work as they complete each stage of the Chapter Project. The completed project may also be added to their portfolios.

Students may choose to survey number of pets *or* type of pet. This decision will help them determine the type of graph to use.

Using the Table After students take their surveys and complete tables similar to the one shown, have them make new tables that contain only the information they will use in their graphs.

Investigations and Projects Masters, **p. 24**

Instructional Resources
- *Study Guide Masters*, p. 8
- *Practice Masters*, p. 8
- *Enrichment Masters*, p. 8
- Transparencies 2-1, A and B
- CD-ROM Program
  - Resource Lesson 2-1
  - Interactive Lesson 2-1

Recommended Pacing	
Standard	Day 2 of 14
Honors	Day 2 of 14
Block	Day 1 of 7

1 FOCUS

5-Minute Check
(Chapter 1)

1. Estimate $378 + 245$ using rounding. $400 + 200 = 600$
2. Evaluate $12 \div 4 + 5 \times 7$. **38**
3. Evaluate $9 + 6a$ if $a = 4$. **33**
4. Write $6^2 \cdot 2^2$ as a product and evaluate. $6 \cdot 6 \cdot 2 \cdot 2 = 144$
5. Solve $m + 21 = 74$ mentally. **53**

 The 5-Minute Check is also available on **Transparency 2-1A** for this lesson.

Motivating the Lesson
Problem Solving Ask students how they would organize a table of the ages of Olympic athletes.

2 TEACH

 Transparency 2-1B contains a teaching aid for this lesson.

Teaching Tip In Example 1, emphasize to students the importance of checking the total number in the frequency column. Discuss what happens if the number is incorrect.

2-1 Frequency Tables

What you'll learn
You'll learn to make and interpret frequency tables.

When am I ever going to use this?
You can use frequency tables to keep track of votes in a class survey.

Word Wise
data
statistics
frequency table

Have you ever purchased something and the cashier asked you for your zip code? Zip codes are **data** that tell the store where its customers live. Data are pieces of information that are usually numerical. **Statistics** involves collecting, organizing, analyzing, and presenting data. In statistics, data are often organized in tables.

One type of table is a **frequency table**. A frequency table shows how many times each piece of data occurs. The frequency table below shows the results of a survey.

Favorite TV Show	Tally	Frequency
Boy Meets World	卌 卌 卌 卌 卌 III	28
Family Matters	卌 卌 卌 卌 卌 I	26
Home Improvement	卌 卌 卌 卌 卌 卌 卌	35
Mad TV	卌 卌 卌 卌 卌 卌	30
The Simpsons	卌 卌 卌 卌 卌 卌 卌 卌 卌 IIII	44
The X–Files	卌 卌 卌	15

There is a tally mark for each response in the corresponding row. The total of each row of marks is given in the last column.

Example 1
APPLICATION

Food Marian's Diner surveyed its customers to determine the favorite lunch item—hamburger (HB), hot-dog (HD), pizza (P), or sub (S). Make a frequency table of the responses shown at the right.

HD	HB	S	P
P	S	HB	HD
S	HB	HD	P
P	S	HB	P
P	S	HD	P
S	P	P	P

Explore What do you know?

You know how each person responded.

You know how many people responded.

Plan Draw a table with three columns and tally the responses.

Solve In the first column, list each lunch item. In the second column, mark the tallies. In the third column, write the frequency or number of tallies.

Lunch Choice	Tally	Frequency
Hamburger	IIII	4
Hot Dog	IIII	4
Pizza	卌 卌	10
Sub	卌 I	6

46 Chapter 2 Statistics: Graphing Data

Multiple Learning Styles

Auditory/Musical Provide students with copies of sheet music of children's songs and identify the notes in the treble clef. Have students make frequency tables of the notes by time value. Based on their results, have them make predictions of the most common notes. Test the predictions on several types of sheet music such as classical, modern, jazz, Top 40, and rock. Evaluate the success of their predictions.

Examine Check to see if the total number of scores in the frequency column matches the number of responses in the original list. If it doesn't, you need to check your work.

You can use the information in a completed frequency table in many problem-solving situations.

Example **2**
CONNECTION

Life Science Mr. Martinez' life science class is taking a field trip to the local zoo to do research for their class project. He decides to ask the students which behind-the-scenes tour they would like to take. They will take the tour receiving the most votes. Mr. Martinez made the frequency table below. Which tour should he schedule for the students?

Tour	Tally	Frequency
reptiles	卌 卌 l	11
large cats	卌 lll	8
pachyderms	lll	3
birds	卌 l	6

Look for the tour that has the most tally marks. Mr. Martinez should schedule a tour about reptiles.

HANDS-ON

MINI-LAB

Work in groups of five. newspaper

Ten of the most commonly used words in written English are *the, and, of, to, in, is, you, that, it,* and *for*.

Try This

1. Select a portion of text from a newspaper article that is at least 4 inches long and 1 column wide.

2. Make a frequency table. Tally each time one of these ten words appears. **1–3. See students' work. Answers will vary.**

Talk About It

3. Which word appeared most often? Do you think it is the most commonly used word in written English?

4. Do you think the results would be the same if you selected a different section of the newspaper? Explain why or why not.

5. Do you think the results would be a lot different if you selected a portion of a novel instead of the newspaper? Explain why or why not. **4–5. Answers will vary.**

In-Class Examples

For Example 1
Students were asked to name their favorite school period—Science (S), English (E), Math (M), Social Studies (H), Phys. Ed. (P), or Lunch (L). Make a frequency table of their responses.

S	S	P	P	L	E	S	H
L	E	L	P	P	L	S	L
M	P	H	S	L	L	E	L

Favorite Period	Tally	Frequency
Science	卌	5
English	lll	3
Math	l	1
Social Studies	ll	2
Phys. Ed.	卌	5
Lunch	卌 lll	8

For Example 2
From a class survey of favorite colors, the frequency table below was made. List the students' favorite colors in order from least to most favorite.

Color	Frequency
red	4
blue	5
black	2
green	3
purple	7
yellow	6

black, green, red, blue, yellow, purple

Using the Mini-Lab After answering Exercises 3–5, discuss how this activity could be applied to writing and solving cryptograms and other codes. As an extension, have students write cryptic messages using what they've learned.

Cross-Curriculum Cue

Inform the teachers on your team that your students are studying graphing and statistics. Suggestions for curriculum integration are:
Health: height/weight charts
Civics: national debt
Earth Science: atmospheric temperature

Check for Understanding

If students need additional practice or instruction after completing Exercises 1–5, one of the following options may be helpful.

- Extra Practice, see p. 559
- Reteaching Activity
- *Study Guide Masters*, p. 8
- *Practice Masters*, p. 8

Assignment Guide

Core: 6, 7, 9, 11–14
Enriched: 7–9, 11–14

Additional Answers

1. Sample answer: to make it easier to study the data.

2. Sample answer: *The Simpsons* was the favorite TV show.

5a.
Age	Tally	Frequency
10	I	1
11	I	1
12	ЖΙ	5
13	III	3
14	II	2

Study Guide Masters, p. 8

Communicating Mathematics

Read and study the lesson to answer each question. 1–2. See margin.

1. *Tell* an advantage of organizing data into a frequency table.

2. *Write* a sentence to tell what you can conclude when you analyze the favorite TV show data in the frequency table on page 46.

HANDS-ON MATH

3. Use the newspaper article from the Mini-Lab on page 47 to make a frequency table of the letters used in the article. **a–d. See students' work.**

   a. Do you think the letter that appeared most often is used most often in the English language? Explain why or why not.

   b. Were any letters in the alphabet not in the article?

   c. What vowel appears most often?

   d. Compare your results with other classmates.

Guided Practice

4. *Physical Education* Coach Estes took a survey of the students in gym class to see which sport she should offer during the spring. She recorded the responses in the frequency table below.

Sport	Tally	Frequency
volleyball	ЖΙ ЖΙ	10
baseball	IIII	4
softball	ЖΙ III	8
basketball	ЖΙ II	7

   a. Copy the table and complete the frequency column.

   b. Which sport should Coach Estes offer? **volleyball**

5. The ages of the guests at Linn's birthday party are shown below.

14	12	13	12	12	14
13	10	11	12	13	12

   a. Make a frequency table of the data. **See margin.**

   b. What was the most common age of the guests? **12**

Applications and Problem Solving

Make a frequency table for each set of data. 6–8. See Answer Appendix.

6. To the nearest hour, how many hours of television did you watch last Saturday?

   1   2   4   3   2   1   1   2   2   3

7. How many times were you absent in the last six weeks?

   1   5   0   1   3   1   1   0   1
   3   5   1   2   0   0   2   0   3

8. What were the daily high temperatures for September?

   89   92   89   89   90   89   89   87   86   87
   89   87   86   90   90   89   90   87   87   86
   87   89   89   87   89   87   87   86   85   86

■ Reteaching the Lesson ■

Activity Have students think of a survey question with a *yes* or *no* answer, such as "Did you eat breakfast today?" Have them survey the class and make a frequency table of their results.

9a. Sample answer: All the shoes sold were between sizes 9 and 11.

9. Business The frequency table shows the sizes of shoes sold in August at the Athlete's Locker.

Size	Tally	Frequency
9	ЖГ ЖГ II	12
9½	ЖГ ЖГ ЖГ I	16
10	ЖГ ЖГ ЖГ ЖГ III	23
10½	ЖГ ЖГ I	11
11	ЖГ ЖГ ЖГ	15

a. Describe the data shown in the table.

b. How many pairs of shoes were sold in August? **77 pairs of shoes**

c. Which size should the Athlete's Locker stock the most? **10**

10. Working on the CHAPTER Project

a. Make a frequency table that shows the number of pets owned by each student on the class list on page 45. **See students' work.**

b. Make a frequency table that shows the number of pets owned by each of your classmates. **See students' work.**

11. Business Refer to the frequency table below.

11a. Sample answer: Based on the data in the table, they should keep potato chips and corn chips well stocked.

11b. Answers will vary. Sample answer: There are too few to tell.

Favorite Snack Food of Food Mart Customers		
Snack Food	Tally	Frequency
potato chips	ЖГ I	6
tortilla chips	III	3
corn chips	ЖГ	5
pretzels	II	2
popcorn	III	3

a. Which snack(s) should Food Mart keep well stocked?

b. Are there any snacks that Food Mart should discontinue stocking? Explain.

12. Critical Thinking Refer to the beginning of the lesson. What are some reasons that a store would like to know the zip codes of its customers? **See margin.**

Mixed Review

13. **Test Practice** Name the number that is the solution of the equation $102 \div q = 17$. *(Lesson 1–7)* **A**

A 6
B 7
C 8
D 9

14. Money Matters Before buying a computer, Jocelyn called several computer stores to compare prices. The highest price was $2,349, and the lowest price was $1,788. Estimate the difference between the highest and lowest prices. *(Lesson 1-3)* **about $500**

Lesson 2-1 Frequency Tables **49**

Extending the Lesson

Enrichment Masters, p. 8

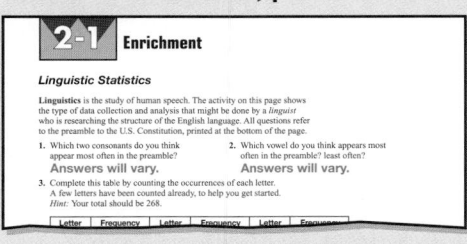

Activity Separate students into groups of three or four. Have them research how frequency tables are used in science, social studies, sports, conservation, or other areas. Have each group report to the class on its findings.

Exercise 10 asks students to advance to the next stage of work on the Chapter Project. You may wish to have students work in groups to complete their frequency tables.

4 ASSESS

Closing Activity

Writing Have pairs of students write a description of a frequency table and how they might use it at school or at home. How might their teacher use one?

Additional Answer

12. Sample answer: The owner of the store could plan a special mailing to residents with the zip code that appears most.

Practice Masters, p. 8

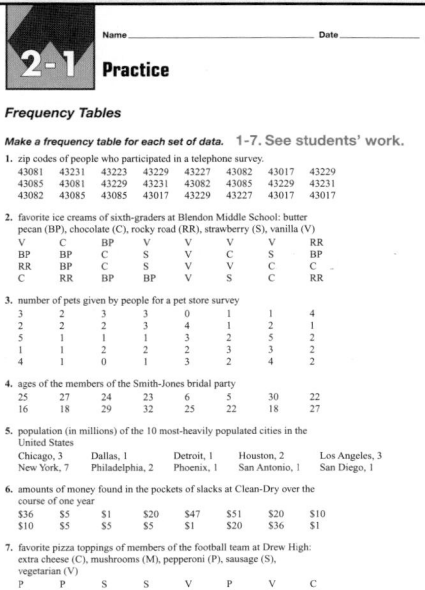

- *Study Guide Masters*, p. 9
- *Practice Masters*, p. 9
- *Enrichment Masters*, p. 9
- Transparencies 2-2, A and B
 CD-ROM Program
  - Resource Lesson 2-2

Recommended Pacing	
Standard	Day 3 of 14
Honors	Day 3 of 14
Block	Day 2 of 7

1 FOCUS

 5-Minute Check
(Lesson 2-1)

Vivian kept track of the time she spent on the telephone for one week.

Telephone Calls		
Minutes	Tally	Frequency
1–14	IIII	4
15–29	IIIII III	8
30–44	IIIII IIIII	10
45–59	IIIII I	6

1. How many calls did Vivian make? **28**
2. How many calls were less than 30 minutes? **12**
3. How many calls were 30 minutes or longer? **16**

 The 5-Minute Check is also available on **Transparency 2-2A** for this lesson.

Motivating the Lesson
Communication Discuss with students how or why the word *scale* may have evolved from the Latin word meaning ladder or staircase. Draw a picture of a ladder on the chalkboard and ask students to arrange a data set of temperatures or weights on the steps. Ask them what they think the word *interval* means.

2-2 Scales and Intervals

What you'll learn
You'll learn to choose appropriate scales and intervals for frequency tables.

When am I ever going to use this?
Knowing how to choose scales and intervals can help you make a bar graph or line graph.

Word Wise
interval
scale

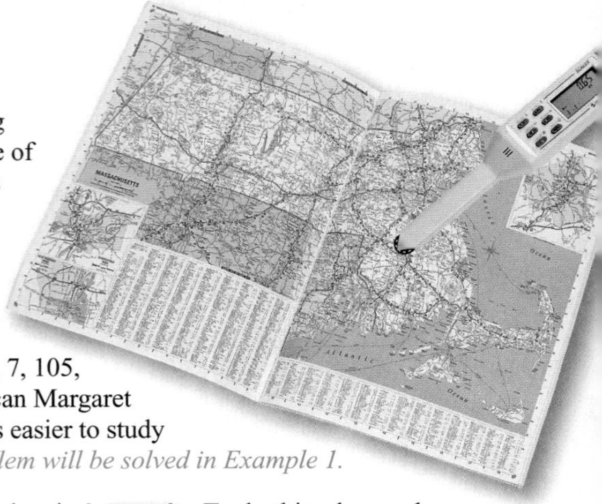

For her history project, Margaret is investigating the early state legislature of Massachusetts. She uses an atlas to find how far each county seat is from the capital, Boston. The distances in miles are 65, 134, 37, 85, 22, 101, 90, 102, 7, 105, 2, 41, 28, and 42. How can Margaret organize this data so it is easier to study the distances? *This problem will be solved in Example 1.*

Margaret can group the data in **intervals**. To do this, she needs to determine the **scale** for her frequency table. The scale must include all of the data.

Example ❶
CONNECTION

History Refer to the beginning of the lesson. Organize Margaret's distance data into a frequency table.

Step 1 Determine the scale. Since the data includes distances from 2 to 134, begin the scale with 0 and end with 140.

Step 2 Determine the interval. To keep the number of intervals manageable, choose an interval of 20.

Step 3 Make the frequency table. Now Margaret can easily see the number of county seats in each interval.

All intervals need to be equal and not overlap.

Distances from County Seats to Boston		
Distance	Tally	Frequency
121–140	I	1
101–120	III	3
81–100	II	2
61–80	I	1
41–60	II	2
21–40	III	3
0–20	II	2

Example ❷
APPLICATION

Sports The scores of the Glencoe Scholarship Golf tournament are shown below.

74, 87, 88, 97, 75, 78, 76, 69, 84, 83, 86, 88, 92, 93, 72, 89, 85, 67, 77, 82, 84, 85, 99, 81

a. Determine an appropriate scale for the data.
b. Make a frequency table for the data.

50 Chapter 2 Statistics: Graphing Data

Investigations for the Special Education Student

This blackline master booklet helps you plan for the needs of your special education students by providing long-term projects along with teacher notes. Investigation 2, *Commercial Success,* may be used with this chapter.

a. A scale from 66 to 100 is appropriate because it includes all of the scores.

b. Since the scale is small, an interval of 5 is used in the table at the right.

Golf Scores		
Score	**Tally**	**Frequency**
96–100	II	2
91–95	II	2
86–90	IIII	5
81–85	IIII II	7
76–80	III	3
71–75	III	3
66–70	II	2

You can use a frequency table to organize data about a group of people, such as your classmates.

 **MINI-LAB**

Work with a partner.

📄 paper ✏️ pencil

Try This 1–3. Answers will vary.

1. Each person should ask ten classmates how many hours a week each of them spend on the telephone. Record each response. Combine your results with your partner.
2. Choose a scale and appropriate intervals for a frequency table.
3. Make a frequency table.

Talk About It 4. the most common responses

4. What does your frequency table tell you?
5. Compare your frequency table to a frequency table from another pair of classmates. Are the scale and intervals the same? Explain why or why not. **Answers will vary.**

Study Hint

Problem Solving When the scale of a set of data is small, intervals can be 1, 2, 3, 4, or 5. When the scale is large, intervals are usually multiples of 10.

CHECK FOR UNDERSTANDING

Communicating Mathematics

1–2. See margin.

Read and study the lesson to answer each question.

1. *Explain* how to find an appropriate scale for the data in the chart.

2. *Write* a sentence or two explaining how to determine an interval for the data in the chart.

Distances Traveled to School by Students in Mr. Alonzo's Class					
18	14	30	62	40	47
34	12	56	57	63	54
32	27	33	46	17	13

Additional Answers

1. Sample answer: The scale must include all of the data. Since the shortest distance is 12 and the longest distance is 63, a scale from 10 to 70 is appropriate.

2. Sample answer: Since the scale is 10 to 70, the best interval would be 10. This will provide a manageable number of intervals. Using 5 would have too many intervals and using 20 would have very few.

---

 2 TEACH

Transparency 2-2B contains a teaching aid for this lesson.

In-Class Examples

For Example 1
Organize this set of average temperatures (°F) in U.S. cities for the month of January into a frequency table.
21, 34, 13, 36, 41, 31, 49, 31, 49, 32, 50, 20, 42, 6, 29, 29, 29, 23, 16, 10, 22, 47, 32, 40, 26, 21, 25, 44, 27, 19, 44, 29

Average Temperatures		
°F	**Tally**	**Frequency**
46–50	IIII	4
41–45	IIII	4
36–40	II	2
31–35	IIII	5
26–30	IIII I	6
21–25	IIII	5
16–20	III	3
11–15	I	1
6–10	II	2

For Example 2
The scores for a social studies test are shown below.
82, 41, 58, 99, 76, 84, 67, 97, 92, 85, 79, 83, 90, 84, 77, 63, 75, 91, 84, 88, 96, 73, 79, 80

a. Determine an appropriate scale for the data. **40 to 100**

b. Make a frequency table for the data.

Test Scores		
Score	**Tally**	**Frequency**
90–100	IIII I	6
80–89	IIII III	8
70–79	IIII I	6
60–69	II	2
50–59	I	1
40–49	I	1

Using the Mini-Lab Encourage students to determine their own responses before beginning their surveys. As surveys are completed, circulate around the room to identify any difficulties in setting up frequency tables. Assign peer tutors as needed.

3 PRACTICE/APPLY

Check for Understanding

If students need additional practice or instruction after completing Exercises 1–7, one of these options may be helpful.
- Extra Practice, see p. 560
- Reteaching Activity
- *Study Guide Masters,* p. 9
- *Practice Masters,* p. 9
-  Interactive Mathematics Tools Software

Assignment Guide

Core: 8, 10, 11, 13, 14, 16, 17, 19, 20–23
Enriched: 9, 10, 12, 13, 15, 16, 17–23

Additional Answer

6.

Amount	Tally	Frequency
20–24	I	1
15–19	II	2
10–14	HHt	5
5–9	IIII	4
0–4		0

Study Guide Masters, p. 9

HANDS-ON MATH

3. Using the data you collected for the Mini-Lab on page 51, make another frequency table to show how the scale and interval would change if you had asked your classmates how many *minutes* they spend on the phone. **See students' work.**

Guided Practice   **Use the set of data below to answer Exercises 4–6.**

$12, $8, $13, $6, $17, $7, $9, $13, $11, $10, $15, $20

4. What scale would you use in making a frequency table for this set of data? **Sample answer: 0 to 20**

5. Would an interval of 2 or 10 be better to use in making a frequency table for this set of data? Explain. **2; 10 is too large.**

6. Make a frequency table of the data. **See margin.**

7a. See Answer Appendix.

7. *Entertainment* Shelby's last eighteen video game scores were: 3,500; 10,200; 30,000; 65,300; 56,000; 28,200; 17,000; 43,200; 23,300; 41,900; 32,600; 29,100; 46,400; 39,500; 25,000; 37,800; 49,700; and 34,300.

   a. Make a frequency table for this set of data.

   b. In what interval are most of Shelby's scores? **Most of her scores were between 30,000 and 39,900.**

EXERCISES

Practice   **Choose the better scale for a frequency table for each set of data.**

8. 3, 17, 9, 6, 2, 5, 12 **b**   **a.** 0 to 10   **b.** 0 to 20

9. 53, 88, 25, 47, 50, 18, 15, 43 **a**   **a.** 0 to 99   **b.** 20 to 80

10. 245, 144, 489, 348, 36, 284, 150, 94, 220 **a**   **a.** 0 to 499   **b.** 0 to 999

Choose the best interval for a frequency table for each set of data.

11. the data in Exercise 8 **a**   **a.** 2   **b.** 10   **c.** 20

12. the data in Exercise 9 **c**   **a.** 4   **b.** 5   **c.** 10

13. the data in Exercise 10 **b**   **a.** 10   **b.** 100   **c.** 1,000

Make a frequency table for each set of data. 14–16. See Answer Appendix.

14. the data in Exercise 8

15. the data in Exercise 9

16. the data in Exercise 10

52   Chapter 2 Statistics: Graphing Data

Reteaching the Lesson

Activity Show the class several small containers containing varying numbers of dried beans. Ask students to count the beans in each container. Make a frequency table with the class. Have students determine the scale and intervals that organize the data.

Error Analysis
Watch for students who have difficulty choosing an appropriate scale.
Prevent by reminding them to use the highest and lowest numbers to determine the scale and to always include both numbers in the table.

17a. No; since Texas is a much larger state, the distances will be larger. If she used the same scale the frequency table would not include all the distances.

17b. Sample answer: Yes; however, the table will have a large number of rows.

17. *Geography* Refer to the beginning of the lesson. Suppose Margaret decided to do her project on the state of Texas instead of Massachusetts.
  a. Could she use the same scale she used for Massachusetts? Explain why or why not.
  b. Could she use the same interval she used for Massachusetts? Explain why or why not.

18. *Health* Read the nutrition label on at least 15 different food items and record the number of milligrams (mg) of sodium per serving.
  a. Make a frequency table of your data. **See students' work.**
  b. What conclusions can you make from your table? **Answers will vary.**

19. *Geography* The highest elevation in states with mountains over 5,000 feet high are shown below.

State	Elevation (ft)	State	Elevation (ft)
Alaska	20,320	New York	5,344
Arizona	12,633	North Carolina	6,684
California	14,494	Oregon	11,239
Colorado	14,433	South Dakota	7,242
Idaho	12,662	Tennessee	6,643
Maine	5,268	Texas	8,749
Montana	12,799	Utah	13,528
Nevada	13,140	Virginia	5,729
New Hampshire	6,288	Washington	14,410
New Mexico	13,161	Wyoming	13,804

  a. What scale would you use in making a frequency table for this set of data? **Sample answer: 5,000–20,999**
  b. Make a frequency table of the data. **See margin.**

20. Sample answer: 21, 24, 34, 32, 31, 22, 23, 36, 35, 33, 32, 39, 34, 32, 23

20. *Critical Thinking* Create a set of at least 15 pieces of data so that a frequency table of the data would need a scale from 20 to 40.

Mixed Review

21. *Sports* The number of hits each player on the little league baseball team had in the last game of the season were 1, 3, 2, 4, 2, 3, 1, 0, and 2. Make a frequency table of the data. *(Lesson 2-1)* **See margin.**

22. *Algebra* Evaluate a^3 if $a = 2$. *(Lesson 1-6)* **8**

23. **Test Practice** A stereo receiver costs $659, speakers cost $472, and a compact disc player costs $326. Which is the best estimate for the total amount for this stereo system? *(Lesson 1-3)* **D**
  A $900
  B $1,100
  C $1,200
  D $1,500
  E $1,900

Lesson 2-2 Scales and Intervals **53**

Extending the Lesson

Enrichment Masters, p. 9

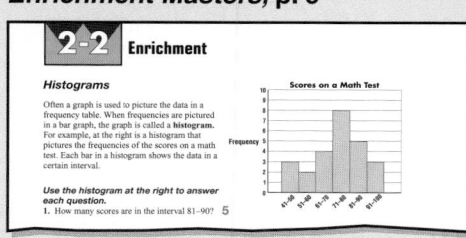

Activity Have students look through newspapers and almanacs for data that would probably not be organized in a frequency table. Ask students to describe the data and explain why a frequency table would not be appropriate.

Closing Activity
Modeling Ask students to measure the distance from their elbows to the tips of their little fingers. Have them gather this information and determine a scale for the data. Then have students separate into groups according to the interval they have selected.

Additional Answers
19b.

Elevation	Tally	Frequency
19,000–20,999	I	1
17,000–18,999		0
15,000–16,999		0
13,000–14,999	ⅢⅡ	7
11,000–12,999	IIII	4
9,000–10,999		0
7,000–8,999	II	2
5,000–6,999	ⅢI	6

21.

Number of Hits	Tally	Frequency
0	I	1
1	II	2
2	III	3
3	II	2
4	I	1

Practice Masters, **p. 9**

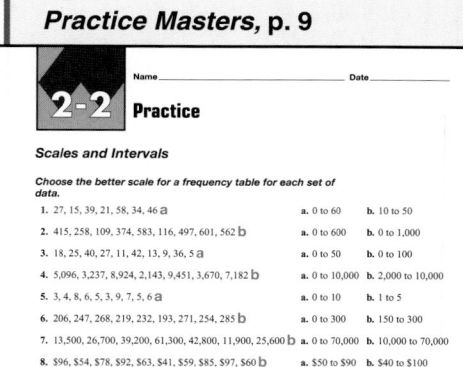

Lesson 2-2 **53**

- *Study Guide Masters*, p. 10
- *Practice Masters*, p. 10
- *Enrichment Masters*, p. 10
- Transparencies 2-3, A and B
- *Assessment and Evaluation Masters*, p. 43
- *Hands-On Lab Masters*, p. 70
- *Technology Masters*, p. 4
- CD-ROM Program
  - Resource Lesson 2-3
  - Interactive Lesson 2-3

Recommended Pacing	
Standard	Day 4 of 14
Honors	Day 4 of 14
Block	Day 2 of 7

1 FOCUS

 5-Minute Check
(Lesson 2-2)

1. Determine an appropriate scale for the following set of data. 24, 52, 44, 41, 87, 79, 43, 88, 63, 82, 20, 15, 50, 39, 69, 22 **10–90**

2. Make a frequency table for the data in Exercise 1.

Data	Tally	Frequency
10–30	IIII	4
31–50	IIII I	5
51–70	III	3
71–90	IIII	4

The 5-Minute Check is also available on **Transparency 2-3A** for this lesson.

Motivating the Lesson

Communication Ask students to read the opening paragraph of the lesson. Then ask these questions.

- Do you think that a graph is a better way to show this information? **yes**
- Why? **Sample answer: It is easier to see the difference in height of bars on a graph than to compare numbers in a table.**

2-3 Bar Graphs and Line Graphs

What you'll learn

You'll learn to construct bar graphs and line graphs.

When am I ever going to use this?

You can use a bar graph to compare the attendance at sporting events and a line graph to show population data.

Word Wise
bar graph
line graph

How do you wake up in the morning? Does someone wake you or do you use an alarm clock? The results of a survey of how 400 parents with children under 18 wake their children is shown in the frequency table.

Method	Frequency
Call Them	172
Alarm Clock	88
Wake On Their Own	64
Other	76

A graph is a more visual representation of data. A **bar graph** is used to compare the frequency of the amount in a category or class.

Example ➊ Use the data from the survey to draw a vertical bar graph.

Step 1 Draw and label both scales.

Step 2 Determine the scale for the vertical axis. Choose an interval that best represents the data and mark off equal spaces on the vertical scale.

> **Scale** The data includes numbers from 76 to 172, so begin the scale with 0 and end with 180.
>
> **Interval** Since the vertical scale needs to go from 0 to 180, use an interval of 20.

Step 3 Mark off equal spaces on the horizontal axis and label each with a method.

Step 4 Draw a bar for each method. The height of each bar represents the number of parents that use that method.

Step 5 Label the graph with a title.

How Parents Wake Up Their Children

Multiple Learning Styles

Visual/Spatial Have small groups use almanacs to find the temperatures over a 12-month period for a city of their choice. Ask groups to show the data in graphs on posterboards. Encourage a variety of cities among the groups, such as Minneapolis, Phoenix, and San Francisco so they can compare graphs.

To show how data changes over a period of time, you could use a **line graph**. A line graph is usually used to show the change and direction of change over time.

2 TEACH

Transparency 2-3B contains a teaching aid for this lesson.

Example 2

APPLICATION

Food Christopher takes a poll to find out how many soft drinks his classmates drink each day. The results are shown at the right. Draw a line graph using the data he collected.

Day	Frequency
Monday	27
Tuesday	22
Wednesday	18
Thursday	26
Friday	57
Saturday	61
Sunday	42

Step 1 Draw and label both scales.

Step 2 Determine the scale for the vertical axis. Choose an interval that best represents the data and mark off equal spaces on the vertical scale.

Scale The data includes numbers from 18 to 61, so begin the scale with 0 and end with 70.

Interval Since the vertical scale needs to go from 0 to 70, use an interval of 10.

Step 3 Mark off equal spaces on the horizontal scale and label them with the days of the week in order.

Step 4 Draw a dot to show the frequency for each day. Draw line segments to connect the dots.

Step 5 Label the graph with a title.

Soft Drinks Consumed in a Week

The graph shows that Christopher's classmates drink many more soft drinks on the weekend.

Study Hint

Technology You can use a TI-83 graphing calculator to make a line graph without a title or labels. There are also many computer software programs available that will construct graphs as you enter the data.

Reading Mathematics After reading the first paragraph on page 55 and discussing Example 2, ask students what changes and direction of change over time they observe.

In-Class Examples

For Example 1
A poll was taken to see how many minutes people read each day. The results showed women read 20 minutes; men, 15; teens, 25; and children, 10. Use the data to draw a vertical bar graph.

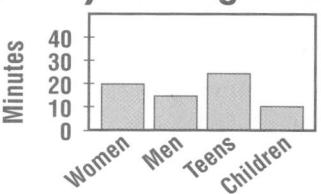

Daily Reading Time

For Example 2
Draw a line graph to show the windchill factors at 35° F.

Thermometer Reading 35°F	
Wind Speed (mph)	**Windchill Factor (°F)**
5	33
10	22
15	16
20	12
25	8
30	6
35	4
40	3
45	2

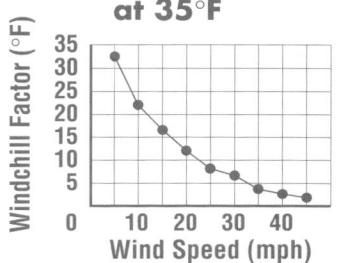

Windchill Factors at 35°F

CHECK FOR UNDERSTANDING

Communicating Mathematics

Read and study the lesson to answer each question.

1. *Examine* the bar graph in Example 1. Tell which method for waking up children was used the most. **Call Them**

2. *Examine* the line graph in Example 2. Tell which day the students drank the least number of soft drinks. **Wednesday**

Classroom Vignette

"I fill a glass jar with color candies and have students estimate the total number of candies and how many are each color. We count the candy by placing them in cups of 10 for each color. We make a pictograph of the cups organized by color, labeling the axes and giving the graph a title."

Dora Brown, Teacher
Awtrey Middle School
Kennesaw, GA

Dora Brown

Check for Understanding

If students need additional practice or instruction after completing Exercises 1–6, you may find one of the following options helpful.
- Extra Practice, see p. 560
- Reteaching Activity
- *Study Guide Masters,* p. 10
- *Practice Masters,* p. 10
- 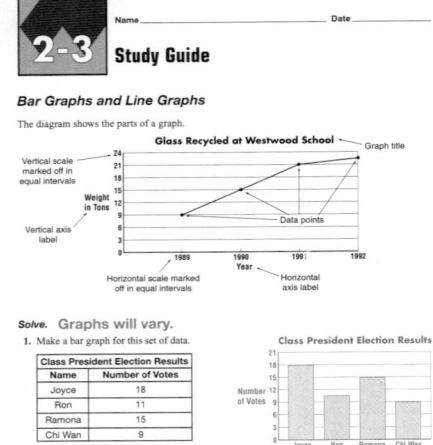 Interactive Mathematics Tools Software

Assignment Guide

Core: 7–15 odd, 16, 18–21
Enriched: 8–14 even, 16, 18–21

Additional Answer

3. Sample answer: If the vertical scale is much higher than the highest value, it makes the graph flatter. Changing the interval does not affect either type of graph.

Math Journal

3. *Write* a sentence or two explaining how the vertical scale and the interval affect the look of a bar graph or line graph. **See margin.**

Guided Practice

Choose a scale and an interval for the vertical axis of a line graph for each set of data.

4. 28, 13, 9, 26, 19 **0–30; 5**

5. 18, 32, 52, 68, 56, 31 **0–70; 10**

6. *Transportation* The number of passengers arriving at and departing from airports are listed in the table.

    a. Choose an interval for a bar graph of the passenger data.

    b. Make a horizontal bar graph for the data. **See Answer Appendix. a. 5,000 or 10,000**

World's Busiest Airports, 1994	
Airport	**Passengers (thousands)**
Chicago-O'Hare	66,400
Atlanta	54,100
Dallas-Ft. Worth	52,600
London Heathrow	51,700
Los Angeles	51,100
Frankfurt	35,100
San Francisco	34,600
Denver	33,100
Miami	30,200

Source: Airports Council International

EXERCISES

Practice

Make a bar graph for each set of data. **7–10. See Answer Appendix.**

7. a vertical bar graph

Favorite Soft Drink	
Soft Drink	**Frequency**
Coke	4
Pepsi	15
Dr. Pepper	9
Mountain Dew	19
Orange Crush	11
A&W Root Beer	8

8. a horizontal bar graph

Average Event-Day Attendance (thousands)	
Sport	**Attendance**
NASCAR Races	83
NFL (football)	62
MLB (baseball)	31
NBA (basketball)	16
NHL (hockey)	14

Source: USA TODAY research

Make a line graph for each set of data.

9.

Average Snowfall for Brighton, Utah	
Month	**Snow (in.)**
Oct.	21
Nov.	50
Dec.	66
Jan.	69
Feb.	63
March	70
April	51
May	16

Source: The USA TODAY Weather Almanac

10.

Sales of Karaoke Players in the U.S. (thousands)	
Year	**Amount**
1990	$77,900
1991	$94,800
1992	$119,000
1993	$116,000
1994	$86,900
1995	$94,300
1996	$82,100

Source: National Association of Music Merchants

56   Chapter 2 Statistics: Graphing Data

Study Guide Masters, p. 10

2-3 **Study Guide**

Name _____ Date _____

Bar Graphs and Line Graphs

The diagram shows the parts of a graph.

Glass Recycled at Westwood School

Graph title
Vertical scale marked off in equal intervals
Weight in Tons
Vertical axis label
Data points
Horizontal scale marked off in equal intervals
Year
Horizontal axis label

Solve. Graphs will vary.

1. Make a bar graph for this set of data.

Class President Election Results	
Name	Number of Votes
Joyce	18
Ron	11
Ramona	15
Chi Wan	9

Class President Election Results
Number of Votes
Name

2. Make a line graph for this set of data.

Evans Family Electric Bill	
Month	Amount
March	$129.90
April	$112.20
May	$105.00
June	$88.50

Evans Family Electric Bill
Amount
Month

© Glencoe/McGraw-Hill T10 *Mathematics: Applications and Connections, Course 1*

Reteaching the Lesson

Activity Take a survey of the color of the shirts, blouses, or sweaters worn by students. Record the results in a frequency table. Have students determine the scale and interval. Draw a vertical bar graph and choose a title and labels for axes. Use masking tape to form the horizontal and vertical axes on the floor. Then have students model the bar graph by standing in the correct place based on the color of the top they are wearing.

Refer to the following tables for Exercises 11–15.

A.

Super Bowl Ticket Prices	
Year	Price
1967	$12
1972	15
1977	20
1982	40
1987	75
1992	150
1997	275

Source: The NFL

B.

Entertainment Expenses per Person in 1994	
Entertainment	Amount
Basic Cable	$110
Books	79
Home Video	73
Recorded Music	56
Daily Newspapers	49
Magazines	36
On-line/Internet	7

Source: Statistical Abstract of the United States, 1996

11. Determine a scale for the prices in Table A. **Sample answer: 0 to 300**

12. What would be the best interval for the amounts in Table B? Explain why you chose it. **12–13. See margin.**

13. Which table could be best represented with a line graph? Explain.

14. Make a bar graph for the data in Table B. **14–15. See Answer Appendix.**

15. Make a line graph for the data in Table A.

Applications and Problem Solving

16a. 18 million

16. *Geography* The line graph shows Australia's population growth.
    a. Estimate the population of Australia in 1991.
    b. Estimate the number of years it took for the population to double. **about 35 years**

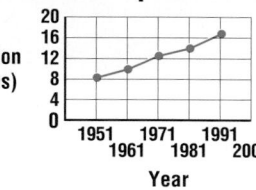

Australia's Population

17. *Working on the* **CHAPTER Project** Refer to your frequency tables from Exercise 10 on page 49. **a–b. See students' work.**
    a. Make a bar graph using your frequency table from Exercise 10a.
    b. Make a bar graph using your frequency table from Exercise 10b.

18. Sample answer: Over half of my friends get a larger allowance.

18. *Critical Thinking* The bar graph shows the allowance of your five best friends. How might you use the information in this graph to argue for a raise in your allowance if you are currently getting $5 a week?

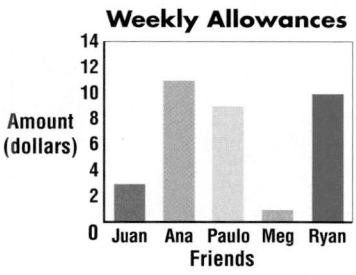

Weekly Allowances

Mixed Review

19. **Test Practice** Brianna's scores on her first six history quizzes were 76, 72, 83, 96, 81, and 85. Choose the best interval for a frequency table for this data. *(Lesson 2-2)* **B**

    A 10   B 5   C 2   D 1

20. *Algebra* Evaluate $ab - c$ if $a = 3$, $b = 16$ and $c = 5$. *(Lesson 1-5)* **43**

21. Evaluate $14 - 2 \times 5$. *(Lesson 1-4)* **4**

Extending the Lesson

Enrichment Masters, p. 10

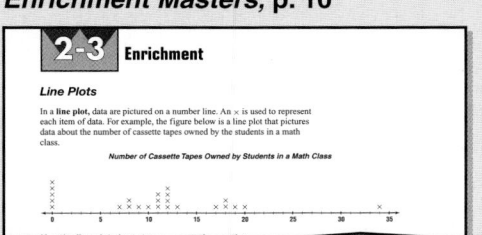

Activity Have students work in groups to look in news and financial magazines for bar and line graphs. Ask groups to compare information and determine whether the proper type of graph was used to display the data.

CHAPTER Project
Exercise 17 asks students to advance to the next stage of work on the Chapter Project. Discuss why they should make bar graphs, not line graphs.

4 ASSESS

Closing Activity

Modeling Have pairs of students contact local companies to obtain sales figures for items such as cars, appliances, and computers. Have students record their information in frequency tables and draw bar or line graphs.

Chapter 2, Quiz A (Lessons 2-1 through 2-3) is available in the *Assessment and Evaluation Masters*, p. 43.

Additional Answers

12. Sample answer: 20; Because the frequency table would have 6 rows.

13. Table A; It shows the change and direction of change over time.

Practice Masters, p. 10

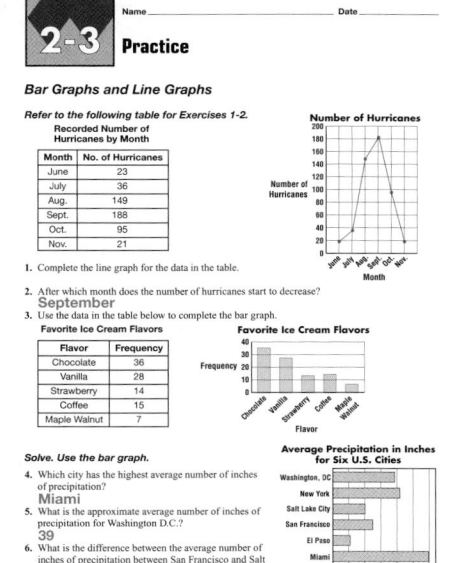

Objective Students solve problems by using a graph.

Recommended Pacing	
Standard	Day 5 of 14
Honors	Day 5 of 14
Block	Day 3 of 7

1 FOCUS

Getting Started Display a bar graph and a table on an overhead transparency. Using a timer, determine how long it takes students to find the greatest number represented by a bar on the graph and by a number in the table. Discuss how each is used.

2 TEACH

Teaching Tip Point out to students that a horizontal bar graph shows the same information as a vertical bar graph. Discuss reasons for making horizontal rather than vertical bar graphs.

In-Class Example
The pictograph shows circulation of four magazines. Refer to it for the following questions.

Magazine Circulation
Homes ❑❑❑❑❑❑
Scene ❑❑◧
Today ❑❑❑❑❑
Video ❑❑❑❑

❑ = 2 million copies

a. What was *Homes* circulation? **12 million**

b. What does the half-magazine for *Scene* mean? **1 million**

c. Which magazine had a circulation of 10 million copies? *Today*

THINKING LAB

PROBLEM SOLVING

2-3B Use a Graph

A Follow-Up of Lesson 2-3

The sixth grade class is planning a field trip to the science center. Their science teacher, Ms. Tai, has asked Steve and Maria to help her decide where they will stop for lunch. The choices are Burger King, Dairy Queen, Hardee's, McDonald's, and Wendy's. Let's listen in!

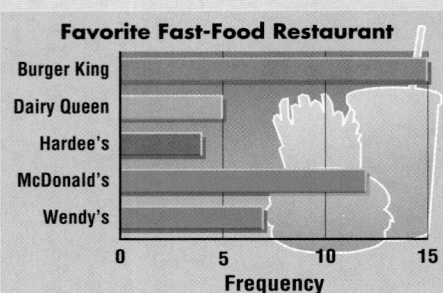
Favorite Fast-Food Restaurant

Steve

> I made a bar graph using our survey results. It looks like we should pick Burger King.

> Yeah, but Ms. Tai told us it had to be healthy. I'm not sure Burger King is our best choice based on the pictograph I made.

> What's a pictograph?

> A pictograph is another type of graph used to compare numbers.

> Well, Hardee's has the least amount of fat, but it's the least favorite.

> How about McDonald's? It was the second most popular choice and the second lowest in fat!

Grams of Fat in a Cheeseburger and Medium Fries

Burger King 34
Dairy Queen 31
Hardee's 28
McDonald's 30
Wendy's 32

🍔 = 8 grams fat

Maria

THINK ABOUT IT

1–2. See students' work.

1. *Compare* the two graphs to see if you agree with Maria's suggestion that they select McDonald's. Explain your answer.

2. *Analyze* the data in the two graphs to see what your second choice would be. Explain your answer.

58 Chapter 2 Statistics: Graphing Data

■ Reteaching the Lesson ■

Activity Use self-adhesive notes to conceal parts of a bar graph. Have students determine what information is missing and why it is needed. For example, without the vertical scale information on the fast-food graph, it is impossible to know the restaurant choices.

3. **Apply** what you have learned about the **use a graph** strategy to solve the following problem. **See margin.**

    Ms. Tai has suggested that students bring a sack lunch. Then they could stop to have dessert on the way home. The bar graph shows the fat grams in a chocolate shake at each of the restaurants. Which restaurant would you recommend and why?

Grams of Fat in a Chocolate Shake

ON YOUR OWN

4–5. See students' work.

4. The second step of the 4-step plan for problem solving is *plan*. **Explain** how the type of graph helps you to make a plan for solving the problem.

5. **Write a Problem** that you could solve more easily if you could use a graph to examine the data in the problem.

6. **Examine** the line graph in Example 2 on page 55. Suppose Christopher's mother is going to buy 2-liter bottles of soft drinks. What day of the week would you recommend that she purchase based on the results shown in the graph? **See margin.**

MIXED PROBLEM SOLVING

STRATEGIES
Look for a pattern.
Solve a simpler problem.
Act it out.
Guess and check.
Draw a diagram.
Make a chart.
Work backward.

Solve. Use any strategy.

7. **Money Matters** Mr. Rivera bought 5 packages of batteries for $5.89 each. About how much did he spend? **$30**

8. **Algebra** Find *x* if the product of *x* and 28 is 252. **9**

9. **Travel** Ogima travels 345 miles per week. How far does he drive in four weeks? **1,380 miles**

10. **Money Matters** Laura collected $2.00 from each student to buy a gift for their teacher. If 27 people contributed, how much money was collected? **$54**

11. **Test Practice** The line graph shows the number of people in Lubbock and Garland from 1960 to 1990.

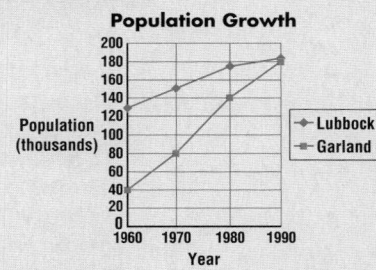
Population Growth

About how many more people were in Lubbock than in Garland in 1980? **C**

A 90,000

B 70,000

C 35,000

D 5,000

3 PRACTICE/APPLY

Check for Understanding
Use the results from Exercise 3 to determine whether students understand how to interpret bar graphs.

Extra Practice If students need additional practice in problem solving, extra practice is available on the following pages.
• Use a Graph, see p. 560
• Mixed Problem Solving, see pp. 593–594

Assignment Guide
All: 4–11

4 ASSESS

Closing Activity
Speaking Display a pictograph. Have small groups of students take turns posing questions about the graph for the rest of the class to answer. Ask why pictographs may be more difficult to interpret. **When there are partial pictures, it may be difficult to give an exact answer.**

Additional Answers
3. Sample answer: McDonald's; It is the second favorite and the healthiest.

6. Sample answer: Thursday, so she'd be prepared for the increased consumption on the weekend.

Extending the Lesson

Activity Have small groups of students determine how many minutes they each spend per day doing household chores. Have each group present its information in a bar graph or pictograph. Graphs can be based on minutes per day, per week, or per month. Then have groups predict how many minutes per year each student spends doing household chores.

Sample problem: *Michelle spends 45 minutes a day doing household chores. How many minutes does she spend per year on household chores?*
16,425 minutes

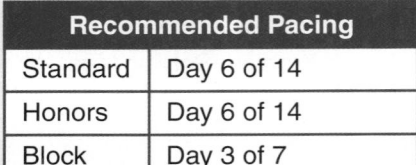

2-4 Lesson Notes

Instructional Resources
- *Study Guide Masters,* p. 11
- *Practice Masters,* p. 11
- *Enrichment Masters,* p. 11
- Transparencies 2-4, A and B
- *Diversity Masters,* p. 2
 CD-ROM Program
  - Resource Lesson 2-4

Recommended Pacing	
Standard	Day 6 of 14
Honors	Day 6 of 14
Block	Day 3 of 7

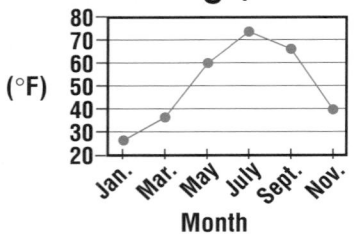 **5-Minute Check**
(Lesson 2-3)

Refer to the graph for Exercises 1–3.

Average Temperatures Pittsburgh, PA

(°F)

Source: *A Geography of Pennsylvania, 1995.*

1. What is the temperature scale? **20°–80°**
2. What is the average temperature in July? **about 72°**
3. How much colder are days in November than in May? **about 20°**

 The 5-Minute Check is also available on **Transparency 2-4A** for this lesson.

Motivating the Lesson

Hands-On Activity Have students trace a circle. Ask them to think of it as a pizza. Have them name toppings. Tell them to divide their "pizza" into "slices" and choose a different topping for each slice. Discuss the results.

60 Chapter 2

2-4 Reading Circle Graphs

What you'll learn
You'll learn to interpret circle graphs.

When am I ever going to use this?
Knowing how to interpret circle graphs can help you analyze a budget.

Word Wise
circle graph

In Lesson 2-3B, you learned how to interpret sets of data displayed in bar graphs and line graphs. In the Mini-Lab below, you will make a **circle graph**.

HANDS-ON MINI-LAB

Work with a partner.   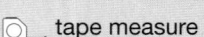 tape measure   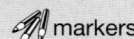 markers

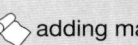 tape   adding machine tape   string

The table shows the age groups of all the people who listen to National Football League radio broadcasts. You can use this data to make a circle graph.

Age Group	Percent
18–24	11%
25–34	27%
35–44	27%
45–54	17%
55–64	9%
65+	9%

Try This 1–4. See students' work.

1. Cut a piece of adding machine tape that is one meter long.
2. For each age group, mark the length in centimeters that corresponds to the percent. For example, the 18–24 age group is 11% so mark off 11 centimeters and label it accordingly.
3. Tape the ends of the adding machine tape together to form a circle. The labels should be facing the inside of the circle.
4. Stand your adding machine tape circle on a flat surface and locate its center. Tape one end of a piece of string to the center. Tape the other end to the point where the adding machine tape is joined together. Repeat with five more pieces of string joining the center to each mark on the tape.

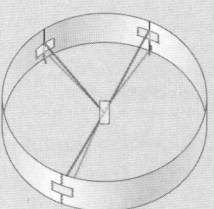

Talk About It 6–7. See margin.

5. What is the total of the percents? **100%**
6. Write a short paragraph describing the circle graph. Describe the sizes of the sections in relation to each other.
7. The circle graph you made represents the same information as the table. Discuss the advantages and disadvantages of each.

60 Chapter 2 Statistics: Graphing Data

Multiple Learning Styles

 Intrapersonal Find a circle graph that contains a topic of interest to students. Ask them which category they would have been in if surveyed. In what categories would their friends have been?

A circle graph is used to compare parts of a whole. The interior of the circle represents a set of data, and the pie-shaped sections are the parts. All the percents given in a circle graph must add up to 100%.

Examples

APPLICATION

1 **Food** The reasons why adults visit a fast food restaurant are shown in the circle graph.

a. What are the three most popular reasons for choosing a fast food restaurant?

b. How does fast service compare to location as a reason for choosing a fast food restaurant?

Choosing A Fast Food Restaurant

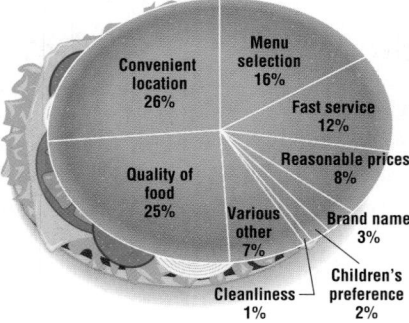

Convenient location 26%
Menu selection 16%
Fast service 12%
Reasonable prices 8%
Quality of food 25%
Various other 7%
Brand name 3%
Children's preference 2%
Cleanliness 1%

Source: *Maritz Marketing Ameripoll, 1995*

a. By looking at the sections of the circle graph, you can see that the three largest parts of the graph are location, food quality, and menu selection. So they are the three most popular reasons for choosing a fast food restaurant.

b. The section relating to location is about twice the size of the section relating to fast service. So, about twice as many people gave location as a reason as gave fast service.

CONNECTION **2** **Physical Science** The circle graph shows what fuels are used to generate electric power in the U.S.

a. Which fuel is used to generate more electricity than the others combined?

b. Which two types of fuel together generate about the same amount of electricity as nuclear fuel?

Electric Power Sources

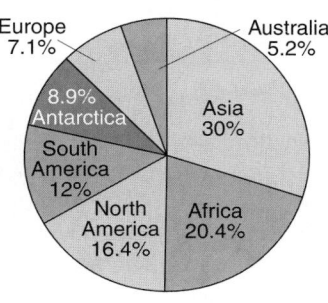

Coal 55%
Petroleum 2%
Hydroelectric 10%
Natural Gas 11%
Nuclear 22%

Source: Energy Information Administration

a. You can see that the section representing coal covers more than half of the circle. Therefore, it is used to generate more than the others combined.

b. If you combine the sections represented by natural gas and hydroelectric fuel, they are about the same size as nuclear fuel. So, they generate about the same amount of electricity as nuclear fuel.

Lesson 2-4 Reading Circle Graphs **61**

Additional Answers for the Mini-Lab

6. **Sample answer: The larger the percent, the larger the part taken in the circle. The 25–34 and 35–44 age groups are each three times larger than each of the 55–64 and 65+ age groups. The 45–54 age group is almost twice as large as each of the 55–64 and 65+ age groups.**

7. **Sample answer: The circle graph allows you to see how all the parts are related.**

 Transparency 2-4B contains a teaching aid for this lesson.

Using the Mini-Lab Students may want to stand their tape on a piece of cardboard and use pins to stabilize it when adding the string.

In-Class Examples

For Example 1

Earth's Land Area

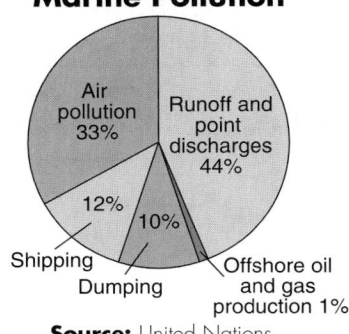

Europe 7.1%
Australia 5.2%
8.9% Antarctica
Asia 30%
South America 12%
North America 16.4%
Africa 20.4%

Source: National Geographic, *Atlas of the World,* 1992.

a. What two continents make up more than half the continental land area? **Asia and Africa**

b. How many continents make up the other half? **5**

For Example 2

Major Sources of Marine Pollution

Air pollution 33%
Runoff and point discharges 44%
12%
10%
Shipping
Dumping
Offshore oil and gas production 1%

Source: United Nations Environment Program, 1990.

a. What are the top two sources of marine pollution? **runoff and point discharges and air pollution**

b. Which two sources of pollution together have about the same result as runoff and point discharges? **air pollution and dumping or air pollution and shipping**

Lesson 2-4 **61**

Check for Understanding

If students need additional practice or instruction after completing Exercises 1–4, one of these options may be helpful.
- Extra Practice, see p. 561
- Reteaching Activity
- *Study Guide Masters*, p. 11
- *Practice Masters*, p. 11

Assignment Guide
Core: 5–10
Enriched: 5–10

Additional Answers

1. **Sample answer: When you want to represent data that look at part to whole relationships.**

2. **The totals should be 100% because 100% represents the whole.**

4a. **5%—Grilled Cheese, Turkey, and Tuna; 11%—Hamburgers, Ham & Cheese; 14%—Bologna; 21%—Other; 28%—Peanut Butter and Jelly;**

5a. **13%—Special; 15%—Egg Noodles; 31%—Short Pasta; 41%—Long Pasta**

Study Guide Masters, p. 11

2-4 Study Guide

Name_____ Date_____

Reading Circle Graphs

Circle graphs show parts of a whole.

For this graph, the whole is all of the dog food sold. The parts are the types of dog food.

How much more of the dog food sold is dry than canned?

To solve, subtract the canned part from the dry part.

70%
−27%
43%

43% more of the dog food sold is dry.

Dog Food Sold

Canned 27%
Semi-moist 3%
Dry 70%

Solve. Use the circle graph.

1. What is the greatest amount of oil used for? **transportation**

2. What is the least amount of oil used for? **electric generation**

3. How much of the oil is used for heating and for electricity generation? **12%**

4. How much more oil is used for transportation than is used in industry? **44%**

5. How much greater is the amount of oil used for transportation than the amount of oil used for other purposes? **32%**

U.S. Oil Consumption

Transportation 66%
Electricity Generation 4%
Heating 8%
Industry 22%

© Glencoe/McGraw-Hill   T11   *Mathematics: Applications and Connections, Course 1*

Communicating Mathematics

Read and study the lesson to answer each question. 1–2. See margin.

1. *Explain* when it is best to use a circle graph.

2. *Write* what the total of the percents in a circle graph should be.

HANDS-ON MATH

3. The table shows the number of hours per week that students ages 12–18 use a computer at school. Use the same procedure you used in the Mini-Lab to make a circle graph of the data in the table. **See students' work.**

Time	Percent
Less than 1 hour	22%
1–2 hours	28%
2–4 hours	16%
4–6 hours	19%
More than 6 hours	11%
Don't know	4%

Source: *Consumer Electronics Manufacturers Association Survey*

Guided Practice

4. *Food* The circle graph shows the favorite sandwiches of children. **a. See margin.**

a. The percents are 5%, 11%, 14%, 21% and 28%. Match each percent with the appropriate section of the graph.

4b. Sample answer: No; the combination of ham and cheese or hamburgers is not $\frac{1}{4}$ of the circle.

b. Suppose hamburgers and ham and cheese were combined. Would the combination be preferred by $\frac{1}{4}$ of children? Explain.

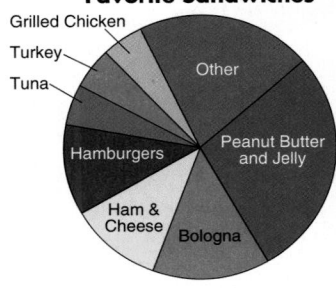

Favorite Sandwiches

Grilled Chicken
Turkey
Tuna
Other
Hamburgers
Peanut Butter and Jelly
Ham & Cheese
Bologna

Source: *Are You Normal?* by Bernice Kanner, St. Martin's, 1995

Practice

5. *Food* The circle graph shows the retail sales by shape at the Pasta Factory Outlet.

5b. special and egg noodles
5d. special and egg noodles

a. The percents are 13%, 15%, 31% and 41%. Match each percent with the appropriate section of the graph. **See margin.**

b. Which two types of pasta have about the same amount of sales?

c. What percent of sales comes from short and long pasta? **72%**

d. Which two types together have about the same amount of sales as short pasta?

e. Which two types together account for about half of the sales? **either long pasta and special (54%) or short pasta and egg noodles (46%)**

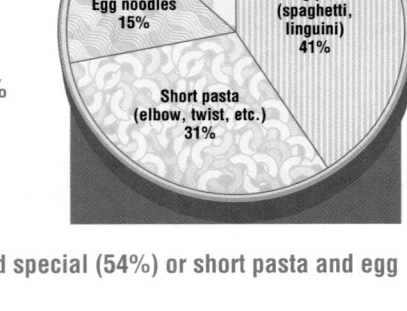

Retail Sales by Pasta Shape

Special (lasagna, jumbo) 13%
Long pasta (spaghetti, linguini) 41%
Egg noodles 15%
Short pasta (elbow, twist, etc.) 31%

62 Chapter 2 Statistics: Graphing Data

Reteaching the Lesson

Activity Have students randomly choose a colored paper scrap from a bag of 10 scraps. Record the results in a frequency table. Then construct a circle graph for the students to interpret. Ask questions such as: *Which color was chosen most often? How do you know?*

Error Analysis

Watch for students who have trouble interpreting a circle graph.

Prevent by reminding students that circle graphs show parts of a whole. If drawn correctly, the greatest numerical value should be the largest section.

Applications and Problem Solving

6a. 14–17 and the under 14 age groups

6c. They are the same so the difference is 0.

7a. LP album, 7–12 in. singles and music videos

6. **Money Matters** The circle graph shows the percent of people who purchase athletic shoes by age group.
   a. What two age groups together make about half of all athletic shoe purchases?
   b. What two age groups together make about one fourth of all athletic shoe purchases? **25–34 and 35–44**
   c. What is the difference between the percent of people 65 and older and the 18–24 age group?

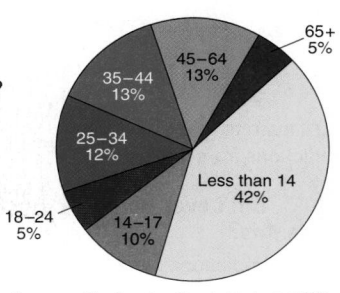

Who Buys Athletic Shoes?

65+ 5%
45–64 13%
35–44 13%
25–34 12%
Less than 14 42%
18–24 5%
14–17 10%

Source: *The Sporting Goods Market in 1995*

7. **Entertainment** The circle graph shows the format of pre-recorded music sales in 1995.
   a. The two formats with the lowest percent of sales each had 1%. What are they?
   b. Cassette album sales were five times cassette single sales and cassette single sales were five times as much as music video sales. The rest are CDs. What are the percents for each of these formats?
   cassette single—5%; cassette album—25%; CDs—68%

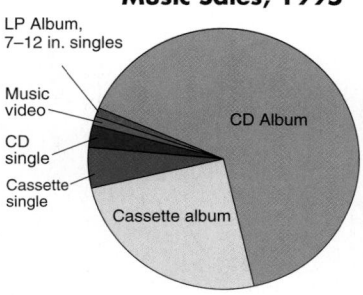

Music Sales, 1995

LP Album, 7–12 in. singles
Music video
CD single
Cassette single
CD Album
Cassette album

Source: RIAA

8. **Critical Thinking** Which would *not* be appropriate data to show in a circle graph? Explain your reasoning.
   a. data showing what teens say would 'most' encourage them to volunteer
   b. data showing how many hours parents help their children with homework each week
   c. data showing the five most wasted foods in the United States
   d. data showing why people dine out
   b; It does not compare parts of a whole.

Mixed Review

9. **Life Science** Janet has a leaf collection that includes 15 birch, 8 willow, 5 oak, 10 maple, and 8 miscellaneous leaves. Make a bar graph for this data. *(Lesson 2-3)* **See Answer Appendix.**

10. **Test Practice** In March, an average of 280 videos were rented per day at Brian's Video Barn. Which is the best estimate for the number of videos rented for the whole month of March? *(Lesson 1-3)* **D**
    A 3,000
    B 6,000
    C 8,000
    D 9,000
    E 10,000

Extending the Lesson

Enrichment Masters, p. 11

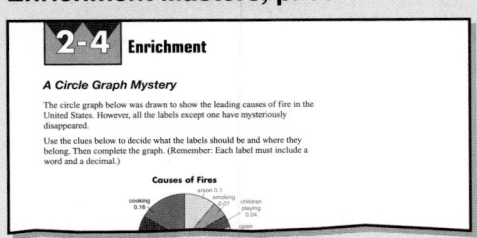

2-4 **Enrichment**

A Circle Graph Mystery

The circle graph below was drawn to show the leading causes of fire in the United States. However, all the labels except one have mysteriously disappeared.

Use the clues below to decide what the labels should be and where they belong. Then complete the graph. (Remember: Each label must include a word and a decimal.)

Causes of Fires

Activity Separate students into groups and have each group choose a feature of weather, such as temperature, wind speed, or humidity, to track and record for a week. Then have them organize the results into a color-coded circle graph. Encourage groups to display their graphs for the class to see.

---

4 ASSESS

Closing Activity

Speaking Have students compare circle graphs with bar and line graphs. Have them discuss how each graph is different from the others. Then display a circle graph from a newspaper or magazine. Have students ask each other questions about the information in the graph.

Practice Masters, p. 11

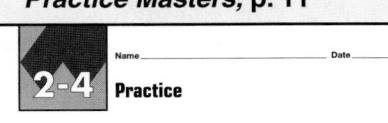

Name_____ Date_____

2-4 **Practice**

Reading Circle Graphs

Solve. Use the circle graph.

Customers' Favorite Pizza Toppings at Pizza Palace

cheese 42%
sausage 11%
pepperoni 25%
onion
mushroom 19%

1. Which pizza topping is the most popular? **cheese**
2. Which pizza topping is the least popular? **onion**
3. Which two toppings together are about as popular as cheese? **mushroom and pepperoni**
4. What percent of the toppings are meat? **36%**

Solve. Use the circle graph.

Student Survey of Extra-Curricular Activities

Sports 30%
Math Club 16%
Debating Team 7%
Drama Club 25%
Chorus 4%
Band 18%

5. Which extracurricular activity is the least popular? **Chorus**
6. Which extracurricular activity is about as popular as Band? **Math Club**
7. Which two activities together are about as popular as Drama Club? **Band and Debating Team**

© Glencoe/McGraw-Hill   T11   *Mathematics: Applications and Connections, Course 1*

Lesson 2-4   63

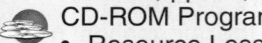

Instructional Resources

- *Study Guide Masters*, p. 12
- *Practice Masters*, p. 12
- *Enrichment Masters*, p. 12
- Transparencies 2-5, A and B
- *Assessment and Evaluation Masters*, pp. 42, 43

CD-ROM Program
- Resource Lesson 2-5
- Extended Activity 2-5

Recommended Pacing	
Standard	Day 7 of 14
Honors	Day 7 of 14
Block	Day 4 of 7

1 FOCUS

5-Minute Check
(Lesson 2-4)

Refer to the following graph for Exercises 1 and 2.

How Micah Spends His Allowance

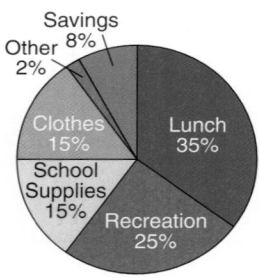

Savings 8%
Other 2%
Clothes 15%
School Supplies 15%
Lunch 35%
Recreation 25%

1. What percent of his allowance does Micah save? **8%**

2. For which two categories does he spend the same amount of money? **school supplies, clothes**

The 5-Minute Check is also available on **Transparency 2-5A** for this lesson.

Motivating the Lesson

Communication Refer to the beginning of the lesson and ask these questions.
- Does the graph show an increase or a decrease? **increase**
- What characteristic of this graph helps you make a prediction? **It shows a steady increase.**

What **you'll learn**

You'll learn to make predictions from line graphs.

When **am I ever going to use this?**

Making predictions can be used to plan housing and highway projects.

The line graph shows the population of the world since 1900. In 1992, the United Nations predicted that world population could increase to over 6 billion by the year 2000.

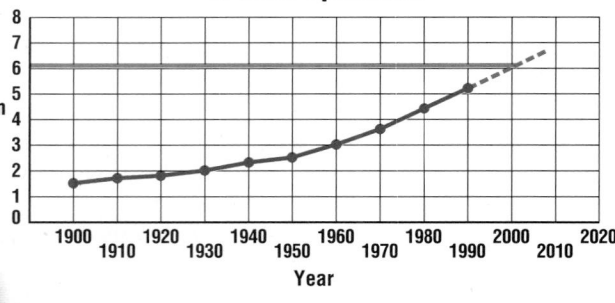

World Population

Source: United Nations Department for Economic and Social Information and Policy Analysis, Population Division.

Line graphs can be used to assist in making predictions as follows.

Step 1 Extend the graph as shown by the dashed line.

Step 2 Draw a horizontal line from the point where the extension intersects the 2000 gridline.

Step 3 You can see that the population of the world could be over 6 billion by the year 2000.

Example 1
CONNECTION

Geography Refer to the graph above.

a. Predict the world population in the year 2020. Identify any assumptions you use to make your prediction.

Extend the dashed line so it intersects the 2020 gridline. If the extension aligns with the segment between 1980 and 1990, world population could be about 8 billion by 2020.

b. Suppose the population growth between 1990 and 2000 matches the rate of growth between 1940 and 1950 instead of between 1980 and 1990. What would the population be in the year 2000?

The line segment between 1940 and 1950 is not as steep as the one between 1980 and 1990. The population would probably be less than 6 billion.

CHECK FOR UNDERSTANDING

Communicating
Mathematics

Read and study the lesson to answer each question.

1. *Predict* the world population in 2010. Identify any assumptions you use to make your prediction. **See margin.**

2. *You Decide* Jamonte predicts that world population will be over 8 billion in 2010. Tashima predicts it will be less than 7 billion. Whose prediction do you agree with? Explain your reasoning. **See students' work.**

Guided Practice

3. *Sports* The line graph shows the winning times in the 3,000 meter steeplechase at the last thirteen Olympics.

Olympic 3,000-Meter Steeplechase, 1948–1996

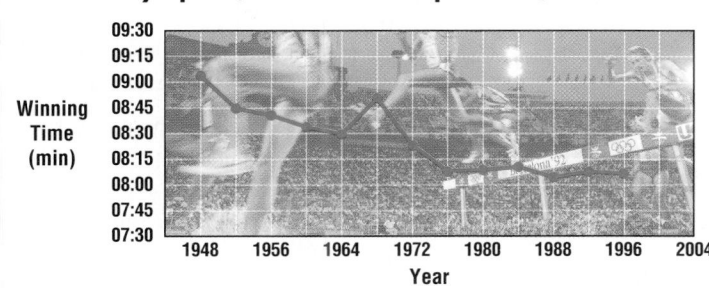

Source: The Universal Almanac, 1997

a. Would you have predicted the winning time in 1968? What might have happened? **Answers will vary.**

b. Would you predict the winning time in 2004 to be under 8 minutes? Explain why or why not. **See students' work.**

EXERCISES

Practice

inter**NET**
CONNECTION
For the latest weather statistics, visit:
www.glencoe.com/sec/
math/mac/mathnet

4a. 60°, 30°, 30°

4b. about 10–15° colder

4c. about 20° colder

4. *Earth Science* The line graph shows the average monthly temperature for three U.S cities. **4a–d. Sample answers given.**

a. Predict the temperature for Galveston in December, Asheville in January, and Minneapolis in November.

b. How much colder would you expect it to be in Asheville than in Galveston in November?

c. How much colder would you expect it to be in Minneapolis than in Asheville in December?

d. How much colder would you expect it to be in Minneapolis than in Galveston in January? **about 45° colder**

Average Monthly Temperature

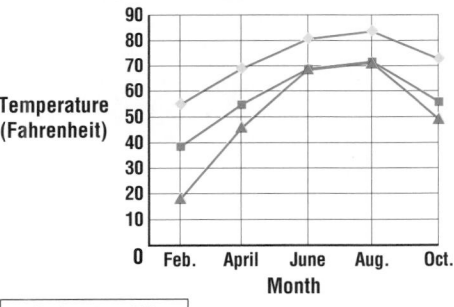

Lesson 2-5 Making Predictions **65**

Thinking Algebraically Analyzing graphs and looking for patterns to predict an unknown value are the basics of the idea of *functions*. Ask students to describe the patterns for the graph on page 64.

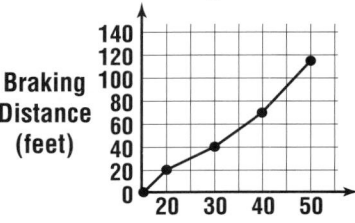

Transparency 2-5B contains a teaching aid for this lesson.

In-Class Example
For the Example
Refer to the graph below.

Braking Distance for Cars on Dry Road

a. Predict the braking distance of a car going 55 miles per hour. **140 feet**

b. A car skidded 100 feet. How fast was it going? **about 46 miles per hour**

3 PRACTICE/APPLY

Check for Understanding
If students need additional practice or instruction after completing Exercises 1–3, one of these options may be helpful.
• Extra Practice, see p. 561
• Reteaching Activity
• *Study Guide Masters,* p. 12
• *Practice Masters,* p. 12

Additional Answer
1. Sample answer: If the extension aligns with the segment between 1980 and 1990, world population could be about 7 billion.

═ Reteaching the Lesson ═

Activity Have students use the following data to draw a line graph on the chalkboard and predict the average tuition and fees at state universities in the year 2000: 1989—$2,035; 1990—$2,159; 1991—$2,349; 1992—$2,410; 1993—$2,537; 1994—$2,689; 1995—$2,811.
Sample answer: $3,250

Average Annual Tuition at U.S. Colleges

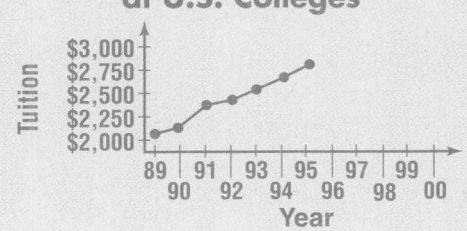

Applications and
Problem Solving

5. *Physical Science* Angelina and Bart each dropped a different ball from
several distances and recorded the highest point of the first bounce. The
graph shows their results. **5a–d. Sample answers given.**

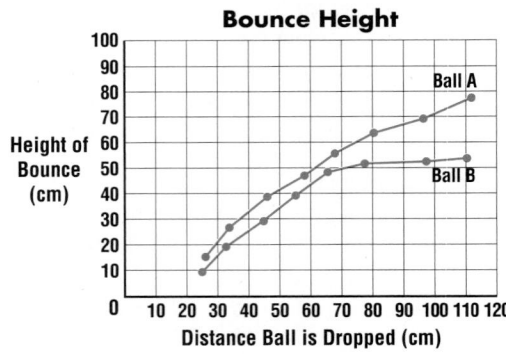

Bounce Height

b. It will bounce about as high as it did when dropped from 110 cm.

a. What would you predict about Ball A if the distance it is dropped
continues to increase? **It will continue to bounce higher.**

b. What would you predict about Ball B if the distance it is dropped
continues to increase?

c. How might you explain these differences? **Ball A might be a Super Ball.**

d. Do you think that each ball will always give a higher bounce with a
higher drop height? Explain. **No; at some point both balls will begin to
level off. If you dropped Ball A from 100 feet it will not bounce 100 feet.**

6. *Sports* Tom and Jim have been running each afternoon preparing for the
track team try-outs. They hope to qualify for the 1,600-meter run. The graph
shows their best times over the last ten weeks.

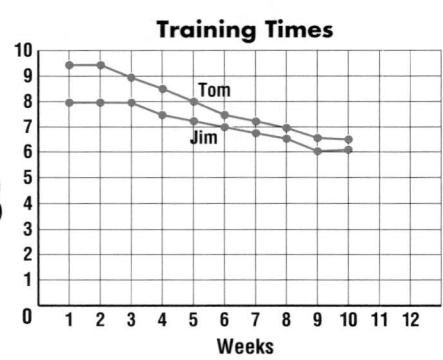

Training Times

a. Who has shown the most improvement? **Tom**

b. Who has the better times? **Jim**

c. Who do you predict will be the faster runner in 15 weeks? What will be
his time? Explain. **Answers will vary.**

**7. The point where
they cross indicates
the times were the
same.**

7. *Critical Thinking* Suppose the lines in Exercise 6 cross after week 12.
What does the point where they cross represent?

2-5 Name_____ Date_____

Study Guide

Making Predictions

With this **line graph**, you can make predictions about the expected number
of calories used while bicycling or playing tennis.

Calories Used in Ten Minutes

Example About how many more calories can a 120-pound person
expect to use bicycling for 10 minutes than playing
tennis for 10 minutes?

From the graph, a 120-pound person will use about 60
calories playing tennis and about 82 calories bicycling for
10 minutes.

82 − 60 = 22

A 120-pound person can expect to use about 22 more
calories in 10 minutes of bicycling.

Refer to the graph.

1. About how many calories can a 150-pound
person expect to use bicycling for 10
minutes?
about 105 calories

2. About how many calories can a 110-
pound person expect to use playing
tennis for 10 minutes?
about 55 calories

3. About how many more calories can a 140-
pound person expect to use bicycling for 10
minutes than playing tennis for 10 minutes?
about 30 more

4. About how many more calories will a
160-pound person use playing tennis for
10 minutes than a 100-pound person will
use?
about 30

© Glencoe/McGraw-Hill   T12   *Mathematics: Applications and Connections, Course 1*

8. *Food* Refer to the circle graph in Example 1 on page 61. Which two reasons together account for half of the responses? *(Lesson 2-4)*

9. | Test Practice | If $m = 6$ and $n = 4$, find the value of the expression $2m - n$. *(Lesson 1-5)* **D**

A 2   **B** 4   **C** 6   **D** 8

CHAPTER 2

Mid-Chapter Self Test

Make a frequency table for the set of data. *(Lesson 2-1)*

1. To the nearest inch, how tall are you? **See Answer Appendix.**

   57   54   58   59   60   57   63   57   55   56   60   57

2. Use the set of data 58, 92, 23, 14, 62, 79, 43, 85, 91, 18, 50, 47, 25, 88. *(Lesson 2-2)*

  a. Choose the better scale for a frequency table for the data: 0 to 99 or 0 to 999. **0 to 99**

  b. Choose the best interval for a frequency table for the data: 1, 10, or 100. **10**

  c. Make a frequency table for this set of data. **See Answer Appendix.**

3. Make a bar graph for the data in the table at the right. *(Lesson 2-3)* **See Answer Appendix.**

Favorite Ice Cream Flavor	
Flavor	**Frequency**
Chocolate	7
Strawberry	9
Vanilla	5

4. The circle graph shows the average family size in the U.S. *(Lesson 2-4)* **a. four persons and three persons**

  a. Which two sizes of families together have about the same percent as the size with the greatest percent?

  b. What percent of all families have five or more members? **14%**

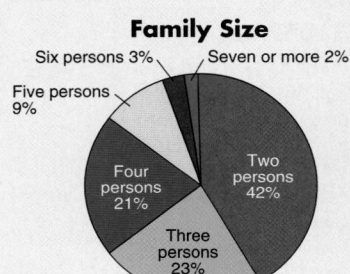

Family Size

Six persons 3%   Seven or more 2%
Five persons 9%
Four persons 21%
Two persons 42%
Three persons 23%

Source: U.S. Census Bureau

5c. Sample answer: No, the horizontal scale doesn't extend far enough.

5. *Economics* Keshia is saving her money to purchase a Nintendo 64. She starts with $40 and adds $10 the first week, $11 the second week, $12 the third week, and so on. The graph at the right shows a record of her savings for the first 10 weeks. *(Lesson 2-5)*

  a. Estimate when Keshia will double her money (have at least $100). **the 5th week.**

  b. Estimate when she will double her money again (have at least $200). **the 11th week.**

  c. Can you predict when she might double her money again (have at least $400)? Explain your reasoning.

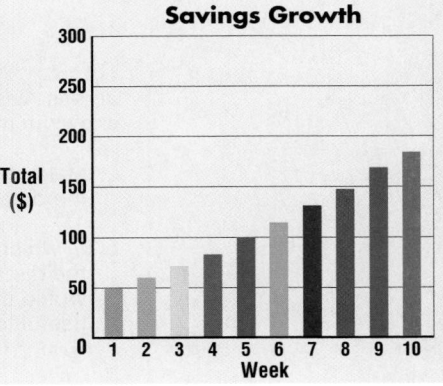

Savings Growth

Total ($) — Week

---

4 ASSESS

Closing Activity

Writing Have students write a few sentences describing how graphs can be used to make predictions.

Chapter 2, Quiz B (Lessons 2-4 and 2-5) is available in the *Assessment and Evaluation Masters,* p. 43.

Mid-Chapter Test (Lessons 2-1 through 2-5) is available in the *Assessment and Evaluation Masters,* p. 42.

Mid-Chapter Self Test

The Mid-Chapter Self Test reviews concepts in Lessons 2-1 through 2-5. Lesson references are given so students can review concepts not yet mastered.

Additional Answer

8. convenient location and quality of food

Practice Masters, p. 12

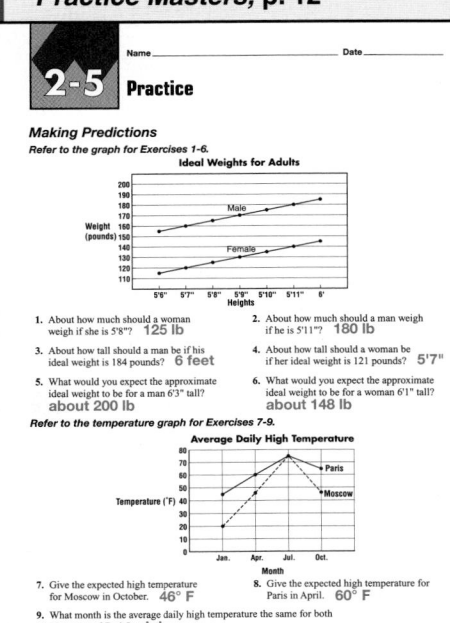

2-5 Practice

Making Predictions
Refer to the graph for Exercises 1-6.

Ideal Weights for Adults

1. About how much should a woman weigh if she is 5'8"? **125 lb**
2. About how much should a man be if he is 5'11"? **180 lb**
3. About how tall should a man be if his ideal weight is 184 pounds? **6 feet**
4. About how tall should a woman be if her ideal weight is 121 pounds? **5'7"**
5. What would you expect the approximate ideal weight to be for a man 6'3" tall? **about 200 lb**
6. What would you expect the approximate ideal weight to be for a woman 6'1" tall? **about 148 lb**

Refer to the temperature graph for Exercises 7-9.

Average Daily High Temperature

7. Give the expected high temperature for Moscow in October. **46° F**
8. Give the expected high temperature for Paris in April. **60° F**
9. What month is the average daily high temperature the same for both Moscow and Paris? **July**

© Glencoe/McGraw-Hill   T12   *Mathematics: Applications and Connections, Course 1*

Extending the Lesson

Enrichment Masters, p. 12

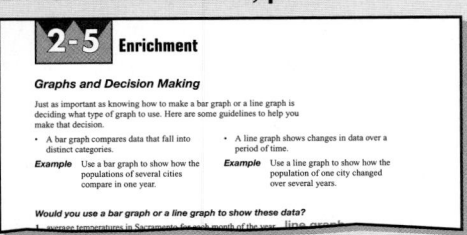

2-5 Enrichment

Graphs and Decision Making

Just as important as knowing how to make a bar graph or a line graph is deciding what type of graph to use. Here are some guidelines to help you make that decision.

- A bar graph compares data that fall into distinct categories.
- A line graph shows changes in data over a period of time.

Example Use a bar graph to show how the populations of several cities compare in one year.

Example Use a line graph to show how the population of one city changed over several years.

Would you use a bar graph or a line graph to show these data?

1. average temperatures in Sacramento for each month of the year **line graph**

Activity Have pairs of students make up a graph of the expected weight gains of a science fiction creature. Suggest that they use a pencil or straw as the line of increase or decrease. Ask them to change the angle of the straightedge. Then have them determine how changing the angle affects the outcome.

Recommended Pacing	
Standard	Day 8 of 14
Honors	Day 8 of 14
Block	Day 4 of 7

1 FOCUS

5-Minute Check
(Lesson 2-5)

Refer to the graph for Exercises 1 and 2.

Rick's Test Results

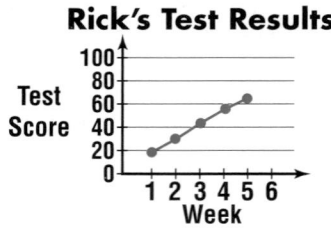

1. Predict what Rick's score will be in week 6. **about 80**
2. At this rate, in what week will he earn a 100 on his test? **week 8**

The 5-Minute Check is also available on **Transparency 2-6A** for this lesson.

Motivating the Lesson

Problem Solving Have students read the first paragraph of the lesson and decide whether most kids watch less than 10, 10 to 20, or more than 20 hours of television in a week. Record the responses.

2 TEACH

Transparency 2-6B contains a teaching aid for this lesson.

Reading Mathematics Ask students to define *stem* and *leaf* in biological terms. Compare these characteristics with those of a stem-and-leaf plot.

***What* you'll learn**

You'll learn to construct stem-and-leaf plots.

***When* am I ever going to use this?**

You can use a stem-and-leaf plot to record a large set of data collected for a science project.

Word Wise

stem-and-leaf plot
stem
leaf

Did you know? Although the length of the Iditarod race varies, the official distance is 1,049 miles—49 symbolizes the fact that Alaska was the 49th state.

D'Ante's brother told him that if he wants to improve his grades, he should watch less television and study more. D'Ante took a survey to find how many hours of television other kids in his grade watch in a week. The results are shown in the chart. What could D'Ante conclude from his data? *This problem will be solved in Example 2.*

2	18	23	14	13
4	5	9	11	13
21	3	19	8	16
9	10	3	3	10
12	15	4	16	19

One way to display a large data set to make it easy to read is to construct a **stem-and-leaf plot**. To display the data 25, 8, 14, 25, 12, and 21 in a stem-and-leaf plot, follow these steps.

Step 1 Find the least and the greatest number. Identify the tens digit in each. The least number, 8, has 0 in the tens place. The greatest number, 25, has 2 in the tens place.

Step 2 Draw a vertical line and write the tens digits from least to greatest to the left of the line. These digits form the **stems**.

```
0 |
1 |
2 |
```

Step 3 Write the units digits to the right of the line, with the corresponding stem. The units digits form the **leaves**.

```
0 | 8
1 | 4 2
2 | 5 5 1
```

Step 4 Order the leaves in each row from least to greatest.

Stem	Leaf
0	8
1	2 4
2	1 5 5

Step 5 Include a key or an explanation. $1|2 = 12$

Example 1

Racing The Iditarod Dog Sled Race is more than 1,000 miles across Alaska. The distances between the checkpoints are shown in the chart.

a. Make a stem-and-leaf plot of the data.

b. In which interval do most of the distances fall? Why would this information be useful to competitors?

20	29	14	52	34
45	30	48	93	48
23	38	90	65	25
18	60	70	90	40
58	48	28	18	55
22				

Source: *USA TODAY* research

a. The least number is 14, and the greatest number is 93. So the stems start with 1 and end with 9.

Stem	Leaf
1	4 8 8
2	0 9 3 5 8 2
3	4 0 8
4	5 8 8 0 8
5	2 8 5
6	5 0
7	0
8	
9	3 0 0

Order the leaves from least to greatest.

→

$1|4 = 14$ miles

Stem	Leaf
1	4 8 8
2	0 2 3 5 8 9
3	0 4 8
4	0 5 8 8 8
5	2 5 8
6	0 5
7	0
8	
9	0 0 3

Include 8 as a stem even though none of the numbers have an 8 in the tens place.

b. Most of the distances are between 20 and 48 miles. This information would be helpful in calculating the amount of supplies needed between checkpoints.

Example **2**
APPLICATION

Surveys Refer to the beginning of the lesson. What could D'Ante conclude from his survey?

Make a stem-and-leaf plot of the survey data to make it easy to summarize. The stem 1 has the greatest number of leaves. D'Ante could conclude that most of the students watch 10-19 hours of television per week.

Stem	Leaf
0	2 3 3 3 4 4 5 8 9 9
1	0 0 1 2 3 3 4 5 6 6 8 9 9
2	1 3

$2|1 = 21$ hours

CHECK FOR UNDERSTANDING

Communicating Mathematics

Read and study the lesson to answer each question.

1. **Tell** when a stem-and-leaf plot would be the best way to represent a set of data. **when there are many numbers in a set of data**

2. **Explain** how a stem-and-leaf plot is like a bar graph. **See margin.**

Math Journal

3. **Find** a list of data in a magazine and make a stem-and-leaf plot. Describe how the plot summarizes the data. **See students' work.**

Guided Practice
4. 0, 1, 2, 3, 4

4. Determine the stems for the data set 9, 42, 33, 21, 11, 6, 40, 5, 5, 9.

5. Make a stem-and-leaf plot for the following set of data.

27, 53, 58, 34, 24, 36, 20, 38, 43, 45, 35, 54, 78, 47, 58, 36 **See Answer Appendix.**

6. **Traffic** Ms. Sangita recorded the number of cars that passed through the intersection in her neighborhood every morning for 3 weeks.

46	53	61	41	70	38	50
33	67	52	69	32	45	62
39	51	66	72	52	59	61

   a. Make a stem-and-leaf plot of her data. **See Answer Appendix.**
   b. In what intervals do most of the data fall? **between 50-59 and 60-69**

Lesson 2-6 Stem–and–Leaf Plots **69**

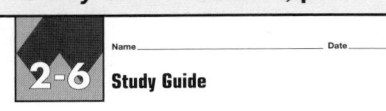

■ Reteaching the Lesson ■

Activity Have students use base-ten blocks to model the data that consist of two-digit numbers that you provide. Have students record the numbers of ten strips they use in the stems column of the stem-and-leaf plot, and the number of ones squares in the leaves column.

Additional Answer

2. The length of each row of numbers is like the length of bars on a graph.

Lesson 2-6 **69**

3 PRACTICE/APPLY

Check for Understanding

If students need additional practice or instruction after completing Exercises 1–6, one of these options may be helpful.

- Extra Practice, see p. 561
- Reteaching Activity, see p. 69
- *Study Guide Masters*, p. 13
- *Practice Masters*, p. 13

Assignment Guide

Core: 7–15 odd, 16–18
Enriched: 8–12 even, 14–18

4 ASSESS

Closing Activity

Speaking Lead a class discussion on the advantages and disadvantages of a stem-and-leaf plot compared to a regular list of data. **Sample answers: Advantages: Stem-and-leaf plot is more concise, displays data in order, easy to interpret large sets of data; Disadvantage: original data must be reconstructed if being used another way.**

Practice Masters, p. 13

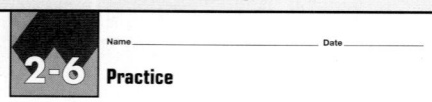

EXERCISES

Practice

Determine the stems for each set of data.

7. 41, 44, 28, 22, 39, 26, 33, 17, 14, 56, 22 **1, 2, 3, 4, 5**

8. 135, 106, 100, 132, 92, 136, 89, 128, 112 **8, 9, 10, 11, 12, 13**

9. 6, 15, 4, 3, 11, 22, 2, 9, 18, 20, 41, 45, 18 **0, 1, 2, 3, 4**

Make a stem-and-leaf plot for each set of data. **10–13. See Answer Appendix.**

10. 35, 29, 27, 28, 27, 30, 44, 33, 32, 45, 38, 28, 22, 21, 31, 31, 38, 24

11. 124, 99, 140, 133, 94, 130, 124, 167, 162, 92, 99, 132

12. 92¢, 83¢, 94¢, 54¢, 85¢, 54¢, 96¢, 89¢, 75¢

13. Make a stem-and-leaf plot of the temperatures (°F) 57°, 34°, 57°, 56°, 27°, 58°, 46°, 53°, 28°, 34°, 9°, 56°, 57°, 45°, 34°, and 29°.

Applications and Problem Solving

14b. Sample answer: He probably will want the equipment around the 12th of October.

14. *Agriculture* Cotton crops are typically harvested when all of the pods on the plants are open. Mr. Smotherman wants to rent harvesting equipment when all of the pods are open. The chart shows the dates in October during the past 19 years when the cotton pods were fully open.

10	19	12	21
18	7	11	16
12	18	20	14
15	16	8	22
9	12	13	

   a. Make a stem-and-leaf plot of the data. **See Answer Appendix.**

   b. Estimate what day Mr. Smotherman will need to rent the harvesting equipment.

15. *Life Science* As part of a science project, Lai'sung recorded the number of Monarch butterflies he saw every afternoon for one month during migrating season.

6	9	26	11	5
7	9	27	10	8
5	9	41	9	11
13	16	37	18	9
19	18	22	22	14
17	23	19	27	29

   a. Make a stem-and-leaf plot of the data. **See Answer Appendix.**

   b. On how many days did Lai'sung see more than 20 butterflies? **9 days**

16. *Critical Thinking* Refer to Exercise 6. The City Council told Ms. Sangita that they would place a traffic light at the intersection in her neighborhood if, on average, there were more than 50 cars that pass through the intersection each morning. Use the stem-and-leaf plot and one other type of graph of the same data to prepare an argument for the City Council meeting. **See students' work.**

Mixed Review
17. Sample answer: 400

17. *Business* Use the line graph to estimate sales in July at First Motors. *(Lesson 2-5)*

Average Car Sales by Month

18. **Test Practice** If a passenger train travels at an average speed of 62 miles per hour, which is a good estimate for the number of hours it will take to travel 293 miles? *(Lesson 1-3)* **C**

   A 3 h   **B** 4 h   **C** 5 h   **D** 7 h   **E** 10 h

Extending the Lesson

Enrichment Masters, p. 13

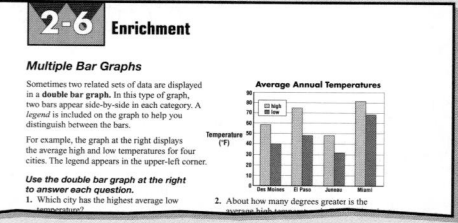

Activity Have students find the high and low temperatures for 15 cities in the U.S. Ask students to make a double stem-and-leaf plot for this data. Tell them that in a double stem-and-leaf plot, one column of stems has two columns of leaves—one on its left for one category and one on its right for the other.

2-7 Mean, Median, and Mode

What you'll learn
You'll learn to find the mean, median, mode, and range to describe a set of data.

When am I ever going to use this?
You can use mean, median, and mode to describe your classmates' allowances.

Word Wise
measures of
  central
  tendency
mean
median
mode
average
range

The table shows the prices of ten compact 35mm cameras.

It is often helpful to find one number to represent one aspect of a set of data. Some of these numbers are known as **measures of central tendency**. Three common measures of central tendency are **mean**, **median**, and **mode**.

Brand	Price
Canon	$160
Kodak	$80
Konica	$230
Minolta	$215
Nikon	$180
Olympus	$220
Pentax	$170
Ricoh	$220
Samsung	$300
Yashica	$185

Source: *Consumer Reports, December, 1996*

Each measure is a different type of **average**. When people use the word *average*, they are usually referring to the mean.

Mean	The mean of a set of data is the sum of the data divided by the number of pieces of data.

The mean of the camera prices is

$$\frac{160 + 80 + 230 + 215 + 180 + 220 + 170 + 220 + 300 + 185}{10} \text{ or } 196.$$

The median is another measure of central tendency.

Median	The median is the middle number when the data are arranged in numerical order.

To find the median camera price, first order the prices. Then find the middle number.

80   160   170   180   185 | 215   220   220   230   300

5 numbers   *5 numbers*

There are two middle numbers, 185 and 215. To find the median, you need to find the mean of these two numbers.

$$\frac{185 + 215}{2} \text{ or } 200$$

The median is 200, so the median price is $200.

Study Hint
Reading Math The median value means that there are just as many values above the median as below it.

Lesson 2-7 Mean, Median, and Mode **71**

Multiple Learning Styles

Interpersonal Give groups of students a price list from a consumer magazine for the top-rated brands of a particular product. Have each group study it and find the mean, median, and mode. Then have students discuss which measure best describes the "average" top-rated products.

2-7 Lesson Notes

Instructional Resources
- *Study Guide Masters,* p. 14
- *Practice Masters,* p. 14
- *Enrichment Masters,* p. 14
- Transparencies 2-7, A and B
- *Assessment and Evaluation Masters,* p. 44
- *School to Career Masters,* p. 2
- *Science and Math Lab Manual,* pp. 5–8
- *Technology Masters,* p. 3
- CD-ROM Program
  - Resource Lesson 2-7

Recommended Pacing

Standard	Days 9 & 10 of 14
Honors	Days 9 & 10 of 14
Block	Day 5 of 7

1 FOCUS

5-Minute Check
(Lesson 2-6)

1. Make a stem-and-leaf plot for the following set of data.
   46, 37, 32, 51, 48, 44, 33, 54, 39, 40, 56, 59

Stem	Leaf
3	2 3 7 9
4	0 4 6 8
5	1 4 6 9 5|1 = 51

2. The stem-and-leaf plot below shows scores on a test.

Stem	Leaf
9	0 3 7 9
8	1 1 2 7 8
7	3 4 5 5
6	2 5
5	
4	3 4|3 = 43

   a. What was the highest score on the test? **99**

   b. In which interval did most of the scores fall? **80–89**

The 5-Minute Check is also available on **Transparency 2-7A** for this lesson.

Hands-On Activity Have students comparison shop in catalogs for one item such as a CD or video game. After they have four or five different prices, ask them how they would determine the average price for each item.

2 TEACH

Transparency 2-7B contains a teaching aid for this lesson.

In-Class Examples

For Example 1
Find the mean, median, and mode of the data in the following graph.

Top 5 Cable T.V. Networks, 1996

- CNN 678
- Discovey Channel 670
- ESPN 679
- TBS 676
- USA Network 672

668 670 672 674 676 678 680

Subscribers (hundred thousands)

675,000; 676,000; no mode

For Example 2
What is the range of data for the graph in Example 1? **19,000**

A third measure of central tendency is the mode.

Mode	The mode of a set of data is the number(s) or item(s) that appear most often.

To find the mode of the camera price data, look for the dollar amount that appears most often. The amount $220 appears twice and is the mode. If the Nikon would have cost $185, then $185 would have been listed twice also, and the data would have two modes. *A set of data may have no mode, one mode, or multiple modes.*

Example ① **APPLICATION**

Entertainment The graph shows the number of channels available from different cable companies. Find the mean, median, and mode of the data.

Nothing's On!

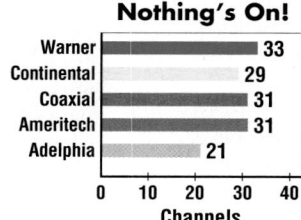

- Warner 33
- Continental 29
- Coaxial 31
- Ameritech 31
- Adelphia 21

0 10 20 30 40
Channels

Did you know? By the time American children reach age 18, they have, on average, watched 17,000 hours of television and seen 360,000 commercials.

Mean $\dfrac{21 + 31 + 31 + 29 + 33}{5}$ or 29

Mode The number 31 appears twice and is the mode.

Median The set of data in order is: 21   29   31   31   33.

The middle number is 31. The median is 31.

The stem-and-leaf plots show the test scores of two sixth grade science classes. Both classes have the same mean score, the same median score, and the same mode score. However, the individual scores are not all the same. How do the two sets of scores differ?

Third Period			Fourth Period	
2			2	0 8
3			3	5 8
4	3 5		4	5
5	2 4 5 6		5	0
6	0 0 0 8		6	0 0
7	2 5 8		7	2 5
8	1 5		8	8 9
9		$8|1 = 81$	9	1 6 8

Another aspect of data is how it is spread out. One way to describe the spread of a set of data is to use the **range**. The range is the difference between the greatest number and the least number in a set of data.

The range of the science test scores are as follows.

Third Period: 85 − 43 or 42   Fourth Period: 98 − 20 or 78

The scores of the fourth period class are more spread out, or dispersed, than those of the third period class.

Classroom Vignette

"We collect age information from the class by routing a chart in which students fill in their name, age in years and months, age in months, and birthday. Then students use the data in various ways determining the most appropriate statistical measures and methods for displaying the data."

Verlene B. DeWitt

Verlene B. DeWitt, Teacher
Lee Burneson Middle School
Westlake, OH

Example ② Refer to the beginning of the lesson. What is the range of the camera prices?

First order the prices.

80 160 170 180 185 215 220 220 230 300

The range is 300 − 80 or 220.

The range of the camera prices is $220.

In this Mini-Lab, you'll use beans to learn more about measures of central tendency.

HANDS-ON MINI-LAB

Work in groups of five. 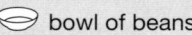 bowl of beans 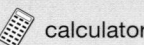 calculator

Try This

- Have one person in your group reach into the bowl and grab a handful of beans. Count and record the number of beans. Replace the beans in the bowl.
- Repeat the previous step for the other four members of your group.
- Find the mean, mode, and median of your group's results.

Talk About It 1–2. See students' work.

1. Suppose you add the number 4 to the set of data so that your group now has six pieces of data.
   a. Find the mean, mode, and median of the new set of data.
   b. Which measure of central tendency is affected the most?
   c. Which measure of central tendency is most representative of the data before adding the 4? after adding the 4?

2. Suppose you add the number 500 to your original set of data.
   a. Find the mean, mode, and median of the new set of data.
   b. Which measure of central tendency is affected the most?
   c. Which measure of central tendency is most representative of the data before adding the 500? after adding the 500?

3. the mean; Sample answer: Adding an extreme value can't affect the mode because it's a new value. It affects the median slightly because the middle value moves one position.

3. In general, which measure of central tendency is most affected by adding an extreme value? Explain.

CHECK FOR UNDERSTANDING

Communicating Mathematics

Read and study the lesson to answer each question. 1–2. See margin.

1. *State* which measure—mean, median, or mode—best represents the original set of data in the Mini-Lab. Explain your reasoning.

2. *Explain* how to find the median of a set of data that has an even number of data in the set.

Lesson 2-7 Mean, Median, and Mode **73**

Reteaching the Lesson

Activity Write a set of data on 3" × 5" cards with one number on each card. Have pairs of students arrange the cards in order from least to greatest. Have them find the mean on their calculators. Then have them find the mode and the median.

Error Analysis
Watch for students who have trouble differentiating mean, median, and mode. **Prevent by** association of common words: median of highway is in the middle; mode is one letter from *more*.

Using the Mini-Lab Ask students what was the most efficient way to count the beans. Ask students how they think their measures would change if they had 100 people in their group.

3 PRACTICE/APPLY

Check for Understanding
If students need additional practice or instruction after completing Exercises 1–6, you may find one of the following options helpful.
- Extra Practice, see p. 562
- Reteaching Activity
- *Study Guide Masters*, p. 14
- *Practice Masters*, p. 14
-  Interactive Mathematics Tools Software

Additional Answers
1. The mean because the numbers do not differ greatly.
2. Find the mean of the two middle pieces of data.

Study Guide Masters, p. 14

2-7 Study Guide

Name _____ Date _____

Mean, Median, and Mode

Mean, median and mode are measures of central tendency.

To find the **mean** of a set of numbers, find the sum of the numbers and divide by the number of addends.

To find the **median** of a set of numbers, arrange the numbers in order and find the middle number.

To find the **mode** of a set of numbers, find the number that appears most often.

To find the **range** of a set of numbers, subtract the least number from the greatest number.

Student Heights in Inches

65	62	66
59	62	60
64	59	66
62	64	67

Example Find the mean, median, and mode of the student heights.

Mean: $\frac{65 + 62 + 66 + 59 + 62 + 60 + 64 + 59 + 66 + 62 + 64 + 67}{12} = 63$

Median: 59 59 60 62 62 64 64 65 66 66 67

$(62 + 64) \div 2 = 63$

Mode: 62

Range: 67 − 59 or 8

Find the mean, median, mode, and range for each set of data.

1. 8, 10, 6, 9, 8, 7 **8; 8; 8; 4**

2. 12, 6, 8, 2, 7, 5, 2 **6; 6; 2; 10**

3. 11, 9, 6, 14, 5, 5, 13 **9; 9; 5; 9**

4. 20, 30, 40, 10, 20, 90, 70 **40; 30; 20; 80**

5. 16, 20, 18, 14, 17 **17; 17; no mode; 6**

6. 17, 31, 29, 42, 17, 36, 24 **28; 29; 17; 25**

7. 152, 148, 150 **150; 150; no mode; 4**

© Glencoe/McGraw-Hill   T14   *Mathematics: Applications and Connections, Course 1*

Lesson 2-7 **73**

Family Activity

Ask students to share with the class the graph they chose. Have them demonstrate how they found the mean, median, and mode and explain what each means.

4 ASSESS

Closing Activity

Writing Have pairs of students write a sentence to explain the method used to find each measure of central tendency.

Chapter 2, Quiz C (Lessons 2-6 and 2-7) is available in the *Assessment and Evaluation Masters,* p. 44.

Practice Masters, p. 14

2-7 Practice

Mean, Median, and Mode

Find the mean, median, mode, and range for each set of data.

1. 6, 9, 2, 4, 3, 6, 5 **5; 5; 6; 7**
2. 25, 18, 14, 27, 25, 14, 18, 25, 23 **21; 23; 25; 13**
3. 453, 345, 543, 345, 534 **444; 453; 345; 198**
4. 13, 6, 7, 13, 6 **9; 7; 6 and 13; 7**
5. 8, 2, 9, 4, 6, 8, 5 **6; 6; 8; 7**
6. 13, 7, 17, 19, 7, 15, 11, 7 **12; 12; 7; 12**
7. 1, 15, 9, 12, 18, 9, 5, 14, 7 **10; 9; 9; 17**
8. 28, 32, 23, 43, 32, 27, 21, 34 **30; 30; 32; 22**
9. 3, 9, 4, 3, 9, 4, 2, 3, 8 **5; 4; 3; 7**
10. 42, 35, 27, 42, 38, 35, 29, 24 **34; 35; 35 and 42; 18**
11. 157, 124, 157, 124, 157, 139 **143; 148; 157; 33**

Use the data below to answer Exercises 12-15.

12. What is the mode? **$5**
13. What is the median? **$10**
14. What is the mean? **$9.00**

After-School Activities	Cost per Hour
Piano	$14.00
Ballet	10.00
Swimming	5.00
Karate	15.00
Skating	12.00
Aerobics	5.00
Arts & Crafts	8.00
Drama	10.00
Scouts	2.00

15. If the price of swimming lessons went up to $6, would it change the
   a. mode? Why? **Yes; only one occurrence of 5.**
   b. median? Why? **No; still 4 numbers below 10.**
   c. mean? Why? **Yes; sum of data is different.**
   d. range? Why? **No; it's not the least or greatest amount.**

© Glencoe/McGraw-Hill   T14   *Mathematics: Applications and Connections, Course 1*

HANDS-ON MATH

Guided Practice

3. Add a number to your original set of data in the Mini-Lab so that the mean and median stay the same. **See students' work.**

List the data in each set from least to greatest. Then find the median, mode, mean, and range. **5. See Answer Appendix.**

4. 1, 2, 3, 5, 5, 7, 9;
   5; 5; ≈ 4.6; 8

4. 3, 7, 1, 5, 5, 9, 2

5. 36, 26, 10, 57, 29, 83, 27, 88, 37

6. **Food** If 18 snacks in a vending machine cost 60¢ each, 9 snacks cost 50¢ each, and 8 snacks cost 80¢ each, what is the mean cost for all 35 snacks? **62¢**

EXERCISES

Practice

7. 44; 44; no mode; 22
8. 28; 27; 27; 10
9. 15; 15; 14 and 16; 4
10. 10; 9; 9; 11
12. 427.5; 430; 377 and 475; 300

Find the mean, median, mode, and range for each set of data.

7. 52, 48, 32, 40, 54, 38

8. 23, 27, 33, 31, 30, 27, 25

9. 13, 17, 16, 16, 14, 14, 16, 14

10. 9, 15, 9, 12, 7, 9, 15, 4

11. 31, 25, 18, 40, 31, 18, 26, 32, 39, 44, 37 **31; 31; 18 and 31; 26**

12. 475, 377, 273, 379, 477, 573, 475, 377, 385, 484

Refer to the beginning of the lesson for Exercises 13–15.

13. Suppose the Samsung camera was removed from the list. How would this affect the mean? **The mean would be lower.**

14. Suppose both the Samsung and the Kodak cameras were removed from the list. How would this affect the median? **It would not affect the median.**

15. If the Konica camera cost as much as the Samsung camera, would this change the mean? Explain why or why not. **See Answer Appendix.**

Applications and Problem Solving

Family Activity

Find a bar graph in a newspaper or magazine in your home. Work with a family member to find the mean, median, and mode of the data in the graph.

16. **Earth Science** The Channel 6 meteorologist made the graph at the right.
   a. What was the mean high temperature for June? **88°**
   b. What was the range of the high temperatures? **7°**

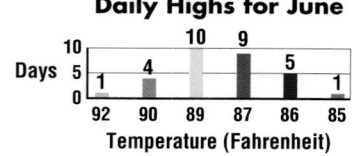

Daily Highs for June

17. **Business Management** The manager of Dairy Land keeps a record of the sizes of the shakes sold. Would the mean, median, or mode be more useful to her? Explain. **See Answer Appendix.**

18. **Critical Thinking** List six numbers such that the mean is 14, the median is 14, the modes are 12 and 14, and the range is 5. **See Answer Appendix.**

Mixed Review

19. **Statistics** Construct a stem-and-leaf plot for 12, 7, 23, 9, 10, 20, 0, 4, 19, 13, 5, 7, 2, 13, 18, and 2. *(Lesson 2-6)* **See Answer Appendix.**

20. **Test Practice** Shamrock, Texas, had a population of 2,206 in 1996. What was the population rounded to the nearest thousand? *(Lesson 1-3)* **A**
   A 2,000   **B** 2,200   **C** 2,020   **D** 2,210

74 Chapter 2 Statistics: Graphing Data

Extending the Lesson

Enrichment Masters, p. 14

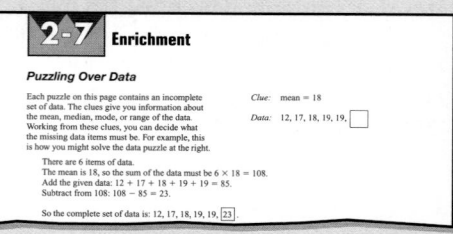

2-7 Enrichment

Puzzling Over Data

Each puzzle on this page contains an incomplete set of data. The clues give you information about the mean, median, mode, or range of the data. Working from these clues, you can decide what the missing data items must be. For example, this is how you might solve the data puzzle at the right.

Clue: mean = 18

Data: 12, 17, 18, 19, 19, ☐

There are 6 items of data.
The mean is 18, so the sum of the data must be 6 × 18 = 108.
Add the given data: 12 + 17 + 18 + 19 + 19 = 85.
Subtract from 108: 108 − 85 = 23.

So the complete set of data is: 12, 17, 18, 19, 19, 23.

Activity Check store catalogs for the range of prices of a favorite item. If you were the store's marketing manager, would you use the range, mean, mode, or median to advertise the items? Explain your reasons.

What's the Average?

Get Ready This game is for four players.

 4 index cards   calculator   2 spinners

Get Set
- Each player should write five whole numbers on an index card. The numbers should be from 1 through 10.
- Label the spinner that has three equal regions with the words *mean, mode,* and *median.*

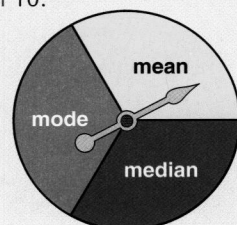

- Label the spinner that has four equal regions with the words *add/increase, add/decrease, remove/increase,* and *remove/decrease.*

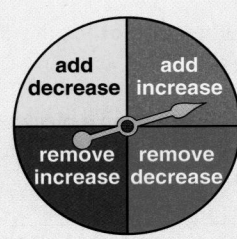

Go
- Mix the index cards and turn them facedown.
- The first player randomly selects a card and spins each spinner once. Then the player adjusts the data set as instructed. For example, if the player gets *mode* and *add/decrease,* the player must add a piece of data to the data set so the mode decreases. If the player gets *median* and *remove/increase,* the player must remove a piece of data from the data set so the median of the set increases.
- The other members then check.
- A player scores two points for each correct solution and loses one point for each incorrect solution.
- The first player to get 10 points wins.

 Visit www.glencoe.com/sec/math/mac/mathnet for more games.

Lesson 2-7 Mean, Median, and Mode **75**

Instructional Resources
Hands-On Lab Masters
- spinners, p. 20

Manipulative Kit
- spinners

Overhead Manipulative Resources
- spinners

You may wish to use an overhead transparency spinner to demonstrate the instructions. Encourage students to devise a pattern that will help them change the mean, median, and mode. For example, to add/increase the median, you would add a number that is larger than the median. *Additional resources for this game can be found on page 42 of the* ***Classroom Games.***

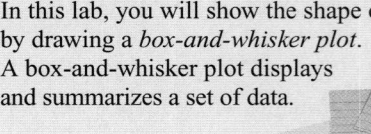

GET READY

Objective Students display and summarize data in a box-and-whisker plot.

Optional Resources
Hands-On Lab Masters
• number lines, p. 16
• worksheet, p. 40

MANAGEMENT TIPS

Recommended Time
45 minutes

Getting Started Have students find examples of line graphs and convert them to vertical bar graphs, using counters to represent the vertical numbers. Then they should practice finding the median of the horizontal axis.

The **Activity** deals with quartiles. Explain that quartiles are just fourths of the data. Compare *quartile* to *quarter*. The upper quartile is the upper fourth of the data. The lower quartile is the lowest fourth of the data.

COOPERATIVE LEARNING

2-7B Box-and-Whisker Plots

A Follow-Up of Lesson 2-7

📇 index cards

📄 cardboard

✏️ markers

📍 push pins

〰️ string

Did you know you can analyze a set of data by looking at its shape? In this lab, you will show the shape of a set of data by drawing a *box-and-whisker plot*. A box-and-whisker plot displays and summarizes a set of data.

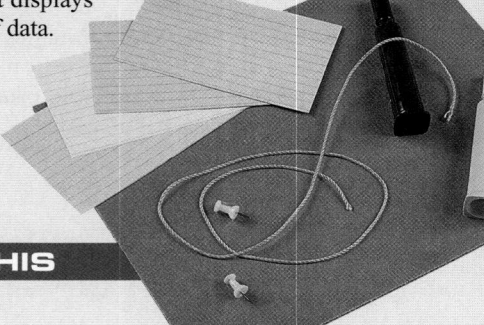

TRY THIS

Work in groups of four.

Make a box-and-whisker plot for the data 20, 10, 22, 25, 18, 22, 16, 26, 23, 27, 14, and 19.

Step 1 Draw a number line on a piece of cardboard. Label it with an appropriate scale. Since the data goes from 10 to 27, begin the scale with 10 and end with 29.

10 11 12 13 14 15 16 17 18 19 20 21 22 23 24 25 26 27 28 29

Step 2 Copy each number from the data set on an index card. Place each number in its appropriate place above the number line.

22
10   14   16   18 19 20   22 23   25 26 27
10 11 12 13 14 15 16 17 18 19 20 21 22 23 24 25 26 27 28 29

Step 3 Mark the median and the *extreme* (least and greatest) *values* in the data by placing push pins on the number line. Label each number. The data is now separated into two halves.

22
10   14   16   18 19 20   22 23   25 26 27
10 11 12 13 14 15 16 17 18 19 20 21 22 23 24 25 26 27 28 29
lower   median   upper
extreme   extreme

Step 4 Now find the median of each half of the data. Mark each by placing a push pin on the number line. The median of the lower half of the data is called the *lower quartile*. The median of the upper half of the data is called the *upper quartile*. Label these on the number line. The data is now separated into fourths or *quartiles,* each having the same number of data.

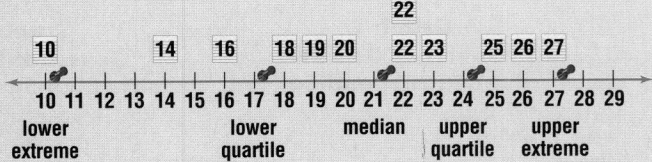

Step 5 Draw a box using the quartiles as the left and right edges. Then draw a line segment through the median.

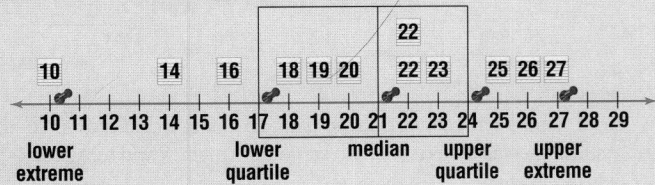

Step 6 Construct the whiskers. Tie a piece of string between the lower quartile and the least number. Tie another piece of string between the upper quartile and the greatest number.

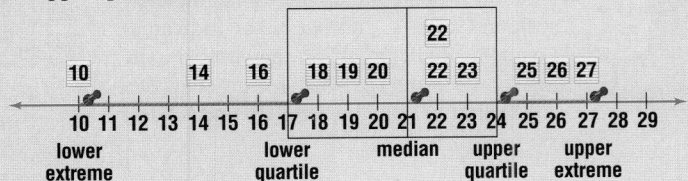

The box-and-whisker plot shows that the bottom fourth of the data is fairly spread out while the top fourth has a small range.

1,3. See Answer Appendix.

ON YOUR OWN

1. **Draw** a box-and-whisker plot for 37, 12, 25, 9, 33, 17, 51, 45, 35, and 39.
2. If the whiskers in a box-and-whisker plot are longer than the box, what does this tell you about the data? **See margin.**
3. **Reflect Back** Make a box-and-whisker plot for the data in the stem-and-leaf plot in Example 2 on page 69.

Lesson 2-7B HANDS-ON LAB **77**

Math Journal Have students summarize what they've learned about box-and-whisker plots and when they might be useful.

Have students complete Exercises 1 and 2. Watch for students who may have difficulty determining the upper and lower quartiles. Remind them to divide the data into fourths to find the quartiles.

Use Exercise 3 to determine whether students understand box-and-whisker plots.

Additional Answer

2. Sample answer: The middle half of the data is tightly clustered and the lower and upper fourths of the data are widely dispersed.

Hands-On Lab Masters, p. 40

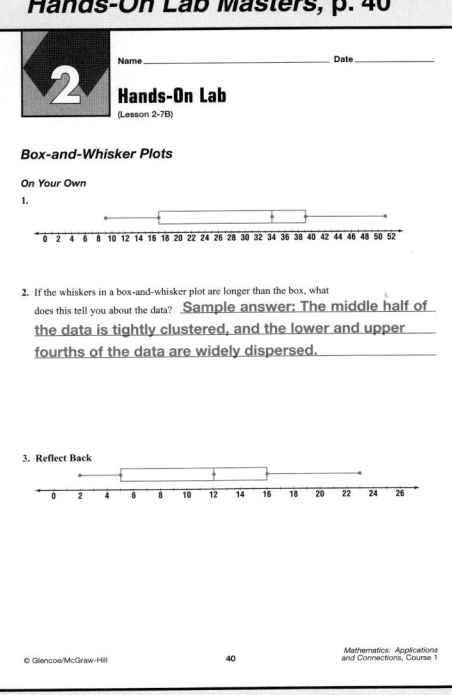

Instructional Resources

- *Study Guide Masters*, p. 15
- *Practice Masters*, p. 15
- *Enrichment Masters*, p. 15
- Transparencies 2-8, A and B
- *Classroom Games*, pp. 3–4
- CD-ROM Program
  - Resource Lesson 2-8

Recommended Pacing	
Standard	Day 11 of 14
Honors	Day 11 of 14
Block	Day 6 of 7

1 FOCUS

5-Minute Check
(Lesson 2-7)

Find the mean, median, mode, and range for each set of data.

1. 2, 5, 6, 8, 4, 7, 8, 9, 5
   6, 6, 5 and 8, 7

2. 13, 80, 23, 45, 13, 81, 25
   40, 25, 13, 68

On a sale rack at a store, 20 CDs cost $5 each, 15 CDs cost $4 each, and 5 CDs cost $7 each. Use this data for Exercises 3–5.

3. What is the mean? **$4.88**

4. What is the mode? **$5**

5. What is the median? **$5**

 The 5-Minute Check is also available on **Transparency 2-8A** for this lesson.

Motivating the Lesson

Problem Solving Ask students to read the opening paragraph of the lesson. Then have them look at the two graphs. Ask the following questions.

- How are the graphs the same?
  Sample answer: They show the same information.
- How are they different?
  Sample answer: There are more intervals in Graph 2; the intervals are larger in Graph 1.

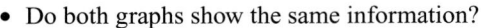

2-8   Misleading Statistics

What you'll learn

You'll learn to recognize when statistics and graphs are misleading.

When am I ever going to use this?

Knowing how to recognize misleading statistics will help you make informed decisions.

During political campaigns, politicians often use statistics to make a point. In a recent election, opposing candidates presented the two graphs shown below.

- Do both graphs show the same information?

- Which graph suggests a drastic cut in government spending? Which graph suggests fairly steady spending?

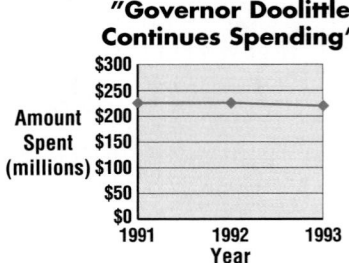

Graph 1
"Governor Doolittle Continues Spending"

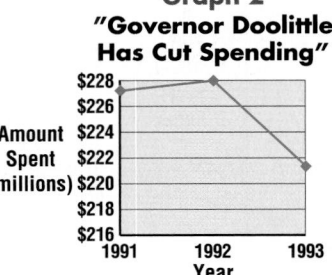

Graph 2
"Governor Doolittle Has Cut Spending"

Both graphs represent the same data. By using a different scale, each candidate was able to take the same information and tell a very different story.

Graph 2 is a result of an expanded vertical scale. For this reason, Graph 2 is visually misleading when compared to the complete scale in Graph 1. This should be indicated using a small "break" in the vertical axis as shown at the right. This shows that the axis is not to scale between $0 and $220.

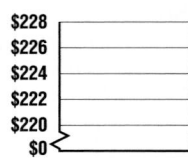

Example ——➊   **Geography** The graphs display the same data.

CONNECTION

a. **Is Graph A misleading? Explain.**

Graph A is misleading because there is no title and there are no labels on either scale. It should also include a break in the vertical axis between 0 and 30.

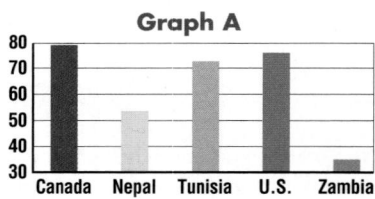
Graph A

b. Is Graph B misleading? Explain.

Graph B is not misleading. The scales are labeled, and the graph includes a title.

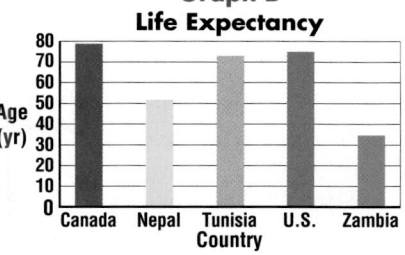

**Graph B
Life Expectancy**

Another way statistics can be misleading is by not using the best measure of central tendency.

 Transparency 2-8B contains a teaching aid for this lesson.

Using Applications For the amounts below, have students determine which measure of central tendency they might choose as a basis for their weekly allowance. **mode** Ask them which measure they might use if they were parents. **mean**

$4 $6 $5 $7 $7

Example 2

APPLICATION

Allowances The results of a class survey on weekly allowances are shown in the table. Felipe told his sister that the average allowance was $7, Megan told her parents the average was $5, and Alba told her brother the average was $4.45.

Allowance	Frequency
$7	6
$6	0
$5	5
$4	1
$3	4
$2	2
$1	2

a. How can all three people think they are correct?

b. Which do you think is the best measure and why?

a. Each person used a different measure of central tendency to describe the set of data. Felipe used the mode, Megan used the median, and Alba used the mean.

b. The mode is misleading because $7 is also the highest allowance. The median is misleading because $5 is the second highest allowance. In this case, the best measure is the mean.

In-Class Examples

For Example 1
The graphs display the same information about life expectancy of women in various countries. Which graph is misleading? Explain. **Graph A; no labels and vertical axis skips from 0 to 50 and 50 to 70**

**Graph A
Life Expectancy**

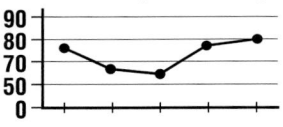

**Graph B
Life Expectancy**

CHECK FOR UNDERSTANDING

Communicating Mathematics

Read and study the lesson to answer each question.

1. *Describe* at least two ways that a graph can be misleading. **See margin.**

2. *Write* a set of eight pieces of data that would best be represented by the median. **Sample answer: 2, 8, 9, 10, 12, 13, 14, 100**

 Math Journal

3. *Write* a few sentences explaining how the mean of a set of data might be used to mislead a consumer. **See Answer Appendix.**

Guided Practice

Tell whether the *mean*, *median*, or *mode* would be best to describe each set of data. Explain each answer.

4. mean — all numbers about the same

4. 500, 400, 390, 405, 390

5. car buyer's favorite color **See margin.**

6. populations of California, Texas, New York, Florida, Pennsylvania, and Idaho **Median — Idaho is much less than the rest.**

For Example 2
The speeds at which animals can travel in a quarter-mile distance are shown below. The mode is 30, the median is 40, and the mean is 38.5. Which average best represents the data? **median**

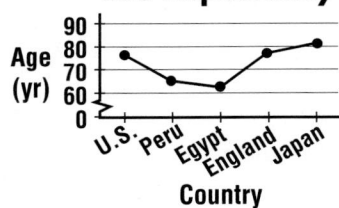

Speed (mph)	Frequency
61	1
50	3
45	2
40	3
35	3
30	4
9	1

Lesson 2-8 Misleading Statistics   **79**

■ Reteaching the Lesson ■

Activity Play a game of "Misleading." Have students roll two number cubes five times, writing down the 10 numbers. Have them find the mean, median, and mode of the numbers. Then have them choose the highest average. If it is misleading, other players may challenge by saying "Misleading."

Additional Answers

1. Sample answer: using different scales and not having a title

5. mode—want to find the most frequent

Check for Understanding

If students need additional practice or instruction after completing Exercises 1–7, one of these options may be helpful.
- Extra Practice, see p. 562
- Reteaching Activity
- *Study Guide Masters*, p. 15
- *Practice Masters*, p. 15

7. *Stock Market* The graphs below display the same information.

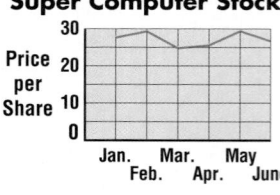

Graph A
Monthly Closing Price of Super Computer Stock

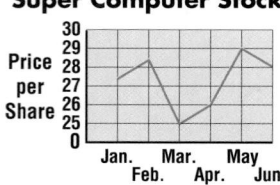

Graph B
Monthly Closing Price of Super Computer Stock

7a. Graph B is misleading because the distance between 0 and 25 is the same as the distance between 25 and 30.

a. Explain why these graphs look different.

b. If a stockbroker wanted to convince a customer to invest in the stock market, which graph would she probably show the customer? Explain.
Graph A because it shows a general rise in price

EXERCISES

Practice

8a. Sample answer: Graph B; it does not go as high giving the impression that the rates are lower.

8b. Sample answer: Graph B; it looks like the calls are cheaper.

8. *Money Matters* A local telephone company employee made two graphs to show the cost of calling Port Arthur.

a. Which graph makes a call to Port Arthur look more economical? Why?

b. Which graph would you show to a telephone customer? Explain.

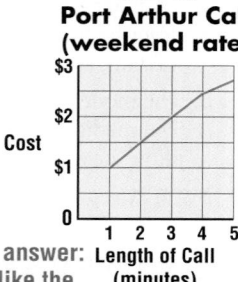
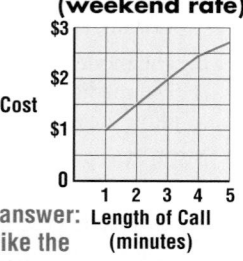

Graph A
Cost of Port Arthur Call (weekend rate)

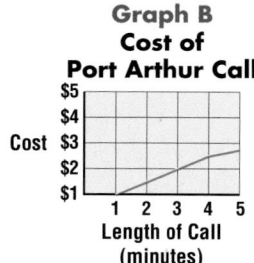

Graph B
Cost of Port Arthur Call

c. Which graph would you show to a company stockholder? Explain. Sample answer: Graph A; it looks like the company is getting more

Determine the measure of central tendency that you think is best for each category. Explain your choice.

Home	Condo	Cape Cod	Ranch	Split Level	2–Story
Bathrooms	1	1	$1\frac{1}{2}$	2	$2\frac{1}{2}$
Bedrooms	2	2	2	4	5
Taxes	$1,200	$2,080	$1,800	$1,740	$2,990
Price	$62,900	$99,450	$124,800	$114,900	$239,400

9. $1\frac{1}{2}$; median

9. number of bathrooms

10. number of bedrooms 2; mode

11. taxes $1962; mean

12. price $114,900; median

Study Guide Masters, p. 15

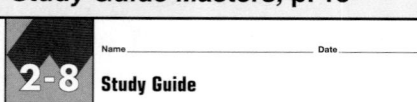

Name _____ Date _____

2-8 Study Guide

Misleading Statistics

Some graphs are misleading. Watch for scales with unequal intervals and missing titles and labels on scales.

Example

This interval is 10.
This interval is 80.

Use the best measure of central tendency.

Measure of Central Tendency	When to Use
Mean	when no numbers are much greater or much less than the others
Median	when there are numbers that are much greater or much less than the others
Mode	when the most frequently occuring number is needed

Use the chart and a measure of central tendency to show each of the following.

1. weight mean, 163
2. age median, 42
3. hair color mode, blond

Name	Bill	Joe	Yi	Tony	Al
Age	43	39	42	45	12
Weight	162	155	168	182	148
Hair	brown	blond	black	blond	blond

Solve. Use the line graph above.

4. Draw a scale that could be used to make a graph that is not misleading. Start at 0 and use equal intervals.

5. Change the title of the graph so that it is misleading. Example: Bicycle sales soar!

© Glencoe/McGraw-Hill   T15   Mathematics: Applications and Connections, Course 1

Applications and Problem Solving

13. *Money Matters* An advertisement for *CD Spins* states that their average price for a CD is $11. How may the owner of *CD Spins* be using average to give the impression of lower prices? **There may be a few very low priced CDs, thus giving a lower average.**

14. *Language Arts* Dora is writing a book report on a book about Henry Aaron entitled *I Had A Hammer*. To show that Aaron is the lifetime home run leader, Dora draws the graphs below.

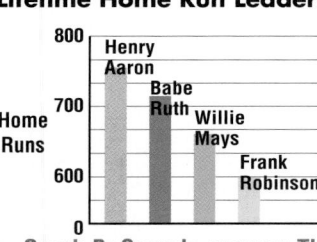
Graph A
Lifetime Home Run Leaders

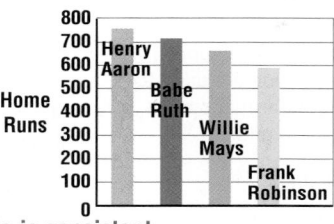
Graph B
Lifetime Home Run Leaders

b. **Graph B; Sample answer: The scale is consistent.**

a. Which graph makes it look like Willie Mays hit about twice as many home runs as Frank Robinson? **Graph A**

b. Which graph should Dora use for a presentation? Explain.

c. How could Graph A be changed to be less misleading? **Show a break in the vertical axis.**

Henry Aaron

15a. mean: 134 ÷ 24 ≈ 5.6 or 6 pets; median: 3 pets; mode: 2 pets

15. *Working on the* CHAPTER Project Refer to your frequency tables from Exercise 10 on page 49.

a. Find the mean, median, and mode for the data in your frequency table from Exercise 10a.

b. Tell which measure of central tendency you think would be best to describe this set of data. Explain your answer. **See margin.**

16. *Critical Thinking* Find a graph in a newspaper or magazine. Redraw the graph so that the data appear to show different results. Which graph describes the data better? Explain. **See students' work.**

Mixed Review

17. [Test Practice] The Chill hockey team played 32 games. They scored a total of 128 goals. What was the mean (average) number of goals scored per game? *(Lesson 2-7)* **A**

A 4
B 5
C 96
D 160

18. Sample answer: 100 to 150

18. *Sports* Erica's bowling scores last weekend were 119, 134, 135, 125, 143, and 130. What is an appropriate scale for her scores? *(Lesson 2-2)*

19. *Patterns* Find the next three numbers in the pattern 1, 4, 9, 16, ... *(Lesson 1-2)* **25, 36, 49**

Lesson 2-8 Misleading Statistics **81**

Extending the Lesson

Enrichment Masters, p. 15

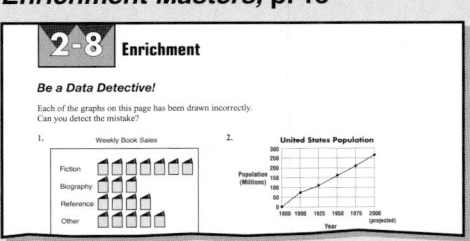

Activity Have small groups of students use the financial page of a daily newspaper to track the price of a stock over a period of one month. Have them determine whether a graph or a measure of central tendency best illustrates the performance of the stock.

Exercise 15 asks students to advance to the next stage of work on the Chapter Project. You may wish to have students work together to discuss the best measure of central tendency. If students can't reach a consensus, have them include all choices and reasons for each.

4 ASSESS

Closing Activity

Speaking Have students supply the words *mean, median,* and *mode* as you read the guidelines from the lesson for the best use of measures of central tendency. Then have them discuss examples of how information can be misleading in a graph.

Additional Answer

15b. Sample answer: The median best represents the data. The student with 24 pets has a big effect on the mean.

Practice Masters, p. 15

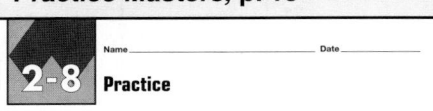
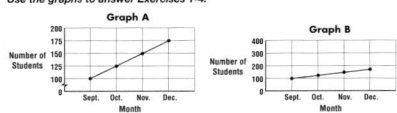

Lesson 2-8 **81**

1 FOCUS

5-Minute Check
(Lesson 2-8)

Ruth's Cookie Sales

Boxes Sold / Day of Week

Ina's Cookie Sales

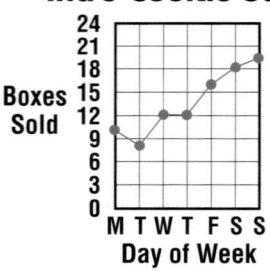

Boxes Sold / Day of Week

Ruth and Ina sold cookies for a week.

1. At first glance, which student appears to be the better salesperson? **Ruth**

2. Which student actually sold the most boxes? **Ina**

3. Which graph is misleading? **Ruth's**

 The 5-Minute Check is also available on **Transparency 2-9A** for this lesson.

2-9

What you'll learn
You'll learn to use ordered pairs to locate points and organize data.

When am I ever going to use this?
In geography, latitude and longitude are used to separate Earth into a coordinate grid.

Word Wise
coordinate system
origin
x-axis
y-axis
ordered pair
x-coordinate
y-coordinate

Integration: Geometry
Graphing Ordered Pairs

In the cartoon, how do the other students know who the machine has caught daydreaming?

Notice how the chart on the front of the machine forms a grid pattern. The grid is used to identify the daydreamer by using a code similar to what is used on road maps. Obviously, everyone knows how to locate the seat designated by 4-B.

CLOSE TO HOME   JOHN McPHERSON

In mathematics, we can locate a point by using a **coordinate system**.

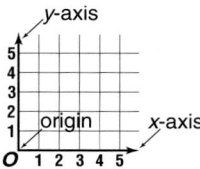

The coordinate system is formed when two number lines intersect at their zero points. This point is called the **origin**. The horizontal number line is called the **x-axis**, and the vertical number line is called the **y-axis**.

You can name any point on a coordinate system by using an **ordered pair** of numbers. The first number in an ordered pair is called the **x-coordinate**, and the second number is called the **y-coordinate**. An ordered pair is written in this form:

$$(4, 2)$$

x-coordinate ↑ ↑ *y-coordinate*

The *x*-coordinate represents the number of horizontal units the point is from 0, and the *y*-coordinate represents the number of vertical units it is from 0.

Motivating the Lesson
Hands-On Activity Have students determine the coordinates of their seat assignments in a manner similar to the one in the opening paragraph of the lesson.

Example ① **Name the ordered pair for point K.**

Start at the origin. Move right along the *x*-axis until you are under point *K*. Since you moved two units to the right, the *x*-coordinate of the ordered pair is 2.

Now move up until you reach point *K*. Since you moved up 4 units, the *y*-coordinate is 4.

The ordered pair for point *K* is (2, 4).

Transparency 2-9B contains a teaching aid for this lesson.

Reading Mathematics Much of mathematics involves symbolism with distinct meanings. Make sure students understand what the expression *B*(4, 2) means.

You can also graph a point on a coordinate system. To graph a point means to place a dot at the point named by an ordered pair.

In-Class Examples

For Example 1
Name the ordered pair for point *M*. **(3, 2)**

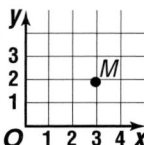

Example ② **Graph each point.**

a. A(6, 4)
Start at the origin. Move 6 units to the right on the *x*-axis. Then move 4 units up to locate the point. Draw a dot and label the dot *A*.

b. B(7, 0)
Start at the origin. Move 7 units to the right on the *x*-axis. Since the *y*-coordinate is 0, stop here. Draw a dot and label the dot *B*.

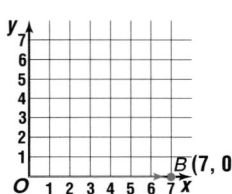

For Example 2
Graph each point.
a. *C* (5, 3)
b. *D* (0, 5)

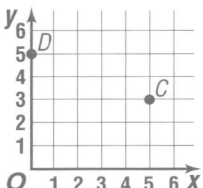

Ordered pairs can also be used to show how data are related.

For Example 3
Rhaghad went to play putt-putt. Games were $3 each. If *x* represents the number of games she played, the expression 3*x* gives the total cost.
a. Make a list of ordered pairs in this form: (number of games played, cost of games).

Example ③ **Algebra** Pazi went bowling at The Bowling Castle. Games were $2 each. If *x* represents the number of games she bowled, the expression 2*x* gives the total cost.

INTEGRATION

a. Make a list of ordered pairs in this form: (number of games bowled, cost of games).

b. Then graph the ordered pairs.

(continued on the next page)

x (games)	3x	y (cost)	(x, y)
1	3(1)	3	(1, 3)
2	3(2)	6	(2, 6)
3	3(3)	9	(3, 9)

b. Graph the ordered pairs.

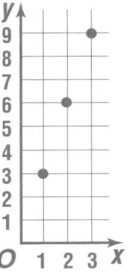

Check for Understanding

If students need additional practice or instruction after completing Exercises 1–10, one of these options may be helpful.
• Extra Practice, see p. 562
• Reteaching Activity
• *Study Guide Masters,* p. 16
• *Practice Masters,* p. 16

Assignment Guide

Core: 11–31 odd, 33–37
Enriched: 12–30 even, 31–37

Additional Answers

3. Sample answer: Raul; to graph (8, 3) you move 8 units right and 3 units up and to graph (3, 8) you move 3 units right and 8 units up.

10.

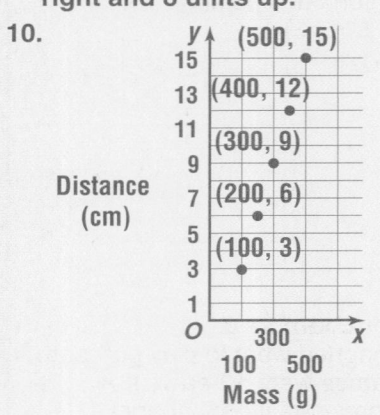

Study Guide Masters, p. 16

2-9 Study Guide

Integration: Geometry
Graphing Ordered Pairs

In mathematics, we lokate points by using a **coordinate system.** This system is formed when two number lines intersect at their zero points. This point is called the **origin,** and is labeled with an O. The horizontal number line is called the **x-axis,** and the vertical number line is called the **y-axis.**

You can name any point on a coordinate system by using an **ordered pair** of numbers. The first number is called the **x-coordinate,** and the second number is called the **y-coordinate.**

Examples 1 Name the ordered pair for point W.

Start at the origin. Move right along the x-axis until you are under point W. Since you moved three units to the right, the x-coordinate of the graphed pair is 3.
Now move up until you reach point W. Since you moved up 1 unit, the y-coordinate is 1. The ordered pair for point W is (3, 1).

2 Graph P(0, 2).

Start at the origin. Since the x-coordinate is 0, do not move any units to the right on the x-axis. Move 2 units up to locate the point. Draw a dot and label the point P.

Use the grid at the right to name the point for each ordered pair.

1. (0, 4) L   2. (3, 6) N
3. (2, 1) J   4. (1, 8) K
5. (7, 5) M   6. (5, 3) H

Graph each point.

7. A(1, 5)   9. X(6, 0)
8. F(9, 2)   10. D(4, 8)

© Glencoe/McGraw-Hill   T16   Mathematics: Applications and Connections, Course 1

Did you know On February 2, 1997, Jeremy Sonnenfeld, a college sophomore, became the first person to bowl a 900 series (3 perfect games in a row) in the 101–year history of the American Bowling Congress.

a. Evaluate the expression $2x$ for 1, 2, and 3 games. A table of ordered pairs that result is shown below.

b.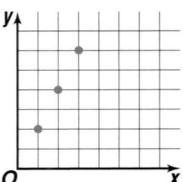

x (games)	2x	v (cost)	(x, y)
1	2 (1)	2	(1, 2)
2	2 (2)	4	(2, 4)
3	2 (3)	6	(3, 6)

(3, 6) means 3 games cost $6.

CHECK FOR UNDERSTANDING

Communicating Mathematics

1. From the origin, go 7 units right, then up 9 units.
2. See students' work.

Read and study the lesson to answer each question.

1. *Tell* how to graph the ordered pair (7, 9).
2. *Draw* a coordinate system and label the origin, x-axis, and y-axis.
3. *You Decide* Lina says that the ordered pairs (8, 3) and (3, 8) name the same point. Raul disagrees. Who is correct and why? **See margin.**

Guided Practice

Use the grid at the right to name the point for each ordered pair.

4. (3, 7) *B*   5. (9, 5) *F*   6. (0, 8) *A*

Use the grid at the right to name the ordered pair for each point.

7. *E* (8, 0)   8. *D* (6, 3)   9. *C* (5, 6)

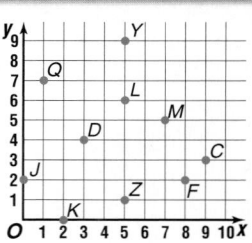

10. See margin for graph. Sample answer: The points appear to be in a straight line.

10. *Physical Science* Odell and Elise conducted an experiment to see how the mass of an object affected the distance a spring stretched. On graph paper, draw a coordinate grid. Then graph the ordered pairs (mass, distance). Use the x-axis for mass and the y-axis for distance. What did they observe?

Stretching of a Spring

Mass	Distance
100 g	3 cm
200 g	6 cm
300 g	9 cm
400 g	12 cm
500 g	15 cm

EXERCISES

Practice

Use the grid at the right to name the point for each ordered pair.

11. (9, 3) *C*   12. (5, 9) *Y*
13. (2, 0) *K*   14. (5, 1) *Z*
15. (1, 7) *Q*   16. (8, 2) *F*
17. (5, 6) *L*   18. (0, 2) *J*
19. (7, 5) *M*   20. (3, 4) *D*

84   Chapter 2 Statistics: Graphing Data

Reteaching the Lesson

Activity Have students use a state map and locate different cities using the letter and number pairs. Discuss the importance of being able to locate places on a map. Have students construct a map of their neighborhood and draw a grid. Have them list ordered pairs that identify neighborhood landmarks.

Error Analysis
Watch for students who go up for the x value and across for the y value.
Prevent by reminding students to go alphabetically. The **x** value is **h**orizontal. The **y** value is **v**ertical.

Use the grid at the right to name the ordered pair for each point.

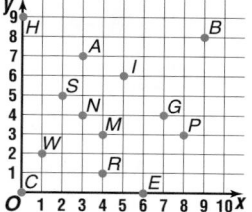

21. G (7, 4)   22. R (4, 1)
23. A (3, 7)   24. P (8, 3)
25. H (0, 9)   26. I (5, 6)
27. C (0, 0)   28. M (4, 3)
29. B (9, 8)   30. S (2, 5)

Applications and Problem Solving

31. **Art** Draw a coordinate grid. Number each axis from 0 to 20. Draw a picture on your grid. List the ordered pairs of points where sides of the drawing meet. Have a classmate draw your picture by graphing your ordered pairs and connecting the dots. **See students' work.**

32. **Cartography** The grid is a simplified version of the map of present-day Washington, D.C.

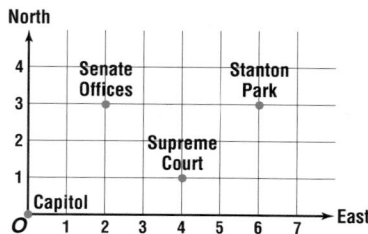

a. Which ordered pair indicates the location of the Supreme Court? **(4, 1)**

b. Which location does (6, 3) indicate? **Stanton Park**

33. **Critical Thinking** Write a short paragraph that tells how graphing ordered pairs compares to graphing statistical data in bar graphs or line graphs. **See students' work.**

Mixed Review

34. **Statistics** Find the mean for the set of Tim's history test scores: 88, 90, 87, 91, 49. Why is this a misleading statistic? *(Lesson 2-8)*

34. Sample answer: 81; All the test scores except 49 are greater than the mean.

35. **Test Practice** Four girls have heights of 152, 158, 164, and 168 centimeters. Which height is a reasonable average height of the girls? *(Lesson 2-7)* **C**

A 120 cm   B 140 cm   C 160 cm
D 180 cm   E 200 cm

36. **Life Science** Make a horizontal bar graph for the set of data in the chart. *(Lesson 2-3)* **See Answer Appendix.**

37. **Algebra** Solve the equation 56 ÷ n = 8. *(Lesson 1-7)* **7**

Top Animal Speeds (mph)	
Antelope	61
Cheetah	70
Coyote	43
Elk	45
Gazelle	50
Gray Fox	42
Hyena	40
Lion	50
Quarterhorse	47.5
Wildebeest	50
Zebra	40

Lesson 2-9 Integration: Geometry   Graphing Ordered Pairs   **85**

═══ Extending the Lesson ═══

Enrichment Masters, p. 16

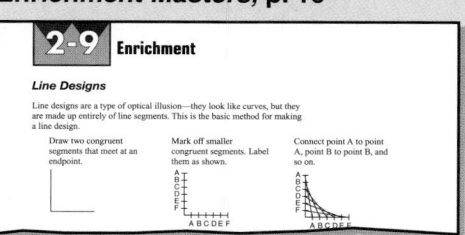

2-9 **Enrichment**

Line Designs

Line designs are a type of optical illusion—they look like curves, but they are made up entirely of line segments. This is the basic method for making a line design.

Activity Display these ordered pairs: (4, 1), (8, 3), (14, 6). Show students that these points lie on the same line. Then display: (6, 4), (10, 4), (20, 8), (11, 4), (2, 0), (26, 12). Have students identify which points also lie on the line. Ask students to identify a "rule" for how the x values are related to the y values.
Sample answer: 2 · y − 2 = x

4 ASSESS

Closing Activity
Writing Have students write a sentence to serve as a memory aid in remembering which direction comes first when writing or locating an ordered pair.

Chapter 2, Quiz D (Lessons 2-8 and 2-9) is available in the *Assessment and Evaluation Masters,* p. 44.

Practice Masters, p. 16

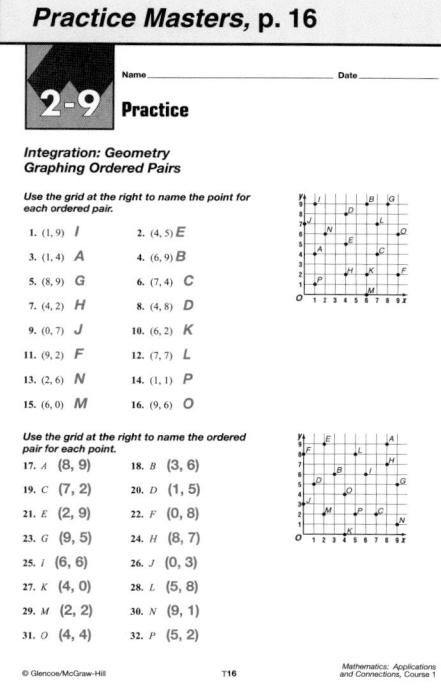

Name_____ Date_____

2-9 **Practice**

Integration: Geometry
Graphing Ordered Pairs

Use the grid at the right to name the point for each ordered pair.

1. (1, 9) *I*   2. (4, 5) *E*
3. (1, 4) *A*   4. (6, 9) *B*
5. (8, 9) *G*   6. (7, 4) *C*
7. (4, 2) *H*   8. (4, 8) *D*
9. (0, 7) *J*   10. (6, 2) *K*
11. (9, 2) *F*   12. (7, 7) *L*
13. (2, 6) *N*   14. (1, 1) *P*
15. (6, 0) *M*   16. (9, 6) *O*

Use the grid at the right to name the ordered pair for each point.

17. A (8, 9)   18. B (3, 6)
19. C (7, 2)   20. D (1, 5)
21. E (2, 9)   22. F (0, 8)
23. G (9, 5)   24. H (8, 7)
25. I (6, 6)   26. J (0, 3)
27. K (4, 0)   28. L (5, 8)
29. M (2, 2)   30. N (9, 1)
31. O (4, 4)   32. P (5, 2)

© Glencoe/McGraw-Hill   T16   Mathematics: Applications and Connections, Course 1

Study Guide and Assessment

CHAPTER 2

Vocabulary

This section provides a listing of the new terms, properties, and phrases that were introduced in this chapter. Have students define each term and provide an example or two of it, if appropriate.

Understanding and Using the Vocabulary

These exercises check students' understanding of the terms by using a variety of verbal formats including matching, completion, and true/false.

Glossaries A complete glossary of terms appears on pages 642–648. The glossary also appears in Spanish on pages 649–656.

Additional Answer

11. Sample answer: A bar graph represents data using rectangles to show the frequency of responses. A line graph uses dots at the frequency points connected by segments to show how data changes over time.

Vocabulary

After completing this chapter, you should be able to define each term, concept, or phrase and give an example or two of each.

Algebra
coordinate system (p. 82)
ordered pair (p. 82)
origin (p. 82)
x-axis (p. 82)
x-coordinate (p. 82)
y-axis (p. 82)
y-coordinate (p. 82)

Statistics and Probability
average (p. 71)
bar graph (p. 54)
box-and-whisker plot (p. 76)
circle graph (p. 60)
data (p. 46)
extreme value (p. 76)
frequency table (p. 46)

interval (p. 50)
leaf (p. 68)
line graph (p. 54)
lower quartile (p. 77)
mean (p. 71)
measures of central tendency (p. 71)
median (p. 71)
mode (p. 71)
range (p. 72)
scale (p. 50)
statistics (p. 46)
stem (p. 68)
stem-and-leaf plot (p. 68)
upper quartile (p. 77)

Problem Solving
use a graph (p. 58)

Understanding and Using the Vocabulary

Choose the letter of the term that best matches each phrase.

1. separates the scale into equal parts **f**
2. the sum of a set of data divided by the number of pieces of data **b**
3. a plot that uses place value to display a large data set **a**
4. the number that appears most often in a set of data **d**
5. the middle number when a set of data is arranged in numerical order **c**
6. the horizontal number line of a coordinate system **g**
7. the vertical number line of a coordinate system **i**
8. the first number in an ordered pair **h**
9. the second number in an ordered pair **j**
10. the name of a point on a coordinate system in the form (*x*-coordinate, *y*-coordinate) **e**

a. stem-and-leaf plot
b. mean
c. median
d. mode
e. ordered pair
f. interval
g. *x*-axis
h. *x*-coordinate
i. *y*-axis
j. *y*-coordinate
k. box-and-whisker plot

In Your Own Words

11. *Explain* the difference between a bar graph and a line graph. See margin.

86 Chapter 2 Statistics: Graphing Data

 MindJogger Videoquizzes

MindJogger Videoquizzes provide an alternative review of concepts presented in this chapter. Students work in teams to answer questions, gaining points for correct answers. The questions are presented in three rounds.
Round 1 Concepts–5 questions
Round 2 Skills–4 questions
Round 3 Problem Solving–4 questions

Objectives & Examples

Upon completing this chapter, you should be able to:

• make and interpret frequency tables *(Lesson 2-1)*

Make a frequency table for this list of favorite fruits.

banana	grapes	apple	banana
apple	apple	grapes	apple

Fruit	Tally	Frequency
apples	IIII	4
bananas	II	2
grapes	II	2

• choose appropriate scales and intervals for frequency tables *(Lesson 2-2)*

Find an appropriate scale and an interval for 7, 11, 24, 9, 16, 19, and 18.

scale = 0 to 25, interval of 5

• construct bar graphs and line graphs *(Lesson 2-3)*

Number of Pets in Animal Shelter	
Season	Pets
Winter	65
Spring	75
Summer	85
Autumn	70

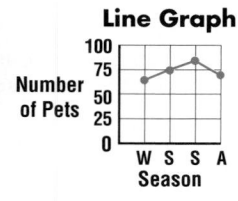

Line Graph

• interpret circle graphs *(Lesson 2-4)*

Magazine Sales by Grade

Sixth graders sold the most subscriptions, earning $906.80.

7th $489.50
6th $906.80
8th $616.75

Review Exercises

Use these exercises to review and prepare for the chapter test.

Make a frequency table for each set of data.

12. How many brothers and sisters do you have? **See Answer Appendix.**

3	1	2	0	4	1
2	1	0	1	3	1
0	2	1	3	1	0

13. What is your favorite color? **See Answer Appendix**

blue	red	pink	blue	orange
red	blue	red	red	yellow
green	blue	brown	pink	green

Choose an appropriate scale and an interval for each set of data. 15–17. See Answer Appendix.

14. 4, 7, 6, 9, 2, 3, 6   **scale = 0 to 10, interval 2**

15. 13, 38, 9, 23, 17, 43, 23

16. 16, 8, 9, 15, 10, 11, 17, 4, 7, 15

17. 138, 152, 112, 127, 173, 136, 98, 125, 145

18. Make a line graph from the data below.

Alma's Grades During Junior High	
Grade	Frequency
A	3
B	9
C	5
D	1
F	0

19. Make a vertical bar graph of the data in Exercise 18. **18–19. See Answer Appendix.**

Use the graph.

20. What is the most common race time? **42.8–54.9 minutes**

10K Race Times (Minutes)

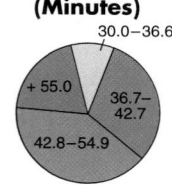

+ 55.0
30.0–36.6
36.7–42.7
42.8–54.9

Objectives & Examples

This section reviews the skills and concepts of the chapter and shows completely worked examples.

Review Exercises

These exercises provide practice for the corresponding objectives.

Assessment and Evaluation Masters, pp. 31–32

Name_____ Date_____

Chapter 2 Test, Form 1B

Refer to the frequency table for Questions 1–2.

Hours Watching Sports Per Week	
Interval	Frequency
0–2	6
3–4	9
5–6	4
7–8	1

1. What is the least common interval of hours of sports watched?
   A. 0–2   B. 3–4
   C. 5–6   D. 7–8   **1. D**

2. How many people spent 3 hours or more per week watching sports events?
   A. 9   B. 20   C. 14   D. 6   **2. C**

3. Determine the stems for this set of data: 15, 23, 18, 7, 2, 16, 12.
   A. 0, 1, 2   B. 1, 2, 3, 4, 5   C. 2, 4, 6, 8, 10   D. 0, 1, 3, 5   **3. A**

4. Choose the best interval for a frequency table for this set of data: 178, 165, 189, 173, 205, 188, 202.
   A. 1   B. 10   C. 100   D. 210   **4. B**

5. Which bar graph below correctly shows the following frequencies: 18 bears, 5 reptiles, 10 fish, and 8 elephants?
   A.   B.   C.   D.   **5. A**

6. Use the bar graph at the right to determine how much more the receipts were for April than for March.
   A. $3,000   B. $7,000
   C. $4,000   D. $11,000   **6. A**

Cafe Receipts

7. Use the pictograph at the right to determine how many more compact discs were sold on Friday than on Thursday.
   A. 5   B. 65
   C. 125   D. 15   **7. D**

Compact Disc Sales

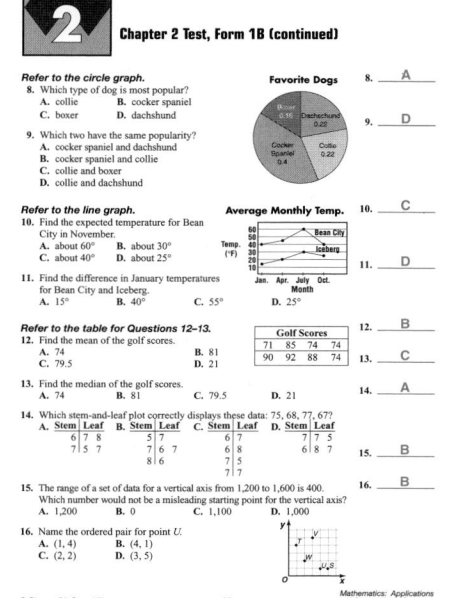

Chapter 2 Test, Form 1B (continued)

Refer to the circle graph.

8. Which type of dog is most popular?
   A. collie   B. cocker spaniel
   C. boxer   D. dachshund   **8. A**

9. Which two have the same popularity?
   A. cocker spaniel and dachshund
   B. cocker spaniel and collie
   C. collie and boxer
   D. collie and dachshund   **9. D**

Favorite Dogs

Refer to the line graph.

10. Find the expected temperature for Bean City in November.
   A. about 60°   B. about 30°
   C. about 40°   D. about 25°   **10. C**

11. Find the difference in January temperatures for Bean City and Iceberg.
   A. 15°   B. 40°   C. 55°   D. 25°   **11. D**

Average Monthly Temp.

Refer to the table for Questions 12–13.

Golf Scores			
71	85	74	74
90	92	88	74

12. Find the mean of the golf scores.
   A. 74   B. 81
   C. 79.5   D. 21   **12. B**

13. Find the median of the golf scores.
   A. 74   B. 81   C. 79.5   D. 21   **13. C**

14. Which stem-and-leaf plot correctly displays these data: 75, 68, 77, 67?
   A.   B.   C.   D.   **14. A**

15. The range of a set of data for a vertical axis from 1,200 to 1,600 is 400. Which number would not be a misleading starting point for the vertical axis?
   A. 1,200   B. 0   C. 1,100   D. 1,000   **15. B**

16. Name the ordered pair for point U.
   A. (1, 4)   B. (4, 1)
   C. (2, 2)   D. (3, 5)   **16. B**

Assessment and Evaluation

Six forms of Chapter 2 Test are available in the *Assessment and Evaluation Masters* as shown in the chart.

Chapter 2 Test, Form 1B, is shown at the right. Chapter 2 Test, Form 2B, is shown on the next page.

1A	Multiple Choice	Honors
1B	Multiple Choice	Average
1C	Multiple Choice	Basic
2A	Free Response	Honors
2B	Free Response	Average
2C	Free Response	Basic

Additional Answer

22.

Stem	Leaf
4	5 8
5	1 3 6
6	0 2 5 6 8
7	2 3 7 9
8	0 4

$4|5 = 45°$

Assessment and Evaluation Masters, pp. 37–38

Name_____ Date_____

2 Chapter 2 Test, Form 2B

Refer to the frequency table.

Passes Completed

Interval	Frequency
0–4	3
5–8	6
9–12	9
13–16	8
17–20	5

1. How many players completed 9 or more passes? **1. 22**
2. Can you tell from the table if any player completed exactly 12 passes? **2. no**
3. Can you tell from the table how many players completed 12 or fewer passes? If so, how many? **3. yes, 18**

Points scored by five basketball players in a game: Majko, 26; Ertl, 32; Min, 11; Naylor, 8; Ben, 17

4. Make a stem-and-leaf plot for the data.

4.
Stem	Leaf
0	8
1	1 7
2	6
3	2

5. What is the range of the data? **5. 24**
6. Make a vertical bar graph showing these data. **6. Points scored**

Refer to the bar graph. **Camera Sales**
7. In what month were sales the least? **7. August**
8. How many cameras were sold in April? **8. 7**

Refer to the circle graph. **Favorite Desserts**
9. Which type of dessert is most popular? **9. pie**
10. Which two types of desserts together are equal to about one half of the desserts? **10. pie and pudding or ice cream and cake**

© Glencoe/McGraw-Hill   37   *Mathematics: Applications and Connections, Course 1*

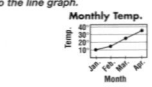

2 Chapter 2 Test, Form 2B (continued)

When Tessa totaled the tips she had earned while working at Badlands Steak-a-Rama, she found she had earned $89 in July, $110 in Aug., $120 in Sept., $100 in Oct., $120 in Nov., and $97 in Dec. Use the data to find the following.
11. mean **11. $106**
12. median **12. $105**
13. mode **13. $120**
14. Use the tip data above to make a line graph. **14. Tips Earned**
15. Name two ways to make your graph misleading. **15. Sample answers: Make scale uneven, omit labels.**
16. Does the mean, median, or mode best describe the data? 4, 20, 26, 29, 30? **16. median**

Refer to the line graph. **Monthly Temp.**
17. Predict the average February temperature. **17. 15°**
18. Predict the average May temperature. **18. 45°**

Refer to the grid.
19. Name the point for the ordered pair (1, 2). **19. F**
20. Name the ordered pair for point J. **20. (4, 1)**

© Glencoe/McGraw-Hill   38   *Mathematics: Applications and Connections, Course 1*

Objectives & Examples

● make predictions from line graphs *(Lesson 2-5)*

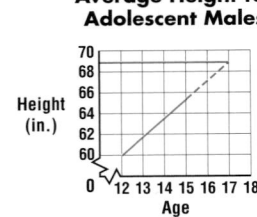

Average Height for Adolescent Males

A male could be about 69 cm tall at age 17.

● construct stem-and-leaf plots *(Lesson 2-6)*

49, 58, 42, 63, 55, 42, 59, 44

Stem	Leaf
4	2 2 4 9
5	5 8 9
6	3

$4|2 = 42$

● find the mean, median, mode, and range of a set of data *(Lesson 2-7)*

117, 98, 104, 108, 104, 111

$$\text{mean} = \frac{117 + 98 + 104 + 108 + 104 + 111}{6} \text{ or } 107$$

$$\text{median} = \frac{104 + 108}{2} \text{ or } 106 \quad \text{mode} = 104$$

$$\text{range} = 117 - 98 \text{ or } 19$$

● recognize when statistics and graphs are misleading *(Lesson 2-8)*

Choose the best measure of central tendency for the set $4, $4, $20, $23, $24, $27, and $30.

The median would be best. The mode is misleading because $4 is the lowest amount.

● use ordered pairs to locate points and organize data *(Lesson 2-9)*

Graph (6, 2).

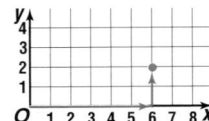

88   Chapter 2 Statistics: Graphing Data

Review Exercises

21. Would you expect the winning time in the year 2000 to be more or less than 2 minutes 15 seconds? **less**

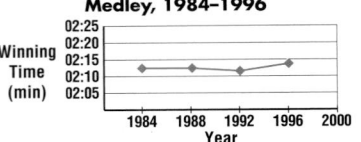

Women's 200-meter Individual Medley, 1984–1996

22. Make a stem-and-leaf plot for the following temperatures (°F). **See margin.**
60°, 53°, 68°, 72°, 66°, 80°, 73°, 51°, 62°, 48°, 56°, 84°, 77°, 45°, 79°, 65°

Use the following set of data for Exercises 23–26. 17, 31, 30, 34, 22, 26, 28, 22, 15

23. Find the mode. **22**
24. Find the median. **26**
25. Find the mean. **25**
26. Find the range. **19**

Tell whether the *mean*, *median*, or *mode* would be best to describe each set of data.

27. $10, $55, $60, $77, $79, $85, $85, $89
28. choices for Homecoming Queen **mode**
29. 81°, 84°, 85°, 87°, 88°, 90°, 91°, 98°

27. median   29. mean

Use the grid to name the point for each ordered pair.
30. (1, 4) **B**   31. (4, 1) **A**

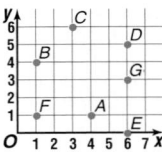

Use the grid to name the ordered pair for each point.
32. E **(6, 0)**   33. C **(3, 6)**

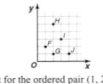 **Test and Review Software**

You may use this software, a combination of an item generator and item bank, to create your own tests or worksheets. Types of items include free response, multiple choice, short answer, and open ended.

CD-ROM Program

The CD-ROM Program contains an Assessment Game whose questions review the concepts in this chapter.

Applications & Problem Solving

34. *Entertainment* The frequency table shows the number of hours per week sixth graders play video games. *(Lesson 2-1)*

a. In what interval did most of the responses fall? **3-4 h**

b. How many people play video games less than 5 hours per week? **21 people**

Hours Playing Video Games per Week	
0–2	10
3–4	11
5–6	5
7–8	3

35. *Television* A survey of the number of minutes a junior high student watches TV on a school day produced the following data.

95, 85, 69, 75, 90, 45,
92, 65, 50, 40, 75

Make a stem-and-leaf plot for the data. *(Lesson 2-6)* **See Answer Appendix.**

36. *Use a Graph* Was the marriage rate in 1980 higher or lower than in 1990? *(Lesson 2-3B)* **higher**

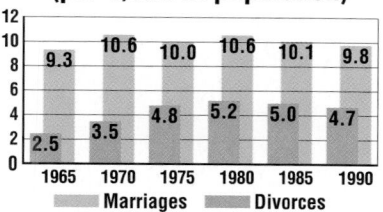

U.S. Marriage and Divorce Rates (per 1,000 of population)

	1965	1970	1975	1980	1985	1990
Marriages	9.3	10.6	10.0	10.6	10.1	9.8
Divorces	2.5	3.5	4.8	5.2	5.0	4.7

Alternative Assessment

Performance Task

Suppose you are the owner of a company. You want to convince a group of investors to contribute money so you can expand your business. Your monthly profits the past year are shown in the table.

January	$2,645	July	$3,200
February	$2,200	August	$2,900
March	$2,205	September	$2,050
April	$2,900	October	$1,950
May	$2,750	November	$2,400
June	$3,050	December	$3,250

Find the mean, median, and mode.

Explain how you would determine whether to present the mean, median, or mode as your "average" monthly profit. **See margin.**

A practice test for Chapter 2 is provided on page 596.

Completing the

Use the following checklist to make sure your presentation is complete.

☑ You have included the frequency tables, bar graphs, and calculations of mean, median, and mode.

☑ The two new bar graphs show the types of pets in both classes.

☑ Your fact sheet about your favorite pet includes a picture or drawing of the pet, the cost of taking care of the pet for one year, and any other interesting facts about the pet.

 Choose a topic of interest to you. Take a survey to collect data about it. Make a frequency table and a bar or line graph. Place all of these materials in your portfolio.

Applications & Problem Solving

This section provides additional practice in solving real-world problems that involve the skills of this chapter.

Alternative Assessment

The **Performance Task** provides students with a performance assessment opportunity to evaluate their work and understanding.

Students should complete the final stages of their project and prepare a class demonstration of their results. A scoring guide for the project is available in the *Investigations and Projects Masters*, p. 23.

 Students should add to their portfolios at this time.

***Assessment and Evaluation Masters*, p. 41**

Name_____   Date_____

2 Chapter 2 Performance Assessment

Instructions: Demonstrate your knowledge by giving a clear, concise solution to each problem. Be sure to include all relevant drawings and to justify your answers. You may show your solutions in more than one way or investigate beyond the requirements of the problems.

1. Use the following table to answer each question.

Planet	Average Distance from Sun
Mercury	36,000,000 miles
Earth	93,003,000 miles
Jupiter	483,900,000 miles
Neptune	2,796,700,000 miles

a. Draw a misleading graph of the average distances from the Sun to the planets listed above.

b. Did you draw a pictograph, bar graph, or line graph? Why?

c. Why is the graph misleading?

d. Tell how to change the graph so that it is not misleading.

e. Tell why a circle graph would be inappropriate for this data.

f. Predict the average distance from the Sun to Venus. Venus is between Mercury and Earth. Check your prediction in an encyclopedia.

2. The scores in the Indianapolis Golf Tournament were 71, 73, 66, 75, 71, 82, 69, 72, 74, 64, 77, 71, 73, 70, 78, and 69.

a. Find the mean, median, and mode for the set of golf scores. Round to the nearest tenth.

b. These measures are said to be measures of central tendency. Explain in your own words what a measure of central tendency is.

c. Tell what a stem-and-leaf plot is. Make one for the golf data.

d. Choose a scale and interval and form a frequency table for the set of golf scores. Label your table.

e. List the ordered pairs you would use to graph the data.

f. Construct a line graph for the data in your table. Label your graph.

g. Tell what the line graph shows.

© Glencoe/McGraw-Hill   41   *Mathematics: Applications and Connections, Course 1*

Additional Answer for the Performance Task

You would use the measure of tendency that shows the greatest profit. $2,625; $2,697.50; $2,900; The mode shows the greatest profit.

Performance Assessment

Additional performance assessment tasks for this chapter are included in the *Assessment and Evaluation Masters* on page 41. A scoring guide is also provided on page 53.

Standardized Test Practice

The Standardized Test Practice may be used to help students prepare for standardized tests. The test items are written in the same style as those in state proficiency tests and standardized tests like CAT, CTBS, ITBS, MAT, SAT, and Terra Nova. The test items cover skills and concepts covered up to this point in the text.

The pages can be used as an overnight assessment. After students have completed the pages, discuss how each problem can be solved, or provide copies of the solutions from the *Solutions Manual.*

Section One: Multiple Choice

There are twelve multiple-choice questions in this section. Choose the best answer. If a correct answer is *not here,* mark the letter for Not Here.

1. How is the product $3 \times 3 \times 5 \times 5 \times 7$ expressed in exponential notation? **D**
   - **A** $3 \times 3 \times 5 \times 5 \times 7$
   - **B** $3 \times 3 \times 5^2 \times 7$
   - **C** $3^2 \times 5 \times 5 \times 7$
   - **D** $3^2 \times 5^2 \times 7$

2. What is the value of $8 + 6 \times 10$? **F**
   - **F** 68
   - **G** 120
   - **H** 140
   - **J** 480

3. If $n = 4$, $m = 3$, and $x = 12$, what is the value of $x \cdot m + n$? **B**
   - **A** 19
   - **B** 40
   - **C** 24
   - **D** 84

4. Which average would be best to describe this data? **H**
   $$195, 188, 70, 185, 190$$
   - **F** range
   - **G** mean
   - **H** median
   - **J** mode

5. Evaluate the expression $c - b$ if $c = 9$ and $b = 3$. **B**
   - **A** 3
   - **B** 6
   - **C** 9
   - **D** 12

6. Minato found 5 seashells. His friend Soto gave him some more, and then he had 10. To find out how many seashells he was given, Minato wrote $x + 5 = 10$. What is the value of x? **G**
   - **F** 2
   - **G** 5
   - **H** 15
   - **J** 35

Please note that Questions 7–12 have five answer choices.

7. Three chimpanzees weigh 62, 75, and 72 pounds. Which weight is a reasonable average weight of the chimpanzees? **C**
   - **A** 80 lb
   - **B** 75 lb
   - **C** 70 lb
   - **D** 65 lb
   - **E** 60 lb

8. The graph shows the number of visitors who toured Gund Arena on Saturday.

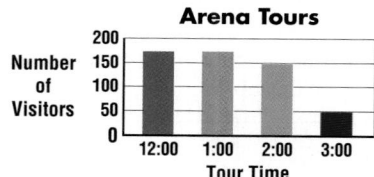

Arena Tours

How many people altogether toured the arena before 3:00 P.M.? **J**
   - **F** 275
   - **G** 350
   - **H** 400
   - **J** 500
   - **K** 550

Assessment and Evaluation Masters, p. 47

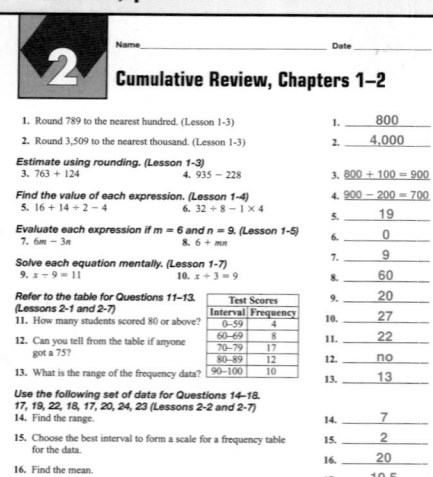

Name_____ Date_____

Cumulative Review, Chapters 1–2

1. Round 789 to the nearest hundred. (Lesson 1-3) 1. _**800**_
2. Round 3,509 to the nearest thousand. (Lesson 1-3) 2. _**4,000**_

Estimate using rounding. (Lesson 1-3)
3. $763 + 124$ 4. $935 - 228$ 3. _**800 + 100 = 900**_
                                  4. _**900 − 200 = 700**_

Find the value of each expression. (Lesson 1-4)
5. $16 + 14 \div 2 - 4$ 6. $32 \div 8 - 1 \times 4$ 5. _**19**_
                                                        6. _**0**_

Evaluate each expression if $m = 6$ and $n = 9$. (Lesson 1-5)
7. $6m - 3n$ 8. $6 + mn$ 7. _**9**_
                             8. _**60**_

Solve each equation mentally. (Lesson 1-7)
9. $x - 9 = 11$ 10. $x + 3 = 9$ 9. _**20**_
                                   10. _**27**_

Refer to the table for Questions 11–13.
(Lessons 2-1 and 2-7)
11. How many students scored 80 or above?   11. _**22**_
12. Can you tell from the table if anyone got a 75?   12. _**no**_
13. What is the range of the frequency data?   13. _**13**_

Test Scores	
Interval	Frequency
0–59	4
60–69	8
70–79	17
80–89	12
90–100	10

Use the following set of data for Questions 14–18.
17, 19, 22, 18, 17, 20, 24, 23 (Lessons 2-2 and 2-7)
14. Find the range.   14. _**7**_
15. Choose the best interval to form a scale for a frequency table for the data.   15. _**2**_
16. Find the mean.   16. _**20**_
17. Find the median.   17. _**19.5**_
18. Find the mode.   18. _**17**_

Use the graph for Questions 19–20.
(Lesson 2-4)
19. Which sport is most popular?   19. _**baseball**_
20. Which two sports add up to half of the sports listed?   20. _**football and soccer**_

Favorite Sport
Football 25%
Soccer 25%
Baseball 50%

© Glencoe/McGraw-Hill   47   *Mathematics: Applications and Connections, Course 1*

◀◀◀Instructional Resources

Another cumulative review is shown at the left and is available in the *Assessment and Evaluation Masters,* p. 47.

9. Jason bought 6 hockey tickets that ranged in price from $7 to $10. Which is a reasonable cost for all 6 tickets? **C**

   A $25

   B $40

   C $54

   D $70

   E $85

10. Serena's scout troop collected 4,230 aluminum cans for a service project. Scott's class collected 3,348 cans. How many more cans did Serena's scout troop collect than Scott's class? **H**

    F 756

    G 880

    H 882

    J 892

    K Not Here

11. The base price of a mini-van is $18,876. With an air conditioner, the mini-van costs $678 more. What is the price of the mini-van with an air conditioner? **E**

    A $19,454

    B $19,444

    C $18,954

    D $18,198

    E Not Here

12. Every class that sells 400 tickets to the school carnival earns a bowling party. Kylie's class sold 209 tickets. How many more tickets must her class sell to earn the bowling party? **H**

    F 101

    G 181

    H 191

    J 211

    K Not Here

Test-Taking Tip

You may be able to eliminate all or most of the answer choices by estimating. Also, look to see which answer choices are not reasonable for the information given in the problem.

Section Two: Free Response

This section contains four questions for which you will provide short answers. Write your answers on your paper.

13. Translate the sentence into an equation and solve. **7 + r = 22; 15**
    The sum of 7 and r is equal to 22.

14. Find the mode, median, and mean for the set of data. 72, 68, 59, 83, 74, 52
    no mode; 70,68

15. If $14 + p = 39$, what is the value of p?
    25

16. The line graph shows the number of houses sold in Mentor and Medina from 1960 to 1990.

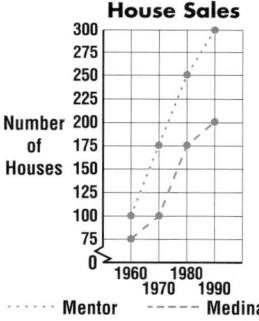

House Sales

About how many more houses were sold in Mentor than in Medina in 1980? **75**

Test-Taking Tip

Encourage students to review Chapter 1, Lesson 3, on how to estimate by rounding. Point out that this skill is very useful when taking a standardized test. It enables students to quickly eliminate certain answers, thereby saving time.

Assessment and Evaluation Masters, pp. 45–46

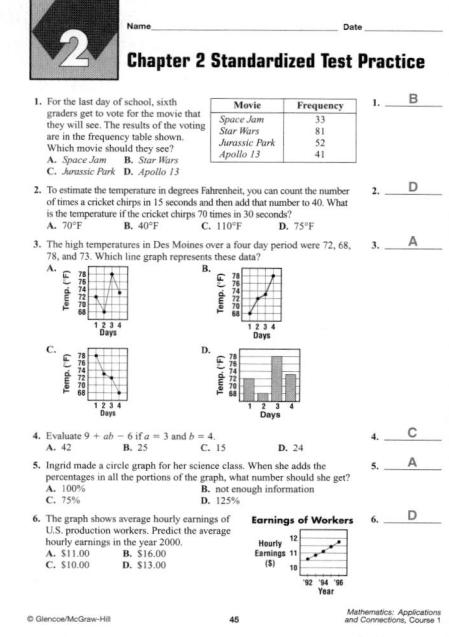

Instructional Resources ▶▶▶

Additional standardized test practice is shown at the right and is available in the *Assessment and Evaluation Masters*, pp. 45–46.

Adding and Subtracting Decimals

Previewing the Chapter

Overview

This chapter explores comparing, ordering, adding, and subtracting decimals through ten-thousandths. Students learn to measure length in metric units and to estimate decimal sums and differences using rounding and clustering. Students also learn to determine whether answers to problems are reasonable.

Lesson (pages)	Lesson Objectives	NCTM Standards	Standardized Tests	State/Local Objectives
3-1A (94)	Model decimals through hundredths.	1–6	CTBS, TN	
3-1 (95–98)	Model, read, and write decimals.	1–6, 13	CTBS, MAT, TN	
3-2A (100–101)	Measure length in metric units.	1–6, 13	CAT	
3-2 (102–104)	Show relationships among metric units of length and measure line segments.	1–6, 12, 13	CAT, MAT	
3-3 (105–108)	Compare decimals and order a set of decimals.	1–6, 10–12	ITBS, MAT, SAT	
3-4 (109–111)	Round decimals.	1–6, 12, 13	CTBS, ITBS, SAT, TN	
3-5 (112–115)	Estimate decimal sums and differences.	1–7, 10, 14	CAT, CTBS, ITBS, SAT, TN	
3-6A (116–117)	Determine whether answers to problems are reasonable.	1–7	CTBS, MAT, SAT, TN	
3-6 (118–121)	Add and subtract decimals.	1–7, 13	CAT, CTBS, ITBS, MAT, SAT, TN	

CAT = California Achievement Tests, CTBS = Comprehensive Tests of Basic Skills, ITBS = Iowa Tests of Basic Skills, MAT = Metropolitan Achievement Tests, SAT = Stanford Achievement Tests, TN = Terra Nova

Organizing the Chapter

CD-ROM

 All of the blackline masters in the Teacher's Classroom Resources are available on the **Electronic Teacher's Classroom Resources** CD-ROM.

LESSON PLANNING GUIDE

Lesson	Extra Practice (Student Edition)	BLACKLINE MASTERS (PAGE NUMBERS)										
		Study Guide	Practice	Enrichment	Assessment & Evaluation	Classroom Games	Diversity	Hands-On Lab	School to Career	Science and Math Lab Manual	Technology	Transparencies A and B
3-1A								41				
3-1	p. 563	17	17	17								3-1
3-2A								42				
3-2	p. 563	18	18	18	71							3-2
3-3	p. 563	19	19	19	70, 71			71			6	3-3
3-4	p. 564	20	20	20			3					3-4
3-5	p. 564	21	21	21	72							3-5
3-6A	p. 564											
3-6	p. 565	22	22	22	72	5–6				3	5	3-6
Study Guide/ Assessment					57–69, 73–75							

OTHER CHAPTER RESOURCES

Student Edition

Chapter Project, pp. 93, 115, 120, 125
Math in the Media, p. 98
School to Career, p. 99
Let the Games Begin, p. 121

Technology
CD-ROM Program
Interactive Mathematics Tools Software

Teacher's Classroom Resources

Applications
Family Letters and Activities, pp. 5–6
Investigations and Projects Masters, pp. 25–28
Meeting Individual Needs
Transition Booklet, pp. 5–8, 11–20, 29–30
Investigations for the Special Education Student, p. 7

Teaching Aids
Answer Key Masters
Block Scheduling Booklet
Lesson Planning Guide
Solutions Manual

Professional Publications
Glencoe Mathematics Professional Series

Planning the Chapter

MindJogger Videoquizzes provide a unique format for reviewing concepts presented in the chapter.

ASSESSMENT RESOURCES

Student Edition
Mixed Review, pp. 98, 104, 108, 111, 115, 121
Mid-Chapter Self Test, p. 111
➡ Math Journal, pp. 114, 119
Study Guide and Assessment, pp. 122–125
Performance Task, p. 125
📎 Portfolio Suggestion, p.125
Standardized Test Practice, pp. 126–127
Chapter Test, p. 597

Assessment and Evaluation Masters
Multiple-Choice Tests (Forms 1A, 1B, 1C), pp. 57–62
Free-Response Tests (Forms 2A, 2B, 2C), pp. 63–68
Performance Assessment, p. 69
Mid-Chapter Test, p. 70
Quizzes A–D, pp. 71–72
Standardized Test Practice, pp. 73–74
Cumulative Review, p. 75

Teacher's Wraparound Edition
5-Minute Check, pp. 95, 102, 105, 109, 112, 118
📎 Building Portfolios, p. 92
➡ Math Journal, pp. 94, 101
Closing Activity, pp. 98, 104, 108, 111, 115, 117, 121

Technology
🖥 Test and Review Software
📹 MindJogger Videoquizzes
💿 CD-ROM Program

MATERIALS AND MANIPULATIVES

Lesson 3-1A
base-ten blocks*

Lesson 3-1
base-ten blocks*

Lesson 3-2A
tape measure*

Lesson 3-2
tape measure*

Lesson 3-3
grid paper†
marker

Lesson 3-6
meterstick
number cubes*
spinners*†
counters*†

*Glencoe Manipulative Kit   †Glencoe Overhead Manipulative Resources

PACING CHART

See pages T25–T27 for the Course Planning Calendar.

COURSE	DAY 1	DAY 2	DAY 3	DAY 4	DAY 5	DAY 6	DAY 7
Standard	Chapter Project	Lessons 3-1A & 3-1		Lessons 3-2A & 3-2		Lesson 3-3	Lesson 3-4
Honors	Chapter Project	Lesson 3-1	Lesson 3-2	Lesson 3-3	Lesson 3-4	Lesson 3-5	Lesson 3-6A
Block	Chapter Project & Lesson 3-1A	Lessons 3-1 & 3-2A	Lessons 3-2 & 3-3	Lessons 3-4 & 3-5	Lessons 3-6A & 3-6	Study Guide and Assessment, Chapter Test	

The *Transition Booklet* (Skills 1–2, 4–8, 13) can be used to practice basic operations with whole numbers and review measuring length.

Interactive Mathematics:
Activities and Investigations

is an activity-based program that may be used as an enhancement for chapters in *Mathematics: Applications and Connections.*

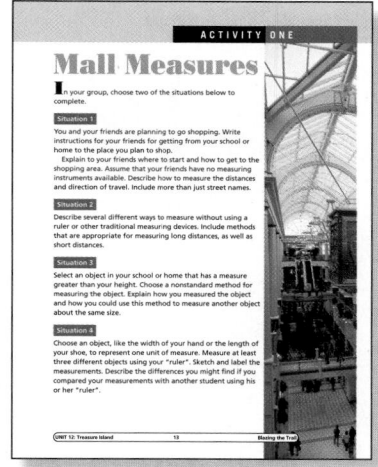

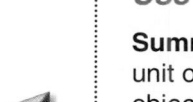

Unit 12, Activity One,
Mall Measures
Use with Lesson 3-2.

Summary Students work individually in groups to develop their own unit of linear measure and a system for using that unit to measure objects.

Math Connection Students use what they know about units of length and systems of measurement to create the "rules" for their new system of measure.

Unit 18, Activity One
Use with Lesson 3-6A.

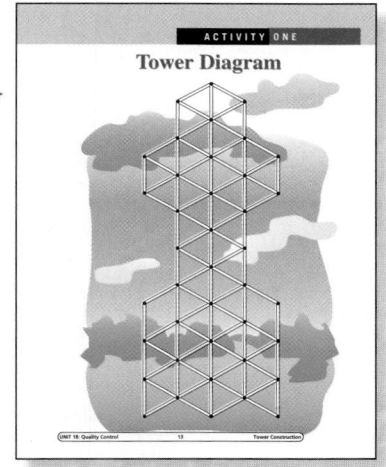

Summary Students work in groups as construction contractors. Each group estimates the time and amount of raw materials needed to build a tower. Then they construct a tower, record the amount of time it took to build the tower, and write a report.

Math Connection Students use estimation, measurement, geometry, spatial visualization, and accuracy to estimate how many construction materials are needed to build a tower. At every stage they need to determine how reasonable their measurements are.

DAY 8	DAY 9	DAY 10	DAY 11	DAY 12	DAY 13	DAY 14	DAY 15
Lesson 3-5	Lesson 3-6A	Lesson 3-6	Study Guide and Assessment	Chapter Test			
Lesson 3-6	Study Guide and Assessment	Chapter Test					

Enhancing the Chapter

APPLICATIONS

Classroom Games, pp. 5–6

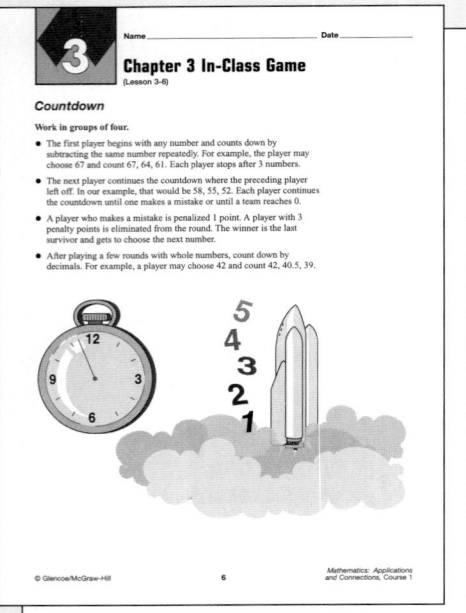

3 Name _____ Date _____

Chapter 3 In-Class Game
(Lesson 3-6)

Countdown

Work in groups of four.

- The first player begins with any number and counts down by subtracting the same number repeatedly. For example, the player may choose 67 and count 67, 64, 61. Each player stops after 3 numbers.

- The next player continues the countdown where the preceding player left off. In our example, that would be 58, 55, 52. Each player continues the countdown until one makes a mistake or until a team reaches 0.

- A player who makes a mistake is penalized 1 point. A player with 3 penalty points is eliminated from the round. The winner is the last survivor and gets to choose the next number.

- After playing a few rounds with whole numbers, count down by decimals. For example, a player may choose 42 and count 42, 40.5, 39.

© Glencoe/McGraw-Hill   6   *Mathematics: Applications and Connections, Course 1*

Diversity Masters, p. 3

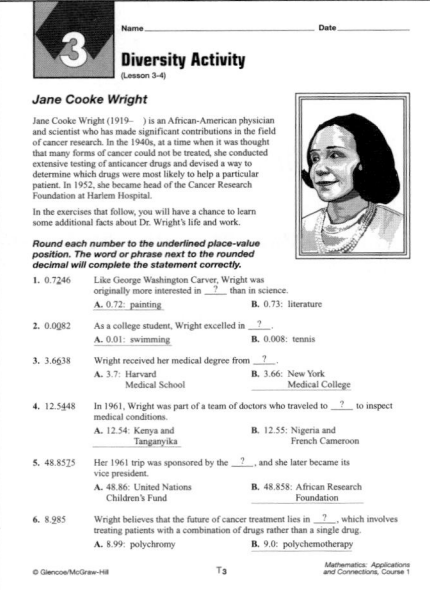

3 Name _____ Date _____

Diversity Activity
(Lesson 3-4)

Jane Cooke Wright

Jane Cooke Wright (1919–) is an African-American physician and scientist who has made significant contributions in the field of cancer research. In the 1940s, at a time when it was thought that many forms of cancer could not be treated, she conducted extensive testing of anticancer drugs and devised a way to determine which drugs were most likely to help a particular patient. In 1952, she became head of the Cancer Research Foundation at Harlem Hospital.

In the exercises that follow, you will have a chance to learn some additional facts about Dr. Wright's life and work.

Round each number to the underlined place-value position. The word or phrase next to the rounded decimal will complete the statement correctly.

1. 0.7246   Like George Washington Carver, Wright was originally more interested in ___?___ than in science.
   A. 0.72: painting   B. 0.73: literature

2. 0.0082   As a college student, Wright excelled in ___?___.
   A. 0.01: swimming   B. 0.008: tennis

3. 3.6638   Wright received her medical degree from ___?___.
   A. 3.7: Harvard   B. 3.66: New York
      Medical School   Medical College

4. 12.5448   In 1961, Wright was part of a team of doctors who traveled to ___?___ to inspect medical conditions.
   A. 12.54: Kenya and   B. 12.55: Nigeria and
      Tanganyika   French Cameroon

5. 48.8575   Her 1961 trip was sponsored by the ___?___, and she later became its vice president.
   A. 48.86: United Nations   B. 48.858: African Research
      Children's Fund   Foundation

6. 8.985   Wright believes that the future of cancer treatment lies in ___?___, which involves treating patients with a combination of drugs rather than a single drug.
   A. 8.99: polychromy   B. 9.0: polychemotherapy

© Glencoe/McGraw-Hill   T3   *Mathematics: Applications and Connections, Course 1*

School to Career Masters, p. 3

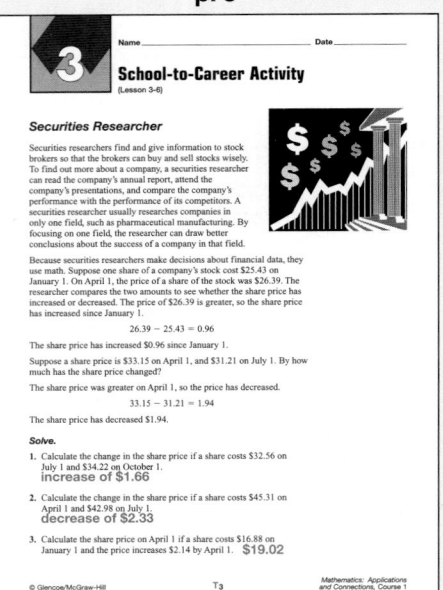

3 Name _____ Date _____

School-to-Career Activity
(Lesson 3-6)

Securities Researcher

Securities researchers find and give information to stock brokers so that the brokers can buy and sell stocks wisely. To find out more about a company, a securities researcher can read the company's annual report, attend the company's presentations, and compare the company's performance with the performance of its competitors. A securities researcher usually researches companies in only one field, such as pharmaceutical manufacturing. By focusing on one field, the researcher can draw better conclusions about the success of a company in that field.

Because securities researchers make decisions about financial data, they use math. Suppose one share of a company's stock cost $25.43 on January 1. On April 1, the price of a share of the stock was $26.39. The researcher compares the two amounts to see whether the share price has increased or decreased. The price of $26.39 is greater, so the share price has increased since January 1.

$$26.39 - 25.43 = 0.96$$

The share price has increased $0.96 since January 1.

Suppose a share price is $33.15 on April 1, and $31.21 on July 1. By how much has the share price changed?

The share price was greater on April 1, so the price has decreased.

$$33.15 - 31.21 = 1.94$$

The share price has decreased $1.94.

Solve.

1. Calculate the change in the share price if a share costs $32.56 on July 1 and $34.22 on October 1.
   increase of $1.66

2. Calculate the change in the share price if a share costs $45.31 on April 1 and $42.98 on July 1.
   decrease of $2.33

3. Calculate the share price on April 1 if a share costs $16.88 on January 1 and the price increases $2.14 by April 1.   **$19.02**

© Glencoe/McGraw-Hill   T3   *Mathematics: Applications and Connections, Course 1*

Family Letters and Activities, pp. 5–6

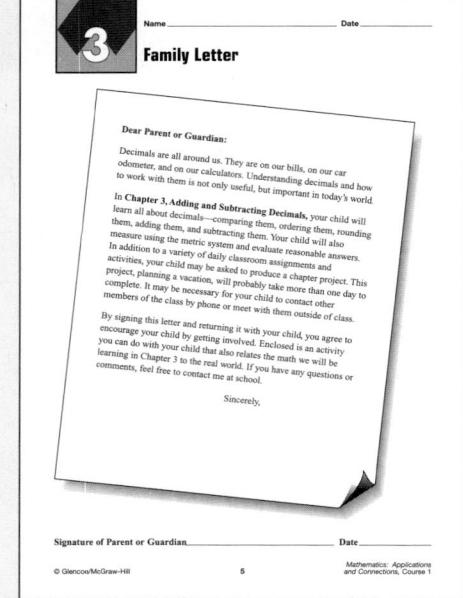

3 Name _____ Date _____

Family Letter

Dear Parent or Guardian:

Decimals are all around us. They are on our bills, on our car odometer, and on our calculators. Understanding decimals and how to work with them is not only useful, but important in today's world.

In **Chapter 3, Adding and Subtracting Decimals,** your child will learn all about decimals—comparing them, ordering them, rounding them, adding them, and subtracting them. Your child will also measure using the metric system and evaluate reasonable answers.

In addition to a variety of daily classroom assignments and activities, your child may be asked to produce a chapter project. This project, planning a vacation, will probably take more than one day to complete. It may be necessary for your child to contact other members of the class by phone or meet with them outside of class.

By signing this letter and returning it with your child, you agree to encourage your child by getting involved. Enclosed is an activity you can do with your child that also relates the math we will be learning in Chapter 3 to the real world. If you have any questions or comments, feel free to contact me at school.

Sincerely,

Signature of Parent or Guardian _____ Date _____

© Glencoe/McGraw-Hill   5   *Mathematics: Applications and Connections, Course 1*

3 Name _____ Date _____

Family Activity

Measuring Length in the Metric System

Work with a family member to answer the following questions. Write the unit of length, millimeter, centimeter, meter, or kilometer, that you would use to measure each of the following items. Then estimate each measure. When you can, use a centimeter ruler or a meterstick to test the accuracy of your estimate.

1. the width of your yard or a neighbor's yard

2. the height of a member of your family

3. the distance from your house to school

4. the height from the floor of your kitchen to the ceiling

5. the thickness of a pen in your home

6. the length of a box of cereal

7. the distance from your house to the grocery store

8. the width of a dollar bill

Estimates will vary: 1. meter 2. meter 3. kilometer 4. meter 5. millimeter 6. centimeter 7. kilometer (or meter) 8. centimeter

© Glencoe/McGraw-Hill   6   *Mathematics: Applications and Connections, Course 1*

Hands-On Lab Masters, p. 71

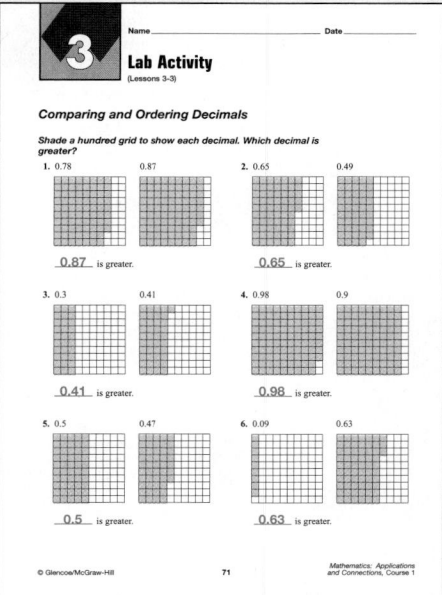

Assessment and Evaluation Masters, pp. 70–72

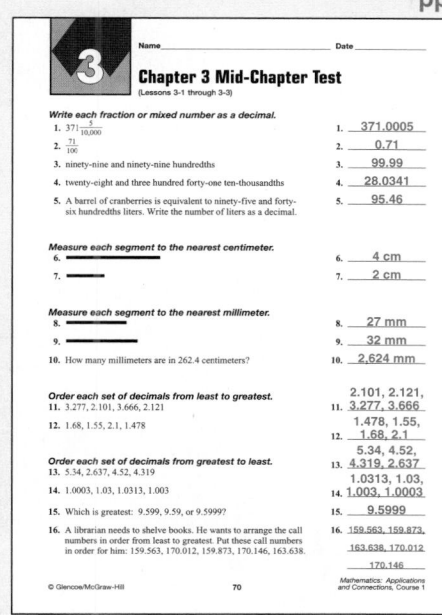

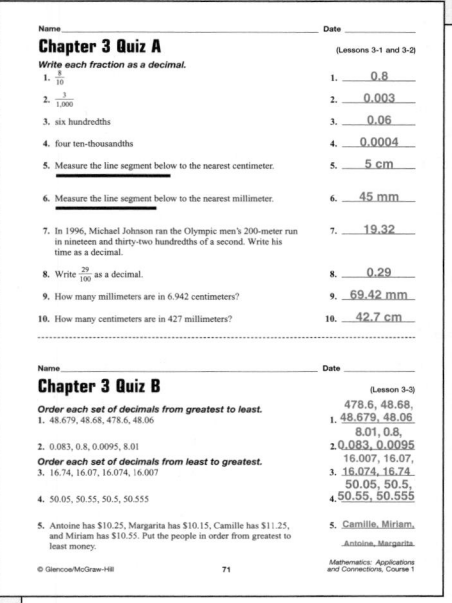

Technology Masters, pp. 5–6

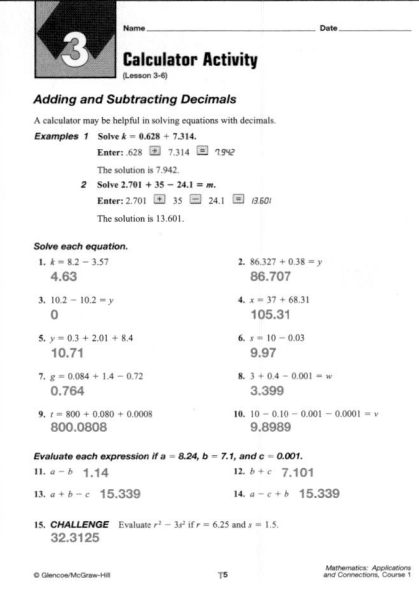

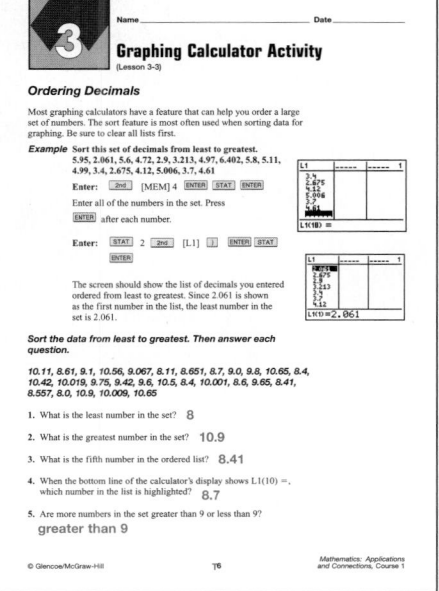

Investigations for the Special Education Student, p. 7

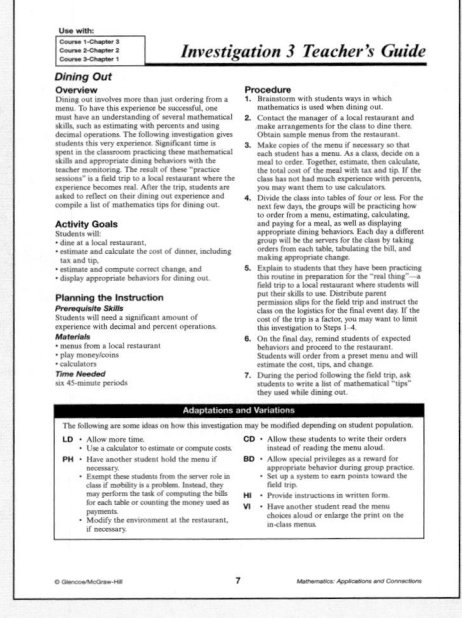

Theme: Travel

One of the most important parts of vacation planning is transportation.

It is a five- or six-hour flight from the West Coast to Honolulu International Airport. The flights may continue on to other islands, but just as often, students will need to arrange separate, inter-island flights. How much money each student spends on transportation depends on many factors, including how many people are in the party and how far in advance the ticket is purchased.

Question of the Day Caitlin, her brother, and her grandparents are flying to Hawaii from New York. Each ticket is $986.50, and additional tax is $47.81 per ticket. Estimate the cost for Caitlin's family to fly to Hawaii. **$4,200** Then use a calculator to find the exact cost. **$4,137.24**

Assess Prerequisite Skills

Ask students to read through the list of objectives presented in "What you'll learn in Chapter 3." You may wish to ask them what each of the objectives means or if they have experienced any of these types of problems before.

 Building Portfolios

Encourage students to revise their portfolios as they study this chapter. They may find some lessons more challenging than others and wish to include work that shows how they have improved their mathematical skills.

 **Math and the Family**

In the *Family Letters and Activities* booklet (pp. 5–6), you will find a letter to the parents explaining what students will study in Chapter 3. An activity appropriate for the whole family is also available.

CHAPTER 3 — Adding and Subtracting Decimals

What you'll learn in Chapter 3

- to model, read, write, compare, order, and round decimals,
- to show relationships between metric units of length and estimate and measure line segments,
- to determine whether answers are reasonable, and
- to estimate and find sums and differences of decimals.

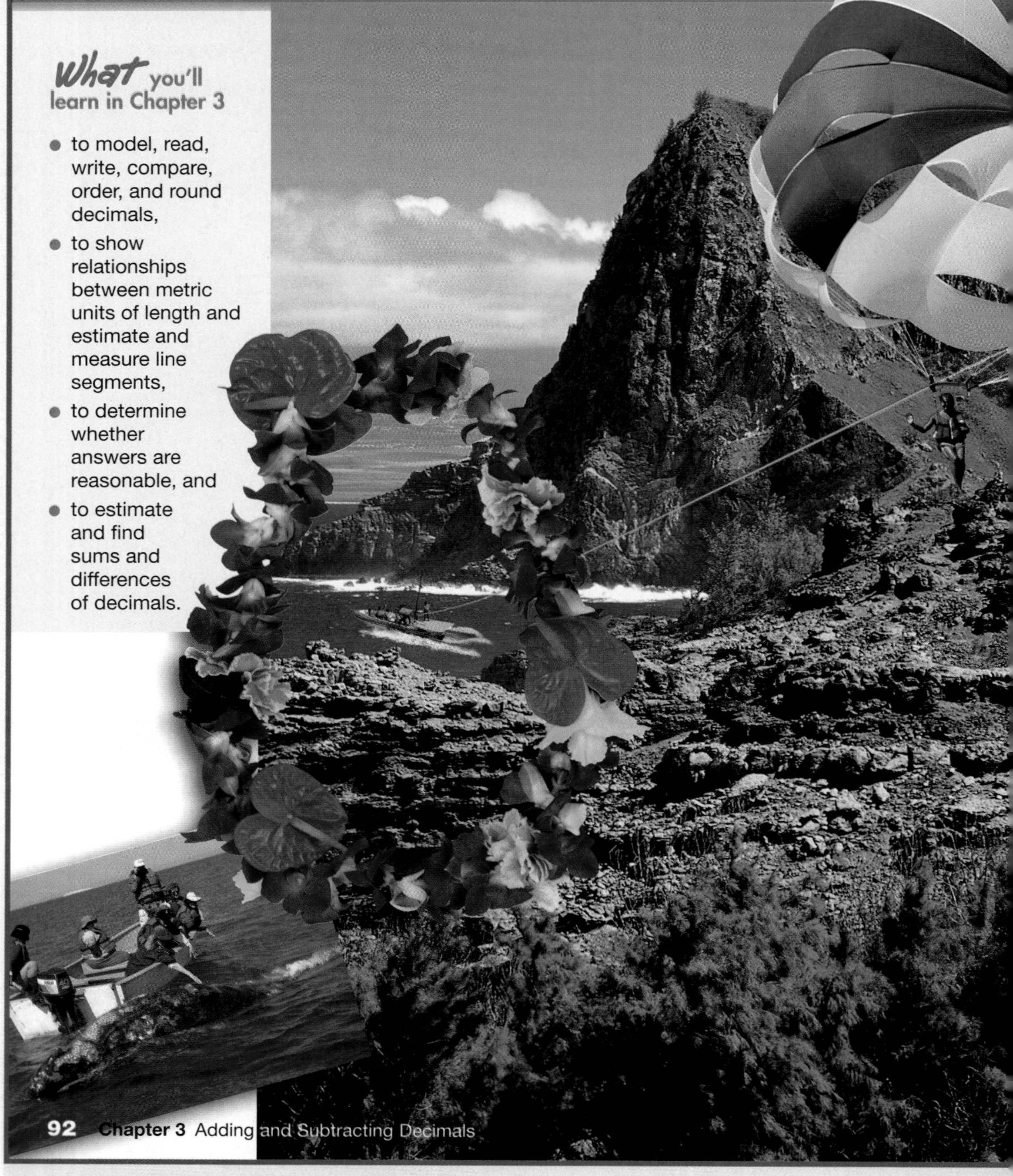

 **CD-ROM Program**

Activities for Chapter 3
- Chapter 3 Introduction
- Interactive Lessons 3-1A, 3-3
- Assessment Game
- Resource Lessons 3-1 through 3-6

CHAPTER Project

VACATION DESTINATION

In this project, you will add and subtract decimals to help you plan a one- or two-week vacation for your family. You will need to research the cost of transportation, lodging, food, and admission to any tourist attractions that you want to visit.

Getting Started

- Look at the table. Estimate the cost for two adults and two children to attend each of the three tours. Make a bar graph to compare your estimates.
- How could you estimate the cost for 4 people to attend the Hawaiian Luau?

Activity	Costs	
Whale Watching Tour	$27.50 (adult, over 12)	$15.00 (children, 12 and under)
Tropical Plantation Tour	$8.50 (adult, over 12)	$3.50 (children, 12 and under)
Bike Tour from Haleakala Crater	$59.00 (basic, all ages)	$99.00 (deluxe, all ages)
Snorkeling Cruise to Molokini	$33.00 (afternoon, all ages)	$39.95 (morning, all ages)
Hawaiian Luau	$44.95 (all ages)	
Parasailing	$29.95 (morning, all ages)	$39.95 (anytime, all ages)

Costs for Selected Activities on Maui, Hawaii

Technology Tips

- Use a **spreadsheet** to find the cost of your vacation.
- Use **computer software** to help you make graphs.
- Use a **word processor**.

interNET CONNECTION For up-to-date information on family vacations, visit:
www.glencoe.com/sec/math/mac/mathnet

Working on the Project

You can use what you'll learn in Chapter 3 to help you make your plans.

Page	Exercise
115	35
120	34
125	Alternative Assessment

---

interNET CONNECTION

Glencoe has made every effort to ensure that the website links for *Mathematics: Applications and Connections* at www.glencoe.com/sec/math/mac/mathnet are current and contain appropriate content. However, these website links are not under Glencoe's control.

Instructional Resources ▶▶▶
A recording sheet to help students organize their data for the Chapter Project is shown at the right and is available in the *Investigations and Projects Masters*, p. 28.

CHAPTER Project
NOTES

Objectives Students should
- understand rounding, estimating, adding, and subtracting decimals.
- be able to solve problems using decimals.

Project Pointer You may suggest that students begin a *Project Folder* to keep their work as they complete each stage of the Chapter Project. The completed project may also be added to their portfolios.

Students who use computer software for their graphs should think carefully about their scale so that the graph is not misleading. You may want them to sketch a graph on paper first, so that they know what a reasonable graph will be for their data.

Using the Table Students need to be careful reading the table. Some of the costs are different by age and some are not. Students may want to explore why that difference exists. Also, they may want to investigate why some activities are less expensive during certain times of the day. Emphasize that all costs are per person.

***Investigations and Projects Masters*, p. 28**

3 Chapter 3 Project

Name _____ Date _____

Vacation Destination

Page 93, Getting Started

Activity	Estimated Costs for 2 Adults and 2 Children	
Whale Watching Tour		
Tropical Plantation Tour		
Bike Tour	(basic)	(deluxe)

You can estimate the cost of 4 people attending the Hawaiian Luau by:

Page 115, Working on the Chapter Project, Exercise 35
a. The estimated total cost for Mandy's family to go on the Tropical Plantation Tour is: Can her family go for less than $35?
b. Meal One Estimates:
   Meal Two Estimates:
   Meal Three Estimates:
   Day Total:

Page 120, Working on the Chapter Project, Exercise 34
Whale Watching Tour:
Hawaiian Luau:
Total:

© Glencoe/McGraw-Hill   28   *Mathematics: Applications and Connections, Course 1*

GET READY

Objective Students model decimals through hundredths.

Optional Resources
Hands-On Lab Masters
• base-ten models, p. 1
• worksheet, p. 41

Overhead Manipulative Resources
• decimal models

Manipulative Kit
• base-ten blocks
CD-ROM Program
• Interactive Lesson 3-1A

MANAGEMENT TIPS

Recommended Time
30 minutes

Getting Started Have students read the first paragraph and identify the base-ten models. Lead them to think of the ones block shown as a ones block that has been divided into 100 equal parts. Then ask the following questions.
• How would you describe the tenths and hundredths blocks?
• When would you use these numbers?

Activities 1 and 2 involve trading tenths blocks and hundredths blocks. When modeling the trade for students, have them describe your actions and justify the trade of tenths for hundredths. Have them determine the result before counting the hundredths.

ASSESS

Before modeling Exercises 1–3, encourage students to predict the answers. Have them describe their prediction methods. Discuss whether their predictions were right or wrong and why.

Use Exercise 4 to determine whether students understand modeling decimals with base-ten blocks.

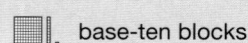

HANDS-ON LAB   COOPERATIVE LEARNING

3-1A Decimals Through Hundredths

A Preview of Lesson 3-1

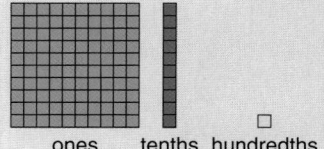 base-ten blocks

You know that base-ten blocks can be used to model whole numbers. Did you know that they can also be used to model decimals? For decimals, the blocks have the meanings shown at the right.

ones   tenths   hundredths

TRY THIS

Work in groups of four.

1 Model three tenths by using base-ten blocks.
• You can also model three tenths by trading the tenths for hundredths. How many hundredths are there?

2 Show four tenths and seven hundredths with base-ten blocks. What decimal have you modeled?

• Trade the tenths for hundredths and count the hundredths blocks.
• There are 47 blocks, so 47 hundredths is the decimal modeled.

ON YOUR OWN

1. Three tenths is the same as how many hundredths? **thirty hundredths**
2. Show six tenths and four hundredths with base-ten blocks. Trade the tenths for hundredths. How many hundredths do you have now? **See margin.**
3. How many tenths are the same as ninety hundredths? Model using base-ten blocks. **9 tenths; See Answer Appendix for model.**
4. **Look Ahead** If you separated a hundredth block into 100 equal parts, what decimal would be modeled by seventeen of the new parts? **17 ten-thousandths**

94   **Chapter 3** Adding and Subtracting Decimals

Additional Answer
2. See students' work; 64 hundredths

 Math Journal

Have students write a description of how base-ten blocks can be used to model decimals. Make sure the description is detailed enough so that they can use it as a reference later.

Decimals Through Ten-Thousandths

3-1 Lesson Notes

What you'll learn

You'll learn to model, read, and write decimals.

When am I ever going to use this?

You can use decimals to record winning times at a track meet.

Word Wise

place value

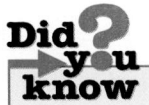

Did you know The duration of the average blink is about three tenths of a second.

Close your eyes and then open and close them as fast as you can. You have just modeled what happens when a camera takes a picture. The time a 35-millimeter camera's shutter stays open can range from a second to less than a millisecond.

A millisecond is $\frac{1}{1,000}$ of a second. What are some other ways to express $\frac{1}{1,000}$?

Decimals are another way to write fractions when the denominators are 10, 100, 1,000, and so on. We use **place-value** positions to name decimals.

In a decimal, the decimal point separates the ones place and the tenths place. The fraction $\frac{1}{1,000}$ can be written as the decimal 0.001 or in words as *one-thousandth*.

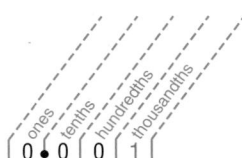

Example 1

INTEGRATION

Statistics Charmaine asked 100 students at Western Hills Middle School to identify their favorite brand name. 42 out of 100 or $\frac{42}{100}$ said their favorite is Nike.

a. Model the decimal for $\frac{42}{100}$.

Display forty-two hundredths blocks. Trade forty blocks for four tenths blocks.

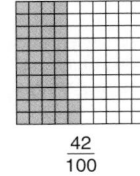

$\frac{42}{100}$

b. Write the decimal in a place-value chart.

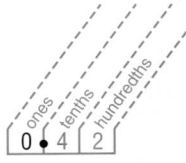

There are four tenths and two hundredths in the model. Write 4 in the tenths place and 2 in the hundredths place.

c. Write the decimal in words.

Notice that the last digit, 2, is in the hundredths place. The decimal in words is *forty-two hundredths*.

Instructional Resources

- *Study Guide Masters*, p. 17
- *Practice Masters*, p. 17
- *Enrichment Masters*, p. 17
- Transparencies 3-1, A and B
- CD-ROM Program
  - Resource Lesson 3-1

Recommended Pacing	
Standard	Days 2 & 3 of 12
Honors	Day 2 of 10
Block	Day 2 of 6

1 FOCUS

5-Minute Check *(Chapter 2)*

Refer to the table for Exercises 1–3.

Camcorder Prices	
1988	$1,025
1989	$750
1990	$880
1991	$1,060
1992	$1,150

1. What did the camcorder cost in 1992? **$1,150**

2. Find the mean, median, and mode of the prices. **$973; $1,025; no mode**

3. Make a line graph for the data in the table.

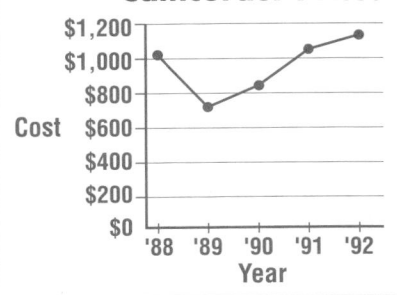

Camcorder Prices

The 5-Minute Check is also available on **Transparency 3-1A** for this lesson.

Cross-Curriculum Cue

Inform the teachers on your team that your students are studying decimals. Suggestions for curriculum integration are:
Health: body temperature
Geography: distance
Earth Science: precipitation
Physical Education: speed

Multiple Learning Styles

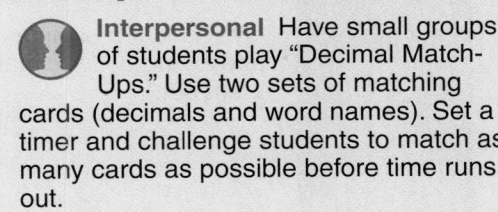

Interpersonal Have small groups of students play "Decimal Match-Ups." Use two sets of matching cards (decimals and word names). Set a timer and challenge students to match as many cards as possible before time runs out.

Motivating the Lesson
Hands-On Activity Have students create designs using centimeter grid paper and colored markers or crayons. Give them directions for coloring such as, *Color six hundredths red.*

2 TEACH

Transparency 3-1B contains a teaching aid for this lesson.

Reading Mathematics When looking at part c of Example 1, point out that when reading a decimal, the last digit gives the place-value name.

In-Class Examples

For Example 1
One hundred birds live in a bird sanctuary. Fifteen are parrots.
a. Write the decimal for the model. **0.15**

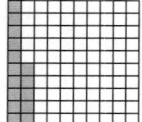

b. Write the decimal in a place-value chart.

c. Write the decimal in words. **fifteen hundredths**

For Example 2
a. Model the decimal for $1\frac{3}{10}$.

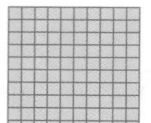 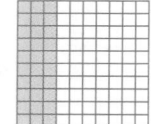

b. Write the decimal in a place-value chart.

c. Write the decimal in words. **one and three tenths**

You can also write mixed numbers, such as $3\frac{7}{10}$, as decimals.

Example 2
a. Model the decimal for $3\frac{7}{10}$.
b. Write the decimal in a place-value chart.
c. Write the decimal in words.

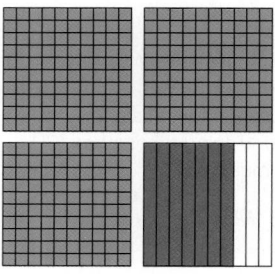

a. Display three ones and seven tenths blocks.

Study Hint
Reading Math The mixed number $3\frac{7}{10}$ is read as *three and seven tenths.* The number 3.7 is also read as *three and seven tenths.*

b. Write 3 in the ones place and 7 in the tenths place.

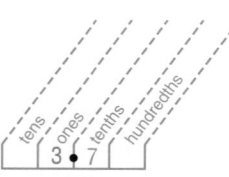

c. The decimal is *three and seven tenths.*

To write very small decimals, you can extend the place-value chart farther to the right.

Example 3
CONNECTION

Life Science The egg of the Vervain hummingbird weighs about $\frac{128}{10,000}$ ounce.

a. Write this fraction as a decimal in a place-value chart.
b. Write the decimal in words.

a.

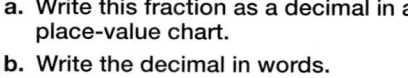

b. The last digit, 8, is in the ten-thousandths place. So the decimal is *one hundred twenty-eight ten-thousandths.*

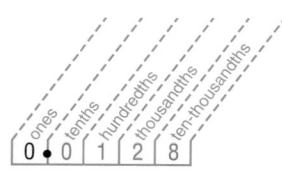

Classroom Vignette

"We have fun using a Koosh ball to represent a decimal point and acting out decimal numbers as I or another student read them out."

Patricia A. Maestranzi, Teacher
Woodbury School
Salem, NH

Patricia A. Maestranzi

CHECK FOR UNDERSTANDING

Communicating Mathematics

Read and study the lesson to answer each question.

3. Mercedes is correct because the last digit, 8, is in the thousandths place.

1. *Write* the fraction, decimal, and word name for the decimal shown by the model. **0.05,** $\frac{5}{100}$**, five hundredths**

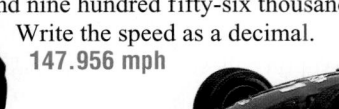

2. *Draw* a model for 2.25. **See Answer Appendix.**

3. *You Decide* Elizabeth says that thirty-eight thousandths is written as 0.0038. Mercedes says that it is written as 0.038. Who is correct? Explain.

Guided Practice

Write each fraction or mixed number as a decimal.

4. $\frac{9}{10}$ **0.9**   5. $\frac{43}{100}$ **0.43**   6. $\frac{563}{1,000}$ **0.563**   7. $4\frac{29}{10,000}$ **4.0029**

8. Write four and three tenths as a decimal. **4.3**

9. Write seven hundred two ten-thousandths as a decimal. **0.0702**

10. *Auto Racing* In 1996, Buddy Lazier of Vail, Colorado, won the Indianapolis 500 with an average speed of one hundred forty-seven and nine hundred fifty-six thousandths miles per hour. Write the speed as a decimal. **147.956 mph**

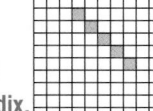
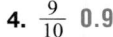

EXERCISES

Practice

19. 47.047
20. 3.0027
21. 172.0001

Write each fraction or mixed number as a decimal.

11. $\frac{3}{10}$ **0.3**   12. $\frac{99}{100}$ **0.99**   13. $\frac{7}{100}$ **0.07**   14. $10\frac{1}{10}$ **10.1**

15. $19\frac{53}{100}$ **19.53**   16. $\frac{375}{1,000}$ **0.375**   17. $\frac{9}{1,000}$ **0.009**   18. $\frac{5,561}{10,000}$ **0.5651**

19. $47\frac{47}{1,000}$ 20. $3\frac{27}{10,000}$ 21. $172\frac{1}{10,000}$ 22. $\frac{2,384}{1,000}$ **2.384**

Write each expression as a decimal.

23. twenty and nine tenths **20.9**

24. eleven hundredths **0.11**

25. three and three thousandths **3.003**

26. one hundred nine and fifteen thousandths **109.015**

27. eight hundred one ten-thousandths **0.0801**

28. 610.0306

28. six hundred ten and three hundred six ten-thousandths

Applications and Problem Solving

29. *Industrial Technology* A micrometer caliper is a device used to measure the thickness of an object. Many metric micrometer calipers can measure to one hundredth of a millimeter. Write one hundredth as a decimal. **0.01**

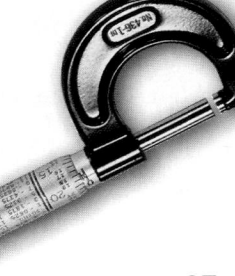

Lesson 3-1 Decimals Through Ten-Thousandths **97**

■ Reteaching the Lesson ■

Activity Display a place-value chart on an overhead transparency. Write a decimal in the place-value chart. Have students model the decimal using base-ten blocks. Students may also work in pairs to challenge each other with decimals to model.

In-Class Example

For Example 3

A hockey rink is about $\frac{379}{10,000}$ of a mile long.

a. Write the fraction as a decimal in a place-value chart.

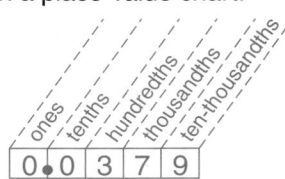

ones	tenths	hundredths	thousandths	ten-thousandths
0	0	3	7	9

b. Write the decimal in words. **three hundred seventy-nine ten-thousandths**

3 PRACTICE/APPLY

Check for Understanding

If students need additional practice or instruction after completing Exercises 1–10, you may find one of the following options helpful.

• Extra Practice, see p. 563
• Reteaching Activity
• *Transition Booklet,* pp. 5–6
• *Study Guide Masters,* p. 17
• *Practice Masters,* p. 17
• Interactive Mathematics Tools Software

Assignment Guide

Core: 11–29 odd, 31–35
Enriched: 12–28 even, 29–35

Study Guide Masters, **p. 17**

3-1 **Study Guide**

Name_____ Date_____

Decimals Through Ten-Thousandths

Fraction: $\frac{5,016}{10,000}$ Decimal: 0.5016

Say: five thousand sixteen ten-thousandths

Ones	Tenths	Hundredths	Thousandths	Ten-thousandths
0	5	0	1	6

Here are some other examples.

Fraction	Decimal	Words
$\frac{924}{1,000}$	0.924	Nine hundred twenty-four thousandths
$5\frac{7}{10}$	5.7	Five and seven tenths

Write each fraction as a decimal.

1. $\frac{31}{100}$ **0.31**   2. $\frac{9}{100}$ **0.09**   3. $\frac{4}{10,000}$ **0.0004**   4. $\frac{35}{1,000}$ **0.035**

5. $\frac{1,654}{10,000}$ **0.1654**   6. $\frac{1}{10}$ **0.1**   7. $\frac{6}{1,000}$ **0.006**   8. $\frac{3}{10}$ **0.3**

Write each expression as a decimal.

9. two hundred fifty-one thousandths **0.251**   10. one and eleven hundredths **1.11**

11. eight hundredths **0.08**   12. seventy and fifty-six thousandths **70.056**

13. five hundred two ten-thousandths **500.0002**   14. thirty-six ten-thousandths **0.0036**

© Glencoe/McGraw-Hill   T17   *Mathematics: Applications and Connections, Course 1*

Closing Activity

Speaking Write several decimals on the chalkboard. Give a clue such as "Find the decimal that has a six in the thousandths place." The student who names the decimal gets to give a clue for another decimal, and so on.

30. four hundred fifty-five and ninety-eight hundredths dollars

Mixed Review

34. An interval of 5 would be best to form a scale because 1 is too small and 100 is too big.

30. **Banking** As a safeguard against error, the dollar amount on a check is written in both standard form and in words. Write $455.98 in words.

31. **Critical Thinking** Use the digits 0, 0, 5, 7 to make the greatest possible decimal and the least possible decimal. **greatest: 0.750; least: 0.057**

32. **Geometry** *(Lesson 2-9)*
   a. Which of the points on the graph has the coordinates (0, 4)? **M**
   b. Write the coordinates of point *N*. **(4, 3)**

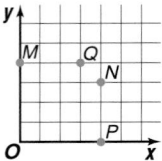

33. **Test Practice** Seven salespeople at Kip's Keyboards sold 13, 14, 10, 6, 3, 25, and 13 keyboards during the month of February. What is the mean for this set of data? *(Lesson 2-7)* **B**

   A 3　　　**B** 12　　　**C** 13　　　**D** 25

34. **Statistics** Tammy took a survey of ten of her classmates to find the number of hours each person reads for pleasure each week. The results were 3, 0, 6, 10, 2, 20, 19, 8, 2, and 4. Would an interval of 1, 5, or 100 be best to form a frequency table of the data? Explain. *(Lesson 2-2)*

35. **Money** Martin collected $38 for the holiday children's fund. Phyllis collected half as much. Use the four-step plan to determine how much money Phyllis collected. *(Lesson 1-1)* **$19**

MATH ⟩ IN THE MEDIA

Tiger *by Bud Blake*

2. No; Ten-thousandths as a decimal is 0.010. The 1 is not in the tenths place.

1. Write one tenth as a decimal. **0.1**

2. Suppose Hugo had said he learned ten thousandths of what he was supposed to. Would he be talking about the same amount? Explain.

Practice Masters, p. 17

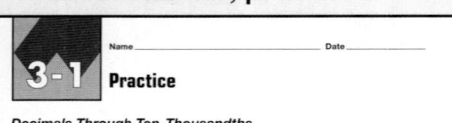

3-1 **Practice**

Decimals Through Ten-Thousandths

Write each fraction or mixed number as a decimal.

1. $\frac{8}{10}$ 0.8　　2. $\frac{19}{100}$ 0.19　　3. $\frac{7}{100}$ 0.07　　4. $\frac{26}{100}$ 0.26

5. $\frac{32}{100}$ 0.32　　6. $\frac{5}{100}$ 0.05　　7. $\frac{4}{10}$ 0.4　　8. $\frac{11}{100}$ 0.11

9. $\frac{6}{100}$ 0.06　　10. $\frac{48}{100}$ 0.48　　11. $\frac{93}{100}$ 0.93　　12. $\frac{2}{10}$ 0.2

13. $\frac{407}{1,000}$ 0.407　　14. $\frac{9}{1,000}$ 0.009　　15. $\frac{2351}{10,000}$ 0.2351

16. $\frac{63}{10,000}$ 0.0063　　17. $\frac{742}{1,000}$ 0.742　　18. $\frac{8}{10,000}$ 0.0008

19. $\frac{914}{10,000}$ 0.0914　　20. $\frac{3,806}{10,000}$ 0.3806　　21. $\frac{59}{1,000}$ 0.059

Write each expression as a decimal.

22. thirteen hundredths 0.13

23. two and forty-nine hundredths 2.49

24. six and eight hundredths 6.08

25. thirty-nine and two tenths 39.2

26. eighty-three hundredths 0.83

27. seven tenths 0.7

28. forty-five and two ten-thousandths 45.0002

29. thirty-one thousandths 0.031

30. four thousandths 0.004

31. twelve and nine hundred five ten-thousandths 12.0905

© Glencoe/McGraw-Hill　　T17　　*Mathematics: Applications and Connections, Course 1*

■ Extending the Lesson ■

Enrichment Masters, p. 17

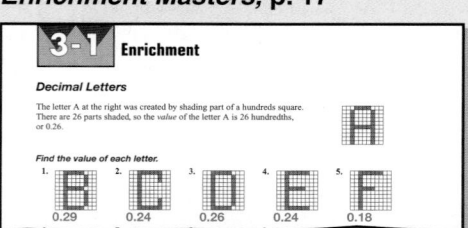

3-1 **Enrichment**

Decimal Letters

The letter A at the right was created by shading part of a hundreds square. There are 26 parts shaded, so the *value* of the letter A is 26 hundredths, or 0.26.

Find the value of each letter.

1. B 0.29　2. C 0.24　3. D 0.26　4. E 0.24　5. F 0.18

MATH ⟩ IN THE MEDIA

Review with students different decimals that are equal to one tenth. For example, ten hundredths or one hundred thousandths.

Aviation

Mayte Greco
PILOT

Mayte Greco took flying lessons as a teenager. Now, the mother of five children, she is a pilot and owns her own air charter company in Florida. She also volunteers one day a week to search for Cuban refugees on rafts. She then radios the Coast Guard and waits until they come to the rescue.

Most companies that employ pilots require at least two years of college. To become a pilot for an airline company, you will need a college degree. Courses in engineering, meteorology, physics and mathematics are helpful in preparing for a pilot's career.

I'd like to fly people to their favorite vacation spots. I could stop there for a day or two myself!

For more information:
Air Line Pilots Association
535 Herndon Parkway
P.O. Box 1169
Herndon, VA 22070

interNET
CONNECTION
www.glencoe.com/sec/
math/mac/mathnet

Your Turn
Research the future of careers in flying. Will more pilots be needed in the next 10 to 20 years? What are the salaries for pilots?

School to Career: Aviation **99**

More About Mayte Greco
- Mayte Greco's parents left Cuba following the revolution led by Fidel Castro in 1958. Mayte was a baby when her parents settled in Miami, Florida. When she was growing up, her father would take her to the airport to watch airplanes coming and going.
- To become a pilot, Mayte Greco had to earn her "wings." A pilot's license requires hours of practice with a trained instructor, passing a written examination, and at least forty hours of flying time.

Motivating Students
As students learned in the chapter opener on pages 92–93, vacations often include a trip on an airplane. Aviation is an important part of recreation and entertainment, as well as business, trade, and government. There are many opportunities to explore in the aviation industry. Students may be interested in a career in that or a related field. To start the discussion, you may ask students questions about aviation history.
- When was the first airplane flight and who achieved it? **December 17, 1903, at Kitty Hawk, North Carolina; Wilbur and Orville Wright**
- When did the first jet airplane fly? **August 1939**
- How fast does the supersonic jet *Concorde* fly? **about 1,200 mph**

Making the Math Connection
When small planes ascend and descend, the barometric pressure changes by small amounts. These changes affect the altimeter, or altitude, setting. The pilot needs to make adjustments by adding and subtracting the pressures, which are measured in decimals, to and from the altimeter readings.

Working on *Your Turn*
Remind students that a career in aviation is not restricted to being a pilot. There are varied jobs associated with aviation that they can check out. These include working for commercial airlines, federal and state aviation boards, the U.S. Armed Forces, and NASA, or working in airport management or aerospace research, design, or manufacturing.

*An additional School to Career activity is available on page 3 of the **School to Career Masters**.*

School to Career **99**

Objective Students measure length in metric units.

Optional Resources
Hands-On Lab Masters
• worksheet, p. 42
Manipulative Kit
• tape measure

MANAGEMENT TIPS

Recommended Time
45 minutes

Getting Started Guide the class in measuring the width of a chair seat. Help them find the answer in meters, centimeters, and millimeters.

HANDS-ON LAB

COOPERATIVE LEARNING

3-2A Measurement

A Preview of Lesson 3-2

tape measure

The basic unit of length in the metric system is the *meter*. All other metric units of length are defined in terms of the meter.

The chart below summarizes the most commonly used metric units of length.

Unit	Symbol	Meaning
millimeter	mm	thousandth
centimeter	cm	hundredth
meter	m	one
kilometer	km	thousand

A metric ruler or tape measure is easy to read. The ruler below is labeled using *centimeters*.

The pencil below is about 12 centimeters long.

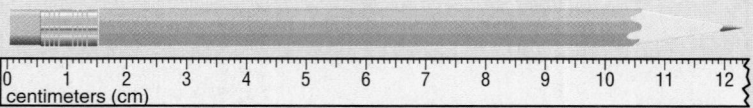

To read *millimeters*, you count each individual unit or mark on the metric ruler. There are ten millimeter marks for each centimeter mark. The pencil is about 124 millimeters long.

$$124 \text{ mm} = 12.4 \text{ cm}$$

There are 100 centimeters in a meter. Since there are 10 millimeters in one centimeter, there are 10×100 or 1,000 millimeters in a meter. The pencil is $\frac{124}{1,000}$ of a meter or 0.124 meters long.

$$124 \text{ mm} = 0.124 \text{ m}$$

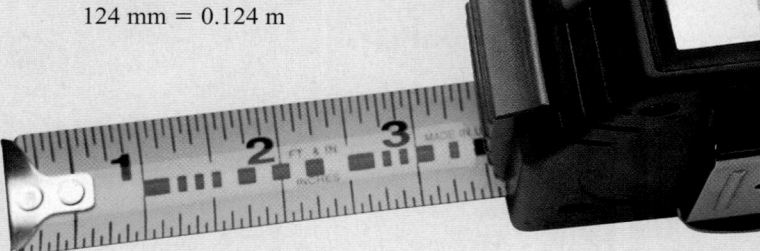

100 Chapter 3 Adding and Subtracting Decimals

TRY THIS

Work in groups of four.

Copy the table. Using a tape measure, measure the objects and complete the table.

See students' work.

Object	Measure		
	m	cm	mm
height of door	X		
width of door		X	
length of classroom	X		
length of math book		X	
length of pencil		X	
length of table or desk		X	
width of hallway	X		
length of sheet of paper		X	
length of chalkboard eraser		X	
length of your hand		X	
width of your little finger			X

ON YOUR OWN

1. On your table, mark the unit of measure that is most appropriate for each item. **See table.**

2. What pattern do you notice in the relationship between the measure you marked and the size of the object?

3. What patterns do you notice in the relationship between the numbers in the columns? **See margin.**

4. Look around your classroom. Select three objects that you think would be best measured in meters, three objects you think would be best measured in millimeters, and three objects you think would be best measured in centimeters. Explain your choices. **See margin.**

5. *Look Ahead* Copy the table below. Write the name of a common object that you think has a length that corresponds to each length in the first column. **Answers will vary.**

Length	Object
5 centimeters	
15 centimeters	
3 meters	
1 meter	
75 centimeters	

2. Answers will vary. Sample answer: The larger the object, the larger the unit that should be used.

Lesson 3-2A HANDS-ON LAB **101**

ASSESS

Have members of each group present some of their measurements to the class and what they chose as their appropriate unit of measure for that object. Discuss the appropriateness of the chosen unit.

Use Exercise 5 to determine whether students understand *appropriate units of measure*.

Additional Answers
3. **Students should notice that the measures are related to powers of 10 and start to notice how decimals are used.**
4. **Answers will vary, but students should select larger objects for using meters and smaller objects when using centimeters and millimeters.**

Hands-On Lab Masters, p. 42

 Math Journal Have students write a paragraph about how to choose the most appropriate unit of measure and why this is important.

Hands-On Lab 3-2A **101**

Instructional Resources

- *Study Guide Masters*, p. 18
- *Practice Masters*, p. 18
- *Enrichment Masters*, p. 18
- Transparencies 3-2, A and B
- *Assessment and Evaluation Masters*, p. 71

 CD-ROM Program
- Resource Lesson 3-2

Recommended Pacing

Standard	Days 4 & 5 of 12
Honors	Day 3 of 10
Block	Day 3 of 6

1 FOCUS

 5-Minute Check
(Lesson 3-1)

Write each fraction or mixed number as a decimal.

1. $\frac{64}{1,000}$ **0.064**

2. $\frac{319}{10,000}$ **0.0319**

3. $5\frac{4}{10}$ **5.4**

Write each expression as a decimal.

4. thirty-eight thousandths
0.038

5. twenty-nine and six ten-thousandths **29.0006**

 The 5-Minute Check is also available on **Transparency 3-2A** for this lesson.

Motivating the Lesson

Communication Ask students how tall they are in centimeters and how they might find this out.

2 TEACH

Transparency 3-2B contains a teaching aid for this lesson.

Using the Mini-Lab Have students estimate heights of things in the classroom and compare these heights in meters and centimeters with their own heights.

What you'll learn

You'll learn to show relationships among metric units of lengths and to measure line segments.

When am I ever going to use this?

Metric units of length are used in cross country running, swimming, and track and field events.

Word Wise

metric system
meter

Integration: Measurement
Length in the Metric System

In the 1996 Summer Olympics, Michael Johnson became the first man to win both the 200-meter and 400-meter races.

Most Olympic events use the **metric system** to measure lengths. The basic unit of length in the metric system is the **meter**. All other metric units of length are defined in terms of the meter.

Decimals are used in the metric system. The chart summarizes the most commonly used metric units of length.

Unit	Symbol	Meaning	Size	Model
millimeter	mm	thousandth	0.001 m	thickness of a dime
centimeter	cm	hundredth	0.01 m	width of large paperclip
meter	m	one	1.0 m	width of doorway
kilometer	km	thousand	1,000 m	six city blocks

Examples

Write the unit of length: millimeter, centimeter, meter or kilometer, that you would use to measure each of the following. Then estimate each measure.

1 thickness of a nickel

Since a nickel is thin, the *millimeter* is the appropriate unit. The thickness of a nickel is a little more than the thickness of a dime, so the thickness of a nickel is about 2 millimeters.

2 length of a baseball bat

Since the length of a baseball bat is close to the width of a doorway, the *meter* is the appropriate unit. The length of a baseball bat is about 1 meter.

3 On the map, what is the distance between San Antonio and Houston in centimeters? in millimeters?

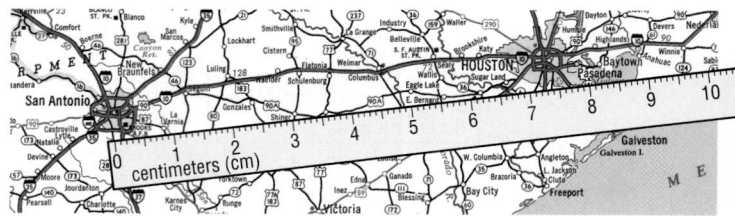

The distance is about 7.3 centimeters, or 73 millimeters.

Multiple Learning Styles

 Kinesthetic Separate the class into groups, and have each group estimate the length of some part of the school. Each group member should measure and record that length using a different body part. Then have them measure that body part with an appropriate metric device. For example, one person might measure the hallway with his or her foot. Another might measure it with an arm. The foot and arm are then measured to the nearest centimeter. Find the length of the hallway using that data to create a chart presenting the various measurements.

Example 4
CONNECTION

Life Science Use a centimeter ruler to measure the length of the body of the chameleon as shown in the photo below.

In the photo, the length of the body of the chameleon is about 9 centimeters.

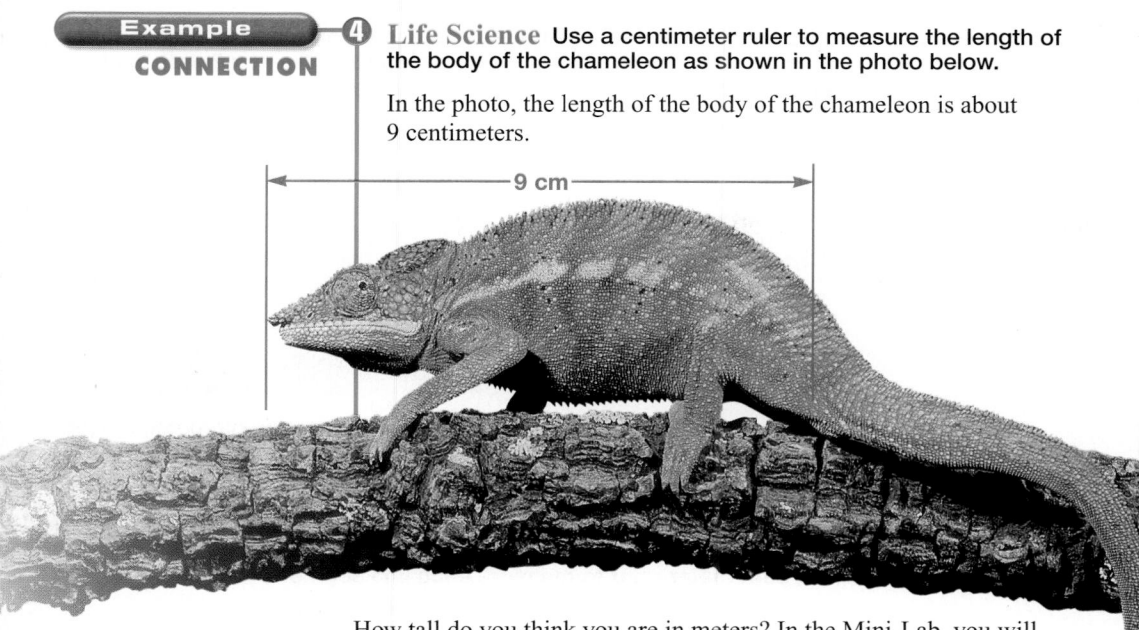

9 cm

How tall do you think you are in meters? In the Mini-Lab, you will estimate and find heights in the metric system.

Did you know? The muscles in a chameleon's throat are so powerful that it can extend its tongue the length of its body in about $\frac{1}{25}$ of a second.

3. Answers will vary. Students should conclude that the larger the object, the larger the unit that should be used.

HANDS-ON MINI-LAB

Work with a partner. ⌇ string ▭ tape measure

Try This
• Cut a length of string that is as long as your partner is tall.
• Estimate each other's height in centimeters and in meters. Then measure your string to find your height.

Talk About It
1. What is your height in centimeters? **Answers will vary.**
2. What is your height in meters? **Answers will vary.**
3. Write a paragraph explaining how you decide what metric unit to use to measure length. Use specific examples in your explanation.

CHECK FOR UNDERSTANDING

Communicating Mathematics

Read and study the lesson to answer each question.

1. *Draw* a line that is 10.5 centimeters long. **See students' work.**

2. *Tell* whether you would use centimeters or meters to measure the length of this book. Explain your answer. **See Answer Appendix.**

HANDS-ON MATH

3. Collect everyone's height data from the Mini-Lab. Organize the data and draw a bar graph to display the data. **See students' work.**

Lesson 3-2 Integration: Measurement Length in the Metric System **103**

Reteaching the Lesson

Activity Place a clear centimeter ruler on the overhead. Place a jumbo paper clip above the ruler. Have students model the activity and determine the lengths of their paper clips in millimeters and in thousandths of a meter. Continue with other items.

Error Analysis
Watch for students who are confused about the placement of the decimal point when converting a fraction to a decimal. **Prevent by** showing them a simple rule. For example, point out that 100 has *two* zeros, and the last digit in centimeters, or hundredths, is *two* places to the right of the decimal point.

In-Class Examples

Write the unit of length that you would use to measure each of the following. Then estimate each measure.

For Example 1
Width of a nickel
cm, about 2 cm

For Example 2
Width of an average window
m, about 1 m

For Example 3
On the map, what is the distance in centimeters between Austin and Houston along U.S. Route 290? in millimeters? **5.7 cm; 57 mm**

For Example 4
Measure the widest part of the chameleon's body, located just behind the foreleg. **about 3.5 cm**

3 PRACTICE/APPLY

Check for Understanding
If students need additional practice or instruction after completing Exercises 1–8, you may find one of the following options helpful.
• Extra Practice, see p. 563
• Reteaching Activity
• *Transition Booklet,* pp. 29–30
• *Study Guide Masters,* p. 18
• *Practice Masters,* p. 18

Study Guide Masters, **p. 18**

3-2 Study Guide

Name _____ Date _____

Integration: Measurement
Length in the Metric System

A dime is about one millimeter (1 mm) thick.
| 1 mm |

A sugar cube is about one centimeter (1 cm) wide.
| 1 cm |

A kitchen counter is about one meter (1 m) high.

Nine times the length of a football field, including end zones, is about 1 kilometer (1 km).

Use a centimeter ruler to measure each line segment.
1. ▬ 1.4 cm
2. ▬▬ 8.9 cm
3. ▬ 2.1 cm
4. ▬▬ 10.5 cm
5. ▪ 0.7 cm
6. ▬ 5.6 cm

Use a centimeter ruler to measure one side of each square.
7. □ 1.2 cm
8. □ 1.7 cm
9. □ 2.5 cm
10. □ 0.8 cm
11. □ 1.7 cm
12. □ 2.0 cm

4 ASSESS

Closing Activity

Modeling Have students demonstrate how the decimal point moves when expressing measurements in meters or in tenths, hundredths, or thousandths of a meter by measuring the same object all four ways.

Chapter 3, Quiz A (Lessons 3-1 and 3-2) is available in the *Assessment and Evaluation Masters*, p. 71.

Additional Answer

22. Sample answer: centigrade, centipede, century

Practice Masters, p. 18

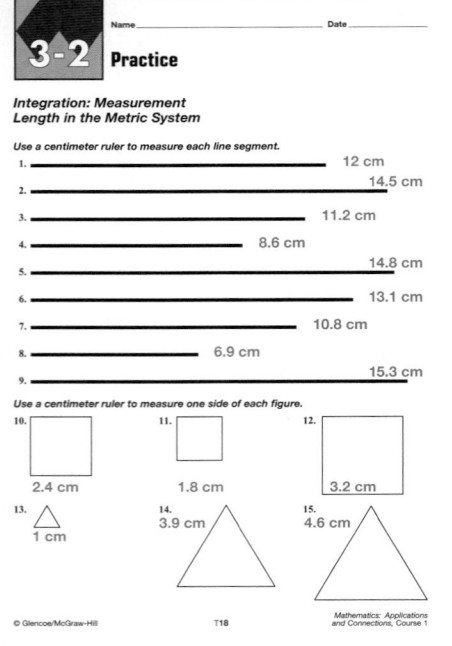

Guided Practice

Write the unit of length: millimeter, centimeter, meter or kilometer, that you would use to measure each of the following. Then estimate each measure.

4. the length of a football field
   m; about 110 m
5. the width of a quarter
   cm; about 2.5 cm

Use a centimeter ruler to measure each line segment.

6. ——— 4.4 cm or 44 mm
7. ——— 3.2 cm or 32 mm

8. **Sports** Runners often participate in races that are 10 kilometers long. How many meters are in 10 kilometers? **10,000 meters**

EXERCISES

Practice

9. cm; about 90 cm
10. mm; 8 mm

Write the unit of length: millimeter, centimeter, meter or kilometer, that you would use to measure each of the following. Then estimate each measure.

9. the length of a skateboard
10. the thickness of a pencil
11. the height of a giraffe
    m; about 5 m
12. the length of a swimming pool
    m; about 50 meters

Use a centimeter ruler to measure each line segment.

13. ——— 1.3 cm or 13 mm
14. ——— 1.9 cm or 19 mm
15. ——— 3.8 cm or 38 mm
16. ——— 5.4 cm or 54 mm

Use a centimeter ruler to measure one side of each figure.

17. 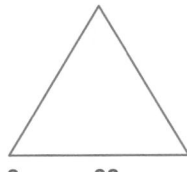 3 cm or 30 mm
18. 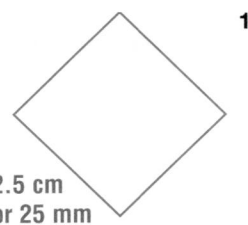 2.5 cm or 25 mm
19. 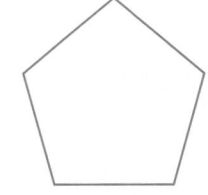 2 cm or 20 mm

20. How many millimeters are in 6.8 centimeters? **68 mm**
21. How many centimeters are in 3 meters? **300 cm**

Applications and Problem Solving

23. See students' work.
24. 0.49 m, 55 cm, 999 mm, 0.0037 km, 3.9 m

Mixed Review

22. **Language Arts** The metric prefix *cent* means *hundred*. Make a list of words that contain this prefix. **See margin.**

23. **Geometry** On a sheet of centimeter grid paper, draw a square such that the perimeter (distance around the square) is 20 centimeters.

24. **Critical Thinking** Order from least to greatest.
    0.0037 km   3.9 m   55 cm   0.49 m   999 mm

25. **Test Practice** Choose the decimal that represents *twelve and sixty-three thousandths*. *(Lesson 3-1)* **C**
    A 1206.3   **B** 120.63   **C** 12.063   **D** 0.12063

26. **Statistics** Construct a horizontal bar graph using the guinea pig weights of 1.8, 1.754, 2.09, 1.91, and 2.1 pounds. *(Lesson 2-3)* **See Answer Appendix.**

27. **Algebra** Find the value of $3x - 2y$ if $x = 6$ and $y = 4$. *(Lesson 1-5)* **10**

104 Chapter 3 Adding and Subtracting Decimals

Extending the Lesson

Enrichment Masters, p. 18

Activity Have students make a scale drawing of the classroom on centimeter grid paper. Then have them construct a cardboard model of the drawing. Display the drawings and models in the classroom.

3-3

Comparing and Ordering Decimals

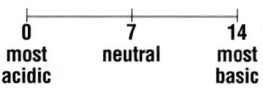 **you'll learn**

You'll learn to compare decimals and order a set of decimals.

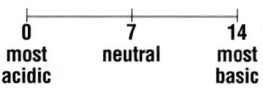 **am I ever going to use this?**

Knowing how to compare and order decimals can help you find library books.

Study Hint

Reading Math
Remember that the symbol < is read as *is less than* and the symbol > is read as *is greater than*. The symbol always points toward the lesser number.

Mrs. Lee's science class is studying acids and bases. Acidity is expressed using the pH scale, which ranges from 0 to 14.

Substance	pH
apple	2.9
baking soda	9.0
carrot	5.1
drinking water	7.0
household ammonia	13.0
lemon	2.1
lye	14.0
rainwater	5.8
tomato	4.2

```
 |--------|--------|
 0        7        14
most    neutral   most
acidic            basic
```

The results of Carmen's group are shown in the table.

Carmen's group needs to compare the pH of the carrot and of rainwater, 5.1 and 5.8, respectively. There are two ways to compare decimals. You can compare the digits in each place-value position, or you can use a number line.

Method 1 Use place value.

Line up the decimal points of the two numbers. Then start at the left, comparing the digits in the same place-value position that are not equal. The decimal with the greater digit is the greater decimal.

carrot: 5.**1**
rainwater: 5.**8**

1 and 8 are not equal. 1 tenth < 8 tenths, so 5.1 < 5.8.

Method 2 Use a number line.

On a number line, numbers to the right are greater than numbers to the left.

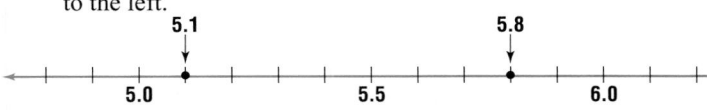

5.1 is left of 5.8, 5.1 < 5.8.

Since 5.1 < 5.8, the carrot is more acidic than the rainwater.

Example ① Which is greater, 2.037 or 2.033?

Method 1 Use place value.

2.03**7**   *Line up decimal points.*
2.03**3**   *Starting at the left, compare each place-value.*

7 and 3 are not equal.

7 thousandths > 3 thousandths, so 2.037 > 2.033.

(continued on the next page)

 Investigations for the Special Education Student

This blackline master booklet helps you plan for the needs of your special education students by providing long-term projects along with teacher notes. Investigation 3, *Dining Out,* may be used with this chapter.

Instructional Resources
- *Study Guide Masters,* p. 19
- *Practice Masters,* p. 19
- *Enrichment Masters,* p. 19
- *Transparencies 3-3, A and B*
- *Assessment and Evaluation Masters,* pp. 70, 71
- *Hands-On Lab Masters,* p. 71
- *Technology Masters,* p. 6
- CD-ROM Program
  - Resource Lesson 3-3
  - Interactive Lesson 3-3

Recommended Pacing	
Standard	Day 6 of 12
Honors	Day 4 of 10
Block	Day 3 of 6

1 FOCUS

 5-Minute Check
(Lesson 3-2)

Use a centimeter ruler to draw a line segment having the given length.
1. 2.8 centimeters
2. 4 centimeters
3. 3.6 centimeters
4. Draw a square with sides 1.5 centimeters long.
   1–4. See students' work.
5. How many millimeters are in 3.2 centimeters? **32**

 The 5-Minute Check is also available on **Transparency 3-3A** for this lesson.

Motivating the Lesson
Problem Solving Ask students how they would determine which item weighs more:
- a turkey that weighs 19.68 pounds
- a turkey that weighs 19.75 pounds

Compare the decimal portion of each weight; 0.75 > 0.68 so 19.75 > 19.68.

2 TEACH

Transparency 3-3B contains a teaching aid for this lesson.

Using the Mini-Lab A brief review with base-ten blocks may be needed by some students. When students are finished, lead them to make a generalization based on their findings.

In-Class Examples

For Example 1
Which is greater, 3.048 or 3.045? **3.048**

For Example 2
Which is the greatest number, 1.32, 1.305, or 1.3? **1.32**

For Example 3
Tanisha is doing research at the library. She has a stack of books to return to the shelves. Help her order the call numbers on the books from least to greatest: 389.426, 389.4, 389.42, 389.342. **389.342, 389.4, 389.42, 389.426**

Teaching Tip While discussing Example 2, stress the importance of lining up the decimal points before comparing.

Method 2 Use a number line.
Compare the decimals on a number line.

2.037 is to the right of 2.033. So 2.037 > 2.033.

You can also use grid paper to compare decimals.

 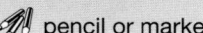

MINI-LAB

Work with a partner. 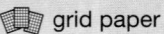 grid paper ✏ pencil or marker

Try This
- Each person should draw a 10-by-10 square on grid paper.
- One person should shade a grid to show 0.4 or four tenths.
- The other person should shade a grid to show 0.40 or forty hundredths.

Talk About It
1. How do the two grids compare?
2. Shade two more grids to show 0.6 and 0.60. Compare the two grids. What can you say about them?

1. The area of the shaded sections are equal.
2. The area of the shaded sections are equal.

You can annex, or place zeros to the right of a decimal so that each decimal has the same number of decimal places.

Example 2

Which is the greatest number, 19.48, 19.481, or 19.4?

Line up the decimal points.	*Annex a zero so that each has the same number of decimal places.*
19.48	19.48**0**
19.481 ➡	19.481
19.4	19.4**00**

Since 8 hundredths is greater than 0 hundredths, 19.480 > 19.400.

Since 1 thousandth is greater than 0 thousandths, 19.481 > 19.480.

The greatest number is 19.481.

Sports The table shows the top rebounder for each team in the Midwest Division of the NBA for the 1997 regular season. Order the rebounding averages from least to greatest.

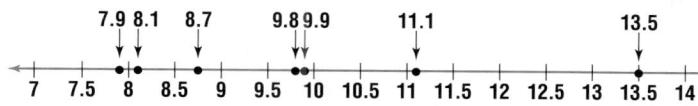

Player	Team	Rebounds/Game
Barkley	Houston	13.5
Green	Dallas	7.9
Malone	Utah	9.9
Gugliotta	Minnesota	8.7
Er. Johnson	Denver	11.1
Perdue	San Antonio	9.8
Reeves	Vancouver	8.1

Use a number line to order the average number of rebounds.

The rebounding averages in order from least to greatest are 7.9, 8.1, 8.7, 9.8, 9.9, 11.1, and 13.5.

CHECK FOR UNDERSTANDING

Communicating Mathematics

Read and study the lesson to answer each question.

1. **Draw** a number line that compares 3.89 and 3.91. **See Answer Appendix.**

2. Sample answer: Since 1.018 is greater, it is to the right of 1.01.

2. **Tell** how the decimals 1.018 and 1.01, graphed on the number line, compare.

HANDS-ON MATH

3. Use base-ten blocks to compare 1.63 and 1.54. **See students' work.**

Guided Practice

State the greatest number in each group.

4. 6.02 or 6.20 **6.20**

5. 0.042, 0.06 or 0.051 **0.06**

6. 198.6, 198.06, 189.5, or 198.067 **198.6**

Order each set of decimals from least to greatest.

7. 13.507; 13.05; 13.9; 13.84 **13.05, 13.507, 13.84, 13.9**

8. 0.002, 0.09, 0.19, 0.2, 0.21, 2.1, 21.9

8. 0.2; 0.09; 0.19; 0.21; 2.1; 21.9; 0.002

9. 389.225, 388.404, 388.246, 385.867, 385.841

9. **Sports** The table shows the scores for the top five teams in women's gymnastics at the 1996 Summer Olympics. Order the scores from greatest to least.

Team	Score
China	385.867
Romania	388.246
Russia	388.404
Ukraine	385.841
USA	389.225

Lesson 3-3 Comparing and Ordering Decimals **107**

Reteaching the Lesson

Activity Have students use a newspaper to find sports statistics such as batting averages. Then ask them to order the averages from greatest to least. Point out that these are actually decimals, so 0.324 is greater than 0.299.

Error Analysis
Watch for students who consistently make errors in comparing decimals. **Prevent by** having these students use base-ten models to shade each decimal before comparing.

3 PRACTICE/APPLY

Check for Understanding
If students need additional practice or instruction after completing Exercises 1–9, you may find one of the following options helpful.
• Extra Practice, see p. 563
• Reteaching Activity
• *Transition Booklet,* pp. 7–8
• *Study Guide Masters,* p. 19
• *Practice Masters,* p. 19

Assignment Guide

Core: 11–27 odd, 28–32
Enriched: 10–24 even, 26–32

Study Guide Masters, p.19

Closing Activity

Modeling Play "So You Think You Have the Greatest Decimal?" Have students write decimals, 1 per card, on 3″ × 5″ index cards. Make 36 cards. Then divide the cards among the players. To play, each player reveals the top card. The player with the greatest decimal takes the cards. Play continues until all cards have been used.

Chapter 3, Quiz B (Lesson 3-3) is available in the *Assessment and Evaluation Masters,* p. 71.

Mid-Chapter Test (Lessons 3-1 through 3-3) is available in the *Assessment and Evaluation Masters,* p. 70.

Practice Masters, p. 19

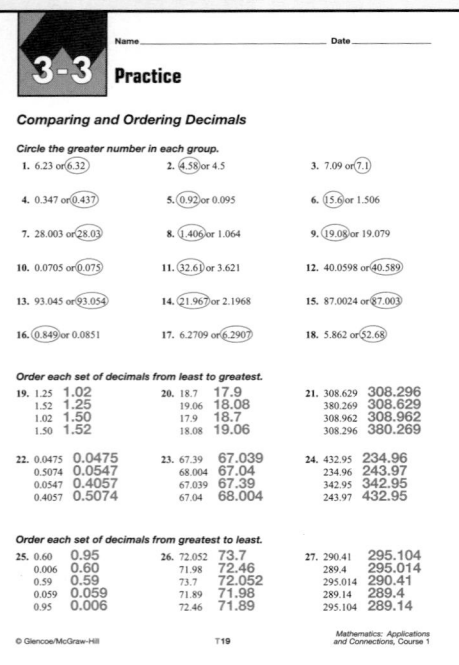

EXERCISES

Practice

State the greatest number in each group.

10. 16.099 or 160.98 **160.98**

11. 0.331 or 0.303 **0.331**

12. 18.607 or 18.06 **18.607**

13. 1.018 or 1.01 **1.018**

14. 0.03, 0.31, or 0.039 **0.31**

15. 47.553, 47.5, or 47.053 **47.553**

16. 547.484 or 547.4843 **547.4843**

17. 0.068, 0.07, or 0.7 **0.7**

Order each set of decimals from least to greatest.

18.
94.7	**94.7**
101.1	**98.5**
99.7	**99.7**
98.5	**101.1**

19.
15	**14.95**
15.8	**15**
14.95	**15.01**
15.01	**15.8**

20.
37.5	**35.7**
35.7	**35.849**
35.849	**36.06**
36.06	**37.5**

Order each set of decimals from greatest to least.

21.
0.025	**0.0316**
0.0316	**0.0306**
0.0306	**0.025**
0.0249	**0.0249**
0.0208	**0.0208**

22.
43.8	**43.8**
42.998	**43.6789**
43.16	**43.16**
42.022	**42.998**
43.6789	**42.022**

23.
379.8778	**397.877**
378.87	**379.9**
397.877	**379.88**
379.9	**379.8778**
379.88	**378.87**

24. Which is the least, 10.59 or 10.599? **10.59**

25. Which is the greatest, 0.0621, 0.603, or 0.06? **0.603**

Applications and Problem Solving

26. 745.2, 745.23, 745.231, 745.325, 745.412

27a. 2.1, 2.9, 4.2, 5.1, 5.8, 7.0, 9.0, 13.0, 14.0

28. $1.70-César; $0.79-Eric; $1.07-Beatriz; $1.18-Antonio; $0.89-Dexter

26. *Library Science* Shannon has to return the stack of books shown at the right to the library shelves. Order the call numbers on the books from least to greatest.

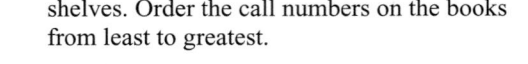
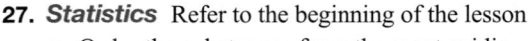

745.231
745.23
745.2
745.412
745.325

27. *Statistics* Refer to the beginning of the lesson.
  a. Order the substances from the most acidic to the most basic.
  b. What is the median pH reading? **5.8**

28. *Critical Thinking* Antonio has more money than Beatriz. Antonio has less money than César. Dexter has $0.10 more than Eric. Use the decimals at the right to determine how much money each person has.

$1.70
$0.79
$1.07
$1.18
$0.89

Mixed Review

29. *Measurement* Measure the line segment. *(Lesson 3-2)*
—————————————— **5.7 cm or 57 mm**

30. Write thirty-seven thousandths as a decimal. *(Lesson 3-1)* **0.037**

31. Sample answer: 95, 91, 93, 89; no number occurs more often than the others

31. *Statistics* Make up a set of data that has no mode and state why it has no mode. *(Lesson 2-7)*

32. **Test Practice** Florida has 67 counties. What is this number rounded to the nearest ten? *(Lesson 1-3)* **C**
  A 60   **B** 65   **C** 70   **D** 75

Extending the Lesson

Enrichment Masters, p. 19

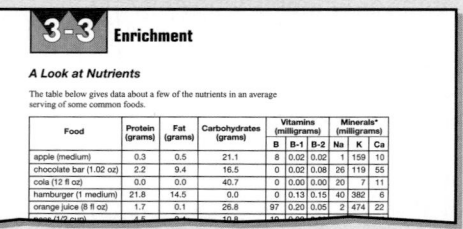

Activity Have students use the catalog system at the local or school library to find a book on any topic of interest. Have them look for the call number and then find the book in the stacks. You may suggest that students research how call numbers are assigned.

Rounding Decimals

What you'll learn

You'll learn to round decimals.

When am I ever going to use this?

Knowing how to round decimals is helpful in estimating with money.

El Pollo Loco created the world's largest burrito in Anaheim, California, on July 31, 1995. It weighed 4,217 pounds and was 3,112.99 feet long. Do you think the weight and length were reported this way in the local papers?

Newspapers often round numbers so it is easier to read. The length of the burrito may have been reported as 3,113 feet.

Decimals may be rounded to any place-value position.

Example ① Round 1.63 to the nearest tenth.

To round 1.63, look at the number line below.

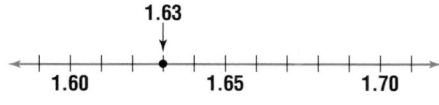

1.63

On the number line, 1.63 is closer to 1.6 than it is to 1.7.

1.63 rounded to the nearest tenth is 1.6.

LOOK BACK
You can refer to Lesson 1-3 for information on rounding whole numbers.

You can also round decimals without using a number line.

Rounding Decimals	• Look at digit to the right of the place being rounded. • The digit remains the same if the digit to the right is 0, 1, 2, 3, or 4. • Round up if the digit to the right is 5, 6, 7, 8, or 9.

Example ②
APPLICATION

Money Matters Ayashe purchased a carbon monoxide detector for $27.44. To the nearest dollar, how much did she spend?

Look at the digit to the right of the ones place.

27.44 *Since 4 < 5, the digit in the ones place stays the same.*
↑
ones place

To the nearest dollar, Ayashe spent $27.

Lesson 3-4 Rounding Decimals **109**

Instructional Resources
• *Study Guide Masters*, p. 20
• *Practice Masters*, p. 20
• *Enrichment Masters*, p. 20
• Transparencies 3-4, A and B
• *Diversity Masters*, p. 3
 CD-ROM Program
 • Resource Lesson 3-4

Recommended Pacing	
Standard	Day 7 of 12
Honors	Day 5 of 10
Block	Day 4 of 6

1 FOCUS

5-Minute Check
(Lesson 3-3)

State the greatest number in each group.
1. 63.08 or 63.18 63.18
2. 0.5214, 0.521, or 0.52 0.5214
3. 75.481, 75.4182, or 75.4818 75.4818
4. Order the decimals from least to greatest: 97.490, 97.4, 97.482, 97.4826. 97.4, 97.482, 97.4826, 97.490
5. Order the decimals from greatest to least: 55.711, 55.701, 55.7, 54.777, 55.071. 55.711, 55.701, 55.7, 55.071, 54.777

The 5-Minute Check is also available on **Transparency 3-4A** for this lesson.

Motivating the Lesson

Communication Have small groups of students construct number lines with decimals in tenths from 74.0 to 76.0. Give each group a penny to slide along the number line. Have them record where the farthest point of the penny stops. Then have them discuss whether it is closer to 74.0, 75.0, or 76.0.

Classroom Vignette

"To practice rounding decimals, we create a human number line. Some students represent increments on the number line and others, numbers to be rounded. We discuss how each number is to be rounded and then the "rounded" student stands behind the appropriate person on the number line."

Joy Donlin, Teacher
Bates Middle School
Annapolis, MD

2 TEACH

Transparency 3-4B contains a teaching aid for this lesson.

Using Number Lines Have students use a number line to round a decimal to the tenth, hundredth, or whole number if they have difficulties.

In-Class Examples

For Example 1
Round 8.46 to the nearest tenth. **8.5**

For Example 2
A bakery manager says the cost per store-made roll is $0.169. To the nearest cent, how much is this? **$0.17**

For Example 3
Round 0.3928 to the nearest hundredth. **0.39**

3 PRACTICE/APPLY

Check for Understanding

If students need additional practice or instruction after completing Exercises 1–8, you may find one of the following options helpful.
- Extra Practice, see p. 564
- Reteaching Activity
- *Study Guide Masters*, p. 20
- *Practice Masters*, p. 20

***Study Guide Masters*, p. 20**

3-4 Study Guide

Name _____ Date _____

Rounding Decimals

Round 34.725 to the nearest tenth.

You can use a number line.

Find the approximate location of 34.725 on the number line.
34.725 is closer to 34.7 than to 34.8
34.725 rounded to the nearest tenth is 34.7.

34.0 34.1 34.2 34.3 34.4 34.5 34.6 34.7 34.8 34.9 35.0

You can also round without a number line.

| Find the place to which you want to round. | Look at the digit to the right. If the digit is less than 5, round down. If the digit is 5 or greater, round up. | 2 is less than 5. Round down. |
| 34.725 | 34.725 | 34.7 |

Use each number line to show how to round the decimal to the nearest tenth.

1. 7.82 → **7.8**   7.0 7.1 7.2 7.3 7.4 7.5 7.6 7.7 7.8 7.9 8.0
2. 0.39 → **0.4**   0.0 0.1 0.2 0.3 0.4 0.5 0.6 0.7 0.8 0.9 1.0
3. 5.071 → **5.1**   5.0 5.1 5.2 5.3 5.4 5.5 5.6 5.7 5.8 5.9 6.0

Round each number to the underlined place-value position.

4. 6.32 → **6.3**
5. 0.4721 → **0.47**
6. 26.444 → **26.4**
7. 1.161 → **1.2**
8. 362.0846 → **362.085**
9. 15.553 → **15.55**
10. 151.391 → **151.39**
11. 0.55 → **0.6**
12. 631.0008 → **631.001**
13. 17.327 → **17.33**
14. 3.09 → **3.1**
15. 1.58 → **1.6**

© Glencoe/McGraw-Hill   T20   *Mathematics: Applications and Connections, Course 1*

Example 3 Round 3.4672 to the nearest hundredth.

Look at the digit to the right of the hundredths place.

3.4672   *Round up since 7 > 5.*
  ↑
hundredths place

3.4672 rounded to the nearest hundredth is 3.47.

CHECK FOR UNDERSTANDING

Communicating Mathematics

Read and study the lesson to answer each question.

1. *Explain* how to round $10.79 to the nearest dollar. **See Answer Appendix.**

2. *Draw* a number line to show why 2.983 rounded to the nearest tenth is 3.0. **See Answer Appendix.**

3. Sharon is correct. Carlos rounded to the nearest hundred.

3. *You Decide* Sharon says that 456.789 rounded to the nearest hundredth is 456.79. Carlos says it is 500. Who is correct? Explain.

Guided Practice

Round each number to the underlined place-value position.

4. 0.2̲8 **0.3**   5. 8.2̲02 **8.20**   6. 0.1̲487 **0.1**   7. 19̲.59 **20**

8. *Money Matters* The unit price of a twenty-ounce bottle of Mountain Dew is $0.0445 per ounce. How much is this to the nearest cent? **$0.04**

EXERCISES

Practice

Round each number to the underlined place-value position.

9. 18̲.44 **18**   10. 0.84̲9 **0.85**   11. 20.4̲5 **20.5**   12. 2.48̲5 **2.49**

13. 49.7̲75 **49.8**   14. 68.8̲8 **68.9**   15. 19.7̲75   16. 48.8̲02 **48.8**

17. 99.9̲8 **100.0**   18. 42.7̲896   19. 6.999̲8   20. 4.00̲4 **4.00**

15. 19.78
18. 42.79
19. 7.000

21. Round $1.69 to the nearest dollar. **$2.00**

22. Round 49,237.1589499 to the nearest ten-thousandth. **49,237.1589**

Applications and Problem Solving

23. *Entertainment* KTXQ in Dallas, Texas, can be found by tuning to 102.1 on the radio. The DJs round the call number to the nearest whole number. What number do the DJs use to refer to KTXQ? **102**

24. *Money Matters* Gasoline prices are usually calculated to the thousandth place. If you purchased gas that cost $1.199 per gallon, what price would you say you paid? **$1.20**

110 Chapter 3 Adding and Subtracting Decimals

■ Reteaching the Lesson ■

Activity Play "Rounds." Use two cubes, one with six different decimals in thousandths (for example, 8.754) and the other with the words "tenths," "hundredths," and "whole number" written twice on opposite faces. Have students take turns rolling the cubes and rounding the decimal to the place indicated.

Multiple Learning Styles

Logical Separate students into groups. Have each student give clues such as the following for others to guess. Encourage students to make their clues complicated. *The decimal rounded to the nearest hundredth is 8.48. The digit in the thousandths place is greater than five. What is the decimal?*

25. _Critical Thinking_ The table shows the average density of the nine planets. If you had to determine the mean, median, and mode of this set of data, would you round to the nearest tenth, nearest whole number, or use the exact numbers? Explain. **See Answer Appendix.**

Planet	Grams per cubic cm
Mercury	5.42
Venus	5.25
Earth	5.52
Mars	3.94
Jupiter	1.33
Saturn	0.69
Uranus	1.27
Neptune	1.71
Pluto	2.03

Mixed Review

27. See Answer Appendix.

26. Order the decimals 2.9, 2.38, 2.474, 2.91, and 2.88 from greatest to least. _(Lesson 3-3)_ **2.91, 2.9, 2.88, 2.474, 2.38**

27. _Statistics_ Is the mean a misleading measure of central tendency for this set of data: 17, 21, 20, 19, 17, 21, 18, 22, and 21? Explain. _(Lesson 2-8)_

28. **Test Practice** The graph shows the number of boys and girls who signed up to play in the Northtown Youth Soccer League. Which is a reasonable conclusion that can be drawn from the information in the graph? _(Lesson 2-3)_ **B**

Interview family members about when they use rounding. Ask how they decide to what place value a number should be rounded.

A More girls than boys signed up for soccer in 1995.
B The number of girls playing soccer has been catching up with the number of boys since 1992.
C The number of boys who joined soccer decreased every year.
D Girls did not want to play soccer before 1992.
E There were more soccer sign-ups in 1994 than any other year.

Northtown Soccer Sign-ups

(bar graph: Boys, Girls for years 1992, 1993, 1994, 1995, 1996; vertical axis 0–45; horizontal axis Year)

CHAPTER 3

Mid-Chapter Self Test

Express each fraction as a decimal. _(Lesson 3-1)_

1. $\frac{7}{10}$ **0.7**
2. $\frac{43}{100}$ **0.43**
3. $\frac{681}{1,000}$ **0.681**
4. $\frac{409}{10,000}$ **0.0409**

Measure each line segment to the nearest centimeter. _(Lesson 3-2)_

5. ———— **3 cm**
6. —————— **5 cm**

7. **_Library_** Miguel Aguilar went to the library to get some information about bees for a project. He is looking for books with the call numbers 638.178 and 638.186. Which number comes first? _(Lesson 3-3)_ **638.178**

Round each number to the underlined place-value position. _(Lesson 3-4)_

8. 6.8 **7**
9. $3.4\underline{0}1$ **3.40**
10. $181.9\underline{8}$ **182.0**

Extending the Lesson

Enrichment Masters, p. 20

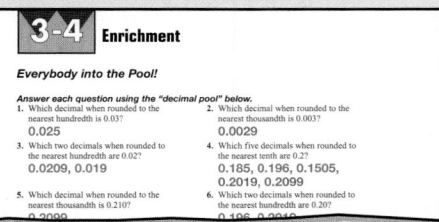

3-4 **Enrichment**

Everybody into the Pool!

Answer each question using the "decimal pool" below.
1. Which decimal when rounded to the nearest hundredth is 0.03?
0.025
2. Which decimal when rounded to the nearest thousandth is 0.003?
0.0029
3. Which two decimals when rounded to the nearest hundredth are 0.02?
0.0209, 0.019
4. Which five decimals when rounded to the nearest tenth are 0.2?
0.185, 0.196, 0.1505, 0.2019, 0.2099
5. Which decimal when rounded to the nearest thousandth is 0.210?
0.2099
6. Which two decimals when rounded to the nearest hundredth are 0.20?
0.196, 0.2049

Activity Have pairs of students divide whole numbers on their calculators and then round the quotients to the nearest whole number, tenth, or hundredth. Have them determine a method for rounding decimals with more than three digits to the right of the decimal point.

1 FOCUS

5-Minute Check
(Lesson 3-4)

1. Draw a number line to show how to round 7.39 to the nearest tenth.

```
                 7.39
                  ↓
 ←──┼──┼──┼──┼──┼──┼──┼──→
   7.3        7.35      7.4
```

Round each number to the underlined place-value position.

2. 32.<u>1</u>09 32.1

3. <u>7</u>.65 8

4. 4.4<u>3</u>8 4.44

5. 6.<u>8</u>611 6.9

The 5-Minute Check is also available on **Transparency 3-5A** for this lesson.

Motivating the Lesson

Problem Solving Ask students to read the opening paragraph of the lesson. Ask the following questions:

- Should you estimate or find the exact answer? Explain.
  Estimate; the question asks for an estimate by using the word *about*.

- Glancing quickly at the graph, would you guess the cost was about $10 billion? $15 billion?
  no; yes

3-5 Estimating Sums and Differences

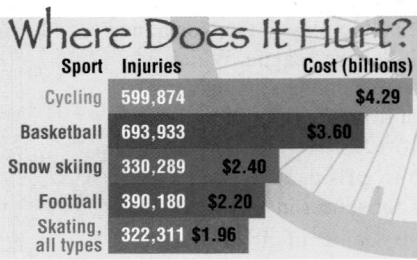

What you'll learn

You'll learn to estimate decimal sums and differences.

When am I ever going to use this?

You'll estimate sums and differences when you need to determine if you have enough money to pay for the groceries in your cart.

Word Wise
clustering

Do you participate in a sport? The graph shows the cost of treating sports injuries in emergency rooms in a recent year. About how much did treating these injuries cost?

Since you only need to know about how much the cost was, you can estimate. One way to estimate is to round the amounts to the same place-value position and then add.

Where Does It Hurt?

Sport	Injuries	Cost (billions)
Cycling	599,874	$4.29
Basketball	693,933	$3.60
Snow skiing	330,289	$2.40
Football	390,180	$2.20
Skating, all types	322,311	$1.96

Source: *Consumer Product Safety Commission, American Academy of Orthopedic Surgeons*

Round each number to the nearest billion.

$4.29	→	$4
$3.60	→	$4
$2.40	→	$2
$2.20	→	$2
+$1.96	→	+$2
		$14

The cost of sports injuries was about $14 billion.

Example
APPLICATION 1

Advertising The table shows the ten companies with the most commercials on prime-time TV between August 26 and September 1 of 1996. About how many households were exposed to McDonald's and Burger King ads?

To estimate, round each number to the nearest hundred million.

Advertiser	Household Exposure (millions)
1. McDonald's	390.4
2. Burger King	216.5
3. Wendy's	183.4
4. Sears	172.7
5. Mazda Protege	172.6
6. Saturn	136.6
7. MCI	131.4
8. Baskin Robbins	126.3
9. Kmart	106.7
10. Olive Garden	106.6

Source: *Nielsen Media Research, Monitor-Plus Service*

390.4	→	400	*9 > 4, so round up.*
+216.6	→	+200	*1 < 5, so the digit stays the same.*
		600	

About 600 million households were exposed to the McDonald's and Burger King ads.

② **Earth Science** The largest cut diamond is the 545.67-carat gem known as the Golden Jubilee. Before it was cut, it weighed 775.50 carats. About how many carats of the uncut diamond were not used?

$$\begin{array}{rcl} 775.50 & \to & 776 \\ -545.67 & \to & -546 \\ & & \overline{230} \end{array}$$ *Round to the nearest carat. Then subtract.*

About 230 carats of the uncut diamond were not used.

APPLICATION ③ **Money Matters** Sega video games are on sale. The sale price for Vectorman is $42.99, and the sale price for Hangtime is $55.88.

a. About how much less is Vectorman than Hangtime?

$$\begin{array}{rcl} \$55.88 & \to & \$56 \\ -\ 42.99 & \to & -\ 43 \\ & & \overline{13} \end{array}$$ *Round to the nearest dollar. Then subtract.*

The Vectorman game is about $13 less than the Hangtime game.

b. If you bought both games while they were on sale, about how much would they cost?

$$\begin{array}{rcl} \$55.88 & \to & \$56 \\ +42.99 & \to & +43 \\ & & \overline{99} \end{array}$$ *Round to the nearest dollar. Then add.*

The games would cost about $99.

Clustering is another way you can estimate sums and differences. Clustering is used when numbers are close to the same number.

Example
APPLICATION ④ **Business** The table shows the charges for business related calls and faxes made by Mrs. Moore last week. About how much were her business related phone charges last week?

Minutes	Amount ($)
1.0	0.31
1.0	0.28
1.0	0.28
1.0	0.26
1.0	0.30

Since each amount is about $0.30, add this amount 5 times.

$$0.30 + 0.30 + 0.30 + 0.30 + 0.30 = 1.50$$

Mrs. Moore spent about $1.50 last week.

Lesson 3-5 Estimating Sums and Differences **113**

3 PRACTICE/APPLY

Check for Understanding

If students need additional practice or instruction after completing Exercises 1–11, you may find one of the following options helpful.
- Extra Practice, see p. 564
- Reteaching Activity
- *Study Guide Masters,* p. 21
- *Practice Masters,* p. 21

Assignment Guide

Core: 13–33 odd, 36–41
Enriched: 12–32 even, 33, 34, 36–41

Additional Answers

1. Sample answer: Round both amounts to the nearest dollar. $17 − $4 = $13

11. Sample answer: Round 0.38 to 0.4 and round 0.21 to 0.2. 0.4 + 0.2 = 0.6. Since 0.6 > 0.5, both samples cannot be stored in the 0.5-liter container.

Study Guide Masters, p. 21

3-5 Study Guide

Name _____ Date _____

Estimating Sums and Differences

Two estimation strategies are **rounding** and **clustering**.
Round each number to the same place.

Examples

Round to the nearest ten dollars.	Round to the nearest tenth.	Round to the nearest one.
$46.90 → $50 + 33.27 → + 30 $80	0.693 → 0.7 − 0.113 → − 0.1 0.6	6.22 → 6 + 0.85 → + 1 7

To use clustering, find a number around which each number in the set "clusters." Then use that number for all of the numbers.

Example $7.62 + $7.89 + $8.01 + $7.99 *These numbers cluster around 8.*
Add $8 four times.
$8 + $8 + $8 + $8 = $32

Estimate using rounding.

| 1. 0.456
 + 0.375
 0.9 | 2. 59.118
 − 17.799
 40 | 3. $6.63
 + 9.29
 $16.00 | 4. 0.0056
 − 0.0028
 0.003 |

5. 8.802 − 6.115 **3**
6. 0.9 − 0.0984 **0.8**

Estimate using clustering.

7. 13.1 + 12.97 + 12.62 + 13.44 **52**
8. 1.01 + 0.67 + 1.39 **3**
9. $19.99 + $20.15 + $19.52 **$60**
10. 5.55 + 6.01 + 5.7 + 6.412 **24**

© Glencoe/McGraw-Hill   T21   *Mathematics: Applications and Connections, Course 1*

CHECK FOR UNDERSTANDING

Communicating Mathematics

Read and study the lesson to answer each question.

1. *Explain* how to use rounding to estimate the difference between $16.98 and $4.29. **See margin.**

2. *Describe* a situation where it makes sense to use the clustering method to estimate a sum. **Answers will vary.**

Math Journal

3. *Write* a paragraph explaining how estimation would help you buy items at a store. **See students' work.**

Guided Practice

Estimate using rounding.

4. 7.75 + 8.95
   8 + 9 = 17

5. 18.52 + 31.3
   20 + 30 = 50

6. $20 − $1.82
   $20 − $2 = $18

Estimate using clustering.

7. 11.36 + 10.84 + 11 + 10.5 11 + 11 + 11 + 11 = 44

8. $3.44 + $3.40 + $3.50 + $3.49 $3 + $3 + $3 + $3 = $12

9. Estimate the sum of 20.09, 20.58, and 19.98. 20 + 20 + 20 = 60

10. About how much is $53.38 minus $32.68? $53 − $33 = $20

11. *Physical Science* Enrique has two samples of the same chemical. He wants to store both samples in a 0.5-liter container. One sample is 0.38 liter. The other sample is 0.21 liter. Can he store both samples in the container? Explain. **See margin.**

EXERCISES

Practice

Estimate using rounding. 12. 0.4 + 0.9 = 1.3   13. 9 − 5 = 4

12. 0.43 + 0.94
13. 8.78 − 5.09
14. 68.69 − 7.43 69 − 7 = 62
15. 31.556 + 17.405
16. 0.612 + 0.185
17. $31.30 − $18.52
18. 0.8 − 0.7383
    0.8 − 0.7 = 0.1
19. $57.98 − $26.95
    $58 − $27 = $31
20. 5.34 + 6.33 + 1.9
    5 + 6 + 2 = 13

15. 32 + 17 = 49
16. 0.6 + 0.2 = 0.8
17. $30 − $20 = $10
22. $11 + $11 + $11 + $11 = $44
23. 1 + 1 + 1 + 1 = 4
24. 6 + 6 + 6 + 6 = 24
29. $200 − $110 = $90
31. 102°F − 99°F = 3°F
32. $7 + $7 + $7 + $7 + $7 = $35

Estimate using clustering. 21. 4 + 4 + 4 = 12

21. 4.38 + 3.68 + 4.42
22. $11.46 + $10.57 + $10.88 + $11
23. 0.805 + 1.006 + 0.64 + 0.9
24. 6.72 + 5.9 + 6.143 + 6.5037
25. $54.45 + $54.07 + $53.99
    $54 + $54 + $54 = $162
26. 95.98 + 98.15 + 104.5 + 104.95
    100 + 100 + 100 + 100 = 400
27. About how much more is $64.50 than $39.95? $65 − $40 = $25
28. Estimate the sum 3.456 + 2.888 + 3.393 + 3.483. 3 + 3 + 3 + 3 = 12
29. About how much above $109.99 is an amount of $199.98?
30. Estimate the difference between 1.685 and 0.454. 2 − 0 = 2
31. About how much above 98.6°F is a body temperature of 102.4°F?
32. Estimate the sum of $7.25, $6.88, $6.75, $7.02, and $6.97.

114 Chapter 3 Adding and Subtracting Decimals

Reteaching the Lesson

Activity Have small groups of students create problems involving estimating decimal sums and differences using rounding or clustering. Provide mail-order catalogs for data. Have groups exchange their problems with others for solving.

Error Analysis

Watch for students who do not identify the specific place to which they will round.

Prevent by encouraging students to underline the place value they will round to.

33. *Measurement* The largest commercial building in terms of floor area is the flower auction building of the Cooperative VBA in Aalsmer, Netherlands. The original floor surface of 3.7 million square feet has now been extended to 5.27 million square feet. By about how many square feet was the floor surface increased? **about 1 million square feet**

34. *Recycling* The table shows the amount received by Mrs. Barsch's class for turning in aluminum cans. About how much money did they receive? **about $28**

Week	Money
1	$6.80
2	$6.60
3	$7.20
4	$7.00

35. *Working on the* CHAPTER Project Refer to the table on page 93.
  a. In Mandy's family, three people are over 12, and one is under 12. Can her whole family go on the Tropical Plantation Tour for less than $35?
  b. Using any strategy, estimate the cost of eating in restaurants for one day for your family or your group. Show the numbers you used to make your estimate. **a–b. See Answer Appendix.**

36. *Critical Thinking* Four same-priced items are purchased. Based on rounding, the estimate of the total was $16.
  a. What is the maximum price each item could have cost? **$4.49**
  b. What is the minimum price each item could have cost? **$3.50**

37. *Sports* Batting averages are rounded to the nearest thousandth. In 1996, Mike Stanley, catcher for the Boston Red Sox, had a batting average of 0.2695 to the nearest ten-thousandth. How will his 1996 batting average be listed on his baseball card? *(Lesson 3-4)* **0.270**

38. *Sports* The chart shows the times for four runners in a 100-meter race. In what order did the participants cross the finish line? *(Lesson 3-3)* **Debbie, Camellia, Fala, Sarah**

Runner	Time
Sarah	14.31 s
Camellia	13.84 s
Fala	13.97 s
Debbie	13.79 s

39. *Statistics* Find the median of the data: 23, 19, 22, 22, 20, 20, 19, 22. *(Lesson 2-7)* **21**

40. Test Practice The circle graph shows pie sales at a local bakery. What part of the total sales is peanut butter and strawberry? *(Lesson 2-4)* **A**
  A 0.20
  B 0.17
  C 0.12
  D 0.08

Pie Sales

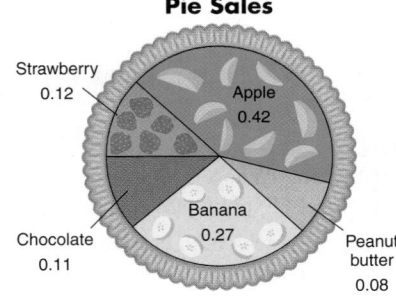
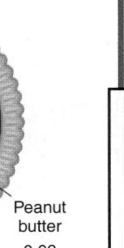

Strawberry 0.12
Apple 0.42
Banana 0.27
Chocolate 0.11
Peanut butter 0.08

41. *Algebra* Evaluate the expression $a + b \div c$ if $a = 5$, $b = 12$, and $c = 4$. *(Lesson 1-5)* **8**

Extending the Lesson

Enrichment Masters, p. 21

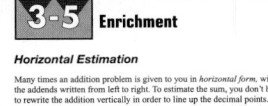

3-5 Enrichment

Horizontal Estimation

Many times an addition problem is given to you in *horizontal* form, with the addends written from left to right. To estimate the sum, you don't have to rewrite the addition vertically in order to line up the decimal points. Just use place value to figure out which digits are most important. Here is an example.

3.11 + 0.4639 + 8.205
The most important digits are in the ones place.
3 + 0 + 8 = 11
The next group of important digits are in the tenths place.
1 tenth + 4 tenths + 2 tenths = 7 tenths

Activity Write items and their prices (under $12) on the chalkboard. Have students roll two number cubes. The sum of the numbers on the cubes is how much "money" they have to spend. Then have students pick an item from the chalkboard and estimate how much change they would receive.

CHAPTER Project

Exercise 35 asks students to advance to the next stage of work on the Chapter Project. You may wish to have students compare costs of different restaurants.

4 ASSESS

Closing Activity

Writing Have pairs of students write sentences describing each estimation method used in the lesson. Guide them to write as if the person reading the description knew nothing about estimation.

Chapter 3, Quiz C (Lessons 3-4 and 3-5) is available in the *Assessment and Evaluation Masters,* p. 72.

Practice Masters, p. 21

3-5 Practice

Estimating Sums and Differences

Estimate using rounding.

1. 5.62 +3.04 **6 + 3 = 9**
2. 18.93 +27.45 **20 + 30 = 50**
3. 9.24 −2.56 **9 − 3 = 6**
4. 16.72 −9.13 **17 − 9 = 8**

5. 0.417 +0.869 **0.4 + 0.9 = 1.3**
6. 42.905 +31.276 **40 + 30 = 70**
7. 0.754 −0.482 **0.8 − 0.5 = 0.3**
8. 87.146 −24.953 **90 − 20 = 70**

9. 0.69 +0.45 **0.7 + 0.5 = 1.2**
10. 0.74 −0.18 **0.7 − 0.2 = 0.5**
11. 12.394 −4.601 **12 − 5 = 7**
12. 24.537 +9.862 **20 + 10 = 30**

13. 6.521 + 4.378 **7 + 4 = 11**
14. 0.932 − 0.485 **0.9 − 0.5 = 0.4**

15. 0.86 + 0.95 **0.9 + 1 = 1.9**
16. 43.058 − 15.726 **40 − 20 = 20**

Estimate using clustering. Sample answers are given.
17. 59.62 + 60.4 + 60 + 61 **240**
18. 8.2 + 7.8 + 7.2 + 7.99 **32**

19. 26.08 + 25.99 + 26.17 **78**
20. 6.73 + 7.01 + 7.53 + 6.91 + 7.1 **35**

21. 3.42 + 3.11 + 2.9 + 2.6 **12**
22. 4.59 + 5.28 + 5.444 **15**

23. 67.24 + 66.905 + 65 + 67.3 **268**
24. 87.04 + 86.55 + 87.101 + 86 **348**

25. $4.79 + $5.29 + $4.99 **$15**
26. 9.634 + 9.9 + 9.46 + 9.91 + 9.7632 **50**

27. 3.604 + 3.918 + 3.342 + 4.1 **16**
28. 2.1 + 2.387 + 2.57 + 1.99 **8**

© Glencoe/McGraw-Hill   T21   *Mathematics: Applications and Connections, Course 1*

Objective Students determine whether an answer is reasonable.

Recommended Pacing	
Standard	Day 9 of 12
Honors	Day 7 of 10
Block	Day 5 of 6

1 FOCUS

Getting Started Discuss the table with students. Round each year of CD shipments. Discuss the differences of rounding to the nearest ones, tens, or hundreds place.

2 TEACH

Teaching Tip Students may not understand what is *reasonable*. Using the CD table, discuss situations when you would round to the tens, hundreds, or thousands place. For example, a distributor would round 662.1 million up to 1 billion to impress buyers. However, an advertiser might round to 660 million to promote more advertising. Then have students work through Exercises 1–4.

In-Class Example
Roberta has $65 to spend on sports equipment. She decides to buy a bat for $29.95, 2 balls for $4.59 each, and a batting glove for $14.89. She thinks she will have enough left over to buy a cap for $15. Does this seem reasonable? **no**

Additional Answer
2. **Sample answer: Brian rounded to the nearest 1,000 and Akira rounded to the nearest 100.**

PROBLEM SOLVING

3-6A Reasonable Answers

A Preview of Lesson 3-6

Akira and Brian were shopping for CDs and noticed there were hundreds of CDs. Let's listen in!

Wow! Look at the selection they have here. Do you think over a billion CDs have been sold yet?

I know where we can find the answer. I have an almanac in my backpack. We were using it today in social studies class. Let's look.

There it is. The table shows the number of CDs that were shipped for sale from 1984 to 1994.

It looks like almost a billion CDs were shipped in 1994 alone!

Akira

How did you figure that? There were less than 700 million CDs shipped that year.

Year	CDs Shipped (millions)
1984	5.8
1985	22.6
1986	53.0
1987	102.1
1988	149.7
1989	207.2
1990	286.5
1991	333.3
1992	407.5
1993	495.4
1994	662.1

Brian

THINK ABOUT IT

1. ***Compare and contrast*** Akira's and Brian's thinking. Whose estimate do you think is correct? Explain why. **Answers will vary.**

2. ***Explain*** why both boys might be considered correct. **2–4. See margin.**

3. ***Choose*** two years when a total of about 200 million CDs were shipped. Explain your reasoning.

4. ***Apply*** what you have learned from Akira and Brian's situation to solve the following problem.

   Kelsey buys a Beanie Baby for $5.54. She pays for the purchase with a $20 bill. Which is a more reasonable estimate for the amount of change she should receive: $12 or $15? Explain why.

116 Chapter 3 Adding and Subtracting Decimals

■ Reteaching the Lesson ■

Activity Have students work in pairs to choose strategies for the exercises. Encourage students to try more than one strategy for each exercise. They should find that some strategies work better on some problems than on others.

Additional Answers

3. **Sample answer: If you round to the nearest hundred, the 1986 and 1987 amounts or the 1987 and 1988 amounts total 200 million. If you round to the nearest ten, the 1986 and 1988 amounts total 200 million.**

4. **$15 is more reasonable because $20 − $6 = $14 and $14 is closer to $15 than $12.**

ON YOUR OWN

5. The last step of the 4-step plan for problem solving asks you to *examine* your solution. *Explain* how the place value to which you round affects how you examine a solution.

6. *Write* a definition of a reasonable answer in your own words.

7. *Explain* how the strategies you used in estimating with large decimal numbers can be applied to estimating with small decimal numbers.

5–7. See students' work.

MIXED PROBLEM SOLVING

STRATEGIES
Look for a pattern.
Solve a simpler problem.
Act it out.
Guess and check.
Draw a diagram.
Make a chart.
Work backward.

Solve. Use any strategy.

8. *Write a Problem* using the following numbers and phrases: $4.59; 3 centimeters; 75° F; 9:00 A.M., and 6 hours.
See students' work.

9. *Statistics* During the fall fund-raising project, Roberta's class sold magazine subscriptions. The pictograph shows how many magazines they sold during a 6-week period.

Subscriptions Sold Per Week

Week 1–6

4 subscriptions = ▢

a. How many subscriptions were sold during week 3? **See margin.**

b. What does the half-magazine shown during week 4 mean? **See margin.**

c. During what week did the class sell 16 subscriptions? **See margin.**

10. *Sports* Suppose 235,532 people attended a 4-game home stand to see the Texas Rangers during the 1997 season. Which is a more reasonable estimate for the number of people that attended each game: 50,000 or 60,000? Explain. **See margin.**

11. *Statistics* Based on the data in the table on page 116, what would you predict about the shipment of CDs from 1995 to 2000? Explain how you made your prediction. **See margin.**

12. *Education* Robert E. Lee High School will graduate 678 seniors on June 8. The ceremony will be held in the gymnasium. The gymnasium holds 2,100 people in addition to the graduates. Is it reasonable to offer each graduate three tickets for family and friends? Explain. **See margin.**

13. **Test Practice** Which diagram does *not* show 0.7? **B**

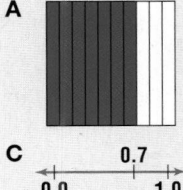

A   B

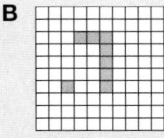

C   0.7
0.0   1.0

D ▵▵▵▵▵
  ▵▵▵▵▵

Lesson 3-6A THINKING **LAB** **117**

Extending the Lesson

Activity Have pairs of students interview the librarian or department heads to determine how they estimate budget figures each year. Have students ask questions such as: *What checking methods are used to determine the accuracy of the figures? What would happen if the figures were not accurate?*

Sample problem: *The art teacher needs to have at least half a ream of paper per student per year. A ream costs $8.95. The teacher has 180 students during the year. What would be a reasonable budget for paper?* **$900**

---

3 PRACTICE/APPLY

Check for Understanding
Use the results from Exercise 4 to determine whether students comprehend how to determine whether an answer is reasonable.

Extra Practice If students need additional practice in problem solving, extra practice is available on the following pages.
- Determine Reasonable Answers, see p. 564
- Mixed-Problem Solving, see pp. 593–594

Assignment Guide
All: 5–13

4 ASSESS

Closing Activity
Writing Have students imagine that during the course of the school year they are going to get 5 hours of math homework each night, for a total of 1,750 hours. Have them write a paragraph describing how they can determine whether that number of hours makes sense. **A more reasonable answer would be 900 hours.**

Additional Answers
9a. They sold 12 subscriptions during week three.

9b. Since each symbol means 4 subscriptions, a half-magazine means 2 subscriptions.

9c. They sold 16 subscriptions during week two.

10. 60,000 is more reasonable because 235,532 is closer to 240,000 than 200,000 and 240,000 ÷ 4 = 60,000.

11. Sample answer: By drawing a line graph and extending it to match the segment between 1993 and 1994, over 1.6 million CDs will be distributed in 2000.

12. Sample answer: yes; 2,100 ÷ 3 = 700. Since there are 678 seniors, this is a reasonable plan.

- *Study Guide Masters*, p. 22
- *Practice Masters*, p. 22
- *Enrichment Masters*, p. 22
- Transparencies 3-6, A and B
- *Assessment and Evaluation Masters*, p. 72
- *Classroom Games*, pp. 5–6
- *School to Career Masters*, p. 3
- *Technology Masters*, p. 5
- CD-ROM Program
  - Resource Lesson 3-6

Recommended Pacing	
Standard	Day 10 of 12
Honors	Day 8 of 10
Block	Day 5 of 6

1 FOCUS

5-Minute Check
(Lesson 3-5)

Estimate using rounding.
1. $0.463 + 0.729$
   $0.5 + 0.7 = 1.2$
2. $29.542 - 11.116$
   $30 - 11 = 19$

Estimate using clustering.
3. $11.75 + 11.89 + 11.60 + 12$
   48
4. $8.75 + 9.22 + 9.48 + 8.99$
   36
5. Mona saw CDs on sale for $8.95 each. About how much money will she need to buy three CDs? **about $27**

The 5-Minute Check is also available on **Transparency 3-6A** for this lesson.

Motivating the Lesson
Hands-On Activity Separate students into pairs and give each pair three number cubes. Have students take turns rolling the cubes, writing down the three numbers in any order to represent dollars and cents. Tell students this figure is the amount of money they can "deposit in their accounts." After three turns each, ask them to add up their own figures to see who has the greater sum. Point out that they are adding decimals.

3-6 Adding and Subtracting Decimals

What you'll learn
You'll learn to add and subtract decimals.

When am I ever going to use this?
You'll add and subtract decimals when you balance your checking account.

Example
APPLICATION

The table shows the top ten movies based on money earned on Thanksgiving weekend in 1996. What was the total earned by the top four movies?

In order to find the total, you need to add the four numbers.

Adding decimals is like adding whole numbers. Make sure that you line up the decimal points before you add or subtract.

Movie	Money Earned (millions)
101 Dalmatians	$45.1
Star Trek: First Contact	25.5
Space Jam	17.4
Ransom	17.3
Jingle All the Way	17.25
The Mirror Has Two Faces	8.1
The English Patient	5.6
Set It Off	4.4
Romeo & Juliet	3.4
Sleepers	1.4

Source: Exhibitor Relations Co. Inc.

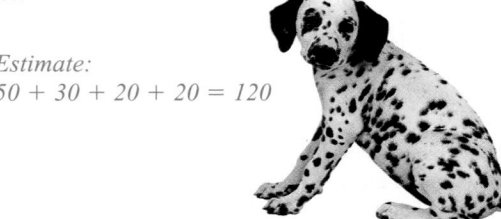

$$\begin{array}{r} \overset{2\ 1}{45.1} \\ 25.5 \\ 17.4 \\ +17.3 \\ \hline 105.3 \end{array}$$ *Estimate:* $50 + 30 + 20 + 20 = 120$

The top four movies earned a total of $105.3 million. The estimate shows that the answer is reasonable.

Entertainment Refer to the table above. How much more money was earned by *101 Dalmatians* than by *Sleepers*?

Explore   What do you know?
You know how much money each movie earned.

What do you need to know?
You need to know how much more money *101 Dalmatians* earned than *Sleepers*.

Plan   You need to subtract the smaller amount earned from the larger amount earned.

Estimate first. $45 - 1 = 44$

Solve $\begin{array}{r} \overset{4\ 11}{4\cancel{5}.\cancel{1}} \\ -\ 1.4 \\ \hline 43.7 \end{array}$ *101 Dalmatians* earned $43.7 million more than *Sleepers*.

Examine   The estimate shows that the answer is reasonable.

Sometimes it is necessary to annex zeros in order to subtract decimals.

 Transparency 3-6B contains a teaching aid for this lesson.

Examples

② Find the difference of 7 and 2.35.

Estimate: 7 − 2 = 5

$$\begin{array}{r} 7.00 \\ -2.35 \\ \hline 4.65 \end{array}$$ *Annex two zeros.*
Rename and subtract.

$$\begin{array}{r} 6\ 9\,10 \\ 7.\cancel{0}\cancel{0} \\ -2.35 \\ \hline 4.65 \end{array}$$

Study Hint

Estimation When you use a calculator to add or subtract decimals, estimate to check that the result is reasonable. If it isn't, check to see if each decimal point was entered correctly.

The estimate shows that the answer is reasonable.

③ Find the sum of 47.68 and 7.8.

Estimate: 50 + 10 = 60

47.68 ⊞ 7.8 ⊟ *55.48*

The estimate shows that the answer is reasonable.

INTEGRATION

④ Algebra Evaluate $x + y$ if $x = 4.56$ and $y = 19.367$.

$x + y = 4.56 + 19.367$ *Replace x with 4.56 and y with 19.367.*

$$\begin{array}{r} 4.560 \\ +19.367 \\ \hline 23.927 \end{array}$$ *Annex a zero and line up the decimal points.*
Add.

The value is 23.927.

Thinking Algebraically When estimating sums and differences of decimals, encourage students to write out their estimate with a variable. For example, in Example 2, they could write $7 − 2 = x$, and then find x.

In-Class Examples

For Example 1
Refer to the table on page 118 and find how much more money was earned by *Space Jam* than by *Jingle All the Way*.
$0.15 million or $150,000

For Example 2
Find the difference of 90 and 9.85. **80.15**

For Example 3
Find the sum of 26.05 and 8.2. **34.25**

For Example 4
Evaluate $m + n$ if $m = 9.24$ and $n = 14.817$. **24.057**

Teaching Tip You may want to have students verify their computations by using a calculator.

CHECK FOR UNDERSTANDING

Communicating Mathematics
2. Estimate: 4 + 0 + 2 = 6. Then line up the decimal points and add.

Math Journal

Guided Practice

Read and study the lesson to answer each question.

1. *Write* directions explaining how to subtract 2.67 from 3. **See margin.**

2. *Explain* how you would find the sum of 3.701, 0.49, and 2.4 using paper and pencil.

3. *Write* a paragraph explaining how adding and subtracting decimals compares to adding and subtracting whole numbers. **See students' work.**

Add or subtract.

4. $\begin{array}{r} 6.4 \\ +3.3 \\ \end{array}$ **9.7**

5. $\begin{array}{r} 1.34 \\ +0.9 \\ \end{array}$ **2.24**

6. $\begin{array}{r} 41.39 \\ -23.17 \\ \end{array}$ **18.22**

7. $\begin{array}{r} 3.7 \\ -2.95 \\ \end{array}$ **0.75**

8. $67.38 − 37.46$ **29.92**

9. $3.702 + 0.49 + 2.4$ **6.592**

10. *Algebra* Solve the equation $c = 0.085 + 2.487$. **2.572**

Lesson 3-6 Adding and Subtracting Decimals **119**

3 PRACTICE/APPLY

Check for Understanding
If students need additional practice or instruction after completing Exercises 1–10, you may find one of the following options helpful.
- Extra Practice, see p. 565
- Reteaching Activity
- *Study Guide Masters,* p. 22
- *Practice Masters,* p. 22
- Interactive Mathematics Tools Software

Additional Answer
1. Sample answer: Annex two zeros. Then line up the decimal points, rename, and subtract.

Reteaching the Lesson

Activity Give groups of three students an addition or subtraction problem. Have one student line up the decimal points and another find the sum or difference. The third student checks the answer on a calculator. The group with the first correct answer makes up the next problem.

Error Analysis
Watch for students who forget to line up the decimal points before they add or subtract.
Prevent by having students turn their papers sideways and place the decimals vertically on the same ruled line.

CHAPTER Project

Exercise 34 asks students to advance to the next stage of work on the Chapter Project. Remind students to estimate an answer before working out the exact cost.

Study Guide Masters, p. 22

EXERCISES

Practice

Add or subtract.

11. 2.3 **6.4**
    +4.1

12. 0.37 **0.92**
    +0.55

13. 0.67 **0.24**
    −0.43

14. 42.76 **11.17**
    −31.59

15. $6.78 **$11.77**
    + 4.99

16.   8 **14.76**
    +6.76

17. 8.267 **1.747**
    −6.52

18. 17.6 **12.87**
    −4.73

19. 6.6 − 4.58 **2.02**

20. 5.77 − 2.374 **3.396**

21. 0.563 + 5.8 + 6.89 **13.253**

22. 23.4 + 9.865 + 18.26 **51.525**

23. Find the sum of 84.34 and 67.235. **151.575**

24. How much is 46 minus 23.78? **22.22**

25. How much more than $102.90 is $115? **$12.10**

26. Find the total of 3.702, 0.49, 77.74, and 62.39. **144.322**

Solve each equation.

27. $29.2 − 2.78 = b$ **26.42**

28. $e = 5.162 + 0.6099$ **5.7719**

29. $c = 4 − 1.9$ **2.1**

30. $478.98 − 46 = k$ **432.98**

31. *Algebra* What is the value of $a − b$ if $a = 34.6$ and $b = 23.88$? **10.72**

32. *Algebra* Evaluate $r + s + t$ if $r = 45.1$, $s = 16$, and $t = 8.091$. **69.191**

Applications and Problem Solving

33a. 2.6 pounds

33. *Food* The graph shows how many pounds of turkey the average person eats during the year.

   a. How many more pounds of turkey does the average person eat from October through December than from January through March?

   b. How many pounds of turkey does the average person eat in one year?
   18 pounds

GOBBLE, GOBBLE

Oct.-Dec. 6.3

July-Sept. 4.1

April-June 3.9

Jan.-March 3.7

Source: National Turkey Federation, 1995

34. *Working on the* **CHAPTER Project** Refer to the table on page 93. In Dawn's family, two people are over 12, and three are under 12. Find the exact cost for her family to take the Whale-Watching Tour and attend the Hawaiian Luau. **$324.75**

35. *Critical Thinking* Arrange the digits 1, 2, 3, 4, 5, 6, 7, and 8 into two decimals so that their difference is as close to 0 as possible. Use each digit only once. **0.2 − 0.1876543 = 0.0123457**

36. *Money Matters* Marta plans to buy a baseball for $6.50, a baseball glove for $37.99, and a baseball cap for $13.79. Estimate the cost of these items before tax is added. *(Lesson 3-5)* **$7 + $38 + $14 = $59**

37. | Test Practice | How long is the pencil in centimeters? *(Lesson 3-2)* **B**

A 9 cm

B 8 cm

C 7 cm

D 6 cm

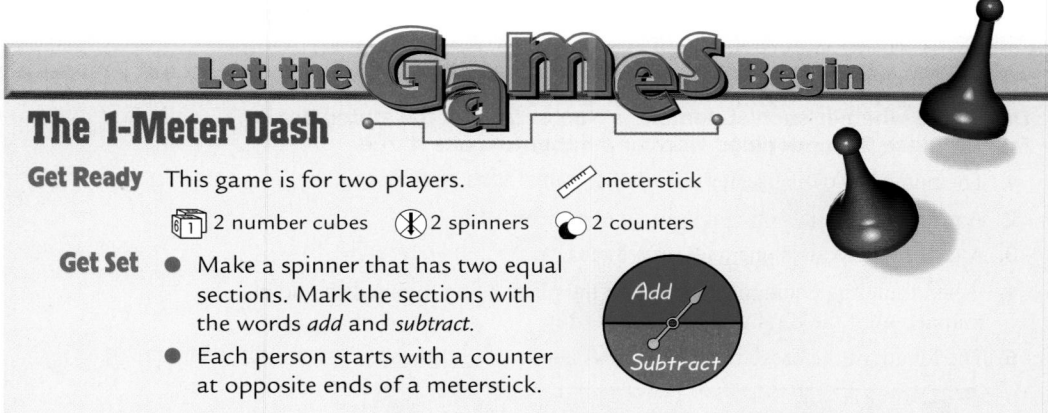

38. The mode is misleading because it is also the highest number in the set.

38. *Statistics* Is the mode a misleading measure of central tendency for this set of data: 21, 20, 19, 13, 21, 18, 12, 21? Explain. *(Lesson 2-8)*

39. *Algebra* Identify the number that is the solution of the equation $42 \div h = 14$; 3, 4, or 5. *(Lesson 1-7)* **3**

Let the Games Begin

The 1-Meter Dash

Get Ready This game is for two players.

🎲 2 number cubes ⊗ 2 spinners 📏 meterstick ◐ 2 counters

Get Set
- Make a spinner that has two equal sections. Mark the sections with the words *add* and *subtract*.

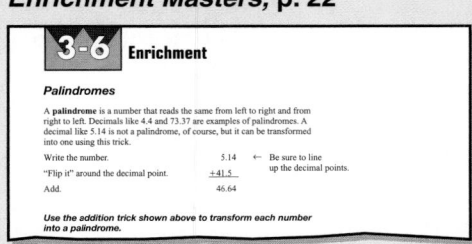

- Each person starts with a counter at opposite ends of a meterstick.

Go
- The first player rolls the number cubes and spins the spinner.
- The player forms a decimal using the numbers on the number cube and then rolls the number cubes again.
- The player forms another decimal, performs the operation shown on the spinner, and then moves the counter that many centimeters along the meterstick. For example, suppose a player rolls a 2 and a 4, spins *subtract*, and then rolls a 5 and 6. That player could move $6.5 - 2.4$ or 4.1 centimeters.
- The winner is the first player to have their counter go beyond the other end of the meterstick.

🖥 **interNET CONNECTION** Visit www.glencoe.com/sec/math/mac/mathnet for more games.

Lesson 3-6 Adding and Subtracting Decimals **121**

■ Extending the Lesson ■

Enrichment Masters, p. 22

3-6 Enrichment

Palindromes

A **palindrome** is a number that reads the same from left to right and from right to left. Decimals like 4.4 and 73.37 are examples of palindromes. A decimal like 5.14 is not a palindrome, of course, but it can be transformed into one using this trick.

Write the number.	5.14	← Be sure to line up the decimal points.
"Flip it" around the decimal point.	+41.5	
Add.	46.64	

Use the addition trick shown above to transform each number into a palindrome.

Let the Games Begin

Point out that students should form decimals that will give them the greatest sum or difference possible. The *Hands-On Lab Masters*, p. 18, contains spinner patterns that can be modified for this activity.

Practice Masters, p. 22

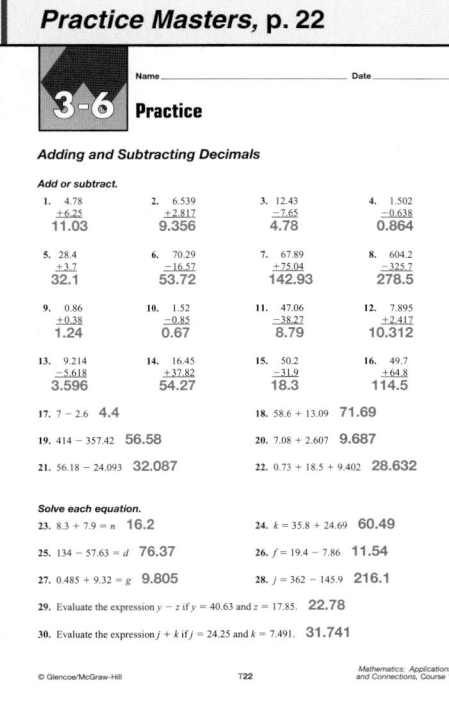

Name _____ Date _____

3-6 Practice

Adding and Subtracting Decimals

Add or subtract.

| 1. | 4.78 +6.25 **11.03** | 2. | 6.539 +2.817 **9.356** | 3. | 12.43 −7.65 **4.78** | 4. | 1.502 −0.638 **0.864** |

| 5. | 28.4 +3.7 **32.1** | 6. | 70.29 −16.57 **53.72** | 7. | 67.89 +75.04 **142.93** | 8. | 604.2 −325.7 **278.5** |

| 9. | 0.86 +0.38 **1.24** | 10. | 1.52 −0.85 **0.67** | 11. | 47.06 −38.27 **8.79** | 12. | 7.895 +2.417 **10.312** |

| 13. | 9.214 −5.618 **3.596** | 14. | 16.45 +37.82 **54.27** | 15. | 50.2 −31.9 **18.3** | 16. | 49.7 +64.8 **114.5** |

17. $7 - 2.6$ **4.4**

18. $58.6 + 13.09$ **71.69**

19. $414 - 357.42$ **56.58**

20. $7.08 + 2.607$ **9.687**

21. $56.18 - 24.093$ **32.087**

22. $0.73 + 18.5 + 9.402$ **28.632**

Solve each equation.

23. $8.3 + 7.9 = n$ **16.2**

24. $k = 35.8 + 24.69$ **60.49**

25. $134 - 57.63 = d$ **76.37**

26. $f = 19.4 - 7.86$ **11.54**

27. $0.485 + 9.32 = g$ **9.805**

28. $j = 362 - 145.9$ **216.1**

29. Evaluate the expression $y - z$ if $y = 40.63$ and $z = 17.85$. **22.78**

30. Evaluate the expression $j + k$ if $j = 24.25$ and $k = 7.491$. **31.741**

© Glencoe/McGraw-Hill   T22   *Mathematics: Applications and Connections, Course 1*

Lesson 3-6 **121**

Study Guide and Assessment

Study Guide and Assessment

Vocabulary

This section provides a listing of the new terms, properties, and phrases that were introduced in this chapter. Have students define each term and provide an example or two of it, if appropriate.

Understanding and Using the Vocabulary

These exercises check students' understanding of the terms by using a variety of verbal formats including matching, completion, and true/false.

Glossaries A complete glossary of terms appears on pages 642–648. The glossary also appears in Spanish on pages 649–656.

Additional Answer

10. Clustering is used to estimate sums and differences when numbers are close to the same numbers.

Vocabulary

After completing this chapter, you should be able to define each term, concept, or phrase and give an example or two of each.

Measurement
centimeter (p. 100)
meter (p. 102)
metric system (p. 102)
millimeter (p. 100)

Number and Operations
clustering (p. 113)
place value (p. 95)

Problem Solving
reasonable answers (pp. 116–117)

Understanding and Using the Vocabulary

Determine whether each statement is *true* or *false*. If the statement is false, replace the underlined word or number to make it true.

1. The number 0.07 is <u>greater</u> than 0.071. **false, less**
2. A millimeter equals <u>one thousandth</u> of a meter. **true**
3. A centimeter equals <u>one tenth</u> of a meter. **false, one hundredth**
4. When rounding decimals, the digit in the place being rounded should be rounded up if the digit to its right is <u>6</u>. **true**
5. The length of the cassette tape below is about 10 <u>millimeters</u>. **false, centimeters**

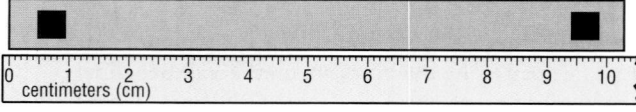

6. In 643.082 the digit 2 names the number two <u>hundredths</u>. **false, thousandths**
7. Six hundred and twelve thousandths written as a decimal is <u>0.612</u>. **false, 600.012**
8. Estimating $3.3 + 2.9 + 3.4 + 3.09$ by computing $3 + 3 + 3 + 3$ is an example of <u>clustering</u>. **true**
9. If the amount of your purchase is $6.74, then a reasonable amount of change from <u>$20</u> is $13.26. **true**

In Your Own Words

10. ***Explain*** when and why the clustering strategy is used to estimate sums. **See margin.**

122 Chapter 3 Adding and Subtracting Decimals

 ### MindJogger Videoquizzes

MindJogger Videoquizzes provide an alternative review of concepts presented in this chapter. Students work in teams to answer questions, gaining points for correct answers. The questions are presented in three rounds.
Round 1 Concepts–5 questions
Round 2 Skills–4 questions
Round 3 Problem Solving–4 questions

Objectives & Examples

Upon completing this chapter, you should be able to:

● model, read, and write decimals *(Lesson 3-1)*

Model $\frac{3}{100}$ using base-ten blocks.

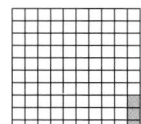

Write $\frac{73}{100}$ as a decimal.

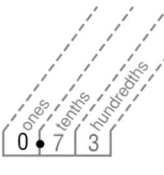

Write 7 in the tenths place and 3 in the hundredths place.

Write 0.024 in words.

The last digit, 4, is in the thousandths place. The decimal in words is *twenty-four thousandths*.

● show relationships among metric units of length and measure line segments *(Lesson 3-2)*

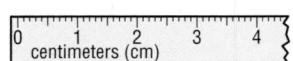

The line segment measures 4.1 centimeters or 41 millimeters.

Review Exercises

Use these exercises to review and prepare for the chapter test.

Write each fraction or mixed number as a decimal.

11. $\frac{8}{100}$ **0.08**   12. $8\frac{9}{10}$ **8.9**

13. $14\frac{17}{1,000}$ **14.017**   14. $\frac{643}{10,000}$ **0.0643**

Write each expression as a decimal.

15. two tenths **0.2**

16. thirty-four hundredths **0.34**

17. fifty-three thousandths **0.053**

18. thirty and twelve ten-thousandths **30.0012**

19. *Life Science* An amoeba is one of the larger creatures of the microscopic world. The length of a typical amoeba is 0.0008 meter. Write 0.0008 in words. **eight ten-thousandths**

Use a centimeter ruler to measure each line segment.

20. ———————————— **4 cm**

21. ———————————————— **6 cm**

22. ——————— **2.5 cm**

Use a centimeter ruler to measure one side of the figure.

23.  **2 cm**

Chapter 3 Study Guide and Assessment **123**

Objectives & Examples

This section reviews the skills and concepts of the chapter and shows completely worked examples.

Review Exercises

These exercises provide practice for the corresponding objectives.

Assessment and Evaluation Masters, pp. 59–60

Name_____ Date_____

3

Chapter 3 Test, Form 1B

1. What is the decimal for $\frac{9}{100}$?
   A. 0.9   B. 0.90   C. 0.09   D. 900   1. __C__

2. What is the decimal for twenty-three and seven tenths?
   A. 0.237   B. 237   C. 23.07   D. 23.7   2. __D__

3. What is the decimal for twenty-eight hundredths?
   A. 2.8   B. 0.28   C. 280   D. 0.028   3. __B__

4. What is the decimal for $\frac{93}{1,000}$?
   A. 0.093   B. 0.93   C. 0.930   D. 930   4. __A__

5. What is the decimal for sixteen and six thousandths?
   A. 16.006   B. 16.6000   C. 16.06   D. 16.060   5. __A__

6. What is the decimal for $\frac{9,563}{10,000}$?
   A. 9,563   B. 9.563   C. 956.3   D. 0.9563   6. __D__

7. What is the length of the segment to the nearest centimeter?
   A. 70 cm   B. 72 cm   C. 7 cm   D. 8 cm   7. __C__

8. What is the length of the segment to the nearest millimeter?
   A. 30 mm   B. 38 mm   C. 3.8 mm   D. 40 mm   8. __B__

9. How many millimeters are in 5.21 centimeters?
   A. 52.1 mm   B. 0.521 mm   C. 521 mm   D. 50 mm   9. __A__

10. Which decimal is greater than 0.073?
    A. 0.07   B. 0.008   C. 0.08   D. 0.072   10. __C__

11. Which decimals are ordered from least to greatest?
    A. 0.006, 0.0005, 0.07   B. 0.054, 3.06, 1.013
    C. 1.013, 0.054, 3.06   D. 0.0005, 0.006, 0.07   11. __D__

12. Which decimals are ordered from greatest to least?
    A. 2.138, 2.158, 1.35   B. 5.0101, 5.01, 5.0103
    C. 5.0103, 5.0101, 5.01   D. 1.35, 2.138, 2.158   12. __C__

© Glencoe/McGraw-Hill   59   Mathematics: Applications and Connections, Course 1

3

Chapter 3 Test, Form 1B (continued)

13. Round 43.628 to the nearest tenth.
    A. 43.6   B. 43.7   C. 43   D. 44   13. __A__

14. Round 1.7968 to the nearest thousandth.
    A. 1.796   B. 1.8   C. 1.80   D. 1.797   14. __D__

15. Round 0.053 to the nearest hundredth.
    A. 0.06   B. 0.05   C. 0.53   D. 0.5   15. __B__

16. Round $72.81 to the nearest dollar.
    A. $73   B. $72   C. $72.80   D. $70   16. __A__

17. Estimate $9.15 + $7.89 using rounding.
    A. $16   B. $17   C. $18   D. $20   17. __B__

18. Estimate 0.93 − 0.19 using rounding.
    A. 0.8   B. 0.7   C. 8   D. 7   18. __B__

19. Use clustering to estimate 3.98 + 4.01 + 4.03 + 4.01.
    A. 16   B. 17   C. 18   D. 20   19. __A__

20. Estimate $29.42 − $16.21.
    A. $13.25   B. $15   C. $13   D. $20   20. __C__

21. Subtract: 76.9 − 52.84.
    A. 24.14   B. 24.16   C. 24.06   D. 23.06   21. __C__

22. Add: 81.1 + 3.04.
    A. 814.40   B. 78.06   C. 111.5   D. 84.14   22. __D__

23. Evaluate $a − b$ if $a = 3.542$ and $b = 0.432$.
    A. 3.111   B. 3.974   C. 3.110   D. 7.862   23. __C__

24. Neal had $32 and earned $8.60. How much money does he have?
    A. $40.60   B. $11.80   C. $41.60   D. $12.80   24. __A__

25. Alice has $10 with which to buy blank video tapes. Each tape costs $2.98. What is a reasonable number of tapes that she can buy?
    A. 2   B. 3   C. 4   D. 5   25. __B__

© Glencoe/McGraw-Hill   60   Mathematics: Applications and Connections, Course 1

Assessment and Evaluation

Six forms of Chapter 3 Test are available in the *Assessment and Evaluation Masters* as shown in the chart.

Chapter 3 Test, Form 1B, is shown at the right. Chapter 3 Test, Form 2B, is shown on the next page.

1A	Multiple Choice	Honors
1B	Multiple Choice	Average
1C	Multiple Choice	Basic
2A	Free Response	Honors
2B	Free Response	Average
2C	Free Response	Basic

Objectives & Examples

● compare decimals and order a set of decimals *(Lesson 3-3)*

Is 5.43 greater than 5.427?

Since 3 hundredths is greater than 2 hundredths, 5.43 is greater than 5.427.

Write 45.93, 46.4, 45.89, and 45.311 in order from least to greatest.

The decimals in order from least to greatest are 45.311, 45.89, 45.93, and 46.4.

● round decimals *(Lesson 3-4)*

Round 4.739 to the nearest tenth.

The digit to the right of the tenths place is 3. So, 7 remains the same.

4.739 rounded to the nearest tenth is 4.7.

● estimate decimal sums and differences *(Lesson 3-5)*

a. rounding

$$7.79 \rightarrow 7.8$$
$$\underline{-2.32} \rightarrow \underline{-2.3}$$
$$5.5$$

b. clustering

$$7.96 + 8.1 + 8.23 + 7.7 \rightarrow 8 + 8 + 8 + 8$$
$$\rightarrow 8 \times 4 = 32$$

● add and subtract decimals *(Lesson 3-6)*

Find the difference of 7.3 and 2.89.

$$\overset{6\ 1210}{7.3\cancel{0}}$$ *Line up the decimal points.*
$$-2.89$$ *Annex a zero.*
$$4.41$$ *Rename and subtract.*

Review Exercises

State the greatest number in each group.

24. 5.218 or 5.207   **25.** 11.6 or 11.13

26. 13.02, 13.022, or 13.21 **13.21**
24. 5.218   25. 11.6

Order each set of decimals from least to greatest. 27. 0.0289, 0.0319, 0.032, 0.31

27. 0.0319, 0.31, 0.032, 0.0289

28. 75.3, 7.598, 7.8, 75.6, 75.09
   7.598, 7.8, 75.09, 75.3, 75.6

Order each set of decimals from greatest to least.

29. 6.32, 6.75, 6.39, 6.02 **6.75, 6.39, 6.32, 6.02**

30. 17.0463, 17.045, 17.023, 17.0201, 17.002
30. 17.0463, 17.045, 17.023, 17.0201, 17.002

Round each number to the underlined place-value position.

31. <u>7</u>.29 **7**   **32.** 76.8<u>0</u>2 **76.80**

33. 13.<u>5</u>81 **13.6**   **34.** 69.<u>9</u>99 **70.0**

Estimate. 35. 5 − 1 = 4   36. $30 + $20 = $50

35. 4.86 − 1.131   **36.** $34.29 + $17.58

37. 6.19 + 5.98 + 5.7 + 6 + 6.3
   6 + 6 + 6 + 6 + 6 = 30
38. *Money Matters* Inali worked three days last week. He earned $19.85 on Monday, $17.75 on Thursday, and $21.30 on Saturday. About how much did he earn in all? **about $60**

Add or subtract.

39.   15.63 **12.912** **40.**   4.63 **9.35**
     − 2.718   +4.72

41. 25.6 + 47.92 + 3.1 + 0.48 **77.1**

42. Solve $x = 12 − 3.45$. **8.55**

Assessment and Evaluation Masters, pp. 65–66

Name_____ Date _____

3   **Chapter 3 Test, Form 2B**

Write each fraction or mixed number as a decimal.
1. $2\frac{5}{10}$ 1.   **2.5**
2. $\frac{29}{1,000}$ 2.   **0.029**
3. $\frac{38}{100}$ 3.   **0.38**
4. $\frac{123}{10,000}$ 4.   **0.0123**
5. forty-four ten-thousandths   5.   **0.0044**
6. nineteen and three thousandths   6.   **19.003**

Measure each line segment to the nearest centimeter.
7. ▬▬▬▬▬▬   7.   **5 cm**
8. ▬▬▬▬   8.   **4 cm**

Measure each line segment to the nearest millimeter.
9. ▬▬▬▬▬   9.   **42 mm**
10. ▬▬▬▬▬▬   10.   **59 mm**
11. Estimate the height of a minivan using the metric system.   11.   **Sample answer: 2 m**

State the greatest number in each group.
12. 0.315 or 0.0325   12.   **0.315**
13. 0.799, 0.08, or 0.8   13.   **0.8**
14. 0.290, 0.30, 0.280   14.   **0.30**
15. Order the set of decimals from least to greatest: 0.038, 0.4, 0.35, 0.0005.   15.   **0.0005, 0.038, 0.35, 0.4**
16. Order the set of decimals from greatest to least: 0.089, 0.090, 0.009.   16.   **0.090, 0.089, 0.009**

© Glencoe/McGraw-Hill   65   *Mathematics: Applications and Connections, Course 1*

3   **Chapter 3 Test, Form 2B (continued)**

Round each number to the underlined place-value position.
17. 0.3<u>4</u>44   17.   **0.344**
18. 38.9<u>4</u>9   18.   **38.9**
19. 4<u>3</u>.61   19.   **44**
20. Round $1.32 to the nearest dollar.   20.   **$1**

Estimate. Use any method.
21. $5.49 + $4.75 + $5.08   21.   **$5 + $5 + $5 = $15**
22. 75.42 − 12.81   22.   **75 − 13 = 62**
23. 3.05 + 2.99 + 3.47 + 2.72   23.   **3 + 3 + 3 + 3 = 12**
24. 70.4 − 35.1   24.   **70 − 35 = 35**
25. 98.2 + 25.46   25.   **100 + 25 = 125**

Add or subtract.
26. 89.4 + 78.56   26.   **167.96**
27. 98.1 − 17.89   27.   **80.21**
28. 50 − 9.3   28.   **40.7**
29. 4.9 + 3.876 + 13.029   29.   **21.805**
30. How much is 48 minus 35.47?   30.   **12.53**
31. Evaluate $r − s$ if $r = 6.792$ and $s = 3.9994$.   31.   **2.7926**

Solve.
32. Emily wants to buy notebooks at $2.95 each. She has $10 to spend. About how many notebooks can she buy?   32.   **3**
33. Four same-priced items were purchased. Based on rounding, the estimate for the total was $20. What is the maximum price each item could have cost?   33.   **$5.49**

© Glencoe/McGraw-Hill   66   *Mathematics: Applications and Connections, Course 1*

Test and Review Software

You may use this software, a combination of an item generator and item bank, to create your own tests or worksheets. Types of items include free response, multiple choice, short answer, and open ended.

CD-ROM Program

The CD-ROM Program contains an Assessment Game whose questions review the concepts in this chapter.

Applications & Problem Solving

43. Inventory Control Cheryl works in a warehouse. She must place items on shelves according to their stock number. Arrange the set of stock numbers in order from least to greatest. *(Lesson 3-3)*

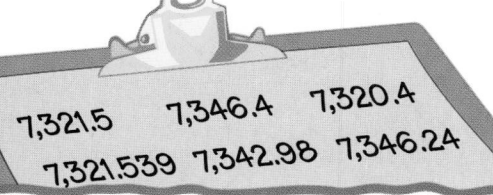

7320.4, 7321.5, 7321.539, 7342.98, 7346.24, 7346.4

44. Money Matters Lamont wrote a check for groceries. On the check, he wrote the amount of purchase as "Seventy-six and $\frac{32}{100}$ dollars." Write the amount as a decimal. *(Lesson 3-1)* **$76.32**

45. Money Matters Mr. Marlin bought three pairs of athletic shoes for his children. The shoes cost $39.99, $75.50, and $89.90. Estimate the amount of money Mr. Marlin spent on shoes. *(Lesson 3-5)* **about $210**

46. Reasonable Answers Jackson bought a pennant at a souvenir shop. He paid for the $7.59 pennant with a $10 bill. Should he expect about $4 or $2 in change? *(Lesson 3-6A)* **about $2**

Alternative Assessment

Performance Task

Suppose you are planning a family picnic. The prices of the items you want are shown in the table.

Item	Cost
chips	2 bags for $4.48
bread	3 loaves for $3.29
cookies	12 for $3.50
apples	4 for $1.25

Estimate the cost to feed your family. **See students' work.**

Suppose you bought the quantities listed in the cost column. Write the total dollar amount in words. If you have a $20 bill, would you expect about $10 or $8 in change?

twelve and $\frac{52}{100}$ dollars; about $8

A practice test for Chapter 3 is provided on page 597.

Completing the CHAPTER Project

Use the following checklist to make sure your plan is complete.

☑ A schedule of travel times and activities.

☑ All travel costs including transportation, lodging, food, souvenirs, and entertainment. You may have to estimate the cost of some items.

☑ A paragraph describing why you chose your vacation spot.

PORTFOLIO Write an example to illustrate using each type of estimation (rounding and clustering), and include them in your portfolio.

Applications & Problem Solving

This section provides additional practice in solving real-world problems that involve the skills of this chapter.

Alternative Assessment

The **Performance Task** provides students with a performance assessment opportunity to evaluate their work and understanding.

CHAPTER Project

Students should complete the final stages of their project and prepare a class demonstration of their results. A scoring guide for the project is available in the *Investigations and Projects Masters*, p. 27.

PORTFOLIO Students should add to their portfolios at this time.

Assessment and Evaluation Masters, p. 69

Name_____ Date_____

3 **Chapter 3 Performance Assessment**

Instructions: Demonstrate your knowledge by giving a clear, concise solution to each problem. Be sure to include all relevant drawings and justify your answers. You may show your solutions in more than one way or investigate beyond the requirements of the problems.

1. Use the table to answer each question.

Amounts of Gases in Dry Air	
Gas	**Amount**
Nitrogen	$\frac{7,802}{10,000}$
Argon	ninety-four ten-thousandths
Oxygen	$\frac{201}{1,000}$
Other	$\frac{94}{10,000}$

a. Write the amount of argon in dry air as a decimal.

b. Write each fraction in the table as a decimal.

c. Order the decimals from least to greatest.

d. Find the total amount of oxygen and nitrogen in dry air.

e. How much more oxygen than argon is there in dry air?

f. Tell how to round a decimal without using a number line.

g. Round each decimal in parts a and b to the nearest hundredth.

2. Refer to the rectangle. The perimeter of a rectangle is $l + l + w + w$.

a. Use a centimeter ruler to measure the length and width of the rectangle.

b. Estimate the perimeter of the rectangle. Explain why you would consider this a reasonable number for the perimeter of the rectangle.

© Glencoe/McGraw-Hill   69   *Mathematics: Applications and Connections, Course 1*

Performance Assessment

Additional performance assessment tasks for this chapter are included in the *Assessment and Evaluation Masters* on page 69. A scoring guide is also provided on page 81.

The Standardized Test Practice may be used to help students prepare for standardized tests. The test items are written in the same style as those in state proficiency tests and standardized tests like CAT, CTBS, ITBS, MAT, SAT, and Terra Nova. The test items cover skills and concepts covered up to this point in the text.

The pages can be used as an overnight assessment. After students have completed the pages, discuss how each problem can be solved, or provide copies of the solutions from the *Solutions Manual.*

Assessment and Evaluation Masters, p. 75

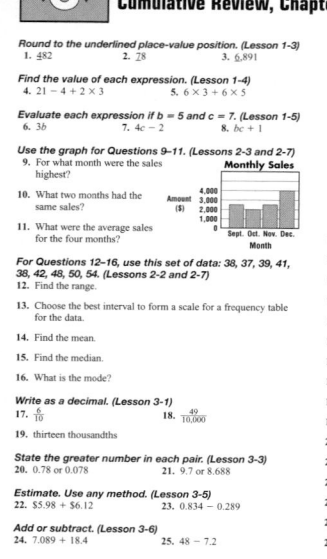

CHAPTERS 1-3

Standardized Test Practice

Assessing Knowledge & Skills

Section One: Multiple Choice

There are twelve multiple-choice questions in this section. Choose the best answer. If a correct answer is *not here*, mark the letter for Not Here.

1. The class scores on a history quiz were: 8, 9, 8, 8, 7, 9, 10, 7, 10, 8, 8, 9, 9, 10, 9, and 8. What was the mode? **A**
  A 8
  B 8.5
  C 8.56
  D 8 and 9

2. Jodi keeps her fish in a tank that has a base 60 centimeters wide and 100 centimeters long. If she gets a tank that is 20 centimeters wider but with the same length as the older tank, what will be the dimensions of the new tank? **H**
  F 40 cm by 100 cm
  G 60 cm by 120 cm
  H 80 cm by 100 cm
  J 80 cm by 120 cm

3. Write $2 \times 3^3 \times 5 \times 7^2$ as the product of factors. **C**
  A $2 \times 3 \times 5 \times 7$
  B $2 \times 3 \times 3 \times 5 \times 7^2$
  C $2 \times 3 \times 3 \times 3 \times 5 \times 7 \times 7$
  D $3^3 \times 7^2$

4. What is the solution of $6a = 42$? **J**
  F 4
  G 5
  H 6
  J 7

5. Which shows the number *five and eighty-six thousandths?* **A**
  A 5.086
  B 5.0086
  C 5.860
  D 5.86

6. The graph shows the number of donuts sold at two stores.

About how many more donuts were sold at The Superstore than at Max's Market on Wednesday? **G**
  F 0
  G 20
  H 40
  J 60

Please note that Questions 7–12 have five answer choices.

7. A TV costs $975, a home theater system costs $395, and a VCR costs $169. Which is the best estimate for the total amount for this entertainment package? **D**
  A $1,000
  B $1,200
  C $1,300
  D $1,600
  E $2,000

8. Four boxes of cereal have weights of 22, 11, 29, and 21 ounces. Which weight is a reasonable average weight of the 4 boxes of cereal? **J**
  F 50 ounces
  G 40 ounces
  H 30 ounces
  J 20 ounces
  K 10 ounces

126   Chapters 1–3 Standardized Test Practice

◀◀◀**Instructional Resources**
Another cumulative review is shown at the left and is available in the *Assessment and Evaluation Masters,* p. 75.

9. A tree farm has 438 seedlings to be planted in 6 rows. How many seedlings need to be planted in each row? **B**

 A 89
 B 73
 C 66
 D 63
 E Not Here

10. Lin paid $275.59 for a 25-inch color TV. The tax was $15.60. What was the price of the TV? **J**

 F $291.19
 G $269.99
 H $260.99
 J $259.99
 K Not Here

11. If an airplane travels 435 miles per hour, how many miles will it travel in 5 hours? **D**

 A 2,055 mi
 B 2,075 mi
 C 2,155 mi
 D 2,175 mi
 E Not Here

12. Each day Toru drives his delivery truck 2.75 kilometers to his first stop, 0.5 kilometer to his second stop, 7.8 kilometers to his third stop, and 5.42 kilometers back to the distribution center. How many kilometers does Toru drive his truck each day? **K**

 F 15.47 km
 G 13.47 km
 H 13.22 km
 J 4.4 km
 K Not Here

Test-Taking Tip

You can prepare for taking standardized tests by working through practice tests like this one. The more you work with different styles of testing, the better you become at test taking.

Section Two: Free Response

This section contains eight questions for which you will provide short answers. Write your answers on your paper.

13. There will be 120 people at the music awards banquet. Each table seats 8 people. How can you find the number of tables needed? **divide 120 by 8**

14. Any number that makes an equation true is a(n) _____. **solution**

15. The lengths in miles of the Great Lakes are Lake Superior, 350; Lake Michigan, 307; Lake Huron, 206; Lake Erie, 241; and Lake Ontario, 193. What is the median of this data? **241 miles**

16. Round 5.99$\underline{9}$8 to the underlined place-value position. **6.000**

17. What is the value of $8 + 6 \times 10$? **68**

18. State the greater of the numbers, 1.089 or 1.09. **1.09**

19. Which measure of central tendency would be best to describe the colors of cars at a used car lot? **mode**

20. Evaluate the expression $a - b$ if $a = 42.2$ and $b = 35.9$. **6.3**

Test-Taking Tip

"Practice makes perfect" makes perfect sense in test-taking. Students should become familiar with the organization and style of standardized tests. The more they practice, the better their chances of scoring well.

Assessment and Evaluation Masters, pp. 73–74

Name_____ Date_____

Chapter 3 Standardized Test Practice

1. Mrs. Moore's math class test scores are 74, 87, 92, 72, 89, 75, 67, 69, 84, 99, 86, and 75. Determine the range of the data. 1. **C**
 A. 75   B. 99   C. 32   D. 30

2. What would be an appropriate scale for the data in Question 1? 2. **A**
 A. 66 to 100   B. 68 to 99   C. 10   D. 70 to 100

3. How many millimeters are in 5.8 centimeters? 3. **C**
 A. 0.58 mm   B. 6 mm   C. 58 mm   D. 5 mm

4. Order the decimals from least to greatest. 9.99, 9.92, 10.25, 9.96 4. **D**
 A. 10.25, 9.99, 9.96, 9.92   B. 9.96, 9.92, 9.99, 10.25
 C. 9.92, 9.99, 9.96, 10.25   D. 9.92, 9.96, 9.99, 10.25

5. A fathom is equal to 1.8288 meters. For easier reference, this number is rounded to the nearest hundredth. What number is used? 5. **B**
 A. 1.8 m   B. 1.83 m   C. 1.829 m   D. 1.82 m

6. A *dram* is a British unit of weight equal to 3.88 grams. A standard aspirin tablet weighs eight hundred thirty six ten-thousandths drams. Write this weight as a decimal. 6. **A**
 A. 0.0836   B. 836.0010   C. 8.3006   D. 0.836

7. As a gift for her father, Yasuaki ordered books for $6.95, $9.95, $7.95, and $6.95. She had a coupon for $15 off her purchase. About what was the cost of her order after the coupon? 7. **D**
 A. $32   B. $16.50   C. $15   D. $17

8. The number of wins for the 1995–96 season for each team in the Central Division of the NBA were 41, 52, 25, 46, 72, 21, 46, 47. Determine the stems for this set of data. 8. **C**
 A. 2, 4, 5, 7   B. 1, 2, 5, 6, 7
 C. 2, 3, 4, 5, 6, 7   D. 1, 3, 5, 7

9. In science class, teams measured how far two frogs jumped in a certain amount of time. Stacey and Moira's frogs jumped 438.762 meters and 521.778 meters. What was the total distance for the two frogs? 9. **A**
 A. 960.540 m   B. 83.016 m
 C. 961 m   D. 959.540 m

10. To reach a friend's house, you have to walk 3 blocks east and 4 blocks north. If your house is at the origin, how could you represent your friend's house with an ordered pair? 10. **B**
 A. (4, 3)   B. (3, 4)   C. 7 blocks   D. You can't.

© Glencoe/McGraw-Hill   73   *Mathematics: Applications and Connections, Course 1*

Chapter 3 Standardized Test Practice (continued)

11. Evaluate $15 \div 3 + 27 \div 3$. 11. **C**
 A. 11   B. 4   C. 14   D. 16

12. A theater has 27 rows of main floor seats. Each row has 112 seats in it. All of the main floor seats are filled and 40 people are seated in the balcony. How many people are in the theater? 12. **B**
 A. 3,024   B. 3,064   C. 2,984   D. 1,192

13. What three types of graphs are commonly seen in newspapers? 13. **A**
 A. bar graphs, line graphs, circle graphs
 B. bar graphs, circle graphs, square graphs
 C. line graphs, calendar graphs, bar graphs
 D. circle graphs, triangle graphs, line graphs

14. Find the next three numbers in the pattern: 28, 32, 36, _?_, _?_, _?_. 14. **D**
 A. 38, 42, 46   B. 44, 48, 52
 C. 42, 44, 46   D. 40, 44, 48

15. Estimate $492 \div 5$. 15. **C**
 A. 98.4   B. 10   C. 100   D. 1,000

16. Solve $54 + a = 72$ using mental math. 16. **B**
 A. 126   B. 18   C. 13   D. 86

17. A movie theater owner calculates the income from a Saturday movie matinee by evaluating the expression $6.50A + 3.50C$, where A = adults and C = children. If A = 25 and C = 105, what is the income? 17. **A**
 A. $530   B. $770   C. $1300   D. $630

18. If you make a frequency chart for the letters in the title "Hail to the Chief," what letter is used most frequently? 18. **A**
 A. H B. A C. I D. T

19. Evaluate $4 \cdot 6^2 + 3^3$. 19. **D**
 A. 181   B. 57   C. 67   D. 171

20. You are making a line graph for the average student ages in your school. The horizontal axis will be grade levels and the vertical axis will be average ages at each grade level. How do you expect the graph to appear? 20. **B**
 A. flat line   B. rising line
 C. falling line   D. cannot determine

© Glencoe/McGraw-Hill   74   *Mathematics: Applications and Connections, Course 1*

Instructional Resources ▶▶▶

Additional standardized test practice is shown at the right and is available in the *Assessment and Evaluation Masters*, pp. 73–74.

Interdisciplinary Investigation

GET READY

This optional investigation is designed to be completed by a group of 4 or 5 students over several days or several weeks.

Mathematical Overview

This investigation utilizes the concepts from Chapters 1–3.
- converting time from minutes to seconds
- adding and subtracting decimals
- drawing statistical graphs

Time Management	
Gathering Data	30 minutes
Calculations	20 minutes
Creating Graphs	30 minutes
Summarizing Data	15 minutes
Presentation	10 minutes

Instructional Resources
- *Investigations and Projects Masters*, pp. 1–4
- *Hands-On Lab Masters*
  - grid paper, p. 11

Investigations and Projects Masters, p. 4

Name_____ Date_____

Interdisciplinary Investigation
(Student Edition, Pages 136–137)

What's for Dinner?

Use these tables to record your data.

Breakfast	Lunch	Dinner

Menu Item	Calories	Grams of Protein	Milligrams of Sodium	Grams of Fat
Total				

Menu Item	Calories	Grams of Protein	Milligrams of Sodium	Grams of Fat
Total				

© Glencoe/McGraw-Hill   4   *Mathematics: Applications and Connections, Course 3*

Interdisciplinary Investigation

"O" IS FOR OLYMPICS

What do Atlanta, Georgia, and Athens, Greece, have in common? Both cities were sites for the modern Olympic Games. In 1896, Athens hosted the first modern Olympic Games with 13 nations participating. Now almost 200 nations compete every four years.

Are athletes today faster? How do the times of runners today compare to runners in 1896? Are swimmers getting faster every year? How do the times for men and women compare?

What You'll Do

In this investigation, you will collect data about an Olympic event. You will graph the data and predict future records in the event.

Materials calculator

 graph paper

Procedure

1. Work in groups of four. Find a source of winning Olympic times for each year that the games were held. Select one timed event in which both women and men participate. The event needs to have been held at least ten times. Two examples are the 100-meter dash and the 100-meter freestyle (swimming).

2. Separate your group into pairs. One pair will work with the men's times, and the other pair with the women's times for the event you selected. Make a table to display the times. Write all times in seconds. For example, 1 m 22.2 s is 60 s + 22.2 s or 82.2 s.

3. Graph your results.

4. Predict the winning times for men and women for the year 2012. Does your prediction seem reasonable?

Technology Tips

- Use a **calculator** to determine the mean and median.

- Use a **graphing calculator** or **graphing software** to display your graphs.

◄◄◄**Instructional Resources**

A recording sheet to help students organize their data for this investigation is shown at the left and is available in the *Investigations and Projects Masters*, p. 4.

 Cooperative Learning

This investigation offers an excellent opportunity for using cooperative learning groups. For more information on cooperative learning strategies and group management, see *Cooperative Learning in the Mathematics Classroom*.

Making the Connection

Use the Olympic data you collected as needed to help in these investigations.

Language Arts

Your friend says, "Men are better than women in sports and always will be." Write a letter to your friend stating whether you agree or disagree with this statement. Use the data from the investigation to support your opinion.

Foreign Language

Select a winner in an Olympic event from a country that speaks a foreign language that you know or are interested in. Write an article about the athlete's victory using that language.

Social Studies

Select a popular sport. Research the history of this sport and write a short report.

Go Further

- Predict when you think men and women will have the same time in the event you used in the investigation. What do you think the fastest time will ever be for that event?

- If possible, time everyone in your class in the 100-meter dash. How do the times compare to the Olympic times?

interNET CONNECTION For more information on the history of the Olympics, visit the following website.
www.glencoe.com/sec/math/mac/mathnet

 You may want to place your work on this investigation in your portfolio.

Interdisciplinary Investigation "O" is for Olympics **129**

Working in Teams Each group of four should research various events that fit the criteria. They should then discuss and agree on what event to investigate. After they separate into pairs, each member of the pair should check the other's conversions, addition, and subtraction. They should work together to create an accurate and colorful graph.

Making the Connection
You may wish to alert other teachers on your team that your students may need their assistance in this investigation.
Language Arts Students need to edit and rewrite their letters until their wording is clear and effective. They need to support their statements with facts from their data.
Foreign Language Encourage students to use a foreign language dictionary and to ask a teacher of that language for help if necessary.
Social Studies Students should include the country of origin and the major countries where the sport is popular today.

Investigations and Projects Masters, p. 3

SCORING GUIDE

Interdisciplinary Investigation
(Student Edition, Pages 136–137)

What's for Dinner?

Level	Specific Criteria
3 Superior	• Shows a thorough understanding of the concepts of *expressing numbers as ratios and converting between fractions and percents.* • Uses appropriate strategies to solve problems. • Computations are correct. • Written explanations are exemplary. • Charts, model, and any statements included are appropriate and sensible. • Goes beyond the requirements of some or all problems.
2 Satisfactory, with minor flaws	• Shows understanding of the concepts of *expressing numbers as ratios and converting between fractions and percents.* • Uses appropriate strategies to solve problems. • Computations are mostly correct. • Written explanations are effective. • Charts, model, and any statements included are appropriate and sensible. • Satisfies the requirements of problems.
1 Nearly Satisfactory, with obvious flaws	• Shows understanding of most of the concepts of *expressing numbers as ratios and converting between fractions and percents.* • May not use appropriate strategies to solve problems. • Computations are mostly correct. • Written explanations are satisfactory. • Charts, model, and any statements included are appropriate and sensible. • Satisfies the requirements of problems.
0 Unsatisfactory	• Shows little or no understanding of the concepts of *expressing numbers as ratios and converting between fractions and percents.* • Does not use appropriate strategies to solve problems. • Computations are incorrect. • Written explanations are not satisfactory. • Charts, model, and any statements included are not appropriate or sensible. • Does not satisfy the requirements of the problems.

© Glencoe/McGraw-Hill   3   *Mathematics: Applications and Connections, Course 3*

You may wish to have students graph their results on posterboard for class presentation. During this presentation, ask students what they learned about time differences in men's and women's events and differences over the years covered.

Instructional Resources ▶▶▶
Sample solutions for this investigation are provided in the *Investigations and Projects Masters* on p. 2. The scoring guide for assessing student performance shown at the right is also available on p. 3.

CHAPTER 4

Multiplying and Dividing Decimals

Previewing the Chapter

Overview

In this chapter, students learn to estimate decimal products, to multiply and divide decimals by whole numbers and decimals, and to round decimal quotients. The distributive property is used to compute products mentally. Finding the perimeter and area of figures and naming metric measurements is integrated into the chapter. Students learn to solve problems by first solving a simpler problem.

Lesson (pages)	Lesson Objectives	NCTM Standards	Standardized Tests	State/Local Objectives
4-1A (132)	Use grid paper to multiply decimals by whole numbers.	1–3, 5–7, 11		
4-1 (133–136)	Estimate and find the products of decimals and whole numbers.	1–10, 13	CAT, CTBS, ITBS, SAT, TN	
4-2 (137–139)	Compute products mentally using the distributive property.	1–10		
4-3A (140)	Use grid paper to multiply decimals.	1–3, 5–7	CTBS, TN	
4-3 (141–143)	Multiply decimals.	1–7, 9, 10, 12	CAT, CTBS, ITBS, MAT, SAT, TN	
4-3B (144)	Explore how data and formulas are entered in a computer spreadsheet.	1–6, 10		
4-4 (145–148)	Find the perimeters and the areas of rectangles and squares.	1–10, 12, 13	MAT, SAT	
4-4B (149)	Explore how perimeter and area are related.	1–4, 8, 12, 13		
4-5A (150–151)	Solve problems by first solving a simpler problem.	1–4, 6, 8–10, 12, 13	ITBS, MAT	
4-5 (152–155)	Divide decimals by whole numbers.	1–9, 12, 13	CAT, CTBS, ITBS, SAT, TN	
4-6A (156)	Use base-ten blocks to divide decimals.	2–7		
4-6 (157–159)	Divide decimals by decimals.	1–7, 10	CTBS, MAT, SAT, TN	
4-7 (161–163)	Divide decimals involving special cases of zero in the quotient.	1–7, 9, 12, 13	CTBS, MAT, TN	
4-8 (164–166)	Use metric units of mass and capacity.	1–5, 13	ITBS, MAT, SAT	
4-9 (167–169)	Change units within the metric system.	1–7, 9	CAT, CTBS, MAT, SAT, TN	

CAT = California Achievement Tests, CTBS = Comprehensive Tests of Basic Skills, ITBS = Iowa Tests of Basic Skills, MAT = Metropolitan Achievement Tests, SAT = Stanford Achievement Tests, TN = Terra Nova

Organizing the Chapter

LESSON PLANNING GUIDE

Lesson	Extra Practice (Student Edition)	BLACKLINE MASTERS (PAGE NUMBERS)											
		Study Guide	Practice	Enrichment	Assessment & Evaluation	Classroom Games	Diversity	Hands-On Lab	School to Career	Science and Math Lab Manual	Technology	Transparencies A and B
4-1A								43				
4-1	p. 565	23	23	23								4-1
4-2	p. 565	24	24	24							7	4-2
4-3A								44				
4-3	p. 566	25	25	25	99							4-3
4-3B												
4-4	p. 566	26	26	26							8	4-4
4-4B								45				
4-5A	p. 566											
4-5	p. 567	27	27	27	98, 99				4	73–76		4-5
4-6A								46				
4-6	p. 567	28	28	28		7–9						4-6
4-7	p. 567	29	29	29	100							4-7
4-8	p. 568	30	30	30								4-8
4-9	p. 568	31	31	31	100		4	72				4-9
Study Guide/ Assessment					85–97, 101–103							

OTHER CHAPTER RESOURCES

Student Edition

Chapter Project, pp. 131, 135, 143, 173
School to Career, p. 160
Let the Games Begin, p. 166

Technology

 CD-ROM Program

Interactive Mathematics Tools Software

Teacher's Classroom Resources

Applications
Family Letters and Activities, pp. 7–8
Investigations and Projects Masters, pp. 29–32
Meeting Individual Needs
Transition Booklet, pp. 21–28, 31–34
Investigations for the Special Education Student, pp. 9–12, 31–34

Teaching Aids
Answer Key Masters
Block Scheduling Booklet
Lesson Planning Guide
Solutions Manual

Professional Publications
Glencoe Mathematics Professional Series

Planning the Chapter

MindJogger Videoquizzes provide a unique format for reviewing concepts presented in the chapter.

ASSESSMENT RESOURCES

Student Edition
Mixed Review, pp. 136, 139, 143, 148, 155, 159, 163, 166, 169
Mid-Chapter Self Test, p. 155
Math Journal, pp. 138, 147, 162
Study Guide and Assessment, pp. 170–173
Performance Task, p. 173
Portfolio Suggestion, p. 173
Standardized Test Practice, pp. 174–175
Chapter Test, p. 598

Assessment and Evaluation Masters
Multiple-Choice Tests (Forms 1A, 1B, 1C), pp. 85–90
Free-Response Tests (Forms 2A, 2B, 2C), pp. 91–96
Performance Assessment, p. 97
Mid-Chapter Test, p. 98
Quizzes A–D, pp. 99–100
Standardized Test Practice, pp. 101–102
Cumulative Review, p. 103

Teacher's Wraparound Edition
5-Minute Check, pp. 133, 137, 141, 145, 152, 157, 161, 164, 167
Building Portfolios, p. 130
Math Journal, pp. 132, 140, 144, 149, 156
Closing Activity, pp. 136, 139, 143, 148, 151, 155, 159, 163, 166, 169

Technology
Test and Review Software
MindJogger Videoquizzes
CD-ROM Program

MATERIALS AND MANIPULATIVES

Lesson 4-1A
grid paper†
markers
scissors*

Lesson 4-3A
grid paper†
markers

Lesson 4-3B
computer
spreadsheet software

Lesson 4-4B
grid paper†
scissors*

Lesson 4-6A
base-ten blocks*

Lesson 4-8
index cards
tape

*Glencoe Manipulative Kit   †Glencoe Overhead Manipulative Resources

PACING CHART

See pages T25–T27 for the Course Planning Calendar.

COURSE	DAY 1	DAY 2	DAY 3	DAY 4	DAY 5	DAY 6	DAY 7
Standard	Chapter Project	Lessons 4-1A & 4-1		Lesson 4-2	Lessons 4-3A & 4-3		Lesson 4-4
Honors	Chapter Project	Lesson 4-1	Lesson 4-2	Lessons 4-3 & 4-3B		Lessons 4-4 & 4-4B	
Block	Chapter Project & Lesson 4-1A	Lessons 4-1 & 4-2	Lessons 4-3A & 4-3	Lessons 4-4 & 4-5A	Lessons 4-5 & 4-6A	Lessons 4-6 & 4-7	Lessons 4-8 & 4-9

The *Transition Booklet* (Skills 9–12, 14–15) can be used to practice multiplying and dividing by 1- and 2-digit numbers and measuring capacity and weight.

Interactive Mathematics:
Activities and Investigations

is an activity-based program that may be used as an enhancement for chapters in *Mathematics: Applications and Connections.*

Unit 2, Activity One, Menu B
Use with Lesson 4-5A.

Summary Students estimate how many beans, popcorn kernels, and rice are in three jars. They may want to solve a simpler problem by measuring out a small amount of the object being counted and use this information to make their estimation. Students write about their findings, post them around the room, and discuss the results.

Math Connection Students use proportional reasoning to solve the estimation problem. A proportion is an equation that shows that two ratios are equivalent. Students may solve a simpler problem as their problem-solving strategy.

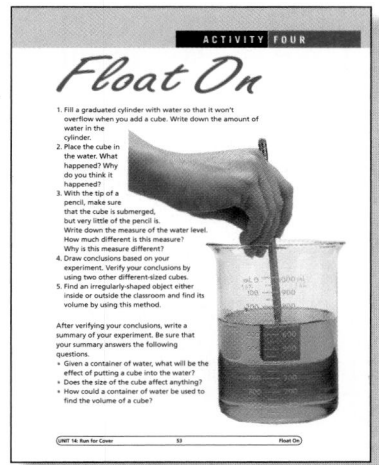

Unit 14, Activity Four
Use with Lesson 4-8.

Summary Students discover that by submerging a cube in water and noticing the difference in the water level, they can find the volume of solids. They then apply this method to find the volumes of irregularly shaped objects. Students also learn about Archimedes' principle and use it to predetermine whether objects will float or sink.

Math Connection Students use measurement and scientific experiments to determine the volume of solids and to determine whether objects will float or sink.

DAY 8	DAY 9	DAY 10	DAY 11	DAY 12	DAY 13	DAYS 14 & 15	DAY 16
Lesson 4-5A	Lesson 4-5	Lessons 4-6A & 4-6		Lesson 4-7	Lesson 4-8	Lesson 4-9, Study Guide & Assessment	Chapter Test
Lesson 4-5A	Lesson 4-5	Lesson 4-6	Lesson 4-7	Lesson 4-8	Lesson 4-9	Study Guide & Assessment, Chapter Test	
Study Guide and Assessment, Chapter Test							

Enhancing the Chapter

APPLICATIONS

Classroom Games, pp. 7–9

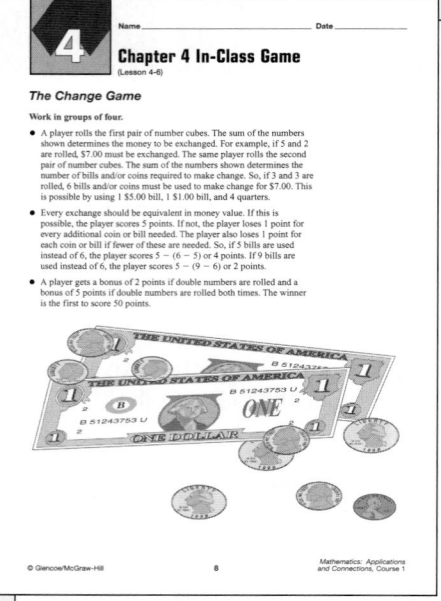

4 Chapter 4 In-Class Game
(Lesson 4-6)

The Change Game

Work in groups of four.

- A player rolls the first pair of number cubes. The sum of the numbers shown determines the money to be exchanged. For example, if 5 and 2 are rolled, $7.00 must be exchanged. The same player rolls the second pair of number cubes. The sum of the numbers shown determines the number of bills and/or coins required to make change. So, if 3 and 3 are rolled, 6 bills and/or coins must be used to make change for $7.00. This is possible by using 1 $5.00 bill, 1 $1.00 bill, and 4 quarters.

- Every exchange should be equivalent in money value. If this is possible, the player scores 5 points. If not, the player loses 1 point for every additional coin or bill needed. The player also loses 1 point for each coin or bill if fewer of these are needed. So, if 5 bills are used instead of 6, the player scores 5 − (6 − 5) or 4 points. If 9 bills are used instead of 6, the player scores 5 − (9 − 6) or 2 points.

- A player gets a bonus of 2 points if double numbers are rolled and a bonus of 5 points if double numbers are rolled both times. The winner is the first to score 50 points.

© Glencoe/McGraw-Hill   8   Mathematics: Applications and Connections, Course 1

Diversity Masters, p. 4

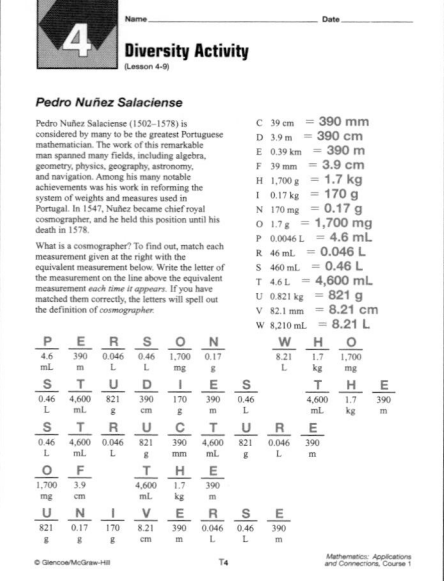

4 Diversity Activity
(Lesson 4-9)

Pedro Nuñez Salaciense

Pedro Nuñez Salaciense (1502–1578) is considered by many to be the greatest Portuguese mathematician. The work of this remarkable man spanned many fields, including algebra, geometry, physics, geography, astronomy, and navigation. Among his many notable achievements was his work in reforming the system of weights and measures used in Portugal. In 1547, Nuñez became chief royal cosmographer, and he held this position until his death in 1578.

What is a cosmographer? To find out, match each measurement at the right with the equivalent measurement below. Write the letter of the measurement on the line above the equivalent measurement *each time it appears*. If you have matched them correctly, the letters will spell out the definition of *cosmographer*.

C 39 cm   = 390 mm
D 3.9 m   = 390 cm
E 0.39 km   = 390 m
F 39 mm   = 3.9 cm
H 1,700 g   = 1.7 kg
I 0.17 kg   = 170 g
N 170 mg   = 0.17 g
O 1.7 g   = 1,700 mg
P 0.0046 L = 4.6 mL
R 46 mL   = 0.046 L
S 460 mL   = 0.46 L
T 4.6 L   = 4,600 mL
U 0.821 kg = 821 g
V 82.1 mm   = 8.21 cm
W 8,210 mL = 8.21 L

```
P    E    R    S    O    N         W    H    O
4.6  390  0.046 0.46 1,700 0.17    8.21 1.7  1,700
mL   m    L    L    mg   g          L   kg   mg

S    T    U    D    I    E    S         T    H    E
0.46 4,600 821 390  170  390  0.46     4,600 1.7  390
L    mL   g   cm   g    m    L          mL   kg   m

S    T    R    U    C    T    U    R    E
0.46 4,600 0.046 821 390 4,600 821 0.046 390
L    mL   L    g   mm   mL   g    L    m

O         F         T    H    E
1,700     3.9       4,600 1.7  390
mg        m         mL   kg   m

U    N    I    V    E    R    S    E
821  0.17 170  8.21 390  0.046 0.46 390
g    g    g    cm   m    L    L    m
```

© Glencoe/McGraw-Hill   T4   Mathematics: Applications and Connections, Course 1

School to Career Masters, p. 4

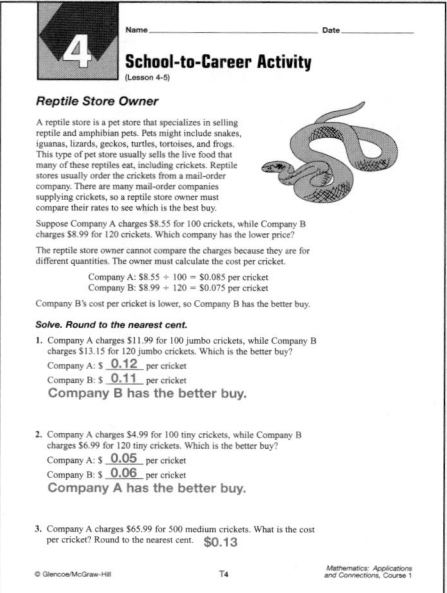

4 School-to-Career Activity
(Lesson 4-5)

Reptile Store Owner

A reptile store is a pet store that specializes in selling reptile and amphibian pets. Pets might include snakes, iguanas, lizards, geckos, turtles, tortoises, and frogs. This type of pet store usually sells the live food that many of these reptiles eat, including crickets. Reptile stores usually order the crickets from a mail-order company. There are many mail-order companies supplying crickets, so a reptile store owner must compare their rates to see which is the best buy.

Suppose Company A charges $8.55 for 100 crickets, while Company B charges $8.99 for 120 crickets. Which company has the lower price?

The reptile store owner cannot compare the charges because they are for different quantities. The owner must calculate the cost per cricket.

Company A: $8.55 ÷ 100 = $0.085 per cricket
Company B: $8.99 ÷ 120 = $0.075 per cricket

Company B's cost per cricket is lower, so Company B has the better buy.

Solve. Round to the nearest cent.

1. Company A charges $11.99 for 100 jumbo crickets, while Company B charges $13.15 for 120 jumbo crickets. Which is the better buy?
   Company A: $ _0.12_ per cricket
   Company B: $ _0.11_ per cricket
   Company B has the better buy.

2. Company A charges $4.99 for 100 tiny crickets, while Company B charges $6.99 for 120 tiny crickets. Which is the better buy?
   Company A: $ _0.05_ per cricket
   Company B: $ _0.06_ per cricket
   Company A has the better buy.

3. Company A charges $65.99 for 500 medium crickets. What is the cost per cricket? Round to the nearest cent. $0.13

© Glencoe/McGraw-Hill   T4   Mathematics: Applications and Connections, Course 1

Family Letters and Activities, pp. 7–8

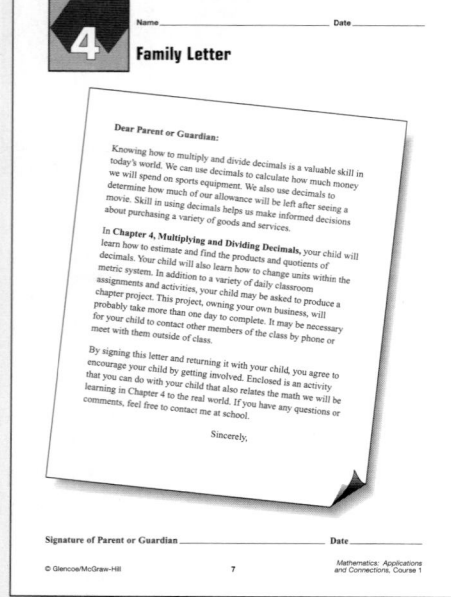

4 Family Letter

Dear Parent or Guardian:

Knowing how to multiply and divide decimals is a valuable skill in today's world. We can use decimals to calculate how much money we will spend on sports equipment. We also use decimals to determine how much of our allowance will be left after seeing a movie. Skill in using decimals helps us make informed decisions about purchasing a variety of goods and services.

In Chapter 4, **Multiplying and Dividing Decimals**, your child will learn how to estimate and find the products and quotients of decimals. Your child will also learn how to change units within the metric system. In addition to a variety of daily classroom assignments and activities, your child may be asked to produce a chapter project. This project, owning your own business, will probably take more than one day to complete. It may be necessary for your child to contact other members of the class by phone or meet with them outside of class.

By signing this letter and returning it with your child, you agree to encourage your child by getting involved. Enclosed is an activity that you can do with your child that also relates the math we will be learning in Chapter 4 to the real world. If you have any questions or comments, feel free to contact me at school.

Sincerely,

Signature of Parent or Guardian _____ Date _____

© Glencoe/McGraw-Hill   7   Mathematics: Applications and Connections, Course 1

4 Family Activity

Calculating Tips and Taxes

Work with a family member to answer the following questions. Visit a favorite restaurant or recall a past visit to a restaurant.

1. Estimate the total cost of the food and beverages.

2. What is your local sales tax rate? Convert the rate to a decimal. For example, a tax rate of $5\frac{1}{2}$% is written as 0.055.

3. Calculate the amount of tax you should be charged for the meal.

4. What is the total amount of the bill (food, beverages, and tax)?

5. Now suppose that you want to leave a 15% tip. Calculate the amount of the tip.

Visit a local retail store. Select three items that you would like to purchase. Use the tax rate from Exercise 2 to estimate the total cost of each item.

6.

7.

8.

© Glencoe/McGraw-Hill   8   Mathematics: Applications and Connections, Course 1

Science and Math Lab Manual, pp. 73–76

19 Science and Mathematics Lab
(Course 1, Lesson 4-5; Course 2, Lesson 3-4; Course 3, Lesson 4-4)

It's Raining, It's Pouring

INTRODUCTION

Rain is very important to plants and animals. Farmers depend on rain to ensure the success of their crops without expensive irrigation. Rain is the source of water for rivers, lakes, and aquifers that provide us with drinking water. Rainfall patterns vary throughout the world and from city to city. In this activity, you will graph your local monthly rainfall totals, find the total annual rainfall, and calculate seasonal rainfall averages.

OBJECTIVES

In this lab, you will:
- graph the monthly rainfall totals.
- find the total annual rainfall and compare it to the average monthly rainfall for the year.

MATERIALS
- local rainfall data for each month of the previous year
- average annual local rainfall
- current local rainfall total for this month

PROCEDURE

1. Fill in the monthly rainfall amounts of the previous year in the Data Table. Be sure to include the unit of measurement.

2. Add the monthly rainfall amounts to find the yearly total.

3. Construct a bar graph from information in the Data Table.

DATA AND OBSERVATIONS

Rainfall Data for _____

Month	Rainfall (in.)
January	
February	
March	
April	
May	
June	
July	
August	
September	
October	
November	
December	
Total	

© Glencoe/McGraw-Hill   75   Mathematics: Applications and Connections

Hands-On Lab Masters, p. 72

4 Lab Activity
(Lesson 4-9)

Name _____ Date _____

Using the Metric System

Use a ruler to measure the lines in centimeters. Then write your measurement in millimeters, meters, and kilometers.

1. ____ **5** centimeters
   50 millimeters **0.05** meters **0.005** kilometers

2. ____ **3** centimeters
   30 millimeters **0.03** meters **0.003** kilometers

3. ____ **7.5** centimeters
   75 millimeters **0.075** meters **0.0075** kilometers

4. ____ **1.5** centimeters
   15 millimeters **0.015** meters **0.0015** kilometers

Find three small objects in your classroom to measure with your ruler. Record what you are measuring. Record its length in centimeters. Write the measurement in millimeters, meters, and kilometers. **Answers will vary.**

5. Object: _____ centimeters
   _____ millimeters _____ meters _____ kilometers

6. Object: _____ centimeters
   _____ millimeters _____ meters _____ kilometers

7. Object: _____ centimeters
   _____ millimeters _____ meters _____ kilometers

© Glencoe/McGraw-Hill   72   *Mathematics: Applications and Connections, Course 1*

Assessment and Evaluation Masters, pp. 98–100

4 Chapter 4 Mid-Chapter Test
(Lessons 4-1 through 4-5)

Name _____ Date _____

Multiply.
1. 0.83×1000 .... 1. **830**
2. 4.38×25 .... 2. **109.5**
3. 0.092×55 .... 3. **5.06**
4. Estimate the product of 2.031 and 21. .... 4. **$2 \times 20 = 40$**

Find each product mentally. Use the distributive property.
5. 6.22×11 .... 5. **68.42**
6. 5.07×21 .... 6. **106.47**
7. Steven bought 12 pairs of socks for $2.96 each. How much did he spend? 7. **$35.52**
8. Brigette jogged around a 440-yard track 5.5 times. How far did she jog? .... 8. **2,420 yd**

Multiply.
9. 0.42×1.25 .... 9. **0.525**
10. 5.1×8.12 .... 10. **41.412**

Evaluate each expression if $a = 3.62$, $b = 0.05$, and $c = 1.5$.
11. ab 11. **0.181**
12. $a(b + c)$ 12. **5.611**
13. Find the perimeter of a square with a side of 3.27 yards. .... 13. **13.08 yd**
14. Find the area of a rectangle 5.9 feet long and 3.8 feet wide. .... 14. **22.42 ft²**
15. The Bakers' living room is 13.5 feet by 11.2 feet. The carpet they like is $2.57 per square foot. How much will it cost to carpet the living room, to the nearest cent? ... 15. **$388.58**

16. Find the area of the figure. 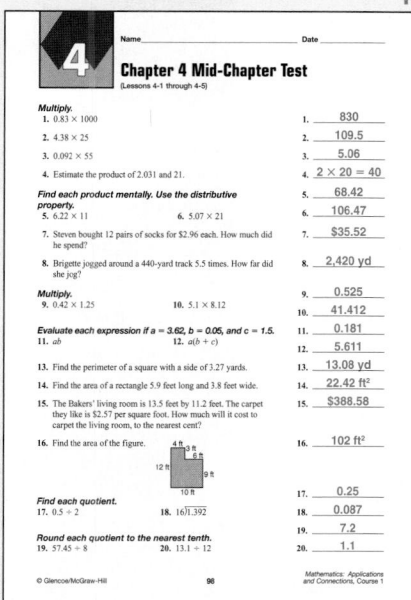 ... 16. **102 ft²**

Find each quotient.
17. $0.5 \div 2$ 17. **0.25**
18. $16\overline{)1.392}$ 18. **0.087**

Round each quotient to the nearest tenth.
19. $57.45 \div 8$ 19. **7.2**
20. $13.1 \div 12$ 20. **1.1**

© Glencoe/McGraw-Hill   98   *Mathematics: Applications and Connections, Course 1*

Chapter 4 Quiz A

Name _____ Date _____
(Lessons 4-1 through 4-3)

1. Estimate 8.79×58 using rounding. .... 1. **$9 \times 60 = 540$**
2. Estimate 248×40.3 using compatible numbers. ... 2. **$250 \times 40 = 10,000$**

Find each product mentally. Use the distributive property.
3. 7×17 4. 62×4 5. 8.1×6 ... 3. **119** 4. **248** 5. **48.6**

Multiply.
6. 9.86×124 7. 88×0.007 .... 6. **1,222.64** 7. **0.616**
8. Five ballpoint pens cost $0.39 each. How much do they cost in all? Solve mentally using the distributive property. ... 8. **$1.95**

Solve each equation.
9. $a = 7.2 \times 7.2$ 10. $c = 0.008 \times 0.9$ ... 9. **51.84** 10. **0.0072**

Chapter 4 Quiz B

Name _____ Date _____
(Lessons 4-4 and 4-5)

Find the perimeter of each figure.
1. 65.5 cm, 38.2 cm, 81.4 cm .... 1. **185.1 cm**
2. rectangle: length, 4.5 in., width, 3.8 in. ..... 2. **16.6 in.**

Find the area of each figure.
3. 12.5 m, 1.5 m .... 3. **18.75 m²**
4. 30 m, 5 m, 18 m, 4 m, 12 m, 9 m .... 4. **198 m²**
5. a square with sides of 9.1 mm .... 5. **82.81 mm²**

Find each quotient.
6. $9\overline{)25.83}$ 7. $64\overline{)492.8}$ 8. $31.74 \div 2$ ... 6. **2.87** 7. **7.7** 8. **15.87**

Round each quotient to the nearest tenth.
9. $7.059 \div 4$ 10. $35.67 \div 15$ 9. **1.8** 10. **2.4**

© Glencoe/McGraw-Hill   99   *Mathematics: Applications and Connections, Course 1*

Technology Masters, pp. 7–8

4 Calculator Activity
(Lesson 4-2)

Name _____ Date _____

The Distributive Property

You can solve word problems by using the distributive property. The parentheses keys on the calculator will help you.

Examples 1 Neil purchased 4 dozen blueberry bagels and 6 dozen cinnamon-raisin bagels for a fund raiser at school. How many bagels did Neil purchase in all?

Enter: 12 ☒ ⎡ 4 ⊞ 6 ⎤ ⩵ 120

Neil purchased 120 bagels in all.

2 Jill has been training to run a marathon for 3 weeks. On the first 7 days, she ran 2.5 miles per day. On the next 7 days, she ran 3 miles each day. On each of the last 7 days, she ran 3.75 miles. How many miles in all did Jill run?

Enter: 7 ☒ ⎡ 2.5 ⊞ 3 ⊞ 3.75 ⎤ ⩵ 64.75

Jill ran 64.75 miles in 3 weeks of training.

Solve each problem using the distributive property.

1. The company assistant put in an order for supplies that included 15 dozen pens and 8 dozen pencils. How many individual pens and pencils were ordered in all? **276**

2. The Music Source is having a sale on CDs and cassettes. They have 140 CDs and 215 cassettes they are selling for $5.29 each. How much money will they earn if all CDs and cassettes are sold? **$1,877.95**

3. If the Music Source decreased the selling price to $4.95, how much money would they earn? What is the difference in earnings from Exercise 2? **$1,757.25; $120.70**

4. Kevin earns $3.15 per hour for each hour he helps Mr. McCready with lawn work. Kevin worked the following hours: Friday: 3.25 hours; Saturday: 4 hours; Sunday: 2.5 hours. How much money did Kevin earn in all? Round to the nearest cent. **$30.71**

© Glencoe/McGraw-Hill   T7   *Mathematics: Applications and Connections, Course 1*

4 Graphing Calculator Activity
(Lesson 4-4)

Name _____ Date _____

Perimeter and Area

You can use graphing calculator to find the perimeter of rectangles. First, enter the expression for the perimeter of a rectangle, $2\ell + 2w$.

Enter: 2 ⎡ L ⎤ ALPHA [L] ⎡ 3 ⎤ ⊞ 2 ⎡ L ⎤ ALPHA [W] ⎡ 3 ⎤ ENTER

Example 1 Find the perimeter of a rectangle whose length is 8 ft and width is 4 ft.

Enter: 2nd [ENTER]

Use the arrow keys to move the cursor to L. Replace the L with 8. Move the cursor to W. Replace the W with 4. Then press ENTER

The perimeter of a rectangle whose length is 8 ft and width is 4 ft is 24 ft.

You can also use a graphing calculator to find the area of rectangles. First, enter the expression for the area of a rectangle, $\ell \times w$.

Enter: ⎡ L ⎤ ALPHA [L] ⎡ 3 ⎤ ☒ ⎡ L ⎤ ALPHA [W] ⎡ 3 ⎤ ENTER

Example 2 Find the area of a rectangle whose length is 9 ft and width is 3 ft.

Enter: 2nd [ENTER]

Use the arrow keys to move the cursor to L. Replace the L with 9. Move the cursor to W. Replace the W with 3. Then press ENTER

The area of a rectangle whose length is 9 ft and width is 3 ft is 27 ft².

Use a graphing calculator to find the perimeter and area of each rectangle described.

1. length = 7 ft, width = 6 ft   **26 ft; 42 ft²**
2. length = 9 ft, width = 4 ft   **26 ft; 36 ft²**
3. length = 8 ft, width = 8 ft   **32 ft; 64 ft²**
4. length = 7 ft, width = 5 ft   **24 ft; 35 ft²**
5. length = 9 ft, width = 2 ft   **22 ft; 18 ft²**
6. length = 6 ft, width = 1 ft   **14 ft; 6 ft²**

© Glencoe/McGraw-Hill   T8   *Mathematics: Applications and Connections, Course 1*

Investigations for the Special Education Student, pp. 9–12, 31–34

Use with:
Course 1–Chapter 4
Course 2–Chapter 7
Course 3–Chapter 3

Investigation 4 Teacher's Guide

The Check's in the Mail

Overview
In this investigation, students simulate how to manage money in the real world. Each student is responsible for choosing a career from the classified ads. Based on the salary earned from that career, students will use the classified ads to find a place to live and develop a monthly budget that is realistic for the salary they earn. In addition, students will keep a checkbook and practice writing checks—another necessary everyday skill. Once budgets are in final form, they are displayed in the classroom for classmates to see and share, so students can learn about many different career possibilities and the interests of their peers.

Activity Goals
Students will:
• write a balanced monthly budget,
• learn to write checks and keep a checkbook, and
• explore different facets of real life.

Planning the Instruction
Prerequisite Skills
Students should have a significant amount of practice computing with decimals and writing numbers in word form.
Materials
• investigation worksheets
• a class set of classified ads
• scissors
• glue
• calculators
Time Needed
five 45-minute periods

Procedure
1. Discuss with the class that managing money and making a budget are necessary skills to have in the real world.
2. First, using the classified ads, students are to choose a job that is of interest to them. Have students glue that career ad onto the Monthly Budget worksheet in the appropriate space. From their selection, each student will use the salary given to compute the monthly paycheck.
3. When the monthly salary is computed, have students set up a checking account using the Keeping a Checkbook worksheet. First, students will mark their first month's salary in the checkbook register. Then, the teacher will show the class the appropriate way to write a check.
4. Once the first month's salary has been deposited, students will search the classifieds for a place to live that is in a reasonable price range. Glue that ad onto the appropriate area of the front of the Monthly Budget worksheet. Have students write and record a check for the first month's rent.
5. With living and working arrangements made, students are now ready to put together the rest of a monthly budget, including transportation, utilities, food, insurance, savings, entertainment, and so on. The work for this step should be done on the back of the Monthly Budget worksheet.
6. When the budget is complete, have students double-check their budget total using a calculator. The final version is then displayed for others to see and learn about various careers, as well as their classmates' interests.

Adaptations and Variations	

The following are some ideas on how this investigation may be modified depending on student population.

LD • Allow more time.
• Offer help when locating items in the classified ads.

PH • Assign a partner or aide to help them search through the classified ads if they are unable to do it themselves.
• Allow a computer to be used.

BD • Monitor students closely during independent work time.
• Isolate students from others if problems arise.

VI • Enlarge certain sections of the classified ads or assign someone to read to them.

© Glencoe/McGraw-Hill   9   *Mathematics: Applications and Connections*

Theme: Business

There are more than 22 million small businesses in the United States today. Not only do they employ more than half of the private workforce, they also account for more than 50 percent of the U.S. Gross Domestic Product. Many businesses begin with one person having an idea for a product or service. With the help of a staff and financial backing, one person's idea can become a thriving company.

Question of the Day Janelle and Lucy make necklaces and sell them for $3.75 each. A neighbor wants to order 125 as gifts for a family reunion. The girls will give him a discount of $0.25 per necklace on the first 100 necklaces and a discount of $0.50 per necklace for any additional. How much will his order cost? **$431.25**

Assess Prerequisite Skills

Ask students to read through the list of objectives presented in "What you'll learn in Chapter 4." You may wish to ask them what each of the objectives means or if they have experienced or used any of these math concepts before.

 Building Portfolios

Encourage students to revise their portfolios as they study this chapter. Have them include specific examples that illustrate their understanding of decimal multiplication and division.

 **Math and the Family**

In the *Family Letters and Activities* booklet (pp. 7–8), you will find a letter to the parents explaining what students will study in Chapter 4. An activity appropriate for the whole family is also available.

Multiplying and Dividing Decimals

What you'll
learn in Chapter 4

● to multiply and divide decimals by whole numbers and decimals,

● to use the distributive property to compute products mentally,

● to find the perimeters and areas of figures,

● to solve problems by first solving a simpler problem,

● to use metric units of mass and capacity, and

● to change units within the metric system

130   Chapter 4 Multiplying and Dividing Decimals

💿 CD-ROM Program

Activities for Chapter 4
• Chapter 4 Introduction
• Interactive Lessons 4-3, 4-4, 4-5, 4-6, 4-9
• Assessment Game
• Resource Lessons 4-1 through 4-9

CHAPTER Project

BE YOUR OWN BOSS!

Owning your own business can be a lot of work. But, it can be very rewarding. It also requires creativity. In this project, you will learn just a little of what goes into making a small business successful. After choosing a product you'd like to sell, you'll explain why you chose it and then determine its selling price. Then you should develop a pricing strategy for large orders. Finally, you need to design an advertisement for your product.

Getting Started

- Decide on a product to sell.
- Determine how much it will cost you to make this product.
- Decide on the price that you want to ask for your product.
- Decide whether you want to do any special promotions such as offering a discount on early orders.

Technology Tips

- Use a **spreadsheet** to organize your price chart and make calculations.
- Use **desktop publishing software** to design your advertisement.

inter NET CONNECTION For up-to-date information on starting a small business, visit:
www.glencoe.com/sec/math/mac/mathnet

Working on the Project

You can use what you'll learn in Chapter 4 to help you sell your product.

Page	Exercise
135	32
143	30
173	Alternative Assessment

inter NET CONNECTION

Glencoe has made every effort to ensure that the website links for *Mathematics: Applications and Connections* at **www.glencoe.com/sec/math/mac/mathnet** are current and contain appropriate content. However, these website links are not under Glencoe's control.

Instructional Resources ▶▶▶

A recording sheet to help students organize their data for the Chapter Project is shown at the right and is available in the *Investigations and Projects Masters,* p. 32.

CHAPTER Project NOTES

Objectives Students should
- develop a pricing strategy for a chosen product.
- design an advertisement.

Project Pointer You may suggest that students begin a *Project Folder* to keep their work as they complete each stage of the Chapter Project. The completed project may also be added to their portfolios.

You may wish to have students organize their information in the form of a business proposal. Explain how they can coordinate their product description, price list, and advertising strategy to create a report potential investors could read.

Investigations and Projects Masters, p. 32

4 Name_____ Date_____

Chapter 4 Project

Be Your Own Boss!

Page 131, Getting Started

Product:

Cost to produce:

Selling price:

Special promotions:

Page 135, Working on the Chapter Project, Exercise 32

Number of Items	Cost
1	
2	
3	
4	
5	
6	
7	
8	
9	
10	

Page 143, Working on the Chapter Project, Exercise 30

a. The after-tax price is:

b. The senior citizen discount price is:

© Glencoe/McGraw-Hill   32   *Mathematics: Applications and Connections, Course 1*

Chapter 4 Project **131**

COOPERATIVE LEARNING

4-1A Multiplying Decimals by Whole Numbers

A Preview of Lesson 4-1

Objective Students use grid paper to multiply decimals by whole numbers.

Optional Resources
Hands-On Lab Masters
• grid paper, p. 11
• worksheet, p. 43

Overhead Manipulative Resources
• decimal models

MANAGEMENT TIPS

Recommended Time
30 minutes

Getting Started Have students read the first paragraph. Then ask questions such as the following.
• How would you represent the number 2? **2 10-by-10 squares**
• What decimal is represented by three columns? **0.3**

The **Activity** shows how to model decimal multiplication. As students are counting squares and finding products, ask what decimal each square represents. Ask how they would write 120 hundredths. Some students may find it helpful to use addition to find the product. For example, 0.60 + 0.60 = 1.20.

ASSESS

Have students complete Exercises 1–4. Make sure students are shading the correct amount.

Use Exercise 5 to determine whether students understand the concept of multiplying decimals by whole numbers.

grid paper

markers

scissors

In this lab, you will use grid paper to draw decimal models. You can use decimal models to multiply a decimal by a whole number. Remember, □ represents one hundredth, ▯ represents one tenth, and 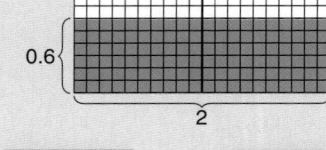 represents one.

In the following activity, the number of rows will represent the first factor of the multiplication, and the number of columns will represent the second factor. The number of shaded rows or columns will represent the product.

TRY THIS

Work in groups of three.

To model 0.6 × 2, follow these steps.

Step 1 Draw two 10-by-10 squares on grid paper to represent the factor 2.

Step 2 Shade six rows to represent 0.6.

Step 3 Cut off the shaded rows and arrange them to form as many 10-by-10 squares as possible.

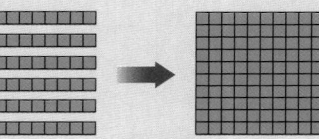

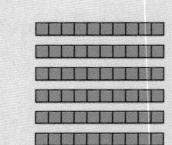

There is one 10-by-10 square, which is 1, and two rows, which is 0.2.

Therefore, 0.6 × 2 = 1.2.

ON YOUR OWN

Draw decimal models to show each product. **1–3. See Answer Appendix.**

1. 4 × 0.6
2. 3 × 0.7
3. 0.5 × 5

4. Write a multiplication problem using decimals for the model shown. **0.7 × 4**

5. *Look Ahead* Find the product of 8 and 0.4 without using models. **3.2**

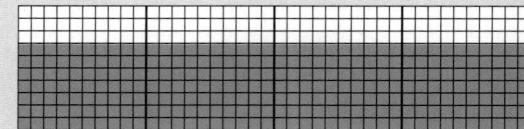

 Math Journal Have students write a paragraph explaining how models show the multiplication of decimals by whole numbers. Students should include an explanation of what one small square, one column, and one large square represent.

Multiplying Decimals by Whole Numbers

What you'll learn

You'll learn to estimate and find the products of decimals and whole numbers.

When am I ever going to use this?

Knowing how to multiply a decimal by a whole number can help you find the amount of money you earn.

Word Wise

compatible numbers

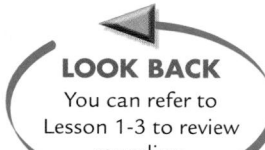

LOOK BACK
You can refer to Lesson 1-3 to review rounding.

Mrs. Sopher spends $2.25 for the Saturday and Sunday edition of her local newspaper. What does this cost per year? *This problem will be solved in Example 2.*

When multiplying a decimal by a whole number, multiply as with whole numbers. One way to determine where to place the decimal point in the product is to use estimation.

There are two methods to estimate products. You can use rounding or **compatible numbers**.

Method 1   Use rounding.

First, round each factor to its greatest place-value position. Then multiply. Do not round 1-digit factors.

Method 2   Use compatible numbers.

It is easy to find the product of compatible numbers mentally.

Examples

1 Find 9 × 78.42.

Estimate using rounding. Round 78.42 to 80. 9 × 80 = 720.

$$
\begin{array}{r}
{\scriptstyle 7\,3\ \ 1} \\
78.42 \\
\times \qquad 9 \\
\hline
705.78
\end{array}
$$

Multiply as with whole numbers.

Since the estimate is 720, place the decimal point after 5.

Check with a calculator. 78.42 9 = 705.78 ✓

APPLICATION

2 Money Matters Refer to the beginning of the lesson. How much does Mrs. Sopher spend per year for newspapers?

Explore You know how much she spends each weekend. You want to know how much she spends in a year.

(continued on the next page)

 Cross-Curriculum Cue

Inform the teachers on your team that your students are studying decimal multiplication and division and metric measures. Suggestions for curriculum integration are:
Health: nutrition labels
Physical Science: formulas

4-1 Lesson Notes

Instructional Resources
- *Study Guide Masters*, p. 23
- *Practice Masters*, p. 23
- *Enrichment Masters*, p. 23
- Transparencies 4-1, A and B
- CD-ROM Program
  - Resource Lesson 4-1

Recommended Pacing	
Standard	Days 2 & 3 of 16
Honors	Day 2 of 15
Block	Day 2 of 8

1 FOCUS

5-Minute Check
(Chapter 3)

1. Write three hundred ten-thousandths as a decimal. **0.0300**

2. Order 980.89, 980.08, 980.189, 890.86, 980.81, and 980.8 from greatest to least. **980.89, 980.81, 980.8, 980.189, 980.08, 890.86**

3. Estimate 18.05 + 39.95 using rounding. **20 + 40 = 60**

4. Subtract: 67.53 − 37.125. **30.405**

5. Add: 30.001 + 14.78. **44.781**

The 5-Minute Check is also available on **Transparency 4-1A** for this lesson.

Motivating the Lesson

Problem Solving Have students read the opening paragraph of the lesson. Ask the following question. *What estimation strategies could be used to find the cost per year for the newspaper?* **Sample answer: rounding**

2 TEACH

Transparency 4-1B contains a teaching aid for this lesson.

Reading Mathematics Have groups of students find the dictionary definitions for *rounding* and *compatible*. Have them discuss how these definitions compare with how the words are used in this lesson.

In-Class Examples

For Example 1
a. Find 6.4×5. **32**
b. Find 3.08×7. **21.56**

For Example 2
During his professional basketball career, Wilt Chamberlain averaged 30.07 points per game for 1,045 games. How many points did he score in his career? **31,423 points**

For Example 3
a. Evaluate the expression $4d$ if $d = 7.2$. **28.8**
b. Evaluate the expression $5s$ if $s = 0.014$. **0.07**

Teaching Tip In Example 3, remind students that $3a$ is the same as $3 \times a$.

3 PRACTICE/APPLY

Check for Understanding
If students need additional practice or instruction after completing Exercises 1–11, one of these options may be helpful.
• Extra Practice, see p. 565
• Reteaching Activity
• *Transition Booklet,* pp. 21–22
• *Study Guide Masters,* p. 23
• *Practice Masters,* p. 23

Cultural Kaleidoscope

If you were passing through the Forum in ancient Rome, you may have seen the *Acta Diurna*, which means "daily events." This was the first newspaper.

Plan There are 52 weeks in a year. Find $\$2.25 \times 52$.
Estimate using compatible numbers.

$\$2.25 \rightarrow \$2$ *Round to the greatest place value.*
$\underline{\times\ 52} \rightarrow \underline{\times 50}$ *Round to 50 since it is easy to multiply 2 and 5 mentally.*

Since $2 \times 5 = 10$, then $2 \times 50 = 100$.

Solve
$$
\begin{array}{r}
\$2.25 \\
\underline{\times\ 52} \\
450 \\
\underline{1125} \\
117.00
\end{array}
$$
Since the estimate is 100, place the decimal point after 7.

Mrs. Sopher spends $\$117.00$ a year for newspapers.

Examine Compared to the estimate, the answer is reasonable.

You can also determine where to place the decimal point in the product by counting the number of decimal places in the decimal factor. The product must have the same number of decimal places. If more decimal places are needed, annex zeros.

Example 3 INTEGRATION

LOOK BACK
You can refer to Lesson 1-5 to review evaluating expressions.

Algebra Evaluate the expression $3a$ if $a = 0.032$.

$3a = 3 \times 0.032$ *Replace a with 0.032.*

$0.032 \leftarrow$ *three decimal places*
$\underline{\times\ \ \ 3}$
$0.096 \leftarrow$ *Annex a zero on the left to make three decimal places.*

Check your answer by adding.
$$
\begin{array}{r}
0.032 \\
0.032 \\
\underline{+\ 0.032} \\
0.096 \ \checkmark
\end{array}
$$
The product is 0.096.

CHECK FOR UNDERSTANDING

Communicating Mathematics

1. Sample answer:
Round 40.32 to 40 and round 251 to 250. You can find 40×250 since 4 and 25 are compatible numbers.

Read and study the lesson to answer each question.

1. *Explain* how you could use compatible numbers to estimate the product of 40.23 and 251.

2. *Write* a multiplication problem for the model shown. **0.9 × 2**

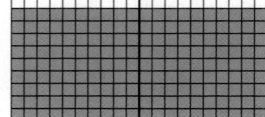

■ Reteaching the Lesson ■

Activity Make multiplication cards that ask students to find the product of a decimal and a whole number. Then have groups of students play "Top of the Tower." Players choose a card, find the product, and add the product to a tower drawn on the chalkboard until they reach the sum of 1,000 (Top of the Tower).

3. *You Decide* Montega uses a calculator to find the product of 34.78 and 452. He gets 15,720.56 for an answer. Is the answer reasonable? Explain why or why not. **See margin.**

Guided Practice

Use estimation to place the decimal point in each product.

4. $0.88 \times 3 = 264$
    2.64

5. $12.6 \times 19 = 2394$
    239.4

6. $254 \times 3.82 = 97028$
    970.28

Multiply.

7. 0.2
  $\times 8$ **1.6**

8. 5.02
  $\times\ 3$ **15.06**

9. 64×0.005 **0.32**

10. *Algebra* Evaluate the expression $12n$ if $n = 5.6$. **67.2**

11. *Astronomy* Pluto, normally the farthest planet from the Sun, is also the slowest. Its average speed around the Sun is 10,604 miles per hour. Earth, by contrast, travels 6.28 times faster. What is the average speed of Earth? **66,593.12 miles per hour**

EXERCISES

Practice

Multiply.

12. 0.6 **3.0**
  $\times 5$

13. 0.28 **1.12**
  $\times\ 4$

14. 2.03 **14.21**
  $\times\ 7$

15. 3.42 **30.78**
  $\times\ 9$

16. 0.007 **0.056**
  $\times\ 8$

17. 10.7 **64.2**
  $\times\ 6$

18. 0.67 **22.11**
  $\times 33$

19.   3.25 **2,606.5**
  $\times 802$

20. $1{,}250 \times 2.5$ **3,125**
21. 0.0125×754 **9.425**
22. $2{,}967 \times 0.071$ **210.657**

Solve each equation.

23. $x = 36 \times 0.007$

24. $4.8 \times 235 = y$

25. $p = 2{,}388 \times 1.65$

26. *Algebra* Evaluate $112d$ if $d = 0.98$. **109.76**

27. What is the solution of $n = 58.002 \cdot 367$? **21,286.734**

28. Will 11.7×5 be closer to 55 or 60? Explain. **See margin.**

29. How does knowing that $5^2 = 25$ help you find the answer to 5×0.5?

23. 0.252
24. 1128
25. 3,940.2
29. Sample answer: Since 0.5 is less than 5, 0.5×5 will be less than 25.

Applications and Problem Solving

30. *Statistics* The graph shows the broadcast television ratings for baseball. In which year were the ratings twice as high as the 1996 ratings? **1995**

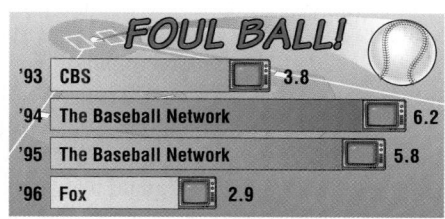

FOUL BALL!

'93	CBS	3.8
'94	The Baseball Network	6.2
'95	The Baseball Network	5.8
'96	Fox	2.9

Sources: Major League Baseball and *USA TODAY* research

31. *Money Exchange* If the Japanese yen (¥) is worth $0.0078, what is the value of ¥ 3,750? **$29.25**

32. *Working on the* CHAPTER Project Determine how much it would cost someone to buy 1 item, 2 items, 3 items, . . . , through 10 items of your product. Organize this data in a table. **See students' work.**

Lesson 4-1 Multiplying Decimals by Whole Numbers **135**

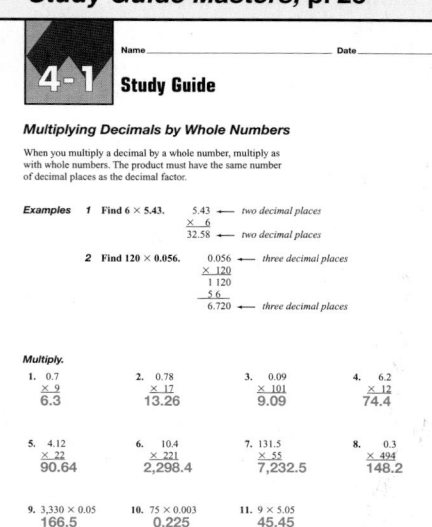

Closing Activity

Writing Have students use information in newspapers or magazines to write real-world problems that involve multiplying a decimal and a whole number.

33. Answers will vary. 33. *Critical Thinking* Create a multiplication problem where the product is
    Sample answer: 45.89 and one of the factors is a whole number.
    5×9.178

Mixed Review   34. **Test Practice** Jaden plans to buy a soccer ball for $21.99, soccer cleats for $45.50, and a soccer shirt for $14.79. What is the cost of these items before tax is added? *(Lesson 3-6)* **A**

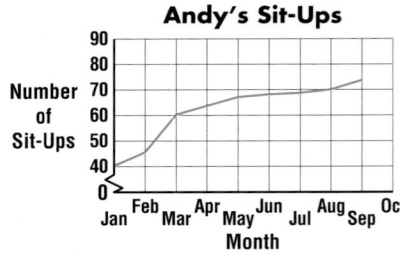

A $82.28

B $81.18

C $80.28

D $80.18

E Not Here

35. *Fitness* The line graph shows Andy's progress doing sit-ups. How many sit-ups do you predict he will be doing in October? *(Lesson 2-5)* **Sample answer: 80**

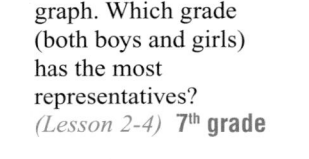

Andy's Sit-Ups

36. *School* At Gillian School, there are 36 student council representatives as shown in the circle graph. Which grade (both boys and girls) has the most representatives? *(Lesson 2-4)* **7ᵗʰ grade**

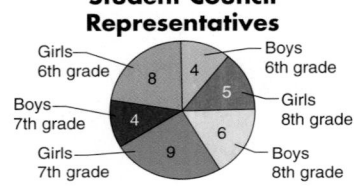

Student Council Representatives

37a. Sample answer: 800 points

37b. Sample answer: Use clustering to total the hundreds and use rounding to estimate to the nearest ten.

37c. Sample answer: You would round to the nearest ten because rounding to the nearest hundred they would all be the same.

37. *Sports* The table shows the total points scored in the eight games in the second round of the midwest region during the 1997 NCAA Division I Women's Basketball Tournament. *(Lesson 1-3)*

a. Estimate the total points scored over the two days.

b. Explain how you could use a combination of methods to estimate the two-day total.

c. Suppose you wanted to make a bar graph comparing the total points scored in each game. Would you round to the nearest hundred or to the nearest ten? Explain.

Total Points Scored	
March 14	March 15
118	106
126	138
141	145
145	148

136 Chapter 4 Multiplying and Dividing Decimals

Practice Masters, p. 23

4-1 Practice

Name_____ Date_____

Multiplying Decimals by Whole Numbers

Use estimation to place the decimal point in each product.

1. $0.73 \times 56 = 4088$ **40.88**
2. $2.7 \times 48 = 1296$ **129.6**
3. $2.94 \times 108 = 31752$ **317.52**
4. $1.035 \times 69 = 71415$ **71.415**
5. $0.8 \times 472 = 3776$ **377.6**
6. $15.06 \times 319 = 480414$ **4,804.14**

Multiply.

7. 0.9×6 **5.4**
8. 3.47×5 **17.35**
9. 0.82×9 **7.38**
10. 27.3×8 **218.4**

11. 0.64×32 **20.48**
12. 5.9×174 **1,026.6**
13. 0.0358×216 **7.7328**
14. 4.76×95 **452.2**

15. 208.7×43 **8,974.1**
16. 0.4×738 **295.2**
17. $1.95 \times 4,620$ **9,009**
18. 0.006×87 **0.522**

19. 89.2×54 **4,816.8**
20. $0.013 \times 2,361$ **30.693**
21. 7.49×105 **786.45**
22. $2.5 \times 3,092$ **7,730**

23. 36×0.07 **2.52**
24. 4.8×235 **1,128**

25. 1.29×614 **792.06**
26. 93×0.57 **53.01**

27. 18×270.9 **4,876.2**
28. 0.006×315 **1.89**

29. Find the product of 58.2 and 67. **3,899.4**

30. What is 1,073 times 2.04? **2,188.92**

© Glencoe/McGraw-Hill   **T23**   Mathematics: Applications and Connections, Course 1

Extending the Lesson

Enrichment Masters, p. 23

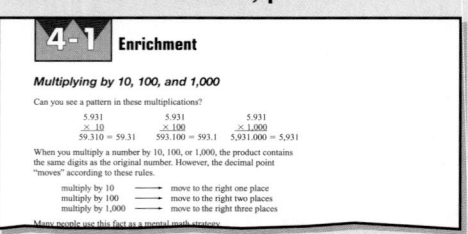

4-1 Enrichment

Multiplying by 10, 100, and 1,000

Can you see a pattern in these multiplications?

5.931	5.931	5.931
× 10	× 100	× 1,000
59.310 = 59.31	593.100 = 593.1	5,931.000 = 5,931

When you multiply a number by 10, 100, or 1,000, the product contains the same digits as the original number. However, the decimal point "moves" according to these rules.

multiply by 10 ⟶ move to the right one place
multiply by 100 ⟶ move to the right two places
multiply by 1,000 ⟶ move to the right three places

Many people use this fact as a mental math strategy.

Activity In the 1930s, a sixth grader could see a movie and buy a candy bar for $0.25. Have students determine the cost of a movie and a candy bar today. Then have them figure out the total cost for the whole class and compare that cost with what it would have been in the '30s.

Using the Distributive Property

What you'll learn

You'll learn to compute products mentally using the distributive property.

When am I ever going to use this?

Knowing how to use the distributive property can help you multiply mentally.

Word Wise

distributive property

Christine's drum teacher charges $15 for a lesson. If Christine had 4 lessons in November and 3 lessons in December, how much did she pay her drum teacher for those two months?

Before you solve this problem, let's look at another way to show multiplication, using grouping symbols such as parentheses. For example, you can write 3×5 as $3(5)$ or $(3)5$. Let's include parentheses with the rules for order of operations.

Order of Operations	1. Do all operations within grouping symbols first.
	2. Do all powers before other operations.
	3. Multiply and divide in order from left to right.
	4. Add and subtract in order from left to right.

Now go back to Christine's problem.

Study Hint

Reading Math

Read the expression $15(4 + 3)$ as fifteen times the quantity four plus three.

Method 1 To find the total cost of the lessons, you can multiply the charge per lesson by the total number of lessons.

$$\underset{\substack{\text{charge per} \\ \text{lesson}}}{15}(\underset{\substack{\text{number of} \\ \text{lessons}}}{4 + 3}) = 15(7) \quad \textit{Use the order of operations. Add inside the parentheses first.}$$

$$= 105$$

Method 2 You can find the cost for each month first, and then add to find the total cost of the lessons.

$$\underset{\substack{\text{money paid} \\ \text{in November}}}{15 \times 4} + \underset{\substack{\text{money paid} \\ \text{in December}}}{15 \times 3} = 60 + 45 \quad \textit{Use the order of operations. Do the multiplication first.}$$

$$= 105$$

The solution is the same using either method. Christine paid the drum teacher $105 for lessons in November and December. So, the following sentence is true.

$$15(4 + 3) = 15 \times 4 + 15 \times 3$$

This is an example of the **distributive property**.

Lesson 4-2 Using the Distributive Property **137**

Investigations for the Special Education Student

This blackline master booklet helps you plan for the needs of your special education students by providing long-term projects along with teacher notes. Investigation 4, *The Check's in the Mail,* and Investigation 11, *Super Star!,* may be used with this chapter.

Instructional Resources
- *Study Guide Masters,* p. 24
- *Practice Masters,* p. 24
- *Enrichment Masters,* p. 24
- Transparencies 4-2, A and B
- *Technology Masters,* p. 7
- CD-ROM Program
  - Resource Lesson 4-2

Recommended Pacing	
Standard	Day 4 of 16
Honors	Day 3 of 15
Block	Day 2 of 8

1 FOCUS

 5-Minute Check *(Lesson 4-1)*

Multiply.
1. $\begin{array}{r} 0.31 \\ \times\ 4 \\ \hline 1.24 \end{array}$ 2. $\begin{array}{r} 5.13 \\ \times\ 2 \\ \hline 10.26 \end{array}$
3. 9.35×47 **439.45**
4. $5,285 \times 0.0261$ **137.9385**
5. In September 1997, a share of Wendy's International stock sold for $22.1875. How much did 30 shares cost? **$665.63**

 The 5-Minute Check is also available on **Transparency 4-2A** for this lesson.

Motivating the Lesson

Communication After reading the opening paragraph of the lesson, ask students:
- How many steps would you use to solve the problem? **two or three steps**
- What operations would you use? **addition and multiplication**

2 TEACH

 Transparency 4-2B contains a teaching aid for this lesson.

Thinking Algebraically Compare the two definitions of the distributive property. Emphasize to students that the algebra definition is a general rule that works for any values of *a, b,* and *c.*

3 PRACTICE/APPLY

Check for Understanding

If students need additional practice or instruction after completing Exercises 1–9, one of these options may be helpful.
- Extra Practice, see p. 565
- Reteaching Activity
- *Study Guide Masters*, p. 24
- *Practice Masters*, p. 24

Study Guide Masters, p. 24

Distributive Property	**Symbols: Arithmetic** $5(3 + 6) = 5 \cdot 3 + 5 \cdot 6$
	Algebra For any numbers a, b, and c, $a(b + c) = ab + ac$.

The distributive property allows us to solve problems in parts. This makes it easy to solve some multiplication problems mentally.

Examples

1 Find 6×45 mentally using the distributive property.

Estimate: $6 \times 50 = 300$

$6 \times 45 = 6(40 + 5)$ *Use 40 + 5 for 45.*
$\quad\quad = 6 \times 40 + 6 \times 5$
$\quad\quad = 240 + 30$
$\quad\quad = 270$ *Compare to the estimate.*

APPLICATION **2** **Money Matters** Adria has a paper route to earn some extra money. She earns $0.18 per customer per week. If she has 80 customers, how much will she make weekly?

Estimate: $80 \times \$0.20 = \$16$

$80 \times 0.18 = 80(0.1 + 0.08)$
$\quad\quad\quad = 80 \times 0.1 + 80 \times 0.08$
$\quad\quad\quad = 8 + 6.4$
$\quad\quad\quad = 14.4$

Adria will make $14.40. *Compare to the estimate.*

CHECK FOR UNDERSTANDING

Communicating Mathematics

Read and study the lesson to answer each question. 1–2. See margin.

1. *Tell* the order of operations you would use to find $9(8^2 + 6)$.

2. *Explain* how to use the distributive property to solve 5.5×8.

Math Journal

3. *Write* a paragraph explaining how the distributive property can help you solve a problem mentally. **See students' work.**

Guided Practice

4. Rewrite $12 \times 5 + 12 \times 8$ using the distributive property.
   $12(5 + 8)$

Find each product mentally. Use the distributive property.

7. $30 + 2 = 32$ 5. 8×18 $80 + 64 = 144$ 6. 52×3 $150 + 6 = 156$ 7. 6.4×5

8. $7 \times 30 + 7 \times 0.9 =$ 8. Find the product of 7 and 30.9 mentally.
   $210 + 6.3 = 216.3$

9. *Money Matters* Emilio is taking Karate lessons. The instructor charges $25.00 for each lesson. If Emilio has 5 lessons in one month, how much does he owe the instructor? $100 + 25 = 125$

■ Reteaching the Lesson ■

Activity Make cards with multiplication problems such as 6×16 and 8.9×5. Highlight the number that can be rewritten using the distributive property ($16 = 10 + 6$; $8.9 = 8 + 0.9$). Have students take turns choosing a card to find the product mentally.

Additional Answers

1. First find 8^2. Then add 64 and 6. Finally, multiply $70(9)$.

2. Sample answer: Find the sum of 8×5 and 8×0.5.

EXERCISES

Practice **Rewrite each expression using the distributive property.**

10. $4(30 + 6)$
$4 \times 30 + 4 \times 6$

11. $3(20 + 7)$
$3 \times 20 + 3 \times 7$

12. $15 \times 20 + 15 \times 0.4$
$15(20 + 0.4)$

Find each product mentally. Use the distributive property.

13. 6×14 **84**

14. 103×7 **721**

15. 3×72 **216**

16. 2.4×11 **26.4**

17. 8.1×6 **48.6**

18. 20×4.3 **86**

19. 20.7×6 **124.2**

20. 14×110 **1,540**

21. 30×3.09 **92.7**

22. Find the product of 12 and 15 mentally. **180**

23. Mentally multiply 60 and 10.5. **630**

24. *Algebra* What is the value of $8(x + 0.7)$ if x is 30? **245.6**

Applications and Problem Solving

25. *Money Matters* The company that employs Mrs. Leshnock pays her 31.5¢ for each mile she drives her personal car for business purposes. How much money will she receive if she drives her car 50 miles on company business? **$15.75**

The Alamo in San Antonio, Texas

26. *Travel* The Garcia family drove from Houston to San Antonio. They filled their car's tank with 12.8 gallons of gas before they left Houston. They filled the tank with 10.2 gallons when they arrived in San Antonio. If gas cost $1.29 per gallon at both gas stations, how much did they spend on gas? **$29.67**

27. *Critical Thinking* Use the distributive property to write two expressions for the figure. $3 \times 5 \rightarrow 3(3 + 2) = (3 \times 3) + (3 \times 2)$

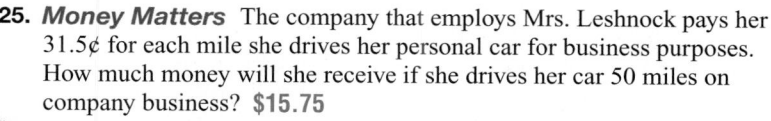

Mixed Review

28. [Test Practice] The sign below was on a sale rack. If Jhan bought 3 items from this sale rack, a reasonable total cost, without tax is — *(Lesson 4-1)* **C**

A $13.00.
B $26.00.
C $75.00.
D $100.00.
E $110.00.

SALE ITEMS! $12.99 to $29.99

29. *Weather* Washington, D.C., has an average annual precipitation of 35.86 inches. Round this amount to the nearest tenth. *(Lesson 3-4)* **35.9 in.**

30. *Statistics* The number of students in Mrs. Jing's history class during the first two weeks of the new semester were: 27, 31, 25, 19, 31, 32, 24, 26, 33, and 31. Construct a stem-and-leaf plot for the data. *(Lesson 2-6)* **See margin.**

Lesson 4-2 Using the Distributive Property **139**

Extending the Lesson

Enrichment Masters, p. 24

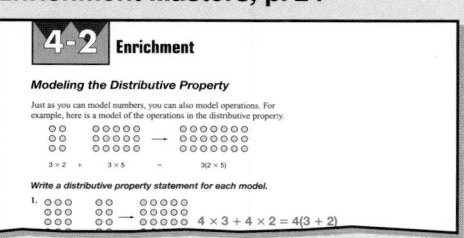

4-2 Enrichment

Modeling the Distributive Property

Just as you can model numbers, you can also model operations. For example, here is a model of the operations in the distributive property.

3×2 + 3×5 → $3(2 \times 5)$

Write a distributive property statement for each model.

1. $4 \times 3 + 4 \times 2 = 4(3 + 2)$

Activity Have pairs of students write problems involving multiplication of whole numbers and decimals. Have them determine whether they can solve the problems mentally using the distributive property. Then have them write generalizations about their strategies.

Right sidebar

Assignment Guide

Core: 11–25 odd, 27–30
Enriched: 10–24 even, 25–30

4 ASSESS

Closing Activity

Speaking Give a multiplication example such as 8.7×30 and have students find the product mentally using the distributive property. Then ask a student to explain the process to the rest of the class and to give another example for the class to discuss and solve.

Additional Answer

30.

Stem	Leaf
1	9
2	4 5 6 7
3	1 1 1 2 3

$3 | 1 = 31$

Practice Masters, p. 24

4-2 Practice

Using the Distributive Property

Rewrite each expression using the distributive property.

1. $4(30 + 6)$
$4 \times 30 +$
4×6

2. $0.5(20 + 9)$
$0.5 \times 20 +$
0.5×9

3. $7(8 + 0.4)$
$7 \times 8 +$
7×0.4

4. $16(50 + 2)$
$16 \times 50 +$
16×2

5. $3(200 + 7)$
$3 \times 200 +$
3×7

6. $2.8(10 + 3)$
$2.8 \times 10 +$
2.8×3

7. $90(60 + 0.5)$
$90 \times 60 +$
90×0.5

8. $4(70 + 0.1)$
$4 \times 70 +$
4×0.1

9. $11(300 + 20)$
$11 \times 300 +$
11×20

Find each product mentally. Use the distributive property.

10. 26×5 **130**

11. 0.7×34 **23.8**

12. 9.8×2 **19.6**

13. 14×3.02 **42.28**

14. 50×6.7 **335**

15. 108×12 **1,296**

16. 11×24 **264**

17. 3×290 **870**

18. 80.4×5 **402**

19. 7×63 **441**

20. 51×0.9 **45.9**

21. 40.2×30 **1,206**

22. 8×2.7 **21.6**

23. 60×5.4 **324**

24. 13×12 **156**

25. 90×20.1 **1,809**

26. 7×508 **3,556**

27. 16×105 **1,680**

28. Find the product of 4.09 and 80 mentally. **327.2**

29. Find the product of 6 and 7.5 mentally. **45**

30. Find the product of 21 and 31 mentally. **651**

© Glencoe/McGraw-Hill   T24   *Mathematics: Applications and Connections, Course 1*

Lesson 4-2   **139**

GET READY

Objective Students use grid paper to multiply decimals.

Optional Resources
Hands-On Lab Masters
• grid paper, p. 11
• worksheet, p. 44
Overhead Manipulative Resources
• decimal models

MANAGEMENT TIPS

Recommended Time
45 minutes

Getting Started Review the activities in Lesson 4-1A by asking questions such as *How did you use grid paper to represent 3 × 0.5?* **3 10-by-10 squares and 5 columns**

Activities 1 and 2 use decimal models to multiply decimals. After students have completed the activities, ask how multiplying two decimals is similar to multiplying a whole number and a decimal. **Sample answer: Both are first multiplied like whole numbers. Then the decimal point is placed in the product according to the number of decimal places in the factors.**

ASSESS

Have students complete Exercises 1–5. Make sure they shade in the correct number of squares.

Use Exercise 6 to determine whether students understand how to multiply decimals.

COOPERATIVE LEARNING

4-3A Multiplying Decimals

A Preview of Lesson 4-3

grid paper

markers

In the Hands-On Lab on page 132, you used decimal models to multiply a whole number and a decimal. In this lab, you will use decimal models to multiply two decimals.

TRY THIS

Work with a partner.

① To model 0.7×0.5, follow these steps.
• Draw a 10-by-10 square. Recall that each small square represents 0.01.
• Color seven rows of the model blue to represent 0.7.
• Color five columns of the model yellow to represent 0.5.

There are 35 small squares that are shaded green.
Therefore, $0.7 \times 0.5 = 0.35$.

② To model 0.9×2.3, you need to draw three 10-by-10 squares.
• Line up the squares as shown.
• Color 9 rows blue to represent 0.9.
• Color 23 columns yellow to represent 2.3.

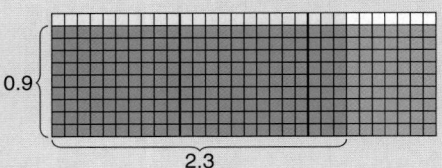

There are 207 small squares that are shaded green. This would cover two 10-by-10 squares, which is 2, and seven small squares, which is 0.07.
Therefore, $0.9 \times 2.3 = 2.07$.

ON YOUR OWN

Write a multiplication problem using decimals for each model. 2. 0.4×3.8

1.  0.6×1.9

2.

Use decimal models to show each product. 3–5. See Answer Appendix for models.

3. 0.4×0.6 **0.24**   **4.** 1.3×0.2 **0.26**   **5.** 2.5×1.1 **2.75**

6. *Look Ahead* Find the product of 1.7 and 2.2 without using models. **3.74**

Math Journal Have students write an explanation of how models can be used to show the multiplication of decimals and how they might use models if they forget the pattern they learned in this lab.

Multiplying Decimals

What you'll learn

You'll learn to multiply decimals.

When am I ever going to use this?

Knowing how to multiply decimals can help you find the amount of tip to leave at a restaurant.

On September 1, 1997, the minimum wage increased to $5.15 an hour. If Domingo works 17.5 hours a week and earns minimum wage, how much does he make each week before taxes?

To answer this question, you need to multiply 5.15 by 17.5. When you multiply decimals, multiply as with whole numbers. One way to place the decimal point is to use estimation.

Estimate: 5 · 20 = 100 Round 5.15 to 5 and 17.5 to 20.

$$
\begin{array}{r}
5.15 \\
\times 17.5 \\
\hline
2\,575 \\
36\,05 \\
51\,5 \\
\hline
90.125
\end{array}
$$

Multiply as with whole numbers.

Since the estimate is 100, place the decimal point after 0.

Rounded to the nearest penny, Domingo makes $90.13 before taxes.

Another way to place the decimal point in the product is by counting the decimal places in each factor. The product will have the same number of decimal places as the sum of the number of decimal places in the factors.

Examples

1 Find **2.4 · 5.9.**   *Estimate: 2 · 6 = 12*

$$
\begin{array}{r}
5.9 \\
\times 2.4 \\
\hline
23\,6 \\
118 \\
\hline
14.16
\end{array}
$$

← *one decimal place*
← *one decimal place*

← *two decimal places*

The product is 14.16.   *Compared to the estimate, the answer is reasonable.*

2 Find **1.45 × 0.7.**   *Estimate: 1 × 1 = 1*

$$
\begin{array}{r}
1.45 \\
\times 0.7 \\
\hline
1.015
\end{array}
$$

← *two decimal places*
← *one decimal place*
← *three decimal places*

The product is 1.015.

Lesson 4-3 Multiplying Decimals   **141**

Multiple Learning Styles

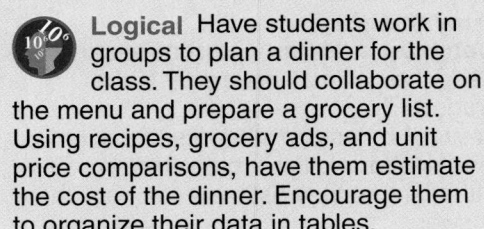

Logical Have students work in groups to plan a dinner for the class. They should collaborate on the menu and prepare a grocery list. Using recipes, grocery ads, and unit price comparisons, have them estimate the cost of the dinner. Encourage them to organize their data in tables.

4-3 Lesson Notes

Instructional Resources

- *Study Guide Masters*, p. 25
- *Practice Masters*, p. 25
- *Enrichment Masters*, p. 25
- Transparencies 4-3, A and B
- *Assessment and Evaluation Masters*, p. 99
- CD-ROM Program
  - Resource Lesson 4-3
  - Interactive Lesson 4-3

Recommended Pacing	
Standard	Days 5 & 6 of 16
Honors	Days 4 & 5 of 15
Block	Day 3 of 8

1 FOCUS

5-Minute Check
(Lesson 4-2)

1. Rewrite 5(20 + 6) using the distributive property.
   5 × 20 + 5 × 6

Find each product mentally. Use the distributive property.

2. 107 × 6 **642**
3. 5 × 6.3 **31.5**
4. 40 × 2.8 **112**
5. 0.6 × 15 **9**

The 5-Minute Check is also available on **Transparency 4-3A** for this lesson.

Motivating the Lesson

Problem Solving Ask students how they would determine which craft material costs more.
- 8.5 yd of ribbon at $0.55 a yard
- 2.8 yd of lace at $1.82 a yard

  Sample answer: Estimate by multiplying; the lace costs more.

2 TEACH

Transparency 4-3B contains a teaching aid for this lesson.

Using Logical Reasoning Have groups of students write riddles such as: *My product is 6.781. One of my factors is a decimal in tenths. How many decimal places does my other factor have?* 2

In-Class Examples

For Example 1
Find 2.5 · 4.3. **10.75**

For Example 2
Find 0.8 × 2.53. **2.024**

For Example 3
Multiply 6.24 and 0.05. **0.312**

For Example 4
Evaluate 7.3*r* if *r* = 15.78.
115.194

3 PRACTICE/APPLY

Check for Understanding

If students need additional practice or instruction after completing Exercises 1–8, one of these options may be helpful.
- Extra Practice, see p. 566
- Reteaching Activity
- *Transition Booklet,* pp. 23–24
- *Study Guide Masters,* p. 25
- *Practice Masters,* p. 25
- Interactive Mathematics Tools Software

Assignment Guide

Core: 9–29 odd, 31–35
Enriched: 10–28 even, 29, 31–35

Study Guide Masters, p. 25

4-3 **Study Guide**

Name_____ Date_____

Multiplying Decimals

Multiply decimals just as you multiply whole numbers. The number of decimal places in the product is equal to the sum of the number of decimal places in the two factors.

Example Multiply 0.16 and 1.025.

```
    1.025  ←——— three decimal places
 ×  0.16   ←——— two decimal places
    6150
    1025
  0.16400  ←——— five decimal places
```

Multiply.

1. 0.5 × 20.2
10.1

2. 1.2 × 2.3
2.76

3. 0.055 × 3.2
0.176

4. 0.014 × 0.4
0.0056

5. 12.4 × 12.4
153.76

6. 3.07 × 1.07
3.2849

Solve each equation.

7. *c* = 15.5 × 3.3
51.15

8. *x* = 202.1 × 1.14
230.394

9. *a* = 0.008 × 65.3
0.5224

Evaluate each expression if m = 0.09, n = 1.2, and p = 8.19.

10. *mn*
0.108

11. *p*(*n* + *m*)
10.5651

12. *pm*
0.7371

© Glencoe/McGraw-Hill   T25   *Mathematics: Applications and Connections, Course 1*

Examples

3 Multiply 0.02 and 1.36.   *Estimate: 0 · 1 = 0*

$$
\begin{array}{r}
1.36 \\
\times 0.02 \\
\hline
0.0272
\end{array}
$$

1.36 ← *two decimal places*
×0.02 ← *two decimal places*
0.0272 ← *To make four decimal places, annex a zero on the left.*

The product is 0.0272.

INTEGRATION **4** **Algebra** Evaluate 4.3*n* if *n* = 10.89.

4.3*n* = 4.3 · 10.89 *Replace n with 10.89.*

Estimate: 4 · 11 = 44

Use a calculator.

4.3 [×] 10.89 [=] *46.827* *Compare to the estimate.*

CHECK FOR UNDERSTANDING

Communicating Mathematics
1. Sample answer: The total decimal places of the factors are 5; 0.40563

Read and study the lesson to answer each question.

1. *Explain* how you know where the decimal point would go in the product 4.507 × 0.09 = 040563.

2. *Write* the multiplication sentence represented by the decimal model. **0.4 × 0.6**

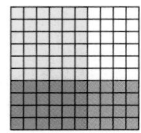

Guided Practice

3. Use estimation to place the decimal point in the product 3.4 · 1.2 = 408. **4.08**

Multiply.

4. 0.4 × 8.3 **3.32**   5. 8.54 · 3.27 **27.9258**   6. 39.6 × 2.417
95.7132

7. Evaluate 0.002*y* if *y* = 3.9. **0.0078**

8. *Money Matters* Helaku wants to buy a new video game. The game he wants costs $41.99. The sales tax is calculated by multiplying the total of the merchandise by 0.0575. If the video game is the only thing Helaku purchases, how much sales tax will he pay? **$2.41**

EXERCISES

Practice

Use estimation to place the decimal point in each product.

9. 5.6 × 12.43 = 69608 **69.608**   10. 0.03 × 1.24 = 00372 **0.0372**

Multiply.

11. 0.35 · 1.4 **0.49**   12. 5.2 × 0.065 **0.338**   13. 3.06 · 4.28 **13.0968**

14. 0.9 × 0.15 **0.135**   15. 18.37 · 908.44   16. 0.003 × 0.012
16,688.0428   **0.000036**

142 Chapter 4 Multiplying and Dividing Decimals

Reteaching the Lesson

Activity Use transparency overlays and grid paper to illustrate multiplication. For example, illustrate 1.2 × 0.8 by covering one whole model and two rows of another model with a yellow overlay. Cover eight rows with a red overlay. Have students count the orange squares to find the product. **0.96**

Error Analysis
Watch for students who consistently place the decimal incorrectly in the product.
Prevent by asking students to explain the rules for multiplying decimals. Suggest that they estimate to check their answer.

Solve each equation.

17. $p = 1.3 \cdot 7.3$ **9.49**

18. $q = 0.3 \cdot 0.012$ **0.0036**

19. $0.6(0.031) = m$ **0.0186**

20. $28.2(4.4) = n$ **124.08**

Evaluate each expression if $a = 1.06$, $b = 0.002$, and $c = 5.5$.

22. 5.83212
23. 5.82788
24. 0.01166
27. 0.00696

21. ab **0.00212**

22. $a(b + c)$

23. $a(c - b)$

24. abc

25. Find the product of 24.8 and 4.389. **108.8472**

26. What is 2.15 multiplied by 3.84? **8.256**

27. *Algebra* What is the product of x and y if $x = 1.16$ and $y = 0.006$?

28. *Algebra* Find the value of prt if $p = \$250$, $r = 0.03$, and $t = 7.5$. **$56.25**

Applications and Problem Solving

29. *Life Science* The graph shows a breakdown by age group of the 26.2 million Americans who were hearing impaired as of January, 1996. The number of people between the ages of 65 and 74 is 3.95 times the number of people in the 0-17 age group. Complete the graph by finding the number of people in the 65-74 age group. **5.41**

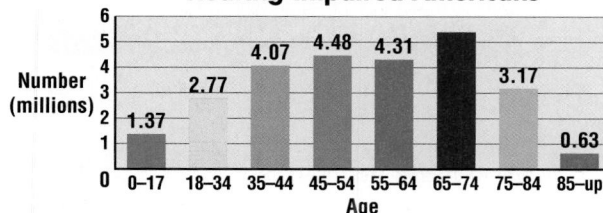

Hearing-Impaired Americans

Source: American Speech–Language–Hearing Association

30. *Working on the*  **CHAPTER Project** Refer to the prices you determined in Exercise 32 on page 135. **a–b. See students' work.**

   a. Suppose your after-tax prices are determined by multiplying the prices by 1.065. Find the after-tax price for your product.

   b. Suppose your senior citizen discount is determined by multiplying the after-tax prices by 0.90. Find the discount prices.

31. *Critical Thinking* Write a multiplication problem where the product of two decimals is between 0.05 and 0.75. **Sample answer: 0.1 × 0.6**

32. 32 + 2 = 34

Mixed Review

32. Find 8.5×4 mentally. Use the distributive property. *(Lesson 4-2)*

33. **Test Practice** Shanee recorded the number of miles she rode her bicycle each day for four days. Which list shows that data in correct order from least to greatest? *(Lesson 3-3)* **D**
   A 7.6, 21.4, 13.7, 9.3
   B 21.4, 13.7, 21.4
   C 13.7, 7.6, 21.4, 9.3
   D 7.6, 9.3, 13.7, 21.4

34. Express $\frac{8}{1,000}$ as a decimal. *(Lesson 3-1)* **0.008**

35. *Algebra* Evaluate y^3 if $y = 6$. *(Lesson 1-6)* **216**

Lesson 4-3 Multiplying Decimals **143**

Extending the Lesson

Enrichment Masters, p. 25

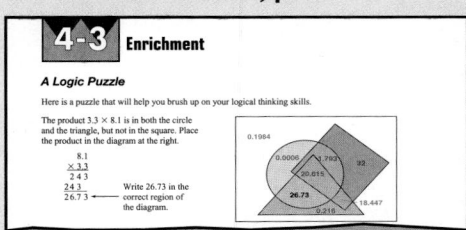

Activity Provide students with brochures for a local theme park, arcade, or game center. Have them use decimal multiplication to determine the cost of hosting a party for the entire class. Assume students will average 2.5 tokens or tickets for each game during the party.

CHAPTER Project

Exercise 30 asks students to advance to the next stage of work on the Chapter Project. You may wish to have them research local tax rates and senior citizen discount rates. Have them apply these rates to their products.

4 ASSESS

Closing Activity

Speaking Have students summarize the rule for calculating the number of decimal places when multiplying decimals. Then have them explain how they can check their decimal point placement.

Chapter 4, Quiz A (Lessons 4-1 through 4-3) is available in the *Assessment and Evaluation Masters*, p. 99.

Practice Masters, p. 25

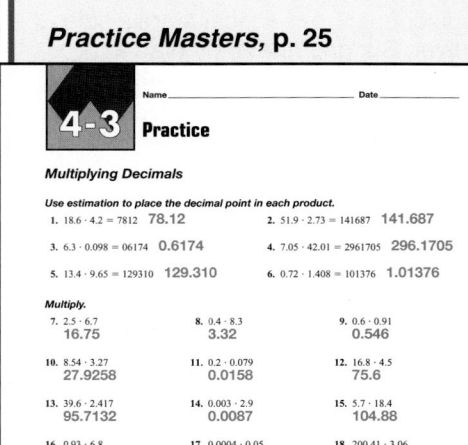

4-3B Reading Spreadsheets

A Follow-Up of Lesson 4-3

 computer

 spreadsheet software

A computer *spreadsheet* arranges data and formulas in a column and row format. A spreadsheet is used for organizing and analyzing data.

A spreadsheet is made up of *cells*. A cell can contain data, labels, or formulas. Each cell has an address. A cell's address is a combination of the letter from the top of a particular column and the number from the left of a particular row. The cell B4 is the cell in column B and row 4. *Cell B4 in the spreadsheet below contains the amount $29.90.*

The spreadsheet below shows a price list for a retail store.

GET READY

Objective Students explore how data and formulas are entered in a computer spreadsheet.

Technology Resources
Suggested spreadsheet software:
- *Lotus 1·2·3*
- *ClarisWorks*
- *Microsoft Excel*

MANAGEMENT TIPS

Recommended Time
45 minutes

Getting Started Make sure students understand the differences between data, labels, and formulas. Also make sure students understand a cell's address and how it is used in formulas.

Even though students have not studied percents yet, explain that the values they are entering in column C are the decimal form of the percent of discount. So $0.2 = 20\%$.

Teaching Tip Have students use a similar setup and enter prices and discounts from a local store's sale flyer. Suggest that they plan a shopping spree using the data.

ASSESS

After students answer Exercises 1–5, ask them how they could change the spreadsheet for different situations, such as inventories, employee discounts, and so on.

TRY THIS

Work with a partner.

The formulas in the cells in column D find the product of the numbers that are entered in columns B and C.

Use the spreadsheet to determine the discounts by making the following substitutions.

C2 = 0.2, C3 = 0.25, C4 = 0.15, C5 = 0.2, C6 = 0.35, C7 = 0.1

	A	B	C	D	E
1	Item	Regular Price	Rate of Discount	Discount	Sale Price
2	Blouse	$35.99		= B2*C2	= B2-D2
3	Skirt	$49.99		= B3*C3	= B3-D3
4	Jeans	$29.90		= B4*C4	= B4-D4
5	Shorts	$15.00		= B5*C5	= B5-D5
6	Sweatshirt	$18.99		= B6*C6	= B6-D6
7	T-Shirts	$19.95		= B7*C7	= B7-D7

ON YOUR OWN

1. Name the cell that holds the data $25.41. **E4**

2. What is stored in cell D5? **$3.00**

3. The formulas in the cells in column E find the difference of the numbers in column B and column D. Use the results from the activity above to determine each sale price. **$28.79, $37.49, $25.41, $12.00, $12.34, $17.95**

4. How could you use the spreadsheet to determine the amount of sales tax on each item? **See margin.**

5. Suppose column C was labeled *Quantity* and column D was labeled *Total*. Would the formulas in column D change? **No; you would still multiply.**

144 Chapter 4 Multiplying and Dividing Decimals

Additional Answer
4. **Sample answer: Add a column with the formula to find the amount of tax.**

 Math Journal Have students write a paragraph explaining how the spreadsheet could be used to make a graph. Would they use a line graph or a bar graph for the given information?

4-4

Integration: Geometry
Perimeter and Area

What you'll learn

You'll learn to find the perimeters and the areas of rectangles and squares.

When am I ever going to use this?

Knowing how to find perimeter can help you buy the right amount of framing. Knowing how to find area can help you buy the right amount of paint or carpet.

Word Wise
perimeter
side
area

Alicia is redecorating her bedroom and wants to add a wallpaper border around the entire room. How much will she need?

Alicia needs to find the **perimeter** of the bedroom so she can know how much border to buy. The perimeter (*P*) of any closed figure is the distance around the figure. You can find the perimeter by adding the measures of the **sides** of the figure.

$P = 15 + 9 + 10 + 3 + 5 + 12$

$P = 54$

The perimeter of Alicia's room is 54 feet. So, Alicia needs 54 feet of wallpaper border.

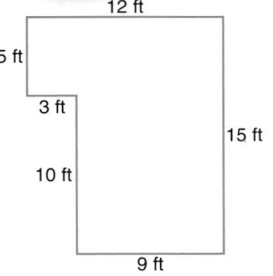

To find the perimeter of a rectangle, add the measures of its four sides. Let *P* represent the measure of the perimeter, ℓ its length, and *w* its width.

$P = \ell + w + \ell + w$

$P = \ell + \ell + w + w$

$P = 2\ell + 2w$

Perimeter of a Rectangle	**Words:** The perimeter of a rectangle is two times the length (ℓ) plus two times the width (*w*).
	Symbols: $P = 2\ell + 2w$ **Model:**

Example ①

Study Hint

Mental Math You can rewrite the formula for perimeter using the distributive property: $2\ell + 2w = 2(\ell + w)$.

Find the perimeter of a rectangle with a length of 18.3 meters and a width of 7.5 meters.

$P = 2\ell + 2w$

$P = 2(18.3) + 2(7.5)$ *Replace ℓ with 18.3 and w with 7.5.*

$P = 36.6 + 15$

$P = 51.6$

The perimeter is 51.6 meters.

Lesson 4-4 Integration: Geometry Perimeter and Area **145**

Multiple Learning Styles

Visual/Spatial The region between Dallas, Texas, and Fort Worth, Texas, forms a rectangle made by Highway 20 on the south, Highway 408 and Loop 12 on the east, Highway 183 on the north, and Highway 183 and I-820 on the west. Have students use maps to determine other such regions that are either square or rectangular in shape. Then have them find the approximate perimeter and area of the region described.

Instructional Resources
- *Study Guide Masters,* p. 26
- *Practice Masters,* p. 26
- *Enrichment Masters,* p. 26
- Transparencies 4-4, A and B
- *Technology Masters,* p. 8
- CD-ROM Program
  - Resource Lesson 4-4
  - Interactive Lesson 4-4

Recommended Pacing	
Standard	Day 7 of 16
Honors	Days 6 & 7 of 15
Block	Day 4 of 8

1 FOCUS

5-Minute Check
(Lesson 4-3)

Multiply.
1. $6.74 \cdot 3.28$ **22.1072**
2. $0.03 \cdot 0.002$ **0.00006**
3. Solve $x = 7.36 \cdot 20.544$. **151.20384**

Evaluate each expression if $a = 12.5$, $b = 3.08$, and $c = 0.009$.
4. $a(b + c)$ **38.6125**
5. abc **0.3465**

The 5-Minute Check is also available on **Transparency 4-4A** for this lesson.

Motivating the Lesson

Hands-On Activity Have small groups of students use toothpicks to make a closed figure. Let 1 toothpick = 10 feet. Have students determine methods for finding the distance around their figure and the amount of space covered by their figure.

Transparency 4-4B contains a teaching aid for this lesson.

Reading Mathematics Draw a rectangle on the chalkboard, including its dimensions. Also, write the words *perimeter, side, area, length,* and *width* on the board. Say each word. Have students use each word in a sentence about the figure. Discuss how the words *length* and *width* might be interpreted for different orientations of a rectangle.

In-Class Examples

For Example 1
Find the perimeter of a rectangle with a length of 22.86 meters and a width of 9.14 meters. **64 m**

For Example 2
Find the perimeter of a square poster whose sides each measure 91.44 centimeters. **365.76 cm**

For Example 3
Sandy is building a garden shed 11.2 feet long and 8.5 feet wide. How much floor material will she need? **95.2 ft²**

Teaching Tip Encourage students to always draw and label the figure (if an illustration is not included) and write the formula they will use before solving a problem.

Since each side of a square has the same length, you can multiply the measure of any of its sides (s) by 4 to find its perimeter. Thus, the formula for the perimeter of a square can be written as $P = 4s$.

Example ② Find the perimeter of a square cement block whose sides each measure 17.875 inches.

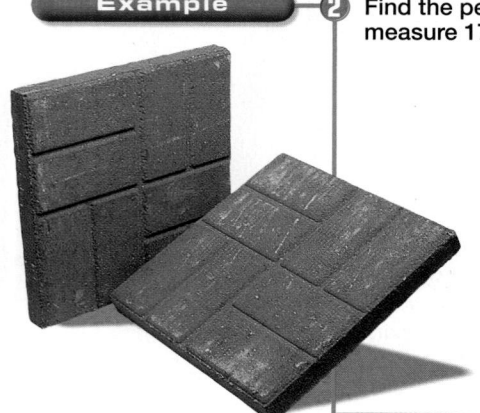

17.875 in.

$P = 4s$
$P = 4 \times 17.875$ *Replace s with 17.875.*
$4 \boxed{\times} 17.875 \boxed{=} 71.5$

The perimeter of the cement block is 71.5 inches.

In addition to the perimeter, we often solve problems by using the **area** of a closed figure. Area is the number of square units needed to cover a surface.

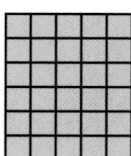

By counting, you can see that the rectangle at the right has an area of 30 square units. You can also find its area by multiplying its length, 6 units, by its width, 5 units.

You can find the area of any rectangle by multiplying its length and its width.

Area of a Rectangle	**Words:** The area of a rectangle is the product of its length (ℓ) and width (w).
	Symbols: $A = \ell \cdot w$ **Model:** 

Area is expressed in square units. For example, if the length and width are measured in feet, the area will be written in square feet, or ft².

Example ③
APPLICATION

Sports International soccer fields are rectangular and measure 100 meters by 73 meters. A soccer field needs to be recovered with sod. How much sod do the groundskeepers need for the field?

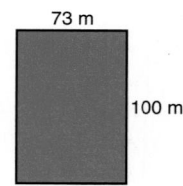

73 m

100 m

$A = \ell \cdot w$
$A = 100 \cdot 73$ *Replace ℓ with 100 and w with 73.*
$A = 7,300$

The groundskeepers need 7,300 square meters of sod.

You can write the formula for the area of a square as $A = s^2$, because each side of a square has the same length.

Example **4** Find the area of a square whose sides measure 4.5 meters.

$A = s^2$

$A = 4.5^2$ *Replace s with 4.5.*

$A = 20.25$

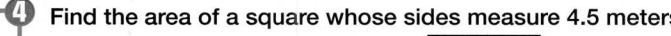

4.5 m

The area of the square is 20.25 square meters.

CHECK FOR UNDERSTANDING

Communicating Mathematics

 Math Journal

Read and study the lesson to answer each question. 1. **See students' work.**

1. *Draw* and label a rectangle that has a perimeter of 36 units. 2. **Square the measure of one of its sides.**

2. *Explain* how to find the area of a square.

3. Use grid paper to draw as many different rectangles as you can with a perimeter of 12 units. *Make a table* listing the length, width, perimeter, and area of each rectangle. Use the relationships between the lengths and widths to create formulas that could be used to find the perimeter and area of any rectangle. **See students' work.**

Guided Practice

Find the perimeter of each figure.

4.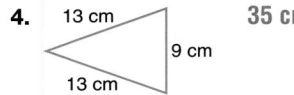
13 cm
9 cm
13 cm
35 cm

5.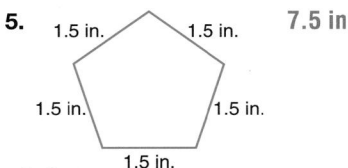
1.5 in.   1.5 in.
1.5 in.   1.5 in.
1.5 in.
7.5 in.

Find the perimeter and the area of each figure.

6.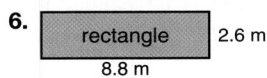
rectangle   2.6 m
8.8 m
$P = 22.8$ m; $A = 22.88$ m²

7.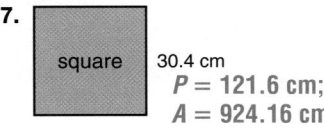
square   30.4 cm
$P = 121.6$ cm; $A = 924.16$ cm²

8. What is the area of a rectangle with a length of 10.25 inches and a width of 8.5 inches? **87.125 in²**

EXERCISES

Practice

Find the perimeter of each figure.

9.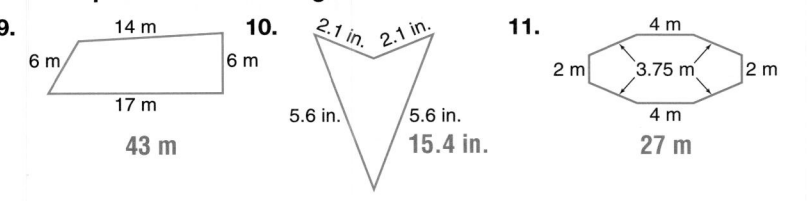
14 m
6 m
17 m
43 m

10.
2.1 in. 2.1 in.
5.6 in. 5.6 in.
15.4 in.

11.
4 m
2 m   3.75 m   2 m
4 m
27 m

Lesson 4-4 Integration: Geometry Perimeter and Area **147**

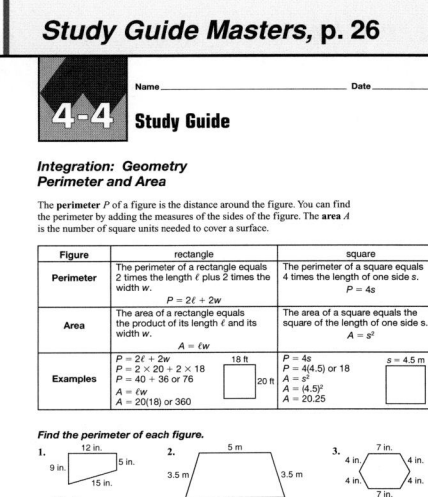

4 ASSESS

Closing Activity
Modeling Have students draw various rectangles and squares on grid paper and determine the length and width in grid units. Then have them find the perimeter and area of each figure.

Additional Answer
23.
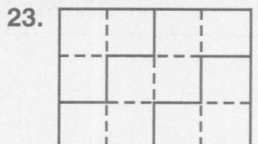

Practice Masters, p. 26

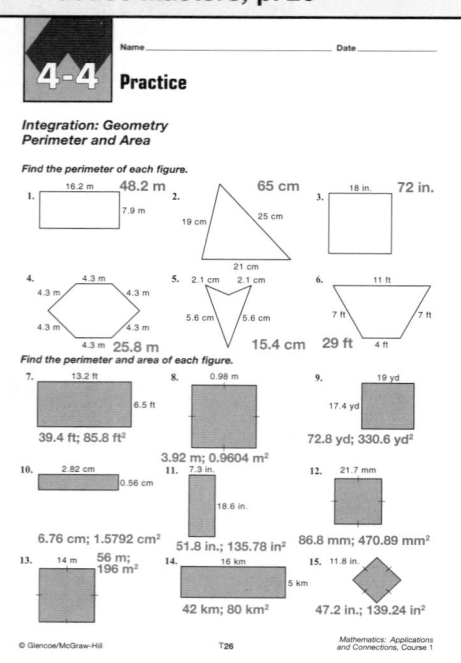

Applications and Problem Solving
14. $P = 64$ ft;
$A = 220$ ft^2
15. $P = 12.8$ m;
$A = 7.68$ m^2
16. $P = 49.2$ cm;
$A = 151.29$ cm^2
17. $P = 46.4$ in.;
$A = 134.56$ in^2
18. See students' work.

22. No; it will increase by two times the original length; see students' work.

Mixed Review

Find the perimeter and the area of each figure.

12.
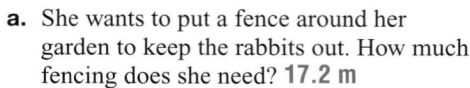
rectangle   7.25 ft
12.5 ft
$P = 39.5$ ft;
$A = 90.625$ ft^2

13.
square   1.5 in.
$P = 6$ in.
$A = 2.25$ in.^{2}

14. rectangle: ℓ, 22 ft; w, 10 ft
15. rectangle: ℓ, 4.8 m; w, 1.6 m
16. square: s, 12.3 cm
17. square: s, 11.6 in.

18. Draw and label a 6-sided figure that has a perimeter of 9 inches.

19. A rectangle is 5.5 meters long, and its perimeter is 20 meters. What is its width? **4.5 m**

20. *Gardening* Mrs. Collins has a rectangular strawberry garden that is 6.1 meters long and 2.5 meters wide.
   a. She wants to put a fence around her garden to keep the rabbits out. How much fencing does she need? **17.2 m**
   b. She needs to cover the garden with a net to keep birds from eating the strawberries. How much netting does she need? **15.25 m^2**

21. *Sports* Badminton was added for the 1992 Summer Olympic games in Barcelona, Spain. A regulation badminton court has the measurements shown in the diagram.
   a. Tape is to be applied to the floor to form a badminton court's boundaries. Do you need to find the area or the perimeter to determine how much tape you need? **perimeter**
   b. What is the area of the court? **880 ft^2**

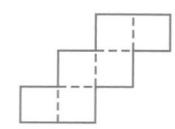
20′
44′

22. *Geometry* Suppose you increase a pair of opposite sides of a square by two units. Will the area of the rectangle be four more square units than the area of the square? Use a model in your explanation.

23. *Critical Thinking* Using the figure shown, draw another figure that has twice the area and still has the same perimeter. **See margin.**

24. *Algebra* Evaluate the expression zwx if $w = 0.2$, $x = 20.7$, and $z = 3.01$. *(Lesson 4-3)* **12.4614**

25. **Test Practice** Ernesto bought 7 spiral notebooks for his classes. Each notebook cost $2.29, including tax. What was the total cost of the notebooks? *(Lesson 4-1)* **C**
   A $8.93   B $16.93   C $16.03   D $17.03   E Not Here

26. Find 52.8 minus 6.75. *(Lesson 3-6)* **46.05**

Extending the Lesson

Enrichment Masters, p. 26

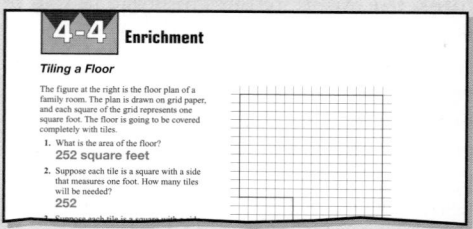

Activity Have students use a Computer Assisted Design (CAD) program to draw closed figures with designated perimeters. Have them determine, for example, how many different figures they can create that have a perimeter of 42 meters or an area of 110 square meters.

COOPERATIVE LEARNING

4-4B Area and Perimeter

A Follow-up of Lesson 4-4

grid paper

In this lab, you will use grid paper to explore how perimeter and area are related.

TRY THIS

Work in groups of three.

Step 1 Copy the table at the right.

Step 2 On centimeter grid paper, draw a rectangle with a length of 6 and a width of 1.

Step 3 Find the perimeter and the area of the rectangle and record them in your table.

Step 4 Repeat Steps 2 and 3 for the remaining dimensions in your table.

Length	Width	Perimeter	Area
6	1		
5	2		
4	3		

ON YOUR OWN

1a–c. See Answer Appendix.

1. Each person in your group should copy the table at the right.

   a. One person should complete the table for all rectangles whose perimeter is 16 with whole number sides.

   b. Another person should complete the table for all rectangles whose perimeter is 18 with whole number sides.

   c. The third person should complete the table for all rectangles whose perimeter is 20 with whole number sides.

Length	Width	Area

2. Suppose you want to enclose a rectangular garden with the greatest area. **a. 13 feet long and 12 feet wide b. 12 feet by 12 feet**

   a. What would be the dimensions of the garden if you have 50 feet of fence?

   b. What would be the dimensions of the garden if you have 48 feet of fence?

3. What can you conclude about the relationship between area and perimeter?

4. *Reflect Back* Find the dimensions of a rectangle with the least perimeter possible if its area is 24 square units. **See margin.**

3. Sample answer: The longer the length and shorter the width, the lesser the area of the rectangular shape.

Lesson 4-4B HANDS-ON **LAB** 149

Math Journal
Have students write a paragraph explaining how perimeter and area are related. Have them describe how they discovered the relationship between perimeter and area by making a table.

Additional Answer
4. 6 units long by 4 units wide

GET READY

Objective Students explore how perimeter and area are related.

MANAGEMENT TIPS

Optional Resources
Hands-On Lab Masters
• grid paper, p. 11
• worksheet, p. 45

Overhead Manipulative Resources
• centimeter grid

Recommended Time
45 minutes

Getting Started Ask students to find the dimensions of a rectangle drawn on centimeter grid paper. Then review the formulas for perimeter and area.

The **Activity** uses grid paper to explore perimeter and area relationships. Some students may wish to use a straightedge for accuracy in drawing lines. Have all students determine the lengths of the sides of their rectangles before they draw them.

ASSESS

Have students complete Exercises 1–3. Watch for students who confuse perimeter and area.

Use Exercise 4 to determine whether students understand the relationship between minimum perimeter and maximum area.

Objective Students solve problems by first solving a simpler problem.

Recommended Pacing	
Standard	Day 8 of 16
Honors	Day 8 of 15
Block	Day 4 of 8

1 FOCUS

Getting Started Have students act out the situation presented at the beginning of the lesson. Then have pairs of students work on Exercises 1–2. After discussing the results as a class, have the pairs solve Exercise 3.

2 TEACH

Teaching Tip Remind students to review what information they are given and what question is being asked before trying to solve the problems.

In-Class Example

A recipe that makes 6 servings of chili calls for 2 pounds of ground beef. How much ground beef is needed for 15 servings?
5 lb

THINKING LAB **PROBLEM SOLVING**

4-5A Solve a Simpler Problem

A Preview of Lesson 4-5

Jemeka and Macy volunteered to plant some flowering bushes along the side of a neighborhood senior center. The wall of the building is 50 feet long. A gardener told them to plant the bushes 2 feet apart to give them room to grow. They are trying to determine how many plants to buy. Let's listen in!

50 divided by 2 is 25. I guess we need 25 plants.

I'm not sure that's right. Let's make it easier. Suppose the wall was 4 feet long. We'd need 3 bushes, not 2. Right?

Let's see. There would be one at the front corner, one in the middle 2 feet from the corner, and one at the back corner. When 3 bushes are planted 2 feet apart, the bushes on the end are 4 feet apart. You're right!

And if the wall was 6 feet long, we'd need one more bush for a total of 4 bushes. Is there a pattern here?

Macy

Yes! The number of plants needed is half the total distance plus one. So for this 50-foot wall, we'll need 26 plants, not 25. It was hard to see that until we worked with a shorter wall.

Jemeka

THINK ABOUT IT

Work with a partner.

1. *Explain* why Macy's first method of solving the problem seemed reasonable. **See margin.**

2. *Think* of another way the girls could have solved the problem. **Sample answer: Draw a diagram and count the plants.**

3. *Apply* what you have learned about the **solve a simpler problem** strategy to solve the following problem.

   Mike's Subs has made a submarine sandwich for eighteen people to share. How many cuts must be made to divide the sandwich equally among the eighteen people? **17 cuts**

150 Chapter 4 Multiplying and Dividing Decimals

■ Reteaching the Lesson ■

Activity Have students solve the problem below. Before they begin, ask how the problem can be simplified. *Find the number of 3-inch strips in 12 yards of string.* **144 strips**

Additional Answer

1. Sample answer: She thought she was dividing the length into 2-foot sections.

ON YOUR OWN

4. The last step of the 4-step plan for problem solving asks you to *examine* your solution. *Explain* how you can use estimation with decimals to help you examine a solution. **See margin.**

5. *Tell* why solving a simpler problem can sometimes help you solve a more difficult problem. **See margin.**

6. *Explain* how using the distributive property is similar to the solving a simpler problem strategy. **See margin.**

MIXED PROBLEM SOLVING

STRATEGIES

Look for a pattern.
Solve a simpler problem.
Act it out.
Guess and check.
Draw a diagram.
Make a chart.
Work backward.

Solve. Use any strategy.

7. *Gardening* Marco wants to fence in his tomato garden that is 8 feet by 6 feet. How much fencing does he need? **28 feet**

8. *Decorating* How many 1-foot square tiles are needed to cover a kitchen floor that is 14 feet by 10 feet? **140 tiles**

9. *Statistics* The bar graph shows revenues for several popular magazines.

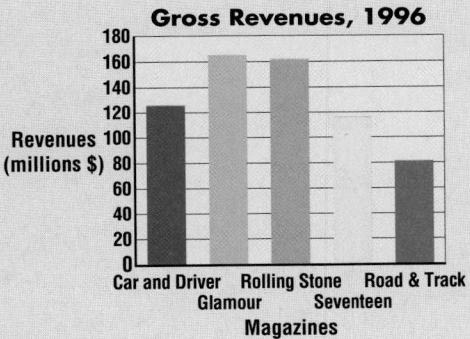

Gross Revenues, 1996

Revenues (millions $)

Car and Driver / Glamour / Rolling Stone / Seventeen / Road & Track

Magazines

Source: *Advertising Age,* June 16, 1997

a. About how much higher were the revenues of *Car and Driver* than *Road & Track*? **$40 million**

b. About how much higher were *Glamour* magazine's revenues than *Seventeen's*? **$50 million**

c. About how much more revenue did the magazine with the highest revenue have than *Rolling Stone*? **$3 million**

10. *Decorating* The diagram shows Paula's dining room and living room area.

a. If she wants to carpet this area, how much carpet will she need? **528 ft²**

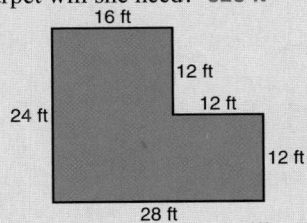

16 ft
12 ft
24 ft / 12 ft
12 ft
28 ft

b. Draw another room that will use the same amount of carpet. **See students' work.**

11. *Sports* Luis bowled a 125, a 148, and a 162 during league competition. Before this competition, his average was 128 pins. Compare his previous average to his competition average. **See margin.**

12. *Entertainment* The Strand Theater seats 536 people. At last night's movie, there was one empty seat for every three people. How many people were in the movie theater? **402 people**

13. **Test Practice** Nick bought 5 packages of strawberries. Each package weighed 1.15 pounds. What was the weight of the 5 packages? **D**

A 4.85 lb
B 5.15 lb
C 5.45 lb
D 5.75 lb
E Not Here

■ Extending the Lesson ■

Activity Most building supply stores have brochures that give detailed descriptions for designing a kitchen. The brochures include cabinet and appliance dimensions, sample layouts, and suggested spacing. Have small groups of students use such a brochure to design a 16.5-foot by 12-foot kitchen. Then have them catalog shop to find the total cost of the items used.

Check for Understanding
Use the results from Exercise 4 to determine whether students understand how to solve a problem by breaking it down into simpler problems.

Extra Practice If students need additional practice in problem solving, extra practice is available on the following pages.
- Solve a Simpler Problem, see p. 566
- Mixed Problem Solving, see pp. 593–594

Assignment Guide

All: 4–13

4 ASSESS

Closing Activity
Writing Guide students to summarize the lesson by asking them to think of a simpler problem to solve the following. *Find the number of hours in 12 years.* **Sample answer: Multiply the number of hours in a day by 365. Then multiply that number by 12. Then have pairs of students create problems that can be solved using a simpler problem.**

Additional Answers
4. Sample answer: By estimating first, you can examine the solution to see if the decimal point is in the correct place.

5. Sample answer: This strategy allows you to break a difficult problem into manageable parts.

6. Sample answer: When you use the distributive property to solve a problem mentally, you solve two simpler problems and add.

11. His competition average was 17 pins higher than his previous average.

- *Study Guide Masters*, p. 27
- *Practice Masters*, p. 27
- *Enrichment Masters*, p. 27
- Transparencies 4-5, A and B
- *Assessment and Evaluation Masters*, pp. 98, 99
- *School to Career Masters*, p. 4
- *Science and Math Lab Manual*, pp. 73–76

 CD-ROM Program
  - Resource Lesson 4-5
  - Interactive Lesson 4-5

Recommended Pacing	
Standard	Day 9 of 16
Honors	Day 9 of 15
Block	Day 5 of 8

1 FOCUS

 5-Minute Check
(Lesson 4-4)
Find the perimeter and the area of each rectangle or square.

1.

2.7 ft

6.3 ft

18 ft, 17.01 ft²

2. rectangle: ℓ, 24 ft; *w*, 16 ft
80 ft, 384 ft²

3. square: *s*, 21 in. **84 in., 441 in²**

4. rectangle: ℓ, 8.22 m; *w*, 3.4 m
23.24 m, 27.948 m²

5. square: *s*, 8.4 mi **33.6 mi, 70.56 mi²**

The 5-Minute Check is also available on **Transparency 4-5A** for this lesson.

Motivating the Lesson

Hands-On Activity Give small groups of students play money (bills and coins). Have them divide different amounts of money equally among themselves.

4-5 Dividing Decimals by Whole Numbers

What you'll learn
You'll learn to divide decimals by whole numbers.

When am I ever going to use this?
Knowing how to divide a decimal by a whole number can help you find your share of a group purchase.

A package that included in-line skates, a helmet, and pads was on sale for $119.50. Josh didn't have enough money to pay for them so, in order to get the sale price, he put them on layaway. The store's layaway policy requires one payment now and a payment equal to that each week for 4 weeks. How much will he pay each time?

You need to divide $119.50 by 5.

First estimate. $100 \div 5 = 20$

When dividing a decimal by a whole number, place the decimal point in the quotient directly above the decimal point in the dividend. Then divide as you do with whole numbers.

```
      23.90    Place the decimal point.
  5)119.50    Then divide as with whole numbers.
   -10
    19
   -15
     4 5
    -4 5
      00
```

He needs to pay $23.90 each week. Checking this answer against the estimate, the answer is reasonable.

Examples

Find each quotient.

1 29.8 ÷ 2

Estimate: 30 ÷ 2 = 15

```
    14.9     Place the
 2)29.8     decimal point.
  -2         Divide as with
   9         whole numbers.
  -8
   1 8
  -1 8
     0
```

29.8 ÷ 2 = 14.9

The estimate shows that the answer is reasonable.

2 8.58 ÷ 12

Estimate: 10 ÷ 10 = 1

```
    0.715    Place the
 12)8.580   decimal point.
   -8 4      Divide as with
    18       whole numbers.
   -12       Annex a zero to
    60       continue dividing.
   -60
     0
```

8.58 ÷ 12 = 0.715

The estimate shows that the answer is reasonable.

Usually when you divide with decimals, the answer does not come out even. You need to round the quotient to a specified place-value position. Always divide to one more place-value position than the place to which you are rounding.

Example ③

APPLICATION

Money Matters Katie and 5 of her friends bought a six-pack of fruit juice after their soccer game. Each friend wants to pay for her share. If the six-pack costs $3.29, how much does each friend owe to the nearest cent?

> **Study Hint**
> **Estimation** When dealing with money, always round the quotient up.

```
    0.548      Estimate: 3.00 ÷ 6 = $0.50
6)3.290        Place the decimal point.
  -3 0         Annex a zero.
     29        Divide to the thousandths place.
    -24
     50
    -48
      2
```

Katie and her five friends cannot pay $0.548 each. $0.548 rounded to the nearest cent is $0.55.

Each person owes $0.55. The estimate shows that the answer is reasonable.

Using Connections Have pairs of students solve the following problem. *Howard has $57 to buy gifts for three people. If he spends an equal amount on each gift, how much can he spend on each person?* $19 Change $57 to $5.70 and have them solve the new problem. $1.90

Teaching Tip Before studying Example 1, review estimation using compatible numbers.

> **In-Class Examples**
> *For Example 1*
> **Find each quotient.**
> **a.** 82.8 ÷ 4 **20.7**
> **b.** 36.4 ÷ 14 **2.6**
>
> *For Example 2*
> **Find each quotient.**
> **a.** 7.56 ÷ 15 **0.504**
> **b.** 2.18 ÷ 4 **0.545**
>
> *For Example 3*
> Mr. Schneider's class went to a movie theater. It cost $85.48 for 23 people. What was the cost per person rounded to the nearest cent? **$3.72**

CHECK FOR UNDERSTANDING

Communicating Mathematics

1. Sample answer: The estimate tells you the answer is close to 1.

Guided Practice

Read and study the lesson to answer each question. 2–3. See margin.

1. **Explain** why you can use an estimate to place the decimal point in the quotient for 34.56 ÷ 25.

2. **Tell** how to round a quotient to a specific place-value position.

3. **You Decide** Payat uses a calculator to find 75.89 ÷ 38. He gets 1.997105263 for an answer. Is the answer reasonable? Explain.

Find each quotient.

4. 13)15.6 **1.2**   5. 6)2.49 **0.415**   6. 55.6 ÷ 8 **6.95**

Round each quotient to the nearest tenth.

7. 55.39 ÷ 7 **7.9**   8. 78.75 ÷ 35 **2.3**   9. 111.37 ÷ 43 **2.6**

10. **Geometry** The perimeter of the square is 30.24 meters. What is the length of one side? **7.56 m**

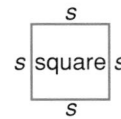

s | square | s
s

3 PRACTICE/APPLY

Check for Understanding
If students need additional practice or instruction after completing Exercises 1–10, one of these options may be helpful.
- Extra Practice, see p. 567
- Reteaching Activity
- *Transition Booklet,* pp. 25–26
- *Study Guide Masters,* p. 27
- *Practice Masters,* p. 27

■ Reteaching the Lesson ■

Activity Play "I Can Estimate That," using division cards. Players reveal a card and challenge each other to get the closest estimate. Then they find the exact answer on a calculator. The player with the closest estimate gets the card. The player with the most cards wins.

Additional Answers

2. Always divide to one more place-value position than the place to which you are rounding.

3. Sample answer: Estimate 80 ÷ 40 = 2; The estimate shows that the answer is reasonable.

EXERCISES

Practice

Find each quotient.

11. $25\overline{)6.25}$ **0.25**   12. $16\overline{)117.44}$ **7.34**   13. $37\overline{)784.4}$ **21.2**

14. $475.2 \div 32$ **14.85**   15. $479.96 \div 52$ **9.23**   16. $256.36 \div 34$ **7.54**

Round each quotient to the nearest tenth.

17. $29.48 \div 11$ **2.7**   18. $28.56 \div 42$ **0.7**   19. $263.25 \div 81$ **3.3**

Round each quotient to the nearest hundredth.

20. $344.736 \div 76$ **4.54** 21. $650.23 \div 29$ **22.42** 22. $4.567 \div 21$ **0.22**

Solve each equation. **24. 0.8956**

23. $x = 29.37 \div 3$ **9.79** 24. $8.956 \div 10 = y$ 25. $z = 302.5 \div 55$ **5.5**

26. Find 72.89 divided by 39 to the nearest thousandth. **1.869**

27. What is the solution of $m = 276.33 \div 61$? **4.53**

28. *Algebra* Evaluate $b \div c$ if $b = 34.56$ and $c = 32$. **1.08**

Applications and Problem Solving

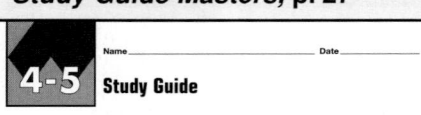

29. *Money Matters* On Monday, Amber was paid $20 for watering her neighbor's plants the previous week. On Tuesday, her mother paid her $14 for raking 14 bags of leaves. She also made $10 baby-sitting on Friday. What was Amber's average pay for these three jobs? **$14.67**

30. *Sports* The Pickerington girls track team ran the 4-by-100 meter relay in 48.9 seconds. What was the average time of each runner? **12.225 seconds**

31. **$0.19, $0.07, $0.06**

31. *Spreadsheets* The spreadsheet shows the unit price for a jar of jelly. To find the unit price, divide the cost of the item by its size. Find the unit price for the next three items. Round to the nearest cent.

	A	B	C	D
1	Item	Cost	Size	Unit Price
2	Jelly	$1.59	12 oz	0.1325
3	Cereal	$3.35	18 oz	
4	Bread	$1.19	16 oz	
5	Ketchup	$0.89	14 oz	

32. *Money Matters* Suppose you wanted to fill a car's 12-gallon gas tank. If you can afford to spend $14.25 on gasoline, what price per gallon should you look for? Round to the nearest cent. **$1.19**

33. *Critical Thinking* Create a division problem that meets all of the following conditions. **Sample answer: 0.9984 ÷ 8**

- the divisor is a whole number
- the dividend is a decimal
- the quotient is 0.125 when rounded to the nearest thousandth
- the quotient is 0.12 when rounded to the nearest hundredth

154 Chapter 4 Multiplying and Dividing Decimals

Study Guide Masters, p. 27

154 **Chapter 4**

34. *Geometry* Find the perimeter of the rectangle. *(Lesson 4-4)* **30.3 m**

11.75 m
3.4 m

35. *Weather* During two weeks of torrential rains, Antwon measured the rainfall. He recorded 11.33 centimeters in the first week and 15.75 centimeters in the second week. About how much rain fell in the two weeks? *(Lesson 3-5)* **about 27 cm**

36. *Careers* Prospective employees were told that the current employees' earnings averaged $475 a week. In the past, 6 people have earned weekly amounts of $200, $400, $410, $260, $320, and $1,260. Were the prospects being misled? Explain. *(Lesson 2-8)* **See margin.**

37. [**Test Practice**] The line graph shows the number of people in Marysville and Plain City from 1960 to 1990. About how many more people were in Marysville than in Plain City in 1980? *(Lesson 2-3)* **B**

A 2,000   B 3,500
C 6,000   D 7,500

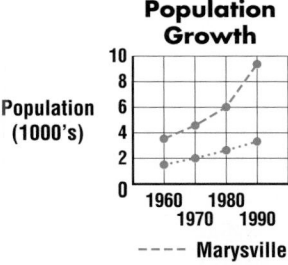

Population Growth

Population (1000's)

- - - - Marysville
.... Plain City

CHAPTER 4

Mid-Chapter Self Test

1. *Money Matters* Mrs. Ortiz has bought a new car. Her payments will be $385.55 a month for 48 months. About how much will she pay in all? *(Lesson 4-1)* **$20,000**

Find each product mentally. Use the distributive property. *(Lesson 4-2)*

2. 6×17 **102**

3. 10.5×40 **420**

Solve each equation. *(Lesson 4-3)*

4. $a = 28.5 \cdot 0.61$ **17.385**

5. $b = 0.006(3.4)$ **0.0204**

6. *Geometry* Find the perimeter of a square if each side is 7.2 inches long. *(Lesson 4-4)* **28.8 in.**

Find the perimeter and the area of each figure. *(Lesson 4-4)*

7. rectangle 0.2 cm
4.7 cm
9.8 cm, 0.94 cm²

8. square **12.4 m, 9.61 m²**
3.1 m

Round each quotient to the nearest hundredth. *(Lesson 4-5)*

9. $597.3 \div 62$ **9.63**

10. $69.597 \div 45$ **1.55**

Lesson 4-5 Dividing Decimals by Whole Numbers **155**

Extending the Lesson

Enrichment Masters, p. 27

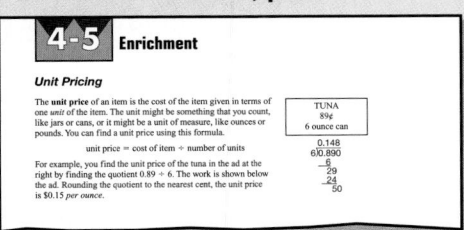

4-5 Enrichment

Unit Pricing

The **unit price** of an item is the cost of the item given in terms of one *unit* of the item. The unit might be something that you count, like jars or cans, or it might be a unit of measure, like ounces or pounds. You can find a unit price using this formula.

unit price = cost of item ÷ number of units

For example, you find the unit price of the tuna in the ad at the right by finding the quotient 0.89 ÷ 6. The work is shown below the ad. Rounding the quotient to the nearest cent, the unit price is $0.15 *per ounce.*

TUNA
89¢
6 ounce can

0.148
6)0.890
—6
 29
 24
 50

Activity Have pairs of students interview independent business owners. Have them ask how the skills taught in this lesson are used in business. Then have students use the information they have gathered to create word problems for classroom sharing.

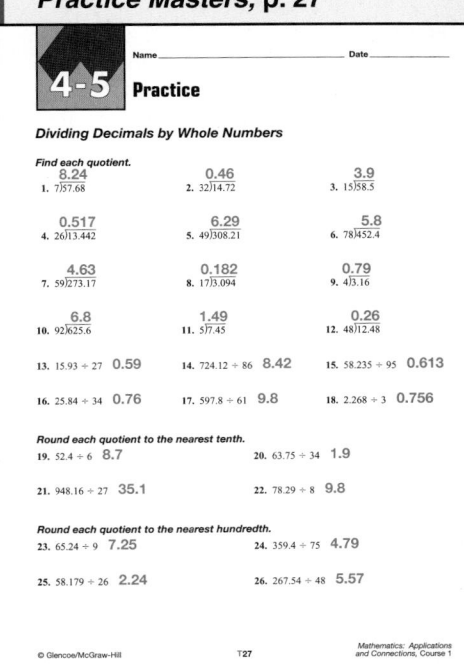

HANDS-ON LAB

4-6A Dividing by Decimals

A Preview of Lesson 4-6

base-ten blocks

In this lab, you will use base-ten blocks to model dividing a decimal by a decimal. Recall that each 10-by-10 block represents 1, each row or column represents 0.1, and each small square represents 0.01.

GET READY

Objective Students use base-ten blocks to divide decimals.

Optional Resources
Hands-On Lab Masters
• base-ten models, p. 1
• worksheet, p. 46

Manipulative Kit
• base-ten blocks

Overhead Manipulative Resources
• decimal models

MANAGEMENT TIPS

Recommended Time
30 minutes

Getting Started Review modeling decimals. Read the first paragraph of the lesson aloud and display each model as you define it. Have students practice representing numbers such as 1.4, 2.3, 0.2, and 0.7.

Activities 1 and 2 involve modeling decimal division. Remind students to carefully count their models when they do the trading.

ASSESS

Have students complete Exercises 1–4. Watch for students who are not trading their blocks correctly or who are not grouping correctly at the end.

Use Exercise 5 to determine whether students understand how to divide decimals.

TRY THIS

Work with a partner.

1 To model $1.2 \div 0.3$, follow these steps.
• Place one and two tenths in front of you.

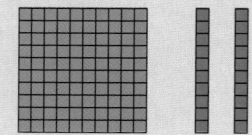

• Trade the ones block for tenths.

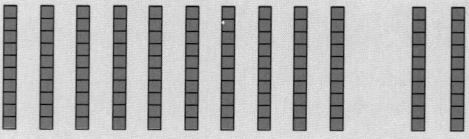

• Separate the tenths into groups of three tenths.

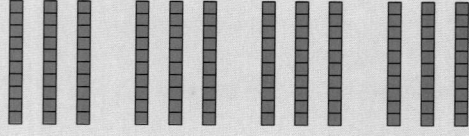

There are four groups of three tenths.
Therefore, $1.2 \div 0.3 = 4$.

2 To model $0.2 \div 0.04$, follow these steps.
• Place two tenths in front of you.

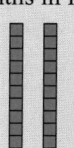

• Trade the tenths blocks for hundredths.

• Separate the hundredths into groups of four hundredths.

There are five groups of four hundredths.
Therefore, $0.2 \div 0.04 = 5$.

ON YOUR OWN

Use base-ten blocks to show each quotient. 1–4. See students' work for models.

1. $1.4 \div 0.7$ **2**   **2.** $2.6 \div 0.2$ **13**   **3.** $0.9 \div 0.03$ **30**   **4.** $1.25 \div 0.25$ **5**

5. *Look Ahead* Find the quotient for $2.4 \div 0.8$ without using models. **3**

156 Chapter 4 Multiplying and Dividing Decimals

Math Journal Have students write a paragraph explaining what expressions like $1.2 \div 0.3$ mean when using models.

4-6 Dividing by Decimals

What you'll learn

You'll learn to divide decimals by decimals.

When am I ever going to use this?

Decimal division can be used to compare the rates of speed over different distances.

The circle graph shows where the United States exported $16.9 billion worth of cars in 1996.

How many times greater was the value of the cars exported to Canada than to Japan?

You need to divide 7.8 by 2.4.

First, estimate the answer.

$$8 \div 2 = 4$$

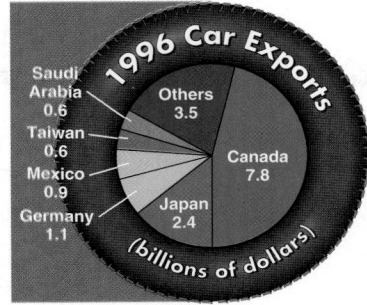

1996 Car Exports

Saudi Arabia 0.6
Taiwan 0.6
Mexico 0.9
Germany 1.1
Japan 2.4
Canada 7.8
Others 3.5

(billions of dollars)

Source: U.S. Census Bureau

The value of the cars exported to Canada was about 4 times greater than the value of the cars exported to Japan.

When dividing decimals by decimals, change the divisor to a whole number. To do this, multiply both the divisor and dividend by the same power of 10. Then divide as with whole numbers.

Multiply 7.8 and 2.4 by 10.

$$2.4\overline{)7.8} \rightarrow 2.4\overline{)7.8} \rightarrow 24\overline{)78.00}$$

$$
\begin{array}{r}
3.25 \\
24\overline{)78.00} \\
-72 \\
\hline
6\ 0 \\
-4\ 8 \\
\hline
1\ 20 \\
-1\ 20 \\
\hline
0
\end{array}
$$

Multiply to check.

$$
\begin{array}{r}
3.25 \\
\times 2.4 \\
\hline
1\ 300 \\
6\ 50 \\
\hline
7.800 \checkmark
\end{array}
$$

Study Hint

Mental Math When multiplying by powers of ten, move the decimal to the right as many places as the number of zeros in the power of ten.

The value of the cars exported to Canada was 3.25 times greater than the value of the cars exported to Japan. The estimate shows that the answer is reasonable.

Example ① Find $8.84 \div 3.4$.

Estimate: $9 \div 3 = 3$

$$3.4\overline{)8.84} \rightarrow 34\overline{)88.4}$$

$$
\begin{array}{r}
2.6 \\
34\overline{)88.4} \\
-68 \\
\hline
20\ 4 \\
-20\ 4 \\
\hline
0
\end{array}
$$

Multiply by 10.
Place the decimal point.
Divide as with whole numbers.

$8.84 \div 3.4 = 2.6$. The estimate shows that the answer is reasonable.

Check: $2.6 \times 3.4 = 8.84$ ✓

Lesson 4-6 Dividing by Decimals **157**

Multiple Learning Styles

Verbal/Linguistic Have students research decimal facts about animal sizes. For example, a 0.1-inch flea can jump 7.75 inches high. Have them use these facts to create word problems about living creatures. Separate students into groups to solve the problems.

In-Class Examples

For Example 1
Find each quotient.
a. 4.352 ÷ 3.2 **1.36**
b. 8.873 ÷ 1.9 **4.67**

For Example 2
Find each quotient.
a. 1.035 ÷ 0.023 **45**
b. 25.16 ÷ 0.068 **370**

For Example 3
Mrs. Sharvy's vegetable garden is 8.25 meters long. She wants to plant tomatoes 0.33 meter apart along the back border. How many plants does she need? **25**

3 PRACTICE/APPLY

Check for Understanding

If students need additional practice or instruction after completing Exercises 1–9, one of these options may be helpful.
- Extra Practice, see p. 567
- Reteaching Activity
- *Transition Booklet*, pp. 27–28
- *Study Guide Masters*, p. 28
- *Practice Masters*, p. 28

Study Guide Masters, p. 28

4-6 Study Guide

Name_____ Date_____

Dividing by Decimals

To divide a decimal by a decimal, first multiply the divisor by a power of ten to make it a whole number. Multiply the dividend by the same power of ten. Then divide as with whole numbers.

Example Find 87.3025 ÷ 3.715.   *Estimate:* 88 ÷ 4 = 22

```
              23.5
    3.715.)87.302.5       Multiply the divisor and the dividend by 1,000.
    − 74 30                Place the decimal point in the quotient.
      13 002              Divide.
    − 11 145
       1 857 5
     − 1 857 5
             0
```

Find each quotient.
1. 1.4)9.8 **7**
2. 2.7)40.5 **15**
3. 0.41)3.69 **9**
4. 2.1)4.41 **2.1**

5. 0.07)2.38 **34**
6. 0.212)1.696 **8**
7. 0.013)0.0208 **1.6**
8. 6.28)87.92 **14**

Solve each equation.
9. a = 27.63 ÷ 0.3 **92.1**
10. 8.652 ÷ 1.2 = z **7.21**
11. 9.594 ÷ 0.06 = h **159.9**
12. s = $1.76 ÷ 32 **$0.055**

© Glencoe/McGraw-Hill   T28   *Mathematics: Applications and Connections, Course 1*

Examples ── ❷ Find 15.176 ÷ 0.028.

```
                                542.
    0.028)15.176  →  28)15176.        Multiply by 1,000.
                      −140            Place the decimal point.
                       117           Divide.
                      −112
                        56
                       −56
                         0
```

15.176 ÷ 0.028 = 542 **Check:** 0.028 × 542 = 15.176 ✓

INTEGRATION ❸ **Measurement** Mr. Henderson's flower garden is 11.25 meters long. He wants to make a border along one side using bricks that are 0.25 meter long. How many bricks does he need?

```
                     45
    0.25)11.25  →  25)1125       Multiply by 100.
                   −100          Place the decimal point.
                    125          Divide.
                   −125
                      0
```

Mr. Henderson needs 45 bricks.

CHECK FOR UNDERSTANDING

Communicating Mathematics

Read and study the lesson to answer each question. 1–2. See margin.

1. *Write* a decimal division problem that would require you to multiply the divisor and dividend by 10 before you divide.

2. *Explain* why the quotient for 2.35 ÷ 0.58 should be about 4.

Guided Practice

Find each quotient.

3. 2.7)51.3 **19**
4. 2.5)43.25 **17.3**
5. 0.18)124.92 **694**

6. 302.5 ÷ 5.5 **55**
7. 70.59 ÷ 1.3 **54.3**
8. 3.965 ÷ 0.065 **61**

9. *Algebra* Evaluate $a ÷ b$ if $a = 7.502$ and $b = 3.41$. **2.2**

EXERCISES

Practice

Find each quotient.

10. 4.9)160.72 **32.8**
11. 0.98)109.76 **112**
12. 3.25)2,606.5 **802**

13. 3.6)1.44 **0.4**
14. 89.6)206.08 **2.3**
15. 0.008)0.0072 **0.9**

17. 2,967

16. 0.32 ÷ 0.005 **64**
17. 210.657 ÷ 0.071
18. 9.425 ÷ 0.0125 **754**

19. 0.0186 ÷ 0.031 **0.6**
20. 40.99524 ÷ 14.3 **2.8668**
21. 5,885.9514 ÷ 703.22 **8.37**

■ Reteaching the Lesson ■

Activity Illustrate 2.5 ÷ 0.5 on the overhead projector using decimal models. Then multiply the divisor and the dividend by ten. Show 25 ÷ 5 with models. Guide students to conclude that multiplying the divisor and the dividend by a power of ten does not change the quotient.

Additional Answers

1. Sample answer: 1.84 ÷ 0.8

2. Sample answer: When you multiply the divisor and dividend by 100, the problem becomes 235 ÷ 58. Estimate 240 ÷ 60 = 4.

Solve each equation.

22. $b = 17.01 \div 0.81$ **21**

23. $p = 94.16 \div 0.88$ **107**

24. $127.6625 \div 102.13 = x$ **1.25**

25. $0.0078 \div 0.002 = y$ **3.9**

26. What is 48.355 divided by 0.095? **509**

27. Find $7.502 \div 2.2$. **3.41**

28. *Algebra* Evaluate $s \div t$ if $s = 1.43$ and $t = 1.3$. **1.1**

29. *Algebra* Evaluate $(p + q) \div r$ if $p = 0.7$, $q = 1.4$, and $r = 2.5$. **0.84**

Applications and Problem Solving

30. *Sports* At the 1996 Olympics, American sprinter Michael Johnson set a world record of 19.32 seconds for the 200-meter dash. A honeybee can fly the same distance in 40.572 seconds. How many times faster than a honeybee is Michael Johnson? **2.1 times faster**

31. *Transportation* The Aero Spacelines Super Guppy, a converted Boeing C-97, can carry 87.5 tons. Tanks that weigh 4.5 tons each are to be loaded onto the Super Guppy. What is the maximum number of tanks it can transport? **19 tanks**

32. *Food* The table shows how many pounds of breakfast cereal are consumed each year by the average person in selected countries.

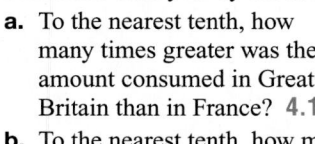

a. To the nearest tenth, how many times greater was the amount consumed in Great Britain than in France? **4.1**

b. To the nearest tenth, how many times greater was the amount consumed in the United States than in Canada? **2.0**

Country	Amount (lb)
Canada	6.02
France	1.78
Great Britain	7.34
South Korea	0.07
United States	11.9

33. *Critical Thinking* Replace each ■ to make a true sentence.

$$\blacksquare.8 \; \blacksquare \; 6 \div 0.35 = 2.5 \blacksquare$$

Sample answer: $0.896 \div 0.35 = 2.56$

Mixed Review

34. Find the quotient when 68.52 is divided by 12. *(Lesson 4-5)* **5.71**

35. Write the number 204.2398 in words. *(Lesson 3-1)* **See margin.**

36. **Test Practice** The stem-and-leaf plot shows the high temperatures for 20 cities on February 23. Which interval had the fewest high temperatures? *(Lesson 2-6)* **C**

A $20° - 29°$

B $30° - 39°$

C $50° - 59°$

D $70° - 79°$

Stem	Leaf
2	5
3	0 2 4 5
4	1 2 6 9
5	
6	0 3 3 4 4 5 5 6
7	0 0 1

$4|2 = 42$

37. *Algebra* Solve $b \times 6 = 42$ mentally. *(Lesson 1-7)* **7**

---

4 ASSESS

Closing Activity

Speaking Write several problems on the chalkboard that involve dividing decimals by decimals. Give students clues to identify a problem such as, "To divide, the divisor and dividend would be multiplied by 100." Have the student who identifies the first problem provide the next clue. Continue for all problems.

Additional Answer

35. two hundred four and two thousand three hundred ninety-eight ten-thousandths

Practice Masters, p. 28

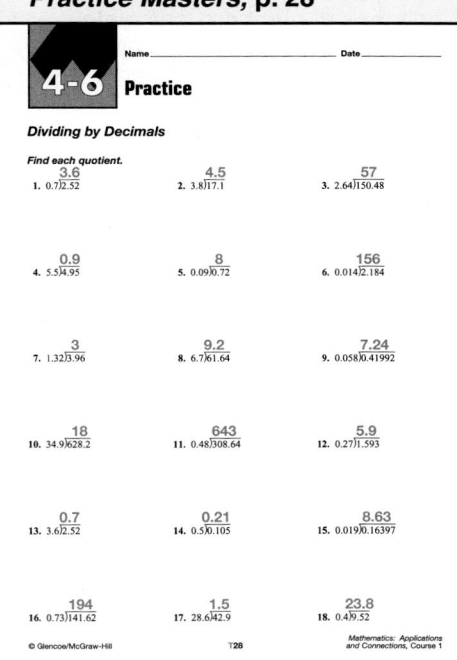

Extending the Lesson

Enrichment Masters, p. 28

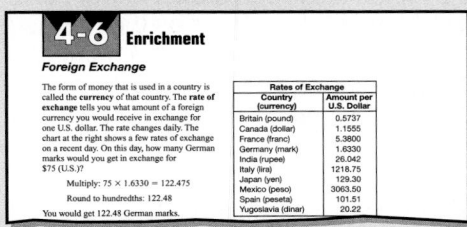

Activity Have groups of students write and solve word problems that involve dividing decimals. Have students use current newspapers, magazines, and catalogs to make the word problems relevant and realistic.

Motivating Students

Owning a small business is the dream of many people. Small business owners have the advantages of choosing the type of company they start, setting their own working hours and conditions, and earning profits from the company in addition to receiving a salary. To begin a discussion of small business ownership, ask students these questions.

- Are there any small businesses in your neighborhood? What are they? **Sample answer: video store, ice cream shop, service station, bakery**
- What would you need to start a small business? **Sample answer: money, a place to work , a product or service to sell, advertising**
- Why do you think some small businesses are more successful than others? **Sample answer: Some products and services are in higher demand than others.**

Making the Math Connection

Small business owners themselves often keep track of expenditures, sales, and inventory. They need to be able to add and subtract columns of numbers, calculate sales tax, and analyze financial data to determine profits and losses.

Working on *Your Turn*

Have students begin by forming groups and brainstorming ideas for small businesses. Have them choose five products or services to include in their survey. After conducting the survey, have groups discuss what their results would mean to someone attempting to start a small business.

*An additional School to Career activity is available on page 4 of the **School to Career Masters.***

SCHOOL to CAREER

BUSINESS

Cedric Walker
Small business owner

Cedric Walker is the co-owner, along with Cal Dupree, of the UniverSoul Circus, the nation's only African-American-owned, operated, and performed circus. After spending three years doing research, Walker opened the circus in 1994.

A person who is interested in owning a small business should take classes in economics, mathematical analysis, business ethics, communications, mathematics of finance, and marketing. A college degree in communications, marketing, business law, business management, or accounting should be considered.

For more information:
The Young Entrepreneurs' Organization
10101 North Glebe Road
Arlington, VA 22207
(703) 519-6700

inter CONNECTION
www.glencoe.com/sec/
math/mac/mathnet

> I can't wait to design my own product and start my own business!

Your Turn
Part of what makes a small business owner successful is offering a product or service that people need. Survey your class or your school and find out what people need that a small business could provide. Write a paragraph describing what you discovered and how you might be able to fill the need with your own small business.

160 Chapter 4 Multiplying and Dividing Decimals

More About Cedric Walker

- When Walker was 18, his father sent him to live with his uncle. He met the musical group the Commodores at his uncle's club and joined their road crew. He went on to a career in music promotion. He produced rap music shows such as the New York City Fresh Festival and gospel plays such as *A Good Man Is Hard to Find*.
- Walker set out to create an African-American variety show, then decided to make it a circus. He stresses that the circus is family entertainment. From a marketing standpoint, he believes the circus will be in business for many years to come.

What you'll learn

You'll learn to divide decimals involving special cases of zero in the quotient.

When am I ever going to use this?

Knowing how to divide all kinds of decimals can help you find the unit cost of a jar of peanut butter.

Did you know?  The men's gymnastic team from Japan won the gold medal in the team combined exercises for five consecutive Olympics, from 1960 through 1976.

On March 30, 1980, Shigeru Iwasaki somersaulted backward 54.68 yards in 10.8 seconds. To the nearest hundredth, how many yards per second did he somersault?

First, estimate the answer.

$$50 \div 10 = 5$$

He went about 5 yards per second.

Method 1 Use a calculator.

54.68 ÷ 10.8 = *5.062962963*

To the nearest hundredth, he went 5.06 yards per second. Compared to the estimate, the answer is reasonable.

Method 2 Use paper and pencil.

You can see why a zero appears in this quotient.

$$
\begin{array}{r}
5.062 \\
108\overline{)546.800} \\
-540 \\
\hline
6\ 8 \\
-0 \\
\hline
6\ 80 \\
-6\ 48 \\
\hline
320 \\
-216 \\
\hline
104
\end{array}
$$

$10.8\overline{)54.68} \rightarrow$ *Divide to the thousandths place to round to the nearest hundredth. 68 < 108 so a zero is written in the quotient.*

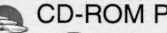

Example ① Find $9.03 \div 0.301$.

$$
\begin{array}{r}
30.0 \\
0.301\overline{)9.030} \\
-9\ 03 \\
\hline
00
\end{array}
$$
Write a zero in the ones place of the quotient.

$9.03 \div 0.301 = 30.0$

Check: $0.301 \times 30 = 9.03$ ✓

Lesson 4-7 Zeros in the Quotient **161**

Instructional Resources

- *Study Guide Masters*, p. 29
- *Practice Masters*, p. 29
- *Enrichment Masters*, p. 29
- Transparencies 4-7, A and B
- *Assessment and Evaluation Masters*, p. 100

CD-ROM Program
- Resource Lesson 4-7

Recommended Pacing	
Standard	Day 12 of 16
Honors	Day 11 of 15
Block	Day 6 of 8

1 FOCUS

5-Minute Check
(Lesson 4-6)

Find each quotient.

1. $4.2\overline{)466.2}$ **111**
2. $5.8\overline{)21.46}$ **3.7**
3. $9.476 \div 1.03$ **9.2**
4. $173.28 \div 45.6$ **3.8**
5. Solve $a = 181.44 \div 15.12$. **12**

The 5-Minute Check is also available on **Transparency 4-7A** for this lesson.

Motivating the Lesson

Communication Have students read the opening paragraphs of the lesson. Ask: *Why is there a zero in the tenths place of the quotient?* **There are no groups of 108 in a group of 68.**

2 TEACH

Transparency 4-7B contains a teaching aid for this lesson.

Using Calculators Have students use a calculator to find quotients in decimal division problems they make up. Have them record any quotients with zeros in them. Then have them make a generalization based on their findings.

In-Class Examples

For Example 1
a. Find $3.80 \div 0.19$. **20**
b. Find $\$2.97 \div 3$. **$0.99**

For Example 2
The area of a rectangular bathroom is 85.012 square feet. The length of the bathroom is 8.02 feet. Find the width of the bathroom. **10.6 ft**

3 PRACTICE/APPLY

Check for Understanding

If students need additional practice or instruction after completing Exercises 1–11, one of these options may be helpful.
• Extra Practice, see p. 567
• Reteaching Activity
• *Study Guide Masters*, p. 29
• *Practice Masters*, p. 29

Additional Answers

2. Sample answer: Multiply the divisor and dividend by 100. Estimate: $1,800 \div 2 = 900$.

3. Sample answer: Multiply the divisor and dividend by 10. The problem becomes $1,561.4 \div 148$. Estimate: $1,500 \div 150 = 10$.

Study Guide Masters, p. 29

Name _____ Date _____

4-7 **Study Guide**

Zeros in the Quotient

Remember to write a zero in the quotient when you need a placeholder.

Examples

Use zero as a placeholder in the ones place.
$$\begin{array}{r} 0.8 \\ 7\overline{)5.6} \\ -5.6 \\ \hline 0 \end{array}$$

There are no 19s in 2. Use zero as a placeholder.
$$\begin{array}{r} 0.012 \\ 19\overline{)0.228} \\ -0 \\ \hline 22 \\ -19 \\ \hline 38 \\ -38 \\ \hline 0 \end{array}$$

There are no 27s in 18. Use zero as a placeholder.
$$\begin{array}{r} 1.07 \\ 2.7\overline{)2.889} \\ -2.7 \\ \hline 18 \\ -0 \\ \hline 189 \\ -189 \\ \hline 0 \end{array}$$

Find each quotient to the nearest hundredth.

1. $22\overline{)6.6}$ **0.3**
2. $14\overline{)1.26}$ **0.09**
3. $7\overline{)28.35}$ **4.05**

4. $2.95 \div 59$ **0.05**
5. $0.0264 \div 2.2$ **0.01**
6. $0.158 \div 7.9$ **0.02**

7. $19.065 \div 9.3$ **2.05**
8. $0.102 \div 34$ **0.003**
9. $72.76 \div 6.8$ **10.7**

© Glencoe/McGraw-Hill   T29   *Mathematics: Applications and Connections, Course 1*

Example **2**
APPLICATION

Money Matters Mr. Alvarez put 11.5 gallons of gas in his van. If he paid $13.90, what was the price per gallon?

$$11.5\overline{)13.90} \rightarrow \begin{array}{r} 1.208 \\ 115\overline{)139.000} \\ -115 \\ \hline 24\,0 \\ -23\,0 \\ \hline 1\,00 \\ -0 \\ \hline 1\,000 \\ -920 \\ \hline 80 \end{array}$$ *Estimate: $10 \div 10 = 1$*

The price of the gasoline was about $1.21 per gallon.

CHECK FOR UNDERSTANDING

Communicating Mathematics

1. Sample answer: $4.53 \div 15$

Math Journal

Read and study the lesson to answer each question. 2–3. See margin.

1. *Write* a decimal division problem that would require a zero in the quotient.
2. *Explain* how you could estimate the quotient for $17.5 \div 0.02$.
3. *Write* a letter to a student explaining why $156.14 \div 14.8$ is not 1.55.

Guided Practice

Find each quotient to the nearest hundredth.

4. $13\overline{)27}$ **2.08**
5. $36\overline{)145}$ **4.03**
6. $61\overline{)2,441}$ **40.02**
7. $75.3 \div 15$ **5.02**
8. $0.68 \div 6$ **0.11**
9. $3.41 \div 0.2$ **17.05**

10. Solve $y = 15.25 \div 0.5$. **30.5**
11. *Algebra* Evaluate $d \div r$ if $d = 51.714$ and $r = 5.07$. **10.2**

EXERCISES

Practice

Find each quotient to the nearest hundredth.

12. $12\overline{)24.32}$ **2.03**
13. $0.25\overline{)75.25}$ **301**
14. $0.45\overline{)0.117}$ **0.26**
15. $24\overline{)125}$ **5.21**
16. $3.98\overline{)4.02}$ **1.01**
17. $7.89\overline{)86.3}$ **10.94**
18. $1.43 \div 13$ **0.11**
19. $83.25 \div 4.11$ **20.26**
20. $234.4 \div 11.2$ **20.93**
21. $2.5 \div 0.24$ **10.42**
22. $0.0125 \div 1.7$ **0.01**
23. $95.23 \div 47.6$ **2.00**

Solve each equation.

24. $g = 6.018 \div 5.9$ **1.02**
25. $d = 59.59 \div 0.59$ **101**
26. $7,576.4 \div 37.6 = f$ **201.5**
27. $3.3894 \div 4.2 = m$ **0.807**

28. What is 0.62524 divided by 30.8? **0.0203**
29. *Algebra* Evaluate $(a + b) \div c$ if $a = 0.016$, $b = 0.008$, and $c = 7.5$. **0.0032**

162   Chapter 4 Multiplying and Dividing Decimals

Reteaching the Lesson

Activity Have pairs of students model $0.53 \div 8$ with base-ten blocks. Ask them how they would write the quotient. Then have them use their calculators to verify the answer. Continue with other examples that have zeros in the quotient.

Error Analysis
Watch for students who omit zeros before the decimal point.
Prevent by reminding students that without a preceding zero, a decimal point is easily overlooked—a $0.79 pen could become a $79 pen!

30. *Music* LeAnn Rimes' *blue*
compact disc has 11 tracks. If the
entire CD takes 34.8 minutes to
play, what is the average time
for each track to the nearest
hundredth of a minute?

31. *Geometry* The area of
the patio in the diagram
is 27.36 square meters.
The width of the patio
is 4.5 meters. Find the
length of the patio. **6.08 m**

ℓ

| 4.5 m | Area = 27.36 sq m | 4.5 m |

ℓ

32. *Physical Science* Sound travels through air at
330 meters per second. How long will it take a bat's cry to reach its prey
and echo back if the prey is 1 meter away? Round to the nearest thousandth
of a second. **0.006 s**

33. *Critical Thinking* If a decimal greater than 0 and less than 1 is divided
by a lesser decimal, would the quotient be always less, sometimes less, or
never less than 1? **never**

Mixed Review

34. Find 5.06 divided by 2.3. *(Lesson 4-6)* **2.2**

35. | Test Practice | Mrs. Zacharias asked her mathematics students to
decide whether 95 ceramic tiles, each measuring 1.5 feet-by-1.5
feet, would be enough to completely cover the floor of her classroom.
What other piece of information would allow the students to answer
the question? *(Lesson 4-4)* **D**

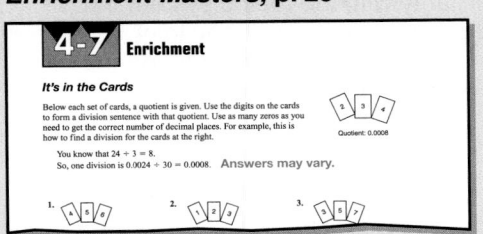

  A the weight of each tile
  B the number of classrooms in the school
  C the number of students in Mrs. Zacharias'
    classroom
  D the area of the classroom floor
  E the number of tiles in a package

36. *Write a Problem* Write a problem about a
real-life situation involving the photo at the
left that can be represented by a mean of
20° F. *(Lesson 2-7)* **See margin.**

37. *Statistics* Choose an appropriate scale for a
frequency table for the following set of data:
24, 67, 11, 9, 52, 38, 114, and 98.
(Lesson 2-2) **Sample answer: 0 to 120**

Lesson 4-7 Zeros in the Quotient **163**

Activity Have students solve the
following problem. *When 6.26 is divided
by a number, the quotient is 2.08 and
there is a remainder of 2. What is the
number?* **3**

Instructional Resources
- *Study Guide Masters*, p. 30
- *Practice Masters*, p. 30
- *Enrichment Masters*, p. 30
- Transparencies 4-8, A and B
 CD-ROM Program
  - Resource Lesson 4-8

Recommended Pacing	
Standard	Day 13 of 16
Honors	Day 12 of 15
Block	Day 7 of 8

1 FOCUS

5-Minute Check
(Lesson 4-7)

Find each quotient to the nearest hundredth.

1. $55\overline{)115}$ **2.09**
2. $4.67 \div 38$ **0.12**
3. $83.7 \div 402.8$ **0.21**
4. Evaluate $r \div s$ if $r = 506.3$ and $s = 12.4$. Round to the nearest hundredth. **40.83**
5. Dinner for three costs $35.40. To the nearest cent, what is the cost for each person? **$11.80**

The 5-Minute Check is also available on **Transparency 4-8A** for this lesson.

Motivating the Lesson
Hands-On Activity Give small groups of students dried beans. Discuss the meanings of "mass." Without using a scale, have them describe how they would measure one kilogram of beans.

2 TEACH

Transparency 4-8B contains a teaching aid for this lesson.

Using a Balance Have students estimate the mass of various items in grams, milligrams, and kilograms. Have them use a pan balance and metric weights to check their work.

4-8

Integration: Measurement
Mass and Capacity in the Metric System

What you'll learn
You'll learn to use metric units of mass and capacity.

When am I ever going to use this?
Knowing how to use metric units of mass and capacity can help you solve problems in science class.

Word Wise
gram (g)
kilogram (kg)
milligram (mg)
liter (L)
milliliter (mL)

Pablo Lara is a weightlifter from Cuba who won the gold medal in the 1996 Olympics in Atlanta, Georgia. Pablo competed in the 76-kilogram division. This means his mass could be no more than 76 kilograms.

In the metric system, all units are defined in terms of a basic unit. The basic unit of mass in the metric system is the **gram (g)**. A **kilogram (kg)** is 1,000 grams, and a **milligram (mg)** is 0.001 gram.

Many items you use every day are measured in grams, kilograms, or milligrams.

- A small paper clip has a mass of about 1 gram.
- Your math textbook has a mass of about 1 kilogram.
- A grain of salt has a mass of about 1 milligram.

Examples

Did you know Pablo Lara lifted 451.75 pounds (204.9 kg) in the clean and jerk and 358 pounds (162.4 kg) in the snatch, yet he weighs only 167.5 pounds.

Write the unit of mass: milligram, gram, or kilogram, that you would use to measure each of the following. Then estimate the mass.

 a compact disc

Since a compact disc has a small mass, the *gram* is the appropriate unit.

The mass of a compact disc is about 50 grams.

 a laptop computer

Since a laptop computer has a mass close to several copies of your textbook, the *kilogram* is the appropriate unit.

The mass of a laptop computer is about 4 kilograms.

The basic unit of capacity in the metric system is the **liter (L)**. A liter is a little more than a quart. The **milliliter (mL)** is 0.001 liter. It takes 1,000 milliliters to make a liter.

- A small pitcher has a capacity of about 1 liter.
- An eyedropper has a capacity of about 1 milliliter.

Classroom Vignette

"To practice students' familiarity with the metric system, I have them work in groups and gather objects that they think will weigh a certain amount, such as 40 grams. You can do the same with length or units of capacity."

Judith J. Cardamone, Teacher
East Middle School
Colorado Springs, CO

Write the unit of capacity: milliliter or liter, that you would use to measure each of the following. Then estimate the capacity.

3 a bathtub

Since a bathtub holds a large amount, the *liter* is the appropriate unit.

An average bathtub has a capacity of about 80 liters.

4 10 drops of food coloring

Since 10 drops of food coloring is a small amount, the *milliliter* is the appropriate unit.

The capacity is about 1 milliliter.

CHECK FOR UNDERSTANDING

Communicating Mathematics

Read and study the lesson to answer each question.

1. *Tell* which unit represents a greater mass, 1 gram or 1 kilogram. **kilogram**

2. *Explain* what a milliliter and a milligram have in common. **See margin.**

Guided Practice

Write the unit that you would use to measure each of the following. Then estimate the mass or capacity. **4. milligram; 500 mg**

3. the mass of a horse **kilogram; 500 kg**   4. an aspirin

5. a large bottle of soft drink **liter; 2 L**   6. a tennis ball **gram; 60 g**

7. *Food* Great Harvest Bread Company makes loaves of bread that have a mass of 1 kilogram. A loaf of honey whole wheat is much larger than a loaf of blueberry apricot. Explain why this happens. **See margin.**

EXERCISES

Practice

Write the unit that you would use to measure each of the following. Then estimate the mass or capacity.

9. milliliter, 200 mL
10. milliliter, 1 mL
11. kilogram,100kg

8. gas in the tank of a car **liter, 50 L**   9. a bottle of cough syrup

10. vanilla used in a cookie recipe   11. the mass of an Olympic boxer

12. a can of soup **milliliter, 300 mL**   13. a bag of sugar **kilogram, 2 kg**

14. a glass of iced tea **milliliter, 350 mL** 15. a dollar bill **milligram, 1 mg**

16. a pitcher of water **liter, 2 L**   17. a canary **gram, 100 g**

18. the liquid in a thermometer **milliliter, 10 mL**   19. a Ping-Pong ball **milligram, 500 mg**

20–22. See students' work.

Name an item that you think has the given measure.

20. about 250 mL   21. about 15 kg

Applications and Problem Solving

22. *Life Science* The human brain averages about 0.025 of a human's total body mass. Find your mass in kilograms and use a calculator to determine the approximate mass of your brain.

23. *Medicine* A pharmacist has 84.73 grams of a prescription medicine. She wants to separate it into 48 capsules. To the nearest thousandth, how many grams will go into each capsule? **1.765 g**

Lesson 4-8 Integration: Measurement Mass and Capacity in the Metric System **165**

■ Reteaching the Lesson ■

Activity Have pairs of students use a scale to determine the mass of common materials such as pencils, books, and scissors. Have them record their findings and then arrange the items in order of increasing mass.

Additional Answers

2. Sample answer: They are both one-thousandth of the basic unit.

7. Sample answer: The fruit is heavier, so the loaf has to be smaller to have the same mass.

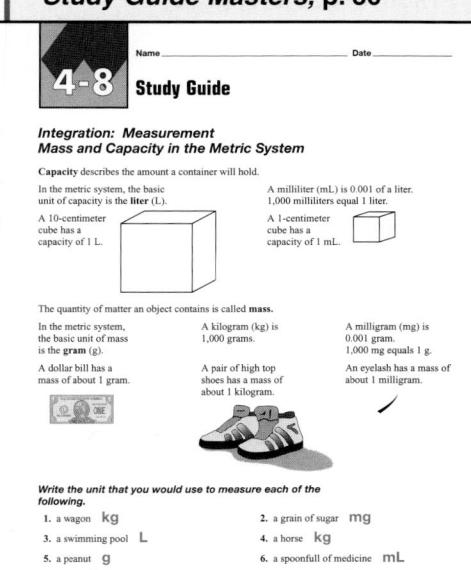

4 ASSESS

Closing Activity

Modeling Have students use clay to form different size balls that estimate various masses. To model capacity, tape pieces of paper together. Compare and discuss the different estimations.

Additional Answer

24b. Sample answer: The 6-pack is the better buy because you get more for the same amount of money.

24. *Money Matters* A can of soft drink contains 355 milliliters.
   a. How many milliliters are in a 6-pack of soft drink? **2,130 mL**
   b. Two liters of soft drink costs the same as a 6-pack of the same soft drink. Which is the better buy? Explain. **See margin.**

25. *Life Science* The table shows part of the recommended daily diet for a gibbon.
   a. Determine how many grams of carrots, bananas, and celery are needed in one week. **4,389 g**
   b. Is the amount of carrots, bananas, and celery in part a more or less than 4 kilograms?
   c. Visit a grocery store and select two or three of the items to weigh. About how many of these items would be needed for one week? **See students' work.**
   b. **more than 4 kg**

380 g lettuce
270 g orange
150 g spinach
143 g sweet potato
270 g banana
147 g carrot
210 g celery
210 g green beans

26. *Critical Thinking* Marva filled a 250-milliliter beaker with sand. She said its mass was 250 milligrams. Is she right? Explain.

Mixed Review

26. No, weight and capacity are not the same.

27. Divide 1,269.45 by 6.3. *(Lesson 4-7)* **201.5**

28. *Physical Science* One cubic meter of carbon dioxide has a mass of 1.977 kilograms. Round this number to the nearest tenth. *(Lesson 3-4)* **2.0**

29. **Test Practice** If a passenger train travels an average of 58 miles per hour, which is a good estimate for the number of hours it will take to travel 286 miles? *(Lesson 1-3)* **C**

   A 3 h   **B** 4 h   **C** 5 h   **D** 6 h   **E** 18 h

Let the GameS Begin

Scavenger Hunt

Get Ready This game is for three to four players.
   5 index cards per player   tape

Get Set Each player is assigned a unit of measure. For example, one player may have centimeters, another player may have liters, and the third player may have kilograms.

Go
   • Each player will write their unit on each of the index cards and find items in the class that would best be measured using that unit.
   • Players should pass their cards to the person on their left.
   • The winner is the first player to tape the cards to the appropriate items.

interNET CONNECTION Visit www.glencoe.com/sec/math/mac/mathnet for more games.

■ Extending the Lesson ■

Let the GameS Begin

You may want to have a fourth player on the team so that three players are matching the cards to the items and the fourth is judging correctness.

What you'll learn

You'll learn to change units within the metric system.

When am I ever going to use this?

You'll often need to change metric units when you are simplifying measurements.

Changing Metric Units

A chemistry experiment requires 3.05 grams of 3% hydrogen peroxide to produce one liter of oxygen. If a scientist wanted 24 liters of oxygen, how many kilograms of 3% hydrogen peroxide will she need? *This problem will be solved in Example 3.*

To change from one unit to another within the metric system, you either multiply or divide by powers of 10. The chart below shows the relationship between the units in the metric system and the powers of 10.

> **Study Hint**
>
> **Mental Math**
> Remember you can multiply or divide by a power of ten by moving the decimal point.

Each place value is ten times the place value to its right.

To change from a larger unit to a smaller unit, you need to multiply. To change from a smaller unit to a larger unit, you need to divide.

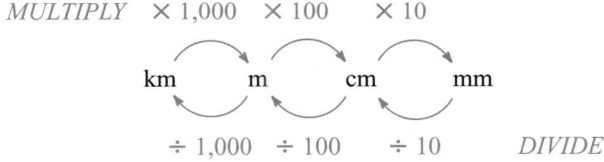

MULTIPLY × 1,000   × 100   × 10

km   m   cm   mm

÷ 1,000   ÷ 100   ÷ 10   *DIVIDE*

Examples

1 345 mL = _?_ L

To change from milliliters to liters, divide by 1,000 since 1 L = 1,000 mL.

$345 \div 1,000 = 0.345$
345 mL = 0.345 L

2 0.9 cm = _?_ mm

To change from centimeters to millimeters, multiply by 10 since 1 cm = 10 mm.

$0.9 \times 10 = 9$
0.9 cm = 9 mm

Lesson 4-9 Integration: Measurement Changing Metric Units **167**

4-9 Lesson Notes

Instructional Resources

- *Study Guide Masters,* p. 31
- *Practice Masters,* p. 31
- *Enrichment Masters,* p. 31
- Transparencies 4-9, A and B
- *Assessment and Evaluation Masters,* p. 100
- *Diversity Masters,* p. 4
- *Hands-On Lab Masters,* p. 72
- CD-ROM Program
  - Resource Lesson 4-9
  - Interactive Lesson 4-9

Recommended Pacing	
Standard	Day 14 of 16
Honors	Day 13 of 15
Block	Day 7 of 8

1 FOCUS

 5-Minute Check
(Lesson 4-8)

Write the unit that you would use to measure each of the following. Then estimate the mass or capacity.

1. a can of soup **milliliter, 350 mL**
2. a rice cake **gram, 2 g**
3. a picnic jug **liter, 2 L**
4. a camera **kilogram, 0.5 kg**
5. an oak leaf **milligram, 5 mg**

 The 5-Minute Check is also available on **Transparency 4-9A** for this lesson.

Motivating the Lesson

Problem Solving Ask students how they would convert one mile to inches. **63,360 inches per mile** Also convert between feet and inches. Ask them how they remember the conversion factors. **memorize** Introduce metric conversion. What is the only conversion factor needed? **10**

Classroom Vignette

"I have students copy the chart for converting within the metric system and show them how to move from what they know (are given) to what they need to know. The number of spaces moved left or right in the chart tells the students how many spaces to move the decimal point to the left or right."

Michelle Hutcheson

Michelle Hutcheson, Teacher
Evans Middle School
Newnan, GA

2 TEACH

Transparency 4-9B contains a teaching aid for this lesson.

Mastering Basic Skills Have students practice multiplying and dividing by multiples of ten to prepare them for changing from one metric unit to another.

In-Class Examples

For Example 1
0.456 L = _?_ mL **456**

For Example 2
5 mm = _?_ cm **0.5**

For Example 3
A saltwater tank requires 11.26 grams of a certain salt and nutrient mixture for one liter of water. If Bob is filling a 45-liter tank, how many kilograms of the salt and nutrient mixture will he need? **0.5067 kg**

For Example 4
760 mg = _?_ g **0.76**

For Example 5
6.57 km = _?_ m **6,570**

Additional Answer

3. Sample answer: Parker and Dinh are both incorrect. Parker multiplied by 10 and Dinh divided by 10. They should have multiplied by 1,000.

Study Guide Masters, p. 31

Name_____ Date_____

4-9 Study Guide

Integration: Measurement
Changing Metric Units

The metric system is a base-10 system. The **meter** is the basic unit of length. The **liter** is the basic unit of capacity. The **gram** is the basic unit of mass.

Prefix	Meaning	Length	Capacity	Mass
kilo-	1,000	kilometer (km)	kiloliter (kL)	kilogram (kg)
hecto-	100	hectometer (hm)	hectoliter (hL)	hectogram (hg)
deka-	10	dekameter (dam)	dekaliter (daL)	dekagram (dag)
basic unit	1	meter (m)	liter (L)	gram (g)
deci-	0.1	decimeter (dm)	deciliter (dL)	decigram (dg)
centi-	0.01	centimeter (cm)	centiliter (cL)	centigram (cg)
milli-	0.001	millimeter (mm)	milliliter (mL)	milligram (mg)

Units may be changed by multiplying or dividing by multiples of 10.

Examples

1 3,500 mL = _____ L
1,000 mL = 1 L
Divide 3,500 by 1,000.
3,500 mL = 3.5 L

2 7.6 km = _____ m
1 km = 1,000 m
Multiply 7.6 by 1,000.
7.6 km = 7,600 m

3 5,000 mg = _____ g
1,000 mg = 1 g
Divide 5,000 by 1,000.
5,000 mg = 5 g

Write whether you multiply or divide to change each measurement. Then complete.

1. 8.74 g = **8,740** mg
   multiply
2. 1,900 g = **1.9** kg
   divide
3. 6.21 L = **6,210** mL
   multiply
4. **5,600** m = 5.6 km
   multiply
5. 226 cm = **2.26** m
   divide
6. **5,114** g = 5.114 kg
   multiply
7. 890 mL = **0.89** L
   divide
8. **0.9** g = 900 mg
   divide
9. **610** cm = 6.1 m
   multiply
10. 0.6 kg = **600** g
    multiply
11. 0.5 km = **500** m
    multiply
12. **1,100** mm = 1.1 m
    multiply
13. **4** cm = 40 mm
    divide
14. 53.7 mL = **0.0537** L
    divide
15. 25 cm = **250** mm
    multiply

© Glencoe/McGraw-Hill   T31   *Mathematics: Applications and Connections, Course 1*

Examples
CONNECTION

3 **Physical Science** Refer to the beginning of the lesson. How many kilograms of 3% hydrogen peroxide will the scientist need?

Explore You know it took 3.05 grams of 3% hydrogen peroxide to produce 1 liter of oxygen. You want to know how many kilograms are needed for 24 liters.

Plan Multiply 3.05 times 24 to find the number of grams needed. Then change this number from grams to kilograms.

Solve 3.05 ⊠ 24 ⊟ **73.2**

Divide to change from smaller units to larger units.

73.2 ÷ 1,000 = 0.0732

She will need 0.0732 kilogram of 3% hydrogen peroxide.

Examine Since 1 g = 0.001 kg, the answer is reasonable.

Study Hint

Reading Math When a measurement is less than 1, the unit of measure stays singular. For example, 0.45 g is read as forty-five hundredths (of a) gram.

4 0.45 g = _?_ mg
To change from grams to milligrams, multiply by 1,000 since 1 g = 1,000 mg.
0.45 × 1,000 = 450
0.45 g = 450 mg

5 8,960 m = _?_ km
To change from meters to kilometers, divide by 1,000 since 1 km = 1,000 m.
8,960 ÷ 1,000 = 8.96
8,960 m = 8.96 km

CHECK FOR UNDERSTANDING

Communicating Mathematics

Read and study the lesson to answer each question.

1. **Explain** how to change milliliters to liters. **divide by 1,000**

2. **Tell** whether you would use grams or kilograms to weigh a large screen television. **kilogram**

3. **You Decide** Parker and Dinh each changed 3.45 kilograms to grams. Parker's answer was 34.5 grams, and Dinh's answer was 0.345 grams. Who is correct? Explain your reasoning. **See margin.**

Guided Practice

Write whether you multiply or divide to change each measurement. Then complete.

4. 328 mL = _?_ L **divide; 0.328**
5. _?_ mm = 0.7 cm **multiply; 7**
6. _?_ g = 150 mg **divide; 0.15**
7. 5.02 kg = _?_ g **multiply; 5,020**
8. _?_ mL = 1.5 L **multiply; 1,500**
9. 150 m = _?_ km **divide; 0.15**

10. **Geography** The circumference of Earth is about 40,000 kilometers. How many meters is it around Earth? **40,000,000**

Reteaching the Lesson

Activity Have students make pairs of cards with equivalent metric measurement units on them. Play "Concentration" with students matching equivalent measures such as 500 mL and 0.5 L.

Error Analysis
Watch for students who move the decimal point the wrong direction.
Prevent by having students check the reasonableness of their answers.

EXERCISES

Practice

Complete. 12. 0.253 14. 6.8 15. 5.24 17. 81.7 18. 0.149 19. 1.953

11. 210 mm = ? cm **21** **12.** ? L = 253 mL **13.** 2.5 L = ? mL **2,500**

14. 0.0068 kg = ? g **15.** ? m = 524 cm **16.** ? m = 2,400 mm **2.4**

17. ? mL = 0.0817 L **18.** 149 mg = ? g **19.** ? g = 1,953 mg

20. 593
21. 3,290
22. 9,750
23. 5,250
24. 0.427
26. 690,800
27. 0.0067

20. 0.593 km = ? m **21.** 3.29 L = ? mL **22.** 975 cm = ? mm

23. 5.25 kg = ? g **24.** ? g = 427 mg **25.** ? mL = 0.01 L **10**

26. ? cm = 6.908 km **27.** 6,700 mg = ? kg **28.** 1.2 km = ? mm

1,200,000

29. How many centimeters are in 0.58 meter? **58 cm**

30. Change 0.6 milligram to grams. **0.0006 g**

31. How many liters are in 213 milliliters? **0.213 L**

Applications and Problem Solving

32. *Earth Science* Seismic waves are waves generated by an earthquake. The focus of an earthquake is the point in Earth's interior where seismic waves originate. The focus can be between 5 and 700 kilometers below the surface of Earth. How many meters can the focus be below the surface? **between 5,000 and 700,000 meters**

33. *Food* A can holds 355 milliliters of soft drink. How many cans would equal 2 liters? **about 5.6 cans**

34. *Critical Thinking* A liter of water at 4°C has a mass of 1 kilogram. What is the mass of 1 milliliter of water at 4°C? **1 gram**

Mixed Review

35. *Measurement* To measure the water in a washing machine, would you use liters or milliliters as your units? *(Lesson 4-8)* **liters**

36. Find 4,507.72 divided by 8.5. *(Lesson 4-6)* **530.32**

37. *Algebra* Evaluate the expression $z + w$ if $z = 6.45$ and $w = 71.2$. *(Lesson 3-6)* **77.65**

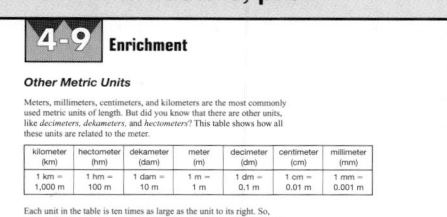

38. **Test Practice** The graph shows the average number of greeting cards purchased by Americans in 1995. Which is a reasonable conclusion that can be drawn from the information in the graph? *(Lesson 2-3)* **D**

Greetings, America!

Source: American Greetings Corporation

A People over 65 buy the most greeting cards.

B People 19 to 24 bought twenty times as many cards as people under 19.

C The number of cards bought increases in every age group.

D People 65 and older bought about five times as many cards as people 19 to 24.

E People under 19 don't buy greeting cards.

Lesson 4-9 Integration: Measurement Changing Metric Units **169**

Extending the Lesson

Enrichment Masters, p. 31

4-9 Enrichment

Other Metric Units

Meters, millimeters, centimeters, and kilometers are the most commonly used metric units of length. But did you know that there are other units, like *decimeters*, *dekameters*, and *hectometers*? This table shows how all these units are related to the meter.

kilometer (km)	hectometer (hm)	dekameter (dam)	meter (m)	decimeter (dm)	centimeter (cm)	millimeter (mm)
1 km = 1,000 m	1 hm = 100 m	1 dam = 10 m	1 m = 1 m	1 dm = 0.1 m	1 cm = 0.01 m	1 mm = 0.001 m

Each unit in the table is ten times as large as the unit to its right. So, 1 km = 10 hm, and 1 hm = 10 dam. It follows that 1 km = (10 × 10) dam.

Activity Help students brainstorm things that may be measured in metric units, such as recorded snowfalls in one month. Ask them to draw a bar graph for their data, changing the metric unit of measure for the scale.

3 PRACTICE/APPLY

Check for Understanding

If students need additional practice or instruction after completing Exercises 1–10, one of these options may be helpful.
- Extra Practice, see p. 568
- Reteaching Activity, see p. 168
- *Study Guide Masters,* p. 31
- *Practice Masters,* p. 31

Assignment Guide

Core: 11–33 odd, 34–38
Enriched: 12–30 even, 32–38

4 ASSESS

Closing Activity

Writing Have students write a paragraph explaining how to change units in the metric system. Make sure they include any pattern they developed for moving the decimal points.

Chapter 4, Quiz D (Lessons 4-8 and 4-9) is available in the *Assessment and Evaluation Masters,* p. 100.

Practice Masters, p. 31

Name_____ Date_____

4-9 Practice

Integration: Measurement
Changing Metric Units

Write whether you multiply or divide to change each measurement. Then complete.

1. 3.72 L = __3,720__ mL multiply
2. __975__ cm = 9.75 m multiply
3. __0.0068__ kg = 6.8 g divide
4. 0.018 kg = __18__ g multiply
5. 149 cm = __1.49__ m divide
6. __5.24__ m = 524 cm divide
7. __560__ g = 0.56 kg multiply
8. 3 mm = __0.3__ cm divide
9. 2.4 m = __2,400__ mm multiply
10. __6,700__ mg = 6.7 g multiply
11. __9,300__ mL = 9.3 L multiply
12. 0.89 m = __89__ cm multiply
13. 0.085 g = __85__ mg multiply
14. __4.6__ m = 4,600 mm divide
15. __7.124__ km = 7,124 m divide
16. 205 g = __0.205__ kg divide
17. 609 mg = __0.609__ g divide
18. __1.9__ mm = 0.0019 m multiply
19. __0.038__ L = 38 mL divide
20. 720 m = __0.72__ km divide
21. 150 cm = __1,500__ mm multiply
22. __4__ cm = 40 mm divide
23. __7,000__ m = 7 km multiply
24. __81.7__ mL = 0.0817 L multiply
25. 480 mL = __0.48__ L divide
26. __530__ 530 mm = 53 cm multiply
27. __3.02__ g = 3,020 mg divide
28. 26 km = __26,000__ m multiply
29. 6.1 mm = __0.0061__ m divide
30. 3.904 L = __3,904__ mL multiply

© Glencoe/McGraw-Hill   T31   *Mathematics: Applications and Connections, Course 1*

Vocabulary

This section provides a listing of the new terms, properties, and phrases that were introduced in this chapter. Have students define each term and provide an example or two of it, if appropriate.

Understanding and Using the Vocabulary

These exercises check students' understanding of the terms by using a variety of verbal formats including matching, completion, and true/false.

Glossaries A complete glossary of terms appears on pages 642–648. The glossary also appears in Spanish on pages 649–656.

Additional Answer

13. **Sample answer: Round 6.9 to 7 and 88 to 90. 7 × 90 = 630**

Vocabulary

After completing this chapter, you should be able to define each term, concept, or phrase and give an example or two of each.

Algebra
distributive property (p. 137)

Geometry
area (p. 146)
perimeter (p. 145)
sides (p. 145)

Number and Operations
compatible numbers (p. 133)

Measurement
gram (p. 164)
kilogram (p. 164)
liter (p. 164)
milligram (p. 164)
milliliter (p. 164)

Problem Solving
solve a simpler problem (pp. 150–151)

Understanding and Using the Vocabulary

State whether each sentence is *true* or *false*. If false, replace the underlined word or number to make a true sentence.

1. You can use compatible numbers to <u>estimate</u> products. **true**
2. Using the distributive property, $4(3 + 2) = 4 \cdot 3 + \underline{3} \cdot 2$. **false, 4**
3. The <u>area</u> of a rectangle is two times the length plus two times the width. **false, perimeter**
4. The perimeter of a rectangle with a length of 4 meters and a width of 1 meter is <u>5</u> meters. **false, 10**
5. The area of a rectangle is the <u>product</u> of its length and width. **true**
6. If each side of a square is 4.2 feet, then its <u>area</u> is 17.64 square feet. **true**
7. The basic unit of mass in the metric system is the <u>gram</u>. **true**
8. A <u>milligram</u> is 1,000 grams. **false, kilogram**
9. The basic unit of capacity in the metric system is the <u>liter</u>. **true**
10. It takes <u>0.001 milliliter</u> to make a liter. **false, 1,000 milliliters**
11. To change from a larger unit to a smaller unit, you need to <u>divide</u>. **false, multiply**
12. To change from a smaller unit to a larger unit, you need to <u>multiply</u>. **false, divide**

In Your Own Words

13. *Explain* how to estimate the product of 6.9 and 88. **See margin.**

170 Chapter 4 Multiplying and Dividing Decimals

MindJogger Videoquizzes

MindJogger Videoquizzes provide an alternative review of concepts presented in this chapter. Students work in teams to answer questions, gaining points for correct answers. The questions are presented in three rounds.
Round 1 Concepts–5 questions
Round 2 Skills–4 questions
Round 3 Problem Solving–4 questions

Objectives & Examples

Upon completing this chapter, you should be able to:

● estimate and find the products of decimals and whole numbers *(Lesson 4-1)*

Find 12.32×12.

12.32 *Estimate: $12 \times 12 = 144$*
$\underline{\times\ 12}$ *Multiply as with whole numbers.*
147.84 *Since the estimate is 144, place the decimal point after 7.*

● compute products mentally using the distributive property *(Lesson 4-2)*

Find 90×3.8.

Estimate: $90 \times 4 = 360$
$90 \times 3.8 = 90(3 + 0.8)$
$= 90 \times 3 + 90 \times 0.8$
$= 270 + 72$
$= 342$

● multiply decimals *(Lesson 4-3)*

Find $38.76 \cdot 4.2$.

38.76 ← *two decimal places*
$\underline{\times\ 4.2}$ ← *one decimal place*
7752
$\underline{15504}$
162.792 ← *three decimal places*

● find the perimeters and the areas of rectangles and squares *(Lesson 4-4)*

Find the perimeter and the area of the figure.

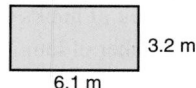
3.2 m
6.1 m

$P = 2\ell + 2w$ $A = \ell \cdot w$
$= 2(6.1) + 2(3.2)$ $= 6.1 \cdot 3.2$
$= 12.2 + 6.4$ $= 19.52\ m^2$
$= 18.6\ m$

Review Exercises

Use these exercises to review and prepare for the chapter test.

Multiply.

14. 17.31 **692.40**
    $\underline{\times\ 40}$

15. 2.15 **488.05**
    $\underline{\times 227}$

16. 38.5×791
    30,453.5

17. 6×7.91 **47.46**

Solve each equation.

18. $m = 201 \times 3.94$ **791.94**

19. $51.2 \times 1,891 = r$ **96,819.2**

Find each product mentally. Use the distributive property.

20. 40×8.9 **356**

21. 30×10.76 **322.8**

22. 9×16 **144**

23. 5×6.8 **34**

Multiply. 24. 19.4902

24. $8.74 \cdot 2.23$

25. 0.04×5.1 **0.204**

26. 0.04×0.0063
    0.000252

27. $11.089 \cdot 5.6$
    62.0984

Solve each equation.

28. $d = 2.6 \cdot 3.9$
    10.14

29. $112.45(4.8) = n$
    539.76

Find the perimeter of each figure.

30.
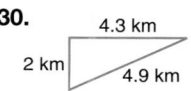
4.3 km
2 km   4.9 km
11.2 km

31.   **27.3 m**
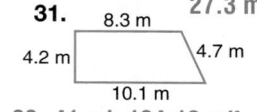
8.3 m
4.2 m   4.7 m
10.1 m
33. 41 mi; 104.16 mi²

Find the perimeter and the area of each figure. 32. 66.2 in.; 202.5 in²

32. rectangle: $\ell = 25$ in.; $w = 8.1$ in.

33. rectangle: $\ell = 11.2$ mi; $w = 9.3$ mi

34. square: $s = 100$ m **400 m; 10,000 m²**

35. square: $s = 7.6$ ft **30.4 ft; 57.76 ft²**

Objectives & Examples

This section reviews the skills and concepts of the chapter and shows completely worked examples.

Review Exercises

These exercises provide practice for the corresponding objectives.

Assessment and Evaluation Masters, pp. 87–88

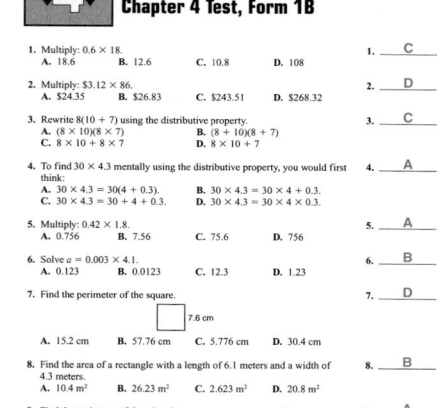

Assessment and Evaluation

Six forms of Chapter 4 Test are available in the *Assessment and Evaluation Masters* as shown in the chart.

Chapter 4 Test, Form 1B, is shown at the right. Chapter 4 Test, Form 2B, is shown on the next page.

1A	Multiple Choice	Honors
1B	Multiple Choice	Average
1C	Multiple Choice	Basic
2A	Free Response	Honors
2B	Free Response	Average
2C	Free Response	Basic

Assessment and Evaluation Masters, pp. 93–94

Chapter 4 Test, Form 2B

1. Estimate: 4.8 × 69.   1. **5 × 70 = 350**

Multiply.
2. 0.72 × 26   2. **18.72**

3. 136 × 4.32   3. **587.52**

4. 8.97 × 6.8.   4. **60.996**

5. Solve *m* = 0.6 × 0.09.   5. **0.054**

Find each product mentally. Use the distributive property.
6. 102 × 8   6. **816**

7. 30.3 × 4   7. **121.2**

Find the perimeter of each figure.
8. 1.7 m, 0.8 m, 1.5 m   8. **4 m**

9. 5.6 cm, 2.9 cm   9. **17 cm**

10. Find the area of the square. 7.5 in.   10. **56.25 in²**

Find each quotient.
11. 8)28.96   11. **3.62**

12. 231.68 ÷ 32   12. **7.24**

13. 19.98 ÷ 0.74   13. **27**

14. 5.9)48.97   14. **8.3**

© Glencoe/McGraw-Hill   93   *Mathematics: Applications and Connections, Course 1*

Chapter 4 Test, Form 2B (continued)

Find each quotient.
15. 73.923 ÷ 12.3   15. **6.01**

16. 495.88 ÷ 98   16. **5.06**

17. Evaluate *a* ÷ *b* if *a* = 38.4 and *b* = 60.   17. **0.64**

Round each quotient to the nearest tenth.
18. 8.6 ÷ 7   18. **1.2**

19. 225.9 ÷ 42   19. **5.4**

Write the unit that you would use to measure each of the following. Then estimate the mass or capacity.
20. capacity of a soup bowl   20. **mL, 300 mL**

21. mass of a sofa   21. **kg, 40 kg**

Complete.
22. 897 mL = _?_ L   22. **0.897**

23. _?_ g = 4.6 kg   23. **4,600**

24. A room is 16.5 feet long by 13.7 feet wide. How many square feet of carpeting are needed to cover the room?   24. **226.05 ft²**

25. Isidro is fencing his yard. How much fencing does he need? 25. **64 m**
24 m, 3 m, 8 m, 10 m, 5 m, 14 m

© Glencoe/McGraw-Hill   94   *Mathematics: Applications and Connections, Course 1*

172   Chapter 4

---

divide decimals by whole numbers
(Lesson 4-5)

Find the quotient of 96.9 and 51.

```
      1.9      Estimate: 100 ÷ 50 = 2
51)96.9        Place the decimal point.
   −51         Divide as with whole numbers.
   45 9
  −45 9
      0
```

divide decimals by decimals *(Lesson 4-6)*

Find the quotient of 166.14 and 21.3.

$$21.3)\overline{166.14} \to 213)\overline{1661.4}$$ with quotient 7.8

Multiply the divisor and dividend by 10.

divide decimals involving special cases of zero in the quotient *(Lesson 4-7)*

Find 571.2 ÷ 56.
```
      10.2
56)571.2
  −56
   11
  − 0
   11 2
  −11 2
      0
```

use metric units of mass and capacity
(Lesson 4-8)

What unit of mass would you use to measure a chicken's egg?

Since a chicken's egg has a small mass, the *gram* is the appropriate unit.

change units within the metric system
(Lesson 4-9)

9.2 g = _?_ mg   *Multiply to change a larger*
9.2 g = 9,200 mg   *unit to a smaller unit.*

Find each quotient.
36. 12.24 ÷ 36 **0.34**
37. 32)203.84 **6.37**
38. 1,000)17.97 **0.01797**
39. 5.2175 ÷ 10 **0.52175**
40. Round 38.86 ÷ 7 to the nearest tenth. **5.6**
41. Round 249.77 ÷ 13 to the nearest hundredth. **19.21**

Find each quotient.
42. 136.5 ÷ 35 **3.9**
43. 0.045)6.345 **141**
44. 0.081)43.254 **534**
45. 18.88 ÷ 3.2 **5.9**

Find each quotient to the nearest hundredth.
46. 7.38 ÷ 36 **0.21**
47. 24.3)0.243 **0.01**
48. 4.2)43.75 **10.42**
49. 8.134 ÷ 98 **0.08**

Solve each equation.
50. *h* = 3,190.95 ÷ 4.5 **709.1**
51. 77.1912 ÷ 14.2 = *w* **5.436**

Write the unit you would use to measure each of the following. Then estimate the mass or capacity. 55. **milliliter, 15 mL**
52. a candy apple **gram, 200 g**
53. a stack of books **kilogram, 5 kg**
54. a pitcher of lemonade **liter, 3 L**
55. a dose of cough medicine

Complete.
56. 300 mL = _?_ L **0.3**
57. _?_ g = 1 mg **0.001**
58. _?_ m = 0.75 km **750**

172   Chapter 4 Multiplying and Dividing Decimals

Test and Review Software

You may use this software, a combination of an item generator and item bank, to create your own tests or worksheets. Types of items include free response, multiple choice, short answer, and open ended.

CD-ROM Program

The CD-ROM Program contains an Assessment Game whose questions review the concepts in this chapter.

Applications & Problem Solving

59. Fitness Tyra exercises every morning by running once around a rectangular neighborhood. The length of the rectangle is 1.7 kilometers, and the width is 0.9 kilometer. How far does she run? *(Lesson 4-4)* **5.2 kilometers**

60. Solve a Simpler Problem The figure below is to be painted on a wall as part of a mural. Find the area of the figure. *(Lesson 4-5A)* **174 square feet**

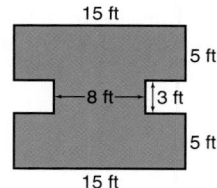

61. Money Matters Drew bought a new car and financed it over a 48-month period. The total of all the payments he will make is $15,336.96. How much is each car payment? *(Lesson 4-5)* **$319.52**

62. Money Matters Mariano wants to collect $131.25 to replace his friend's CD player that was lost in a fire. He is asking everyone to contribute $1.25. How many people will he need to collect from? *(Lesson 4-7)* **105 people**

Alternative Assessment

Performance Task

Suppose you run a landscape company. The label on a bag of fertilizer states that it will cover 3 square meters. How can you determine how many bags to bring with you to fertilize your client's yard? **See margin.**

Suppose one client has a rectangular yard with dimensions of 7.3 meters by 5.2 meters. For how many bags of fertilizer will you charge the client?

It will take about 12.65 bags, so charge them for 13 bags.

 Select an item from this chapter that you feel shows your best work. Place it in your portfolio. Explain why you selected it.

Completing the **CHAPTER Project**

Use the following checklist to make sure your project is complete.

☑ You have included your price chart for multiple items, your prices with tax included, and your senior citizen discount prices.

☑ Your paragraph describes why you chose the product that you did and how you determined its price.

☑ You have designed an advertisement for your product. Be sure to include your business name, a picture of the product, the price of the item, and anything else that you think would help sell the product.

A practice test for Chapter 4 is provided on page 598.

Chapter 4 Study Guide and Assessment **173**

Applications & Problem Solving

This section provides additional practice in solving real-world problems that involve the skills of this chapter.

Alternative Assessment

The **Performance Task** provides students with a performance assessment opportunity to evaluate their work and understanding.

 **CHAPTER Project**

Students should complete the final stages of their project and prepare a class demonstration of their results. A scoring guide for the project is available in the *Investigations and Projects Masters*, p. 31.

 Students should add to their portfolios at this time.

Assessment and Evaluation Masters, p. 97

Additional Answer for the Performance Task

First, find the area of the yard in square meters. Then divide the area of the yard by 3 to find the number of bags.

 **Performance Assessment**

Additional performance assessment tasks for this chapter are included in the *Assessment and Evaluation Masters* on page 97. A scoring guide is also provided on page 109.

The Standardized Test Practice may be used to help students prepare for standardized tests. The test items are written in the same style as those in state proficiency tests and standardized tests like CAT, CTBS, ITBS, MAT, SAT, and Terra Nova. The test items cover skills and concepts covered up to this point in the text.

The pages can be used as an overnight assessment. After students have completed the pages, discuss how each problem can be solved, or provide copies of the solutions from the *Solutions Manual*.

Section One: Multiple Choice

There are twelve multiple-choice questions in this section. Choose the best answer. If a correct answer is *not here*, mark the letter for Not Here.

1. Kenda found 18 crayons. Her classmate Marissa gave her some more, and then she had 42. To find out how many crayons she was given, Kenda wrote $18 + c = 42$. What is the value of c? **D**
   A 60
   B 52
   C 34
   D 24

2. The exponential form of $3 \times 3 \times 7 \times 7 \times 7 \times 11$ is $3^2 \times 7^? \times 11$. What should replace the question mark? **H**
   F 1
   G 2
   H 3
   J 7

3. Julian is 1.54 meters tall. How many centimeters tall is Julian? **C**
   A 0.0154 cm
   B 15.4 cm
   C 154 cm
   D 1,540 cm

4. Fairfield has an average precipitation of 35.22 inches in June and July. Round this amount to the nearest tenth. **G**
   F 40
   G 35.2
   H 35.22
   J 35.3

Please note that Questions 5–12 have five answer choices.

5. A sign in a store window reads as follows:

   ALL ITEMS
   $5.99
   to
   $13.99

   If Manny bought 3 items in the store, what would be a reasonable total cost? **C**
   A $6.00
   B $12.00
   C $24.00
   D $45.00
   E $60.00

6. 8×3.21 is about— **F**
   F 24.
   G 32.
   H 240.
   J 320.
   K 2,568.

7. Jesse wants to buy ice-cream cones for himself and 3 friends. Each cone costs $1.19. Which is the best estimate of the amount of money Jesse will need in order to buy the cones? **C**
   A $1
   B $3
   C $5
   D $7
   E $9

Assessment and Evaluation Masters, p. 103

Cumulative Review, Chapters 1–4

Name_____ Date_____

Perform the indicated operation. (Lessons 4-3, 3-6, 4-5, and 4-6)
1. 3.8×2.5 1. 9.5
2. $7.66 + 1.5$ 2. 9.16
3. $38\overline{)53.2}$ 3. 1.4
4. $0.27\overline{)0.999}$ 4. 37

Evaluate. (Lessons 1-4, 1-6, and 1-5)
5. $7 + 6 \times 3 + 6$ 6. 7^4 5. 31
7. mn if $m = 8$ and $n = 25$ 8. x^6 if $x = 10$ 6. 2,401
9. Order the following decimals from least to greatest: 0.003, 0.95, 0.2, 0.0056. (Lesson 3-3)   7. 200
                                                       8. 1,000,000
Golf Scores: 78, 136, 92, 100, 134, 92   9. 0.003, 0.0056, 0.2, 0.95
10. What is the range of the golf scores? (Lesson 2-7) 10. 58
11. What is the median of the golf scores? (Lesson 2-7) 11. 96
12. Refer to the circle graph at the right. Which two flavors received about the same number of votes? (Lesson 2-4)   12. chocolate and vanilla
    Favorite Flavors
13. The equation for finding the perimeter of a hexagon with six equal sides is $P = 6s$. If the length (s) of each side of a hexagon is 18 centimeters, find the perimeter. (Lesson 4-4)   13. 108 cm
Estimate. (Lessons 3-5 and 1-3)
14. $0.716 + 0.855$ 15. 49×86 14. $0.7 + 0.9 = 1.6$
16. $355 \div 7$ 17. $166 + 299$ 15. $50 \times 90 = 4,500$
                                                       16. $350 \div 7 = 50$
18. Change 1.9 kilograms to grams. (Lesson 4-9)   17. $170 + 300 = 470$
19. Find the area of the figure at the right. (Lesson 4-4)   18. 1,900 g
20. Round $18.65 \div 3.8$ to the nearest hundredth. (Lesson 4-7)   19. 330 mm²
                                                       20. 4.91

© Glencoe/McGraw-Hill   103   *Mathematics: Applications and Connections, Course 1*

◄◄◄Instructional Resources

Another cumulative review is shown at the left and is available in the *Assessment and Evaluation Masters*, p. 103.

8. Aubrey charges $1.75 per hour for walking her neighbor's dogs. If she works for 8 hours, how much will she earn? **H**
- **F** $17.50
- **G** $16.00
- **H** $14.00
- **J** $7.50
- **K** Not Here

9. Three watermelons have weights of 18, 17, and 22 pounds. Which weight is a reasonable average weight of the three watermelons? **B**
- **A** 10 lb
- **B** 20 lb
- **C** 25 lb
- **D** 45 lb
- **E** 60 lb

10. Bill's Boy Scout troop collected $439 for the Muscular Dystrophy Association. Kelly's Girl Scout troop collected $672. How much more did Kelly's troop collect than Bill's? **G**
- **F** $133
- **G** $233
- **H** $243
- **J** $247
- **K** Not Here

11. Rashid cut a ribbon into 5 equal pieces. If each piece was 2.4 meters long, how long was the ribbon before he cut it? **C**
- **A** 0.48 m
- **B** 1.2 m
- **C** 12 m
- **D** 120 m
- **E** Not Here

12. Solve the equation $y = 43.1 + 7.256$. **G**
- **F** 51.456
- **G** 50.356
- **H** 11.566
- **J** 7.687
- **K** Not Here

Test-Taking Tip

Instead of waiting until the night before a test, allow yourself plenty of time to review the basic skills and formulas. If you prepare early, you will have time to find your weaknesses and ask for help.

19. Sample answer: The 539-gram can of soup is a better buy. It costs 0.135¢ per gram while the 306-gram can costs 0.137¢ per gram.

Section Two: Free Response

This section contains seven questions for which you will provide short answers. Write your answers on your paper.

13. Five students took Mr. Castillo's history test. The scores were 93, 82, 95, 95, and 75. Find the mean, median and mode of the scores. **88, 93, 95**

14. Maria plans to buy dog food for $15.99, a bone for $5.79, and a dog toy for $2.99. What is the cost of these items before tax is added? **$24.77**

15. To measure the water in a swimming pool, would you use liters or milliliters? **liters**

16. Evaluate $(t + w) \div z$ if $t = 0.007$ $w = 0.014$, and $z = 2.5$. **0.0084**

17. 23.5 mL = _?_ L **0.0235**

18. Without computing, explain why 312.28 divided by 29.6 is not 105.5. See margin.

19. Which is a better buy, a 306-gram can of soup for $0.42 or a 539-gram can for $0.73? Explain.

Test-Taking Tip

Encourage students to set aside 15 to 30 minutes each night to review basic skills. Formulas can be written on flash cards to aid in self-testing.

Assessment and Evaluation Masters, pp. 101–102

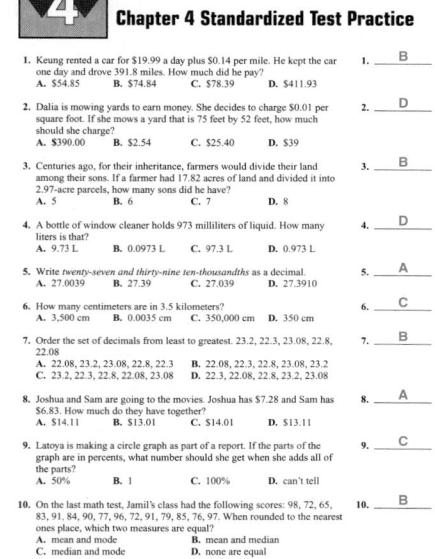

Additional Answer

18. Sample answer: If you round to compatible numbers, you can estimate that the quotient is $300 \div 30 = 10$. This is not close to 105.5.

Instructional Resources ▶▶▶

Additional standardized test practice is shown at the right and is available in the *Assessment and Evaluation Masters*, pp. 101–102.

Using Number Patterns, Fractions, and Ratios

Previewing the Chapter

Overview

This chapter helps students develop number sense and an understanding of fractions and ratios by investigating number patterns. Students learn to use the greatest common factor to simplify fractions and the least common multiple to order and compare them. They solve problems by making an organized list and use customary units of length. Students convert between decimals and fractions and explore terminating and repeating decimals.

Lesson (pages)	Lesson Objectives	NCTM Standards	Standardized Tests	State/Local Objectives
5-1 (178–180)	Use divisibility rules for 2, 3, 5, 6, 9, and 10.	1–9, 12	CTBS, ITBS, MAT, SAT, TN	
5-2A (181)	Identify prime and composite numbers by using rectangular arrays.	1–3, 8, 12		
5-2 (182–184)	Find the prime factorization of a composite number.	1–10	SAT	
5-3A (186–187)	Solve problems by making an organized list.	1–4, 6, 7		
5-3 (188–190)	Find the greatest common factor of two or more numbers.	1–7, 10, 12, 13	CAT, CTBS, ITBS, MAT, SAT, TN	
5-4A (191–192)	Use models to represent fractions and equivalent fractions.	1–3, 5, 6	CAT, CTBS, TN	
5-4 (193–196)	Express fractions and ratios in simplest form.	1–6, 7, 10	CTBS, ITBS, MAT, SAT, TN	
5-4B (197)	Determine the experimental probability for a given set of data.	1–6, 9, 13	CTBS, SAT, TN	
5-5 (198–201)	Express mixed numbers as improper fractions and vice versa.	1–6, 9, 12, 13	CTBS, SAT, TN	
5-6 (202–205)	Measure line segments and objects with a ruler divided in halves, fourths, and eighths.	1–6, 13		
5-7 (206–209)	Find the least common multiple of two or more numbers.	1–6, 10, 13	CAT, CTBS, ITBS, MAT, SAT, TN	
5-8 (210–213)	Compare and order fractions.	1–6, 10, 13	CAT, CTBS, ITBS, MAT, SAT, TN	
5-9 (214–216)	Express terminating decimals as fractions in simplest form.	1–6, 9, 10	CTBS, SAT, TN	
5-10 (217–219)	Express fractions as terminating and repeating decimals.	1–8, 10		

CAT = California Achievement Tests, CTBS = Comprehensive Tests of Basic Skills, ITBS = Iowa Tests of Basic Skills, MAT = Metropolitan Achievement Tests, SAT = Stanford Achievement Tests, TN = Terra Nova

Organizing the Chapter

CD-ROM

All of the blackline masters in the Teacher's Classroom Resources are available on the **Electronic Teacher's Classroom Resources** CD-ROM.

LESSON PLANNING GUIDE

Lesson	Extra Practice (Student Edition)	BLACKLINE MASTERS (PAGE NUMBERS)										Transparencies A and B
		Study Guide	Practice	Enrichment	Assessment & Evaluation	Classroom Games	Diversity	Hands-On Lab	School to Career	Science and Math Lab Manual	Technology	
5-1	p. 568	32	32	32				73		77–80		5-1
5-2A								47				
5-2	p. 569	33	33	33			5					5-2
5-3A	p. 569											
5-3	p. 569	34	34	34	127							5-3
5-4A								48				
5-4	p. 570	35	35	35								5-4
5-4B								49				
5-5	p. 570	36	36	36	126, 127				5			5-5
5-6	p. 570	37	37	37								5-6
5-7	p. 571	38	38	38							10	5-7
5-8	p. 571	39	39	39	128	11–13						5-8
5-9	p. 571	40	40	40								5-9
5-10	p. 572	41	41	41	128					9–12	9	5-10
Study Guide/ Assessment					113–125, 129–131							

OTHER CHAPTER RESOURCES

Student Edition

Chapter Project, pp. 177, 196, 219, 223
Math in the Media, p. 205
School to Career, p. 185
Let the Games Begin, p. 209

Technology

 CD-ROM Program

Interactive Mathematics Tools Software

Teacher's Classroom Resources

Applications
Family Letters and Activities, pp. 9–10
Investigations and Projects Masters, pp. 33–36
Meeting Individual Needs
Transition Booklet, pp. 21–22, 25–26, 29–30
Investigations for the Special Education Student, pp. 13–14, 35–38

Teaching Aids
Answer Key Masters
Block Scheduling Booklet
Lesson Planning Guide
Solutions Manual

Professional Publications
Glencoe Mathematics Professional Series

Planning the Chapter

MindJogger Videoquizzes provide a unique format for reviewing concepts presented in the chapter.

ASSESSMENT RESOURCES

Student Edition
Mixed Review, pp. 180, 184, 190, 196, 201, 205, 209, 213, 216, 219
Mid-Chapter Self Test, p. 201
Math Journal, pp. 179, 212
Study Guide and Assessment, pp. 220–223
Performance Task, p. 223
Portfolio Suggestion, p. 223
Standardized Test Practice, pp. 224–225
Chapter Test, p. 599

Assessment and Evaluation Masters
Multiple-Choice Tests (Forms 1A, 1B, 1C), pp. 113–118
Free-Response Tests (Forms 2A, 2B, 2C), pp. 119–124
Performance Assessment, p. 125
Mid-Chapter Test, p. 126
Quizzes A–D, pp. 127–128
Standardized Test Practice, pp. 129–130
Cumulative Review, p. 131

Teacher's Wraparound Edition
5-Minute Check, pp. 178, 182, 188, 193, 198, 202, 206, 210, 214, 217
Building Portfolios, p. 176
Math Journal, pp. 181, 192, 197
Closing Activity, pp. 180, 184, 187, 190, 196, 201, 205, 209, 213, 216, 219

Technology
Test and Review Software
MindJogger Videoquizzes
CD-ROM Program

MATERIALS AND MANIPULATIVES

Lesson 5-2A
square tiles*

Lesson 5-4A
ruler*

Lesson 5-4B
number cube*

Lesson 5-5
ruler*

Lesson 5-6
tape measure*
string

Lesson 5-7
ruler*
scissors*
spinners†

Lesson 5-10
calculator

*Glencoe Manipulative Kit   †Glencoe Overhead Manipulative Resources

PACING CHART

See pages T25–T27 for the Course Planning Calendar.

COURSE	DAY 1	DAY 2	DAY 3	DAY 4	DAY 5	DAY 6	DAY 7
Standard	Chapter Project	Lesson 5-1	Lessons 5-2A & 5-2		Lesson 5-3A	Lesson 5-3	Lessons 5-4A & 5-4
Honors	Chapter Project	Lesson 5-1	Lesson 5-2	Lesson 5-3A	Lesson 5-3	Lessons 5-4 & 5-4B	
Block	Chapter Project & Lesson 5-1	Lessons 5-2A & 5-2	Lessons 5-3A & 5-3	Lessons 5-4A & 5-4	Lessons 5-5 & 5-6	Lessons 5-7 & 5-8	Lessons 5-9 & 5-10

The *Transition Booklet* (Skills 9, 11, 13) can be used to practice multiplying and dividing by 1-digit numbers and measuring length.

Interactive Mathematics:
Activities and Investigations

is an activity-based program that may be used as an enhancement for chapters in *Mathematics: Applications and Connections.*

Activities and Investigations

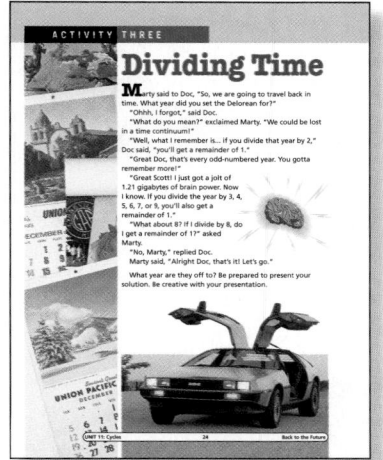

Unit 11, Activity Three
Dividing Time
Use with Lesson 5-1.

Summary Students work in groups to solve a problem involving different clock systems. They may solve the problem using various methods, including using common multiples, variables, and a guess-and-check process. Then the groups present their solutions to the class.

Math Connection Students solve a problem using multiple clock sizes. If students know their divisibility rules, they can use them to determine quickly for which year +1 is correct.

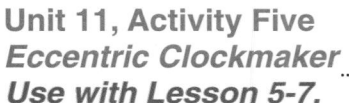

Unit 11, Activity Five
Eccentric Clockmaker
Use with Lesson 5-7.

Summary Students work in groups to determine when, if ever, all three clocks, each of different size and set to sound at different intervals, will sound simultaneously. The groups are then invited to share their solution and problem-solving methods with the class.

Math Connection Students explore cyclical patterns involving clocks of different sizes. They may use the least common multiple to solve the clock problem.

DAY 8	DAY 9	DAY 10	DAY 11	DAYS 12 & 13	DAY 14	DAY 15	DAY 16
(continue from Day 7)	Lesson 5-5	Lesson 5-6	Lesson 5-7	Lessons 5-8 & 5-9	Lesson 5-10	Study Guide and Assessment	Chapter Test
Lesson 5-5	Lesson 5-6	Lesson 5-7	Lesson 5-8	Lessons 5-9 & 5-10	Study Guide and Assessment	Chapter Test	
Study Guide and Assessment, Chapter Test							

Enhancing the Chapter

APPLICATIONS

Classroom Games, pp. 11–13

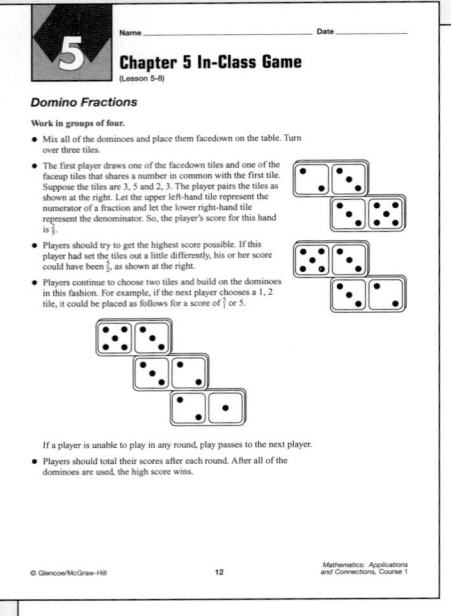

5 Chapter 5 In-Class Game
(Lesson 5-8)

Name _____ Date _____

Domino Fractions

Work in groups of four.

- Mix all of the dominoes and place them facedown on the table. Turn over three tiles.
- The first player draws one of the facedown tiles and one of the faceup tiles that shares a number in common with the first tile. Suppose the tiles are 3, 5 and 2, 3. The player pairs the tiles as shown at the right. Let the upper left-hand tile represent the numerator of a fraction and let the lower right-hand tile represent the denominator. So, the player's score for this hand is $\frac{5}{2}$.
- Players should try to get the highest score possible. If this player had set the tiles out a little differently, his or her score could have been $\frac{2}{5}$, as shown at the right.
- Players continue to choose two tiles and build on the dominoes in this fashion. For example, if the next player chooses a 1, 2 tile, it could be placed as follows for a score of $\frac{1}{5}$ or 5.

If a player is unable to play in any round, play passes to the next player.

- Players should total their scores after each round. After all of the dominoes are used, the high score wins.

© Glencoe/McGraw-Hill   12   Mathematics: Applications and Connections, Course 1

Diversity Masters, p. 5

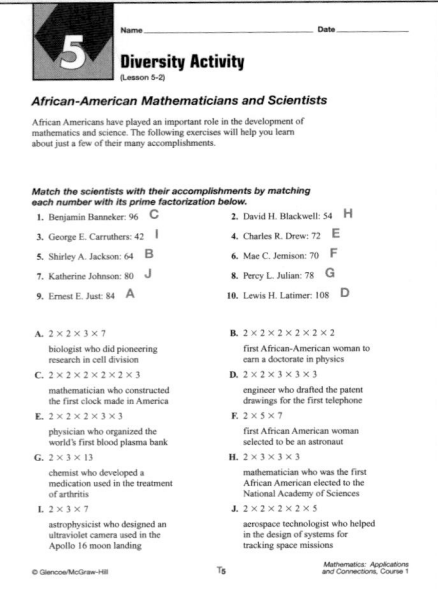

5 Diversity Activity
(Lesson 5-2)

Name _____ Date _____

African-American Mathematicians and Scientists

African Americans have played an important role in the development of mathematics and science. The following exercises will help you learn about just a few of their many accomplishments.

Match the scientists with their accomplishments by matching each number with its prime factorization below.

1. Benjamin Banneker: 96   C
2. David H. Blackwell: 54   H
3. George E. Carruthers: 42   I
4. Charles R. Drew: 72   E
5. Shirley A. Jackson: 64   B
6. Mae C. Jemison: 70   F
7. Katherine Johnson: 80   J
8. Percy L. Julian: 78   G
9. Ernest E. Just: 84   A
10. Lewis H. Latimer: 108   D

A. $2 \times 2 \times 3 \times 7$
biologist who did pioneering research in cell division

B. $2 \times 2 \times 2 \times 2 \times 2 \times 2$
first African-American woman to earn a doctorate in physics

C. $2 \times 2 \times 2 \times 2 \times 2 \times 3$
mathematician who constructed the first clock made in America

D. $2 \times 2 \times 3 \times 3 \times 3$
engineer who drafted the patent drawings for the first telephone

E. $2 \times 2 \times 2 \times 3 \times 3$
physician who organized the world's first blood plasma bank

F. $2 \times 5 \times 7$
first African American woman selected to be an astronaut

G. $2 \times 3 \times 13$
chemist who developed a medication used in the treatment of arthritis

H. $2 \times 3 \times 3 \times 3$
mathematician who was the first African American elected to the National Academy of Sciences

I. $2 \times 3 \times 7$
astrophysicist who designed an ultraviolet camera used in the Apollo 16 moon landing

J. $2 \times 2 \times 2 \times 2 \times 5$
aerospace technologist who helped in the design of systems for tracking space missions

© Glencoe/McGraw-Hill   T5   Mathematics: Applications and Connections, Course 1

School to Career Masters, p. 5

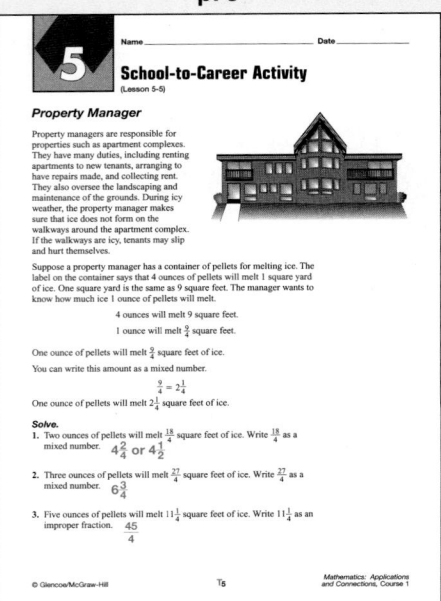

5 School-to-Career Activity
(Lesson 5-5)

Name _____ Date _____

Property Manager

Property managers are responsible for properties such as apartment complexes. They have many duties, including renting apartments to new tenants, arranging to have repairs made, and collecting rent. They also oversee the landscaping and maintenance of the grounds. During icy weather, the property manager makes sure that ice does not form on the walkways around the apartment complex. If the walkways are icy, tenants may slip and hurt themselves.

Suppose a property manager has a container of pellets for melting ice. The label on the container says that 4 ounces of pellets will melt 9 square yard of ice. One square yard is the same as 9 square feet. The manager wants to know how much ice 1 ounce of pellets will melt.

4 ounces will melt 9 square feet.

1 ounce will melt $\frac{9}{4}$ square feet.

One ounce of pellets will melt $\frac{9}{4}$ square feet of ice. You can write this amount as a mixed number.

$$\frac{9}{4} = 2\frac{1}{4}$$

One ounce of pellets will melt $2\frac{1}{4}$ square feet of ice.

Solve.

1. Two ounces of pellets will melt $\frac{18}{4}$ square feet of ice. Write $\frac{18}{4}$ as a mixed number. $4\frac{2}{4}$ or $4\frac{1}{2}$

2. Three ounces of pellets will melt $\frac{27}{4}$ square feet of ice. Write $\frac{27}{4}$ as a mixed number. $6\frac{3}{4}$

3. Five ounces of pellets will melt $11\frac{1}{4}$ square feet of ice. Write $11\frac{1}{4}$ as an improper fraction. $\frac{45}{4}$

© Glencoe/McGraw-Hill   T5   Mathematics: Applications and Connections, Course 1

Family Letters and Activities, pp. 9–10

5 Family Letter

Name _____ Date _____

Dear Parent or Guardian:

Using fractions, ratios, and decimals is part of our daily lives. Knowing how to represent and interpret these types of numbers helps us analyze statistics, measure lengths, and compare prices when shopping. In order to emphasize the importance of math skills, we try to relate classroom concepts to everyday events.

In Chapter 5, Using Number Patterns, Fractions, and Ratios, your child will learn to find the prime factorization of a number, to solve problems by making an organized list, and to find the greatest common factor and least common multiple of numbers to simplify numbers as improper fractions and to write decimals as fractions and vice versa. In addition to a variety of daily classroom assignments, your child may be asked to produce a chapter project. This project, investigating magazines, will probably take more than one day to complete. It may be necessary for your child to contact other members of the class by phone or meet with them outside of class.

By signing this letter and returning it with your child, you agree to encourage your child by getting involved. Enclosed is an activity you can do with your child that also relates the math we will be learning in Chapter 5 to the real world. If you have any questions or comments, feel free to contact me at school.

Sincerely,

Signature of Parent or Guardian _____ Date _____

© Glencoe/McGraw-Hill   9   Mathematics: Applications and Connections, Course 1

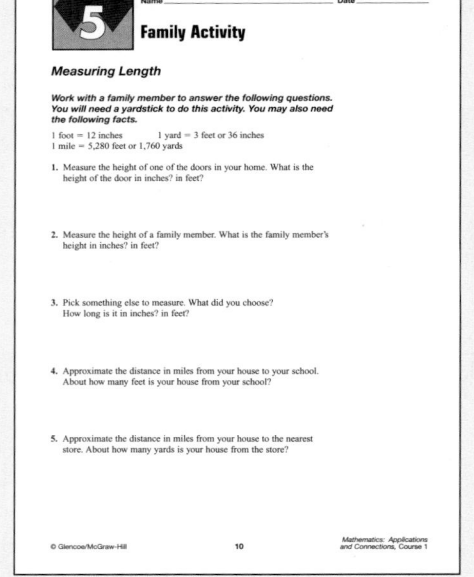

5 Family Activity

Name _____ Date _____

Measuring Length

Work with a family member to answer the following questions. You will need a yardstick to do this activity. You may also need the following facts.

1 foot = 12 inches   1 yard = 3 feet or 36 inches
1 mile = 5,280 feet or 1,760 yards

1. Measure the height of one of the doors in your home. What is the height of the door in inches? in feet?

2. Measure the height of a family member. What is the family member's height in inches? in feet?

3. Pick something else to measure. What did you choose? How long is it in inches? in feet?

4. Approximate the distance in miles from your house to your school. About how many feet is your house from your school?

5. Approximate the distance in miles from your house to the nearest store. About how many yards is your house from the store?

© Glencoe/McGraw-Hill   10   Mathematics: Applications and Connections, Course 1

Science and Math Lab Manual, pp. 9–12

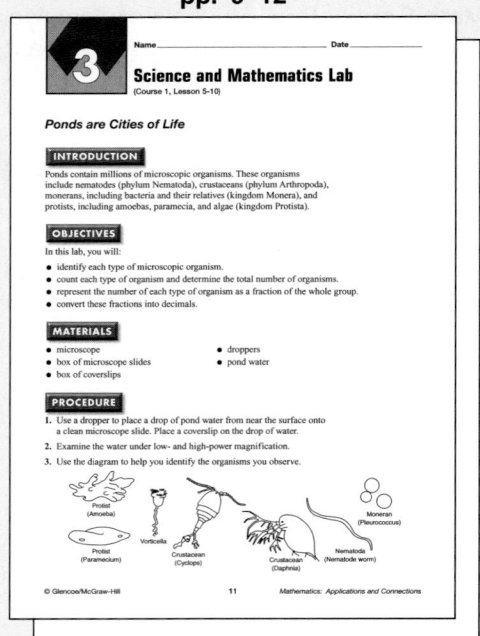

3 Science and Mathematics Lab
(Course 1, Lesson 5-10)

Name _____ Date _____

Ponds are Cities of Life

INTRODUCTION

Ponds contain millions of microscopic organisms. These organisms include nematodes (phylum Nematoda), crustaceans (phylum Arthropoda), monerans, including bacteria and their relatives (kingdom Monera), and protists, including amoebas, paramecia, and algae (kingdom Protista).

OBJECTIVES

In this lab, you will:
- identify each type of microscopic organism.
- count each type of organism and determine the total number of organisms.
- represent the number of each type of organism as a fraction of the whole group.
- convert these fractions into decimals.

MATERIALS
- microscope
- box of microscope slides
- box of coverslips
- droppers
- pond water

PROCEDURE

1. Use a dropper to place a drop of pond water from near the surface onto a clean microscope slide. Place a coverslip on the drop of water.
2. Examine the water under low- and high-power magnification.
3. Use the diagram to help you identify the organisms you observe.

Protist (Amoeba)   Moneran (Pleurococcus)   Protist (Paramecium)   Vorticella   Crustacean (Cyclops)   Crustacean (Daphnia)   Nematode (Nematode worm)

© Glencoe/McGraw-Hill   11   Mathematics: Applications and Connections

MANIPULATIVES/MODELING

Hands-On Lab Masters, p. 73

5 Name _____ Date _____

Lab Activity
(Lesson 5-1)

Divisibility Patterns

1. List the numbers that divide 18 evenly. Use the rules for divisibility. **1, 2, 3, 6, 9, 18**

2. Take 18 beans. Separate them into groups evenly. The numbers you found in Exercise 1 will work as group numbers. Try to find all of the possible arrangements. Sketch each arrangement you find. **Sample answer:**

3. How many different numbers divide 18 evenly? **6** List each way.
   $18 ÷ 1 = 18; 18 ÷ 2 = 9; 18 ÷ 3 = 6; 18 ÷ 6 = 3;$
   $18 ÷ 9 = 2; 18 ÷ 18 = 1$

4. Put your 18 beans into a jar with everyone else's beans. Grab some more beans without counting. When you have the beans you want, count them. How many beans do you have? **Sample answer: 15**

5. What numbers will divide the number of beans you have evenly? Use your rules for divisibility. **Sample answer: 1, 3, 5, 15**

6. Separate your beans into groups evenly. Find all of the arrangements you can. Sketch each one.
Sample answer:

7. How many different ways can you divide your beans? **Sample answer: 4**
   List the numbers represented by each grouping. **Sample answer:**
   $15 ÷ 1 = 15; 15 ÷ 3 = 5; 15 ÷ 5 = 3; 15 ÷ 15 = 1$

© Glencoe/McGraw-Hill   73   Mathematics: Applications and Connections, Course 1

ASSESSMENT/EVALUATION

Assessment and Evaluation Masters, pp. 126–128

5 Name _____ Date _____

Chapter 5 Mid-Chapter Test
(Lessons 5-1 through 5-5)

Determine whether the first number is divisible by the second number.
1. 2,233; 9
2. 134; 2

1. _____ no
2. _____ yes

State whether each number is divisible by 2, 3, 5, 6, 9, or 10.
3. 945
4. 324

3. _____ 3, 5, 9
4. _____ 2, 3, 6, 9

Tell whether each number is prime, composite, or neither.
5. 221
6. 0

5. _____ composite
6. _____ neither

Find the prime factorization of each number for Questions 7–8.
7. 162
8. 140

9. Fernando's grocery store has powdered, glazed, and iced donuts. These come in holes, mini, and regular sizes. How many different options are available to the customers?

7. _____ $2 × 3^4$
8. _____ $2^2 × 5 × 7$
9. _____ 9

Find the GCF of each set of numbers.
10. 77, 35
11. 88, 104, 136
12. 17, 153, 187

10. _____ 7
11. _____ 8
12. _____ 17

Replace each ▦ with a number so that the fractions are equivalent.
13. $\frac{3}{9} = \frac{▦}{162}$
14. $\frac{10}{11} = \frac{▦}{66}$

13. _____ 54
14. _____ 60

State whether each fraction or ratio is in simplest form. If not, write each fraction or ratio in simplest form.
15. 39 to 270
16. $\frac{2}{21}$

15. _____ no, 13 to 90
16. _____ yes

Express each mixed number as an improper fraction.
17. $15\frac{2}{7}$
18. $6\frac{1}{17}$

17. _____ $\frac{107}{7}$
18. _____ $\frac{101}{15}$

Express each improper fraction as a mixed number.
19. $\frac{155}{20}$
20. $\frac{72}{5}$

19. _____ $7\frac{3}{4}$
20. _____ $14\frac{2}{5}$

© Glencoe/McGraw-Hill   126   Mathematics: Applications and Connections, Course 1

Name _____ Date _____

Chapter 5 Quiz A
(Lessons 5-1 through 5-3)

State whether each number is divisible by 2, 3, 5, 6, 9, or 10 for Questions 1–2.
1. 72
2. 315
3. Find the prime factorization of 150.
4. Is 51 prime, composite, or neither?
5. Find the prime factorization of 40.
6. A sweatshirt is available in white, black, or red and in three sizes: small, medium, and large. How many different types of sweatshirts are available?

1. _____ 2, 3, 6, 9
2. _____ 3, 5, 9
3. _____ $2 × 3 × 5 × 5$
4. _____ composite
5. _____ $2^3 × 5$
6. _____ 9

Find the GCF of each set of numbers.
7. 35, 45
8. 52, 72
9. 15, 40
10. 27, 105, 126

7. _____ 5
8. _____ 4
9. _____ 5
10. _____ 3

Name _____ Date _____

Chapter 5 Quiz B
(Lessons 5-4 and 5-5)

State whether each fraction or ratio is in simplest form. If not, write each fraction or ratio in simplest form.
1. $\frac{9}{35}$
2. 54 to 60
3. 23:69
4. 8 out of 128

1. _____ yes
2. _____ no, 9 to 10
3. _____ no, 1:3
4. _____ no, 1 out of 16

Express each mixed number as an improper fraction.
5. $6\frac{3}{8}$
6. $22\frac{1}{4}$
7. $2\frac{7}{9}$
8. $40\frac{2}{3}$

5. _____ $\frac{51}{8}$
6. _____ $\frac{89}{4}$
7. _____ $\frac{25}{9}$
8. _____ $\frac{122}{3}$

Express each improper fraction as a mixed number.
9. $\frac{53}{9}$
10. $\frac{165}{40}$

9. _____ $5\frac{8}{9}$
10. _____ $4\frac{1}{8}$

© Glencoe/McGraw-Hill   127   Mathematics: Applications and Connections, Course 1

TECHNOLOGY/MULTIMEDIA

Technology Masters, pp. 9–10

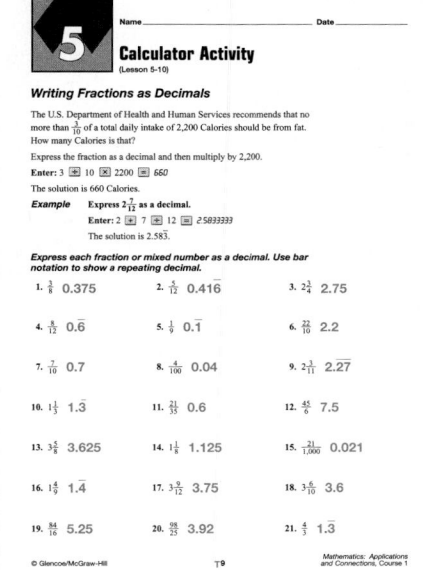

5 Name _____ Date _____

Calculator Activity
(Lesson 5-10)

Writing Fractions as Decimals

The U.S. Department of Health and Human Services recommends that no more than $\frac{3}{10}$ of a total daily intake of 2,200 Calories should be from fat. How many Calories is that?

Express the fraction as a decimal and then multiply by 2,200.

Enter: 3 ÷ 10 × 2200 = 660

The solution is 660 Calories.

Example   Express $2\frac{7}{12}$ as a decimal.
   Enter: 2 + 7 ÷ 12 = 2.5833333
   The solution is 2.583.

Express each fraction or mixed number as a decimal. Use bar notation to show a repeating decimal.

1. $\frac{3}{8}$ 0.375
2. $\frac{5}{12}$ 0.416
3. $2\frac{3}{4}$ 2.75
4. $\frac{8}{12}$ 0.6
5. $\frac{1}{9}$ 0.1
6. $2\frac{2}{10}$ 2.2
7. $\frac{7}{10}$ 0.7
8. $\frac{4}{100}$ 0.04
9. $2\frac{3}{11}$ 2.27
10. $1\frac{1}{3}$ 1.3
11. $\frac{21}{35}$ 0.6
12. $\frac{45}{6}$ 7.5
13. $3\frac{5}{8}$ 3.625
14. $1\frac{1}{8}$ 1.125
15. $\frac{21}{1,000}$ 0.021
16. $1\frac{4}{9}$ 1.4
17. $3\frac{9}{12}$ 3.75
18. $3\frac{6}{10}$ 3.6
19. $\frac{84}{16}$ 5.25
20. $\frac{98}{25}$ 3.92
21. $\frac{4}{3}$ 1.3

© Glencoe/McGraw-Hill   T9   Mathematics: Applications and Connections, Course 1

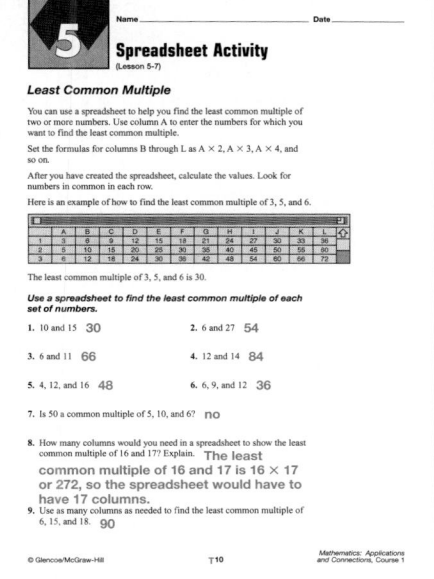

5 Name _____ Date _____

Spreadsheet Activity
(Lesson 5-7)

Least Common Multiple

You can use a spreadsheet to help you find the least common multiple of two or more numbers. Use column A to enter the numbers for which you want to find the least common multiple.

Set the formulas for columns B through L as A × 2, A × 3, A × 4, and so on.

After you have created the spreadsheet, calculate the values. Look for numbers in common in each row.

Here is an example of how to find the least common multiple of 3, 5, and 6.

	A	B	C	D	E	F	G	H	I	J	K	L
1	3	6	9	12	15	18	21	24	27	30	33	36
2	5	10	15	20	25	30	35	40	45	50	55	60
3	6	12	18	24	30	36	42	48	54	60	66	72

The least common multiple of 3, 5, and 6 is 30.

Use a spreadsheet to find the least common multiple of each set of numbers.

1. 10 and 15 30
2. 6 and 27 54
3. 6 and 11 66
4. 12 and 14 84
5. 4, 12, and 16 48
6. 6, 9, and 12 36

7. Is 50 a common multiple of 5, 10, and 6? no

8. How many columns would you need in a spreadsheet to show the least common multiple of 16 and 17? Explain. **The least common multiple of 16 and 17 is 16 × 17 or 272, so the spreadsheet would have to have 17 columns.**

9. Use as many columns as needed to find the least common multiple of 6, 15, and 18. 90

© Glencoe/McGraw-Hill   T10   Mathematics: Applications and Connections, Course 1

MEETING INDIVIDUAL NEEDS

Investigations for the Special Education Student, pp. 13–14, 35–38

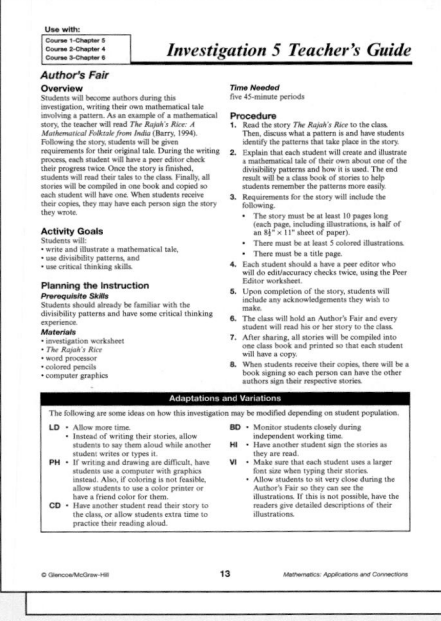

Use with:
Course 1-Chapter 5
Course 2-Chapter 4
Course 3-Chapter 6

Investigation 5 Teacher's Guide

Author's Fair

Overview
Students will become authors during this investigation, writing their own mathematical tale involving a pattern. As an example of a mathematical story, the teacher will read *The Rajah's Rice: A Mathematical Folktale from India* (Barry, 1994). Following the story, students will be given requirements for their original tale. During the writing process, each student will have a peer editor check their progress twice. Once the story is finished, students will read their tales to the class. Finally, all stories will be compiled in one book and copied so each student will have one. When students receive their copies, they may have each person sign the story they wrote.

Activity Goals
Students will:
• write and illustrate a mathematical tale,
• use divisibility patterns, and
• use critical thinking skills.

Planning the Instruction
Prerequisite Skills
Students should already be familiar with the divisibility patterns and have some critical thinking experience.
Materials
• investigation worksheet
• *The Rajah's Rice*
• word processor
• colored pencils
• computer graphics

Time Needed
five 45-minute periods

Procedure
1. Read the story *The Rajah's Rice* to the class. Then, discuss what a pattern is and have students identify the patterns that take place in the story.
2. Explain that each student will create and illustrate a mathematical tale of their own about one of the divisibility patterns and how it is used. The end result will be a class book of stories to help students remember the patterns more easily.
3. Requirements for the story will include the following.
   • The story must be at least 10 pages long (each page, including illustrations, is half of an $8\frac{1}{2}" × 11"$ sheet of paper).
   • There must be at least 5 colored illustrations.
   • There must be a title page.
4. Each student should have a peer editor who will do edit/accuracy checks twice, using the Peer Editor worksheet.
5. Upon completion of the story, students will include any acknowledgements they wish to make.
6. The class will hold an Author's Fair and every student will read his or her story to the class.
7. After sharing, all stories will be compiled into one class book and printed so that each student will have a copy.
8. When students receive their copies, there will be a book signing so each person can have the other authors sign their respective stories.

Adaptations and Variations

The following are some ideas on how this investigation may be modified depending on student population.

LD • Allow more time.
• Instead of writing their stories, allow students to say them aloud while another student writes or types it.

PH • If writing and drawing are difficult, have students use a computer with graphics instead. Also, if coloring is not feasible, allow students to use a color printer or have a friend color for them.

CD • Have another student read their story to the class, or allow students extra time to practice their reading aloud.

BD • Monitor students closely during independent working time.

HI • Have another student sign the stories as they are read.

VI • Make sure that each student uses a larger font size when typing their stories.
• Allow students to sit very close during the Author's Fair so they can see the illustrations. If this is not possible, have the readers give detailed descriptions of their illustrations.

© Glencoe/McGraw-Hill   13   Mathematics: Applications and Connections

Theme: Publishing

Advertising did not become common in print until the 18th century although printing began in the 15th century. In the 19th century, advertising agencies were started as brokers for space in printed media. It was not until the 20th century that agencies actually produced the advertising messages.

Question of the Day

Use the data in the table to write ratios of ads to articles for each magazine listed. Which has the highest ratio? *Nickelodeon / August 1997* Which has the lowest ratio? *International Wildlife / March/April 1997* Discuss reasons for the numbers of ads in different magazines.

Assess Prerequisite Skills

Ask students to read through the list of objectives presented in "What you'll learn in Chapter 5." You may wish to ask them what each of the objectives means or if they have experienced or used any of these math concepts before.

Building Portfolios

Encourage students to revise their portfolios as they study this chapter. Encourage them to write examples of real-life situations that require knowledge of each skill.

 Math and the Family

In the *Family Letters and Activities* booklet (pp. 9–10), you will find a letter to the parents explaining what students will study in Chapter 5. An activity appropriate for the whole family is also available.

What you'll learn in Chapter 5

- to use divisibility patterns to find the prime factorization and greatest common factor,
- to solve problems by making an organized list,
- to simplify fractions and to express mixed numbers as improper fractions and vice versa,
- to measure length in the customary system,
- to find the least common multiple of two or more numbers and compare and order fractions, and
- to write decimals as fractions and vice versa.

176 Chapter 5 Using Number Patterns, Fractions, and Ratios

CD-ROM Program

Activities for Chapter 5
- Chapter 5 Introduction
- Interactive Lessons 5-3, 5-4, 5-7
- Extended Activity 5-2
- Assessment Game
- Resource Lessons 5-1 through 5-10

CHAPTER Project

HOT OFF THE PRESS

Imagine that you have been hired as a reviewer for a magazine publisher. Your job is to investigate competitors' magazines to determine what parts of the magazines have articles and what parts have advertisements. In this project, you will use fractions, ratios, and decimals to help you prepare a report that summarizes your findings. You can choose any four magazines for your project.

Getting Started

- Look at the table below. Find the total number of pages in each issue of the given magazines.
- Use the table to determine which magazine issue has the least number of pages of articles and advertisements. Which issue has the greatest number of pages of articles and advertisements?

Magazine/Issue	Pages		
	Articles	Ads	Table of Contents
Sports Illustrated for Kids/June, 1997	58	20	1
Nickelodeon/ August, 1997	47	19	1
International Wildlife/ March/April, 1997	48	2	1
Disney Adventures/ August 30, 1997	80	29	2

Technology Tips

- Use a **spreadsheet** to express fractions as decimals.
- Use computer **software** to make graphs.

 **interNET**
CONNECTION For up-to-date information on magazines, visit:
www.glencoe.com/sec/math/mac/mathnet

Working on the Project

You can use what you'll learn in Chapter 5 to help you investigate magazines.

Page	Exercise
196	36
219	27
223	Alternative Assessment

CHAPTER Project
NOTES

Objectives Students should
- use fractions and ratios to show a comparison between two numbers.
- express fractions as decimals and compare.

Project Pointer You may suggest that students begin a *Project Folder* to keep their work as they complete each stage of the Chapter Project. The completed project may also be added to their portfolios.

Students may want to investigate why different ads are used in different magazines. Have them determine what the target audience is for the ad and for the magazine.

Using the Table Students may want to copy the table and add a column for the total number of pages. Make sure that they add all three categories to get the total.

***Investigations and Projects Masters*, p. 36**

Instructional Resources ▶▶▶
A recording sheet to help students organize their data for the Chapter Project is shown at the right and is available in the *Investigations and Projects Masters*, p. 36.

---

5 Chapter 5 Project

Name_____ Date_____

Hot Off the Press

Page 177, Getting Started

Magazine/Issue	Total Number of Pages
Sports Illustrated for Kids/June, 1997	
Nickelodeon/August, 1997	
International Wildlife/March/April, 1997	
Disney Adventures/August 30, 1997	

Page 196, Working on the Chapter Project, Exercise 36

Magazine/Issue	Articles	Ads	Table of Contents	Articles/ Total Pages	Ads/ Total Pages	Contents/ Total Pages

Page 219, Working on the Chapter Project, Exercise 27

a.
Magazine/Issue	Articles/Total Pages	Ads/Total Pages	Contents/Total Pages

b.
Page Breakdown for Four Magazines
Decimal Portion
0.8
0.6
0.4
0.2
0
Articles Ads Contents

© Glencoe/McGraw-Hill   36   Mathematics: Applications and Connections, Course 1

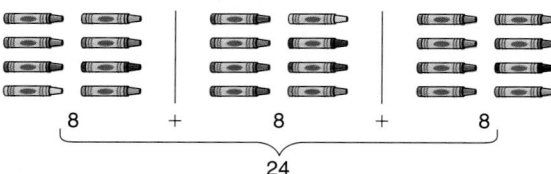

Instructional Resources

- *Study Guide Masters*, p. 32
- *Practice Masters*, p. 32
- *Enrichment Masters*, p. 32
- Transparencies 5-1, A and B
- *Science and Math Lab Masters*, pp. 77–80
- *Hands-On Lab Masters*, p. 73

 CD-ROM Program
- Resource Lesson 5-1

Recommended Pacing	
Standard	Day 2 of 16
Honors	Day 2 of 15
Block	Day 1 of 8

1 FOCUS

5-Minute Check
(Chapter 4)

Multiply.
1. 92.26 × 8 **738.08**
2. 12.35 × 0.76 **9.386**

Divide.
3. 712.5 ÷ 15 **47.5**
4. 46.8 ÷ 2.5 **18.72**
5. Complete: 41.6 kg = _?_ g.
   41,600

 The 5-Minute Check is also available on **Transparency 5-1A** for this lesson.

Motivating the Lesson

Problem Solving Ask students how they would determine whether each statement is true or false without using division or estimation.
- There is no remainder in the quotient of 2,364 ÷ 2. **true; ones digit is divisible by 2**
- The quotient of 98,633 ÷ 6 is a whole number. **false; number is not divisible by 2 and 3**

What **you'll learn**
You'll learn to use divisibility rules for 2, 3, 5, 6, 9, and 10.

When **am I ever going to use this?**
Knowing how to use divisibility rules can help you form groups when going on a field trip.

Crayola crayons were first introduced in 1903. Each box contained eight crayons. Today, Crayola crayons are sold in boxes of 8, 16, 24, 32, 48, 64, and 96 crayons. Suppose you have a box of 24 crayons. Can you divide the crayons evenly among three children?

To solve this problem, you can divide 24 by 3. 24 ÷ 3 means to separate 24 into 3 equal groups.

8 + 8 + 8 = 24

Since $8 + 8 + 8 = 24$, $24 ÷ 3 = 8$. The quotient, 8, is a whole number, so you can say that 24 is *divisible* by 3. The 24 crayons can be divided evenly among 3 children.

You can test for divisibility mentally by using divisibility rules. The divisibility rules for 2, 3, 5, 6, 9, and 10 are as follows.

A number is divisible by:

- 2 if the ones digit is divisible by 2.
- 3 if the sum of the digits is divisible by 3.
- 5 if the ones digit is 0 or 5.
- 6 if the number is divisible by both 2 and 3.
- 9 if the sum of the digits is divisible by 9.
- 10 if the ones digit is 0.

Examples

1 Is 46 divisible by 2?
The ones digit is 6. Since $6 ÷ 2 = 3$, 6 is divisible by 2. So, 46 is divisible by 2.

2 Is 428 divisible by 3?
The sum of the digits is $4 + 2 + 8$, or 14. Since 14 is not divisible by 3, 428 is not divisible by 3.

Check by using a calculator.

428 ÷ 3 = *142.66667* *Since the result is not a whole number, the remainder is not zero.*

428 is not divisible by 3.

178 Chapter 5 Using Number Patterns, Fractions, and Ratios

 ## Cross-Curriculum Cue

Inform the other teachers on your team that your students are studying number patterns, fractions, and ratios. Suggestions for curriculum integration are:

Economics: stock market
Statistics: ratio comparisons
Drafting: scale drawings

Examples

3 Is 2,736 divisible by 2, 3, 5, 6, 9, or 10?

2: Yes. The ones digit, 6, is divisible by 2.

3: Yes. The sum of the digits, 18, is divisible by 3.

5: No. The ones digit is not 0 or 5.

6: Yes. The number is divisible by both 2 and 3.

9: Yes. The sum of the digits, 18, is divisible by 9.

10: No. The ones digit, 6, is not 0.

So, of the given numbers, 2,736 is divisible by 2, 3, 6, and 9.

APPLICATION **4** **Gardening** Lyn has 4 dozen marigolds. She wants to plant the flowers in rows so that each row has the same number of flowers. Can she plant the flowers in 3 equal rows?

Explore   You know that Lyn has 4 dozen marigolds. You need to find out whether she can plant the flowers in 3 equal rows.

Plan   There are 12 in a dozen, so 4 dozen equals 48. Use divisibility rules to see whether 48 is divisible by 3.

Solve   The sum of the digits, 12, is divisible by 3. Thus, she can plant the flowers in 3 equal rows.

Examine   By using a model, you find that 48 flowers can be planted in 3 rows of 16.

CHECK FOR UNDERSTANDING

Communicating Mathematics

Read and study the lesson to answer each question. 2–3. See margin.

1. *Draw* a picture showing how a box of 24 crayons can be equally divided among 2, 6, or 12 children. **See Answer Appendix.**

2. *Explain* how to determine whether 129 is divisible by 3.

3. *Write* the divisibility rules for 2, 3, 5, 6, 9, and 10.

Guided Practice

Determine whether the first number is divisible by the second number.

4. 243; 3 **yes**   5. 2,081; 2 **no**   6. 963; 9 **yes**

State whether each number is divisible by 2, 3, 5, 6, 9, or 10.

7. 17 **none**   8. 104 **2**   9. 1,032 **2, 3, 6**

Lesson 5-1 Divisibility Patterns **179**

■ **Reteaching the Lesson** ■

Activity Have students work in pairs. Ask one partner to enter a random number on a calculator. Ask the other partner to determine whether it is divisible by 2, 3, 5, 6, 9, or 10. Continue the activity with students reversing roles.

Additional Answer

3. 2: if the ones digit is divisible by 2.
   3: if the sum of the digits is divisible by 3.
   5: if the ones digit is 5 or 0.
   6: if the number is divisible by 2 and 3.
   9: if the sum of the digits is divisible by 9.
   10: if the ones digit is zero.

2 TEACH

Transparency 5-1B contains a teaching aid for this lesson.

Using Charts Separate the class into six groups. Give each group a divisibility rule. Write a number on the chalkboard and have each group apply its rule to the number. Record the findings in a chart on the board. Ask students to look for patterns in numbers that work for more than one divisibility rule.

In-Class Examples

For Example 1
a. Is 45 divisible by 5? **yes**
b. Is 27 divisible by 6? **no**

For Example 2
a. Is 355 divisible by 9? **no**
b. Is 726 divisible by 10? **no**

For Example 3
Is 216 divisible by 2, 3, 5, 6, 9, or 10? **2, 3, 6, and 9**

For Example 4
If Lyn has 5 dozen zinnias, can she plant the flowers in six even rows? **yes**

Additional Answer

2. Sample answer: The sum of the digits is divisible by 3.

Study Guide Masters, p. 32

Name _____ Date _____

5-1 **Study Guide**

Divisibility Patterns

The following rules will help you determine if a number is divisible by 2, 3, 5, 6, 9, or 10.

A number is divisible by:
- 2 if the ones digit is divisible by 2.
- 3 if the sum of the digits is divisible by 3.
- 5 if the ones digit is 0 or 5.
- 6 if the number is divisible by 2 and 3.
- 9 if the sum of the digits is divisible by 9.
- 10 if the ones digit is 0.

Example   Is 1,120 divisible by 2, 3, 5, 6, 9, or 10?

2:   Yes. The ones digit 0, is divisible by 2.
3:   No. The sum of the digits, 1 + 1 + 2 + 0 or 4, is not divisible by 3.
5:   Yes. The ones digit is 0.
6:   No. The number is divisible by 2, but not by 3.
9:   No. The sum of the digits, 1 + 1 + 2 + 0 or 4, is not divisible by 9.
10: Yes. The ones digit is 0. 1,120 is divisible by 10.

1,120 is divisible by 2, 5, and 10.

Determine whether the first number is divisible by the second number.

1. 785; 3   2. 655; 5   3. 415; 10
   no   yes   no

4. 819; 9   5. 772; 2   6. 652; 5
   yes   yes   no

7. 1,180; 10   8. 8,764; 9   9. 669; 3
   yes   no   yes

10. 4,655; 3   11. 6,202; 9   12. 3,704; 2
    no   no   yes

© Glencoe/McGraw-Hill   T32   Mathematics: Applications and Connections, Course 1

Lesson 5-1   179

3 PRACTICE/APPLY

Check for Understanding

If students need additional practice or instruction after completing Exercises 1–10, one of these options may be helpful.
- Extra Practice, see p. 568
- Reteaching Activity, see p. 179
- *Transition Booklet,* pp. 25–26
- *Study Guide Masters,* p. 32
- *Practice Masters,* p. 32

Assignment Guide

Core: 11–35 odd, 36–40
Enriched: 12–32 even, 34–40

4 ASSESS

Closing Activity

Speaking Have students work in groups. Ask one member to call out a 3-digit number. Have them discuss whether the number is divisible by 2, 3, 5, 6, 9, or 10 and give reasons to support their answers. Continue until each group member has called out a number.

Additional Answer

35. Yes; each chaperone will have 15 students.

Practice Masters, p. 32

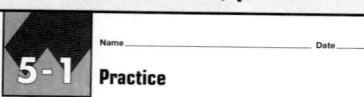

5-1 Practice

Divisibility Patterns

Determine whether the first number is divisible by the second number.

1. 527; 3 no	2. 1,048; 6 no	3. 693; 9 yes	4. 1,974; 2 yes
5. 305; 10 no	6. 860; 5 yes	7. 4,672; 9 no	8. 2,310; 6 yes
9. 816; 3 yes	10. 13,509; 5 no	11. 2,847; 2 no	12. 192; 6 yes

State whether each number is divisible by 2, 3, 5, 6, 9 or 10.

13. 36 2, 3, 6, 9	14. 450 2, 3, 5, 6, 9, 10	15. 192 2, 3, 6	16. 87 3
17. 264 2, 3, 6	18. 1,251 3, 9	19. 890 2, 5, 10	20. 738 2, 3, 6, 9
21. 2,601 3, 9	22. 675 3, 5, 9	23. 498 2, 3, 6	24. 2,019 3

Use mental math skills, paper and pencil, or a calculator to find a number that satisfies the given conditions. Sample answers are given.

25. a three-digit number divisible by both 2 and 9 198
26. a number divisible by both 5 and 6 60
27. a four-digit number divisible by 2, 3, and 9 3,096
28. a three-digit number divisible by 5, 6, and 9 630
29. a four-digit number divisible by 9 and 10 1,980
30. a five-digit number divisible by 5 and 9 14,625

© Glencoe/McGraw-Hill   T32   *Mathematics: Applications and Connections, Course 1*

---

10. **World Records** On February 28, 1898, in Milton, Massachusetts, Henry Helm Clayton and A.E. Sweetland set a record when their kite reached an altitude of 12,471 feet. Determine whether 12,471 is divisible by 2, 3, 5, 6, 9, or 10. **3**

EXERCISES

Practice   Determine whether the first number is divisible by the second number.

11. 435; 5 yes	**12.** 240; 10 yes	**13.** 624; 6 yes
14. 333; 9 yes	**15.** 1,330; 3 no	**16.** 1,752; 2 yes
17. 5,514; 3 yes	**18.** 507; 5 no	**19.** 11,112; 6 yes

State whether each number is divisible by 2, 3, 5, 6, 9, or 10.

20. 208 2	**21.** 576 2, 3, 6, 9	**22.** 175 5
23. 397 none	**24.** 403 none	**25.** 4,077 3, 9
26. 918 2, 3, 6, 9	**27.** 2,860 2, 5, 10	**28.** 9,910 2, 5, 10

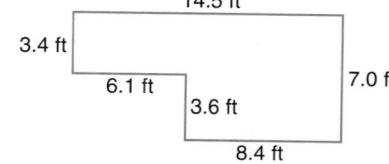

29. Is 747 divisible by 9? **yes**

30. Is 8,760 divisible by 5? **yes**

31. Determine whether 3,517 is divisible by 6. **no**

32. Find a number that is divisible by both 3 and 5.

32. Sample answer: 30

33. Find a number that is divisible by 2, 9, and 10.

33. Sample answer: 90

Applications and Problem Solving

34. **Games** Reynaldo and two of his friends are going to play Old Maid. To play the card game, they need to deal out all of the cards. If there are 51 cards in the deck, will each player get the same number of cards? **Yes; each player will get 17 cards.**

35. **School** The sixth grade students at East Middle School are planning a trip to Grand Canyon National Park. There are 135 students and 9 chaperones. Will each chaperone have the same number of students in his or her group? **See margin.**

36. **Critical Thinking** How can you determine whether a number is divisible by 15? Explain. **Sample answer: if the number is divisible by 3 and 5.**

Mixed Review

37. **Measurement** How many grams are there in 0.05 kilogram? *(Lesson 4-9)* **50**

38. **Geometry** Find the perimeter of the figure. *(Lesson 4-4)* **43.0 ft**

39. Find the product of 17.241 and 16. *(Lesson 4-1)* **275.856**

[figure: 14.5 ft, 3.4 ft, 6.1 ft, 3.6 ft, 7.0 ft, 8.4 ft]

40. [Test Practice] The Clothes Depot is having a sale on denim shirts. The prices of the shirts range from $19.99 to $35.99. If Zachary bought 4 denim shirts, what would be a reasonable total cost for the shirts? *(Lesson 3-5)* **C**

   A $60   **B** $70   **C** $120   **D** $160   **E** $180

Extending the Lesson

Enrichment Masters, p. 32

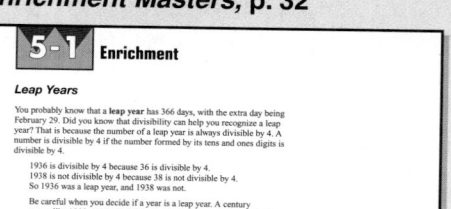

5-1 Enrichment

Leap Years

You probably know that a **leap year** has 366 days, with the extra day being February 29. Did you know that divisibility can help you recognize a leap year? That is because the number of a leap year is always divisible by 4. A number is divisible by 4 if the number formed by its tens and ones digits is divisible by 4.

  1936 is divisible by 4 because 36 is divisible by 4.
  1938 is not divisible by 4 because 38 is not divisible by 4.
  So 1936 was a leap year, and 1938 was not.

Be careful when you decide if a year is a leap year. A century year—like 1800, 1900, or 2000—is a leap year only if its number is divisible by 400.

Activity Have small groups of students determine whether divisibility rules could be written for other numbers such as 4, 7, 8, and 11. Have them write out any rules that they find.

HANDS-ON LAB

COOPERATIVE LEARNING

5-2A Rectangular Arrays

A Preview of Lesson 5-2

square tiles

A *composite number* is a number that has more than two factors. A *prime number* is a number that has exactly two factors. You can use rectangular arrays to find out whether a number is composite or prime. *You can refer to Lesson 1-6 to review factors.*

TRY THIS

Work with a partner.

1 Determine whether 6 is a prime number or a composite number.

- Use 6 square tiles to build as many different-shaped rectangles as possible.

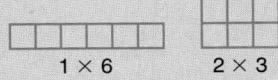

1 × 6   2 × 3

- There is a rectangle that is 1 unit by 6 units and one that is 2 units by 3 units. So, the factors of 6 are 1, 2, 3, and 6.

Since 6 has four factors, it is a composite number.

2 Determine whether 11 is a prime number or a composite number.

- Use 11 square tiles to build as many different-shaped rectangles as possible.

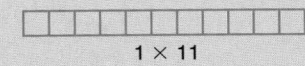

1 × 11

- There is one rectangle whose dimensions are 1 unit by 11 units. So, the factors of 11 are 1 and 11.

Since 11 has exactly two factors, it is a prime number.

ON YOUR OWN

1. Repeat the process with areas of 2 square units through 12 square units. **See Answer Appendix.**
2. Which numbers of square tiles had only one arrangement? **2, 3, 5, 7, 11**
3. Which numbers of square tiles had more than one arrangement? **4, 6, 8, 9, 10, 12**
4. Is there a relationship between a multiplication table and the number of arrangements you have? If so, describe it. **See margin.**
5. Make a guess about which numbers between 12 and 25 square units can have more than one rectangular shape. Explain why you selected those numbers. **See margin.**
6. **Look Ahead** Write a sentence describing the characteristics of prime and composite numbers. **See margin.**

Lesson 5-2A HANDS-ON **LAB** **181**

Additional Answers

4. The greater the number of arrangements, the more times the number appears on the multiplication table.

5. 12, 14, 15, 16, 18, 20, 21, 22, 24; these numbers have several factors.

6. Sample answer: A prime number has exactly 2 factors, 1 and the number itself. A composite number has more than 2 factors.

GET READY

Objective Students identify prime and composite numbers by using rectangular arrays.

Optional Resources
Hands-On Lab Masters
- grid paper, pp. 11, 12
- worksheet, p. 47

Overhead Manipulative Resources
- algebra/area tiles

Manipulative Kit
- algebra/area tiles

MANAGEMENT TIPS

Recommended Time
30 minutes

Getting Started Have students practice making rectangles with the tiles or on grid paper. Make sure that students form complete rectangles. Show students how to model a rectangle with an area of 13 square units and one of 15 square units.

Activities 1 and 2 model prime and composite numbers. Explain that while rectangles can be positioned to look different, the shapes in this activity must have different widths and lengths. Tell students that a rectangle measuring 1 unit by 8 units is the same as a rectangle measuring 8 units by 1 unit.

ASSESS

Have students take turns modeling the rectangles and recording the findings. Also, to help students record all rectangles, have them start with a length of 1 unit. Then go to a length of 2 units, 3 units, and so on.

Math Journal Have students write a paragraph that describes the patterns they discovered while modeling with the tiles.

- *Study Guide Masters*, p. 33
- *Practice Masters*, p. 33
- *Enrichment Masters*, p.33
- Transparencies 5-2, A and B
- *Diversity Masters*, p. 5

 CD-ROM Program
- Resource Lesson 5-2
- Extended Activity 5-2

Recommended Pacing	
Standard	Days 3 & 4 of 16
Honors	Day 3 of 15
Block	Day 2 of 8

1 FOCUS

 5-Minute Check
(Lesson 5-1)

Determine whether the first number is divisible by the second number.

1. 342; 6 **yes**

2. 795; 9 **no**

State whether each number is divisible by 2, 3, 5, 6, 9, or 10.

3. 113 **none**

4. 6,291 **3, 9**

5. Students at the Cassady School raised $98 for two homeless shelters in their town. Will they be able to give the same amount to each shelter? **yes**

The 5-Minute Check is also available on **Transparency 5-2A** for this lesson.

2 TEACH

 Transparency 5-2B contains a teaching aid for this lesson.

Reading Mathematics Display the words *prime, composite, factor tree,* and *prime factorization.* Have students give an example of each. Write the examples below the words. Then have students use the words and the examples in a sentence.

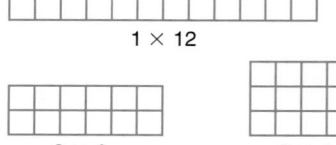

Prime Factorization

What **you'll learn**

You'll learn to find the prime factorization of a composite number.

When **am I ever going to use this?**

Knowing how to find the prime factorization of a number can help you determine the possible dimensions of a box when given the box's volume.

Word Wise

composite number
prime number
prime factorization
factor tree

Each rectangle shown below has an area of 12 square units. The dimensions of the rectangles are 1 by 12, 2 by 6, and 3 by 4. So, the factors of 12 are 1, 2, 3, 4, 6, and 12.

1 × 12

2 × 6   3 × 4

Since 12 has six factors, it is a **composite number**. A composite number is any whole number greater than one with more than two factors.

In Lesson 5-2A, you found that there was only one rectangle you could make with 2, 3, 5, 7, and 11 square tiles. Numbers such as these are **prime numbers**. A prime number has exactly two factors, 1 and the number itself.

The numbers 0 and 1 are neither prime nor composite. Zero has an endless number of factors. The number 1 has only one factor, itself.

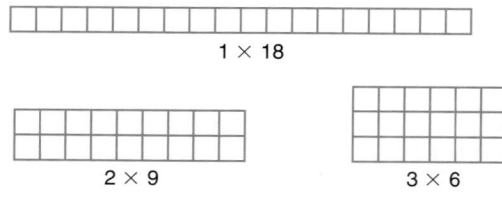 **Example** **1** Determine whether 18 is a prime or composite number. Draw rectangles to explain your answer.

1 × 18

2 × 9   3 × 6

The factors of 18 are 1, 2, 3, 6, 9, and 18. Since 18 has more than two factors, it is a composite number.

Motivating the Lesson

Hands-On Activity Give pairs of students 40 small items, such as dried beans or paper clips. Have them separate the items into equal groups and record their findings. Challenge them to find as many equal groups as possible.

Multiple Learning Styles

 Visual/Spatial Have pairs of students use the geoboard's circular pattern. With colored rubber bands or string, students should hook the various intervals that represent different factors. Have students chart the information depicted on the geoboard and discuss. Repeat with a 24-peg board or a different composite number.

Every composite number can be expressed as a product of prime numbers. This is called the **prime factorization** of a number. To find the prime factorization of a number, you can use a **factor tree** or division.

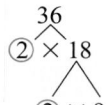**Examples** ➋ Find the prime factorization of 36.

Method 1 Use a factor tree.

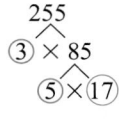

Factor 36. 36 is divisible by 2.
Circle the prime number 2.
How do you know 2 is prime?
Factor 18. Circle the prime number 2.
Factor 9. Circle the prime numbers 3.

Method 2 Use division.

$$\begin{array}{r} 3 \\ 3\overline{)9} \\ 2\overline{)18} \\ 2\overline{)36} \end{array}$$

Begin with the least prime that is a factor.
Then divide the quotient by the least possible prime factor. In this case, it is 2.
Repeat until the quotient is prime.

The prime factorization of 36 is $2 \times 2 \times 3 \times 3$, or $2^2 \times 3^2$.

LOOK BACK
You can refer to Lesson 1-6 to review exponents.

INTEGRATION ➌ **Geometry** The volume of a box is found by multiplying its height, length, and width. If the measure of the volume of a box is 255, what could its dimensions be?

Find the prime factorization of 255.

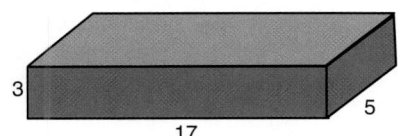

The prime factorization of 255 is $3 \times 5 \times 17$. So, 3, 5, and 17 are factors of 255. Thus, the dimensions of the box could be 3 by 5 by 17.

CHECK FOR UNDERSTANDING

Communicating Mathematics

Read and study the lesson to answer each question. 1–3. Answer Appendix.

1. ***Draw*** a picture to determine whether 10 is a prime or composite number.

2. ***Explain*** the difference between a prime number and a composite number.

3. ***You Decide*** Michelle says that the prime factorization of 120 is $2 \times 2 \times 2 \times 3 \times 5$. Lorenzo says that it is $2 \times 5 \times 3 \times 2 \times 2$. Who is correct? Explain your reasoning.

Lesson 5-2 Prime Factorization **183**

Reteaching the Lesson

Activity Have students use prime number cards to physically build a factor tree. Have them use self-adhesive notes to write composite numbers. As they factor each composite number, have them replace that note with a pair of factors (cards and/or other notes). Continue until the tree is composed totally of cards.

Error Analysis
Watch for students who rely on products in multiplication facts to determine whether a number is prime.
Prevent by reminding students that numbers such as 51 are not products in basic multiplication facts.

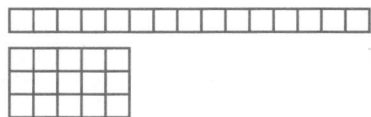

In-Class Examples

For Example 1
Determine whether 15 is a prime or composite number. Draw rectangles to explain your answer. **15 is a composite number.**

For Example 2
Find the prime factorization of 48. $2 \times 2 \times 2 \times 2 \times 3$ or $2^4 \times 3$

For Example 3
If the measure of the volume of a box is 609, what could its dimensions be? **3 by 7 by 29**

---

3 PRACTICE/APPLY

Check for Understanding
If students need additional practice or instruction after completing Exercises 1–12, one of these options may be helpful.
- Extra Practice, see p. 569
- Reteaching Activity
- *Transition Booklet,* pp. 21–22
- *Study Guide Masters,* p. 33
- *Practice Masters,* p. 33

***Study Guide Masters,* p. 33**

Name _____ Date _____

5-2 Study Guide

Prime Factorization

A **prime number** has exactly two factors, 1 and the number itself.

Example 1 13 is a prime number. It has only 1 and 13 as factors.

A **composite number** has more than two factors.
A composite number may be written as the product of prime numbers.

Example 2 **Find the prime factorization of 48.**

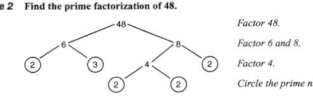

Factor 48.
Factor 6 and 8.
Factor 4.
Circle the prime numbers.

Write the prime numbers in order from least to greatest.
Check to see if the product is 48.
$2 \times 2 \times 2 \times 2 \times 3 = 48$
The numbers 0 and 1 are neither prime nor composite.
0 has an endless number of factors. 1 has only one factor, itself.

Find the prime factorization of each number.

1. 24 $2 \times 2 \times 2 \times 3$
2. 25 5×5
3. 70 $7 \times 5 \times 2$

4. 55 5×11
5. 44 $2 \times 2 \times 11$
6. 84 $2 \times 2 \times 3 \times 7$

7. 68 $2 \times 2 \times 17$
8. 121 11×11
9. 92 $2 \times 2 \times 23$

© Glencoe/McGraw-Hill   T33   Mathematics: Applications and Connections, Course 1

Lesson 5-2 183

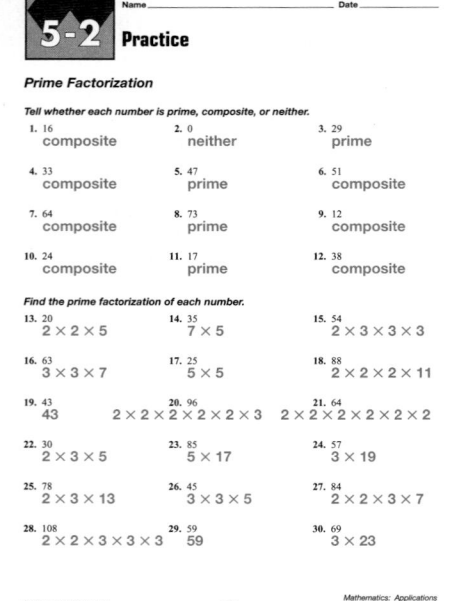
Guided Practice

Tell whether each number is *prime*, *composite*, or *neither*.

4. 17 prime   **5.** 0 neither   **6.** 23 prime   **7.** 57 composite

Find the prime factorization of each number.

8. 49 7^2 **9.** 75 3×5^2 **10.** 32 2^5 **11.** 104 $2^3 \times 13$

12. *Geography* The state of North Carolina is made up of 100 counties. Express 100 as a product of primes. $2^2 \times 5^2$

EXERCISES

Practice

Tell whether each number is *prime*, *composite*, or *neither*.

13. 1 neither   **14.** 15   **15.** 29 prime   **16.** 44 composite

17. 45   **18.** 53 prime   **19.** 56   **20.** 31 prime

21. 87 composite   **22.** 110 composite   **23.** 93 composite   **24.** 114 composite

14. composite
17. composite
19. composite
25. $2 \times 3 \times 7$
33. $2 \times 3 \times 17$
35. $2^3 \times 3 \times 5$

Find the prime factorization of each number.

25. 42   **26.** 81 3^4 **27.** 65 5×13 **28.** 38 2×19

29. 17 prime   **30.** 24 $2^3 \times 3$ **31.** 18 2×3^2 **32.** 40 $2^3 \times 5$

33. 102   **34.** 97 prime   **35.** 120   **36.** 48 $2^4 \times 3$

37. What is the least prime number that is greater than 80? **83**

38. Express 126 as a product of prime numbers. $2 \times 3^2 \times 7$

39. What is the only even prime number? Explain. **See margin.**

Applications and Problem Solving

40. *Math History* In 1742, Christian Goldbach of Russia suggested that any even number greater than two could be expressed as the sum of two prime numbers. For example, 28 = 11 + 17 and 30 = 13 + 17. Find two prime numbers whose sum is the given number. **a–c. Sample answers given.**
   a. 32 **13 + 19**   **b.** 40 **17 + 23**   **c.** 58 **41 + 17**

41. *Number Patterns* Twin primes are two prime numbers that are consecutive odd integers such as 3 and 5, 5 and 7, and 11 and 13. Find all of the twin primes that are less than 100. **See margin.**

42. *Critical Thinking* Every odd number greater than 7 can be expressed as the sum of three prime numbers. Which three prime numbers have a sum of 57? **Sample answer: 7 + 19 + 31**

Mixed Review

43. Determine whether or not 462 is divisible by 6. *(Lesson 5-1)* **yes**

44. ▮ **Test Practice** ▮ Seth purchased 3 CDs for $51.12. If each CD sold for the same amount, what was the price of each CD? *(Lesson 4-5)* **D**
   A $15.06   **B** $13.10   **C** $10.98   **D** $17.04   **E** Not Here

45. *Algebra* Evaluate $a - b$ if $a = 7.3$ and $b = 2.94$. *(Lesson 3-6)* **4.36**

46. *Statistics* Find the mean of the numbers 120, 112, 88, 100, 141, and 147. *(Lesson 2-7)* **118**

47. Write 17^7 as a product. *(Lesson 1-6)* $17 \times 17 \times 17 \times 17 \times 17 \times 17 \times 17$

184   Chapter 5   Using Number Patterns, Fractions, and Ratios

Extending the Lesson

Enrichment Masters, p. 33

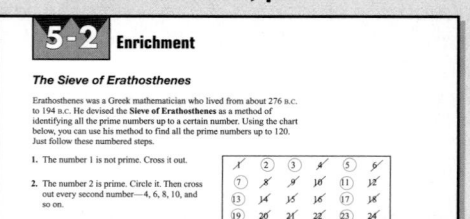

Activity Have students research *perfect, deficient*, and *abundant numbers*. Have them determine how they are related to factors of a number.

GRAPHIC DESIGN

CYBER GRAFIX

Kathryn Sharar Prusinski
GRAPHIC DESIGNER

Kathryn Sharar Prusinski is co-owner of CyberGraphix, a company that designs online magazines and other computer-related products. She holds a degree in advertising design and a certificate in advanced computer graphics. Ms. Prusinski designed several online materials for the Smithsonian and Sesame Street.

Most graphic designers have a bachelor's degree in graphic design, art, or art history. Graphic designers should have an understanding of computer technology, especially painting and graphic design tools. Computer and mathematics courses are an essential part of preparing for a career in graphics design.

For more information:
Graphic Arts Technical Foundation
200 Deer Run Road
Sewickley, PA 15143-2600

*inter*NET CONNECTION
www.glencoe.com/sec/
math/mac/mathnet

Your Turn
Write and design a brochure that describes the advantages of a career in the graphic design industry.

I'd like to design graphics for a website.

School to Career: Graphic Design **185**

Motivating Students
Anyone who has created a poster or newsletter has tried graphic design. Graphic design can be employed for both commercial and artistic purposes. Its most popular use today is for the creation of websites on the Internet. To begin a discussion of graphic design, ask students these questions.
- What does the term *graphic design* mean? **a format that involves written words, drawings, engravings, photographs, or any combination of these items**
- What examples of graphic design are in the classroom? **Sample answer: bulletin board display, posters, textbooks**
- Why would someone creating a website need to understand graphic design? **to include text and illustrations in an interesting manner so that surfers will stop to read the site**

Making the Math Connection
Graphic designers need to understand angles, geometric shapes, area, and perimeter to lay out a design. Knowledge of scale and the ability to add, subtract, multiply, and divide using units of measure are also important.

Working on *Your Turn*
Have students separate into groups to brainstorm ideas for their brochures. Have them sketch their design ideas before creating the finished product. If possible, have students create their brochures on a computer.

More About Kathryn Sharar Prusinski
- Kathryn Sharar Prusinski has a degree in Advertising Design from the Fashion Institute of Technology and a certificate in Advanced Computer Graphics from the Center for Media Arts. She also studied fine arts at Rutgers University and film animation at the School of Visual Arts in New York.
- Ms. Prusinski's on-line design credits include the following: the on-line magazine *National Geographic;* "NASA Space Challenge," a voyage with the astronauts on the space shuttle *Atlantis;* Disney's Gargoyles; and Saban's Mighty Morphin Power Rangers.

*An additional School to Career activity is available on page 5 of the **School to Career Masters.***

Objective Students solve problems by making an organized list.

Recommended Pacing	
Standard	Day 5 of 16
Honors	Day 4 of 15
Block	Day 3 of 8

1 FOCUS

Getting Started Ask students how they would determine which situation offers a greater variety of items.
- T-shirts available in 6 colors and 4 sizes
- 7 varieties of nuts, each in 3 package sizes

Suggest that students begin by listing the options.

2 TEACH

Teaching Tip An organized list will help students find a pattern and also determine whether all possibilities have been exhausted.

In-Class Example
How many 3-digit numbers can you make using the digits 4, 6, 8, and 9 if no repeated digits are allowed? **24 3-digit numbers**

THINKING LAB

PROBLEM SOLVING

5-3A Make a List

A Preview of Lesson 5-3

Harvi and Kendra are making a banner for spirit week using three sheets of paper. Harvi is trying to figure out how many different banners they can make using two colors of paper. Let's listen in!

Harvi

Our school colors are red and blue so we should use red and blue paper. How many different banners can we make using three sheets of paper?

Well, let's arrange the sheets of paper in different ways and see.

It's going to be hard to remember each arrangement. While you're arranging the paper, I'll list the possibilities.

Arrangements

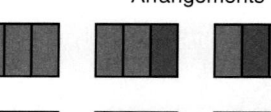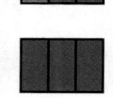

That's a good idea.

Well, there are eight different possibilities.

RRR	RRB	RBR	RBB
BRR	BBR	BRB	BBB

Kendra

THINK ABOUT IT

Work with a partner.

1. *Analyze* the eight different possibilities. Do you agree or disagree with the possibilities? Explain your reasoning. **See students' work.**

2. *Think* of another way that Harvi and Kendra could have solved the problem. **See students' work.**

3. *Apply* the **make a list** strategy to solve the following problem.

   Molly's Muffin Shop sells apple, cranberry, blueberry, and strawberry muffins in three sizes: mini, regular, and jumbo. How many different options does a customer have to buy a muffin? **12 possible options**

186 Chapter 5 Using Number Patterns, Fractions, and Ratios

■ Reteaching the Lesson ■

Activity Give small groups of students a take-out pizza menu. Have them list all of the possible options available.

ON YOUR OWN

4. The third step of the 4-step plan for problem solving asks you to *solve* the problem. *Explain* the advantages of making an organized list over a random list when solving the problem. **See margin.**

5. *Write a Problem* that can be solved by making a list. Then ask a classmate to solve the problem. **See students' work.**

6. *Look Back* Explain how you can use the make a list strategy to answer Exercise 41 on page 184. **See margin.**

MIXED PROBLEM SOLVING

Solve. Use any strategy.

STRATEGIES
Look for a pattern.
Solve a simpler problem.
Act it out.
Guess and check.
Draw a diagram.
Make a chart.
Work backward.

7. *Manufacturing* A sweater company offers 6 different styles of sweaters in 5 different colors. How many combinations of style and color are possible? **30**

8. *Money Matters* Luisa saved $125 in May and $175 in June. How much does she need to save in July to average $150 per month in savings? **$150**

9. *Number Patterns* Colleen is planning to travel to Dallas, San Antonio, and Houston. She cannot decide in which order to schedule her visits. How many choices does she have? **6**

10. *Food* The graph shows that strawberry and grape jelly account for more than half of the yearly jelly sales in the United States. *About* how much more money is spent on strawberry and grape jelly than the other types of jelly? **Sample answer: $370 million–$300 million or $70 million.**

Yearly Jelly Sales

$366.2 million

$291.5 million

Strawberry and Grape

All others

Source: Nielsen Marketing Research

11. *Travel* The Guerrero's vacation lasted 12 days. They budgeted $75 per night for lodging and $65 per day for food. How much did they budget for these two items for the entire trip? **$1,680**

12. *Number Sense* How many different 2-digit numbers can you make using the digits 1, 3, 5, and 7 if no number is repeated? **12**

13. **Test Practice** Samantha is going on vacation with her father. She has asked her neighbor to feed her 2 dogs while she is gone. Each dog eats 8.5 cans of dog food per week. What else do you need to know to find how many cans of food Samantha should buy before she goes on vacation? **C**

  A the number of ounces in each can of dog food

  B the price of one can of dog food

  C the number of weeks Samantha will be on vacation

  D the breed of each dog

  E the weight of each dog

14. *Manufacturing* How many license plate combinations can you make with the letters M, N, and P and the numbers 1, 2, and 3 if each license plate starts with three different letters and ends with three different numbers? **36**

Lesson 5-3A THINKING 187

■ Extending the Lesson ■

Activity Have small groups of students contact local automobile dealers to obtain new-car brochures. Have them choose a model and list the types of options. Then have them determine how many cars the dealers would need to keep on the lot if they wanted to have all of the possible combinations available for their customers.

Check for Understanding
Use the results from Exercise 3 to determine whether students understand how to make a list to solve a problem.

Extra Practice If students need additional practice in problem solving, extra practice is available on the following pages.
• Make a List, see p. 569
• Mixed Problem Solving, see pp. 593–594

Assignment Guide

All: 4–14

4 ASSESS

Closing Activity
Writing Have students list situations that are most easily answered by making a list. Suggest that they compare lists.

Additional Answers
4. Sample answer: An organized list allows you to make sure you have all the possibilities.

6. Sample answer: Make a list of all consecutive integers. Then circle each pair of consecutive odd integers that are both prime.

Instructional Resources
- *Study Guide Masters*, p. 34
- *Practice Masters*, p. 34
- *Enrichment Masters*, p. 34
- Transparencies 5-3, A and B
- *Assessment and Evaluation Masters*, p. 127

 CD-ROM Program
- Resource Lesson 5-3
- Interactive Lesson 5-3

Recommended Pacing	
Standard	Day 6 of 16
Honors	Day 5 of 15
Block	Day 3 of 8

1 FOCUS

5-Minute Check
(Lesson 5-2)

Tell whether each number is *prime, composite,* or *neither.*
1. 49 composite
2. 67 prime

Find the prime factorization of each number.
3. 118 2×59
4. 90 $2 \times 3 \times 3 \times 5$ or $2 \times 3^2 \times 5$
5. The first prime number is 2. What is the fourteenth prime number? 43

 The 5-Minute Check is also available on **Transparency 5-3A** for this lesson.

Motivating the Lesson
Communication Ask a student to read aloud the opening paragraph of the lesson. Discuss whether the greatest number of people is a product, a factor, or a sum, and why. **a factor because the larger groups are being separated into smaller groups**

2 TEACH

 Transparency 5-3B contains a teaching aid for this lesson.

What you'll learn
You'll learn to find the greatest common factor of two or more numbers.

When am I ever going to use this?
Knowing how to find the greatest common factor of two or more numbers can help you coordinate a parade.

Word Wise
greatest common factor (GCF)

Suppose that 27 flag line and 45 drill team members were selected to participate in the Flag Day parade. The parade director wants all members to line up in rows that have the same number of people. Flag line and drill team members will be in different rows. What is the greatest number of people that can be in each row?

To answer this question, you can use the factors of 27 and 45.

factors of 27: 1, 3, 9, 27
factors of 45: 1, 3, 5, 9, 15, 45

Notice that 1, 3, and 9 are common factors of both 27 and 45. The greatest of the common factors of two or more numbers is called the **greatest common factor (GCF)** of the numbers. So, the greatest common factor of 27 and 45 is 9.

The greatest number of people that can be in each row is 9.

To find the GCF of two or more numbers, you can make a list or use prime factorization.

Method 1 Make a list.
- List all of the factors of each number.
- Identify the common factors.
- The greatest of the common factors is the GCF.

Method 2 Use prime factorization.
- Write the prime factorization of each number.
- Identify all of the common prime factors.
- The product of the common prime factors is the GCF.

Example ① Find the GCF of 24 and 32 by making a list.

factors of 24: 1, 2, 3, 4, 6, 8, 12, 24   *List all of the factors*
factors of 32: 1, 2, 4, 8, 16, 32   *of each number.*

The common factors are 1, 2, 4, and 8.
The GCF of 24 and 32 is 8.

188   Chapter 5 Using Number Patterns, Fractions, and Ratios

Classroom Vignette

"I assign a color to each of the first eight prime numbers. Students make up factor trees to find the prime factorization of a number. When they finish each tree, they circle each prime number with its appropriate color. They can then spot the factors that numbers have in common to find the GCF."

Frances S. Gattis, Teacher
Ware County Middle School
Waycross, GA

2 Find the GCF of 36 and 54 by using prime factorization.

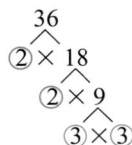 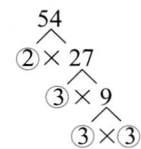

Write the prime factorization of each number.

The common prime factors are 2, 3, and 3.
The GCF of 36 and 54 is $2 \times 3 \times 3$, or 18.

CONNECTION **3** **Life Science** Kaja has 32 fir, 48 cedar, and 80 maple tree seedlings to sell at the annual Agriculture Fair. He wants to display them in rows with the same number of seedlings in each row with no mixing of rows.

a. What is the greatest number of seedlings that he could put in each row?

b. How many rows of each tree seedlings will there be?

a. factors of 32: 1, 2, 4, 8, 16, 32
factors of 48: 1, 2, 3, 4, 6, 8, 12, 16, 24, 48
factors of 80: 1, 2, 4, 5, 8, 10, 16, 20, 40, 80

List all of the factors of each number.

The common factors are 1, 2, 4, 8, and 16.
The GCF of 32, 48, and 80 is 16.
So, Kaja could put 16 tree seedlings in a row.

b. Since $2 \times 16 = 32$, there would be 2 rows of fir seedlings.
Since $3 \times 16 = 48$, there would be 3 rows of cedar seedlings.
Since $5 \times 16 = 80$, there would be 5 rows of maple seedlings.

CHECK FOR UNDERSTANDING

Communicating Mathematics

Read and study the lesson to answer each question. 1–2. Answer Appendix.

1. *Explain* how to find the GCF of 12 and 18.

2. *Tell* how the GCF of two numbers could be one of the numbers. Give two examples.

Guided Practice

Find the GCF of each set of numbers by making a list.

3. 10, 15 **5**   **4.** 8, 30 **2**   **5.** 14, 35, 84 **7**

Find the GCF of each set of numbers by using prime factorization.

6. 15, 45 **15**   **7.** 8, 88 **8**   **8.** 16, 24, 72 **8**

9. *School* Carena has two rolls of streamers to use in decorating the school gym for a pep rally. One roll is 64 feet long, and the other roll is 72 feet long. If she wants to make all of the streamers the same length, what is the greatest length each streamer can be? **8 ft**

Lesson 5-3 Greatest Common Factor **189**

■ Reteaching the Lesson ■

Activity Display these numbers and their factors on an overhead transparency. Ask leading questions to help students find the GCF.
36: 1, 2, 3, 4, 6, 9, 12, 18, 36
90: 1, 2, 3, 5, 6, 9, 10, 15, 18, 30, 45, 90
GCF: 18

Using Technology The TI-83 calculator has a GCD function (greatest common divisor), which calculates the GCF for any pair of numbers. To use the function, press [MATH], select NUM, and [ENTER] 9. Enter the first number, a comma, the second number, and the closing parenthesis. Press [ENTER] to obtain the GCF.

---

3 PRACTICE/APPLY

Check for Understanding
If students need additional practice or instruction after completing Exercises 1–9, one of these options may be helpful.
• Extra Practice, see p. 569
• Reteaching Activity
• *Study Guide Masters,* p. 34
• *Practice Masters,* p. 34
•  Interactive Mathematics Tools Software

***Study Guide Masters,* p. 34**

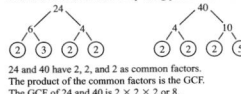

Name_____ Date_____

5-3 **Study Guide**

Greatest Common Factor

The greatest of the common factors of two or more numbers is called the **greatest common factor (GCF)** of the numbers.

Example 1 Find the GFC of 30 and 42 by making a list.
factors of 30: 1, 2, 3, 5, 6, 10, 15, 30
factors of 42: 1, 2, 3, 6, 7, 14, 21, 42

The common factors of 30 and 42 are: 1, 2, 3, and 6.
The greatest common factor of 30 and 42 is 6.

You can also use prime factorization to find the GCF.

Example 2 Find the GCF of 24 and 40 by using prime factorization.

24 and 40 have 2, 2, and 2 as common factors.
The product of the common factors is the GCF.
The GCF of 24 and 40 is $2 \times 2 \times 2$ or 8.

Find the GCF of each set of numbers using either method.

1. 12, 16   2. 18, 24   3. 20, 16
   4   6   4

4. 30, 36   5. 35, 49   6. 32, 40
   6   7   8

7. 72, 36, 54   8. 63, 42, 21   9. 28, 42, 56
   18   21   14

10. 12, 60, 24   11. 35, 28, 21   12. 16, 64, 48
    12   7   16

© Glencoe/McGraw-Hill   134   *Mathematics: Applications and Connections, Course 1*

4 ASSESS

Closing Activity

Speaking Ask students how knowing divisibility rules can help them when finding the GCF of two numbers.

Chapter 5, Quiz A (Lessons 5-1 through 5-3) is available in the *Assessment and Evaluation Masters,* p. 127.

Practice Masters, p. 34

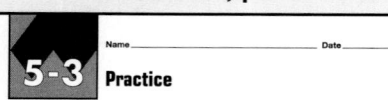

EXERCISES

Practice

Find the GCF of each set of numbers by making a list.

10. 8, 32 **8**
11. 6, 44 **2**
12. 42, 56 **14**
13. 75, 30 **15**
14. 9, 18, 42 **3**
15. 22, 55, 88 **11**

Find the GCF of each set of numbers by using prime factorization.

16. 16, 56 **8**
17. 17, 34 **17**
18. 35, 65 **5**
19. 96, 108 **12**
20. 16, 52, 76 **4**
21. 12, 18, 60 **6**

Find the GCF of each set of numbers using either method.

22. 15, 36 **3**
23. 21, 45 **3**
24. 42, 90 **6**
25. 34, 85 **17**
26. 6, 57, 99 **3**
27. 19, 95, 152 **19**

28. What is the GCF of 42, 77, and 63? **7**
29. Find the GCF of 39, 48, and 51. **3**
30. Write two numbers whose GCF is 18. **Sample answer: 18 and 36**

Applications and Problem Solving

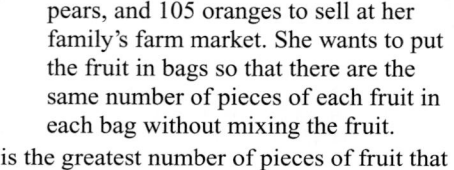

31. *Remodeling* Shawn is covering a portion of his bathroom wall with equal-sized ceramic square tiles. The portion of the wall to be tiled measures 16 inches wide and 72 inches long.
   a. What is the largest square tile that can be used so that no tiles will need to be cut? **8 in. by 8 in.**
   b. What is the total number of tiles Shawn will need? **18**

32. *Business* Marsha has 45 apples, 75 pears, and 105 oranges to sell at her family's farm market. She wants to put the fruit in bags so that there are the same number of pieces of each fruit in each bag without mixing the fruit.

32b. 3 bags of apples, 5 bags of pears, 7 bags of oranges

   a. What is the greatest number of pieces of fruit that can be put in each bag? **15**
   b. How many bags of each kind of fruit will there be?

33. *Critical Thinking* Two composite numbers that have 1 as their only common factor are said to be *relatively prime*. Find two composite numbers less than 25 that are relatively prime. **Sample answer: 8 and 9**

Mixed Review

34. **Test Practice** Which expression represents the prime factorization of 378? *(Lesson 5-2)* **C**
   A $2 \times 3 \times 7^3$ **B** $2^3 \times 3 \times 7$ **C** $2 \times 3^3 \times 7$ **D** $2^3 \times 3^3 \times 7^3$

35. *Decorating* Antonieta is buying new carpet for her bedroom. If her bedroom measures 5.3 yards by 4.7 yards, how many square yards of carpeting will she need? *(Lesson 4-4)* **24.9 yd²**

36. Sample answer: $4 + 4 + 4 + 4 + 4 = 20$

36. Estimate $4.231 + 3.98 + 4 + 4.197 + 3.76$. *(Lesson 3-5)*

37. Write *nine and sixteen thousandths* as a decimal. *(Lesson 3-1)* **9.016**

Extending the Lesson

Enrichment Masters, p. 34

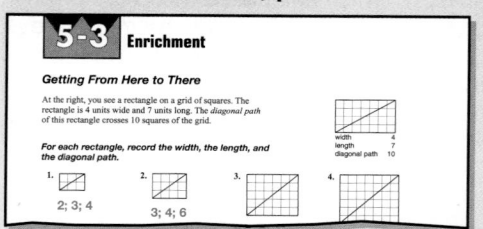

Activity Have pairs of students make generalizations about the GCF of two prime numbers. Then have them compare these generalizations with generalizations about the GCF of two composite numbers.

HANDS-ON LAB

COOPERATIVE LEARNING

5-4A Fractions and Ratios

A Preview of Lesson 5-4

ruler

You have used models to help you understand decimals. You can also use models to help you understand ratios and fractions. A *ratio* is a comparison of two numbers by division. The ratio comparing 3 to 4 can be stated as 3 out of 4, 3 to 4, 3:4, or $\frac{3}{4}$.

TRY THIS

Work with a partner.

1 Model the ratio 1 *out of* 4.
- Copy the rectangle shown.
- Separate the rectangle into four equal parts.
- Shade 1 part.

One out of four parts is shaded.

Therefore, the model represents the ratio *1 out of 4*. Note that the model also represents the fraction $\frac{1}{4}$.

2 Model the fraction $\frac{3}{5}$.
- Copy the rectangle shown.
- Separate the rectangle into five equal parts.
- Shade 3 parts.

Three out of five of the parts are shaded.

Therefore, the model represents the fraction $\frac{3}{5}$ as well as the ratio *3 out of 5*.

ON YOUR OWN

Write a ratio and a fraction for each model shown.   3. 5 out of 9, 5 to 9, or 5:9; $\frac{5}{9}$

1.

5 out of 6, 5 to 6, or 5:6; $\frac{5}{6}$

2.
3 out of 8,
3 to 8, or 3:8; $\frac{3}{8}$

3.
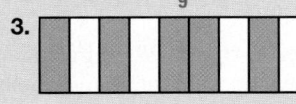

Lesson 5-4A HANDS-ON LAB 191

HANDS-ON
5-4A LAB Notes

GET READY

Objective Students use models to represent fractions and equivalent fractions.

Optional Resources
Hands-On Lab Masters
- fraction models, pp. 3, 4
- worksheet, p. 48

Manipulative Kit
- ruler

MANAGEMENT TIPS

Recommended Time
45 minutes

Getting Started Have students read the opening paragraph of the lesson. Draw a variety of shapes and ask whether they can be divided into equal parts. Have students determine how they would divide a rectangle into equal parts.

Activities 1 and 2 model a single ratio or fraction. Emphasize to students that a fraction is just a way of writing a ratio.

Activity 3 models two fractions and compares them. Make sure that the two initial rectangles are identical. For students having difficulty drawing identical rectangles and dividing them into appropriate sections, you may opt to use the premade fraction models in the *Hands-On Lab Masters,* which are based on identical rectangles or circles.

ASSESS

Have students work in pairs. Ask them to represent a pair of equivalent fractions with a model. Then have them tell which fractions they have modeled.

Use Exercise 8 to determine whether students understand how to use models to represent and compare fractions and ratios.

Hands-On Lab Masters, p. 48

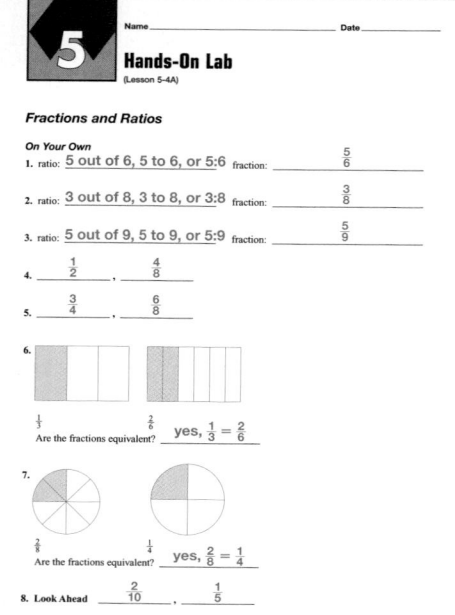

TRY THIS

Work with a partner.

③ Compare $\frac{4}{6}$ and $\frac{2}{3}$.

• Draw two identical rectangles as shown.

• Separate the upper rectangle into 6 equal parts. Separate the lower rectangle into 3 equal parts.

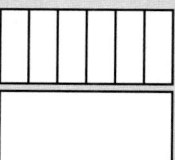

• Shade 4 parts of the upper rectangle and 2 parts of the lower rectangle.

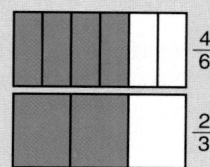

$\frac{4}{6}$

$\frac{2}{3}$

Since the areas shaded are the same, the fractions are equivalent.

ON YOUR OWN

Write the pair of fractions represented by each model.

4. 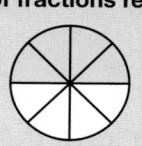 $\frac{1}{2}, \frac{4}{8}$

5. 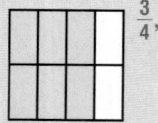 $\frac{3}{4}, \frac{6}{8}$

Draw models for each pair of fractions. Then tell whether the pair are equivalent.

6. $\frac{1}{3}, \frac{2}{6}$ **See Answer Appendix.**

7. $\frac{2}{8}, \frac{1}{4}$ **See Answer Appendix.**

8. ***Look Ahead*** What fraction is represented by the model shown? Name another fraction that is represented by the model. $\frac{2}{10}, \frac{1}{5}$

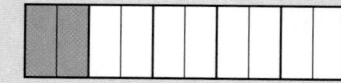

192 Chapter 5 Using Number Patterns, Fractions, and Ratios

Math Journal Have students write a description of how they used models to compare fractions and ratios and why models are helpful. Encourage them to include drawings.

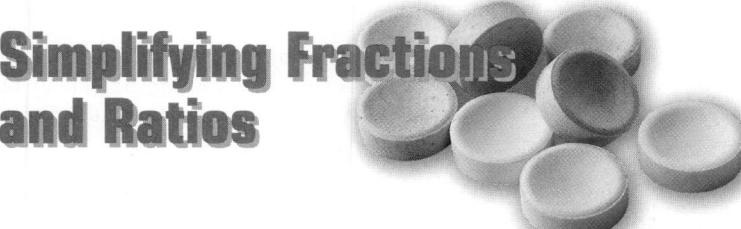

5-4 Simplifying Fractions and Ratios

What you'll learn

You'll learn to express fractions and ratios in simplest form.

When am I ever going to use this?

Knowing how to write fractions and ratios in simplest form can help you analyze baseball statistics.

Word Wise

ratio
equivalent fractions
simplest form

A roll of Smarties contains about 16 candies. Tyra has a roll in which 4 of the 16 candies are pink. You can compare the number of pink candies to the total number of candies using a **ratio**.

A ratio is a comparison of two numbers by division. The ratio that compares 4 to 16 can be written in several ways.

    4 to 16   4 out of 16   4:16

Ratios can be expressed as fractions. In this case, the fraction is $\frac{4}{16}$.

You can write the fraction $\frac{4}{16}$ as $\frac{2}{8}$ and also as $\frac{1}{4}$. Fractions that name the same number are called **equivalent fractions**.

Method 1 Use a model.

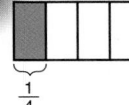 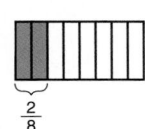

$\frac{1}{4}$ $\frac{2}{8}$

The rectangles are the same size and the same part or fraction of each rectangle is shaded. So, the fractions are equivalent. That is, $\frac{1}{4} = \frac{2}{8}$.

Method 2 Use paper and pencil.

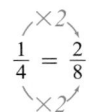 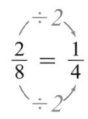

Multiply or divide the numerator and the denominator of a fraction by the same nonzero number.

Examples

Replace each ■ with a number so that the fractions are equivalent.

❶ $\frac{3}{8} = \frac{■}{24}$

Since $8 \times 3 = 24$, multiply the numerator and denominator by 3.

$\frac{3}{8} = \frac{■}{24}$, so $\frac{3}{8} = \frac{9}{24}$.

❷ $\frac{15}{25} = \frac{3}{■}$

Since $15 \div 3 = 5$, divide the numerator and denominator by 5.

$\frac{15}{25} = \frac{3}{■}$, so $\frac{15}{25} = \frac{3}{5}$.

Lesson 5-4 Simplifying Fractions and Ratios   **193**

Multiple Learning Styles

 Kinesthetic Use volunteers to form two groups of students in two different sizes such as 6 and 12. Ask one group to arrange themselves so they show $\frac{2}{6}$. Ask the second group to show an equivalent amount. Ask what other groups they could form to show fractions equivalent to $\frac{2}{6}$ and $\frac{4}{12}$.

Instructional Resources

- *Study Guide Masters,* p. 35
- *Practice Masters,* p. 35
- *Enrichment Masters,* p. 35
- Transparencies 5-4, A and B
- CD-ROM Program
  - Resource Lesson 5-4
  - Interactive Lesson 5-4

Recommended Pacing	
Standard	Days 7 & 8 of 16
Honors	Days 6 & 7 of 15
Block	Day 4 of 8

1 FOCUS

 5-Minute Check
(Lesson 5-3)

Find the GCF of each set of numbers.

1. 22, 46 **2**
2. 36, 54 **18**
3. 72, 39 **3**
4. 48, 56 **8**
5. 12, 45, 60 **3**

 The 5-Minute Check is also available on **Transparency 5-4A** for this lesson.

Motivating the Lesson

Hands-On Activity Have pairs of students hold their hands behind their backs. On the count of three, have them put both hands out in front with any number of fingers outstretched. Have them express in simplest form the total number of fingers as a fraction of the total number possible (20). Have students do ten rounds and compare results.

Transparency 5-4B contains a teaching aid for this lesson.

Using Patterns As students examine Examples 1 and 2, have them identify whether they are seeking a number greater or less than the given numerator (or denominator). This determines whether they should divide or multiply. Expand this pattern to show that when simplifying, they always divide.

In-Class Examples

For Example 1
Replace ■ with a number so that $\frac{3}{8} = \frac{■}{24} \cdot \frac{9}{24}$

For Example 2
Replace ■ with a number so that $\frac{21}{28} = \frac{■}{4} \cdot \frac{3}{4}$

For Example 3
Write $\frac{3}{9}$ in simplest form. $\frac{1}{3}$

For Example 4
Write $\frac{35}{84}$ in simplest form. $\frac{5}{12}$

For Example 5
82 of the first 100 elements in the periodic table were discovered before 1900. Express the ratio 82:100 in simplest form. **41:50**

Teaching Tip After Example 4, stress that both the numerator and the denominator must be divided by the GCF when writing a fraction in simplest form. However, point out that a second division can be performed in case they did not use the GCF.

A fraction is in **simplest form** when the GCF of the numerator and denominator is 1. To write a fraction in simplest form, find the GCF of the numerator and the denominator. Then divide the numerator and denominator by the GCF.

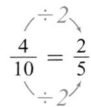
Examples

3 Write $\frac{4}{10}$ in simplest form.

factors of 4: 1, 2, 4   *Find the GCF of the numerator*
factors of 10: 1, 2, 5, 10   *and the denominator.*

The GCF of 4 and 10 is 2.

$$\overset{\div 2}{\frac{4}{10}} = \frac{2}{5} \quad \underset{\div 2}{}$$
Divide the numerator and denominator by the GCF, 2.

Since the GCF of 2 and 5 is 1, the fraction $\frac{2}{5}$ is in simplest form.

4 Write $\frac{18}{21}$ in simplest form.

factors of 18: 1, 2, 3, 6, 9, 18   *Find the GCF of the numerator*
factors of 21: 1, 3, 7, 21   *and the denominator.*

The GCF of 18 and 21 is 3.

$$\overset{\div 3}{\frac{18}{21}} = \frac{6}{7} \quad \underset{\div 3}{}$$
Divide the numerator and denominator by the GCF, 3.

Since the GCF of 6 and 7 is 1, the fraction $\frac{6}{7}$ is in simplest form.

INTEGRATION **5**

Statistics Approximately 26 out of 100 American households have two or more VCR's. Express the ratio 26:100 in simplest form.

factors of 26: 1, 2, 13, 26   *Find the GCF.*
factors of 100: 1, 2, 4, 5, 10, 20, 25, 50, 100

The GCF of 26 and 100 is 2.

$$\frac{\overset{13}{\cancel{26}}}{\underset{50}{\cancel{100}}} = \frac{13}{50} \quad \text{Simplify. Divide both the numerator and the denominator by the GCF, 2.}$$

In simplest form, the ratio is 13:50.

Study Hint

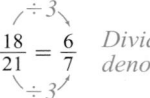

Technology You can use a fraction calculator to simplify fractions. For example, to simplify $\frac{4}{16}$, enter

4 [/] 16 [SIMP] [=].

Repeat [SIMP] [=]

until the N/D → n/d does not appear on the screen.

Classroom Vignette

"I use colored candies to introduce fractions. Each group of students sorts and counts the pieces by color and finds the total number of pieces. Then they write the fraction representing each color's part of the whole bag. We post the results and students look for patterns in the product."

Sharon E. Eaton

Sharon E. Eaton, Teacher
Thunderbolt Middle School
Lake Havasu City, AZ

CHECK FOR UNDERSTANDING

Communicating Mathematics

Read and study the lesson to answer each question. 2. See margin.

1. *Examine* the figure shown.
   a. Write the fraction that represents the shaded part of the figure. $\frac{8}{12}$
   b. Find the GCF of the numerator and denominator of the fraction. 4
   c. Write the fraction in simplest form. $\frac{2}{3}$
   d. Draw a model that represents the fraction in simplest form. **See margin.**

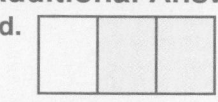

2. *Explain* how you can tell whether a fraction is in simplest form.

Guided Practice

Replace each ■ with a number so that the fractions are equivalent.

3. $\frac{1}{3} = \frac{■}{27}$ 9   4. $\frac{12}{16} = \frac{3}{■}$ 4   5. $\frac{■}{5} = \frac{9}{15}$ 3

State whether each fraction or ratio is in simplest form. If not, write each fraction or ratio in simplest form.

6. $\frac{5}{8}$ yes   7. 4 to 36 no; 1 to 9

8. $\frac{42}{49}$ no; $\frac{6}{7}$ 9. 79:100 yes

10. **Statistics** Approximately 36 of every 100 people in the United States listen to compact discs on portable CD players. Express the ratio 36:100 as a fraction in simplest form. $\frac{9}{25}$

EXERCISES

Practice

Replace each ■ with a number so that the fractions are equivalent.

11. $\frac{1}{2} = \frac{■}{8}$ 4   12. $\frac{10}{15} = \frac{2}{■}$ 3   13. $\frac{3}{4} = \frac{■}{12}$ 9

14. $\frac{14}{18} = \frac{■}{9}$ 7   15. $\frac{8}{9} = \frac{■}{27}$ 24   16. $\frac{30}{35} = \frac{6}{■}$ 7

17. $\frac{13}{78} = \frac{1}{■}$ 6   18. $\frac{4}{5} = \frac{■}{40}$ 32   19. $\frac{57}{60} = \frac{19}{■}$ 20

State whether each fraction or ratio is in simplest form. If not, write each fraction or ratio in simplest form.

23. no; 3 out of 4
24. no; 4 to 11
27. no; 3:5
31. no; 1 out of 4

20. $\frac{4}{20}$ no; $\frac{1}{5}$ 21. $\frac{10}{38}$ no; $\frac{5}{19}$ 22. 8:25 yes   23. 18 out of 24

24. 28 to 77   25. $\frac{15}{100}$ no; $\frac{3}{20}$ 26. $\frac{27}{54}$ no; $\frac{1}{2}$ 27. 21:35

28. 41:85 yes   29. $\frac{42}{50}$ no; $\frac{21}{25}$ 30. $\frac{11}{67}$ yes   31. 12 out of 48

32. Express the ratio 12 out of 144 in simplest form. **1 out of 12**

33. Write a fraction that can be expressed as $\frac{2}{3}$ in simplest form.
    Sample answer: $\frac{4}{6}$

Lesson 5-4 Simplifying Fractions and Ratios **195**

Reteaching the Lesson

Activity Have small groups of students play "Fraction Pairs." Make pairs of equivalent fraction cards. Place all cards facedown. Have players take turns picking two cards and determining whether they are matching pairs. One player may challenge another by drawing a model to prove that a match is incorrect.

Error Analysis

Watch for students who divide by a common factor that is not the greatest common factor and assume the fraction is simplified.

Prevent by asking students to examine the resulting fraction to see whether its numerator and denominator can be divided again.

---

3 PRACTICE/APPLY

Check for Understanding

If students need additional practice or instruction after completing Exercises 1–10, one of these options may be helpful.
- Extra Practice, see p. 570
- Reteaching Activity
- *Study Guide Masters*, p. 35
- *Practice Masters*, p. 35

Assignment Guide
Core: 11–35 odd, 37–42
Enriched: 12–32 even, 34, 35, 37–42

Additional Answers

1d.

2. **Sample answer:** The numerator and the denominator do not have any common factors other than 1.

Study Guide Masters, p. 35

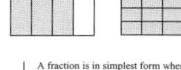

5-4 Study Guide

Simplifying Fractions and Ratios

Fractions that name the same number are called **equivalent fractions.**

Example 1 The same part of each rectangle is shaded. The fractions $\frac{3}{4}$ and $\frac{9}{12}$ are equivalent.

Multiply or divide the numerator and the denominator of a fraction by the same number (not zero) to find an equivalent fraction.

A fraction is in simplest form when the GCF of the numerator and denominator is 1.

Examples 2 Solve $\frac{5}{6} = \frac{?}{18}$.

$\frac{5}{6} = \frac{?}{18}$ Multiply by 3.

$\frac{5}{6} = \frac{15}{18}$

3 Write $\frac{24}{36}$ in simplest form.

$\frac{24}{36} = \frac{2}{3}$ The GCF of 24 and 36 is 12. Divide by 12.

Refer to the figure at the right.

1. What fraction does the shaded part of the figure show? $\frac{12}{15}$
2. Find the GCF of the numerator and denominator of the fraction. 3
3. Write the fraction in simplest form. $\frac{4}{5}$
4. Draw a picture of the fraction in simplified form.

State whether each fraction or ratio is in simplest form. If not, write each fraction or ratio in simplest form.

5. $\frac{9}{21}$ no, $\frac{3}{7}$ 6. $\frac{13}{15}$ yes   7. 49:56 no, 7:8

8. 48 out of 64 no, 3 out of 4   9. $\frac{12}{25}$ no, $\frac{3}{5}$ 10. $\frac{10}{21}$ yes

11. $\frac{23}{33}$ yes   12. 32:42 no, 16:21   13. $\frac{24}{54}$ no, $\frac{4}{9}$

© Glencoe/McGraw-Hill   T35   Mathematics: Applications and Connections, Course 1

Lesson 5-4 **195**

CHAPTER Project

Exercise 36 asks students to advance to the next stage of work on the Chapter Project. Students should determine how to count pages that include both articles and ads.

4 ASSESS

Closing Activity

Modeling Have students model fractions such as $\frac{9}{30}$. Then have them model the simplified equivalent. Have students trade models to check each others' work.

Additional Answers

35a. Gwynn, $\frac{2}{5}$; Ripken, $\frac{1}{3}$;
Gonzalez, $\frac{11}{36}$; Williams, $\frac{8}{21}$;
Bonds, $\frac{1}{3}$; Rodriguez, $\frac{5}{14}$

35b. Ripken and Bonds

Applications and Problem Solving

34. Games Scrabble, a crossword board game, contains 100 tiles, most of which are labeled with a letter. The table shows that the letter O appears on 8 of the tiles.

  a. Write a ratio comparing the number of times the letter O appears to the total number of wooden tiles. $\frac{8}{100}$

  b. Write this ratio in simplest form. $\frac{2}{25}$

A-9	J-1	S-4
B-2	K-1	T-6
C-2	L-4	U-4
D-4	M-2	V-2
E-12	N-6	W-2
F-2	O-8	X-1
G-3	P-2	Y-2
H-2	Q-1	Z-1
I-9	R-6	
BLANK-2		

35. Sports The chart shows the number of at-bats and the number of hits for certain baseball players during the first few games of the season.

Player	At-Bats	Hits
Gwynn	40	16
Ripken	45	15
Gonzalez	36	11
Williams	42	16
Bonds	36	12
Rodriguez	28	10

  a. For each player, write a fraction in simplest form that shows the number of at-bats compared to the total number of hits. **See margin.**

  b. Which two players had the same "batting average"? **See margin.**

interNET CONNECTION
For the latest major league statistics, visit: www.glencoe.com/sec/math/mac/mathnet

36. Working on the CHAPTER Project Refer to page 177.

  a. Count the number of pages of articles, advertisements, and table of contents in each issue of your four chosen magazines. Record the data in a table. **See students' work.**

  b. For each issue, write a ratio in simplest form that compares the number of pages of articles to the total number of pages, the number of pages of ads to the total number of pages, and the number of pages of table of contents to the total number of pages. **See students' work.**

37. Critical Thinking A fraction is equivalent to $\frac{3}{4}$ and the sum of the numerator and denominator is 84. What is the fraction? $\frac{36}{48}$

Mixed Review

38. Find the greatest common factor of 40 and 36. *(Lesson 5-3)* **4**

39. **Test Practice** Which expression is equivalent to $(2.3 \times 4) + (6.7 \times 4)$? *(Lesson 4-2)* **C**
  A $4(2.3 \times 6.7)$
  B $6.7(2.3 + 4)$
  C $4(2.3 + 6.7)$
  D $2.3(4 + 6.7)$

40. Money Matters If Andreina works Monday through Friday baby-sitting and makes $10.75 each day she works, how much will she make in three weeks? *(Lesson 4-1)* **$161.25**

41. Order the decimals 27.025, 26.98, 27.13, 27.9, and 27.131 from least to greatest. *(Lesson 3-3)* **26.98, 27.025, 27.13, 27.131, 27.9**

42. Find $45 \div 3 \times 3 - 7 + 12$. *(Lesson 1-4)* **50**

Practice Masters, p. 35

5-4 Practice

Simplifying Fractions and Ratios

Refer to the figure at the right.

1. What fraction does the shaded part of the figure describe? $\frac{12}{20}$

2. What is the GCF of the numerator and denominator of the fraction? **4**

3. Write the fraction in simplest form. $\frac{3}{5}$

4. Draw a picture of the fraction in simplest form.

State whether each fraction or ratio is in simplest form. If not, write each fraction or ratio in simplest form.

5. $\frac{7}{14}$ no; $\frac{1}{2}$ 6. $\frac{6}{21}$ no; $\frac{2}{7}$ 7. $\frac{45}{72}$ no; $\frac{5}{8}$ 8. 36:45 no; 4:5

9. $\frac{15}{60}$ no; $\frac{1}{4}$ 10. 70:84 no; 5:6 11. $\frac{24}{81}$ no; $\frac{2}{3}$ 12. 31 out of 61 yes

13. 28:36 no; 7:9 14. $\frac{81}{90}$ no; $\frac{9}{10}$ 15. 8 to 21 yes 16. $\frac{14}{35}$ no; $\frac{2}{5}$

17. $\frac{23}{46}$ no; $\frac{1}{2}$ 18. 45 to 48 no; 15 to 16 19. $\frac{12}{27}$ no; $\frac{4}{9}$ 20. 17:51 no; 1:3

21. $\frac{42}{56}$ no; $\frac{3}{4}$ 22. $\frac{66}{121}$ no; $\frac{6}{11}$ 23. 10:25 no; 2:5 24. $\frac{13}{25}$ yes

25. 20:28 no; 5:7 26. $\frac{49}{56}$ no; $\frac{7}{8}$ 27. 13 out of 57 yes 28. $\frac{16}{96}$ no; $\frac{1}{6}$

29. $\frac{63}{108}$ no; $\frac{7}{12}$ 30. 18:31 yes 31. $\frac{49}{70}$ no; $\frac{7}{10}$ 32. $\frac{24}{64}$ no; $\frac{3}{8}$

© Glencoe/McGraw-Hill T35 Mathematics: Applications and Connections, Course 1

Extending the Lesson

Enrichment Masters, p. 35

5-4 Enrichment

Fraction Mysteries

Here is a set of mysteries that will help you sharpen your thinking skills. In each exercise, use the clues to discover the identity of the mystery fraction.

1. My numerator is 6 less than my denominator.
   I am equivalent to $\frac{1}{4}$. $\frac{18}{24}$

2. My denominator is 5 more than twice my numerator.
   I am equivalent to $\frac{1}{3}$. $\frac{5}{15}$

3. The GCF of my numerator and denominator is 3.

4. The GCF of my numerator and denominator is 5.

Activity Fractions have been replaced by decimals in many places today. For example, supermarket scales and gas pumps are now decimal-based. Have students list at least two everyday uses of fractions. Then have them predict how fractions will be used in a hundred years.

COOPERATIVE LEARNING

5-4B Experimental Probability

A Follow-Up of Lesson 5-4

number cube

A number cube is marked with 1, 2, 3, 4, 5, and 6. If you roll it 60 times, what is the *probability*, or chance, it will show a 2? a 4? a 6? *Experimental probability* is a ratio that compares the number of ways a certain outcome occurs to the total number of outcomes. You can conduct an experiment to find the experimental probability of rolling a 2, a 4, and a 6.

TRY THIS

Work with a partner.

To find the probability that the number cube will show a 2, 4, or 6, follow these steps.

- Copy the chart shown.
- Estimate the number of times the number cube will show a 2, 4, and 6 if it is rolled 60 times.
- Roll the number cube 60 times. Make a tally of each roll.
- Suppose the number cube shows a 4 eleven times out of 60 rolls. The ratio, or experimental probability, of showing a 4 is 11 out of 60 or $\frac{11}{60}$. Based on your results, what is the experimental probability of showing a 2? a 4? a 6? Write a ratio to represent the experimental probability.

	2	4	6
estimate			
actual			

1. $\frac{7}{48}$ 2. $\frac{5}{48}$ 3. $\frac{17}{48}$ 4. $\frac{1}{4}$

ON YOUR OWN

Six index cards are labeled L, O, C, K, E, and R. Without looking, Jackie chooses a card, records its letter, and replaces it. She repeats the activity 48 times. The chart shows the results of her experiment.

1. What is the experimental probability of choosing an O?

2. What is the experimental probability of choosing an E?

3. What is the experimental probability of choosing an O or a C?

4. *Reflect Back* Find the experimental probability of choosing a vowel. Write the answer in simplest form.

L	O	C	K	E	R
卌 卌 Ⅲ	卌 Ⅱ	卌 卌 卌	卌 卌	卌 卌	卌 Ⅲ

Lesson 5-4B HANDS-ON **LAB** **197**

Math Journal Have pairs of students write a brief paragraph describing their observations and how experimental probability is affected by the number of trials.

GET READY

Objective Students determine the experimental probability for a given set of data.

Optional Resources
Hands-On Lab Masters
- number cube patterns, p. 21
- worksheet, p. 49
Manipulative Kit
- number cube

MANAGEMENT TIPS

Recommended Time
30 minutes

Getting Started Have students read the opening paragraph. Ask them why experimental probability is being used to solve this problem. Have them describe other problems that could be solved in a similar way.

The **Activity** works through the experimental probability of rolling a number cube. Have students check their tallies after 10, 20, 30, 40, and 50 rolls. Make sure they notice that the more rolls there are, the more equal the tallies become. Experimental probability becomes more accurate with a higher number of outcomes.

Teaching Tip The CD-ROM Program contains a tool for simulating large numbers of rolls of number cubes.

ASSESS

Have students complete Exercises 1–3. You may want to have groups of students perform the same activity and compare results. Combine the results for the whole class, and discuss how the probabilities are affected.

Use Exercise 4 to determine whether students understand the basics of experimental probability.

Instructional Resources

- *Study Guide Masters*, p. 36
- *Practice Masters*, p. 36
- *Enrichment Masters*, p. 36
- Transparencies 5-5, A and B
- *Assessment and Evaluation Masters*, pp. 126, 127
- *School to Career Masters*, p. 5

 CD-ROM Program
- Resource Lesson 5-5

Recommended Pacing

Standard	Day 9 of 16
Honors	Day 8 of 15
Block	Day 5 of 8

1 FOCUS

 5-Minute Check
(Lesson 5-4)

State whether each fraction or ratio is in simplest form. If not, write it in simplest form.

1. $\frac{7}{32}$ **simplest form**

2. $\frac{6}{33}$ **$\frac{2}{11}$**

3. 35:40 **7:8**

4. $\frac{17}{85}$ **$\frac{1}{5}$**

5. Ten Americans in every 100 are hearing impaired. Write this as a fraction in simplest form. **$\frac{1}{10}$**

The 5-Minute Check is also available on **Transparency 5-5A** for this lesson.

Motivating the Lesson

Problem Solving You would like to bring a fruit punch to class to celebrate your birthday. If you triple the recipe, $\frac{5}{8}$ cup grape juice will be $\frac{15}{8}$ cups, $\frac{3}{4}$ cup orange juice will be $\frac{9}{4}$ cups, and $\frac{2}{3}$ cup apple juice will be $\frac{6}{3}$ cups. How would you measure these amounts?

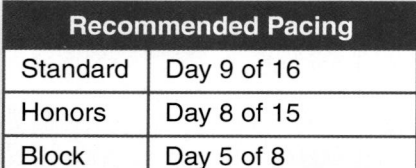

5-5 Mixed Numbers and Improper Fractions

***What* you'll learn**

You'll learn to express mixed numbers as improper fractions and vice versa.

***When* am I ever going to use this?**

Knowing how to express numbers in different forms can help you determine the amounts of the ingredients when increasing a recipe.

Word Wise

mixed number
improper fraction

Tennis anyone? Tennis racquets come in a variety of grip sizes. Three of these are $4\frac{3}{8}$ inches, $4\frac{1}{4}$ inches, and $3\frac{7}{8}$ inches.

The numbers $4\frac{3}{8}$, $4\frac{1}{4}$, and $3\frac{7}{8}$ are examples of **mixed numbers**.

A mixed number shows the sum of a whole number and a fraction.

In the following Mini-Lab, you'll learn that mixed numbers can be written as fractions.

HANDS-ON MINI-LAB

Work with a partner.   ruler

Try This

- Draw a rectangle like the one shown. Shade the rectangle to represent the whole number 1.
- Draw an identical rectangle beside the first one. Separate this rectangle into four equal parts to show fourths. Shade one part to represent $\frac{1}{4}$.
- Separate the whole number portion into $\frac{1}{4}$s.

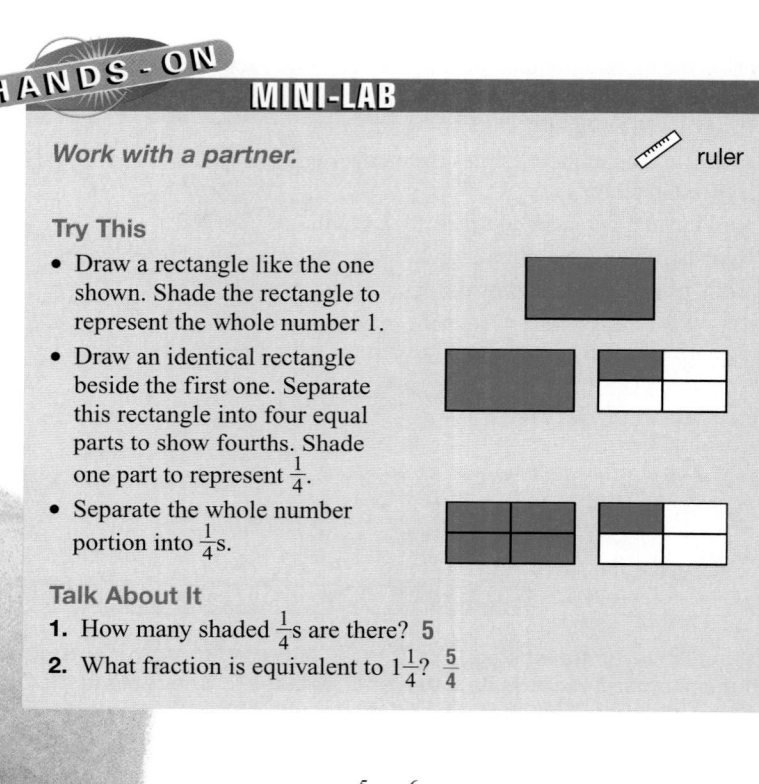

Talk About It

1. How many shaded $\frac{1}{4}$s are there? **5**
2. What fraction is equivalent to $1\frac{1}{4}$? **$\frac{5}{4}$**

A fraction, like $\frac{5}{4}$ or $\frac{6}{5}$, with a numerator that is greater than or equal to the denominator is called an **improper fraction**. To express a mixed number as an improper fraction, you can use a model.

Investigations for the Special Education Student

This blackline master booklet helps you plan for the needs of your special education students by providing long-term projects along with teacher notes. Investigation 5, *Author's Fair*, and Investigation 12, *Yard Sale*, may be used with this chapter.

Express $3\frac{1}{2}$ as an improper fraction.

Change the whole number into halves. Then count the total number of halves.

There are seven $\frac{1}{2}$s in $3\frac{1}{2}$. So, $3\frac{1}{2}$ can be expressed as $\frac{7}{2}$.

Study Hint
Reading Math
A mixed number like $3\frac{1}{2}$ is read as *three and one half*. $3\frac{1}{2}$ means $3 + \frac{1}{2}$.

A shortcut to writing a mixed number as an improper fraction is to first multiply the whole number by the denominator and add the numerator. Then write this sum over the denominator.

$$1\frac{1}{4} = \frac{(1 \times 4) + 1}{4} = \frac{5}{4}$$

Example ➋
CONNECTION

Life Science The body of the vampire bat measures $2\frac{3}{4}$ inches. Express the body length of a vampire bat as an improper fraction.

$2\frac{3}{4} = \frac{(2 \times 4) + 3}{4}$ *Multiply 2 by 4 and add 3.*

$= \frac{11}{4}$ *Then write the result over 4.*

The body length of a vampire bat can be expressed as $\frac{11}{4}$ inches.

You can also express an improper fraction as a mixed number. To do this, divide the numerator by the denominator.

Example ➌

Express $\frac{7}{3}$ as a mixed number.

Divide 7 by 3.

$\begin{array}{r} 2 \\ 3\overline{)7} \\ -6 \\ \hline 1 \end{array}$ *Write the remainder in the numerator of a fraction that has the divisor as the denominator.*

$\frac{7}{3} = 2\frac{1}{3}$

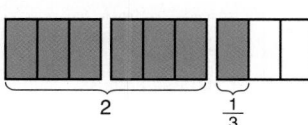

2          $\frac{1}{3}$

 Transparency 5-5B contains a teaching aid for this lesson.

Teaching Tip Before the Mini-Lab, show students that mixed numbers can be written as the sum of a whole number and a fraction by writing $1\frac{1}{4} = 1 + \frac{1}{4}$, $1\frac{3}{16} = 1 + \frac{3}{16}$, and $1\frac{1}{2} = 1 + \frac{1}{2}$.

Using the Mini-Lab Have students model other mixed numbers such as $\frac{9}{4}$, $\frac{6}{5}$, or $\frac{5}{3}$. Make sure students separate the whole number portion into the same number of parts as they separate the extra rectangle.

In-Class Examples
For Example 1
Express $3\frac{5}{8}$ as an improper fraction. $\frac{29}{8}$

For Example 2
The distance between the center of the pitcher's mound and home plate on a baseball diamond is $60\frac{1}{2}$ feet. Express this distance as an improper fraction. $\frac{121}{2}$ feet

For Example 3
a. Express $\frac{9}{8}$ as a mixed number. $1\frac{1}{8}$
b. Express $\frac{7}{2}$ as a mixed number. $3\frac{1}{2}$

Check for Understanding

If students need additional practice or instruction after completing Exercises 1–10, one of these options may be helpful.
- Extra Practice, see p. 570
- Reteaching Activity
- *Study Guide Masters*, p. 36
- *Practice Masters*, p. 36

Assignment Guide

Core: 11–33 odd, 35–39
Enriched: 12–32 even, 33–39
All: Self Test, 1–10

Additional Answers

1. Sample answer: a fraction with a numerator greater than or equal to the denominator

2. $2\frac{3}{4}$ and $1\frac{1}{4}$

3.

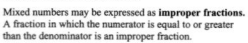
$\frac{9}{4}$; $2\frac{1}{4}$

Study Guide Masters, p. 36

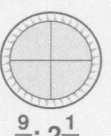
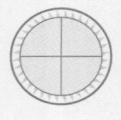
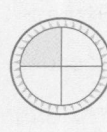

5-5 Study Guide

Name_____ Date_____

Mixed Numbers and Improper Fractions

The figure shows 2 whole circles plus $\frac{1}{3}$ of a circle. You can write the mixed number $2\frac{1}{3}$ to describe the number of circles.

Mixed numbers may be expressed as **improper fractions**. A fraction in which the numerator is equal to or greater than the denominator is an improper fraction.

Example 1 Express $2\frac{1}{3}$ as an improper fraction.

Multiply the whole number by the denominator.	Add the numerator to the product.	Write the sum over the denominator.
$2 \times 3 = 6$	$6 + 1 = 7$	$\frac{7}{3}$

$2\frac{1}{3} = \frac{(2 \times 3) + 1}{3} = \frac{7}{3}$

An improper fraction may be written as a mixed number.

Example 2 Express $\frac{14}{5}$ as a mixed number.

Divide the numerator by the denominator.	Write the quotient as the whole number as the fraction.	Write the remainder over the denominator
$14 \div 5 = 2 \text{ R } 4$	2	$\frac{4}{5}$

$\frac{14}{5} = 2\frac{4}{5}$

Express each mixed number as an improper fraction.

1. $1\frac{2}{5}$ $\frac{7}{5}$ 2. $3\frac{1}{2}$ $\frac{7}{2}$ 3. $7\frac{2}{3}$ $\frac{23}{3}$ 4. $1\frac{7}{8}$ $\frac{15}{8}$

5. $4\frac{3}{4}$ $\frac{19}{4}$ 6. $2\frac{5}{6}$ $\frac{17}{6}$ 7. $5\frac{1}{9}$ $\frac{46}{9}$ 8. $1\frac{2}{7}$ $\frac{9}{7}$

Express each improper fraction as a mixed number.

9. $\frac{12}{7}$ $1\frac{5}{7}$ 10. $\frac{17}{9}$ $1\frac{8}{9}$ 11. $\frac{17}{5}$ $3\frac{2}{5}$ 12. $\frac{25}{6}$ $4\frac{1}{6}$

13. $\frac{13}{3}$ $4\frac{1}{3}$ 14. $\frac{15}{2}$ $7\frac{1}{2}$ 15. $\frac{30}{5}$ 6 16. $\frac{33}{4}$ $8\frac{1}{4}$

© Glencoe/McGraw-Hill   T36   Mathematics: Applications and Connections, Course 1

Communicating Mathematics

Read and study the lesson to answer each question. 1–3. See margin.

1. *Define* improper fraction.

2. *Write* a mixed number and an improper fraction for the model shown.

HANDS-ON MATH

3. *Fold* each of three paper plates into four equal parts to show fourths. Shade nine parts. What improper fraction do the shaded parts represent? What mixed number is equivalent to this improper fraction?

Guided Practice

Express each mixed number as an improper fraction.

4. $2\frac{1}{2}$ $\frac{5}{2}$ 5. $1\frac{2}{5}$ $\frac{7}{5}$ 6. $4\frac{3}{4}$ $\frac{19}{4}$

Express each improper fraction as a mixed number.

7. $\frac{11}{4}$ $2\frac{3}{4}$ 8. $\frac{13}{2}$ $6\frac{1}{2}$ 9. $\frac{24}{6}$ 4

10. Express *nine-fourths* as a mixed number. $2\frac{1}{4}$

Practice

Express each mixed number as an improper fraction.

11. $3\frac{1}{5}$ $\frac{16}{5}$ 12. $2\frac{3}{8}$ $\frac{19}{8}$ 13. $1\frac{1}{8}$ $\frac{9}{8}$ 14. $4\frac{3}{4}$ $\frac{19}{4}$ 15. $1\frac{7}{9}$ $\frac{16}{9}$

16. $4\frac{5}{6}$ $\frac{29}{6}$ 17. $8\frac{2}{3}$ $\frac{26}{3}$ 18. $5\frac{8}{9}$ $\frac{53}{9}$ 19. $7\frac{3}{5}$ $\frac{38}{5}$ 20. $12\frac{4}{7}$ $\frac{88}{7}$

Express each improper fraction as a mixed number.

21. $\frac{19}{3}$ $6\frac{1}{3}$ 22. $\frac{13}{8}$ $1\frac{5}{8}$ 23. $\frac{15}{8}$ $1\frac{7}{8}$ 24. $\frac{17}{5}$ $3\frac{2}{5}$ 25. $\frac{29}{4}$ $7\frac{1}{4}$

26. $\frac{23}{6}$ $3\frac{5}{6}$ 27. $\frac{19}{9}$ $2\frac{1}{9}$ 28. $\frac{32}{8}$ 4 29. $\frac{25}{7}$ $3\frac{4}{7}$ 30. $\frac{48}{11}$ $4\frac{4}{11}$

31. Express *six and seven-eighths* as an improper fraction. $\frac{55}{8}$

32. Find the mixed number that is equivalent to *thirty-eight ninths*. $4\frac{2}{9}$

Applications and Problem Solving
33. See margin.

33. *Sports* To win the U.S. Triple Crown, a horse must win the Kentucky Derby, the Preakness Stakes, and the Belmont Stakes. The table shows the distance of each of these races. Express each distance as an improper fraction.

Race	Distance
Kentucky Derby	$1\frac{1}{4}$ mi
Preakness Stakes	$1\frac{3}{16}$ mi
Belmont Stakes	$1\frac{1}{2}$ mi

200 Chapter 5 Using Number Patterns, Fractions, and Ratios

■ Reteaching the Lesson ■

Activity Have students take turns rolling two number cubes, once for the numerator and again for the denominator. Have them record the fraction. Then, if the fraction is improper, have them express it as a mixed number.

Additional Answer

33. Kentucky Derby—$\frac{5}{4}$; Preakness Stakes—$\frac{19}{16}$; Belmont Stakes—$\frac{3}{2}$

34. Entertainment *Snow White and the Seven Dwarfs,* created in 1937, was Walt Disney's first full-length film. Over the years, other all-time classics were made. The table shows the running time of a few of Disney's movies.

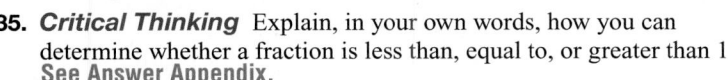

Movie	Running Time (min)
The Lion King	88
Cinderella	76
Snow White and the Seven Dwarfs	84
Bambi	69

a. For each movie, write a fraction that compares the running time to the number of minutes in an hour.

b. Express each fraction as a mixed number in simplest form. **a–b. See Answer Appendix.**

35. Critical Thinking Explain, in your own words, how you can determine whether a fraction is less than, equal to, or greater than 1. **See Answer Appendix.**

Mixed Review

36. Express $\frac{35}{42}$ in simplest form. *(Lesson 5-4)* $\frac{5}{6}$

37. Find the prime factorization of 204. *(Lesson 5-2)* $2^2 \times 3 \times 17$

38. **Test Practice** Jalisa is planning to buy baseball cards that cost $2.27 per pack including tax. How many packs of baseball cards can she buy with $32? *(Lesson 4-6)* **C**

    A 12   **B** 13   **C** 14   **D** 15   **E** Not Here

39. Write *thirteen hundredths* as a decimal. *(Lesson 3-1)* **0.13**

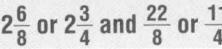

CHAPTER 5 — Mid-Chapter Self Test

State whether each number is divisible by 2, 3, 5, 6, 9, or 10. *(Lesson 5-1)*

1. 435 **3, 5**   **2.** 827 **none**   **3.** 1,090 **2, 5, 10**

Find the prime factorization of each number. *(Lesson 5-2)*

4. 36 $2^2 \times 3^2$ **5.** 88 $2^3 \times 11$ **6.** 105 $3 \times 5 \times 7$

7. Find the GCF of 90 and 36. *(Lesson 5-3)* **18**

8. Express $\frac{12}{28}$ in simplest form. *(Lesson 5-4)* $\frac{3}{7}$

9. Geography Eight out of 50 states in the United States are located in the South Atlantic region. Express the ratio 8:50 in simplest form. *(Lesson 5-4)* **4:25**

10. Write a mixed number and an improper fraction for the model shown. *(Lesson 5-5)*

$2\frac{6}{8}$ or $2\frac{3}{4}$ and $\frac{22}{8}$ or $\frac{11}{4}$

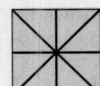 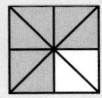

Lesson 5-5 Mixed Numbers and Improper Fractions **201**

Extending the Lesson

Enrichment Masters, p. 36

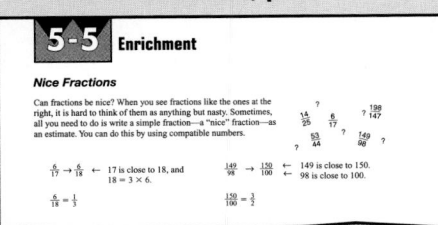

Activity Ask students to bring in recipes that include mixed numbers as measurements. Then have students practice writing the ingredients as improper fractions. Ask them why this skill might be necessary when making the recipe.

4 ASSESS

Closing Activity

Modeling Give the class a stopwatch. At the beginning of the period, have a student start the watch, stop it at some point after one minute, and record the time passed in minutes and seconds. Have the students pass the watch to others, who will also start and stop the watch and record the time. At the end of the period, have students write their times on the chalkboard as both mixed numbers and improper fractions.

Chapter 5, Quiz B (Lessons 5-4 and 5-5) is available in the *Assessment and Evaluation Masters,* p. 127.

Mid-Chapter Test (Lessons 5-1 through 5-5) is available in the *Assessment and Evaluation Masters,* p. 126.

Mid-Chapter Self Test

The Mid-Chapter Self Test reviews the concepts in Lessons 5-1 through 5-5. Lesson references are given so students can review concepts not yet mastered.

Practice Masters, p. 36

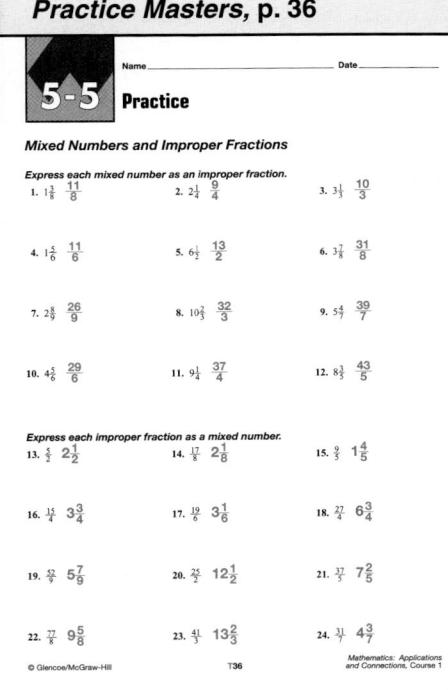

- *Study Guide Masters*, p. 37
- *Practice Masters*, p. 37
- *Enrichment Masters*, p. 37
- Transparencies 5-6, A and B
   CD-ROM Program
  - Resource Lesson 5-6

Recommended Pacing	
Standard	Day 10 of 16
Honors	Day 9 of 15
Block	Day 5 of 8

1 FOCUS

 5-Minute Check
(Lesson 5-5)

Express each mixed number as an improper fraction.

1. $3\frac{2}{7}$ $\frac{23}{7}$
2. $1\frac{5}{6}$ $\frac{11}{6}$

Express each improper fraction as a mixed number.

3. $\frac{14}{3}$ $4\frac{2}{3}$
4. $\frac{38}{12}$ $3\frac{1}{6}$
5. What improper fraction represents the mixed number six and five ninths? $\frac{59}{9}$

The 5-Minute Check is also available on **Transparency 5-6A** for this lesson.

Motivating the Lesson
Hands-On Activity Have students use inch rulers to measure things in the classroom such as doors, desks, books, paper, chalk, and pencils. Have them determine whether it is best to express their measurements in inches, feet, or yards and then measure to the nearest whole unit.

5-6

What you'll learn
You'll learn to measure line segments and objects with a ruler divided in halves, fourths, and eighths.

When am I ever going to use this?
Knowing how to use the customary system can help you measure everyday objects.

Word Wise
inch
foot
yard
mile

Integration: Measurement
Length in the Customary System

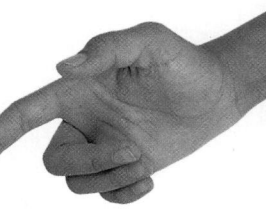

Did you know that, for many people, the length of their arm is about 8 times the length of their index finger?

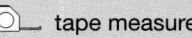
MINI-LAB

Work with a partner. 🔲 tape measure 〰 string

Try This
- On a long piece of string, mark off eight segments using the length of your index finger. Measure this length to the nearest inch. Record.
- Measure the length of your arm from your shoulder to the end of your index finger to the nearest inch. Record.

Talk About It 1–2. See students' work.
1. How does the total length of the eight segments on the string compare to the actual length of your arm?
2. Divide the actual length of your arm by 8 to get an estimate of the length of your index finger. How does this estimate compare to the actual length of your index finger?

One of the most commonly used customary units of length is the **inch**. Some others are the **foot**, **yard**, and **mile**.

$$1 \text{ foot (ft)} = 12 \text{ inches (in.)}$$

$$1 \text{ yard (yd)} = 3 \text{ feet or } 36 \text{ inches}$$

$$1 \text{ mile (mi)} = 5{,}280 \text{ feet or } 1{,}760 \text{ yards}$$

To change from one unit to another unit in the customary system, you can use either multiplication or division.

Examples

1 4 ft = _?_ in.
Since 1 ft = 12 in., it follows that 4 feet equals 4×12, or 48 inches.

4 ft = 48 in.

2 15 ft = _?_ yd
Since 1 yd = 3 ft, it follows that 15 feet equals $15 \div 3$, or 5 yards.

15 ft = 5 yd

Sometimes you need to measure objects using units less than an inch. Most rulers are separated into eighths.

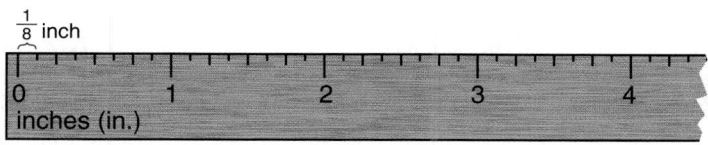
$\frac{1}{8}$ inch

inches (in.)

Transparency 5-6B contains a teaching aid for this lesson.

Using the Mini-Lab You may suggest that once students have marked off one index length, they can fold the string to mark off the other seven lengths instead of laying their finger down each time.

Examples

③ Draw a line segment measuring $1\frac{3}{4}$ inches.

Find $1\frac{3}{4}$ on the ruler.
Draw a line segment
from 0 to $1\frac{3}{4}$.

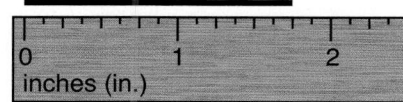
inches (in.)

INTEGRATION ④ **Geometry**

a. Find the length of $\overline{AD}$ to the nearest fourth inch.
b. Find the length of $\overline{AE}$ to the nearest eighth inch.
c. Find the length of $\overline{BC}$ to the nearest inch.

Study Hint

Reading Math The notation $\overline{AB}$ is read as line segment AB.

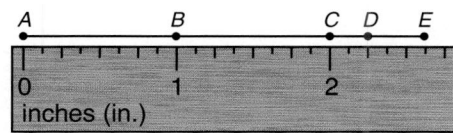
inches (in.)

a. To the nearest fourth inch, the length of $\overline{AD}$ is $2\frac{1}{4}$ inches.

b. To the nearest eighth inch, the length of $\overline{AE}$ is $2\frac{5}{8}$ inches.

c. To the nearest inch, the length of $\overline{BC}$ is $2 - 1$, or 1 inch.

In-Class Examples

For Example 1
6 ft = _?_ in. **72**

For Example 2
36 ft = _?_ yd **12**

For Example 3
Draw a line segment measuring $2\frac{1}{4}$ inches. **Check students' drawings.**

For Example 4
Have students draw three line segments. Have them measure one to the nearest half inch, one to the nearest fourth inch, and the third to the nearest eighth inch. **Check students' drawings.**

Teaching Tip Before Example 3, if your students' rulers are divided into sixteenths, use an overhead transparency to show how to read it.

CHECK FOR UNDERSTANDING

Communicating
Mathematics

Read and study the lesson to answer each question.

1. **Tell** how you would change 24 inches to feet. **Divide 24 by 12.**

2. **Explain** why $\frac{4}{16}$ inch is the same measure as $\frac{1}{4}$ inch. **See margin.**

HANDS-ON
MATH

3. A cubit is about 18 in. A span is about 8 to 9 in.

3. A *cubit* and a *span* are examples of *nonstandard units* of measurement. A cubit is the measure from a person's elbow to the end of the middle finger. A span is the measure from the end of the thumb to the end of the little finger as the hand is outstretched. **Use** a ruler or tape measure to find the approximate length of your cubit and your span in the customary system.

Lesson 5-6 Integration: Measurement   Length in the Customary System   **203**

---

3 PRACTICE/APPLY

Check for Understanding

If students need additional practice or instruction after completing Exercises 1–10, one of these options may be helpful.
- Extra Practice, see p. 570
- Reteaching Activity
- *Transition Booklet,* pp. 29–30
- *Study Guide Masters,* p. 37
- *Practice Masters,* p. 37

Additional Answer
2. In simplest form, $\frac{4}{16} = \frac{1}{4}$.

Reteaching the Lesson

Activity Write lengths, such as $1\frac{3}{8}$ inches, $2\frac{1}{4}$ inches, and so on, on the chalkboard. Have pairs of students create a design by taking turns drawing line segments of the given lengths. Have them use all of the segments at least once.

Error Analysis
Watch for students who read a ruler incorrectly.
Prevent by showing them how the lengths of the notches correspond to the fractional unit of measure.

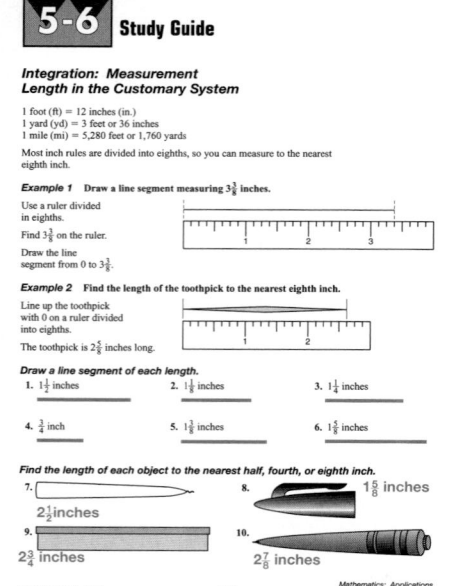
Guided Practice

Complete.

4. 7 ft = _?_ in. **84**

5. 36 in. = _?_ ft **3**

Draw a line segment of each length.

6. $1\frac{1}{8}$ inches

7. $2\frac{1}{4}$ inches

Find the length of each line segment or object to the nearest half, fourth, or eighth inch.

8. ━━━━━━━━━━━━
$2\frac{1}{4}$ in.

9.
$\frac{7}{8}$ in.

10. **Auto Mechanics** Mrs. Huang uses wrenches of different sizes when working on cars. She needs to choose a wrench that will tighten the bolt shown. What size wrench will she need: $\frac{1}{2}$ inch, $\frac{5}{8}$ inch, or $\frac{3}{4}$ inch? $\frac{1}{2}$ in.

EXERCISES

Practice

Complete.

11. 60 in. = _?_ ft **5**

12. 3 ft = _?_ yd **1**

13. 3 yd = _?_ in. **108**

14. 78 in. = _?_ ft $6\frac{1}{2}$

15. 2 mi = _?_ ft **10,560**

16. 225 ft = _?_ yd **75**

Draw a line segment of each length. 17–22. See Answer Appendix.

17. $1\frac{1}{4}$ inches

18. $1\frac{1}{2}$ inches

19. $\frac{3}{8}$ inch

20. 2 inches

21. $1\frac{5}{8}$ inches

22. $2\frac{1}{2}$ inches

Find the length of each line segment or object to the nearest half, fourth, or eighth inch.

23. ━━━━
$\frac{1}{2}$ in.

24.
$\frac{3}{4}$ in.

25. ⌐─┐
$\frac{1}{2}$ in.

26.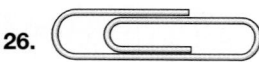
$1\frac{3}{8}$ in.

27. ━━━━━━━━━━
$1\frac{1}{4}$ in.

28. ━━━━━━━━━━━━━━━
$2\frac{1}{2}$ in.

29. Which is greater: 16 inches or $1\frac{1}{2}$ feet? $1\frac{1}{2}$ ft

30. Order the measurements 1 yard, 24 inches, 4 feet, and 40 inches from least to greatest. **24 in., 1 yd, 40 in., 4 ft**

31. Order the measurements $\frac{1}{2}$ inch, $\frac{1}{8}$ inch, 1 inch, and $\frac{1}{4}$ inch from greatest to least. **See margin.**

Applications and Problem Solving

32. Postal Service The first animated character to be featured on a 32-cent stamp issued by the United States Postal Service was Bugs Bunny. What is the measure of the width and height of the stamp to the nearest eighth inch? **width: 1 in., height: $1\frac{5}{8}$ in. or $1\frac{1}{2}$ in.**

33. Geometry Find the measure of each line segment.

a. $\overline{MN}$ **1 in.**   b. $\overline{MO}$ **$1\frac{3}{4}$ in.**   c. $\overline{MP}$ **$2\frac{1}{8}$ in.**

d. $\overline{MR}$ **$3\frac{7}{8}$ in.**   e. $\overline{NS}$ **3 in.**   f. $\overline{NO}$ **$\frac{3}{4}$ in.**

34. Critical Thinking How many eighth inches are in a foot? How many fourth inches are in a yard? **96 eighths; 144 fourths**

Mixed Review

35. Express $5\frac{3}{8}$ as an improper fraction. *(Lesson 5-5)* **$\frac{43}{8}$**

36. Find the greatest common factor of 45 and 75. *(Lesson 5-3)* **15**

37. **Test Practice** The sides of a regulation football field measure 120 yards and about 53.3 yards. About how far does a person have to walk to go all the way around a regulation football field? *(Lesson 4-4)* **C**

A 173.3 yd   B 246.6 yd   C 346.6 yd   D 6,396 yd

38. Solve the equation $n = 3.569 + 781.2$. *(Lesson 3-6)* **784.769**

MATH IN THE MEDIA

Peanuts

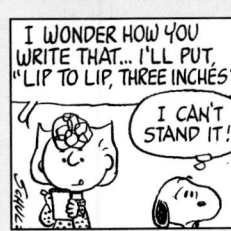

1. Do you agree or disagree with the method Sally used to measure Snoopy's mouth? Explain. **See students' work.**

2. Tell how Sally got a measurement of 3 inches. **Subtract 6 in. from 9 in.**

Lesson 5-6 Integration: Measurement   Length in the Customary System   **205**

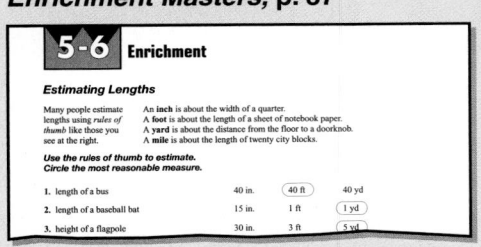

Instructional Resources
- *Study Guide Masters*, p. 38
- *Practice Masters*, p. 38
- *Enrichment Masters*, p. 38
- Transparencies 5-7, A and B
- *Technology Masters*, p. 10

 CD-ROM Program
- Resource Lesson 5-7
- Interactive Lesson 5-7

Recommended Pacing	
Standard	Day 11 of 16
Honors	Day 10 of 15
Block	Day 6 of 8

1 FOCUS

 5-Minute Check
(Lesson 5-6)

Complete.
1. $\frac{1}{2}$ mi = _?_ ft **2,640**
2. 3 yd = _?_ in. **108**

Draw a line segment of each length.

3. $1\frac{1}{4}$ inches **See students' drawings.**

4. $\frac{5}{8}$ inch **See students' drawings.**

5. Find the length of a new pencil to the nearest eighth inch. **Sample answer: $7\frac{1}{2}$ in.**

The 5-Minute Check is also available on Transparency 5-7A for this lesson.

Motivating the Lesson

Problem Solving Ask students how they would solve this problem. *Alec is making a design with 50 wooden squares in a row. He is painting every other square blue and every fifth square red. Which squares will have two layers of paint?* **10, 20, 30, 40, 50**

5-7 Least Common Multiple

What **you'll learn**
You'll learn to find the least common multiple of two or more numbers.

When **am I ever going to use this?**
You can use the least common multiple to help you lay tile on a floor or wall.

Word Wise
multiple
common multiples
least common multiple (LCM)

José is shopping for party supplies. He finds a package of 10 plates, a package of 16 napkins, and a package of 8 cups. What is the least number of packages of plates, napkins, and cups José can buy so that he has the same number of plates, napkins, and cups? *This problem will be solved in Example 3.*

To answer this question, you can use multiples. A **multiple** of a number is the product of the number and any whole number.

HANDS-ON MINI-LAB

Work with a partner. scissors ruler

Try This
- Cut six strips of paper 2 inches long by 1 inch wide.
- Cut six strips of paper 3 inches long by 1 inch wide.
- Place the 2-inch strips of paper end to end to make a train.
- Repeat the process with the 3-inch strips of paper. Place this train below the 2-inch strip train.

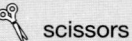 ←2 in.→
 ← 3 in.→

Talk About It

1. See students' work;
2: 0, 2, 4, 6, 8, 10;
3: 0, 3, 6, 9, 12, 15

1. Sketch a diagram of the trains. What are the first six multiples of 2 and of 3?
2. Describe where the ends of the strips of paper are lined up. What are the measures of these lengths? **0 in., 6 in., and 12 in.**
3. At what measurement, other than zero, do the strips of paper line up for the first time? **6 in.**

Notice that 0, 6, and 12 are multiples of both 2 and 3. They are called **common multiples**. The least of the common multiples of two or more numbers, other than zero, is called the **least common multiple (LCM)**. The least common multiple of 2 and 3 is 6.

Example ❶ Determine whether 65 is a multiple of 13.

multiples of 13: 0, 13, 26, 39, 52, 65, 78,...   *$13 \times 0 = 0$*
                                                  $13 \times 1 = 13$
So, 65 is a multiple of 13.   *$13 \times 2 = 26$*
                                                  $\vdots$

206 Chapter 5 Using Number Patterns, Fractions, and Ratios

To find the LCM, you can use either one of the following methods.

Method 1 Make a list.

- List several multiples of each number.
- Identify the common multiples.
- The least of the common multiples is the LCM.

Method 2 Use a calculator.

- List the multiples of the greater number.
- Divide the multiples of the greater number by the lesser number until you get a whole number quotient.

Method 3 Use prime factorization.

- Write the prime factorization of each number.
- Identify all common prime factors. Then find the product of the prime factors using each common prime factor only once and any remaining factors. This product is the LCM.

Examples

2 Find the LCM of 4 and 7.

Method 1 Make a list.

multiples of 4:
0, 4, 8, 12, 16, 20, 24, 28,...

multiples of 7:
0, 7, 14, 21, 28, 35, 42,...

Method 2 Use a calculator.

multiples of 7:
0, 7, 14, 21, 28, 35, 42,...

7 ÷ 4 = 1.75

14 ÷ 4 = 3.5

21 ÷ 4 = 5.25

28 ÷ 4 = 7 ✓

The LCM of 4 and 7 is 28.

Study Hint
Problem Solving You can make a list to find the multiples.

APPLICATION **3** **Shopping** Refer to the beginning of the lesson. What is the least number of packages of plates, napkins, and cups José can buy so that he has the same number of plates, napkins, and cups?

Explore You need to find the LCM of 10, 16, and 8.

Plan Use prime factorization.

Solve
$10 = 2 \times 5$
$16 = 2 \times 2 \times 2 \times 2$
$8 = 2 \times 2 \times 2$

The LCM of 10, 16, and 8 is $2 \times 2 \times 2 \times 2 \times 5$, or 80.

Since $10 \times 8 = 80$, he will need 8 packages of plates.
Since $16 \times 5 = 80$, he will need 5 packages of napkins.
Since $8 \times 10 = 80$, he will need 10 packages of cups.

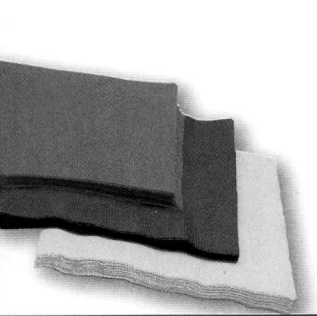

(continued on the next page)

Lesson 5-7 Least Common Multiple **207**

Transparency 5-7B contains a teaching aid for this lesson.

Using the Mini-Lab It is possible to cut all of the strips from one $8\frac{1}{2} \times 11$-inch sheet of paper. Have students begin by cutting eight 1-inch-wide strips as long as the paper. Then have them cut each strip to the lengths given in the instructions. There will be paper leftover.

Reading Mathematics
Emphasize to students that the word *multiple* is from the same root as the word *multiply*. It is easy to confuse factors and multiples.

In-Class Examples

For Example 1
Determine whether 98 is a multiple of 14. **yes**

For Example 2
a. Find the LCM of 5 and 10.
10
b. Find the LCM of 4 and 6. **12**

For Example 3
The sixth grade is decorating the gym for a school dance. They plan to put blue balloons on every second chair, pink balloons on every fourth chair, and purple balloons on every fifth chair. What is the first chair that will have three balloons?
20th chair

Teaching Tip Before Example 2, explain that the LCM can be, but is not necessarily, the product of the two numbers. Also, the LCM can be the same as the higher number.

Teaching Tip Students can use a calculator to find multiples of a number. Remember that multiplication is repeated addition. On the TI-Explorer Plus, press ⊕, the number, and [OP₁]. Then press 0 ⊕ [OP₁] to get the first multiple. Continue pressing [OP₁] to get additional multiples.

Check for Understanding

If students need additional practice or instruction after completing Exercises 1–11, one of these options may be helpful.
- Extra Practice, see p. 571
- Reteaching Activity
- *Study Guide Masters,* p. 38
- *Practice Masters,* p. 38

Assignment Guide
Core: 13–35 odd, 37–42
Enriched: 12–34 even, 35–42

Additional Answers

1. The least common multiple is the least number, other than zero, that is a multiple of two or more numbers.

2. List the multiples of both 8 and 12. Then identify the least common multiple.

11. 4 packages of hot dogs, 5 packages of hot dog buns, 2 packages of plates

Study Guide Masters, p. 38

5-7 **Study Guide**

Least Common Multiple

A **multiple** of a number is the product of the number and any whole number. The least of the common multiples of two or more numbers, other than zero, is called the **least common multiple (LCM).**

Examples **1 Find the LCM of 9 and 12.**

multiples of 9: 0, 9, 18, 27, 36, 45, 54, 63, 72, 81, . . .
multiples of 12: 0, 12, 24, 36, 48, 60, 72, 84, . . .
0, 36, and 72 are common multiples of 9 and 12.
The LCM is 36.

2 Find the LCM of 4, 6, and 8.

multiples of 4: 0, 4, 8, 12, 16, 20, 24, 28, 32, 36, 40, 44, 48, . . .
multiples of 6: 0, 6, 12, 18, 24, 30, 36, 42, 48, . . .
multiples of 8: 0, 8, 16, 24, 32, 40, 48, . . .

The LCM of 4, 6, and 8 is 24.

Determine whether the first number is a multiple of the second number.

| 1. 48, 7 no | 2. 32, 6 no | 3. 144, 4 yes | 4. 60, 12 yes |
| 5. 140, 8 no | 6. 90, 18 yes | 7. 98, 4 no | 8. 54, 6 yes |

Find the LCM for each set of numbers.

9. 12, 20 60	10. 15, 6 30	11. 7, 9 63	12. 15, 9 45
13. 6, 10 30	14. 4, 13 52	15. 18, 27 54	16. 30, 20 60
17. 4, 8, 10 40	18. 3, 5, 7 105	19. 2, 6, 8 24	20. 5, 15, 75 75

© Glencoe/McGraw-Hill   T38   Mathematics: Applications and Connections, Course 1

208 Chapter 5

Examine Each package of plates contains 10 plates. So, eight packages will contain 10 × 8, or 80 plates.

Each package of napkins contains 16 napkins. So, five packages will contain 16 × 5, or 80 napkins.

Each package of cups contains 8 cups. So, ten packages will contain 8 × 10, or 80 cups.

Since he will have exactly 80 of each item, the answer is correct.

CHECK FOR UNDERSTANDING

Communicating Mathematics

Read and study the lesson to answer each question.

1. *Define* least common multiple. **See margin.**

2. *Explain* to a classmate how to find the LCM of 8 and 12. **See margin.**

3. Using paper strips, make a model to find the LCM of 5 and 6. What is the LCM of 5 and 6? **See student's work; 30**

Guided Practice

Determine whether the first number is a multiple of the second number.

4. 40; 5 **yes**   5. 32; 8 **yes**   6. 133; 3 **no**

Find the LCM for each set of numbers.

7. 3, 9 **9**   8. 5, 12 **60**   9. 2, 11 **22**   10. 9, 12, 15 **180**

11. *Party Planning* Refer to the chart shown. What is the least number of packages of each item that should be purchased so that there are the same number of hot dogs, hot dog buns, and plates?

The Food Shop	
Item	**Quantity**
Hot Dogs	10/pkg
Hot Dog Buns	8/pkg
Plates	20/pkg

EXERCISES

Practice

Determine whether the first number is a multiple of the second number.

12. 15; 9 **no**   13. 30; 6 **yes**   14. 14; 8 **no**

15. 32; 16 **yes**   16. 27; 4 **no**   17. 35; 3 **no**

18. 84; 7 **yes**   19. 115; 3 **no**   20. 142; 7 **no**

Find the LCM for each set of numbers. 31. **280**   32. **224**

21. 6, 12 **12**   22. 5, 8 **40**   23. 3, 7 **21**   24. 18, 24 **72**

25. 16, 20 **80**   26. 12, 21 **84**   27. 13, 16 **208**   28. 21, 28 **84**

29. 4, 6, 9 **36**   30. 15, 25, 75 **75**   31. 10, 35, 40   32. 14, 28, 32

33. What is the least common multiple of 3, 9, and 18? **18**

34. Find the LCM of 7, 8, and 28. **56**

■ Reteaching the Lesson ■

Activity Have students play "Gotcha LCM!" using the following rules. (1) Roll two number cubes. Record the numbers. (2) Use a calculator to find the first twelve multiples of the greater number. Record them. (3) Find the multiples of the other number until you have a match. Say, "Gotcha LCM!"

35. *Mechanics* Two interlocking gears have 48 teeth and 28 teeth, respectively. The lead teeth of each gear are opposite each other. How many complete rotations must the smaller gear make before the lead teeth are lined up again? **12**

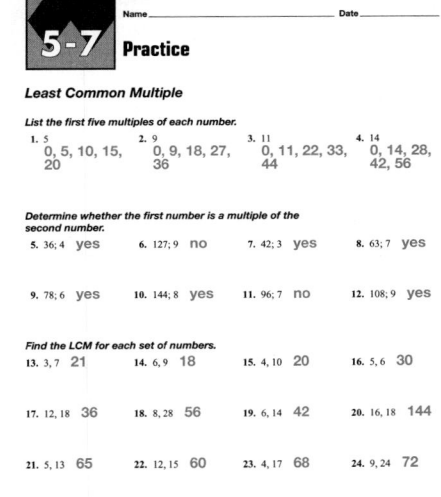

36. *Remodeling* Mrs. Guzman is replacing the tile on the wall in her bathroom. She places 3-inch tiles in the first row, 4-inch tiles in the second row, and 5-inch tiles in the third row. At what point will all three tiles be lined up? **60 inches**

37. *Critical Thinking* The LCM of two numbers is $2^3 \times 3^2$. Find two pairs of numbers that fit this description. **Sample answer: 18 and 24, 36 and 72**

Mixed Review

38. *Measurement* Draw a line segment measuring $\frac{5}{8}$ inch. *(Lesson 5-6)*
See Answer Appendix.

39. Express the fraction $\frac{47}{5}$ as a mixed number. *(Lesson 5-5)* $9\frac{2}{5}$

40. Test Practice Dom is planning a trip to England and wishes to exchange his U.S. currency for British pound. If one U.S. dollar equals 0.623 pounds, about how many pounds will Dom get for $126?
(Lesson 4-1) **B**

A 86   **B** 79   **C** 75   **D** 57   **E** Not Here

41. Estimate $6.291 + 234.38$. *(Lesson 3-5)* **Sample answer: 6 + 234 = 240**

42. *Statistics* Find the mean for the scores 19, 17, 28, 32, and 23. *(Lesson 2-7)* **23.8**

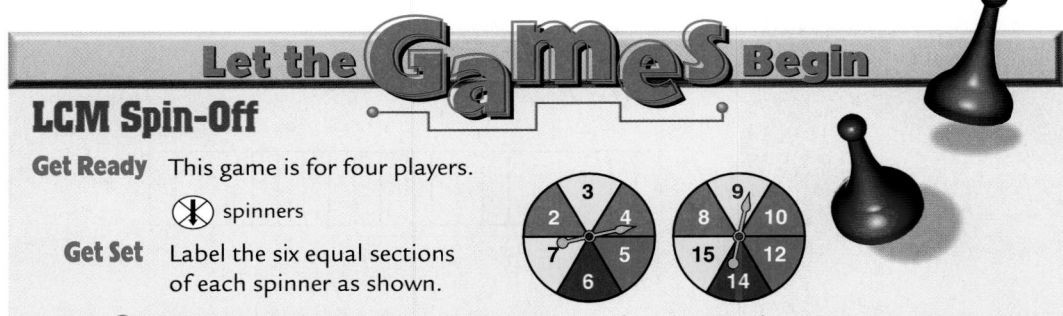

LCM Spin-Off

Get Ready   This game is for four players.

   ✦ spinners

Get Set   Label the six equal sections of each spinner as shown.

Go   Form two pairs of players. The members of one pair each spin a spinner. The members in the other pair compete to be the first to name the LCM of the two numbers. The first member to correctly name the LCM gets 1 point. Pairs take turns spinning the spinner and guessing the LCM. The first pair to get 5 points wins.

🖳 **inter**NET
CONNECTION   Visit www.glencoe.com/sec/math/mac/mathnet for more games.

■ **Extending the Lesson** ■

Enrichment Masters, **p. 38**

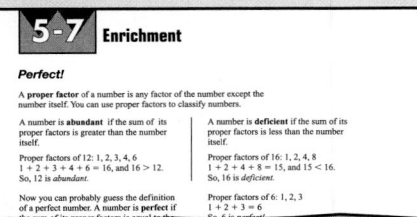

5-7 Enrichment

Perfect!

A **proper factor** of a number is any factor of the number except the number itself. You can use proper factors to classify numbers.

A number is **abundant** if the sum of its proper factors is greater than the number itself.	A number is **deficient** if the sum of its proper factors is less than the number itself.
Proper factors of 12: 1, 2, 3, 4, 6 $1 + 2 + 3 + 4 + 6 = 16$, and $16 > 12$. So, 12 is *abundant*.	Proper factors of 16: 1, 2, 4, 8 $1 + 2 + 4 + 8 = 15$, and $15 < 16$. So, 16 is *deficient*.
Now you can probably guess the definition of a perfect number. A number is **perfect** if the sum of its proper factors is equal to the	Proper factors of 6: 1, 2, 3 $1 + 2 + 3 = 6$ So, 6 is *perfect*.

The students who spin the spinner need to check the answers of the other pair. Have students make a table to keep track of each turn.

Teaching Tip As an extension of Exercise 35, have students experiment with a *Spirograph* to determine its gears' LCM and GCF.

4 ASSESS

Closing Activity

Writing Have students write a sentence or two explaining how to find the LCM of two or more numbers.

Practice Masters, **p. 38**

Name _____   Date _____

5-7 Practice

Least Common Multiple

List the first five multiples of each number.

1. 5   **2.** 9   **3.** 11   **4.** 14
0, 5, 10, 15, 20   0, 9, 18, 27, 36   0, 11, 22, 33, 44   0, 14, 28, 42, 56

Determine whether the first number is a multiple of the second number.

5. 36; 4 **yes**   **6.** 127; 9 **no**   **7.** 42; 3 **yes**   **8.** 63; 7 **yes**

9. 78; 6 **yes**   **10.** 144; 8 **yes**   **11.** 96; 7 **no**   **12.** 108; 9 **yes**

Find the LCM for each set of numbers.

13. 3, 7 **21**   **14.** 6, 9 **18**   **15.** 4, 10 **20**   **16.** 5, 6 **30**

17. 12, 18 **36**   **18.** 8, 28 **56**   **19.** 6, 14 **42**   **20.** 16, 18 **144**

21. 5, 13 **65**   **22.** 12, 15 **60**   **23.** 4, 17 **68**   **24.** 9, 24 **72**

25. 8, 13 **104**   **26.** 15, 18 **90**   **27.** 12, 14 **84**   **28.** 7, 13 **91**

29. 3, 5, 12 **60**   **30.** 6, 16, 24 **48**   **31.** 12, 18, 24 **72**   **32.** 7, 10, 14 **70**

© Glencoe/McGraw-Hill   T38   *Mathematics: Applications and Connections, Course 1*

Lesson 5-7 **209**

1 FOCUS

5-Minute Check
(Lesson 5-7)

Determine whether the first number is a multiple of the second number.

1. 42; 6 yes
2. 83; 3 no

Find the LCM for each set of numbers.

3. 7, 12 84
4. 12, 15 60
5. 2, 3, 5 30

The 5-Minute Check is also available on **Transparency 5-8A** for this lesson.

Motivating the Lesson

Communication Ask students to discuss the following scenario.

A class survey showed that $\frac{1}{3}$ of the class participates in sports and $\frac{2}{5}$ of the class participates in after-school clubs. How could we determine which activity has more participants? Sample answer: Compare them by finding the least common denominator.

Comparing and Ordering Fractions

***What* you'll learn**
You'll learn to compare and order fractions.

***When* am I ever going to use this?**
Knowing how to order fractions can help you determine better buys when grocery shopping.

Word Wise
least common denominator (LCD)

The graph shows that $\frac{3}{5}$ of smog is caused by cars and $\frac{3}{20}$ is caused by power plant emissions. Is more smog caused by cars or by power plants?

Source: Environmental Protection Agency

You can solve the problem by comparing the fractions $\frac{3}{5}$ and $\frac{3}{20}$.

One way to compare fractions is to express them as fractions with the same denominator. Any common denominator could be used. But the **least common denominator (LCD)** makes the computation easier.

The least common denominator is the LCM of the denominators. To find the LCD of $\frac{3}{5}$ and $\frac{3}{20}$, you need to find the LCM of 5 and 20.

multiples of 5: 0, 5, 10, 15, 20,…

multiples of 20: 0, 20, 40, 60, 80, …

The LCM of the denominators, 5 and 20, is 20.

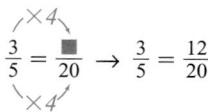

$$\frac{3}{5} = \frac{\blacksquare}{20} \rightarrow \frac{3}{5} = \frac{12}{20}$$

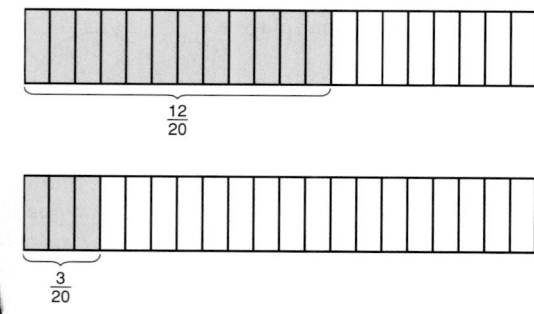

$$\frac{12}{20}$$

$$\frac{3}{20}$$

Since $12 > 3$, $\frac{12}{20} > \frac{3}{20}$. Therefore, $\frac{3}{5} > \frac{3}{20}$. So, more smog is caused by cars than by power plants.

Replace each ● with < , > , or = to make a true sentence.

1 $\frac{3}{5}$ ● $\frac{2}{3}$

The LCM of 5 and 3 is 15. Express $\frac{3}{5}$ and $\frac{2}{3}$ as fractions with a denominator of 15.

$$\frac{3}{5} = \frac{\blacksquare}{15}, \text{ so } \frac{3}{5} = \frac{9}{15}. \qquad \frac{2}{3} = \frac{\blacksquare}{15}, \text{ so } \frac{2}{3} = \frac{10}{15}.$$

(×3)   (×5)

Since $9 < 10$, $\frac{9}{15} < \frac{10}{15}$. Therefore, $\frac{3}{5} < \frac{2}{3}$.

2 $\frac{9}{10}$ ● $\frac{4}{5}$

The LCM of 10 and 5 is 10. Express $\frac{4}{5}$ as a fraction with a denominator of 10.

$$\frac{4}{5} = \frac{\blacksquare}{10} \rightarrow \frac{4}{5} = \frac{8}{10}$$

(×2)

Since $9 > 8$, $\frac{9}{10} > \frac{8}{10}$. Therefore, $\frac{9}{10} > \frac{4}{5}$.

APPLICATION **3** **Travel** The graph shows how people prefer to pay for travel expenses. Do more people prefer to pay for travel expenses with credit cards or with cash?

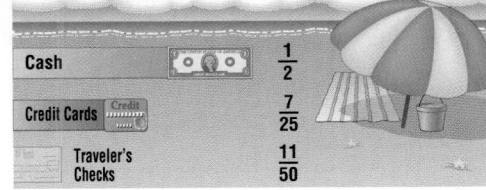

Paying for Travel Expenses

Cash — $\frac{1}{2}$

Credit Cards — $\frac{7}{25}$

Traveler's Checks — $\frac{11}{50}$

Source: Cirrus System, Inc.

You need to compare the fractions $\frac{7}{25}$ and $\frac{1}{2}$. The LCM of 25 and 2 is 50. So, express each fraction with a denominator of 50.

$$\frac{7}{25} = \frac{\blacksquare}{50}, \text{ so } \frac{7}{25} = \frac{14}{50}. \qquad \frac{1}{2} = \frac{\blacksquare}{50}, \text{ so } \frac{1}{2} = \frac{25}{50}.$$

(×2)   (×25)

Since $14 < 25$, $\frac{14}{50} < \frac{25}{50}$. Therefore, $\frac{7}{25} < \frac{1}{2}$.

So, more people prefer to pay with cash than with credit cards.

Lesson 5-8 Comparing and Ordering Fractions **211**

Thinking Algebraically When students are writing equivalent fractions, encourage them to think of the missing numerator as a variable. For example, $\frac{3}{5} = \frac{x}{20}$. This way they are actually solving an algebraic equation.

Teaching Tip Be sure students understand the greater than (>) and less than (<) symbols.

In-Class Examples

For Example 1
Replace each ● with < , > , or = to make a true sentence.
a. $\frac{4}{5}$ ● $\frac{3}{4}$ >
b. $\frac{5}{6}$ ● $\frac{15}{18}$ =
c. $\frac{5}{9}$ ● $\frac{7}{8}$ <

For Example 2
Replace each ● with < , > , or = to make a true sentence.
a. $\frac{3}{4}$ ● $\frac{7}{8}$ <
b. $\frac{13}{15}$ ● $\frac{2}{3}$ >

For Example 3
Refer to the graph in Example 3 of the Student Edition. Do more people prefer to pay with credit cards or traveler's checks?
credit cards

Teaching Tip After students understand how to compare fractions written with like denominators, you may wish to present the means-extremes property. That is, for $\frac{a}{b}$ and $\frac{c}{d}$ if $ad > bc$, then $\frac{a}{b} > \frac{c}{d}$.

3 PRACTICE/APPLY

Check for Understanding

If students need additional practice or instruction after completing Exercises 1–9, one of these options may be helpful.
- Extra Practice, see p. 571
- Reteaching Activity
- *Study Guide Masters,* p. 39
- *Practice Masters,* p. 39

Assignment Guide
Core: 11–29 odd, 30–36 **Enriched:** 10–26 even, 28–36

Additional Answers

1. Use the LCM of the denominators to rename fractions so they will have the same denominators. Then compare the numerators.

3. The LCD of $\frac{2}{5}$ and $\frac{4}{9}$ is 45. So, rename the fractions with a denominator of 45. $\frac{2}{5} = \frac{18}{45}$ and $\frac{4}{9} = \frac{20}{45}$. Since $18 < 20$, $\frac{2}{5} < \frac{4}{9}$.

Study Guide Masters, p. 39

CHECK FOR UNDERSTANDING

Communicating Mathematics

Read and study the lesson to answer each question. 1, 3. See margin.

1. *Explain* how the LCD can be used to compare fractions.

2. *State* the LCD of $\frac{3}{8}$ and $\frac{5}{6}$. **24**

3. *Write* one or two sentences describing how to compare $\frac{2}{5}$ and $\frac{4}{9}$.

Guided Practice

Find the LCD for each pair of fractions.

4. $\frac{2}{3}, \frac{1}{6}$ **6**

5. $\frac{3}{5}, \frac{3}{4}$ **20**

Replace each ● with $<$, $>$, or $=$ to make a true sentence.

6. $\frac{5}{6}$ ● $\frac{7}{8}$ **<**

7. $\frac{6}{9}$ ● $\frac{2}{3}$ **=**

8. $\frac{5}{12}$ ● $\frac{3}{4}$ **<**

9. **Food** The graph shows where people usually eat dessert in their homes. Do more people eat dessert in their kitchen or in their living room? **kitchen**

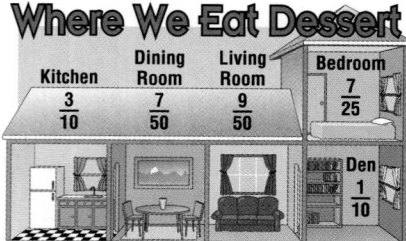

Source: The Alden Group

EXERCISES

Practice

Find the LCD for each pair of fractions.

10. $\frac{1}{2}, \frac{3}{8}$ **8**
11. $\frac{3}{4}, \frac{5}{6}$ **12**
12. $\frac{7}{12}, \frac{5}{8}$ **24**
13. $\frac{3}{5}, \frac{4}{9}$ **45**
14. $\frac{1}{6}, \frac{5}{9}$ **18**

Replace each ● with $<$, $>$, or $=$ to make a true sentence.

15. $\frac{1}{3}$ ● $\frac{2}{9}$ **>**
16. $\frac{3}{5}$ ● $\frac{3}{4}$ **<**
17. $\frac{10}{16}$ ● $\frac{5}{8}$ **=**

18. $\frac{1}{2}$ ● $\frac{3}{8}$ **>**
19. $\frac{2}{3}$ ● $\frac{1}{5}$ **>**
20. $\frac{9}{15}$ ● $\frac{11}{15}$ **<**

21. $\frac{9}{28}$ ● $\frac{5}{14}$ **<**
22. $\frac{3}{10}$ ● $\frac{1}{4}$ **>**
23. $\frac{5}{7}$ ● $\frac{15}{21}$ **=**

24. Which is greater, $\frac{2}{5}$ or $\frac{3}{7}$? $\frac{3}{7}$

25. Which is less, $\frac{1}{3}$ or $\frac{2}{7}$? $\frac{2}{7}$

26. Order the fractions $\frac{1}{2}, \frac{3}{5}, \frac{5}{6}, \frac{2}{3}$ from least to greatest. $\frac{1}{2}, \frac{3}{5}, \frac{2}{3}, \frac{5}{6}$

27. Order the fractions $\frac{1}{6}, \frac{2}{5}, \frac{3}{7}, \frac{3}{5}$ from greatest to least. $\frac{3}{5}, \frac{3}{7}, \frac{2}{5}, \frac{1}{6}$

212 Chapter 5 Using Number Patterns, Fractions, and Ratios

Reteaching the Lesson

Activity Write the following steps for comparing fractions on the chalkboard. (1) Find the LCD. (2) Rewrite the fractions with common denominators. (3) Compare the fractions. Ask students which step they would start with to compare $\frac{1}{5}$ and $\frac{3}{5}$; $\frac{2}{9}$ and $\frac{3}{6}$. Compare the fractions.

Error Analysis

Watch for students who compare numerators before checking the denominators.

Prevent by reminding students that fractions cannot be compared unless they have the same denominator.

Applications and Problem Solving

28. *Entertainment* According to Nielson Media Research, during an average week from 8:00 P.M. until 11:00 P.M., male teens spend $\frac{1}{3}$ of this time watching TV, and males over the age of 18 years of age spend $\frac{3}{7}$ of this time watching TV. Who spends more time watching TV in the evening, male teens or males over the age of 18? **See margin.**

29. *Money Matters* Hiroshi needs to purchase a can of garbanzo beans to make a salad for his family picnic. The table shows the cost of garbanzo beans at his neighborhood market. Write a fraction that compares the cost of each can to its weight. Which can is the better buy? Explain. **See margin.**

Isley's Market Garbanzo Beans	
Weight	Cost
8 oz	$0.45
16 oz	$0.90

30. *Critical Thinking* I am a fraction in simplest form. I am not improper. My numerator and denominator are twin primes. The sum of my numerator and denominator is equal to a dozen. Who am I? (*Hint*: Twin primes are prime numbers that have a difference of two.) $\frac{5}{7}$

Mixed Review

31. Find the least common multiple of 15, 20, and 25. *(Lesson 5-7)* **300**

32. *Measurement* How many yards are in 138 feet? *(Lesson 5-6)* **46 yards**

33. Write the fraction $\frac{100}{112}$ in simplest form. *(Lesson 5-4)* $\frac{25}{28}$

34. | Test Practice | Kenny drives his dairy truck 7.5 miles to his first stop. 8.4 miles to his second stop, 9.2 miles to his third stop, and 10.5 miles back to the dairy company. How many miles does he travel in his dairy truck altogether? *(Lesson 3-6)* **C**

A 36.5 mi
B 38.4 mi
C 35.6 mi
D 39.2 mi
E Not Here

35. *Algebra* Which of the numbers 21, 23, or 27 is the solution of $41 - m = 18$? *(Lesson 1-7)* **23**

36. *Patterns* Draw the next three figures in the pattern. *(Lesson 1-2)*

Lesson 5-8 Comparing and Ordering Fractions **213**

Extending the Lesson

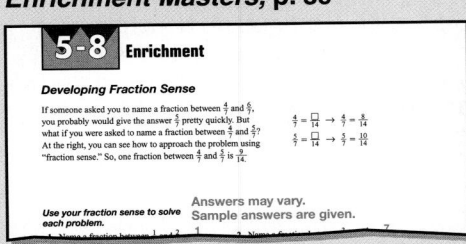
Activity Have students create math puzzlers such as the following. *We are two fractions in simplest form. Our LCD is 15. Our numerators are twin primes. Who are we?* $\frac{1}{3}$ and $\frac{3}{5}$ Have students illustrate their puzzles and display them on the bulletin board.

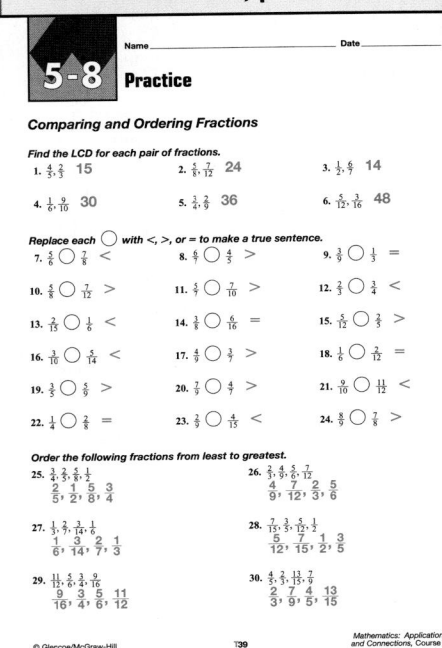

- *Study Guide Masters,* p. 40
- *Practice Masters,* p. 40
- *Enrichment Masters,* p. 40
- Transparencies 5-9, A and B CD-ROM Program
  - Resource Lesson 5-9

Recommended Pacing	
Standard	Day 13 of 16
Honors	Day 12 of 15
Block	Day 7 of 8

1 FOCUS

5-Minute Check
(Lesson 5-8)

1. Find the LCD for $\frac{1}{4}$ and $\frac{3}{8}$. **8**

Replace each ● with <, >, or = to make a true sentence.

2. $\frac{2}{7}$ ● $\frac{1}{5}$ **>**

3. $\frac{4}{9}$ ● $\frac{2}{3}$ **<**

4. $\frac{3}{4}$ ● $\frac{5}{6}$ **<**

5. Order $\frac{1}{2}$, $\frac{3}{5}$, $\frac{9}{10}$, and $\frac{1}{4}$ from least to greatest. $\frac{1}{4}, \frac{1}{2}, \frac{3}{5}, \frac{9}{10}$

The 5-Minute Check is also available on **Transparency 5-9A** for this lesson.

2 TEACH

Transparency 5-9B contains a teaching aid for this lesson.

Using Connections Bring in food boxes that have the items' weight listed as a decimal. Have students write the weight of each item as a fraction or a mixed number. Then have students discuss why the makers of the food products chose to list the weights in decimals rather than in fractions.

5-9

Writing Decimals as Fractions

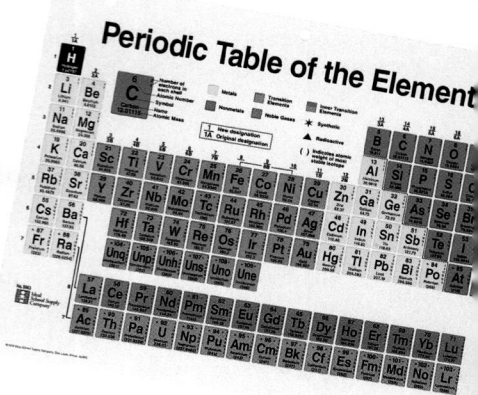

Periodic Table of the Element

What you'll learn
You'll learn to express terminating decimals as fractions in simplest form.

When am I ever going to use this?
Decimals and fractions are used interchangeably in situations where weight and length are measured.

Word Wise
terminating decimal

The periodic table gives the symbol, name, atomic number, and atomic mass of each element. The atomic mass of lead (Pb) is 207.2.

The number 207.2 is an example of a **terminating decimal**. Terminating decimals can be written as fractions with denominators of 10, 100, 1,000, and so on. For example, 207.2 can be written as the mixed number $207\frac{2}{10}$ or $207\frac{1}{5}$ in simplest form.

Cultural Kaleidoscope

In 1869, Russian chemist Dmitri Mendeleev devised and published the periodic table of the elements.

Express each decimal as a fraction or mixed number in simplest form.

① 0.8

$0.8 = \frac{8}{10}$ *Write the decimal as a fraction. 0.8 means eight tenths.*

$= \frac{\overset{4}{8}}{\underset{5}{10}}$ *Simplify. Divide the numerator and denominator each by the GCF, 2.*

$= \frac{4}{5}$

② 0.28

$0.28 = \frac{28}{100}$ *Write the decimal as a fraction. 0.28 means twenty-eight hundredths.*

$= \frac{\overset{7}{28}}{\underset{25}{100}}$ *Simplify. Divide by the GCF, 4.*

$= \frac{7}{25}$

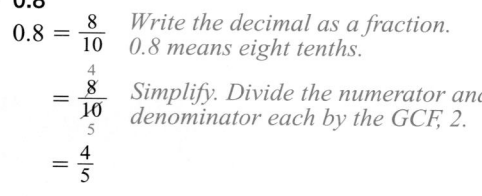

Study Hint

Mental Math Here are some commonly used decimal-fraction equivalencies.

$0.5 = \frac{1}{2}$ | $0.25 = \frac{1}{4}$

$0.2 = \frac{1}{5}$ | $0.125 = \frac{1}{8}$

③ 15.125

$15.125 = 15\frac{125}{1,000}$ *Write the decimal as a mixed number. 15.125 means fifteen and one hundred twenty-five thousandths.*

$= 15\frac{\overset{1}{125}}{\underset{}{1,000}}$ *Simplify. Divide by the GCF, 125.*

$= 15\frac{1}{8}\overset{8}{}$

214 **Chapter 5** Using Number Patterns, Fractions, and Ratios

---

Motivating the Lesson
Communication Ask students to read the opening paragraphs of the lesson. Ask the following questions.
- How would you say 207.2 in words? two hundred seven and two tenths
- To write 36.61 as a mixed number, would you use a denominator of 10, 100, or 1,000? 100

Multiple Learning Styles

Intrapersonal Have students write a paragraph describing which measurements are easier for them to understand—those written as fractions or those written as decimals. Have them explain why.

Example 4

APPLICATION

Transportation The F-16C Fighting Falcon can reach a maximum speed of mach 2.05 or 1,320 miles per hour. Express its mach speed as a mixed number in simplest form.

$$2.05 = 2\frac{5}{100}$$ *Write the decimal as a mixed number. 2.05 means two and five hundredths*

$$= 2\frac{\overset{1}{\cancel{5}}}{\underset{20}{\cancel{100}}}$$ *Simplify. Divide by the GCF, 5.*

$$= 2\frac{1}{20}$$

The mach speed can be expressed as $2\frac{1}{20}$.

CHECK FOR UNDERSTANDING

Communicating Mathematics

Read and study the lesson to answer each question.

1. *Explain* how to express 0.36 as a fraction in simplest form. **See Answer Appendix.**

2. *Tell* what decimal is represented by the model shown. Then write the decimal as a fraction in simplest form. **0.55; $\frac{11}{20}$**

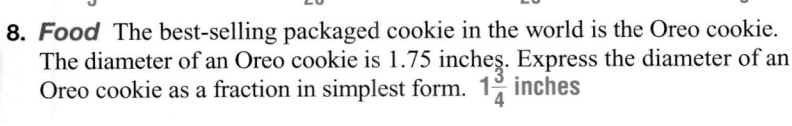

3. *You Decide* Ann says that to express 3.55 as a fraction, you use a denominator of 100. Aisha says to use a denominator of 1,000. Who is correct? Explain your reasoning. **See Answer Appendix.**

Guided Practice

Express each decimal as a fraction or mixed number in simplest form.

4. 0.6 $\frac{3}{5}$ 5. 0.45 $\frac{9}{20}$ 6. 2.08 $2\frac{2}{25}$ 7. 4.375 $4\frac{3}{8}$

8. *Food* The best-selling packaged cookie in the world is the Oreo cookie. The diameter of an Oreo cookie is 1.75 inches. Express the diameter of an Oreo cookie as a fraction in simplest form. $1\frac{3}{4}$ inches

EXERCISES

Practice

Express each decimal as a fraction or mixed number in simplest form.

9. 3.2 $3\frac{1}{5}$ 10. 5.26 $5\frac{13}{50}$ 11. 8.65 $8\frac{13}{20}$ 12. 0.04 $\frac{1}{25}$

13. 5.64 $5\frac{16}{25}$ 14. 6.018 $6\frac{9}{500}$ 15. 2.4 $2\frac{2}{5}$ 16. 4.303 $4\frac{303}{1,000}$

17. 13.009 $13\frac{9}{1,000}$ 18. 1.234 $1\frac{117}{500}$ 19. 7.89 $7\frac{89}{100}$ 20. 9.82 $9\frac{41}{50}$

21. Write *thirty-eight hundredths* as a decimal and as a fraction in simplest form. **0.38; $\frac{19}{50}$**

22. Express *twelve and sixteen thousandths* as a decimal and as a fraction in simplest form. **12.016; $12\frac{2}{125}$**

═ Reteaching the Lesson ═

Activity Have small groups of students model pairs of decimals and fractions using base-ten models. Have them compare the results. Then give them decimals such as 0.7 and 2.25. Have them use the models to determine how to express each decimal as a fraction in simplest form.

Error Analysis
Watch for students who use the incorrect denominator when writing decimals as fractions.
Prevent by reminding students that the number of decimal places equals the number of zeros in the denominator.

In-Class Examples

For Example 1
Express 0.6 as a fraction in simplest form. $\frac{3}{5}$

For Example 2
Express 0.75 as a fraction in simplest form. $\frac{3}{4}$

For Example 3
Express 42.425 as a mixed number in simplest form. $42\frac{17}{40}$

For Example 4
A 21.5-foot mirror has replaced six smaller mirrors at the Multiple Mirror Telescope in Arizona. Express the new mirror's size as a fraction in simplest form. $21\frac{1}{2}$ feet

---

3 PRACTICE/APPLY

Check for Understanding
If students need additional practice or instruction after completing Exercises 1–8, one of these options may be helpful.
- Extra Practice, see p. 571
- Reteaching Activity
- *Study Guide Masters,* p. 40
- *Practice Masters,* p. 40
-  Interactive Mathematics Tools Software

---

Study Guide Masters, p. 40

5-9 **Study Guide**

Name _____ Date _____

Writing Decimals as Fractions

A **terminating decimal** can be written as a fraction with a denominator of 10, 100, 1,000, and so on.

Examples 1 Express 0.5 as a fraction in simplest form.
$0.5 = \frac{5}{10}$ *Write the decimal as a fraction. Use 10 as the denominator since 0.5 is 5 tenths.*
$\frac{5}{10} = \frac{1}{2}$ *Simplify.*

2 Express 7.005 as a fraction in simplest form.
$7.005 = 7\frac{5}{1,000}$ *Write the decimal as a mixed number. Use 1,000 as the denominator since 0.005 is 5 thousandths.*
$7\frac{5}{1,000} = 7\frac{1}{200}$ *Simplify.*

Express each decimal as a fraction or mixed number in simplest form.

1. 0.4 $\frac{2}{5}$
2. 0.15 $\frac{3}{20}$
3. 0.125 $\frac{1}{8}$
4. 0.88 $\frac{22}{25}$
5. 5.008 $5\frac{1}{125}$
6. 7.8 $7\frac{4}{5}$
7. 11.25 $11\frac{1}{4}$
8. 3.525 $3\frac{21}{40}$
9. 25.2 $25\frac{1}{5}$
10. 6.65 $6\frac{13}{20}$
11. 10.475 $10\frac{19}{40}$
12. 7.75 $7\frac{3}{4}$
13. 12.002 $12\frac{1}{500}$
14. 5.5 $5\frac{1}{2}$
15. 8.34 $8\frac{17}{50}$
16. 2.1 $2\frac{1}{10}$

17. Write *fourteen thousandths* as a decimal and as a fraction in simplest form.
0.014, $\frac{7}{500}$

18. Write *one and eighty-five hundredths* as a decimal and as a fraction in simplest form.
1.85, $1\frac{17}{20}$

© Glencoe/McGraw-Hill   T40   Mathematics: Applications and Connections, Course 1

4 ASSESS

Closing Activity

Speaking Say a decimal and ask one student to express it as a fraction in simplest form. Have another student check to see if the fraction is correct and, if not, give the correct fraction. Continue until all students have had a turn.

Additional Answers

25. krypton: $83\frac{4}{5}$; selenium: $78\frac{24}{25}$; sulfur: $32\frac{3}{50}$; carbon: $12\frac{11}{1,000}$

26. True; a terminating decimal can have a denominator of 10, 100, 1,000, ... Since 10 is divisible by 2 and 5, the denominator of every terminating decimal is divisible by 2 and 5.

Practice Masters, p. 40

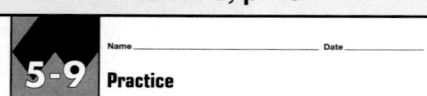

Applications and Problem Solving

23. *Travel* The traffic sign shown tells drivers the distance and direction in which a landmark is located from an exit ramp. What fraction of a mile is each landmark located from the exit?

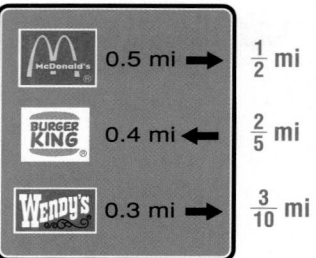

24. *History* Greenbacks, or paper money, were issued during the Civil War to help the North and South pay for the war. Today, a dollar bill is 6.14 inches long, 2.61 inches wide, and weighs 0.033 ounce. Express the length, width, and weight of a dollar bill as fractions or mixed numbers in simplest form.
length: $6\frac{7}{50}$ in., width: $2\frac{61}{100}$ in., weight: $\frac{33}{1,000}$ oz

25. *Physical Science* The table shows the atomic mass of certain elements. Express the atomic mass of each element as a mixed number in simplest form. **See margin.**

Element	Atomic Mass
Krypton (Kr)	83.8
Selenium (Se)	78.96
Sulfur (S)	32.06
Carbon (C)	12.011

26. *Critical Thinking* *True* or *False*? Every terminating decimal can be written as a fraction with a denominator that is divisible by 2 and 5. Explain your reasoning. **See margin.**

Mixed Review

27. Which fraction is greater, $\frac{13}{40}$ or $\frac{3}{7}$? *(Lesson 5-8)* $\frac{3}{7}$

28. *Geometry* Find the perimeter of the rectangle if $h = 3.5$ and $g = 2.1$. *(Lesson 4-4)* **11.2 feet**

h feet

g feet

29. **Test Practice** Madison wants to buy banana splits for herself and 3 friends. Each banana split costs $2.49. Which is the best estimate of the amount of money Madison will need in order to buy the banana splits? *(Lesson 1-3)* **D**

A $4   B $6   C $8   D $10   E $12

30. *Entertainment* For the film *101 Dalmations*, Disney animators drew a total of 6,469,952 spots on the animated dalmations. Round this number to the nearest thousand. *(Lesson 1-3)* **6,470,000**

31. *Patterns* Draw the next figure in the pattern shown. *(Lesson 1-2)*

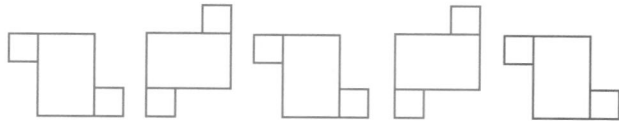

Extending the Lesson

Activity Brad, Janna, and Sven are on a hiking trip. Brad's water bottle holds $31\frac{5}{8}$ ounces of liquid, Janna's bottle holds 31.42 ounces, and Sven's bottle holds 31.425 ounces. Have students use fractions to determine whose bottle holds the most water. **Brad's bottle**

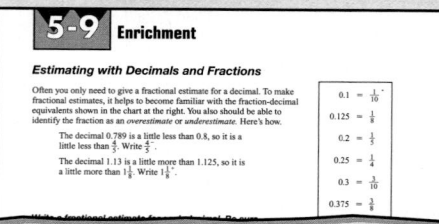

Writing Fractions as Decimals

What you'll learn

You'll learn to express fractions as terminating and repeating decimals.

When am I ever going to use this?

Knowing how to express numbers in different forms can help you interpret weather statistics.

Word Wise

repeating decimal
bar notation

Home gardening has become a popular hobby. The top five vegetables grown in home gardens are tomatoes, peppers, onions, cucumbers, and beans. An estimated $\frac{2}{5}$ of home gardeners grow beans.

Any fraction can be written as a decimal using division.

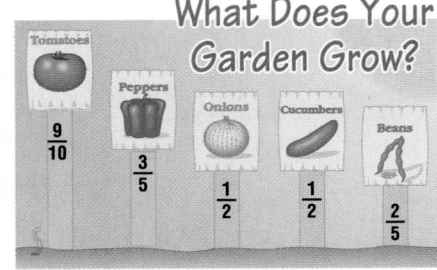

What Does Your Garden Grow?

Tomatoes — $\frac{9}{10}$
Peppers — $\frac{3}{5}$
Onions — $\frac{1}{2}$
Cucumbers — $\frac{1}{2}$
Beans — $\frac{2}{5}$

Source: National Gardening Association, Gallup Organization

Method 1 Use paper and pencil.

$\frac{2}{5}$ means $2 \div 5$.

$$
\begin{array}{r}
0.4 \\
5\overline{)2.0} \\
-20 \\
\hline
0
\end{array}
$$
Write 2 as 2.0.
Place the decimal point in the quotient. Divide as with whole numbers.

The fraction $\frac{2}{5}$ can be written as 0.4. A decimal like 0.4 is called a *terminating decimal* because the division ends, or terminates, when the remainder is zero.

Method 2 Use a calculator.

2 5 ⊟ *0.4*

Example ➊ **Express $\frac{1}{8}$ as a decimal.**

Method 1 Use paper and pencil.

$$
\begin{array}{r}
0.125 \\
8\overline{)1.000} \\
-8 \\
\hline
20 \\
-16 \\
\hline
40 \\
-40 \\
\hline
0
\end{array}
$$
Divide 1 by 8.

Therefore, $\frac{1}{8} = 0.125$.

Method 2 Use a calculator.

1 8 ⊟ *0.125*

Lesson 5-10 Writing Fractions as Decimals **217**

Motivating the Lesson

Problem Solving Ask students how they would determine which basket of vegetables weighs more than $5\frac{1}{2}$ pounds.
- a 5.46-lb basket of tomatoes
- a 5.64-lb basket of peppers

---

1 FOCUS

5-Minute Check
(Lesson 5-9)

Express each decimal as a fraction or mixed number in simplest form.

1. 0.4 $\frac{2}{5}$
2. 0.45 $\frac{9}{20}$
3. 3.675 $3\frac{27}{40}$
4. 22.52 $22\frac{13}{25}$
5. 17.8 $17\frac{4}{5}$

The 5-Minute Check is also available on **Transparency 5-10A** for this lesson.

2 TEACH

Transparency 5-10B contains a teaching aid for this lesson.

Reading Mathematics Have students look up the word "terminate" in the dictionary. Ask them to think of antonyms of the word. What other terms could be used for "repeating" decimal? Point out that a decimal equivalent of a fraction will always terminate or repeat. However, there are decimals such as π (3.14159265...) that do neither.

3 PRACTICE/APPLY

Check for Understanding

Not all fractions are terminating decimals. Decimals like 0.5555555... are called **repeating decimals** because the digits repeat. The **bar notation** 0.5̄ can be used to indicate that the 5 digit repeats forever. Several repeating decimals are shown.

0.433333333... = 0.43̄	*The digit 3 repeats.*
2.121212121... = 2.12̄	*The digits 12 repeat.*
13.567567567... = 13.567̄	*The digits 567 repeat.*

Examples

2 Express $\frac{3}{11}$ as a decimal. Use bar notation to show a repeating decimal.

Method 1 Use paper and pencil.

```
     0.2727...
11)3.0000        Divide 3 by 11.
   −22
     80
    −77
     30
    −22
     80      The pattern will
    −77      continue.
      3
```

Method 2 Use a calculator.

3 ÷ 11 = *0.27272727...*

Therefore, $\frac{3}{11}$ = 0.2727... or 0.27̄.

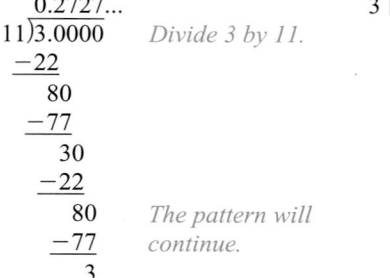

CONNECTION 3 Earth Science The average annual precipitation in Georgia is about $48\frac{3}{5}$ inches. How many inches of precipitation does Georgia average each month?

Divide $48\frac{3}{5}$ by 12. First, express $48\frac{3}{5}$ as a decimal.

48 + 3 ÷ 5 = *48.6* $48\frac{3}{5} = 48 + \frac{3}{5}$

Then divide by 12.

48.6 ÷ 12 = *4.05*

Georgia averages about 4.05 inches of precipitation each month.

CHECK FOR UNDERSTANDING

Communicating Mathematics

1–2. See Answer Appendix.

Read and study the lesson to answer each question.

1. ***Explain*** to a classmate how to express a fraction as a decimal.

2. ***Give*** an example of a terminating decimal and a repeating decimal.

Guided Practice

Write each repeating decimal using bar notation.

3. 0.4444444... **0.4̄**

4. 10.34343434... **10.34̄**

218 Chapter 5 Using Number Patterns, Fractions, and Ratios

Express each fraction or mixed number as a decimal. Use bar notation to show a repeating decimal.

5. $\frac{3}{8}$ **0.375**

6. $2\frac{4}{11}$ **2.$\overline{36}$**

7. $\frac{7}{12}$ **0.58$\overline{3}$**

8. Express the fraction $\frac{7}{8}$ as a decimal. **0.875**

EXERCISES

Practice

Write each repeating decimal using bar notation.

9. 0.77777777... **0.$\overline{7}$**
10. 1.33333333... **1.$\overline{3}$**
11. 2.45454545... **2.$\overline{45}$**
12. 17.0909090... **17.$\overline{09}$**
13. 0.83183183... **0.$\overline{831}$**
14. 5.01289289... **5.01$\overline{289}$**

Express each fraction or mixed number as a decimal. Use bar notation to show a repeating decimal.

15. $\frac{3}{4}$ **0.75**
16. $4\frac{1}{5}$ **4.2**
17. $\frac{2}{9}$ **0.$\overline{2}$**
18. $2\frac{5}{8}$ **2.625**

19. $\frac{5}{11}$ **0.$\overline{45}$**
20. $7\frac{1}{3}$ **7.$\overline{3}$**
21. $\frac{11}{12}$ **0.91$\overline{6}$**
22. $\frac{7}{10}$ **0.7**

23. Express *four and one-eleventh* as a decimal. **4.$\overline{09}$**
24. Find the decimal equivalent of the fraction *four fifteenths*. **0.2$\overline{6}$**

Applications and Problem Solving

25. *Food* A Hershey Kiss weighs $\frac{1}{6}$ ounce. Find the decimal equivalent of the weight of a Hershey Kiss. **0.1$\overline{6}$ oz**

26. *Geography* Japan is a group of islands in the Pacific Ocean. Tokyo is Japan's largest city. The city receives about $5\frac{1}{8}$ inches of rain each month. How many inches of rain does Tokyo receive in a year? **61.5 in.**

27. *Working on the* **CHAPTER Project** Refer to the table you made on page 196 in Exercise 36.
   a. Express each of your fractions as a decimal. If necessary, round to the nearest hundredth. Which of your magazines has the highest decimal portion devoted to articles? to advertisements? Which has the lowest decimal portion devoted to articles? to advertisements?
   b. Make a bar graph for the set of data. **a–b. See students' work.**

28. *Critical Thinking* Tell how you can determine whether a fraction in simplest form will be expressed as a terminating or repeating decimal by looking at the denominator. **See margin.**

Mixed Review
29. $29\frac{3}{4}$

29. *Earth Science* Mercury moves in its orbit at a speed of 29.75 miles per second. Write this speed as a mixed number in simplest form. *(Lesson 5-9)*

30. Find the LCM of 24 and 30. *(Lesson 5-7)* **120**

31. **Test Practice** Express 24 as a product of primes *(Lesson 5-2)* **D**
   A $2 \times 4 \times 3$
   B $2 \times 2 \times 2 \times 6$
   C $2 \times 2 \times 8$
   D $2 \times 2 \times 2 \times 3$

32. Find $2 + 17 - 16 \div 4$. *(Lesson 1-4)* **15**

Lesson 5-10 Writing Fractions as Decimals **219**

Extending the Lesson

Enrichment Masters, p. 41

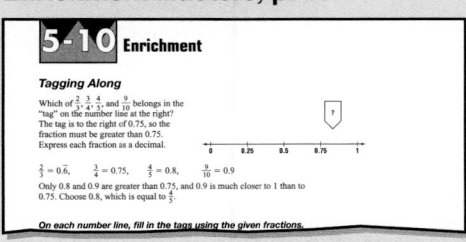

Activity Have small groups of students create a table of fractions and their decimal equivalents. Have students look for patterns in families of repeating decimals, such as $\frac{1}{9} = 0.1\overline{1}$, $\frac{2}{9} = 0.2\overline{2}$, $\frac{3}{9} = 0.3\overline{3}$.

Assignment Guide
Core: 9–25 odd, 28–32
Enriched: 10–24 even, 25, 26, 28–32

CHAPTER Project

Exercise 27 asks students to advance to the next stage of work on the Chapter Project. You may require students to place their work in their Project Folders. Ask students to determine which sets of data—fractions or decimals—are easier to interpret.

4 ASSESS

Closing Activity
Writing Have students use paper and pencil to determine whether $\frac{14}{15}$ is a repeating or terminating decimal. **0.9$\overline{3}$, repeating**

Chapter 5, Quiz D (Lessons 5-9 and 5-10) is available in the *Assessment and Evaluation Masters*, p. 128.

Additional Answer
28. If the prime factorization of the denominator contains only 2's or 5's, the fraction is a terminating decimal.

Practice Masters, p. 41

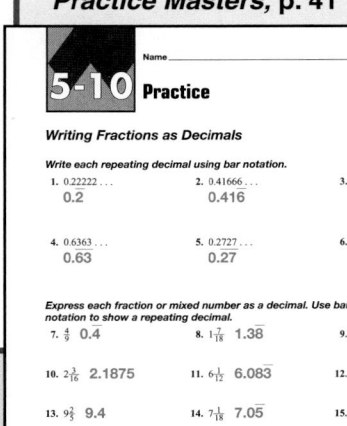

Vocabulary

This section provides a listing of the new terms, properties, and phrases that were introduced in this chapter. Have students define each term and provide an example or two of it, if appropriate.

Understanding and Using the Vocabulary

These exercises check students' understanding of the terms by using a variety of verbal formats including matching, completion, and true/false.

Glossaries A complete glossary of terms appears on pages 642–648. The glossary also appears in Spanish on pages 649–656.

Additional Answer

8. An inch is the most commonly used unit of length. 1 foot = 12 inches; 1 yard = 3 feet, or 36 inches; 1 mile = 5,280 feet, or 1,760 yards.

Vocabulary

After completing this chapter, you should be able to define each term, concept, or phrase and give an example or two of each.

Number and Operations
bar notation (p. 218)
common multiples (p. 206)
composite number (pp. 181, 182)
equivalent fractions (p. 193)
factor tree (p. 183)
greatest common factor (GCF) (p. 188)
improper fraction (p. 198)

least common denominator (LCD) (p. 210)
least common multiple (LCM) (p. 206)
mixed number (p. 198)
multiple (p. 206)
prime factorization (p. 183)
prime number (pp. 181, 182)
ratio (pp. 191, 193)
repeating decimal (p. 218)
simplest form (p. 194)
terminating decimal (p. 214)

Measurement
foot (p. 202)
inch (p. 202)
mile (p. 202)
yard (p. 202)

Probability
experimental (p. 197)

Problem Solving
make a list (p. 186)

Understanding and Using the Vocabulary

Choose the letter of the term that best matches each phrase.
1. a comparison of two numbers by division **g**
2. the product of a number and any whole number **a**
3. a number having more than two factors **e**
4. numbers showing the sum of a whole number and a fraction **h**
5. a number having exactly two factors, 1 and itself **b**
6. the LCM of 15 and 9 **c**
7. the greatest common factor of 56 and 70 **f**

a. multiple
b. prime number
c. 45
d. improper fraction
e. composite number
f. 14
g. ratio
h. mixed number

In Your Own Words
8. *Explain* the relationship between inch, foot, yard, and mile. **See margin.**

Objectives & Examples

Upon completing this chapter, you should be able to:

● use divisibility rules for 2, 3, 5, 6, 9, and 10 *(Lesson 5-1)*

Is 630 divisible by 2, 3, 5, 6, 9, or 10?
It is divisible by 2, 3, 5, 6, 9, and 10.

Review Exercises

Use these exercises to review and prepare for the chapter test.

State whether each number is divisible by 2, 3, 5, 6, 9, or 10.
9. 51 **3**
10. 300 **2, 3, 5, 6, 10**
11. 423 **3, 9**
12. 1,250 **2, 5, 10**

220 Chapter 5 Using Number Patterns, Fractions, and Ratios

MindJogger Videoquizzes

MindJogger Videoquizzes provide an alternative review of concepts presented in this chapter. Students work in teams to answer questions, gaining points for correct answers. The questions are presented in three rounds.
Round 1 Concepts–5 questions
Round 2 Skills–4 questions
Round 3 Problem Solving–4 questions

Objectives & Examples

find the prime factorization of a composite number *(Lesson 5-2)*

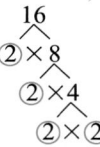

```
        16
       ╱  ╲
      ②× 8
         ╱ ╲
        ②× 4
           ╱ ╲
          ②× ②
```

The prime factorization of 16 is $2 \times 2 \times 2 \times 2$ or 2^4.

find the greatest common factor of two or more numbers *(Lesson 5-3)*

Find the GCF of 12 and 18.

factors of 12: 1, 2, 3, 4, 6, 12

factors of 18: 1, 2, 3, 6, 9, 18

The GCF of 12 and 18 is 6.

express fractions and ratios in simplest form *(Lesson 5-4)*

Write $\frac{12}{36}$ in simplest form.

factors of 12: 1, 2, 3, 4, 6, 12

factors of 36: 1, 2, 3, 4, 6, 9, 12, 18, 36

The GCF of 12 and 36 is 12.

$$\frac{12}{36} = \frac{1}{3}$$ (÷12)

In simplest form, the fraction is $\frac{1}{3}$.

express mixed numbers as improper fractions and vice versa *(Lesson 5-5)*

Express $1\frac{2}{3}$ as an improper fraction.

$$1\frac{2}{3} = \frac{(1 \times 3) + 2}{3}$$
$$= \frac{5}{3}$$

Review Exercises

Tell whether each number is *prime*, *composite*, or *neither*.

13. 37 prime   **14.** 78 composite

15. 1 neither   **16.** 47 prime

Find the prime factorization of each number.

17. 54 2×3^3 **18.** 75 3×5^2

19. 124 $2^2 \times 31$ **20.** 36 $2^2 \times 3^2$

Find the GCF of each set of numbers by making a list.

21. 30, 36 6   **22.** 39, 26 13

Find the GCF of each set of numbers by using prime factorization.

23. 18, 28 2   **24.** 12, 24, 30 6

Replace each ■ with a number so that the fractions are equivalent.

25. $\frac{5}{6} = \frac{■}{24}$ 20   **26.** $\frac{15}{35} = \frac{3}{■}$ 7

State whether each fraction or ratio is in simplest form. If not, write each fraction or ratio in simplest form.

27. $\frac{15}{18}$ no; $\frac{5}{6}$ **28.** 2 out of 9 yes

29. 14 to 16 no; 7 to 8   **30.** $\frac{24}{28}$ no; $\frac{6}{7}$

Express each mixed number as an improper fraction.

31. $3\frac{3}{5}$ $\frac{18}{5}$ **32.** $4\frac{2}{7}$ $\frac{30}{7}$

Express each improper fraction as a mixed number.

33. $\frac{19}{5}$ $3\frac{4}{5}$ **34.** $\frac{36}{7}$ $5\frac{1}{7}$

Chapter 5 Study Guide and Assessment   **221**

Objectives & Examples

This section reviews the skills and concepts of the chapter and shows completely worked examples.

Review Exercises

These exercises provide practice for the corresponding objectives.

***Assessment and Evaluation Masters,* pp. 115–116**

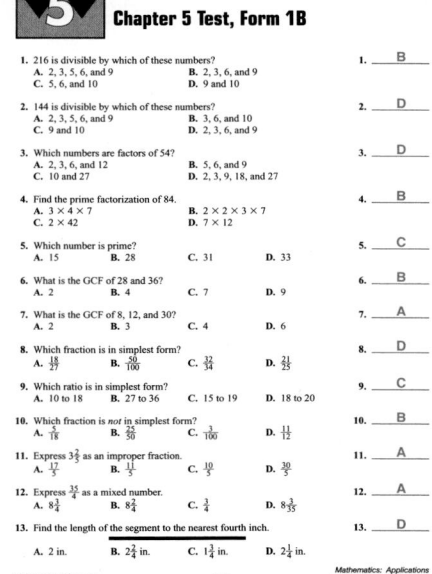

Assessment and Evaluation

Six forms of Chapter 5 Test are available in the *Assessment and Evaluation Masters* as shown in the chart.

Chapter 5 Test, Form 1B, is shown at the right. Chapter 5 Test, Form 2B, is shown on the next page.

1A	Multiple Choice	Honors
1B	Multiple Choice	Average
1C	Multiple Choice	Basic
2A	Free Response	Honors
2B	Free Response	Average
2C	Free Response	Basic

Assessment and Evaluation Masters, pp. 121–122

Name_____   Date_____

Chapter 5 Test, Form 2B

State whether each number is divisible by 2, 3, 5, 6, 9, or 10 for Questions 1–3.
1. 225   1. ___3, 5, 9___
2. 351   2. ___3, 9___
3. 135   3. ___3, 5, 9___
4. How many 3-digit numbers can you make using the digits 3, 6, and 9 if digits can be repeated?   4. ___27___

Find the prime factorization of each number for Questions 5–7.
5. 70   5. ___2 × 5 × 7___
6. 192   6. ___2⁶ × 3___
7. 54   7. ___2 · 3³___
8. List all of the factors of 60.   8. ___1, 2, 3, 4, 5, 6, 10, 12, 15, 20, 30, 60___

Find the GCF of each pair of numbers.
9. 32, 36   9. ___4___
10. 33, 55   10. ___11___

State whether each fraction or ratio is in simplest form. If not, write each fraction or ratio in simplest form.
11. 45/80   11. ___no, 9/16___
12. 5 to 32   12. ___yes___
13. 15:35   13. ___no, 3:7___
14. 9/12   14. ___no, 3/4___

Express each fraction as a mixed number.
15. 32/7   15. ___4 4/7___
16. 50/9   16. ___5 5/9___

© Glencoe/McGraw-Hill   121   Mathematics: Applications and Connections, Course 1

Chapter 5 Test, Form 2B (continued)

Express each mixed number as an improper fraction for Questions 17–18.
17. 5 7/12   17. ___67/12___
18. 10 3/20   18. ___203/20___
19. Find the length of the segment to the nearest eighth inch.   19. ___1 5/8 in.___
20. Complete: 76 inches = _?_ feet.   20. ___6 1/3___

Find the LCM for each pair of numbers.
21. 18, 10   21. ___90___
22. 21, 35   22. ___105___
23. 25, 30   23. ___150___

Replace each ● with , <, >, or = to make a true sentence.
24. 4/7 ● 3/8   24. ___>___
25. 20/30 ● 4/6   25. ___=___
26. 73/100 ● 23/50   26. ___>___

Express each decimal as a fraction in simplest form.
27. 0.82   28. 2.4   29. 0.45   27. ___41/50___
                                          28. ___2 2/5___
                                          29. ___9/20___

Express each fraction as a decimal for Questions 30–32. Use bar notation to show a repeating decimal.
30. 39/50   30. ___0.78___
31. 7/10   31. ___0.7___
32. 2/9   32. ___0.2̄___
33. One ribbon is 80 inches long, and another ribbon is 32 inches long. The ribbons are to be cut into pieces of equal length. What is the longest length possible for the pieces if there is to be no leftover ribbon?   33. ___16 in.___

© Glencoe/McGraw-Hill   122   Mathematics: Applications and Connections, Course 1

Objectives & Examples

● measure line segments and objects with a ruler divided in halves, fourths, and eighths *(Lesson 5-6)*

Draw a line segment measuring $1\frac{3}{8}$ inches.

_____

● find the least common multiple of two or more numbers *(Lesson 5-7)*

Find the LCM of 8 and 12.
multiples of 8: 0, 8, 16, 24, 32, …
multiples of 12: 0, 12, 24, 36, …
The LCM of 8 and 12 is 24.

● compare and order fractions *(Lesson 5-8)*

$\frac{3}{7}$ ● $\frac{4}{9}$ *The LCM of 7 and 9 is 63.*

$$\frac{3}{7} = \frac{27}{63} \qquad \frac{4}{9} = \frac{28}{63}$$

(×9)   (×7)

Since $\frac{27}{63} < \frac{28}{63}, \frac{3}{7} < \frac{4}{9}$.

● express terminating decimals as fractions in simplest form *(Lesson 5-9)*

Express 1.72 as a mixed number in simplest form.

$$1.72 = 1\frac{72}{100} \text{ or } 1\frac{18}{25}$$

● express fractions as terminating and repeating decimals *(Lesson 5-10)*

Express $\frac{5}{6}$ as a decimal.

5 ÷ 6 = *0.833333…*

$\frac{5}{6} = 0.8\overline{3}$

Review Exercises

Draw a line segment of each length.
35. $1\frac{3}{4}$ inches   36. $2\frac{7}{8}$ inches
35–36. See Answer Appendix.
37. Find the length of the line segment to the nearest half, fourth, or eighth inch.
_____ $1\frac{3}{4}$ in.

38. Determine whether 82 is a multiple of 12. **no**

Find the LCM for each set of numbers.
39. 15, 25 **75**   40. 28, 35 **140**
41. 7, 12 **84**   42. 12, 15, 20 **60**

Find the LCD for each pair of fractions.
43. $\frac{3}{5}, \frac{1}{4}$ **20**   44. $\frac{5}{6}, \frac{7}{8}$ **24**

Replace each ● with < , > , or = to make a true sentence.
45. $\frac{5}{9}$ ● $\frac{6}{11}$ **>**   46. $\frac{8}{12}$ ● $\frac{6}{9}$ **=**

Express each decimal as a fraction or mixed number in simplest form.
47. 0.8   48. 0.04
49. 3.65   50. 7.36
47. $\frac{4}{5}$ 48. $\frac{1}{25}$ 49. $3\frac{13}{20}$ 50. $7\frac{9}{25}$

Express each fraction or mixed number as a decimal. Use bar notation to show a repeating decimal.
51. $\frac{5}{8}$ **0.625**   52. $1\frac{5}{11}$ **1.$\overline{45}$**

Test and Review Software

You may use this software, a combination of an item generator and item bank, to create your own tests or worksheets. Types of items include free response, multiple choice, short answer, and open ended.

CD-ROM Program

The CD-ROM Program contains an Assessment Game whose questions review the concepts in this chapter.

Applications & Problem Solving

53. Make a List When school is cancelled because of snow, the superintendent has to call four radio stations to let them know. How many ways can the superintendent make the phone calls? *(Lesson 5-3A)*

54. Landscaping Brandon used $10\frac{1}{4}$ bags of mulch around the flower beds in his yard. Write $10\frac{1}{4}$ as an improper fraction. *(Lesson 5-5)* $\frac{41}{4}$

55. Money Matters Zu-Wang worked $6\frac{2}{5}$ hours last week. If he earns $4.75 per hour, how much did he earn last week? *(Lesson 5-10)* **$30.40**

53. 24 ways

56. School The graph shows the results of a survey on field trip preferences. Where do more students prefer to go on their field trip? *(Lesson 5-8)* **ZOO**

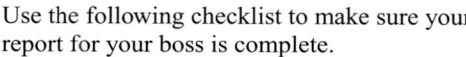

Field Trip

Zoo	$\frac{9}{20}$
Museum	$\frac{1}{4}$
State Capitol	$\frac{3}{10}$

Alternative Assessment

Performance Task

Suppose you are buying party supplies for a younger sibling's birthday party. You find a package of 6 decks of cards, a package of 8 candy bars, and a package of 4 cans of modeling dough. How can you determine the least number of packages of cards, candy, and modeling dough to buy so that you have the same number of each? **Find the LCM of 4, 6, and 8.**

Find the least number of packages to buy so that you will have the same number of each item. If there will be 15 children at the party, what is the least number of packages of each item you would have to buy? **See margin.**

A practice test for Chapter 5 is provided on page 599.

Completing the CHAPTER Project

Use the following checklist to make sure your report for your boss is complete.

☑ The table showing the number of pages of articles, ads, and table of contents in each issue of your magazines is included.

☑ The data comparing the number of pages of articles, ads, and table of contents to the total number of pages in each magazine is correct.

☑ The data showing what portion of each magazine is devoted to articles and ads is correct.

☑ The bar graphs of the data are included.

 PORTFOLIO Select one of the assignments from this chapter and place it in your portfolio. Attach a note to it explaining why you selected it.

Chapter 5 Study Guide and Assessment   **223**

Applications & Problem Solving

This section provides additional practice in solving real-world problems that involve the skills of this chapter.

Alternative Assessment

The **Performance Task** provides students with a performance assessment opportunity to evaluate their work and understanding.

CHAPTER Project

Students should complete the final stages of their project and prepare a class demonstration of their results. A scoring guide for the project is available in the *Investigations and Projects Masters*, p. 35.

 PORTFOLIO Students should add to their portfolios at this time.

Assessment and Evaluation Masters, p. 125

Name_____ Date_____

5 **Chapter 5 Performance Assessment**

Instructions: Demonstrate your knowledge by giving a clear, concise solution to each problem. Be sure to include all relevant drawings and justify your answers. You may show your solutions in more than one way or investigate beyond the requirements of the problems.

1. Mr. Berkowitz is planning the half-time show for the first football game of the season. He expects 120 band members this year and needs to determine possible marching formations.
   a. Tell how to find the prime factorization of a number.
   b. Find the prime factorization of 120. Show your work.
   c. Give all possible rectangular formations the band can make.
   d. At one point in the show, the woodwind and brass sections of the band will march toward each other from the end zones in rows across the field of equal length. If there are 30 woodwinds and 50 in the brass section, what is the longest row possible? What is this number called?
   e. In the finale, band members in rows of 10 perform with members of the flag corp in rows of 6. What is the LCM of 6 and 10? If he wishes to use the same number of band members as flag corp members, what is the least number of rows of each that he can use?

2. A restaurant owner is planning a menu for the next day.
   a. She has $2\frac{2}{3}$ apple pies left. If each serving is one-sixth of a pie, how many servings does she have left? Write the answer as an improper fraction and as a decimal.
   b. She sells about 50 servings of pecan pie a day. How many pecan pies does she usually sell a day?
   c. If she is sold out of pecan pie, how many pies should she order for the next day?
   d. Write a word problem that uses mixed numbers. Solve and give the meaning of the answer.

3. Dominique is trimming a picture to fit in a frame. The frame measures 0.75 foot by 0.5 foot.
   a. Change the decimals to fractions.
   b. What are the measurements of the frame in inches? Show you how you determine this.

© Glencoe/McGraw-Hill   125   *Mathematics: Applications and Connections, Course 1*

Performance Assessment

Additional performance assessment tasks for this chapter are included in the *Assessment and Evaluation Masters* on page 125. A scoring guide is also provided on page 137.

Additional Answer for the Performance Task

The LCM is 24. So, 4 packages of cards, 3 packages of candy, and 6 packages of modeling dough should be purchased. If 15 children attend the party, 3 packages of cards, 2 packages of candy, and 4 packages of modeling dough should be purchased.

Standardized Test Practice

The Standardized Test Practice may be used to help students prepare for standardized tests. The test items are written in the same style as those in state proficiency tests and standardized tests like CAT, CTBS, ITBS, MAT, SAT, and Terra Nova. The test items cover skills and concepts covered up to this point in the text.

The pages can be used as an overnight assessment. After students have completed the pages, discuss how each problem can be solved, or provide copies of the solutions from the *Solutions Manual.*

Assessment and Evaluation Masters, p. 131

Cumulative Review, Chapters 1–5

Perform the indicated operation.

1. 56)190.4 ..... 1. ___3.4___
2. 18 × 3.52 ..... 2. ___63.36___
3. 7.284 ÷ 1.2 ..... 3. ___6.07___

Evaluate.

4. 7 × 12 + 6 × 12 ..... 4. ___156___
5. n^3 if $n = 13$ ..... 5. ___2,197___

Estimate.

6. $3.98 + $5.13 ..... 6. ___$4 + $5 = $9___
7. 9.2 × 28.3 ..... 7. ___9 × 30 = 270___

8. Find the best interval for a frequency table for these scores: 91, 88, 83, 83, 90. ..... 8. ___2___
9. Find the median of the following scores: 17, 13, 25, 22, 13. ..... 9. ___17___
10. A United States one-dollar bill is 15.7 centimeters long and 6.6 centimeters wide. Find the perimeter. ..... 10. ___44.6 cm___
11. Complete: 63 in. = _?_ ft. ..... 11. ___$5\frac{1}{4}$ ft___
12. Change 3.8 kilograms to grams. ..... 12. ___3,800 g___
13. Express $\frac{31}{5}$ as a mixed number. ..... 13. ___$6\frac{1}{5}$___

Write as a decimal.

14. $\frac{81}{10,000}$ ..... 14. ___0.0081___
15. $\frac{15}{16}$ ..... 15. ___0.9375___

Order from least to greatest for Questions 16–17.

16. 0.006, 0.2, 0.03, 0.0008 ..... 16. ___0.0008, 0.006, 0.03, 0.2___
17. $\frac{3}{4}, \frac{1}{8}, \frac{15}{16}, \frac{1}{2}$ ..... 17. ___$\frac{1}{4}, \frac{3}{8}, \frac{1}{2}, \frac{15}{16}$___
18. Find the GCF of 18 and 30. ..... 18. ___6___
19. Express 0.06 as a fraction in simplest form. ..... 19. ___$\frac{3}{50}$___
20. Find the LCM for 14, 15, and 21. ..... 20. ___210___

© Glencoe/McGraw-Hill ..... 131 ..... *Mathematics: Applications and Connections, Course 1*

Section One: Multiple Choice

There are thirteen multiple-choice questions in this section. Choose the best answer. If a correct answer is *not here*, choose the letter for Not Here.

1. Which is a true sentence? **A**

   A $0.5 > 0.48$

   B $0.48 > 0.5$

   C $1.4 > 2.5$

   D $5.01 < 4.08$

2. Find the area of the rectangle. **H**

   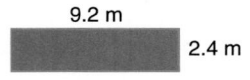

   9.2 m

   2.4 m

   F 11.8 m^2 **G** 23.28 m^2

   H 22.08 m^2 **J** 90.4 m^2

3. Shalena's birthday is 7 weeks and 5 days away. How many days away is her birthday? **B**

   A 49 days

   B 54 days

   C 35 days

   D 12 days

4. Simplify the fraction $\frac{18}{144}$. **G**

   F $\frac{1}{6}$ **G** $\frac{1}{8}$

   H $\frac{1}{14}$ **J** $\frac{1}{12}$

5. Xavier High School has a student enrollment of 2,155 students. What is this number rounded to the nearest thousand? **A**

   A 2,000

   B 2,100

   C 2,150

   D 2,200

Please note that Questions 6–13 have five answer choices.

6. Greg purchased 4 paperback books. Each book costs between $4.99 and $8.99. What is a reasonable total cost for the books? **G**

   F $10   **G** $30

   H $15   **J** $45

   K $50

7. Mitena is buying 3 binders that cost $2.89 each and 4 pens that cost $0.49 each. Which expression can be used to find the total cost of the items? **B**

   A $3 \times 2.89 \times 4 \times 0.49$

   B $3 \times 2.89 + 4 \times 0.49$

   C $3 + 2.89 + 4 + 0.49$

   D $3 + 2.89 \times 4 + 0.49$

   E $3 \times 2.89 \div 4 \times 0.49$

8. The graph shows the number of teams who entered the annual 3-on-3-basketball tournament from 1994 to 1998.

   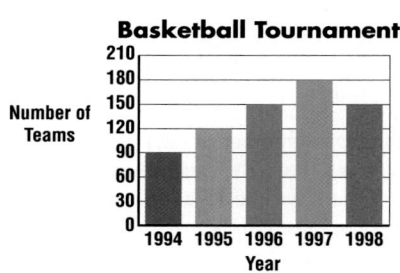

   Basketball Tournament

   How many more teams entered the tournament in 1997 than in 1995? **F**

   F 60   **G** 30

   H 90   **J** 120

   K 150

◀◀◀ **Instructional Resources**

Another cumulative review is shown at the left and is available in the *Assessment and Evaluation Masters,* p. 131.

9. If $n = 4.35 \div 1.5$, what is the value of n to the nearest tenth? **A**

A 2.9

B 0.09

C 0.29

D 1.2

E Not Here

10. Nashala wants to buy movie tickets for herself and three friends. Each ticket costs $2.95. Which is the best estimate for the total cost of the tickets? **J**

F $3

G $6

H $9

J $12

K Not Here

11. Find the value of $23 - 6 + 2 \times 5$. **C**

A 195

B 7

C 27

D 29

E Not Here

12. The Framing Center charges $12.95 to frame one 3-inch by 5-inch picture. How much would it cost to frame three 3-inch by 5-inch pictures? **H**

F $35.95

G $42.99

H $38.85

J $45.79

K Not Here

13. Solve $11.05 \div 0.65$. **D**

A 9

B 15

C 14

D 17

E Not Here

Test-Taking Tip

It is a good idea to review formulas before taking a standardized test. Usually the formulas are given in the test booklet, but reviewing the formulas before taking a test may give you an advantage.

Section Two: Free Response

This section contains six questions for which you will provide short answers. Write your answers on your paper. **16.** $4\frac{57}{125}$

14. Erika's bowling scores were 119, 134, 135, 125, 143, and 135. What is the mode of her scores? **135**

15. What is the value of $y + 42$ if $y = 87$? **129**

16. Express the decimal 4.456 as a mixed number in simplest form.

17. Levon is shopping for groceries. He buys a 12-pack of cola for $3.98, a bottle of laundry detergent for $4.79, and 2 frozen pizzas for $5.86 each. What is the total cost of the items before tax is added? **$20.49**

18. What is the perimeter of a rectangular bulletin board that is 4 feet wide and 6 feet long? **20 ft**

19. Angel purchased a red, a black, and a blue T-shirt. The total cost for the shirts was $14.37. How much change should he receive from a $20 bill? **$5.63**

Chapters 1–5 Standardized Test Practice   **225**

Instructional Resources ▶▶▶
Additional standardized test practice is shown at the right and is available in the *Assessment and Evaluation Masters*, pp. 129–130.

Test-Taking Tip

Not having to look up formulas can save students time during a test. Encourage them to review the formulas by putting them on flash cards so that they can test themselves.

Assessment and Evaluation Masters, pp. 129–130

5 Chapter 5 Standardized Test Practice

1. The volume of a box is found by multiplying its height, length, and width. If the measure of the volume of a box is 175, what could its dimensions be?   **1. C**
   A. $3 \times 5 \times 9$ B. $3 \times 5 \times 7$ C. $5 \times 5 \times 7$ D. $5 \times 5 \times 9$

2. Find the GCF of 16, 36, and 40.   **2. A**
   A. 4   B. 2   C. 6   D. 1

3. Ethan is training for a race. He ran $\frac{17}{4}$ miles. Express this distance as a mixed number.   **3. B**
   A. $1\frac{7}{4}$ B. $4\frac{1}{4}$ C. $3\frac{5}{4}$ D. $4\frac{1}{17}$

4. Taro feeds his kitten 0.625 of a can of food. What fraction is this in simplest form?   **4. D**
   A. $6\frac{2}{5}$ B. $\frac{3}{5}$ C. $62\frac{1}{2}$ D. $\frac{5}{8}$

5. Rosario is taking trombone lessons for $22 per week. If he takes 8 weeks of lessons, how much do they cost?   **5. C**
   A. $17.60   B. $30   C. $176   D. $200

6. Jacqueline bought 8 yards of ribbon for $4.96. How much was the ribbon per yard?   **6. A**
   A. $0.62   B. $0.40   C. $0.71   D. $0.58

7. Mistyville received 24.72 inches of rain in 24 hours. How much did it rain per hour?   **7. A**
   A. 1.03 in.   B. 1.3 in.   C. 0.103 in.   D. 10.3 in.

8. The distance horses run in the Kentucky Derby is $1\frac{1}{4}$ miles. Write this distance as a decimal.   **8. C**
   A. 1.025   B. 1.40   C. 1.25   D. 0.0125

9. One mile equals 1.609 kilometers. Round this measurement to the nearest tenth.   **9. A**
   A. 1.6 km   B. 1.61 km   C. 1.7 km   D. 2 km

10. A photo is 3.5 inches high. The frame is 7.77 inches high. Estimate how much total extra space there will be between the photo and the frame.   **10. C**
    A. 3.5 in.   B. 3 in.   C. 4 in.   D. 5 in.

© Glencoe/McGraw-Hill   129   Mathematics: Applications and Connections, Course 1

---

5 Chapter 5 Standardized Test Practice (continued)

11. Choose the best interval for a frequency table for the data shown.   **11. B**
    51, 87, 42, 53, 28, 12, 17, 72, 43
    A. 4   B. 10   C. 5   D. 20

12. Bobbie's science project requires her to track her plant's growth. Refer to the graph and predict the plant's height at 12 weeks.   **12. D**
    A. 15 in.   B. 25 in.   C. 18 in.   D. 20 in.
    Plant Growth

13. Determine the stems for the data shown.   **13. A**
    5, 1, 22, 18, 7, 24, 41
    A. 0, 1, 2, 3, 4   B. 1, 2, 4
    C. 1, 2, 3, 4   D. 0, 1, 2, 4

14. The distance from Berlin to Moscow is 996 miles. The distance from Moscow to Tokyo is 4,650 miles. About how far is it from Berlin to Tokyo if you go through Moscow?   **14. C**
    A. 1,000 miles   B. 4,700 miles
    C. 5,700 miles   D. 6,000 miles

15. Evaluate $3a + 2b - 4 \div c$ if $a = 4$, $b = 5$, and $c = 2$.   **15. B**
    A. 9   B. 20   C. 33   D. 28

16. Joanne orders a T-shirt for a total cost of $15. Shipping and handling is $2.50. If t stands for the cost of the T-shirt, the equation $15 = 2.50 + t$ results. Find the cost of the T-shirt.   **16. D**
    A. $17.50   B. $10.00   C. $2.50   D. $12.50

17. A bank is paying 5% interest. Use prt to find the interest on $500 invested for 3.5 years. Use $p = 500$, $r = 0.5$, and $t = 3.5$.   **17. A**
    A. $87.50   B. $875   C. $8.75   D. $75

18. A pygmy shrew weighs 4.53 grams and a mouse weighs 22.40 grams. What is the difference in their weights?   **18. B**
    A. 22.9 g   B. 17.87 g   C. 67.7 g   D. 93 g

19. Shayla purchased $7\frac{7}{8}$ yards of material. Find the decimal equivalent for the amount of material purchased.   **19. D**
    A. 7.75 yd   B. 7.85 yd   C. 7.78 yd   D. 7.875 yd

20. Find the mean distance from the Sun in millions of miles if Venus is 67 million miles, Earth is 93 million miles, and Mars is 142 million miles from the Sun.   **20. A**
    A. 100.67 mi   B. 116 mi   C. 96.5 mi   D. 93 mi

© Glencoe/McGraw-Hill   130   Mathematics: Applications and Connections, Course 1

Adding and Subtracting Fractions

Previewing the Chapter

Overview

This chapter explores adding and subtracting fractions. Students round fractions and mixed numbers and estimate sums and differences. They learn to add and subtract fractions with like denominators, fractions with unlike denominators, and mixed numbers. Connections include problem solving by eliminating possibilities and adding and subtracting measures of time.

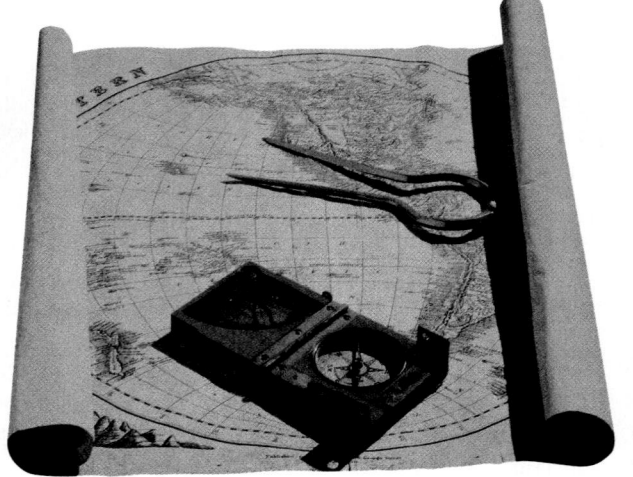

Lesson (pages)	Lesson Objectives	NCTM Standards	Standardized Tests	State/Local Objectives
6-1 (228–231)	Round fractions and mixed numbers.	1–7, 12, 13	SAT	
6-2 (232–234)	Estimate sums and differences of fractions and mixed numbers.	1–7, 12, 13	CTBS, TN	
6-2B (236–237)	Solve problems by eliminating possibilities.	1–7	MAT	
6-3 (238–241)	Add and subtract fractions with like denominators.	1–7, 12, 13	CAT, CTBS, ITBS, MAT, SAT, TN	
6-4A (242)	Find common unit names for adding different objects.	1–3, 10		
6-4 (243–245)	Add and subtract fractions with unlike denominators.	1–7, 9, 10	CAT, CTBS, ITBS, MAT, SAT, TN	
6-5 (246–249)	Add and subtract mixed numbers.	1–7, 12, 13	CAT, CTBS, SAT, TN	
6-6 (250–253)	Subtract mixed numbers involving renaming.	1–7, 9, 10	CTBS, SAT, TN	
6-7 (254–257)	Add and subtract measures of time.	1–7, 13		

CAT = California Achievement Tests, CTBS = Comprehensive Tests of Basic Skills, ITBS = Iowa Tests of Basic Skills,
MAT = Metropolitan Achievement Tests, SAT = Stanford Achievement Tests, TN = Terra Nova

Organizing the Chapter

CD-ROM

 All of the blackline masters in the Teacher's Classroom Resources are available on the **Electronic Teacher's Classroom Resources** CD-ROM.

LESSON PLANNING GUIDE

Lesson	Extra Practice (Student Edition)	Blackline Masters (Page Numbers)										
		Study Guide	Practice	Enrichment	Assessment & Evaluation	Classroom Games	Diversity	Hands-On Lab	School to Career	Science and Math Lab Manual	Technology	Transparencies A and B
6-1	p. 572	42	42	42					6			6-1
6-2	p. 572	43	43	43	155							6-2
6-2B	p. 573											
6-3	p. 573	44	44	44			6	74				6-3
6-4A								50				
6-4	p. 573	45	45	45	154, 155						12	6-4
6-5	p. 574	46	46	46		15–18						6-5
6-6	p. 574	47	47	47	156							6-6
6-7	p. 574	48	48	48	156						11	6-7
Study Guide/ Assessment					141–153, 157–159							

OTHER CHAPTER RESOURCES

Student Edition
Chapter Project, pp. 227, 231, 252, 261
School to Career, p. 235
Let the Games Begin, p. 253

Technology
 CD-ROM Program
 Interactive Mathematics Tools Software

Teacher's Classroom Resources

Applications
Family Letters and Activities, pp. 11–12
Investigations and Projects Masters, pp. 37–40
Meeting Individual Needs
Investigations for the Special Education Student, pp. 15, 17–18

Teaching Aids
Answer Key Masters
Block Scheduling Booklet
Lesson Planning Guide
Solutions Manual

Professional Publications
Glencoe Mathematics Professional Series

Planning the Chapter

MindJogger Videoquizzes provide a unique format for reviewing concepts presented in the chapter.

ASSESSMENT RESOURCES

Student Edition
Mixed Review, pp. 231, 234, 241, 245, 249, 253, 257
Mid-Chapter Self Test, p. 249
Math Journal, pp. 230, 255
Study Guide and Assessment, pp. 258–261
Performance Task, p. 261
Portfolio Suggestion, p. 261
Standardized Test Practice, pp. 262–263
Chapter Test, p. 600

Assessment and Evaluation Masters
Multiple-Choice Tests (Forms 1A, 1B, 1C), pp. 141–146
Free-Response Tests (Forms 2A, 2B, 2C), pp. 147–152
Performance Assessment, p. 153
Mid-Chapter Test, p. 154
Quizzes A–D, pp. 155–156
Standardized Test Practice, pp. 157–158
Cumulative Review, p. 159

Teacher's Wraparound Edition
5-Minute Check, pp. 228, 232, 238, 243, 246, 250, 254
Building Portfolios, p. 226
Math Journal, p. 242
Closing Activity, pp. 231, 234, 237, 241, 245, 249, 253, 257

Technology
Test and Review Software
MindJogger Videoquizzes
CD-ROM Program

MATERIALS AND MANIPULATIVES

Lesson 6-3
grid paper†
colored pencils

Lesson 6-4A
pennies
nickels
pencils
pens

Lesson 6-5
paper plates
scissors*

Lesson 6-6
number cubes*
spinner*†

*Glencoe Manipulative Kit   †Glencoe Overhead Manipulative Resources

PACING CHART

See pages T25–T27 for the Course Planning Calendar.

COURSE	DAY 1	DAY 2	DAY 3	DAY 4	DAY 5	DAY 6	DAY 7
Standard	Chapter Project	Lesson 6-1	Lesson 6-2	Lesson 6-2B	Lesson 6-3	Lessons 6-4A & 6-4	
Honors	Chapter Project	Lesson 6-1	Lesson 6-2	Lesson 6-2B	Lesson 6-3	Lesson 6-4	Lesson 6-5
Block	Chapter Project & Lesson 6-1	Lessons 6-2 & 6-2B	Lessons 6-3 & 6-4A	Lesson 6-4 & 6-5	Lesson 6-6	Lesson 6-7	Study Guide and Assessment, Chapter Test

Interactive Mathematics:
Activities and Investigations

is an activity-based program that may be used as an enhancement for chapters in *Mathematics: Applications and Connections.*

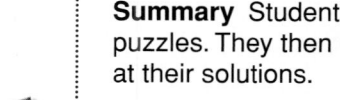

Unit 5, Activity One, Menu A
Use with Lesson 6-2B.

Summary Students work in groups to solve two cooperative logic puzzles. They then explain in writing the methods they used to arrive at their solutions.

Math Connection Students work cooperatively in groups to determine their own strategies in order to solve two logic puzzles. Each group is given two envelopes, one per puzzle, with the instructions for the puzzle on the outside and clue cards on the inside.

Unit 6, Pre-Assessment Activity
Use with Lesson 6-7.

Summary Students are given a map and a problem situation. They work in pairs or groups to determine the best route to travel and at which time they should begin their trip to ensure on-time arrival. Each group gives an oral presentation of their decisions.

Math Connection Students will be using their knowledge of discrete mathematics. Discrete mathematics is the study of objects and ideas that can be divided into distinct parts.

DAY 8	DAY 9	DAY 10	DAY 11	DAY 12	DAY 13	DAY 14	DAY 15
Lesson 6-5	Lesson 6-6		Lesson 6-7	Study Guide and Assessment	Chapter Test		
	Lesson 6-6		Lesson 6-7	Study Guide and Assessment	Chapter Test		

Enhancing the Chapter

APPLICATIONS

Classroom Games, pp. 15–18

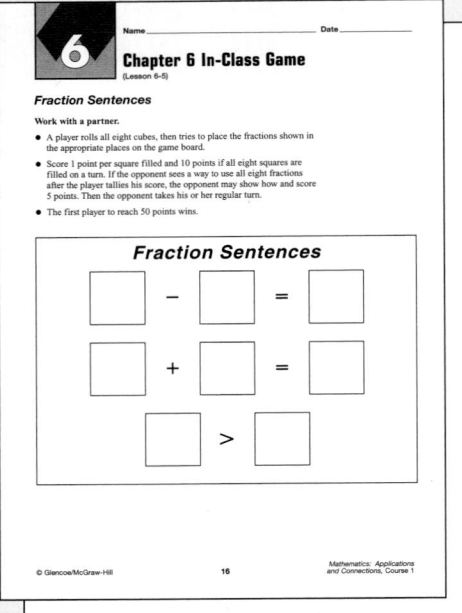

6 Name _____ Date _____

Chapter 6 In-Class Game
(Lesson 6-5)

Fraction Sentences

Work with a partner.
- A player rolls all eight cubes, then tries to place the fractions shown in the appropriate places on the game board.
- Score 1 point per square filled on a turn. If the opponent sees a way to use all eight fractions after the player tallies his score, the opponent may show how and score 5 points. Then the opponent takes his or her regular turn.
- The first player to reach 50 points wins.

Fraction Sentences

□ − □ = □

□ + □ = □

□ > □

© Glencoe/McGraw-Hill   16   Mathematics: Applications and Connections, Course 1

Diversity Masters, p. 6

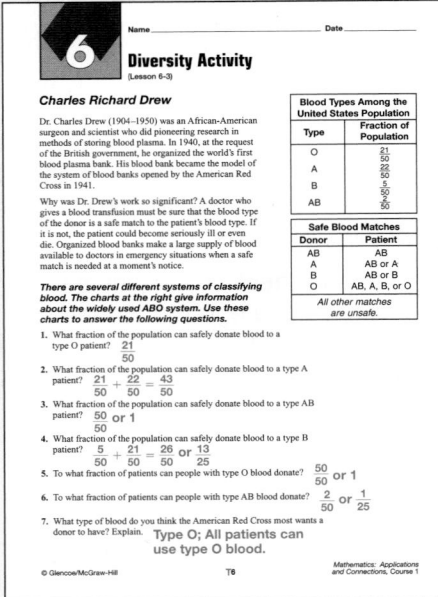

6 Name _____ Date _____

Diversity Activity
(Lesson 6-3)

Charles Richard Drew

Dr. Charles Drew (1904–1950) was an African-American surgeon and scientist who did pioneering research in methods of storing blood plasma. In 1940, at the request of the British government, he organized the world's first blood plasma bank. His blood bank became the model of the system of blood banks opened by the American Red Cross in 1941.

Why was Dr. Drew's work so significant? A doctor who gives a blood transfusion must be sure that the blood type of the donor is a safe match to the patient's blood type. If it is not, the patient could become seriously ill or even die. Organized blood banks make a large supply of blood available to doctors in emergency situations when a safe match is needed at a moment's notice.

There are several different systems of classifying blood. The charts at the right give information about the widely used ABO system. Use these charts to answer the following questions.

Blood Types Among the United States Population	
Type	Fraction of Population
O	$\frac{21}{50}$
A	$\frac{22}{50}$
B	$\frac{5}{50}$
AB	$\frac{2}{50}$

Safe Blood Matches	
Donor	Patient
AB	AB
A	AB or A
B	AB or B
O	AB, A, B, or O

All other matches are unsafe.

1. What fraction of the population can safely donate blood to a type O patient? $\frac{21}{50}$

2. What fraction of the population can safely donate blood to a type A patient? $\frac{21}{50} + \frac{22}{50}$ $\frac{43}{50}$

3. What fraction of the population can safely donate blood to a type AB patient? $\frac{50}{50}$ or 1

4. What fraction of the population can safely donate blood to a type B patient? $\frac{5}{50} + \frac{21}{50}$ $\frac{26}{50}$ or $\frac{13}{25}$

5. To what fraction of patients can people with type O blood donate? $\frac{50}{50}$ or 1

6. To what fraction of patients can people with type AB blood donate? $\frac{2}{50}$ or $\frac{1}{25}$

7. What type of blood do you think the American Red Cross most wants a donor to have? Explain. Type O; All patients can use type O blood.

© Glencoe/McGraw-Hill   T6   Mathematics: Applications and Connections, Course 1

School to Career Masters, p. 6

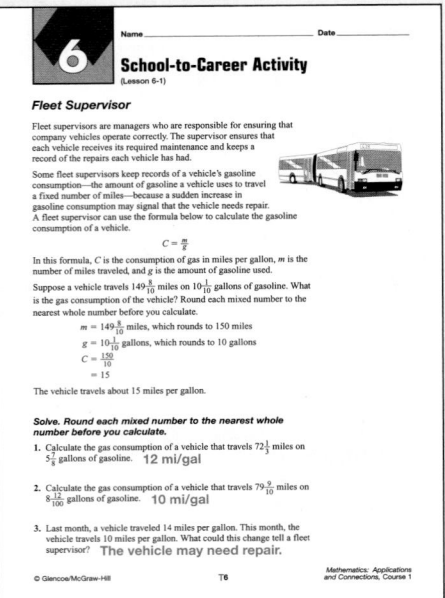

6 Name _____ Date _____

School-to-Career Activity
(Lesson 6-1)

Fleet Supervisor

Fleet supervisors are managers who are responsible for ensuring that company vehicles operate correctly. The supervisor ensures that each vehicle receives its required maintenance and keeps a record of the repairs each vehicle has had.

Some fleet supervisors keep records of a vehicle's gasoline consumption—the amount of gasoline a vehicle uses to travel a fixed number of miles—because a sudden increase in gasoline consumption may signal that the vehicle needs repair. A fleet supervisor can use the formula below to calculate the gasoline consumption of a vehicle.

$$C = \frac{m}{g}$$

In this formula, C is the consumption of gas in miles per gallon, m is the number of miles traveled, and g is the amount of gasoline used.

Suppose a vehicle travels $149\frac{8}{10}$ miles on $10\frac{1}{10}$ gallons of gasoline. What is the gas consumption of the vehicle? Round each mixed number to the nearest whole number before you calculate.

$m = 149\frac{8}{10}$ miles, which rounds to 150 miles
$g = 10\frac{1}{10}$ gallons, which rounds to 10 gallons
$C = \frac{150}{10}$
$= 15$

The vehicle travels about 15 miles per gallon.

Solve. Round each mixed number to the nearest whole number before you calculate.

1. Calculate the gas consumption of a vehicle that travels $72\frac{1}{8}$ miles on $5\frac{5}{8}$ gallons of gasoline. 12 mi/gal

2. Calculate the gas consumption of a vehicle that travels $79\frac{9}{10}$ miles on $8\frac{12}{100}$ gallons of gasoline. 10 mi/gal

3. Last month, a vehicle traveled 14 miles per gallon. This month, the vehicle travels 10 miles per gallon. What could this change tell a fleet supervisor? The vehicle may need repair.

© Glencoe/McGraw-Hill   T6   Mathematics: Applications and Connections, Course 1

Family Letters and Activities, pp. 11–12

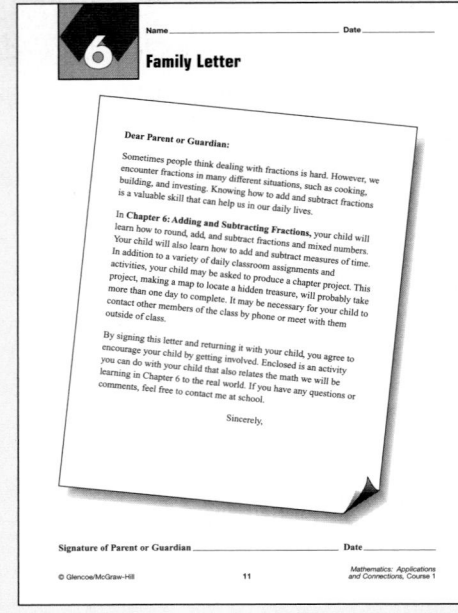

6 Name _____ Date _____

Family Letter

Dear Parent or Guardian:

Sometimes people think dealing with fractions is hard. However, we encounter fractions in many different situations, such as cooking, building, and investing. Knowing how to add and subtract fractions is a valuable skill that can help us in our daily lives.

In **Chapter 6: Adding and Subtracting Fractions**, your child will learn how to round, add, and subtract fractions and mixed numbers. Your child will also learn how to add and subtract measures of time. In addition to a variety of daily classroom assignments and activities, your child may be asked to produce a chapter project. This project, making a map to locate a hidden treasure, will probably take more than one day to complete. It may be necessary for your child to contact other members of the class by phone or meet with them outside of class.

By signing this letter and returning it with your child, you agree to encourage your child by getting involved. Enclosed is an activity you can do with your child that also relates the math we will be learning in Chapter 6 to the real world. If you have any questions or comments, feel free to contact me at school.

Sincerely,

Signature of Parent or Guardian _____ Date _____

© Glencoe/McGraw-Hill   11   Mathematics: Applications and Connections, Course 1

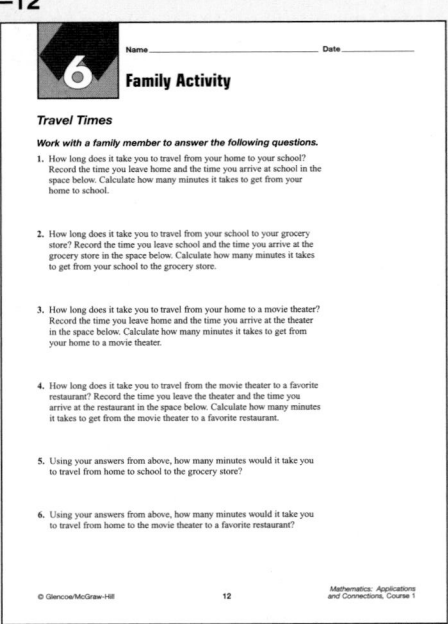

6 Name _____ Date _____

Family Activity

Travel Times

Work with a family member to answer the following questions.

1. How long does it take you to travel from your home to your school? Record the time you leave home and the time you arrive at school in the space below. Calculate how many minutes it takes to get from your home to school.

2. How long does it take you to travel from your school to your grocery store? Record the time you leave school and the time you arrive at the grocery store in the space below. Calculate how many minutes it takes to get from your school to the grocery store.

3. How long does it take you to travel from your home to a movie theater? Record the time you leave home and the time you arrive at the theater in the space below. Calculate how many minutes it takes to get from your home to a movie theater.

4. How long does it take you to travel from the movie theater to a favorite restaurant? Record the time you leave the theater and the time you arrive at the restaurant in the space below. Calculate how many minutes it takes to get from the movie theater to a favorite restaurant.

5. Using your answers from above, how many minutes would it take you to travel from home to school to the grocery store?

6. Using your answers from above, how many minutes would it take you to travel from home to the movie theater to a favorite restaurant?

© Glencoe/McGraw-Hill   12   Mathematics: Applications and Connections, Course 1

Hands-On Lab Masters, p. 74

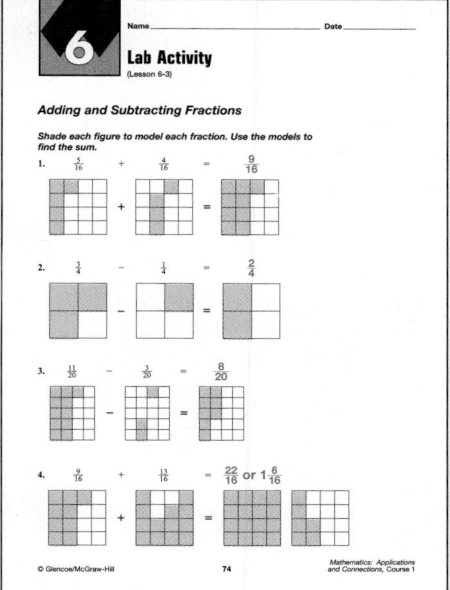

Assessment and Evaluation Masters, pp. 154–156

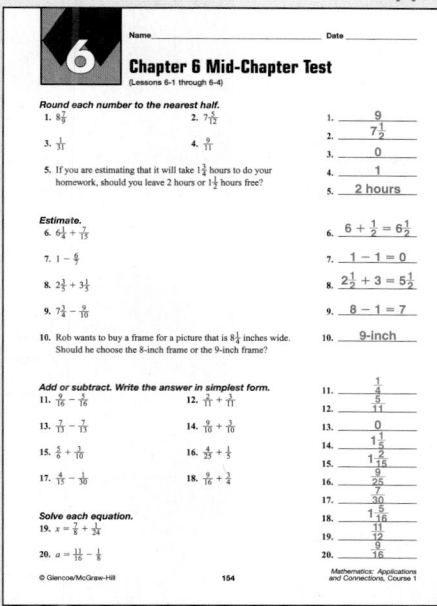

Technology Masters, pp. 11–12

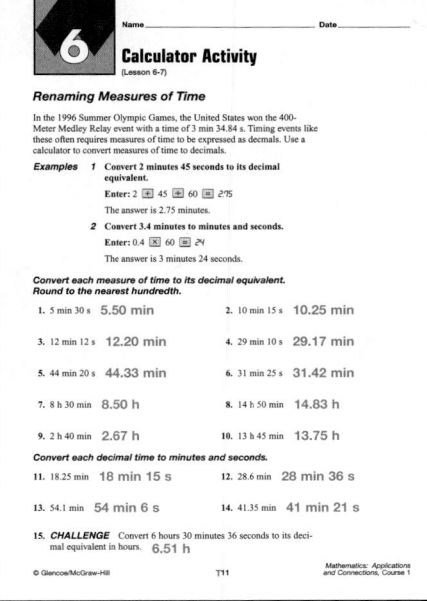

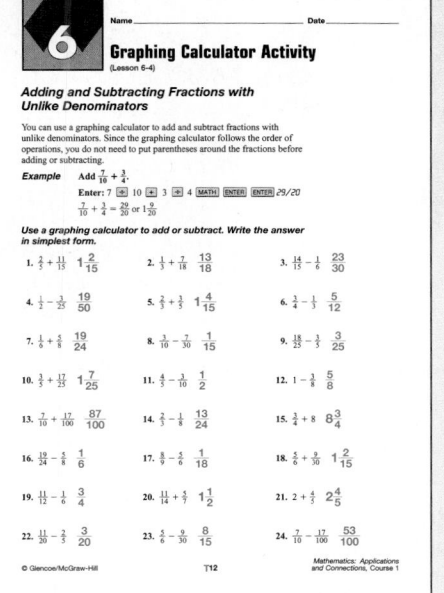

Investigations for the Special Education Student, pp. 15, 17–18

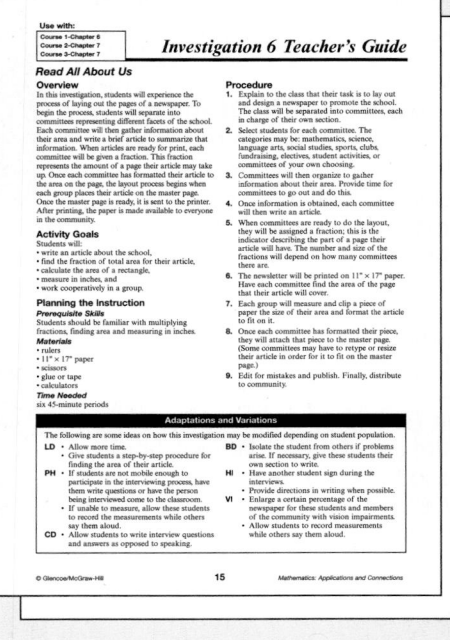

Theme: Maps

Buried treasures can be found with thorough research and modern technology. In 1988, Tommy Thompson and his crew found the remains of the *Central America* off the coast of South Carolina. Their researchers used first-hand narratives and newspaper accounts to map the location of the sunken ship. Today, remote-controlled vehicles and nuclear submarines are used to plumb the depths of the oceans looking for lost ships and their treasures.

Question of the Day Jeanne is following directions to Raphael's house. She travels $1\frac{1}{2}$ miles and turns left, $\frac{3}{4}$ mile and turns right, $\frac{1}{4}$ mile and turns left, and $2\frac{5}{8}$ miles and turns right. Then she continues $\frac{1}{3}$ mile to Raphael's house. How far has Jeanne traveled?
$5\frac{11}{24}$ miles

Assess Prerequisite Skills

Ask students to read through the list of objectives presented in "What you'll learn in Chapter 6." You may wish to ask them what each of the objectives means or if they have experienced or used any of these math concepts before.

 Building Portfolios

Encourage students to revise their portfolios as they study this chapter. In addition to their maps, you may wish to have them include examples of the work they did to create the maps.

 **Math and the Family**

In the *Family Letters and Activities* booklet (pp. 11–12), you will find a letter to the parents explaining what students will study in Chapter 6. An activity appropriate for the whole family is also available.

Adding and Subtracting Fractions

What you'll learn in Chapter 6

- to round fractions and mixed numbers,
- to estimate and find sums and differences of fractions and mixed numbers,
- to solve problems by eliminating possibilities, and
- to add and subtract measures of time.

226 Chapter 6 Adding and Subtracting Fractions

 CD-ROM Program

Activities for Chapter 6
- Chapter 6 Introduction
- Interactive Lessons 6-3, 6-5, 6-6
- Assessment Game
- Resource Lessons 6-1 through 6-7

CHAPTER Project

TRAIL TO THE TREASURE TROVE

Countless stories and movies have been written about people in search of hidden or lost treasure. Often, these people found or bought maps describing the location of the treasure. In this project, you will create and draw your own map for locating a hidden treasure. You will then provide instructions so that someone can find your treasure by using your map.

Getting Started

● On your map, the distance between the starting point and the treasure, in a straight line, must be $5\frac{3}{4}$ inches.
● You will use the landmarks and distances shown in the table when you draw your map.

Landmarks	Distance
starting point to first landmark	$\frac{7}{8}$ in.
center of fountain to tallest oak tree	$1\frac{3}{8}$ in.
tallest oak tree to bench near pond	$2\frac{1}{2}$ in.
bench near pond to seal-shaped rock	$1\frac{5}{8}$ in.
last landmark to treasure	$1\frac{1}{8}$ in.

Technology Tips

● Use a **word processor** to list your instructions.
● Use **computer software** to draw your map.
● Use an **electronic encyclopedia** to learn more about maps and map making.

 *inter*NET CONNECTION For up-to-date information on maps and map making, visit:
www.glencoe.com/sec/math/mac/mathnet

Working on the Project

You can use what you'll learn in Chapter 6 to help you make your treasure map.

Page	Exercise
231	34
252	37
261	Alternative Assessment

 *inter*NET CONNECTION

Glencoe has made every effort to ensure that the website links for *Mathematics: Applications and Connections* at **www.glencoe.com/sec/math/mac/mathnet** are current and contain appropriate content. However, these website links are not under Glencoe's control.

CHAPTER Project NOTES

Objectives Students should
● create and draw a map for locating hidden treasure.
● provide instructions to find the treasure using the map.

Project Pointer You may suggest that students begin a *Project Folder* to keep their work as they complete each stage of the Chapter Project. The completed project may also be added to their portfolios.

Have students develop a plan for creating a map that meets all of the distance requirements. Encourage them to see how many different ways they can place the landmarks and still have a total distance of $5\frac{3}{4}$ inches from the starting point to the treasure. Also point out that several requirements for the map will be given in the chapter.

Using the Table Make sure students understand that in addition to using the distances given for each landmark, they must also follow the guideline given for the total distance from the starting point to the treasure.

***Investigations and Projects Masters*, p. 40**

6 | Name_____ Date_____
Chapter 6 Project

Trail to the Treasure Trove

© Glencoe/McGraw-Hill   40   *Mathematics: Applications and Connections, Course 1*

Instructional Resources ▶▶▶
A recording sheet to help students organize their data for the Chapter Project is shown at the right and is available in the *Investigations and Projects Masters,* p. 40.

Chapter 6 Project   **227**

- *Study Guide Masters*, p. 42
- *Practice Masters*, p. 42
- *Enrichment Masters*, p. 42
- Transparencies 6-1, A and B
- *School to Career Masters*, p. 6

CD-ROM Program
- Resource Lesson 6-1

Recommended Pacing	
Standard	Day 2 of 13
Honors	Day 2 of 12
Block	Day 1 of 7

1 FOCUS

5-Minute Check
(Chapter 5)

1. Find the GCF of 12 and 42. **6**

2. Express $\frac{21}{5}$ as a mixed number. **$4\frac{1}{5}$**

3. Find the LCM of 6, 8, and 12. **24**

4. Order $\frac{3}{10}$, $\frac{1}{5}$, $\frac{3}{4}$, and $\frac{1}{2}$ from least to greatest. **$\frac{1}{5}, \frac{3}{10}, \frac{1}{2}, \frac{3}{4}$**

5. Express $4\frac{1}{3}$ as a decimal. Use bar notation to show a repeating decimal. **$4.\overline{3}$**

The 5-Minute Check is also available on **Transparency 6-1A** for this lesson.

Motivating the Lesson
Hands-On Activity Use transparency overlays to illustrate rounding fractions to the nearest whole number or half. For example, show $\frac{3}{7}$ rounded to $\frac{1}{2}$ by drawing a rectangle, shading $\frac{3}{7}$, and then using a dashed line to separate the model into halves. Repeat for $\frac{4}{5}$ and $\frac{1}{6}$.

6-1 Rounding Fractions and Mixed Numbers

What you'll learn
You'll learn to round fractions and mixed numbers.

When am I ever going to use this?
Knowing how to round fractions and mixed numbers can help you order enough food for a party.

Sophia is making a skirt to wear in her class play. The pattern for the skirt calls for $2\frac{7}{8}$ yards of fabric. She rounds the amount of fabric needed to the nearest whole number because she can use the extra material to make a matching headband.

To round fractions and mixed numbers to the nearest unit, you can use the following guidelines. A number line can help you decide how to round.

LOOK BACK
You can refer to Lesson 1-3 to review rounding.

- If the numerator is almost as large as the denominator, round the number up to the next whole number.

    $\frac{7}{8}$ rounds to 1.   *7 is almost as large as 8.*

- If the numerator is about half of the denominator, round the fraction to $\frac{1}{2}$.

    $\frac{3}{5}$ rounds to $\frac{1}{2}$.   *3 is about half of 5.*

- If the numerator is much smaller than the denominator, round the number down to the next whole number.

    $2\frac{3}{16}$ rounds to 2.   *3 is much smaller than 16.*

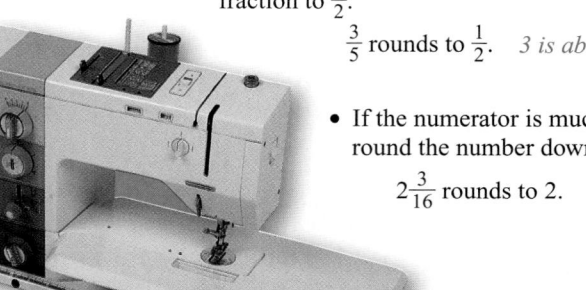

Example 1 Round $3\frac{4}{5}$ to the nearest half.

$3\frac{4}{5}$ is closer to 4 than $3\frac{1}{2}$.

The numerator is almost as large as the denominator. $3\frac{4}{5}$ rounds to 4.

Cross-Curriculum Cue

Inform the teachers on your team that your students are studying addition and subtraction of fractions. Suggestions for curriculum integration are:

Health: growth charts
Family and Consumer Science: cooking and sewing measurements
Industrial Technology: measurements of materials

2 Round $\frac{3}{8}$ to the nearest half.

$\frac{3}{8}$ is closer to $\frac{1}{2}$ than 0.

The numerator is about half of the denominator. $\frac{3}{8}$ rounds to $\frac{1}{2}$.

INTEGRATION **3** **Measurement** **Find the length of the line segment to the nearest one-half inch.**

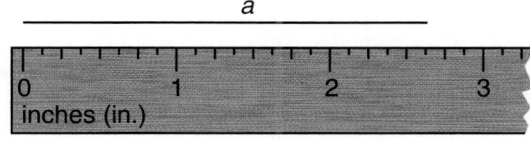

$2\frac{5}{8}$ inches is closer to $2\frac{1}{2}$ inches than to 3 inches. Therefore, to the nearest one-half inch, the length of the line segment is $2\frac{1}{2}$ inches.

As shown in the following example, you should round a number down when it is better for a measure to be too small than too large.

Example **4** **Cooking** **Suppose you are camping and you only have one cup of water left to use for cooking. You think the recipe for a pancake mix needs $\frac{2}{3}$ cup of water. Should you round down to one-half cup or round up to one cup?**

APPLICATION

To avoid making the mix too thin, you should round down to one-half cup of water. You can always add more.

Sometimes it is necessary to round a number up despite what the rule says.

Example **5** **Measurement** **Ahn is being fitted for a tuxedo to wear in his sister's wedding. His neck measured $14\frac{3}{4}$ inches around. Shirt collars are measured in half inch increments. What size shirt should Ahn order?**

INTEGRATION

If Ahn rounds down, his shirt collar will be too tight. Ahn should round up and order a shirt with a 15-inch collar.

Lesson 6-1 Rounding Fractions and Mixed Numbers **229**

Investigations for the Special Education Student

This blackline master booklet helps you plan for the needs of your special education students by providing long term projects along with teacher notes. Investigation 6, *Read All About Us,* and Investigation 7, *Wall Street Week,* may be used with this chapter.

---

2 TEACH

Transparency 6-1B contains a teaching aid for this lesson.

Reading Mathematics Another definition of the word *numerator* is "one that numbers." The word *denominator* can also mean "a common trait." Show a fraction on the chalkboard, and point out the numerator and the denominator. Have students think about these alternate meanings and examine the mathematical role of the numerator and denominator when rounding fractions and mixed numbers.

In-Class Examples

Round each number to the nearest half.

For Example 1

a. $2\frac{7}{9}$ 3

b. $3\frac{2}{5}$ $3\frac{1}{2}$

For Example 2

a. $\frac{5}{11}$ $\frac{1}{2}$

b. $\frac{6}{7}$ 1

For Example 3

Find the width of your math textbook to the nearest one-half inch. $8\frac{1}{2}$ in.

For Example 4

A recipe for a stir-fry dish calls for $1\frac{1}{4}$ pounds of chicken. Should you buy a $1\frac{1}{8}$-pound package or a $1\frac{1}{2}$-pound package?

$1\frac{1}{2}$-pound package

For Example 5

Mila is buying wood for a bookcase shelf. The width of the bookcase is 3 feet 4 inches. Wood for shelving is sold in half-foot increments. What length of wood should Mila buy? $3\frac{1}{2}$ ft

---

Check for Understanding

If students need additional practice or instruction after completing Exercises 1–11, one of these options may be helpful.
- Extra Practice, see p. 572
- Reteaching Activity
- *Study Guide Masters*, p. 42
- *Practice Masters*, p. 42
- 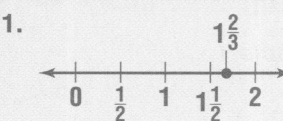 Interactive Mathematics Tools Software

Assignment Guide

Core: 13–33 odd, 35–39
Enriched: 12–32 even, 33, 35–39

Additional Answers

1.

$$1\frac{2}{3}$$

$0 \quad \frac{1}{2} \quad 1 \quad 1\frac{1}{2} \quad 2$

3. Sample answer: If the numerator is close in value to the denominator, round up. If the numerator is about half the denominator, round to $\frac{1}{2}$. If the numerator is much smaller than the denominator, round down.

Study Guide Masters, p. 42

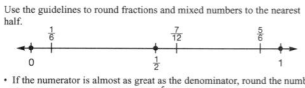

CHECK FOR UNDERSTANDING

Communicating Mathematics

Read and study the lesson to answer each question. 1. See margin.

1. **Draw** a number line that shows how to round $1\frac{2}{3}$ to the nearest half.

2. **Describe** a situation where it would make sense to round a fraction up to the nearest unit.

3. **Write**, in your own words, how you know whether to round a fraction to $0, \frac{1}{2}$, or 1 when rounding to the nearest half. **See margin.**

Guided Practice
2. Sample answer: When estimating the amount of wood needed to build a bookshelf.

Round each number to the nearest half.

4. $\frac{5}{8}$ **$\frac{1}{2}$**
5. $3\frac{15}{16}$ **4**
6. $\frac{4}{10}$ **$\frac{1}{2}$**
7. $\frac{7}{10}$ **$\frac{1}{2}$**
8. $5\frac{2}{5}$ **$5\frac{1}{2}$**

Tell whether each number should be rounded up or down.

9. the weight limit of a bridge **down**

10. the capacity of a container needed to hold $4\frac{3}{4}$ liters of gasoline **up**

11. **Measurement** Find the amount of water in the measuring cup to the nearest one-half cup.
$2\frac{1}{2}$ **cups**

EXERCISES

Practice

Round each number to the nearest half.

12. $1\frac{1}{10}$ **1**
13. $\frac{5}{6}$ **1**
14. $\frac{3}{8}$ **$\frac{1}{2}$**
15. $6\frac{2}{3}$ **$6\frac{1}{2}$**
16. $\frac{1}{5}$ **0**

17. $\frac{9}{16}$ **$\frac{1}{2}$**
18. $2\frac{4}{5}$ **3**
19. $12\frac{1}{6}$ **12**
20. $7\frac{3}{10}$ **$7\frac{1}{2}$**
21. $4\frac{2}{9}$ **4**

22. $\frac{1}{8}$ **0**
23. $10\frac{7}{10}$ **$10\frac{1}{2}$**
24. $\frac{1}{9}$ **0**
25. $5\frac{3}{7}$ **$5\frac{1}{2}$**
26. $\frac{27}{32}$ **1**

Tell whether each number should be rounded up or down.

27. the amount of chili pepper needed for a pot of chili **down**

28. a patch for a $2\frac{1}{4}$-inch tear in a pair of jeans **up**

29. the depth of the stream where you are fishing while wearing hip boots **up**

30. the width of blinds to fit in a window $63\frac{3}{4}$ inches wide **down**

31. the weight of cargo on an airplane **up**

■ Reteaching the Lesson ■

Activity You will need measuring cups, spoons, and a variety of containers. Have pairs of students fill the containers with water or a dry material such as uncooked oatmeal, beans, popcorn, or rice. Then have them measure the contents to the nearest half-unit.

Applications and Problem Solving

32. $1\frac{1}{2}$ pounds, 1 pound isn't enough.

33a. $\frac{23}{28}, \frac{2}{7}, \frac{5}{28}, \frac{1}{2},$ $\frac{11}{14}, \frac{17}{28}, \frac{5}{7}, \frac{3}{4}$

33b. salads, turkey

Mixed Review

32. **Cooking** A recipe to make tacos calls for $1\frac{1}{4}$ pounds of ground beef. Should you buy a $1\frac{1}{2}$-pound package or a 1-pound package? Explain.

33. **Food** The graph shows the results of a survey of the food served at 28 major league baseball stadiums.

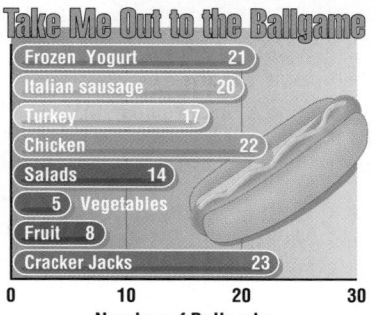

Take Me Out to the Ballgame

Frozen Yogurt	21
Italian sausage	20
Turkey	17
Chicken	22
Salads	14
Vegetables	5
Fruit	8
Cracker Jacks	23

Number of Ballparks: 0 10 20 30

   a. Write a fraction to represent the number of stadiums that sell each type of food.
   b. Which foods are sold in about half of the stadiums?
   c. Which foods are sold in almost all of the stadiums? **Cracker Jacks, chicken, frozen yogurt**

34. **Working on the CHAPTER Project** Refer to the table on page 227. Begin drawing your map. Choose the first two landmarks so that the distance between them is $1\frac{1}{2}$ inches when rounded to the nearest half inch. However, on your map, use the actual distance. **See margin.**

35. **Critical Thinking** Name three mixed numbers that round to $5\frac{1}{2}$. **Sample answer:** $5\frac{3}{8}, 5\frac{7}{16}, 5\frac{3}{5}$

36. Write $1\frac{5}{18}$ as a decimal using bar notation. *(Lesson 5-10)* **1.2$\overline{7}$**

37. **Measurement** One acre is about 0.0016 square mile. Write this decimal in words. *(Lesson 3-1)* **sixteen ten-thousandths**

38. **Statistics** Make a horizontal bar graph and one other type of graph for the set of data. *(Lesson 2-3)*

Raul's Earnings				
Monday	Tuesday	Wednesday	Thursday	Friday
$21.50	$13.75	$19.15	$20.00	$25.50

See Answer Appendix.

39. **Test Practice** A telephone operator answered 500 calls during the first three weeks of a month. She answered 137 calls the fourth week. A reasonable conclusion would be that the telephone operator answered — *(Lesson 1-3)* **D**

   A less than 100 calls per week.
   B between 100 and 125 calls per week.
   C between 126 and 150 calls per week.
   D between 151 and 175 calls per week.
   E more than 175 calls per week.

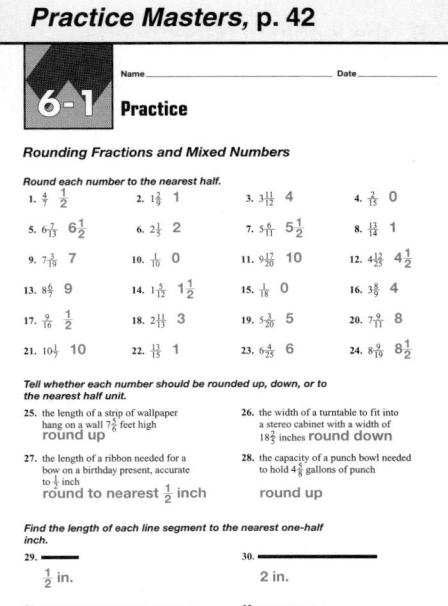

Lesson 6-1 Rounding Fractions and Mixed Numbers **231**

Extending the Lesson

Enrichment Masters, p. 42

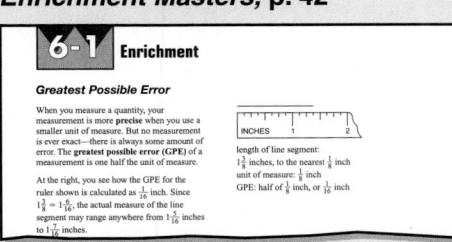

6-1 Enrichment

Greatest Possible Error

When you measure a quantity, your measurement is more precise when you use a smaller unit of measure. But no measurement is ever exact—there is always some amount of error. The **greatest possible error (GPE)** of a measurement is one half the unit of measure.

At the right, you see how the GPE for the ruler shown is calculated as $\frac{1}{16}$ inch. Since $1\frac{3}{8} = 1\frac{6}{16}$, the actual measure of the line segment may range anywhere from $1\frac{5}{16}$ inches to $1\frac{7}{16}$ inches.

length of line segment:
$1\frac{3}{8}$ inches, to the nearest $\frac{1}{8}$ inch
unit of measure: $\frac{1}{8}$ inch
GPE: half of $\frac{1}{8}$ inch, or $\frac{1}{16}$ inch

Activity Have pairs of students contrast examples of appropriate and inappropriate units of measurement for rounding. For example, the time spent mowing a lawn could be rounded to the nearest one-half hour, but even the nearest one-half minute would not be exact enough to record the finishing time in an Olympic race.

CHAPTER Project

Exercise 34 asks students to advance to the next stage of work on the Chapter Project. You may wish to provide students with blank tables to help them organize their information.

4 ASSESS

Closing Activity

Speaking Name a fraction and ask a student whether it should be rounded up, down, or to the nearest half unit. Then have that student name a fraction for someone else, and continue around the room until everyone has had a turn.

Additional Answer

34. **Sample answer:** The two landmarks can be either the center of the fountain and the tallest oak tree or the bench near the pond and the seal-shaped rock.

Practice Masters, p. 42

6-1 Practice

Rounding Fractions and Mixed Numbers

Round each number to the nearest half.

1. $4\frac{7}{9}$ $\frac{1}{2}$
2. $1\frac{1}{9}$ 1
3. $3\frac{11}{12}$ 4
4. $\frac{2}{15}$ 0
5. $6\frac{7}{13}$ $6\frac{1}{2}$
6. $2\frac{1}{5}$ 2
7. $5\frac{6}{11}$ $5\frac{1}{2}$
8. $\frac{13}{14}$ 1
9. $7\frac{3}{19}$ 7
10. $\frac{1}{10}$ 0
11. $9\frac{17}{20}$ 10
12. $4\frac{13}{25}$ $4\frac{1}{2}$
13. $8\frac{6}{9}$ 9
14. $1\frac{5}{12}$ $1\frac{1}{2}$
15. $\frac{1}{18}$ 0
16. $3\frac{3}{8}$ 4
17. $\frac{9}{16}$ $\frac{1}{2}$
18. $2\frac{11}{13}$ 3
19. $5\frac{3}{20}$ 5
20. $7\frac{9}{11}$ 8
21. $10\frac{1}{7}$ 10
22. $\frac{11}{13}$ 1
23. $6\frac{4}{25}$ 6
24. $8\frac{9}{19}$ $8\frac{1}{2}$

Tell whether each number should be rounded up, down, or to the nearest half unit.

25. the length of a strip of wallpaper to hang on a wall $7\frac{2}{5}$ feet high **round up**
26. the width of a turntable to fit into a stereo cabinet with a width of $18\frac{2}{3}$ inches **round down**
27. the length of a ribbon needed for a bow on a birthday present, accurate to $\frac{1}{4}$ inch **round to nearest $\frac{1}{2}$ inch**
28. the capacity of a punch bowl needed to hold $4\frac{3}{8}$ gallons of punch **round up**

Find the length of each line segment to the nearest one-half inch.

29. $\frac{1}{2}$ in.
30. 2 in.
31. $2\frac{1}{2}$ in.
32. 1 in.

© Glencoe/McGraw-Hill   T42   Mathematics: Applications and Connections, Course 1

Lesson 6-1 231

- *Study Guide Masters*, p. 43
- *Practice Masters*, p. 43
- *Enrichment Masters*, p. 43
- Transparencies 6-2, A and B
- *Assessment and Evaluation Masters*, p. 155
- CD-ROM Program
  - Resource Lesson 6-2

Recommended Pacing	
Standard	Day 3 of 13
Honors	Day 3 of 12
Block	Day 2 of 7

1 FOCUS

5-Minute Check
(Lesson 6-1)

Round each number to the nearest half.

1. $\frac{1}{8}$ 0
2. $3\frac{9}{16}$ $3\frac{1}{2}$
3. $\frac{9}{11}$ 1

Tell whether each number should be rounded up or down.

4. the number of bricks to load in a $\frac{3}{4}$-ton pickup truck
   down
5. the length of paper needed to wrap several gifts **up**

The 5-Minute Check is also available on **Transparency 6-2A** for this lesson.

2 TEACH

Transparency 6-2B contains a teaching aid for this lesson.

Using Modeling Have students work in groups of three. Ask them to measure the lengths of their outstretched arms (fingertip to fingertip). Then have them hold hands with arms outstretched and estimate their combined length. Have them compare their estimates with others and with their actual combined length.

6-2 Estimating Sums and Differences

What you'll learn

You'll learn to estimate sums and differences of fractions and mixed numbers.

When am I ever going to use this?

You'll estimate sums and differences when planning what songs to record on a cassette tape.

Maxine is making a cassette tape of some of her favorite songs from her CD collection so that she can listen to them while she jogs. She has already recorded about $49\frac{3}{4}$ minutes of music on her 60-minute cassette tape. The songs she wants to record are about $3\frac{3}{4}$ minutes, $4\frac{1}{6}$ minutes, and $3\frac{1}{3}$ minutes long. Is there enough tape left to record all three songs? *This problem will be solved in Example 3.*

A good way to estimate the sum or difference of fractions is to round each fraction to the nearest half and then add or subtract.

Examples

Estimate.

1 $\frac{9}{16} - \frac{1}{8}$

$\frac{9}{16}$ rounds to $\frac{1}{2}$.

$\frac{1}{8}$ rounds to 0.

Subtract: $\frac{1}{2} - 0 = \frac{1}{2}$

$\frac{9}{16} - \frac{1}{8}$ is about $\frac{1}{2}$.

2 $\frac{5}{6} + \frac{7}{12}$

$\frac{5}{6}$ rounds to 1.

$\frac{7}{12}$ rounds to $\frac{1}{2}$.

Add: $1 + \frac{1}{2} = 1\frac{1}{2}$

$\frac{5}{6} + \frac{7}{12}$ is about $1\frac{1}{2}$.

To estimate sums and differences of mixed numbers, round each number to the nearest whole number.

Example
APPLICATION

3 **Music** Refer to the beginning of the lesson. Is there enough tape left to record all three songs?

Explore You know how long the tape is, how much has been used, and the length of each of the three songs Maxine wants to record.

Plan Subtract to estimate the amount of time left on the tape. Then compare it to an estimate of the total length of the three songs.

Solve $49\frac{3}{4}$ rounds to 50.

Subtract: $60 - 50 = 10$

There are about 10 minutes left on the tape.

$3\frac{3}{4}$ rounds to 4.

$4\frac{1}{6}$ rounds to 4.

$3\frac{1}{3}$ rounds to 3.

Motivating the Lesson

Problem Solving Ask students how to estimate which pair of chores would take longer to finish.

- washing floors for $\frac{1}{2}$ hour and cleaning the garage for $2\frac{1}{4}$ hours
- washing the car for $1\frac{3}{4}$ hours and raking leaves for $\frac{1}{2}$ hour

Round each mixed number to the nearest whole number.

Add: $4 + 4 + 3 = 11$

The three songs need about 11 minutes of tape.

The time needed is greater than the time left on the tape. Maxine cannot record all three songs.

Examine   Check by adding the total time for the three songs to the time already used.

Sometimes when estimating sums and differences of fractions and mixed numbers, you need to round all fractions up.

Example ④

APPLICATION

Crafts   Leon is making a wallet in a craft class. He plans to trim the wallet with a different color of leather. The wallet is $6\frac{3}{4}$ inches long and $3\frac{1}{2}$ inches wide. About how much leather trim does he need?

Leon wants to make sure he buys enough leather. He rounds up.

Since he wants to trim the entire wallet, he needs to estimate the perimeter of the wallet.

$6\frac{3}{4}$ rounds to 7, and $3\frac{1}{2}$ rounds to 4.

Estimate: $7 + 4 + 7 + 4 = 22$

Leon needs about 22 inches of leather trim.

$6\frac{3}{4}$ in.

$3\frac{1}{2}$ in.

CHECK FOR UNDERSTANDING

Communicating Mathematics
1. See Answer Appendix.

Read and study the lesson to answer each question.

1. **Draw** a number line that shows about where $2\frac{3}{4}$ is located.

2. **Explain** how you would estimate $3\frac{3}{4}$ minus $1\frac{1}{3}$. **Sample answer: $4 - 1 = 3$**

3. **You Decide** Nina wants to make a square picture frame that is $4\frac{3}{8}$ inches on each side. Should she buy 16 inches of framing material or 20 inches of framing material? Explain. **See Answer Appendix.**

Guided Practice

Estimate.   4–9. Sample answers are given.

4. $\frac{7}{8} - \frac{5}{16}$ $1 - \frac{1}{2} = \frac{1}{2}$ 5. $6\frac{7}{10} + 3\frac{5}{8}$ $7 + 4 = 11$ 6. $\frac{1}{2} + \frac{4}{5}$ $\frac{1}{2} + 1 = 1\frac{1}{2}$

7. $2\frac{5}{12} - \frac{1}{3}$ $2\frac{1}{2} - \frac{1}{2} = 2$ 8. $7\frac{1}{4} - 2\frac{3}{16}$ $7 - 2 = 5$ 9. $5\frac{7}{12} + 9\frac{5}{6}$ $6 + 10 = 16$

10. **Write a Problem** in which you would need to estimate the difference between $5\frac{2}{3}$ and $1\frac{1}{4}$. **See Answer Appendix.**

Lesson 6-2 Estimating Sums and Differences   **233**

■ Reteaching the Lesson ■

Activity Give pairs of students catalog pages that include measurements of curtains, furniture, or rugs. Have them use the information to create word problems involving estimating sums and differences of fractions and mixed numbers. Have them exchange their problems with others to solve.

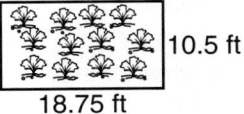

Check for Understanding
If students need additional practice or instruction after completing Exercises 1–10, one of these options may be helpful.
- Extra Practice, see p. 572
- Reteaching Activity, see p. 233
- *Study Guide Masters*, p. 43
- *Practice Masters*, p. 43

Assignment Guide
Core: 11–29 odd, 31–36
Enriched: 12–26 even, 28–36

4 ASSESS

Closing Activity
Modeling Have pairs of students use fraction bars to estimate sums and differences. For example, have them use bars for $\frac{7}{12} + \frac{7}{8}$. To determine whether they need to round up, down, or to the nearest half, have them compare the $\frac{1}{2}$ model with $\frac{7}{12}$ and $\frac{7}{8}$. Then have them write $\frac{1}{2} + 1 = 1\frac{1}{2}$.

Chapter 6, Quiz A (Lessons 6-1 and 6-2) is available in the *Assessment and Evaluation Masters*, p. 155.

Practice Masters, p. 43

6-2 Practice
Name_____ Date_____

Estimating Sums and Differences

Round each fraction or mixed number to the nearest whole number.
1. $3\frac{3}{9}$ 4
2. $2\frac{2}{5}$ 2
3. $\frac{9}{16}$ 1
4. $5\frac{1}{4}$ 6
5. $\frac{3}{8}$ 0
6. $7\frac{4}{12}$ 7
7. $4\frac{7}{10}$ 5
8. $\frac{6}{7}$ 1

Estimate. Sample answers are given.
9. $3\frac{4}{9} + 2\frac{2}{3}$
   $3 + 3 = 6$
10. $\frac{5}{8} - \frac{1}{7}$
    $1 - 0 = 1$
11. $4\frac{7}{12} - \frac{11}{20}$
    $5 - \frac{1}{2} = 4\frac{1}{2}$
12. $\frac{9}{10} + \frac{7}{8}$
    $1 + 1 = 2$
13. $6\frac{3}{7} - 1\frac{9}{14}$
    $6 - 2 = 4$
14. $\frac{11}{16} - \frac{7}{16}$
    $1 - \frac{1}{2} = \frac{1}{2}$
15. $7\frac{3}{7} + \frac{6}{9}$
    $8 + \frac{1}{2} = 8\frac{1}{2}$
16. $\frac{5}{9} + \frac{4}{5}$
    $\frac{1}{2} + 1 = 1\frac{1}{2}$
17. $9\frac{11}{16} - 3\frac{1}{4}$
    $10 - 3 = 7$
18. $8\frac{1}{10} + 5\frac{13}{15}$
    $8 + 6 = 14$
19. $6\frac{7}{20} - \frac{1}{7}$
    $6 - 0 = 6$
20. $\frac{7}{15} + 2\frac{2}{9}$
    $\frac{1}{2} + 2 = 2\frac{1}{2}$
21. $\frac{3}{8} + \frac{15}{16}$
    $0 + 1 = 1$
22. $12\frac{11}{14} - 7\frac{4}{7}$
    $13 - 8 = 5$
23. $\frac{8}{9} - \frac{8}{15}$
    $1 - \frac{1}{2} = \frac{1}{2}$
24. $8\frac{1}{6} + \frac{14}{15}$
    $8 + 1 = 9$
25. $2\frac{3}{7} + 1\frac{4}{14}$
    $2 + 1 = 3$
26. $4\frac{1}{3} - \frac{1}{10}$
    $4 - 0 = 4$
27. Estimate the sum of $1\frac{5}{9}$ and $4\frac{1}{2}$.
    $2 + 5 = 7$
28. Estimate the difference of $\frac{9}{10}$ and $\frac{11}{20}$.
    $1 - \frac{1}{2} = \frac{1}{2}$

© Glencoe/McGraw-Hill   T43   *Mathematics: Applications and Connections, Course 1*

Practice

Estimate. 11–27. Sample answers are given.

11. $\frac{7}{8} + \frac{5}{16}$ $1 + \frac{1}{2} = 1\frac{1}{2}$
12. $5\frac{1}{3} - 4\frac{3}{4}$ $5 - 5 = 0$
13. $8\frac{1}{4} - \frac{9}{16}$ $8 - 1 = 7$
14. $7\frac{4}{5} + 3\frac{1}{3}$ $8 + 3 = 11$
15. $\frac{9}{10} + \frac{1}{2}$ $1 + \frac{1}{2} = 1\frac{1}{2}$
16. $3\frac{3}{4} - 2\frac{7}{8}$ $4 - 3 = 1$
17. $\frac{2}{3} + 6\frac{3}{8}$ $1 + 6 = 7$
18. $2\frac{3}{10} - 1\frac{7}{8}$ $2 - 2 = 0$
19. $8\frac{5}{6} + \frac{1}{12}$ $9 + 0 = 9$

20. $11 - 1 = 10$
21. $22 + 5 = 27$
23. $5 \text{ min} - 1 \text{ min} = 4 \text{ min}$
25. $20 \text{ cups} - 11 \text{ cups} = 9 \text{ cups}$

20. $11\frac{7}{16} - \frac{5}{9}$
21. $21\frac{5}{8} + 4\frac{3}{4}$
22. $12\frac{4}{5} + \frac{2}{3}$ $13 + 1 = 14$

23. About how much longer than $\frac{5}{6}$ minute is $4\frac{1}{2}$ minutes?
24. Estimate the sum $3\frac{3}{10} + 2\frac{4}{5} + 3\frac{1}{3}$. $3 + 3 + 3 = 9$
25. About how much more is $19\frac{3}{4}$ cups than $10\frac{7}{8}$ cups?
26. Estimate the difference between $1\frac{3}{5}$ and $\frac{1}{6}$. $2 - 0 = 2$
27. Estimate the sum of $7\frac{1}{3}$, $6\frac{4}{5}$, $6\frac{3}{4}$, $7\frac{1}{10}$, and $6\frac{15}{16}$. $7 + 7 + 7 + 7 + 7 = 35$

Applications and Problem Solving

28. **Geometry** Estimate the perimeter of the rectangle.
    $12 \text{ in.} + 3 \text{ in.} + 12 \text{ in.} + 3 \text{ in.} = 30 \text{ in.}$
    $2\frac{13}{16}$ in. $12\frac{1}{8}$ in.

29. **Carpentry** A board that is $63\frac{5}{8}$ inches long is about how much longer than a board that is $62\frac{1}{4}$ inches long? $64 \text{ in.} - 62 \text{ in.} = 2 \text{ in.}$

interNET CONNECTION
For the latest immigration statistics, visit:
www.glencoe.com/sec/math/mac/mathnet

30. **Geography** In 1994, about $\frac{3}{8}$ of the U.S. immigrants came from Latin America, about $\frac{1}{5}$ came from Canada and Europe, and about $\frac{1}{25}$ came from Africa and Australia. Asians made up the rest of the immigrant population.
    a. About what fraction of immigrants were from Latin America, Canada, and Europe? $\frac{1}{2} + 0 = \frac{1}{2}$
    b. Estimate to find the fraction of immigrants from Asia.

30b. $1 - \left(\frac{1}{2} + 0 + 0\right)$
     $= \frac{1}{2}$

31. **Critical Thinking** The estimate for the sum of two fractions is 1. If 2 is added to each fraction, the estimate for the sum of the two mixed numbers is 4. What are examples of the fractions? Sample answer: $\frac{5}{8}$ and $\frac{7}{16}$

Mixed Review

32. **Test Practice** What is $6\frac{4}{7}$ rounded to the nearest half? *(Lesson 6-1)* **B**
    A 6 **B** $6\frac{1}{2}$ **C** 7 **D** $7\frac{1}{2}$

33. **Physical Science** Helium has a mass of 0.17 kilogram per cubic meter. Express this mass as a fraction in simplest form. *(Lesson 5-9)* $\frac{17}{100}$

34. Find the least common multiple of 7 and 21. *(Lesson 5-7)* **21**

35. **Measurement** Change 4 kilometers to centimeters. *(Lesson 4-9)* **400,000 cm**

36. Estimate $39.7 - 28.561$ using rounding. *(Lesson 3-4)* **10**

Extending the Lesson

Enrichment Masters, p. 43

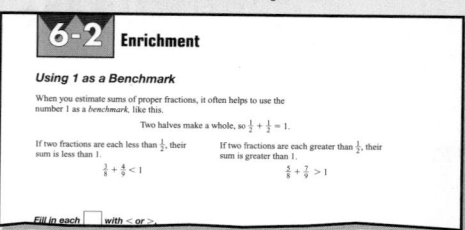

6-2 Enrichment

Using 1 as a Benchmark

When you estimate sums of proper fractions, it often helps to use the number 1 as a *benchmark*, like this.

Two halves make a whole, so $\frac{1}{2} + \frac{1}{2} = 1$.

If two fractions are each less than $\frac{1}{2}$, their sum is less than 1.
$\frac{2}{5} + \frac{4}{9} < 1$

If two fractions are each greater than $\frac{1}{2}$, their sum is greater than 1.
$\frac{5}{8} + \frac{7}{9} > 1$

Fill in each □ with < or >.

Activity Choose a favorite recipe. Estimate the amount of ingredients needed to make enough for the whole class. Ask the cafeteria staff if there are any special requirements for adapting the recipe for a large group.

CARTOGRAPHY

Richard Jimenez
CARTOGRAPHER

Richard Jimenez is a cartographer for the U.S. Geological Survey. He creates maps of different areas of the United States. Mr. Jimenez is very skilled in accurately measuring distances and determining the location of any given place on a map by using coordinates.

A person who is interested in becoming a cartographer should take classes in geography, mathematics, mechanical drawing, and computer science. The mathematics courses should include algebra and trigonometry. A college degree in engineering or physical science is often required. College courses should include technical mathematics, drafting, and mapping.

For more information:
American Congress on Surveying and Mapping
5410 Grosvenor Lane, Suite 210
Bethesda, MD 20814

www.glencoe.com/sec/math/mac/mathnet

I use maps when I go hiking. Someday I'd like to create maps that people can use to travel around the world!

Your Turn

Plan a one-week vacation. Write a report about where you will travel, the sites you will visit along the way, and the distance and driving time for your trip. Include in your report a map to refer to while reporting your plan to the class.

School to Career: Cartography **235**

Motivating Students

People have been making maps for thousands of years. The oldest example that has been found is a clay tablet map created by a Babylonian in 2300 B.C. Maps grew more scientific beginning in the eighteenth century, when decorative illustrations were replaced by factual representations. To begin a discussion of cartography, ask students these questions.
- How were maps drawn before inventions were created to aid the process? **Cartographers used their own observations and explorers' accounts.**
- What do cartographers use today to help them draw maps? **aerial and satellite photographs, computer graphics**

Making the Math Connection

Measuring angles is extremely important in cartography. Also of importance are the abilities to add and subtract degrees of latitude and longitude, determine distances according to scale, plot points using coordinates, and depict a sphere on a plane.

Working on *Your Turn*

Encourage students to draw their maps based on written descriptions and photographs of the area, rather than relying on professionally created maps. Have them include the sites they are most interested in visiting. As they report their plan to the class, have them draw the route they plan to take on their map.

More About Richard Jimenez
- Richard Jimenez received a B.S. degree in Surveying and Mapping from Metropolitan State College in Denver, Colorado. As a cartographer, he has done analytical mapping, field work, digital mapping, and photogrammetry (making measurements using aerial photographs).
- Mr. Jimenez became a cartographer because he enjoyed relating mathematical measurements of land to nature. He believes it is important not to spend too much of one's time in front of a computer—after all, maps represent the real world.

*An additional School to Career activity is available on page 6 of the **School to Career Masters.***

Objective Students solve problems by eliminating possibilities.

Recommended Pacing	
Standard	Day 4 of 13
Honors	Day 4 of 12
Block	Day 2 of 7

1 FOCUS

Getting Started Ask students how they can determine which movie to see on a Saturday afternoon, considering they only have $2.00 each.
- a drama rated PG-13 at a second-run theater, which charges $2.00
- a spy thriller rated R at a first-run theater, which charges $3.50
- a comedy rated G at a second-run theater, which charges $1.50
- a rock music documentary rated PG at a first-run theater, which charges $6.00

eliminate the higher-priced movies and those with adult ratings

2 TEACH

Teaching Tip Suggest that students list the possible solutions to problems such as Exercise 12, crossing out the ones that are not reasonable.

In-Class Example

Carlito has $60 to buy clothes. Looking at the prices of the items below, what can he buy?
- shoes that cost $39.95
- a jacket that costs $69.95
- jeans that cost $29.95
- a shirt that costs $18.95
- a suit that costs $249.95

shoes, jeans, shirt, shirt and jeans, shirt and shoes, 2 jeans, 3 shirts

---

PROBLEM SOLVING

6-2B Eliminate Possibilities

A Follow-Up of Lesson 6-2

Cesar and Franco are planning to take the bus to the City Center Mall. They have a map and timetables for the bus routes through their neighborhood. Let's listen in.

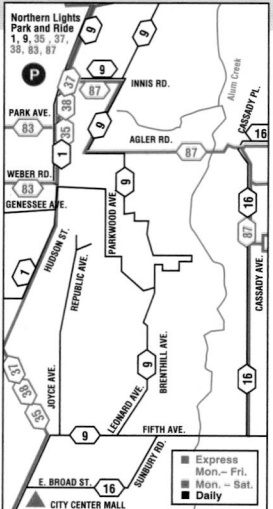

There are seven different bus routes that leave the Northern Lights Park and Ride. Which one do we want?

Since it's Saturday, we can eliminate the three express routes. They only run on weekdays.

And the number 1 and number 83 are east-west routes. We want to go south.

Cesar

Number 87 doesn't go downtown, but we could pick up number 16 at Cassady Avenue.

But then we would have to pay twice. That leaves us with the number 9 bus. Let's check the timetables.

Franco

THINK ABOUT IT

Work with a partner.

1. **Tell** how Cesar and Franco eliminated possibilities to solve the problem.
2. **Explain** how estimating could help you to **eliminate possibilities.** See margin.

1. Sample answer: Franco and Cesar eliminated the routes that they could not take to see which route they could.

3. **Apply** the eliminate possibilities strategy to solve the following problem.

Mrs. Danko bought a ham that weighed $5\frac{1}{4}$ pounds and a turkey that weighed $13\frac{1}{2}$ pounds. About how many pounds of meat did she buy? **C**

A 8 lb   B 10 lb   C 19 lb   D 20 lb

236 Chapter 6 Adding and Subtracting Fractions

■ Reteaching the Lesson ■

Activity Have students work in small groups and brainstorm as many different sports as they can. Then have them refine their lists to only the sports which they can play on the school grounds on that day. Tell them to consider the weather, available equipment, and surroundings, and to give a reason for why they eliminated each sport from their lists.

Additional Answer

2. Sample answer: Estimating gets you close to the actual answer. You could eliminate choices that aren't close to your estimate.

Check for Understanding
Use the discussion from Exercise 3 to determine whether students understand the eliminate possibilities strategy.

Extra Practice If students need additional practice in problem solving, extra practice is available on the following pages.
- Eliminate Possibilities, see p. 573
- Mixed Problem Solving, see pp. 593–594

Assignment Guide
All: 4–15

ON YOUR OWN

4. The fourth step of the 4-step plan for problem solving asks you to *examine*. *Explain* how you can use the strategy of eliminating possibilities to examine a solution.
4–6. See margin.

5. *Write a Problem* that could be solved by eliminating possibilities.

6. *Reflect Back* Tell how the eliminating possibilities strategy would help you on a multiple-choice test.

MIXED PROBLEM SOLVING

10. about 44 feet

> **STRATEGIES**
> Look for a pattern.
> Solve a simpler problem.
> Act it out.
> Guess and check.
> Draw a diagram.
> Make a chart.
> Work backward.

Solve. Use any strategy.

7. Viho's father is 4 times as old as Viho. His grandfather is twice as old as Viho's father. The sum of their three ages is 104. How old is Viho, his father, and his grandfather? **See margin.**

8. <u>Test Practice</u> Last year, Fred's car odometer read 45,500.4. A year later the odometer reads 57,200.9. About how many miles did Fred drive over the past year? **C**

A 1,000,000
B 100,000
C 10,000
D 1,000
E 100

9. *Money Matters* Candy bars are priced at 2 for 99¢. What will one candy bar cost? **50¢**

10. *Geometry* Mrs. Coe wants to buy fence to put around her rectangular flower bed. About how much fence will she need?

$13\frac{1}{4}$ ft
$8\frac{1}{2}$ ft

11. *Food* Ani is preparing a meal for her friends. She wants to make three desserts — a cake, some cookies, and an apple pie. To make the cake, she needs $1\frac{1}{2}$ cups of flour. She needs $1\frac{1}{4}$ cups for the cookies, and $2\frac{3}{4}$ cups for the pie. About how much flour does she need to make the three desserts? **about 7 cups**

12. *Sports* Jill, Kai, Lyndsi, and Mykia are friends. Each of them is on one of the following school teams: basketball, golf, soccer, or tennis. Use the following information to determine who plays on each team. **See margin.**

> Jill is shorter than the girl who plays basketball. Kai only likes to play games played on a court. Lyndsi has a problem with her knee and cannot run. Mykia practices kicking a ball as part of her training.

13. *Fashion* A manufacturer offers four different styles of tennis shoes in white, black, and blue. How many combinations of style and color are possible? **12**

14. *Statistics* Is the mean of 123.9, 43.6, 120.89, 502.9, 12.7, and 72.34 about 150 or 15? Explain. **See margin.**

15. <u>Test Practice</u> Anita is making curtains for her room. She needs $12\frac{3}{4}$ yards of material for a larger window and $7\frac{1}{2}$ yards for a smaller window. She bought 30 yards of material. About how much will she have left after making the curtains? **D**

A 21 yd
B 20 yd
C 19 yd
D 10 yd
E 5 yd

Lesson 6-2B THINKING **LAB** **237**

4 ASSESS

Closing Activity
Writing Have students write a paragraph explaining how eliminating possibilities can be used to solve a problem. Have them include a situation where they could apply this strategy.

Additional Answers
4. Estimate before getting the exact answer and then compare the exact answer to the estimate.

5. Sample answer: Is 15,840 ÷ 18 equal to 80, 88, 880, or 8,800?

6. Sample answer: By eliminating possibilities, especially obviously wrong choices, you can narrow down the choices until you arrive at the correct answer.

7. Viho is 8 years old, his father is 32, and his grandfather is 64.

12. Jill—tennis; Kai—basketball; Lyndsi—golf; Mykia—soccer

14. about 150; Sample answer: An estimate of the total is 800, and 800 ÷ 5 is greater than 100.

■ Extending the Lesson ■

Activity Have small groups of students write a multiple-choice quiz on estimating sums and differences of fractions and mixed numbers. Ask them to prepare an answer key. Then have the groups exchange their work with other groups to critique the answer key in terms of being able to use the eliminating possibilities strategy.

- *Study Guide Masters*, p. 44
- *Practice Masters*, p. 44
- *Enrichment Masters*, p. 44
- *Transparencies 6-3, A and B*
- *Diversity Masters*, p. 6
- *Hands-On Lab Masters*, p. 74

 CD-ROM Program
- Resource Lesson 6-3
- Interactive Lesson 6-3

Recommended Pacing

Standard	Day 5 of 13
Honors	Day 5 of 12
Block	Day 3 of 7

1 FOCUS

 5-Minute Check
(Lesson 6-2)

Estimate.

1. $\frac{5}{6} + \frac{1}{5}$ $1 + 0 = 1$

2. $7\frac{1}{8} - 4\frac{3}{5}$ $7 - 5 = 2$

3. $3\frac{6}{7} + 8\frac{3}{4}$ $4 + 9 = 13$

4. $5\frac{2}{3} - \frac{7}{13}$ $6 - \frac{1}{2} = 5\frac{1}{2}$

5. Estimate the amount of framing material needed for a picture that is $18\frac{3}{8}$ inches long and $10\frac{1}{2}$ inches wide.
   $19 + 11 + 19 + 11 = 60$ in.

The 5-Minute Check is also available on **Transparency 6-3A** for this lesson.

Motivating the Lesson

Communication Ask students to read the opening paragraph of the lesson. Ask the following questions.

- If everyone in this class wanted to plant one square of a garden, how many squares would we need?
- What fraction of the garden would be planted if we put in four squares of tomatoes and five squares of peppers?

6-3 Adding and Subtracting Fractions with Like Denominators

What **you'll learn**

You'll learn to add and subtract fractions with like denominators.

When **am I ever going to use this?**

You'll add and subtract fractions with like denominators when you work with equal parts of a whole, such as sections of a garden.

Word Wise

like fractions

John prepared an area for a vegetable garden. He used rope to make a 3-by-5 grid of squares. He planted green beans in the back three squares and corn in two squares. How much of the garden has he planted so far?

You can use grid paper to model this problem.

HANDS-ON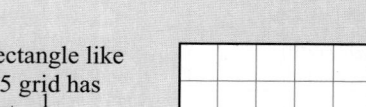

MINI-LAB

Work with a partner. 🪟 grid paper 🖍 colored pencils

Try This

- On your grid paper, draw a rectangle like the one shown. Since a 3-by-5 grid has 15 squares, each one represents $\frac{1}{15}$.
- With a colored pencil, color three squares to represent the green beans.
- With a different colored pencil, color two more squares to represent the corn.

Talk About It

1. How many squares are colored? **5 squares**
2. What fraction represents the number of colored squares inside the rectangle? $\frac{5}{15}$ or $\frac{1}{3}$
3. If you color four more squares, what fraction would that represent? $\frac{9}{15}$ or $\frac{3}{5}$

Fractions with the same denominator are called **like fractions**. You add and subtract the numerators of like fractions the same way you add and subtract whole numbers. The denominator of the fraction names the units being added or subtracted.

From the Mini-Lab, we know that

3 fifteenths	*plus*	*2 fifteenths*	*equals*	*5 fifteenths.*
$\frac{3}{15}$	$+$	$\frac{2}{15}$	$=$	$\frac{5}{15}$ or $\frac{1}{3}$

John has planted $\frac{1}{3}$ of the garden so far.

Classroom Vignette

"I have students begin writing their addition and subtraction problems vertically in Lesson 6-3. This makes finding common denominators easier."

Doris Jones, Teacher
Creekland Middle School
Lawrenceville, GA

Doris Jones

Adding Like Fractions	To add fractions with like denominators, add the numerators. Use the same denominator in the sum.

Example

1 Find the sum of $\frac{3}{5}$ and $\frac{4}{5}$.

Estimate: $\frac{1}{2} + 1 = 1\frac{1}{2}$

$$\frac{3}{5} + \frac{4}{5} = \frac{3+4}{5}$$
$$= \frac{7}{5}$$
$$= 1\frac{2}{5} \quad \textit{Compared to the estimate, the answer is reasonable.}$$

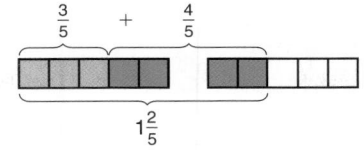

$$\frac{3}{5} \quad + \quad \frac{4}{5}$$
$$1\frac{2}{5}$$

Study Hint

Estimation Use rounding to estimate your answer before adding or subtracting. Then compare your answer to your estimate to see if it is reasonable.

Subtracting Like Fractions	To subtract fractions with like denominators, subtract the numerators. Use the same denominator in the difference.

Subtraction has three meanings. Each of these is shown in one of the following examples.

- to take away part of a set
- to find a missing addend
- to compare the size of two sets

Examples

2 Reiko opened a carton of milk and drank $\frac{1}{4}$ of it. How much of the carton of milk is left?

Since 1 is equal to $\frac{4}{4}$, you need to find $\frac{4}{4} - \frac{1}{4}$.

$$\frac{4}{4} - \frac{1}{4} = \frac{4-1}{4}$$
$$= \frac{3}{4}$$

$\frac{3}{4}$ of the carton is left.

3 Patrice was refilling her watering can. The can holds $\frac{7}{8}$ gallon of water. If it took $\frac{5}{8}$ gallon to fill it, how much water was already in the can?

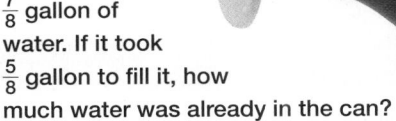

You can use mental math to solve this problem.

$x + \frac{5}{8} = \frac{7}{8}$ *Think: What plus* $\frac{5}{8} = \frac{7}{8}$*?*

$\frac{2}{8} + \frac{5}{8} = \frac{7}{8}$, so the watering can already contained $\frac{2}{8}$ or $\frac{1}{4}$ gallon.

LOOK BACK
Refer to Lesson 1-7 to review solving equations using mental math.

Lesson 6-3 Adding and Subtracting Fractions with Like Denominators **239**

---

2 TEACH

 Transparency 6-3B contains a teaching aid for this lesson.

Using the Mini-Lab You may wish to model the activity on an overhead transparency. Remind students that all fractions representing a portion of the garden will have a denominator of 15 unless simplified. Continue with other examples as needed.

In-Class Examples

For Example 1
Find the sum of $\frac{8}{9}$ and $\frac{5}{9}$. $1\frac{4}{9}$

For Example 2
Mario and his friends ate $\frac{4}{7}$ of a giant submarine sandwich. How much of the sub is left? $\frac{3}{7}$

For Example 3
The route from Ramon's house to the public library and then to school is $\frac{9}{10}$ mile. It is $\frac{3}{10}$ mile from the library to the school. What is the distance from Ramon's house to the library? $\frac{3}{5}$ mile

Teaching Tip Before Example 3, remind students to write fractions in their simplest form when answering a problem.

In-Class Example

For Example 4
A large orange weighs $\frac{11}{16}$ pound. A small orange weighs $\frac{5}{16}$ pound. How much more does the large orange weigh?
$\frac{3}{8}$ **pound**

3 PRACTICE/APPLY

Check for Understanding

If students need additional practice or instruction after completing Exercises 1–10, one of these options may be helpful.
- Extra Practice, see p. 573
- Reteaching Activity
- *Study Guide Masters*, p. 44
- *Practice Masters*, p. 44
- Interactive Mathematics Tools Software

Assignment Guide

Core: 11–31 odd, 32–36
Enriched: 12–28 even, 29–36

Study Guide Masters, p. 44

6-3 **Study Guide**

Name_____ Date_____

Adding and Subtracting Fractions with Like Denominators

To add fractions with like denominators, add the numerators.

Example 1 Find the sum of $\frac{7}{8} + \frac{5}{8}$.
$\frac{7}{8} + \frac{5}{8} = \frac{7+5}{8}$
$= \frac{12}{8}$
$= 1\frac{4}{8}$ or $1\frac{1}{2}$

To subtract fractions with like denominators, subtract the numerators.

Example 2 Find the difference between $\frac{9}{10}$ and $\frac{7}{10}$.
$\frac{9}{10} - \frac{7}{10} = \frac{9-7}{10}$
$= \frac{2}{10}$ or $\frac{1}{5}$

Add or subtract. Write each answer in simplest form.

1. $\frac{7}{12} + \frac{2}{12}$ $\frac{3}{4}$
2. $\frac{9}{10} - \frac{3}{10}$ $\frac{3}{5}$
3. $\frac{7}{9} + \frac{8}{9}$ $1\frac{1}{3}$
4. $\frac{7}{16} - \frac{3}{16}$ $\frac{1}{4}$
5. $\frac{5}{11} + \frac{6}{11}$ 1
6. $\frac{7}{8} - \frac{5}{8}$ $\frac{1}{4}$
7. $\frac{2}{3} + \frac{2}{3}$ $1\frac{1}{3}$
8. $\frac{11}{12} - \frac{5}{12}$ $\frac{1}{2}$
9. $\frac{3}{4} + \frac{3}{4}$ $1\frac{1}{2}$
10. $\frac{4}{5} - \frac{1}{5}$ $\frac{3}{5}$
11. $\frac{5}{6} + \frac{1}{6}$ 1
12. $\frac{7}{10} - \frac{1}{10}$ $\frac{3}{5}$
13. $\frac{3}{7} + \frac{4}{7}$ 1
14. $\frac{15}{16} - \frac{3}{16}$ $\frac{3}{4}$
15. $\frac{5}{8} + \frac{3}{8}$ 1

© Glencoe/McGraw-Hill   T44   *Mathematics: Applications and Connections, Course 1*

Example ——④ **Geography** According to the 1990 census, about $\frac{12}{100}$ of the population of the United States lives in California. Another $\frac{7}{100}$ of the population lives in Texas. How much more of the population lives in California than in Texas?

CONNECTION

Did you know According to 1994 Census Bureau estimates, Texas passed New York to become the second most populous state.

$\frac{12}{100} - \frac{7}{100} = \frac{12-7}{100}$ *Subtract the numerators.*

$= \frac{5}{100}$ or $\frac{1}{20}$ *Simplify.*

About $\frac{1}{20}$ more of the population of the United States lives in California than in Texas.

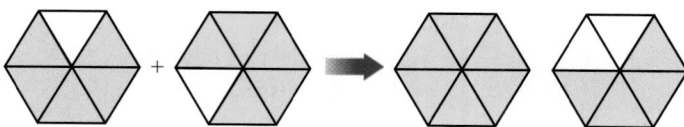
CHECK FOR UNDERSTANDING

Communicating Mathematics
1. Sample answer: Add numerators and keep the denominators the same.

Read and study the lesson to answer each question.

1. *Tell* a simple rule for adding and subtracting like fractions.
2. *Write* the addition sentence shown by the model. $\frac{5}{6} + \frac{5}{6} = 1\frac{2}{3}$

HANDS-ON MATH

3. *Make a model* to show the sum of $\frac{5}{8}$ and $\frac{5}{8}$. Write the sum as a mixed number. **See Answer Appendix.**

Guided Practice

Add or subtract. Write the answer in simplest form.

4. $\frac{3}{8} + \frac{1}{8}$ $\frac{1}{2}$
5. $\frac{3}{5} - \frac{2}{5}$ $\frac{1}{5}$
6. $\frac{7}{10} - \frac{3}{10}$ $\frac{2}{5}$
7. $\frac{5}{6} + \frac{5}{6}$ $1\frac{2}{3}$

8. Find the sum of $\frac{2}{3}$ and $\frac{2}{3}$. $1\frac{1}{3}$
9. What is $\frac{11}{12}$ minus $\frac{5}{12}$? $\frac{1}{2}$

10. *Food* Angie is making a punch mixture that calls for $\frac{3}{4}$ quart of grapefruit juice, $\frac{3}{4}$ quart of orange juice, and $\frac{3}{4}$ quart of pineapple juice. How much punch does the recipe make? $2\frac{1}{4}$ **quarts**

EXERCISES

Practice

Add or subtract. Write the answer in simplest form.

11. $\frac{7}{8} + \frac{5}{8}$ $1\frac{1}{2}$
12. $\frac{9}{16} - \frac{5}{16}$ $\frac{1}{4}$
13. $\frac{3}{4} - \frac{3}{4}$ 0
14. $\frac{3}{5} + \frac{4}{5}$ $1\frac{2}{5}$
15. $\frac{1}{3} + \frac{2}{3}$ 1
16. $\frac{3}{10} + \frac{5}{10}$ $\frac{4}{5}$
17. $\frac{5}{6} - \frac{1}{6}$ $\frac{2}{3}$
18. $\frac{11}{12} - \frac{7}{12}$ $\frac{1}{3}$
19. $\frac{5}{9} - \frac{2}{9}$ $\frac{1}{3}$
20. $\frac{5}{8} - \frac{1}{8}$ $\frac{1}{2}$
21. $\frac{11}{16} + \frac{13}{16}$ $1\frac{1}{2}$
22. $\frac{7}{15} + \frac{8}{15}$ 1

Reteaching the Lesson

Activity Have pairs of students roll two number cubes. The greater number is the denominator, and the other number is the numerator. Ask students to think of the greatest fraction that could be added to it so the sum is less than 1 and the difference is greater than 0. For example, if the fraction is $\frac{4}{6}$, $\frac{1}{6}$ could be added or subtracted.

Error Analysis
Watch for students who add both the numerator and the denominator.
Prevent by reminding students that denominators are never added or subtracted.

Solve each equation mentally. Write the solution in simplest form.

23. $a + \frac{2}{5} = \frac{4}{5}$ $\frac{2}{5}$ **24.** $b = \frac{7}{8} + \frac{3}{8}$ $1\frac{1}{4}$ **25.** $\frac{8}{9} - c = \frac{4}{9}$ $\frac{4}{9}$

26. How much longer than $\frac{7}{16}$ inch is $\frac{15}{16}$ inch? $\frac{1}{2}$ **inch**

27. Find the sum of $\frac{3}{10}$, $\frac{9}{10}$, and $\frac{7}{10}$. $1\frac{9}{10}$

28. How much more is $\frac{3}{4}$ cup than $\frac{1}{4}$ cup? $\frac{1}{2}$ **cup**

Applications and Problem Solving

29. *Carpentry* In industrial technology class, Namid made a plaque by gluing a piece of $\frac{3}{8}$-inch oak to a piece of $\frac{5}{8}$-inch poplar. What was the total thickness of the plaque? **1 inch**

30. *Food* Chad found $\frac{5}{8}$ of a pizza in the refrigerator. He ate $\frac{3}{8}$ of the original pizza. How much of the original pizza is left? $\frac{1}{4}$

31. *Life Science* The inner organs of an electric eel are located in the first $\frac{1}{5}$ of its body. The rest of the eel contains the organs that produce an electric current. How much of an eel produces an electric current? $\frac{4}{5}$

32. *Critical Thinking* Find the sum $\frac{1}{20} + \frac{19}{20} + \frac{2}{20} + \frac{18}{20} + \frac{3}{20} + \frac{17}{20} + \ldots + \frac{10}{20}$. Look for a pattern to help you. **10**

Mixed Review

33. Estimate $4\frac{1}{5} + 1\frac{7}{8}$. *(Lesson 6-2)* **4 + 2 = 6**

34. ▉ **Test Practice** Order the fractions $\frac{7}{9}$, $\frac{2}{3}$, $\frac{2}{6}$, $\frac{5}{2}$, and $\frac{4}{9}$ from greatest to least. *(Lesson 5-8)* **C**

A $\frac{7}{9}, \frac{5}{2}, \frac{4}{9}, \frac{2}{6}, \frac{2}{3}$

B $\frac{7}{9}, \frac{4}{9}, \frac{2}{6}, \frac{2}{3}, \frac{4}{9}$

C $\frac{5}{2}, \frac{7}{9}, \frac{2}{3}, \frac{4}{9}, \frac{2}{6}$

D $\frac{2}{6}, \frac{4}{9}, \frac{2}{3}, \frac{7}{9}, \frac{5}{2}$

35. Name a fraction to describe what part of the figure is shaded. *(Lesson 5-4)* $\frac{6}{8}$ **or** $\frac{3}{4}$

36. *Statistics* The number of days that it rained each month in Cincinnati, Ohio, was recorded for a complete year with the following results: 12, 17, 9, 21, 15, 7, 14, 7, 15, 22, 14, 19. Construct a stem-and-leaf plot for this data. *(Lesson 2-6)* **See margin.**

Lesson 6-3 Adding and Subtracting Fractions with Like Denominators **241**

Extending the Lesson

Enrichment Masters, p. 44

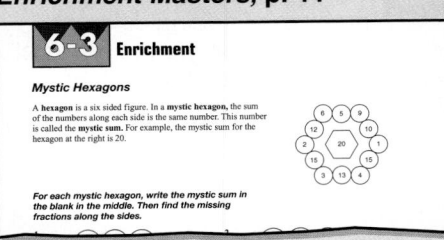

6-3 Enrichment

Mystic Hexagons

A **hexagon** is a six sided figure. In a **mystic hexagon**, the sum of the numbers along each side is the same number. This number is called the **mystic sum**. For example, the mystic sum for the hexagon at the right is 20.

For each mystic hexagon, write the mystic sum in the blank in the middle. Then find the missing fractions along the sides.

Activity Have students measure the tops of their desks or tables to the nearest $\frac{1}{2}$ inch and find the perimeter. Have students also measure the portion they actually use when their books and notebooks are out. Then have them subtract the perimeter of their work space from the perimeter of the entire surface.

---

4 ASSESS

Closing Activity

Modeling Have students use egg cartons and counters to model $\frac{3}{5} + \frac{1}{5} = \frac{4}{5}$ and $\frac{6}{7} - \frac{3}{7} = \frac{3}{7}$. Cut the cartons so there are 5 cups and 7 cups and have students place the counters in the cups.

Additional Answer

36.

Stem	Leaf
0	7 7 9
1	2 4 4 5 5 7 9
2	1 2 2|1 = 21

Practice Masters, p. 44

6-3 Practice

Adding and Subtracting Fractions with Like Denominators

Add or subtract. Write the answer in simplest form.

1. $\frac{2}{9} + \frac{4}{9}$ $\frac{2}{3}$
2. $\frac{13}{16} - \frac{7}{16}$ $\frac{3}{8}$
3. $\frac{5}{8} + \frac{7}{8}$ $1\frac{1}{2}$
4. $\frac{17}{18} - \frac{9}{18}$ $\frac{4}{9}$
5. $\frac{13}{15} - \frac{4}{15}$ $\frac{3}{5}$
6. $\frac{3}{4} + \frac{2}{4}$ $1\frac{1}{4}$
7. $\frac{11}{12} - \frac{7}{12}$ $\frac{1}{3}$
8. $\frac{19}{20} - \frac{11}{20}$ $\frac{2}{5}$
9. $\frac{8}{14} + \frac{7}{14}$ $1\frac{1}{7}$
10. $\frac{9}{10} - \frac{4}{10}$ $\frac{1}{2}$
11. $\frac{4}{5} + \frac{1}{5}$ 1
12. $\frac{6}{7} + \frac{5}{7}$ $1\frac{4}{7}$
13. $\frac{10}{11} - \frac{7}{11}$ $\frac{8}{11}$
14. $\frac{17}{18} + \frac{4}{18}$ $1\frac{1}{6}$
15. $\frac{5}{6} + \frac{4}{6}$ $1\frac{1}{2}$
16. $\frac{12}{13} - \frac{12}{13}$ 0
17. $\frac{9}{16} + \frac{11}{16}$ $1\frac{1}{4}$
18. $\frac{14}{15} - \frac{9}{15}$ $\frac{1}{3}$
19. $\frac{13}{20} - \frac{7}{20}$ $\frac{3}{10}$
20. $\frac{11}{14} - \frac{5}{14}$ $\frac{3}{7}$
21. $\frac{11}{18} + \frac{7}{18}$ $\frac{7}{9}$
22. $\frac{9}{11} + \frac{7}{11}$ $1\frac{5}{11}$
23. $\frac{10}{13} - \frac{4}{13}$ $\frac{6}{13}$
24. $\frac{7}{10} + \frac{9}{10}$ $1\frac{3}{5}$

Tell whether you would add or subtract to solve. Then solve.

25. Amber spent $\frac{2}{5}$ of an hour on her math assignment and $\frac{4}{5}$ of an hour studying for her science test. How much time did she spend doing her homework?
add; $\frac{2}{5} + \frac{4}{5} = 1\frac{1}{5}$ hours

26. Mr. Garcia planted $\frac{11}{16}$ of his fields with corn and $\frac{5}{16}$ of his fields with wheat. How much more of his fields were planted with corn than with wheat?
subtract; $\frac{11}{16} - \frac{5}{16} = \frac{3}{8}$ More of his fields were planted with corn.

© Glencoe/McGraw-Hill   T44   *Mathematics: Applications and Connections, Course 1*

Lesson 6-3 241

HANDS-ON LAB Notes

GET READY

Objective Students find common unit names for adding different objects.

Optional Resources
Hands-On Lab Masters
• worksheet, p. 50

MANAGEMENT TIPS

Recommended Time
15 minutes

Getting Started Have students read the opening paragraph of the lesson. Ask them to think of items in a department store that could have a common unit name. Ask students to think of reasons for a store to group objects into common units.

Some students may need encouragement when thinking of names for the sum of pennies and nickels. Guide them with questions such as the following. *What are they made of? What shape are they? What do we use them for? What is their value?*

The **Activity** has students make a list of common unit names for pennies and nickels and then for pencils and pens. Encourage the class to agree on the best unit names and to discuss why the names they chose are appropriate.

ASSESS

Have students complete Exercises 1–5. Make sure they see the connection between the idea of common unit names and common denominators. Students need to understand that $\frac{5}{10}$, $\frac{3}{6}$, $\frac{2}{4}$, and $\frac{4}{8}$ are equivalent fractions.

HANDS-ON LAB

COOPERATIVE LEARNING

6-4A Renaming Sums

A Preview of Lesson 6-4

 pennies and nickels

pencils

pens

If you had 7 cassette tapes and 5 CDs, how would you tell someone how many you have all together? Since you can't add tapes and CDs, you need to find a common unit name for them. You could name them 12 things, 12 objects, 12 recordings, or 12 albums. Probably the best unit name would be albums.

In this lab, you will find a common unit name for other common objects.

TRY THIS

Work with a partner.

• Choose one person to be the recorder.
• Put 4 pennies and 3 nickels together. Write as many unit names as you can think of to describe the sum of the pennies and nickels.
• Look at your list. Choose the best unit name for the pennies and nickels.
• Repeat the steps using pencils and pens.

1. **Sample answer: you rename the units so that the answer makes sense.**
4. **Sample answer: paper clips, staples, and rubber bands – paper holders**

ON YOUR OWN

1. Explain why you need a common unit name to find the sum.
2. Did you find that some unit names fit better than others? Explain why or why not. **See students' work.**
3. Make a graph showing the different units' names used by your class for the pennies and nickels. **See students' work.**
4. Make a list of different objects that could have a common unit name.
5. *Look Ahead* What do you think you need to do to find the sum of $\frac{1}{2}$ and $\frac{3}{4}$? **See margin.**

242 Chapter 6 Adding and Subtracting Fractions

Additional Answer
5. Sample answer: Find a common denominator.

 Math Journal Have students write a paragraph explaining how a common unit name solves the problem of adding different objects. Include one or two examples of common unit names and what they describe.

Adding and Subtracting Fractions with Unlike Denominators

What you'll learn

You'll learn to add and subtract fractions with unlike denominators.

When am I ever going to use this?

You'll add and subtract fractions with unlike denominators when you analyze graphs.

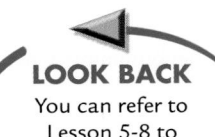

LOOK BACK
You can refer to Lesson 5-8 to review LCD.

What would you do if you won $100,000? The graph shows how adults ages 35–50 said they would spend a $100,000 prize. What fraction of the money did the people surveyed say they would spend on a new car and a new home? *This problem will be solved in Example 1.*

To find the sum, you need a common unit name. In Lesson 6-4A, you came up with common unit names for a group of different objects. When you work with fractions with different, or unlike, denominators, you do the same thing.

To find the sum or difference of two fractions with unlike denominators, rename the fractions using the least common denominator (LCD). Then add or subtract and simplify.

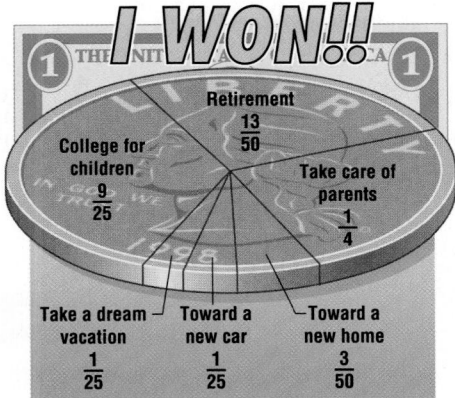

I WON!!

Retirement $\frac{13}{50}$

College for children $\frac{9}{25}$

Take care of parents $\frac{1}{4}$

Take a dream vacation $\frac{1}{25}$

Toward a new car $\frac{1}{25}$

Toward a new home $\frac{3}{50}$

Source: Market Research Institute Survey

Examples

APPLICATION

1 **Money Matters** Refer to the beginning of the lesson. What part of the $100,000 prize did people say they would spend on a new home and a new car?

Add $\frac{3}{50}$ and $\frac{1}{25}$.   *The LCD of $\frac{3}{50}$ and $\frac{1}{25}$ is 50.*

$\frac{3}{50} + \frac{1}{25} = \frac{3}{50} + \frac{2}{50}$ *Rename $\frac{1}{25}$ as $\frac{2}{50}$.*

$= \frac{3+2}{50}$

$= \frac{5}{50}$ or $\frac{1}{10}$

The people surveyed said they would spend $\frac{1}{10}$ of the prize on a new home and a new car.

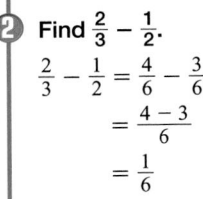

2 Find $\frac{2}{3} - \frac{1}{2}$.

$\frac{2}{3} - \frac{1}{2} = \frac{4}{6} - \frac{3}{6}$ *The LCD of $\frac{2}{3}$ and $\frac{1}{2}$ is 6. Rename $\frac{2}{3}$ as $\frac{4}{6}$ and $\frac{1}{2}$ as $\frac{3}{6}$.*

$= \frac{4-3}{6}$ *Subtract the numerators.*

$= \frac{1}{6}$

Lesson 6-4 Adding and Subtracting Fractions with Unlike Denominators **243**

Multiple Learning Styles

Auditory/Musical Give each pair of students several measures of music that contain notes with different rhythmic values. Have them label each note with its fractional value and find the sum. Encourage students who read music to help the others with this activity.

Motivating the Lesson

Communication Have students discuss how they would determine which cake has more sweetener.
- a chocolate chip cake that has $\frac{3}{4}$ cup of sugar and $\frac{1}{3}$ cup of brown sugar
- a black walnut cake that has $\frac{1}{2}$ cup of brown sugar and $\frac{2}{3}$ cup of corn syrup

---

6-4 Lesson Notes

Instructional Resources
- *Study Guide Masters*, p. 45
- *Practice Masters*, p. 45
- *Enrichment Masters*, p. 45
- Transparencies 6-4, A and B
- *Assessment and Evaluation Masters*, pp. 154, 155
- *Technology Masters*, p. 12
- CD-ROM Program
  - Resource Lesson 6-4

Recommended Pacing

Standard	Days 6 & 7 of 13
Honors	Day 6 of 12
Block	Day 4 of 7

1 FOCUS

5-Minute Check
(Lesson 6-3)

Add or subtract. Write the answer in simplest form.

1. $\frac{3}{8} + \frac{1}{8}$ $\frac{1}{2}$

2. $\frac{7}{12} - \frac{4}{12}$ $\frac{1}{4}$

3. $\frac{9}{14} - \frac{6}{14}$ $\frac{3}{14}$

4. $\frac{4}{7} + \frac{6}{7}$ $1\frac{3}{7}$

5. How much more is $\frac{5}{9}$ gallon than $\frac{2}{9}$ gallon? $\frac{1}{3}$ **gallon**

 The 5-Minute Check is also available on **Transparency 6-4A** for this lesson.

2 TEACH

 Transparency 6-4B contains a teaching aid for this lesson.

Using Graphs Work with students to find the LCD for the circle graph at the beginning of the lesson. Add the fractions and remind students that circle graphs compare parts of a whole.

244   Chapter 6

In-Class Examples

For Example 1
At 8 years old, most people have $\frac{3}{8}$ of their permanent teeth. Between the ages of 8 and 13, people get another $\frac{1}{2}$ of their permanent teeth. What fraction of their permanent teeth do 13-year-olds have? **$\frac{7}{8}$ of their teeth**

For Example 2
Find $\frac{4}{5} - \frac{3}{4}$. **$\frac{1}{20}$**

For Example 3
Evaluate $b - f$ if $b = \frac{2}{3}$ and $f = \frac{5}{8}$. **$\frac{1}{24}$**

3 PRACTICE/APPLY

Check for Understanding

If students need additional practice or instruction after completing Exercises 1–9, one of these options may be helpful.
- Extra Practice, see p. 573
- Reteaching Activity
- *Study Guide Masters*, p. 45
- *Practice Masters*, p. 45

Assignment Guide

Core: 11–33 odd, 35–40
Enriched: 10–30 even, 32–40

Study Guide Masters, p. 45

6-4   Study Guide

Name_____ Date_____

Adding and Subtracting Fractions with Unlike Denominators

To find the sum or difference of two fractions with unlike denominators, write equivalent fractions with a common denominator. Then add or subtract.

Examples 1 Find $\frac{3}{4} + \frac{5}{6}$.
$\frac{3}{4} + \frac{5}{6} = \frac{9}{12} + \frac{10}{12}$ The LCM of 4 and 6 is 12. Rename the fractions with 12 as the denominator.
$= \frac{9 + 10}{12}$
$= \frac{19}{12}$ or $1\frac{7}{12}$

2 Find $\frac{2}{3} - \frac{3}{5}$.
$\frac{2}{3} - \frac{3}{5} = \frac{10}{15} - \frac{9}{15}$ The LCM of 3 and 6 is 15. Rename the fractions with 15 as the denominator.
$= \frac{10 - 9}{15}$
$= \frac{1}{15}$

Find the LCD for each pair of fractions.

1. $\frac{1}{6}, \frac{2}{3}$ **6**   2. $\frac{1}{2}, \frac{2}{5}$ **10**   3. $\frac{7}{8}, \frac{5}{6}$ **24**   4. $\frac{4}{9}, \frac{1}{3}$ **9**

5. $\frac{5}{6}, \frac{1}{3}$ **24**   6. $\frac{7}{10}, \frac{4}{15}$ **30**   7. $\frac{5}{12}, \frac{1}{2}$ **12**   8. $\frac{11}{20}, \frac{2}{5}$ **20**

Add or subtract. Write each answer in simplest form.

9. $\frac{1}{6} + \frac{1}{2}$ **$\frac{2}{3}$**   10. $\frac{2}{3} - \frac{1}{2}$ **$\frac{1}{6}$**   11. $\frac{1}{4} + \frac{7}{8}$ **$1\frac{1}{8}$**   12. $\frac{9}{10} - \frac{3}{5}$ **$\frac{3}{10}$**

13. $\frac{4}{5} + \frac{7}{12}$ **$\frac{53}{60}$**   14. $\frac{11}{15} - \frac{1}{3}$ **$\frac{2}{5}$**   15. $\frac{1}{9} + \frac{1}{6}$ **$\frac{5}{18}$**   16. $\frac{1}{2} - \frac{7}{16}$ **$\frac{1}{16}$**

17. $\frac{3}{10} + \frac{4}{5}$ **$1\frac{1}{10}$**   18. $\frac{5}{6} - \frac{1}{5}$ **$\frac{19}{30}$**   19. $\frac{2}{3} + \frac{1}{2}$ **$1\frac{1}{6}$**   20. $\frac{7}{8} - \frac{4}{9}$ **$\frac{31}{72}$**

© Glencoe/McGraw-Hill   T45   *Mathematics: Applications and Connections, Course 1*

In Chapter 1, you evaluated expressions when the variables were whole numbers. Now you can also evaluate expressions when the variables are fractions.

Example INTEGRATION

3 Algebra Evaluate $p - q$ if $p = \frac{5}{6}$ and $q = \frac{3}{4}$.

$p - q = \frac{5}{6} - \frac{3}{4}$ *Replace p with $\frac{5}{6}$ and q with $\frac{3}{4}$.*

$= \frac{10}{12} - \frac{9}{12}$ *Rename $\frac{5}{6}$ as $\frac{10}{12}$ and $\frac{3}{4}$ as $\frac{9}{12}$.*

$= \frac{10 - 9}{12}$ *Subtract the numerators.*

$= \frac{1}{12}$

CHECK FOR UNDERSTANDING

Communicating Mathematics

Read and study the lesson to answer each question.

1. ***Explain*** why you must rename fractions with unlike denominators when you add or subtract them. **See margin.**

2. ***Tell*** how to find the LCD for $\frac{2}{3}$ and $\frac{5}{8}$. **See margin.**

Guided Practice

Add or subtract. Write the answer in simplest form.

3. $\frac{3}{8} + \frac{1}{16}$ **$\frac{7}{16}$**   4. $\frac{3}{5} - \frac{1}{10}$ **$\frac{1}{2}$**   5. $\frac{1}{3} - \frac{1}{4}$ **$\frac{1}{12}$**   6. $\frac{5}{6} + \frac{3}{4}$ **$1\frac{7}{12}$**

7. What is $\frac{2}{3}$ minus $\frac{5}{12}$? **$\frac{1}{4}$**   8. Find the sum of $\frac{1}{2}$ and $\frac{4}{5}$. **$1\frac{3}{10}$**

9. ***Algebra*** Evaluate $c + d$ if $c = \frac{1}{4}$ and $d = \frac{5}{8}$. **$\frac{7}{8}$**

EXERCISES

Practice

Add or subtract. Write the answer in simplest form.

10. $\frac{5}{8} + \frac{1}{4}$ **$\frac{7}{8}$**   11. $\frac{9}{16} - \frac{1}{2}$ **$\frac{1}{16}$**   12. $\frac{7}{8} - \frac{3}{4}$ **$\frac{1}{8}$**   13. $\frac{3}{5} + \frac{1}{2}$ **$1\frac{1}{10}$**

14. $\frac{2}{3} - \frac{1}{5}$ **$\frac{7}{15}$**   15. $\frac{7}{10} - \frac{1}{6}$ **$\frac{8}{15}$**   16. $\frac{1}{6} + \frac{3}{4}$ **$\frac{11}{12}$**   17. $\frac{11}{12} - \frac{5}{8}$ **$\frac{7}{24}$**

18. $\frac{1}{4} + \frac{2}{3}$ **$\frac{11}{12}$**   19. $\frac{3}{4} - \frac{2}{5}$ **$\frac{7}{20}$**   20. $\frac{5}{8} + \frac{5}{6}$ **$1\frac{11}{24}$**   21. $\frac{5}{9} - \frac{1}{12}$ **$\frac{17}{36}$**

22. Find the sum of $\frac{3}{8}$ and $\frac{5}{6}$. **$1\frac{5}{24}$**

23. How much is $\frac{7}{8}$ minus $\frac{1}{2}$? **$\frac{3}{8}$**

24. How much more is $\frac{3}{4}$ cup than $\frac{2}{3}$ cup? **$\frac{1}{12}$ cup**

25. How much longer than $\frac{9}{16}$ inch is $\frac{7}{8}$ inch? **$\frac{5}{16}$ inch**

26. What is the sum of $\frac{2}{3}$, $\frac{5}{8}$, and $\frac{7}{12}$? **$1\frac{7}{8}$**

244   Chapter 6 Adding and Subtracting Fractions

■ Reteaching the Lesson ■

Activity Write examples such as $\frac{6}{12} - \frac{1}{3}$, $\frac{2}{5} + \frac{3}{10}$, and $\frac{7}{9} - \frac{3}{4}$ on the chalkboard. Have pairs of students decide what steps need to be performed to find each sum or difference. Then have them write and solve an equation of their own.

Additional Answers

1. Sample answer: You rename fractions with unlike denominators to get a common unit name.

2. Sample answer: Find the least common multiple of 3 and 8.

Solve each equation. Write the solution in simplest form.

27. $x = \frac{3}{10} + \frac{9}{20}$ $\frac{3}{4}$ 28. $\frac{4}{5} + \frac{1}{2} = y$ $1\frac{3}{10}$ 29. $\frac{5}{6} - \frac{3}{4} = d$ $\frac{1}{12}$

30. *Algebra* Evaluate $j - k$ if $j = \frac{3}{4}$ and $k = \frac{1}{3}$. $\frac{5}{12}$

31. *Algebra* Evaluate $r + s$ if $r = \frac{3}{5}$ and $s = \frac{7}{10}$. $1\frac{3}{10}$

Applications and Problem Solving

32. *Money Matters* Refer to the graph at the beginning of the lesson. How much bigger is the part that would be spent on taking care of parents than on a dream vacation? $\frac{21}{100}$

33. *Life Science* Proper diet is essential to raising champion horses. Their diet usually consists of grain and hay. The table shows the prescribed amounts that three horses are fed each time, two times a day. How many cans of grain are needed to feed all three horses per feeding? $1\frac{1}{2}$

Horse's Name	Amount of Grain	Amount of Hay
Penny	$\frac{3}{4}$ can	2 slices
Fancy Free	$\frac{1}{4}$ can	3 slices
Max	$\frac{1}{2}$ can	2 slices

34. *Carpentry* The Cabinet Shop made a desktop by gluing a sheet of oak veneer to a sheet of $\frac{3}{4}$-inch plywood. The total thickness of the desktop is $\frac{13}{16}$ inch. What was the thickness of the oak veneer? $\frac{1}{16}$

35. *Critical Thinking* A piece of wire was cut into thirds. One-third was used. One-fifth of the remaining wire was used. The piece that remains is 16 feet long. What was the original length of the wire? **30 feet**

Mixed Review

36. Find $\frac{3}{5} - \frac{1}{5}$. *(Lesson 6-3)* $\frac{2}{5}$

37. *Measurement* Find the width of your pencil to the nearest eighth inch. *(Lesson 5-6)* **Sample answer:** $\frac{2}{8}$ or $\frac{1}{4}$ in.

38. *Geometry* Find the perimeter of the figure. *(Lesson 4-4)* **44 m**

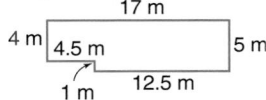
17 m, 4 m, 4.5 m, 5 m, 1 m, 12.5 m

39. **Test Practice** Alphonse Bielevich caught an Atlantic codfish in New Hampshire that weighed 98.75 pounds. Donald Vaughn caught a Pacific codfish in Alaska that weighed 30 pounds. What is the difference in the weights of the two fish? *(Lesson 3-6)* **D**

A 98.45 lb
B 95.75 lb
C 70 lb
D 68.75 lb
E Not Here

40. *Algebra* Solve the equation $17 - m = 4$. *(Lesson 1-7)* **13**

Lesson 6-4 Adding and Subtracting Fractions with Unlike Denominators **245**

Family Activity

Find the recipes for two of your family's favorite desserts. Make a list of the total amount of each ingredient you would need to make both recipes.

Extending the Lesson

Enrichment Masters, p. 45

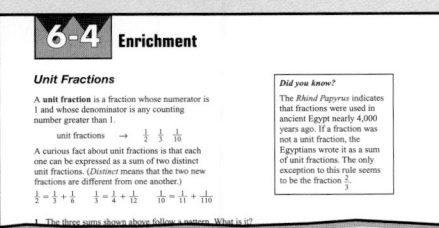

6-4 Enrichment

Unit Fractions

A **unit fraction** is a fraction whose numerator is 1 and whose denominator is any counting number greater than 1.

unit fractions → $\frac{1}{2}$ $\frac{1}{3}$ $\frac{1}{10}$

A curious fact about unit fractions is that each one can be expressed as a sum of two distinct unit fractions. (*Distinct* means that the two new fractions are different from one another.)

$\frac{1}{2} = \frac{1}{3} + \frac{1}{6}$ $\frac{1}{3} = \frac{1}{4} + \frac{1}{12}$ $\frac{1}{10} = \frac{1}{11} + \frac{1}{110}$

Did you know?

The *Rhind Papyrus* indicates that fractions were used in ancient Egypt nearly 4,000 years ago. If a fraction was not a unit fraction, the Egyptians wrote it as a sum of unit fractions. The only exception to this rule seems to be the fraction $\frac{2}{3}$.

1. The three sums shown above follow a pattern. What is it?

Activity Have students survey parents or other adults about the use of fractions and measurements in their work. Invite a parent or other adult to speak to the class about the use of fractions and the importance of accurate measurements in his or her career.

Family Activity

Have students bring recipes to class and explain how they determined the total amount of each ingredient needed. Suggest that students make one of the desserts with their family. Ask them how their knowledge of fractions helped them make the dessert.

4 ASSESS

Closing Activity

Writing Have students work in small groups to create word problems that can be solved by using equations that involve addition or subtraction of fractions. Ask them to use fractions that relate to the group; for example, the fraction of the group that has curly hair or blue eyes.

Chapter 6, Quiz B (Lessons 6-3 and 6-4) is available in the *Assessment and Evaluation Masters*, p. 155.

Mid-Chapter Test (Lessons 6-1 through 6-4) is available in the *Assessment and Evaluation Masters*, p. 154.

Practice Masters, p. 45

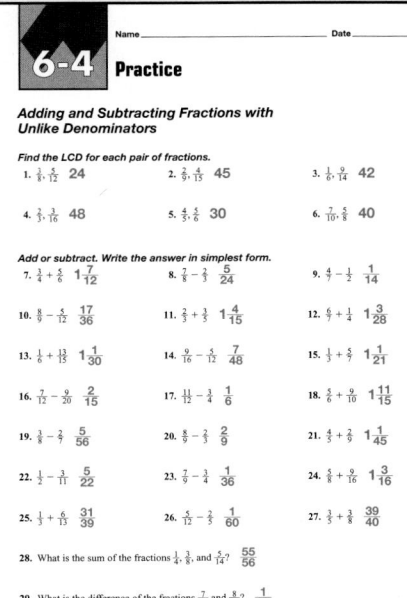

6-4 Practice

Adding and Subtracting Fractions with Unlike Denominators

Find the LCD for each pair of fractions.

1. $\frac{2}{3}, \frac{5}{12}$ 24
2. $\frac{2}{9}, \frac{4}{15}$ 45
3. $\frac{1}{6}, \frac{9}{14}$ 42
4. $\frac{2}{3}, \frac{7}{16}$ 48
5. $\frac{4}{5}, \frac{1}{6}$ 30
6. $\frac{7}{10}, \frac{5}{8}$ 40

Add or subtract. Write the answer in simplest form.

7. $\frac{3}{4} + \frac{4}{6}$ $1\frac{7}{12}$
8. $\frac{7}{8} - \frac{3}{6}$ $\frac{5}{24}$
9. $\frac{4}{7} - \frac{1}{2}$ $\frac{1}{14}$
10. $\frac{8}{9} - \frac{4}{12}$ $\frac{17}{36}$
11. $\frac{2}{3} + \frac{3}{5}$ $1\frac{4}{15}$
12. $\frac{4}{9} + \frac{1}{4}$ $1\frac{3}{28}$
13. $\frac{1}{6} + \frac{13}{15}$ $1\frac{1}{30}$
14. $\frac{9}{16} - \frac{4}{12}$ $\frac{7}{48}$
15. $\frac{1}{3} + \frac{5}{7}$ $1\frac{1}{21}$
16. $\frac{7}{12} - \frac{9}{20}$ $\frac{2}{15}$
17. $\frac{11}{12} - \frac{1}{4}$ $\frac{1}{6}$
18. $\frac{5}{6} + \frac{9}{10}$ $1\frac{11}{15}$
19. $\frac{3}{8} - \frac{2}{7}$ $\frac{5}{56}$
20. $\frac{8}{9} - \frac{2}{3}$ $\frac{2}{9}$
21. $\frac{4}{5} + \frac{2}{9}$ $1\frac{1}{45}$
22. $\frac{1}{2} - \frac{3}{11}$ $\frac{5}{22}$
23. $\frac{7}{9} - \frac{3}{4}$ $\frac{1}{36}$
24. $\frac{5}{8} + \frac{9}{16}$ $1\frac{3}{16}$
25. $\frac{1}{3} + \frac{6}{13}$ $\frac{31}{39}$
26. $\frac{5}{12} - \frac{2}{3}$ $\frac{1}{60}$
27. $\frac{3}{5} + \frac{3}{8}$ $\frac{39}{40}$

28. What is the sum of the fractions $\frac{1}{4}$, $\frac{3}{8}$, and $\frac{5}{14}$? $\frac{55}{56}$

29. What is the difference of the fractions $\frac{7}{12}$ and $\frac{8}{15}$? $\frac{1}{20}$

© Glencoe/McGraw-Hill T45 Mathematics: Applications and Connections, Course 1

Lesson 6-4 245

Instructional Resources
- *Study Guide Masters*, p. 46
- *Practice Masters*, p. 46
- *Enrichment Masters*, p. 46
- Transparencies 6-5, A and B
- *Classroom Games*, pp. 15–18
- CD-ROM Program
  - Resource Lesson 6-5
  - Interactive Lesson 6-5

Recommended Pacing	
Standard	Day 8 of 13
Honors	Day 7 of 12
Block	Day 4 of 7

1 FOCUS

5-Minute Check
(Lesson 6-4)

Add or subtract. Write the answer in simplest form.

1. $\frac{2}{5} - \frac{1}{3}$ $\frac{1}{15}$
2. $\frac{3}{8} + \frac{2}{5}$ $\frac{31}{40}$
3. $\frac{9}{10} - \frac{1}{4}$ $\frac{13}{20}$
4. $\frac{1}{8} + \frac{1}{6}$ $\frac{7}{24}$
5. Evaluate $p - q$ if $p = \frac{2}{3}$ and $q = \frac{2}{5}$. $\frac{4}{15}$

 The 5-Minute Check is also available on **Transparency 6-5A** for this lesson.

Motivating the Lesson

Problem Solving Ask students how they would solve the following problem. *Part of the daily diet of polar bears at the Bronx Zoo is $1\frac{1}{4}$ pounds of apples and a $1\frac{1}{2}$-pound mixture of oats and barley. What is the combined weight of these items?* $2\frac{3}{4}$ pounds

---

Adding and Subtracting Mixed Numbers

What you'll learn

You'll learn to add and subtract mixed numbers.

When am I ever going to use this?

You'll add and subtract mixed numbers when you work with lumber dimensions.

Sandy is taking a flight from Atlanta to San Antonio. Travel time is about $2\frac{3}{4}$ hours. When she reports to the gate, they tell her the flight has a $1\frac{1}{2}$-hour delay. In how many hours will she arrive in San Antonio?

You need to find $2\frac{3}{4} + 1\frac{1}{2}$. You can use models to solve the problem.

 HANDS-ON **MINI-LAB**

Work with a partner.   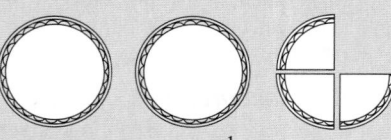 paper plates   scissors

Try This

- Place two paper plates in front of you to show the whole number 2 in the mixed number $2\frac{3}{4}$. Cut another paper plate into fourths. Place three fourths of the plate in front of you to show the fraction $\frac{3}{4}$ in the mixed number $2\frac{3}{4}$.

- Use more paper plates to show $1\frac{1}{2}$.

- Combine the pieces to make as many whole paper plates as you can.

Talk About It

1. How many whole paper plates do you have? **4**
2. What fraction is represented by the pieces of paper plate that you have left over? $\frac{1}{4}$

3. $4\frac{1}{4}$ hours
4. $1\frac{1}{4}$; See students' work.

3. How long will it be before Sandy arrives in San Antonio?
4. Use paper plates to find $2\frac{3}{4} - 1\frac{1}{2}$.

Multiple Learning Styles

 Naturalist Have students bring in samples from nature such as sticks, leaves, feathers, nuts, or pinecones. Have them measure the lengths of the items using fractional parts of customary measures. Then have students write problems that require adding or subtracting the lengths of the sticks, leaves, and so on.

The Mini-Lab suggests the following rule.

Adding and Subtracting Mixed Numbers	1. Add or subtract the fractions. 2. Then add or subtract the whole numbers. 3. Rename and simplify if necessary.

Examples

1 Find $4\frac{2}{3} - 2\frac{1}{3}$. *Estimate: 5 − 2 = 3*

Subtract the fractions.   *Subtract the whole numbers.*

$$
\begin{array}{r} 4\frac{2}{3} \\ -2\frac{1}{3} \\ \hline \frac{1}{3} \end{array}
\qquad \rightarrow \qquad
\begin{array}{r} 4\frac{2}{3} \\ -2\frac{1}{3} \\ \hline 2\frac{1}{3} \end{array}
$$

2 Find $12\frac{1}{4} + 3\frac{5}{6}$. *Estimate: 12 + 4 = 16*

                Add the fractions.   *Add the whole numbers.*

$$
\begin{array}{r} 12\frac{1}{4} \\ +3\frac{5}{6} \\ \hline \end{array}
\rightarrow
\begin{array}{r} 12\frac{3}{12} \\ +3\frac{10}{12} \\ \hline \frac{13}{12} \end{array}
\rightarrow
\begin{array}{r} 12\frac{3}{12} \\ +3\frac{10}{12} \\ \hline 15\frac{13}{12} \end{array}
$$

Rename $\frac{13}{12}$ as $1\frac{1}{12}$.

$15 + 1\frac{1}{12} = 16\frac{1}{12}$

INTEGRATION **3** **Algebra** **Evaluate** $x - y$ **if** $x = 5\frac{9}{10}$ **and** $y = 2\frac{1}{2}$.

$x - y = 5\frac{9}{10} - 2\frac{1}{2}$ *The LCM of 2 and 10 is 10.*

$ = 5\frac{9}{10} - 2\frac{5}{10}$ *Rename $2\frac{1}{2}$ as $2\frac{5}{10}$.*

$ = 3\frac{4}{10}$ or $3\frac{2}{5}$

INTEGRATION **4** **Geometry** **Find the perimeter of the triangle.**

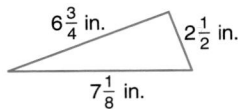

$6\frac{3}{4}$ in. $2\frac{1}{2}$ in. $7\frac{1}{8}$ in.

LOOK BACK
You can refer to Lesson 4-4 to review perimeter.

$P = 2\frac{1}{2} + 6\frac{3}{4} + 7\frac{1}{8}$ *The LCM of 2, 4, and 8 is 8.*

$P = 2\frac{4}{8} + 6\frac{6}{8} + 7\frac{1}{8}$ *Rename $\frac{1}{2}$ as $\frac{4}{8}$ and $\frac{3}{4}$ as $\frac{6}{8}$.*

$P = \frac{4 + 6 + 1}{8} + 2 + 6 + 7$ *Add the fractions. Then add the whole numbers.*

$P = \frac{11}{8} + 15$

$P = 15\frac{11}{8}$ or $16\frac{3}{8}$ *Rename $\frac{11}{8}$ as $1\frac{3}{8}$. $15 + 1\frac{3}{8} = 16\frac{3}{8}$.*

The perimeter of the triangle is $16\frac{3}{8}$ inches.

 Transparency 6-5B contains a teaching aid for this lesson.

Using the Mini-Lab Each group will need at least four paper plates. If paper plates are not available, students can trace circles on sheets of paper and then cut them out.

Teaching Tip Before Example 1, stress the importance of adding and subtracting the fractions before adding and subtracting the whole numbers. This will encourage good habits for later work with renaming mixed numbers.

In-Class Examples

For Example 1
Find $12\frac{3}{8} - 5\frac{1}{8}$. $7\frac{1}{4}$

For Example 2
Find $3\frac{7}{9} + 15\frac{1}{6}$. $18\frac{17}{18}$

For Example 3
Evaluate $f + g$ if $f = 21\frac{4}{5}$ and $g = 6\frac{1}{10}$. $27\frac{9}{10}$

For Example 4
Find the perimeter of the trapezoid. $25\frac{7}{12}$ ft

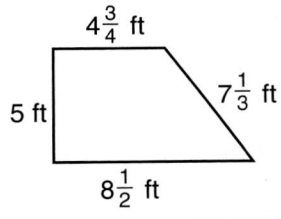

$4\frac{3}{4}$ ft $7\frac{1}{3}$ ft   5 ft $8\frac{1}{2}$ ft

Check for Understanding
If students need additional practice or instruction after completing Exercises 1–11, one of these options may be helpful.
- Extra Practice, see p. 574
- Reteaching Activity
- *Study Guide Masters*, p. 46
- *Practice Masters*, p. 46

Assignment Guide
Core: 13–31 odd, 32–36
Enriched: 12–28 even, 29–36
All: Self Test, 1–10

Additional Answer
2. Sample answer: The length of a rectangle is $3\frac{5}{8}$ inches, and the width is $1\frac{1}{2}$ inches. How much longer is the length than the width?

CHECK FOR UNDERSTANDING

Communicating Mathematics

Read and study the lesson to answer each question.

1. **Tell** the first step you should take when adding or subtracting mixed numbers. **Sample answer: Find the LCM of the denominators.**

2. **Write a Problem** where you need to subtract $1\frac{1}{2}$ from $3\frac{5}{8}$. **See margin.**

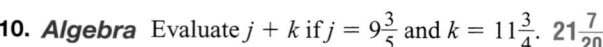
HANDS-ON MATH

3. **Make a model** to show $2\frac{1}{4} + 2\frac{3}{4}$. **See Answer Appendix.**

Guided Practice

Add or subtract. Write the answer in simplest form.

4. $6\frac{3}{4} - 2\frac{1}{4}$ $4\frac{1}{2}$ 5. $7\frac{5}{8} + 3\frac{1}{8}$ $10\frac{3}{4}$ 6. $8\frac{9}{10} + 6\frac{1}{4}$ $15\frac{3}{20}$

7. $21\frac{4}{5} + 6\frac{3}{10}$ $28\frac{1}{10}$ 8. $11\frac{1}{3} - 8\frac{1}{6}$ $3\frac{1}{6}$ 9. $7\frac{2}{3} - \frac{3}{5}$ $7\frac{1}{15}$

10. **Algebra** Evaluate $j + k$ if $j = 9\frac{3}{5}$ and $k = 11\frac{3}{4}$. $21\frac{7}{20}$

11. **Life Science** A rare African giant frog was captured in 1989 on the Sanaga River, in Cameroon. The body of the frog measured $14\frac{1}{2}$ inches. With its legs fully extended, the frog was $34\frac{1}{2}$ inches long. How much longer was the frog when its legs were extended? **20 inches**

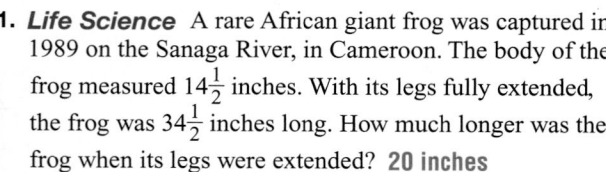

EXERCISES

Practice

Add or subtract. Write the answer in simplest form.

12. $4\frac{3}{5} + 2\frac{2}{5}$ 7 13. $3\frac{5}{6} - 1\frac{1}{6}$ $2\frac{2}{3}$ 14. $2\frac{1}{2} + 3\frac{2}{3}$ $6\frac{1}{6}$

15. $13\frac{1}{3} + 5\frac{3}{4}$ $19\frac{1}{12}$ 16. $8\frac{9}{10} - 6\frac{1}{4}$ $2\frac{13}{20}$ 17. $5\frac{3}{4} - 3\frac{1}{2}$ $2\frac{1}{4}$

18. $17\frac{3}{10} + 9\frac{1}{4}$ $26\frac{11}{20}$ 19. $7\frac{4}{5} + \frac{3}{5}$ $8\frac{2}{5}$ 20. $8\frac{1}{3} - 2\frac{1}{6}$ $6\frac{1}{6}$

21. $18\frac{5}{6} - 18\frac{5}{8}$ $\frac{5}{24}$ 22. $11\frac{3}{4} + 9\frac{3}{5}$ $21\frac{7}{20}$ 23. $14\frac{3}{5} - 5\frac{7}{16}$ $9\frac{13}{80}$

24. How much longer is $28\frac{1}{2}$ seconds than $23\frac{3}{10}$ seconds? $5\frac{1}{5}$ **seconds**

25. Find the sum of $4\frac{1}{5}$, $8\frac{7}{8}$, and $1\frac{7}{10}$. $14\frac{31}{40}$

Evaluate each expression if $a = 2\frac{3}{4}$, $b = 5\frac{1}{6}$, and $c = 7\frac{2}{3}$.

26. $a + b$ $7\frac{11}{12}$ 27. $c - b$ $2\frac{1}{2}$ 28. $a + c$ $10\frac{5}{12}$

Applications and Problem Solving

29. **Health** Dakota's baby sister weighed $7\frac{1}{2}$ pounds at birth. She weighed $8\frac{3}{4}$ pounds at her one-month checkup. How much weight did the baby gain? $1\frac{1}{4}$ **pounds**

Study Guide Masters, p. 46

Name _____ Date _____

6-5 Study Guide

Adding and Subtracting Mixed Numbers

Follow these steps to add and subtract mixed numbers.

Examples	Write equivalent fractions with a common denominator. Add or subtract the fractions.	Add or subtract the whole numbers. Rename and simplify. If necessary.

1. $4\frac{7}{8}$ $+ 3\frac{1}{2}$ → $4\frac{7}{8}$ $+ 3\frac{4}{8}$ → $4\frac{7}{8}$ $+ 3\frac{4}{8}$ $7\frac{11}{8}$ $7 + 1\frac{3}{8} = 8\frac{3}{8}$

2. $5\frac{5}{6}$ $- 3\frac{1}{2}$ → $5\frac{5}{6}$ $- 3\frac{3}{6}$ → $5\frac{5}{6}$ $- 3\frac{3}{6}$ $2\frac{2}{6}$ $2\frac{2}{6} = 2\frac{1}{3}$

Add or subtract. Write each answer in simplest form.

1. $6\frac{1}{4} + 2\frac{1}{4}$ $8\frac{1}{2}$ 2. $7\frac{7}{9} - 4\frac{2}{9}$ $3\frac{5}{9}$ 3. $8\frac{2}{3} + 2\frac{1}{3}$ 11

4. $6\frac{6}{7} - 5\frac{3}{7}$ $1\frac{3}{7}$ 5. $10\frac{1}{2} + 4\frac{1}{8}$ $14\frac{5}{8}$ 6. $12\frac{5}{6} - 3\frac{1}{3}$ $9\frac{1}{2}$

7. $7\frac{1}{10} + 2\frac{1}{5}$ $9\frac{3}{10}$ 8. $9\frac{1}{2} - 5\frac{1}{6}$ $4\frac{1}{3}$ 9. $5\frac{1}{4} + 2\frac{5}{8}$ $8\frac{3}{8}$

10. $18\frac{3}{4} - 6\frac{3}{4}$ 12 11. $5\frac{6}{9} + 4\frac{5}{9}$ $10\frac{11}{21}$ 12. $9\frac{3}{4} - 2\frac{1}{6}$ $7\frac{7}{12}$

■ Reteaching the Lesson ■

Activity Write examples such as $2\frac{3}{4} + 5\frac{1}{8}$ on index cards. (Include subtraction examples also.) Have groups of three students add the fractions first, then the whole numbers, and simplify the answer. Have students take turns completing each part.

30a. $68\frac{2}{5}$ million acres

30. Agriculture The table shows the acreage of crops planted in the United States in 1996 and estimates for 1997.

  a. In 1996, how many more acres of wheat were planted than barley?

  b. Find the total acreage used to raise corn, wheat, cotton, and rice in 1996. $172\frac{3}{5}$ **million acres**

  c. How many more acres of soybeans were projected to be planted in 1997? $4\frac{3}{5}$ **million acres**

  d. Were more or fewer acres of rice projected to be planted in 1997? Explain.

Acreage (millions)		
Crop	1996	1997 (est.)
corn	$79\frac{1}{2}$	$81\frac{1}{2}$
wheat	$75\frac{3}{5}$	$69\frac{1}{5}$
soybeans	$64\frac{1}{5}$	$68\frac{4}{5}$
cotton	$14\frac{7}{10}$	$14\frac{1}{2}$
sorghum	$13\frac{1}{5}$	$10\frac{9}{10}$
barley	$7\frac{1}{5}$	7
oats	$4\frac{7}{10}$	$5\frac{3}{10}$
rice	$2\frac{4}{5}$	$2\frac{9}{10}$

Source: Agriculture Department Economic Research Service

31. Write a Problem that involves the two amounts of acreage of cotton given in the table in Exercise 30. **See margin.**

32. Critical Thinking Use the numbers 1, 2, 3, 4, 1, and 2 to create an addition of mixed numbers problem so that the sum is $4\frac{1}{4}$.

Mixed Review

33.  **Test Practice** What is the perimeter of the figure? *(Lesson 6-4)* **D**

  A $\frac{16}{24}$ in.   **B** $\frac{16}{8}$ in.

  C 2 in.   **D** $2\frac{1}{2}$ in.

$\frac{1}{4}$ in. $\frac{3}{4}$ in. $\frac{7}{8}$ in. $\frac{5}{8}$ in.

30d. more; Sample answer: $2\frac{9}{10}$ is $\frac{1}{10}$ more than $2\frac{4}{5}$

34. Express $\frac{100}{6}$ as a mixed number. *(Lesson 5-5)* $16\frac{2}{3}$

32. Sample answer: $1\frac{3}{4} + 2\frac{1}{2} = 4\frac{1}{4}$

35. Measurement If a rabbit has a mass of 800 grams, what is its mass in kilograms? *(Lesson 4-9)* **0.8 kg**

36. Estimate $21.1 + 19 + 20 + 20.3 + 20.1 + 18.8$. *(Lesson 3-5)*
Sample answer: $20 + 20 + 20 + 20 + 20 + 20 = 120$

CHAPTER 6

Mid-Chapter Self Test

1. Food You are not sure how much salt to put in a pot of chili. Should you put one-half teaspoon or one teaspoon of salt in the pot? Explain. *(Lesson 6-1)* **See margin.**

Estimate. *(Lesson 6-2)*

2. $4\frac{3}{4} - 2\frac{2}{5}$ $5 - 2 = 3$

3. $22\frac{3}{8} - 14\frac{9}{10}$ $22 - 15 = 7$

4. $9\frac{1}{3} + 5\frac{5}{6}$ $9 + 6 = 15$

Add or subtract. Write the answer in simplest form. *(Lessons 6-3, 6-4, and 6-5)*

5. $\frac{13}{20} - \frac{7}{20}$ $\frac{3}{10}$

6. $\frac{4}{5} + \frac{3}{5}$ $1\frac{2}{5}$

7. $\frac{2}{5} - \frac{1}{6}$ $\frac{7}{30}$

8. $\frac{3}{4} + \frac{7}{10}$ $1\frac{9}{20}$

9. $13\frac{2}{3} + 6\frac{1}{2}$ $20\frac{1}{6}$

10. $11\frac{5}{8} - 9\frac{2}{5}$ $2\frac{9}{40}$

Lesson 6-5 Adding and Subtracting Mixed Numbers **249**

Extending the Lesson

Enrichment Masters, p. 46

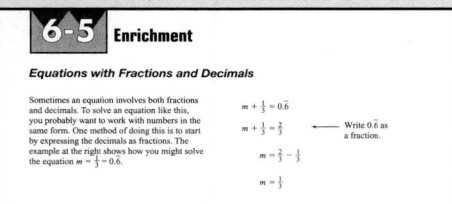

6-5 Enrichment

Equations with Fractions and Decimals

Activity Challenge students to use what they know about decimals and fractions to solve the following problem. *Winona buys stationery for $0.99 a pound. She bought $3\frac{2}{5}$ pounds of pink paper and $1\frac{1}{2}$ pounds of purple paper. Not including tax, how much did Winona spend?* **$4.85**

Subtracting Mixed Numbers with Renaming

Instructional Resources

- *Study Guide Masters,* p. 47
- *Practice Masters,* p. 47
- *Enrichment Masters,* p. 47
- Transparencies 6-6, A and B
- *Assessment and Evaluation Masters,* p. 156
- CD-ROM Program
  - Resource Lesson 6-6
  - Interactive Lesson 6-6

Recommended Pacing	
Standard	Days 9 & 10 of 13
Honors	Days 8 & 9 of 12
Block	Day 5 of 7

1 FOCUS

 5-Minute Check
(Lesson 6-5)

Add or subtract. Write the answer in simplest form.

1. $7\frac{5}{6} - 3\frac{3}{4}$ $4\frac{1}{12}$
2. $5\frac{2}{5} + 1\frac{3}{10}$ $6\frac{7}{10}$
3. $15\frac{5}{7} - 7\frac{1}{2}$ $8\frac{3}{14}$
4. Aaron bought $3\frac{5}{6}$ pounds of bananas and $4\frac{1}{2}$ pounds of oranges. How many pounds of fruit did he buy? $8\frac{1}{3}$ lb
5. Evaluate $a - b$ if $a = 3\frac{5}{12}$ and $b = 1\frac{3}{4}$. $1\frac{2}{3}$

The 5-Minute Check is also available on **Transparency 6-6A** for this lesson.

Motivating the Lesson

Hands-On Activity Have students cut models from centimeter grid paper to illustrate renaming mixed numbers. The models below illustrate renaming $2\frac{1}{4}$ as $1\frac{5}{4}$.

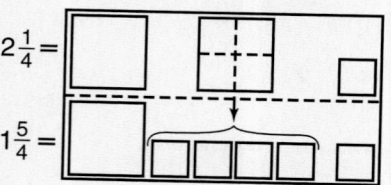

$2\frac{1}{4} =$

$1\frac{5}{4} =$

What you'll learn

You'll learn to subtract mixed numbers involving renaming.

When am I ever going to use this?

You'll subtract mixed numbers involving renaming when you work with stock market prices.

LOOK BACK
You can refer to Lesson 5-5 to review renaming mixed numbers.

Example

① Find $4 - 1\frac{2}{3}$. *Estimate: $4 - 2 = 2$*

$\begin{array}{r} 4 \\ -1\frac{2}{3} \end{array} \rightarrow \begin{array}{r} 3\frac{3}{3} \\ -1\frac{2}{3} \\ \hline 2\frac{1}{3} \end{array}$ *Rename 4 as $3\frac{3}{3}$.*

$4 - 1\frac{2}{3} = 2\frac{1}{3}$

Brenda belongs to a cycling club. The club meets and rides every weekend. Brenda rides during the week to stay in shape. On Tuesday she rode her bike $4\frac{1}{2}$ miles, and on Thursday she rode $6\frac{1}{4}$ miles. How much farther did she ride on Thursday? You need to subtract to compare the mileage.

$\begin{array}{r} 6\frac{1}{4} \\ -4\frac{1}{2} \end{array} \rightarrow \begin{array}{r} 6\frac{1}{4} \\ -4\frac{2}{4} \end{array}$

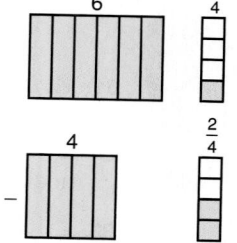

Notice that you cannot subtract $\frac{2}{4}$ from $\frac{1}{4}$. Rename $6\frac{1}{4}$ as $5\frac{5}{4}$.

$\begin{array}{r} 6\frac{1}{4} \\ -4\frac{1}{2} \end{array} \rightarrow \begin{array}{r} 6\frac{1}{4} \\ -4\frac{2}{4} \end{array} \rightarrow \begin{array}{r} 5\frac{5}{4} \\ -4\frac{2}{4} \\ \hline 1\frac{3}{4} \end{array}$

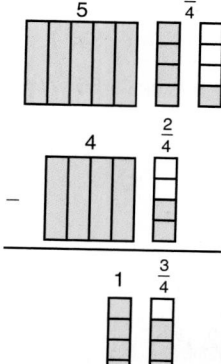

Brenda rode $1\frac{3}{4}$ miles farther on Thursday.

As shown above, sometimes it is necessary to rename the fraction of a mixed number as an improper fraction in order to subtract.

Multiple Learning Styles

 Verbal/Linguistic Have each pair of students create a large map on a piece of paper, labeling fractional lengths on it. Have them write a series of directions to travel a path on their map that would be achieved by adding or subtracting fractional lengths. Have groups exchange maps and directions and complete.

2 Find $17\frac{1}{4} - 3\frac{5}{8}$. *Estimate: 17 − 4 = 13*

Step 1 $17\frac{1}{4}$ → $17\frac{2}{8}$ *The LCM of 4 and 8 is 8.*

$\quad\quad\quad -3\frac{5}{8}$ → $-3\frac{5}{8}$

Step 2 $17\frac{2}{8}$ → $16\frac{10}{8}$ *Since $\frac{5}{8}$ is greater than $\frac{2}{8}$, you must*

$\quad\quad\quad -3\frac{5}{8}$ → $\dfrac{-3\frac{5}{8}}{13\frac{5}{8}}$ *rename $17\frac{2}{8}$ as $16\frac{10}{8}$.*

$17\frac{1}{4} - 3\frac{5}{8} = 13\frac{5}{8}$

CONNECTION **3** **Life Science** A male California sea lion grows to be between $6\frac{1}{2}$ and 8 feet long. Find the difference between the greatest and least lengths.

Subtract $8 - 6\frac{1}{2}$.   *Estimate: 8 − 7 = 1*

$\quad 8$ → $7\frac{2}{2}$ *Rename 8 as $7\frac{2}{2}$.*

$-6\frac{1}{2}$ → $\dfrac{-6\frac{1}{2}}{1\frac{1}{2}}$

The difference between the greatest and least lengths is $1\frac{1}{2}$ feet.

CHECK FOR UNDERSTANDING

Communicating Mathematics

Read and study the lesson to answer each question.

1. *Draw* a model or use paper plates to show how to subtract $4\frac{1}{2} - 2\frac{3}{8}$. **See Answer Appendix.**

2. *Tell* why you might rename $3\frac{5}{8}$ as $2\frac{13}{8}$ in a subtraction problem.

2. Sample answer: The number being subtracted is $\frac{6}{8}$ or $\frac{7}{8}$.

3. *You Decide* Gail subtracted $10\frac{1}{5} - 3\frac{4}{5}$. Her answer was $6\frac{7}{5}$. Does Gail's answer make sense? How do you think Gail got her answer? **See Answer Appendix.**

Guided Practice

Complete.

4. $4\frac{1}{8} = 3\frac{\blacksquare}{8}$ **9**

5. $\blacksquare = 29\frac{4}{4}$ **30**

Subtract. Write the answer in simplest form.

6. $\begin{array}{r} 6\frac{3}{10} \\ -1\frac{4}{5} \end{array}$ **$4\frac{1}{2}$**

7. $11\frac{1}{6} - 8\frac{1}{3}$ **$2\frac{5}{6}$**

8. $7\frac{1}{4} - \frac{3}{5}$ **$6\frac{13}{20}$**

9. $3\frac{5}{8} - 2\frac{3}{4}$ **$\frac{7}{8}$**

10. *Transportation* The U.S. Department of Transportation regulations prohibit a truck driver from driving more than 70 hours in any 8-day period. Mr. Galvez has driven $53\frac{3}{4}$ hours in the last 6 days. How many more hours is he allowed to drive during the next 2 days? **$16\frac{1}{4}$ hours**

Lesson 6-6 Subtracting Mixed Numbers with Renaming   **251**

Reteaching the Lesson

Activity Have pairs of students make and use drawings to model examples such as $12 - 3\frac{1}{4}$. Ask them to make enough shapes to show 12. Have them illustrate renaming by cutting one of the shapes. Then have them remove $3\frac{1}{4}$ to find the difference.

Error Analysis
Watch for students who forget to subtract the whole numbers.
Prevent by encouraging students to develop a habit of estimating and checking each answer.

Exercise 37 asks students to advance to the next stage of work on the Chapter Project. You may wish to suggest to the students that they follow their instructions on their maps. This will ensure that their maps will make sense to others.

Study Guide Masters, p. 47

6-6 Study Guide

Subtracting Mixed Numbers with Renaming

Sometimes it is necessary to rename a mixed number as an improper fraction before you can subtract.

Examples 1 $6\frac{1}{4}$ $6\frac{1}{4}$ You cannot subtract $5\frac{1}{4}$ *Rename* $6\frac{1}{4}$
                $-2\frac{3}{4} \rightarrow -2\frac{3}{4}$ $\frac{3}{4}$ *from* $\frac{1}{4}$. $\rightarrow -2\frac{3}{4}$ *as* $5\frac{5}{4}$.
                                                                                      $\overline{3\frac{2}{4}}$ *Then subtract.*

            2 $8 \rightarrow 7\frac{8}{8}$ *Rename 8 as* $7\frac{8}{8}$.
               $-4\frac{5}{8} \rightarrow -4\frac{5}{8}$ *Then subtract.*
                          $\overline{3\frac{3}{8}}$

Complete.

1. $7\frac{5}{6} = \square\frac{11}{6}$ **6**   2. $4\frac{2}{4} = 3\frac{\square}{4}$ **7**   3. $2\frac{3}{8} = 1\frac{\square}{8}$ **11**

4. $9\frac{3}{8} = \square\frac{3}{8}$ **8**   5. $10\frac{1}{3} = 9\frac{\square}{3}$ **4**   6. $15 = 14\frac{\square}{2}$ **2**

7. $20\frac{5}{12} = 19\frac{\square}{12}$ **17**   8. $13 = 12\frac{\square}{7}$ **7**   9. $6\frac{3}{5} = \square\frac{\square}{5}$ **5**

Subtract. Write each answer in simplest form.

10. $5\frac{1}{3} - 3\frac{2}{3}$ **1$\frac{2}{3}$**   11. $12\frac{1}{8} - 7\frac{5}{8}$ **4$\frac{4}{8}$**   12. $8\frac{3}{8} - 3\frac{5}{8}$ **4$\frac{3}{4}$**

13. $9\frac{1}{2} - 4\frac{3}{4}$ **4$\frac{3}{4}$**   14. $12 - 1\frac{2}{5}$ **10$\frac{3}{5}$**   15. $8\frac{1}{2} - \frac{7}{8}$ **7$\frac{5}{8}$**

16. $15\frac{1}{2} - 9\frac{5}{6}$ **5$\frac{1}{2}$**   17. $7\frac{1}{2} - 3\frac{11}{12}$ **3$\frac{7}{12}$**   18. $22 - 10\frac{8}{9}$ **11$\frac{1}{9}$**

© Glencoe/McGraw-Hill   T47   Mathematics: Applications and Connections, Course 1

252   Chapter 6

EXERCISES

Practice   **Complete.**

11. $3\frac{5}{8} = \blacksquare\frac{13}{8}$ **2**   12. $\blacksquare\frac{3}{10} = 5\frac{13}{10}$ **6**   13. $9\frac{2}{5} = 8\frac{\blacksquare}{5}$ **7**

14. $7\frac{\blacksquare}{3} = 6\frac{4}{3}$ **1**   15. $7\frac{13}{16} = 6\frac{\blacksquare}{16}$ **29**   16. $\blacksquare\frac{11}{12} = 7\frac{23}{12}$ **8**

Subtract. Write the answer in simplest form.

17. $30\frac{1}{4}$ **5$\frac{1}{2}$**   18. $20\frac{1}{5}$ **15$\frac{3}{5}$**   19. $5\frac{2}{5}$ **3$\frac{7}{10}$**   20. $18\frac{3}{8}$ **12$\frac{5}{8}$**
     $-24\frac{3}{4}$ $-4\frac{3}{5}$ $-1\frac{7}{10}$ $-5\frac{3}{4}$

21. $7\frac{1}{3} - 2\frac{3}{4}$ **4$\frac{7}{12}$**   22. $15\frac{3}{8} - 9\frac{5}{6}$ **5$\frac{13}{24}$**   23. $16\frac{1}{6} - 8\frac{3}{4}$ **7$\frac{5}{12}$**

24. $13 - 8\frac{7}{8}$ **4$\frac{1}{8}$**   25. $30 - 20\frac{1}{4}$ **9$\frac{3}{4}$**   26. $6\frac{4}{5} - 1\frac{5}{6}$ **4$\frac{29}{30}$**

27. Find the difference between $11\frac{4}{7}$ and $3\frac{3}{5}$. **7$\frac{34}{35}$**

28. What is $2\frac{1}{2}$ less than $8\frac{1}{5}$? **5$\frac{7}{10}$**

Solve each equation. Write the solution in simplest form.

29. $f = 9\frac{1}{3} - 8\frac{2}{3}$ **$\frac{2}{3}$**   30. $p = 18\frac{5}{8} - 8\frac{3}{4}$ **9$\frac{7}{8}$**   31. $12\frac{3}{5} - 10\frac{5}{6} = t$ **1$\frac{23}{30}$**

Evaluate each expression if $a = 4\frac{1}{2}$**,** $b = 3\frac{2}{3}$**, and** $c = 2\frac{5}{6}$**.**

32. $a - b$ **$\frac{5}{6}$**   33. $a - c$ **1$\frac{2}{3}$**   34. $b - c$ **$\frac{5}{6}$**

Applications and Problem Solving

35a. softball; $2\frac{5}{8}$ to $3\frac{1}{8}$ in. larger

35. **Sports** Some of the differences between Olympic softball and Olympic baseball are listed in the table.

a. Which sport's ball has the larger circumference? How much larger is it compared to the other sport's ball?

35b. softball; 1 to 2 ounces heavier

b. Which ball is heavier? How much heavier is it?

	Olympic Softball	Olympic Baseball
Size	$11\frac{7}{8}$ to $12\frac{1}{8}$ inches in circumference	9 to $9\frac{1}{4}$ inches in circumference
Weight	$6\frac{1}{4}$ to 7 ounces	5 to $5\frac{1}{4}$ ounces

36. $8\frac{1}{2}$ in. by $2\frac{1}{2}$ in.

36. **Geometry** Suppose you fold the short side of an $8\frac{1}{2}$ inch-by-11 inch piece of paper to align with the long side. What are the dimensions of the rectangle at the side of the paper?

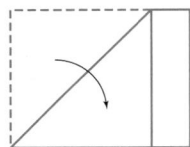

37. **Working on the CHAPTER Project** Refer to the table on page 227.

a. Complete your map using the remaining landmarks. **See students' work.**

b. How much longer is the path from the start, through the landmarks, to the treasure than the shortest distance between the start and the treasure? **1$\frac{3}{4}$ in.**

Classroom Vignette

"You may want to allow students to use the paper plate models they made in the Mini-Lab on page 246 to illustrate the renaming process used in subtracting mixed numbers."

Billie Pikur, Teacher
West Middle School
Rochester Hills, MI

Billie E. Pikur

38. Critical Thinking Write a problem where you must subtract with renaming and the difference is between $\frac{1}{3}$ and $\frac{1}{2}$.
Sample answer: **What is $5\frac{1}{4} - 4\frac{7}{8}$?**

Mixed Review

39. Find the sum of $\frac{7}{8}$ and $1\frac{5}{6}$. Write the answer in simplest form. *(Lesson 6-5)* $2\frac{17}{24}$

40. ▐ Test Practice ▐ Train A runs every 8 minutes and Train B runs every 6 minutes. If they both leave the station at 9:00 A.M., at what time will they next leave the station together? *(Lesson 5-7)* **A**

   A 9:24 A.M.   **B** 9:48 A.M.
   C 10:48 A.M.   **D** 12:00 P.M.

41. Find $586.1 \div 0.58$ to the nearest tenth. *(Lesson 4-7)* **1,010.5**

42. Statistics Find the mean and the median for the following set of low temperatures for a city: 45, 41, 28, 42, 44, 40, 40. *(Lesson 2-7)* **40; 41**

43. Algebra Evaluate $3x - y$ if $x = 4$ and $y = 1$. *(Lesson 1-5)* **11**

Let the Games Begin

What Difference Does It Make?

Get Ready   This game is for two players.

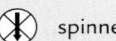 2 number cubes   ✖ spinner

Get Set   Make the spinner shown.

Go
- Each player creates a mixed number by rolling the number cubes and spinning the spinner. For example, suppose a player rolls a 1 and a 6 and spins $\frac{2}{3}$. The player would record the mixed number $7\frac{2}{3}$.

- Each player creates a second mixed number using the same method.

- If the second mixed number is less than the first, the player should subtract it from the first.

- If the second mixed number is greater than the first, the player should add it to the first.

- Players continue to create mixed numbers and either add to or subtract from the previous result.

- The winner is the player with the smallest number after six rounds.

▐ inter**NET** ▐
CONNECTION Visit www.glencoe.com/sec/math/mac/mathnet for more games.

Lesson 6-6 Subtracting Mixed Numbers with Renaming   **253**

■ Extending the Lesson ■

Enrichment Masters, p. 47

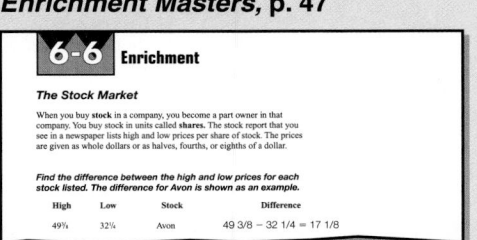

6-6 Enrichment

The Stock Market

When you buy **stock** in a company, you become a part owner in that company. You buy stock in units called **shares**. The stock report that you see in a newspaper lists high and low prices per share of stock. The prices are given as whole dollars or as halves, fourths, or eighths of a dollar.

Find the difference between the high and low prices for each stock listed. The difference for Avon is shown as an example.

High	Low	Stock	Difference
49⅜	32¼	Avon	49 3/8 − 32 1/4 = 17 1/8

Let the Games Begin

Suggest to students before they start that finding the LCD of the fractions on the spinner may help them add or subtract more quickly during play. You may also choose to have students play in teams instead of in pairs.

4 ASSESS

Closing Activity

Modeling Have small groups of students make sets of cards with subtraction problems and answers to use in a "Concentration" game. A sample set is shown below.

$$15 - 6\frac{3}{4}$$ $$8\frac{1}{4}$$

Chapter 6, Quiz C (Lessons 6-5 and 6-6) is available in the *Assessment and Evaluation Masters*, p. 156.

Practice Masters, p. 47

6-6 Practice

Name_____ Date_____

Subtracting Mixed Numbers with Renaming

Complete.
1. $6\frac{3}{5} = 5\frac{□}{5}$ **11**   2. $9\frac{1}{10} = \boxed{}\frac{□}{10}$ **8**   3. $8\frac{1}{12} = \boxed{}\frac{17}{12}$ **7**   4. $3\frac{5}{9} = 2\frac{□}{9}$ **16**
5. $10\frac{7}{14} = 9\frac{□}{14}$ **23**   6. $7\frac{4}{11} = 6\frac{□}{11}$ **14**   7. $12\frac{7}{8} = \boxed{}\frac{15}{8}$ **11**   8. $14\frac{13}{15} = 13\frac{□}{15}$ **28**

Subtract. Write the answer in simplest form.

9.	10.	11.	12.
$4\frac{2}{7}$	$7\frac{1}{6}$	10	$12\frac{5}{8}$
$-1\frac{4}{7}$	$-3\frac{5}{6}$	$-5\frac{1}{4}$	$-3\frac{3}{4}$
$2\frac{6}{7}$	$3\frac{1}{3}$	$4\frac{3}{4}$	$8\frac{7}{8}$

13.	14.	15.	16.
$15\frac{1}{2}$	$17\frac{1}{3}$	$15\frac{1}{7}$	$13\frac{1}{2}$
$-9\frac{4}{9}$	$-7\frac{3}{5}$	$-8\frac{4}{7}$	$-7\frac{4}{5}$
$5\frac{13}{18}$	$9\frac{8}{15}$	$6\frac{11}{14}$	$5\frac{7}{10}$

17.	18.	19.	20.
$10\frac{1}{8}$	$2\frac{1}{7}$	$14\frac{1}{12}$	$15\frac{5}{16}$
$-2\frac{5}{8}$	$-\frac{2}{3}$	$-3\frac{5}{9}$	$-6\frac{3}{8}$
$7\frac{19}{24}$	$1\frac{10}{21}$	$10\frac{7}{36}$	$8\frac{15}{16}$

21.	22.	23.	24.
$3\frac{2}{5}$	$16\frac{2}{3}$	$18\frac{1}{9}$	$9\frac{7}{15}$
$-2\frac{3}{4}$	$-12\frac{11}{12}$	$-8\frac{5}{9}$	$-6\frac{4}{5}$
$\frac{13}{20}$	$3\frac{3}{4}$	$9\frac{5}{9}$	$2\frac{2}{3}$

25.	26.	27.	28.
12	$14\frac{2}{7}$	$16\frac{1}{3}$	$8\frac{1}{20}$
$-5\frac{7}{11}$	$-9\frac{1}{4}$	$-5\frac{5}{6}$	$-7\frac{3}{10}$
$6\frac{4}{11}$	$4\frac{15}{28}$	$10\frac{11}{12}$	$\frac{3}{4}$

29. If you subtract $4\frac{2}{5}$ from $8\frac{5}{8}$, what is the result? $3\frac{39}{40}$

30. Find the difference of $5\frac{1}{6}$ and $2\frac{5}{12}$. $2\frac{3}{4}$

© Glencoe/McGraw-Hill   T47   *Mathematics: Applications and Connections, Course 1*

6-7 Lesson Notes

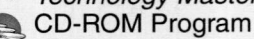

Instructional Resources

- *Study Guide Masters*, p. 48
- *Practice Masters*, p. 48
- *Enrichment Masters*, p. 48
- Transparencies 6-7, A and B
- *Assessment and Evaluation Masters*, p. 156
- *Technology Masters*, p. 11
- CD-ROM Program
  - Resource Lesson 6-7

Recommended Pacing

Standard	Day 11 of 13
Honors	Day 10 of 12
Block	Day 6 of 7

1 FOCUS

5-Minute Check
(Lesson 6-6)

Complete.

1. $8\frac{2}{6} = 7\frac{\blacksquare}{6}$ 8

2. $10\frac{\blacksquare}{7} = 9\frac{12}{7}$ 5

Subtract. Write the answer in simplest form.

3. $20\frac{1}{4} - 15\frac{3}{4}$ $4\frac{1}{2}$

4. $11\frac{2}{5} - 8\frac{3}{10}$ $3\frac{1}{10}$

5. $22 - 13\frac{2}{7}$ $8\frac{5}{7}$

The 5-Minute Check is also available on **Transparency 6-7A** for this lesson.

Motivating the Lesson

Problem Solving Have students read the introduction to the lesson and calculate how long they spend in math class each week.

6-7

Integration: Measurement
Adding and Subtracting Measures of Time

***What* you'll learn**
You'll learn to add and subtract measures of time.

***When* am I ever going to use this?**
You'll add and subtract measures of time when you're trying to determine how long a flight will last.

Demetria's school uses block scheduling. She has an A schedule and a B schedule. For example, she goes to math class this week on Monday, Wednesday, and Friday (A Schedule). Next week she will go to math class on Tuesday and Thursday (B Schedule). How much time will she spend in math class during the week when she only goes two days if each class is 1 hour and 30 minutes?

To add or subtract measures of time:
1. Add or subtract the seconds.
2. Add or subtract the minutes.
3. Add or subtract the hours.
Rename if necessary in each step.

1 hour (h) = 60 minutes (min)	
1 minute = 60 seconds (s)	

Tuesday: 1 h 30 min
Thursday: + 1 h 30 min
                2 h 60 min

Since 2 hours 60 minutes equals 3 hours, Demetria spends 3 hours in math class during weeks when she only goes two days.

Example **APPLICATION** 1

Aerospace The countdown for the launching of a spacecraft was stopped at 12 hours, 7 minutes, 36 seconds before launch. Another stop occurred 1 hour, 28 minutes, 10 seconds before launch. How far did the countdown progress between stops?

Estimate: 12 hours − 1 hour = 11 hours

12 h 7 min 36 s   *Since you cannot subtract 28 min from 7 min,*
− 1 h 28 min 10 s   *you must rename 12 h 7 min as 11 h 67 min.*
        26 s

12 h 7 min 36 s   →   11 h 67 min 36 s
− 1 h 28 min 10 s   − 1 h 28 min 10 s
        26 s   10 h 39 min 26 s

The countdown progressed 10 hours, 39 minutes, 26 seconds between the two stops.

Travel Josefina's flight is scheduled to leave Chicago at 11:14 A.M. and arrive in Houston at 1:36 P.M. How long will the flight last?

To answer this question, you need to find how much time has elapsed. Think of a clock.

11:14 A.M. to 12:00 noon   12:00 noon to 1:36 P.M.
        is 46 minutes.   is 1 hour 36 minutes.

The elapsed time is 46 min + 1 h 36 min or 1 h 82 min. Now rename 82 min as 1 h and 22 min.

1 h + 1 h 22 min = 2 h 22 min

Josefina's flight will last 2 hours and 22 minutes.

APPLICATION 3

Sports Joan Benoit of the United States won the first women's Summer Olympic Games marathon in 1984. Her time was 2 hours, 24 minutes, 52 seconds. In the 1996 Summer Olympic Games, Fatuma Roba of Ethiopia won the marathon with a time of 2 hours, 26 minutes, 5 seconds. How much faster was Benoit's time?

Roba's time: 2 h 26 min 5 s   *Since you cannot subtract 52 s*
Benoit's time: − 2 h 24 min 52 s   *from 5 s, you must rename*
                                        26 min 5 s as 25 min 65 s.

  2 h 26 min 5 s   →   2 h 25 min 65 s
− 2 h 24 min 52 s   − 2 h 24 min 52 s
                                1 min 13 s

Benoit ran the marathon 1 minute, 13 seconds faster than Roba.

CHECK FOR UNDERSTANDING

Communicating Mathematics

Read and study the lesson to answer each question.

1. *Write* the number of minutes in 660 seconds. **11 minutes**

2. *Draw* clocks to show how much time elapses between when you leave home to go to school and when you return home from school. Then determine how much time has elapsed. **See Answer Appendix.**

3. *Write*, in your own words, how adding and subtracting measures of time is similar to adding and subtracting mixed numbers. **See margin.**

Additional Answer

3. Sample answer: Adding and subtracting measures of time is similar to adding and subtracting mixed numbers because you have to rename.

---

 2 TEACH

Transparency 6-7B contains a teaching aid for this lesson.

Reading Mathematics Be sure students can read time numbers and labels from left to right correctly. Careful alignment of labels is necessary before adding or subtracting time.

In-Class Examples

For Example 1
In 1973, Secretariat won the Kentucky Derby in 1 minute, 59 seconds, and also won the Belmont Stakes in 2 minutes, 24 seconds. How much slower was Secretariat's time in the Belmont? **25 seconds**

For Example 2
If a 13-pound turkey is put in the oven at 11:20 A.M. and taken out at 4:30 P.M., how long did the turkey cook? **5 h 10 min**

For Example 3
Wesley Paul set an age group world record in the 1977 New York City Marathon. He ran the race in 3 hours 31 seconds. He was 8 years old at the time. Suppose he ran 2 hours 58 minutes 48 seconds in the practice before the race. How much faster was Wesley's practice time? **1 min 43 s**

Teaching Tip Compare adding or subtracting measures of time to adding and subtracting three-digit numbers. Remind students that renaming is not a new idea.

Check for Understanding

If students need additional practice or instruction after completing Exercises 1–13, one of these options may be helpful.
- Extra Practice, see p. 574
- Reteaching Activity
- *Study Guide Masters*, p. 48
- *Practice Masters*, p. 48

Assignment Guide

Core: 15–43 odd, 44–49
Enriched: 14–40 even, 41–49

Study Guide Masters, p. 48

6-7 Study Guide

Name_____ Date_____

Integration: Measurement
Adding and Subtracting Measures of Time

To add measures of time, add the seconds, add the minutes, and add the hours. Rename if necessary.

Remember: 1 hour (h) = 60 minutes (min)
            1 minute (min) = 60 seconds (s)

Example 1   4 h 25 min 40 s
              + 5 h 30 min 25 s
              _____
               9 h 55 min 65 s   *Rename 65 s as 1 min 5 s.*
              9 h 55 min + 1 min 5 s = 9 h 56 min 5 s

To subtract measures of time, rename if necessary. Then subtract seconds, subtract minutes, and subtract hours.

Example 2   7 h 15 min 40 s   *You cannot subtract 20 min from 15 min.*
              − 3 h 20 min 10 s
                   ↓
               6 h 75 min 40 s   *Rename 7 h 15 min as 6 h 75 min.*
              − 3 h 20 min 10 s   *Then subtract.*
               3 h 55 min 30 s

Complete.

1. 14 min 85 s = _15_ min 25 s   2. 9 h 5 min = 8 h _65_ min

3. 3 h 20 min 7 s = 3 h _19_ min _67_ s   4. 7 h 9 min 25 s = _6_ h _69_ min 25 s

Add or subtract. Rename if necessary.

5.   6 h 20 min   6.   35 min 45 s   7.   12 h 15 s
   − 3 h 17 min   + 12 min 12 s   + 10 h 55 s
   _____   _____   _____
    3 h 3 min   47 min 57 s   22 h 1 min 10 s

8. 9 h 45 min 10 s   9. 1 h 55 min 12 s   10. 7 h 20 min
   − 3 h 30 min 50 s   + 3 h 25 min 34 s   − 2 h 9 min 10 s
   _____   _____   _____
    6 h 14 min 20 s   5 h 20 min 46 s   5 h 10 min 50 s

© Glencoe/McGraw-Hill   †48   *Mathematics: Applications and Connections, Course 1*

Guided Practice

Complete.

4. 8 h 18 min = 7 h _?_ min **78**   5. 9 h 93 min 8 s = _?_ h 33 min 8 s **10**

Add or subtract. Rename if necessary.

6.   5 h 34 min   7.   2 min 35 s   8.   9 h 20 min
   − 4 h 9 min   + 8 min 10 s   − 3 h 45 min
   _____   _____   _____
    1 h 25 min   10 min 45 s   5 h 35 min

9.   6 h   20 s   10. 12 h 40 min 30 s
   − 2 h 9 min 40 s   + 4 h 30 min 50 s
   _____   _____
    3 h 50 min 40 s   17 h 11 min 20 s

Find the elapsed time.

11. 7:50 A.M. to 2:35 P.M.   12. 10:30 P.M. to 6:15 A.M.
    6 h 45 min   **7 h 45 min**

13. *Money Matters* Kenny worked for the Moore family by baby-sitting this weekend. On Friday, he worked 3 hours and 15 minutes. On Saturday, he worked 4 hours and 45 minutes. If he gets paid $3.50 an hour, how much money did he make this weekend? **$28**

EXERCISES

Practice   **Complete.**

14. 17 min 15 s = 16 min _?_ s **75**

15. 4 h 8 min = 3 h _?_ min **68**

16. 2 min 85 s = _?_ min 25 s **3**

17. 42 h 5 min 87 s = 42 h _?_ min _?_ s **6; 27**

18. 2 h 23 min 28 s = 2 h 22 min _?_ s **88**

19. 26 h 83 min 8 s = _?_ h _?_ min 8 s **27; 23**

Add or subtract. Rename if necessary.

20.   15 min 54 s   21.   7 h 20 min   22.   19 h 30 min
    − 9 min 50 s   + 3 h 18 min   − 12 h 40 min
    _____   _____   _____
     6 min 4 s   10 h 38 min   6 h 50 min

23.   5 h 25 min   24. 12 h 38 min   25.   15 h 8 min
    + 10 h 36 min   − 3 h 46 min   − 14 h 48 min
    _____   _____   _____
     16 h 1 min   8 h 52 min   20 min

26.   4 h 54 min   27.   4 min   28.   4 h 20 s
    + 12 h 6 min   + 12 min 55 s   − 3 h 45 s
    _____   _____   _____
     17 h   16 min 55 s   59 min 35 s

29. 18 h 25 min   30.   7 h 20 min   31.   3 h 35 min 40 s
    − 2 h 9 min 40 s   + 2 h 48 min 10 s   + 6 h 50 min 40 s
    _____   _____   _____
     16 h 15 min 20 s   10 h 8 min 10 s   10 h 26 min 20 s

32.   8 h   33.   6 h   20 s   34.   2 h 9 min 23 s
    − 3 h 20 min 15 s   − 2 h 9 min 40 s   − 1 h 10 min 32 s
    _____   _____   _____
     4 h 39 min 45 s   3 h 50 min 40 s   58 min 51 s

Reteaching the Lesson

Activity Have students work in small groups to develop a chart of "A Day in the Life of a Sixth Grader." Have them record beginning and ending times for daily activities. Then have them determine how long each activity takes. Have them use their calculators to check their answers by determining if the total time is 24 hours.

Error Analysis
Watch for students who have difficulty renaming units of time.
Prevent by providing examples such as 85 seconds renamed as 1 minute 25 seconds.

Find the elapsed time.

35. 10:20 A.M. to 11:19 A.M. **59 min**   **36.** 6:30 A.M. to 12:35 P.M. **6 h 5 min**

37. 8:15 P.M. to 6:45 A.M. **10 h 30 min**   **38.** 8:05 A.M. to 4:15 P.M. **8 h 10 min**

39. 21 h 20 min 10 s **39.** Find the sum of 7 hours, 25 minutes, 10 seconds and 13 hours 55 minutes.

40. How much time is there between 6:15 A.M. and 7:15 P.M.? **13 h**

Applications and Problem Solving

41. *Music* Orlando's piano lesson started at 4:45 P.M. and ended at 5:30 P.M. How long was his lesson? **45 min**

42. *Cooking* Rosalinda is making biscuits. She put the biscuits in the oven at 8:45 A.M. They need to bake for about 20 minutes. When should she check them? **9:05 A.M.**

43. *Travel* Peter flew from Tampa, Florida, to New York City. His plane left Tampa at 6:34 A.M., and the flight took 3 hours 55 minutes. What time did he arrive in New York City? **10:29 A.M.**

44. *Critical Thinking* Bridgette is flying from Philadelphia to San Francisco to visit her grandmother. Her flight leaves at 8:15 A.M. The non-stop flight takes about 6 hours. About what time will it be in San Francisco when Bridgette arrives? (*Hint:* Remember that there is a time difference between Philadelphia and San Francisco.) **11:15 A.M.**

Mixed Review

45. *Health* Joaquin is $65\frac{1}{2}$ inches tall. Jesús is $61\frac{3}{4}$ inches tall. How much taller is Joaquin than Jesús? *(Lesson 6-6)* $3\frac{3}{4}$ **in.**

46. *Cooking* Paul needs $2\frac{1}{4}$ cups of flour for making cookies, $1\frac{2}{3}$ cups for almond bars, and $3\frac{1}{2}$ cups for cinnamon rolls. How much flour does he need in all? *(Lesson 6-5)* $7\frac{5}{12}$ **cups**

47. | Test Practice | What is the greatest 3-digit number that is not divisible by 3 or 7? *(Lesson 5-1)* **B**
   A 999
   B 998
   C 997
   D 996

48. *Measurement* Which is the better estimate for the capacity of a glass of milk: 360 liters or 360 milliliters? *(Lesson 4-8)* **360 milliliters**

49. *Astronomy* The distance from Earth to the Sun is close to 10^8 miles. How many miles is this? *(Lesson 1-6)* **100,000,000 miles**

4 ASSESS

Closing Activity

Writing Have students explain in writing how they would solve the following problem. *While Adam was on his way home from school, there was a power outage. When he got home from school, his watch read 3:30 P.M. The kitchen clock read 2:55 P.M. How long was the power out?* **35 minutes**

Chapter 6, Quiz D (Lesson 6-7) is available in the *Assessment and Evaluation Masters,* p. 156.

Practice Masters, p. 48

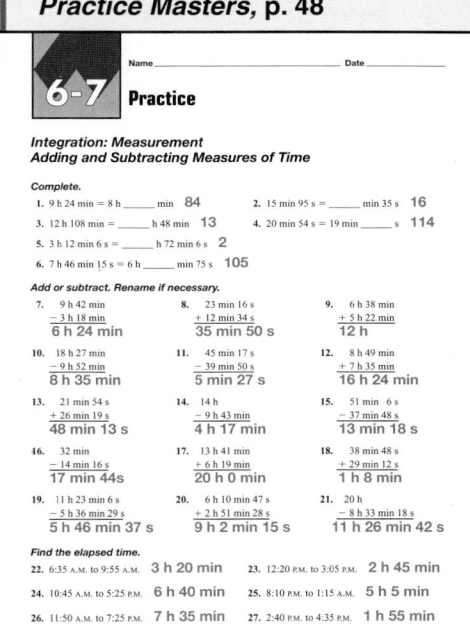

Enrichment Masters, p. 48

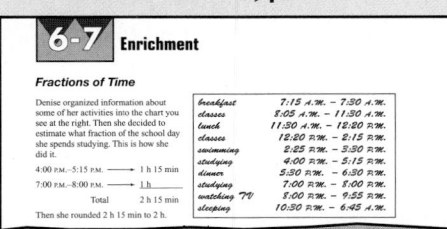

Extending the Lesson

Activity Provide students with copies of pages from a bus or airline schedule. Ask them to plan a trip and use the schedules to calculate the amount of travel time.

Vocabulary

This section provides a listing of the new terms, properties, and phrases that were introduced in this chapter. Have students define each term and provide an example or two of it, if appropriate.

Understanding and Using the Vocabulary

These exercises check students' understanding of the terms by using a variety of verbal formats including matching, completion, and true/false.

Glossaries A complete glossary of terms appears on pages 642–648. The glossary also appears in Spanish on pages 649–656.

Additional Answer

10. Sample answer: 8 hours 15 minutes should be renamed as 7 hours 75 minutes. Subtract 50 minutes from 75 minutes to get 25 minutes, and subtract 5 hours from 7 hours to get 2 hours.

Vocabulary

After completing this chapter, you should be able to define each term, concept, or phrase and give an example or two of each.

Number and Operations
like fractions (p. 238)

Problem Solving
eliminate possibilities (pp. 236–237)

Understanding and Using the Vocabulary

Choose the correct term or number to complete each sentence.

1. Rounded to the nearest half, $5\frac{1}{5}$ rounds to $\left(\underline{5}, 5\frac{1}{2}\right)$.

2. An estimate for $\frac{9}{20} + \frac{8}{9}$ is $\left(\frac{1}{2}, 1\frac{1}{2}\right)$.

3. The sum of $\frac{2}{5}$ and $\left(\frac{3}{5}, \frac{4}{5}\right)$ is $1\frac{1}{5}$.

4. (Denominators, <u>Numerators</u>) are subtracted when you subtract fractions with like denominators.

5. The $\left(\underline{\text{LCD}}, \text{GCF}\right)$ is the least common multiple of the denominators of unlike fractions.

6. The LCD of $\frac{1}{8}$ and $\frac{3}{10}$ is $\left(80, \underline{40}\right)$.

7. The solution of $x = 6\frac{7}{10} - 1\frac{1}{5}$ is $\left(5\frac{1}{2}, 5\frac{3}{5}\right)$.

8. The mixed number $9\frac{1}{4}$ can be renamed as $\left(8\frac{3}{4}, 8\frac{5}{4}\right)$.

9. The number of minutes in one hour and twenty minutes is $\left(120, \underline{80}\right)$.

In Your Own Words

10. **Explain** how to subtract 5 hours 50 minutes from 8 hours 15 minutes. See margin.

 MindJogger Videoquizzes

MindJogger Videoquizzes provide an alternative review of concepts presented in this chapter. Students work in teams to answer questions, gaining points for correct answers. The questions are presented in three rounds.
Round 1 Concepts–5 questions
Round 2 Skills–4 questions
Round 3 Problem Solving–4 questions

Objectives & Examples

Upon completing this chapter, you should be able to:

● round fractions and mixed numbers *(Lesson 6-1)*

Round to the nearest half.

$\frac{13}{16}$ → 1   *13 is close to 16.*

$\frac{5}{9}$ → $\frac{1}{2}$ *5 is about half of 9.*

$\frac{2}{11}$ → 0   *2 is much smaller than 11.*

● estimate sums and differences of fractions and mixed numbers *(Lesson 6-2)*

$\frac{13}{16} - \frac{5}{12}$ → $1 - \frac{1}{2} = \frac{1}{2}$

$\frac{7}{8} + 8\frac{1}{5}$ → $1 + 8 = 9$

● add and subtract fractions with like denominators *(Lesson 6-3)*

$\frac{7}{8} + \frac{5}{8} = \frac{7+5}{8}$

$= \frac{12}{8}$

$= 1\frac{4}{8}$ or $1\frac{1}{2}$

Review Exercises

Use these exercises to review and prepare for the chapter test.

Round each number to the nearest half.

11. $\frac{9}{16}$ $\frac{1}{2}$

12. $9\frac{2}{9}$ 9

13. $11\frac{4}{7}$ $11\frac{1}{2}$

14. $\frac{9}{11}$ 1

Tell whether each number should be rounded up or down.

15. the diameter of a plant to put in an $11\frac{7}{8}$-inch pot **down**

16. the weight limit of an elevator **down**

17. the capacity of a pitcher needed to hold $\frac{7}{8}$ gallon of lemonade **up**

18. the height of a truck that can go under a bridge **up**

Estimate.

19. $8\frac{3}{16} - 4\frac{1}{4}$ $8 - 4 = 4$

20. $\frac{5}{8} + \frac{15}{16}$ $\frac{1}{2} + 1 = 1\frac{1}{2}$

21. $\frac{1}{8} + 4\frac{2}{3}$ $0 + 5 = 5$

22. $\frac{7}{8} - \frac{5}{12}$ $1 - \frac{1}{2} = \frac{1}{2}$

Add or subtract. Write the answer in simplest form.

23. $\frac{8}{15} + \frac{11}{15}$ $1\frac{4}{15}$

24. $\frac{4}{7} - \frac{1}{7}$ $\frac{3}{7}$

25. $\frac{7}{8} - \frac{1}{8}$ $\frac{3}{4}$

26. $\frac{5}{6} + \frac{5}{6}$ $1\frac{2}{3}$

Chapter 6 Study Guide and Assessment **259**

Objectives & Examples

This section reviews the skills and concepts of the chapter and shows completely worked examples.

Review Exercises

These exercises provide practice for the corresponding objectives.

Assessment and Evaluation Masters, pp. 143–144

Assessment and Evaluation

Six forms of Chapter 6 Test are available in the *Assessment and Evaluation Masters* as shown in the chart.

Chapter 6 Test, Form 1B, is shown at the right. Chapter 6 Test, Form 2B, is shown on the next page.

1A	Multiple Choice	Honors
1B	Multiple Choice	Average
1C	Multiple Choice	Basic
2A	Free Response	Honors
2B	Free Response	Average
2C	Free Response	Basic

Objectives & Examples

add and subtract fractions with unlike denominators *(Lesson 6-4)*

$$\frac{7}{12} - \frac{11}{24} = \frac{14}{24} - \frac{11}{24}$$
$$= \frac{14 - 11}{24}$$
$$= \frac{3}{24} \text{ or } \frac{1}{8}$$

add and subtract mixed numbers *(Lesson 6-5)*

$$11\frac{2}{3} \;\rightarrow\; 11\frac{16}{24}$$
$$+ 2\frac{3}{8} \;\rightarrow\; +2\frac{9}{24}$$
$$13\frac{25}{24} = 13 + 1\frac{1}{24} \text{ or } 14\frac{1}{24}$$

subtract mixed numbers involving renaming *(Lesson 6-6)*

Find $4\frac{1}{4} - \frac{2}{3}$.

$$4\frac{1}{4} \rightarrow 4\frac{3}{12} \rightarrow 3\frac{15}{12}$$
$$-\frac{2}{3} \rightarrow -\frac{8}{12} \rightarrow -\frac{8}{12}$$
$$3\frac{7}{12}$$

$$4\frac{1}{4} - \frac{2}{3} = 3\frac{7}{12}$$

add and subtract measures of time *(Lesson 6-7)*

$$\begin{array}{r} 3\text{ h } 50\text{ min} \\ + 2\text{ h } 15\text{ min} \\ \hline 5\text{ h } 65\text{ min} \end{array}$$

Rename 65 min as 1 h and 5 min.

5 h + 1 h 5 min = 6 h 5 min

Review Exercises

Add or subtract. Write the answer in simplest form.

27. $\frac{7}{10} - \frac{1}{5}$ **$\frac{1}{2}$**   28. $\frac{2}{3} + \frac{1}{4}$ **$\frac{11}{12}$**

29. $\frac{4}{9} + \frac{5}{6}$ **$1\frac{5}{18}$**   30. $\frac{2}{3} - \frac{1}{5}$ **$\frac{7}{15}$**

Add or subtract. Write the answer in simplest form.

31. $7\frac{3}{7} - 2\frac{1}{7}$ **$5\frac{2}{7}$**   32. $8\frac{2}{3} + 1\frac{2}{3}$ **$10\frac{1}{3}$**

33. $9\frac{4}{5} - 8\frac{1}{2}$ **$1\frac{3}{10}$**   34. $11\frac{5}{8} - 3\frac{1}{6}$ **$8\frac{11}{24}$**

Subtract. Write the answer in simplest form.

35. $\begin{array}{r} 8\frac{1}{8} \\ -3\frac{1}{4} \\ \hline \end{array}$ **$4\frac{7}{8}$**   36. $\begin{array}{r} 7\frac{1}{6} \\ -4\frac{1}{3} \\ \hline \end{array}$ **$2\frac{5}{6}$**

37. $8 - 2\frac{2}{3}$ **$5\frac{1}{3}$**   38. $18\frac{7}{16} - 8\frac{3}{4}$ **$9\frac{11}{16}$**

Solve each equation. Write the solution in simplest form.

39. $f = 6\frac{1}{8} - \frac{1}{4}$ **$5\frac{7}{8}$**   40. $5\frac{2}{5} - 1\frac{7}{10} = h$ **$3\frac{7}{10}$**

Complete.

41. 6 h 8 min = 5 h ? min **68**
42. 27 min 75 s = ? min ? s **28, 15**

Add or subtract. Rename if necessary.

43. $\begin{array}{r} 5\text{ h } 20\text{ min} \\ + 2\text{ h } 16\text{ min} \\ \hline \textbf{7 h 36 min} \end{array}$ 44. $\begin{array}{r} 7\text{ h } 45\text{ min} \\ - 4\text{ h } 32\text{ min} \\ \hline \textbf{3 h 13 min} \end{array}$

45. $\begin{array}{r} 9\text{ h } 7\text{ min} \\ - 8\text{ h } 7\text{ min } 8\text{ s} \\ \hline \textbf{59 min 52 s} \end{array}$ 46. $\begin{array}{r} 2\text{ h } 35\text{ min} \\ + 6\text{ h } 41\text{ min} \\ \hline \textbf{9 h 16 min} \end{array}$

260   Chapter 6 Adding and Subtracting Fractions

Assessment and Evaluation Masters, pp. 149–150

Name _____ Date _____

6 Chapter 6 Test, Form 2B

Round each number to the nearest half.
1. $\frac{7}{11}$ 1. $\frac{1}{2}$
2. $\frac{19}{20}$ 2. 1
3. $3\frac{1}{16}$ 3. 3
4. $12\frac{1}{6}$ 4. 12
5. Jermaine is buying a lava lamp that he will store in a bookcase with shelves that are $11\frac{1}{4}$ inches apart. When looking for the lamp, should he round up or round down to determine its proper height?   5. down

Estimate.
6. $7\frac{1}{10} - 4\frac{19}{20}$ 6. 7 − 5 = 2
7. $\frac{24}{25} - \frac{1}{9}$ 7. 1 − 0 = 1
8. $\frac{17}{18} + \frac{9}{10}$ 8. 1 + 1 = 2
9. $3\frac{2}{3} + 2\frac{1}{5}$ 9. 4 + 2 = 6

Add or subtract. Write the answer in simplest form.
10. $\frac{5}{14} + \frac{3}{14}$ 10. $\frac{4}{7}$
11. $\frac{7}{10} - \frac{1}{5}$ 11. $\frac{1}{2}$
12. $\frac{1}{10} + \frac{2}{3}$ 12. $\frac{23}{30}$
13. $\frac{5}{6} - \frac{2}{15}$ 13. $\frac{7}{10}$
14. $7\frac{9}{16} - 3\frac{5}{16}$ 14. $4\frac{1}{4}$
15. $5\frac{4}{5} + \frac{7}{8}$ 15. $6\frac{27}{40}$
16. $6\frac{3}{16} + 6\frac{3}{16}$ 16. $12\frac{3}{8}$
17. $12 - 3\frac{1}{15}$ 17. $8\frac{14}{15}$
18. $18\frac{2}{5} - 14\frac{11}{20}$ 18. $3\frac{17}{20}$

© Glencoe/McGraw-Hill   149   *Mathematics: Applications and Connections, Course 1*

---

6 Chapter 6 Test, Form 2B (continued)

Add or subtract. Write the answer in simplest form.
19. $5\frac{4}{5} - 2\frac{5}{6}$ 19. $2\frac{23}{30}$
20. $10\frac{15}{32} - 3\frac{1}{4}$ 20. $7\frac{7}{32}$
21. $3\frac{1}{2} + 2\frac{3}{10} + 1\frac{1}{3}$ 21. 7
22. $5\frac{3}{4} + 2\frac{5}{6}$ 22. $8\frac{7}{12}$
23. Monica is going to buy a shirt for $15.99. Should she take $15 or $20 with her?   23. $20

Solve each equation. Write the solution in simplest form.
24. $x = \frac{11}{24} + \frac{5}{24}$ 24. $\frac{2}{3}$
25. $a = \frac{9}{14} - \frac{5}{14}$ 25. $\frac{2}{7}$
26. $m = 9\frac{3}{5} + 6\frac{1}{10}$ 26. $15\frac{7}{10}$
27. $y = 9\frac{2}{3} - 3\frac{3}{4}$ 27. $5\frac{11}{12}$
28. A recipe calls for $\frac{1}{2}$ cup of pecans, $\frac{3}{8}$ cup of almonds, and $\frac{1}{4}$ cup of raisins. How many total cups of pecans, almonds, and raisins are in the recipe?   28. $1\frac{1}{8}$ cups
29. A train is scheduled to leave at 6:40 A.M. and arrive at 9:10 A.M. that same day. How long is the scheduled trip?   29. 2 h 30 min

Add or subtract.
30. $\begin{array}{r} 4\text{ h } 39\text{ min } 22\text{ s} \\ + 2\text{ h } 50\text{ min } 19\text{ s} \end{array}$ 30. 7 h 29 min 41 s
31. $\begin{array}{r} 13\text{ h } 25\text{ min} \\ - 11\text{ h } 50\text{ min} \end{array}$ 31. 1 h 35 min
32. $\begin{array}{r} 3\text{ h } 18\text{ min} \\ - 1\text{ h } 50\text{ min} \end{array}$ 32. 1 h 28 min
33. Nicholas and Sarah watched two movies. One lasted 2 hours and 17 minutes, and the other lasted 3 hours and 21 minutes. How long were they watching the movies?   33. 5 h 38 min

© Glencoe/McGraw-Hill   150   *Mathematics: Applications and Connections, Course 1*

Test and Review Software

You may use this software, a combination of an item generator and item bank, to create your own tests or worksheets. Types of items include free response, multiple choice, short answer, and open ended.

CD-ROM Program

The CD-ROM Program contains an Assessment Game whose questions review the concepts in this chapter.

9. Which is a reasonable remainder when a number is divided by 5? **E**

   A 8

   B 7

   C 6

   D 5

   E 4

10. Pedro bought 6 batteries for his radio. Each battery cost $1.79, including tax. What was the total cost of the batteries? **H**

   F $6.24

   G $10.24

   H $10.74

   J $11.74

   K Not Here

11. Each day Matias drives his delivery truck 3.25 miles to his first stop, 4.23 miles to his second stop, 8.8 miles to his third stop, and 6 miles back to headquarters. How many miles does Matias drive the truck each day? **A**

   A 22.28 miles

   B 16.28 miles

   C 8.42 miles

   D 7.48 miles

   E Not Here

12. Ron was making curtains for his room. He had 7 yards of material. It took $5\frac{1}{2}$ yards to make the curtains. How much material was not used for the curtains? **J**

   F $12\frac{1}{2}$ yd

   G $2\frac{1}{2}$ yd

   H 2 yd

   J $1\frac{1}{2}$ yd

   K Not Here

Test-Taking Tip

Most standardized tests have a time limit, so you must use your time wisely. Some questions will be easier than others. If you cannot answer a question quickly, go on to another. If there is time remaining when you are finished, go back to the questions that you skipped.

Section Two: Free Response

This section contains seven questions for which you will provide short answers. Write your answers on your paper.

13. When is the product of two decimals less than both factors? **when both factors are less than one**

14. *Geometry* Find the perimeter of the rectangle. **38 in.**

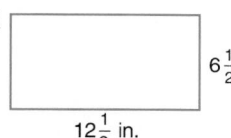

$6\frac{1}{2}$ in.

$12\frac{1}{2}$ in.

15. Write $\frac{35}{63}$ in simplest form. $\frac{5}{9}$

16. Express $\frac{18}{11}$ as a mixed number. $1\frac{7}{11}$

17. *Algebra* Evaluate $a + b$ if $a = \frac{7}{8}$ and $b = \frac{3}{16}$. $1\frac{1}{16}$

18. What is $12\frac{3}{15}$ minus $\frac{2}{5}$? $11\frac{4}{5}$

19. Find the elapsed time from 8:30 P.M. to 10:40 A.M.   **14 h 10 min**

Test-Taking Tip

One of the important skills in test taking is time management. Remind students that they can answer several easier questions in the time they might otherwise spend trying to answer one question they find difficult.

Assessment and Evaluation Masters, pp. 157–158

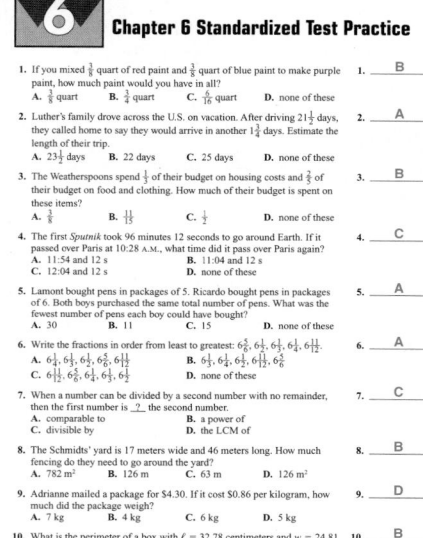

Instructional Resources ▶▶▶

Additional standardized test practice is shown at the right and is available in the *Assessment and Evaluation Masters*, pp. 157–158.

GET READY

This optional investigation is designed to be completed by a group of 4 or 5 students over several days or several weeks.

Mathematical Overview

This investigation utilizes the concepts from Chapters 1–6.
- rounding whole numbers
- writing fractions
- ordering whole numbers
- writing comparison ratios
- calculating perimeter

Time Management	
Gathering Data	30 minutes
Calculations	30 minutes
Summarizing Data	20 minutes
Presentation	15 minutes

Instructional Resources
- *Investigations and Projects Masters,* pp. 5–8

Investigations and Projects Masters, p. 8

Name _____ Date _____

Interdisciplinary Investigation
(Student Edition, Pages 264–265)

Just "State" The Facts!

Use this table to record your data.

2.

Cities (Metropolitan Areas) – 1990 Population ___			
City (Metropolitan Area)	Population	Population (Rounded)	Fraction
Other			
State Total			

3. Perimeter: _____

© Glencoe/McGraw-Hill   8   *Mathematics: Applications and Connections, Course 1*

264   Chapter 6

JUST "STATE" THE FACTS!

Do you know the population of the largest city in the United States? According to the 1990 census, New York City had a population of 7,322,564. That is more than twice as many people as Los Angeles, the second largest city. In 1990, eight U. S. cities had populations over 1,000,000.

What You'll Do

In this investigation, you will choose a state and investigate what part of that state's population live in its largest cities.

Materials almanac   calculator

Procedure

1. Work in groups of four. Each member of your group should choose a different state. Find a resource book, map, or website that lists the populations of the largest cities.

2. Make a table, similar to the one below, for your chosen state. List the ten largest cities in order from greatest to least population. Determine how many people live in the rest of the state and call that category "other."

Montana Cities – 1990 Population			
City	Population	Population (rounded)	Fraction
Billings	81,125	81,000	$\frac{81}{799}$
Great Falls	55,125	55,000	$\frac{55}{799}$
⋮	⋮	⋮	⋮
Other	498,357	498,000	$\frac{498}{799}$
State Total	799,065	799,000	

In column 2, write the population of each city. In column 3, round the population to the nearest thousand. In column 4, write a fraction to show the population of each city compared to the total population of the state. (Round the total population to the nearest thousand.)

Technology Tips
- Use a **spreadsheet** to record populations.
- Use a **graphing calculator** or **graphing software** to make graphs.
- Use a **word processor** to prepare your display.
- Surf the **Internet** for more information.

◄◄◄**Instructional Resources**
A recording sheet to help students organize their data for this investigation is shown at the left and is available in the *Investigations and Projects Masters,* p. 8.

 Cooperative Learning

This investigation offers an excellent opportunity for using cooperative learning groups. For more information on cooperative learning strategies and group management, see *Cooperative Learning in the Mathematics Classroom.*

3. For your chosen state, use a map and the scale for the map to find the perimeter in miles. Place a string along each side and then add the lengths of string.

4. Prepare a display of the facts about your group's four states. Include some similarities and differences between the states.

Making the Connection

Use the data you collected about state population as needed to help in these investigations.

Language Arts

Imagine that the people of your chosen state stand on the perimeter and hold hands. Write directions for finding out whether there are enough people in the state to form a human border.

Science

Research the climate of your chosen state. Make a poster describing the climate. Include the high and low temperatures and draw a graph showing the amount of precipitation for each month.

Social Studies

Research the geography of your chosen state. Write a short report about the major geographic features of the state.

Go Further

- Make two or more different types of graphs comparing the populations of the cities in your chosen state and another student's state.

- Suppose that the state you chose has a random drawing for a prize. The names of all people in the state are included. What is the probability that the winner lives in the largest city?

*inter*NET **CONNECTION** For current information from the U.S. Bureau of the Census, visit:

www.glencoe.com/sec/math/mac/mathnet

You may want to place your work on this investigation in your portfolio.

Working in Teams Be sure students use the most current information available. Each student should prepare his or her own table. Encourage students to choose states with fairly regular shapes to simplify finding the perimeter. Suggest that students check each other's perimeter measurements.

Making the Connection
You may wish to alert other teachers on your team that your students may need their assistance in this investigation. *Language Arts* Students should begin by figuring out how many people can span one mile. Tell them that the average arm span is $4\frac{1}{2}$ feet.
Science Students may wish to investigate how the climate affects population movement in their state. *Social Studies* Students may wish to include a discussion of how the geographic features affect the population distribution in their state. Have them make maps of their states including the major geographic features.

Investigations and Projects Masters, p. 7

SCORING GUIDE

Interdisciplinary Investigation
(Student Edition, Pages 264–265)

Just "State" The Facts!

Level	Specific Criteria
3 Superior	• Shows a thorough understanding of the concepts of *rounding whole numbers, writing and simplifying fractions*, and *constructing bar graphs*. • Uses appropriate strategies to solve problems. • Computations are correct. • Written explanations are exemplary. • Charts, model, and any statements included are appropriate and sensible. • Goes beyond the requirements of some or all problems.
2 Satisfactory, with minor flaws	• Shows understanding of the concepts of *rounding whole numbers, writing and simplifying fractions*, and *constructing bar graphs*. • Uses appropriate strategies to solve problems. • Computations are mostly correct. • Written explanations are effective. • Charts, model, and any statements included are appropriate and sensible. • Satisfies the requirements of problems.
1 Nearly Satisfactory, with obvious flaws	• Shows understanding of most of the concepts of *rounding whole numbers, writing and simplifying fractions*, and *constructing bar graphs*. • May not use appropriate strategies to solve problems. • Computations are mostly correct. • Written explanations are satisfactory. • Charts, model, and any statements included are appropriate and sensible. • Satisfies the requirements of problems.
0 Unsatisfactory	• Shows little or no understanding of the concepts of *rounding whole numbers, writing and simplifying fractions*, and *constructing bar graphs*. • Does not use appropriate strategies to solve problems. • Computations are incorrect. • Written explanations are not satisfactory. • Charts, model, and any statements included are not appropriate or sensible. • Does not satisfy the requirements of the problems.

© Glencoe/McGraw-Hill 7 Mathematics: *Applications and Connections, Course 1*

ASSESS

You may have groups finish their presentations with a population comparison of their chosen states to their home state. Encourage groups to include visuals such as posters.

Instructional Resources ▶▶▶
Sample solutions for this investigation are provided in the *Investigations and Projects Masters* on p. 6. The scoring guide for assessing student performance shown at the right is also available on p. 7.

Previewing the Chapter

Overview

In this chapter, students learn to multiply and divide fractions and mixed numbers. Compatible numbers are used to estimate fraction products. Connections include finding the circumference of a circle and changing units within the customary system of measurement. Students learn to solve problems by finding and extending a pattern. The look-for-a-pattern strategy and spreadsheets are used to help recognize and extend sequences.

Lesson (pages)	Lesson Objectives	NCTM Standards	Standardized Tests	State/Local Objectives
7-1 (268–270)	Estimate fraction products using compatible numbers and rounding.	1–7, 10, 12, 13	CTBS, TN	
7-2A (271–272)	Multiply fractions by using a geoboard.	1–3, 5–7, 12, 13		
7-2 (273–276)	Multiply fractions.	1–7, 9, 13	CAT, CTBS, ITBS, MAT, SAT, TN	
7-3 (277–279)	Multiply mixed numbers.	1–7, 9, 12, 13	SAT	
7-4 (280–283)	Find the circumference of circles.	1–9, 12, 13		
7-5A (284)	Divide fractions by using a geoboard.	2, 5–7		
7-5 (285–288)	Divide fractions.	1–9	CAT, CTBS, MAT, SAT, TN	
7-6 (289–291)	Divide mixed numbers.	1–7, 9, 10	CTBS, SAT, TN	
7-7 (292–294)	Change units within the customary system.	1–7, 9, 13	CAT, CTBS, MAT, SAT, TN	
7-7B (295)	Compare weights.	1–7, 13		
7-8A (296–297)	Solve problems by finding and extending a pattern.	1–8, 12, 13	CAT, CTBS, ITBS, MAT, SAT, TN	
7-8 (298–300)	Recognize and extend sequences.	1–10, 12, 13		
7-8B (301)	Use a spreadsheet to extend sequences.	1–10, 12, 13		

CAT = California Achievement Tests, CTBS = Comprehensive Tests of Basic Skills, ITBS = Iowa Tests of Basic Skills, MAT = Metropolitan Achievement Tests, SAT = Stanford Achievement Tests, TN = Terra Nova

Organizing the Chapter

CD-ROM

All of the blackline masters in the Teacher's Classroom Resources are available on the **Electronic Teacher's Classroom Resources** CD-ROM.

LESSON PLANNING GUIDE

Lesson	Extra Practice (Student Edition)	BLACKLINE MASTERS (PAGE NUMBERS)										
		Study Guide	Practice	Enrichment	Assessment & Evaluation	Classroom Games	Diversity	Hands-On Lab	School to Career	Science and Math Lab Manual	Technology	Transparencies A and B
7-1	p. 575	49	49	49								7-1
7-2A								51				
7-2	p. 575	50	50	50	183				7			7-2
7-3	p. 575	51	51	51			7					7-3
7-4	p. 576	52	52	52	182, 183			75				7-4
7-5A								52				
7-5	p. 576	53	53	53		19–20					14	7-5
7-6	p. 576	54	54	54	184							7-6
7-7	p. 577	55	55	55								7-7
7-7B								53				
7-8A	p. 577											
7-8	p. 577	56	56	56	184						13	7-8
7-8B												
Study Guide/ Assessment					169–181, 185–187							

OTHER CHAPTER RESOURCES

Student Edition
Chapter Project, pp. 267, 279, 291, 305
Math in the Media, p. 279
Let the Games Begin, p. 276

Technology
 CD-ROM Program

Interactive Mathematics Tools Software

Teacher's Classroom Resources

Applications
Family Letters and Activities, pp. 13–14
Investigations and Projects Masters, pp. 41–44
Meeting Individual Needs
Transition Booklet, pp. 29–34
Investigations for the Special Education Student, pp. 19–20

Teaching Aids
Answer Key Masters
Block Scheduling Booklet
Lesson Planning Guide
Solutions Manual

Professional Publications
Glencoe Mathematics Professional Series

Planning the Chapter

MindJogger Videoquizzes
provide a unique format for reviewing concepts presented in the chapter.

ASSESSMENT RESOURCES

Student Edition
Mixed Review, pp. 270, 276, 279, 283, 288, 291, 294, 300
Mid-Chapter Self Test, p. 283
Math Journal, pp. 275, 290
Study Guide and Assessment, pp. 302–305
Performance Task, p. 305
Portfolio Suggestion, p. 305
Standardized Test Practice, pp. 306–307
Chapter Test, p. 601

Assessment and Evaluation Masters
Multiple-Choice Tests (Forms 1A, 1B, 1C), pp. 169–174
Free-Response Tests (Forms 2A, 2B, 2C), pp. 175–180
Performance Assessment, p. 181
Mid-Chapter Test, p. 182
Quizzes A–D, pp. 183–184
Standardized Test Practice, pp. 185–186
Cumulative Review, p. 187

Teacher's Wraparound Edition
5-Minute Check, pp. 268, 273, 277, 280, 285, 289, 292, 298
Building Portfolios, p. 266
Math Journal, pp. 272, 284, 295, 301
Closing Activity, pp. 270, 276, 279, 283, 288, 291, 294, 297, 300

Technology
Test and Review Software
MindJogger Videoquizzes
CD-ROM Program

MATERIALS AND MANIPULATIVES

Lesson 7-2A
geoboard*†
geobands*†
dot paper

Lesson 7-2
poster board
number cubes*
calculator

Lesson 7-4
tape measure*
calculator
circular objects

Lesson 7-5A
geoboard*†
geobands*†
dot paper

Lesson 7-7B
measuring cups*
containers
water

Lesson 7-8B
computer
spreadsheet software

*Glencoe Manipulative Kit   †Glencoe Overhead Manipulative Resources

PACING CHART

See pages T25–T27 for the Course Planning Calendar.

COURSE	DAY 1	DAY 2	DAY 3	DAY 4	DAY 5	DAY 6	DAY 7
Standard	Chapter Project	Lesson 7-1	Lessons 7-2A & 7-2		Lesson 7-3	Lesson 7-4	Lessons 7-5A & 7-5
Honors	Chapter Project	Lesson 7-1	Lesson 7-2	Lesson 7-3	Lesson 7-4	Lesson 7-5	Lesson 7-6
Block	Chapter Project & Lesson 7-1	Lessons 7-2A & 7-2	Lesson 7-3	Lesson 7-4	Lessons 7-5A & 7-5	Lessons 7-6 & 7-7	Lessons 7-8A & 7-8

The *Transition Booklet* (Skills 13–15) can be used to practice measuring length, capacity, and weight.

Interactive Mathematics:
Activities and Investigations

is an activity-based program that may be used as an enhancement for chapters in *Mathematics: Applications and Connections.*

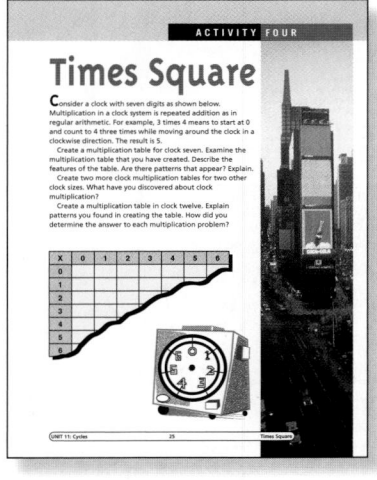

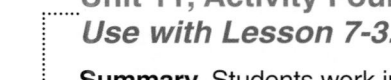

Unit 11, Activity Four
Use with Lesson 7-3.

Summary Students work in groups to solve multiplication problems in three different clock systems. Then they write about their observations. Each group presents its findings to the class.

Math Connection Students explore the characteristics of multiplication in clock systems, which parallels multiplication of real numbers. Clock multiplication tables allow some students to observe symmetry, patterns, sequences, and the special cases of zero and one. They also may make generalizations about the properties of real numbers.

Unit 12, Pre-Assessment Activity
Use with Lesson 7-7.

Summary Students work in pairs to determine a route and a description using nonstandard units for getting from school to another location. They are to write a brief report and are invited to share their presentations with the class.

Math Connection Students use their knowledge of direction and distance to write instructions for getting from school to another location. Knowledge of customary units of measure is used to complete this activity.

DAY 8	DAY 9	DAY 10	DAY 11	DAY 12	DAY 13	DAY 14	DAY 15
(continue from Day 7)	Lesson 7-6	Lesson 7-7		Lesson 7-8A	Lesson 7-8	Study Guide and Assessment	Chapter Test
Lessons 7-7 & 7-7B		Lesson 7-8A	Lessons 7-8 & 7-8B		Study Guide and Assessment	Chapter Test	
Study Guide and Assessment, Chapter Test							

Enhancing the Chapter

APPLICATIONS

Classroom Games,
pp. 19–20

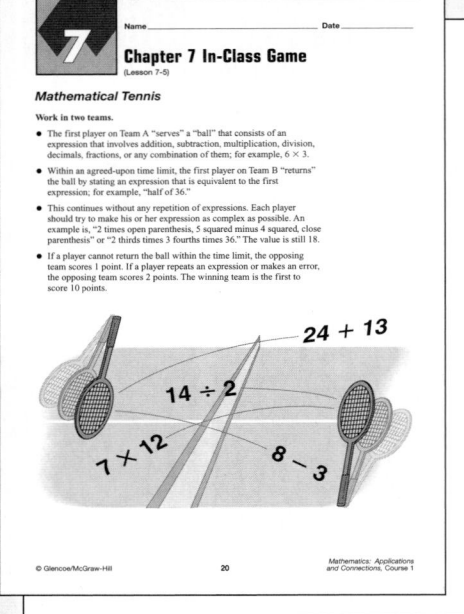

Diversity Masters,
p. 7

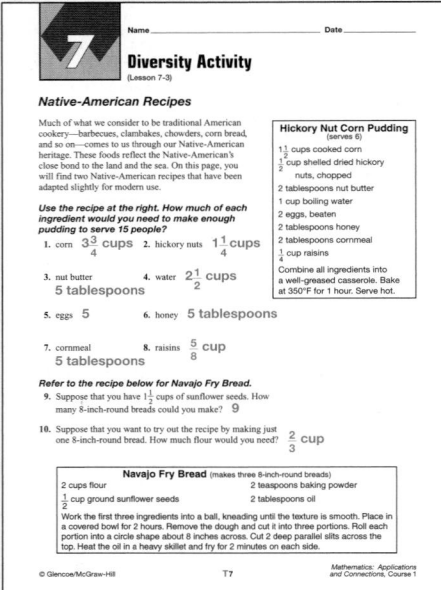

School to Career Masters,
p. 7

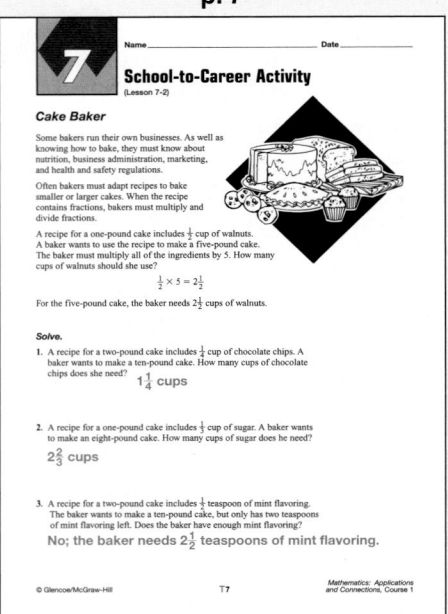

Family Letters and Activities,
pp. 13–14

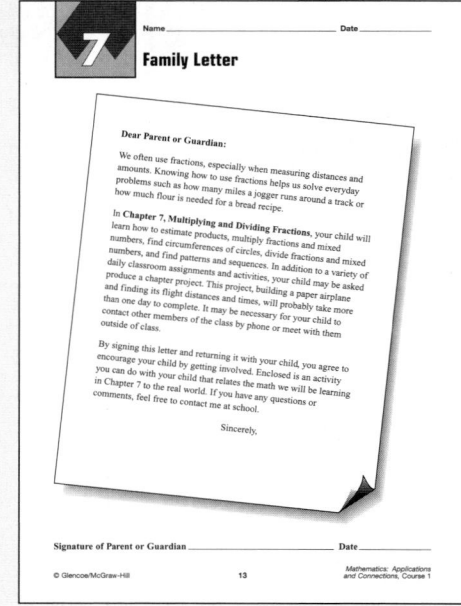

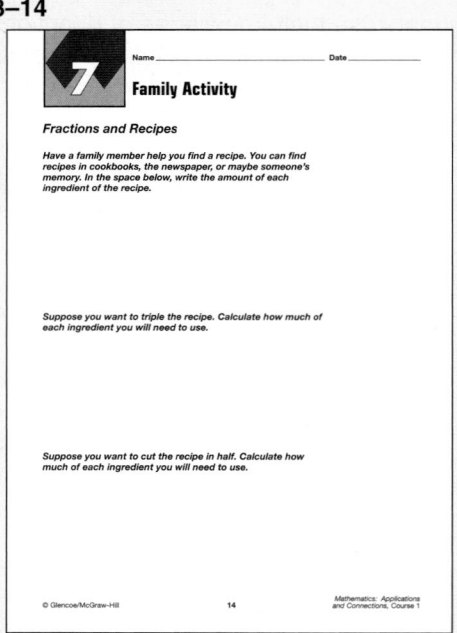

MANIPULATIVES/MODELING

Hands-On Lab Masters, p. 75

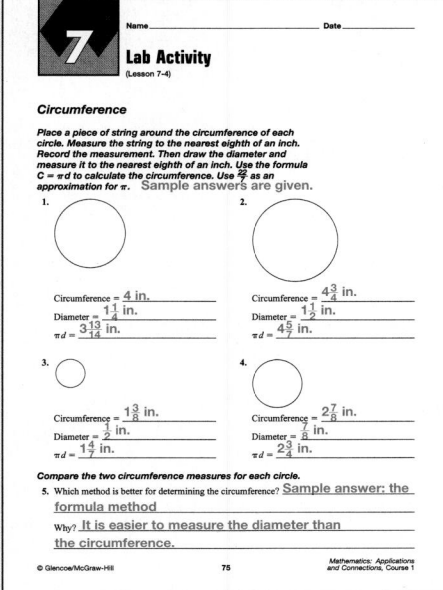

7 Name_____ Date_____
Lab Activity
(Lesson 7-4)

Circumference

Place a piece of string around the circumference of each circle. Measure the string to the nearest eighth of an inch. Record the measurement. Then draw the diameter and measure it to the nearest eighth of an inch. Use the formula $C = \pi d$ to calculate the circumference. Use $\frac{22}{7}$ as an approximation for π. **Sample answers are given.**

1.   2.

Circumference = __4 in.__   Circumference = __$4\frac{3}{4}$ in.__
Diameter = __$1\frac{1}{4}$ in.__   Diameter = __$1\frac{1}{2}$ in.__
πd = __$3\frac{13}{14}$ in.__ πd = __$4\frac{5}{7}$ in.__

3.   4.

Circumference = __$1\frac{3}{8}$ in.__   Circumference = __$2\frac{7}{8}$ in.__
Diameter = __$\frac{1}{2}$ in.__   Diameter = __$\frac{7}{8}$ in.__
πd = __$1\frac{4}{7}$ in.__ πd = __$2\frac{3}{4}$ in.__

Compare the two circumference measures for each circle.
5. Which method is best for determining the circumference? __Sample answer: the formula method__
Why? __It is easier to measure the diameter than the circumference.__

© Glencoe/McGraw-Hill   75   *Mathematics: Applications and Connections, Course 1*

ASSESSMENT/EVALUATION

Assessment and Evaluation Masters, pp. 182–184

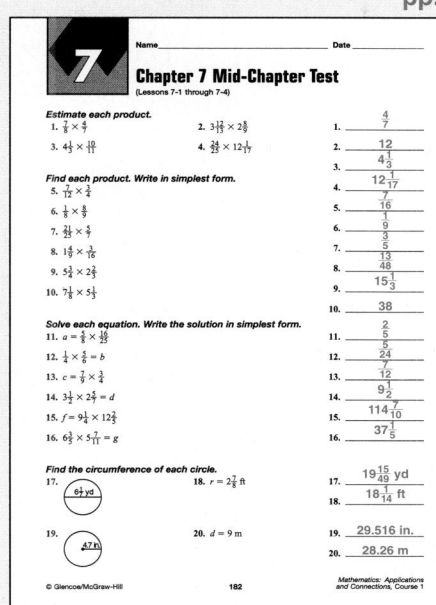

7 Name_____ Date_____
Chapter 7 Mid-Chapter Test
(Lessons 7-1 through 7-4)

Estimate each product.
1. $\frac{1}{8} \times \frac{4}{7}$ 2. $3\frac{13}{14} \times 2\frac{5}{8}$
3. $4\frac{1}{3} \times \frac{10}{11}$ 4. $4\frac{5}{9} \times 12\frac{1}{17}$

Find each product. Write in simplest form.
5. $\frac{7}{12} \times \frac{4}{7}$
6. $\frac{1}{8} \times \frac{8}{9}$
7. $\frac{9}{21} \times \frac{7}{8}$
8. $1\frac{8}{9} \times \frac{7}{16}$
9. $5\frac{3}{4} \times 2\frac{3}{8}$
10. $7\frac{1}{8} \times 5\frac{1}{3}$

Solve each equation. Write the solution in simplest form.
11. $a = \frac{4}{8} \times \frac{13}{19}$
12. $\frac{1}{4} \times \frac{8}{9} = b$
13. $c = \frac{3}{7} \times \frac{3}{4}$
14. $3\frac{1}{4} \times 2\frac{4}{9} = d$
15. $f = 9\frac{1}{4} \times 12\frac{2}{3}$
16. $6\frac{2}{3} \times 5\frac{7}{11} = g$

Find the circumference of each circle.
17.   18. $r = 2\frac{2}{8}$ ft
19.   20. $d = 9$ m

1. __$\frac{4}{7}$__
2. __12__
3. __$4\frac{1}{3}$__
4. __$12\frac{7}{17}$__
5. __$\frac{16}{7}$__
6. __$\frac{1}{9}$__
7. __$\frac{3}{5}$__
8. __$\frac{13}{48}$__
9. __$15\frac{1}{3}$__
10. __38__
11. __$\frac{5}{6}$__
12. __$\frac{5}{24}$__
13. __$\frac{12}{9}$__
14. __$9\frac{1}{6}$__
15. __$114\frac{7}{10}$__
16. __$37\frac{1}{5}$__
17. __$19\frac{15}{49}$ yd__
18. __$18\frac{1}{14}$ ft__
19. __29.516 in.__
20. __28.26 m__

© Glencoe/McGraw-Hill   182   *Mathematics: Applications and Connections, Course 1*

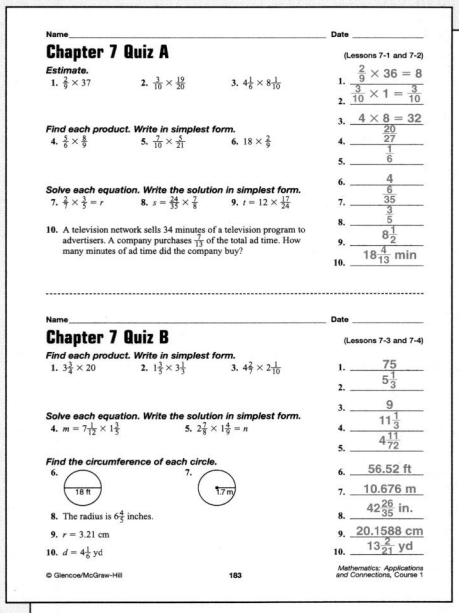

Name_____ Date_____
Chapter 7 Quiz A   (Lessons 7-1 and 7-2)
Estimate.
1. $\frac{2}{9} \times 37$ 2. $\frac{7}{10} \times \frac{10}{20}$ 3. $4\frac{1}{6} \times 8\frac{7}{10}$

Find each product. Write in simplest form.
4. $\frac{4}{5} \times \frac{6}{9}$ 5. $\frac{7}{10} \times \frac{5}{21}$ 6. $18 \times \frac{4}{9}$

Solve each equation. Write the solution in simplest form.
7. $\frac{4}{9} \times \frac{3}{2} = r$ 8. $s = \frac{43}{55} \times \frac{7}{8}$ 9. $r = 12 \times \frac{13}{14}$

10. A television network sells 34 minutes of a television program to advertisers. A company purchases $\frac{7}{17}$ of the total ad time. How many minutes of ad time did the company buy?

1. __$\frac{2}{9} \times 36 = 8$__
2. __$\frac{3}{10} \times 1 = \frac{3}{10}$__
3. __$4 \times 8 = 32$__
4. __$\frac{20}{27}$__
5. __$\frac{1}{6}$__
6. __$\frac{4}{6}$__
7. __$\frac{6}{35}$__
8. __$\frac{5}{9}$__
9. __$8\frac{1}{2}$__
10. __$18\frac{4}{13}$ min__

Name_____ Date_____
Chapter 7 Quiz B   (Lessons 7-3 and 7-4)
Find each product. Write in simplest form.
1. $3\frac{1}{4} \times 20$ 2. $1\frac{5}{9} \times 3\frac{1}{3}$ 3. $4\frac{3}{7} \times 2\frac{1}{10}$

Solve each equation. Write the solution in simplest form.
4. $m = 7\frac{1}{12} \times 1\frac{3}{5}$ 5. $2\frac{3}{8} \times 1\frac{4}{9} = n$

Find the circumference of each circle.
6.   7.
8. The radius is $6\frac{2}{7}$ inches.
9. $r = 3.21$ cm
10. $d = 4\frac{1}{9}$ yd

1. __75__
2. __$5\frac{1}{3}$__
3. __9__
4. __$11\frac{1}{3}$__
5. __$4\frac{11}{72}$__
6. __56.52 ft__
7. __10.676 m__
8. __$42\frac{26}{35}$ in.__
9. __20.1588 cm__
10. __$13\frac{2}{21}$ yd__

© Glencoe/McGraw-Hill   183   *Mathematics: Applications and Connections, Course 1*

TECHNOLOGY/MULTIMEDIA

Technology Masters, pp. 13–14

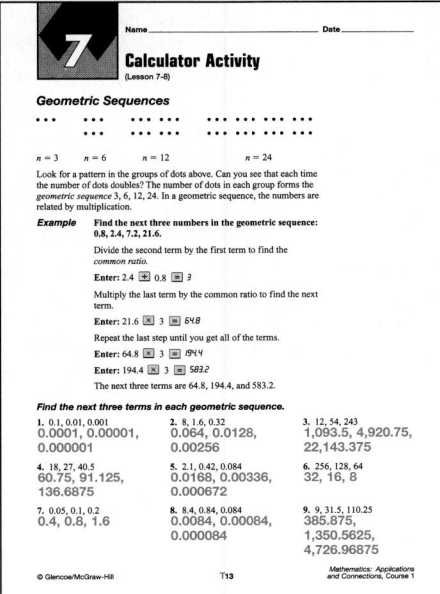

7 Name_____ Date_____
Calculator Activity
(Lesson 7-8)

Geometric Sequences

• • •   • • •   • • • • • •   • • • • • • • • • • • •
$n = 3$ $n = 6$ $n = 12$ $n = 24$

Look for a pattern in the groups of dots above. Can you see that each time the number of dots doubles? The number of dots in each group forms the *geometric sequence* 3, 6, 12, 24. In a geometric sequence, the numbers are related by multiplication.

Example   Find the next three numbers in the geometric sequence: 0.8, 2.4, 7.2, 21.6.
Divide the second term by the first term to find the *common ratio*.
Enter: 2.4 ÷ 0.8 = 3
Multiply the last term by the common ratio to find the next term.
Enter: 21.6 × 3 = 64.8
Repeat the last step until you get all of the terms.
Enter: 64.8 × 3 = 194.4
Enter: 194.4 × 3 = 583.2
The next three terms are 64.8, 194.4, and 583.2.

Find the next three terms in each geometric sequence.
1. 0.1, 0.01, 0.001
   0.0001, 0.00001, 0.000001
2. 8, 1.6, 0.32
   0.064, 0.0128, 0.00256
3. 12, 54, 243
   1,093.5, 4,920.75, 22,143.375
4. 18, 27, 40.5
   60.75, 91.125, 136.6875
5. 2.1, 0.42, 0.084
   0.0168, 0.00336, 0.000672
6. 256, 128, 64
   32, 16, 8
7. 0.05, 0.1, 0.2
   0.4, 0.8, 1.6
8. 8.4, 0.84, 0.084
   0.0084, 0.00084, 0.000084
9. 9, 31.5, 110.25
   385.875, 1,350.5625, 4,726.96875

© Glencoe/McGraw-Hill   T13   *Mathematics: Applications and Connections, Course 1*

7 Name_____ Date_____
Graphing Calculator Activity
(Lesson 7-5)

Dividing Fractions

You cannot enter fractions on most graphing calculators, but you can still perform calculations and have the calculator convert the answer into fraction form. Under the MATH button you will find a menu of formats for numbers.

Examples   1 Find $\frac{1}{3} \div \frac{2}{5}$.
Enter: (1 ÷ 3) ÷ (2 ÷ 5) ENTER 0.8333333333
Now express the answer in fraction form.
Enter: MATH ENTER ENTER $\frac{5}{6}$
So, $\frac{1}{3} \div \frac{2}{5} = \frac{5}{6}$.
2 Find $\frac{5}{9} \div 6$.
Enter: (5 ÷ 9) ÷ 6 MATH ENTER ENTER $\frac{5}{54}$
So, $\frac{5}{9} \div 6 = \frac{5}{54}$.

Find each quotient. Write in simplest form.
1. $\frac{1}{8} \div \frac{2}{5}$ $\frac{5}{16}$ 2. $\frac{2}{9} \div \frac{1}{4}$ $\frac{8}{9}$ 3. $\frac{3}{10} \div \frac{2}{3}$ $\frac{9}{20}$
4. $\frac{4}{3} \div 6$ $\frac{2}{15}$ 5. $\frac{8}{9} \div 5$ $\frac{8}{45}$ 6. $\frac{3}{4} \div 8$ $\frac{3}{32}$

Solve each equation. Write the solution in simplest form.
7. $g = \frac{1}{2} \div \frac{3}{5}$ $\frac{5}{6}$ 8. $\frac{6}{7} \div 7 = j$ $\frac{6}{49}$ 9. $\frac{3}{8} \div 3 = m$ $\frac{1}{8}$
10. $7 \div \frac{1}{2} = y$ 14   11. $k = \frac{2}{9} \div \frac{7}{10}$ $\frac{20}{63}$ 12. $d = \frac{5}{6} \div 2$ $\frac{5}{12}$

© Glencoe/McGraw-Hill   T14   *Mathematics: Applications and Connections, Course 1*

MEETING INDIVIDUAL NEEDS

Investigations for the Special Education Student, pp. 19–20

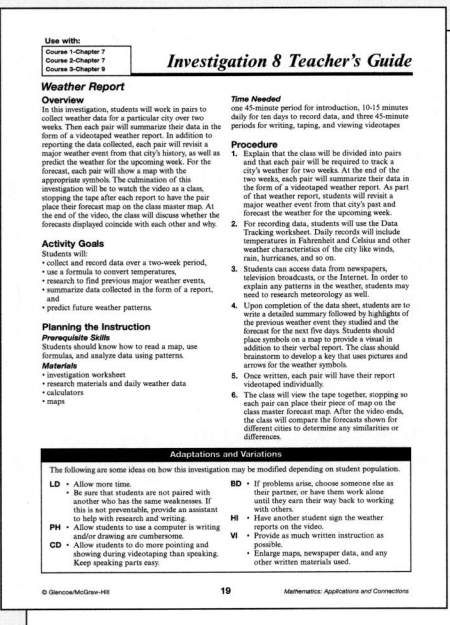

Use with:
Course 1–Chapter 7
Course 2–Chapter 7
Course 3–Chapter 9

Investigation 8 Teacher's Guide
Weather Report

Overview
In this investigation, students will work in pairs to collect weather data for a particular city over two weeks. Then each pair will summarize their data in the form of a videotaped weather report. In addition to reporting the data collected, each pair will revisit a major weather event from that city's history, as well as predict the weather for the upcoming week. For the forecast, each pair will show a map with the appropriate symbols. The culmination of this investigation will be to watch the video as a class, stopping the tape after each report to have the pair place their forecast map on the class master map. At the end of the video, the class will discuss whether the forecasts displayed coincide with each other and why.

Activity Goals
Students will:
• collect and record data over a two-week period,
• use a formula to convert temperatures,
• research to find previous major weather events,
• summarize data collected in the form of a report, and
• predict future weather patterns.

Planning the Instruction
Prerequisite Skills
Students should know how to read a map, use formulas, and analyze data using patterns.
Materials
• investigation worksheet
• research materials and daily weather data
• calculators
• maps

Time Needed
one 45-minute period for introduction, 10-15 minutes daily for ten days to record data, and three 45-minute periods for writing, taping, and viewing videotapes

Procedure
1. Explain that the class will be divided into pairs and that each pair will be required to track a city's weather for two weeks. At the end of the two weeks, each pair will summarize their data in the form of a videotaped weather report. As part of that weather report, students will revisit a major weather event from that city's past and forecast the weather for the upcoming week.
2. For recording data, students will use the Data Tracking worksheet. Daily records will include temperatures in Fahrenheit and Celsius and other weather characteristics of the city like winds, rain, hurricanes, and so on.
3. Students can access data from newspapers, television broadcasts, or the Internet. In order to explain any patterns in the weather, students may need to research meteorology as well.
4. Upon completion of the data sheet, students are to write a detailed summary followed by highlights of the previous weather event they studied and the forecast for the next five days. Students should place symbols on a map to provide a visual in addition to their verbal report. The class should brainstorm to develop a key that uses pictures and arrows for the weather symbols.
5. Once written, each pair will have their report videotaped individually.
6. The class will view the tape together, stopping so each pair can place their piece of map on the class master forecast map. After the video ends, the class will compare the forecasts shown for different cities to determine any similarities or differences.

Adaptations and Variations
The following are some ideas on how this investigation may be modified depending on student population.

LD • Allow more time.
• Be sure that students are not paired with another who has the same weaknesses. If this is not prevented, provide an assistant to help with research and writing.
PH • Allow students to use a computer is writing and/or drawing are cumbersome.
CD • Allow students to do more pointing and showing during videotaping than speaking. Keep speaking parts easy.
BD • If problems arise, choose someone else as their partner, or have them work alone until they earn their way back to working with others.
HI • Have another student sign the weather reports on the video.
VI • Provide as much written instruction as possible.
• Enlarge maps, newspaper data, and any other written materials used.

© Glencoe/McGraw-Hill   19   *Mathematics: Applications and Connections*

Theme: Flight

Ken Blackburn began to experiment with paper airplanes when he was a child. At age 13, he designed a square-shaped glider that quickly became his favorite plane. Seven years and countless improvements later, Blackburn became the world-record holder for the longest flight by a paper airplane, with a flight of 16.89 seconds. He has since broken his own record twice, most recently in 1994 when his glider stayed airborne for 18.8 seconds. Blackburn has every intention of breaking the 20-second barrier.

Question of the Day Speed equals distance divided by time. If Mitiku's first flight traveled $11\frac{1}{4}$ feet in 2 seconds, how fast did it fly? $5\frac{5}{8}$ ft/s

Assess Prerequisite Skills

Ask students to read through the list of objectives presented in "What you'll learn in Chapter 7." You may wish to ask them what each of the objectives means or if they have used or experienced any of these math concepts before.

 Building Portfolios

Encourage students to revise their portfolios as they study this chapter. They may want to include summaries of what they did in the labs to remind them of relevant applications of fractions.

 **Math and the Family**

In the *Family Letters and Activities* booklet (pp. 13–14), you will find a letter to the parents explaining what students will study in Chapter 7. An activity appropriate for the whole family is also available.

What you'll learn in Chapter 7

- to multiply and divide fractions and mixed numbers,
- to find the circumference of circles,
- to change units within the customary system,
- to recognize and extend sequences, and
- to solve real-world problems by finding and extending a pattern.

266 **Chapter 7** Multiplying and Dividing Fractions

 **CD-ROM Program**

Activities for Chapter 7
- Chapter 7 Introduction
- Interactive Lesson 7-2
- Extended Activity 7-4
- Assessment Game
- Resource Lessons 7-1 through 7-8

CHAPTER Project

BUILD A PAPER AIRPLANE

In this project, you will use multiplication and division of fractions to help you find flight distances and flight times of a paper airplane. You can create any type of airplane you wish, such as a bomber or a fighter plane.

Getting Started

- Look at the flight table. In which flight did Mitiku's paper airplane travel the farthest?
- Use the table to determine how far Mitiku's plane could have traveled in flight 3 if it had remained in the air for 2 seconds.

Mitiku's Flight and Distance Table		
Flight Number	Distance (ft)	Time (s)
1	$11\frac{1}{4}$	2
2	$13\frac{1}{2}$	3
3	$5\frac{3}{4}$	1

- Use a sheet of notebook paper and tape to construct a paper airplane.
  - Throw your airplane three times. Use a tape measure to measure the distance of each flight, in feet, your airplane travels. Use a stopwatch to time each flight. Round to the nearest second. Record your flight results in a table similar to the one above.

Technology Tips

- Use a **spreadsheet** to record your flight results and to find the speed of your airplane.

 **inter NET CONNECTION** For up-to-date information on paper airplanes, visit:
www.glencoe.com/sec/math/mac/mathnet

Working on the Project

You can use what you'll learn in Chapter 7 to help you find flight distances and flight times of your paper airplane.

Page	Exercise
279	42
291	42
305	Alternative Assessment

Chapter 7 **267**

 **inter NET CONNECTION**

Glencoe has made every effort to ensure that the website links for *Mathematics: Applications and Connections* at **www. glencoe.com/sec/math/mac/mathnet** are current and contain appropriate content. However, these website links are not under Glencoe's control.

Instructional Resources ▶▶▶
A recording sheet to help students organize their data for the Chapter Project is shown at the right and is available in the *Investigations and Projects Masters*, p. 44.

Instructional Resources

- *Study Guide Masters,* p. 49
- *Practice Masters,* p. 49
- *Enrichment Masters,* p. 49
- Transparencies 7-1, A and B
- CD-ROM Program
  - Resource Lesson 7-1

Recommended Pacing

Standard	Day 2 of 15
Honors	Day 2 of 14
Block	Day 1 of 8

1 FOCUS

5-Minute Check
(Chapter 6)

Add or Subtract.

1. $13\frac{2}{5} - 8\frac{1}{3}$ $5\frac{1}{15}$
2. $\frac{1}{8} + \frac{9}{8}$ $1\frac{1}{4}$
3. $\frac{3}{4} + \frac{1}{6}$ $\frac{11}{12}$
4. $\begin{array}{r} 9\text{ h }22\text{ min }12\text{ s} \\ -\ 6\text{ h }43\text{ min }54\text{ s} \\ \hline 2\text{ h }38\text{ min }18\text{ s} \end{array}$
5. Solve $10\frac{7}{9} = 15 - n$. $4\frac{2}{9}$

The 5-Minute Check is also available on **Transparency 7-1A** for this lesson.

Motivating the Lesson

Hands-On Activity Give groups of students a handful of small items such as plastic cubes or beans. Ask them to find $\frac{1}{2}$ of the items, then $\frac{1}{5}$ of the items. Have students record their answers and then discuss their methods.

2 TEACH

Transparency 7-1B contains a teaching aid for this lesson.

Using Manipulatives Each pair of students will need two whole number cubes and two teacher-made fraction cubes with fractions such as $\frac{1}{3}, \frac{3}{8},$ and $\frac{6}{7}$. Have them use the cubes to generate two numbers and estimate their product.

268 Chapter 7

What you'll learn

You'll learn to estimate fraction products using compatible numbers and rounding.

When am I ever going to use this?

You'll use estimation to approximate the discount on a sale item.

Word Wise

compatible numbers

Where do you like to study? The circle graph shows where students prefer to study. If there are 16 students in Taylor's class, about how many students prefer to study in their bedrooms?

To solve the problem, you can estimate the product of $\frac{1}{3}$ and 16.

One way to estimate the product is to use **compatible numbers**. Compatible numbers are easy to divide mentally.

Where Students Prefer to Study

$\frac{1}{3}$ Bedroom

$\frac{2}{3}$

Kitchen, Dining Room, or Family Room

Source: Federal National Mortgage Association

$\frac{1}{3} \times 16 = ?$ *$\frac{1}{3} \times 16$ means $\frac{1}{3}$ of 16.*

$\frac{1}{3} \times 15 = ?$ *Think: For 16, the nearest multiple of 3 is 15. 3 and 15 are compatible numbers since $15 \div 3 = 5$.*

$\frac{1}{3} \times 15 = 5$ *Multiplying by $\frac{1}{3}$ is the same as dividing by 3.*

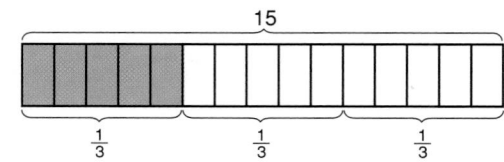

15

$\frac{1}{3}$ $\frac{1}{3}$ $\frac{1}{3}$

So, the product of $\frac{1}{3}$ and 16 is about 5.

About 5 of the 16 students prefer to study in their bedrooms.

Example ① Estimate $\frac{5}{8} \times 25$.

$\frac{1}{8} \times 24 = 3$ *For 25, the nearest multiple of 8 is 24. $\frac{1}{8}$ of 24 is 3.*

$\frac{5}{8} \times 24 = 15$ *Since $\frac{1}{8}$ of 24 is 3, it follows that $\frac{5}{8}$ of 24 is 5×3 or 15.*

So, $\frac{5}{8} \times 25$ is about 15.

Cross-Curriculum Cue

Inform the other teachers on your team that your students are studying how to multiply and divide fractions. Suggestions for curriculum integration are:
Music: musical notes
Demographics: statistics
Poetry: verse

You can also estimate products by rounding fractions to 0, $\frac{1}{2}$, or 1.

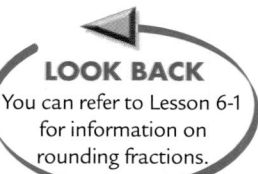

Example 2 Estimate $\frac{2}{9} \times \frac{5}{6}$.

Think: $\frac{5}{6}$ is close to 1.

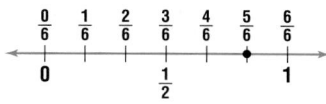

$$\frac{2}{9} \times 1 = \frac{2}{9} \quad \text{Round } \frac{5}{6} \text{ to 1. Any number multiplied by 1 is the number.}$$

So, $\frac{2}{9} \times \frac{5}{6}$ is about $\frac{2}{9}$.

LOOK BACK
You can refer to Lesson 6-1 for information on rounding fractions.

To estimate the product of mixed numbers, you can round each mixed number to the nearest whole number and then multiply.

Example 3
APPLICATION

Cooking Kim needs $3\frac{1}{2}$ batches of cookies. If one recipe calls for $2\frac{1}{4}$ cups of flour, about how many cups of flour are needed?

You need to estimate $3\frac{1}{2} \times 2\frac{1}{4}$.

$4 \times 2 = 8$ *Round $3\frac{1}{2}$ to 4. Round $2\frac{1}{4}$ to 2.*

So, $3\frac{1}{2} \times 2\frac{1}{4}$ is about 8.

Kim will need about 8 cups of flour.

CHECK FOR UNDERSTANDING

Communicating Mathematics

1–3. See Answer Appendix.

Read and study the lesson to answer each question.

1. **Explain** how you would use rounding to estimate $3\frac{2}{3} \times 9\frac{1}{8}$.

2. **Tell** how the model shows the use of compatible numbers to estimate $\frac{3}{4} \times 9$.

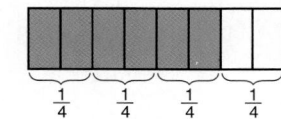

$\frac{1}{4} \quad \frac{1}{4} \quad \frac{1}{4} \quad \frac{1}{4}$

3. **You Decide** Juan says that $8\frac{1}{2} \times 6\frac{1}{4}$ is about 54. Odina says that the product is about 48. Whose estimate is better? Explain your reasoning.

Guided Practice

Round each fraction to 0, $\frac{1}{2}$, or 1 and each mixed number to the nearest whole number.

4. $\frac{1}{8}$ 0

5. $\frac{5}{9}$ $\frac{1}{2}$

6. $7\frac{2}{3}$ 8

Estimate each product.

7. $\frac{1}{8} \times 15$ $\frac{1}{8} \times 16 = 2$

8. $\frac{2}{5} \times \frac{6}{7}$ $\frac{2}{5} \times 1 = \frac{2}{5}$

9. $3\frac{1}{4} \times 8\frac{6}{7}$ $3 \times 9 = 27$

10. **Geometry** Estimate the area of the rectangle. **See margin.**

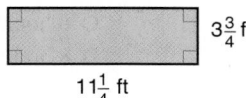 $3\frac{3}{4}$ ft

$11\frac{1}{4}$ ft

Lesson 7-1 Estimating Products **269**

In-Class Examples
Estimate each product.

For Example 1
a. $\frac{1}{5} \times 11$ $\frac{1}{5} \times 10 = 2$
b. $\frac{3}{7} \times 33$ $\frac{3}{7} \times 35 = 15$

For Example 2
a. $\frac{3}{4} \times \frac{8}{9}$ $\frac{3}{4} \times 1 = \frac{3}{4}$
b. $\frac{7}{8} \times \frac{1}{9}$ $1 \times \frac{1}{9} = \frac{1}{9}$

For Example 3
Takeo painted a mural that was $9\frac{3}{4}$ feet long by $5\frac{1}{3}$ feet wide. Estimate the mural's area, $\ell \times w$. 10 ft $\times$ 5 ft = 50 ft²

3 PRACTICE/APPLY

Check for Understanding
If students need additional practice or instruction after completing Exercises 1–10, you may find one of the following options helpful.
• Extra Practice, see p. 575
• Reteaching Activity
• *Study Guide Masters,* p. 49
• *Practice Masters,* p. 49

***Study Guide Masters,* p. 49**

7-1 Name_____ Date_____
Study Guide

Estimating Products

One way to estimate products is using **compatible numbers.** Compatible numbers are easy to divide mentally.

Example 1 Estimate $\frac{3}{5} \times 9$.

$\frac{1}{5} \times 10 = 2$ *For 9, the nearest multiple of 5 is 10. $\frac{1}{5}$ of 10 is 2.*
$\frac{3}{5} \times 10 = 6$ *Since $\frac{1}{5} \times 10 = 2$, it follows that $\frac{3}{5}$ of 10 is 3 × 2 or 6.*

So, $\frac{3}{5} \times 9$ is about 6.

You can also estimate products by rounding. Round fractions to 0, $\frac{1}{2}$ or 1. Round mixed numbers to the nearest whole number.

Examples 2 Estimate $\frac{5}{6} \times \frac{3}{7}$.

$\frac{5}{6} \times \frac{3}{7}$ *Think of a number line. $\frac{5}{6}$ rounds to 1.*
$1 \times \frac{3}{7} = \frac{3}{7}$ *So, $\frac{5}{6} \times \frac{3}{7}$ is about $\frac{3}{7}$.*

3 Estimate $9\frac{9}{11} \times 5\frac{1}{8}$.

$9\frac{9}{11} \times 5\frac{1}{8}$. *Round $9\frac{9}{11}$ to 10. Round $5\frac{1}{8}$ to 5.*
$10 \times 5 = 50$ *So, $9\frac{9}{11} \times 5\frac{1}{8}$ is about 50.*

Round each fraction to 0, $\frac{1}{2}$ or 1.
1. $\frac{6}{13}$ $\frac{1}{2}$
2. $\frac{3}{8}$ $\frac{1}{2}$
3. $\frac{9}{10}$ 1
4. $\frac{1}{9}$ 0

Estimate each product.
5. $\frac{1}{5} \times 24$ 5
6. $7\frac{2}{3} \times 5\frac{3}{4}$ 42
7. $\frac{2}{3} \times 19$ 12
8. $\frac{9}{10} \times \frac{1}{3}$ $\frac{1}{2}$
9. $15 \times \frac{1}{4}$ 4
10. $8\frac{7}{8} \times 2\frac{9}{10}$ 27
11. $\frac{1}{9} \times \frac{1}{12}$ 0
12. $\frac{2}{5} \times 27$ 12

© Glencoe/McGraw-Hill   T49   Mathematics: Applications and Connections, Course 1

■ Reteaching the Lesson ■

Activity Have students work in small groups, with each student in the group writing examples of multiplying a fraction and a whole number and multiplying two fractions. For each example, groups should use an appropriate estimation method and then work to find the best estimate.

Additional Answer
10. about 4 × 11, or 44 ft²

4 ASSESS

Family Activity

Have students share their daily situations that involve fractions or mixed numbers. Examples may include cooking, telling time, reading music, or timing sports events. Keep a list on the chalkboard to tally repeats.

Practice Masters, p. 49

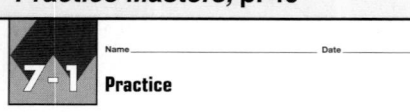

EXERCISES

Practice

Round each fraction to 0, $\frac{1}{2}$, or 1 and each mixed number to the nearest whole number.

11. $\frac{6}{11}$ $\frac{1}{2}$ **12.** $\frac{4}{5}$ 1   **13.** $\frac{4}{9}$ $\frac{1}{2}$ **14.** $\frac{2}{15}$ 0

15. $6\frac{1}{8}$ 6   **16.** $12\frac{3}{4}$ 13   **17.** $3\frac{1}{2}$ 3 or 4   **18.** $15\frac{2}{9}$ 15

Estimate each product. 19–27. See Answer Appendix.

19. $\frac{1}{4} \times 31$ **20.** $\frac{3}{5} \times \frac{1}{2}$ **21.** $\frac{1}{7} \times 22$

22. $2\frac{1}{6} \times 5\frac{1}{2}$ **23.** $\frac{1}{6} \times 40$ **24.** $\frac{7}{8} \times 3\frac{1}{4}$

25. $\frac{4}{9} \times 46$ **26.** $\frac{1}{6} \times \frac{7}{8}$ **27.** $7\frac{3}{4} \times 3\frac{1}{8}$

28. Estimate $4\frac{5}{6}$ multiplied by $16\frac{7}{8}$. 5 × 17 = 85

29. Estimate $14\frac{8}{9} \times 6\frac{3}{5}$. 15 × 7 = 105

Applications and Problem Solving

30. $\frac{1}{20} \times 100$, or about 5 people

Family Activity

Ask a family member to help you make a list of situations in your daily life that involve a fraction or a mixed number. Then estimate each number by rounding.

30. *Travel* The circle graph shows when people pack for a vacation. Suppose 96 people were surveyed. About how many people pack the day they leave?

31. *Recreation* There are about 7 million pleasure boats in the United States. About $\frac{2}{3}$ of these boats are motorboats. About how many motorboats are in the United States? $\frac{2}{3} \times 6$, or about 4 million

When Vacationers Pack

1 day before leaving $\frac{1}{2}$
2 or more days before leaving $\frac{9}{20}$
The day they leave $\frac{1}{20}$

Source: Carlson Wagonlit Travel Survey

32. *Critical Thinking* Which point on the number line could be the graph of the product of the numbers graphed at *C* and *D*? **point *N***

0 M N C R D 1

Mixed Review

33. 1 hr 3 min 51 s

35. 0.498

33. *Measurement* Add 26 min 37 s and 37 min 14 s. *(Lesson 6-7)*

34. Find the prime factorization of 300. *(Lesson 5-2)* 2 · 2 · 3 · 5 · 5

35. *Measurement* How many kilograms are in 498 grams? *(Lesson 4-9)*

36. **Test Practice**   A bus travels at a speed of 47 miles per hour. About how long will it take the bus to travel 316 miles? *(Lesson 1-3)* **C**

A 4 hours
B 5 hours
C 6 hours
D 8 hours
E 10 hours

270   **Chapter 7** Multiplying and Dividing Fractions

Extending the Lesson

Enrichment Masters, p. 49

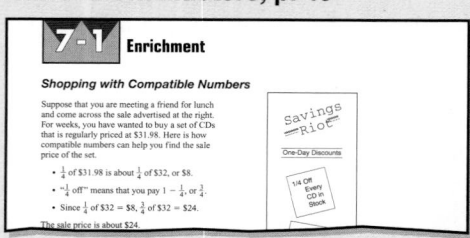

Activity Provide home furnishings catalogs or ask students to bring them from home. Have small groups of students use the catalogs to estimate the cost of window coverings for their classroom windows.

COOPERATIVE LEARNING

7-2A Multiplying Fractions

A Preview of Lesson 7-2

You can multiply fractions by using geoboards.

 geoboard
or dot paper

geobands

TRY THIS

Work with a partner.

1 Model $\frac{1}{3} \times \frac{1}{2}$ using a geoboard.

Step 1 Place one geoband in a straight line along the bottom of the geoboard to show thirds.

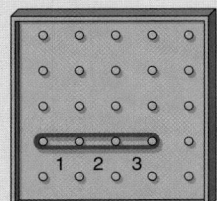

Step 2 Place one geoband perpendicular to the first geoband to show halves.

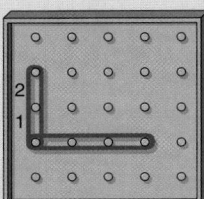

Step 3 Use geobands to make a rectangle.

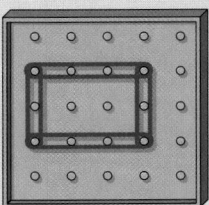

Step 4 Place one geoband on the peg as shown. This represents $\frac{1}{3}$.

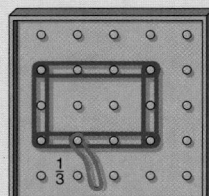

Step 5 Place one geoband on the peg as shown. This represents $\frac{1}{2}$.

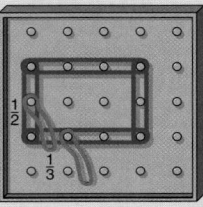

Step 6 Connect the geobands to show a small rectangle.

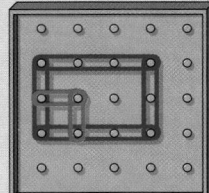

ON YOUR OWN

1. The area of the small rectangle has an area of 1 square unit. What is the area of the large rectangle? **6 square units**

2. What fraction of the area of the large rectangle is the area of the small rectangle? $\frac{1}{6}$

3. What is $\frac{1}{3} \times \frac{1}{2}$? $\frac{1}{6}$

Lesson **7-2A** HANDS-ON **271**

GET READY

Objective Students multiply fractions by using a geoboard.

Optional Resources
Hands-On Lab Masters
• square dot paper, p. 13
• worksheet, p. 51

Overhead Manipulative Resources
• geoboard
• geobands

Manipulative Kit
• geoboard
• geobands

MANAGEMENT TIPS

Recommended Time
30 minutes

Teaching Tip Each pair of students will need a geoboard and six geobands. If geoboards and geobands are not available, students may draw the figures on square dot paper from the *Hands-On Lab Masters*.

Getting Started Ask students how they would use geoboards to show fractions. Have them illustrate thirds and sixths on their geoboards. Then ask them to determine how they would show $\frac{1}{3} \times \frac{1}{6}$.

Activity 1 uses geobands to represent fractions. Make sure students are counting spaces between pegs, not the pegs themselves. Remind students how to find the area defined by the geobands.

Activity 2 repeats Activity 1 with the problem $\frac{1}{4} \times \frac{1}{3}$. Have students count the 12 squares outlined on the geoboard and relate the resulting small square to the answer of $\frac{1}{12}$.

ASSESS

After students complete Exercises 1–7, have pairs of students write an example of multiplying fractions and then model it on their geoboards.

Use Exercise 8 to determine whether students understand how to multiply fractions.

Additional Answer
7. **To multiply fractions, multiply the numerators and then multiply the denominators.**

Hands-On Lab Masters, p. 51

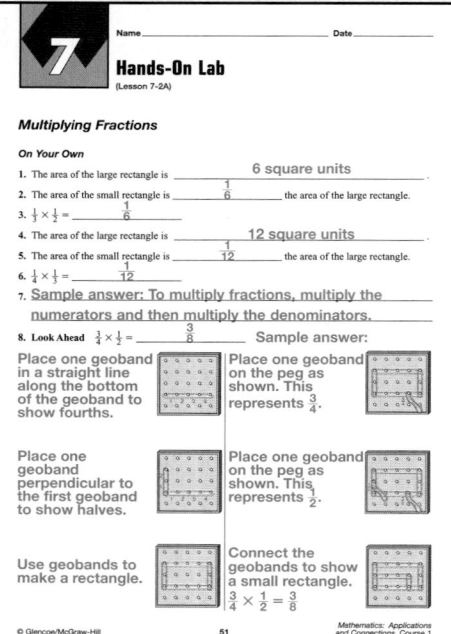

7

Name _____ Date _____

Hands-On Lab
(Lesson 7-2A)

Multiplying Fractions

On Your Own

1. The area of the large rectangle is _____ 6 square units _____.
2. The area of the small rectangle is _____ $\frac{1}{6}$ _____ the area of the large rectangle.
3. $\frac{1}{2} \times \frac{1}{3} =$ _____ $\frac{1}{6}$ _____
4. The area of the large rectangle is _____ 12 square units _____.
5. The area of the small rectangle is _____ $\frac{1}{12}$ _____ the area of the large rectangle.
6. $\frac{1}{4} \times \frac{1}{3} =$ _____ $\frac{1}{12}$ _____
7. Sample answer: To multiply fractions, multiply the numerators and then multiply the denominators.
8. **Look Ahead** $\frac{3}{4} \times \frac{1}{2} =$ _____ $\frac{3}{8}$ _____ Sample answer:

Place one geoband in a straight line along the bottom of the geoboard to show fourths. | Place one geoband on the peg as shown. This represents $\frac{3}{4}$.

Place one geoband perpendicular to the first geoband to show halves. | Place one geoband on the peg as shown. This represents $\frac{1}{2}$.

Use geobands to make a rectangle. | Connect the geobands to show a small rectangle. $\frac{3}{4} \times \frac{1}{2} = \frac{3}{8}$

© Glencoe/McGraw-Hill   51   Mathematics: Applications and Connections, Course 1

TRY THIS

Work with a partner.

2 Model $\frac{1}{4} \times \frac{1}{3}$ using a geoboard.

Step 1 Place one geoband in a straight line along the bottom of the geoboard to show fourths.

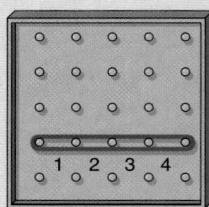

Step 4 Place one geoband on the peg as shown. This represents $\frac{1}{4}$.

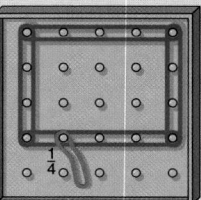

Step 2 Place one geoband perpendicular to the first geoband to show thirds.

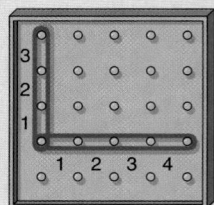

Step 5 Place one geoband on the peg as shown. This represents $\frac{1}{3}$.

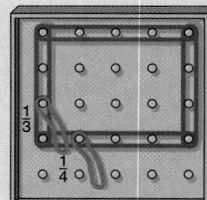

Step 3 Use geobands to make a rectangle.

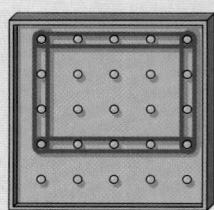

Step 6 Connect the geobands to show a small rectangle.

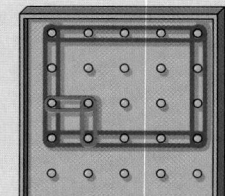

ON YOUR OWN

4. The small rectangle has an area of 1 square unit. What is the area of the large rectangle? **12 square units**

5. What fraction of the area of the large rectangle is the area of the small rectangle? $\frac{1}{12}$

6. What is $\frac{1}{4} \times \frac{1}{3}$? $\frac{1}{12}$

7. Write a rule you can use to multiply fractions. **See margin.**

8. **Look Ahead** Use your rule to find $\frac{3}{4} \times \frac{1}{2}$. Check your answer using a geoboard. $\frac{3}{8}$; **See Answer Appendix.**

272   **Chapter 7** Multiplying and Dividing Fractions

Math Journal Have students write a summary of how they modeled multiplying fractions on their geoboards. Make sure they include any patterns they discovered that will help them multiply fractions in future lessons.

Multiplying Fractions

What you'll learn

You'll learn to multiply fractions.

When am I ever going to use this?

Knowing how to multiply fractions can help you make half a batch of chocolate chip cookies.

Did you know? If all of the blood vessels in your body could be laid end to end, they would be about 60,000 miles long.

Do you know your blood type? The human body can contain any one of four different blood types. Suppose the employees of a very large company have the fractions of blood types shown in the table. If $\frac{1}{2}$ of the company's employees donate blood, what fraction will donate type A blood?

Blood Type	Fraction of People With Each Blood Type
A	$\frac{2}{5}$
B	$\frac{1}{10}$
AB	$\frac{1}{20}$
O	$\frac{9}{20}$

You need to find $\frac{2}{5}$ of $\frac{1}{2}$, which means $\frac{2}{5} \times \frac{1}{2}$.

Method 1 Use a model.

Separate a rectangle into halves. Then color $\frac{1}{2}$ of the rectangle blue.

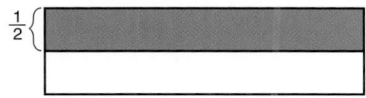

Next, separate the rectangle into fifths, and color $\frac{2}{5}$ of the rectangle yellow.

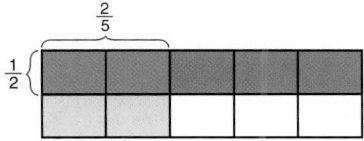

2 parts are green. 10 parts in all.

Since 2 out of 10 parts are green, the product is $\frac{2}{10}$ or $\frac{1}{5}$.

Method 2 Use a rule.

Find the product by multiplying the numerators and multiplying the denominators.

$\frac{2}{5} \times \frac{1}{2} = \frac{2 \cdot 1}{5 \cdot 2}$ *Multiply the numerators.*
Multiply the denominators.

$= \frac{2}{10}$ or $\frac{1}{5}$ *Simplify.*

So, $\frac{1}{5}$ of the company's employees will donate type A blood.

Lesson 7-2 Multiplying Fractions **273**

Instructional Resources

- *Study Guide Masters,* p. 50
- *Practice Masters,* p. 50
- *Enrichment Masters,* p. 50
- Transparencies 7-2, A and B
- *Assessment and Evaluation Masters,* p. 183
- *School to Career Masters,* p. 7
- CD-ROM Program
  - Resource Lesson 7-2
  - Interactive Lesson 7-2

Recommended Pacing	
Standard	Days 3 & 4 of 15
Honors	Day 3 of 14
Block	Day 2 of 8

1 FOCUS

 5-Minute Check *(Lesson 7-1)*

Estimate each product.

1. $\frac{3}{5} \times 31$ $\frac{3}{5} \times 30 = 18$
2. $\frac{2}{3} \times 44$ $\frac{2}{3} \times 45 = 30$
3. $\frac{3}{4} \times \frac{1}{3}$ $1 \times \frac{1}{3} = \frac{1}{3}$
4. $\frac{1}{8} \times \frac{25}{28}$ $\frac{1}{8} \times 1 = \frac{1}{8}$
5. $14\frac{1}{8} \times 6\frac{9}{10}$ $14 \times 7 = 98$

 The 5-Minute Check is also available on **Transparency 7-2A** for this lesson.

Motivating the Lesson

Problem Solving Ask students how they would determine which group would get the longer length of submarine sandwich.

- Group A will get $\frac{1}{6}$ of the $\frac{3}{4}$-yard submarine sandwich.
- Group B will get $\frac{1}{8}$ of the $\frac{4}{5}$-yard submarine sandwich. **Multiply the fractions; Group A**

Multiple Learning Styles

Logical Have groups of students devise a coded message that requires multiplying by fractions to decode. Let A through Z equal 1 through 26, respectively. Then write a code such as $\frac{2}{3}$X, $\frac{6}{7}$U, 3C, $\frac{13}{2}$D, $\frac{1}{4}$T. This word

decodes as "prize." Have students post their coded phrases on a bulletin board. The phrases could be changed daily and the decoding could include additional operations as they are learned.

Transparency 7-2B contains a teaching aid for this lesson.

Using Models You may wish to use pieces of transparencies to create fraction models based on the same rectangular form. Use a permanent black pen to create the whole divided into a particular fractional unit and then use overhead markers to color in the appropriate area. By overlapping the pieces, you can model each multiplication shown in the Examples and Guided Practice.

In-Class Examples

For Example 1
Find $\frac{1}{2} \times \frac{3}{4}$. $\frac{3}{8}$

For Example 2
Find $\frac{1}{8} \times 6$. $\frac{3}{4}$

For Example 3
Find $\frac{2}{3} \times \frac{9}{10}$. $\frac{3}{5}$

For Example 4
Solve $\frac{3}{5} \times \frac{10}{21} = n$. $\frac{2}{7}$

Teaching Tip In Examples 3 and 4, emphasize that the numerator and the denominator must be divided by the same number when simplifying fractions.

3 PRACTICE/APPLY

Check for Understanding
If students need additional practice or instruction after completing Exercises 1–11, you may find one of the following options helpful.
- Extra Practice, see p. 575
- Reteaching Activity, see p. 275
- *Study Guide Masters,* p. 50
- *Practice Masters,* p. 50
- Interactive Mathematics Tools Software

You can use the following rule to multiply fractions.

Multiplying Fractions	To multiply fractions, multiply the numerators. Then multiply the denominators. Simplify if necessary.

Examples ① Find $\frac{1}{4} \times \frac{2}{3}$.

Method 1 Use a model.

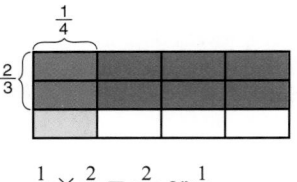

$\frac{1}{4} \times \frac{2}{3} = \frac{2}{12}$ or $\frac{1}{6}$

Method 2 Use a rule.

$\frac{1}{4} \times \frac{2}{3} = \frac{1 \cdot 2}{4 \cdot 3}$ *Multiply.*

$= \frac{2}{12}$ or $\frac{1}{6}$ *Simplify.*

Study Hint
Technology You can use a fraction calculator to multiply fractions. For example, to find $\frac{1}{2} \times \frac{3}{4}$, enter

1 / 2 × 3 / 4 =

② Find $\frac{2}{5} \times 4$. *Estimate:* $\frac{1}{2} \times 4 = 2$

$\frac{2}{5} \times 4 = \frac{2}{5} \times \frac{4}{1}$ *Express 4 as an improper fraction, $\frac{4}{1}$.*

$= \frac{2 \cdot 4}{5 \cdot 1}$ *Multiply.*

$= \frac{8}{5}$ or $1\frac{3}{5}$ *Simplify. Compare to the estimate.*

If the numerator of one fraction and the denominator of another fraction have a common factor, you can simplify *before* you multiply.

Examples ③ Find $\frac{5}{9} \times \frac{3}{4}$. *Estimate:* $\frac{1}{2} \times 1 = \frac{1}{2}$

Simplify before multiplying.

$\frac{5}{9} \times \frac{3}{4} = \frac{5 \cdot \overset{1}{3}}{\underset{3}{9} \cdot 4}$ *The GCF of 3 and 9 is 3. So, divide both the numerator and the denominator by 3.*

$= \frac{5}{12}$

INTEGRATION ④ **Algebra** Solve $\frac{3}{10} \times \frac{4}{9} = n$.

LOOK BACK
Refer to Lesson 5-3 for information on GCF.

$\frac{3}{10} \times \frac{4}{9} = n$

$\frac{\overset{1}{3} \cdot \overset{2}{4}}{\underset{5}{10} \cdot \underset{3}{9}} = n$ *The GCF of 4 and 10 is 2. The GCF of 3 and 9 is 3. So, divide both the numerator and the denominator by 2 and then by 3.*

$\frac{2}{15} = n$

The solution is $\frac{2}{15}$.

Communicating Mathematics

Read and study the lesson to answer each question. **1–3. See margin.**

1. *Explain* how the model shows the product of $\frac{1}{3}$ and $\frac{1}{2}$.

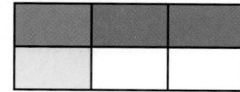

2. *Draw* a model to show that $\frac{2}{3} \times \frac{1}{5} = \frac{2}{15}$.

3. *Write* how to find the product of any two fractions.

Guided Practice

Find each product. Write in simplest form.

4. $\frac{1}{6} \times \frac{1}{2}$ $\frac{1}{12}$

5. $\frac{3}{5} \times \frac{1}{4}$ $\frac{3}{20}$

6. $\frac{1}{5} \times \frac{5}{7}$ $\frac{1}{7}$

7. $\frac{2}{3} \times \frac{1}{5}$ $\frac{2}{15}$

Solve each equation. Write the solution in simplest form.

8. $14 \times \frac{2}{7} = a$ **4**

9. $w = \frac{3}{4} \times \frac{4}{9}$ $\frac{1}{3}$

10. $\frac{6}{7} \times \frac{5}{12} = x$ $\frac{5}{14}$

11. **Life Science** About $\frac{7}{10}$ of the human body is water. If a person weighs 150 pounds, how many pounds are water? **about 105 pounds**

Practice

Find each product. Write in simplest form.

12. $\frac{1}{3} \times \frac{3}{5}$ $\frac{1}{5}$

13. $\frac{8}{9} \times \frac{1}{3}$ $\frac{8}{27}$

14. $\frac{7}{8} \times \frac{3}{4}$ $\frac{21}{32}$

15. $\frac{2}{3} \times \frac{1}{6}$ $\frac{1}{9}$

16. $\frac{2}{3} \times 5$ $3\frac{1}{3}$

17. $\frac{3}{5} \times \frac{2}{9}$ $\frac{2}{15}$

18. $6 \times \frac{5}{8}$ $3\frac{3}{4}$

19. $\frac{7}{9} \times \frac{6}{7}$ $\frac{2}{3}$

20. $\frac{1}{2} \times \frac{8}{9}$ $\frac{4}{9}$

21. $\frac{2}{3} \times \frac{5}{9}$ $\frac{10}{27}$

22. $\frac{3}{4} \times \frac{8}{9}$ $\frac{2}{3}$

23. $\frac{1}{9} \times \frac{2}{3}$ $\frac{2}{27}$

Solve each equation. Write the solution in simplest form.

24. $x = \frac{5}{12} \times \frac{3}{7}$ $\frac{5}{28}$

25. $10 \times \frac{2}{3} = h$ $6\frac{2}{3}$

26. $y = \frac{5}{8} \times \frac{2}{3}$ $\frac{5}{12}$

27. $m = \frac{5}{16} \times \frac{4}{15}$ $\frac{1}{12}$

28. $\frac{1}{2} \times \frac{4}{9} = b$ $\frac{2}{9}$

29. $d = \frac{5}{6} \times \frac{8}{15}$ $\frac{4}{9}$

30. $c = \frac{7}{10} \times \frac{5}{14}$ $\frac{1}{4}$

31. $18 \times \frac{7}{12} = a$ $10\frac{1}{2}$

32. $p = \frac{9}{13} \times \frac{8}{9}$ $\frac{8}{13}$

33. **Algebra** Evaluate mn if $m = \frac{1}{3}$ and $n = \frac{12}{15}$. $\frac{4}{15}$

Applications and Problem Solving

34. *Broadcasting* A television network sells 16 minutes of air time to advertisers. A company purchases $\frac{3}{8}$ of the total ad time. How many minutes of ad time did the company buy? **6 minutes**

■ Reteaching the Lesson ■

Activity Have students use counters to model problems such as $\frac{1}{4} \times \frac{3}{5}$. Give students 20 counters. Have them divide the counters into piles representing $\frac{3}{5}$ and $\frac{2}{5}$. Have them divide the $\frac{3}{5}$ pile into fourths. How many counters are in one-fourth? Write this as a fraction. $\frac{3}{20}$

Additional Answers

1. The rectangle is separated into thirds. Then the rectangle is separated into halves. The shaded portion shows overlap between $\frac{1}{3}$ of the rectangle and $\frac{1}{2}$ of the rectangle. This area represents the product, $\frac{1}{6}$.

2.

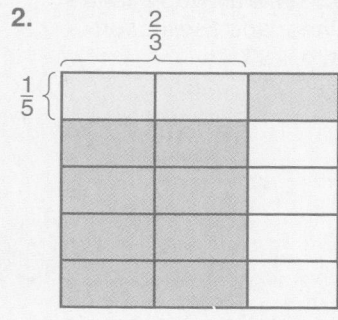

3. To find the product of fractions, you need to multiply the numerators and then multiply the denominators. If necessary, simplify the answer.

Study Guide Masters, p. 50

7-2 Study Guide

Multiplying Fractions

To multiply fractions:	Multiply the numerators.	Then multiply the denominators. Write the product in simplest form.
$\frac{5}{7} \times \frac{3}{5}$	$\frac{5}{7} \times \frac{3}{5} = \frac{15}{}$	$\frac{5}{7} \times \frac{3}{5} = \frac{15}{35} = \frac{3}{7}$
To multiply whole numbers and fractions:	Rename the whole number as an improper fraction. Multiply the numerators.	Multiply the denominators. Write the product in simplest form.
$\frac{3}{8} \times 7$	$\frac{3}{8} \times \frac{7}{1} = \frac{21}{}$	$\frac{3}{8} \times \frac{7}{1} = \frac{21}{8} = 2\frac{5}{8}$

If the numerator of one fraction and the denominator of the other fraction have a common factor, you can simplify before you multiply.

Example Find $\frac{8}{11} \times \frac{3}{4}$. The GCF of 8 and 4 is 4.

$\frac{8^{2}}{11} \times \frac{3}{4_{1}} = \frac{6}{11}$ Divide the numerator and denominator by 4. Then multiply.

Find each product. Write in simplest form.

1. $\frac{1}{3} \times \frac{1}{5}$ $\frac{1}{15}$

2. $\frac{5}{8} \times \frac{1}{2}$ $\frac{5}{16}$

3. $\frac{4}{9} \times \frac{3}{4}$ $\frac{1}{3}$

4. $6 \times \frac{2}{3}$ **4**

5. $\frac{3}{5} \times 10$ **6**

6. $\frac{2}{3} \times \frac{3}{8}$ $\frac{1}{4}$

7. $\frac{1}{7} \times \frac{1}{7}$ $\frac{1}{49}$

8. $\frac{2}{9} \times \frac{1}{2}$ $\frac{1}{9}$

9. $12 \times \frac{5}{6}$ **10**

Solve each equation. Write the solution in simplest form.

10. $m = 8 \times \frac{1}{4}$ **2**

11. $\frac{3}{7} \times \frac{6}{} = n$ $\frac{2}{7}$

12. $c = \frac{2}{7} \times \frac{1}{3}$ $\frac{2}{21}$

13. $\frac{5}{8} \times 24 = a$ **15**

14. $k = \frac{5}{12} \times \frac{1}{3}$ $\frac{1}{12}$

15. $\frac{1}{2} \times \frac{1}{5} = t$ $\frac{1}{10}$

16. $e = \frac{4}{9} \times \frac{8}{15}$ $\frac{16}{35}$

17. $\frac{4}{12} \times 10 = t$ $4\frac{1}{6}$

18. $h = \frac{8}{9} \times \frac{9}{10}$ $\frac{4}{5}$

© Glencoe/McGraw-Hill   T50   Mathematics: Applications and Connections, Course 1

Closing Activity

Writing Have students determine how they would use fraction multiplication to solve the following problem. *What fraction of your classmates have both a first and a last name that begins with a letter from A to M?* Ask them to write out the steps and provide an answer to the problem.

Chapter 7, Quiz A (Lessons 7-1 and 7-2) is available in the *Assessment and Evaluation Masters, p. 183.*

---

35. *Algebra* The expression 3^2 means 3×3, or 9.

   a. If $m = \frac{3}{8}$, what is the value of m^2? $\frac{9}{64}$

   b. Evaluate h^2 if $h = \frac{4}{5}$. $\frac{16}{25}$

36. *Critical Thinking* Find $\frac{1}{2} \times \frac{2}{3} \times \frac{3}{4} \times \frac{4}{5} \times \ldots \times \frac{99}{100}$. $\frac{1}{100}$

Mixed Review

37. Estimate the product of $\frac{3}{4}$ and 37. *(Lesson 7-1)* $\frac{3}{4} \times 36 = 27$

38. *Algebra* Solve the equation $x - \frac{11}{12} = \frac{5}{12}$. *(Lesson 6-3)* $\frac{16}{12}$ or $1\frac{1}{3}$

39. Determine whether 786 is divisible by 2, 3, 5, 6, 9, or 10. **2, 3, 6** *(Lesson 5-1)*

40. **Test Practice** Hayley works 8 hours a week baby-sitting and earns $22.40. How much does she earn per hour? *(Lesson 4-5)* **D**

   A $3.20   **B** $2.50   **C** $3.40   **D** $2.80   **E** Not Here

---

Let the Games Begin

Multiplication Mania

Get Ready   This game is for any number of players.

   ▢ poster board   🎲 2 number cubes   🖩 calculator

Get Set   On a poster board, draw a large game board similar to the one shown at the right.

Go
- Place the game board on the floor.
- Each player rolls the number cubes. The person with the highest total starts.
- The first player rolls the number cubes onto the game board. If a number cube rolls off the game board or lands on a line, roll it again. The player then multiplies the two numbers on which the number cubes land and simplifies the product. Use a calculator to determine whether the answer is correct. Each correct answer is worth 1 point.
- The first player to score 10 points wins.

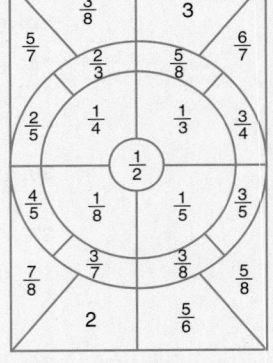

interNET CONNECTION Visit www.glencoe.com/sec/math/mac/mathnet for more games.

276   **Chapter 7** Multiplying and Dividing Fractions

■ Extending the Lesson ■

Enrichment Masters, p. 50

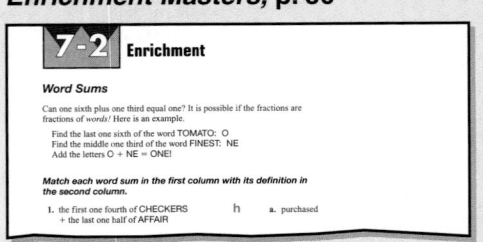

Let the Games Begin

Another variation for Multiplication Mania is to use 3 number cubes. To also use this game in Lesson 7-5 for division of fractions, students could drop the number cubes one at a time and divide the first fraction by the second.

*Additional resources for this game can be found on page 43 of the **Classroom Games**.*

Practice Masters, p. 50

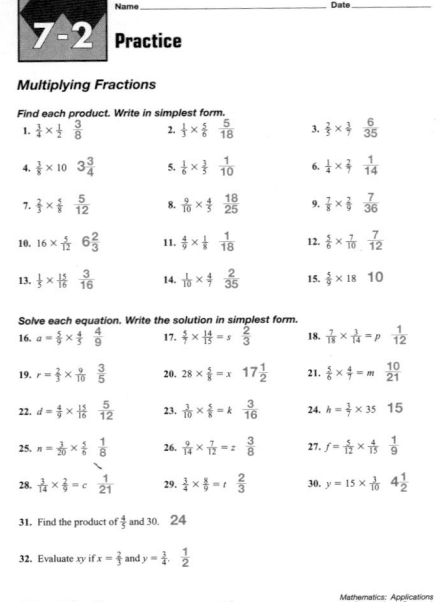

Name _____ Date _____

7-2 Practice

Multiplying Fractions

Find each product. Write in simplest form.

1. $\frac{1}{4} \times \frac{1}{2}$ $\frac{3}{8}$
2. $\frac{1}{3} \times \frac{5}{6}$ $\frac{5}{18}$
3. $\frac{2}{3} \times \frac{3}{5}$ $\frac{6}{35}$
4. $\frac{3}{8} \times 10$ $3\frac{3}{4}$
5. $\frac{1}{3} \times \frac{1}{3}$ $\frac{1}{10}$
6. $\frac{1}{4} \times \frac{4}{7}$ $\frac{1}{14}$
7. $\frac{2}{3} \times \frac{4}{8}$ $\frac{5}{12}$
8. $\frac{9}{10} \times \frac{4}{5}$ $\frac{18}{25}$
9. $\frac{7}{8} \times \frac{2}{9}$ $\frac{7}{36}$
10. $16 \times \frac{5}{12}$ $6\frac{2}{3}$
11. $\frac{4}{9} \times \frac{1}{8}$ $\frac{1}{18}$
12. $\frac{5}{6} \times \frac{7}{10}$ $\frac{7}{12}$
13. $\frac{1}{3} \times \frac{15}{16}$ $\frac{3}{16}$
14. $\frac{1}{10} \times \frac{4}{7}$ $\frac{2}{35}$
15. $\frac{5}{9} \times 18$ 10

Solve each equation. Write the solution in simplest form.

16. $a = \frac{2}{3} \times \frac{2}{3}$ $\frac{4}{9}$
17. $\frac{4}{7} \times \frac{11}{13} = s$ $\frac{2}{3}$
18. $\frac{7}{18} \times \frac{3}{14} = p$ $\frac{1}{12}$
19. $r = \frac{2}{3} \times \frac{9}{10}$ $\frac{3}{5}$
20. $28 \times \frac{5}{8} = x$ $17\frac{1}{2}$
21. $\frac{5}{6} \times \frac{4}{7} = m$ $\frac{10}{21}$
22. $d = \frac{4}{9} \times \frac{15}{16}$ $\frac{5}{12}$
23. $\frac{7}{10} \times \frac{5}{8} = k$ $\frac{3}{16}$
24. $h = \frac{5}{7} \times 35$ 15
25. $n = \frac{3}{20} \times \frac{5}{6}$ $\frac{1}{8}$
26. $\frac{9}{14} \times \frac{7}{12} = z$ $\frac{3}{8}$
27. $f = \frac{5}{12} \times \frac{4}{15}$ $\frac{1}{9}$
28. $\frac{3}{14} \times \frac{2}{9} = c$ $\frac{1}{21}$
29. $\frac{1}{3} \times \frac{4}{9} = t$ $\frac{2}{3}$
30. $y = 15 \times \frac{3}{10}$ $4\frac{1}{2}$

31. Find the product of $\frac{4}{5}$ and 30.   24

32. Evaluate xy if $x = \frac{2}{3}$ and $y = \frac{3}{4}$. $\frac{1}{2}$

© Glencoe/McGraw-Hill   T50   *Mathematics: Applications and Connections, Course 1*

7-2 Enrichment

Word Sums

Can one sixth plus one third equal one? It is possible if the fractions are fractions of *words!* Here is an example.

  Find the last one sixth of the word TOMATO: O
  Find the middle one third of the word FINEST: NE
  Add the letters O + NE = ONE!

Match each word sum in the first column with its definition in the second column.

1. the first one fourth of CHECKERS + the last one half of AFFAIR   h   a. purchased

What you'll learn

You'll learn to multiply mixed numbers.

When am I ever going to use this?

Knowing how to multiply mixed numbers can help you determine the ingredients needed to make a cake.

LOOK BACK

You can refer to Lesson 5-5 for information on improper fractions.

A dot following a musical note ($\mathbf{o}\cdot$) means that the note gets $1\frac{1}{2}$ times as many beats as the same note without a dot ($\mathbf{o}$). If a whole note gets four beats, how many beats would a dotted whole note get?

You need to find $1\frac{1}{2}$ of 4, or $1\frac{1}{2} \times 4$. To multiply mixed numbers, express the mixed numbers as improper fractions and then multiply as with fractions.

$$1\frac{1}{2} \times 4 = \frac{3}{2} \times \frac{4}{1} \quad \textit{Express the mixed numbers as improper fractions.}$$

$$= \frac{3 \cdot 4}{2 \cdot 1} \quad \textit{Multiply the numerators. Multiply the denominators.}$$

$$= \frac{12}{2} \text{ or } 6 \quad \textit{Simplify.}$$

A dotted whole note gets 6 beats.

Examples

1 Find $\frac{1}{3} \times 3\frac{3}{8}$. *Estimate: $\frac{1}{3} \times 3 = 1$*

$$\frac{1}{3} \times 3\frac{3}{8} = \frac{1}{3} \times \frac{27}{8} \quad \textit{Express } 3\frac{3}{8} \textit{ as an improper fraction.}$$

$$= \frac{1 \cdot \overset{9}{27}}{\underset{1}{3} \cdot 8} \quad \textit{Divide 27 and 3 by the GCF, 3. Then multiply.}$$

$$= \frac{9}{8} \text{ or } 1\frac{1}{8} \quad \textit{Compare with your estimate.}$$

INTEGRATION

2 **Algebra** If $m = 2\frac{2}{5}$ and $n = 1\frac{7}{8}$, what is the value of mn?

$$mn = 2\frac{2}{5} \cdot 1\frac{7}{8} \quad \textit{Replace m with } 2\frac{2}{5} \textit{ and n with } 1\frac{7}{8}.$$

$$\textit{Estimate: } 2 \times 2 = 4$$

$$= \frac{12}{5} \cdot \frac{15}{8} \quad \textit{Express the mixed numbers as improper fractions.}$$

$$= \frac{\overset{3}{12} \cdot \overset{3}{15}}{\underset{1}{5} \cdot \underset{2}{8}} \quad \textit{Divide 12 and 8 by the GCF, 4. Divide 15 and 5 by the GCF, 5.}$$

$$\textit{Then multiply.}$$

$$= \frac{9}{2} \text{ or } 4\frac{1}{2} \quad \textit{Compare with your estimate.}$$

Lesson 7–3 Multiplying Mixed Numbers **277**

7-3 Lesson Notes

Instructional Resources

- *Study Guide Masters*, p. 51
- *Practice Masters*, p. 51
- *Enrichment Masters*, p. 51
- Transparencies 7-3, A and B
- *Diversity Masters, p. 7*
- CD-ROM Program
  - Resource Lesson 7-3

Recommended Pacing	
Standard	Day 5 of 15
Honors	Day 4 of 14
Block	Day 3 of 8

1 FOCUS

5-Minute Check
(Lesson 7-2)

Find each product. Write in simplest form.

1. $\frac{1}{4} \times \frac{3}{7}$ $\frac{3}{28}$

2. $8 \times \frac{3}{4}$ 6

3. $\frac{2}{3} \times \frac{9}{14}$ $\frac{3}{7}$

Solve each equation. Write the solution in simplest form.

4. $m = \frac{3}{5} \times \frac{1}{9}$ $\frac{1}{15}$

5. $\frac{4}{9} \times \frac{27}{32} = t$ $\frac{3}{8}$

The 5-Minute Check is also available on **Transparency 7-3A** for this lesson.

2 TEACH

Transparency **7-3B** contains a teaching aid for this lesson.

Using Models Represent mixed numbers with models as we did fractions. For example, model $3\frac{3}{8} \times \frac{1}{3}$ as shown and write out as $\frac{8}{24} + \frac{8}{24} + \frac{8}{24} + \frac{3}{24} = \frac{27}{24}$, or $1\frac{1}{8}$.

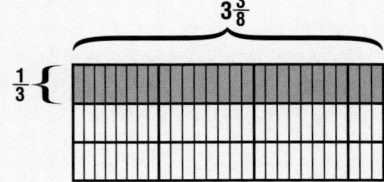

Motivating the Lesson

Communication Bring in a blueberry muffin recipe that makes 12 muffins. Discuss with the class how they might increase the recipe to make enough muffins for everybody in the class. $2\frac{1}{2}$ times for a class of 30 students

For Example 1
Find each product.
a. $\frac{2}{5} \times 3\frac{1}{6}$ $1\frac{4}{15}$
b. $6\frac{6}{11} \times \frac{4}{9}$ $2\frac{10}{11}$

For Example 2
If $c = 3\frac{1}{2}$ and $d = 4\frac{2}{5}$, what is the value of cd? $15\frac{2}{5}$

3 PRACTICE/APPLY

Check for Understanding
If students need additional practice or instruction after completing Exercises 1–13, you may find one of the following options helpful.
• Extra Practice, see p. 575
• Reteaching Activity
• *Study Guide Masters*, p. 51
• *Practice Masters*, p. 51

Assignment Guide

Core: 15–41 odd, 43–48
Enriched: 14–40 even, 41, 43–48

Study Guide Masters, p. 51

7-3 **Study Guide**

Name_____ Date_____

Multiplying Mixed Numbers

To multiply mixed numbers, express each mixed number as an improper fraction. Then multiply the fractions.

Example Find $7\frac{1}{2} \times 3\frac{1}{3}$.   Estimate: $8 \times 3 = 24$

$7\frac{1}{2} \times 3\frac{1}{3} = \frac{15}{2} \times \frac{10}{3}$ Express the mixed numbers as improper fractions.

$= \frac{{}^5 15 \cdot 10 {}^5}{{}_1 2 \cdot 3 {}_1}$ Divide 15 and 3 by the GCF 3.
Divide 10 and 2 by the GCF 2.

$= \frac{25}{1}$ or 25

Express each mixed number as an improper fraction.
1. $5\frac{3}{4}$ 2. $3\frac{7}{9}$ 3. $2\frac{4}{5}$ 4. $1\frac{15}{16}$
   $\frac{23}{4}$ $\frac{34}{9}$ $\frac{14}{5}$ $\frac{31}{16}$

Find each product. Write in simplest form.
5. $\frac{2}{3} \times 3\frac{1}{2}$ 6. $5\frac{1}{4} \times \frac{2}{3}$ 7. $9 \times 1\frac{5}{6}$ 8. $2\frac{4}{9} \times \frac{4}{11}$
   $2\frac{1}{3}$ $3\frac{5}{6}$ $16\frac{1}{2}$ $\frac{8}{9}$

9. $1\frac{1}{8} \times \frac{2}{3}$ 10. $2\frac{1}{2} \times 1\frac{1}{5}$ 11. $\frac{1}{9} \times 1\frac{1}{2}$ 12. $8 \times 1\frac{1}{4}$
   $\frac{3}{4}$ 3 $\frac{1}{6}$ 10

Solve each equation. Write the solution in simplest form.
13. $2\frac{1}{2} \times 4 = n$ 14. $k = 4\frac{2}{3} \times 1\frac{1}{2}$ 15. $\frac{4}{5} \times 1\frac{1}{4} = p$
    10   7   1

16. $y = 6 \times 3\frac{1}{3}$ 17. $4\frac{1}{2} \times \frac{8}{9} = a$ 18. $8\frac{1}{3} \times \frac{3}{5} = r$
    20   4   5

© Glencoe/McGraw-Hill   T51   Mathematics: Applications and Connections, Course 1

CHECK FOR UNDERSTANDING

Communicating Mathematics

Read and study the lesson to answer each question.
1. *Tell* how to multiply mixed numbers. **1–2. See Answer Appendix.**
2. *Explain* to a classmate how to find the product of $3\frac{1}{2}$ and $1\frac{1}{2}$. What is the product?

Guided Practice

Express each mixed number as an improper fraction.
3. $6\frac{1}{3}$ $\frac{19}{3}$ 4. $2\frac{5}{8}$ $\frac{21}{8}$ 5. $5\frac{3}{4}$ $\frac{23}{4}$

Find each product. Write in simplest form.
6. $6 \times 2\frac{1}{4}$ $13\frac{1}{2}$ 7. $\frac{2}{3} \times 5\frac{2}{5}$ $3\frac{3}{5}$ 8. $2\frac{4}{5} \times 3\frac{3}{4}$ $10\frac{1}{2}$

Solve each equation. Write the solution in simplest form.
9. $12 \times 3\frac{1}{6} = n$ **38**   10. $a = 2\frac{4}{5} \times 3\frac{1}{8}$ $8\frac{3}{4}$ 11. $1\frac{1}{4} \times \frac{8}{9} = x$ $1\frac{1}{9}$

12. *Algebra* Evaluate the expression st if $s = 2\frac{2}{3}$ and $t = 4\frac{2}{3}$. $12\frac{4}{9}$

13. *Music* Refer to the beginning of the lesson.
   a. If an eighth note gets half of a beat, how many beats would a dotted eighth note get? $\frac{3}{4}$
   b. A sixteenth note gets one-fourth of a beat. How many beats would a dotted sixteenth note get? $\frac{3}{8}$

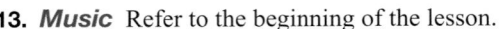
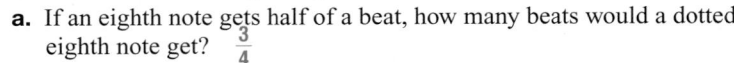

EXERCISES

Practice

Express each mixed number as an improper fraction.
14. $4\frac{1}{2}$ $\frac{9}{2}$ 15. $7\frac{2}{3}$ $\frac{23}{3}$ 16. $3\frac{1}{4}$ $\frac{13}{4}$ 17. $4\frac{2}{3}$ $\frac{14}{3}$
18. $5\frac{5}{7}$ $\frac{40}{7}$ 19. $6\frac{3}{5}$ $\frac{33}{5}$ 20. $1\frac{5}{8}$ $\frac{13}{8}$ 21. $8\frac{3}{4}$ $\frac{35}{4}$

Find each product. Write in simplest form.
22. $1\frac{4}{5} \times 3$ $5\frac{2}{5}$ 23. $1\frac{7}{9} \times 3\frac{3}{4}$ $6\frac{2}{3}$ 24. $12\frac{1}{2} \times \frac{1}{5}$ $2\frac{1}{2}$
25. $4\frac{3}{8} \times 2\frac{2}{5}$ $10\frac{1}{2}$ 26. $4 \times \frac{2}{5}$ $1\frac{3}{5}$ 27. $3\frac{3}{8} \times 1\frac{4}{9}$ $4\frac{7}{8}$
28. $\frac{5}{11} \times 6\frac{2}{7}$ $2\frac{6}{7}$ 29. $7\frac{7}{8} \times 5\frac{1}{3}$ **42**   30. $3\frac{5}{8} \times 4$ $14\frac{1}{2}$

Solve each equation. Write the solution in simplest form.
31. $6 \times 3\frac{1}{2} = y$ **21**   32. $h = \frac{1}{3} \times 2\frac{1}{4}$ $\frac{3}{4}$ 33. $k = 3\frac{1}{6} \times 2\frac{4}{7}$ $8\frac{1}{7}$
34. $j = \frac{5}{12} \times 1\frac{4}{5}$ $\frac{3}{4}$ 35. $2\frac{3}{4} \times 2\frac{2}{3} = d$ $7\frac{1}{3}$ 36. $8 \times 1\frac{5}{6} = c$ $14\frac{2}{3}$

37. *Algebra* What is the value of $\frac{2}{3}a$ if $a = 45$? **30**

38. *Algebra* Evaluate ab if $a = \frac{2}{3}$ and $b = 3\frac{1}{2}$. $2\frac{1}{3}$

Reteaching the Lesson

Activity Have students use grid paper to model mixed numbers. $5\frac{7}{8}$ would use 6 columns of 8 boxes. Shade in 5 whole columns and 7 boxes in the last column. How many boxes are shaded? **47** Since each box represents $\frac{1}{8}$, there are $\frac{47}{8}$ shaded.

Error Analysis
Watch for students who combine whole numbers and then fractions when multiplying mixed numbers.
Prevent by reinforcing the fact that mixed numbers are rewritten as improper fractions *before* multiplying.

39. *Algebra* If $x = \frac{9}{10}$ and $y = 8\frac{1}{3}$, find the value of xy. $7\frac{1}{2}$

Applications and Problem Solving

40. *Cooking* A recipe for a two-layer, 8-inch cake calls for a box of cake mix, 2 eggs, and $1\frac{1}{3}$ cups of water. How much of each ingredient is needed to make a three-layer, 8-inch cake? $1\frac{1}{2}$ **boxes of cake mix, 3 eggs, 2 cups of water**

41. *Geometry* To find the area of a parallelogram, you can use the formula, $A = b \times h$, where b is the length of the base and h is the height. Find the area of the parallelogram. $8\frac{1}{4}$ **sq ft**

$2\frac{3}{4}$ ft

3 ft

42. *Working on the* **CHAPTER Project** Refer to the table on page 267.

   a. In flight 2, Mitiku's airplane flew $13\frac{1}{2}$ feet in 3 seconds. How far would his airplane travel if it remained in the air for 1 minute? Find the distance the other two airplanes would travel in 1 minute. **270 ft; $337\frac{1}{2}$ ft; 345 ft**

   b. For each of your flights, find the distance the plane would travel in 1 minute. **See students' work.**

43. *Critical Thinking* Is $2\frac{2}{3} \times 4\frac{1}{2}$ more or less than 10? Explain how you know without actually multiplying. **more; $2\frac{2}{3} \times 4\frac{1}{2} \approx 3 \times 4$ or 12**

Mixed Review

44. **Test Practice** Evaluate ab if $a = \frac{7}{8}$ and $b = \frac{2}{3}$. *(Lesson 7-2)* **D**

   A $\frac{72}{83}$ **B** $\frac{21}{16}$ **C** $\frac{9}{11}$ **D** $\frac{7}{12}$ **E** Not Here

45. Find $7\frac{1}{6} - 3\frac{3}{4}$. Write the answer in simplest form. *(Lesson 6-6)* $3\frac{5}{12}$

46. Round $\frac{8}{11}$ to the nearest half. *(Lesson 6-1)* **1**

47. *Money Matters* Mr. Maldonado kept track of the phone calls each of his four children made during the month of September. Make a bar graph for the set of data. *(Lesson 2-3)* **See margin.**

Name	Number of Calls
Luis	45
Lorena	40
Diana	25
Mirna	25

48. Evaluate the expression $12 - 8 \div 2 + 1$. *(Lesson 1-4)* **9**

MATH IN THE MEDIA

1. Explain why the comic is funny. **See students' work.**
2. Suppose Sally multiplies $4\frac{1}{2}$ by $6\frac{5}{8}$. What is the product? $29\frac{13}{16}$

Peanuts

■ Extending the Lesson ■

Enrichment Masters, p. 51

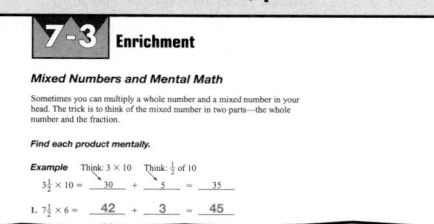

7-3 **Enrichment**

Mixed Numbers and Mental Math

Sometimes you can multiply a whole number and a mixed number in your head. The trick is to think of the mixed number in two parts—the whole number and the fraction.

Find each product mentally.

Example Think: 3×10 Think: $\frac{1}{2}$ of 10

$3\frac{1}{2} \times 10 = \underline{30} + \underline{5} = \underline{35}$

1. $7\frac{1}{2} \times 6 = \underline{42} + \underline{3} = \underline{45}$

MATH IN THE MEDIA

Begin a class discussion/debate by asking the same question Sally asks: "Why should I learn that?" Encourage students to look for or to create other comics that involve fractions.

CHAPTER Project

Exercise 42 asks students to advance to the next stage of work on the Chapter Project. You may ask the class to set up an algebraic expression to answer Exercise 42b.

4 ASSESS

Closing Activity

Modeling Bring in or obtain a variety of boxes that have mixed numbers as lengths and widths. Have pairs of students choose a box and determine the area of one of its sides by multiplying length × width.

Additional Answer

47. **Phone Calls**

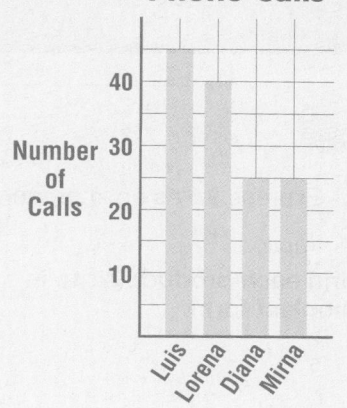

Practice Masters, p. 51

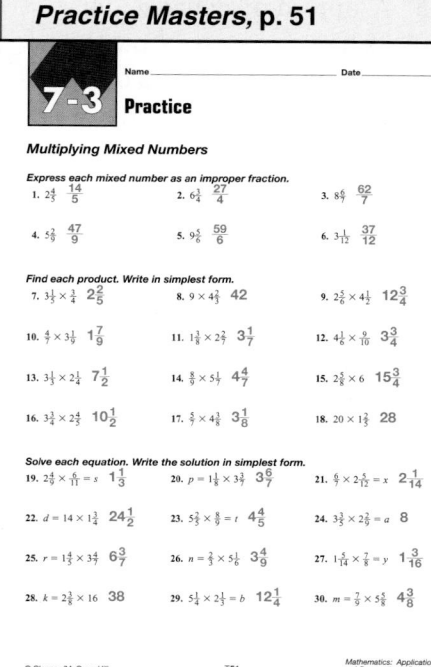

Name _____ Date _____

7-3 **Practice**

Multiplying Mixed Numbers

Express each mixed number as an improper fraction.

1. $2\frac{4}{5}$ $\frac{14}{5}$ 2. $6\frac{3}{4}$ $\frac{27}{4}$ 3. $8\frac{6}{7}$ $\frac{62}{7}$

4. $5\frac{2}{9}$ $\frac{47}{9}$ 5. $9\frac{5}{6}$ $\frac{59}{6}$ 6. $3\frac{1}{12}$ $\frac{37}{12}$

Find each product. Write in simplest form.

7. $3\frac{1}{3} \times \frac{1}{4}$ $2\frac{2}{5}$ 8. $9 \times 4\frac{2}{3}$ 42   9. $2\frac{5}{6} \times 4\frac{1}{2}$ $12\frac{3}{4}$

10. $\frac{4}{5} \times 3\frac{1}{9}$ $1\frac{7}{9}$ 11. $1\frac{3}{8} \times 2\frac{2}{7}$ $3\frac{1}{7}$ 12. $4\frac{1}{6} \times \frac{9}{10}$ $3\frac{3}{4}$

13. $3\frac{1}{3} \times 2\frac{1}{4}$ $7\frac{1}{2}$ 14. $\frac{8}{9} \times 5\frac{1}{4}$ $4\frac{4}{7}$ 15. $2\frac{5}{8} \times 6$ $15\frac{3}{4}$

16. $3\frac{3}{4} \times 2\frac{4}{5}$ $10\frac{1}{2}$ 17. $\frac{5}{6} \times 4\frac{1}{2}$ $3\frac{3}{8}$ 18. $20 \times 1\frac{2}{5}$ 28

Solve each equation. Write the solution in simplest form.

19. $2\frac{4}{9} \times \frac{6}{11} = s$ $1\frac{1}{3}$ 20. $p = 1\frac{1}{8} \times 3\frac{1}{3}$ $3\frac{6}{7}$ 21. $\frac{6}{7} \times 2\frac{5}{12} = x$ $2\frac{1}{14}$

22. $d = 14 \times 1\frac{3}{4}$ $24\frac{1}{2}$ 23. $5\frac{2}{5} \times \frac{8}{9} = t$ $4\frac{4}{5}$ 24. $3\frac{1}{2} \times 2\frac{2}{7} = a$ 8

25. $r = 1\frac{4}{5} \times 3\frac{4}{9}$ $6\frac{3}{7}$ 26. $n = \frac{5}{7} \times 5\frac{1}{6}$ $3\frac{4}{9}$ 27. $1\frac{1}{14} \times \frac{7}{8} = y$ $1\frac{3}{16}$

28. $k = 2\frac{3}{8} \times 16$ 38   29. $5\frac{1}{4} \times 2\frac{1}{3} = b$ $12\frac{1}{4}$ 30. $m = \frac{7}{9} \times 5\frac{5}{8}$ $4\frac{3}{8}$

© Glencoe/McGraw-Hill   T51   Mathematics: Applications and Connections, Course 1

Instructional Resources

- *Study Guide Masters*, p. 52
- *Practice Masters*, p. 52
- *Enrichment Masters*, p. 52
- Transparencies 7-4, A and B
- *Assessment and Evaluation Masters*, pp. 182, 183
- *Hands-On Lab Masters*, p. 75
- CD-ROM Program
  - Resource Lesson 7-4
  - Extended Activity 7-4

Recommended Pacing

Standard	Day 6 of 15
Honors	Day 5 of 14
Block	Day 4 of 8

1 FOCUS

5-Minute Check
(Lesson 7-3)

1. Express $3\frac{5}{7}$ as an improper fraction. $\frac{26}{7}$

Find each product. Write in simplest form.

2. $\frac{2}{5} \times 6\frac{1}{4}$ $2\frac{1}{2}$

3. $6 \times 3\frac{5}{12}$ $20\frac{1}{2}$

4. Solve $n = 4\frac{1}{2} \times 3\frac{4}{9}$. $15\frac{1}{2}$

5. A farm co-op donates $\frac{1}{8}$ of its harvest to a local pantry. How much of $2\frac{2}{3}$ tons of vegetables will go to the pantry? $\frac{1}{3}$ ton

The 5-Minute Check is also available on **Transparency 7-4A** for this lesson.

Motivating the Lesson

Problem Solving Ask students how they would determine the dimensions of a basketball.

Integration: Geometry
Circles and Circumference

What you'll learn

You'll learn to find the circumference of circles.

When am I ever going to use this?

Knowing how to find the circumference of a circle can help you determine the amount of fencing needed to surround a circular swimming pool.

Word Wise

circle
center
radius
diameter
circumference

The first African-American woman to win an Olympic gold medal was Alice Coachman. She received a gold medal for the high jump in the 1948 Olympics. An Olympic medal is a model of a circle.

A **circle** is a set of points in a plane, all of which are the same distance from a fixed point in the plane called the **center**.

The distance from the center to any point on the circle is called the **radius (r)**. The distance across the circle through its center is called the **diameter (d)**. The diameter of a circle is twice the length of its radius. The **circumference (C)** is the distance around the circle.

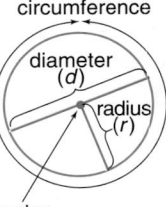

Circumference and diameter are related in a special way.

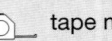

HANDS-ON MINI-LAB

Work with a partner.   tape measure   calculator   circular objects

Try This

- Wrap a tape measure around the circumference of one of the circular objects. Record the measurement. It is the circumference of the object.
- Measure and then record the diameter of the object.
- Using a calculator, divide the circumference by the diameter. Record the results.
- Repeat the steps for each circular object.

Talk About It

1. Compare the quotients you obtained when you divided each circle's circumference by its diameter. What do you notice?
2. What can you say about the diameter of a circle and its circumference?

1. The quotients are close to 3.

2. The circumference of a circle is always a little more than three times the diameter.

280   Chapter 7   Multiplying and Dividing Fractions

Classroom Vignette

"Each group of students gets a circle, a piece of string, a ruler, and a calculator. They use the string to find a length representing the circumference. They measure the string and the diameter of the circle in both metric and customary units. We calculate the ratio of circumference to diameter to estimate π."

Marilyn L. McNamara

Marilyn L. McNamara, Teacher
Walter T. Bergen Middle School
Bloomingdale, NJ

In the Mini-Lab, you discovered that the circumference of a circle is always a little more than three times its diameter. The exact number of times is represented by the Greek letter π (pi).

Circumference	**Words:** The circumference of a circle is equal to π times its diameter or π times twice its radius.	
	Symbols: $C = \pi d$ or $C = 2\pi r$	**Model:** 

The decimal 3.14 and the fraction $\frac{22}{7}$ are used as approximations for π.

Examples

(1) Find the circumference of a circle with a diameter of $1\frac{1}{2}$ feet.

$C = \pi d$

$\approx \frac{22}{7} \cdot 1\frac{1}{2}$ *Replace π with $\frac{22}{7}$ and d with $1\frac{1}{2}$.*

$\approx \frac{22}{7} \cdot \frac{3}{2}$ *Write $1\frac{1}{2}$ as an improper fraction.*

$\approx \frac{\overset{11}{\cancel{22}}}{7} \cdot \frac{3}{\underset{1}{\cancel{2}}}$ *Divide 2 and 22 by the GCF, 2.*

$\approx \frac{33}{7}$ or $4\frac{5}{7}$ *Simplify.*

The circumference of the circle is $4\frac{5}{7}$ feet.

APPLICATION

(2) **Inventions** 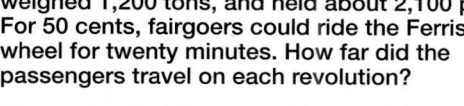 The first Ferris wheel was built for the 1893 World's Columbian Exposition in Chicago. It had a radius of 125 feet, weighed 1,200 tons, and held about 2,100 people. For 50 cents, fairgoers could ride the Ferris wheel for twenty minutes. How far did the passengers travel on each revolution?

You need to find the circumference of the Ferris wheel. The radius of the wheel is 125 feet.

$C = 2\pi r$

$\approx 2 \cdot 3.14 \cdot 125$ *Replace π with 3.14 and r with 125.*

2 ✕ 3.14 ✕ 125 = 785

The passengers traveled about 785 feet on each revolution.

Lesson 7-4 Integration: Geometry Circles and Circumference **281**

Transparency 7-4B contains a teaching aid for this lesson.

Using the Mini-Lab Review the importance of accuracy when using measuring instruments. Suggest that students record their measurements in a four-column table with headings *Object, Circumference, Diameter,* and $\frac{Circumference}{Diameter}$.

In-Class Examples

For Example 1
Find the circumference of a circle with a diameter of $2\frac{3}{4}$ feet. $8\frac{9}{14}$ ft

For Example 2
A radio telescope, located at Jodrell Bank in Great Britain, has a radius of 38.1 meters. Find the circumference of the Jodrell Bank radio dish rounded to the nearest tenth. **about 239.3 m**

Teaching Tip Show students how $\pi d = 2\pi r$ by including an extra step. Write out $C = \pi d = \pi(2r) = 2\pi r$ while reminding them that diameter is twice the radius.

Teaching Tip If students use the π key on their calculators, their answers may differ from those found when using 3.14.

Check for Understanding

If students need additional practice or instruction after completing Exercises 1–8, you may find one of the following options helpful.
- Extra Practice, see p. 576
- Reteaching Activity
- *Study Guide Masters*, p. 52
- *Practice Masters*, p. 52

Assignment Guide

Core: 9–21 odd, 23–26
Enriched: 10–20 even, 21–26
All: Self Test, 1–10

Additional Answers

1. Sample answer: Multiply 2 by π by 3.

2. If the diameter or radius of a circle is a fraction, use $\frac{22}{7}$ for π. If the diameter or radius is a whole number or decimal, use 3.14 for π.

Study Guide Masters, p. 52

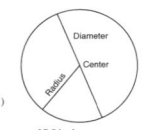

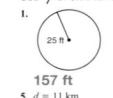 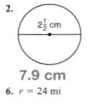  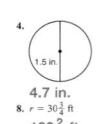

CHECK FOR UNDERSTANDING

Communicating Mathematics

Read and study the lesson to answer each question. 1–2. See margin.

1. *Write*, in your own words, how to find the circumference of a circle with a radius of 3 centimeters.

2. *Tell* how you know when to use either $\frac{22}{7}$ or 3.14 for π.

HANDS-ON MATH

3. *Locate* two circular objects in your home. Measure the circumference and diameter of each object. How does the quotient of each circumference and its diameter compare to π? **See students' work.**

Guided Practice

Find the circumference of each circle shown or described. Use $\frac{22}{7}$ or 3.14 for π. Round decimal answers to the nearest tenth.

4. 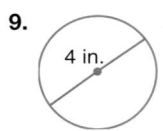 87.9 ft 14 ft

5.  $2\frac{1}{2}$ yd $\quad \pi = \frac{22}{7}$; $7\frac{6}{7}$ yd

6. $d = \frac{7}{8}$ yd $\quad \pi = \frac{22}{7}$; $2\frac{3}{4}$ yd

7. Find the circumference of a circle with a radius of 0.95 meter. **6.0 m**

8. *Earth Science* Scientists in Goldstone, California, use a radio telescope to search for signs of life in space. This radio telescope has a circular dish with a diameter of 112 feet. What is the circumference of the radio dish? $\pi = \frac{22}{7}$; **352 ft**

EXERCISES

Practice

Find the circumference of each circle shown or described. Use $\frac{22}{7}$ or 3.14 for π. Round decimal answers to the nearest tenth.

Answers are calculated using 3.14 and then rounded unless otherwise noted.

9. 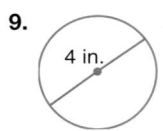 12.6 in. 4 in.

10.  $1\frac{3}{4}$ ft $\quad \pi = \frac{22}{7}$; **11 ft**

11. 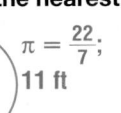 $2\frac{1}{3}$ in. $\quad \pi = \frac{22}{7}$; $7\frac{1}{3}$ in.

12. 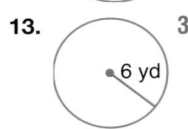 13.5 ft 4.3 ft

13. 37.7 yd 6 yd

14. 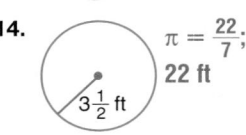 $\quad \pi = \frac{22}{7}$; 22 ft $3\frac{1}{2}$ ft

16. $\pi = \frac{22}{7}$; $29\frac{1}{3}$ in.

15. $d = 8$ ft **25.1 ft**

16. $r = 4\frac{2}{3}$ in.

17. $d = 3.1$ yd **9.7 yd**

18. Find the circumference of a circle with a radius of 9.6 inches. **60.3 in.**

19. Find the measure of the circumference of a circle that has a diameter of 16.8 centimeters. **52.8 cm**

20. The radius of a circle measures $5\frac{1}{4}$ feet. What is the measure of its circumference? **33.0 ft**

Reteaching the Lesson

Activity Provide small groups of students with objects such as paper plates, plastic lids, or CDs. First they should measure the diameter to figure out the circumference of each object. Then they should check their results by measuring the circumference.

Error Analysis
Watch for students who confuse the radius and diameter in a circle.
Prevent by reminding them that the word *radius* is shorter than the word *diameter*. So the radius is the shorter segment.

Applications and Problem Solving

21. $\pi = \frac{22}{7}$; $15\frac{5}{7}$ ft

21. *Transportation* A certain model of a "penny-farthing" bicycle had a large front wheel with a radius of $2\frac{1}{2}$ feet. How far would the bicycle travel on each rotation of the wheel?

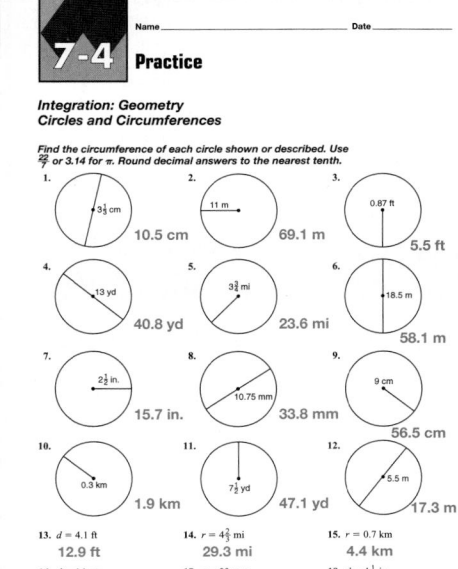

22. *Entertainment* The tallest Ferris wheel in the United States today is the Texas Star Ferris wheel at Fair Park in Dallas, Texas. Its diameter is about 212 feet. What is the difference between the circumference of this Ferris wheel and the Ferris wheel described in Example 2 on page 281? **about 119 feet**

23. The circumference is twice as long.

23. *Critical Thinking* Tell how the circumference of two circles compare if the diameter of one is twice as long as the diameter of the other.

Mixed Review

24. Find $1\frac{5}{7} \times 2\frac{5}{8}$. Write in simplest form. *(Lesson 7-3)* $4\frac{1}{2}$

25. *Algebra* Solve $x - \frac{1}{12} = \frac{5}{12}$. Write the solution in simplest form. *(Lesson 6-3)* $\frac{1}{2}$

26. **Test Practice** In July, an average of 209 books were checked out each day at the Strongville Public Library. Which is the best estimate for the number of books checked out for the whole month of July? *(Lesson 1-3)* **D**
   A 2,000   **B** 4,000   **C** 5,000   **D** 6,000   **E** 8,000

CHAPTER 7

Mid-Chapter Self Test

Estimate each product. *(Lesson 7-1)*

1. $17 \times \frac{3}{4}$ $16 \times \frac{3}{4} = 12$

2. $3\frac{4}{5} \times 6\frac{1}{7}$ $4 \times 6 = 24$

3. $2\frac{3}{4} \times \frac{2}{5}$ $3 \times \frac{1}{2} = 1\frac{1}{2}$

Find each product. *(Lessons 7-2 and 7-3)*

4. $\frac{1}{6} \times \frac{4}{5}$ $\frac{2}{15}$

5. $\frac{3}{10} \times \frac{5}{6}$ $\frac{1}{4}$

6. $\frac{1}{3} \times 4\frac{2}{7}$ $1\frac{3}{7}$

7. $2\frac{5}{8} \times 1\frac{2}{7}$ $3\frac{3}{8}$

Find the circumference of each circle shown or described. Use $\frac{22}{7}$ or 3.14 for π. Round decimal answers to the nearest tenth. *(Lesson 7-4)*

8. 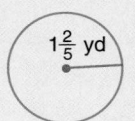 $1\frac{2}{5}$ yd $\pi = \frac{22}{7}$; $8\frac{4}{5}$ yd

9. $d = 13.6$ cm **42.7 cm**

10. *Pets* If a hamster wheel has a radius of 4 inches, how far does the hamster run in one rotation of the wheel? *(Lesson 7-4)* **25.1 in.**

Extending the Lesson

Enrichment Masters, p. 52

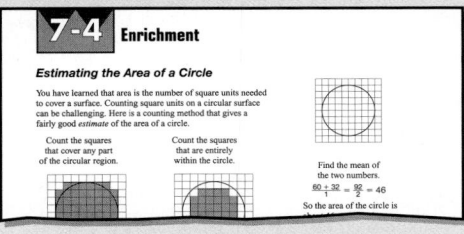

7-4 Enrichment

Estimating the Area of a Circle

You have learned that area is the number of square units needed to cover a surface. Counting square units on a circular surface can be challenging. Here is a counting method that gives a fairly good *estimate* of the area of a circle.

Count the squares that cover any part of the circular region.

Count the squares that are entirely within the circle.

Find the mean of the two numbers.
$\frac{60 + 32}{2} = \frac{92}{2} = 46$
So the area of the circle is

Activity Have pairs of students research the history of π. Have them find out, for example, how π came to be used and how the introduction of decimals in the 1600s affected its value.

Practice Masters, p. 52

Name _____ Date _____

7-4 **Practice**

Integration: Geometry
Circles and Circumferences

Find the circumference of each circle shown or described. Use $\frac{22}{7}$ or 3.14 for π. Round decimal answers to the nearest tenth.

1. $3\frac{1}{3}$ cm **10.5 cm**
2. 11 m **69.1 m**
3. 0.87 ft **5.5 ft**
4. 13 yd **40.8 yd**
5. $3\frac{3}{4}$ mi **23.6 mi**
6. 18.5 m **58.1 m**
7. $2\frac{1}{2}$ in. **15.7 in.**
8. 10.75 mm **33.8 mm**
9. 9 cm **56.5 cm**
10. 0.3 km **1.9 km**
11. $7\frac{1}{2}$ yd **47.1 yd**
12. 5.5 m **17.3 m**

13. $d = 4.1$ ft **12.9 ft**
14. $r = 4\frac{2}{3}$ mi **29.3 mi**
15. $r = 0.7$ km **4.4 km**
16. $d = 16$ cm **50.2 cm**
17. $r = 22$ mm **138.2 mm**
18. $d = 1\frac{1}{4}$ in. **3.9 in.**

© Glencoe/McGraw-Hill T52 Mathematics: Applications and Connections, Course 1

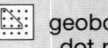
GET READY

Objective Students divide fractions by using a geoboard.

Optional Resources
Hands-On Lab Masters
• square dot paper, p. 13
• worksheet, p. 52

Overhead Manipulative Resources
• geoboard
• geobands

Manipulative Kit
• geoboard
• geobands

MANAGEMENT TIPS

Recommended Time
30 minutes

Getting Started Review using geoboards to multiply fractions by having students model $\frac{1}{8} \times \frac{1}{2}$. Then ask them how they would model $\frac{1}{2} \div \frac{1}{8}$.

The **Activity** uses a geoboard and geobands to model dividing fractions. After students have modeled the activity, ask the following questions.
• What is the area of the rectangle? **16 square units**
• How does the area relate to the fractions being divided? **16 is a multiple of 4 and 8.**
• Can you use other shapes to model division with fractions? **A rectangle with an area of 8 square units also could be used.**

ASSESS

After students complete Exercises 1–4, have pairs of students write a sentence or two explaining how geoboards can be used to model division with fractions.

Use Exercise 5 to determine whether students understand dividing fractions.

COOPERATIVE LEARNING

7-5A Dividing Fractions

A Preview of Lesson 7-5

 geoboard or dot paper

geobands

In Lesson 7-2A, you learned how to use a geoboard to multiply fractions. You can also use a geoboard to divide fractions.

TRY THIS

Work with a partner.

Step 1 Use geobands to make a rectangle as shown at the right. This rectangle represents 1.

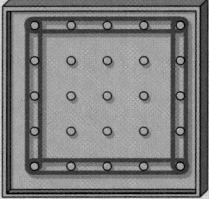

Step 2 Use geobands to show fourths. There are four $\frac{1}{4}$'s in 1. Therefore, $1 \div \frac{1}{4} = 4$.

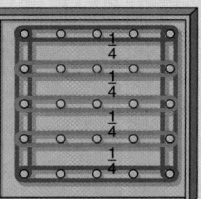

Step 3 Use geobands to show eighths. There are eight $\frac{1}{8}$'s in 1. Therefore, $1 \div \frac{1}{8} = 8$.

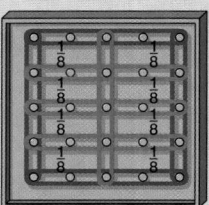

Examine the geoboard. What is $\frac{1}{4} \div \frac{1}{8}$? *Think: $\frac{1}{4} \div \frac{1}{8}$ means: How many $\frac{1}{8}$'s are in $\frac{1}{4}$?*

ON YOUR OWN

Use a geoboard and geobands to find each quotient.

1. $1 \div \frac{1}{2}$ **2**
2. $\frac{1}{2} \div \frac{1}{8}$ **4**
3. $\frac{2}{3} \div \frac{1}{6}$ **4**
4. $\frac{3}{4} \div \frac{1}{8}$ **6**

1–4. See Answer Appendix for geoboard models.

5. *Look Ahead* Find the quotient of $\frac{1}{3}$ and $\frac{1}{6}$ without using a geoboard. **2**

284 Chapter 7 Multiplying and Dividing Fractions

Math Journal Have students write a paragraph explaining how to divide fractions using a geoboard.

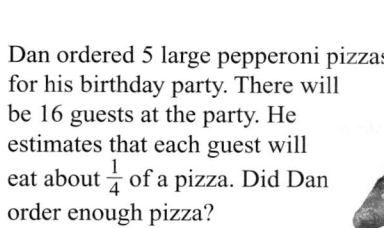

7-5 Dividing Fractions

What you'll learn

You'll learn to divide fractions.

When am I ever going to use this?

Knowing how to divide fractions can help you decide how much food to order for a party.

Word Wise
reciprocal

Dan ordered 5 large pepperoni pizzas for his birthday party. There will be 16 guests at the party. He estimates that each guest will eat about $\frac{1}{4}$ of a pizza. Did Dan order enough pizza?

You need to find the number of $\frac{1}{4}$ servings that are in 5 pizzas.

The model shows $5 \div \frac{1}{4}$.

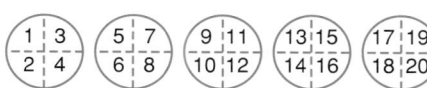

The dashed lines show that each pizza contains four $\frac{1}{4}$-pizza servings. Five pizzas contain 5 times 4, or 20 servings of pizza. That is, $5 \div \frac{1}{4} = 20$. So, Dan will have enough pizza.

$$5 \div \frac{1}{4} = 20 \Leftrightarrow 5 \times 4 = 20 \quad \textit{Dividing by } \frac{1}{4} \textit{ gives the same result as multiplying by 4.}$$

The numbers $\frac{1}{4}$ and 4 have a special relationship. Their product is 1.

$$\frac{1}{4} \times 4 = 1$$

Any two numbers whose product is 1 are called **reciprocals**.

Examples

Find the reciprocal of each number.

1 6

Since $6 \times \frac{1}{6} = 1$,

the reciprocal of 6 is $\frac{1}{6}$.

2 $\frac{3}{8}$

Since $\frac{3}{8} \times \frac{8}{3} = 1$,

the reciprocal of $\frac{3}{8}$ is $\frac{8}{3}$.

Study Hint

Mental Math The reciprocal of a number is found by "inverting" the fraction. That is, the numerator and the denominator are interchanged.

You can use reciprocals to divide fractions.

| **Dividing Fractions** | To divide by a fraction, multiply by its reciprocal. |

Lesson 7-5 Dividing Fractions **285**

Multiple Learning Styles

 Interpersonal Make a list of division examples and corresponding multiplication examples. Give the multiplication examples to small groups of students. Write the division examples, one at a time, on the overhead projector. Have the group with the corresponding multiplication example find the product.

7-5 Lesson Notes

Instructional Resources

- *Study Guide Masters*, p. 53
- *Practice Masters*, p. 53
- *Enrichment Masters*, p. 53
- Transparencies 7-5, A and B
- *Classroom Games*, pp. 19–20
- *Technology Masters*, p. 14

 CD-ROM Program
- Resource Lesson 7-5

Recommended Pacing	
Standard	Days 7 & 8 of 15
Honors	Day 6 of 14
Block	Day 5 of 8

1 FOCUS

 5-Minute Check
(Lesson 7-4)

Find the circumference of each circle described. Use $\frac{22}{7}$ or 3.14 for π. Round decimal answers to the nearest tenth.

1. $d = 4.2$ ft **13.2 ft**
2. $r = 3\frac{7}{8}$ in. **$24\frac{5}{14}$ in.**
3. $d = 12.8$ cm **40.2 cm**
4. $r = 5.7$ yd **35.8 yd**
5. $d = 7\frac{3}{11}$ ft **$22\frac{6}{7}$ ft**

The 5-Minute Check is also available on **Transparency 7-5A** for this lesson.

Motivating the Lesson

Hands-On Activity Have students work in small groups using measuring cups and water to determine:

- the number of $\frac{1}{2}$-cup servings in 3 cups of punch, **6**
- the number of $\frac{1}{4}$-cup servings in $\frac{7}{8}$-cup of sour cream. **$3\frac{1}{2}$**

Transparency 7-5B contains a teaching aid for this lesson.

Mastering Basic Skills Division of fractions requires mastery of multiplying fractions. While practicing division, students are also practicing multiplication and simplifying fractions.

Teaching Tip Before beginning the examples, remind students that all whole numbers can be written as fractions with a denominator of 1. Notice that as numbers get larger, their reciprocals get smaller, and vice versa. Emphasize that the reciprocal is the fraction flipped upside down.

In-Class Examples

For Example 1
Find the reciprocal of 13. $\frac{1}{13}$

For Example 2
Find the reciprocal of $\frac{7}{9}$. $\frac{9}{7}$

For Example 3
Find $\frac{3}{5} \div \frac{9}{10}$. $\frac{2}{3}$

For Example 4
Mr. Newman made a giant calzone that was $\frac{1}{2}$ of a yard long. He wants to give each of his four children an equal portion of the calzone. How long will each child's portion be? $4\frac{1}{2}$ in.

For Example 5
Solve the equation $\frac{2}{3} \div \frac{7}{11} = a$. $1\frac{1}{21}$

Example 3

Find $\frac{5}{8} \div \frac{3}{4}$.

Study Hint

Reading Math The expression $\frac{5}{8} \div \frac{3}{4}$ is read as five-eighths divided by three-fourths.

$$\frac{5}{8} \div \frac{3}{4} = \frac{5}{8} \times \frac{4}{3} \quad \textit{Multiply by the reciprocal of } \frac{3}{4}.$$
$$= \frac{5}{8} \times \frac{\overset{1}{4}}{3} \quad \textit{Divide 4 and 8 by the GCF, 4.}$$
$$= \frac{5}{2} \times \frac{1}{3} \quad \begin{array}{l}\textit{Multiply the numerators.}\\\textit{Multiply the denominators.}\end{array}$$
$$= \frac{5}{6}$$

Recall that division is related to multiplication. Sometimes it is helpful to write a multiplication sentence first and then write the related division sentence.

Example 4

APPLICATION

Food Mr. Tadashi had $\frac{3}{4}$ of a pan of lasagna left for dinner. He decided to divide the remaining lasagna into 6 equal pieces for his family. What part of the pan of lasagna will each person get?

Explore You know the number of pieces, 6, and the total amount of lasagna remaining, $\frac{3}{4}$ of a pan.

Plan

$$6 \quad \times \quad ? \quad = \quad \frac{3}{4} \quad \Leftrightarrow \quad \frac{3}{4} \quad \div \quad 6 \quad = \quad ?$$

| number of pieces | size of each piece | product or total | product or total | number of pieces | size of each piece |

Solve
$$\frac{3}{4} \div 6 = \frac{3}{4} \times \frac{1}{6} \quad \textit{Multiply by the reciprocal of 6.}$$
$$= \frac{\overset{1}{3}}{4} \times \frac{1}{\underset{2}{6}} \quad \textit{Divide 3 and 6 by the GCF, 3.}$$
$$= \frac{1}{4} \times \frac{1}{2} \quad \begin{array}{l}\textit{Multiply the numerators.}\\\textit{Multiply the denominators.}\end{array}$$
$$= \frac{1}{8}$$

Each person will get $\frac{1}{8}$ of a whole pan of lasagna.

Examine The number of pieces of lasagna times the size of each piece should equal $\frac{3}{4}$.

$$6 \times \frac{1}{8} = \frac{3}{4}$$
$$\frac{6}{8} = \frac{3}{4}$$
$$\frac{3}{4} = \frac{3}{4} \quad \checkmark$$

There are 8 parts and 6 are shaded. So, $\frac{6}{8}$ or $\frac{3}{4}$ of the lasagna was left.

Classroom Vignette

"I like to use geoboards when teaching multiplication and division of fractions. Use different colors of rubber bands to show the fractional parts and the whole. Visual displays, like the geoboard, help students with understanding and picturing fractions."

Mrs. Trudy C. Boyd, Teacher
Chapel Hill Middle School
Douglasville, GA

Trudy C. Boyd

$$\frac{3}{4} \div \frac{2}{5} = m$$

$$\frac{3}{4} \times \frac{5}{2} = m \qquad \textit{Multiply by the reciprocal of } \frac{2}{5}.$$

$$\frac{15}{8} = m \qquad \begin{array}{l}\textit{Multiply the numerators.}\\ \textit{Multiply the denominators.}\end{array}$$

$$1\frac{7}{8} = m$$

The solution is $1\frac{7}{8}$.

CHECK FOR UNDERSTANDING

Communicating Mathematics

Read and study the lesson to answer each question. 1–2. See margin.

1. **Explain** how you would use the reciprocal to find $\frac{1}{2} \div \frac{2}{3}$.

2. **Write** a sentence that tells how the model shows $2 \div \frac{1}{3} = 6$.

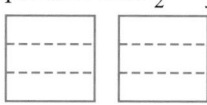

Guided Practice

Find the reciprocal of each number.

3. $\frac{1}{2}$ 2

4. $\frac{3}{5}$ $\frac{5}{3}$

5. 8 $\frac{1}{8}$

Find each quotient. Write in simplest form.

6. $\frac{1}{5} \div \frac{1}{4}$ $\frac{4}{5}$

7. $\frac{2}{3} \div 4$ $\frac{1}{6}$

8. $\frac{5}{8} \div \frac{1}{2}$ $1\frac{1}{4}$

9. $\frac{3}{4} \div \frac{2}{3}$ $1\frac{1}{8}$

Solve each equation. Write the solution in simplest form.

10. $\frac{1}{6} \div \frac{1}{2} = y$ $\frac{1}{3}$

11. $h = 8 \div \frac{2}{5}$ 20

12. $\frac{8}{9} \div \frac{1}{3} = a$ $2\frac{2}{3}$

13. **Algebra** What is the value of $m \div n$ if $m = \frac{2}{3}$ and $n = \frac{2}{5}$? $1\frac{2}{3}$

14. **Food** Each serving of a peach pie is $\frac{1}{12}$ of the pie. If $\frac{1}{2}$ of the peach pie is left, how many servings are left? **6 servings**

EXERCISES

Practice

Find the reciprocal of each number.

15. 5 $\frac{1}{5}$

16. $\frac{2}{5}$ $\frac{5}{2}$

17. $\frac{1}{3}$ 3

18. $\frac{5}{6}$ $\frac{6}{5}$

19. $\frac{1}{7}$ 7

20. 4 $\frac{1}{4}$

21. 1 1

22. $\frac{3}{8}$ $\frac{8}{3}$

Find each quotient. Write in simplest form.

23. $\frac{1}{3} \div \frac{3}{5}$ $\frac{5}{9}$

24. $\frac{5}{6} \div \frac{5}{8}$ $1\frac{1}{3}$

25. $\frac{3}{5} \div \frac{3}{4}$ $\frac{4}{5}$

26. $2 \div \frac{1}{6}$ 12

27. $\frac{2}{5} \div 4$ $\frac{1}{10}$

28. $\frac{1}{5} \div \frac{1}{6}$ $1\frac{1}{5}$

29. $\frac{2}{9} \div \frac{2}{3}$ $\frac{1}{3}$

30. $\frac{1}{2} \div \frac{1}{3}$ $1\frac{1}{2}$

31. $\frac{5}{8} \div 2$ $\frac{5}{16}$

---

3 PRACTICE/APPLY

Check for Understanding

If students need additional practice or instruction after completing Exercises 1–14, you may find one of the following options helpful.

- Extra Practice, see p. 576
- Reteaching Activity
- *Study Guide Masters,* p. 53
- *Practice Masters,* p. 53

Assignment Guide

Core: 15–43 odd, 45–50
Enriched: 16–42 even, 43–50

Additional Answers

1. Multiply $\frac{1}{2}$ by $\frac{3}{2}$.

2. Sample answer: There are 2 boxes and each box is divided into three equal parts. There is a total of 6 smaller sections. So, $2 \div \frac{1}{3} = 6$.

Study Guide Masters, p. 53

7-5 Study Guide

Dividing Fractions

Two numbers are **reciprocals** if their product is 1.

$\frac{1}{3}$ and 3 are reciprocals. $\frac{17}{18}$ and $\frac{18}{17}$ are reciprocals.

$\frac{1}{3} \times 3 = 1$ $\frac{17}{18} \times \frac{18}{17} = 1$

You use reciprocals to divide fractions.
To divide by a fraction, multiply by its reciprocal.

Example Find $\frac{5}{6} \div \frac{2}{3}$.

$\frac{5}{6} \div \frac{2}{3} = \frac{5}{6} \times \frac{3}{2}$ *Multiply by the reciprocal of $\frac{2}{3}$.*
$= \frac{5}{6} \times \frac{3}{2}$ *Divide 3 and 6 by the GCF, 3.*
$= \frac{5}{2} \times \frac{1}{4} = \frac{5}{4}$ *Multiply the numerators. Multiply the denominators.*
$= 1\frac{1}{4}$

Find the reciprocal of each number.

1. $\frac{3}{4}$ $\frac{4}{3}$

2. $\frac{5}{8}$ $\frac{8}{5}$

3. 9 $\frac{1}{9}$

4. $\frac{12}{13}$ $\frac{13}{12}$

Find each quotient. Write in simplest form.

5. $\frac{1}{2} \div \frac{3}{4}$ $\frac{2}{3}$

6. $\frac{4}{5} \div \frac{1}{10}$ 8

7. $\frac{3}{8} \div \frac{3}{4}$ $\frac{1}{2}$

8. $\frac{7}{9} \div \frac{1}{3}$ $2\frac{1}{3}$

9. $\frac{14}{15} \div 7$ $\frac{2}{15}$

10. $\frac{5}{12} \div \frac{5}{6}$ $\frac{1}{2}$

11. $\frac{9}{10} \div 3$ $\frac{3}{10}$

12. $\frac{12}{13} \div \frac{1}{3}$ $3\frac{9}{13}$

Solve each equation. Write the solution in simplest form.

13. $8 \div \frac{1}{2} = a$ 16

14. $x = \frac{3}{5} \div \frac{9}{10}$ $\frac{2}{3}$

15. $\frac{5}{9} \div \frac{5}{6} = w$ $\frac{2}{3}$

16. $m = \frac{11}{12} \div 6$ $\frac{11}{72}$

© Glencoe/McGraw-Hill T53 Mathematics: Applications and Connections, Course 1

■ Reteaching the Lesson ■

Activity Have students draw four rectangles of the same size on centimeter grid paper. Ask them to divide each rectangle into thirds. Then have them color pairs of thirds in different colors—red, blue, green, and so on—until all thirds are colored. Ask how many two-thirds there are in the 4 rectangles. **6** Have students write the division problem.

Error Analysis

Watch for students who forget to multiply by the reciprocal of the divisor.

Prevent by using whole numbers to demonstrate that multiplication can be used to solve a division problem such as $6 \div 3 = 2$ only if the reciprocal of the divisor is used ($6 \times \frac{1}{3} = 2$).

Closing Activity

Writing Have students write directions for a person who needs to divide $\frac{3}{4}$ of a pound of cheese into $\frac{1}{16}$-pound slices.

---

32. If you divide $\frac{5}{6}$ by $\frac{1}{4}$, what is the quotient? $3\frac{1}{3}$

Solve each equation. Write the solution in simplest form.

33. $\frac{2}{3} \div \frac{3}{4} = m$ $\frac{8}{9}$

34. $k = \frac{3}{4} \div 12$ $\frac{1}{16}$

35. $\frac{1}{5} \div \frac{3}{10} = z$ $\frac{2}{3}$

36. $d = \frac{1}{4} \div \frac{1}{8}$ 2

37. $\frac{3}{4} \div \frac{5}{6} = v$ $\frac{9}{10}$

38. $a = \frac{1}{3} \div \frac{1}{2}$ $\frac{2}{3}$

39. $\frac{8}{9} \div 4 = h$ $\frac{2}{9}$

40. $x = \frac{5}{9} \div \frac{2}{3}$ $\frac{5}{6}$

41. $3 \div \frac{6}{7} = w$ $3\frac{1}{2}$

42. *Algebra* Find the value of $a \div b$ if $a = \frac{1}{2}$ and $b = 9$. $\frac{1}{18}$

Applications and Problem Solving

43. *Money Matters* Mr. Barojas parked his car at a 4-hour parking meter. If each half-hour of parking costs 25 cents, how many quarters does he need for 4 hours of parking? **8 quarters**

44. *Geography* The table lists the continents' approximate sizes relative to Earth's total landmass.

a. About how many times larger is North America than South America? $1\frac{1}{3}$

b. About how many times larger is Asia than North America? $1\frac{4}{5}$

c. About how many times larger is Asia than Africa? $1\frac{1}{2}$

45. *Critical Thinking* If $\frac{14}{15} \div \frac{x}{9} = \frac{3}{10}$, what is the value of x? **28**

Continent	Fraction of Earth's Landmass
Asia	$\frac{3}{10}$
Africa	$\frac{1}{5}$
North America	$\frac{1}{6}$
South America	$\frac{1}{8}$
Antarctica	$\frac{1}{10}$
Europe	$\frac{1}{16}$
Australia	$\frac{1}{20}$

Mixed Review

46. *Geometry* Find the circumference of a circle with a radius of 4 meters. *(Lesson 7-4)* **25.1 m**

47. Express $\frac{7}{9}$ as a decimal. Use bar notation to show a repeating decimal. *(Lesson 5-10)* $0.\overline{7}$

48. **Test Practice** Warren's dog weighs 25.9 kilograms. How many grams does the dog weigh? *(Lesson 4-9)* **A**

A 25,900 g
B 0.0259 g
C 1025.9 g
D 259,000 g

interNET CONNECTION
For the latest weather statistics, visit:
www.glencoe.com/sec/math/mac/mathnet

49. about 4 inches

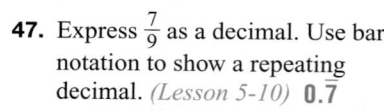

49. *Statistics* The average annual precipitation in Georgia is 48.61 inches. About how many inches of precipitation does Georgia average each month? *(Lesson 4-5)*

50. Add 15.783 and 390.81. *(Lesson 3-6)* **406.593**

288 Chapter 7 Multiplying and Dividing Fractions

Practice Masters, p. 53

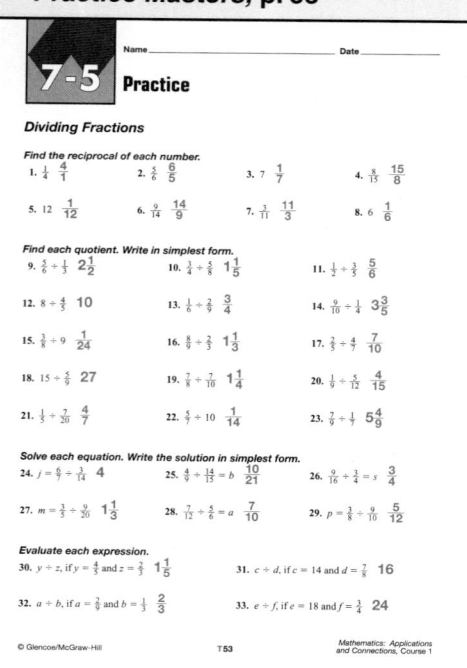

Extending the Lesson

Enrichment Masters, p. 53

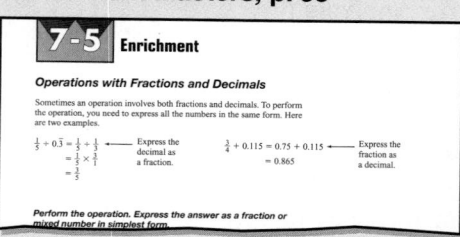

Activity Using two different colors of fraction cubes, have pairs of students take turns rolling the cubes. Using one color as the divisor and the other color as the dividend, students should write out and solve the expression. They should continue until they get a whole number quotient or until a predetermined time runs out.

What you'll learn

You'll learn to divide mixed numbers.

When am I ever going to use this?

Knowing how to divide mixed numbers can help you find the average speed you travel on a trip.

A tsunami, (soo-NAH-mee), or tidal wave, can travel from one side of the Pacific Ocean to the other in less than a day. If a tsunami traveled 1,400 miles from a point in the Pacific Ocean to the Alaskan coastline in $2\frac{1}{2}$ hours, how many miles per hour did it travel?

You need to find $1,400 \div 2\frac{1}{2}$.
To divide mixed numbers, express each mixed number as an improper fraction. Then divide as with fractions.

$$1,400 \div 2\frac{1}{2} = \frac{1,400}{1} \div \frac{5}{2} \quad \textit{Express each mixed number as an improper fraction.}$$

$$= \frac{1,400}{1} \times \frac{2}{5} \quad \textit{Multiply by the reciprocal.}$$

$$= \frac{\overset{280}{\cancel{1,400}}}{1} \times \frac{2}{\underset{1}{\cancel{5}}} \quad \textit{Divide 5 and 1,400 by the GCF, 5.}$$

$$= 560$$

The tsunami traveled 560 miles per hour.

Examples

1 Find $4\frac{1}{2} \div 3\frac{3}{4}$. *Estimate: $4 \div 4 = 1$*

$$4\frac{1}{2} \div 3\frac{3}{4} = \frac{9}{2} \div \frac{15}{4} \quad \textit{Express each mixed number as an improper fraction.}$$

$$= \frac{9}{2} \times \frac{4}{15} \quad \textit{Multiply by the reciprocal.}$$

$$= \frac{\overset{3}{\cancel{9}}}{\underset{1}{2}} \times \frac{\overset{2}{\cancel{4}}}{\underset{5}{\cancel{15}}} \quad \begin{array}{l}\textit{Divide 9 and 15 by the GCF, 3.}\\ \textit{Divide 2 and 4 by the GCF, 2.}\end{array}$$

$$= \frac{6}{5} \text{ or } 1\frac{1}{5} \quad \textit{Simplify. Compare with your estimate.}$$

INTEGRATION **2** **Algebra** Solve $w = 2\frac{4}{5} \div \frac{7}{8}$. *Estimate: $3 \div 1 = 3$*

$$w = \frac{14}{5} \div \frac{7}{8} \quad \textit{Express each mixed number as an improper fraction.}$$

$$w = \frac{\overset{2}{\cancel{14}}}{5} \times \frac{8}{\underset{1}{\cancel{7}}} \quad \begin{array}{l}\textit{Multiply by the reciprocal.}\\ \textit{Divide 7 and 14 by the GCF, 7.}\end{array}$$

$$w = \frac{16}{5} \text{ or } 3\frac{1}{5} \quad \textit{Simplify. Compare with your estimate.}$$

Multiple Learning Styles

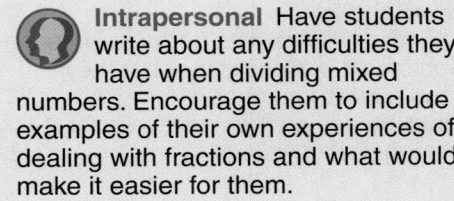 **Intrapersonal** Have students write about any difficulties they have when dividing mixed numbers. Encourage them to include examples of their own experiences of dealing with fractions and what would make it easier for them.

Motivating the Lesson

Problem Solving Ask students how they would solve the following problem. *A square brownie has an area of $1\frac{1}{2}$ square inches. How many brownies could be made in a pan having an area of $114\frac{3}{4}$ square inches?* $76\frac{1}{2}$ **brownies**

Instructional Resources
- *Study Guide Masters*, p. 54
- *Practice Masters*, p. 54
- *Enrichment Masters*, p. 54
- Transparencies 7-6, A and B
- *Assessment and Evaluation Masters*, p. 184
- CD-ROM Program
  - Resource Lesson 7-6

Recommended Pacing

Standard	Day 9 of 15
Honors	Day 7 of 14
Block	Day 6 of 8

1 FOCUS

 5-Minute Check
(Lesson 7-5)

1. Find the reciprocal of $\frac{4}{5}$. $\frac{5}{4}$

Find each quotient. Write in simplest form.

2. $\frac{3}{4} \div \frac{1}{5}$ $3\frac{3}{4}$

3. $\frac{1}{6} \div \frac{2}{3}$ $\frac{1}{4}$

4. $5 \div \frac{1}{4}$ **20**

5. Evaluate $r \div s$, if $r = \frac{3}{5}$ and $s = 6$. $\frac{1}{10}$

 The 5-Minute Check is also available on **Transparency 7-6A** for this lesson.

2 TEACH

 Transparency 7-6B contains a teaching aid for this lesson.

Thinking Algebraically Some students may make the connection that dividing and multiplying are inverse operations. Give them a simple equation such as $\frac{a}{2} = 6$ and ask them to find a. Show them that $\frac{a}{2} = 6$ is the same as $a \cdot \frac{1}{2} = 6$, which can also be expressed as the related equation $a = 6 \div \frac{1}{2}$.

For Example 1

a. Find $4\frac{1}{8} \div 1\frac{3}{4}$. $2\frac{5}{14}$

b. Find $5\frac{5}{6} \div 2\frac{5}{12}$. $2\frac{12}{29}$

For Example 2

a. Solve $q = 12\frac{1}{2} \div 3\frac{3}{5}$. $3\frac{17}{36}$

b. Solve $j = 8\frac{3}{4} \div 2\frac{1}{7}$. $4\frac{1}{12}$

Teaching Tip If students have difficulty dividing common factors before multiplying, they can simplify the fraction after multiplying.

3 PRACTICE/APPLY

Check for Understanding

If students need additional practice or instruction after completing Exercises 1–13, you may find one of the following options helpful.
- Extra Practice, see p. 576
- Reteaching Activity
- *Study Guide Masters*, p. 54
- *Practice Masters*, p. 54

Assignment Guide

Core: 15–41 odd, 43–50
Enriched: 14–38 even, 40, 41, 43–50

Study Guide Masters, p. 54

Name_____ Date_____

7-6 Study Guide

Dividing Mixed Numbers

To divide mixed numbers, express each mixed number as an improper fraction. Then divide as with fractions.

Example Solve $m = 2\frac{5}{8} \div 1\frac{3}{4}$. *Estimate:* $2 \div 2 = 1$

$m = \frac{21}{8} \div \frac{7}{4}$ Express each mixed number as an improper fraction.

$m = \frac{^3\cancel{21}}{\cancel{8}_2} \times \frac{\cancel{4}^1}{\cancel{7}_1}$ Multiply by the reciprocal. Divide 21 and 7 by the GCF, 7. Divide 4 and 8 by the GCF, 4.

$m = \frac{3}{2} \times \frac{1}{1}$ Simplify.

$m = \frac{3}{2}$ or $1\frac{1}{2}$ Compare with your estimate.

Find each quotient. Write in simplest form.

1. $2\frac{1}{2} \div \frac{4}{5}$ $3\frac{1}{8}$
2. $1\frac{2}{3} \div 1\frac{1}{4}$ $1\frac{1}{3}$
3. $5 \div 1\frac{3}{7}$ $3\frac{1}{2}$
4. $2\frac{1}{3} \div \frac{7}{9}$ 3
5. $5\frac{2}{3} \div \frac{9}{10}$ 6
6. $7\frac{1}{2} \div 1\frac{3}{5}$ $4\frac{1}{2}$

Solve each equation. Write the solution in simplest form.

7. $n = 10\frac{1}{2} \div \frac{7}{10}$ 15
8. $3\frac{1}{3} \div 10 = p$ $\frac{9}{25}$
9. $r = 6\frac{3}{4} \div 2\frac{1}{2}$ 3
10. $15 \div 3\frac{1}{2} = t$ $4\frac{2}{7}$
11. $6\frac{2}{3} \div 1\frac{2}{3} = c$ 4
12. $h = 2\frac{11}{12} \div 5$ $\frac{5}{12}$
13. $m = 18 \div \frac{9}{11}$ 22
14. $r = 4\frac{4}{5} \div \frac{8}{15}$ 9
15. $6\frac{3}{4} \div 1\frac{1}{8} = k$ 6

© Glencoe/McGraw-Hill T·54 Mathematics: Applications and Connections, Course 1

290 Chapter 7

CHECK FOR UNDERSTANDING

Communicating
Mathematics
1–3. See Answer
Appendix.

Math Journal

Read and study the lesson to answer each question.

1. **Explain** to a classmate how to find the reciprocal of $4\frac{5}{8}$.

2. **Tell** whether $4 \div \frac{1}{2}$ is more or less than 4. Explain your reasoning.

3. **Write**, in your own words, the steps to follow when finding the quotient of 12 and $2\frac{2}{3}$.

Guided Practice

Write each mixed number as an improper fraction. Then write its reciprocal.

4. $2\frac{1}{4}$ $\frac{9}{4}, \frac{4}{9}$

5. $7\frac{2}{3}$ $\frac{23}{3}, \frac{3}{23}$

6. $6\frac{1}{3}$ $\frac{19}{3}, \frac{3}{19}$

Find each quotient. Write in simplest form.

7. $8 \div 2\frac{1}{2}$ $3\frac{1}{5}$

8. $9\frac{1}{2} \div 4\frac{3}{4}$ 2

9. $3\frac{3}{4} \div \frac{5}{6}$ $4\frac{1}{2}$

Solve each equation. Write the solution in simplest form.

10. $6\frac{1}{4} \div \frac{1}{2} = g$ $12\frac{1}{2}$

11. $5\frac{1}{3} \div 4\frac{2}{3} = n$ $1\frac{1}{7}$

12. If $9\frac{1}{2}$ is divided by $\frac{1}{2}$, what is the quotient? 19

13. **Food** Mrs. Golubec needs to cut a zucchini into slices that measure $\frac{3}{8}$ inch thick. If the zucchini measures $13\frac{1}{2}$ inches long, how many slices can she cut? **36 slices**

EXERCISES

Practice

Write each mixed number as an improper fraction. Then write its reciprocal.

14. $4\frac{1}{2}$ $\frac{9}{2}, \frac{2}{9}$

15. $6\frac{2}{5}$ $\frac{32}{5}, \frac{5}{32}$

16. $1\frac{3}{8}$ $\frac{11}{8}, \frac{8}{11}$

17. $2\frac{4}{5}$ $\frac{14}{5}, \frac{5}{14}$

18. $5\frac{5}{6}$ $\frac{35}{6}, \frac{6}{35}$

19. $3\frac{3}{4}$ $\frac{15}{4}, \frac{4}{15}$

20. $7\frac{1}{2}$ $\frac{15}{2}, \frac{2}{15}$

21. $9\frac{1}{4}$ $\frac{37}{4}, \frac{4}{37}$

Find each quotient. Write in simplest form.

22. $7\frac{1}{3} \div 6$ $1\frac{2}{9}$

23. $4\frac{3}{4} \div \frac{5}{8}$ $7\frac{3}{5}$

24. $5\frac{1}{4} \div 3\frac{1}{2}$ $1\frac{1}{2}$

25. $10\frac{1}{2} \div \frac{7}{8}$ 12

26. $1\frac{5}{9} \div 2\frac{1}{3}$ $\frac{2}{3}$

27. $5 \div 6\frac{1}{4}$ $\frac{4}{5}$

28. $2\frac{4}{5} \div 5\frac{3}{5}$ $\frac{1}{2}$

29. $10 \div 2\frac{2}{7}$ $4\frac{3}{8}$

30. $13 \div 2\frac{3}{5}$ 5

Solve each equation. Write the solution in simplest form.

31. $3 \div 4\frac{1}{2} = t$ $\frac{2}{3}$

32. $a = 1\frac{3}{8} \div 2\frac{3}{4}$ $\frac{1}{2}$

33. $6\frac{1}{2} \div \frac{1}{4} = m$ 26

34. $y = 4\frac{4}{7} \div 2\frac{2}{3}$ $1\frac{5}{7}$

35. $\frac{4}{5} \div 1\frac{1}{5} = q$ $\frac{2}{3}$

36. $c = 5\frac{1}{3} \div 4$ $1\frac{1}{3}$

290 Chapter 7 Multiplying and Dividing Fractions

Reteaching the Lesson

Activity Have two pairs of students play "Check." Have Team 1 give Team 2 a mixed number division problem to solve. Have Team 1 use estimation and a fraction calculator to check Team 2's answer. When the answer checks, Team 2 then gives Team 1 a problem, and so on.

Error Analysis

Watch for students who forget to write the reciprocal of the divisor after they rename a mixed number.

Prevent by having them estimate the quotient before dividing.

37. *Algebra* What is the value of $m \div n$ if $m = 6\frac{2}{3}$ and $n = \frac{4}{5}$? $8\frac{1}{3}$

38. *Algebra* Evaluate $p \div q$ if $p = 9$ and $q = 3\frac{3}{5}$. $2\frac{1}{2}$

39. *Algebra* If $y = 1\frac{1}{6}$ and $z = 4\frac{3}{8}$, what is the value of $y \div z$? $\frac{4}{15}$

Applications and Problem Solving

40. *Statistics* The table shows how many hours Tiarri watched cartoons each day after school. What is the mean number of hours Tiarri watched cartoons in the 5 days? $1\frac{1}{10}$ **hours**

Day	Hours
Monday	$1\frac{1}{4}$
Tuesday	$\frac{1}{2}$
Wednesday	$1\frac{1}{2}$
Thursday	$\frac{1}{2}$
Friday	$1\frac{3}{4}$

Sylvester and Tweety Bird and other Looney Tunes characters and names and related indicia are registered trademarks of Warner Bros., a Time Warner Entertainment Company, ™ and © 1998. All rights reserved.

41. *Measurement* Marnee is working on her history project. She wants to place photographs of people from various countries in one vertical row on a poster board that is $17\frac{1}{2}$ inches long. If each photograph is $2\frac{3}{4}$ long, how many photographs can Marnee place on the poster board? **6 photographs**

42. *Working on the* CHAPTER Project Refer to the table on page 267.
   a. Use the information from Mitiku's first flight to determine how long it would take his airplane to fly 100 miles. **about 26 hours**
   b. Use the information from your first flight to determine how long it would take your airplane to fly 100 miles. **See students' work.**

43. *Critical Thinking* Tell whether $\frac{8}{10} \div \frac{2}{3}$ is greater than or less than $\frac{8}{10} \div \frac{3}{4}$ without solving. Explain your reasoning. **Greater than; $\frac{2}{3}$ is less than $\frac{3}{4}$.**

Mixed Review

44. Find $\frac{3}{7} \div \frac{3}{5}$. Write in simplest form. *(Lesson 7-5)* $\frac{5}{7}$

45. Find $\frac{5}{7} \times \frac{2}{3}$. Write in simplest form. *(Lesson 7-2)* $\frac{10}{21}$

46. *Algebra* Solve $y - \frac{7}{12} = \frac{5}{12}$. *(Lesson 6-3)* $\frac{12}{12}$ or 1

47. **Test Practice** What is the GCF of 96 and 172? *(Lesson 5-3)* **B**
   A 2   **B** 4   **C** 24   **D** 43

48. *Geometry* What is the area of a rectangle with a length of 12.8 feet and a width of 7.4 feet? *(Lesson 4-4)* **94.7 sq ft**

49. *Life Science* A 150-pound human body contains about 0.2 ounce of iron, 0.07 ounce of zinc, and 0.004 ounce of copper. Write each of these decimals in words. *(Lesson 3-1)* **See margin.**

50. *Technology* Bob's answering machine can record fifty 30-second messages. How many minutes of tape does his machine have? *(Lesson 1-1)* **25 minutes**

Lesson 7-6 Dividing Mixed Numbers **291**

Extending the Lesson

Enrichment Masters, p. 54

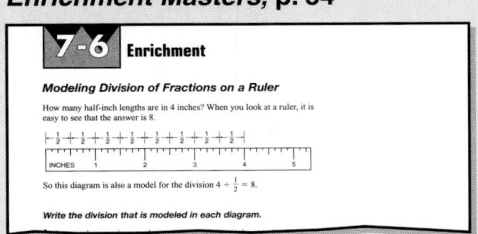

7-6 Enrichment

Modeling Division of Fractions on a Ruler

How many half-inch lengths are in 4 inches? When you look at a ruler, it is easy to see that the answer is 8.

So this diagram is also a model for the division $4 \div \frac{1}{2} = 8$.

Write the division that is modeled in each diagram.

Activity Ask students to design a shipping crate for a floppy disk case that is $4\frac{1}{8}$ in. long by 2 in. wide by $4\frac{5}{8}$ in. high. Use these guidelines: (1) The wood used for the crate will be $5\frac{1}{4}$ in. wide, $\frac{1}{4}$ in. thick, and 8 ft long; (2) The design must have a minimum amount of waste; (3) The crate must hold 12 cases.

CHAPTER Project

Exercise 42 asks students to advance to the next stage of work on the Chapter Project. You may wish to review converting miles to feet.

4 ASSESS

Closing Activity

Modeling Have students work in pairs to illustrate a division problem with blocks, a geoboard, or dot paper and colored pencils.

Chapter 7, Quiz C (Lessons 7-5 and 7-6) is available in the *Assessment and Evaluation Masters,* p. 184.

Additional Answer

49. two tenths; seven hundredths; four thousandths

Practice Masters, p. 54

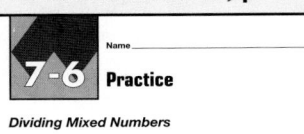

Name _____ Date _____

7-6 **Practice**

Dividing Mixed Numbers

Write each mixed number as an improper fraction. Then write its reciprocal.

1. $8\frac{3}{4}$ $\frac{35}{4}$, $\frac{4}{35}$ 2. $9\frac{6}{7}$ $\frac{69}{7}$, $\frac{7}{69}$ 3. $7\frac{5}{6}$ $\frac{47}{6}$, $\frac{6}{47}$

4. $3\frac{5}{12}$ $\frac{41}{12}$, $\frac{12}{41}$ 5. $1\frac{7}{16}$ $\frac{23}{16}$, $\frac{16}{23}$ 6. $6\frac{7}{8}$ $\frac{55}{8}$, $\frac{8}{55}$

Find each quotient. Write in simplest form.

7. $4 \div 2\frac{2}{5}$ $1\frac{2}{3}$ 8. $3\frac{1}{4} \div 1\frac{3}{8}$ $2\frac{4}{11}$ 9. $\frac{8}{9} \div 5\frac{1}{3}$ $\frac{1}{6}$

10. $2\frac{1}{3} \div 4\frac{4}{7}$ $\frac{7}{12}$ 11. $3\frac{1}{9} \div 7$ $\frac{4}{9}$ 12. $6\frac{2}{3} \div 4\frac{4}{9}$ $1\frac{7}{18}$

13. $2\frac{1}{7} \div \frac{3}{14}$ 10   14. $3\frac{3}{5} \div 2\frac{2}{5}$ $1\frac{2}{5}$ 15. $9 \div 3\frac{1}{2}$ $2\frac{5}{8}$

16. $1\frac{5}{9} \div 1\frac{5}{6}$ $\frac{2}{3}$ 17. $\frac{7}{10} \div 2\frac{5}{8}$ $\frac{4}{15}$ 18. $3\frac{1}{3} \div 1\frac{7}{9}$ $1\frac{4}{5}$

19. $1\frac{3}{4} \div 14$ $\frac{1}{8}$ 20. $2\frac{7}{15} \div 3\frac{5}{9}$ $\frac{3}{5}$ 21. $2\frac{1}{10} \div \frac{7}{8}$ $2\frac{2}{5}$

22. $6\frac{1}{4} \div 1\frac{7}{20}$ 5   23. $18 \div 1\frac{1}{8}$ 16   24. $4\frac{1}{6} \div 1\frac{3}{4}$ $2\frac{11}{12}$

Solve each equation. Write the solution in simplest form.

25. $\frac{5}{12} \div 2\frac{1}{2} = p$ $\frac{1}{6}$ 26. $s = 2\frac{2}{3} \div 1\frac{5}{6}$ $1\frac{5}{11}$ 27. $a = 1\frac{4}{5} \div 6$ $\frac{3}{10}$

28. $k = 2\frac{2}{3} \div 1\frac{7}{9}$ $1\frac{7}{20}$ 29. $1\frac{1}{6} \div \frac{5}{18} = d$ $4\frac{1}{5}$ 30. $1\frac{3}{5} \div 3\frac{3}{5} = x$ $\frac{56}{125}$

Evaluate each expression.

31. $f \div g$, if $f = 5\frac{1}{4}$ and $g = 1\frac{5}{9}$ $3\frac{3}{8}$ 32. $w \div z$, if $w = 9$ and $x = 2\frac{1}{5}$ $4\frac{1}{5}$

33. $t \div v$, if $t = 6\frac{1}{2}$ and $v = 1\frac{7}{8}$ $3\frac{7}{15}$ 34. $j \div k$, if $j = 12$ and $k = 2\frac{4}{5}$ $4\frac{2}{7}$

© Glencoe/McGraw-Hill   54   Mathematics: Applications and Connections, Course 1

Lesson 7-6 **291**

Instructional Resources

- *Study Guide Masters*, p. 55
- *Practice Masters*, p. 55
- *Enrichment Masters*, p. 55
- Transparencies 7-7, A and B

CD-ROM Program
- Resource Lesson 7-7

Recommended Pacing	
Standard	Days 10 & 11 of 15
Honors	Days 8 & 9 of 14
Block	Day 6 of 8

1 FOCUS

5-Minute Check
(Lesson 7-6)

Find each quotient. Write in simplest form.

1. $5 \div 2\frac{1}{3}$ $2\frac{1}{7}$
2. $\frac{14}{15} \div 1\frac{2}{3}$ $\frac{14}{25}$
3. $6\frac{2}{5} \div 3\frac{1}{3}$ $1\frac{23}{25}$

Solve each equation. Write the solution in simplest form.

4. $3\frac{5}{6} \div \frac{1}{3} = v$ $11\frac{1}{2}$
5. $x = 4\frac{2}{3} \div 5\frac{1}{4}$ $\frac{8}{9}$

The 5-Minute Check is also available on **Transparency 7-7A** for this lesson.

Motivating the Lesson

Communication Ask students to read the opening paragraph of the lesson. Ask the following questions:
- What units of measure are mentioned? **quarts and gallons**
- Can you think of goods around your house that come in quarts or gallons? **Sample answers: ice cream, laundry detergent, oil, milk, paint**

7-7

What you'll learn

You'll learn to change units within the customary system.

When am I ever going to use this?

You'll often need to change customary units when you are comparing labels in the grocery store.

Word Wise

fluid ounce	gallon
cup	ounce
pint	pound
quart	ton

Cultural Kaleidoscope

Various Asian countries use trained elephants in the logging industry to carry heavy loads on their back or with their trunk.

Integration: Measurement
Changing Customary Units

Kijana (meaning "little boy" in Swahili) is the first African elephant since 1984 to survive birth in captivity. His mother rejected him so the keepers at the Oakland Zoo in California decided to raise him. Kijana drinks about 25 quarts of formula a day. How many gallons of formula does he drink each day?

You need to find out how many gallons are in 25 quarts. It takes 4 quarts to make 1 gallon. Since you need to find out how many sets of 4 quarts there are in 25 quarts, divide:

$$25 \div 4 = 6\frac{1}{4}$$

$$\begin{array}{r} 6 \\ 4\overline{)25} \\ -24 \\ \hline 1 \end{array}$$

Kijana drinks $6\frac{1}{4}$ gallons of formula a day.

The most commonly used customary units of capacity are the **fluid ounce**, **cup**, **pint**, **quart**, and **gallon**.

1 cup (c) = 8 fluid ounces (fl oz)
1 pint (pt) = 2 cups
1 quart (qt) = 2 pints
1 gallon (gal) = 4 quarts

The most commonly used customary units of weight are **ounce**, **pound**, and **ton**.

1 pound (lb) = 16 ounces (oz)
1 ton (T) = 2,000 pounds

To change customary units of capacity and weight,

1. Determine whether you are changing from smaller to larger units or from larger to smaller units.
2. To change from smaller to larger units, divide. To change from larger to smaller units, multiply.

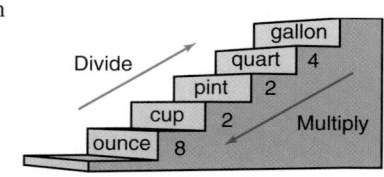

To solve Example 1, you can use the ratio of 2 cups to 1 pint.

1 5 pt = _?_ c   *Think: Each pint equals 2 cups.*

5 × 2 = 10   *Multiply to change from a larger unit (pt) to smaller unit (c)*

5 pt = 10 c

2 5,000 lb = _?_ T   *Think: It takes 2,000 pounds to make 1 ton.*

$5,000 ÷ 2,000 = 2\frac{1}{2}$ *Divide to change from smaller units (lb) to larger units (T).*

$5,000 \text{ lb} = 2\frac{1}{2} \text{ T}$

3 9 gal = _?_ pt   *Think: Each gallon equals 4 quarts. Each quart equals 2 pints. You need to multiply twice.*

9 × 4 = 36   *Multiply to change from gallons to quarts.*

36 × 2 = 72   *Multiply to change from quarts to pints.*

9 gal = 72 pt

CONNECTION **4** **Life Science** Miquel estimates that the finches eat 8 ounces of bird seed a day at his feeder. If he buys a 25-pound bag of bird seed, about how many days will it last?

Explore   You know how much the finches eat each day. You need to find how many days a 25-pound bag of bird seed will last.

Plan   First, find the total number of ounces in 25 pounds. Then find how many sets of 8 ounces there are in the 25-pound bag of bird seed.

Solve   25 lb = _?_ oz   *Think: Each pound equals 16 ounces.*

25 × 16 = 400   *Multiply to change from pounds to ounces.*

There are 400 ounces in a 25-pound bag of bird seed.

400 ÷ 8 = 50

So, the bag of bird seed will last about 50 days.

Examine   To see if your answer makes sense, think 8 × 50 = 400. So, 50 days is a reasonable answer.

CHECK FOR UNDERSTANDING

Communicating Mathematics

Read and study the lesson to answer each question.

1. *List* the common customary units of capacity and weight. **See margin.**

2. *Tell* how to change 3 tons to pounds. **Multiply 3 by 2,000.**

Lesson 7-7 Integration: Measurement   Changing Customary Units   **293**

■ Reteaching the Lesson ■

Activity Have pairs of students fill containers with water and measure the contents in fluid ounces, cups, pints, quarts, or gallons as appropriate. Have them record their findings in a chart. Use a scale and a similar recording method to weigh various sized objects in ounces and pounds.

Additional Answer

1. fluid ounce, cup, pint, quart, gallon, ounce, pound, and ton

---

2 TEACH

Transparency 7-7B contains a teaching aid for this lesson.

Reading Mathematics Students need to readily recognize the symbol for each unit of measurement when reading problems. Part of the understanding of these units should be to determine which unit is smaller than the other.

In-Class Examples

For Example 1
3 gal = _?_ qt **12**

For Example 2
96 oz = _?_ lb **6**

For Example 3
5 qt = _?_ c **20**

For Example 4
Lamont is bringing refreshments to a reception for 200 people. He estimates one cup of drink for each person. How many gallons of juice should Lamont buy? $12\frac{1}{2}$ **gal**

Study Guide Masters, p. 55

7-7 Study Guide

Integration: Measurement
Changing Customary Units

Customary Units	
Weight	**Liquid Capacity**
1 pound (lb) = 16 ounces (oz)	1 cup (c) = 8 fluid ounces (fl oz)
1 ton (T) = 2,000 pounds	1 pint (pt) = 2 cups
	1 quart (qt) = 2 pints
	1 gallon (gal) = 4 quarts

To change from larger units to smaller units, multiply.

Example 1 $4\frac{1}{2}$ lb = _____ oz   *Think: Each pound equals 16 ounces.*

  $4\frac{1}{2}$ × 16 = 72   *Multiply to change from pounds to ounces.*

  $4\frac{1}{2}$ lb = 72 oz

To change from smaller units to larger units, divide.

Example 2 700 qt = _____ gal   *Think: It takes 4 quarts to make 1 gallon.*

  700 ÷ 4 = 175   *Divide to change from quarts to gallons.*

  700 qt = 175 gal

Complete.

1. 2 lb = _____ oz **32**   2. 3 T = _____ lb **6,000**   3. 5 c = _____ fl oz **40**

4. 1.5 lb = _____ oz **24**   5. 10 pt = _____ qt **5**   6. 12 qt = _____ gal **3**

7. 12 fl oz = _____ c $1\frac{1}{2}$ 8. 24 oz = _____ lb $1\frac{1}{2}$ 9. 7 c = _____ fl oz **56**

10. 2.5 pt = _____ c **5**   11. 1.5 gal = _____ qt **6**   12. 3.5 qt = _____ pt **7**

13. 48 oz = _____ lb **3**   14. 8,000 lb = _____ T **4**   15. 2 pt = _____ c **4**

© Glencoe/McGraw-Hill   155   *Mathematics: Applications and Connections, Course 1*

Check for Understanding

If students need additional practice or instruction after completing Exercises 1–9, you may find one of the following options helpful.
- Extra Practice, see p. 577
- Reteaching Activity, see p. 293
- *Transition Booklet,* pp. 29–34
- *Study Guide Masters,* p. 55
- *Practice Masters,* p. 55

Assignment Guide

Core: 11–27 odd, 29–33
Enriched: 10–26 even, 27–33

4 ASSESS

Closing Activity

Writing Have students work in pairs to create word problems that involve changing units of measure within the customary system. Have them exchange their problems with others to solve.

Practice Masters, p. 55

7-7 Practice

Integration: Measurement
Changing Customary Units

Complete.

1. 6 pt = **12** c   2. 20 qt = **5** gal   3. 64 oz = **4** lb

4. 7 c = **56** fl oz   5. 12 qt = **24** pt   6. 12,000 lb = **6** T

7. 8 gal = **32** qt   8. 9 lb = **144** oz   9. 72 fl oz = **9** c

10. 4 T = **8,000** lb   11. 14 c = **7** pt   12. 26 pt = **13** qt

13. 10 qt = **2½** gal   14. 2 T = **64,000** oz   15. 18 pt = **36** c

16. 3½ c = **28** fl oz   17. 128 c = **8** gal   18. 96 oz = **6** lb

19. 21 qt = **42** pt   20. 750 lb = **0.375** T   21. 3 qt = **96** fl oz

22. 15 pt = **7½** qt   23. 7 T = **14,000** lb   24. 2 gal = **32** c

25. 19 c = **9½** pt   26. 4 qt = **16** c   27. 5¼ lb = **84** oz

28. 6 gal = **24** qt   29. 104 fl oz = **13** c   30. 64 oz = **8** gal

© Glencoe/McGraw-Hill   T55   *Mathematics: Applications and Connections, Course 1*

Guided Practice **Complete.**

3. 2 gal = _?_ qt **8**   **4.** 14 c = _?_ pt **7**   **5.** 64 oz = _?_ lb **4**

6. $3\frac{1}{2}$ T = _?_ lb **7,000**   **7.** 32 fl oz = _?_ pt **2**   **8.** 4 T = _?_ oz **128,000**

9. *Life Science* Kijana, the baby elephant, will weigh about $6\frac{1}{2}$ tons when fully grown. About how many pounds will he weigh? **about 13,000 lb**

EXERCISES

Practice **Complete.**

10. 8 qt = _?_ pt **16**   **11.** 24 qt = _?_ gal **6**   **12.** 7 c = _?_ pt $3\frac{1}{2}$

13. 9 lb = _?_ oz **144**   **14.** 6 gal = _?_ qt **24**   **15.** 16 fl oz = _?_ c **2**

16. 10 pt = _?_ qt **5**   **17.** 5 c = _?_ fl oz **40**   **18.** 500 lb = _?_ T $\frac{1}{4}$

19. $2\frac{1}{4}$ T = _?_ lb **4,500**   **20.** $4\frac{1}{2}$ pt = _?_ c **9**   **21.** $2\frac{1}{2}$ qt = _?_ c **10**

22. 32 c = _?_ gal **2**   **23.** 12 pt = _?_ gal $1\frac{1}{2}$ **24.** 12 c = _?_ qt **3**

25. How many tons are in 80,000 pounds? **40**

26. Find how many quarts there are in 40 fluid ounces. $1\frac{1}{4}$ qt

Applications and Problem Solving

27. *Geography* Giant clams live along the Great Barrier Reef off the coast of Australia. They measure up to 6 feet long and weigh as much as 0.25 ton. How many pounds is this? **500 pounds**

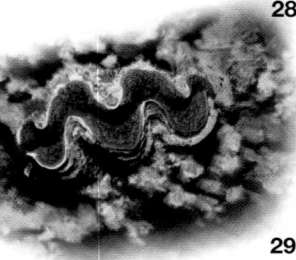

28. *Money Matters* Tashauna rides her moped from her home in Connecticut to her ballet lessons in New York. She uses about 3 quarts of gasoline a week. How much will she save a year in taxes if she buys her gasoline in New York rather than Connecticut? **$6.75**

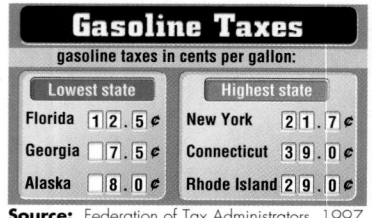

Gasoline Taxes

gasoline taxes in cents per gallon:

Lowest state		Highest state	
Florida	12.5¢	New York	21.7¢
Georgia	7.5¢	Connecticut	39.0¢
Alaska	8.0¢	Rhode Island	29.0¢

Source: Federation of Tax Administrators, 1997

29. *Critical Thinking* What can you divide by to change 256 cups directly to gallons? **16**

Mixed Review

30. Find $2\frac{3}{5} \div 1\frac{2}{3}$. Write the quotient in simplest form. *(Lesson 7-6)* $1\frac{14}{25}$

31. State whether 405 is divisible by 2, 3, 5, 6, 9, or 10. *(Lesson 5-1)* **3, 5, 9**

32. 12.237 liters

32. *Measurement* Change 12,237 milliliters to liters. *(Lesson 4-9)*

33. **Test Practice** Frank purchased $13.72 worth of gasoline. He gives the gas station attendant a $20 bill. How much change should he receive? *(Lesson 3-6)* **E**

A $7.72   B $7.58   C $7.28   D $6.72   E Not Here

Extending the Lesson

Enrichment Masters, p. 55

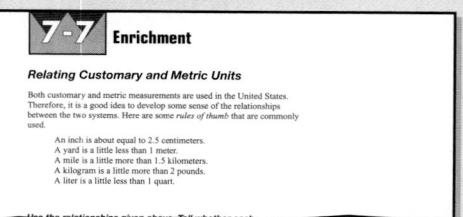

7-7 Enrichment

Relating Customary and Metric Units

Both customary and metric measurements are used in the United States. Therefore, it is a good idea to develop some sense of the relationships between the two systems. Here are some *rules of thumb* that are commonly used.

An inch is about equal to 2.5 centimeters.
A yard is a little less than 1 meter.
A mile is a little more than 1.5 kilometers.
A kilogram is a little more than 2 pounds.
A liter is a little less than 1 quart.

Activity Have small groups of students interview professionals such as chemists, building contractors, chefs, or meteorologists to determine the system of measurement they use in their work. Encourage students to ask why they use that system and how it is used in their profession.

HANDS-ON LAB

COOPERATIVE LEARNING

7-7B Measurement

A Follow-Up of Lesson 7-7

Have you ever dreamed of visiting another planet or the moon? You would discover many things to be different, even your weight!

- measuring cups
- containers
- water

Planets' Weight Factors

| Sun 28 | Mercury $\frac{1}{3}$ | Venus $\frac{9}{10}$ | Earth 1 | Moon $\frac{1}{6}$ | Mars $\frac{3}{8}$ | Jupiter 3 |

TRY THIS

Work with a partner.

Step 1 Choose a work station. Record the name of the planet and its weight factor relative to Earth.

Step 2 Fill the 1-cup measuring cup with water. This represents the weight of one cup of water on Earth. Pour this amount into the container labeled *Earth*.

Step 3 Use the planet's weight factor to fill as many measuring cups with water as needed to represent the weight of one cup of water on this planet. For example, Jupiter's weight factor is 3 times that of the Earth's. So, empty 3 cups of water into the container labeled *Jupiter*. Record your results.

Step 4 Repeat Steps 1–3 at each work station.

4. Sample answer: A 126-pound student would weigh 21 pounds.

ON YOUR OWN

1. Analyze your results. Is the weight of 1 cup of water on each planet more or less than the weight of one cup of water on Earth? more or less than one cup of water on Jupiter? Record your results. 1–2. See Answer Appendix.

2. Which planet's container weighs the most? Explain your reasoning.

3. How much would a 22-pound dog weigh on Jupiter? 66 pounds

4. How much would you weigh on the moon?

5. *Reflect Back* On Earth, a certain object weighs 1 pound. How many ounces would the same object weigh on Mars? 6 ounces

Lesson 7-7B HANDS-ON LAB **295**

Math Journal Have students write a paragraph about comparing weights on different planets and how astronauts are affected by these changes. Have them include a description of how working with fractions applies to this activity.

HANDS-ON

7-7B LAB Notes

GET READY

Objective Students compare weights.

Optional Resources
Hands-On Lab Masters
- worksheet, p. 53

Manipulative Kit
- measuring cups

MANAGEMENT TIPS

Recommended Time
45 minutes

Getting Started Ask students to read the opening sentences of the lesson. Then ask: *How are astronauts affected by changes in weight in our solar system?*

Prepare work stations that have a measuring cup, water, a container labeled Earth, and one labeled Moon or one of the other planets. Ask students to pour the water back into its original container before moving on to the next station.

Teaching Tip Encourage students to use approximate measures at the Venus and Moon stations. Ask questions such as: *Of all the containers at each station, which one weighs the least? How do you know?*

ASSESS

After students complete Exercises 1–4, have pairs of students write and solve word problems about the weight of an object on a different planet.

Use Exercise 5 to determine whether students understand how to compare weights.

THINKING LAB

Objective
Students solve problems by finding and extending a pattern.

Recommended Pacing

Standard	Day 12 of 15
Honors	Day 10 of 14
Block	Day 7 of 8

1 FOCUS

Getting Started Ask students how they would solve this problem.
Once Jenny started making pizzas, she couldn't stop. She made one pizza on Monday, two pizzas on Tuesday, four pizzas on Wednesday, and eight pizzas on Thursday. If this pattern continues, how many pizzas will Jenny make on Sunday? **64 pizzas**

2 TEACH

Teaching Tip To help students understand how patterns are relevant to their everyday lives, have them list three activities in their daily lives that follow a pattern. Then ask them to create a problem using one of the activities.

In-Class Example
The Mayan numbering system used symbols that were combined to give different values. The chart shows some of the numbers. Draw the Mayan symbols for 4 and 13.

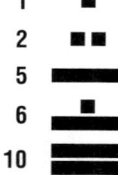

PROBLEM SOLVING

7-8A Look for a Pattern

A Preview of Lesson 7-8

Kerri and her brother Eric are at the museum. They are learning about the history and culture of the Hopi Indians. Let's listen in!

Kerri

It says here that the Hopi Indians began making pottery about 1,500 years ago. Wow, that's a long time ago!

Here is one of the designs they used to decorate the pottery. It appears that they liked to use geometric shapes in the designs.

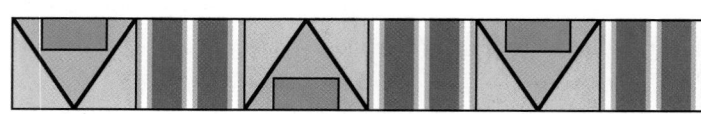

I wonder what the design would look like if the pattern was continued to the right.

Well, since the Hopi Indians liked to use geometric shapes, maybe the design would include circles and triangles.

I disagree. The beauty of the design is in the pattern. I think that the same pattern is probably continued to the right.

Eric

THINK ABOUT IT

Work with a partner. 1–2. See students' work.

1. *Compare and contrast* Kerri's and Eric's thinking. Whose thinking do you think is accurate? Explain your reasoning.

2. *Describe* the pattern the Hopi Indians used in the design.

3. *Draw* the next three segments to the right in the pattern. **See Answer Appendix.**

4. *Apply* the **look for a pattern** strategy to find the next two numbers in each pattern below.
   a. 3, 6, 9, 12, _?_ , _?_ **15,18**

   b. 5, 10, 20, 40, _?_ , _?_ **80,160**

296 Chapter 7 Multiplying and Dividing Fractions

▪ Reteaching the Lesson ▪

Activity Have students create a pattern with fractions. Have them write the beginning of their pattern on an overhead transparency. Then have the class ask yes-and-no questions such as "Did you use addition?" to try to guess the pattern.

ON YOUR OWN

5. The third step of the 4-step plan for problem solving asks you to *solve* the problem. ***Tell*** how you can use the look for a pattern strategy to help you solve a problem.
See margin.
6–7. See students' work.

6. ***Write a Problem*** in which you would use the look for a pattern strategy to solve. Then ask a classmate to describe the pattern and solve the problem.

7. ***Look Ahead*** Explain how you could use the look for a pattern strategy to find the next number in the sequence in Exercise 17 on page 300.

MIXED PROBLEM SOLVING

STRATEGIES
Look for a pattern.
Solve a simpler problem.
Act it out.
Guess and check.
Draw a diagram.
Make a chart.
Work backward.

Solve. Use any strategy.

8. ***Fashion*** Matthew is choosing a shirt and a pair of jeans to wear. He has black, blue, and stonewashed jeans, and blue, gray, and white shirts. How many different combinations of jeans and shirts are possible? **9 combinations**

9. ***Food*** Richard eats half of a ham sandwich at 1:00 P.M. At 3:00 P.M., he eats half of what was left of his sandwich from lunch. At 5:00 P.M., he eats half of what was left from his ham sandwich.

  a. If Richard continues eating at this rate for four more hours, how much of the sandwich has been eaten? **See margin.**
  b. At this rate, will Richard ever eat the entire sandwich? Explain. **No; see margin.**

10. ***Measurement*** What is the missing measurement in the pattern?
$\ldots, \underline{\ ?\ }, \frac{1}{4}$ in., $\frac{1}{8}$ in., $\frac{1}{16}$ in., $\ldots$
See margin.

11. **1,476 sq cm**

11. ***Geometry*** Find the area of the figure.
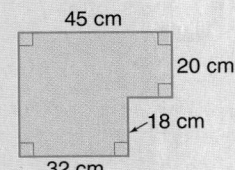
45 cm
20 cm
18 cm
32 cm

12. ***Money Matters*** Roger and Celina began working for the same company in 1997. Celina earned $18,000 per year, and Roger earned $14,500. Each year Roger received a $1,500 raise, and Celina received a $1,000 raise.
  a. In what year will they earn the same amount of money? **2004**
  b. What will be their annual salary in that year? **$25,000**

13. **Test Practice** At Hershal Middle School, the bell rings at 8:55, 9:40, 10:25, and 11:10 each morning. If this pattern continues, when would the next bell ring? **C**
  A 11:50 A.M.
  B 11:45 A.M.
  C 11:55 A.M.
  D 10:40 A.M.

Lesson 7-8A THINKING **LAB** **297**

Check for Understanding
Have students work through Exercises 1–4. Use the results from Exercise 4 to determine whether students comprehend how to extend patterns.

Extra Practice If students need additional practice in problem solving, extra practice is available on the following pages.
• Look for a Pattern, see p. 577
• Mixed Problem Solving, see pp. 593–594.

Assignment Guide

All: 5–13

4 ASSESS

Closing Activity
Modeling Have groups of students design new patterns using manipulatives or pictures. Then ask them to trade the designs with other groups and look for patterns.

Additional Answers
5. Sample answer: Once you identify the pattern, then you can continue the pattern to solve the problem.

9a. $\frac{31}{32}$

9b. Richard will not finish the sandwich, but the amount will be so small that it will be like he did eat it all.

10. $\frac{1}{2}$ in.

Extending the Lesson

Activity Have small groups of students look for patterns in history. For example, they could research the population of their community for the past 50 years. Ask them to record their findings in a graph. Have them determine whether they can make a prediction for the future.

Sample problem: *U.S. presidential elections follow a pattern. If there were elections in 1988, 1992, and 1996, when are the next three elections?* **2000, 2004, 2008**

Instructional Resources
- *Study Guide Masters*, p. 56
- *Practice Masters*, p. 56
- *Enrichment Masters*, p. 56
- Transparencies 7-8, A and B
- *Assessment and Evaluation Masters*, p. 184
- *Technology Masters*, p. 13

CD-ROM Program
- Resource Lesson 7-8

Recommended Pacing	
Standard	Day 13 of 15
Honors	Days 11 & 12 of 14
Block	Day 7 of 8

1 FOCUS

5-Minute Check
(Lesson 7-7)

Complete.
1. 6 gal = _?_ qt **24**
2. 14 c = _?_ pt **7**
3. 4 T = _?_ lb **8,000**
4. 2 T = _?_ oz **64,000**
5. A jar of salsa weighs 12 oz. How many pounds will 48 jars of salsa weigh? **36 lb**

The 5-Minute Check is also available on **Transparency 7-8A** for this lesson.

2 TEACH

Transparency 7-8B contains a teaching aid for this lesson.

Reading Mathematics Write the words *sequence* and *reciprocal* on the chalkboard and pronounce them. Give an example of a sequence such as $\frac{1}{2}, \frac{1}{4}, \frac{1}{8}$, and $\frac{1}{16}$. Then write $\frac{2}{1}, \frac{4}{1}, \frac{8}{1}$, and $\frac{16}{1}$, and have students match the reciprocals. Ask students to use the words in sentences.

7-8

Integration: Patterns and Functions
Sequences

What you'll learn
You'll learn to recognize and extend sequences.

When am I ever going to use this?
You can use sequences to determine the cost of a long-distance phone call.

Word Wise
sequence

> **Study Hint**
> **Mental Math**
> Add 32 because
> $32 - 0 = 32$,
> $64 - 32 = 32$,
> $96 - 64 = 32$,
> and $128 - 96 = 32$.

On Earth, gravity causes all falling objects to accelerate at 32 ft/s². This means that if gravity is the only force acting on a falling object, its speed will increase about 32 feet per second each second.

The table shows the effect of gravity. What will the speed of the object be after 5 seconds?

Seconds After Object is Dropped	0	1	2	3	4	5
Speed of Object (ft/s)	0	32	64	96	128	?

The numbers 0, 32, 64, 96, and 128 form a **sequence**. A sequence is a list of numbers in a specific order.

In the sequence, notice that 32 is added to each number.

$$0, \quad 32, \quad 64, \quad 96, \quad 128$$
$$+32 \quad +32 \quad +32 \quad +32$$

The next number in the sequence is 128 + 32, or 160. So, after 5 seconds, the speed of the object will be 160 ft/s.

Examples

Find the next number in each sequence.

1 $18, \quad 24, \quad 30, \quad 36, \ldots$
$$+6 \quad +6 \quad +6$$

In this sequence, 6 is added to each number. The next number is 36 + 6, or 42.

2 $21, \quad 18\frac{1}{2}, \quad 16, \quad 13\frac{1}{2}, \ldots$
$$-2\frac{1}{2} \quad -2\frac{1}{2} \quad -2\frac{1}{2}$$

In this sequence, $2\frac{1}{2}$ is subtracted from each number. The next number is $13\frac{1}{2} - 2\frac{1}{2}$ or 11.

Another type of sequence is one where the numbers are found by multiplying by the same number.

Motivating the Lesson
Hands-On Activity Bring in or obtain copies of simple sheet music. Have students work in small groups to look for patterns in the musical arrangements.

Find the next number in each sequence.

③ 2,   8,   32,   128,...

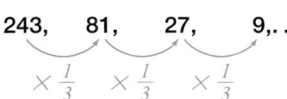

× 4   × 4   × 4

Each number in the sequence is multiplied by 4. The next number is 128 × 4, or 512.

④ 243,   81,   27,   9,...

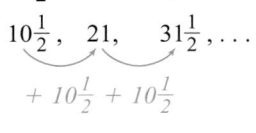

$\times \frac{1}{3}$ $\times \frac{1}{3}$ $\times \frac{1}{3}$

Each number in the sequence is multiplied by $\frac{1}{3}$. The next number is $9 \times \frac{1}{3}$, or 3.

Study Hint

Problem Solving
You may want to use the look for a pattern strategy to help you with these sequences.

APPLICATION ⑤ **Carpentry** A roof rafter of a building is to be braced as shown. The length of braces *A*, *B*, and *C* are $10\frac{1}{2}$ inches, 21 inches, and $31\frac{1}{2}$ inches, respectively. Find the lengths of braces *D* and *E*.

$10\frac{1}{2}$,   21, $31\frac{1}{2}$,...

$+ 10\frac{1}{2}$ $+ 10\frac{1}{2}$

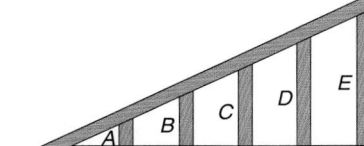

In the sequence, $10\frac{1}{2}$ is added to each number.
The length of brace *D* is $31\frac{1}{2} + 10\frac{1}{2}$, or 42 inches.
The length of brace *E* is $42 + 10\frac{1}{2}$, or $52\frac{1}{2}$ inches.

CHECK FOR UNDERSTANDING

Communicating Mathematics

Read and study the lesson to answer each question. 1–2. See margin.

1. *Write*, in your own words, a definition for sequence.

2. *Tell* how the numbers are related in the sequence 16, 4, 1, $\frac{1}{4}$.

3. *You Decide* Janet says that the next number in the sequence 5, $7\frac{1}{2}$, 10, $12\frac{1}{2}$, is 14. Joshua says the next number is 15. Who is correct? Explain your reasoning. **Joshua; see Answer Appendix for explanations.**

Guided Practice

Find the next two numbers in each sequence.

4. 6, 18, 54, 162, . . . **486, 1,458**   5. 45, 38, 31, 24, . . . **17, 10**

Find the missing number in each sequence.

6. 2, _?_, 9, $12\frac{1}{2}$, . . . $5\frac{1}{2}$ 7. _?_, 25, 5, 1, . . . **125**

8. *Geometry* Draw the next two figures in the sequence.

Lesson 7-8 Integration: Patterns and Functions Sequences **299**

■ Reteaching the Lesson ■

Activity Give a group of students a "rule" by which they will create a sequence. Have them roll a number cube or select a random number to begin the sequence. Each person in the group names the next number in the sequence.

Additional Answers

1. Sample answer: A sequence is a list of numbers in a specific order.
2. Each number in the sequence is multiplied by $\frac{1}{4}$.

In-Class Examples
Find the next number in each sequence.

For Example 1
3, 10, 17, 24, . . . **31**

For Example 2
28, 23, 18, 13, . . . **8**

For Example 3
4, 6, 9, $13\frac{1}{2}$, . . . **$20\frac{1}{4}$**

For Example 4
567, 189, 63, 21, . . . **7**

For Example 5
Jacob put in three fence posts measured from the corner of a shed at 8 ft, $20\frac{1}{2}$ ft, and 33 ft, respectively. Where would he dig a hole for a fourth post? **$45\frac{1}{2}$ feet**

3 PRACTICE/APPLY

Check for Understanding
If students need additional practice or instruction after completing Exercises 1–8, you may find one of the following options helpful.
• Extra Practice, see p. 577
• Reteaching Activity
• *Study Guide Masters,* p. 56
• *Practice Masters,* p. 56
• Interactive Mathematics Tools Software

Study Guide Masters, p. 56

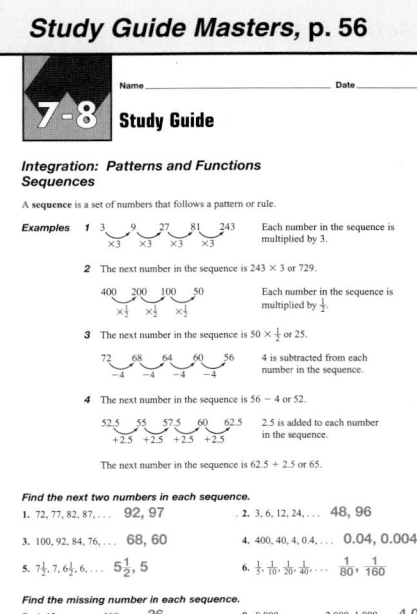

300   Chapter 7

Assignment Guide

Core: 9–23 odd, 25–28
Enriched: 10–22 even, 23–28

4 ASSESS

Closing Activity

Modeling Have students work in small groups to create sequences using grid paper or other small items. Have them challenge other groups to find the next two items in the sequence.

Chapter 7, Quiz D (Lessons 7-7 and 7-8) are available in the *Assessment and Evaluation Masters,* p. 184.

Additional Answers

23. $\frac{1}{16}$ (sixteenth note), $\frac{1}{32}$ (thirty-second note), and $\frac{1}{64}$ (sixty-fourth note)

25a. $\frac{1}{2}, \frac{1}{4}, \frac{1}{8}, \frac{1}{16}, \frac{1}{32}, \frac{1}{64}, \frac{1}{128}, \frac{1}{256}, \frac{1}{512}, \frac{1}{1,024}$

Practice Masters, p. 56

Name_____ Date_____

7-8 Practice

Integration: Patterns and Functions
Sequences

Find the next two numbers in each sequence.
1. 2, 8, 14, 20, ... **26, 32**
2. 31, 27, 23, 19, ... **15, 11**
3. $\frac{1}{3}$, 1, 3, 9, ... **27, 81**
4. 108, 36, 12, 4, ... **$1\frac{1}{3}, \frac{4}{9}$**
5. 43, 38, 33, 28, ... **23, 18**
6. 1.2, 2.4, 3.6, 4.8, ... **6, 7.2**
7. 3, 6, 12, 24, ... **48, 96**
8. 63, 56, 49, 42, ... **35, 28**
9. 1, $1\frac{2}{3}, 2\frac{1}{3}$, 3, ... **$3\frac{2}{3}, 4\frac{1}{3}$**
10. 4, 12, 36, 108, ... **324, 972**
11. 81, 72, 63, 54, ... **45, 36**
12. 6, 11, 16, 21, ... **26, 31**
13. 5, 20, 35, 50, ... **65, 80**
14. 27, 22.5, 18, 13.5, ... **9, 4.5**
15. 18, 36, 54, 72, ... **90, 108**
16. 1, 5, 25, 125, ... **625, 3,125**

Find the missing number in each sequence.
17. $\frac{1}{6}, \frac{1}{3}$, _____, $\frac{2}{3}$, ... **$\frac{1}{2}$**
18. 54, _____, 42, 36, ... **48**
19. _____, 11, 14, 17, ... **8**
20. 1.7, 3.4, 5.1, _____, ... **6.8**
21. ..., 200, 100, _____, 25 **50**
22. $1\frac{3}{4}, 3\frac{1}{2}$, _____, 7, ... **$5\frac{1}{4}$**
23. _____, 12, 48, 192, ... **3**
24. ..., 91, 78, _____, 52, ... **65**
25. 9, _____, 23, 30, ... **16**
26. 0.4, 0.8, 1.2, _____, ... **1.6**
27. ..., _____, $\frac{3}{5}$, _____, $1\frac{4}{5}, 2\frac{2}{5}$ **$1\frac{1}{5}$**
28. 30, 26, _____, 18, ... **22**
29. _____, 7.5, 22.5, 67.5, ... **2.5**
30. ..., 0.004, 0.04, _____, 4 **0.4**

© Glencoe/McGraw-Hill   T56   *Mathematics: Applications and Connections, Course 1*

EXERCISES

Practice   Find the next two numbers in each sequence.

9. 15, 30, 45, 60, ... **75, 90**
10. 4, 12, 36, 108, ... **324, 972**
11. 12, 6, 3, $1\frac{1}{2}$, ... **$\frac{3}{4}, \frac{3}{8}$**
12. 19, $18\frac{1}{2}$, 18, $17\frac{1}{2}$, ... **17, $16\frac{1}{2}$**
13. $\frac{1}{3}$, 2, 12, 72, ... **432, 2,592**
14. $28\frac{1}{2}$, 30, $31\frac{1}{2}$, 33, ... **$34\frac{1}{2}$, 36**

Find the missing number in each sequence.

15. 49, 41, _?_, 25, ... **33**
16. 64, 16, _?_, 1, ... **4**
17. $\frac{2}{3}$, _?_, $1\frac{1}{3}$, $1\frac{2}{3}$, ... **1**
18. _?_, $56\frac{1}{2}$, 53, $49\frac{1}{2}$, ... **60**
19. ..., 4, 40, _?_, 4,000, ... **400**
20. 1, $\frac{3}{4}$, _?_, $\frac{27}{64}$, ... **$\frac{9}{16}$**

21. What is the next term in the sequence $x, x^2, x^3, x^4, \ldots$? **x^5**

22. Find the missing term in the sequence $x + 5, x + 4, \underline{\ ?\ }, x + 2, x + 1$. **$x + 3$**

Applications and Problem Solving

23. **Music** The diagram shows the most common notes used in music. The names of the first four notes are whole note, half note, quarter note, and eighth note. What are the names of the next three notes? **See margin.**

Notes

whole (1) $\frac{1}{2}$ $\frac{1}{4}$ $\frac{1}{8}$

24. **Money Matters** Roberta Salgado rents an apartment for $565 a month. Each year, the monthly rent is expected to increase $12. What will be the monthly rent at the end of four years? **$601**

25. **Critical Thinking** The large square shown represents 1.

   a. Find the first ten numbers of the sequence represented by the model. The first number is $\frac{1}{2}$. **See margin.**

   b. Estimate the sum of the first ten numbers without actually adding. **1**

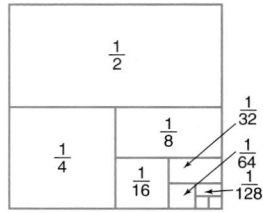

$\frac{1}{2}$ $\frac{1}{8}$ $\frac{1}{32}$ $\frac{1}{4}$ $\frac{1}{16}$ $\frac{1}{64}$ $\frac{1}{128}$

Mixed Review

26. **Measurement** How many pints are in 64 fluid ounces? *(Lesson 7-7)* **4**

27. **Test Practice** Mrs. Matthews bought $3\frac{1}{4}$ pounds of caramels and $2\frac{1}{2}$ pounds of chocolate. How many pounds of candy did she buy altogether? *(Lesson 6-5)* **A**

   A $5\frac{3}{4}$ **B** $5\frac{2}{6}$ **C** $1\frac{1}{2}$ **D** $\frac{3}{4}$ **E** Not Here

28. **Geometry** Find the perimeter of a triangle whose sides are each 23 centimeters long. *(Lesson 4-4)* **69 cm**

300   Chapter 7 Multiplying and Dividing Fractions

Extending the Lesson

Enrichment Masters, p. 56

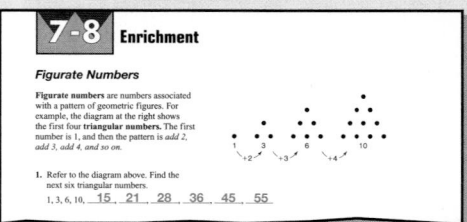

7-8 Enrichment

Figurate Numbers

Figurate numbers are numbers associated with a pattern of geometric figures. For example, the diagram at the right shows the first four **triangular numbers.** The first number is 1, and then the pattern is add 2, add 3, add 4, and so on.

1. Refer to the diagram above. Find the next six triangular numbers.
1, 3, 6, 10, **15**, **21**, **28**, **36**, **45**, **55**

Activity Have students investigate different styles and types of poetry. Then have them find examples of different patterns or sequences, such as Haiku. Have groups of students work together to write a poem that follows one of these patterns or sequences. Ask the groups to read their poems to the class.

SPREADSHEETS

7-8B Sequences

A Follow-Up of Lesson 7-8

- computer
- spreadsheet software

As you learned in Lesson 4-3B, you can use a spreadsheet to prepare tables easily. You can use a spreadsheet to project results, make calculations, and print almost anything that can be arranged in a table. Spreadsheets can also be used to simulate experiments when the variables can be described algebraically.

TRY THIS

Work with a partner.

Use a spreadsheet to solve the following problem.

On each bounce, a ball rebounds and goes back up 0.8 of the way to its starting point. If the ball is dropped from a height of 20 meters, how high will it rebound on the 5th bounce?

Input the initial height of the ball in cell B1 and the rebound ratio in cell B2. The values of the remaining cells are determined by the formulas. The computer does the calculations.

The screen shows the results of running the spreadsheet. Cell B9 shows that after the 5th bounce, the height is approximately 6.554 meters.

B4*B2 means multiply the value in cell B4 by the value in cell B2.

	A	B
	BALL BOUNCE	
1	Initial Ht (m) =	20
2	Rebound Ratio =	0.8
3	Number of bounces	RETURN HT
4	zero → 0	= B1
5	A4 + 1	= B4*B2
6	A5 + 1	= B5*B2
7	A6 + 1	= B6*B2

A4 + 1 means add 1 to the value in cell A4.

	A	B
	BALL BOUNCE	
1	Initial Ht (m) =	20
2	Rebound Ratio =	0.8
3	Number of bounces	RETURN HT
4	0	20.000
5	1	16.000
6	2	12.800
7	3	10.240
8	4	8.192
9	5	6.554

ON YOUR OWN

1. Explain the meaning of the formula in cell B6. **Multiply the value in cell B5 by the value in cell B2.**

2. Explain the meaning of the formula in cell A7. **Add 1 to the value in cell A6.**

3. If the ball rebounds 0.4 of the way to its starting point, what are the heights of its first six bounces? **8.000m, 3.200m, 1.280m, 0.512m, 0.2048m, 0.08192m**

4. Suppose a certain ball rebounds $\frac{1}{2}$ of the way to its starting point. How would you change the spreadsheet to show the heights of the bounces? **Change the number in cell B2 to 0.5.**

Lesson 7-8B TECHNOLOGY **LAB** 301

GET READY

Objective Students use a spreadsheet to extend sequences.

Technology Resources
Suggested spreadsheet software:
- *Lotus 1·2·3*
- *ClarisWorks*
- *Microsoft Excel*

MANAGEMENT TIPS

Recommended Time
30 minutes

Getting Started Remind students how each cell of a spreadsheet is named. Also review how formulas using cell names are entered into the spreadsheet.

After entering the data and having the computer generate the values, have students look for patterns in each column.

Have students work through the exercises using the data they generated. What would be an easy way to change the spreadsheet for Exercises 3 and 4? **change the value in B2**

ASSESS

After students answer Exercises 1–4, ask them to determine how they would use the patterns they found to extend sequences when a spreadsheet is not available.

Math Journal Have students write a paragraph about how a spreadsheet could be helpful in tracking population growth and decline.

Vocabulary

This section provides a listing of the new terms, properties, and phrases that were introduced in this chapter. Have students define each term and provide an example or two of it, if appropriate.

Understanding and Using the Vocabulary

These exercises check students' understanding of the terms by using a variety of verbal formats including matching, completion, and true/false.

Glossaries A complete glossary of terms appears on pages 642–648. The glossary also appears in Spanish on pages 649–656.

Vocabulary

After completing this chapter, you should be able to define each term, concept, or phrase and give an example or two of each.

Number and Operations
compatible numbers (p. 268)
reciprocals (p. 285)

Geometry
center (p. 280)
circle (p. 280)
circumference (p. 280)
diameter (p. 280)
radius (p. 280)

Problem Solving
look for a pattern (p. 296)

Measurement
cup (p. 292)
fluid ounce (p. 292)
gallon (p. 292)
ounce (p. 292)
pint (p. 292)
pound (p. 292)
quart (p. 292)
ton (p. 292)

Patterns and Functions
sequence (p. 298)

10. When changing from smaller units to larger units, divide. When changing from larger units to smaller units, multiply.

Understanding and Using the Vocabulary

Choose the letter of the term that best matches each phrase.

1. 2,000 pounds k
2. the distance across a circle through the center e
3. 2 cups i
4. the distance from the center of a circle to any point on the circle c
5. 8 fluid ounces h
6. a list of numbers in a specific order g
7. the distance around a circle f
8. any two numbers whose product is 1 a
9. the set of all points in a plane that are the same distance from a given point d

a. reciprocals
b. center
c. radius
d. circle
e. diameter
f. circumference
g. sequence
h. cup
i. pint
j. gallon
k. ton

In Your Own Words

10. *Explain* how you would change customary units of capacity such as cups to gallons or tons to pounds.

302 Chapter 7 Multiplying and Dividing Fractions

 **MindJogger Videoquizzes**

MindJogger Videoquizzes provide an alternative review of concepts presented in this chapter. Students work in teams to answer questions, gaining points for correct answers. The questions are presented in three rounds.
Round 1 Concepts–5 questions
Round 2 Skills–4 questions
Round 3 Problem Solving–4 questions

Objectives & Examples

Upon completing this chapter, you should be able to:

- estimate fraction products using compatible numbers and rounding *(Lesson 7-1)*

  Estimate $\frac{4}{5} \times 11$.

  For 11, the nearest multiple of 5 is 10.
  $\frac{4}{5}$ *of 10 is 8.*

  So, $\frac{4}{5} \times 11$ is about 8.

- multiply fractions *(Lesson 7-2)*

  $\frac{5}{9} \times \frac{3}{10} = \frac{\overset{1}{5} \cdot \overset{1}{3}}{\underset{3}{9} \cdot \underset{2}{10}}$ *Divide by the GCF.*

  $= \frac{1}{6}$

- multiply mixed numbers *(Lesson 7-3)*

  $3\frac{1}{2} \times 4\frac{2}{3} = \frac{7}{2} \times \frac{\overset{7}{14}}{\underset{1}{3}}$

  $= \frac{49}{3}$ or $16\frac{1}{3}$

- find the circumference of circles *(Lesson 7-4)*

  $C = 2\pi r$
  $C \approx 2 \cdot 3.14 \cdot 2.4$
  $C \approx 15.072$

  The circumference of the circle is about 15.1 meters.

Review Exercises

Use these exercises to review and prepare for the chapter test.

Estimate each product.

11. $10 \times 2\frac{3}{4}$ **12.** $4\frac{1}{7} \times 5\frac{4}{5}$

13. $\frac{5}{6} \times 13$ **14.** $\frac{7}{8} \times \frac{1}{2}$

15. $\frac{2}{3} \times 19$ **16.** $7\frac{3}{4} \times \frac{1}{4}$

11–16. See Answer Appendix.

Find each product. Write in simplest form.

17. $\frac{7}{8} \times \frac{4}{5}$ $\frac{7}{10}$ **18.** $\frac{14}{15} \times \frac{10}{21}$ $\frac{4}{9}$

Solve each equation. Write the solution in simplest form.

19. $y = 12 \times \frac{5}{8}$ $7\frac{1}{2}$ **20.** $\frac{4}{5} \times \frac{3}{8} = t$ $\frac{3}{10}$

Find each product. Write in simplest form.

21. $3\frac{3}{4} \times 1\frac{1}{5}$ $4\frac{1}{2}$ **22.** $3\frac{2}{3} \times 6$ 22

Solve each equation. Write the solution in simplest form.

23. $7\frac{1}{5} \times 1\frac{7}{8} = x$ $13\frac{1}{2}$ **24.** $w = 3\frac{1}{8} \times 2\frac{2}{5}$ $7\frac{1}{2}$

26. 51.5 cm   **27.** $\pi = \frac{22}{7}$; $5\frac{1}{2}$ yd

Find the circumference of each circle shown or described. Use $\frac{22}{7}$ or 3.14 for π. Round decimal answers to the nearest tenth.

25.   **26.** $r = 8.2$ cm

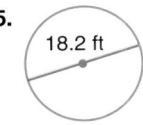

18.2 ft

57.1 ft

27. $d = 1\frac{3}{4}$ yd

Objectives & Examples

This section reviews the skills and concepts of the chapter and shows completely worked examples.

Review Exercises

These exercises provide practice for the corresponding objectives.

Assessment and Evaluation Masters, pp. 171–172

Chapter 7 Test, Form 1B

1. Estimate $\frac{1}{3} \times 17$.
   A. 8   B. 60   C. 5   D. 6   1. __D__

2. Estimate $\frac{9}{10} \times \frac{5}{16}$.
   A. $\frac{9}{10}$ B. $\frac{5}{16}$ C. 0   D. 1   2. __B__

3. Estimate $5\frac{7}{8} \times 3\frac{1}{6}$.
   A. 15   B. 20   C. 18   D. 24   3. __C__

4. Find $\frac{12}{16} \times \frac{8}{21}$. Write in simplest form.
   A. $\frac{23}{37}$ B. $\frac{2}{7}$ C. $\frac{5}{14}$ D. $\frac{120}{336}$ 4. __C__

5. Solve $18 \times \frac{5}{6} = n$. Write the solution in simplest form.
   A. 15   B. $\frac{5}{108}$ C. $\frac{5}{6}$ D. $21\frac{3}{5}$ 5. __A__

6. A foot of metal pipe sells for $9. How much will $\frac{2}{3}$ foot cost?
   A. $\$\frac{2}{27}$ B. $13.50   C. $3   D. $6   6. __D__

7. Find $7\frac{1}{2} \times 4\frac{1}{2}$. Write in simplest form.
   A. $28\frac{7}{8}$ B. 30   C. $\frac{1}{30}$ D. $\frac{7}{12}$ 7. __B__

8. Solve $2\frac{1}{16} \times 1\frac{5}{8} = x$. Write the solution in simplest form.
   A. $3\frac{1}{2}$ B. $2\frac{9}{80}$ C. $2\frac{5}{16}$ D. $1\frac{47}{128}$ 8. __A__

9. Find the area of a room that is $5\frac{1}{2}$ yards long by $3\frac{1}{3}$ yards wide.
   A. $8\frac{5}{6}$ yd²   B. $18\frac{1}{3}$ yd²   C. $15\frac{5}{6}$ yd²   D. $3\frac{1}{2}$ yd²   9. __B__

10. Find the circumference of a circle with a radius of 18 feet. Use 3.14 for π.
    A. 113.04 ft   B. 56.52 ft   C. 36 ft   D. 9 ft   10. __A__

11. Find the circumference of the circle.
    A. 40.82 m   B. 163.28 m
    C. 8,164 m   D. 81.64 m   26 m   11. __D__

12. Find the circumference of a circle with a diameter of $24\frac{1}{2}$ inches.
    A. $77\frac{1}{35}$ in.   B. $152\frac{5}{35}$ in.   C. $76\frac{6}{35}$ in.   D. $75\frac{3}{35}$ in.   12. __C__

13. Find $18 \div \frac{1}{3}$. Write in simplest form.
    A. $\frac{1}{54}$ B. 54   C. 6   D. $\frac{1}{6}$ 13. __B__

14. Solve $\frac{2}{9} \div \frac{3}{5} = x$. Write the solution in simplest form.
    A. $1\frac{1}{9}$ B. $2\frac{1}{2}$ C. $\frac{9}{10}$ D. $\frac{2}{5}$ 14. __A__

© Glencoe/McGraw-Hill   171   *Mathematics: Applications and Connections, Course 1*

Chapter 7 Test, Form 1B (continued)

15. Evaluate $a \div b$ if $a = \frac{5}{18}$ and $b = 3$.
    A. $\frac{5}{21}$ B. $\frac{5}{54}$ C. $10\frac{4}{5}$ D. $\frac{5}{6}$ 15. __B__

16. Find $3\frac{5}{16} \div 2$. Write in simplest form.
    A. $6\frac{5}{8}$ B. $\frac{8}{53}$ C. $1\frac{21}{32}$ D. $\frac{32}{53}$ 16. __C__

17. Solve $2\frac{3}{16} \times 1\frac{3}{4} = m$. Write the solution in simplest form.
    A. $\frac{64}{245}$ B. $3\frac{53}{64}$ C. $\frac{4}{7}$ D. $1\frac{1}{4}$ 17. __D__

18. Divide $5\frac{1}{4}$ by $1\frac{1}{6}$.
    A. $\frac{4}{9}$ B. $\frac{87}{24}$ C. $4\frac{1}{2}$ D. $5\frac{1}{24}$ 18. __C__

19. Complete: 5 T = __?__ lb.
    A. 5,000   B. 10,000   C. 80   D. 1,000   19. __B__

20. Complete: 80 oz = __?__ lb.
    A. 5   B. 1,280   C. 8   D. $6\frac{2}{3}$ 20. __A__

21. Find the next number in the sequence 200, 100, 50, 25, ...
    A. 10   B. 20   C. 12.5   D. 12   21. __C__

22. Find the next number in the sequence 640, 565, 490, 415, ...
    A. 365   B. 75   C. 340   D. 400   22. __C__

23. Find the next number in the sequence $\frac{1}{5}$, $\frac{1}{15}$, $\frac{1}{45}$, $\frac{1}{135}$, ...
    A. $\frac{1}{405}$ B. $\frac{1}{270}$ C. $\frac{1}{180}$ D. 405   23. __A__

24. Keisha needs 12 gallons of water for a science project. The largest container she can find is a quart milk carton. How many times will she need to fill the quart carton to get 12 gallons?
    A. 3   B. 48   C. 24   D. 6   24. __B__

25. Rashaan was measuring the speed of a dropped ball for science class. At 1 second, the speed was 32.16 ft/s. At 2 seconds, it was 64.32 ft/s. At 3 seconds, it was 96.48 ft/s. What will it be after 7 seconds?
    A. 192.96 ft/s   B. 32.16 ft/s
    C. 225.12 ft/s   D. 160.8 ft/s   25. __C__

© Glencoe/McGraw-Hill   172   *Mathematics: Applications and Connections, Course 1*

Assessment and Evaluation

Six forms of Chapter 7 Test are available in the *Assessment and Evaluation Masters* as shown in the chart.

Chapter 7 Test, Form 1B, is shown at the right. Chapter 7 Test, Form 2B, is shown on the next page.

1A	Multiple Choice	Honors
1B	Multiple Choice	Average
1C	Multiple Choice	Basic
2A	Free Response	Honors
2B	Free Response	Average
2C	Free Response	Basic

Objectives & Examples

Review Exercises

● divide fractions *(Lesson 7-5)*

$$\frac{3}{8} \div \frac{2}{3} = \frac{3}{8} \times \frac{3}{2} \quad \textit{Multiply by the reciprocal of } \frac{2}{3}.$$

$$= \frac{9}{16}$$

Find each quotient. Write in simplest form.

28. $\frac{4}{9} \div 8$ $\frac{1}{18}$ **29.** $\frac{5}{6} \div \frac{3}{4}$ $1\frac{1}{9}$

Solve each equation. Write the solution in simplest form.

30. $14 \div \frac{7}{8} = m$ 16   **31.** $\frac{4}{9} \div \frac{2}{3} = d$ $\frac{2}{3}$

● divide mixed numbers *(Lesson 7-6)*

$$5\frac{1}{2} \div 1\frac{5}{6} = \frac{11}{2} \div \frac{11}{6}$$

$$= \frac{\overset{1}{\cancel{11}}}{\underset{1}{\cancel{2}}} \times \frac{\overset{3}{\cancel{6}}}{\underset{1}{\cancel{11}}}$$

$$= \frac{3}{1} \text{ or } 3$$

Find each quotient. Write in simplest form.

32. $1\frac{7}{8} \div 7\frac{1}{2}$ $\frac{1}{4}$ **33.** $3\frac{1}{3} \div 10$ $\frac{1}{3}$

Solve each equation. Write the solution in simplest form.

34. $x = 6 \div 2\frac{2}{3}$ $2\frac{1}{4}$ **35.** $2\frac{1}{4} \div 4\frac{2}{7} = d$ $\frac{21}{40}$

● change units within the customary system *(Lesson 7-7)*

a. 5 qt = _?_ pt   *2 pt = 1 qt*
       larger to smaller → multiply

   $5 \times 2 = 10$

   5 qt = 10 pt

b. 64 oz = _?_ lb   *16 oz = 1 lb*
       smaller to larger → divide

   $64 \div 16 = 4$

   64 oz = 4 lb

Complete.

36. 45 c = _?_ qt $11\frac{1}{4}$

37. $5\frac{1}{2}$ T = _?_ lb   11,000

38. 2.75 gal = _?_ pt   22

39. 3 pt = _?_ fl oz   48

40. 12 oz = _?_ lb $\frac{3}{4}$

● recognize and extend sequences *(Lesson 7-8)*

Find the next number in the sequence
24, 28, 32, 36, ...

In this sequence, 4 is added to each number. The next number is 36 + 4, or 40.

Find the next two numbers in each sequence.

41. 256, 236, 216, 196, . . . **176, 156**

42. $\frac{1}{4}$, $\frac{1}{2}$, 1, 2, 4, . . . **8, 16**

Find the missing number in each sequence.

43. $\frac{2}{5}$, _?_ , 10, 50, . . . **2**

44. $6, 7\frac{1}{3}$, _?_ , 10, . . . $8\frac{2}{3}$

304 Chapter 7 Multiplying and Dividing Fractions

Assessment and Evaluation Masters, pp. 177–178

Test and Review Software

You may use this software, a combination of an item generator and item bank, to create your own tests or worksheets. Types of items include free response, multiple choice, short answer, and open ended.

CD-ROM Program

The CD-ROM Program contains an Assessment Game whose questions review the concepts in this chapter.

Applications & Problem Solving

45. *Statistics* Yana conducted a survey to find out how many students would participate in a school play. Out of 76 students, $\frac{3}{4}$ said they would participate in a play. How many students would participate in the play? *(Lesson 7-2)* **57 students**

46. *Geometry* To find the perimeter of a square, you can use the formula $P = 4s$, where s is the length of one side of the square. Find the perimeter of the square. *(Lesson 7-3)* **22 ft**

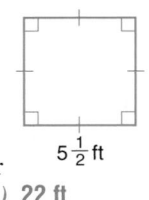
$5\frac{1}{2}$ ft

47. *Food* Adwoa bought 9 gallons of apple cider for the school party. How many 1-cup servings will he be able to serve? *(Lesson 7-7)* **144 one-cup servings**

48. *Look for a Pattern* Matsuko is using a computer to type her term paper. The table shows how many pages she has typed each day. If the pattern continues, how many pages long is the term paper? (Assume that she completes the term paper on the fourth day). *(Lesson 7-8A)* **9 pages**

Day	1	2	3	4
Number of Pages Typed	3	5	7	?

Alternative Assessment

Performance Task

Suppose you use the following ingredients to make one batch of modeling dough for your science project.

1 T oil	1 c salt
$2\frac{1}{2}$ c water	$2\frac{1}{2}$ c flour
2 pkg. soft drink mix	$1\frac{1}{2}$ t cream of tartar

How will you decide the amount of each ingredient needed to make 3 batches of the dough? **See margin.**

Suppose you need to make $2\frac{1}{2}$ batches of the dough. How much of each ingredient will you need? **See margin.**

A practice test for Chapter 7 is provided on page 601.

Completing the CHAPTER Project

Use the following checklist to complete your project.

☑ Your paper airplane is included.

☑ Directions for making the paper airplane are included.

☑ The flight table containing the flight distances and flight times of your paper airplane is included.

Add any finishing touches that you would like to make your project attractive.

 PORTFOLIO   Select one of the assignments from this chapter and place it in your portfolio. Attach a note to it explaining why you selected it.

Applications & Problem Solving

This section provides additional practice in solving real-world problems that involve the skills of this chapter.

Alternative Assessment

The **Performance Task** provides students with a performance assessment opportunity to evaluate their work and understanding.

CHAPTER Project

Students should complete the final stages of their project and prepare a class demonstration of their results. A scoring guide for the project is available in the *Investigations and Projects Masters*, p. 43.

 PORTFOLIO   Students should add to their portfolios at this time.

Assessment and Evaluation Masters, p. 181

Additional Answers for the Performance Task

- Multiply each ingredient by 3.
- To make $2\frac{1}{2}$ batches of modeling dough, you will need 5 pkg. of soft drink mix, $2\frac{1}{2}$ T oil, $6\frac{1}{4}$ c water, $2\frac{1}{2}$ c salt, $6\frac{1}{4}$ c flour, and $3\frac{3}{4}$ tsp cream of tartar.

 ## Performance Assessment

Additional performance assessment tasks for this chapter are included in the *Assessment and Evaluation Masters* on page 181. A scoring guide is also provided on page 193.

Standardized Test Practice
Assessing Knowledge & Skills

The Standardized Test Practice may be used to help students prepare for standardized tests. The test items are written in the same style as those in state proficiency tests and standardized tests like CAT, CTBS, ITBS, MAT, SAT, and Terra Nova. The test items cover skills and concepts covered up to this point in the text.

The pages can be used as an overnight assessment. After students have completed the pages, discuss how each problem can be solved, or provide copies of the solutions from the *Solutions Manual*.

Assessment and Evaluation Masters, p. 187

Section One: Multiple Choice

There are twelve multiple-choice questions in this section. Choose the best answer. If a correct answer is *not here*, mark the letter for Not Here.

1. What are the next two numbers in the sequence 1,024, 256, 64, 16, . . . ? **C**
   - **A** 8, 4
   - **B** 8, 2
   - **C** 4, 1
   - **D** 4, 0

2. What is the area of the rectangle? **F**

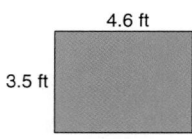

4.6 ft

3.5 ft

   - **F** 16.1 sq ft
   - **G** 14.0 sq ft
   - **H** 12.3 sq ft
   - **J** 12.0 sq ft

3. Jamonte's room is 8 feet wide. How many inches is this? **B**
   - **A** 92 inches
   - **B** 96 inches
   - **C** 20 inches
   - **D** 44 inches

4. Evaluate $p \div m - 4$ if $p = 10$ and $m = 2$. **F**
   - **F** 1
   - **G** 0
   - **H** 5
   - **J** 2

5. Express $3 \times 4 \times 4 \times 4 \times 5$ in exponential notation. **B**
   - **A** $3^2 \times 4^2 \times 5$
   - **B** $3 \times 4^3 \times 5$
   - **C** $3 \times 4^2 \times 5$
   - **D** $3 \times 4^4 \times 5$

Please note that Questions 6–12 have five answer choices.

6. How many 1-foot square tiles are needed to cover the floor of a kitchen that is 13 feet by 10 feet? **G**
   - **F** 13
   - **G** 130
   - **H** 1,300
   - **J** 13,000
   - **K** 130,000

7. A sign at The Gift Gallery reads:

Clearance Sale
$2.99
to
$9.99

   Suppose a customer purchases four items that are on sale. What is a reasonable total cost of the items, without tax? **B**
   - **A** $5
   - **B** $30
   - **C** $50
   - **D** $10
   - **E** $75

8. A car travels at an average speed of 64 miles per hour. At this rate, how many hours will it take to travel 384 miles? **J**
   - **F** 2 hours
   - **G** 4 hours
   - **H** 3 hours
   - **J** 6 hours
   - **K** 5 hours

◄◄◄**Instructional Resources**

Another cumulative review is shown at the left and is available in the *Assessment and Evaluation Masters*, p. 187.

---

9. Suppose a car is traveling at a speed of 51 miles per hour. Which is a good estimate for the number of hours it will take to travel 248 miles? **D**

  A 2 h

  B 3 h

  C 4 h

  D 5 h

  E 6 h

10. Mrs. Alvarez bought groceries for $19.58. How much change should she receive from a $20 bill? **G**

  F $1.42

  G $0.42

  H $1.52

  J $0.52

  K Not Here

11. Mr. Meuser is planning to travel from Dallas to New York. The flight will take about 3 hours. It is time to leave, but the flight has been delayed for $2\frac{1}{2}$ hours. About how long will it be before he arrives in New York? **A**

  A $5\frac{1}{2}$ hours

  B $6\frac{1}{2}$ hours

  C $4\frac{1}{2}$ hours

  D 5 hours

  E Not Here

12. Mrs. Miles purchased a ham that weighed $6\frac{1}{4}$ pounds and a turkey that weighed $11\frac{1}{2}$ pounds. How many pounds of meat did she purchase? **H**

  F $17\frac{1}{2}$ pounds **G** $18\frac{1}{4}$ pounds

  H $17\frac{3}{4}$ pounds **J** $16\frac{1}{4}$ pounds

  K Not Here

Test-Taking Tip

If you're working on a group of questions and find that the questions are getting too difficult, quickly read through the rest of the questions in the section and answer the ones that you know. Then come back to the ones you skipped.

Section Two: Free Response

This section contains six questions for which you will provide short answers. Write your answers on your paper.

13. Jamaal works 22 hours a week and earns $111.10. How much money does he earn per hour? **$5.05**

14. Order the fractions $\frac{2}{7}, \frac{1}{3}, \frac{2}{9}, \frac{2}{5}, \frac{4}{1}$ from greatest to least. **See margin.**

15. Andrea had purchased $6\frac{1}{2}$ yards of material to make curtains for her dining room. If she used 6 yards of material to make the curtains, how much of the material was not used? $\frac{1}{2}$ **yd**

16. How many fluid ounces are in 3 quarts? **96**

17. Draw the next figure in the sequence.

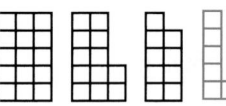

18. Four dogs have weights of 17, 24, 18, and 25 pounds. What is the average weight of the dogs? **21 pounds**

Chapters 1–7 Standardized Test Practice **307**

Assessment and Evaluation Masters, pp. 185–186

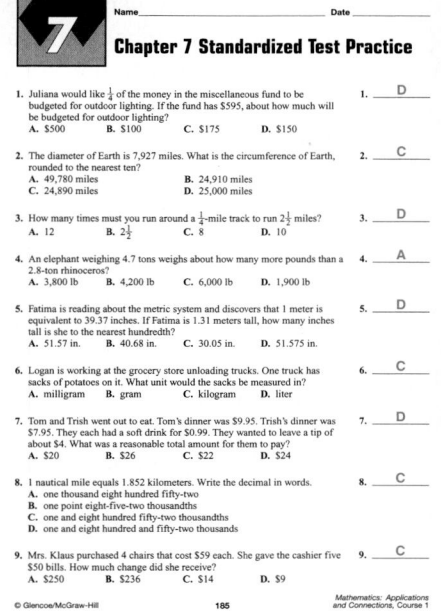

Additional Answer

14. $\frac{4}{1}, \frac{2}{5}, \frac{1}{3}, \frac{2}{7}, \frac{2}{9}$

Instructional Resources ►►►

Additional standardized test practice is shown at the right and is available in the *Assessment and Evaluation Masters*, pp. 185–186.

Exploring Ratio, Proportion, and Percent

Overview

In this chapter, students learn to express ratios and rates as fractions and to solve proportions by using cross products. Then, using their knowledge of ratios, students find actual lengths from scale drawings. Students explore the relationship among fractions, decimals, and percents as well as find the percent of a number. Students also learn to solve problems by drawing a diagram.

Lesson (pages)	Lesson Objectives	NCTM Standards	Standardized Tests	State/Local Objectives
8-1A (310–311)	Use tangram pieces to explore ratios and the relationship between ratio and area.	1–3, 5, 12		
8-1 (312–315)	Express ratios and rates as fractions.	1–7, 12, 13	CAT, ITBS, MAT	
8-1B (316)	Explore ratios and probability.			
8-2 (317–320)	Solve proportions by using cross products.	1–7, 9, 12, 13	MAT	
8-2B (321)	Use spreadsheets to solve problems involving proportions.			
8-3A (322–323)	Solve problems by drawing diagrams.	1–4, 7, 8, 12, 13	CTBS, MAT, SAT, TN	
8-3 (324–327)	Find actual length from a scale drawing.	1–9, 12, 13	CTBS, MAT, SAT, TN	
8-4A (329)	Illustrate the meaning of percent using models.	1–7	CTBS, TN	
8-4 (330–333)	Express percents as fractions and vice versa.	1–7, 9, 10	CTBS, ITBS, TN	
8-5 (334–336)	Express percents as decimals and vice versa.	1–8, 10, 12, 13	CTBS, ITBS, TN	
8-6 (337–339)	Estimate the percent of a number.	1–7, 9, 10	CTBS, ITBS, TN	
8-7 (340–343)	Find the percent of a number.	1–7, 9, 10, 12, 13	CTBS, ITBS, TN	

CAT = California Achievement Tests, CTBS = Comprehensive Tests of Basic Skills, ITBS = Iowa Tests of Basic Skills, MAT = Metropolitan Achievement Tests, SAT = Stanford Achievement Tests, TN = Terra Nova

Organizing the Chapter

CD-ROM

All of the blackline masters in the Teacher's Classroom Resources are available on the **Electronic Teacher's Classroom Resources** CD-ROM.

LESSON PLANNING GUIDE

Lesson	Extra Practice (Student Edition)	BLACKLINE MASTERS (PAGE NUMBERS)										
		Study Guide	Practice	Enrichment	Assessment & Evaluation	Classroom Games	Diversity	Hands-On Lab	School to Career	Science and Math Lab Manual	Technology	Transparencies A and B
8-1A								54				
8-1	p. 578	57	57	57								8-1
8-1B								55				
8-2	p. 578	58	58	58	211				8			8-2
8-2B												
8-3A	p. 578											
8-3	p. 579	59	59	59							16	8-3
8-4A								56				
8-4	p. 579	60	60	60	210, 211							8-4
8-5	p. 579	61	61	61						81–84		8-5
8-6	p. 580	62	62	62	212		8	76		85–88		8-6
8-7	p. 580	63	63	63	212	21–22					15	8-7
Study Guide/ Assessment					197–209, 213–215							

OTHER CHAPTER RESOURCES

Student Edition

Chapter Project, pp. 309, 315, 327, 333, 347
School to Career, p. 328
Let the Games Begin, p. 343

Technology

 CD-ROM Program

 Interactive Mathematics Tools Software

Teacher's Classroom Resources

Applications
Family Letters and Activities, pp. 15–16
Investigations and Projects Masters, pp. 45–48
Meeting Individual Needs
Investigations for the Special Education Student, pp. 21–28

Teaching Aids
Answer Key Masters
Block Scheduling Booklet
Lesson Planning Guide
Solutions Manual

Professional Publications
Glencoe Mathematics Professional Series

Planning the Chapter

MindJogger Videoquizzes provide a unique format for reviewing concepts presented in the chapter.

ASSESSMENT RESOURCES

Student Edition

Mixed Review, pp. 315, 320, 327, 333, 336, 339, 343
Mid-Chapter Self Test, p. 327
Math Journal, pp. 313, 338
Study Guide and Assessment, pp. 344–347
Performance Task, p. 347
Portfolio Suggestion, p. 347
Standardized Test Practice, pp. 348–349
Chapter Test, p. 602

Assessment and Evaluation Masters

Multiple-Choice Tests (Forms 1A, 1B, 1C), pp. 197–202
Free-Response Tests (Forms 2A, 2B, 2C), pp. 203–208
Performance Assessment, p. 209
Mid-Chapter Test, p. 210
Quizzes A–D, pp. 211–212
Standardized Test Practice, pp. 213–214
Cumulative Review, p. 215

Teacher's Wraparound Edition

5-Minute Check, pp. 312, 317, 324, 330, 334, 337, 340
Building Portfolios, p. 308
Math Journal, pp. 311, 316, 321, 329
Closing Activity, pp. 315, 320, 323, 327, 333, 336, 339, 343

Technology

Test and Review Software
MindJogger Videoquizzes
CD-ROM Program

MATERIALS AND MANIPULATIVES

Lesson 8-1A
patty paper
scissors*

Lesson 8-1
grid paper†
ruler*

Lesson 8-1B
spinner*

Lesson 8-2
pattern blocks*

Lesson 8-2B
computer
spreadsheet software

Lesson 8-4A
grid paper†
colored pencils

Lesson 8-4
wastebasket
coin
grid paper†
colored pencils

Lesson 8-7
index cards
scissors*

*Glencoe Manipulative Kit   †Glencoe Overhead Manipulative Resources

PACING CHART

See pages T25–T27 for the Course Planning Calendar.

COURSE	DAY 1	DAY 2	DAY 3	DAY 4	DAY 5	DAY 6	DAY 7
Standard	Chapter Project	Lessons 8-1A & 8-1			Lesson 8-2	Lesson 8-3A	Lesson 8-3
Honors	Chapter Project	Lessons 8-1 & 8-1B		Lessons 8-2 & 8-2B		Lesson 8-3A	Lesson 8-3
Block	Chapter Project & Lessons 8-1A & 8-1		Lessons 8-2 & 8-3A	Lessons 8-3 & 8-4A	Lessons 8-4 & 8-5	Lessons 8-6 & 8-7	Study Guide and Assessment, Chapter Test

Interactive Mathematics:
Activities and Investigations

is an activity-based program that may be used as an enhancement for chapters in *Mathematics: Applications and Connections.*

Activities and Investigations

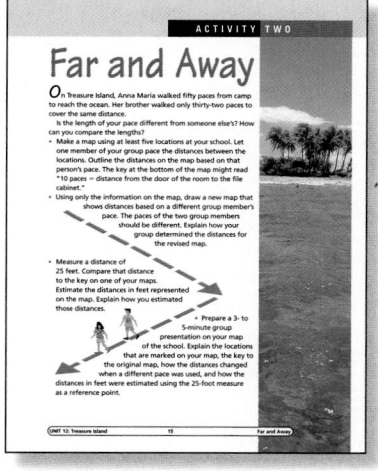

Unit 12, Activity Two
Use with Lesson 8-2.

Summary Students work in groups to make a map of a variety of locations in the school, using the paces of two members of the group as unit measures. They develop a key comparing a number of paces with a specific distance.

Math Connection Students compare units of measure and use proportions. They draw maps based on nonstandard units of measure. They then determine distances on the map using a standard unit of measure.

Unit 2, Activity Five
Use with Lesson 8-7.

Summary Students use spreadsheets to investigate how to spend one billion dollars under certain regulations. Students write a report on their findings, explaining their expenditures.

Math Connection Students use a computer and spreadsheet software to organize their spending into categories. Because students must stay within 12% of the entire budget in each category, they need to find percents of whole numbers.

DAY 8	DAY 9	DAY 10	DAY 11	DAY 12	DAY 13	DAY 14	DAY 15
Lessons 8-4A & 8-4		Lesson 8-5	Lesson 8-6	Lesson 8-7	Study Guide and Assessment	Chapter Test	
Lesson 8-4	Lesson 8-5	Lesson 8-6	Lesson 8-7	Study Guide and Assessment	Chapter Test		

Enhancing the Chapter

APPLICATIONS

Classroom Games, pp. 21–22

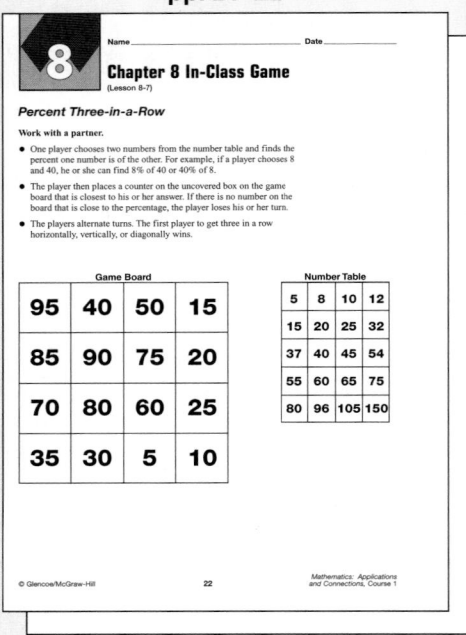

Diversity Masters, p. 8

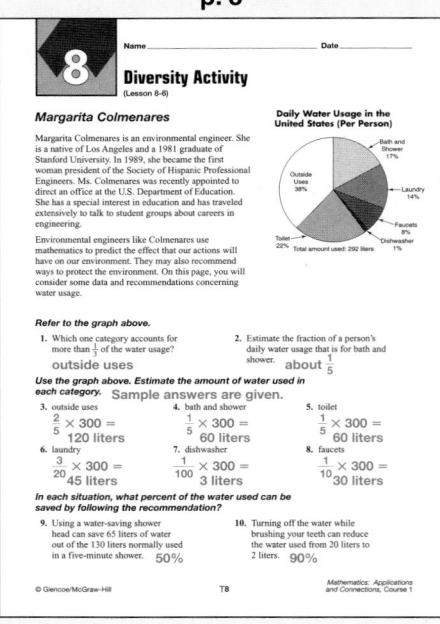

School to Career Masters, p. 8

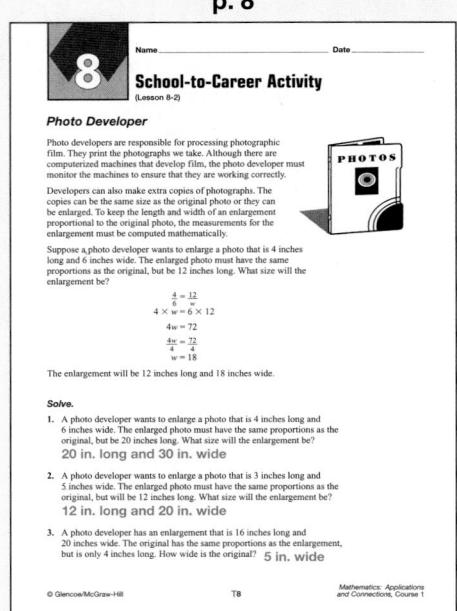

Family Letters and Activities, pp. 15–16

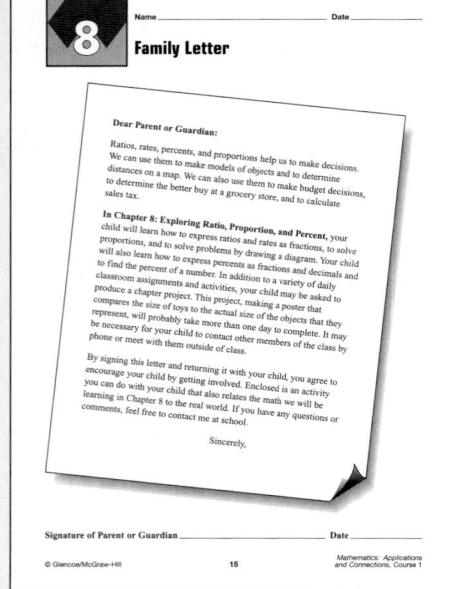

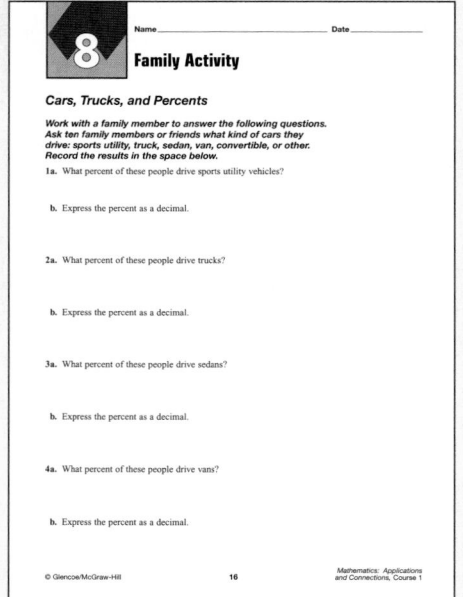

Science and Math Lab Manual, pp. 81–88

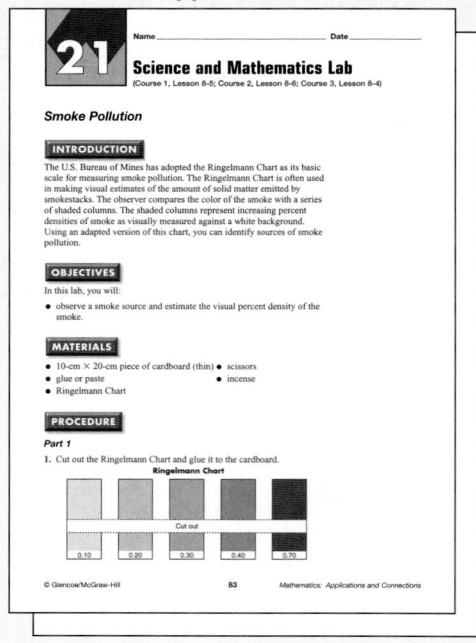

Hands-On Lab Masters, p. 76

8
Name _____ Date _____

Lab Activity
(Lesson 8-6)

Percent and Estimation

Estimate the percent of each figure that is shaded by just looking. Then count the grid squares to find the actual percent shaded. **All estimates are sample answers.**

1.
Estimate: 75%
Actual: 60%

2.
Estimate: 50%
Actual: 50%

3.
Estimate: 35%
Actual: 35%

4.
Estimate: 50%
Actual: 51%

5. How did your estimates compare with the actual percents? **Sample answer: Estimates are close to the actual percents, but not exact.**

6. Shade your own grid. Estimate the percent shaded and count to find the exact percent. **Sample answer in red.**
Estimate: 25%
Actual: 20%

© Glencoe/McGraw-Hill   76   Mathematics: Applications and Connections, Course 1

Assessment and Evaluation Masters, pp. 210–212

8
Name _____ Date _____

Chapter 8 Mid-Chapter Test
(Lessons 8-1 through 8-4)

1. Write the ratio in three different ways: 7 out of 23 M & M's are brown.

1. $\frac{7}{23}$; 7:23; 7 out of 23

Express each ratio as a fraction in simplest form.
2. 22 games won out of 44 games played

2. $\frac{1}{2}$

3. 15 sheep out of 55 animals

3. $\frac{3}{11}$

Express each ratio as a rate.
4. 360 miles in 6 hours   5. 4 tapes for $23.96

4. 60 mph

5. $5.99 per tape

Use cross products to determine whether each pair of ratios forms a proportion.
6. $\frac{9}{20}, \frac{18}{43}$ 7. $\frac{5}{10}, \frac{4}{8}$

6. no

7. yes

Solve each proportion.
8. $\frac{2}{20} = \frac{6}{w}$ 9. $\frac{6}{c} = \frac{18}{30}$

8. 60

9. 10

10. $\frac{7}{36} = \frac{6}{8}$

10. 27

11. Leslie is setting up a singles tennis tournament. She wants each player to play every other player one time. If 4 players enter, how many games will Leslie need to set up?

11. 6 games

Scale drawings were made of four famous structures. The scale for each drawing is 1 cm = 30 m. The given measurement is the size of the drawing to the nearest centimeter. Find the actual height of each structure.
12. Statue of Liberty, New York City, 3 cm

12. 90 m

13. Washington Monument, Washington, D.C., 5 cm

13. 150 m

14. Gateway Arch, St. Louis, Missouri, 6 cm

14. 180 m

If you were to make a scale drawing for the following structures with a scale of 1 in. = 50 ft, what would be the measurement of the drawing?
15. a skyscraper 500 ft tall   16. a space shuttle 125 ft long

15. 10 in.

16. 2.5 in.

Express each percent as a fraction in simplest form.
17. 56%   18. 143%

17. $\frac{14}{25}$

18. $1\frac{43}{100}$

Express each fraction as a percent.
19. $\frac{2}{5}$ 20. $\frac{9}{50}$

19. 40%

20. 18%

© Glencoe/McGraw-Hill   210   Mathematics: Applications and Connections, Course 1

8
Name _____ Date _____

Chapter 8 Quiz A
(Lessons 8-1 and 8-2)

Express each ratio as a fraction in simplest form.
1. 45 people out of 60

1. $\frac{3}{4}$

2. 12 red marbles out of 28 marbles

2. $\frac{3}{7}$

Express each ratio as a rate.
3. 178 feet in 2 seconds

3. 89 ft per s

4. 12 muffins for $3

4. $0.25 per muffin

5. 24 cans of soft drink for $4.99

5. $0.21 per can

Solve each proportion.
6. $\frac{18}{m} = \frac{50}{100}$ 7. $\frac{27}{36} = \frac{x}{16}$ 8. $\frac{45}{60} = \frac{60}{y}$

6. 36

7. 12

8. 80

9. Use cross products to determine if $\frac{9}{19}$ and $\frac{12}{42}$ form a proportion.

9. no

10. Use cross products to determine if $\frac{7}{12}$ and $\frac{14}{24}$ form a proportion.

10. yes

Chapter 8 Quiz B
(Lessons 8-3 and 8-4)

Solve by drawing a diagram.
1. The Hys choose mirror squares that are 16 inches by 16 inches to cover a wall that is 8 feet tall and 4 feet wide. How many mirror squares are needed?

1. 18

2. The distance between two cities on a map is $3\frac{3}{4}$ inches. The scale reads "1 in. = 24 mi." What is the actual distance between the cities?

2. 90 miles

Express each fraction as a percent.
3. $\frac{23}{25}$ 4. $\frac{13}{20}$

3. 92%

4. 65%

5. $\frac{7}{8}$ 6. $\frac{7}{20}$

5. 87.5%

6. 35%

Express each percent as a fraction.
7. 95%   8. 15%

7. $\frac{19}{20}$

8. $\frac{3}{20}$

9. 24%   10. 58%

9. $\frac{6}{25}$

10. $\frac{29}{50}$

© Glencoe/McGraw-Hill   211   Mathematics: Applications and Connections, Course 1

Technology Masters, pp. 15–16

8
Name _____ Date _____

Calculator Activity
(Lesson 8-7)

The Percent Key

Most calculators have a special key to work with percents.

Example   Find 25.5% of 82.
Enter: 82 × 25.5 2nd [%] = 20.91
The solution is 20.91.

Find the percent of each number.
1. 80% of 40   32
2. 16% of 18   2.88
3. 7% of 35   2.45
4. 12% of 42   5.04
5. 1.8% of 14   0.252
6. 0.5% of 30   0.15
7. 120% of 18   21.6
8. 33% of 300   99
9. 99% of 200   198
10. 25% of 412   103
11. 8.4% of 17   1.428
12. 213% of 241   513.33
13. 0.01% of 120   0.012
14. 0.12% of 310   0.372
15. 0.15% of 114   0.171
16. 114% of 18.2   20.748
17. 3.1% of 30.5   0.9455
18. 5.2% of 92.4   4.8048
19. 9.25% of 12.4   1.147
20. 122% of 0.055   0.0671

© Glencoe/McGraw-Hill   T15   Mathematics: Applications and Connections, Course 1

8
Name _____ Date _____

Graphing Calculator Activity
(Lesson 8-3)

Scale Drawings

You can use a graphing calculator to draw geometric shapes and scale drawings. First, clear all lists by pressing 2nd [MEM] 4 ENTER.

Example   An architect was given a floor plan with corners at $A(0, 0)$, $B(2, 0)$, $C(2, 1)$, $D(3, 1)$, $E(3, 4)$, and $F(0, 4)$. Graph the scale drawing of the floor plan that has a scale of 1 to 2.

Step 1   Press STAT ENTER and enter the x-coordinates in column L1 and the y-coordinates in column L2. Repeat the first entry in each column at the end of each column.

Step 2   To graph a scale drawing with a scale factor of 2, multiply each coordinate in the ordered pairs by 2. Enter the new coordinates in columns L3 and L4.

Step 3   You must turn on the two point plots.
Enter: 2nd [STAT PLOT] ENTER ENTER ▼ ►
ENTER ▲ ◄ ENTER ENTER ▼ ►
ENTER ▼ 2nd [L3] ▼ 2nd [L4]

Step 4   Press WINDOW and make entries to match the settings shown.

Step 5   Press GRAPH.

Use a graphing calculator to draw the polygons and their scale drawings. (Reset WINDOW so that Xmax = 20 and Ymax = 20.) Sketch the drawings. Label the vertices and coordinates of each polygon.
1. $A(1, 2)$; $B(3, 4)$; $C(4, 3)$ scale: 1 to 5
2. $A(4, 8)$; $B(16, 16)$; $C(12, 4)$ scale: 4 to 1
3. $A(1, 6)$; $B(6, 4)$; $C(4, 2)$ scale: 1 to 4
4. $A(1, 1)$; $B(1, 3)$; $C(3, 3)$; $D(3, 1)$ scale: 1 to 4
5. $A(1.8, 9)$; $B(12, 9)$; $C(16, 18)$; $D(21, 18)$ scale: 3 to 1
6. Why did the upper right-hand corner of Exercise 5 not show on the screen?
   The x-coordinate of a vertex is 21. This is greater than the Xmax setting.

© Glencoe/McGraw-Hill   T16   Mathematics: Applications and Connections, Course 1

Investigations for the Special Education Student, pp. 21–28

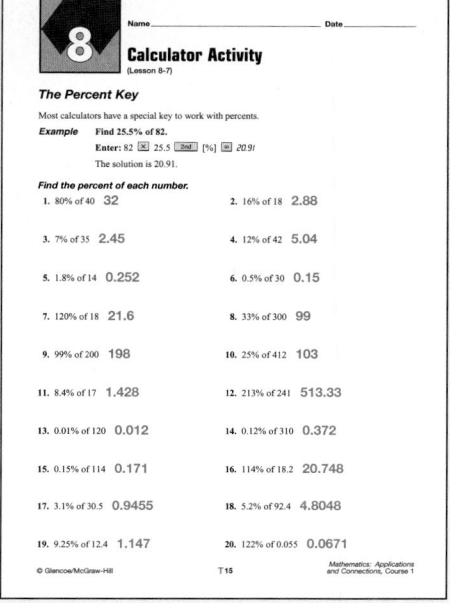

Use with:
Course 1-Chapter 10
Course 2-Chapter 11
Course 3-Chapter 9

Investigation 9 Teacher's Guide

Park It!

Overview
Students will work in groups of four to design a park according to preset criteria. During this investigation, students will experience firsthand how measurement, proportions, and decimal operations are necessary for careers that involve design. Each student is responsible for a certain job in the group: engineer, draftsperson, finance officer, or public relations person. The culminating point of the investigation is a videotaped, oral presentation of the designs.

Activity Goals
Students will:
• design a park to scale using proportions,
• use decimal operations for budgeting, and
• work cooperatively and responsibly with a group.

Planning the Instruction
Prerequisite Skills
Students should have a significant amount of practice measuring, solving proportions, using scales, and computing with decimals.
Materials
• investigation worksheets
• calculators
• rulers, measuring tape
• drawing paper
• crayons, colored pencils, markers
• video camera and videotape

Time Needed
ten 45-minute periods

Procedure
1. Discuss the investigation with the class. Explain that their task is to design a park to scale that will fit on a 200-foot by 300-foot piece of land. The investigation will end with a videotaped, oral presentation to the class.
2. Brainstorm with the entire class a list of features they would like to include in their designs.
3. Set up student groups.
4. Have each group meet and compile their own list of features, using ideas from the class list.
5. Describe the different jobs to the class. Then have the groups discuss and sign the Group Job Contract worksheet according to their responsibilities.
6. Have students begin their designated jobs according to their Job Specification worksheets.
7. Monitor student progress through observations and group meetings. In addition, be available to answer questions when necessary.
8. Set a date for the presentations.
9. Conduct and videotape presentations while students evaluate each other.
10. Collect evaluations and have students view the recorded presentations.

Adaptations and Variations
The following are some ideas on how this investigation may be changed depending on student population.

LD • Allow more time.
• If measuring is difficult, allow students to use nonstandard units such as their hands.
• If converting measurements is difficult using scale factors and proportions, allow students to convert using nonstandard scale units such as one hand in real length equals one paper clip in drawing length.

PH • For students who have trouble drawing, allow them to use computer graphics.

CD • If a communicatively-troubled student should choose to be the public relations person, have each group member say several lines of the speech.

BD • Assign job specification tasks one at a time, with a reward system in place for each one completed in an appropriate manner.

HI • Have another student sign each presentation.

VI • Make sure students give a detailed description of their designs when presenting.

© Glencoe/McGraw-Hill   21   Mathematics: Applications and Connections

Theme: Toys

For many years toys were created by individual craftspeople. However, in the twentieth century, manufacturing processes made it possible for millions of children to own the exact same toy. One of the most popular manufactured toys is the Matchbox car, created in 1947 by Leslie and Rodney Smith. Another popular toy is the Barbie doll, created by Ruth Handler in 1959.

Question of the Day Original Matchbox cars have a 1:75 scale. This means 1 inch on the model equals 75 inches on the actual car. If an original Matchbox London bus is $2\frac{1}{4}$ inches long, how long is an actual London bus, to the nearest foot? **14 ft**

Assess Prerequisite Skills

Ask students to read through the list of objectives presented in "What you'll learn in Chapter 8." You may wish to ask them what each of the objectives means or if they have experienced or used any of these math concepts before.

 Building Portfolios

Encourage students to revise their portfolios as they study this chapter. Have them include work that illustrates their growing understanding of ratios, proportions, and percents.

 **Math and the Family**

In the *Family Letters and Activities* booklet (pp. 15–16), you will find a letter to the parents explaining what students will study in Chapter 8. An activity appropriate for the whole family is also available.

What you'll learn in Chapter 8

- to express ratios and rates as fractions,
- to solve proportions by using cross products,
- to solve problems by drawing a diagram,
- to find actual length from a scale drawing,
- to express percents as fractions and as decimals, and
- to find the percent of a number.

308 Chapter 8 Exploring Ratio, Proportion, and Percent

 CD-ROM Program

Activities for Chapter 8
- Chapter 8 Introduction
- Interactive Lessons 8-1, 8-2, 8-6, 8-7
- Extended Activity 8-4
- Assessment Game
- Resource Lessons 8-1 through 8-7

CHAPTER Project

THE WONDERFUL WORLD OF TOYS

Many toys are small replicas of actual objects. For example, Matchbox cars are small replicas of cars. In this project, you will use ratios and proportions to find and compare the size of toys and the actual size of the objects that they represent. You will then make a poster to illustrate your findings.

Getting Started

- Look at the drawings of the two race cars. What is the length of stock car 88? What is the height of stock car 16?
- How could you use these dimensions to calculate the actual length and height of the stock cars that they represent?

Stock Car 88 Scale:
1 cm = 144 cm

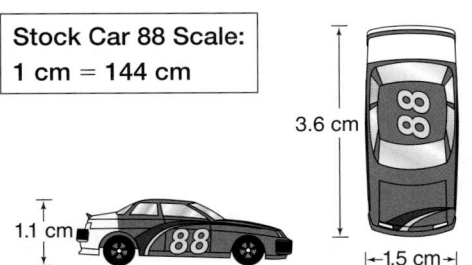

3.6 cm

1.1 cm

|←1.5 cm→|

Stock Car 16 Scale:
1 cm = 64 cm

2.3 cm

3.0 cm

|←———— 7.7 cm ————→|

Technology Tips

- Use a **spreadsheet** to record the measurements of toys and actual objects and to calculate measurements.
- Use **publishing software** to make parts of your poster such as graphs, tables, and drawings.
- Use a **calculator** to help you find the actual dimensions of the stock cars.

inter NET CONNECTION For up-to-date information on toy cars, visit: **www.glencoe.com/sec/math/mac/mathnet**

Working on the Project

You can use what you'll learn in Chapter 8 to help you make your poster.

Page	Exercise
315	35
327	11
333	37
347	Alternative Assessment

Instructional Resources ▶▶▶
A recording sheet to help students organize their data for the Chapter Project is shown at the right and is available in the *Investigations and Projects Masters*, p. 48.

CHAPTER Project
NOTES

Objectives Students should
- use ratios and proportions to find and compare the size of toys and the actual size of the objects that they represent.
- make a poster to illustrate their findings.

Project Pointer You may suggest that students begin a *Project Folder* to keep their work as they complete each stage of the Chapter Project. The completed project may also be added to their portfolios.

Suggest that students give an oral presentation explaining the results of their comparison. Then have them display their posters in the classroom.

Using the Diagrams Have students explain in their own words what the diagrams represent. Ask them why they think it would be important to draw diagrams to scale. Then have them describe how ratios and proportions can be used to make toy cars.

Investigations and Projects Masters, p. 48

HANDS-ON
LAB

GET READY

Objective Students use tangram pieces to explore ratios and the relationship between ratio and area.

Optional Resources
Hands-On Lab Masters
• worksheet, p. 54
• tangram, p. 15

Manipulative Kit
• scissors

MANAGEMENT TIPS

Recommended Time
30 minutes

Getting Started Patty paper can be purchased at restaurant supply stores. If patty paper is not available, squares of waxed paper can be used.
Prepare a tangram to be displayed on the overhead. Show how the pieces may be reconfigured to form many creative, different shapes. Show that a tangram always has seven pieces—five triangles, a square, and a rhombus.

The **Activity** leads students through five steps of cutting and folding to make a tangram. At Steps 1, 3, 4, and 5, students can use their tangram pieces to verify the given ratios.

COOPERATIVE LEARNING

8-1A Ratios

A Preview of Lesson 8-1

◇ 2 sheets of patty paper

✂ scissors

Have you ever put together a tangram? The tangram was first developed in China and was very popular in the 1800s. In this lab, you will construct a tangram and use ratios to compare the areas of the tangram pieces. Recall that a ratio is a comparison of two numbers by division.

TRY THIS

Work with a partner.

Step 1 Using one sheet of patty paper, fold the top left corner to the bottom right corner. Crease the paper and unfold the square. Cut along the fold.

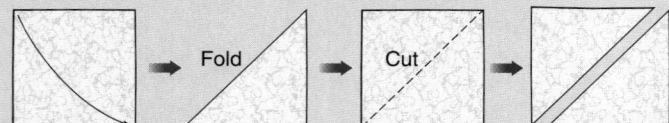

The ratio 1 to 2 compares the area of one of the triangles to the area of the uncut square.

Step 2 Using one of the triangles, fold the bottom left corner to the bottom right corner. Make a crease and unfold the triangle. Cut along the fold. Label the triangles A and B.

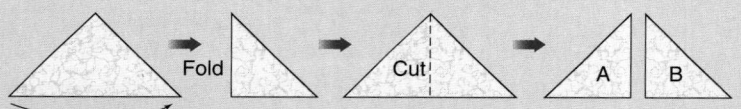

Step 3 Using the other triangle, fold the bottom left corner to the bottom right corner as shown. Make a crease and unfold the triangle. Then fold the top corner to the bottom edge along the crease. Crease and cut on this second crease line. Label the small triangle C.

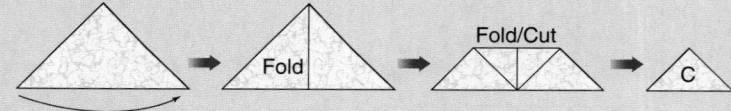

The ratio 1 to 3 compares the area of triangle C to the area of the remaining piece.

310 Chapter 8 Exploring Ratio, Proportion, and Percent

Step 4 Cut on the crease of the remaining piece as shown. Using the piece on the left, fold the bottom left corner to the bottom right corner. Make a crease and unfold. Cut on this fold. Label the triangle D and the square E.

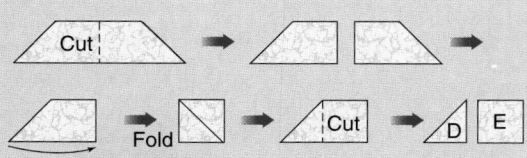

The ratio 1 to 6 compares the area of triangle D to the area of the original figure in this step. The ratio 1 to 3 compares the area of square E to the area of the original figure.

Step 5 Using the remaining piece, fold the bottom left corner to the top right corner. Make a crease and unfold. Cut along the fold. Label the triangle F and the other figure G.

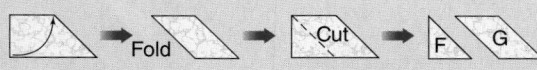

The ratio 1 to 3 compares the area of triangle F to the area of the original figure in this step.

1a. $\frac{1}{2}$ square unit   **1b.** Triangle C is $\frac{1}{2}$ the size of triangle B.

2a. $\frac{1}{4}$ square unit   **2b.** Triangle F is $\frac{1}{4}$ the size of triangle B.

3a. $\frac{1}{2}$ **3b.** $\frac{1}{4}$ **3c.** 1   **3d.** $\frac{2}{1}$ **3e.** $\frac{1}{2}$ **3f.** $\frac{1}{2}$

ON YOUR OWN

1a. If the area of triangle B is 1 square unit, what is the area of triangle C?

 b. How does the area of triangle C compare to the area of triangle B?

2a. If the area of triangle B is 1 square unit, what is the area of triangle F?

 b. How does the area of triangle F compare to the area of triangle B?

3. Write a ratio that compares the areas of the tangram pieces. Express each ratio as a fraction.

a. C to A	**b.** F to A	**c.** A to B
d. B to C	**e.** D to E	**f.** F to G

4. *Look Ahead* Write a ratio that compares the number of yellow squares to the total number of squares. Express the ratio as a fraction in simplest form. $\frac{1}{3}$

ASSESS

Have students complete Exercises 1–3. Encourage students to experiment with their tangram pieces to find the areas and ratios.

Use Exercise 4 to determine whether students understand that a ratio is the comparison of a part to a whole.

***Hands-On Lab Masters,* p. 54**

Name _____ Date _____

Hands-On Lab
(Lesson 8-1A)

Ratios

On Your Own

1a. The area of triangle C is ____$\frac{1}{2}$ square unit____.

 b. Triangle C is ____$\frac{1}{2}$____ the size of triangle B.

2a. The area of triangle F is ____$\frac{1}{4}$ square unit____.

 b. Triangle F is ____$\frac{1}{4}$____ the size of triangle B.

3a. C to A = ____$\frac{1}{2}$____   **b.** F to A = ____$\frac{1}{4}$____

 c. A to B = ____$\frac{1}{1}$____   **d.** B to C = ____$\frac{2}{1}$____

 e. D to E = ____$\frac{1}{2}$____   **f.** F to G = ____$\frac{1}{2}$____

4. Look Ahead The ratio that compares the number of yellow squares to the total number of squares written as a fraction in simplest form is

____$\frac{1}{3}$____.

© Glencoe/McGraw-Hill   54   *Mathematics: Applications and Connections, Course 1*

Math Journal Suggest that students create interesting figures using tangrams and include them, or drawings of them, in their journals.

Hands-On Lab 8-1A **311**

Instructional Resources

- *Study Guide Masters*, p. 57
- *Practice Masters*, p. 57
- *Enrichment Masters*, p. 57
- Transparencies 8-1, A and B

 CD-ROM Program
  - Resource Lesson 8-1
  - Interactive Lesson 8-1

Recommended Pacing	
Standard	Days 2–4 of 14
Honors	Days 2 & 3 of 13
Block	Days 1 & 2 of 7

1 FOCUS

 5-Minute Check
(Chapter 7)

1. Write the product of $2\frac{1}{3}$ and $\frac{3}{4}$ in simplest form. $1\frac{3}{4}$

2. Find the circumference of an apple pie with an 8-inch diameter. Use 3.14 for π. **25.12 in.**

3. Find the next two numbers in the sequence 15, $12\frac{3}{4}$, $10\frac{1}{2}$, $8\frac{1}{4}$. **6, $3\frac{3}{4}$**

4. Write the quotient of $3\frac{1}{5}$ and $\frac{3}{10}$ in simplest form. $10\frac{2}{3}$

5. Complete: $3\frac{1}{2}$ lb = ___?___ oz. **56**

The 5-Minute Check is also available on **Transparency 8-1A** for this lesson.

Motivating the Lesson

Hands-On Activity Have students choose a short paragraph of reading material. Then ask questions such as:
- How many total words are there?
- How many of the words contain the letter r?
- How would you express the relationship of the words with the letter r to the total number of words?

What you'll learn
You'll learn to express ratios and rates as fractions.

When am I ever going to use this?
Knowing how to express ratios as fractions can help you determine better buys in a grocery store.

Word Wise
rate

Have you ever heard of the game UNO Hearts? UNO Hearts is a card game created by Mattel that is similar to UNO. The table shows the cards in the game. Notice that 24 out of the 108 cards are purple.

Type of Card	Amount
Heart	24
Yellow	24
Purple	24
Green	24
Wild Draw	10
Extra Cards	2

You can compare these two numbers by using a ratio. Recall that the ratio that compares 24 to 108 can be written in several ways.

24 to 108   24:108

24 out of 108 $\frac{24}{108}$

A common way to express a ratio is as a fraction in simplest form.

$$\overset{\div 12}{\underset{\div 12}{\frac{24}{108}}} = \frac{2}{9}$$

The GCF of 24 and 108 is 12. Divide the numerator and the denominator by the GCF, 12.

So, $\frac{2}{9}$ of the cards are purple.

The ratio can also be expressed as 2 to 9, 2:9, or 2 out of 9.

Example  Express the ratio that compares the number of strawberries to the total number of berries as a fraction in simplest form.

$$\begin{array}{c} strawberries \to \\ berries \to \end{array} \quad \overset{\div 2}{\underset{\div 2}{\frac{4}{10}}} = \frac{2}{5}$$

The GCF of 4 and 10 is 2. Divide the numerator and the denominator by the GCF, 2.

The ratio in simplest form is $\frac{2}{5}$.

If the two quantities that you are comparing have different units of measure, the ratio is called a **rate**. For example, $\frac{125 \text{ miles}}{2 \text{ hours}}$ compares the number of miles traveled to the number of hours the trip took.

Rate A rate is a ratio of two measurements that have different units.

Investigations for the Special Education Student

This blackline master booklet helps you plan for the needs of your special education students by providing long-term projects along with teacher notes. Investigation 9, *Park It!*, may be used with this chapter.

Rates are usually expressed in a *per unit* form, where the number in the denominator is 1.

Examples — **②** **Express the ratio *7 inches of rain in 28 days* as a rate.**

$$\underbrace{\frac{7 \text{ inches}}{28 \text{ days}}}_{\div 28} = \underbrace{\frac{0.25 \text{ inch}}{1 \text{ day}}}_{\div 28}$$

Divide the numerator and denominator by 28 to get a denominator of 1.

The total rainfall is equivalent to 0.25 inch each day.

APPLICATION **③** **Money Matters** Dhara purchased a 14.5-ounce bag of chips for $3.19. What was the cost per ounce?

Explore   You know that the 14.5-ounce bag of chips costs $3.19. You need to find the cost per ounce.

Plan   Use a rate to compare the cost to the weight.

Solve
$$\underbrace{\frac{\$3.19}{14.5 \text{ ounces}}}_{\div 14.5} = \underbrace{\frac{\$0.22}{1 \text{ ounce}}}_{\div 14.5}$$

Divide the numerator and denominator by 14.5 to get a denominator of 1.
$3.19 \div 14.5 = 0.22$

The cost is 22 cents per ounce.

Examine $\$0.22 \times 14.5 = \$3.19$

So, the answer is correct.

CHECK FOR UNDERSTANDING

Communicating Mathematics

Read and study the lesson to answer each question. 1–3. See margin.

1. ***Explain*** the difference between a ratio and a rate.

2. ***Draw*** a picture in which the ratio of blue circles to the total number of circles is $\frac{3}{5}$.

3. ***Write*** about a situation in which rates would be helpful.

Guided Practice
4–7. See margin.

Write each ratio in three different ways.

4. 7 carnations out of 24 flowers

5. 15 mint chip cookies in a bag of 34 cookies

Express each ratio as a fraction in simplest form.

6. 16 beagles out of 24 dogs   7. 10 emeralds out of 25 gems

Express each ratio as a rate.

8. 120 words in 3 minutes   9. 5 soft drinks for $3.25

8. 40 words per minute
9. $0.65 per soft drink
10. 40 miles per hour

10. ***Life Science*** The fastest running bird is the ostrich. An ostrich can run 240 miles in 6 hours. What is the average speed of an ostrich?

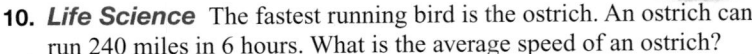

Lesson 8-1 Ratios and Rates   **313**

Reteaching the Lesson

Activity Give small groups of students a set of assorted items. Then have them make and complete a chart such as the following:

	Red	Blue
Number of each type	2	3
Total number	14	14
Ratio	1:7	3:14

Error Analysis

Watch for students who write ratios as number of items of one type compared to remaining items.

Prevent by reminding students that the denominator in a ratio is the *total* number of items being considered.

2 TEACH

 Transparency 8-1B contains a teaching aid for this lesson.

Thinking Algebraically The equals sign between two ratios indicates that the two fractions have the same relationship between numerator and denominator. One-half represents the same value as $\frac{16}{32}$.

In-Class Examples

For Example 1
Kisha has 16 nickels, 4 dimes, and 5 quarters in her pocket. Express the ratio that compares the number of nickels to the total number of coins as a fraction in simplest form. $\frac{16}{25}$

For Example 2
Express the ratio of 130 miles to 4.2 gallons of gas as a rate. **30.95 mpg**

For Example 3
Maria purchased a 24-ounce jar of salsa for $2.88. What was the cost per ounce? **$0.12**

Additional Answers

1. A ratio is a comparison of two numbers by division. A rate is a ratio of two measurements that have different units.

2. Sample answer:

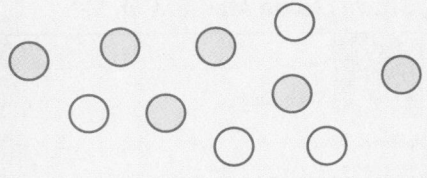

3. Sample answer: Rates would be useful when shopping in the grocery store. They allow you to determine better buys.

4. Sample answer: 7 out of 24, 7 to 24, 7:24

5. Sample answer: 15 to 34, $\frac{15}{34}$, 15 out of 34

6. $\frac{2}{3}$

7. $\frac{2}{5}$

3 PRACTICE/APPLY

Check for Understanding

If students need additional practice or instruction after completing Exercises 1–10, one of these options may be helpful.

- Extra Practice, see p. 578
- Reteaching Activity, see p. 313
- *Study Guide Masters*, p. 57
- *Practice Masters*, p. 57
- Interactive Mathematics Tools Software

Assignment Guide

Core: 11–33 odd, 36–40
Enriched: 12–30 even, 31–34, 36–40

Study Guide Masters, p. 57

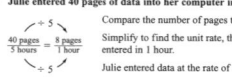

8-1 Study Guide

Ratios and Rates

You can compare numbers with the same unit of measure using a **ratio**. A ratio is a comparison of two numbers by division.

Example 1 There were 52 animals in the ugly pet show. 24 of them were dogs.

The ratio can be written as follows.
24 to 52 24:52 24 out of 52 $\frac{24}{52}$

24 and 52 have 4 as a common factor.
The ratio can be simplified as $\frac{6}{13}$.

A **rate** is a ratio that compares two different units of measure.

Example 2 Julie entered 40 pages of data into her computer in 5 hours.

$\frac{40 \text{ pages}}{5 \text{ hours}} = \frac{8 \text{ pages}}{1 \text{ hour}}$

Compare the number of pages to the number of hours. Simplify to find the unit rate, the number of pages entered in 1 hour.
Julie entered data at the rate of 8 pages per hour.

Express each ratio as a fraction in simplest form.

1. 12 out of 25 people $\frac{12}{25}$
2. 8 out of 10 bicycles $\frac{4}{5}$
3. 9 wins in 12 games $\frac{3}{4}$
4. 14 station wagons out of 40 vehicles $\frac{7}{20}$
5. 10 out of 15 days $\frac{2}{3}$
6. 18 baseball caps out of 36 hats $\frac{1}{2}$

Express each ratio as a rate.

7. $18 for 6 tickets $3 per ticket
8. 30 km in 6 hours 5 km per hour
9. $42 for 7 books $6 per book
10. $28 for 4 hours $7 per hour

Use the letters in the word "WORLDWIDE." Write the ratios comparing the numbers of letters in simplest form.

11. W to D 1:1
12. R to W 1:2
13. vowels to consonants 1:2

© Glencoe/McGraw-Hill   T57   *Mathematics: Applications and Connections, Course 1*

314 Chapter 8

EXERCISES

Practice

11–16. Sample answers are given.

11. 9 out of 16, 9 to 16, 9:16
12. 13 to 16, 13:16, $\frac{13}{16}$
13. 18 out of 29, 18 to 29, 18:29
14. $\frac{11}{15}$, 11:15, 11 out of 15
15. 23 out of 25, 23 to 25, 23:25
16. 7:19, 7 out of 19, $\frac{7}{19}$

Write each ratio in three different ways.

11. 9 pairs of boots out of 16 are yellow.
12. 13 bikes out of 16 bikes have combination locks.
13. 18 out of 29 centimeter cubes are pink.
14. 11 brownies out of 15 brownies contain walnuts.
15. 23 out of 25 ants are red.
16. 7 out of 19 days were cloudy.

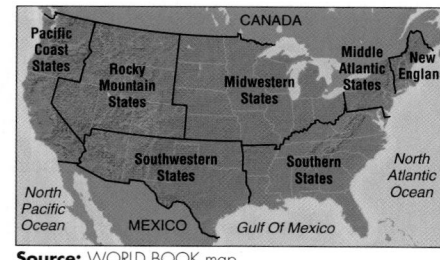

Express each ratio as a fraction in simplest form.

17. 19 games won out of 57 games played $\frac{1}{3}$
18. 25 brick houses out of 45 houses $\frac{5}{9}$
19. 14 tabby cats out of 18 cats $\frac{7}{9}$
20. 16 states visited out of 50 states $\frac{8}{25}$
21. 32 cows out of 72 animals $\frac{4}{9}$
22. 10 drums out of 75 instruments $\frac{2}{15}$

Express each ratio as a rate. 23–29. See margin.

23. 395 kilometers in 5 hours
24. 3 CDs for $41.91
25. 4 tickets for $35
26. 63 million albums in 7 years
27. $1.32 for a dozen eggs
28. 79.8 miles on 3 gallons of gas

29. Write the ratio *13 quarters out of 24 coins* in three different ways.

30. Express the ratio *ten oranges for two dollars* as a rate. **$0.20 per orange**

Applications and Problem Solving

31a. $\frac{2}{25}$

32d. $\frac{1}{3}, \frac{1}{3}, \frac{1}{9}$;
The ratio that compares the length of the sides times the ratio that compares the perimeters equals the ratio that compares the areas.

31. *Geography* Refer to the map.
   a. Write the ratio that compares the number of states in the Southwestern States region to the total number of states.
   b. Write the ratio that compares the number of states in the Midwestern States to the total number of states. $\frac{6}{25}$

Source: WORLD BOOK map

32. *Geometry* Draw a 1-by-1 unit square and a 2-by-2 unit square on grid paper. Then write the ratio that compares each of the following.
   a. the length of the side of the small square to the length of the side of the large square $\frac{1}{2}$
   b. the perimeter of the small square to the perimeter of the large square $\frac{1}{2}$
   c. the area of the small square to the area of the large square $\frac{1}{4}$
   d. Draw a 3-by-3 unit square. Compare the same measures between the 1-by-1 unit square and the 3-by-3 unit square. Describe the relationship that exists between the ratios of the sides, perimeters, and areas.

314 Chapter 8 Exploring Ratio, Proportion, and Percent

Additional Answers

23. 79 kilometers per hour
24. $13.97 per CD
25. $8.75 per ticket
26. 9 million albums per year
27. $0.11 per egg
28. 26.6 miles per gallon
29. Sample answer: 13 out of 24; 13 to 24; 13:24

33. Life Science Scientists say that a pterodactyl could fly 75 miles in three hours. How far could a pterodactyl travel in one hour? **25 miles**

34. The 16-ounce bag costs $0.16 per ounce. The 32-ounce bag costs $0.11 per ounce. So, the better buy is the 32-ounce bag.

35b. $\frac{1}{64}$

34. Money Matters At the supermarket, the cost of a 16-ounce bag of gum balls is $2.56. A 32-ounce bag costs $3.52. Which is the better buy? Explain.

35. Working on the CHAPTER Project
Refer to page 309.

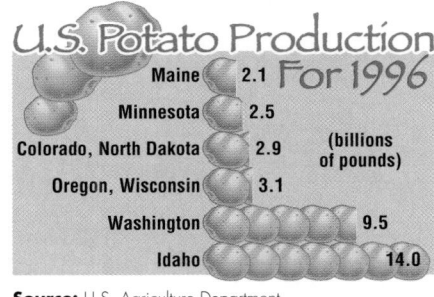

a. Write the ratio that compares the size of the toy stock car 88 to the actual size of the race car as a fraction. $\frac{1}{144}$

b. Write the ratio that compares the size of the toy stock car 16 to the actual size of the race car as a fraction.

c. Find three toys that represent real-life objects. Measure the height and width of each toy to the nearest centimeter. Then find out the actual height and width of the objects that the toys represent. Write the ratios that compare the height, width, and length of each toy to the actual height, width, and length of each real object as a fraction in simplest form. **See students' work.**

36. Critical Thinking Mr. Sarmiento stated that 18 out of 24 students in his class scored 85% or higher on the last test. He instructed the class to write a ratio comparing the number of students scoring below 85% to the total number of students. Explain why each answer is incorrect. **See Answer Appendix.**

a. 18:24   b. 24:6   c. 24:18

Mixed Review

37. Patterns Find the missing number in the sequence $3\frac{7}{8}$, $4\frac{1}{4}$, $\underline{?}$, 5, $5\frac{3}{8}$.
(Lesson 7-8) $4\frac{5}{8}$

38. Geometry What is the circumference of a circle that has a diameter of 13.4 centimeters? *(Lesson 7-4)* **42.1 cm**

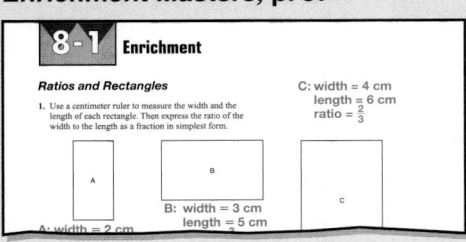

39. Agriculture Refer to the graph. *(Lesson 3-6)*

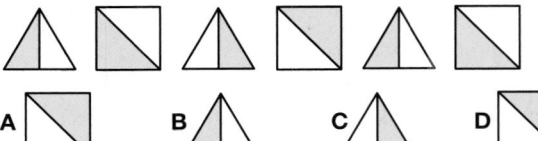

a. How many more pounds of potatoes were harvested in Washington than in Colorado and North Dakota? **6.6 billion lb**

b. Find the total amount of potatoes harvested in these eight states. **34.1 billion lb**

U.S. Potato Production For 1996	
Maine	2.1
Minnesota	2.5
Colorado, North Dakota	2.9
Oregon, Wisconsin	3.1
Washington	9.5
Idaho	14.0

(billions of pounds)

Source: U.S. Agriculture Department

40. Test Practice Find the next figure in the pattern. *(Lesson 1-2)* **C**

A   B   C   D

Lesson 8-1 Ratios and Rates **315**

Extending the Lesson

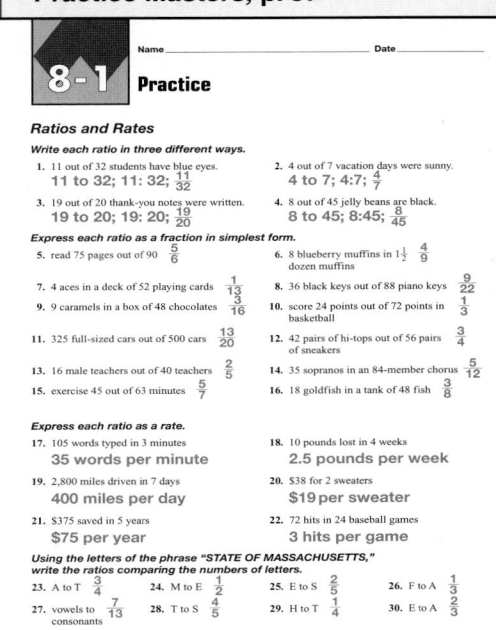
Activity Food coloring, water, cotton swabs, paper, paper cups, and droppers or measuring spoons will be needed. Using these materials, have students develop ratios of food coloring and water for creating certain colors. Ask them to make a chart that shows their colors and the required ratios.

CHAPTER Project

Exercise 35 asks students to advance to the next stage of work on the Chapter Project. You may wish to have students bring in small toys and work in pairs to find measurements.

4 ASSESS

Closing Activity

Writing Have students explain in writing how a rate is different from a ratio in lowest terms.

COOPERATIVE LEARNING

8-1B Ratios and Probability

A Follow-Up of Lesson 8-1

✴ spinner

Suppose the probability that a basketball player makes a basket from the free throw line is 0.75. Since the decimal 0.75 means seventy-five hundredths, the probability can also be expressed as the ratio $\frac{75}{100}$. This means that the player makes a free throw 75 out of 100 times. In this lab, you will explore ratios and probability. *You can refer to Lesson 5-4B to review experimental probability.*

GET READY

Objective Students explore ratios and probability.

Optional Resources
Hands-On Lab Masters
• spinners, p. 20
• worksheet, p. 55

Overhead Manipulative Resources
• spinners

Manipulative Kit
• spinners

MANAGEMENT TIPS

Recommended Time
30 minutes

Getting Started Have students read the introduction to the lab. Display a spinner divided into 4 sections. Ask students what the possibilities are for labeling sections as "make" or "miss." Ask them whether the order of the sections would make any difference in the results of spinning.

Activities 1 and 2 use spinners to generate experimental data for 100 free throws. In both activities, the resulting number of baskets is written as a fraction over 100 and as a decimal. Point out to students that if a player makes 1 out of every 2 free throws, half of the throws are good. Thus, the probability is $\frac{1}{2}$.

ASSESS

Have students complete Exercises 1–4. Make sure students realize that the fraction is formed as the number of "makes" over the number of tries or spins.

Use Exercise 5 to check that students set up ratios correctly and can express the probability as a decimal.

TRY THIS

Work with a partner.

1 • Make a spinner with four equal sections.

• Mark one section of the spinner "make" and the other sections "miss". Each section represents the player making or missing a basket.

• Spin the spinner 100 times. Record the number of baskets made as the ratio $\frac{baskets\ made}{100}$. Write the ratio as a decimal. What is the probability that the player makes a basket from the free throw line?

2 • Mark three sections of the spinner "make" and one section "miss". Again, each section represents the player making or missing a basket.

• Spin the spinner 100 times. Record the number of baskets made as the ratio $\frac{baskets\ made}{100}$. Write the ratio as a decimal. What is the probability that the player makes a basket from the free throw line?

ON YOUR OWN

1. Write a ratio that compares the area of each region marked "make" to the total area of the circle for each spinner. $\frac{1}{4}$, $\frac{3}{4}$

2. What is the probability of making a basket on each spinner? Express the probability as a decimal. **0.25; 0.75**

3. Compare your results from the activity to each probability in Exercise 2. **See margin.**

4. Describe how ratio and probability are related. **See margin.**

5. *Look Back* A player makes a basket every 50 out of 80 times. What is the probability that he will make a basket on his next attempt? Express the probability as a decimal. **0.625**

316 Chapter 8 Exploring Ratio, Proportion, and Percent

Additional Answers

3. Sample answer: The results are about the same.

4. Sample answer: Probability is the ratio of the number of times an event occurs to the total number of possible outcomes.

Math
Journal

Have students write a paragraph about why their results with the spinners might not be exactly the same as the expected probabilities.

What you'll learn

You'll learn to solve proportions by using cross products.

When am I ever going to use this?

Knowing how to solve proportions can help you make scale models.

Word Wise

proportion
cross products

Have you ever played Dominoes? In the standard set of Dominoes, 7 out of 28 tiles, or one-fourth of the tiles, are doubles.

The ratios $\frac{7}{28}$ and $\frac{1}{4}$ are equivalent. That is, $\frac{7}{28} = \frac{1}{4}$. The equation $\frac{7}{28} = \frac{1}{4}$ is an example of a **proportion**.

| **Proportion** | **Words:** | A proportion is an equation that shows that two ratios are equivalent. |
| | **Symbols:** | $\frac{a}{b} = \frac{c}{d}$, $b \neq 0$, $d \neq 0$ |

In the following Mini-Lab, you will use pattern blocks to explore ratios that are equivalent and ratios that are not equivalent.

Cultural Kaleidoscope

Historians believe that dominoes were brought from China to Italy in the 14th century.

3. Sample answer: write a proportion.

4. Sample answer: If you first make the ratios equivalent, then add or subtract a triangle to one numerator or denominator, the resulting ratios will not be equivalent.

HANDS-ON

MINI-LAB

Work with a partner.   pattern blocks

Try This

Use the triangle pattern block as the unit of measure.

1. Build a shape that has a ratio equivalent to each ratio shown.

    a.

    See margin.

    b.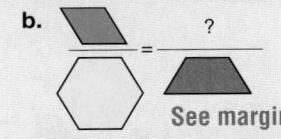

    See margin.

2. Build a shape that has a ratio *not* equivalent to each ratio shown.

    a.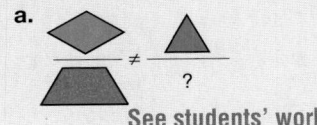

    See students' work.

    b.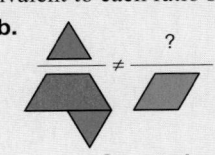

    See students' work.

Talk About It

3. How did you find the equivalent ratios?

4. How did you know when two ratios were not equivalent?

Lesson 8-2 Solving Proportions **317**

Additional Answers for the Mini-Lab

1a.

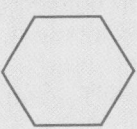

1b.

8-2 Lesson Notes

Instructional Resources

- *Study Guide Masters*, p. 58
- *Practice Masters*, p. 58
- *Enrichment Masters*, p. 58
- Transparencies 8-2, A and B
- *Assessment and Evaluation Masters*, p. 211
- *School to Career Masters*, p. 8
- CD-ROM Program
  - Resource Lesson 8-2
  - Interactive Lesson 8-2

Recommended Pacing	
Standard	Day 5 of 14
Honors	Days 4 & 5 of 13
Block	Day 3 of 7

1 FOCUS

 5-Minute Check
(*Lesson 8-1*)

Write each ratio in three different ways.

1. 7 out of 15 cakes are chocolate $\frac{7}{15}$, 7:15, 7 to 15

2. 6 out of 49 people have red hair **6 out of 49; 6 to 49; 6:49**

Express each ratio as a fraction in simplest form.

3. 5 red cars out of 20 cars $\frac{1}{4}$

4. 2 days absent out of 185 days of school $\frac{2}{185}$

5. Express the ratio *220 Calories in 4 ounces of fish* as a rate. **55 Calories in each ounce**

 The 5-Minute Check is also available on **Transparency 8-2A** for this lesson.

Motivating the Lesson

Problem Solving Ask students how far they would travel in a minute if they were in a car moving at 45 mph. **0.75 or $\frac{3}{4}$ mile**

Transparency 8-2B contains a teaching aid for this lesson.

Using the Mini-Lab To get started, have pairs of students use the pattern blocks to make other geometric figures. Ask them to find all the fractional parts of each figure.

In-Class Examples

Use cross products to determine whether each pair of ratios forms a proportion.

For Example 1
$\frac{3}{8}, \frac{8}{24}$ no

For Example 2
$\frac{5}{6}, \frac{40}{48}$ yes

Solve each proportion.

For Example 3
$\frac{2}{5} = \frac{n}{25}$ 10

For Example 4
$\frac{3.5}{x} = \frac{1.4}{2}$ 5

Teaching Tip Tell students that in some cases, they can solve a proportion mentally by using equivalent fractions. For example, $\frac{3}{4} = \frac{x}{16}$. By thinking $4 \times 4 = 16$, and $3 \times 4 = 12$, x must be 12.

One way to determine whether two ratios form a proportion is to find their **cross products**. If the cross products of two ratios are equal, then the ratios form a proportion. In the proportion shown, notice that the cross products, 6×3 and 9×2, are equal.

$$\frac{6}{9} \bowtie \frac{2}{3} \quad \begin{array}{l} 6 \times 3 = 18 \\ 9 \times 2 = 18 \end{array}$$

Property of Proportions	**Words:**	The cross products of a proportion are equal.
	Symbols:	If $\frac{a}{b} = \frac{c}{d}$, then $ad = bc$.

Examples

Use cross products to determine whether each pair of ratios forms a proportion.

① $\frac{2}{13}, \frac{4}{26}$

$$\frac{2}{13} \bowtie \frac{4}{26}$$

$2 \times 26 \stackrel{?}{=} 13 \times 4$ *Write cross products.*

$52 = 52$ *Multiply.*

The cross products are equal. So, the pair of ratios forms a proportion.

② $\frac{4}{9}, \frac{16}{38}$

$$\frac{4}{9} \bowtie \frac{16}{38}$$

$4 \times 38 \stackrel{?}{=} 9 \times 16$

$152 \neq 144$

The cross products are not equal. So, the pair of ratios does not form a proportion.

> **Study Hint**
> **Mental Math** In some cases, you can solve a proportion mentally by using equivalent fractions.
> $\frac{3}{4} = \frac{x}{16}$
> Think: $4 \times 4 = 16$
> $3 \times 4 = 12$
> So, $x = 12$.

If one value in a proportion is unknown, you can use cross products to solve the proportion.

Examples

Solve each proportion.

③ $\frac{5}{9} = \frac{z}{54}$

$5 \times 54 = 9 \times z$ *Write cross products.*

$270 = 9z$ *Multiply.*

$\frac{270}{9} = \frac{9z}{9}$ *Divide.*

$30 = z$

The solution is 30.

④ $\frac{1.4}{1.8} = \frac{3.5}{w}$

$1.4 \times w = 1.8 \times 3.5$

$1.4w = 6.3$

$\frac{1.4w}{1.4} = \frac{6.3}{1.4}$

$w = 4.5$

The solution is 4.5.

Multiple Learning Styles

Kinesthetic Have groups of students mark off a distance of 10 meters. Have each student walk the distance three times and find the average number of steps taken. Compute the averages to the nearest whole number of steps. After averages are found, have students compute their ratio of steps per meter and then compare their leg lengths with these ratios.

You can use proportions to make predictions.

Example 5
APPLICATION

Technology According to the results of a test conducted by *Zillions* magazine, 12 out of 17 kids prefer the Sony PlayStation to Sega Saturn. Suppose there are 3,400 kids in your community. Predict how many will prefer the PlayStation.

Set up a proportion that compares the number of kids who preferred the PlayStation in the test to the number of kids in your community. Let p represent the number of kids out of 3,400 who will prefer the PlayStation.

Test			Your Community

$$\begin{array}{ccc}
\textit{prefer PlayStation} \rightarrow & \dfrac{12}{17} = \dfrac{p}{3{,}400} & \leftarrow \textit{prefer PlayStation} \\
\textit{total} \rightarrow & & \leftarrow \textit{total}
\end{array}$$

The ratios are equivalent so the cross products are equal.

$$\dfrac{12}{17} \rlap{\diagup}{\diagup} \dfrac{p}{3{,}400}$$

$12 \times 3{,}400 = 17 \times p$ *Write the cross products.*

$40{,}800 = 17p$ *Multiply.*

$\dfrac{40{,}800}{17} = \dfrac{17p}{17}$ *Divide each side by 17.*

$2{,}400 = p$

You can predict that 2,400 will prefer PlayStation.

CHECK FOR UNDERSTANDING

Communicating Mathematics

Read and study the lesson to answer each question. 1–2. See margin.

1. *Explain*, in your own words, the meaning of proportion.

2. *Tell* how you can determine whether two ratios form a proportion.

HANDS-ON  3. *Use* pattern blocks to build a proportion. Draw a picture of your model and explain why the two ratios are proportional. **See students' work.**

Guided Practice

Use cross products to determine whether each pair of ratios forms a proportion.

4. $\dfrac{1}{8}, \dfrac{8}{64}$ **yes**

5. $\dfrac{7}{12}, \dfrac{8}{15}$ **no**

Solve each proportion.

6. $\dfrac{6}{7} = \dfrac{y}{42}$ **36**

7. $\dfrac{7}{12} = \dfrac{42}{h}$ **72**

8. $\dfrac{2.4}{a} = \dfrac{1.2}{1.5}$ **3**

9. *Advertising* According to an advertisement for Crest toothpaste, 8 out of 10 dentists prefer Crest. There are 150 dentists in a certain city. Predict how many of them prefer Crest. **120 dentists**

Lesson 8-2 Solving Proportions **319**

■ **Reteaching the Lesson** ■

Activity Have pairs of students play "Proportion Pro." Write numbers such as 8, 25, 2, and 100 on index cards. Give each pair four cards and see if they can form all the combinations of proportions in a given time limit.

Additional Answers

1. A proportion is an equation stating that two ratios are equivalent.

2. Find their cross products.

4 ASSESS

Closing Activity

Modeling Give groups of students 20 jelly beans. Ask them to count and sort the jelly beans by color. Ask them to predict how many of each color would be in a bag of 200 beans.

Chapter 8, Quiz A (Lessons 8-1 and 8-2) is available in the *Assessment and Evaluation Masters*, p. 211.

Practice Masters, p. 58

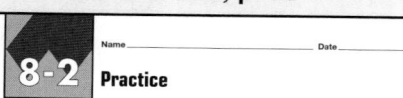

EXERCISES

Practice

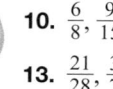

Use cross products to determine whether each pair of ratios forms a proportion.

10. $\frac{6}{8}, \frac{9}{15}$ no   11. $\frac{4}{5}, \frac{16}{20}$ yes   12. $\frac{5}{6}, \frac{30}{36}$ yes

13. $\frac{21}{28}, \frac{3}{7}$ no   14. $\frac{24}{75}, \frac{8}{25}$ yes   15. $\frac{1.3}{3.5}, \frac{2.8}{1.6}$ no

Solve each proportion.

16. $\frac{d}{5} = \frac{24}{40}$ 3   17. $\frac{w}{10} = \frac{4}{5}$ 8   18. $\frac{6}{9} = \frac{r}{72}$ 48   19. $\frac{5}{q} = \frac{25}{55}$ 11

20. $\frac{1.7}{3} = \frac{85}{c}$ 150   21. $\frac{23}{20} = \frac{115}{m}$ 100   22. $\frac{1.6}{2.4} = \frac{2.8}{k}$ 4.2   23. $\frac{25}{n} = \frac{12}{48}$ 100

24. Suppose a car travels 174 miles in 3 hours. How long will it take to travel 290 miles? **5 hours**

25. Suppose you can buy 3 pairs of sunglasses for $38.91. How many pairs can you buy for $77.82? **6 pairs**

Applications and Problem Solving

26. **School** At Market Street Middle School, the teacher to student ratio is 3 to 85. If there are 510 students enrolled at the school, how many teachers are there at the school? **18 teachers**

27. **Health** According to the results of a survey, 27 out of 50 people exercise regularly. Suppose there are 2,600 people in a community. How many people can be expected to exercise regularly? **1,404**

28. **Geometry** A series of rectangles are cut so that the ratio of the length of the short side to the long side is 3:5.
    a. Suppose the short side of one of the rectangles measures 6 units. What is the measure of the long side? **10 units**
    b. Suppose the long side of one of the rectangles measures 25 units. What is the measure of the short side? **15 units**

29. **Critical Thinking** Suppose 24 out of 180 people said they liked hiking, and 5 out of every 12 hikers buy Acme hiking shoes. In a group of 270 people, how many would you expect to have Acme hiking shoes? **15**

Mixed Review

30. **$0.21 per picture**

30. **Money Matters** The Shutter Bug Camera Shop charges $5.04 to develop 24 pictures. What is the cost of developing each picture? *(Lesson 8-1)*

31. **Algebra** Evaluate $w \div v$ if $w = 5\frac{3}{8}$ and $v = \frac{3}{4}$. *(Lesson 7-6)* $7\frac{1}{6}$

32. Find the value of m if $m = 15.64 \div 0.34$. *(Lesson 4-6)* **46.0**

33. **Test Practice** Which shows the decimal twenty-two and four hundred five ten-thousandths? *(Lesson 3-1)* **D**
    A 22.00405   **B** 2.2405   **C** 22.405   **D** 22.0405

Extending the Lesson

Enrichment Masters, p. 58

Activity Have students contact local businesses to ask whether they use ratios or proportions in any way and, if so, to collect examples. Have students use the information they have gathered to create word problems using proportions.

TECHNOLOGY LAB

SPREADSHEETS

8-2B Proportions

A Follow-Up of Lesson 8-2

- computer
- spreadsheet software

Cara and Jeff are going to cater the Nielson family reunion. Their recipe for potato salad serves 10 people. How much of each ingredient will they need to make enough potato salad for 75 people?

You can find the amount of each ingredient needed by using a spreadsheet.

Mustard Potato Salad

$\frac{1}{2}$ c light mayonnaise $\frac{1}{4}$ t salt

2 T Dijon mustard $\frac{1}{8}$ t pepper

2 T sweet pickle relish   5 c cooked potatoes

1 T white vinegar $\frac{1}{4}$ c minced parsley

Combine the first six ingredients in a large bowl. Add potatoes and toss. Cover and chill overnight. Garnish with parsley. Serves 10.

TRY THIS

Work with a partner.

B1 is the number of people, and the number of servings per batch is 10. So, the number of batches needed is $\frac{B1}{10}$, which is given in cell B2. The formula in cell B4 takes the number of batches needed and multiplies it by $\frac{1}{2}$ or 0.5, the amount of mayonnaise needed for one batch. The result is the total amount of mayonnaise needed to make enough potato salad for 75 people.

The printout shows the results of entering the number of servings needed, 75, in cell B1.

POTATO SALAD RECIPE

	A	B	C
1	People To Serve	= B1	
2	Batches needed	= B1/10	
3	INGREDIENT	NUMBER	
4	mayonnaise	= B2 * 0.5	cups
5	mustard	= B2 * 2	T
6	relish	= B2 * 2	T
7	vinegar	= B2 * 1	T
8	salt	= B2 * 0.25	t
9	pepper	= B2 * 0.125	t
10	potatoes	= B2 * 5	cups
11	parsley	= B2 * 0.25	cups

POTATO SALAD RECIPE

	A	B	C
1	People To Serve	75	
2	Batches needed	7.5	
3	INGREDIENT	NUMBER	
4	mayonnaise	3.75	cups
5	mustard	15.0	T
6	relish	15.0	T
7	vinegar	7.5	T
8	salt	1.875	t
9	pepper	0.9375	t
10	potatoes	37.5	cups
11	parsley	1.875	cups

ON YOUR OWN

1. Explain the formula in cell B6. **1–4. See margin.**
2. Using the results of the spreadsheet, write each ingredient amount needed as a fraction in simplest form.
3. Use the spreadsheet to find the amount of each ingredient needed to make enough potato salad for 120 people. Write each ingredient amount as a fraction in simplest form.
4. How could you change the spreadsheet if one batch of potato salad served 12 people?

Lesson 8-2B TECHNOLOGY LAB   **321**

GET READY

Objective Students use spreadsheets to solve problems involving proportions.

Technology Resources
Suggested spreadsheet software:
- *Lotus 1·2·3*
- *ClarisWorks*
- *Microsoft Excel*

MANAGEMENT TIPS

Recommended Time
45 minutes

Getting Started Review reading spreadsheets by referring to Technology Lab 4-3B on page 144. Make sure students know how to name the cells of the spreadsheet and how values are generated.

Using a Calculator If spreadsheet software is not available, you can have groups of students create a table similar to a spreadsheet and use a calculator to perform the operation. Have one group member act as the recorder and let the others be responsible for each column of division.

ASSESS

Have students answer Exercises 1–4. Use Exercise 4 to determine whether students understand the use of proportions and the setup of spreadsheets.

Math Journal Have students write a paragraph about how a spreadsheet could be helpful to a food store manager.

Additional Answers

1. Multiply the contents of cell B2 by 2.

2. $3\frac{3}{4}$ cups of mayonnaise, 15 T of mustard, 15 T of relish, $7\frac{1}{2}$ T of vinegar, $1\frac{7}{8}$ t of salt, $\frac{15}{16}$ t of pepper, $37\frac{1}{2}$ cups of potatoes, $1\frac{7}{8}$ cups of parsley

3. 6 cups of mayonnaise, 24 T of mustard, 24 T of relish, 12 T of vinegar, 3 t of salt, $1\frac{1}{2}$ t of pepper, 60 cups of potatoes, 3 cups of parsley

4. Change the formula in cell B2 to $B\frac{1}{12}$.

Objective Students solve problems by drawing diagrams.

Recommended Pacing	
Standard	Day 6 of 14
Honors	Day 6 of 13
Block	Day 3 of 7

1 FOCUS

Getting Started Ask students to discuss how a drawing or diagram can help to solve a problem. Ask for eight volunteers to perform handshakes in the manner described at the beginning of the lab. As they do so, ask another student to keep a record of the handshakes on the chalkboard.

2 TEACH

Teaching Tip Some students may be hesitant to draw diagrams. Encourage them to draw diagrams to help them understand and picture the information before attempting to solve the problem.

In-Class Example

The stops on Toya, Amy, and Zoe's subway line are 1st St., 2nd St., and so on. Amy gets off 1 stop before Toya. Toya gets off 3 stops before Zoe. Zoe gets off 6 stops after Amy. If Toya gets off on 3rd St., what are Amy's and Zoe's stops? **2nd St., 8th St.**

Additional Answers

1.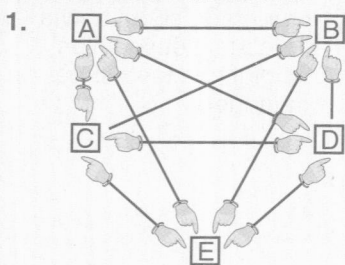

2. The number of handshakes increases by 2, 3, 4, and so on.

THINKING LAB — PROBLEM SOLVING

8-3A Draw a Diagram

A Preview of Lesson 8-3

Alison and Ben are the first guests to arrive at Cindy's party. They are discussing how many handshakes there will be if every guest at the party shakes hands with everyone else once. Let's listen in!

If we shake hands, then there would be one handshake.

Alison

If Cindy joined us, then she would shake hands with you and me. That's two, a total of three.

Let's draw a diagram to picture the situation.

Ben

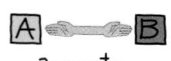
2 guests
1 handshake

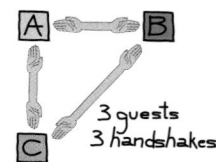
3 guests
3 handshakes

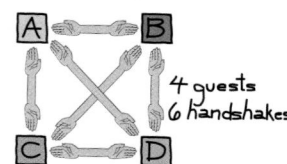
4 guests
6 handshakes

For five people, we can add the handshakes for the fifth person. Add four more handshakes. A total of 10!

Wow! We can use this method to find the number of handshakes for any number of people.

THINK ABOUT IT

Work with a partner. 1–2. See margin.

1. ***Draw a diagram*** to illustrate the number of handshakes with five guests.

2. ***Describe*** the pattern that exists between the number of guests and the number of handshakes.

3. ***Extend*** the pattern to predict how many handshakes there will be with six guests. **15 handshakes**

4. ***Apply*** the **draw a diagram** strategy to solve the following problem.

   Kayla lives in Littleton and works in Parker. There is no direct route from Littleton to Parker, so Kayla goes through either Stoney Creek or Castle Rock. There is one road between Stoney Creek and Castle Rock. How many different ways can Kayla drive to work? **4 ways**

322 Chapter 8 Exploring Ratio, Proportion, and Percent

■ Reteaching the Lesson ■

Activity Have students work with a partner to draw a diagram that helps find the solution to the following problem. *If tiles measuring 2 feet by 2 feet are used to cover a 16 foot by 12 foot floor, how many tiles are needed?* **48**

ON YOUR OWN

5. The third step of the 4-step plan for problem solving asks you to *solve* the problem. *Explain* how the draw a diagram strategy can help you solve a problem. **A diagram will help you understand and picture the information more clearly.**

6. *Write a Problem* that can be solved by using the draw a diagram strategy. Then ask a classmate to determine the answer by drawing a diagram. **See margin.**

7. *Look Ahead* Draw a diagram that you can use to answer Exercise 8 on page 326. **See margin.**

MIXED PROBLEM SOLVING

9. See margin.

STRATEGIES
Look for a pattern.
Solve a simpler problem.
Act it out.
Guess and check.
Draw a diagram.
Make a chart.
Work backward.

Solve. Use any strategy.

8. *Number Sense* A number multiplied by itself is 676. What is the number? **26**

9. *School* At the Science Festival's bridge-building competition, Juliet came in second, and Daniel finished behind Pedro. Keela finished ahead of Pedro, and Reynelda won first place. In what order did they finish?

10. *Recreation* Robin purchased a tent for camping. Each side of the four sides of the tent needs 3 stakes to secure properly to the ground. How many stakes should there be in the box? **8**

11. *Life Science* The chart shows the average weight in kilograms of certain bears. What is the mean of these weights? **417.5 kg**

Weight of Bears

Polar Bear 410 kg
Grizzly Bear 360 kg
Asiatic Black Bear 120 kg
Kodiak Bear 780 kg

Source: Encyclopedia Americana

12. *Patterns* What are the next two figures in the pattern?

13. *Construction* India has a piece of lumber that is 140 inches long. She wants to cut it into 20-inch pieces. How many cuts does she need to make? **6 cuts**

14. **Test Practice** The graph shows the number of students who signed up to volunteer at the Rush Creek Middle School's annual spaghetti dinner.

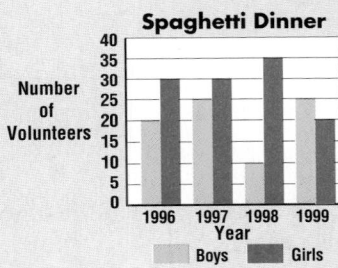

Spaghetti Dinner

Number of Volunteers (y-axis: 0 to 40)
Year (x-axis: 1996, 1997, 1998, 1999)
Boys, Girls

Which is a conclusion that can be drawn from the data on the graph? **D**

A There were more volunteers in 1998 than in 1996.

B The number of volunteers increased each year.

C More boys signed up to volunteer in 1997 than in any other year.

D There were fewer volunteers in 1999 than in 1996.

E More girls signed up to volunteer in 1997 than in 1998.

■ Extending the Lesson ■

Activity Ask students to determine how many routes they could take from one building directly to another. **one** from one building to two others? **two** Then have them draw diagrams to determine all of the routes between 4 points, 5 points, and 8 points if they must pass through each point exactly once. **6; 24; 5,040**

 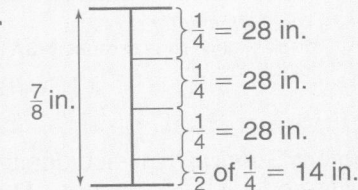

8-3 Lesson Notes

Instructional Resources

- *Study Guide Masters,* p. 59
- *Practice Masters,* p. 59
- *Enrichment Masters,* p. 59
- Transparencies 8-3, A and B
- *Technology Masters,* p. 16
- CD-ROM Program
  - Resource Lesson 8-3

Recommended Pacing	
Standard	Day 7 of 14
Honors	Day 7 of 13
Block	Day 4 of 7

1 FOCUS

5-Minute Check
(Lesson 8-2)

Use cross products to determine whether each pair of ratios forms a proportion.

1. $\frac{3}{5}, \frac{9}{15}$ yes
2. $\frac{4}{7}, \frac{8}{21}$ no

Solve each proportion.

3. $\frac{7}{10} = \frac{n}{100}$ 70
4. $\frac{x}{75} = \frac{1}{3}$ 25
5. $\frac{5}{p} = \frac{12}{96}$ 40

The 5-Minute Check is also available on **Transparency 8-3A** for this lesson.

Motivating the Lesson

Problem Solving Have students read the opening paragraphs of the lesson. Ask: *Would a scale of 1 inch = 2 feet work for a drawing of your school?* no *the doorway?* yes *your math book?* no

8-3 Integration: Geometry
Scale Drawings

What you'll learn

You'll learn to find actual length from a scale drawing.

When am I ever going to use this?

Knowing how to find actual length from a scale drawing can help you determine mileage between cities on a map.

Word Wise

scale drawing

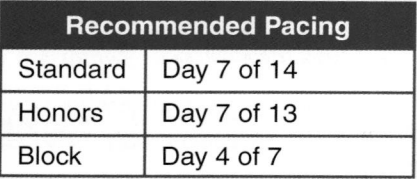

One of the largest meat-eating dinosaurs to have ever lived was the Tyrannosaurus Rex. Since it would be difficult to draw the dinosaur actual size, you can make a **scale drawing**. A scale drawing shows an object exactly as it looks, but it is generally larger or smaller. The scale gives the ratio that compares the lengths on the drawing to the actual lengths of the object.

Suppose a drawing of a dinosaur had a scale of 1 inch = 20 feet. If the length of the dinosaur on the drawing is 2 inches, what is the actual length?

Study Hint

Reading Math The scale 1 inch = 20 feet can be read as one inch is equivalent to twenty feet.

You can write a proportion to find the actual length, ℓ.

Scale		Dinosaur
length in drawing → *actual length* →	$\frac{1 \text{ inch}}{20 \text{ feet}} = \frac{2 \text{ inches}}{\ell \text{ feet}}$	← *length in drawing* ← *actual length*
	$1 \times \ell = 20 \times 2$	*Find the cross products.*
	$\ell = 40$	*Multiply.*

The actual length of the dinosaur was 40 feet or 480 inches.

Example CONNECTION

1 **Life Science** **Suppose a drawing of a dinosaur had a scale of 1 inch = 18 feet. If the height of the dinosaur at the hips is $\frac{1}{2}$ inch, what is the actual height at the hips?**

Let h represent the height of the dinosaur at the hips.

Write and solve a proportion.

Scale		Dinosaur
length in drawing → *actual length* →	$\frac{1 \text{ inch}}{18 \text{ feet}} = \frac{\frac{1}{2} \text{ inch}}{h \text{ feet}}$	← *length in drawing* ← *actual length*
	$1 \times h = 18 \times \frac{1}{2}$	*Find the cross products.*
	$h = 9$	*Multiply.*

The height of the dinosaur at the hips is 9 feet or 108 inches.

Cross-Curriculum Cue

Inform the other teachers on your team that your students are studying scale drawings. Suggestions for curriculum integration are:

Earth Science: geological measurements
Geography: map reading
Family and Consumer Sciences: fashion design
Industrial Technology: blueprints

Example — **2**
CONNECTION

History In 1936, the first Oscar Mayer Wienermobile was built. Since then six different Wienermobiles have been built. The 1995 Oscar Mayer Wienermobile is 8 feet wide and 27 feet long. If a scale model of the Wienermobile is built with a width of 4 inches, what will be its length?

Let ℓ represent the length of the model.
Write and solve a proportion.

	Model		Weinermobile	
width of model →	$\frac{4 \text{ inches}}{8 \text{ feet}}$	$= \frac{\ell \text{ inches}}{27 \text{ feet}}$	← length of model	
actual width →			← actual length	

$4 \times 27 = 8 \times \ell$ *Find the cross products.*

$108 = 8\ell$ *Multiply.*

$\frac{108}{8} = \frac{8\ell}{8}$ *Divide each side by 8.*

$13.5 = \ell$

The length of the model is 13.5 inches or $1\frac{1}{8}$ feet.

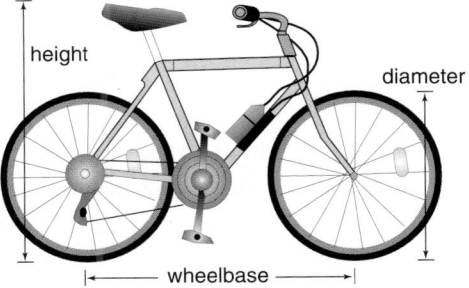

CHECK FOR UNDERSTANDING

Communicating Mathematics

Read and study the lesson to answer each question. 1–2. See margin.

1. **Explain** why an architect would make a scale drawing or a scale model of a building.

2. **Tell** whether a scale drawing of a ladybug would be smaller or larger than the actual ladybug. Explain your reasoning.

Guided Practice

3. The drawing of the bicycle has a scale of 1 inch = 2 feet. Use a ruler to measure each dimension to the nearest eighth inch. Then find the actual measure of each dimension.

3a. $2\frac{1}{4}$ feet

3b. $3\frac{1}{2}$ feet

3c. $3\frac{1}{2}$ feet

   a. the diameter of the wheels
   b. the height of the bicycle
   c. the distance between the centers of the wheels

height
diameter
wheelbase

4. **Architecture** An architect's drawing of a building has a height of $15\frac{3}{4}$ inches. If the scale on the drawing is $\frac{1}{2}$ inch = 1 foot, how tall is the building? $31\frac{1}{2}$ feet

Lesson 8-3 Integration: Geometry Scale Drawings **325**

---

2 TEACH

Transparency 8-3B contains a teaching aid for this lesson.

Reading Mathematics It is important for students to recognize ratio or proportion when reading word problems. Tell students that ratio or proportion problems, such as scale drawings, often include words or phrases such as *compared to, actual, smaller than,* and *larger than.* Tell students to also look for measurement terms such as *height, length, width, depth,* or *weight.*

In-Class Examples

For Example 1
A dinosaur is drawn to a scale of 10 centimeters = 25 meters. If the drawn height at the head is 9 centimeters, what is the actual height at the head? **22.5 m**

For Example 2
A carpenter is making a desk with a top 5 feet long and 3 feet wide. If a scale drawing of the desk is made with a width of 6 inches, what will be its length on the scale drawing? **10 inches**

3 PRACTICE/APPLY

Check for Understanding
If students need additional practice or instruction after completing Exercises 1–4, one of these options may be helpful.
• Extra Practice, see p. 579
• Reteaching Activity
• *Study Guide Masters,* p. 59
• *Practice Masters,* p. 59

Additional Answers
1. An architect would make a scale drawing or a model in order to show a building exactly as it looks, but smaller.

2. Larger; A ladybug is small; therefore, the scale drawing should be larger.

Reteaching the Lesson

Activity Have students work in small groups. Ask them to construct similar figures using toothpicks. Then have them determine a scale for each of their figures.

Error Analysis
Watch for students who multiply the numerators and divide by the denominators to solve a proportion. **Prevent by** reviewing proportions and cross products.

EXERCISES

Practice

5. The drawing of the pup tent has a scale of $\frac{3}{4}$ inch = 1 yard. Use a ruler to measure each dimension to the nearest fourth inch. Then find the actual dimensions.

5a. $1\frac{2}{3}$ yards
5b. $1\frac{1}{3}$ yards
5c. 2 yards

    a. the height
    b. the width
    c. the length

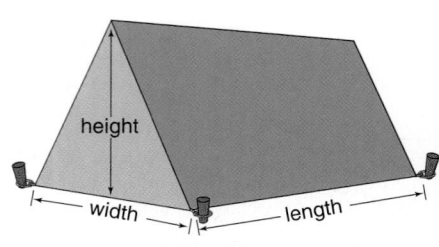

6. The floor plan shown has a scale of $\frac{1}{2}$ inch = 5 feet. Use a ruler to measure each room to the nearest eighth of an inch. Then find the actual measurements. **See margin.**

Rooms	Drawing		Actual	
	Length	Width	Length	Width
a. porch				
b. kitchen				
c. bath				
d. dining				
e. living				

7. A map of the northern portion of Arkansas is shown. It has a scale of 1 inch = 19 miles. Use a ruler to measure each map distance. Then find the actual distance.

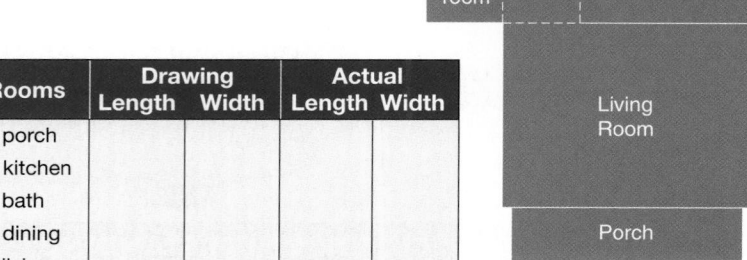

7a. $14\frac{1}{4}$ miles
7b. $23\frac{3}{4}$ miles
7c. $16\frac{5}{8}$ miles
7d. $45\frac{1}{8}$ miles
7e. $8\frac{5}{16}$ miles

    a. Wideman to Horseshoe Bend
    b. Camp to Newburg
    c. Sturkie to Saddle
    d. Mountain Home to Mount Pleasant
    e. Salem to Byron

Applications and Problem Solving

8. *Transportation* On a drawing of a truck, the truck has a height of $\frac{7}{8}$ inch. What is the actual height of the truck if the drawing has a scale of $\frac{1}{4}$ inch = 28 inches? **98 inches**

326 Chapter 8 Exploring Ratio, Proportion, and Percent

Additional Answer

6.

Rooms	Drawing		Actual	
	Length	Width	Length	Width
a. porch	$1\frac{1}{2}$ in.	$\frac{3}{8}$ in.	15 ft	$3\frac{3}{4}$ ft
b. kitchen	1 in.	1 in.	10 ft	10 ft
c. bath	$\frac{1}{2}$ in.	$\frac{1}{2}$ in.	5 ft	5 ft
d. dining	$1\frac{1}{8}$ in.	$1\frac{1}{2}$ in.	$11\frac{1}{4}$ ft	15 ft
e. living	$1\frac{5}{8}$ in.	$1\frac{1}{4}$ in.	$16\frac{1}{4}$ ft	$12\frac{1}{2}$ ft

9. *History* During World War II, America's main strategic weapon was the B-17 Flying Fortress. This aircraft has a length of about 80 feet and a wingspan of about 104 feet. If a scale model of the airplane is built with a wingspan of 26 feet, what will be its length? **20 feet**

10. *School* An original 5-inch by 8-inch photograph must be reduced to $1\frac{1}{4}$ by 2 inches to fit in the school yearbook. What is the scale of the reduced photograph to the original in simplest form? **1 in. = 4 in.**

11a. length: 5.18 m;
width: 2.16 m;
height: 1.58 m
11b. length: 4.93 m;
width: 1.92 m;
height: 1.47 m

11. *Working on the* **CHAPTER Project** Refer to the drawings on page 309.
   a. The drawing of stock car 88 has a scale of 1 cm = 144 cm. Find the length, width, and height of the actual stock car 88 race car.
   b. The drawing of stock car 16 has a scale of 1 cm = 64 cm. Find the length, width, and height of the actual stock car 16 race car.
   c. For each of the three toys that you selected, write a scale that can be used to compare the size of the toy to the actual size of the object. **See students' work.**

12. *Critical Thinking* If you were asked to make a scale drawing of the Statue of Liberty, which scale would you use: 1 cm = 1 m or 1 cm = 1 mm? Explain. **1 cm = 1 m; The other scale would make the statue smaller than the drawing.**

Mixed Review

13. Solve $\frac{7.3}{h} = \frac{14.6}{10.8}$. *(Lesson 8-2)* **5.4**

14. **Test Practice** Valerie purchased $40.06 worth of groceries. She gives the cashier a $50 bill. How much change should she receive? *(Lesson 3-6)* **C**

   A $9.84   **B** $9.34   **C** $9.94   **D** $10.84   **E** Not Here

Mid-Chapter Self Test

Write each ratio in three different ways. *(Lesson 8-1)* **1–2. See margin.**

1. 11 rulers out of 19 are blue.
2. 7 students out of 39 are boys.

Express each ratio as a fraction in simplest form. *(Lesson 8-1)*

3. 12 parrots out of 28 birds $\frac{3}{7}$
4. 15 green apples out of 48 apples $\frac{5}{16}$

Express each ratio as a rate. *(Lesson 8-1)*

5. 270 miles in 4.5 hours **60 mph**
6. 8 kiwis for $1.00 **12.5 cents per kiwi**

Solve each proportion. *(Lesson 8-2)*

7. $\frac{8}{9} = \frac{w}{108}$ **96**
8. $\frac{m}{17} = \frac{35}{42.5}$ **14**
9. $\frac{1.3}{h} = \frac{5.2}{11.2}$ **2.8**

10. *Geography* On a map, the distance between the towns of Gorden and Florin measures $\frac{3}{4}$ inch. Suppose the map has a scale of $\frac{1}{4}$ inch = 10 miles. What is the actual distance between the two towns? *(Lesson 8-3)* **30 miles**

Extending the Lesson

Enrichment Masters, p. 59

8-3 Enrichment

Planning a Room

Before moving furniture into a room, many people plan an arrangement by making a scale drawing. This makes it possible to find the best arrangement for the room without actually moving heavy furniture.

For each piece of furniture, actual measurements are given. Compute scale measurements using the scale $\frac{1}{2}$ inch = 1 foot.

1. bed: $6\frac{1}{2}$ feet long, 3 feet wide $3\frac{1}{4}$ inches long, $1\frac{1}{2}$ inches wide
2. bedside table: $1\frac{1}{2}$ feet long, $1\frac{1}{2}$ feet wide $\frac{3}{4}$ inch long, $\frac{3}{4}$ inch wide
3. bookcase: $3\frac{1}{2}$ feet long, 1 foot wide $1\frac{3}{4}$ inches long, $\frac{1}{2}$ inch wide

Activity Have students work in pairs to draw a scale drawing of the school playground, the basketball court, or the athletic field. They may wish to use centimeter grid paper.

Exercise 11 asks students to advance to the next stage of work on the Chapter Project. You may wish to provide students with a list of reference materials in which they may find actual sizes of objects.

4 ASSESS

Closing Activity

Speaking Tell students that the wingspan of a Boeing 747-400 is 213 feet. Ask: *How long would the wingspan be in a scale drawing if the scale were 1 inch = 71 feet?* **3 in.**

Mid-Chapter Self Test

The Mid-Chapter Self Test reviews concepts in Lessons 8-1 through 8-3. Lesson references are given so students can review concepts not yet mastered.

Additional Answers for the Self Test

1–2. Sample answers are given.
1. 11 to 19, $\frac{11}{19}$, 11:19
2. 7 out of 39, 7:39, $\frac{7}{39}$

Practice Masters, p. 59

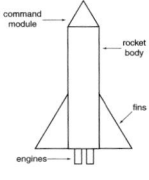

Name _____ Date _____

8-3 Practice

Integration: Geometry
Scale Drawings

1. The drawing of the rocket has a scale of $\frac{1}{2}$ inch = 12 feet. Use a ruler to measure the drawing to the nearest $\frac{1}{2}$ inch and compute the actual measurements of the rocket.

	Drawing	Actual
a. height of command module	$\frac{1}{2}$ in.	12 ft
b. height of rocket body	2 in.	48 ft
c. height of engines	$\frac{1}{4}$ in.	6 ft
d. width of rocket body and fins	$1\frac{1}{2}$ in.	36 ft
e. total length	$2\frac{3}{4}$ in.	66 ft

2. The proposed plan for a new amusement park at the right has a scale of 1 centimeter = 5 feet. Use a centimeter ruler to measure each section of the park to the nearest centimeter. Then compute the actual measurements.

Attraction	Drawing length (cm)	Drawing width (cm)	Actual length (ft)	Actual width (ft)
a. rides	3	5	15	25
b. roller coaster	2	6	10	30
c. snack bar	1	3	5	15
d. picnic area	3	2	15	10
e. arcades	6	1	30	5
f. zoo	2	3	10	15

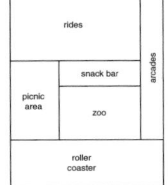

© Glencoe/McGraw-Hill   T59   *Mathematics: Applications and Connections, Course 1*

Motivating Students

Students learned about toy scale models in the Chapter Project on page 309. Students may be interested in toymaking or a related career. Professional doll makers strive for precision in proportions and carefully crafted details. To begin a discussion of toy making, ask students these questions.

- Have you ever owned a homemade toy? What made it different from a manufactured toy? **Answers will vary.**
- What should a toy maker consider when designing a new toy? **Sample answer: the age for which the toy is intended, the educational value of the toy, safety**

Making the Math Connection

To make dolls or other toys, a craftsperson needs a thorough knowledge of scale, the ability to measure accurately and perform calculations involving measurements, and an understanding of geometric principles such as perimeter, area, and volume.

Working on *Your Turn*

Have students prepare questions in advance and interview craftspeople in groups. Then have groups write a paragraph summarizing the information they gathered during the interview.

*An additional School to Career activity is available on page 8 of the **School to Career Masters.***

SCHOOL to CAREER

Arts and Crafts

Carol Larsen
DOLL MAKER

Carol Larsen, mother of nine children, has turned her hobby of doll collecting into a career. In addition to designing and making her own dolls, she also sews the clothing for the dolls, and teaches classes on doll making.

There are various types of crafts in the arts and crafts industry. The skills that you will need depend on the type of craft that you would like to make. To design and make dolls or any other toys, you need to have a good understanding of mathematics. Having a good knowledge of proportions and scale drawings will enable you to create finished products that have the correct dimensions.

For more information:
Toy Manufacturers of America, Inc.
200 5th Avenue, Suite 740
New York, NY 10010

interNET CONNECTION
www.glencoe.com/sec/math/mac/mathnet

Someday, I'd like to design and create toys.

Your Turn
Interview a person in your community that is involved in the arts and crafts industry. Find out how they began their career and how they use mathematics when making their craft.

328 Chapter 8 Exploring Ratio, Proportion, and Percent

More About Carol Larsen

- Carol Larsen lives on a farm in west central Iowa. She has been sewing since she was a young girl, but she did not take up doll making until her own children were grown.

- Ms. Larsen makes porcelain dolls by pouring the liquid porcelain into molds and firing it in a kiln. Then she paints the faces and sews each doll a special outfit. She enjoys designing dolls for meaningful occasions and often dresses her dolls in wedding or christening gowns.

COOPERATIVE LEARNING

8-4A Modeling Percents

A Preview of Lesson 8-4

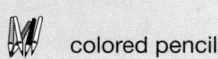 grid paper

 colored pencils

You know that a 10 × 10 grid can be used to represent hundredths. Since the word *percent* means *out of one hundred,* you can also use a 10 × 10 grid to model percents.

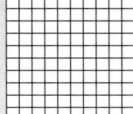

TRY THIS

Work with a partner.

1 Model 25%.

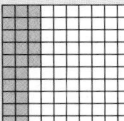

Shade 25 of the 100 squares.

25% means "25 out of 100". By shading 25 out of 100 squares, you can show 25%. What decimal does the model represent?

2 Shade two fifths of the 10 × 10 grid. What percent have you modeled?

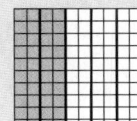

Separate the model into fifths. Shade two of the fifths.

There are 40 squares shaded, so 40% is modeled. Two fifths is equivalent to 40%.

ON YOUR OWN

1. Draw five 10 × 10 squares on a sheet of grid paper. Shade a 10 × 10 grid to represent each percent. **1–3. See Answer Appendix.**

   a. 1%   **b.** 75%   **c.** 30%   **d.** 100%   **e.** 60%

2. Draw five more 10 × 10 squares on another sheet of grid paper. Shade a 10 × 10 grid to represent each fraction or number.

   a. 0.3   **b.** $\frac{1}{100}$ **c.** 0.75   **d.** 1   **e.** $\frac{3}{5}$

3. How do the two sets of shaded 10 × 10 grids compare?

4. **Look Ahead** Express 36% as a fraction in simplest form. Use a model if necessary. $\frac{9}{25}$

 Math Journal Have students write a paragraph explaining how $\frac{3}{4}$ and 75% are related. Have them include a grid to model this percent and fraction.

GET READY

Objective Students illustrate the meaning of percent using models.

Optional Resources
Hands-On Lab Masters
- grid paper, p. 11
- worksheet, p. 56

Overhead Manipulative Resources
- decimal models
- grid paper

MANAGEMENT TIPS

Recommended Time
30 minutes

Getting Started Review using 10 × 10 grids to write decimals as fractions. For example, ask students to write 0.83 as a fraction by showing it first on a grid.

Activity 1 produces a model for 25% by shading 25 out of 100 squares on a 10 × 10 grid. What fraction does the model represent? $\frac{1}{4}$

Activity 2 requires students to divide the 10 × 10 grid into 5 sections, shade 2 sections, and then recognize that 40 out of 100 squares have been shaded. Follow up by asking how students would model $\frac{1}{25}$.

ASSESS

Have students complete Exercises 1–3. Watch for students who have difficulty comparing the percents in Exercise 1 with the numbers in Exercise 2. Remind students what *percent* means.

Use Exercise 4 to determine whether students comprehend the concept of percent.

Instructional Resources

- *Study Guide Masters*, p. 60
- *Practice Masters*, p. 60
- *Enrichment Masters*, p. 60
- Transparencies 8-4, A and B
- *Assessment and Evaluation Masters*, pp. 210, 211

CD-ROM Program

- Resource Lesson 8-4
- Extended Activity 8-4

Recommended Pacing

Standard	Days 8 & 9 of 14
Honors	Day 8 of 13
Block	Day 5 of 7

1 FOCUS

5-Minute Check
(Lesson 8-3)

The floor plan shown has a scale of 1 cm = 2 m. Compute the actual measurement of each room.

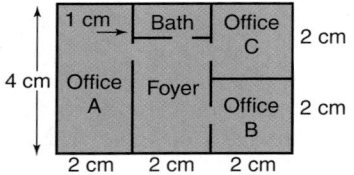

	Drawing (cm)		Actual (m)	
	ℓ	w	ℓ	w
Foyer	3	2	6	4
Office A	4	2	8	4
Office B	2	2	4	4
Office C	2	2	4	4
Bathroom	2	1	4	2

The 5-Minute Check is also available on **Transparency 8-4A** for this lesson.

***What* you'll learn**

You'll learn to express percents as fractions and vice versa.

***When* am I ever going to use this?**

Knowing how to express a number in a different form can help you interpret a monthly budget.

Word Wise

percent

> **Study Hint**
> Reading Math The symbol % means percent.

Did you know that the word pasta is an Italian term meaning dough? Pasta comes in more than 100 shapes and sizes. What is your favorite type of pasta: spaghetti, ravioli, vermicelli, or possibly gnocchi? According to the results of a survey conducted by the National Pasta Association, 10% of people eat lasagna at least once every two weeks.

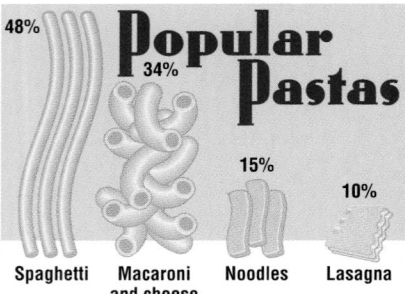

Popular Pastas

48%   34%   15%   10%

Spaghetti   Macaroni and cheese   Noodles   Lasagna

Source: National Pasta Association

A **percent** is a ratio that compares a number to 100. All of the percents in the graph can be expressed as fractions. To express a percent as a fraction, follow these steps.

- Express the percent as a fraction with a denominator of 100.
- Simplify.

 **Examples**

Express each percent as a fraction in simplest form.

1 28%

28% means "28 out of 100."

$28\% = \dfrac{28}{100}$ *Express the percent as a fraction with a denominator of 100.*

$= \dfrac{\overset{7}{\cancel{28}}}{\underset{25}{\cancel{100}}}$ *Simplify. Divide the numerator and the denominator by 4, the GCF of 28 and 100.*

$= \dfrac{7}{25}$

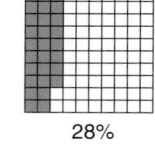

28%

2 118%

118% means "118 out of 100."

$118\% = \dfrac{118}{100}$ *Express the percent as a fraction with a denominator of 100.*

$= 1\dfrac{18}{100}$ or $1\dfrac{9}{50}$ *Simplify.*

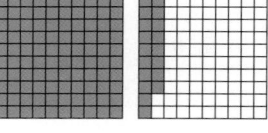

118%

Classroom Vignette

"I have groups of students count, by color, the number of M&Ms in a bag. They write fractions for each color and then change those fractions to percents. We apply the percents to making circle graphs and display the results around the room. It's a big and fun project!"

Mary Schirm

Mary Schirm, Teacher
Summerour Middle School
Norcross, GA

Example
CONNECTION

3 **Health** Refer to the graph. What fraction of teen-age boys always or often use sunscreen before going out into the sun?

From the graph, you can see that 30% of teen-age boys use sunscreen before going into the sun.

$$30\% = \frac{30}{100}$$
$$= \frac{3}{10}$$

So, $\frac{3}{10}$ of teen-age boys always or often use sunscreen.

How regularly teenagers use sunscreen

36% 21% 34% 33% 30% 46%

Boys Girls Boys Girls Boys Girls
Rarely/never **Sometimes** **Always/often**

Source: Sovereign Research for Seventeen, Nivea, and the American Academy of Dermatology

You can also express fractions as percents.

HANDS-ON
MINI-LAB

Work in groups of four. 📄 paper 🗑 wastebasket

Try This
• Stand behind a line 15 feet from a wastebasket.
• Have each person throw a wadded paper ball into the basket. Each person will have five tries.
• Have one group member record a ratio for the total number of baskets made out of the total number of tries.
• Calculate the percent of baskets made for the group.

Talk About It **2. See students' work.**
1. How did you find the percent? **See margin.**
2. How does your group's percent compare to the other groups?

To express a fraction as a percent, write a proportion and solve it.

Examples

Study Hint
Mental Math When changing $\frac{1}{5}$ to a percent, think of finding $\frac{1}{5}$ of 100 parts. $\frac{1}{5}$ = 20%.

4 Express $\frac{1}{5}$ as a percent.

$$\frac{1}{5} = \frac{n}{100}$$ *Set up a proportion.*
$$1 \times 100 = 5 \times n$$ *Find the cross products.*
$$100 = 5n$$
$$100 \div 5 = 5n \div 5$$ *Divide.*
$$20 = n$$

So, $\frac{1}{5}$ is equivalent to 20%.

5 Express $\frac{9}{4}$ as a percent.

$$\frac{9}{4} = \frac{c}{100}$$
$$9 \times 100 = 4 \times c$$
$$900 = 4c$$
$$900 \div 4 = 4c \div 4$$
$$225 = c$$

So, $\frac{9}{4}$ is equivalent to 225%.

Lesson 8-4 Percents and Fractions **331**

Additional Answer for the Mini-Lab
1. number of shots made ÷ number of tries × 100

Motivating the Lesson
Communication Have students work in small groups. Give each group $100 in play money to budget on entertainment, food, savings, and so on. Ask students how they could use fractions and percents to describe their budgets.

2 TEACH

 Transparency 8-4B contains a teaching aid for this lesson.

In-Class Examples
Express each percent as a fraction in simplest form.

For Example 1
40% $\frac{2}{5}$

For Example 2
125% $\frac{5}{4}$

For Example 3
Refer to the graph on page 331. What fraction of teenage girls always or often uses sunscreen before going out into the sun? $\frac{23}{50}$

Teaching Tip Before Example 2, tell students that percents can be greater than 100%.

Using the Mini-Lab You may wish to have the groups complete this activity simultaneously by having each group use a different wastebasket.

In-Class Examples
Express each fraction as a percent.

For Example 4
$\frac{3}{10}$ 30%

For Example 5
$\frac{7}{2}$ 350%

Check for Understanding

If students need additional practice or instruction after completing Exercises 1–10, one of these options may be helpful.
- Extra Practice, see p. 579
- Reteaching Activity
- *Study Guide Masters,* p. 60
- *Practice Masters,* p. 60

Assignment Guide
Core: 11–35 odd, 38–43
Enriched: 12–34 even, 35, 36, 38–43

CHECK FOR UNDERSTANDING

Communicating Mathematics

Read and study the lesson to answer each question.

1. *Identify* the percent that is represented by each model shown.

a.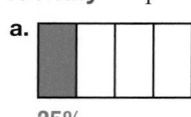
25%

b.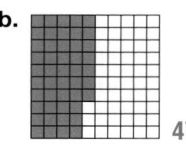
47%

c.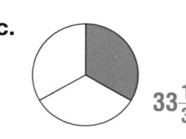
$33\frac{1}{3}$%

d.
80%

e.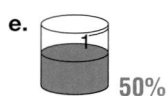
50%

f. ▲▲△ / ▲▲△ $66\frac{2}{3}$%

2. *Tell* what fraction is equivalent to 75%. $\frac{3}{4}$

HANDS-ON MATH

3. Toss a coin 50 times. Record the number of times the coin shows heads. Then calculate the percent that represents the number of times the coin showed heads. **See students' work.**

Guided Practice

Express each percent as a fraction in simplest form.

4. 2% $\frac{1}{50}$ 5. 55% $\frac{11}{20}$ 6. 120% $1\frac{1}{5}$

Express each fraction as a percent.

7. $\frac{34}{100}$ 34%   8. $\frac{13}{20}$ 65%   9. $\frac{5}{4}$ 125%

10. *Technology* According to a survey, 36% of consumers prefer to buy a personal computer at an electronics store. What fraction of consumers is this? $\frac{9}{25}$

EXERCISES

Practice

11. $\frac{7}{50}$ 12. $\frac{9}{10}$
13. $\frac{1}{100}$ 14. $\frac{13}{20}$
15. $1\frac{3}{10}$ 16. $\frac{24}{25}$
17. $1\frac{1}{20}$ 18. $\frac{33}{50}$
19. $\frac{17}{100}$ 20. $1\frac{3}{25}$

Express each percent as a fraction in simplest form.

11. 14%   12. 90%   13. 1%   14. 65%   15. 130%
16. 96%   17. 105%   18. 66%   19. 17%   20. 112%

Use a 10 × 10 grid to shade the amount stated in each fraction. Then express each fraction as a percent. 21–24. For grids, see Answer Appendix.

21. $\frac{4}{5}$ 80%   22. $\frac{4}{10}$ 40%   23. $\frac{9}{20}$ 45%   24. $\frac{13}{50}$ 26%

Express each fraction as a percent.

25. $\frac{63}{100}$ 63%   26. $\frac{1}{20}$ 5%   27. $\frac{3}{2}$ 150%   28. $\frac{12}{25}$ 48%

29. $\frac{37}{50}$ 74%   30. $\frac{17}{20}$ 85%   31. $\frac{19}{25}$ 76%   32. $\frac{17}{10}$ 170%

33. How is forty-five hundredths written as a percent? 45%

34. Express eighty-five percent as a fraction in simplest form. $\frac{17}{20}$

332 Chapter 8 Exploring Ratio, Proportion, and Percent

Study Guide Masters, p. 60

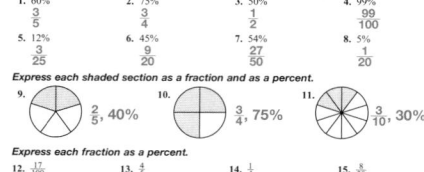

Name_____ Date_____

8-4 Study Guide

Percents and Fractions

A **percent** is a ratio that compares a number to 100.
To express a percent as a fraction, write it as a fraction with a denominator of 100 and then simplify.

Example 1 48% = $\frac{48}{100}$
= $\frac{12}{25}$ *Divide the numerator and denominator by 4.*

To express a fraction as a percent, first set up a proportion. Then solve the proportion using cross products.

Example 2 Express $\frac{13}{20}$ as a percent.
$\frac{13}{20} = \frac{k}{100}$ *Set up a proportion.*
$13 \times 100 = 20 \times k$ *Find the cross products.*
$1,300 = 20k$
$1,300 \div 20 = 20k \div 20$ *Divide each side by 20.*
$65 = k$
$\frac{13}{20} = \frac{65}{100}$ or 65%

Express each percent as a fraction in simplest form.

1. 60% $\frac{3}{5}$ 2. 75% $\frac{3}{4}$ 3. 50% $\frac{1}{2}$ 4. 99% $\frac{99}{100}$

5. 12% $\frac{3}{25}$ 6. 45% $\frac{9}{20}$ 7. 54% $\frac{27}{50}$ 8. 5% $\frac{1}{20}$

Express each shaded section as a fraction and as a percent.

9. $\frac{2}{5}$, 40%   10. $\frac{3}{4}$, 75%   11. $\frac{3}{10}$, 30%

Express each fraction as a percent.

12. $\frac{17}{100}$ 17%   13. $\frac{4}{5}$ 80%   14. $\frac{1}{4}$ 25%   15. $\frac{8}{20}$ 40%

16. $\frac{1}{50}$ 2%   17. $\frac{7}{10}$ 70%   18. $\frac{6}{25}$ 24%   19. $\frac{1}{10}$ 10%

© Glencoe/McGraw-Hill   T60   Mathematics: Applications and Connections, Course 1

Reteaching the Lesson

Activity Have students separate into groups according to the months in which they were born. Then have each group figure out what percent of the class their group represents. Have students express their results in both fraction and percent form.

Error Analysis

Watch for students who are confused by percents greater than 100.

Prevent by reviewing the steps in changing improper fractions to mixed numbers and vice versa.

Applications and Problem Solving

35. *Travel* The graph shows why RV owners like to travel in an RV.

  a. What percent of those surveyed believe that traveling in a recreational vehicle is the best way to see the United States? **48%**

  b. What percent of those surveyed have other reasons why they like to travel in an RV? **9%**

Traveling in an RV

Best way to see the USA	$\frac{12}{25}$
Cost-efficient travel	$\frac{7}{25}$
Spend time with family/friends	$\frac{3}{20}$
Other	$\frac{9}{100}$

Source: Allstate Motor Club Survey of 400 RV owners

36. *Money Matters* The circle graph shows the monthly budget for the Balint family.

36a. $\frac{7}{20}$ 36b. $\frac{7}{25}$

  a. What fraction, in simplest form, represents the portion of the family budget that is spent on rent?

  b. What fraction, in simplest form, represents the portion of the family budget that is spent on food and utilities?

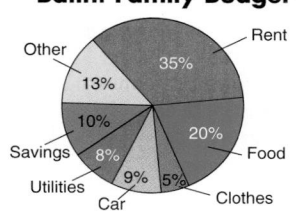
Balint Family Budget

Rent 35%, Food 20%, Clothes 5%, Car 9%, Utilities 8%, Savings 10%, Other 13%

37. *Working on the* **CHAPTER Project** Refer to Exercise 35 on page 315.

37a. $\frac{1}{144}$, 0.01, 1%

37b. $\frac{1}{64}$, 0.02, 2%

37c. See students' work.

  a. Express the ratio in part a as a fraction, decimal, and as a percent.

  b. Express the ratio in part b as a fraction, decimal, and as a percent.

  c. Express each ratio in part c as a fraction, decimal, and as a percent.

38. *Critical Thinking* A woman made a will leaving $\frac{1}{2}$ of her fortune to her sister, $\frac{1}{3}$ to her brother, $\frac{1}{7}$ to her nephew, and the remainder to her cat. What percent of the woman's fortune was left to the cat? Round to the nearest whole percent. **about 2%**

Mixed Review

39. *Geography* A map has a scale of 1 inch = 20 miles. The distance from Hartville to Dixon is $2\frac{3}{4}$ inches. What is the actual distance between these two cities? *(Lesson 8-3)* **55 miles**

40. *Measurement* Painters used 170 gallons of white topcoat to paint the famous Hollywood sign. How many quarts are in 170 gallons? *(Lesson 7-7)* **680 quarts**

41. Find $\frac{3}{8} + \frac{7}{8}$. *(Lesson 6-3)* $1\frac{1}{4}$

42. **Test Practice** What is the perimeter of the figure? *(Lesson 4-4)* **B**

  A 74.3 m
  B 73.2 m
  C 72.2 m
  D 75.3 m

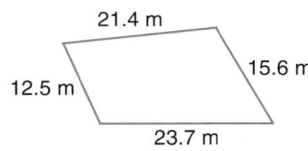
21.4 m, 15.6 m, 12.5 m, 23.7 m

43. *Algebra* Evaluate $gh + j$ if $g = 4$, $h = 7$, and $j = 2$. *(Lesson 1-5)* **30**

Lesson 8-4 Percents and Fractions **333**

Extending the Lesson

Enrichment Masters, p. 60

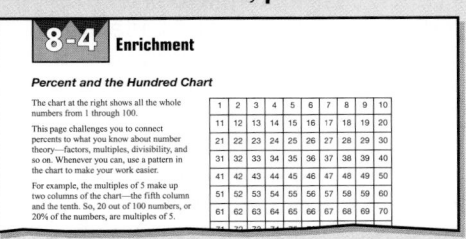

8-4 Enrichment

Percent and the Hundred Chart

The chart at the right shows all the whole numbers from 1 through 100.

This page challenges you to connect percents to what you know about number theory—factors, multiples, divisibility, and so on. Whenever you can, use a pattern in the chart to make your work easier.

For example, the multiples of 5 make up two columns of the chart—the fifth column and the tenth. So, 20 out of 100 numbers, or 20% of the numbers, are multiples of 5.

Activity Have students use spreadsheet software to create a bar graph. Ask them to use the graph to create problems about writing percents as fractions. Then have them display the graph and the problems on the bulletin board for others to solve.

Exercise 37 asks students to advance to the next stage of work on the Chapter Project. You may wish to have students work in pairs to express the ratios as fractions, decimals, and percents.

4 ASSESS

Closing Activity

Act It Out Have students work in groups to determine what fraction and percent of their group is left-handed.

Chapter 8, Quiz B (Lessons 8-3 and 8-4) is available in the *Assessment and Evaluation Master*, p. 211.

Mid-Chapter Test (Lessons 8-1 through 8-4) is available in the *Assessment and Evaluation Masters*, p. 210.

Practice Masters, p. 60

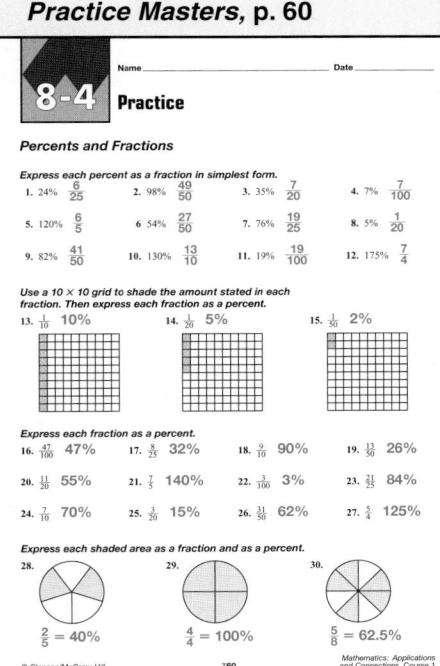

8-4 Practice

Percents and Fractions

Express each percent as a fraction in simplest form.

1. 24% $\frac{6}{25}$ 2. 98% $\frac{49}{50}$ 3. 35% $\frac{7}{20}$ 4. 7% $\frac{7}{100}$

5. 120% $\frac{6}{5}$ 6. 54% $\frac{27}{50}$ 7. 76% $\frac{19}{25}$ 8. 5% $\frac{1}{20}$

9. 82% $\frac{41}{50}$ 10. 130% $\frac{13}{10}$ 11. 19% $\frac{19}{100}$ 12. 175% $\frac{7}{4}$

Use a 10 × 10 grid to shade the amount stated in each fraction. Then express each fraction as a percent.

13. $\frac{1}{10}$ 10%   14. $\frac{1}{20}$ 5%   15. $\frac{1}{50}$ 2%

Express each fraction as a percent.

16. $\frac{47}{100}$ 47%   17. $\frac{8}{25}$ 32%   18. $\frac{9}{10}$ 90%   19. $\frac{13}{50}$ 26%

20. $\frac{11}{20}$ 55%   21. $\frac{7}{5}$ 140%   22. $\frac{3}{100}$ 3%   23. $\frac{21}{25}$ 84%

24. $\frac{7}{10}$ 70%   25. $\frac{3}{20}$ 15%   26. $\frac{31}{50}$ 62%   27. $\frac{5}{4}$ 125%

Express each shaded area as a fraction and as a percent.

28. $\frac{2}{5}$ = 40%   29. $\frac{4}{4}$ = 100%   30. $\frac{5}{8}$ = 62.5%

© Glencoe/McGraw-Hill   T60   *Mathematics: Applications and Connections, Course 1*

Instructional Resources

- *Study Guide Masters*, p. 61
- *Practice Masters*, p. 61
- *Enrichment Masters*, p. 61
- Transparencies 8-5, A and B
- *Science and Math Lab Manual*, pp. 81–84

- CD-ROM Program
  - Resource Lesson 8-5

Recommended Pacing	
Standard	Day 10 of 14
Honors	Day 9 of 13
Block	Day 5 of 7

1 FOCUS

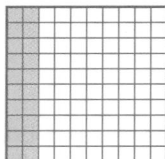

5-Minute Check
(Lesson 8-4)

Express each percent as a fraction in simplest form.

1. 55% $\frac{11}{20}$
2. 72% $\frac{18}{25}$

3. Use a 10 × 10 grid to shade $\frac{1}{5}$ of the model. Then express the fraction as a percent. **20%**

Express each fraction as a percent.

4. $\frac{3}{5}$ **60%**
5. $\frac{24}{21}$ **114%**

 The 5-Minute Check is also available on **Transparency 8-5A** for this lesson.

Motivating the Lesson

Hands-On Activity Have groups of students divide 100 pennies equally, putting any extra pennies aside. Then have them express the amount they received as a percent and as a decimal of the whole.

***What* you'll learn**
You'll learn to express percents as decimals and vice versa.

***When* am I ever going to use this?**
Knowing how to express a percent as a decimal can help you determine the amount of sales tax paid on a dollar.

In 1996, five years after it opened, Eurodisney became France's top tourist attraction. The graph shows where the 11.7 million visitors came from. What part of the visitors came from Italy? *This problem will be solved in Example 3.*

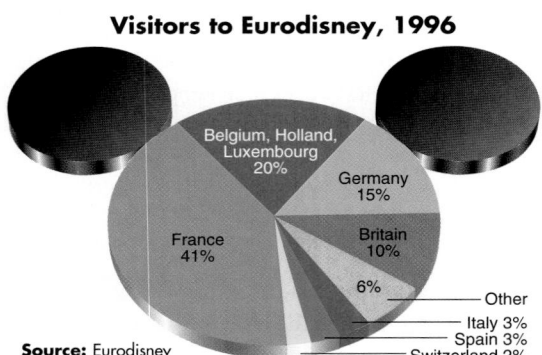

Visitors to Eurodisney, 1996

Belgium, Holland, Luxembourg 20%
Germany 15%
France 41%
Britain 10%
6%
Other
Italy 3%
Spain 3%
Switzerland 2%

Source: Eurodisney

Notice that the sections of the graph are labeled as percents. In Lesson 8-4, you learned how to express percents as fractions. Percents can also be written as decimals. To express a percent as a decimal, follow these steps.

- Rewrite the percent as a fraction with a denominator of 100.
- Express the fraction as a decimal.

Examples

Express each percent as a decimal.

Study Hint

Mental Math To express a percent as a decimal, you can use this shortcut. Move the decimal point two places to the left, which is the same as dividing by 100.

1 65%

$65\% = \frac{65}{100}$ *Rewrite the percent as a fraction with a denominator of 100.*

$= 0.65$ *Express the fraction as a decimal.*

2 0.5%

$0.5\% = \frac{0.5}{100}$ *Rewrite the percent as a fraction with a denominator of 100.*

$= \frac{5}{1,000}$ $\frac{0.5}{100} \times \frac{10}{10} = \frac{5}{1,000}$

$= 0.005$ *Express the fraction as a decimal.*

Classroom Vignette

"After distributing baseball cards, I ask students to find averages and ratios that aren't listed on them and write them as decimals and percents. Students would find slugging and on-base averages for hitters and win and strikeout ratios for pitchers."

Betty Fulton, Teacher
Clearmount Elementary School
North Canton, OH

Example ③ **INTEGRATION**

Statistics Refer to the beginning of the lesson. What part of the visitors came from Italy? Express the answer as a decimal.

From the graph, you can see that 3% of the visitors came from Italy.

$$3\% = \frac{3}{100}$$
$$= 0.03$$

The graph shows that 0.03 of the visitors came from Italy.

To express a decimal as a percent, follow these steps.

- Express the decimal as a fraction.
- Express the fraction as a percent.

Examples

Express each decimal as a percent.

④ **0.72**

$$0.72 = \frac{72}{100} \quad \textit{Express the decimal as a fraction.}$$
$$= 72\% \quad \textit{Express the fraction as a percent.}$$

⑤ **0.257**

$$0.257 = \frac{257}{1,000} \quad \textit{Express the decimal as a fraction.}$$
$$= \frac{25.7}{100} \quad \frac{257 \div 10}{1,000 \div 10} = \frac{25.7}{100}$$
$$= 25.7\% \quad \textit{Express the fraction as a percent.}$$

> **Study Hint**
> **Mental Math** To express a decimal as a percent you can use another shortcut. Move the decimal point two places to the right, which is the same as multiplying by 100.

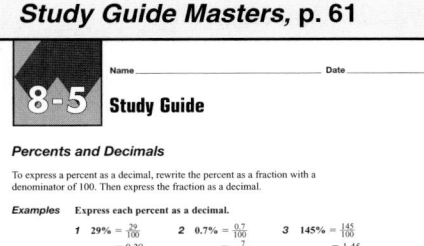

CHECK FOR UNDERSTANDING

Communicating Mathematics

Read and study the lesson to answer each question. 1–3. See Answer Appendix.

1. *Draw* a model that shows the decimal equivalent of 36%.

2. *Explain* how to express 0.008 as a percent.

3. *You Decide* Kellile says that the decimal 3.78 is less than 100%. Betty says that the decimal is more than 100%. Who is correct? Explain your reasoning.

Guided Practice

Express each percent as a decimal.

4. 46% **0.46**   5. 81% **0.81**   6. 7% **0.07**   7. 0.8% **0.008**

Express each decimal as a percent.

8. 0.52 **52%**   9. 0.9 **90%**   10. 0.175 **17.5%**   11. 0.02 **2%**

12. *Life Science* About 95% of all species of fish have skeletons made of bone. Express 95% as a decimal. **0.95**

Lesson 8-5 Percents and Decimals **335**

Reteaching the Lesson

Activity Have students solve this problem. *George Brett of the Kansas City Royals was the American League batting champion in 1980 and again in 1990. His batting average was 39% in 1980 and 32.8% in 1990. How would you express Brett's batting averages in decimals?* **0.390; 0.328**

Error Analysis
Watch for students who remove the percent symbol when expressing a decimal percent as a decimal.
Prevent by reviewing the definition of percent.

---

2 TEACH

Transparency 8-5B contains a teaching aid for this lesson.

Reading Mathematics Reading a fraction or decimal correctly can help students understand place value. Fractions and decimals are read the same way. For example, $19\frac{3}{10}$ and 19.3 are both read as *nineteen and three tenths*.

In-Class Examples

Express each percent as a decimal.

For Example 1
23% **0.23**

For Example 2
0.8% **0.008**

For Example 3
Refer to the graph on page 334. What part of the visitors came from Germany? Express the answer as a decimal. **0.15**

Express each decimal as a percent.

For Example 4
0.64 **64%**

For Example 5
0.519 **51.9%**

Study Guide Masters, p. 61

Check for Understanding
If students need additional practice or instruction after completing Exercises 1–12, one of these options may be helpful.
- Extra Practice, see p. 579
- Reteaching Activity, see p. 335
- *Study Guide Masters*, p. 61
- *Practice Masters*, p. 61

Assignment Guide
Core: 13–41 odd, 42–47
Enriched: 14–38 even, 40–47

4 ASSESS

Closing Activity
Modeling Ask students to look through magazine ads or newspaper articles to find uses of percents. Have them use 10 × 10 grids to express each percent as a decimal.

Practice Masters, p. 61

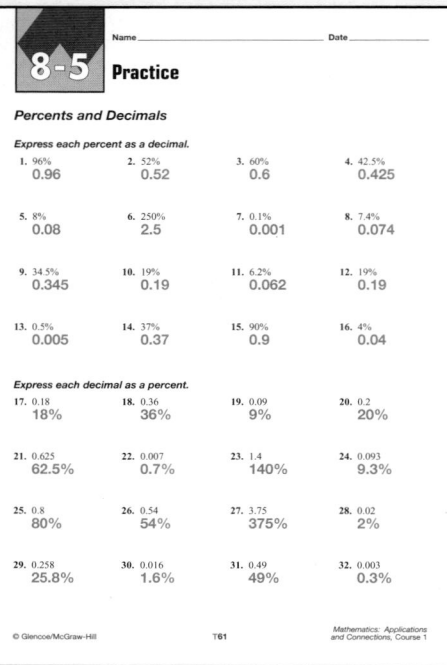

EXERCISES

Practice

Express each percent as a decimal.

13. 32% **0.32** **14.** 84% **0.84** **15.** 1% **0.01** **16.** 0.9% **0.009**

17. 6% **0.06** **18.** 17% **0.17** **19.** 0.3% **0.003** **20.** 63% **0.63**

21. 39% **0.39** **22.** 3.5% **0.035** **23.** 4% **0.04** **24.** 26% **0.26**

Express each decimal as a percent.

25. 0.96 **96%** **26.** 0.1 **10%** **27.** 0.364 **36.4%** **28.** 0.27 **27%**

29. 0.716 **71.6%** **30.** 0.66 **66%** **31.** 0.07 **7%** **32.** 0.5 **50%**

33. 0.08 **8%** **34.** 0.104 **10.4%** **35.** 0.2 **20%** **36.** 0.03 **3%**

37. How is eighteen thousandths written as a percent? **1.8%**

38. Write two and four tenths percent as a decimal. **0.024**

39. *Write a Problem* that can be solved by expressing a decimal as a percent. **See students' work.**

Applications and Problem Solving

40. *Taxes* In 1997, the residents of the state of Texas paid a 8.25% sales tax on purchases.
   a. Express 8.25% as a decimal. **0.0825**
   b. On each dollar, how much money did Texans pay in sales tax? **8.25 cents**

41. *Media* Refer to the graph.
   a. What percent of those surveyed spend more than an hour reading the Sunday paper? **18%**
   b. What percent of those surveyed do *not* spend more than an hour reading the Sunday paper? **82%**

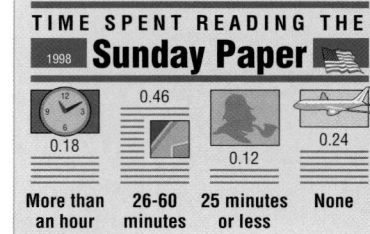

TIME SPENT READING THE
Sunday Paper

Source: Impact Resources

42. *Critical Thinking* Order 23.4%, 2.34, 0.0234, 20.34% from greatest to least. **2.34, 23.4%, 20.34%, 0.0234**

Mixed Review

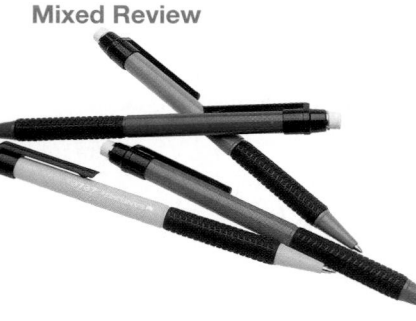

43. Express 156% as a fraction in simplest form. *(Lesson 8-4)* $1\frac{14}{25}$

44. *Money Matters* Ahmik purchased a package of four mechanical pencils for $6.48. How much does each pencil cost? *(Lesson 8-1)* **$1.62**

45. **Test Practice** What number is missing from the sequence 14, 56, _?_, 896, 3,584? *(Lesson 7-8)* **D**
   A 284   **B** 194
   C 334   **D** 224

46. *Measurement* How many inches are in $2\frac{1}{2}$ yards? *(Lesson 5-6)* **90 inches**

47. Round 64.35 ÷ 12 to the nearest tenth. *(Lesson 4-5)* **5.4**

Extending the Lesson

Enrichment Masters, p. 61

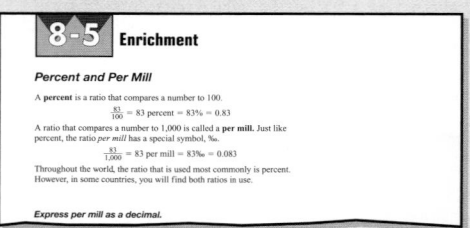

Activity Have students in small groups discuss how changing percents to decimals and vice versa might be useful in a job setting. Have students make a list of professions they think might use this skill. You may wish to have the groups compare their lists to see how similar they are.

Estimating with Percents

What you'll learn

You'll learn to estimate the percent of a number.

When am I ever going to use this?

Knowing how to estimate the percent of a number can help you determine the tip at a restaurant.

The Watch Store is having a sale on Armitron watches. Every watch is on sale for 60% of the regular price. About how much would you pay for a Marvin Martian watch that normally sells for $34.99? *This problem will be solved in Example 3.*

The word *about* tells you that an exact answer is not needed. So, you can estimate the answer. When you estimate with percents, you should round the percent to a fraction that is easy to multiply. The chart shows some commonly-used percents and their fraction equivalents.

$20\% = \frac{1}{5}$	$25\% = \frac{1}{4}$	$12\frac{1}{2}\% = \frac{1}{8}$	$16\frac{2}{3}\% = \frac{1}{6}$
$40\% = \frac{2}{5}$	$50\% = \frac{1}{2}$	$37\frac{1}{2}\% = \frac{3}{8}$	$33\frac{1}{3}\% = \frac{1}{3}$
$60\% = \frac{3}{5}$	$75\% = \frac{3}{4}$	$62\frac{1}{2}\% = \frac{5}{8}$	$66\frac{2}{3}\% = \frac{2}{3}$
$80\% = \frac{4}{5}$	$100\% = 1$	$87\frac{1}{2}\% = \frac{7}{8}$	$83\frac{1}{3}\% = \frac{5}{6}$

Examples

Study Hint

Problem Solving You can also make a model to help you estimate 22% of 197.

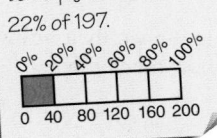

Estimate each percent.

❶ 22% of 197

22% is close to 20% or $\frac{1}{5}$.

Round 197 to 200.

$\frac{1}{5} \times 200 = \frac{1}{5} \times \frac{\overset{40}{200}}{1}$

$= 40$

So, 22% of 197 is about 40.

❷ 50% of 1,512

50% is $\frac{1}{2}$.

Round 1,512 to 1,500.

$\frac{1}{2} \times 1,500 = \frac{1}{2} \times \frac{\overset{750}{1,500}}{1}$

$= 750$

So, 50% of 1,512 is about 750.

APPLICATION **❸ Money Matters** Refer to the beginning of the lesson. About how much would you pay for a Marvin Martian watch that normally sells for $34.99?

Express 60% as a fraction. $60\% = \frac{60}{100} = \frac{3}{5}$ *Simplify.*

Then estimate the product of the fraction and the price.

$\frac{3}{5} \times 35 = \frac{3}{5} \times \frac{35}{1}$ *Think: $34.99 is close to $35.*

$= \frac{3}{\underset{1}{5}} \times \frac{\overset{7}{35}}{1}$ *Divide 5 and 35 by their GCF, 5.*

$= 21$ *Multiply.*

You will pay about $21 for the Marvin Martian watch.

Lesson 8-6 Estimating with Percents   **337**

Multiple Learning Styles

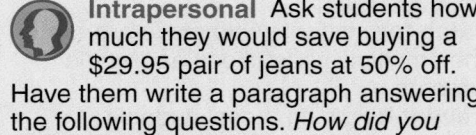

Intrapersonal Ask students how much they would save buying a $29.95 pair of jeans at 50% off. Have them write a paragraph answering the following questions. *How did you figure the savings? Are you more comfortable working with decimals or fractions?*

Motivating the Lesson

Problem Solving Ask students how they would determine which pair of boots costs less.
- a $69 pair at 35% off
- a $79 pair at 50% off
- a $79 pair at 50% off

Instructional Resources
- *Study Guide Masters,* p. 62
- *Practice Masters,* p. 62
- *Enrichment Masters,* p. 62
- Transparencies 8-6, A and B
- *Assessment and Evaluation Masters,* p. 212
- *Diversity Masters,* p. 8
- *Hands-On Lab Masters,* p. 76
- *Science and Math Lab Manual,* pp. 85–88
- CD-ROM Program
  - Resource Lesson 8-6
  - Interactive Lesson 8-6

Recommended Pacing	
Standard	Day 11 of 14
Honors	Day 10 of 13
Block	Day 6 of 7

1 FOCUS

 5-Minute Check (Lesson 8-5)

Express each percent as a decimal.
1. 14% **0.14** 2. 1.36% **0.0136**

Express each decimal as a percent.
3. 0.74 **74%** 4. 0.249 **24.9%**

5. The annual interest on Sam's savings account is 5%. How much money does he earn each year in interest for every dollar he saves? **$0.05**

 The 5-Minute Check is also available on **Transparency 8-6A** for this lesson.

2 TEACH

 Transparency 8-6B contains a teaching aid for this lesson.

Using Applications Have students use newspaper ads to estimate the sale price of merchandise. Discuss how being able to estimate discounts quickly can help them bargain shop.

In-Class Examples

Estimate each percent.

For Example 1

24% of 411 $\frac{1}{4}$ of 400 = 100

For Example 2

78% of 1,187 $\frac{3}{4}$ of 1,200 = 900

For Example 3

Justin is buying a radio that is on sale for 80% of the regular price. If it normally sells for $39.99, about how much will he pay? $\frac{4}{5} \times \$40 = \$32$

For Example 4

Estimate the percent of the figure that is shaded. **25%**

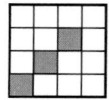

3 PRACTICE/APPLY

Check for Understanding

If students need additional practice or instruction after completing Exercises 1–9, one of these options may be helpful.

- Extra Practice, see p. 580
- Reteaching Activity
- *Study Guide Masters*, p. 62
- *Practice Masters*, p. 62

Study Guide Masters, p. 62

8-6 Study Guide

Estimating with Percents

You can use rounding to estimate with percents.

Examples **1 Estimate 75% of 788.**

$75\% = \frac{3}{4}$ *Express 75% as a fraction.*

Round 788 to 800.

$\frac{3}{4} \times 800 = 600$ *Multiply.*

75% of 788 is about 600.

2 Estimate 65% of 90.

65% is close to $66\frac{2}{3}\%$ or $\frac{2}{3}$.   *Express 65% as a compatible fraction.*

$\frac{2}{3} \times 90 = 60$ *Estimate the product using the fraction.*

65% of 90 is about 60.

3 Estimate 48% of 192.

48% is close to 50% or $\frac{1}{2}$.   *Express 48% as a fraction.*

Round 192 to 200.

$\frac{1}{2} \times 200 = 100$ *Multiply.*

48% of 192 is about 100.

Estimate each percent.

1. 31% of 150 about 50	2. 50% of 27 about 15	3. 79% of 102 about 80
4. 9% of 450 about 45	5. 23% of 60 about 15	6. 30% of 96 about 30
7. 53% of 82 about 40	8. 21% of 200 about 40	9. 68% of 89 about 60
10. 10.7% of 160 about 16	11. 79% of 21 about 16	12. 0.6% of 201 about 1

© Glencoe/McGraw-Hill   T62   *Mathematics: Applications and Connections, Course 1*

Sometimes you need to estimate a percent.

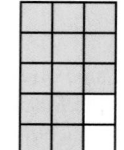
Example 4

Estimate the percent of the figure that is shaded.

13 out of 15 squares are shaded.

$\frac{13}{15}$ is about $\frac{12}{15}$ or $\frac{4}{5}$.

$\frac{4}{5} = 80\%$

So, about 80% of the figure is shaded.

CHECK FOR UNDERSTANDING

Communicating Mathematics

Read and study the lesson to answer each question.

1. *Explain* how you would estimate 75% of 1,976. **See Answer Appendix.**

2. *Draw* a figure in which the percent of the figure that is shaded is about 60%. **See students' work.**

Math Journal

3. *Write* about a situation in which you would use estimating to find the percent of a number. **See students' work.**

Guided Practice

Estimate each percent. 4–6. Sample answers are given.

4. $\frac{1}{5} \times 35 = 7$

5. $\frac{1}{4} \times 200 = 50$

6. $\frac{1}{10} \times 15 = 1.5$

4. 17% of 34   5. 25% of 208   6. 8% of 15

Estimate the percent of each figure that is shaded.

7.

Sample answer: 50%

8.
Sample answer: 25%

9. *Money Matters* When the Chou family went out to dinner, their bill was $35.98. Mrs. Chou wants to leave a 20% tip. About how much money should she leave? **Sample answer: $\frac{1}{5} \times \$35 = \$7$**

EXERCISES

Practice

10. $\frac{3}{4} \times 40 = 30$

11. $\frac{1}{3} \times 60 = 20$

12. $\frac{1}{10} \times 90 = 9$

13. $\frac{1}{2} \times 50 = 25$

14. $\frac{4}{5} \times 125 = 100$

15. $\frac{2}{5} \times 100 = 40$

Estimate each percent.   10–24. Sample answers are given.

10. 77% of 39   11. 34% of 60   12. 9% of 91

13. 47% of 52   14. 80% of 123   15. 38% of 104

16. 66% of 89   17. 97% of 302   18. 24% of 276

16–18. See Answer Appendix.

Estimate the percent of each figure that is shaded.

19. 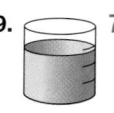 75%

20. 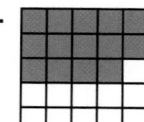 60%

21.  $66\frac{2}{3}\%$

Reteaching the Lesson

Activity Write the example 21% of 249 on an overhead transparency. Then ask the following questions. *What is 21% expressed as a fraction? What is 249 rounded to the nearest ten? Is 21% of 249 about $\frac{1}{3}$ of 200 or about $\frac{1}{5}$ of 250?* Continue with other examples. $\frac{1}{5}$, 250, $\frac{1}{5}$ of 250

Error Analysis

Watch for students who make errors when multiplying fractions and whole numbers.

Prevent by giving these students additional practice exercises and reviewing multiplying fractions and whole numbers.

22.  25%

23. 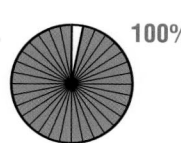 100%

24. 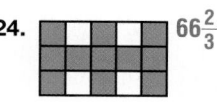 $66\frac{2}{3}$%

25. Estimate 37.5% of 50. **Sample answer:** $\frac{3}{8} \times 48 = 18$

26. *True* or *false*? 26% of 1,400 is about 350. Explain. **See margin.**

27. About how much is one and two tenths percent of ten?
Sample answer: $\frac{1}{100} \times 10 = \frac{1}{10}$

Applications and Problem Solving

28. Sample answer: $\frac{1}{8} \times 2,400 = 300$

28. *Life Science* According to a survey, 13% of gardeners plant geraniums in their flowerbed. If 2,450 gardeners participated in the survey, about how many can be expected to plant geraniums in their flowerbed?

29. *Food* Refer to the graph. Suppose 5,012 people participated in the fast food survey. About how many more people can be expected to eat takeout food in their home than in their car? **See margin.**

 Where do you eat takeout food?
53% Home   14% Work   14% Other   19% Car

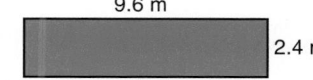
Source: Beef Industry Council

30. *Money Matters* The Bike Shop is having their annual bike sale. Every 15-speed bicycle is on sale for 75% of the regular price, and every mountain bicycle is on sale for 80% of the regular price.

30b. Sample answer: $\frac{4}{5} \times \$200 = \$160$; $\$200 - \$160 = \$40$

a. About how much would you pay for a 15-speed bicycle that usually sells for $125? **Sample answer:** $\frac{3}{4} \times \$120$ or $90

b. About how much would you save if you purchased a mountain bike that normally sells for $196?

31. *Critical Thinking* Which percent problem does not belong? Explain.

a. 50% of 22   **b.** 22% of 50   **c.** $\frac{1}{2}$% of 22

c; because a and b equal 11 while c equals 0.11.

Mixed Review

32. *Statistics* According to a survey, 56% of people brush their teeth after eating snacks. Express 56% as a decimal. *(Lesson 8-5)* **0.56**

33. *Geometry* Find the area of the rectangle. *(Lesson 4-4)* **23.04 m²**
9.6 m   2.4 m

34. **Test Practice** Shalana is buying a jean shirt for $35.95, a pair of jeans for $25.98, and a T-shirt for $11.50. Find the total cost of the items not including tax. *(Lesson 3-6)* **B**
A $68.93   **B** $73.43   **C** $70.63   **D** $72.83   **E** Not Here

35. *Statistics* Find the mean of 14, 73, 25, 25, and 53. *(Lesson 2-7)* **38**

Extending the Lesson

Enrichment Masters, p. 62

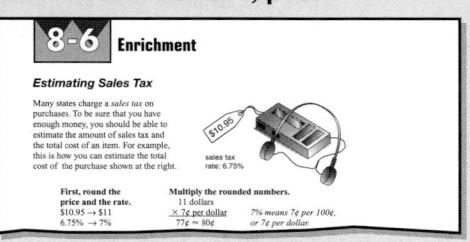

8-6 Enrichment

Estimating Sales Tax

Activity Ask students how percents are used in developing a budget. Discuss what may be involved in developing a budget for their school. What items need to be included, and from where does the funding come? Compare this budget to the state and national budget.

---

4 ASSESS

Closing Activity

Writing Have students work in pairs to write an explanation of how to estimate the percent of a number for a new employee in a department store.

Chapter 8, Quiz C (Lessons 8-5 and 8-6) is available in the *Assessment and Evaluation Masters,* p. 212.

Additional Answers

26. True; 26% is close to 25%. $25\% = \frac{1}{4}$; $\frac{1}{4} \times 1,400 = 350$

29. Sample answer:
$50\% \times 5,000 = 2,500$
$20\% \times 5,000 = 1,000$
$2,500 - 1,000 = 1,500$

Practice Masters, p. 62

8-6 Practice

Estimating with Percents

Estimate each percent. Sample answers are given.

1. 18% of 35
$\frac{1}{5} \times 35 = 7$
2. 26% of 48
$\frac{1}{4} \times 48 = 12$
3. 79% of 40
$\frac{4}{5} \times 40 = 32$
4. 52% of 110
$\frac{1}{2} \times 110 = 55$
5. 34% of 120
$\frac{1}{3} \times 120 = 40$
6. 68% of 82
$\frac{7}{10} \times 80 = 56$
7. 43% of 59
$\frac{2}{5} \times 60 = 24$
8. 77% of 160
$\frac{3}{4} \times 160 = 120$
9. 28% of 142
$\frac{9}{10} \times 140 = 42$
10. 66% of 91
$\frac{2}{3} \times 90 = 60$
11. 59% of 25
$\frac{3}{5} \times 25 = 15$
12. 11% of 178
$\frac{1}{10} \times 180 = 18$
13. 4.8% of 40
$\frac{1}{20} \times 40 = 2$
14. 92% of 29
$\frac{9}{10} \times 30 = 27$
15. 49% of 56
$\frac{1}{2} \times 56 = 28$
16. 62% of 10
$\frac{3}{5} \times 10 = 6$
17. 32% of 270
$\frac{1}{3} \times 270 = 90$
18. 24% of 203
$\frac{1}{4} \times 200 = 50$
19. 9% of 98
$\frac{1}{10} \times 100 = 10$
20. 81% of 45
$\frac{4}{5} \times 45 = 36$
21. 69% of 122
$\frac{7}{10} \times 120 = 84$
22. 7.7% of 50
$\frac{2}{25} \times 50 = 4$
23. 38% of 109
$\frac{2}{5} \times 110 = 44$
24. 89% of 61
$\frac{9}{10} \times 60 = 54$
25. 76% of 43
$\frac{3}{4} \times 40 = 30$
26. 21% of 95
$\frac{1}{5} \times 95 = 19$
27. 67% of 240
$\frac{2}{3} \times 240 = 160$

Use the chart at the right to estimate each percent.

28. percent of first twenty prime numbers from 2 through 71 that contain the digit 4
$\frac{3}{20} \approx \frac{4}{20} = 20\%$

29. percent of first twenty prime numbers from 11 through 71 whose sum of the digits is an even number
$\frac{9}{20} \approx \frac{10}{20} = 50\%$

30. percent of first twenty prime numbers from 2 through 71 that contain the digit 7
$\frac{6}{20} \approx \frac{5}{20} = 25\%$

First Twenty Prime Numbers

2	3	5	7
11	13	17	19
23	29	31	37
41	43	47	53
59	61	67	71

© Glencoe/McGraw-Hill   T 62   Mathematics: Applications and Connections, Course 1

- *Study Guide Masters*, p. 63
- *Practice Masters*, p. 63
- *Enrichment Masters*, p. 63
- Transparencies 8-7, A and B
- *Assessment and Evaluation Masters*, p. 212
- Classroom Games, pp. 21–22
- *Technology Masters*, p. 15
- CD-ROM Program
  - Resource Lesson 8-7
  - Interactive Lesson 8-7

Recommended Pacing	
Standard	Day 12 of 14
Honors	Day 11 of 13
Block	Day 6 of 7

1 FOCUS

5-Minute Check
(Lesson 8-6)

Estimate each percent.

1. 18% of 34 $\frac{1}{5} \times 35 = 7$

2. 72% of 95 $\frac{3}{4} \times 100 = 75$

3. 43% of 68 $\frac{2}{5} \times 70 = 28$

4. 17% of 41 $\frac{1}{5} \times 40 = 8$

5. Of the 675 types of toys in the Fun-R-We toystore, 17% are games. About how many of the toys are games? **135 toys**

The 5-Minute Check is also available on **Transparency 8-7A** for this lesson.

Motivating the Lesson

Communication Have students analyze and discuss the graph at the beginning of the lesson. Ask them these questions.

- Why are percents used instead of whole numbers? **easier to compare statistics**
- Why would retailers be interested in graphs like this one? **get an idea of consumer needs**
- What other information can be represented as a percent? **any measure that is part of a whole unit**

8-7 Percent of a Number

What **you'll learn**

You'll learn to find the percent of a number.

When **am I ever going to use this?**

Knowing how to find the percent of a number can help you determine the sale price of a skateboard.

According to the graph, 20% of parents buy shoes for their children every four to five months. If 2,500 parents were surveyed, how many said that they purchase shoes for their children every four to five months?

How often parents buy shoes for their children

1%	Don't know
6%	Once a month
36%	Every 2-3 months
20%	Every 4-5 months
27%	2 times a year
10%	Once a year or less

Source: Opinion Research for Payless Shoe Source

To find 20% of 2,500, you can change the percent to a fraction or to a decimal, and then multiply it by the number. You can also use a model or a calculator.

Method 1 Change the percent to a fraction.

$20\% = \frac{20}{100}$ or $\frac{1}{5}$

$\frac{1}{5}$ of 2,500 = $\frac{1}{5} \times 2,500$

$= 500$

Method 2 Change the percent to a decimal.

$20\% = \frac{20}{100}$ or 0.2

0.2 of 2,500 = 0.2 × 2,500

$= 500$

Method 3 Use a model.

Since $20\% = \frac{1}{5}$, separate a rectangle into fifths. Label the top and bottom in equal intervals as shown.

0%	20%	40%	60%	80%	100%
0	500	1,000	1,500	2,000	2,500

20% of 2,500 is 500.

Method 4 Use a calculator.

20 [2nd] [%] [×] 2,500 [=] *500*

500 out of 2,500 parents surveyed buy shoes for their children every four to five months.

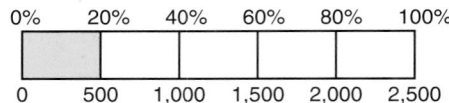

Example ① Find 24% of 250 by changing the percent to a fraction.

$24\% = \frac{24}{100}$ or $\frac{6}{25}$ *Change the percent to a fraction.*

$\frac{6}{25} \times 250 = \frac{6}{25} \times \frac{\overset{10}{250}}{1}$ *Divide 25 and 250 by the GCF, 25.*

$= 60$

24% of 250 is 60.

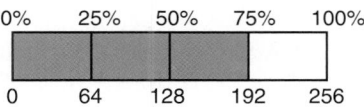

2 Find 9% of 105 by changing the percent to a decimal.

$9\% = \frac{9}{100}$ or 0.09   *Change the percent to a decimal.*

$0.09 \times 105 = 9.45$ *Multiply 0.09 by 105.*

9% of 105 is 9.45.

3 Find 75% of 256 by using a model.

Since $75\% = \frac{3}{4}$, separate a rectangle into fourths.

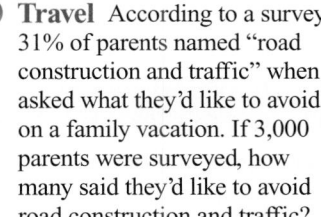

75% of 256 is 192.

4 Find 0.5% of 188 by using a calculator.

0.5 [2nd] [%] [×] 188 [=] *0.94*

0.5% of 188 is 0.94.

APPLICATION

5 **Travel** According to a survey, 31% of parents named "road construction and traffic" when asked what they'd like to avoid on a family vacation. If 3,000 parents were surveyed, how many said they'd like to avoid road construction and traffic?

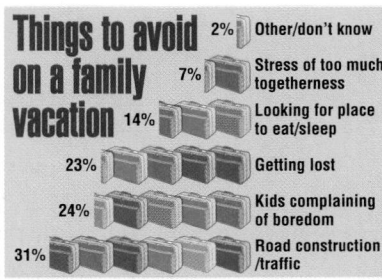

Source: Travel Industry Association, Amtrak

Explore You need to know the number of parents, out of 3,000, who said to avoid road construction and traffic.

Plan Multiply 31% by the number of parents surveyed, 3,000.

Estimate: 31% is close to 30% or $\frac{3}{10}$.

$\frac{3}{10} \times 3,000 = 900$

Solve Change 31% to a decimal. Then multiply it by 3,000.

$31\% = \frac{31}{100}$ or 0.31

0.31 of 3,000 = 0.31 × 3,000
                   = 930

930 out of 3,000 parents said they'd like to avoid road construction and traffic.

Examine Check the answer by comparing it to the estimate. 930 is close to the estimate. So, the answer is correct.

Lesson 8-7 Percent of a Number   **341**

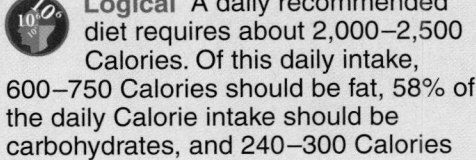

Multiple Learning Styles

Logical A daily recommended diet requires about 2,000–2,500 Calories. Of this daily intake, 600–750 Calories should be fat, 58% of the daily Calorie intake should be carbohydrates, and 240–300 Calories should be proteins. What percent of the daily caloric intake should be fat? **30%** How many Calories should be carbohydrates? **1,160–1,450** What percent of the daily caloric intake should be proteins? **12%**

Check for Understanding

If students need additional practice or instruction after completing Exercises 1–8, one of these options may be helpful.
- Extra Practice, see p. 580
- Reteaching Activity
- *Study Guide Masters*, p. 63
- *Practice Masters*, p. 63

Assignment Guide

Core: 9–25 odd, 27–30
Enriched: 10–24 even, 25–30

Additional Answer

1. Sample answer:
   Method 1: Change the percent to a fraction. First, change the percent to a fraction. Then multiply the number by the fraction.

   Method 2: Change the percent to a decimal. First, change the percent to a decimal. Then multiply the number by the decimal.

   Method 4: Use a calculator. First, enter the percent by entering the number [2nd] [%]. Then press [×] and the number by which you are multiplying. Press [=] .

Study Guide Masters, p. 63

8-7 Study Guide

Name _____ Date _____

Percent of a Number

One way to find the percent of a number is to change the percent to a fraction and then multiply.

Example 1 Find 40% of 90 by changing the percent to a fraction.

$40\% = \frac{40}{100}$ or $\frac{2}{5}$ *Change the percent to a fraction.*
$90 \times \frac{2}{5} = 36$ *Multiply the number by the fraction.*
40% of 90 is 36.

Another way to find the percent of a number is to change the percent to a decimal and then multiply.

Example 2 Find 8% of 68 by changing the percent to a decimal.

$8\% = 0.08$ *Change the percent to a decimal.*
$0.08 \times 68 = 5.44$ *Multiply the number by the decimal.*
8% of 68 is 5.44.

Find the percent of each number.

1. 12% of 140 16.8	2. 25% of 164 41	3. 65% of 125 81.25
4. 77% of 90 69.3	5. 16% of 48 7.68	6. 55% of 96 52.8
7. 105% of 62 65.1	8. 340% of 91 309.4	9. 0.5% of 180 0.9
10. 2% of 84 1.68	11. 200% of 13 26	12. 5% of 80 4

© Glencoe/McGraw-Hill   T 63   Mathematics: Applications and Connections, Course 1

Communicating Mathematics

3. Kosey is correct.
   125% = 1.25.
   1.25 × 150 = 187.5. So, 125% of 150 is 187.5.

Read and study the lesson to answer each question.

1. *Describe* the steps in three of the four methods for finding the percent of a number. **See margin.**

2. *Tell* what number the shaded portion of the model shows. **570**

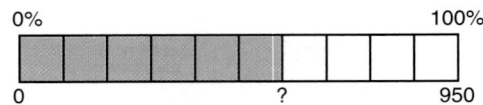

3. *You Decide* Arturo says that 125% of 150 is 18.75. Kosey disagrees. He says that 125% of 150 is 187.5. Who is correct? Explain your reasoning.

Guided Practice

Find the percent of each number.

4. 40% of 65 **26**   5. 8% of 84 **6.72**   6. 0.3% of 500 **1.5**

7. What is 98% of 6? **5.88**

8. *Sports* The Fitch High School softball team won 88% of their games. If they played 25 games, how many games did they win? **22**

EXERCISES

Practice

Find the percent of each number.

9. 25% of 72 **18**	10. 7% of 7 **0.49**	11. 80% of 115 **92**
12. 3% of 156 **4.68**	13. 15% of 40 **6**	14. 33% of 390 **128.7**
15. 101% of 98 **98.98**	16. 0.5% of 85 **0.425**	17. 125% of 145 **181.25**
18. 0.4% of 20 **0.08**	19. 100% of 137 **137**	20. 0.1% of 250 **0.25**

21. What is 60% of 365? **219**

22. Find 22% of 55. **12.1**

23. What is eight-tenths percent of eight hundred? **6.4**

24. Find one hundred fifty percent of ninety-eight. **147**

Applications and Problem Solving

25. *Money Matters* A skateboard is on sale for 85% of the regular price. If it is regularly priced at $40, how much is the sale price? **$34**

26. *Technology* Rachel is using her computer to decode a secret message. The table shows the results of the scan. Suppose the secret message contains 1,500 vowels.

   a. How many of the vowels are a? **375**
   b. How many of the vowels are o? **300**
   c. How many of the vowels are e? **450**

Vowel	Occurred (%)
A	25%
E	30%
I	20%
O	20%
U	5%

342 Chapter 8 Exploring Ratio, Proportion, and Percent

■ Reteaching the Lesson ■

Activity Have students mark off 10 squares on centimeter grid paper strips. Ask them to shade 30%. Ask: *Could you use the strip to show 30% of 100? of 45?* Have students practice finding percents of different numbers.

27. Critical Thinking Rey purchased an electronic dartboard for $210. Manuel purchased the same dartboard at another store for 105% of the amount Rey paid.

a. Who paid more? **Manuel**

b. How much more did that person pay than the other? **$10.50**

Mixed Review
28. Sample answer:
$\frac{2}{5} \times 900$ or 360

28. Hobbies The graph shows how much readers are willing to spend on a book. Suppose 932 readers participated in the survey. About how many of those surveyed are willing to spend between $5 and $14.99? *(Lesson 8-6)*

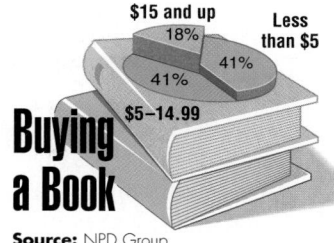

Buying a Book

$15 and up 18%
Less than $5 41%
41%
$5–14.99

Source: NPD Group

29. $\frac{3}{8} \times 64 = 24$

29. Estimate $\frac{3}{8} \times 65$. *(Lesson 7-1)*

30. **Test Practice** What is the greatest common factor of 35, 56, and 63? *(Lesson 5-3)* **D**

A 6   **B** 9   **C** 8   **D** 7

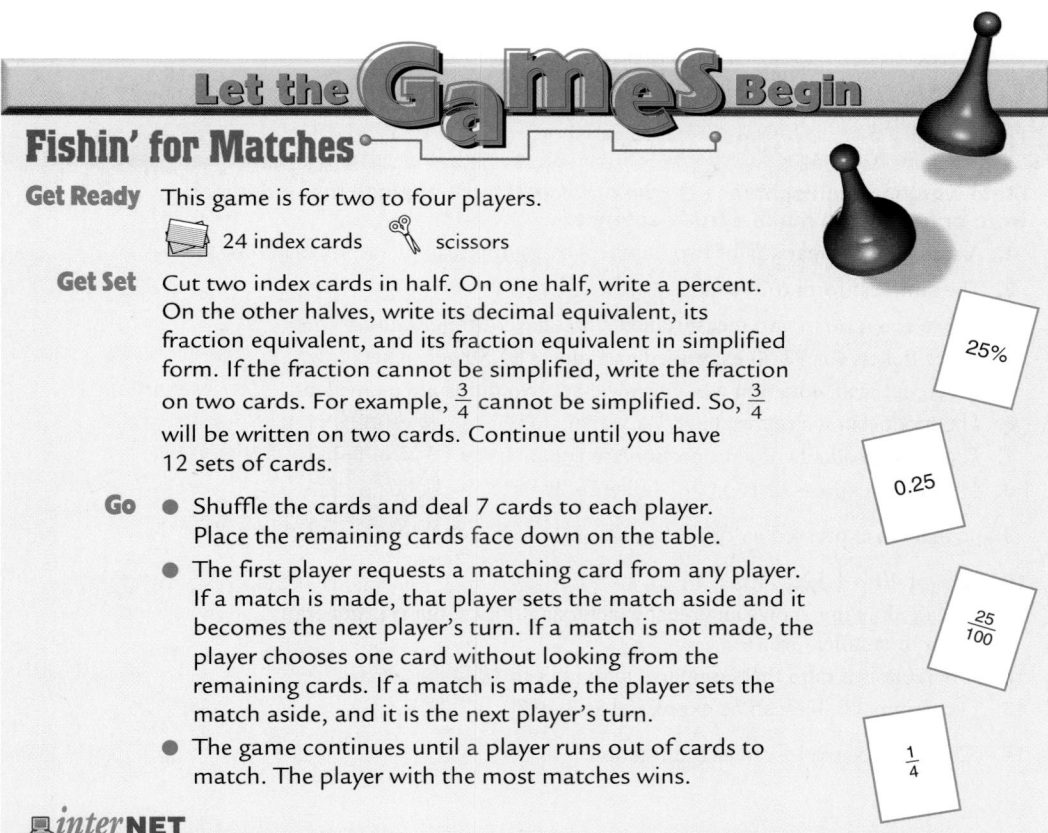

Let the Games Begin

Fishin' for Matches

Get Ready This game is for two to four players.

24 index cards   scissors

Get Set Cut two index cards in half. On one half, write a percent. On the other halves, write its decimal equivalent, its fraction equivalent, and its fraction equivalent in simplified form. If the fraction cannot be simplified, write the fraction on two cards. For example, $\frac{3}{4}$ cannot be simplified. So, $\frac{3}{4}$ will be written on two cards. Continue until you have 12 sets of cards.

Go
● Shuffle the cards and deal 7 cards to each player. Place the remaining cards face down on the table.

● The first player requests a matching card from any player. If a match is made, that player sets the match aside and it becomes the next player's turn. If a match is not made, the player chooses one card without looking from the remaining cards. If a match is made, the player sets the match aside, and it is the next player's turn.

● The game continues until a player runs out of cards to match. The player with the most matches wins.

25%

0.25

$\frac{25}{100}$

$\frac{1}{4}$

interNET CONNECTION Visit www.glencoe.com/sec/math/mac/mathnet for more games.

■ Extending the Lesson ■

Enrichment Masters, p. 63

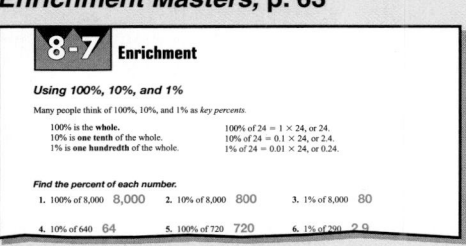

8-7 Enrichment

Using 100%, 10%, and 1%

Many people think of 100%, 10%, and 1% as *key percents*.

100% is **the whole.**   100% of 24 = 1 × 24, or 24.
10% is **one tenth** of the whole.   10% of 24 = 0.1 × 24, or 2.4.
1% is **one hundredth** of the whole.   1% of 24 = 0.01 × 24, or 0.24.

Find the percent of each number.
1. 100% of 8,000   8,000   2. 10% of 8,000   800   3. 1% of 8,000   80
4. 10% of 640   64   5. 100% of 720   720   6. 1% of 290   2.9

Let the Games Begin

The cards could be used to practice writing fractions, decimals, and percents in other forms. Players could challenge each other to draw a card and write that number in as many equivalent ways as they can in a set amount of time.

4 ASSESS

Closing Activity

Speaking Have students give examples of problems they would solve by changing a percent to a fraction or to a decimal, or by using a calculator.

Chapter 8, Quiz D (Lesson 8-7) is available in the *Assessment and Evaluation Masters*, p. 212.

Practice Masters, p. 63

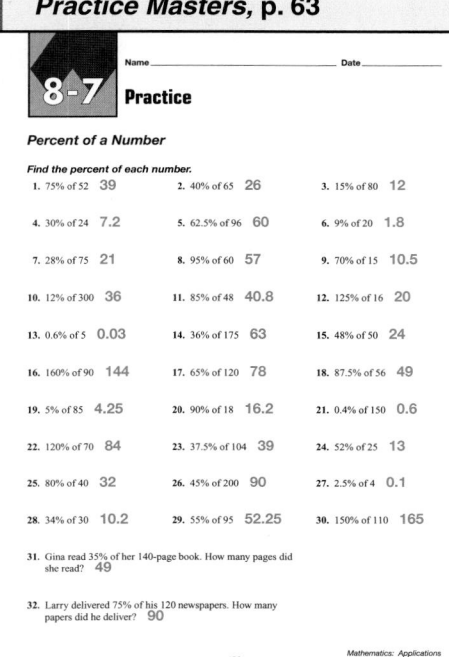

8-7 Practice

Percent of a Number

Find the percent of each number.

1. 75% of 52 39	2. 40% of 65 26	3. 15% of 80 12
4. 30% of 24 7.2	5. 62.5% of 96 60	6. 9% of 20 1.8
7. 28% of 75 21	8. 95% of 60 57	9. 70% of 15 10.5
10. 12% of 300 36	11. 85% of 48 40.8	12. 125% of 16 20
13. 0.6% of 5 0.03	14. 36% of 175 63	15. 48% of 50 24
16. 160% of 90 144	17. 65% of 120 78	18. 87.5% of 56 49
19. 5% of 85 4.25	20. 90% of 18 16.2	21. 0.4% of 150 0.6
22. 120% of 70 84	23. 37.5% of 104 39	24. 52% of 25 13
25. 80% of 40 32	26. 45% of 200 90	27. 2.5% of 4 0.1
28. 34% of 30 10.2	29. 55% of 95 52.25	30. 150% of 110 165

31. Gina read 35% of her 140-page book. How many pages did she read?   49

32. Larry delivered 75% of his 120 newspapers. How many papers did he deliver?   90

© Glencoe/McGraw-Hill   T63   Mathematics: Applications and Connections, Course 1

Study Guide and Assessment

Vocabulary

This section provides a listing of the new terms, properties, and phrases that were introduced in this chapter. Have students define each term and provide an example or two of it, if appropriate.

Understanding and Using the Vocabulary

These exercises check students' understanding of the terms by using a variety of verbal formats including matching, completion, and true/false.

Glossaries A complete glossary of terms appears on pages 642–648. The glossary also appears in Spanish on pages 649–656.

Additional Answers

15. Sample answer: Express the percent as a fraction with a denominator of 100 and then simplify.

16. $\frac{3}{5}$

17. $\frac{11}{20}$

Vocabulary

After completing this chapter, you should be able to define each term, concept, or phrase and give an example or two of each.

Number and Operations
cross products (p. 318)
percent (p. 330)
proportion (p. 317)
rate (p. 312)

Geometry
scale drawing (p. 324)

Problem Solving
draw a diagram (p. 322)

Understanding and Using the Vocabulary

State whether each sentence is *true* or *false*. If false, replace the underlined word or number to make a true sentence.

1. A ratio is a comparison of two numbers by <u>multiplication</u>. **false; division**
2. The simplest form of the ratio $\frac{12}{28}$ is $\frac{3}{7}$. **true**
3. A <u>rate</u> is a ratio of two measurements that have different units. **true**
4. Three tickets for $7.50 expressed as a rate is <u>$1.50</u> per ticket. **false; $2.50**
5. A <u>percent</u> is an equation which shows that two ratios are equivalent. **false; proportion**
6. The model shown represents <u>85%</u>. **false; 73%**
7. The cross products of a proportion are <u>equal</u>. **true**
8. 8% can be expressed as <u>0.008</u>. **false; 0.08**
9. $\frac{3}{25}$ can be expressed as <u>12%</u>. **true**
10. 38% of 40 is <u>1,520</u>. **false; 15.2**
11. A <u>scale drawing</u> shows an object exactly as it looks, but it is generally larger or smaller. **true**
12. A percent is a ratio that compares a number to <u>10</u>. **false; 100**
13. The decimal 0.346 can be expressed as <u>3.46%</u>. **false; 34.6%**
14. $62\frac{1}{2}\%$ can be expressed as the fraction $\frac{5}{8}$. **true**

In Your Own Words

15. *Explain* how to change a percent to a fraction. **See margin.**

344 Chapter 8 Exploring Ratio, Proportion, and Percent

MindJogger Videoquizzes

MindJogger Videoquizzes provide an alternative review of concepts presented in this chapter. Students work in teams to answer questions, gaining points for correct answers. The questions are presented in three rounds.
Round 1 Concepts–5 questions
Round 2 Skills–4 questions
Round 3 Problem Solving–4 questions

Objectives & Examples

Upon completing this chapter, you should be able to:

● express ratios and rates as fractions *(Lesson 8-1)*

Express the ratio 3 winners out of 12 competitors as a fraction in simplest form.

$$\frac{3 \text{ winners}}{12 \text{ competitors}} \overset{\div 3}{\underset{\div 3}{=}} \frac{1 \text{ winner}}{4 \text{ competitors}}$$

● solve proportions by using cross products *(Lesson 8-2)*

Solve the proportion $\frac{9}{12} = \frac{g}{8}$.

$9 \times 8 = 12 \times g$ *Write cross products.*
$72 = 12g$ *Multiply.*
$\frac{72}{12} = \frac{12g}{12}$ *Divide.*
$6 = g$

● find actual length from a scale drawing *(Lesson 8-3)*

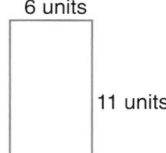
6 units

11 units

The drawing of the room has a scale of 1 unit = 2 feet. What is the actual length of the room?

drawing → $\frac{1 \text{ unit}}{2 \text{ feet}} = \frac{11 \text{ units}}{d \text{ feet}}$ ← *drawing*
actual →   ← *actual*

$1 \times d = 2 \times 11$
$d = 22$

The actual length is 22 feet.

Review Exercises

Use these exercises to review and prepare for the chapter test.

Express each ratio as a fraction in simplest form. 16–17. See margin.

16. 18 out of 30 babies are girls.

17. 11 potatoes out of 20 are rotten.

Express each ratio as a rate.

18. 3 inches of rain in 6 months

19. 189 pounds of garbage in 12 weeks

18. **0.5 inch per month**

19. **15.75 pounds per week**

Use cross products to determine whether each pair of ratios forms a proportion.

20. $\frac{4}{11}, \frac{7}{22}$ no

21. $\frac{3}{9}, \frac{12}{36}$ yes

Solve each proportion.

22. $\frac{7}{11} = \frac{m}{33}$ 21

23. $\frac{12}{20} = \frac{15}{k}$ 25

24. $\frac{g}{20} = \frac{9}{12}$ 15

25. $\frac{10}{12} = \frac{25}{h}$ 30

Refer to the scale drawing at the left.

26. What is the actual width of the room? **12 feet**

27. Suppose that the doorway to the room is actually 3 feet wide. How wide will the doorway be on the drawing? **1.5 units**

28. **Geography** On a map, the measure from Baxter to Sidney is 2 inches. Suppose that the map has a scale of $\frac{3}{4}$ inch = 30 miles. What is the actual distance between the two cities? **80 miles**

Chapter 8 Study Guide and Assessment **345**

Objectives & Examples

This section reviews the skills and concepts of the chapter and shows completely worked examples.

Review Exercises

These exercises provide practice for the corresponding objectives.

Assessment and Evaluation Masters, pp. 199–200

Assessment and Evaluation

Six forms of Chapter 8 Test are available in the *Assessment and Evaluation Masters* as shown in the chart.

Chapter 8 Test, Form 1B, is shown at the right. Chapter 8 Test, Form 2B, is shown on the next page.

1A	Multiple Choice	Honors
1B	Multiple Choice	Average
1C	Multiple Choice	Basic
2A	Free Response	Honors
2B	Free Response	Average
2C	Free Response	Basic

Assessment and Evaluation Masters, pp. 205–206

8

Name_____   Date_____

Chapter 8 Test, Form 2B

Write each ratio as a fraction in simplest form.
1. 16 blue-eyed people out of 50 people
   1. $\frac{8}{25}$

2. $25 out of every $500 collected
   2. $\frac{1}{20}$

Express each ratio as a rate.
3. 424 kilometers in 8 hours
   3. 53 km/h

4. $60 for 12 months
   4. $5/month

5. 40 feet in 5 seconds
   5. 8 ft per s

Solve each proportion.
6. $\frac{10}{15} = \frac{18}{z}$
   6. 27

7. $\frac{18}{m} = \frac{27}{30}$
   7. 20

8. $\frac{40}{100} = \frac{y}{20}$
   8. 8

9. $\frac{c}{25} = \frac{16}{200}$
   9. 2

10. $\frac{62}{d} = \frac{93}{168}$
    10. 112

11. The scale drawing of a window is shown at the right. The scale is $\frac{1}{4}$ in. = 2 ft. What is the actual width labeled w?
    11. 7 ft

12. On a map, the scale is 1 cm = 20 km. If the map distance between Ironton and Bedford is 1.6 centimeters, what is the actual distance?
    12. 32 km

13. On a map, the scale is 1 in. = 10 mi. If the distance on the map between two cities is 4 inches, what is the actual distance?
    13. 40 miles

Express each fraction as a percent.
14. $\frac{28}{50}$
    14. 56%

15. $\frac{17}{20}$
    15. 85%

16. $\frac{4}{5}$
    16. 80%

© Glencoe/McGraw-Hill   205   *Mathematics: Applications and Connections, Course 1*

8

Chapter 8 Test, Form 2B (continued)

Express each percent as a fraction in simplest form and as a decimal.
17. 82%
    17. $\frac{41}{50}$, 0.82

18. 6%
    18. $\frac{3}{50}$, 0.06

19. 3.4%
    19. $\frac{17}{500}$, 0.034

Express each decimal as a percent.
20. 0.77
    20. 77%

21. 0.046
    21. 4.6%

22. 4.7
    22. 470%

Estimate each percent.
23. 48% of 61
    23. $\frac{1}{2}$ of 60 = 30

24. 74% of 21
    24. $\frac{3}{4}$ of 20 = 15

25. 19% of 203
    25. $\frac{1}{5}$ of 200 = 40

26. 32% of 62
    26. $\frac{1}{3}$ of 60 = 20

27. 52% of 143
    27. $\frac{1}{2}$ of 140 = 70

Find the percent of each number.
28. 30% of 80
    28. 24

29. 51% of 63
    29. 32.13

30. 3% of 500
    30. 15

31. 20% of 160
    31. 32

32. 60% of 240
    32. 144

Solve by drawing a diagram.
33. Ricardo has a piece of lumber that is 105 inches long. He wants to cut it into 15-inch pieces. How many cuts does he need to make?
    33. 6

© Glencoe/McGraw-Hill   206   *Mathematics: Applications and Connections, Course 1*

Objectives & Examples

• express percents as fractions and vice versa *(Lesson 8-4)*

Express $\frac{3}{10}$ as a percent.

$$\frac{3}{10} \times \frac{n}{100}$$
$$3 \times 100 = 10 \times n$$
$$300 = 10n$$
$$300 \div 10 = 10n \div 10$$
$$30 = n$$

So, $\frac{3}{10}$ is equivalent to 30%.

• express percents as decimals and vice versa *(Lesson 8-5)*

Express 21% as a decimal.

$$21\% = \frac{21}{100}$$
$$= 0.21$$

• estimate the percent of a number *(Lesson 8-6)*

Estimate 33% of 60.

33% is close to $33\frac{1}{3}\%$ or $\frac{1}{3}$.

$$\frac{1}{3} \times 60 = \frac{1}{3} \times \frac{\overset{20}{\cancel{60}}}{1}$$
$$= 20$$

So, 33% of 60 is about 20.

• find the percent of a number *(Lesson 8-7)*

Find 20% of 50.

$$20\% = \frac{20}{100} \text{ or } \frac{1}{5}$$
$$\frac{1}{5} \times 50 = 10$$

20% of 50 is 10.

Review Exercises

Express each percent as a fraction in simplest form.
29. 3% $\frac{3}{100}$
30. 17% $\frac{17}{100}$
31. 150% $1\frac{1}{2}$
32. 48% $\frac{12}{25}$

Express each fraction as a percent.
33. $\frac{3}{5}$ 60%
34. $\frac{9}{10}$ 90%
35. $\frac{11}{20}$ 55%
36. $\frac{5}{25}$ 20%
37. How is five hundredths written as a percent? 5%

Express each percent as a decimal.
38. 2.2% 0.022
39. 38% 0.38
40. 140% 1.4
41. 66% 0.66

Express each decimal as a percent.
42. 0.003 0.3%
43. 1.3 130%
44. 0.65 65%
45. 0.591 59.1%

46–50. Sample answer given.
Estimate each percent.
46. 19% of 99 $\frac{1}{5} \times 100 = 20$
47. 27% of 82 $\frac{1}{4} \times 80 = 20$
48. 48% of 48 $\frac{1}{2} \times 50 = 25$
49. 41% of 243 $\frac{2}{5} \times 240 = 96$
50. Estimate the percent of the figure that is shaded. 30%

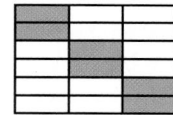

Find the percent of each number.
51. 18% of 89 16.02
52. 40% of 150 60
53. 5% of 340 17
54. 0.8% of 132 1.056
55. What is one-tenth percent of 162? 0.162
56. Find two hundred thirty-six percent of 42. 99.12

346   Chapter 8 Exploring Ratio, Proportion, and Percent

Test and Review Software

You may use this software, a combination of an item generator and item bank, to create your own tests or worksheets. Types of items include free response, multiple choice, short answer, and open ended.

CD-ROM Program

The CD-ROM Program contains an Assessment Game whose questions review the concepts in this chapter.

Applications & Problem Solving

57. *Travel* Tatanka drove 300 miles in 6 hours. How many miles did he drive per hour? *(Lesson 8-1)* **50 mph**

58. *Statistics* According to a survey, 37% of Americans feel that public libraries will become the place to go to use computers. If 1,500 people participated in the survey, how many said that libraries would become the place to use computers? *(Lesson 8-2)* **555**

59. *Draw a Diagram* A store manager displays 36 cans of cat food in a triangular shape. The display is one can deep. How many cans are on the bottom row? *(Lesson 8-3A)* **8 cans**

60. *School* Nick's class is planning a fund-raiser. The results of a class vote show that 75% of the students want to sell candy, 15% of the students want to sell wrapping paper, and 10% of the students want to sell calendars. *(Lessons 8-4 and 8-7)*

   a. What fraction of the class wants to sell calendars? $\frac{1}{10}$

   b. What fraction of the class wants to sell candy? $\frac{3}{4}$

   c. If there are 40 students in the class, how many voted to sell candy? **30**

Alternative Assessment

Performance Task

Suppose your teacher assigns you to make a scale drawing of your school. What information would you need to collect? What decisions would you need to make? **See margin.**

Suppose your school is rectangular-shaped with a length of 220 feet and a width of 90 feet. You want to make the length of the school on the scale drawing 11 inches. Find the scale and the width of the school on the map.

1 inch = 20 feet; 4.5 inches

A practice test for Chapter 8 is provided on page 602.

Completing the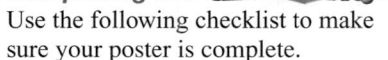

Use the following checklist to make sure your poster is complete.

☑ A sketch of each toy is included.

☑ The dimensions of each toy are labeled, and the dimensions of the actual objects are listed.

☑ For each sketch, include a scale that can be used to compare the size of the toy to the actual size of the object.

Add any finishing touches to make your poster unique.

 Select an item from the chapter that you found to be challenging. Place it in your portfolio and write a paragraph that explains why you found the item to be challenging.

Applications & Problem Solving

This section provides additional practice in solving real-world problems that involve the skills of this chapter.

Alternative Assessment

The *Performance Task* provides students with a performance assessment opportunity to evaluate their work and understanding.

Students should complete the final stages of their project and prepare a class demonstration of their results. A scoring guide for the project is available in the *Investigations and Projects Masters*, p. 47.

 Students should add to their portfolios at this time.

Assessment and Evaluation Masters, p. 209

Name _____ Date _____

8 Chapter 8 Performance Assessment

Instructions: Demonstrate your knowledge by giving a clear, concise solution to each problem. Be sure to include all relevant drawings and justify your answers. You may show your solutions in more than one way or investigate beyond the requirements of the problems.

1. a. Tell in your own words the meaning of *ratio.*

   b. Give an example of a ratio. Write the ratio in four ways.

   c. Tell in your own words the meaning of *rate.* Give two examples of rates.

   d. Tell in your own words the meaning of *proportion.*

   e. Write a word problem that uses proportions. Include a scale drawing.

   f. Solve the word problem in part e. Explain each step.

2. Matthew is in charge of marking merchandise down for the back-to-school sale. Clothing is to be marked down 30%, School supplies are to be on sale for $\frac{1}{4}$ off, and sporting goods prices will be reduced 20%.

   a. Tell how to express a percent as a fraction and as a decimal.

   b. Find the sale price of a $35 basketball in two ways.

   c. Estimate the sale price of a jacket regularly priced $40. Explain your reasoning.

   d. Find the sale price of the jacket in part c.

   e. Tell how to express a fraction as a percent.

   f. Express $\frac{1}{4}$ as a percent. Show your work.

   g. On which item would a customer save more money, the $35 basketball or a $32 book for school? Explain your reasoning.

© Glencoe/McGraw-Hill   **209**   *Mathematics: Applications and Connections, Course 1*

Additional Answer for the Performance Task

You would need to know the dimensions of the school. You would need to decide how much detail to show, such as the parking lot, playground, classroom locations, etc. You would also have to decide on what scale to use.

Performance Assessment

Additional performance assessment tasks for this chapter are included in the *Assessment and Evaluation Masters* on page 209. A scoring guide is also provided on page 221.

The Standardized Test Practice may be used to help students prepare for standardized tests. The test items are written in the same style as those in state proficiency tests and standardized tests like CAT, CTBS, ITBS, MAT, SAT, and Terra Nova. The test items cover skills and concepts covered up to this point in the text.

The pages can be used as an overnight assessment. After students have completed the pages, discuss how each problem can be solved, or provide copies of the solutions from the *Solutions Manual*.

Assessment and Evaluation Masters, p. 215

Section One: Multiple Choice

There are thirteen multiple-choice questions in this section. Choose the best answer. If a correct answer is *not here*, choose the letter for Not Here.

1. How could you estimate a 20% tip for a $23.25 restaurant bill? **C**

   A Find $\frac{1}{2}$ of $23.25.

   B Find $\frac{1}{4}$ of $23.00.

   C Find $\frac{1}{5}$ of $23.00.

   D Find $\frac{1}{20}$ of $24.00.

2. Rogelio is 1.6 meters tall. How many centimeters is this? **H**

   F 0.016 cm   G 16 cm

   H 160 cm   J 1,600 cm

3. The stem-and-leaf plot shows the speeds in miles per hour for the fastest birds in the world. What is the mean speed? **A**

Stem	Leaf
6	5 5 8
7	0 2 7
8	0 8
9	5
10	6

   $8|8 = 88$ mph

   A 78.6 mph   B 74.5 mph

   C 75.6 mph   D 77.8 mph

4. Round $3\frac{3}{5}$ to the nearest half. **H**

   F $3\frac{1}{5}$ G 3

   H $3\frac{1}{2}$ J 4

5. Find the value of $132 \div 6 + 4 \times 3$. **D**

   A 36   B 78

   C 68   D 34

Please note that Questions 6–13 have five answer choices.

6. The table shows the cost for 4 brands of tennis shoes at two different stores.

Brand	Sporting Goods Store	Department Stores
R	$49.99	$45.65
S	$67.98	$65.00
T	$37.50	$37.95
U	$75.97	$72.98

   At the department store, how much more is brand S than brand R? **J**

   F $20.45

   G $19.45

   H $20.35

   J $19.35

   K $19.25

7. Use the table in Question 6. Connie purchases brand T and brand R from the sporting goods store. What is the total cost before sales tax is added? **D**

   A $87.39

   B $90.59

   C $88.99

   D $87.49

   E $86.59

8. Tickets to Splash City cost $10 for adults and $8 for children. If 12 people paid a total of $102, how many were adults, and how many were children? **H**

   F 7 adults, 4 children

   G 5 adults, 6 children

   H 3 adults, 9 children

   J 6 adults, 6 children

   K 4 adults, 8 children

◀◀◀ **Instructional Resources**
Another cumulative review is shown at the left and is available in the *Assessment and Evaluation Masters*, p. 215.

8 Cumulative Review, Chapters 1–8

Name_____ Date_____

Estimate. (Lessons 7-1 and 8-6)
1. $8\frac{7}{8} \times 10\frac{1}{6}$ 2. 32% of 152

Solve each equation. (Lessons 4-3 and 4-7)
3. $3.12 \times 1.8 = m$ 4. $x = 73.44 \div 3.6$

5. Find the LCM of 50 and 80. (Lesson 5-7)

Order from greatest to least. (Lessons 3-3 and 5-8)
6. 1.008, 1.7, 1.09   7. $\frac{5}{16}, \frac{17}{48}, \frac{5}{24}$

Complete. (Lessons 4-9 and 7-7)
8. 8,500 mg = ___ g   9. 12 gal = ___ c

10. Subtract 3 h 52 min from 7 h 12 min. (Lesson 6-7)

Perform the indicated operation. Write answers in simplest form. (Lessons 6-4, 6-5, 7-2, and 7-6)
11. $\frac{7}{20} + \frac{1}{30}$ 12. $3\frac{1}{2} - 1\frac{9}{10}$

13. $\frac{1}{18} \times \frac{3}{4}$ 14. $5\frac{1}{3} \div 2\frac{1}{4}$

15. A stadium has 15,000 seats. If 12,892 people attend a game in the stadium, how many seats are left? Solve using the four-step plan. (Lesson 1-1)

16. List the stems for a stem-and-leaf plot for this set of data: 88, 93, 74, 68, 97, 82. (Lesson 2-6)

17. Solve the proportion $\frac{26}{40} = \frac{39}{x}$. (Lesson 8-2)

Express each percent as a fraction in simplest form and as a decimal. (Lessons 8-4 and 8-5)
18. 38%   19. 5.4%

20. Find 40% of 800. (Lesson 8-7)

1. $\frac{9 \times 11 = 99}{}$
2. $\frac{1}{3} \times 150 = 50$
3. 5.616
4. 20.4
5. 400
6. 1.7, 1.09, 1.008
7. $\frac{17}{48}, \frac{5}{16}, \frac{5}{24}$
8. 8.5 g
9. 192 c
10. 3 h 20 min
11. $\frac{23}{60}$
12. $1\frac{3}{5}$
13. $\frac{1}{24}$
14. $2\frac{10}{27}$
15. 2,108
16. 6, 7, 8, 9
17. 60
18. $\frac{19}{50}$, 0.38
19. $\frac{27}{500}$, 0.054
20. 320

© Glencoe/McGraw-Hill   215   Mathematics: Applications and Connections, Course 1

9. Solve $\frac{16}{25} = \frac{x}{100}$. **D**

   A 4

   B 10

   C 16

   D 64

   E Not Here

10. Yoruba bought 6 packages of hot dogs for a picnic. Each package weighs 1.15 pounds. What is the total weight of the packages? **G**

   F 7.15 lb

   G 6.90 lb

   H 8.40 lb

   J 6.55 lb

   K Not Here

11. Find the sum of 12.675 and 7.081. **C**

   A 18.996

   B 19.566

   C 19.756

   D 18.796

   E Not Here

12. A bicycle costs $159, a helmet costs $59, and bike shorts cost $24. What is the cost of these items before tax is added? **G**

   F $280

   G $242

   H $220

   J $275

   K Not Here

13. $3\frac{3}{4} \times 2\frac{3}{5} =$ **D**

   A $6\frac{9}{20}$ **B** $8\frac{1}{2}$

   C $7\frac{2}{5}$ **D** $9\frac{3}{4}$

   E Not Here

Test-Taking Tip

Remember that, on most tests, you get as much credit for correctly answering the easy questions as you do for the difficult ones. Answer the easy ones first and then spend time on the more challenging questions.

Section Two: Free Response

This section contains seven questions for which you will provide short answers. Write your answers on your paper.

14. What is the value of $3c - 4d$ if $c = 8$ and $d = 3$? **12**

15. Two-fifths of the registered voters voted in the town election. What percent of the voters did not vote? **60%**

16. Estimate 39% of 70. **See margin.**

17. To the nearest eighth of an inch, what is the length of the line segment shown? **See margin.**

   |———————————|

18. Find the value of the expression $76 - 16 \div 4 \times 3 + 12$. **76**

19. Dawit sells small bouquets of flowers at his flower shop for $2.75 each. On Saturday, he made $57.75 in sales from the bouquets. How many bouquets of flowers did he sell? **21**

20. Add a number to the following set of data so that the mode, median, and mean of the new set of data are the same as those for the original set of data: 91, 93, 93, 95, 95, 98, 100. **95**

Test-Taking Tip

Answering easier questions will help students determine how much time they can devote to the more difficult questions. Any question for which they cannot remember how to find an answer should be saved for last. This will prevent students from missing points for easier problems.

Assessment and Evaluation Masters, pp. 213–214

Chapter 8 Standardized Test Practice

1. An average camel can drink 114 liters in 10 minutes. Express this ratio as a rate. **1. B**
   A. 114 liters   B. 11.4 liters per minute
   C. 10 minutes per liter   D. none of these

2. Alex used a proportion to convert between feet and inches. He knows that 1 foot = 12 inches. He is 56 inches tall. Solve the proportion $\frac{12}{1} = \frac{56}{x}$ to determine Alex's height in feet. **2. C**
   A. 5 ft 6 in.   B. 5.6 ft   C. $4\frac{2}{3}$ ft   D. $4\frac{1}{2}$ ft

3. Oxygen makes up 65% of the chemical elements in the human body. Express this percent as a decimal. **3. C**
   A. $\frac{65}{100}$ B. 65.0   C. 0.65   D. 6.5

4. Five of the first 25 presidents were born in March. What percent of the first 25 presidents were born in March? **4. A**
   A. 20%   B. $\frac{1}{5}$ C. $\frac{5}{25}$ D. 25%

5. In music, putting a flag on the stem of a note halves the value of the note. If you start with a quarter note and add a flag, what note results? **5. C**
   A. half note   B. quarter note
   C. eighth note   D. sixteenth note

6. Josiah is making dip for a party. The recipe calls for $\frac{1}{2}$ teaspoon of lemon juice. He wants to make $2\frac{1}{2}$ times the recipe. How much lemon juice does Josiah need? **6. A**
   A. $1\frac{1}{4}$ tsp   B. $2\frac{1}{4}$ tsp   C. $\frac{1}{4}$ tsp   D. none of these

7. Mrs. Smith is ordering pizzas for her class as a reward. She has decided that she needs $4\frac{1}{4}$ pizzas to feed her class. How many pizzas should she order? **7. C**
   A. 4   B. $4\frac{1}{4}$ C. 5   D. none of these

8. If $\frac{2}{9}$ of the marbles in a bag are red and $\frac{5}{9}$ of the marbles are white, what portion of the marbles are neither red nor white? **8. D**
   A. $\frac{4}{9}$ B. $\frac{1}{3}$ C. 2   D. $\frac{2}{9}$

9. Eight presidents out of 41 were born in Virginia. Express this ratio as a decimal rounded to the nearest thousandth. **9. B**
   A. $\frac{8}{41}$ B. 0.195   C. 5.125   D. 19.5

© Glencoe/McGraw-Hill   213   Mathematics: Applications and Connections, Course 1

Chapter 8 Standardized Test Practice (continued)

10. Joy wanted to bundle by weight the newspapers collected for recycling by each of the three classes in the sixth grade. The classes collected 22 pounds, 55 pounds, and 44 pounds. Joy used the greatest common factor of these numbers for the weight of each newspaper bundle. That way no class would have unbundled newspapers left over. How many pounds did each bundle weigh? **10. A**
    A. 11 lb   B. 5 lb   C. 7 lb   D. none of these

11. An average mountain lion weighs 77 kilograms. What is this weight in grams? **11. C**
    A. 0.077 g   B. 77 g   C. 77,000 g   D. 7,700 g

12. Solve $3 \times 12,541 = a$ using the distributive property. **12. B**
    A. 4,180   B. 37,623   C. 3,762   D. none of these

13. The cubit is an old unit of length. The Roman cubit is 0.444 meter. The Greek cubit is 0.463 meter. The Assyrian cubit is 0.469 meter. The Egyptian short cubit is 0.450 meter. Put these decimals in order from least to greatest. **13. A**
    A. 0.444, 0.450, 0.463, 0.469   B. 0.469, 0.463, 0.450, 0.444
    C. 0.450, 0.469, 0.463, 0.444   D. none of these

14. One square meter is equivalent to 1.196 square yards. Round the square yards to the nearest hundredth. **14. B**
    A. 2.0   B. 1.20   C. 1.0   D. none of these

15. Mrs. Overbeck's math class had the following scores on a test: 75, 90, 85, 90, 80, 52, 95, 83, 90. What is the best measure of central tendency to describe this data? **15. D**
    A. range   B. mode   C. mean   D. median

16. Susie started taking piano lessons. The first week she practiced 2 hours, the second week she practiced $2\frac{1}{2}$ hours, the third week she practiced 3 hours, and the fourth week she practiced $3\frac{1}{2}$ hours. If she continues this pattern, how many hours will she practice the sixth week? **16. B**
    A. 4 hours   B. $4\frac{1}{2}$ hours   C. 5 hours   D. $5\frac{1}{2}$ hours

© Glencoe/McGraw-Hill   214   Mathematics: Applications and Connections, Course 1

Additional Answers

16. Sample answer: 40% of 70 = 28

17. $1\frac{3}{8}$ inches

Instructional Resources ►►►

Additional standardized test practice is shown at the right and is available in the *Assessment and Evaluation Masters*, pp. 213–214.

Geometry: Investigating Patterns

Previewing the Chapter

Overview

This chapter explores geometry and its many applications. Students learn to classify, measure, and draw angles; to construct congruent segments and angles; and to bisect line segments and angles. Students investigate two-dimensional figures by describing and defining lines of symmetry and by determining congruence and similarity. Students learn to solve problems by using logical reasoning.

Lesson (pages)	Lesson Objectives	NCTM Standards	Standardized Tests	State/Local Objectives
9-1 (352–355)	Classify and measure angles.	1–7, 12, 13	CAT, CTBS, ITBS, SAT, TN	
9-1B (356–357)	Solve problems by using logical reasoning.	1–8, 12		
9-2 (358–361)	Draw angles and estimate measures of angles.	1–7, 10, 12, 13		
9-3A (362–363)	Construct congruent segments and angles.	1–3, 5–7, 12, 13	CTBS, TN	
9-3 (364–366)	Bisect line segments and angles.	1–7, 12, 13	CAT	
9-4A (368–369)	Classify triangles and quadrilaterals.	1–3, 12	CAT, ITBS	
9-4 (370–373)	Name two-dimensional figures.	1–7, 9, 12, 13	CAT, ITBS	
9-4B (374)	Make a net and use it to wrap a cube.	1–3, 12		
9-5 (375–378)	Describe and define lines of symmetry.	1–7, 12	CTBS, SAT, TN	
9-6 (379–383)	Determine congruence and similarity.	1–7, 12, 13	CTBS, TN	
9-6B (384–385)	Create Escher-like drawings by using translations.	1–3, 8, 12		

CAT = California Achievement Tests, CTBS = Comprehensive Tests of Basic Skills, ITBS = Iowa Tests of Basic Skills, MAT = Metropolitan Achievement Tests, SAT = Stanford Achievement Tests, TN = Terra Nova

CD-ROM

All of the blackline masters in the Teacher's Classroom Resources are available on the **Electronic Teacher's Classroom Resources** CD-ROM.

LESSON PLANNING GUIDE

Lesson	Extra Practice (Student Edition)	BLACKLINE MASTERS (PAGE NUMBERS)										
		Study Guide	Practice	Enrichment	Assessment & Evaluation	Classroom Games	Diversity	Hands-On Lab	School to Career	Science and Math Lab Manual	Technology	Transparencies A and B
9-1	p. 580	64	64	64							18	9-1
9-1B	p. 581											
9-2	p. 581	65	65	65	239				9		17	9-2
9-3A								57				
9-3	p. 581	66	66	66	238, 239							9-3
9-4A								58				
9-4	p. 582	67	67	67		23–26	9					9-4
9-4B								59				
9-5	p. 582	68	68	68	240			77				9-5
9-6	p. 582	69	69	69	240					13–16		9-6
9-6B								60				
Study Guide/ Assessment					225–237, 241–243							

OTHER CHAPTER RESOURCES

Student Edition

Chapter Project, pp. 351, 361, 373, 382, 389
Math in the Media, p. 382
School to Career, p. 367
Let the Games Begin, p. 383

Technology

 CD-ROM Program

 Interactive Mathematics Tools Software

Teacher's Classroom Resources

Applications
Family Letters and Activities, pp. 17–18
Investigations and Projects Masters, pp. 49–52
Meeting Individual Needs
Investigations for the Special Education Student, p. 29

Teaching Aids
Answer Key Masters
Block Scheduling Booklet
Lesson Planning Guide
Solutions Manual

Professional Publications
Glencoe Mathematics
Professional Series

Planning the Chapter

MindJogger Videoquizzes
provide a unique format for reviewing concepts presented in the chapter.

ASSESSMENT RESOURCES

Student Edition
Mixed Review, pp. 355, 361, 366, 373, 378, 382
Mid-Chapter Self Test, p. 373
➡ Math Journal, p. 360
Study Guide and Assessment, pp. 386–389
Performance Task, p. 389
📖 Portfolio Suggestion, p. 389
Standardized Test Practice, pp. 390–391
Chapter Test, p. 603

Assessment and Evaluation Masters
Multiple-Choice Tests (Forms 1A, 1B, 1C), pp. 225–230
Free-Response Tests (Forms 2A, 2B, 2C), pp. 231–236
Performance Assessment, p. 237
Mid-Chapter Test, p. 238
Quizzes A–D, pp. 239–240
Standardized Test Practice, pp. 241–242
Cumulative Review, p. 243

Teacher's Wraparound Edition
5-Minute Check, pp. 352, 358, 364, 370, 375, 379
📖 Building Portfolios, p. 350
✏ Math Journal, pp. 363, 369, 374, 385
Closing Activity, pp. 355, 357, 361, 366, 373, 378, 382

Technology
💾 Test and Review Software
📹 MindJogger Videoquizzes
💿 CD-ROM Program

MATERIALS AND MANIPULATIVES

Lesson 9-1
protractor*†

Lesson 9-2
protractor*†
straightedge*†
paper plate
scissors*

Lesson 9-3A
straightedge*†
compass*†

Lesson 9-3
straightedge*†
ruler*†
protractor*†
compass*†

Lesson 9-4A
geoboards*†
geobands*†
dot paper†
scissors*

Lesson 9-4B
scissors*
cube*

Lesson 9-5
geomirror*
tracing paper
scissors*

Lesson 9-6
scissors*
tracing paper
grid paper†
geoboard*†
geobands*†

Lesson 9-6B
grid paper†
colored pencils

*Glencoe Manipulative Kit   †Glencoe Overhead Manipulative Resources

PACING CHART

See pages T25–T27 for the Course Planning Calendar.

COURSE	DAY 1	DAY 2	DAY 3	DAY 4	DAY 5	DAY 6	DAY 7
Standard	Chapter Project	Lesson 9-1	Lesson 9-1B	Lesson 9-2	Lessons 9-3A & 9-3		Lessons 9-4A & 9-4
Honors	Chapter Project	Lesson 9-1	Lesson 9-1B	Lesson 9-2	Lesson 9-3	Lessons 9-4 & 9-4B	
Block	Chapter Project & Lesson 9-1	Lessons 9-1B & 9-2	Lessons 9-3A & 9-3	Lessons 9-4A & 9-4	Lesson 9-5	Lesson 9-6	Study Guide and Assessment, Chapter Test

Interactive Mathematics:
Activities and Investigations

is an activity-based program that may be used as an enhancement for chapters in *Mathematics: Applications and Connections.*

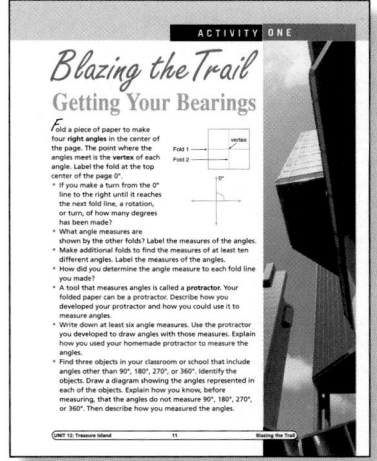

Unit 12, Activity One
Use before Lesson 9-1.

Summary Students make their own protractors by folding paper and labeling the angles. Using their protractors, they draw angles of varying sizes and justify their measurements. Then they use nonstandard measuring devices to measure length. Students then design a map.

Math Connection Students identify angles of 90°, 180°, 270°, and 360° and then use them to find the measures of other angles. They also measure angles and lengths without the use of standard measuring tools, and use map scale and angle measurements to give directions.

Unit 4, Activity Two

Use with Lesson 9-5.

Summary Students work in groups using a geomirror to test if each logo found in the Data Bank has symmetry and where the line of symmetry is located. Then they complete six drawings on a recording sheet by using symmetry.

Math Connection Students investigate line symmetry using a geomirror.

DAY 8	DAY 9	DAY 10	DAY 11	DAY 12	DAY 13	DAY 14	DAY 15
(continue from Day 7)		Lesson 9-5	Lesson 9-6		Study Guide and Assessment	Chapter Test	
Lesson 9-5	Lessons 9-6 & 9-6B		Study Guide and Assessment	Chapter Test			

Enhancing the Chapter

APPLICATIONS

Classroom Games, pp. 23–26

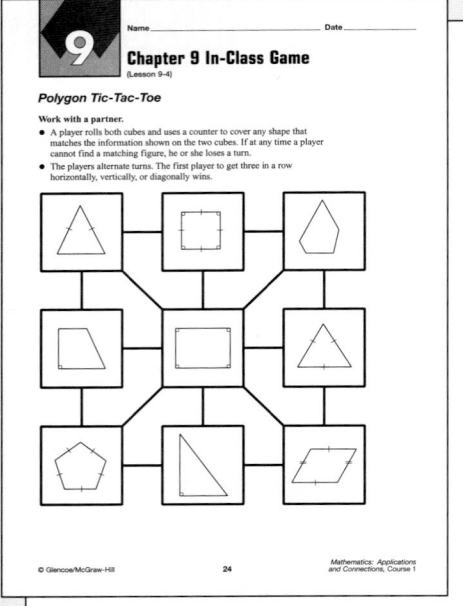

Diversity Masters, p. 9

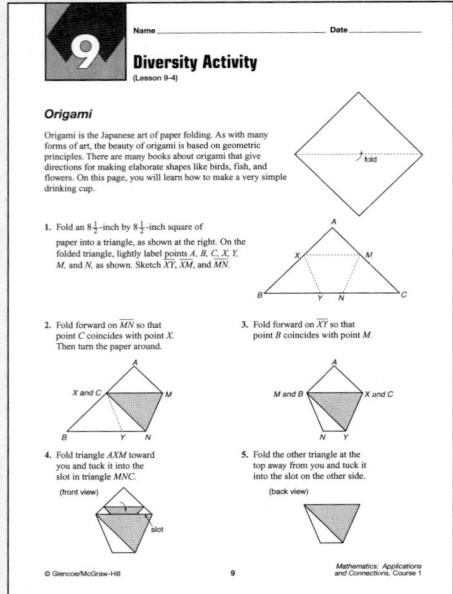

School to Career Masters, p. 9

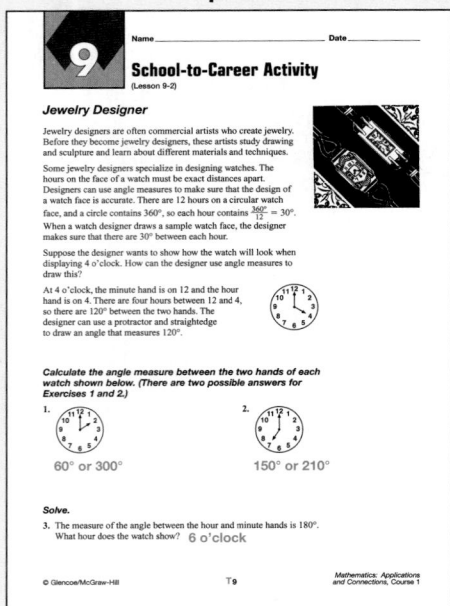

Family Letters and Activities, pp. 17–18

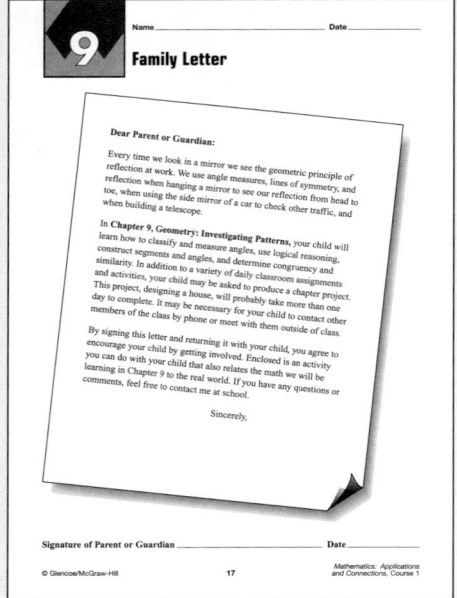

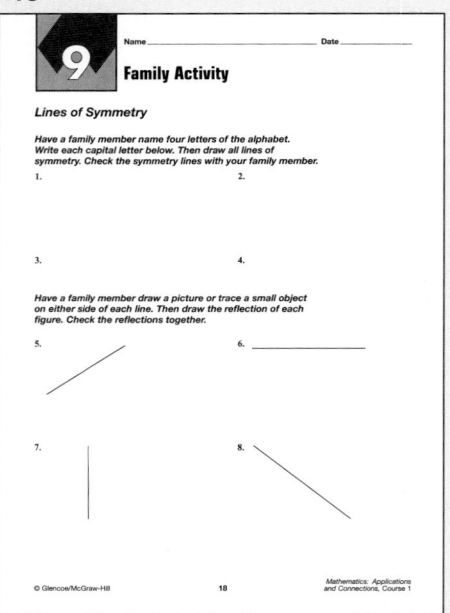

Science and Math Lab Manual, pp. 13–16

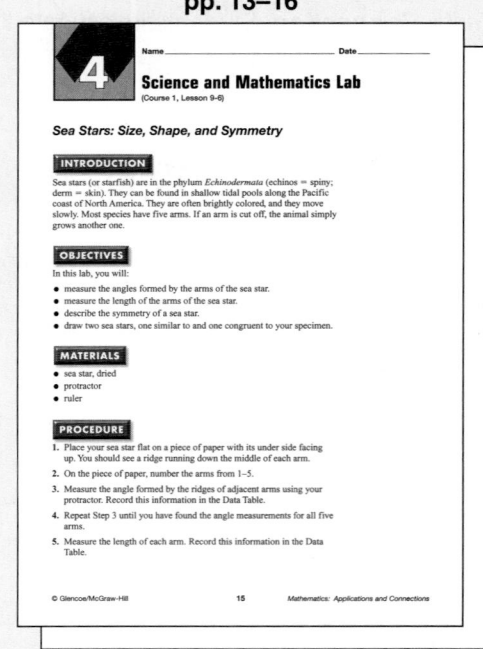

Hands-On Lab Masters, p. 77

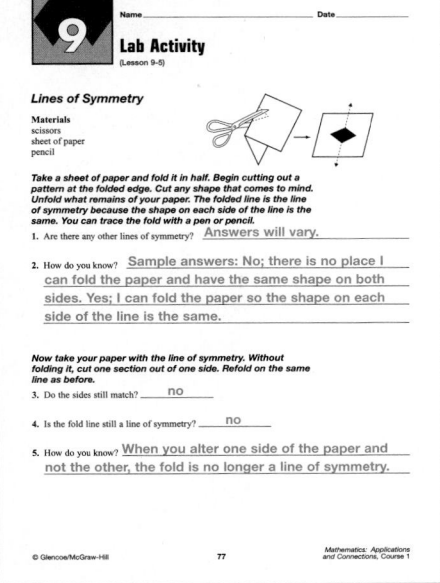

Assessment and Evaluation Masters, pp. 238–240

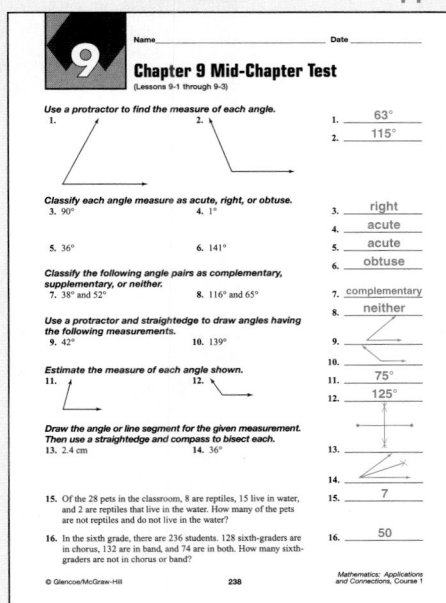

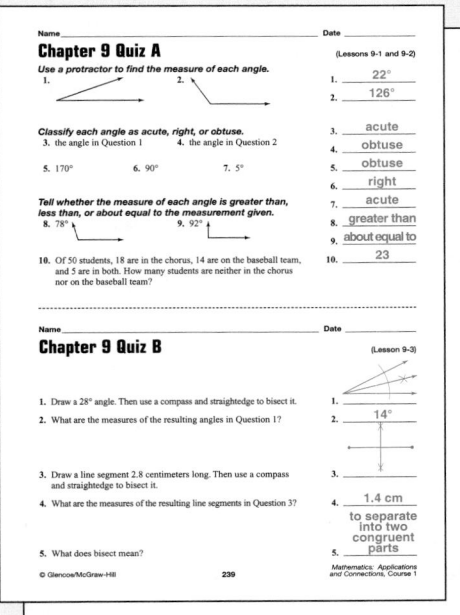

Technology Masters, pp. 17–18

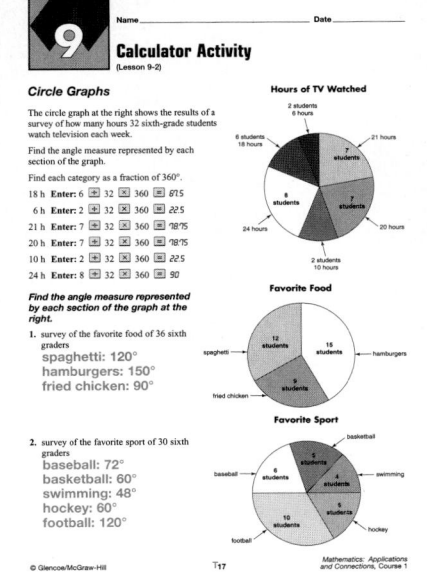

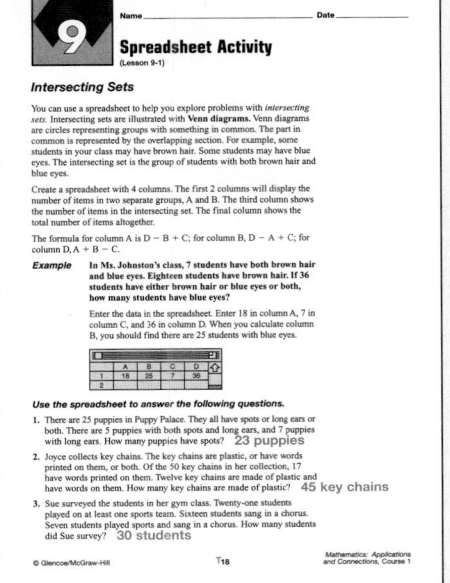

Investigations for the Special Education Student, p. 29

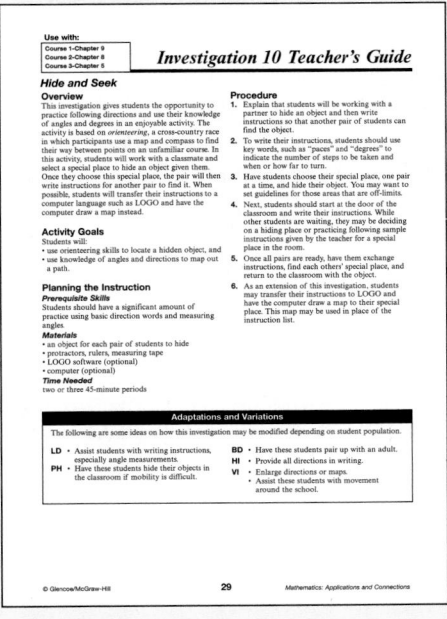

Theme: Architecture

Architects often redefine the American "dream" house. In the mid-twentieth century, simple designs with plenty of rooms and modern appliances were considered the ideal. By the end of the twentieth century, "dream" was defined as luxury with many specialized rooms added to houses built both for comfort and for the spectacular views they afforded.

Question of the Day How many lines of symmetry does the shape formed by the exterior walls of the dream house have? 8

Assess Prerequisite Skills

Ask students to read through the list of objectives presented in "What you'll learn in Chapter 9." You may wish to ask them what each of the objectives means or if they have experienced or used any of these math concepts before.

 Building Portfolios

Encourage students to revise their portfolios as they study this chapter. Suggest that students add work to their portfolios that demonstrates progression as their math knowledge increases.

 **Math and the Family**

In the *Family Letters and Activities* booklet (pp. 17–18), you will find a letter to the parents explaining what students will study in Chapter 9. An activity appropriate for the whole family is also available.

CHAPTER 9 Geometry: Investigating Patterns

What you'll learn in Chapter 9

- to classify and measure angles,
- to solve problems by using logical reasoning,
- to bisect segments and angles,
- to name two-dimensional figures, and
- to determine congruence and similarity.

CD-ROM Program

Activities for Chapter 9
- Chapter 9 Introduction
- Interactive Lessons 9-2, 9-5
- Extended Activity 9-6
- Assessment Game
- Resource Lessons 9-1 through 9-6

CHAPTER Project

DESIGNING A DREAM HOUSE

In this project, you will use geometry to find the actual angle measures and the actual lengths of the walls of the dream house shown. You will need a ruler and a protractor to complete this project.

Getting Started

- Look at the floor plan shown. What shape is the house?

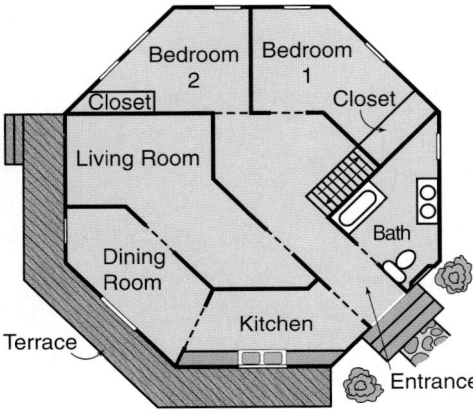

$\frac{1}{8}$ in. = $2\frac{1}{4}$ ft

Technology Tips

- Use geometry or drawing **software** to design a dream house.
- Surf the **Internet** for more information on designing houses.
- Use a **spreadsheet** to convert from inches to feet.

🖳 **inter**NET
CONNECTION For up-to-date information on houses, visit:
www.glencoe.com/sec/math/mac/mathnet

Working on the Project

You can use what you'll learn in Chapter 9 to help you design a dream house.

Page	Exercise
361	27
373	27
382	19
389	Alternative Assessment

Chapter 9 **351**

Instructional Resources ▶▶▶
A recording sheet to help students organize their data for the Chapter Project is shown at the right and is available in the *Investigations and Projects Masters*, p. 52.

CHAPTER Project
N O T E S

Objectives Students use a scale drawing to find
- angle measures.
- wall lengths.

Project Pointer You may suggest that students begin a *Project Folder* to keep their work as they complete each stage of the Chapter Project. The completed project may also be added to their portfolios.

In addition to the Chapter Project, have students add their own design for a dream house to their portfolios.

Using the Diagram Make sure students understand which lines represent walls and how doorways are indicated by line breaks. Have them look at the diagram from the entry point of view and visualize what the dream house looks like.

Investigations and Projects Masters, p. 52

9 Name _____ Date _____
Chapter 9 Project

Designing a Dream House

Page 361, Working on the Chapter Project, Exercise 27
Number each angle in the bathroom, dining room, and second bedroom.

Bathroom	Dining room	Bedroom 2
m∠1:	m∠5:	m∠10:
m∠2:	m∠6:	m∠11:
m∠3:	m∠7:	m∠12:
m∠4:	m∠8:	m∠13:
	m∠9:	

$\frac{1}{8}$ in. = $2\frac{1}{4}$ ft

Page 373, Working on the Chapter Project, Exercise 27
Polygons: Sum of measures of the angles:
                    Bathroom:
                    Dining room:
                    Bedroom 2:
                    Octagon:

Page 382, Working on the Chapter Project, Exercise 19
Bathroom: Dining room: Bedroom 2:

The actual perimeter of the house is:

© Glencoe/McGraw-Hill   52   *Mathematics: Applications and Connections, Course 1*

Instructional Resources
- *Study Guide Masters*, p. 64
- *Practice Masters*, p. 64
- *Enrichment Masters*, p.64
- Transparencies 9-1, A and B
- *Technology Masters*, p. 18

 CD-ROM Program
- Resource Lesson 9-1

Recommended Pacing	
Standard	Day 2 of 14
Honors	Day 2 of 12
Block	Day 1 of 7

1 FOCUS

5-Minute Check
(Chapter 8)

1. Solve the proportion $\frac{5}{6} = \frac{x}{18}$.
   $\frac{5}{6} = \frac{15}{18}$

2. Express $\frac{14}{25}$ as a percent.
   56%

3. Express 2.8% as a fraction and as a decimal. $\frac{7}{250}$, 0.028

4. Express 7.3 as a percent.
   730%

5. Find 36% of 92. **33.12**

 The 5-Minute Check is also available on **Transparency 9-1A** for this lesson.

Motivating the Lesson
Hands-On Activity Have students read the beginning of the lesson and work in pairs to identify angles in the classroom. Then have them order the angles from least measure to greatest.

2 TEACH

 Transparency 9-1B contains a teaching aid for this lesson.

Reading Mathematics Have students discuss how the non-mathematical meanings of *acute* and *obtuse* relate to the mathematical meanings. Also, point out that *complementary* is spelled with an "e," not an "i."

9-1 Angles

What you'll learn
You'll learn to classify and measure angles.

When am I ever going to use this?
Knowing how to classify and measure angles helps a carpenter build houses.

Word Wise
edges
vertex
angle
degree
protractor
acute angle
obtuse angle
right angle
complementary
supplementary

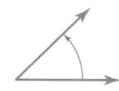
Study Hint
Reading Math The measurement 0° is read as zero degrees.

Notice the arrows on the sides of the angles. These tell you that you can extend the sides.

Have you ever watched a magician place a person in a box and then "saw the box in half?" The **edges** of the front of the box look like two lines that meet at a point called the **vertex**.

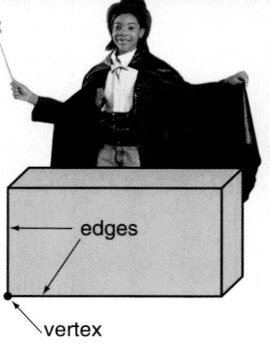
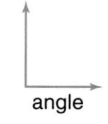

edges
vertex

Vertices and edges of the box form **angles**.

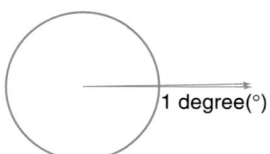
angle

The most common unit of measure for angles is the **degree**. A circle can be separated into 360 equal-sized parts. Each part would make up a one-degree (1°) angle as shown.

1 degree(°)

You can use a **protractor** to measure angles.

Step 1 Place the center of the protractor on the vertex of the angle with the straightedge along one side.

Step 2 Use the scale that begins with 0° on the side of the angle. Read the angle measure where the other side crosses the same scale. Extend the sides if needed.

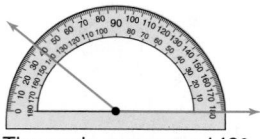
The angle measures 140°.

Angles can be classified according to their measure.

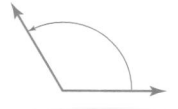
Acute angles
measure between 0° and 90°.

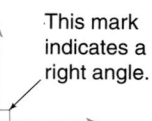
Obtuse angles
measure between 90° and 180°.

This mark indicates a right angle.
Right angles
measure 90°.

352 Chapter 9 Geometry: Investigating Patterns

Cross-Curriculum Cue

Inform the teachers on your team that your students are studying shapes and angles. Suggestions for curriculum integration are:
History: historical monuments
Industrial Technology: molding, trim
Physical Education: playing fields
Art: sculpture, painting, pottery

Examples

Study Hint

Problem Solving Lay a corner of your notebook paper on top of the given angle to determine whether an angle is acute, right, or obtuse.

Use a protractor to find the measure of each angle. Then classify the angle as acute, right, or obtuse.

①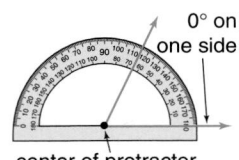

0° on one side

center of protractor

The angle measures 65°. It is an acute angle.

②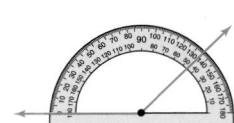

The angle measures 135°. It is an obtuse angle.

Some pairs of angles are **complementary** or **supplementary**. If the sum of the measures of two angles is 90°, the angles are complementary. If the sum of the measures of two angles is 180°, the angles are supplementary.

Study Hint

Reading Math The symbol $m\angle 1$ is read as *the measure of angle 1.*

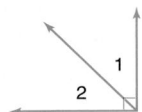

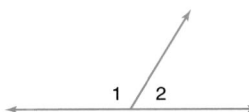

complementary angles
$m\angle 1 + m\angle 2 = 90°$

supplementary angles
$m\angle 1 + m\angle 2 = 180°$

Example

INTEGRATION

③ **Algebra** The angles shown are complementary. If $m\angle X = 36°$, what is $m\angle Y$?

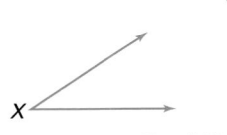

$m\angle X + m\angle Y = 90°$ *complementary angles*
$36 + m\angle Y = 90$ *Replace $m\angle X$ with 36.*
$m\angle Y = 54$ *You know that $36 + 54 = 90$.*

The measure of the angle is 54°.

CHECK FOR UNDERSTANDING

Communicating Mathematics

Read and study the lesson to answer each question.

1. *Draw* an obtuse angle. Then write the steps describing how to use a protractor to measure your obtuse angle. **See Answer Appendix.**

2. *Tell* how you can determine whether two angles are complementary or supplementary. **See margin.**

Lesson 9-1 Angles **353**

Reteaching the Lesson

Activity Open the classroom door to form a right angle. Have pairs of students use masking tape on the floor to mark the right angle formed. Then have them open and close the door to show an acute angle and an obtuse angle. Have students measure the angles with a large classroom protractor.

Error Analysis
Watch for students who make errors measuring angles with protractors.
Prevent by reminding them to align one side with 0° and follow along the scale to see where the other side lands.

Teaching Tip To help students remember the types of angles, encourage them to develop mnemonic devices, such as "a cute" animal is usually small, such as a puppy or a kitten.

In-Class Examples

Use a protractor to find the measure of each angle. Classify each angle as *acute*, *right*, or *obtuse*.

For Example 1

80°; acute

For Example 2

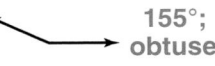

155°; obtuse

For Example 3
The angles shown are supplementary. If $m\angle X = 48°$, what is $m\angle Y$? **132°**

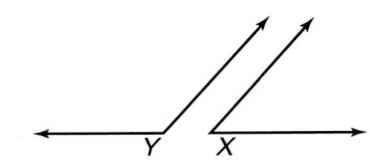

Teaching Tip Before Example 3, remind students that the sum of two numbers minus one of the numbers equals the other number.

3 PRACTICE/APPLY

Check for Understanding
If students need additional practice or instruction after completing Exercises 1–9, one of these options may be helpful.
• Extra Practice, see p. 580
• Reteaching Activity
• *Study Guide Masters,* p. 64
• *Practice Masters,* p. 64

Additional Answer

2. If the sum of the two angles is 90°, the angles are complementary. If the sum of the angles is 180°, the angles are supplementary.

Assignment Guide

Core: 11–31 odd, 33–38
Enriched: 10–30 even, 31–38

Guided Practice Use a protractor to find the measure of each angle. Then classify the angle as *acute, right,* or *obtuse.*

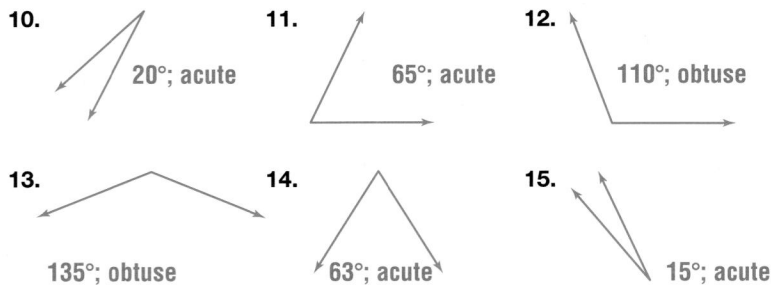

3.   50°; acute

4.   135°; obtuse

Classify each angle measure as *acute, right,* or *obtuse.*

5. 140° obtuse   6. 17° acute   7. 86° acute

8. Classify the angles shown as complementary, supplementary, or neither. **complementary**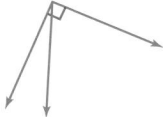

9. *Algebra* Angles *A* and *B* are supplementary. If $m\angle B = 106°$, find $m\angle A$. **74°**

EXERCISES

Practice Use a protractor to find the measure of each angle. Then classify the angle as *acute, right,* or *obtuse.*

10.   20°; acute

11.   65°; acute

12.   110°; obtuse

13.   135°; obtuse

14.   63°; acute

15.   15°; acute

Classify each angle measure as *acute, right,* or *obtuse.*

16. 74° acute   17. 90° right   18. 28° acute   19. 174° obtuse

20. 168° obtuse   21. 54.6° acute   22. 137.5° obtuse   23. 4° acute

Classify each pair of angles as *complementary, supplementary,* or *neither.*

24.   complementary

25.   neither

26.   supplementary

27. An angle measures 89.9°. Is it an acute angle or a right angle? **acute**

28. If an angle measures 90.4°, is it an obtuse angle or a right angle? **obtuse**

29. *Algebra* Angles *J* and *K* are complementary angles. Find $m\angle K$ if $m\angle J = 72°$. **18°**

354 **Chapter 9** Geometry: Investigating Patterns

Study Guide Masters, p. 64

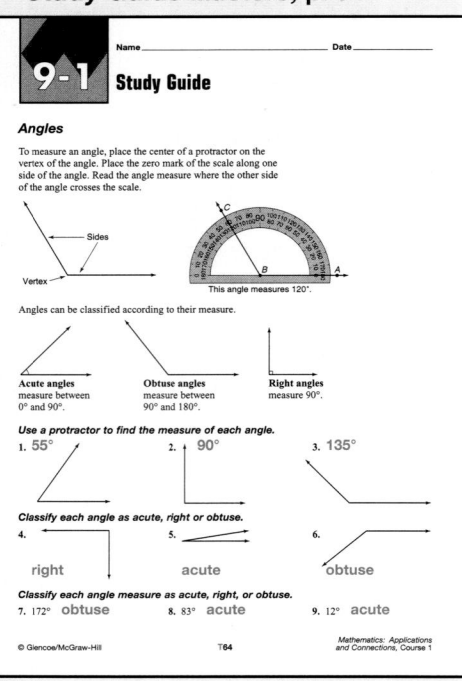

9-1 Study Guide

Name_____ Date_____

Angles

To measure an angle, place the center of a protractor on the vertex of the angle. Place the zero mark of the scale along one side of the angle. Read the angle measure where the other side of the angle crosses the scale.

Sides

Vertex

This angle measures 120°.

Angles can be classified according to their measure.

Acute angles measure between 0° and 90°.   **Obtuse angles** measure between 90° and 180°.   **Right angles** measure 90°.

Use a protractor to find the measure of each angle.

1. 55°   2. 90°   3. 135°

Classify each angle as *acute, right* or *obtuse.*

4. right   5. acute   6. obtuse

Classify each angle measure as *acute, right,* or *obtuse.*

7. 172° obtuse   8. 83° acute   9. 12° acute

© Glencoe/McGraw-Hill   T64   Mathematics: Applications and Connections, Course 1

💡 **Investigations for the Special Education Student**

This blackline master booklet helps you plan for the needs of special education students by providing long-term projects along with teacher notes. Investigation 10, *Hide and Seek,* may be used with this chapter.

354 **Chapter 9**

30. Algebra Angles M and N are supplementary angles. If angle M measures 119°, what is the measure of angle N? **61°**

Applications and Problem Solving

31. Carpentry A carpenter joins two pieces of molding as shown. What types of angles can you identify and where? Explain. **See margin.**

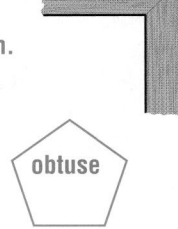

32. Geometry The geometric figure shown is a regular pentagon. Are the angles inside a regular pentagon acute, right, or obtuse?

obtuse

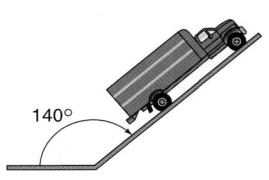

33. Critical Thinking Truck drivers find it difficult to accelerate when the grade of a hill is steep.

140°

a. How would you change the grade so that it is not so steep? **See margin.**

b. Would the obtuse angle with the ground become larger or smaller? **larger**

Mixed Review

35. $0.76 = \frac{19}{25}$;
$0.15 = \frac{3}{20}$;
$0.09 = \frac{9}{100}$

34. Estimate 1.5% of $127,500. *(Lesson 8-6)* **See margin.**

35. Statistics Eight out of 10 mountain bicyclists have been injured at least once while riding their mountain bike. The graph shows where accidents occur. Express each decimal as a fraction in simplest form. *(Lesson 5-9)*

Mountain Bike Accidents

Downhill 0.76
Flat 0.15
Uphill 0.09

Source: *Acute Injuries from Mountain Biking,* by T. Chow, M. Bracker, K. Patrick, M.D.'s

36. | Test Practice | The table shows the results of a cost comparison for three compact discs at different stores. Which statement is true? *(Lesson 3-3)* **A**

Compact Disc	Music Store	Department Store	Discount Store
X	$13.95	$18.95	$11.95
Y	$15.95	$14.50	$13.50
Z	$16.50	$14.98	$15.98

A Compact disc X costs the most at the department store.
B The discount store has the best price for any of the three compact discs.
C Compact disc Y costs the most at the department store.
D The least expensive place to get compact disc Z is at the discount store.

37. Algebra Evaluate $3 + xy - 5$ if $x = 6$ and $y = 9$. *(Lesson 1-5)* **52**

38. Round 108 to the nearest ten. *(Lesson 1-3)* **110**

Lesson 9-1 Angles **355**

Extending the Lesson

Enrichment Masters, p. 64

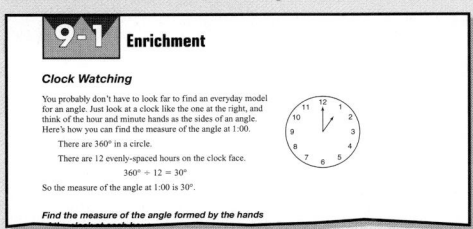

Activity Have students work in pairs to determine how they can use Computer Aided Design (CAD) to measure and copy angles. Have them produce angles of various sizes if the software is available.

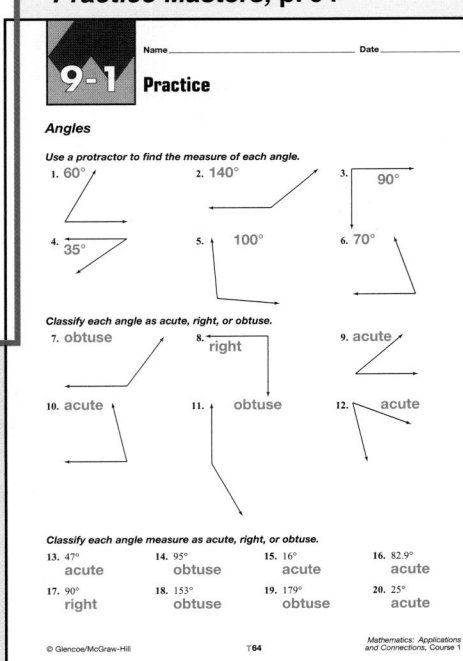

Objective Students solve problems by using logical reasoning.

Recommended Pacing	
Standard	Day 3 of 14
Honors	Day 3 of 12
Block	Day 2 of 7

1 FOCUS

Getting Started Have students act out the situation presented at the beginning of the lesson using smaller numbers. Have them put the Venn diagram on the chalkboard. Then have them work in pairs to answer Exercises 1–3. After the class discusses the results, have them solve Exercise 4.

2 TEACH

Teaching Tip To help students solve Exercise 4b, suggest that they draw 24 boxes to represent students. Write *LB* in 7 boxes for both lunch bags and backpacks. Then mark *L* in enough boxes so $L + LB = 12$ and *B* in enough boxes so $B + LB = 10$. The number of boxes with no letters answers Exercise 4b.

In-Class Example

A survey of 83 students showed that 56 chose books by author, 23 chose books by subject, and 15 chose books both by author and by subject. How many students used other methods to choose books? **19 students**

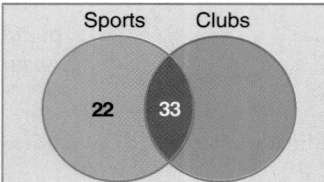

THINKING LAB

9-1B Use Logical Reasoning

A Follow-Up of Lesson 9-1

Jesse and Estella are standing in front of the sign-up board for extra-curricular activities. There are 55 students who play sports, 60 who are involved in clubs, and 33 who are involved in both sports and clubs. They are trying to find out how many students participate only in sports. Let's listen in!

I made a Venn diagram using the information from the sign-up sheets.

Jesse

What's a Venn diagram?

Sports Clubs

22 33

You can use a Venn diagram to solve the problem. The circle on the left represents sports. The circle on the right represents clubs. The overlapping region represents both sports and clubs and the region outside the circles represents neither sports nor clubs.

Estella

I see! There are 55 students who play sports and 33 who are involved in both sports and clubs. 55 − 33 = 22. So, 22 students participate only in sports.

THINK ABOUT IT

Work with a partner. 1. See margin. 2. 27 students

1. **Analyze** the information in the *Venn diagram*. Do you agree or disagree with Estella's and Jesse's thinking? Explain your reasoning.

2. **Use logical reasoning** to find the number of students who are involved only in clubs.

3. **Find** the total number of students who participate in extracurricular activities by **using logical reasoning**. 82 students

4. **Apply** the logical reasoning strategy to solve the following problem.

   In a line of 24 students waiting to go on a field trip, 12 have lunch bags, 10 have backpacks, and 7 have both lunch bags and backpacks.

   a. *How many students have only lunch bags?* 5 students

   b. *How many have neither a lunch bag nor a backpack?* 9 students

■ Reteaching the Lesson ■

Activity Have students draw a Venn diagram outline and use it to group various manipulatives. For example, use colored three-dimensional solids and group red figures, circular figures, and red and circular figures. Have students write out the situation as a word problem.

Additional Answer

1. Sample answer: Jesse's and Estella's thinking does make sense because $22 + 33 = 55$, the number of students who play sports.

ON YOUR OWN

5. The last step of the 4-step plan for problem solving asks you to *examine* the solution. *Explain* how you can use the solution in Exercise 4 to verify the number of students who are going on the field trip. See margin.

6. *Write a Problem* that can be solved by using logical reasoning. Then ask a classmate to solve the problem. See Answer Appendix.

7. *Reflect back* Explain how you can use logical reasoning to answer Exercise 28 on page 354. See margin.

MIXED PROBLEM SOLVING

STRATEGIES
Look for a pattern.
Solve a simpler problem.
Act it out.
Guess and check.
Draw a diagram.
Make a chart.
Work backward.

Solve. Use any strategy.

8. *Communication* Lisa places a long distance phone call to Jason and talks for 33 minutes at a rate of 17 cents per minute. Did she spend about $4 or $6 for the call? about $6

9. *Patterns* What is the missing number in the pattern? ..., 234, 345, _?_, 567, ... 456

10. *School* Of the 150 students at Washington Middle School, 55 are in the orchestra, 75 are in marching band, and 25 are in both orchestra and marching band. How many students are in neither orchestra nor marching band? 45 students

11. *Patterns* The Sweepstakes Prize Company is having a contest. To enter the contest, you need to draw the next figure in the pattern.

a. Draw a sample entry. See margin.
b. Explain why your entry would be the next figure in the pattern. See margin.

12. Enrico

12. *Careers* Enrico, Angela, and Marcus are all engineers. Their specialties are electrical, mechanical, and civil engineering, but not in that particular order. The civil engineer and Enrico played golf together on Friday. Marcus, who works for the power company, is married to the civil engineer. Who is the mechanical engineer?

13. **Test Practice** The Venn diagram shows the relationship between the members in Moesha's scout troop.

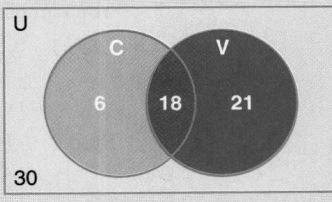

U = set of all members in Moesha's scout troop
C = set of members who have a camping badge
V = set of members who have a volunteer badge
Which of the following is *not* true? C

A 30 scouts do not have a camping badge or a volunteer badge.

B 21 scouts have only a volunteer badge.

C 63 students have at least one badge.

D There are more scouts with a camping badge than those who have only a volunteer badge.

E 6 students have only a camping badge.

Lesson 9-1B THINKING **LAB** 357

■ Extending the Lesson ■

Activity Have students look for logic problems in various sources, such as crossword puzzle books and juvenile magazines. Have them solve the problems and then present their favorites for the class to solve.

Check for Understanding
Use the results from Exercise 4 to determine whether students comprehend how to solve problems by using logical thinking.

Extra Practice If students need additional practice in problem solving, extra practice is available on the following pages.
- Use Logical Reasoning, see p. 581
- Mixed Problem Solving, see pp. 593–594

Assignment Guide
All: 5–13

Closing Activity
Writing Have small groups of students develop a quiz on logical reasoning for the rest of the class. Have them include at least one of each type of problem used in the lesson.

Additional Answers

5. You found that 5 students have only lunch bags, 10 have only backpacks, and 9 have neither a lunch bag nor a backpack. 5 + 10 + 9 = 24. So, there are 24 students going on the trip.

7. A right angle measures 90°. The measure of the angle given is 90.4°. Since 90.4° is greater than 90°, it is only logical that the angle is an obtuse angle.

11a. Sample answer:

11b. Each figure has one more side than the previous figure, so the next figure must have six sides.

Instructional Resources

- *Study Guide Masters*, p. 65
- *Practice Masters*, p. 65
- *Enrichment Masters*, p. 65
- Transparencies 9-2, A and B
- *Assessment and Evaluation Masters*, p. 239
- *School to Career Masters*, p. 9
- *Technology Masters*, p. 17
- CD-ROM Program
  - Resource Lesson 9-2
  - Interactive Lesson 9-2

Recommended Pacing	
Standard	Day 4 of 14
Honors	Day 4 of 12
Block	Day 2 of 7

1 FOCUS

5-Minute Check
(*Lesson 9-1*)

Use a protractor to find the measure of the angle. Then classify the angle as *acute*, *right*, or *obtuse*.

1.

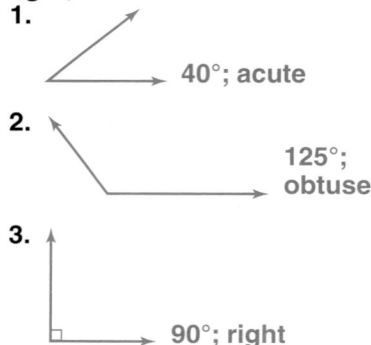

40°; acute

2.

125°; obtuse

3.

90°; right

Classify each angle measure as *acute*, *right*, or *obtuse*.
4. 176° obtuse
5. 34° acute

The 5-Minute Check is also available on **Transparency 9-2A** for this lesson.

Motivating the Lesson

Communication Ask students if they think a protractor could be used to draw an angle with a certain measure. Have them suggest ways they could draw a specific angle.

9-2 Using Angle Measures

What you'll learn
You'll learn to draw angles and estimate measures of angles.

When am I ever going to use this?
Knowing how to use angle measures helps a sailor navigate a boat.

Word Wise
straightedge

> **Study Hint**
> Problem Solving You can use the draw-a-diagram strategy to help you with angle measures.

Many people enjoy the relaxation of sailing. Others enjoy the excitement of sailboat racing. To determine their direction when sailing, sailors can use a compass.

If a sailboat traveling east increases its directional angle by 45°, in what direction is it now sailing? (*Hint*: The directional angle is measured clockwise from magnetic north.)

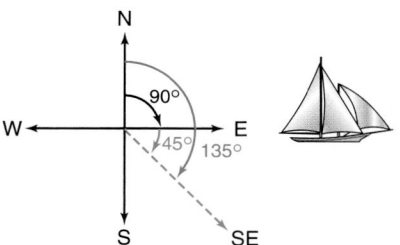

East (E) has a directional angle of 90° measured from the north (N). By increasing this angle by 45°, the new directional angle is 90° + 45°, or 135°. The sailboat is now traveling southeast (SE).

You can use a protractor and a straightedge to draw an angle.

Step 1 Draw one side of the angle. Then mark the vertex and draw an arrow.

Step 2 Place the protractor along the side as you would to measure an angle. On the protractor, find the number of degrees needed for the angle you are drawing and make a pencil mark as shown.

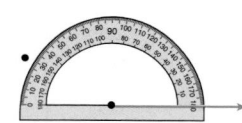

Step 3 A **straightedge** is a ruler or any object with a straight side, which can be used to draw a line. With a straightedge, draw the side that connects the vertex and the pencil mark. Draw an arrow on the end of the other side.

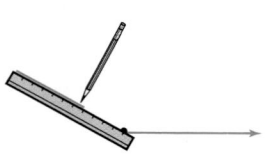

The angle drawn is a 150° angle.

Example 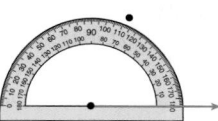 **1** Draw a 68° angle.

Draw one side. Mark the vertex and draw an arrow.

Find 68° on the appropriate scale. Make a pencil mark.

Draw the side that connects the vertex and the pencil mark.

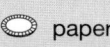 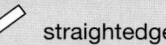

  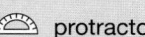

 Transparency 9-2B contains a teaching aid for this lesson.

In-Class Example
For Example 1
Draw a 75° angle.

You can estimate the measure of an angle by comparing it to an angle whose measure you know.

HANDS-ON

MINI-LAB

Work with a partner. ⬭ paper plate ▱ straightedge

✂ scissors ⌓ protractor

Try This

- Find and mark the center of a paper plate by folding it in half twice.
- Measure and cut wedges of 180°, 90°, and 45° from the plate. Write the measure on each wedge.
- Have one partner draw any angle on a piece of paper. The other partner should use the wedges to estimate the measure of the angle.
- Repeat for several different angles.

Talk About It

1. Explain how you could estimate the measure of an angle without using the wedges. **Compare the angle with a right angle.**
2. What is the measure of the angle of the remaining wedge? **45°**

Using the Mini-Lab You may wish to use wedges cut from color transparencies to model the activity with students on the overhead projector. Encourage students to use several different wedges to determine the closest estimate. Students can also form 45° angles by folding a 90° wedge in half.

In-Class Example
For Example 2
Estimate the measure of the angle shown.

Sample answer: about 80°

Example 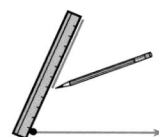 **2** Estimate the measure of the angle.

You can estimate the measure of the angle by using the angle wedges from the Mini-Lab. The angle shown is about the same as the 90° angle wedge and the 45° angle wedge. So, the measure of the angle is about 90° + 45°, or 135°.

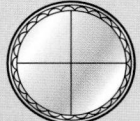

Lesson 9-2 Using Angle Measures **359**

Multiple Learning Styles

 Verbal/Linguistic Have pairs of students use geoboards and geobands to estimate the measure of angles. Have one student in each pair tell the other student an angle measure or a type of angle. The other student then creates the angle on the geoboard. Students should measure the angle with a protractor for accuracy. Have them switch roles and repeat.

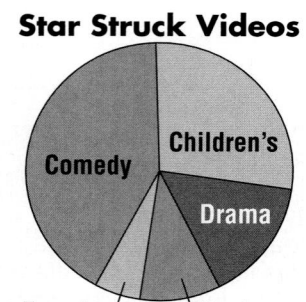
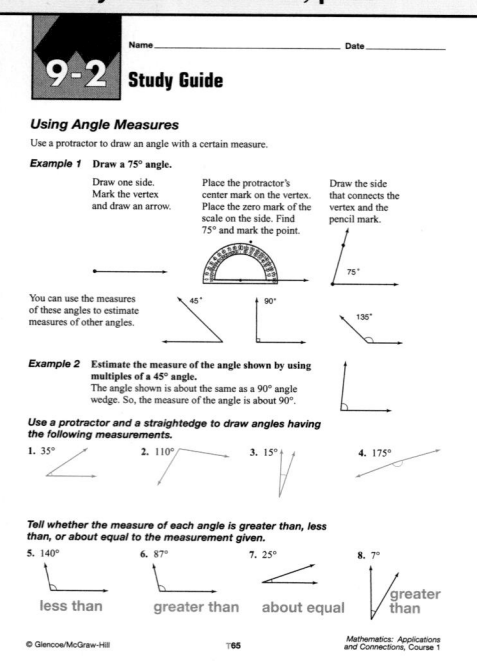
360 Chapter 9

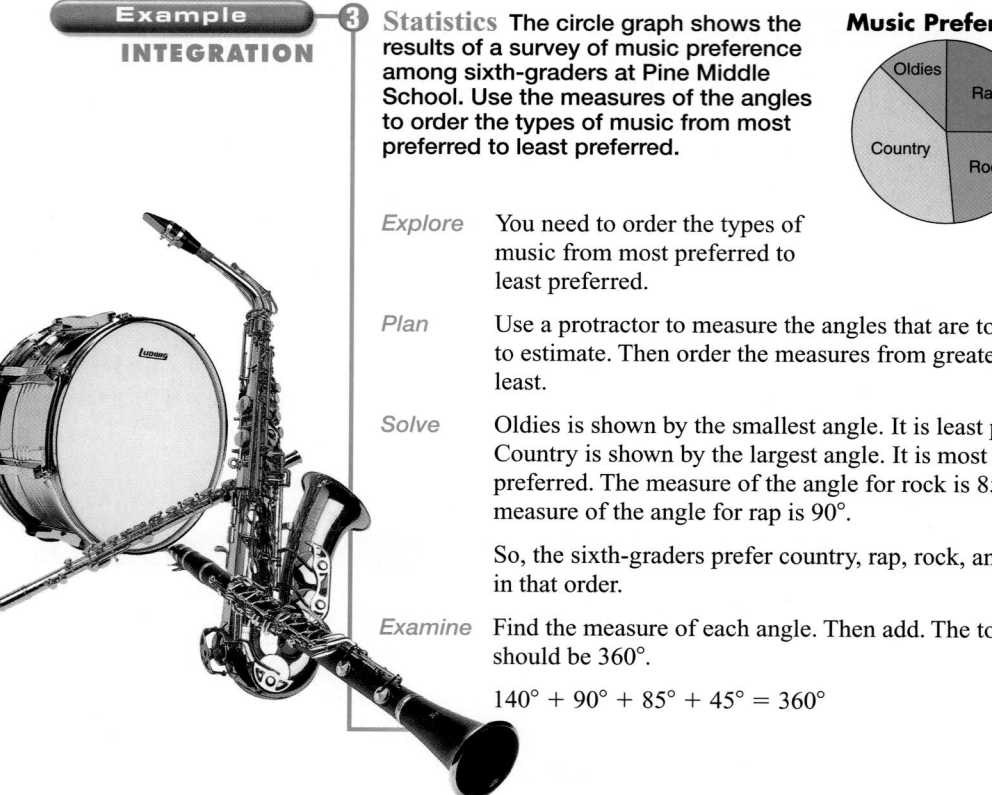

Example —③ **Statistics** The circle graph shows the results of a survey of music preference among sixth-graders at Pine Middle School. Use the measures of the angles to order the types of music from most preferred to least preferred.

INTEGRATION

Explore You need to order the types of music from most preferred to least preferred.

Plan Use a protractor to measure the angles that are too close to estimate. Then order the measures from greatest to least.

Solve Oldies is shown by the smallest angle. It is least preferred. Country is shown by the largest angle. It is most preferred. The measure of the angle for rock is 85°. The measure of the angle for rap is 90°.

So, the sixth-graders prefer country, rap, rock, and oldies, in that order.

Examine Find the measure of each angle. Then add. The total should be 360°.

$$140° + 90° + 85° + 45° = 360°$$

CHECK FOR UNDERSTANDING

Communicating Mathematics

Read and study the lesson to answer each question.   1–2. See students' wor[k]

1. *Explain* to a classmate how to draw a 145° angle.
2. *Show* an angle of about 120° using two pencils as the sides.
3. *Write* a sentence explaining how to use paper folding to show an angle of 45°. **See margin.**

Guided Practice

Use a protractor and a straightedge to draw angles having the following measurements. 4–6. See Answer Appendix.

4. 120°   5. 25°   6. 43°

Estimate the measure of each angle. 7–8. Sample answers are given.

7. about 120°   8. about 90°

9. Is the measure of angle *ABC* greater than, less than, or about equal to 155°? **about equal to**

360   Chapter 9 Geometry: Investigating Patterns

EXERCISES

Practice

Use a protractor and a straightedge to draw angles having the following measurements. 10–17. See Answer Appendix.

10. 165°   **11.** 90°   **12.** 85°   **13.** 8°

14. 95°   **15.** 145°   **16.** 32°   **17.** 66°

Estimate the measure of each angle. 18–23. Sample answers are given.

18. about 60°   **19.** about 40°   **20.** about 90°

21. about 55°   **22.** about 135°   **23.** about 70°

24. Is the angle shown greater than, less than, or about equal to 79°? **less than**

Applications and Problem Solving

25. *Navigation* A pilot is flying northeast (45° east of north). He increases the directional angle of the plane by 135°.
    a. In which direction is the plane now traveling? **south**
    b. How many degrees from north is the plane now traveling? **180°**

26. *Life Science* The branches on young trees should be spread to form angles of at least 60° with the tree trunk. This strengthens the branches and allows for more air circulation and light. Which branches on the tree shown need to be spread? **A, D, E**

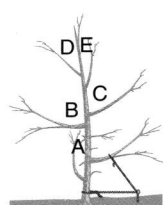

27. *Working on the* CHAPTER Project   Refer to page 351. Trace the floor plan onto your paper. Use a protractor to measure each angle of the bathroom, the dining room, and the second bedroom. Then label the measures of the angles on your diagram. Save your diagram for use in other lessons. **See Answer Appendix.**

28. *Critical Thinking* If it is 10:00 A.M., at approximately what times during the next hour will the hands of the clock show an acute angle?

28. between 10:00 and 10:05, and between 10:40 and 11:00

Mixed Review

29. *Geometry* Classify the angle shown as *acute, right,* or *obtuse.* *(Lesson 9-1)* **acute**

30. Test Practice   If an airplane travels 438 miles per hour, how many miles will it travel in 5 hours? *(Lesson 8-2)* **D**
    A 2,050 mi   **B** 2,090 mi   **C** 2,150 mi   **D** 2,190 mi   **E** Not Here

31. Find the LCM of 24 and 30. *(Lesson 5-7)* **120**

32. *Measurement* How many milliliters are in 2.14 liters? *(Lesson 4-9)* **2,140 mL**

Lesson 9-2 Using Angle Measures   **361**

Extending the Lesson

Enrichment Masters, p. 65

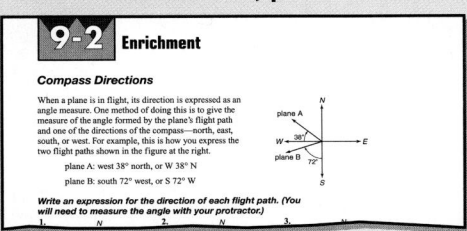

Activity Have small groups of students determine how pilots use their knowledge of angles to fly an airplane. Have them look for information about the angle required for takeoff and how angles are used in navigation.

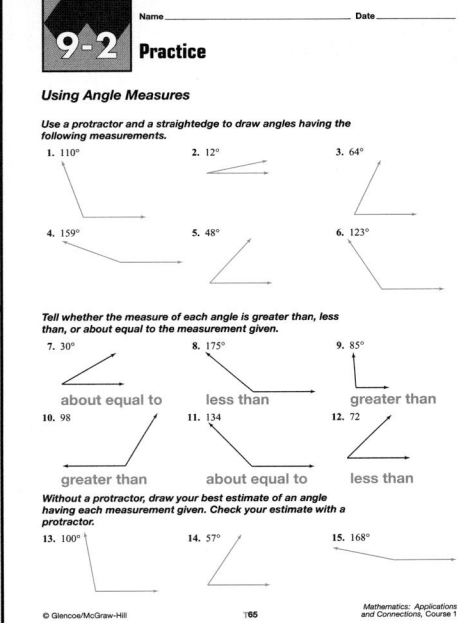

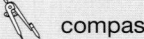

COOPERATIVE LEARNING

GET READY

Objective Students construct congruent segments and angles.

Optional Resources
Hands-On Lab Masters
• worksheet, p. 57
Overhead Manipulative Resources
• compass
Manipulative Kit
• compass

MANAGEMENT TIPS

Recommended Time
45 minutes

Getting Started Have students work in small groups. Ask them to identify and sketch classroom objects having the same length or angle measure.

Activity 1 has students construct a line segment congruent to a given line segment. Make <u>sure</u> students choose a length for *LM* that is easy to measure with a compass.

Teaching Tip The compasses provided in the Manipulative Kit are Triman safety compasses. Students may need extra time to practice using these and how to adapt them to the ball-bearing-compass drawings shown in the Student Edition.

/ straightedge

compass

9-3A Constructing Congruent Segments and Angles

A Preview of Lesson 9-3

A *line segment* is a straight path between two endpoints. To indicate line segment *JK*, write $\overline{JK}$. Line segments that have the same length are called *congruent segments*.

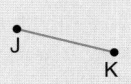

You can construct congruent segments using a straightedge and a compass. A compass is used to draw circles or circular arcs and to measure distances.

TRY THIS

Work with a partner.

1 To construct a line segment congruent to $\overline{JK}$, follow these steps.
• Draw $\overline{JK}$. Then use a straightedge to draw a line segment longer than $\overline{JK}$. Label it $\overline{LM}$.

• Place the compass point at *J* and adjust the compass setting so that the pencil is on point *K*. The compass setting equals the length of $\overline{JK}$.

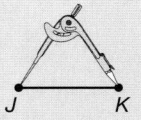

• Use this compass setting and place the compass point at *L*. Draw an arc that intersects $\overline{LM}$ at *P*. $\overline{LP}$ is congruent to $\overline{JK}$.

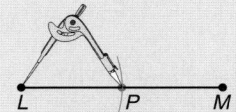

ON YOUR OWN

Trace each segment. Then construct a segment congruent to it. 1–3. See Answer Appendix.

1. *A*   *B*   2. *S*   *T*   3. *X*   *Y*

4. *Look Ahead* The length of $\overline{MN}$ is 38 millimeters. If $\overline{MN}$ is separated into two congruent parts, what will be the length of each part? **19 mm**

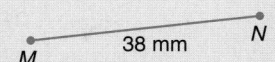
M 38 mm N

The symbol ∠ is used to indicate an angle. The angle shown can be named in two ways, ∠*JKL* or ∠*LKJ*. The middle letter is always the vertex.

You can also use a straightedge and a compass to construct congruent angles. *Congruent angles* are angles that have the same measure.

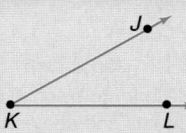

ASSESS

Have students complete Exercises 1–3 and 5–7. After students construct congruent segments and angles, have them check their answers by cutting one out and laying it on top of the other one. Students may hold their papers up to the light to check their accuracy.

Use Exercises 4 and 8 to determine whether students understand constructing congruent segments and angles.

TRY THIS

Work with a partner.

❷ To construct an angle congruent to ∠*JKL*, follow these steps.

- Draw ∠*JKL*. Then use a straightedge to draw side $\overrightarrow{BA}$. $\overrightarrow{BA}$ *means ray BA. A ray is a path that extends infinitely from one point in a certain direction.*

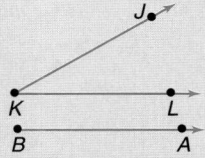

- With the compass point at *K*, draw an arc that intersects the sides of ∠*JKL*. Label the intersections *M* and *N*.

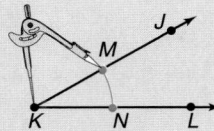

- With the same compass setting, move to side *BA*. Place the compass point at *B* and draw an arc that intersects side $\overrightarrow{BA}$. Label this intersection *P*.

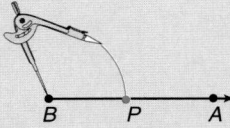

- On ∠*JKL*, set the compass at points *M* and *N* as shown. With that setting, go to side $\overrightarrow{BA}$. Place the compass point at *P*, and draw an arc that intersects the larger arc you drew before. Label the intersection *Q*.

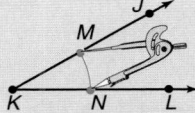

- Use a straightedge to draw side $\overrightarrow{BQ}$. ∠*QBA* is congruent to ∠*JKL*.

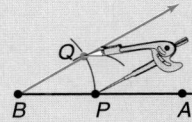

ON YOUR OWN

Trace each angle. Then construct an angle congruent to it. 5–7. See Answer Appendix.

5.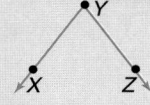

6.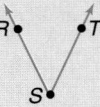

7.

8. **Look Ahead** The measure of an angle is 113°. If the angle is divided into two congruent angles, what will be the measure of each angle? **56.5°**

Lesson 9-3A HANDS-ON **LAB** 363

Have students write the procedures for constructing congruent line segments and angles. Suggest that they include an illustration of each.

Hands-On Lab Masters, **p. 57**

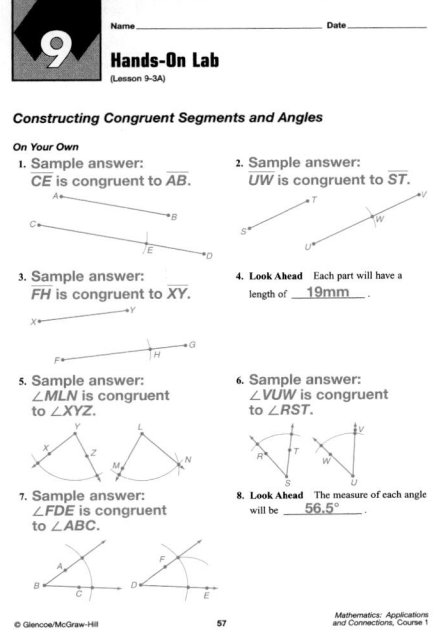

Hands-On Lab 9-3A **363**

- *Study Guide Masters,* p. 66
- *Practice Masters,* p. 66
- *Enrichment Masters,* p. 66
- Transparencies 9-3, A and B
- *Assessment and Evaluation Masters,* pp. 238, 239

 CD-ROM Program
- Resource Lesson 9-3

Recommended Pacing	
Standard	Days 5 & 6 of 14
Honors	Day 5 of 12
Block	Day 3 of 7

1 FOCUS

5-Minute Check
(Lesson 9-2)

Use a protractor and a straightedge to draw angles having the following measurements.

1. 160°

2. 25°

3. 95°

Estimate the measure of each angle.

4.

Sample answer: about 65°

5.

Sample answer: about 135°

 The 5-Minute Check is also available on **Transparency 9-3A** for this lesson.

Motivating the Lesson
Problem Solving Ask students how they would determine the midpoint of a piece of ribbon that is $3\frac{11}{16}$ inches long.

9-3 Constructing Bisectors

What you'll learn
You'll learn to bisect line segments and angles.

When am I ever going to use this?
Knowing how to find bisectors helps a carpenter make miter cuts.

Word Wise
bisect

The Greek mathematician Euclid answered the following problem more than 2,000 years ago.

Given a line segment of any length, find a geometric method for dividing the line segment into two equal parts.

To **bisect** something means to separate it into two congruent parts.

 **MINI-LAB**

Work with a partner.  ruler

Try This
- Draw $\overline{AB}$ using a straightedge.
- Fold point A onto point B and make a crease as shown. The crease bisects $\overline{AB}$. Label the intersection point C.

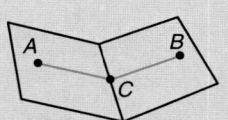

Talk About It It is halfway between point A and B.
Measure $\overline{AC}$ and $\overline{CB}$. What can you say about point C?

You can also use a straightedge and a compass to bisect a line segment.

Example

When segments meet to form right angles, they are perpendicular. You can say that $\overline{WV}$ is a perpendicular bisector of $\overline{YZ}$.

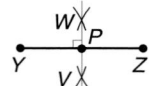

1 Use a straightedge and a compass to bisect $\overline{YZ}$.
- Use a straightedge to draw $\overline{YZ}$.
- Place the compass point at Y. Set the compass to more than half the length of $\overline{YZ}$. Draw two arcs as shown.
- With the same compass setting, place the compass point at Z and draw two arcs as shown. These arcs should intersect the first arc at W and V.
- With a straightedge, draw $\overline{WV}$. $\overline{WV}$ bisects $\overline{YZ}$ at P. So, $\overline{PY}$ and $\overline{PZ}$ are congruent.

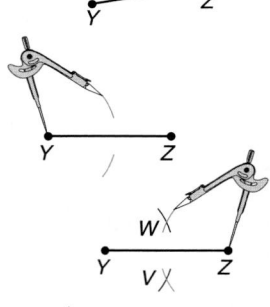

In the previous Mini-Lab and in Example 1, you learned how to bisect a line segment. You can also bisect an angle.

MINI-LAB

Work with a partner. 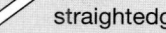 straightedge 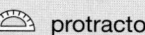 protractor

Try This
- Draw ∠ABC using a straightedge.
- Fold side $\overline{BA}$ onto side $\overline{BC}$. Make a crease. The crease bisects ∠ABC.

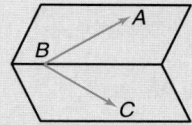

Talk About It They have the same measure.
Use a protractor to compare the measures of the two angles formed by the crease. What can you say about the two angles?

You can also use a straightedge and a compass to bisect an angle.

Example ② Use a straightedge and a compass to bisect ∠DEF.
- Draw ∠DEF using a straightedge.

- Place the compass point at E and draw an arc that intersects both sides of the angle. Label these points G and H.

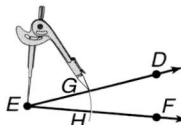

- Place the compass point at G and draw an arc as shown.

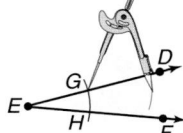

- With the same compass setting, place the compass point at H and draw an arc that intersects the one drawn in the previous step. Label the intersection J.

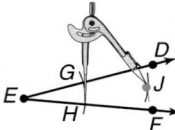

- Using a straightedge, draw $\overrightarrow{EJ}$. Side $\overrightarrow{EJ}$ bisects ∠DEF. So, ∠DEJ and ∠JEF are congruent.

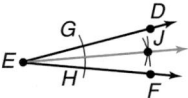

Lesson 9-3 Constructing Bisectors **365**

2 TEACH

Transparency 9-3B contains a teaching aid for this lesson.

Using the Mini-Lab For both of the labs, you may find it helpful for students to use patty paper or tracing paper to draw their segments and angles. They can also hold their papers up to a window to help in matching the endpoints or sides.

In-Class Examples
For Example 1
Use a straightedge and a compass to bisect $\overline{XY}$.

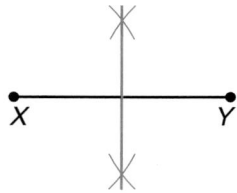

For Example 2
Use a straightedge and a compass to bisect ∠MNP.

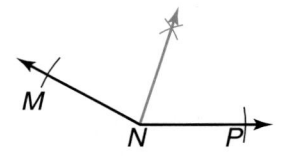

Study Guide Masters, p. 66

9-3 **Study Guide**

Constructing Bisectors

A **bisector** divides a line segment or an angle into two equal parts.

Examples 1 Use a straightedge and a compass to bisect $\overline{AB}$.
- Use the straightedge to draw $\overline{AB}$.
- Place the compass point at A. Set the compass for more than half the length of $\overline{AB}$. Draw two arcs as shown.
- With the same compass setting, place the compass point at B and draw two arcs as shown. These arcs should intersect the first arc at C and D.
- With a straightedge, draw $\overline{CD}$. $\overline{CD}$ bisects $\overline{AB}$.

2 Use a straightedge and a compass to bisect ∠ABC.
- Draw ∠ABC using a straightedge.
- Place the compass point at B and draw an arc that intersects both sides of the angle. Label these points D and E.
- Place the compass point at D and draw an arc as shown.
- With the same compass setting, place the compass point at E and draw an arc that intersects the one drawn in the previous step. Label the intersection F.
- Using a straightedge, draw $\overrightarrow{BF}$. $\overrightarrow{BF}$ bisects ∠ABC.

Draw the angle or line segment with the given measurement. Then use a straightedge and a compass to bisect each angle or line segment. See students' work.
1. 72° 2. 50 mm 3. 152°
4. $1\frac{1}{2}$ in. 5. 49° 6. 6 cm

■ Reteaching the Lesson ■

Activity Draw a line segment *ST* on an overhead transparency. As you bisect it, discuss the procedure. Then ask: *Why is it necessary to draw arcs above and below the line segment?* The distance from each end of the segment must be equal to locate the midpoint.

Error Analysis
Watch for students who have difficulty remembering how to bisect a line segment or an angle.
Prevent by having those students work with peer tutors until they remember the methods.

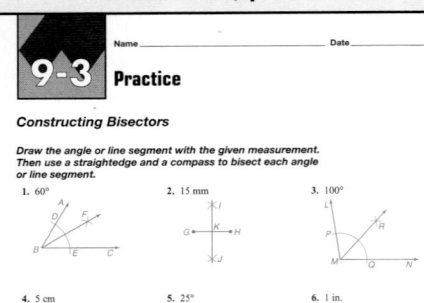
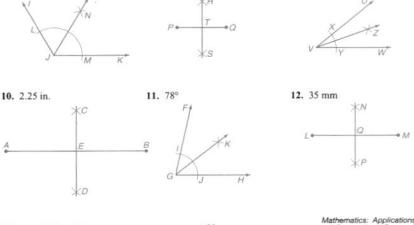

CHECK FOR UNDERSTANDING

Communicating Mathematics

Read and study the lesson to answer each question. 1–3. See Answer Appendix.

1. *Explain* the meaning of the word *bisect*.

2. *Tell* how to bisect a line segment using a straightedge and a compass.

HANDS-ON MATH

3. *Draw* an angle on a sheet of paper. Then fold the paper through the vertex of the angle so that the sides of the angle match. Make a crease and unfold the paper. What can you say about the crease?

Guided Practice

4–7. See Answer Appendix.

Draw the angle or line segment with the given measurement. Then use a straightedge and a compass to bisect each angle or line segment.

4. 4 in.   5. 65°   6. 7 cm

7. *Geometry* Draw a 119° angle. Then bisect the angle. What is the measure of each angle formed?

EXERCISES

Practice

8–15. See Answer Appendix.

Draw the angle or line segment with the given measurement. Then use a straightedge and a compass to bisect each angle or line segment.

8. 120°   9. 75°   10. 5 cm   11. 30°

12. 25 mm   13. 9 cm   14. 108°   15. 3 in.

16. In the figure shown, name the side that appears to bisect ∠*MNP*.   **side** $\overrightarrow{NB}$

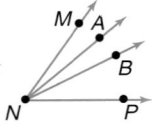

Applications and Problem Solving

17–19. See Answer Appendix.

17. *Geometry* The line segments connecting opposite corners of the square are called *diagonals*. Copy the figure. Use a protractor to measure each angle formed by one side of the square and a diagonal. What can you say about the diagonals of a square?

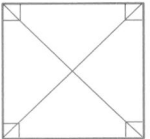

18. *Design* Draw a circle that is 6 inches in diameter. Draw a diameter. Construct a second diameter that bisects the first one. Construct two more diameters that bisect the angles of the intersection.

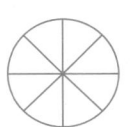

19. *Critical Thinking* Draw a 7-inch line segment. Then use a straightedge and a compass to separate the segment into four congruent parts. How can you tell that the four parts are congruent?

Mixed Review

20. *Geometry* Estimate the measure of the angle. *(Lesson 9-2)*

20. Sample answer: about 30°

21. Express 0.74 as a fraction in simplest form. *(Lesson 5-9)* $\frac{37}{50}$

22. **Test Practice**   Tom, Nina, and Sue bought sandwiches for $11.55. How much did each person pay if they shared the price equally? *(Lesson 4-5)* **D**

    A $3.25   **B** $3.45   **C** $3.65   **D** $3.85   **E** $3.55

Extending the Lesson

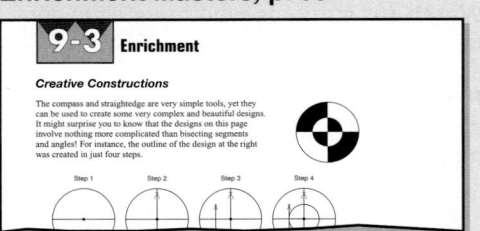
Activity Have small groups of students use a protractor to draw a triangle whose three angles are all 60°. Then have them bisect each angle using their compasses. Ask students to discuss what they have observed.

Architecture

Maya Ying Lin
Architect

When Maya Ying Lin designed the *Vietnam Veterans Memorial* in 1980, she became the first woman to design a major Washington, D.C., monument. Since then she has designed other memorials including the *Civil Rights Memorial* in Montgomery, Alabama.

To be an architect, you'll need at least a bachelor's degree in architecture and three years of practical experience. In addition to organizational skills and computer skills, math skills are needed to determine the dimensions of a design and to draw blue prints. Architects design shopping centers, schools, homes, museums, office buildings, and sports arenas, to name a few.

For more information:
American Institute of Architects
1735 New York Avenue, NW
Washington, DC 20006

*inter*NET
CONNECTION
www.glencoe.com/sec/
math/mac/mathnet

Someday, I'd like to design memorials and monuments like Maya Lin.

Your Turn
Research architecture as a career. Then write a report that describes the benefits of a career in architecture.

School to Career: Architecture **367**

More About Maya Lin
- Maya Lin was a 21-year-old student at Yale University when she won the architectural competition to design the *Vietnam Veterans Memorial.*
- Originally from Athens, Ohio, Maya Lin now works as an architect in New York City.

SCHOOL to CAREER

Motivating Students
When visitors arrive in a city, the first thing they see is the city's skyline—its tallest buildings. Ideally, these structures are built artistically to please the eye; however, they must also be practical. Architects must consider the safety, comfort, and efficiency of each structure they design. To start a discussion about this profession, ask students questions about architectural designs.
- What three shapes are most often used for buildings? **squares, rectangles, and spheres**
- How can an architect design a building to make use of solar energy? **Include many large windows or walls made of glass.**

Making the Math Connection
As students learned in the Chapter Project on page 351, designing a house requires a floor plan. An architect must draw the building plans to scale, which means the size of each part of the drawing must be directly proportional to the size of that part of the actual building.

Working on *Your Turn*
Remind students that most architects work for an architectural firm. Many specialize in designing certain types of buildings, such as commercial offices or houses.

*An additional School to Career activity is available on page 9 of the **School to Career Masters.***

GET READY

Objective Students classify triangles and quadrilaterals.

Optional Resources
Hands-On Lab Masters
- square dot paper p. 13
- worksheet, p. 58

Overhead Manipulative Resources
- geoboards
- geobands
- rectangular dot paper

Manipulative Kit
- geoboards
- geobands
- scissors

MANAGEMENT TIPS

Recommended Time
45 minutes

Getting Started Provide small groups of students with magazines and catalogues. Ask them to record items that have surfaces or designs that are triangles or quadrilaterals. Ask them what they have observed about how the shapes are used. Make sure students can differentiate between triangles and quadrilaterals.

In **Activity 1** students make and sort acute, right, and obtuse triangles. Discuss with students why a triangle cannot have more than one obtuse angle.

In **Activity 2** on page 369, students make and sort quadrilaterals using student-chosen criteria. Discuss with students what criteria they could use for their sorting.

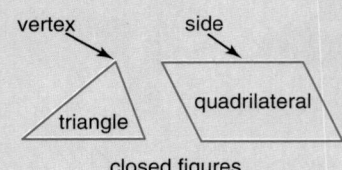

HANDS-ON
LAB

COOPERATIVE LEARNING

9-4A Triangles and Quadrilaterals

A Preview of Lesson 9-4

- geoboard
- geobands
- dot paper
- scissors

Polygons are simple closed figures formed by three or more line segments. The line segments, or sides, intersect at their endpoints. These points of intersection are called *vertices* (plural of vertex).

vertex → side → triangle quadrilateral

closed figures

In this lab, you will classify *triangles* and *quadrilaterals*. A triangle is a polygon with three sides. A quadrilateral is a polygon with four sides.

TRY THIS

Work with a partner.

1 Use geobands to make a triangle. A sample triangle is shown.
- Draw the triangle on dot paper and cut it out.
- Repeat these steps until you have ten different triangles.
- Every triangle has at least two acute angles. The triangle shown has two acute angles. Since the third angle is obtuse, the triangle is an obtuse triangle.
- Sort your triangles into three groups, based on the third angle. Name the groups acute, right, and obtuse triangles.

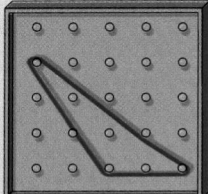

ON YOUR OWN

Classify each triangle as *acute*, *right*, or *obtuse*.

1.
acute

2.
obtuse

3.
right

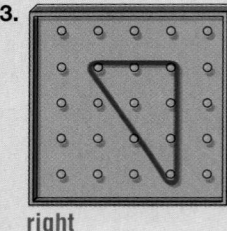

4. *Look Ahead* Write a definition for each type of triangle. **See margin.**

368 Chapter 9 Geometry: Investigating Patterns

Additional Answer

4. An acute triangle is a triangle that has 3 acute angles. A right triangle is a triangle that has one right angle. An obtuse triangle is a triangle that has one obtuse angle.

Triangles can be classified by the type of angles they have while quadrilaterals are classified by the characteristics of their sides and angles.

Have students complete Exercises 1–3 and 5–7. Make sure students correctly differentiate between the shapes.

Use Exercises 4 and 8 to determine whether students can recognize types of triangles and quadrilaterals.

Additional Answers

5. Sample answer: The quadrilateral has two sides that are congruent.

6. Sample answer: The opposite sides of the quadrilateral are congruent.

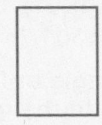

7. Sample answer: The opposite sides of the quadrilateral are parallel.

TRY THIS

Work with a partner.

➋ Use geobands to make a quadrilateral. Several samples are shown.

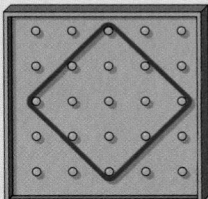 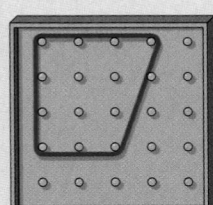 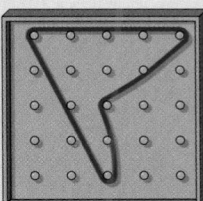

- Draw the quadrilateral on dot paper and cut it out.
- Repeat these steps until you have ten different quadrilaterals.
- You can classify quadrilaterals according to the characteristics of their sides and angles. The first quadrilateral shown can be classified as a quadrilateral with four congruent sides and four right angles.
- Sort your quadrilaterals into three groups, based on any characteristic you choose. Write a description of the quadrilaterals in each group.

ON YOUR OWN

Write a sentence that describes one characteristic of each quadrilateral. Then draw a different quadrilateral that has the same characteristic. 5–8. See margin.

5.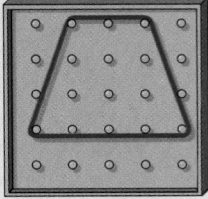
6.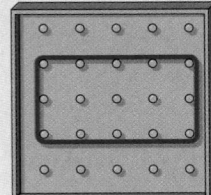
7.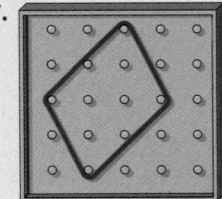

8. **Look Ahead** Draw a quadrilateral that has two obtuse angles and exactly one pair of opposite sides congruent.

Lesson 9-4A HANDS-ON **LAB** 369

Hands-On Lab Masters, p. 58

Math Journal Have students write a paragraph about the different types of triangles and quadrilaterals. Have them include illustrations and examples of everyday objects that fit the categories.

Additional Answer
8. Sample answer:

Hands-On Lab 9-4A **369**

Instructional Resources

- *Study Guide Masters*, p. 67
- *Practice Masters*, p. 67
- *Enrichment Masters*, p. 67
- Transparencies 9-4, A and B
- *Classroom Games*, pp. 23–26
- *Diversity Masters*, p. 9

 CD-ROM Program
- Resource Lesson 9-4

Recommended Pacing	
Standard	Days 7–9 of 14
Honors	Days 6 & 7 of 12
Block	Day 4 of 7

1 FOCUS

 5-Minute Check
(Lesson 9-3)

Draw the angle or line segment with the given measurement. Then use a straightedge and compass to bisect each angle or line segment. See students' work.
1. 35 mm   2. 5 in.
3. 74°   4. 128°

 The 5-Minute Check is also available on **Transparency 9-4A** for this lesson.

Motivating the Lesson

Hands-On Activity Have students read the introduction to the lesson and the discussion of polygons. Then have students choose items in the room and determine what types of polygons they represent.

2 TEACH

 Transparency 9-4B contains a teaching aid for this lesson.

Reading Mathematics
The word *polygon* and each of the names of polygons have their meanings hidden in their root words from Latin and Greek. *Poly-* means many, and *-gon* means sides. So a polygon has many sides. Have students use a dictionary to look up the root meaning of each of the polygon names.

9-4

What you'll learn
You'll learn to name two-dimensional figures.

When am I ever going to use this?
Knowing how to name two-dimensional figures can help you identify the shapes of traffic signs.

Word Wise
polygon
triangle
quadrilateral
pentagon
hexagon
octagon
decagon
regular polygon
parallel
parallelogram
equilateral triangle

Two-Dimensional Figures

Powwows are American Indian celebrations of heritage. Dancing, games, and parades are all part of the festivities. The colorful headdress shown is an example of what an American Indian may wear to a powwow. The figures on the headdress are examples of polygons.

A **polygon** is a simple closed figure formed by three or more sides. The number of sides determines the name of the polygon.

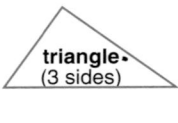

triangle (3 sides)   quadrilateral (4 sides)   pentagon (5 sides)

hexagon (6 sides)   octagon (8 sides)   decagon (10 sides)

Any polygon with all sides congruent and all angles congruent is called a **regular polygon**. Some examples of common regular polygons are shown below.

Examples

Study Hint
Reading Math The red slash marks show congruent sides. The red arcs show congruent angles.

Name each polygon. Then tell if the polygon is a regular polygon.

❶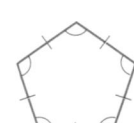

❷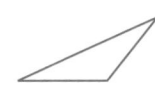

❸

❶ The polygon is a pentagon. The sides and angles are congruent so it is a regular polygon.

❷ The polygon is a triangle. The sides and angles are not congruent so it is not a regular polygon.

❸ The polygon is a quadrilateral. The sides are not all congruent so it is not a regular polygon.

Certain types of quadrilaterals have special characteristics.

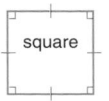
square

- All sides are congruent.
- All angles are right angles.
- Opposite sides are **parallel**. If you extend the lengths of the sides, the opposite sides will never meet.

parallelogram

- Opposite sides are congruent.
- Opposite sides are parallel.

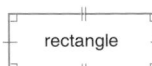
rectangle

- Opposite sides are congruent.
- All angles are right angles.
- Opposite sides are parallel.

Example 4

Explain how a rectangle and a parallelogram are alike and how they are different.

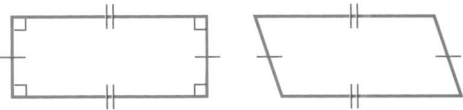

A rectangle and a parallelogram are alike because they
- have four sides.
- have opposite sides parallel.
- have opposite sides congruent.

They are different because a rectangle has four right angles and a parallelogram does not necessarily have four right angles.

A triangle with three congruent sides is called an **equilateral triangle**.

Example 5

APPLICATION

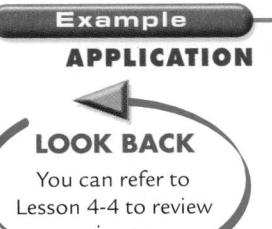

LOOK BACK

You can refer to Lesson 4-4 to review perimeter.

Public Safety The traffic sign shown is a yield sign.
a. What shape is a yield sign?
b. What is the perimeter of the sign?

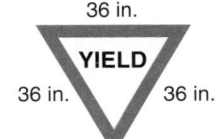

36 in.

YIELD

36 in.   36 in.

a. A yield sign is shaped like an equilateral triangle.
b. The perimeter is the sum of the lengths of the sides.

$$36 + 36 + 36 = 108$$

So, the perimeter of the yield sign is 108 inches.

Lesson 9-4 Two-Dimensional Figures   **371**

In-Class Examples

Name each polygon. Then tell if the polygon is a regular polygon.

For Example 1

hexagon, regular

For Example 2

pentagon, not regular

For Example 3

pentagon, regular

For Example 4
Explain how the square and the quadrilateral are alike and how they are different.

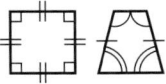

Alike: 4 sides, 4 angles
Different: square has 4 equal sides and angles, quadrilateral does not; both sets of opposite sides of square are parallel, one set of opposite sides of quadrilateral is parallel.

For Example 5

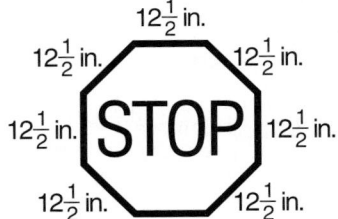

$12\frac{1}{2}$ in.

$12\frac{1}{2}$ in. $12\frac{1}{2}$ in.

$12\frac{1}{2}$ in. $12\frac{1}{2}$ in.

STOP

$12\frac{1}{2}$ in. $12\frac{1}{2}$ in.

The traffic sign shown is a stop sign.
a. What shape is a stop sign? **octagon**
b. What is the perimeter of the sign? **100 in.**

Teaching Tip You may want to have students investigate the names of seven-, nine-, or twelve-sided figures. **heptagons, nonagons, and dodecagons**

Check for Understanding

If students need additional practice or instruction after completing Exercises 1–10, one of these options may be helpful.
- Extra Practice, see p. 582
- Reteaching Activity
- *Study Guide Masters*, p. 67
- *Practice Masters*, p. 67

Assignment Guide

Core: 11–25 odd, 28–32
Enriched: 12–24 even, 25, 26, 28–32
All: Self Test, 1–10

Additional Answers

3. Jonathan: all squares are parallelograms since both sets of opposite sides of a square are parallel. Victoria is incorrect. Some parallelograms are not squares since some parallelograms do not have right angles.

6. Alike: quadrilaterals, opposite sides parallel, four right angles, opposite sides congruent
Different: a square must have four congruent sides; a rectangle may not have four congruent sides.

Study Guide Masters, p. 67

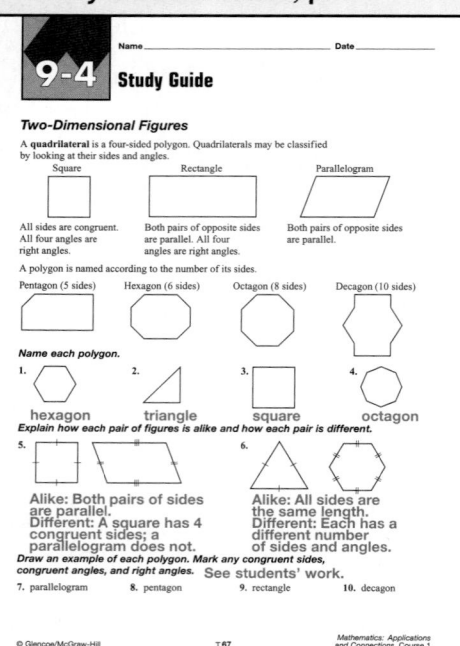

CHECK FOR UNDERSTANDING

Communicating Mathematics

1–2. See Answer Appendix.

Read and study the lesson to answer each question.

1. *Identify* two different polygons that you see in your classroom.

2. *Draw* a triangle that has exactly one pair of congruent sides.

3. *You Decide* Victoria says that all parallelograms are squares. Jonathan says that all squares are parallelograms. Who is correct? Explain your reasoning. **See margin.**

Guided Practice

Name each polygon. Then tell if the polygon is a regular polygon.

4. hexagon; no
5. equilateral triangle; yes

Explain how each pair of figures is alike and how each pair is different.

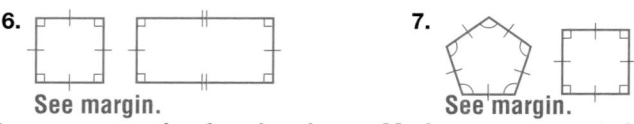

6. See margin.
7. See margin.

Draw an example of each polygon. Mark any congruent sides, congruent angles, and right angles. 8–9. See Answer Appendix.

8. parallelogram
9. regular hexagon

10. *Design* Winona needs to design a table that will seat six people with equal comfort and work space. What shape should she make the table? **hexagon**

EXERCISES

Practice

11. pentagon; yes
12. hexagon; no
13. square; yes
14. quadrilateral; no

15–17. See Answer Appendix.

25. hexagon, pentagon

Applications and Problem Solving

Name each polygon. Then tell if the polygon is a regular polygon.

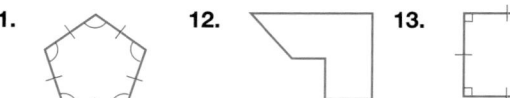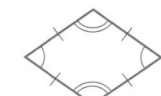

11.   12.   13.   14.

Explain how each pair of figures is alike and how each pair is different.

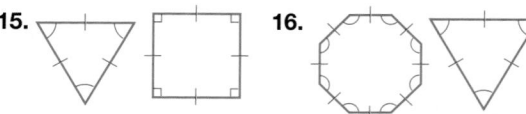

15.   16.   17.

Draw an example of each polygon. Mark any congruent sides, congruent angles, and right angles. 18–23. See Answer Appendix.

18. quadrilateral
19. pentagon
20. equilateral triangle
21. regular octagon
22. triangle
23. rectangle

24. How many sides does a regular decagon have? **10**

25. *Sports* The most popular game in the world is soccer. A soccer ball is shown. Which two polygons make up the pattern on a soccer ball?

Study Guide Masters, p. 67

9-4 **Study Guide**

Name _____ Date _____

Two-Dimensional Figures

A **quadrilateral** is a four-sided polygon. Quadrilaterals may be classified by looking at their sides and angles.

Square | Rectangle | Parallelogram

All sides are congruent. All four angles are right angles. | Both pairs of opposite sides are parallel. All four angles are right angles. | Both pairs of opposite sides are parallel.

A polygon is named according to the number of its sides.

Pentagon (5 sides)   Hexagon (6 sides)   Octagon (8 sides)   Decagon (10 sides)

Name each polygon.

1. hexagon   2. triangle   3. square   4. octagon

Explain how each pair of figures is alike and how each pair is different.

5. Alike: Both pairs of sides are parallel. Different: A square has 4 congruent sides; a parallelogram does not.

6. Alike: All sides are the same length. Different: Each has a different number of sides and angles.

Draw an example of each polygon. Mark any congruent sides, congruent angles, and right angles. See students' work.

7. parallelogram   8. pentagon   9. rectangle   10. decagon

© Glencoe/McGraw-Hill   T 67   *Mathematics: Applications and Connections, Course 1*

■ Reteaching the Lesson ■

Activity Have students make a chart listing all of the polygons named in this lesson. For each polygon, they should identify the number of angles and tell whether the sides are parallel or congruent.

Additional Answer

7. Alike: all sides congruent, all angles congruent
Different: pentagon has five sides and square has four sides.

26. **Carpentry** A clubhouse is shaped like a regular pentagon. If the perimeter of the floor is 30 feet long, how long is each wall? **6 feet**

27a. quadrilaterals, pentagons, hexagons, octagons

27b. 360°; 540°; 360°; 1,080°; See students' explanations.

27. **Working on the** CHAPTER Project **Refer to page 351.**
   a. Name the different types of polygons in the floor plan.
   b. Find the sum of the measures of the angles of the bathroom, the dining room, and the second bedroom. What is the sum of the measures of the angles of an octagon? Explain.

28. **Critical Thinking** A point is in the *interior* of a polygon if it lies on the inside of the polygon and does not lie on the polygon itself. A point is in the *exterior* if it lies outside the polygon and does not lie on the polygon itself. Tell whether each point is in the interior of the polygon, the exterior of the polygon, or neither.

a.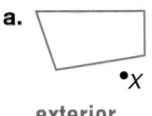

exterior

b.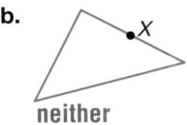

neither

c.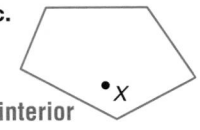

interior

Mixed Review

29. **Geometry** Trace $\overline{CD}$. Use a compass and straightedge to bisect it. *(Lesson 9-3)* **See Answer Appendix.**

30. Find 84% of 24. *(Lesson 8-7)* **20.16**

31. **Test Practice** Dr. Rodriguez drove 384.2 miles on 17 gallons of fuel. How many miles per gallon did his car get? *(Lesson 8-1)* **B**
   A 22.5 mpg   B 22.6 mpg   C 126 mpg   D 226 mpg   E Not Here

32. Express 2.375 as a fraction. *(Lesson 5-9)* $2\frac{3}{8}$

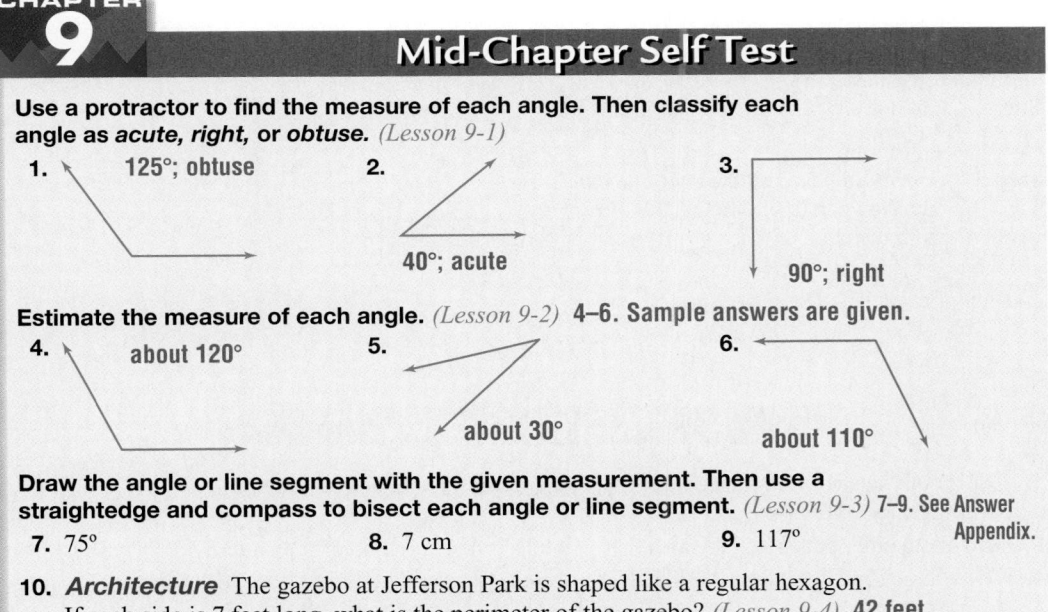

CHAPTER 9

Mid-Chapter Self Test

Use a protractor to find the measure of each angle. Then classify each angle as *acute, right,* or *obtuse.* *(Lesson 9-1)*

1. 125°; obtuse

2. 40°; acute

3. 90°; right

Estimate the measure of each angle. *(Lesson 9-2)* **4–6. Sample answers are given.**

4. about 120°

5. about 30°

6. about 110°

Draw the angle or line segment with the given measurement. Then use a straightedge and compass to bisect each angle or line segment. *(Lesson 9-3)* **7–9. See Answer Appendix.**

7. 75°

8. 7 cm

9. 117°

10. **Architecture** The gazebo at Jefferson Park is shaped like a regular hexagon. If each side is 7 feet long, what is the perimeter of the gazebo? *(Lesson 9-4)* **42 feet**

Extending the Lesson

Enrichment Masters, p. 67

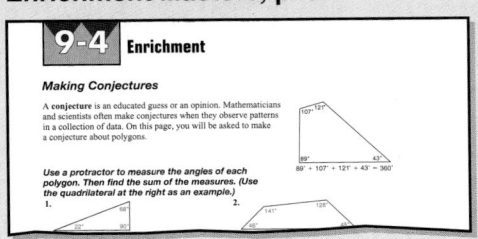

Activity Have students design a creature with squares, parallelograms, pentagons, hexagons, and equilateral triangles so that at least one side of each polygon touches another polygon. Use cardboard models of the polygons to construct the creatures.

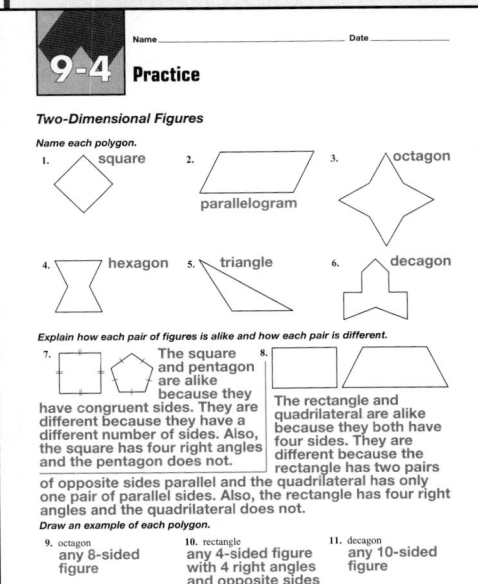

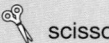
GET READY

Objective Students make a net and use it to wrap a cube.

Optional Resources
Hands-On Lab Masters
• cube pattern, p. 24
• worksheet, p. 59

Manipulative Kit
• centimeter cube
• scissors

MANAGEMENT TIPS

Recommended Time
30 minutes

Getting Started Have students read the opening paragraph of the lesson and then ask the following questions. *Have you ever wrapped a box? How did you do it? How do you think a net is different from the material you used for wrapping?*

The **Activity** has students make a net by tracing the faces of a cube. Ask students if it matters which face they trace. **No; all six sides are congruent.** Make sure that students' nets have six squares.

ASSESS

Have students complete Exercises 1–3. Make sure students check their nets for coverability. Save any nets that don't work to discuss with the class.

Use Exercise 4 to determine whether students have discovered what is required to make a net.

 Have students use their findings to write a conclusion about the requirements of a net for wrapping a box. Encourage them to include diagrams of various nets.

HANDS-ON LAB

COOPERATIVE LEARNING

9-4B Using Nets to Wrap a Cube

A Follow-Up of Lesson 9-4

scissors

cube

In this lab, you will make a figure called a *net* and use it to wrap a cube. The faces of the cube are shaped like squares.

TRY THIS

Work with a partner.

Step 1 Count the number of faces of the cube.

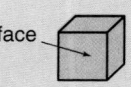
face

Step 2 Place the cube on the paper as shown and draw a square.

Step 3 Continue drawing squares to make a figure like the one shown. This figure is called a net.

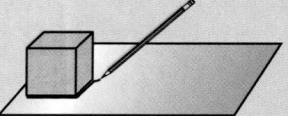

Step 4 Cut out the net and try to wrap the cube.

Step 5 Repeat Step 3 to make a net like the one shown. Cut out the net and try to wrap the cube.

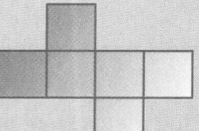

ON YOUR OWN

1. *Tell* if both nets covered the cube. If not, explain why the net or nets did not cover the cube. **See margin.**
2. *Find* three other nets of six squares that will cover the cube. Look for a pattern. **See Answer Appendix.**
3. *Write* three sentences describing the net patterns you found. **See Answer Appendix.**
4. *Reflect Back* Give an example of a net that can be used to cover a box whose faces are shaped like rectangles. **See margin.**

374 Chapter 9 Geometry: Investigating Patterns

Additional Answers

1. **The second net did not cover the cube. The "top" and "bottom" covers are on the same side of the cube.**

4. **Sample answer:**

9-5 Lines of Symmetry

What you'll learn
You'll learn to describe and define lines of symmetry.

When am I ever going to use this?
Knowing how to describe lines of symmetry can help you design and build a kite.

Word Wise
line of symmetry
reflection

More than 2,000 years ago, the Chinese military used kites for military purposes. They attached bamboo pipes to kites and flew them over the enemy at night. As wind passed through the pipes, the whistling noise caused the enemy to flee.

If you draw a line down the center of a diamond-shaped kite, the two halves match. When this happens, the line is called a **line of symmetry**. Lines of symmetry can be found in certain figures.

Examples

Draw all lines of symmetry for each figure.

❶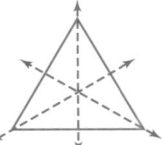

An equilateral triangle has 3 lines of symmetry.

❷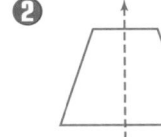

This quadrilateral has 1 line of symmetry.

❸

The letter F has no lines of symmetry.

CONNECTION ❹ **Geography** The flag of Jamaica is shown. How many lines of symmetry does the flag have?

Explore You know the design of the Jamaican flag. You need to find all of the lines of symmetry.

Plan To find the lines of symmetry, you can draw them.

Solve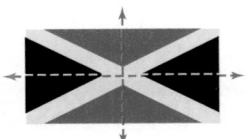

There are 2 lines of symmetry.

Examine Imagine folding the flag in half vertically, horizontally, and diagonally. Since the sides and patterns match exactly when it is folded vertically and horizontally, it has 2 lines of symmetry.

Multiple Learning Styles

Interpersonal Separate students into small cooperative groups. Each group should use geoboards and geobands to design pictures or figures that have symmetry. When all of the designs are complete, have groups present their designs to each other.

Motivating the Lesson

Communication Display a diamond-shaped kite that has a non-symmetric pattern on it. Ask students how many lines of symmetry the kite has. Point out that the figure itself has one line of symmetry. However, by looking at the figure with the design, the two halves do not match.

9-5 Lesson Notes

Instructional Resources
- *Study Guide Masters*, p. 68
- *Practice Masters*, p. 68
- *Enrichment Masters*, p. 68
- Transparencies 9-5, A and B
- *Assessment and Evaluation Masters*, p. 240
- *Hands-On Lab Masters*, p. 77
- CD-ROM Program
  - Resource Lesson 9-5
  - Interactive Lesson 9-5

Recommended Pacing

Standard	Day 10 of 14
Honors	Day 8 of 12
Block	Day 5 of 7

1 FOCUS

5-Minute Check
(Lesson 9-4)

Name each polygon. Then tell if the polygon is a regular polygon.

1. 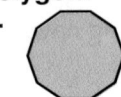 2.

decagon, yes; triangle, yes

3. In what ways are the triangle and the pentagon alike? In what ways are they different?

Alike: polygons with congruent sides
Different: number of sides

Draw an example of each polygon. Mark any congruent sides, congruent angles, and right angles. See students' work.

4. square
5. octagon

 The 5-Minute Check is also available on **Transparency 9-5A** for this lesson.

Transparency 9-5B contains a teaching aid for this lesson.

In-Class Examples

Draw all lines of symmetry for each figure.

For Example 1

For Example 2

For Example 3

For Example 4
A Texas flag (1510–1685) is shown. How many lines of symmetry does the flag have? Explain.

None; the design would not match exactly when folded horizontally, vertically, or diagonally.

Tell whether the figure shows a reflection. Write *yes* or *no*.

For Example 5

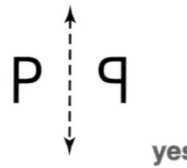

yes

For Example 6

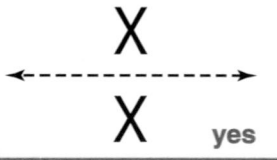

yes

The capital letter M has one line of symmetry. Notice that the right half is a **reflection** of the left half. A reflection is a mirror image of a figure across a line of symmetry. You can use a geomirror to draw reflections.

MINI-LAB

Work with a partner. ▱ geomirror 🧩 tracing paper

Try This

1a–c. See students' work.

1. Trace each figure shown. Then place the geomirror on the dashed line of each figure and draw its reflection.

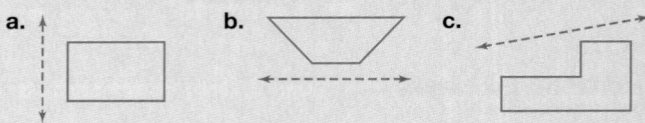

Talk About It 2. mirror image

2. How did the reflection look on the other side of the geomirror?

3. What would you call the dashed line with respect to the figure and its reflection? **line of reflection**

Examples

Tell whether the figure shows a reflection.

⑤

Since the figure on the right is a mirror image of the figure on the left, it is a reflection.

⑥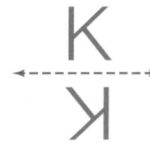

The figure on the bottom is not a mirror image of the figure on the top. So, it is not a reflection.

CHECK FOR UNDERSTANDING

Communicating Mathematics

Read and study the lesson to answer each question.

1. *Draw* a figure that has exactly four lines of symmetry. **See students' work.**

2. *Tell* which figure does not have at least one line of symmetry. **C**

Classroom Vignette

"I have my students construct an art project using line symmetry. I have had them use both the computer and manual drawing techniques. This helps them to see and appreciate mathematics in artwork and other designs."

Deanna Gallagher, Teacher
Nathan Hale Middle School
Northvale, NJ

Deanna Gallagher

3. *Draw* and cut out a rectangle and a square. How many ways can you fold each figure so that one half matches the other half? How many lines of symmetry does each figure have? **See students' work.**

Guided Practice

Tell whether the dashed line is a line of symmetry. Write *yes* or *no*.

4.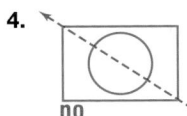
no

5.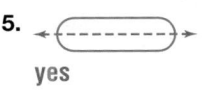
yes

6.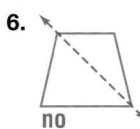
no

Trace each figure. Draw all lines of symmetry.

7.

8.
none

9.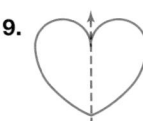

Tell whether the figure shows a reflection. Write *yes* or *no*.

10.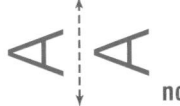
no

11.
yes

12. *Life Science* How many lines of symmetry does the starfish shown have?
5

EXERCISES

Practice

Tell whether the dashed line is a line of symmetry. Write *yes* or *no*.

13.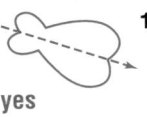
yes

14.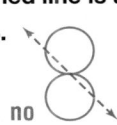
no

15.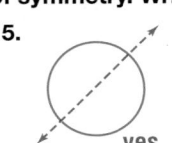
yes

16.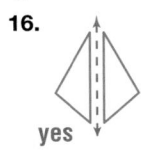
yes

17.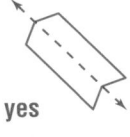
yes

18.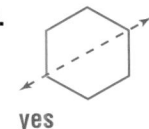
yes

19.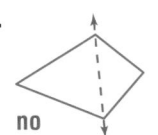
no

20.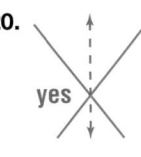
yes

Trace each figure. Draw all lines of symmetry.

21.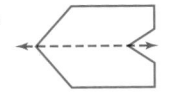

22.
none

23.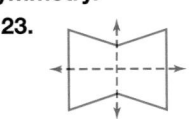

24.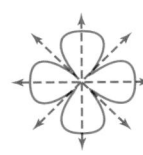

Lesson 9-5 Lines of Symmetry **377**

Using the Mini-Lab If a geomirror or a similar reflector is not available, any reflective surface may be used, for example, mirrored glass tiles or pancake turners.

3 PRACTICE/APPLY

Check for Understanding
If students need additional practice or instruction after completing Exercises 1–12, one of these options may be helpful.
• Extra Practice, see p. 582
• Reteaching Activity
• *Study Guide Masters,* p. 68
• *Practice Masters,* p. 68

Assignment Guide

Core: 13–37 odd, 38–43
Enriched: 14–34 even, 36–43

Study Guide Masters, p. 68

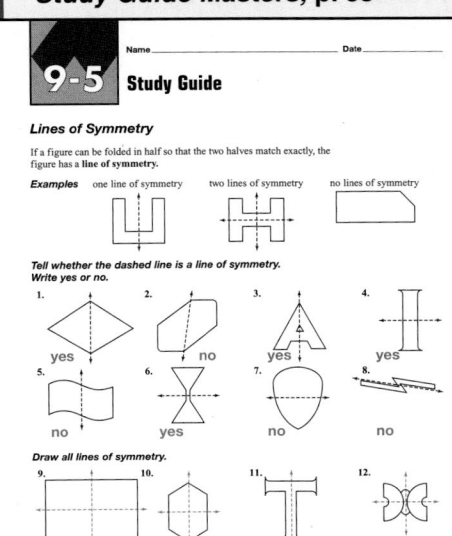

Reteaching the Lesson

Activity Have students play a game like "Concentration." Use sets of cards with one-half of a symmetrical shape on each card. Have players place all of the cards in front of them. Then have them take turns trying to match the halves of each shape.

Error Analysis
Watch for students who draw lines of symmetry between all of the vertices in a multi-sided figure.
Prevent by reminding students that a symmetrical shape may have only one line of symmetry.

Closing Activity

Modeling Have students work in pairs to draw an object that has three lines of symmetry. Have them exchange their drawings and draw the lines of symmetry.

Chapter 9, Quiz C (Lessons 9-4 and 9-5) is available in the *Assessment and Evaluation Masters,* p. 240.

Trace each figure. Draw all lines of symmetry.

25.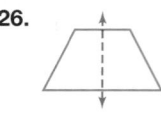
26.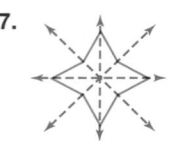
27.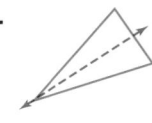
28.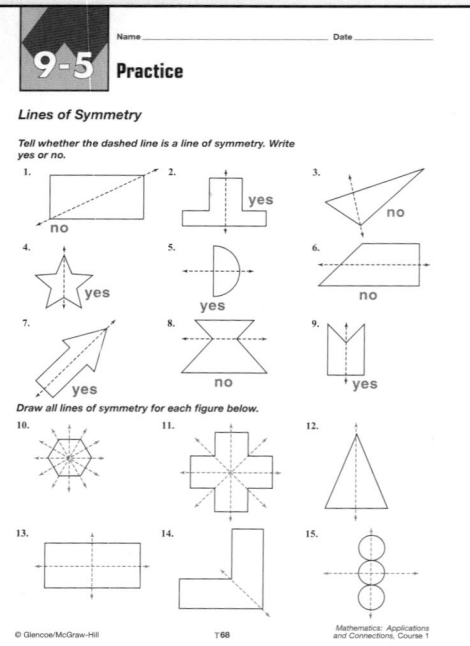

Tell whether the figure shows a reflection. Write *yes* or *no*.

29. 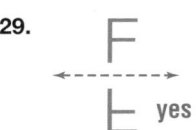 yes
30.  no
31. 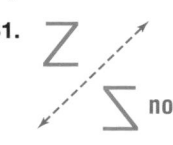 no

32. 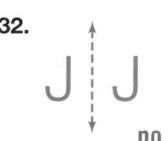 no
33. 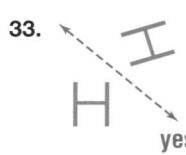 yes
34. 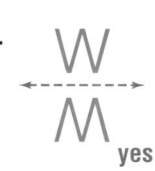 yes

35. How many lines of symmetry does a stop sign have? **8**

Applications and Problem Solving

36. ***Geography*** Switzerland is located in the Alps mountains in central Europe. How many lines of symmetry does the flag of Switzerland have? **4**

37. ***Life Science*** There are more than 20,000 different types of bees. The mining bee is known for making its nest in loose ground. How many lines of symmetry does the mining bee shown at the left have? **1**

38. ***Critical Thinking*** Copy the figure shown. Then shade enough squares so that the **Sample** figure has a diagonal line of symmetry. **answer:**

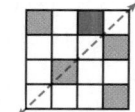

Mixed Review

39. ***Geometry*** Draw an example of an octagon. *(Lesson 9-4)*

39. See students' work.

40. | **Test Practice** | Three-fifths of the sixth-grade students went on the class trip. What percent of the sixth-grade students did *not* go on the class trip? *(Lesson 8-4)* **A**

    A 40%   **B** 45%   **C** 50%   **D** 60%

41. ***Algebra*** If $x = 1\frac{2}{3}$, and $y = 3\frac{4}{5}$, what is the value of xy? *(Lesson 7-3)* $6\frac{1}{3}$

42. Express 0.84 as a fraction in simplest form. *(Lesson 5-9)* $\frac{21}{25}$

43. ***Entertainment*** The graph shows the number of visitors who toured the Rock and Roll Hall of Fame in Cleveland on a Saturday. How many visitors altogether toured the Rock and Roll Hall of Fame before 1:00 P.M.? *(Lesson 2-3)* **525 people**

Rock and Roll Hall of Fame

Number of Visitors (50–450) vs. Time (10:00, 11:00, 12:00, 1:00, 2:00, 3:00)

Practice Masters, p. 68

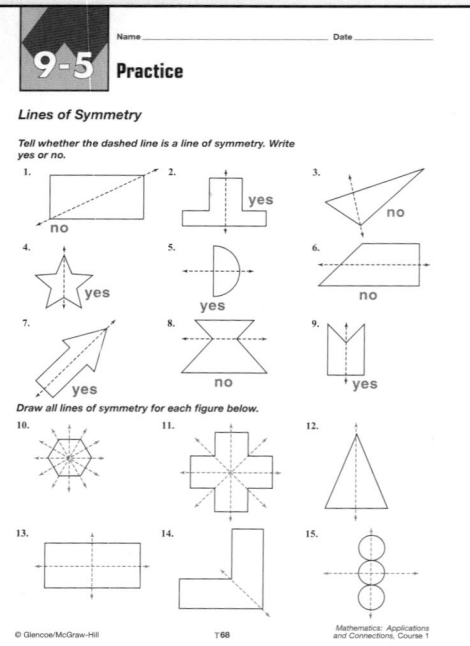

9-5 Practice

Lines of Symmetry

Tell whether the dashed line is a line of symmetry. Write yes or no.

1. no 2. yes 3. no
4. yes 5. yes 6. no
7. yes 8. no 9. yes

Draw all lines of symmetry for each figure below.

10. 11. 12.
13. 14. 15.

© Glencoe/McGraw-Hill   T68   *Mathematics: Applications and Connections, Course 1*

═══ Extending the Lesson ═══

Enrichment Masters, p. 68

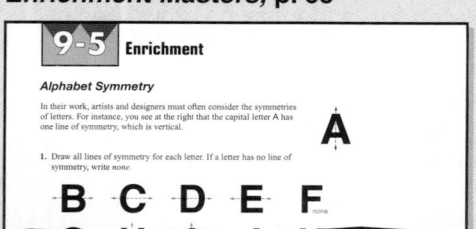

9-5 Enrichment

Alphabet Symmetry

In their work, artists and designers must often consider the symmetries of letters. For instance, you see at the right that the capital letter A has one line of symmetry, which is vertical.

A

1. Draw all lines of symmetry for each letter. If a letter has no line of symmetry, write *none*.

B C D E F none

Activity Ask small groups of students to go to the library to find pictures of works of art that have symmetry or pictures of state or country flags that show symmetry. Have groups present their findings to the class.

9-6

Size and Shape

What you'll learn

You'll learn to determine congruence and similarity.

When am I ever going to use this?

Knowing how to determine congruence and similarity helps an architect design statues.

Word Wise

similar figures
congruent figures

The New York-New York Hotel in Las Vegas, Nevada, has a 150-foot replica of the Statue of Liberty. The height of the actual Statue of Liberty is 305 feet.

Figures that have the same shape and angles, but different size are called **similar figures**. The symbol ~ means *is similar to.*

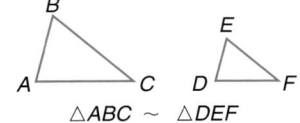

$\triangle ABC \sim \triangle DEF$

Suppose the replica was 305 feet tall. The statues would be the same size and shape.

Figures that are the same size and shape are called **congruent figures**. The symbol ≅ means *is congruent to.*

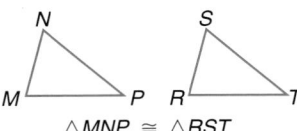

$\triangle MNP \cong \triangle RST$

In the following lab, you will explore some characteristics of congruent figures and similar figures.

> **Study Hint**
> **Reading Math** The symbol △ABC is read as *triangle ABC.*

1. yes; The figures in choice b.
2. In choice a, the angles are congruent. In choice b, the angles and sides are congruent.
3. yes; the figures in choice a

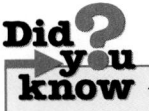
Did you know The length of the Statue of Liberty's nose is about 0.04 of her height.

Instructional Resources

- *Study Guide Masters,* p. 69
- *Practice Masters,* p. 69
- *Enrichment Masters,* p. 69
- Transparencies 9-6, A and B
- *Assessment and Evaluation Masters,* p. 240
- *Science and Math Lab Manual,* pp. 13–16
- CD-ROM Program
  - Resource Lesson 9-6
  - Extended Activity 9-6

Recommended Pacing		
Standard	Days 11 & 12 of 14	
Honors	Days 9 & 10 of 12	
Block	Day 6 of 7	

1 FOCUS

5-Minute Check
(Lesson 9-5)

Tell whether the dashed line is a line of symmetry. Write *yes* or *no.*

1. yes

2. 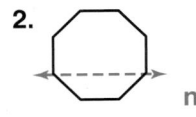 no

Copy each figure. Draw all lines of symmetry.

3.

4.

5. Does the figure show a reflection?

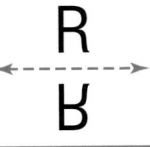 yes

HANDS-ON MINI-LAB

Work with a partner.   ✂ scissors   tracing paper

Try This

- Trace each pair of figures and cut them out.

a.
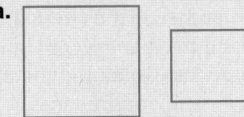

b.

- Try to match each pair of figures.

Talk About It

1. Congruent figures are exact matches. Did either pair of figures match? If so, name the congruent figures.

2. Compare the angles and segments of each pair of figures. How are they alike? How are they different?

3. Did either pair of figures not match? A pair of figures may be similar without being congruent. Name the similar figures.

Lesson 9-6 Size and Shape **379**

The 5-Minute Check is also available on **Transparency 9-6A** for this lesson.

Motivating the Lesson

Problem Solving Have students read the opening paragraphs of the lesson. Then hand out magazines and ask students to look for similar and congruent shapes. Have them cut out the shapes and identify them as similar, congruent, or neither.

Multiple Learning Styles

 Intrapersonal Have students write a paragraph about ways to estimate angle measures and determine if figures are similar or congruent. Have them include a description of reminders or shortcuts that help them.

Lesson 9-6 **379**

Transparency 9-6B contains a teaching aid for this lesson.

Using the Mini-Lab Point out that all congruent figures are similar but not all similar figures are congruent. You may use an overhead projector to model the activity.

In-Class Examples

Tell whether each triangle is congruent or similar to the triangle shown.

For Example 1

congruent

For Example 2

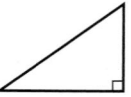

neither congruent nor similar

For Example 3

similar

For Example 4
△*ABC* is congruent to △*DEF*.

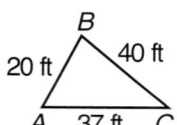

a. What side of △*DEF* corresponds to side $\overline{AC}$? $\overline{DF}$

b. What is the perimeter of △*DEF*? 97 ft

Teaching Tip Explain to students that congruent figures are like exact photocopies. Similar figures are like photocopies made using the zoom feature.

3 PRACTICE/APPLY

Check for Understanding
If students need additional practice or instruction after completing Exercises 1–7, one of these options may be helpful.
• Extra Practice, see p. 582
• Reteaching Activity
• *Study Guide Masters*, p. 69
• *Practice Masters*, p. 69

380 Chapter 9

Examples

Tell whether each quadrilateral is congruent or similar to the quadrilateral shown.

①  The quadrilaterals have the same shape and angles, but not the same size. They are similar.

② The quadrilaterals are the same size and shape. They are congruent.

③  The quadrilaterals are not the same shape or size. They are neither congruent nor similar.

INTEGRATION ④ **Geometry** △*MNP* is congruent to △*RST*.

a. What side of △*RST* corresponds to side $\overline{MN}$?

b. What is the perimeter of △*RST*?

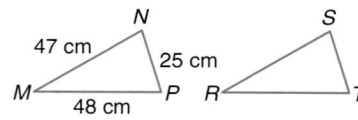

a. Side *RS* corresponds to side $\overline{MN}$.

b. The perimeter of △*MNP* is 25 + 48 + 47, or 120 centimeters. Since △*MNP* and △*RST* are congruent, they have the same size and shape. So, the perimeter of △*RST* is also 120 centimeters.

CHECK FOR UNDERSTANDING

Communicating Mathematics

Read and study the lesson to answer each question. 2–3. See students' work.

1. ***Tell*** why the two figures shown are similar. **See margin.**

2. ***Draw*** two similar triangles.

HANDS-ON MATH **3.** ***Use*** a geoboard and geobands to make two similar figures and two congruent figures. Then write a sentence explaining the difference between similarity and congruence.

Guided Practice

Tell whether each pair of polygons is *congruent*, *similar*, or *neither*.

4. 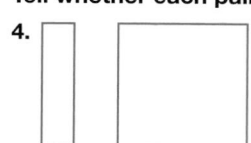 neither

5. congruent

6. ***Geometry*** Quadrilateral *ABCD* is congruent to quadrilateral *EFGH*.
 a. What is the measure of side $\overline{AD}$? **4 m**
 b. What side corresponds to side $\overline{BC}$? $\overline{FG}$

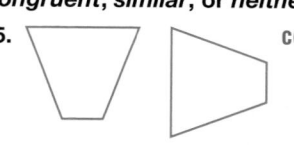

380 Chapter 9 Geometry: Investigating Patterns

■ Reteaching the Lesson ■

Activity Have students use geoboards and geobands to make similar and congruent figures. If geoboards are unavailable, dot paper may be substituted. Have students take turns making figures and asking others to make a similar or congruent figure.

Additional Answer

1. The angles are congruent, they have the same shape, and they are different sizes.

7. Geography The flag of the state of North Carolina is shown. How many pairs of congruent quadrilaterals are there? **3**

EXERCISES

Practice

Tell whether each pair of polygons is *congruent, similar,* or *neither.*

8. 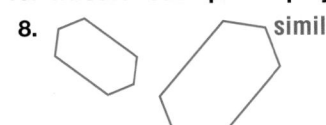 similar

9.  neither

10.  congruent

11.  neither

12. 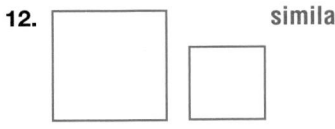 similar

13. 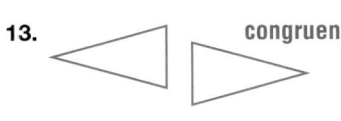 congruent

14. Which pair of polygons are neither similar nor congruent? **b.**

a.   **b.**   **c.**

15. The triangles shown are acute triangles and △*DEF* ≅ △*LMN*. **a. 2.5 m** **b.** $\overline{NM}$

a. What is the measure of side $\overline{DF}$?

b. What side corresponds to side $\overline{FE}$?

16. Quadrilateral *MNOP* is congruent to quadrilateral *WXYZ*.

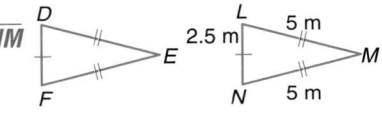

a. Find the perimeter of quadrilateral *WXYZ*. **18 in.**

b. What side corresponds to side $\overline{MP}$? $\overline{WZ}$

Applications and Problem Solving

17. Puzzles A tangram is a Chinese puzzle consisting of seven geometric shapes. The diagram shows how the pieces of the puzzle can form a square. Which of the geometric shapes in a tangram are similar? **See margin.**

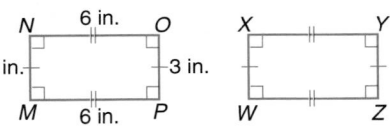

Lesson 9-6 Size and Shape **381**

Classroom Vignette

"To help my students understand congruence, I explain that the congruence symbol (≅) is made up of a similar sign (~) and an equals sign (=). *Equals* means the same size and *similar* means the same shape. Congruent figures are the same size and shape. This makes sense to students."

Richie Berman, Ph.D., Consultant
University of California, Santa Barbara
Santa Barbara, CA

Additional Answer

17. A and D; A and F; A and G; B and D; B and F; B and G; F and D; A and B; D and G; F and G

Study Guide Masters, p. 69

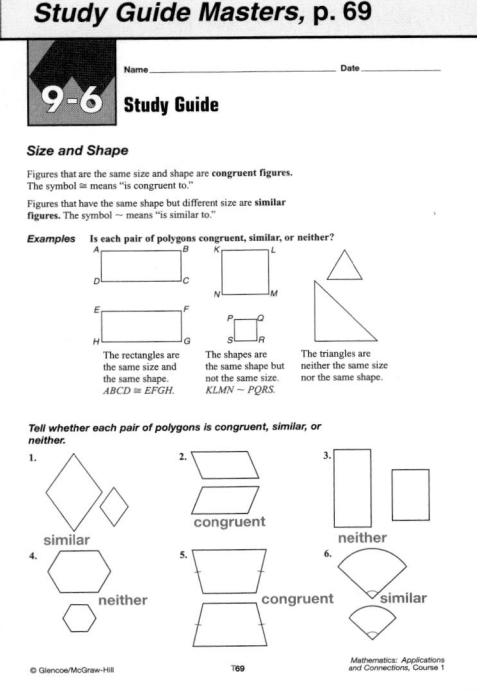

Exercise 19 asks students to advance to the next stage of work on the Chapter Project. You may wish to have students make a scale drawing of the floor plan on grid paper and estimate the area of each room and the house by counting grid squares.

4 ASSESS

Closing Activity

Speaking Have students explain how to tell whether two figures are congruent or similar. Suggest that they demonstrate their explanations by using toothpicks or drinking straws to make congruent and similar figures.

Chapter 9, Quiz D (Lesson 9-6) is available in the *Assessment and Evaluation Masters*, p. 240.

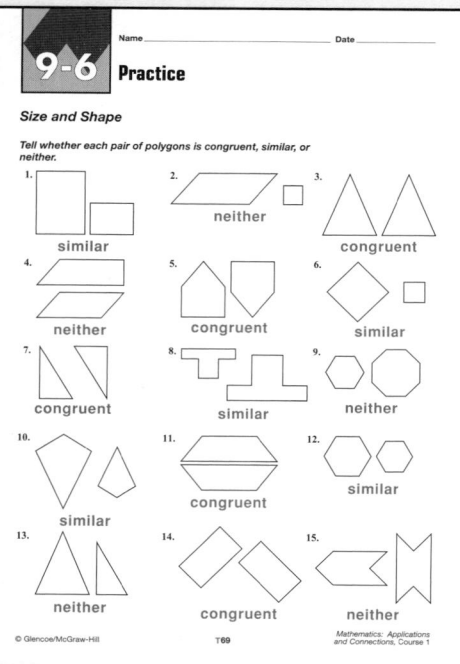

18. **Technology** A photocopier can make copies that are reduced or enlarged. At 50%, a photocopier makes a copy that is 50%, or one-half, of the original length and width. Nina needs to photocopy a picture that is 7 inches long and 5 inches wide. **b. similar**

    a. Draw a diagram that shows the dimensions of the original picture and the dimensions of the copy reduced by 50%. **See Answer Appendix**
    b. Are the two pictures similar or congruent?

19. **Working on the CHAPTER *Project***
    Refer to the floor plan on page 351. Use the given scale to find the actual length of each wall of the bathroom, dining room, and the second bedroom. Then find the actual perimeter of the house. Label each length on your diagram. **See Answer Appendix.**

20. yes; They all have 4 right angles and 4 congruent sides.

Mixed Review

20. **Critical Thinking** Are all squares similar? Explain your reasoning.

21. **Test Practice** Which figure does *not* have at least one line of symmetry? *(Lesson 9-5)* **B**

    A    B    C 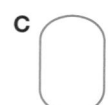   D

22. **Money Matters** One fourth of the 417 students in the sixth grade owe the library for fines. About how many sixth grade students owe the library for fines? *(Lesson 7-2)* **about 100 students**

23. Find $7\frac{1}{2} - 1\frac{3}{4}$. *(Lesson 6-6)* $5\frac{3}{4}$

MATH) IN THE MEDIA

Peanuts

1. What does it mean to be congruent to? **equal in size and shape**

2. *Draw* two octagons that are congruent. **See students' work.**

382 Chapter 9 Geometry: Investigating Patterns

■ Extending the Lesson ■

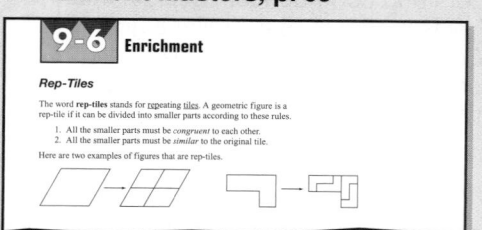
MATH) IN THE MEDIA

Have students practice drawing the symbols for *congruent* and *similar*. Discuss meanings of other symbols that students use in math or in other areas, such as addition or subtraction symbols and recycling or traffic symbols.

Let the Games Begin

Pento

Get Ready This game is for two players.

🗒 grid paper

Get Set A *pento* is a game piece made of five squares. Some sample pentos are shown. Two *pentos* are alike if they can be rotated so that they exactly match each other when stacked. They are different if they do not match when stacked or if they must be flipped over in order to match.

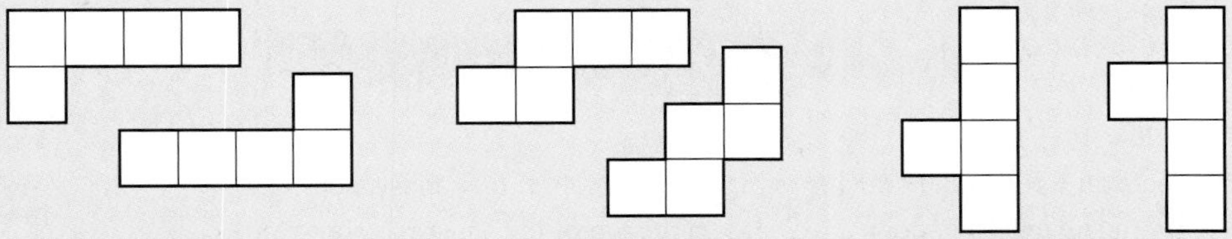

alike   **different**   **different**

Go ● **Round 1:** Each player has 1 minute to draw as many different *pentos* as possible on grid paper. Each player scores 1 point for every *pento* drawn. Two points are scored for each *pento* found by only one of the players.

● **Round 2:** Cut out all of the *pentos* from round 1. Then choose one *pento* as a model. Using any four different *pentos*, each player builds a similar figure where the length of each side is twice the length of the side in the model. Sketch the results on grid paper.

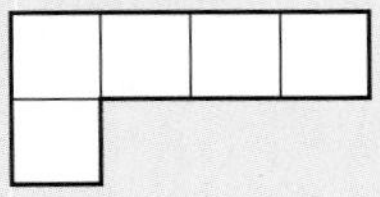

model 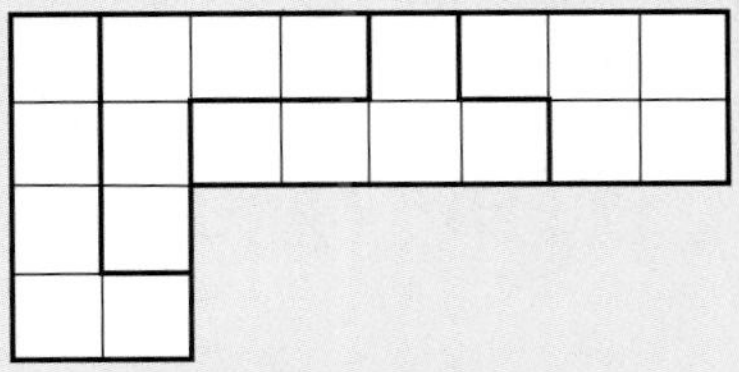   **similar figure**

Continue this round for 2 minutes. Each player scores 3 points for each similar figure drawn. The player with the most total points wins.

*inter*NET
CONNECTION Visit www.glencoe.com/sec/math/mac/mathnet for more games.

Let the Games Begin

In Round 1, make sure students check carefully for duplicates. Some additional original pentos are shown.

Another possible solution for Round 2 is below.

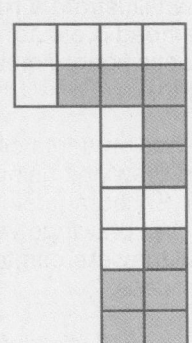

GET READY

Objective Students create Escher-like drawings by using translations.

Optional Resources

Hands-On Lab Masters
• centimeter grid, p. 12
• worksheet, p. 60

Overhead Manipulative Resources
• centimeter grid

MANAGEMENT TIPS

Recommended Time
45 minutes

Getting Started Ask students to read the opening paragraph of the lesson. Then illustrate a translation on the overhead projector using colored plastic figures on a hundred grid.

Activity 1 has students create Escher-like drawings using one translation. Encourage students to keep their first drawings very simple until they are comfortable with the process.

Teaching Tip Tell students that covering a surface with interlocking figures is also called *tessellating*. They can research more Escher tessellations in the library.

COOPERATIVE LEARNING

9-6B Translations and Escher Drawings

A Follow-Up of Lesson 9-6

 grid paper

 colored pencils

M. C. Escher (1898-1972), a Dutch artist, was famous for his unusual artwork. His most famous pieces were created by using congruent figures shaped as birds, reptiles, or fish that fit together to cover the entire surface of the artwork. You can create an Escher-like drawing by using a *translation*. When you do a translation, you slide a figure from one location to another.

TRY THIS

Work with a partner.

1 To create an Escher-like drawing using a translation, follow these steps.
• Draw a square. Then draw a triangle on the top of the square as shown.
• Translate, or slide, the triangle to the opposite side of the square. The pattern unit for our Escher-like drawing is formed.
• Repeat the pattern unit you made in the steps above to create an Escher-like drawing on grid paper. Use colored pencils to decorate your drawing.

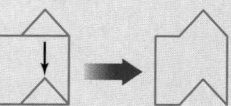

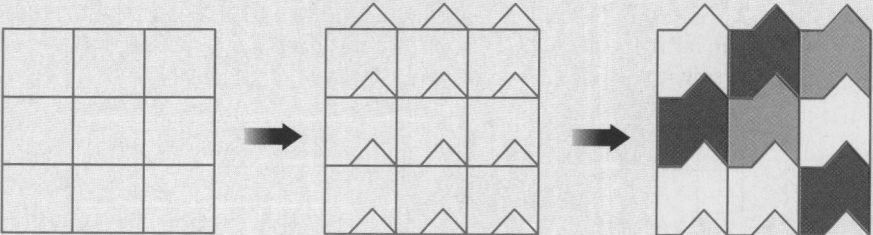

ON YOUR OWN

Create an Escher-like drawing for each pattern unit shown. 1–3. See Answer Appendix.

1.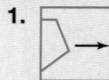

2.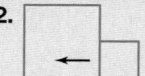

3.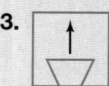

4. *Reflect Back* Are the pattern units in your Escher-like drawing congruent? **yes**

384 **Chapter 9** Geometry: Investigating Patterns

You can also create an Escher-like drawing using two translations.

TRY THIS

Work with a partner.

2 To create an Escher-like drawing using two translations, follow these steps.

- Draw a square. Then draw a rectangle on the top of the square as shown. Next, translate the rectangle to the opposite side of the square.

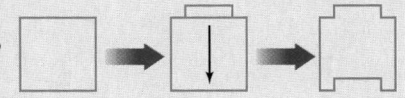

- Draw a triangle on the left side and translate the triangle to the right side.

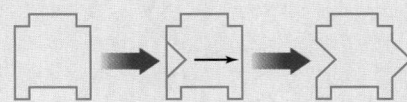

- Repeat the pattern unit you made above to create an Escher-like drawing on grid paper. Use colored pencils to decorate your drawing.

ON YOUR OWN

Create an Escher-like drawing for each pattern unit shown. 5–7. See Answer Appendix.

5.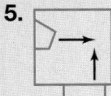

6.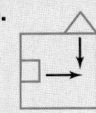

7.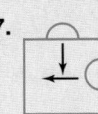

8. **Reflect Back** Why do the pieces fit together exactly? **They are congruent.**

Lesson 9-6B HANDS-ON **LAB** **385**

Math Journal Have students write a few sentences describing how translations can be used to make an Escher-like drawing. Encourage them to include designs they have created.

Activity 2 has students use two translations to create the drawings. Emphasize to students that the order of translations doesn't matter as long as they perform both.

ASSESS

Have students complete Exercises 1–3 and 5–7. Watch for students who forget to make the translation before repeating the pattern.

Use Exercises 4 and 8 to determine whether students understand how to create Escher-like drawings by using translations.

Hands-On Lab Masters, p. 60

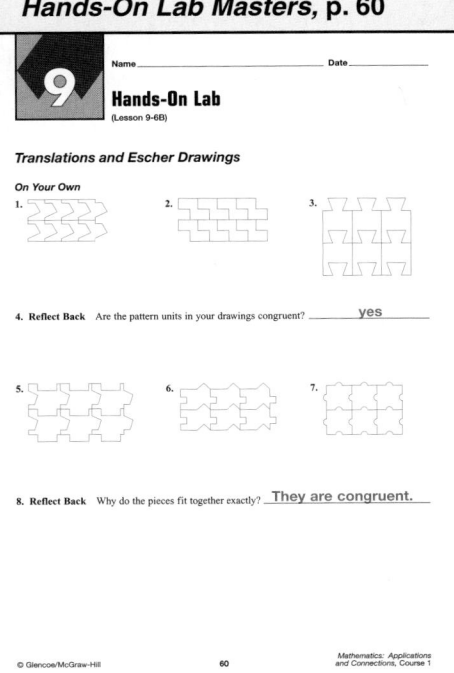

Hands-On Lab 9-6B **385**

Vocabulary

This section provides a listing of the new terms, properties, and phrases that were introduced in this chapter. Have students define each term and provide an example or two of it, if appropriate.

Understanding and Using the Vocabulary

These exercises check students' understanding of the terms by using a variety of verbal formats including matching, completion, and true/false.

Glossaries A complete glossary of terms appears on pages 642–648. The glossary also appears in Spanish on pages 649–656.

Additional Answer

11. An acute angle is less than 90°, a right angle is 90°, and an obtuse angle is greater than 90°.

Vocabulary

After completing this chapter, you should be able to define each term, concept, or phrase and give an example or two of each.

Geometry
acute angle (p. 352)
angle (p. 352)
bisect (p. 364)
complementary (p. 353)
congruent angles (p. 363)
congruent figures (p. 379)
congruent segments (p. 362)
decagon (p. 370)
degree (p. 352)
edges (p. 352)
equilateral triangle (p. 371)

hexagon (p. 370)
line of symmetry (p. 375)
line segment (p. 362)
net (p. 374)
obtuse angle (p. 352)
octagon (p. 370)
parallel (p. 371)
parallelogram (p. 371)
pentagon (p. 370)
polygon (pp. 368, 370)
protractor (p. 352)
quadrilateral (pp. 368, 370)
ray (p. 363)

rectangle (p. 371)
reflection (p. 376)
regular polygon (p. 370)
right angle (p. 352)
similar figures (p. 379)
square (p. 371)
straightedge (p. 358)
supplementary (p. 353)
translation (p. 384)
triangle (pp. 368, 370)
vertex (p. 352)

Problem Solving
use logical reasoning (p. 356)

Understanding and Using the Vocabulary

Choose the letter of the term that best matches each phrase.

1. a line dividing a shape into two matching halves f
2. a six-sided figure i
3. a polygon with all sides and all angles congruent a
4. a quadrilateral with opposite sides parallel j
5. a simple closed figure formed by three or more sides e
6. a ruler or any object with a straight side c
7. the point where two edges of a polygon intersect b
8. the most common unit of measure for an angle h
9. an angle whose measure is between 0° and 90° d
10. an angle whose measure is between 90° and 180° g

a. regular polygon
b. vertex
c. straightedge
d. acute angle
e. polygon
f. line of symmetry
g. obtuse angle
h. degree
i. hexagon
j. parallelogram

In Your Own Words

11. *Explain* how to classify an angle as acute, right, or obtuse. See margin.

386   Chapter 9 Geometry: Investigating Patterns

MindJogger Videoquizzes

MindJogger Videoquizzes provide an alternative review of concepts presented in this chapter. Students work in teams to answer questions, gaining points for correct answers. The questions are presented in three rounds.
Round 1 Concepts–5 questions
Round 2 Skills–4 questions
Round 3 Problem Solving–4 questions

Objectives & Examples

Upon completing this chapter, you should be able to:

● classify and measure angles *(Lesson 9-1)*

The angle is an acute angle since it measures between 0° and 90°.

The angle is a right angle since it measures 90°.

The angle is an obtuse angle since it measures between 90° and 180°.

● draw angles and estimate measures of angles *(Lesson 9-2)*

You can use a protractor and a straightedge to draw an angle.

The angle drawn is a 47° angle.

● bisect line segments and angles *(Lesson 9-3)*

To *bisect* something means to divide it into two congruent parts. You can use paper folding or a straightedge and a compass to bisect a line segment or an angle.

Review Exercises

Use these exercises to review and prepare for the chapter test.

Use a protractor to find the measure of each angle. Then classify the angle as *acute*, *right*, or *obtuse*.

12.

90°; right

13.
42°; acute

Classify each angle measure as *acute*, *right*, or *obtuse*.

14. 147° obtuse   15. 38.5° acute

16. 93° obtuse   17. 12.2° acute

Use a protractor and a straightedge to draw angles having the following measurements. 18–19. See Answer Appendix.

18. 36°   19. 127°

Estimate the measure of each angle.

20.

about 120°

21.
about 90°

Draw the angle or line segment with the given measurement. Then use a straightedge and a compass to bisect each angle or line segment. 22–25. See Answer Appendix.

22. 5 in.   23. 100°

24. 35°   25. 7 cm

Chapter 9 Study Guide and Assessment **387**

Objectives & Examples

This section reviews the skills and concepts of the chapter and shows completely worked examples.

Review Exercises

These exercises provide practice for the corresponding objectives.

Assessment and Evaluation Masters, pp. 227–228

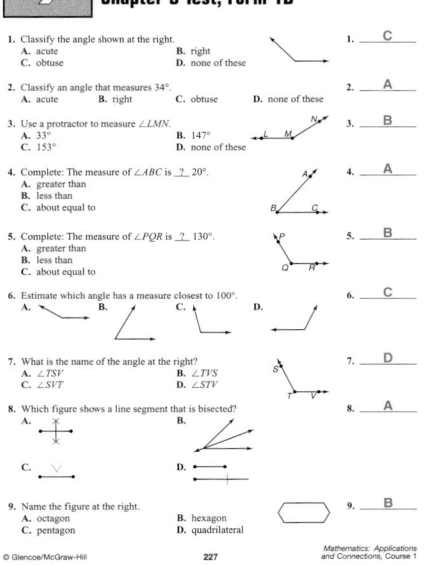

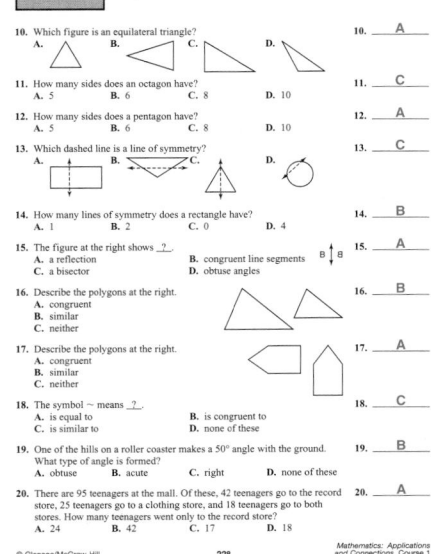

Assessment and Evaluation

Six forms of Chapter 9 Test are available in the *Assessment and Evaluation Masters* as shown in the chart.

Chapter 9 Test, Form 1B, is shown at the right. Chapter 9 Test, Form 2B, is shown on the next page.

1A	Multiple Choice	Honors
1B	Multiple Choice	Average
1C	Multiple Choice	Basic
2A	Free Response	Honors
2B	Free Response	Average
2C	Free Response	Basic

Objectives & Examples

name two-dimensional figures *(Lesson 9-4)*

Name each polygon. Then tell if the polygon is a regular polygon.

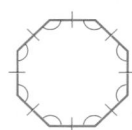 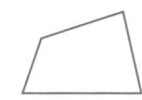

The polygon is an octagon. The sides and angles are congruent so it is a regular polygon.

The polygon is a quadrilateral. The sides and angles are not congruent so it is not a regular polygon.

describe and define lines of symmetry *(Lesson 9-5)*

Draw all lines of symmetry for each figure.

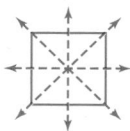

A square has 4 lines of symmetry.

The letter R has no lines of symmetry.

determine congruence and similarity *(Lesson 9-6)*

Tell whether each pair of polygons is congruent, similar, or neither.

congruent   similar   neither

Review Exercises

Name each polygon. Then tell if the polygon is a regular polygon.

26.
square; yes

27.
pentagon; no

Draw an example of each polygon. Mark any congruent sides and right angles.

28. pentagon

29. equilateral triangle

28–29. See Answer Appendix.

Tell whether the dashed line is a line of symmetry. Write *yes* or *no*.

30.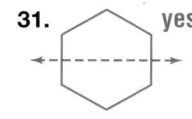
no

31. yes

Trace each figure. Draw all lines of symmetry.

32.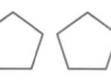

33.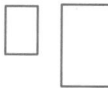

Tell whether each pair of polygons is *congruent, similar*, or *neither*.

34.
congruent

35.
similar

388   Chapter 9 Geometry: Investigating Patterns

Assessment and Evaluation Masters, pp. 233–234

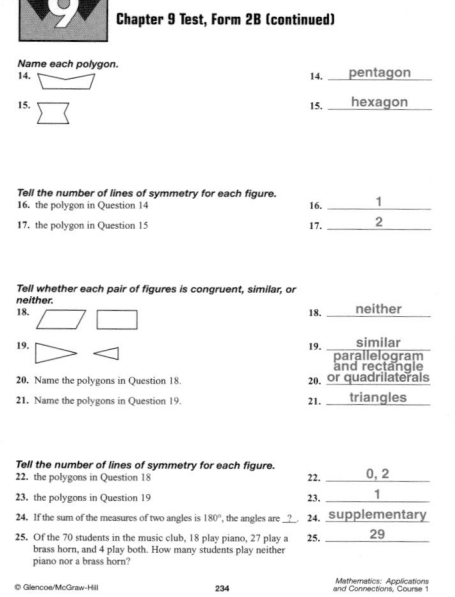

Test and Review Software

You may use this software, a combination of an item generator and item bank, to create your own tests or worksheets. Types of items include free response, multiple choice, short answer, and open ended.

CD-ROM Program

The CD-ROM Program contains an Assessment Game whose questions review the concepts in this chapter.

Applications & Problem Solving

36. Use Logical Reasoning There are 32 students in Mr. Miyar's literature class. Of these, 12 take Spanish, 15 take tennis, and 8 take both Spanish and tennis. How many take neither Spanish nor tennis? *(Lesson 9-1B)* **13 students**

37. Geometry Kurano drew the figure shown as a design on a round tablecloth. She wants to draw three more stripes from the center that will bisect the three angles. Copy her design and bisect the angles to show what the new design will look like. *(Lesson 9-3)* **See Answer Appendix.**

38. Remodeling A kitchen floor is to be covered with vinyl flooring. The pattern of the flooring is a square surrounded by four regular octagons. Draw a section of the flooring. Mark any congruent sides and angles. *(Lesson 9-4)* **See Answer Appendix.**

39. School The student council voted to have a contest to design a school flag. The flag may be of any shape but must have at least one line of symmetry. Use any shapes, letters, objects, and colors to design a flag. *(Lesson 9-5)* **See students' work.**

Alternative Assessment

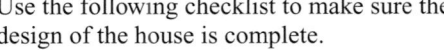

● **Performance Task**

Suppose that you are a graphic designer. A client has asked you to make a sign for the grand opening of her store, Sports Emporium. Design a logo that displays line symmetry. Use two-dimensional figures in the design of your sign.

Identify the types of angles in your design. Explain how your design meets all of the requirements. **See students' work.**

A practice test for Chapter 9 is provided on page 603.

● **Completing the** CHAPTER Project

Use the following checklist to make sure the design of the house is complete.

☑ The angle measures of the dining room, bathroom, and second bedroom are correct.

☑ The measurements of the walls of the dining room, bathroom, and second bedroom are correct.

☑ The floor plans of the house's actual angle measures and actual lengths are included.

 PORTFOLIO Select one of the assignments from this chapter and place it in your portfolio. Attach a note to it explaining why you selected it.

Applications & Problem Solving

This section provides additional practice in solving real-world problems that involve the skills of this chapter.

Alternative Assessment

The **Performance Task** provides students with a performance assessment opportunity to evaluate their work and understanding.

CHAPTER Project

Students should complete the final stages of their project and prepare a class demonstration of their results. A scoring guide for the project is available in the *Investigations and Projects Masters*, p. 51.

 PORTFOLIO Students should add to their portfolios at this time.

Assessment and Evaluation Masters, p. 237

● **Performance Assessment**

Additional performance assessment tasks for this chapter are included in the *Assessment and Evaluation Masters* on page 237. A scoring guide is also provided on page 249.

The Standardized Test Practice may be used to help students prepare for standardized tests. The test items are written in the same style as those in state proficiency tests and standardized tests like CAT, CTBS, ITBS, MAT, SAT, and Terra Nova. The test items cover skills and concepts covered up to this point in the text.

The pages can be used as an overnight assessment. After students have completed the pages, discuss how each problem can be solved, or provide copies of the solutions from the *Solutions Manual*.

Assessment and Evaluation Masters, p. 243

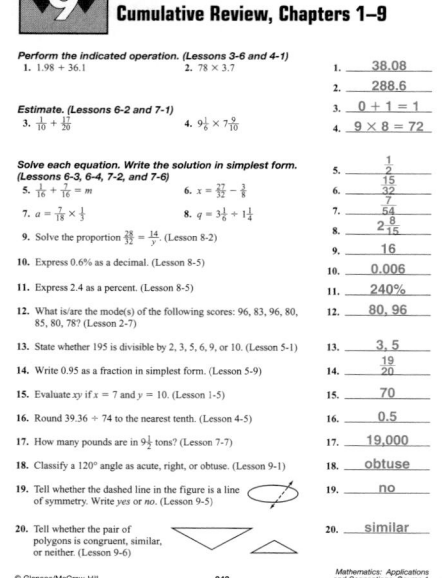

Section One: Multiple Choice

There are twelve multiple-choice questions in this section. Choose the best answer. If a correct answer is *not here*, mark the letter for Not Here.

1. What is the measure of the line segment shown? **C**

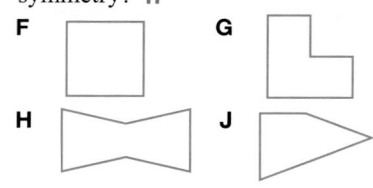

  A 60 km
  B 60 m
  C 60 mm
  D 60 cm

2. Which figure has only two lines of symmetry? **H**

  F   G

  H   J

3. A pedometer measures the number of miles a person has walked. Which list shows the pedometer readings in order from least to greatest? **D**

  A 4.75, 4.7, 4.04, 4.17
  B 4.17, 4.75, 4.7, 4.04
  C 4.04, 4.7, 4.17, 4.75
  D 4.04, 4.17, 4.7, 4.75

4. Which point is in the exterior of quadrilateral *WXYZ*? **H**

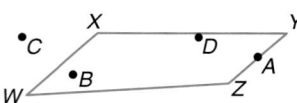

  F point *A*
  G point *B*
  H point *C*
  J point *D*

Please note that Questions 5–12 have five answer choices.

5. Four students are 15, 11, 12, and 14 years old. What is an average age of the four students? **A**

  A 13   B 10
  C 14   D 12
  E 15

6. $\triangle MNP \cong \triangle RST$. Which of the following is a true statement? **J**

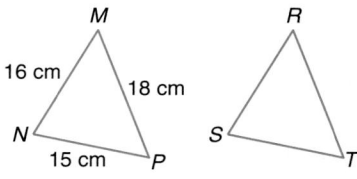

  F The perimeter of $\triangle RST$ is 59 centimeters.
  G The measure of side *RT* is 16 centimeters.
  H The measures of angles *M* and *T* are equal.
  J The measures of sides *RS* and *MN* are equal.
  K The measure of side *RS* is equal to 18 centimeters.

7. The lights on a street sign blink 5 times every 6 seconds. How many times do they blink in 3 minutes? **B**

  A 900   B 150
  C 75   D 30
  E 450

8. Which is a reasonable remainder when a number is divided by 8? **J**

  F 10   G 8
  H 11   J 7
  K 9

◄◄◄ **Instructional Resources**
Another cumulative review is shown at the left and is available in the *Assessment and Evaluation Masters*, p. 243.

9. Find the product of $\frac{3}{4}$ and $\frac{4}{5}$. **A**

   A $\frac{3}{5}$

   B $\frac{15}{16}$

   C $1\frac{2}{5}$

   D 12

   E Not Here

10. Teisha works 17 hours a week and earns $99.45. How much money does she earn each hour? **H**

    F $5.65

    G $4.75

    H $5.85

    J $4.95

    K Not Here

11. The students at Glenwood Middle School sold 1,808 magazine subscriptions. The students at Center Middle School sold 2,046 subscriptions. How many more subscriptions did the students at Center Middle School sell than the students at Glenwood Middle School? **C**

    A 148

    B 242

    C 238

    D 762

    E Not Here

12. Solve $m = 26 \div 3.25$. **J**

    F 0.008

    G 0.08

    H 0.8

    J 8

    K Not Here

Test-Taking Tip

You can improve your chances of doing well on a test by keeping your test anxiety under control. Getting a good night's sleep will help. Right before the test, take deep breaths and focus on doing well. Panicking or getting overly worried about a test can keep you from doing your best.

Section Two: Free Response

This section contains six questions for which you will provide short answers. Write your answers on your paper.

13. A dog is on a 20-foot leash attached to a stake in the ground. If the dog walks in a circle at the end of the leash, how far can he walk before returning to his starting point? Round to the nearest foot. **126 feet**

14. Determine whether 708 is divisible by 2, 3, 5, 6, 9, or 10. **2, 3, 6**

15. Write the decimal 0.125 as a fraction in simplest form. $\frac{1}{8}$

16. Evaluate $a + b$ if $a = 4\frac{1}{2}$ and $b = 7\frac{1}{4}$. $11\frac{3}{4}$

17. Brenda keeps her turtle in an aquarium that is 14 inches wide and 18 inches long. She buys a new aquarium that is 12 inches wider, but is the same length as the other cage. What are the dimensions of the new aquarium? **26 inches by 18 inches**

18. Angles C and D are supplementary angles. If angle C measures $101.3°$, what is the measure of angle D? **78.7°**

Test-Taking Tip

Tell students that other ways to reduce test anxiety include waking up early to avoid being rushed, eating a healthy breakfast for energy, and making a checklist to remember what to bring to the test.

Assessment and Evaluation Masters, pp. 241–242

Instructional Resources ▶▶▶

Additional standardized test practice is shown at the right and is available in the *Assessment and Evaluation Masters*, pp. 241–242.

Interdisciplinary Investigation

Interdisciplinary Investigation

GET READY

This optional investigation is designed to be completed by a group of 3 or 4 students over several days or several weeks.

Mathematical Overview

This investigation utilizes the mathematical skills and concepts presented in Chapters 7–9.
- multiplying and dividing fractions and mixed numbers
- changing customary units
- exploring ratios, percents, and fractions

Time Management	
Gathering Data	45 minutes
Calculations	30 minutes
Creating Graphs	30 minutes
Summarizing Data	30 minutes
Presentation	10 minutes

Instructional Resources
- *Investigations and Projects Masters*, pp. 9–12

Investigations and Projects Masters, p. 12

Name _____ Date _____

Interdisciplinary Investigation
(Student Edition, Pages 392–393)

That Is One Humongous Pie!

1. A record-setting food was:
2. The original recipe would have to be increased: times
   Ratio:
   This ratio was determined by:

Ingredient	Original Recipe	Giant-Sized Recipe	Cost in Recipe	Cost in Giant-Sized Food
Total				

3. In this recipe, the most expensive item is:
   The total cost of the ingredients for the giant-sized food is:
4. Graph:

© Glencoe/McGraw-Hill   12   *Mathematics: Applications and Connections, Course 1*

THAT IS ONE HUMONGOUS PIE!

Did you know that the largest pie ever made weighed almost 38,000 pounds? The largest milkshake ever made was 2,000 gallons. Imagine the amount of each ingredient needed to make a record-setting food.

What You'll Do

In this investigation, you will write a recipe for a giant-sized food and determine the cost of making this food.

Materials
  a copy of *The Guinness Book of Records*

  cookbooks

  supermarket advertisements

  calculator

Procedure

1. Work in groups of 3 or 4. Find a record-setting food in *The Guinness Book of Records*. Then find a recipe for that food in a cookbook. Make sure that the recipe has at least 6 ingredients and that at least 3 of the ingredients have measurements that are given as fractions or mixed numbers.

2. Tell how many times the recipe would need to be increased to make a record-setting food. Express this relationship as a ratio. Explain your reasoning. Then find the amount of each ingredient needed to make the record-setting food. Organize the information in a table.

3. Research the cost of each ingredient in your recipe. Which ingredient costs the most? Find the total cost of the ingredients.

4. Use your data to find the percent of the total cost that each ingredient represents. Make a graph of your data.

5. Make a booklet or brochure that contains your recipe, the costs, the graphs, and any other information on your project.

Technology Tips

- Use a **spreadsheet** to calculate the amount of each ingredient needed and the total cost of the ingredients.

- Use **graphing software** to display the data.

- Use **publishing software** to help make your booklet or brochure.

◄◄◄ **Instructional Resources**

A recording sheet to help students organize their data for this investigation is shown at the left and is available in the *Investigations and Projects Masters*, p. 12.

 Cooperative Learning

This investigation offers an excellent opportunity for using cooperative learning groups. For more information on cooperative learning strategies and group management, see *Cooperative Learning in the Mathematics Classroom*.

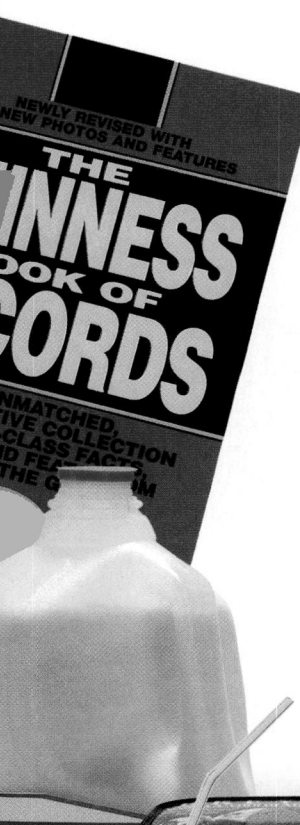

Making the Connection

Use the data collected about your record-setting food as needed to help in these investigations.

Language Arts

Suppose you want to make a record-setting food to put your school in *The Guinness Book of Records*. Write a newspaper article convincing people to donate to your project. Use the data that you gathered in the investigation. Be creative and describe the cost involved.

Health

Find the number of Calories in one serving of your record-setting food, and also in the whole food.

Social Studies

Research the history of *The Guinness Book of Records*. When and why was the first book published?

Go Further

- Estimate the area and perimeter of your record-setting food. Mark off an area in your school or on the school grounds to demonstrate the size of the food. How does the perimeter of the record-setting food compare to its area?

  - How many people could have a serving of your record-setting food? Determine how much money you should charge for a serving so that enough is made to pay for the project.

 interNET
CONNECTION For current information on recipes, visit:
www.glencoe.com/sec/math/mac/mathnet

 PORTFOLIO
You may want to place your work on this investigation in your portfolio.

Interdisciplinary Investigation That is One Humongous Pie! **393**

MANAGEMENT TIPS

Working in Teams Have the teams decide which recipe to use and the amount of increase. Each team member should be responsible for a specific ingredient and should present a different part of the investigation if they do an oral presentation.

Making the Connection
You may wish to alert other teachers on your team that your students may need their assistance in this investigation.
Language Arts Students may need to research different styles of newspaper articles and reporting.
Health Students need to know about how many Calories their bodies need in a day.
Social Studies Students may wish to research the importance of record keeping in society.

ASSESS

You may wish to have students do an oral presentation. If so, they should make large graphs and cost tables on posterboard to accompany their discussion.

Investigations and Projects Masters, p. 11

Instructional Resources ▶▶▶
Sample solutions for this investigation are provided in the *Investigations and Projects Masters* on p. 10. The scoring guide for assessing student performance shown at the right is also available on p. 11.

SCORING GUIDE

Interdisciplinary Investigation
(Student Edition, Pages 392–393)

That Is One Humongous Pie!

Level	Specific Criteria
3 Superior	• Shows a thorough understanding of the concepts of *multiplying fractions, multiplying mixed numbers, and changing customary units.* • Uses appropriate strategies to solve problems. • Computations are correct. • Written explanations are exemplary. • Charts, model, and any statements included are appropriate and sensible. • Goes beyond the requirements of some or all problems.
2 Satisfactory, with minor flaws	• Shows understanding of the concepts of *multiplying fractions, multiplying mixed numbers, and changing customary units.* • Uses appropriate strategies to solve problems. • Computations are mostly correct. • Written explanations are effective. • Charts, model, and any statements included are appropriate and sensible. • Satisfies the requirements of problems.
1 Nearly Satisfactory, with obvious flaws	• Shows understanding of most of the concepts of *multiplying fractions, multiplying mixed numbers, and changing customary units.* • May not use appropriate strategies to solve problems. • Computations are mostly correct. • Written explanations are satisfactory. • Charts, model, and any statements included are appropriate and sensible. • Satisfies the requirements of problems.
0 Unsatisfactory	• Shows little or no understanding of the concepts of *multiplying fractions, multiplying mixed numbers, and changing customary units.* • Does not use appropriate strategies to solve problems. • Computations are incorrect. • Written explanations are not satisfactory. • Charts, model, and any statements included are not appropriate or sensible. • Does not satisfy the requirements of the problems.

© Glencoe/McGraw-Hill   11   *Mathematics: Applications and Connections, Course 1*

Geometry: Understanding Area and Volume

Previewing the Chapter

Overview

In this chapter, students learn to find the areas of parallelograms, triangles, and circles and to construct circle graphs. Students identify and draw three-dimensional figures. They find the surface area and volume of rectangular prisms, both with and without using a spreadsheet. Students learn to solve problems by making models.

Lesson (pages)	Lesson Objectives	NCTM Standards	Standardized Tests	State/Local Objectives
10-1A (396–397)	Find the areas of irregular shapes.	1–4, 7, 12, 13	CAT, MAT, SAT	
10-1 (398–401)	Find the area of parallelograms.	1–7, 9, 12, 13	CAT, MAT, SAT	
10-2 (402–405)	Find the area of triangles.	1–9, 12, 13	CAT, MAT, SAT	
10-3 (406–409)	Find the area of circles.	1–9, 12, 13	CAT, MAT, SAT	
10-3B (410–411)	Construct circle graphs.	1–7, 10, 12, 13	CTBS, TN	
10-4 (412–414)	Identify three-dimensional figures.	1–7, 12, 13	MAT	
10-4B (415)	Draw three-dimensional figures.	1–3, 12		
10-5A (416–417)	Solve problems by making models.	1–9, 12, 13	CTBS, MAT, TN	
10-5 (418–420)	Find the volume of rectangular prisms.	1–7, 9, 12, 13	ITBS, MAT	
10-6 (421–424)	Find the surface area of rectangular prisms.	1–7, 12, 13	MAT	
10-6B (425)	Use spreadsheets to find the surface area and volume of rectangular prisms.	1–4, 7–9, 12, 13		

CAT = California Achievement Tests, CTBS = Comprehensive Tests of Basic Skills, ITBS = Iowa Tests of Basic Skills, MAT = Metropolitan Achievement Tests, SAT = Stanford Achievement Tests, TN = Terra Nova

Organizing the Chapter

CD-ROM

 All of the blackline masters in the Teacher's Classroom Resources are available on the **Electronic Teacher's Classroom Resources** CD-ROM.

LESSON PLANNING GUIDE

Lesson	Extra Practice (Student Edition)	Blackline Masters (page numbers)										
		Study Guide	Practice	Enrichment	Assessment & Evaluation	Classroom Games	Diversity	Hands-On Lab	School to Career	Science and Math Lab Manual	Technology	Transparencies A and B
10-1A								61				
10-1	p. 583	70	70	70							20	10-1
10-2	p. 583	71	71	71	267			78		17–20		10-2
10-3	p. 583	72	72	72	266, 267	27–28					19	10-3
10-3B								62				
10-4	p. 584	73	73	73			10					10-4
10-4B								63				
10-5A	p. 584											
10-5	p. 584	74	74	74	268							10-5
10-6	p. 585	75	75	75	268				10			10-6
10-6B												
Study Guide/ Assessment					253–265, 269–271							

Other Chapter Resources

Student Edition
Chapter Project, pp. 395, 420, 424, 429
Let the Games Begin, p. 409

Technology
 CD-ROM Program

 Interactive Mathematics Tools Software

Teacher's Classroom Resources

Applications
Family Letters and Activities, pp. 19–20
Investigations and Projects Masters, pp. 53–56
Meeting Individual Needs
Investigations for the Special Education Student, pp. 43–44

Teaching Aids
Answer Key Masters
Block Scheduling Booklet
Lesson Planning Guide
Solutions Manual

Professional Publications
Glencoe Mathematics Professional Series

Planning the Chapter

MindJogger Videoquizzes provide a unique format for reviewing concepts presented in the chapter.

ASSESSMENT RESOURCES

Student Edition
Mixed Review, pp. 401, 405, 409, 414, 420, 424
Mid-Chapter Self Test, p. 414
Math Journal, pp. 413, 423
Study Guide and Assessment, pp. 426–429
Performance Task, p. 429
Portfolio Suggestion, p. 429
Standardized Test Practice, pp. 430–431
Chapter Test, p. 604

Assessment and Evaluation Masters
Multiple-Choice Tests (Forms 1A, 1B, 1C), pp. 253–258
Free-Response Tests (Forms 2A, 2B, 2C), pp. 259–264
Performance Assessment, p. 265
Mid-Chapter Test, p. 266
Quizzes A–D, pp. 267–268
Standardized Test Practice, pp. 269–270
Cumulative Review, p. 271

Teacher's Wraparound Edition
5-Minute Check, pp. 398, 402, 406, 412, 418, 421
Building Portfolios, p. 394
Math Journal, pp. 397, 411, 415, 425
Closing Activity, pp. 401, 405, 409, 414, 417, 420, 424

Technology
Test and Review Software
MindJogger Videoquizzes
CD-ROM Program

MATERIALS AND MANIPULATIVES

Lesson 10-1A
centimeter grid paper*†
ruler*

Lesson 10-1
grid paper*†
scissors*
ruler*

Lesson 10-2
plain paper
scissors*
ruler*

Lesson 10-3
paper plate
scissors*
counters*†
felt squares
markers
compass*†

Lesson 10-3B
colored pencils
ruler*
compass*†
protractor*†
calculator

Lesson 10-4B
isometric dot paper†
ruler*

Lesson 10-5
centimeter cubes*

Lesson 10-6
centimeter grid paper*†
scissors*
ruler*
tape
calculator

Lesson 10-6B
computer
spreadsheet software

*Glencoe Manipulative Kit　　　　†Glencoe Overhead Manipulative Resources

PACING CHART

See pages T25–T27 for the Course Planning Calendar.

COURSE	DAY 1	DAY 2	DAY 3	DAY 4	DAY 5	DAY 6	DAY 7
Standard	Chapter Project	Lessons 10-1A & 10-1		Lesson 10-2	Lesson 10-3		Lesson 10-4
Honors	Chapter Project	Lesson 10-1	Lesson 10-2	Lessons 10-3 & 10-3B		Lessons 10-4 & 10-4B	
Block	Chapter Project & Lesson 10-1A	Lessons 10-1 & 10-2	Lesson 10-3	Lesson 10-4	Lessons 10-5A & 10-5	Lesson 10-6	Study Guide and Assessment, Chapter Test

Interactive Mathematics:
Activities and Investigations

is an activity-based program that may be used as an enhancement for chapters in **Mathematics: Applications and Connections.**

Activities and Investigations

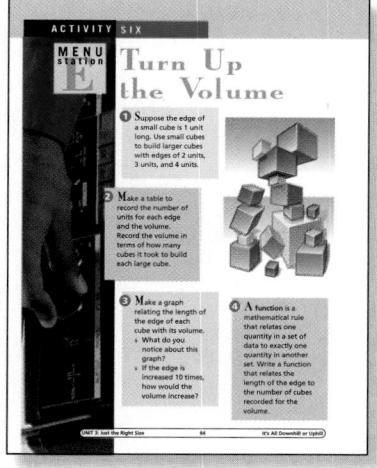

Unit 3, Activity Six, Menu E
Use with Lesson 10-5.

Summary Students use cubes to build various larger cubes for the given edge lengths. They record the number of cubes for each figure and graph the relationship between the edge length and the volume. They look for a pattern among the graphed points and write a function describing the graphed relationship.

Math Connection Students graph data and look for patterns created by modeling cubes of different edge lengths. They express the relationship between the edge length of a cube and its volume in terms of a function.

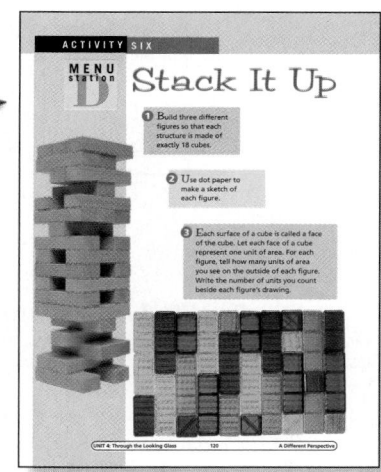

Unit 4, Activity Six, Menu D
Use with Lesson 10-6.

Summary Students use blocks to build three 3-dimensional structures composed of 18 cubes. They make a sketch of each one using dot paper. Then students determine the surface area of each figure.

Math Connection Students build 3-dimensional structures and explore the concept of surface area. They create isometric drawings of their structures. Isometric dot paper is dot paper in which each dot is equidistant from the dots around it, thus allowing for lengths to be preserved. Angle measures are not preserved in such drawings.

DAY 8	DAY 9	DAY 10	DAY 11	DAY 12	DAY 13	DAY 14	DAY 15
(continue from Day 7)	Lesson 10-5A	Lesson 10-5	Lesson 10-6	Study Guide and Assessment	Chapter Test		
Lesson 10-5A	Lesson 10-5	Lessons 10-6 & 10-6B		Study Guide and Assessment	Chapter Test		

Enhancing the Chapter

APPLICATIONS

Classroom Games, pp. 27–28

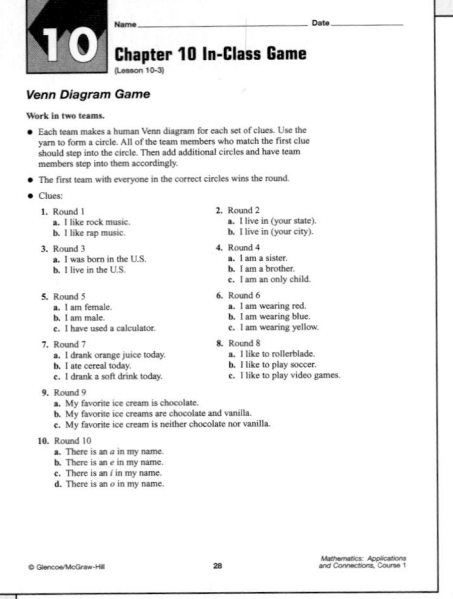

Diversity Masters, p. 10

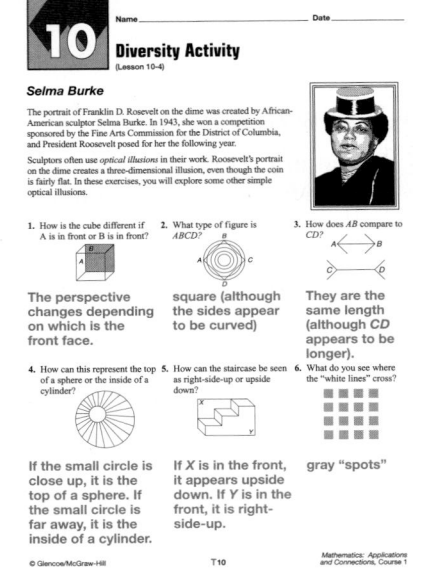

School to Career Masters, p. 10

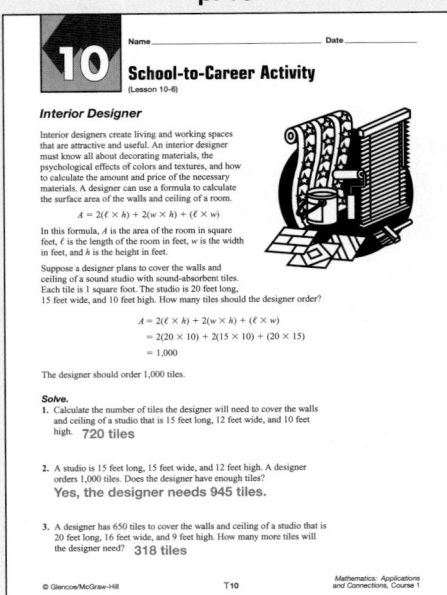

Family Letters and Activities, pp. 19–20

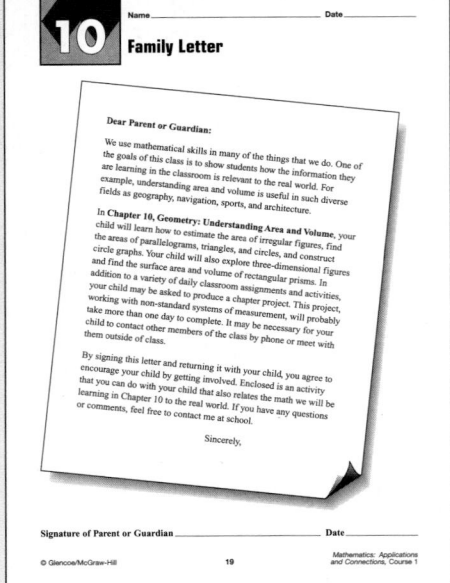

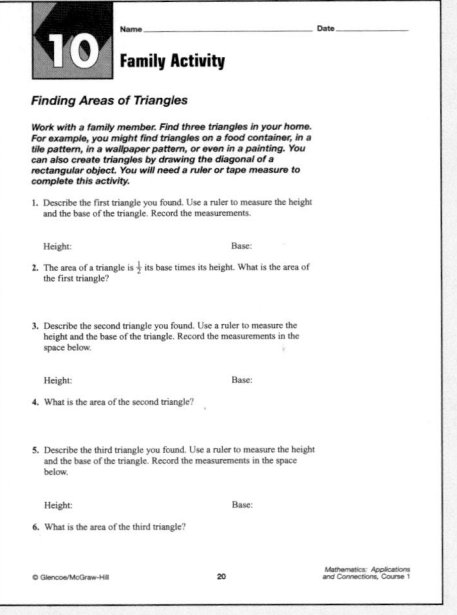

Science and Math Lab Manual, pp. 17–20

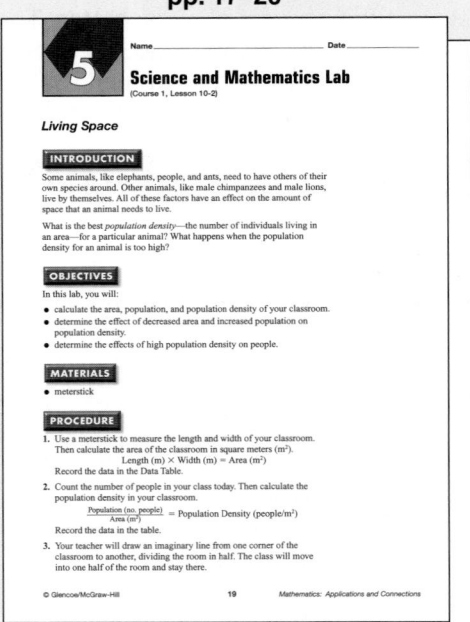

Hands-On Lab Masters, p. 78

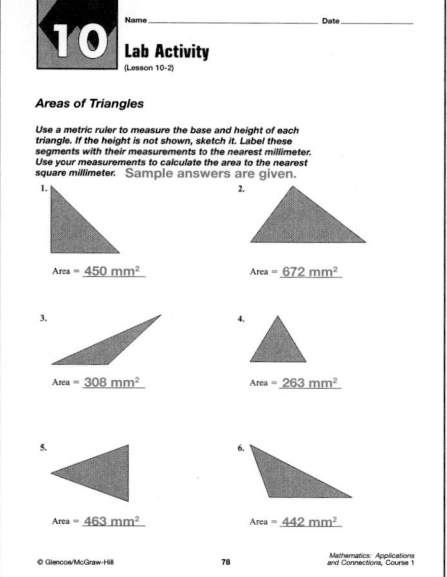

10 Lab Activity
(Lesson 10-2)

Areas of Triangles

Use a metric ruler to measure the base and height of each triangle. If the height is not shown, sketch it. Label these segments with their measurements to the nearest millimeter. Use your measurements to calculate the area to the nearest square millimeter. **Sample answers are given.**

1. Area = 450 mm²
2. Area = 672 mm²
3. Area = 308 mm²
4. Area = 263 mm²
5. Area = 463 mm²
6. Area = 442 mm²

Assessment and Evaluation Masters, pp. 266–268

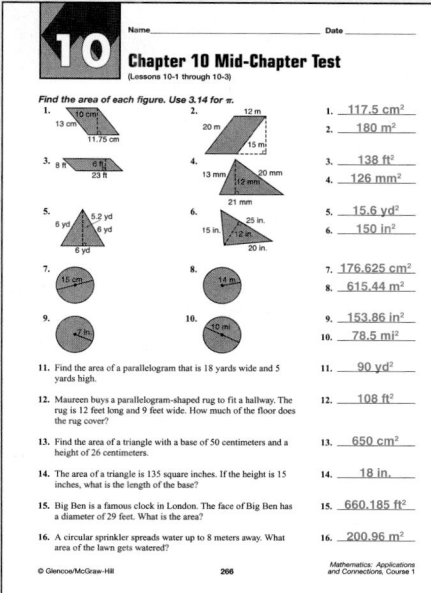

10 Chapter 10 Mid-Chapter Test
(Lessons 10-1 through 10-3)

Find the area of each figure. Use 3.14 for π.

1. 117.5 cm²
2. 180 m²
3. 138 ft²
4. 126 m²
5. 15.6 yd²
6. 150 in²
7. 176.625 cm²
8. 615.44 m²
9. 153.86 in²
10. 78.5 mi²
11. Find the area of a parallelogram that is 18 yards wide and 5 yards high. — 90 yd²
12. Maureen buys a parallelogram-shaped rug to fit a hallway. The rug is 12 feet long and 9 feet wide. How much of the floor does the rug cover? — 108 ft²
13. Find the area of a triangle with a base of 50 centimeters and a height of 26 centimeters. — 650 cm²
14. The area of a triangle is 135 square inches. If the height is 15 inches, what is the length of the base? — 18 in.
15. Big Ben is a famous clock in London. The face of Big Ben has a diameter of 29 feet. What is the area? — 660.185 ft²
16. A circular sprinkler spreads water up to 8 meters away. What area of the lawn gets watered? — 200.96 m²

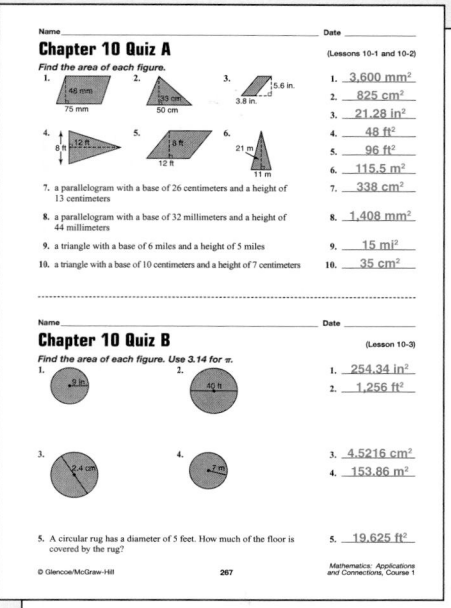

Chapter 10 Quiz A
(Lessons 10-1 and 10-2)

Find the area of each figure.

1. 3,600 mm²
2. 825 cm²
3. 21.28 in²
4. 48 ft²
5. 96 ft²
6. 115.5 m²
7. a parallelogram with a base of 26 centimeters and a height of 13 centimeters — 338 cm²
8. a parallelogram with a base of 32 millimeters and a height of 44 millimeters — 1,408 mm²
9. a triangle with a base of 6 miles and a height of 5 miles — 15 mi²
10. a triangle with a base of 10 centimeters and a height of 7 centimeters — 35 cm²

Chapter 10 Quiz B
(Lesson 10-3)

Find the area of each figure. Use 3.14 for π.

1. 254.34 in²
2. 1,256 ft²
3. 4.5216 cm²
4. 153.86 m²
5. A circular rug has a diameter of 5 feet. How much of the floor is covered by the rug? — 19.625 ft²

Technology Masters, pp. 19–20

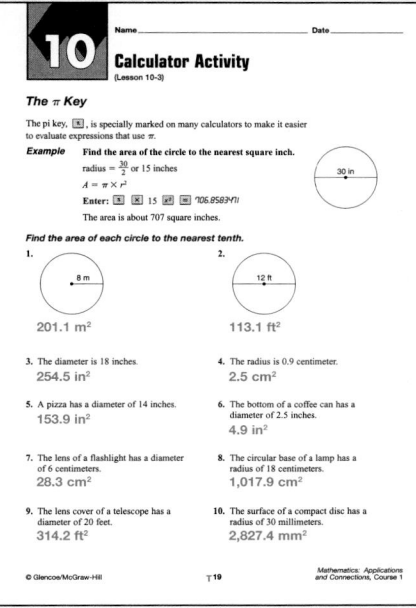

10 Calculator Activity
(Lesson 10-3)

The π Key

The pi key, is specially marked on many calculators to make it easier to evaluate expressions that use π.

Example Find the area of the circle to the nearest square inch.

radius = 30/2 or 15 inches
A = π × r²
Enter: π × 15 x² 706.8583471

The area is about 707 square inches.

Find the area of each circle to the nearest tenth.

1. 201.1 m²
2. 113.1 ft²
3. The diameter is 18 inches. — 254.5 in²
4. The radius is 0.9 centimeter. — 2.5 cm²
5. A pizza has a diameter of 14 inches. — 153.9 in²
6. The bottom of a coffee can has a diameter of 2.5 inches. — 4.9 in²
7. The lens of a flashlight has a diameter of 6 centimeters. — 28.3 cm²
8. The circular base of a lamp has a radius of 18 centimeters. — 1,017.9 cm²
9. The lens cover of a telescope has a diameter of 20 feet. — 314.2 ft²
10. The surface of a compact disc has a radius of 30 millimeters. — 2,827.4 mm²

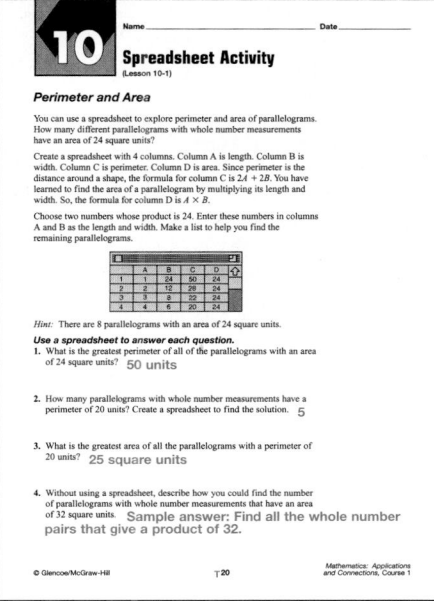

10 Spreadsheet Activity
(Lesson 10-1)

Perimeter and Area

You can use a spreadsheet to explore perimeter and area of parallelograms. How many different parallelograms with whole number measurements have an area of 24 square units?

Create a spreadsheet with 4 columns. Column A is length. Column B is width. Column C is perimeter. Column D is area. Since perimeter is the distance around a shape, the formula for column C is 2A + 2B. You have learned to find the area of a parallelogram by multiplying its length and width. So, the formula for column D is A × B.

Choose two numbers whose product is 24. Enter these numbers in columns A and B as the length and width. Make a list to help you find the remaining parallelograms.

	A	B	C	D
1		24	50	24
2	2	12	28	24
3	3	8	22	24
4	4	6	20	24

Hint: There are 8 parallelograms with an area of 24 square units.

Use a spreadsheet to answer each question.

1. What is the greatest perimeter of all of the parallelograms with an area of 24 square units? — 50 units
2. How many parallelograms with whole number measurements have a perimeter of 20 units? Create a spreadsheet to find the solution. — 5
3. What is the greatest area of all the parallelograms with a perimeter of 20 units? — 25 square units
4. Without using a spreadsheet, describe how you could find the number of parallelograms with whole number measurements that have an area of 32 square units. — Sample answer: Find all the whole number pairs that give a product of 32.

Investigations for the Special Education Student, pp. 43–44

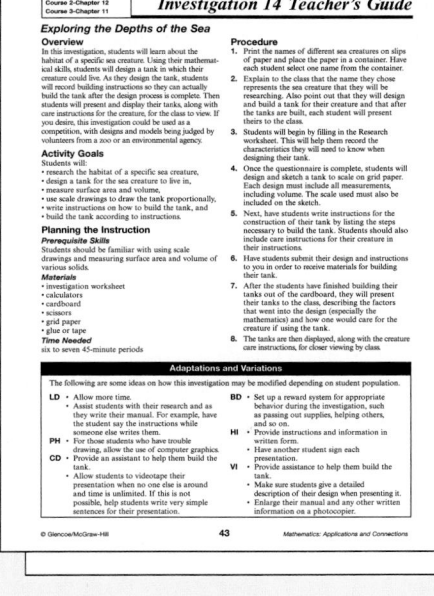

Use with:
Course 1–Chapter 10
Course 2–Chapter 12
Course 3–Chapter 11

Investigation 14 Teacher's Guide

Exploring the Depths of the Sea

Overview
In this investigation, students will learn about the habitat of a specific sea creature. Using their mathematical skills, students will design a tank in which their creature could live. As they design the tank, students will record building instructions so they can actually build the tank after the design process is complete. Then students will research and display their tanks, along with care instructions for the creature, for the class to view. If you desire, this investigation could be used as a competition, with designs and models being judged by volunteers from a zoo or an environmental agency.

Activity Goals
Students will:
- research the habitat of a specific sea creature,
- design a tank for the sea creature to live in,
- measure surface area and volume,
- use scale drawings to draw the tank proportionally,
- write instructions on how to build the tank, and
- build the tank according to instructions.

Planning the Instruction
Prerequisite Skills
Students should be familiar with using scale drawings and measuring surface area and volume of various solids.
Materials
- investigation worksheet
- calculators
- cardboard
- scissors
- grid paper
- glue or tape
Time Needed
six to seven 45-minute periods

Procedure
1. Print the names of different sea creatures on slips of paper and place the paper in a container. Have each student select one name from the container.
2. Explain to the class that the name they chose represents the sea creature that they will be researching. Also point out that they will design and build a tank for their creature and that after the tanks are built, each student will present theirs to the class.
3. Students will begin by filling in the Research worksheet. This will help them record the characteristics they will need to know when designing their tank.
4. Once the questionnaire is complete, students will design and sketch a tank to scale on grid paper. Each design must include all measurements, including volume. The scale used must also be included on the sketch.
5. Next, have students write instructions for the construction of their tank by listing the steps necessary to build the tank. Students should also include care instructions for their creature in their instructions.
6. Have students submit their design and instructions to you in order to receive materials for building their tank.
7. After the students have finished building their tanks out of the cardboard, they will present their tanks to the class, describing the factors that went into the design (especially the mathematics) and how one would care for the creature if using the tank.
8. The tanks are then displayed, along with the creature care instructions, for closer viewing by class.

Adaptations and Variations

The following are some ideas on how this investigation may be modified depending on student population.

LD • Allow more time.	**BD** • Set up a reward system for appropriate behavior during the investigation, such as passing out supplies, helping others, and so on.
• Assist students with their research and as they write their manual. For example, have the student say the instructions while someone else writes them.	**HI** • Provide instructions and information in written form.
PH • For those students who have trouble drawing, allow the use of computer graphics.	• Have another student sign each presentation.
CD • Provide an assistant to help them build the tank.	**VI** • Provide assistance to help them build the tank.
• Allow students to videotape their presentation when no one else is around and time is unlimited. If this is not possible, help students write very simple sentences for their presentation.	• Make sure students give a detailed description of their design when presenting it.
	• Enlarge their manual and any other written information on a photocopier.

CHAPTER 10

Geometry: Understanding Area and Volume

Theme: Measurement

The United States does not use the metric system. The Metric Program at the National Institute of Standards and Technology is encouraging voluntary conversion from the customary system to the metric system.

Question of the Day What is the volume in cubic inches of a cube whose edge is 1 foot long? **1,728 in³**

Assess Prerequisite Skills

Ask students to read through the list of objectives presented in "What you'll learn in Chapter 10." You may wish to ask them what each of the objectives means or if they have experienced or used any of these math concepts before.

 Building Portfolios

Encourage students to revise their portfolios as they study this chapter. Have them choose examples of their work that show improvement. Have students include an explanation of how each example demonstrates increased understanding of math concepts.

 **Math and the Family**

In the *Family Letters and Activities* booklet (pp. 19–20), you will find a letter to the parents explaining what students will study in Chapter 10. An activity appropriate for the whole family is also available.

What you'll learn in Chapter 10

- to estimate the areas of irregular figures,
- to find the areas of parallelograms, triangles, and circles,
- to construct circle graphs,
- to identify and draw three-dimensional figures,
- to solve problems by making a model, and
- to find the surface area and volume of rectangular prisms.

394 Chapter 10 Geometry: Understanding Area and Volume

 CD-ROM Program

Activities for Chapter 10
- Chapter 10 Introduction
- Interactive Lessons 10-1, 10-2, 10-5, 10-6
- Extended Activity 10-4
- Assessment Game
- Resource Lessons 10-1 through 10-6

CHAPTER Project

MEASURING UP

You are familiar with inches, feet, yards, centimeters, and meters. In this project, you will design your own system of measure. You will take measurements using your system and compare your system to a standard system. You will present your findings in a poster.

Getting Started

- Early measurement systems in many cultures were based on readily available objects. For example, the cubit is one of the oldest units. A cubit is the length of your forearm from the point of your elbow to the tip of your middle finger. The table shows some other early units of measure.

Early Units of Measure	
Unit	**Measure**
girth	distance around the waist
span	length from tip of thumb to little finger of outstretched hand
fathom	distance between finger tips when arms are outstretched
grain	weight of one seed of grain
Sun or moon	time in one day

Select a unit of length to use for your project. You may choose a historical unit such as your span, or make your own unit from something you have close by. Some units could be the length of a pen, your little brother's height, or the length of your foot.

Technology Tips

- Use a **spreadsheet** to calculate areas and volumes using your own measurement system.
- Use an **electronic encyclopedia** to research historical or unusual measurements.
- Use a **word processor** to create all or part of your poster.

 **inter NET**
C O N N E C T I O N For up-to-date information on the history of mathematics, visit:
www.glencoe.com/sec/math/mac/mathnet

Working on the Project

You can use what you'll learn in Chapter 10 to help you make a poster about your system of measurement.

Page	Exercise
420	17
424	22
429	Alternative Assessment

CHAPTER Project
NOTES

Objectives Students should
- develop a new system of measure.
- compare their system with a standard system.

Project Pointer You may suggest that students begin a *Project Folder* to keep their work as they complete each stage of the Chapter Project. The completed project may also be added to their portfolios.

Make sure students choose a unit of length that is accessible and easy to use. Encourage them to use their system to measure many objects before comparing it to a standard system of measure.

Using the Table Make sure students realize that the units listed were used because no standard system of measure existed. Point out that measurements based on the human body would be defined differently by each person using them. Demonstrate how your fathom compares to a student's fathom.

***Investigations and Projects Masters*, p. 56**

10 Name_____ Date_____
Chapter 10 Project

Measuring Up

Page 420, Working on the Chapter Project, Exercise 17

Object Chosen:	Length	Width	Height	Volume
Standard Unit				
Nonstandard Unit				

Page 424, Working on the Chapter Project, Exercise 22

Object Chosen:	Surface Area
Standard Unit	
Nonstandard Unit	

© Glencoe/McGraw-Hill   56   *Mathematics: Applications and Connections, Course 1*

Instructional Resources ▶▶▶
A recording sheet to help students organize their data for the Chapter Project is shown at the right and is available in the *Investigations and Projects Masters*, p. 56.

Objective Students find the areas of irregular shapes.

Optional Resources
Hands-On Lab Masters
• centimeter grid, p. 12
• worksheet, p. 61

Overhead Manipulative Resources
• centimeter grid

Manipulative Kit
• ruler

MANAGEMENT TIPS

Recommended Time
45 minutes

Getting Started Use a centimeter grid overhead transparency to review finding the area of a rectangle. Ask students to determine how they would use centimeter grid paper to find the area of a leaf.

Activity 1 on page 396 directs students to trace their hand onto a grid. Have students make two columns to record their numbers—one for whole squares and the other for partial squares. They could then add a third column for total number of squares. Students may want to mark or shade squares they have counted so they don't miss any.

Activity 2 on page 397 directs students to draw freehand an irregular shape. Encourage them to make the shape very irregular. Have them use the same table as in Activity 1. Suggest that they redo each activity using the other method and compare results.

HANDS-ON LAB

COOPERATIVE LEARNING

10-1A Area of Irregular Shapes

A Preview of Lesson 10-1

centimeter grid paper

ruler

In Chapter 4, you learned how to find the area of a rectangle. You can also find the areas of irregular shapes. There are two different methods you might use.

TRY THIS

Work with a partner.

1 Find the mean.
• Place your hand on a sheet of centimeter grid paper so that your fingers and thumb are close together. Then draw an outline.

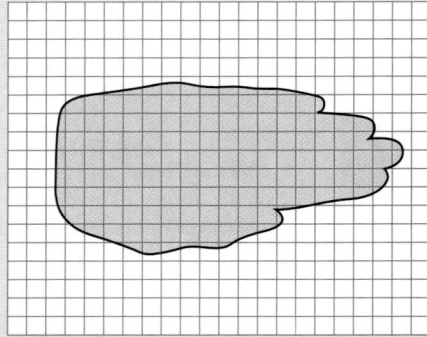

• Count the number of whole squares inside of your outline. Record the number.
• Count the number of squares that have the outline running through them. Add this number to the number of whole squares that you recorded. Then record this number.

ON YOUR OWN

1. Find the mean of the two numbers that you recorded to estimate the area of your hand. *Refer to Lesson 2-7 for information on finding the mean.* **Answers will vary.**

2. Can you think of another way to find the approximate area of your hand? Explain your answer. **See margin.**

Additional Answer
2. Sample answer: Draw two rectangles to enclose the hand outline. Find the area of each rectangle. Then add the two areas to find the total area of the hand.

TRY THIS

Work with a partner.

2 Use a rectangle.

- Draw an outline of an irregular shape like the one at the right on a piece of centimeter grid paper.

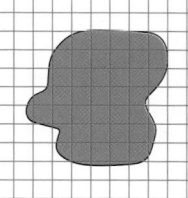

- Use the grid lines to draw a rectangle that encloses most of the figure.
- Count squares to find the length of the rectangle.
- Then count squares to find the width of the rectangle.

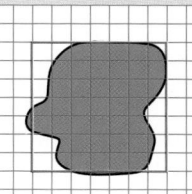

ON YOUR OWN

3–5. See students' work.

3. Find the area of the rectangle you drew to estimate the area of the irregular figure.

4. Use the method in the first activity to estimate the area of the irregular figure.

5. Which estimation method do you think is more accurate? Explain your reasoning.

Estimate the area of each figure. Use whichever method you prefer. 6–10. Sample answers given.

6.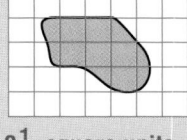

$9\frac{1}{2}$ square units

7. 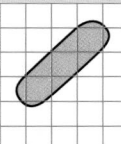 5 square units

8. 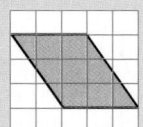 9 square units

9.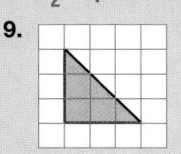

$4\frac{1}{2}$ square units

10.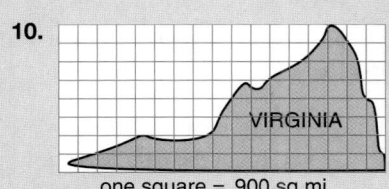

48,600 square miles

VIRGINIA

one square = 900 sq mi

11. Look Ahead Compare the area of the figure in Exercise 8 to the area of a square with sides 3 units long. What do you observe? **The areas are the same.**

Lesson 10-1A HANDS-ON LAB 397

 Have students write a paragraph about the accuracy of each method and how to determine when to use each method. Have them describe how to find the area of the top surface of a skateboard.

Have students complete Exercises 1–10. You may want to have students work in pairs for Exercises 6–10 and have each partner use a different method. Have students compare the methods and the answers. Ask if it is possible to get an exact answer using either method.

Use Exercise 11 to determine whether students understand finding the area of an irregular shape.

Hands-On Lab Masters, p. 61

10 Name _____ Date _____

Hands-On Lab
(Lesson 10-1A)

Area of Irregular Shapes

1 Try This
The number of whole squares inside the outline of my hand = See students' work.
The number of squares that have the outline running through them = See students' work.
Sum of the above numbers of squares = See students' work.

On Your Own
1. Answers will vary.
2. Sample answer: Draw two rectangles to enclose the hand outline. Find the area of each rectangle. Then add the two areas to find the total area of the hand.

2 Try This
Length of the rectangle = See students' work.
Width of the rectangle = See students' work.

On Your Own
3. Estimate of the area of the figure = See students' work.
4. Answers will vary.
5. See students' work. Sample answers are given for Exercises 6-11.
6. $9\frac{1}{2}$ square units 7. 5 square units 8. 9 square units
9. $4\frac{1}{2}$ square units 10. 48,600 square miles
11. Look Ahead The areas are the same.

© Glencoe/McGraw-Hill 61 Mathematics: Applications and Connections, Course 1

Hands-On Lab 10-1A **397**

Instructional Resources

- *Study Guide Masters*, p. 70
- *Practice Masters*, p. 70
- *Enrichment Masters*, p. 70
- Transparencies 10-1, A and B
- *Technology Masters*, p. 20

 CD-ROM Program
- Resource Lesson 10-1
- Interactive Lesson 10-1

Recommended Pacing	
Standard	Days 2 & 3 of 13
Honors	Day 2 of 13
Block	Day 2 of 7

1 FOCUS

5-Minute Check
(Chapter 9)

1. Draw a 74° angle. Then use a straightedge and compass to bisect it.

2. Tell whether the polygons are *congruent*, *similar*, or *neither*. **neither**

 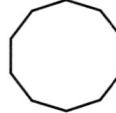

3. Name the polygons in Question 2. **octagon, decagon**

4. Draw an equilateral triangle and show any lines of symmetry.

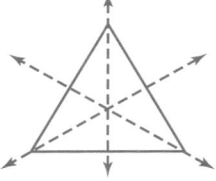

 The 5-Minute Check is also available on **Transparency 10-1A** for this lesson.

10-1 Area of Parallelograms

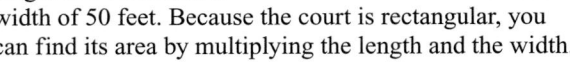

What you'll learn

You'll learn to find the area of parallelograms.

When am I ever going to use this?

Knowing how to find the area of a parallelogram can help you determine the amount of wood needed to build a deck.

Word Wise

base
height

LOOK BACK
Refer to Lesson 4-4 for information on the area of rectangles.

If you watched the 1996 Olympic Games in Atlanta, you probably remember the way the United States women's basketball team rolled over the competition to win the gold medal.

A basketball court is rectangular, with a length of 94 feet and a width of 50 feet. Because the court is rectangular, you can find its area by multiplying the length and the width.

$$A = \ell \times w$$
$$A = 94 \times 50 \quad \textit{Replace } \ell \textit{ with 94 and w with 50.}$$
$$A = 4{,}700$$

The area of a basketball court is 4,700 square feet.

A rectangle is a special type of parallelogram. A parallelogram is a quadrilateral with two pairs of parallel sides. The **base** of a parallelogram is any one of its sides. The shortest distance from the base to the opposite side is the **height** of the parallelogram.

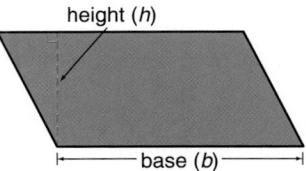

In the Mini-Lab, you will compare the area of a parallelogram to the area of a rectangle with the same height and base.

HANDS-ON

MINI-LAB

Work with a partner. grid paper 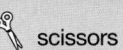 scissors   ruler

Try This
- Draw a parallelogram on a sheet of grid paper.
- Cut out the parallelogram.
- Draw a straight line to represent the height of the parallelogram. Then cut along the line.

• Reassemble the pieces to form a rectangle as shown.

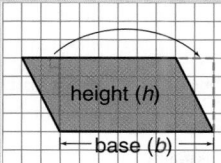

height (h)

base (b)

Talk About It

1. How do the areas of the parallelogram and the rectangle compare? **They are the same.**

2. What part of the rectangle is the same as the height of the parallelogram? What part is the same as the parallelogram's base? **height; base**

3. Use what you observed to write a formula for the area of a parallelogram. **$A = bh$**

Area of a Parallelogram	**Words:** The area (A) of a parallelogram equals the product of its base (b) and height (h).	
	Symbols: $A = bh$	**Model:** 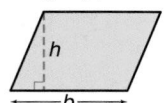

Examples

Find the area of each parallelogram to the nearest tenth.

①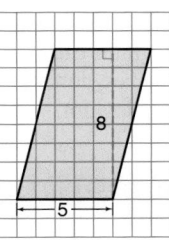

8

5

$A = bh$

$A = 5 \cdot 8$

$A = 40$

The area is 40 square units.

②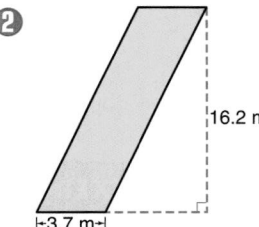

16.2 m

3.7 m

$A = bh$

$A = 3.7 \cdot 16.2$

3.7 ⊠ 16.2 ⊟ *59.94*

The area is 59.9 square meters.

Lesson 10-1 Area of Parallelograms **399**

Multiple Learning Styles

Kinesthetic Have pairs of students cut a rectangle out of grid paper. The area of the rectangle should be 20 square units. Have one partner draw a diagonal through the rectangle to form two triangles and then cut along the line. Then have the other partner arrange the two pieces to form a parallelogram. The area is still 20 square units. Repeat the activity several times with students trading roles each time. Challenge students to see how many different parallelograms they can model that have an area of 20 square units.

Lesson 10-1 **399**

---

2 TEACH

Transparency 10-1B contains a teaching aid for this lesson.

Using the Mini-Lab Have students share tasks. Ask them whether the method used would work for all parallelograms. Have students explain their answers, drawing additional parallelograms if necessary.

In-Class Examples
Find the area of each parallelogram to the nearest tenth.

For Example 1

6

5

30 square units

For Example 2

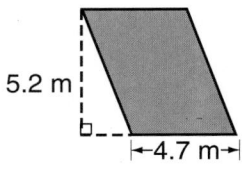

5.2 m

4.7 m

24.4 m²

Teaching Tip For Example 3, remind students how to multiply mixed numbers by changing them to improper fractions and how to change the answer back to a mixed number.

3 PRACTICE/APPLY

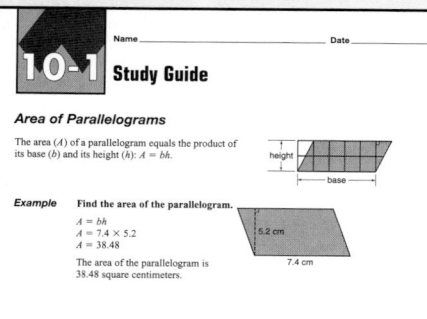
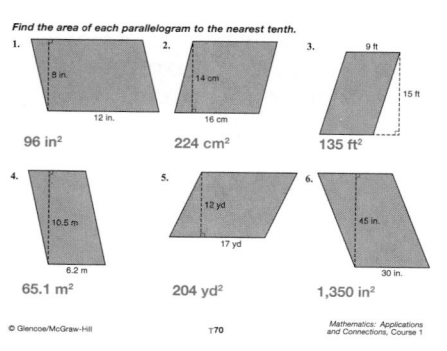

Example 3

APPLICATION

Manufacturing F.E. Beigert Company of Dallas, Texas, makes ceramic tile that is used to cover floors or walls. Find the area of a ceramic tile that needs to be glazed if it is shaped like a parallelogram with a height of $4\frac{1}{2}$ inches and a base of $6\frac{3}{4}$ inches.

Cultural Kaleidoscope

Tile was used on floors and walls in Persia before the 13ᵗʰ century. Islamic architecture is famous for its tile work.

Use the formula for the area of a parallelogram to find the area of the top of the tile.

$A = bh$

$A = \left(6\frac{3}{4}\right)\left(4\frac{1}{2}\right)$ *Replace b with $6\frac{3}{4}$ and h with $4\frac{1}{2}$.*

$A = \left(\frac{27}{4}\right)\left(\frac{9}{2}\right)$ $6\frac{3}{4} = \frac{27}{4}$, and $4\frac{1}{2} = \frac{9}{2}$

$A = \frac{243}{8}$ or $30\frac{3}{8}$ The area of the tile is $30\frac{3}{8}$ square inches.

CHECK FOR UNDERSTANDING

Communicating Mathematics

Read and study the lesson to answer each question. 1–3. See margin.

1. *State* why the formula for finding the area of a rectangle is related to the formula for finding the area of a parallelogram.

2. *Explain* why a rectangle is a parallelogram, but a parallelogram may not be a rectangle.

HANDS-ON MATH

3. *Draw* a rectangle and a parallelogram that have the same area. Explain how you know they have the same area.

Guided Practice

Find the area of each parallelogram to the nearest tenth. 4. 48 square units

4.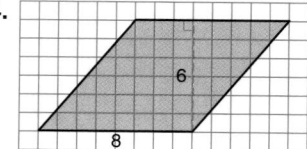

5. 13.6 m²

interNET CONNECTION

To investigate the shapes in other flags, visit: www.glencoe.com/sec/math/mac/mathnet

6. *Flags* The flag of Brunei Darussalam has white and black parallelograms on a field of yellow. If a flag is 40 inches wide and the white and black parallelograms each have an area of 240 square inches, how long is the base of each parallelogram? **6 inches**

EXERCISES

Practice

Find the area of each parallelogram. Round decimal answers to the nearest tenth.

7.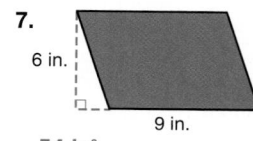
6 in.
9 in.
54 in²

8.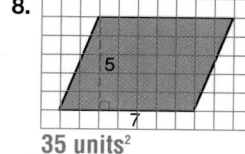
5
7
35 units²

9.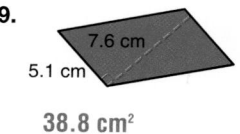
7.6 cm
5.1 cm
38.8 cm²

10. $6\frac{7}{8}$ in. $48\frac{7}{8}$ in²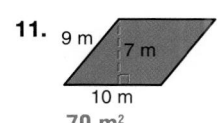
$5\frac{3}{4}$ in. $8\frac{1}{2}$ in.

11. 9 m, 7 m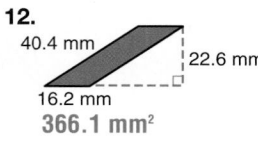
10 m
70 m²

12. 40.4 mm, 22.6 mm
16.2 mm
366.1 mm²

13. Find the area of a parallelogram that is 5.4 centimeters wide and 7.9 centimeters high to the nearest tenth. **42.7 cm²**

14. The area of a parallelogram is $20\frac{7}{12}$ square yards. What is the length of the parallelogram if the base is $3\frac{1}{4}$ yards long? $6\frac{1}{3}$ **yards**

Applications and Problem Solving

15. *Sports* Some alpine snowboards resemble parallelograms. Estimate the area of a snowboard if it is 29 centimeters wide and 159 centimeters long. **4,611 cm²**

16. *Crafts* The quilt pattern at the right is called the Lone Star pattern.
  a. Find the area of one of the smallest parallelograms. **1.65 in²**
  b. If the fabric chosen is 36 inches wide, how long should the piece be to have enough to make all of the pieces for a quilt square? **about 3 inches**

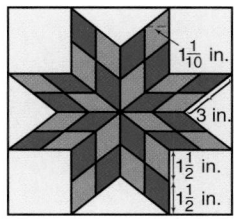
$1\frac{1}{10}$ in.
3 in.
$1\frac{1}{2}$ in.
$1\frac{1}{2}$ in.

17. *Critical Thinking* Suppose you double the height of a parallelogram.
  a. How does the area change? **The area doubles.**
  b. What happens to the area if you double the base as well as the height? **The area is increased four times.**

Mixed Review

18. **Test Practice** Which best describes the two triangles? *(Lesson 9-6)* **C**
  A symmetrical
  B congruent
  C similar
  D regular
  E reflections

2 in. / 4 in. / 4 in.
6 in. / 3 in. / 6 in.

19. State whether 708 is divisible by 2, 3, 5, 6, 9, or 10 using divisibility rules. *(Lesson 5-1)* **2, 3, 6**

Extending the Lesson

Enrichment Masters, p. 70

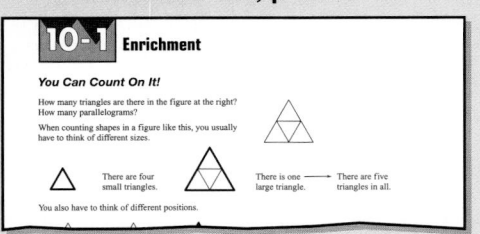

Activity Have students look for everyday items that have parallelograms as faces. Remind them that rectangles are special parallelograms. Ask students to make a scale drawing of one item.

---

4 ASSESS

Closing Activity

Modeling Have each student draw a parallelogram and its dimensions on one index card and write its area on another index card. Mix up the cards and have students match the drawing cards to the area cards.

Practice Masters, p. 70

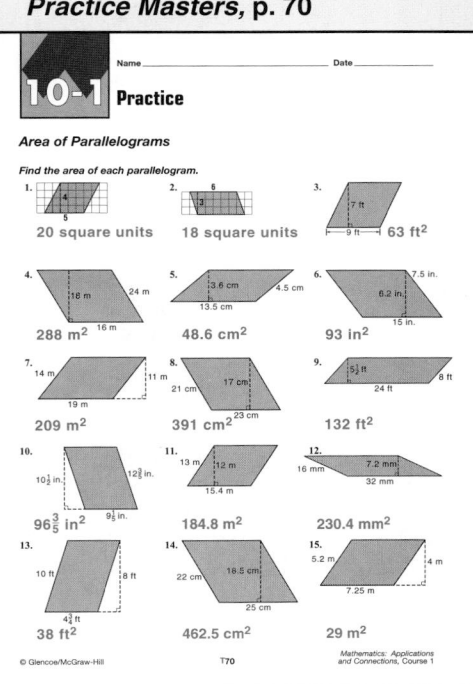

Instructional Resources

- *Study Guide Masters*, p. 71
- *Practice Masters*, p. 71
- *Enrichment Masters*, p. 71
- Transparencies 10-2, A and B
- *Assessment and Evaluation Masters*, p. 267
- *Hands-On Lab Masters*, p. 78
- *Science and Math Lab Manual*, pp. 17–20

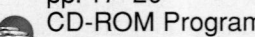 CD-ROM Program
- Resource Lesson 10-2
- Interactive Lesson 10-2

Recommended Pacing

Standard	Day 4 of 13
Honors	Day 3 of 13
Block	Day 2 of 7

1 FOCUS

5-Minute Check
(Lesson 10-1)

Find the area of each parallelogram.

1. base: 13 cm
   height: 8 cm **104 cm²**
2. base: 7 ft
   height: 9 ft **63 ft²**
3.

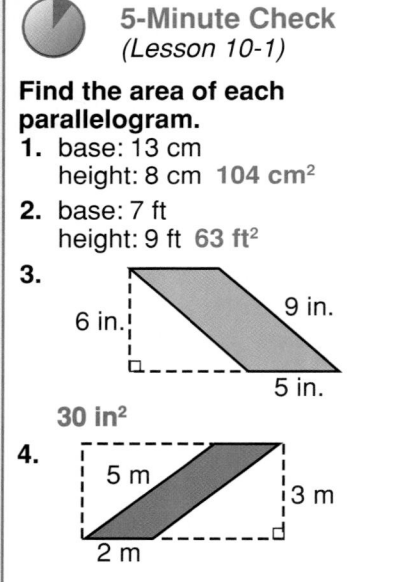

6 in. 9 in. 5 in.

30 in²

4.

5 m 3 m 2 m

6 m²

 The 5-Minute Check is also available on **Transparency 10-2A** for this lesson.

Motivating the Lesson

Communication Ask students to find the area of a triangle using the method they learned in Hands-On Lab 10-1A. Ask them if this method would work to get an exact answer.

10-2 Area of Triangles

What you'll learn

You'll learn to find the area of triangles.

When am I ever going to use this?

You can use the area of a triangle to solve problems in navigation and architecture.

Sailors have told stories of unusual occurrences in the area known as The Bermuda Triangle. This imaginary triangle has Melbourne, Florida; Bermuda; and Puerto Rico as its vertices as shown in the diagram. What is the area enclosed by the Bermuda Triangle? *You will solve this problem in Example 3.*

To answer this and other questions, you need to discover a formula for the area of a triangle. In the Mini-Lab, you will compare the area of a triangle to the area of a parallelogram.

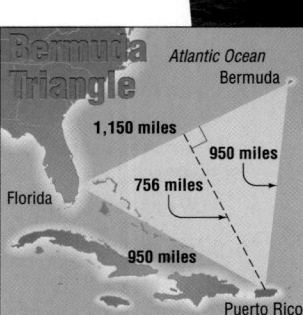

Bermuda Triangle

Atlantic Ocean
Bermuda
1,150 miles
950 miles
756 miles
Florida
950 miles
Puerto Rico

HANDS-ON MINI-LAB

Work with a partner.   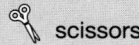 plain paper    scissors   ruler

Try This

- Trace two triangles like the one at the right on a sheet of paper.
- Cut out both triangles.
- Place the triangles together to form a parallelogram.

X Y Z

Talk About It 1–3. See margin.

1. How does the area of one of the triangles compare to the area of the parallelogram?
2. What part of the parallelogram is the same as the height of the triangle? What part is the same as the triangle's base?
3. Write a formula for finding the area of a triangle.
4. Rearrange the triangles into a different parallelogram. Does your formula still hold true? **See students' work.; yes**

402 Chapter 10 Geometry: Understanding Area and Volume

Cross-Curriculum Cue

Inform the other teachers on your team that your students are studying triangles. Suggestions for curriculum integration are:
Health: food pyramid
Social Studies: sails and navigation
Physical Science: velocity

A parallelogram can be formed by two congruent triangles. You know that the area of a parallelogram is $A = bh$. The triangles have the same area. So, the area of a triangle is one-half the area of the parallelogram, or $A = \frac{1}{2} bh$.

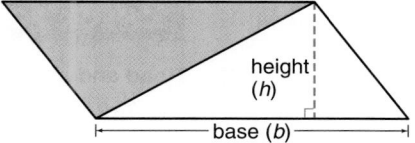

The base of a triangle is any one of its sides. The height of the triangle is the distance from a base to the opposite vertex.

Area of a Triangle	**Words:** The area (A) of a triangle equals half of the product of the length of the base (b) and the height (h).
	Symbols: $A = \frac{1}{2}bh$ **Model:** 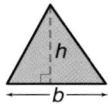

Examples

Find the area of each triangle.

1

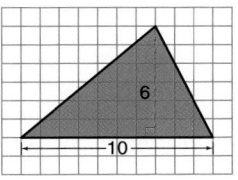

$A = \frac{1}{2} bh$

$A = \frac{1}{2} \cdot 10 \cdot 6$ *Replace b with 10 and h with 6.*

$A = 5 \cdot 6$ or 30

The area is 30 square units.

2

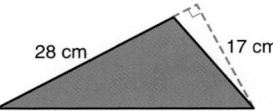

The base is 28 cm, and the height is 17 cm.

$A = \frac{1}{2} bh$

$A = \frac{1}{2} \cdot 28 \cdot 17$ *Use a calculator.*

0.5 ☒ 28 ☒ 17 🟰 *238*

$A = 238$

The area is 238 square centimeters.

CONNECTION

3 **Geography** Refer to the beginning of the lesson. Find the area of the Bermuda Triangle.

Let the base be the distance from Melbourne to Bermuda, which is 1,150 miles long. Then the height will be 756 miles. So, $b = 1,150$ and $h = 756$.

$A = \frac{1}{2} bh$

$A = \frac{1}{2} \cdot 1,150 \cdot 756$

$A = 434,700$ *0.5* ☒ *1150* ☒ *756* 🟰 *434700*

The area of the Bermuda Triangle is 434,700 square miles.

Lesson 10-2 Area of Triangles   **403**

---

2 TEACH

Transparency 10-2B contains a teaching aid for this lesson.

Using the Mini-Lab If necessary, review the formula for finding the area of a parallelogram. Ask students whether this formula works for any two triangles. Have students repeat the activity with triangles they draw.

Teaching Tip Before Example 1, point out that the formula $A = \frac{1}{2} bh$ is the same as $A = \frac{bh}{2}$.

In-Class Examples

Find the area of each triangle.

For Example 1

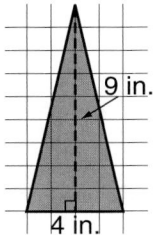

18 in²

For Example 2

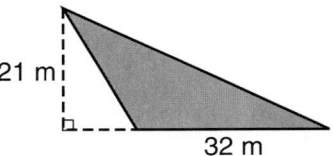

336 m²

For Example 3
Twila, Julie, and Renata are making school pennants. The height of each pennant is 18 inches. The base is 6 inches. Find the area of each pennant.
54 in²

Investigations for the Special Education Student

This blackline master booklet helps you plan for the needs of your special education students by providing long-term projects along with teacher notes. Investigation 14, *Exploring the Depths of the Sea,* may be used with this chapter.

Additional Answers for the Mini-Lab

1. **The area of the parallelogram is twice the area of the triangle.**

2. **height; base**

3. $A = \frac{1}{2}bh$

Check for Understanding

If students need additional practice or instruction after completing Exercises 1–8, one of these options may be helpful.
- Extra Practice, see p. 583
- Reteaching Activity
- *Study Guide Masters,* p. 71
- *Practice Masters,* p. 71

Assignment Guide
Core: 9–23 odd, 24–27   **Enriched:** 10–18 even, 20–27

Additional Answer

3. Ebony is correct. If 5.8 is used as the height, then the base is the length of the side to which the height is drawn, the 7-inch side.

Study Guide Masters, p. 71

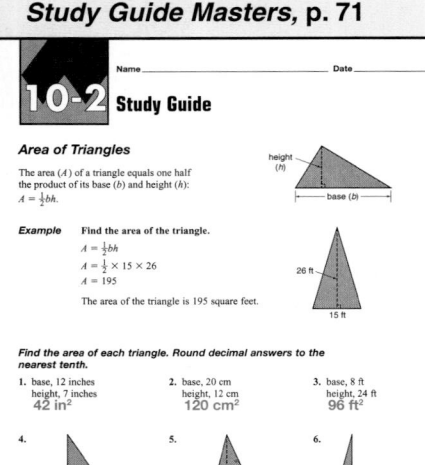

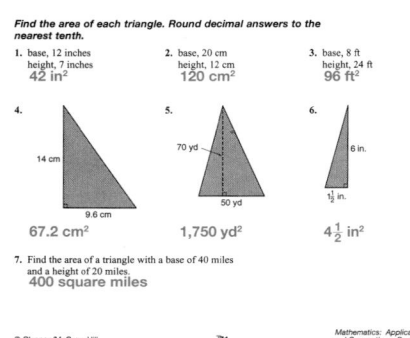

Read and study the lesson to answer these questions.

Communicating Mathematics

1. ***Tell*** why the area of a triangle can be defined as half the area of a parallelogram. **Two congruent triangles form a parallelogram.**

2. ***Develop*** a way to remember the formulas for finding the area of a parallelogram and a triangle. **See students' work.**

3. ***You Decide*** Ebony says that for the triangle at the right, $A = \frac{1}{2} \cdot 7 \cdot 5.8$. Dália says that $A = \frac{1}{2} \cdot 8 \cdot 5.8$. Who is correct and why? **See margin.**

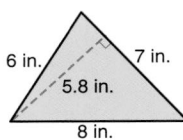

6 in.   7 in.   5.8 in.   8 in.

Guided Practice

Find the area of each triangle.

4.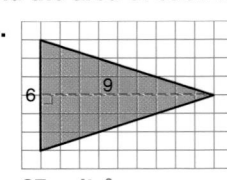
6   9
27 units²

5.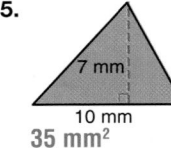
7 mm   10 mm
35 mm²

6.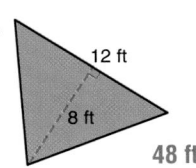
12 ft   8 ft
48 ft²

7. What is the area of a triangle with a height of 40 meters and a base 56.8 meters long? **1,136 m²**

8. ***Architecture*** The Rock and Roll Hall of Fame in Cleveland opened to fans in 1995. Part of the hall is a pyramid covered in glass. Each of the four sides of the pyramid is a triangle with a base of 241 feet and a height of 165 feet. How much glass was used to cover the entire pyramid? **79,530 ft²**

Practice

Find the area of each triangle. Round decimal answers to the nearest tenth.

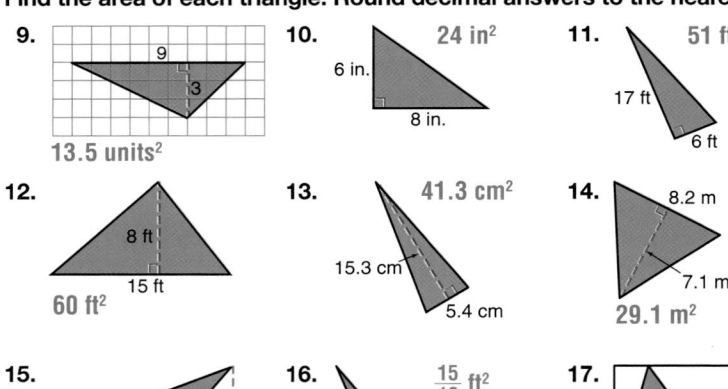

9. 9   3
13.5 units²

10. 6 in.   8 in.
24 in²

11. 17 ft   6 ft
51 ft²

12. 8 ft   15 ft
60 ft²

13. 15.3 cm   5.4 cm
41.3 cm²

14. 8.2 m   7.1 m
29.1 m²

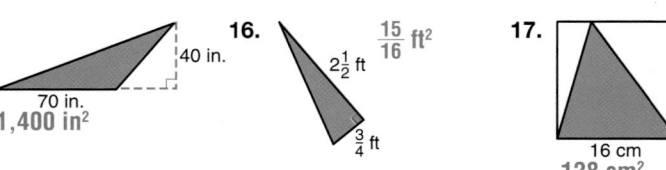

15. 40 in.   70 in.
1,400 in²

16. $\frac{15}{16}$ ft² $2\frac{1}{2}$ ft $\frac{3}{4}$ ft

17. 16 cm   16 cm
128 cm²

Reteaching the Lesson

Activity Have each partner of a student pair draw a triangle on centimeter grid paper. Then have the partners estimate the area of each other's triangle. Have them use the formula learned in this lesson to find the area of the triangles.

Error Analysis
Watch for students who multiply both the base and the height by $\frac{1}{2}$.
Prevent by reminding students that in a multiplication problem, each factor is used only once.

18. Find the area of a triangle with a base of 42 inches and a height of 35 inches. **735 in²**

19. The area of a triangle is 320 square centimeters. If the height is 20 centimeters, what is the length of the base? **32 cm**

Applications and Problem Solving
20. 355.3 square feet

20. ***Sailing*** When comparing sailboats, sailors often consider the sail area. The sail area is the total of the areas of all the sails used on the boat. On a Precision 28, the jib sail is a triangle with a base of 10 feet 6 inches and a height of 31 feet. The base of the triangular mainsail is 11 feet 7 inches and the height is 33 feet 3 inches. What is the sail area of the boat?

21. ***Life Science*** African elephants have large ears that are roughly triangular in shape. They help the elephant to keep cool in the hot climate. Research the ways that the ears cool an elephant in a reference book or on the Internet. **See margin.**

22. ***Sports*** The triangular wing of a hang glider has an area of 27 square feet. If the wingspread is 9 feet, what is the height of the wing? **6 ft**

23. ***Write a Problem*** that can be solved by finding the area of a triangle. **See students' work.**

24. ***Critical Thinking*** Draw two different triangles that have areas of 18 square feet. **See margin.**

Mixed Review

25. ***Geometry*** Find the area of the parallelogram at the right. *(Lesson 10-1)* **320 mm²**

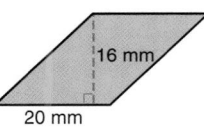
16 mm
20 mm

26. **Test Practice** At Langley High School, 19% of the 2,200 students walk to school. How many students walk to school? *(Lesson 8-7)* **B**
   A 400
   B 418
   C 428
   D 476
   E Not Here

27. ***Waste*** Every person in the U.S. is responsible for about 2.9 pounds of waste that is disposed of per day. How many ounces is that? *(Lesson 7-7)* **46.4 ounces**

Extending the Lesson

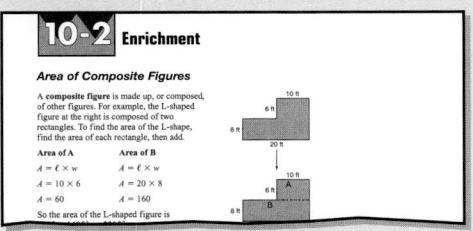

Activity Give students the area and one dimension of a triangle and have them find the other dimension. For example, ask: *What is the base of a triangle with an area of 32 square meters and a height of 16 meters?* **4 meters**

Closing Activity

Modeling Have pairs of students use geoboards and geobands to make triangles of different sizes. Ask them to find the area of each triangle in square units.

Chapter 10, Quiz A (Lessons 10-1 and 10-2) is available in the *Assessment and Evaluation Masters,* p. 267.

Additional Answers

21. **Sample answer: An elephant's ears expand its overall surface area, allowing it more space to release heat and cool itself.**

24. **Sample answers: a triangle with a base 9 feet long and a height of 4 feet; a triangle with a base of 6 feet and a height of 6 feet**

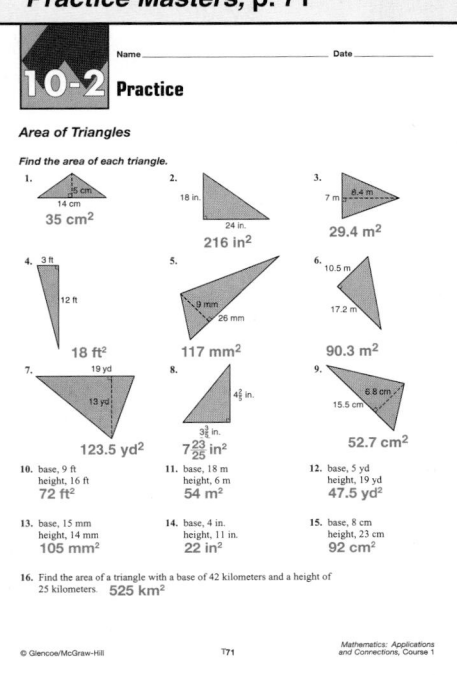

Instructional Resources

- *Study Guide Masters*, p. 72
- *Practice Masters*, p. 72
- *Enrichment Masters*, p. 72
- Transparencies 10-3, A and B
- *Assessment and Evaluation Masters*, pp. 266, 267
- *Classroom Games*, pp. 27–28
- *Technology Masters*, p. 19

 CD-ROM Program
- Resource Lesson 10-3

Recommended Pacing

Standard	Days 5 & 6 of 13
Honors	Days 4 & 5 of 13
Block	Day 3 of 7

1 FOCUS

5-Minute Check
(Lesson 10-2)

Find the area of each triangle. Round decimal answers to the nearest tenth.

1. base = 14 cm
   height = 22 cm **154 cm²**
2. base = 9 m
   height = 6.3 m **28.4 m²**

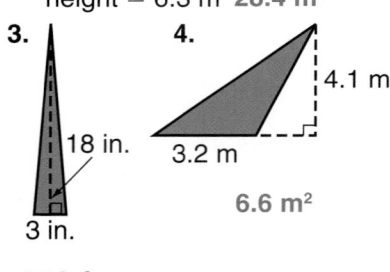

3. **27 in²**   4. **6.6 m²**
   18 in., 3 in.   4.1 m, 3.2 m

 The 5-Minute Check is also available on **Transparency 10-3A** for this lesson.

Motivating the Lesson

Problem Solving Ask students how they would solve this problem. *Antonio's Pizzeria sells a 12-inch pizza for $8.75 and a 14-inch pizza for $10.25. Which pizza costs less per square inch?*

10-3   Area of Circles

***What* you'll learn**
You'll learn to find the area of circles.

***When* am I ever going to use this?**
You can use the area of a circle to find the area of your yard covered by a sprinkler.

The people of Kobe, Japan, were awakened at 5:46 the morning of January 17, 1995 by one of the most destructive earthquakes in history. The epicenter of the quake was near the island of Awaji-shima. Damage extended as far as Osaka, 27 miles away. What is the area of the region affected by the earthquake? *This problem will be solved in Example 3.*

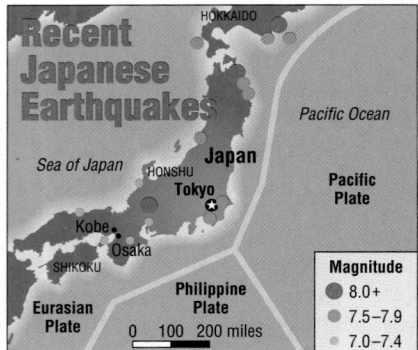

Recent Japanese Earthquakes

Magnitude
- 8.0+
- 7.5–7.9
- 7.0–7.4

Source: U.S. Geological Survey, Agence France Presse

You need a formula for the area of a circle to answer this question. In the Mini-Lab, you will find the area of a circle by forming a parallelogram.

1. the radius of the circle
2. half of the circumference of the circle
3. base × height

 LOOK BACK
Refer to Lesson 7-4 for information on the formula for circumference of a circle.

HANDS-ON   MINI-LAB

Work with a partner.   paper plate   scissors

Try This
- Fold your paper plate into eighths.
- Unfold the plate and cut along the creases.
- Arrange the pieces to form a "parallelogram" as shown below.

Talk About It
1. What is the height of the parallelogram?
2. What is the length of the parallelogram's base?
3. How would you find the area of the parallelogram?

The shape of the figure you formed in the Mini-Lab is nearly a parallelogram. The length of its base is one-half of the circumference of the circle, $\frac{1}{2}C$. So $b = \frac{1}{2}C$. Its height is the radius of the circle, r. So $h = r$.

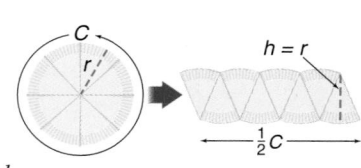

406   Chapter 10 Geometry: Understanding Area and Volume

Multiple Learning Styles

Auditory/Musical Drums come in many sizes, each with a circular head. Have pairs of students make three drums of different sizes. Have them find the area of each drumhead. After sounding each drum, discuss how the sound is affected by the size of the drum.

$A = bh$ *Formula for the area of a parallelogram*

$A = \left(\frac{1}{2}C\right)r$ *Replace b with $\frac{1}{2}C$ and h with r.*

$A = \frac{1}{2}(2\pi r)r$ *Replace C with $2\pi r$.*

$A = \pi \cdot r \cdot r$ *Simplify. $\frac{1}{2} \cdot 2 = 1$*

$A = \pi r^2$ *Simplify. $r \cdot r = r^2$*

Study Hint

Estimation Estimate the area of a circle by multiplying the square of the radius by 3.

Area of a Circle	**Words:**	The area (*A*) of a circle equals the product of π and the square of the radius (*r*).
	Symbols: $A = \pi r^2$	**Model:** 

Examples

Find the area of each circle. Use 3.14 for π.

1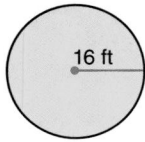

16 ft

$A = \pi r^2$
$A = \pi(16)^2$
3.14 ⊠ 16 $\boxed{x^2}$ ⊟ *803.84*
The area is about 804 ft².

2

1.4 m

The diameter is 1.4 meters. So the radius is 1.4 ÷ 2 or 0.7 m.

$A = \pi(0.7)^2$
$A \approx 1.5386$ *Use a calculator.*
The area is about 1.54 m².

CONNECTION **3** **Earth Science** Refer to the beginning of the lesson. Find the area of the region affected by the earthquake.

The radius of the circular region affected was 27 miles.

$A = \pi r^2$
$A = \pi(27)^2$
3.14 ⊠ 27 $\boxed{x^2}$ ⊟ *2289.06*
$A \approx 2,289$

The area of the region is about 2,289 square miles.

Study Hint

Technology You may want to use the $\boxed{\pi}$ key on the calculator for a more accurate calculation.

CHECK FOR UNDERSTANDING

Communicating Mathematics

Read and study the lesson to answer each question.

1. *Explain* how the area of a parallelogram is related to the area of a circle. **See margin.**

2. *Tell* why you can estimate the area of a circle by multiplying the square of the radius by 3. **π is close to 3, so $3r^2$ is close to the area of a circle.**

Additional Answer

1. A circle can be cut and arranged into a figure like a parallelogram to find the area.

---

2 TEACH

 Transparency 10-3B contains a teaching aid for this lesson.

Using the Mini-Lab Lightweight paper plates will be needed for folding and cutting. Students may want to use a ruler to compare the height of the "parallelogram" with the radius of the plate. Repeat the activity with each wedge cut into two congruent wedges (16 slices). Does this figure look more like a parallelogram?

In-Class Examples

Find the area of each circle. Use 3.14 for π.

For Example 1

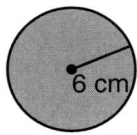

6 cm

113.0 cm²

For Example 2

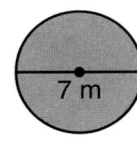

7 m

38.5 m²

For Example 3
The world's largest cylindrical sundial is at Walt Disney World in Orlando, Florida. Arata Isozaki of Tokyo, Japan, designed it. The face of the sundial has a diameter of 122 feet. What is the area of the face? Round to the nearest whole number. **11,690 ft²**

Teaching Tip Remind students to notice whether the measurement given is the radius or the diameter of the circle before finding the area.

Check for Understanding

If students need additional practice or instruction after completing Exercises 1–8, one of these options may be helpful.
- Extra Practice, see p. 583
- Reteaching Activity
- *Study Guide Masters*, p. 72
- *Practice Masters*, p. 72

Assignment Guide

Core: 9–23 odd, 25–28
Enriched: 10–22 even, 23–28

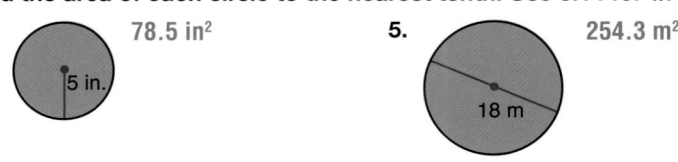

HANDS-ON MATH **3.** *Find* an object with a circular face such as a jar lid or the bottom of a mug. Then find the area of the circular region. **See students' work.**

Guided Practice **Find the area of each circle to the nearest tenth. Use 3.14 for π.**

4. 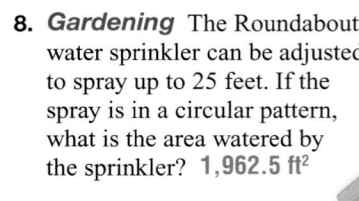 78.5 in²   5 in.

5. 254.3 m²   18 m

6. radius, 3.7 millimeters 43.0 mm²   **7.** diameter, $1\frac{1}{3}$ feet 1.4 ft²

8. *Gardening* The Roundabout water sprinkler can be adjusted to spray up to 25 feet. If the spray is in a circular pattern, what is the area watered by the sprinkler? **1,962.5 ft²**

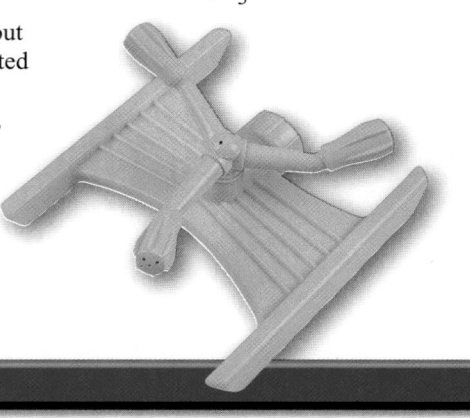

EXERCISES

Practice **Find the area of each circle to the nearest tenth. Use 3.14 for π.**

9. 28.3 in²   3 in.

10. 201.0 m²   8 m

11. 615.4 ft²   28 ft

12. 907.5 cm²   34 cm

13. 95.0 in² $5\frac{1}{2}$ in.

14. 72.3 m²   9.6 m

15. radius, 11 inches 379.9 in²   **16.** radius, 23 centimeters 1,661.1 cm²
17. radius, 4.5 meters 63.6 m²   **18.** diameter, 16 inches 201.0 in²
19. diameter, 22 millimeters 379.9 mm² **20.** diameter, $6\frac{2}{3}$ yards 34.9 yd²

21. A circle has a radius 5.5 meters long. Find the area of the circle. 95.0 m²

22. What is the area of a circle with a diameter of $32\frac{1}{2}$ feet? 829.2 ft²

Applications and Problem Solving **23.** *Physical Therapy* A therapy pool in the Central High School training room is in the shape of a circle. The diameter is 9 meters. The coach would like to have a cover made to conserve energy when the pool is not in use. How much material is needed to cover the pool? about 63.6 m²

408 Chapter 10 Geometry: Understanding Area and Volume

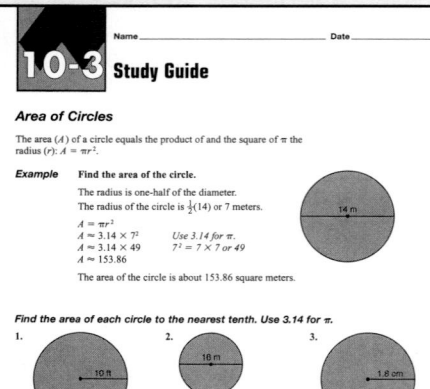
Reteaching the Lesson

Activity Have students use a compass to draw circles of different diameters. Then have them measure each diameter, find the radius, and use a calculator to find the area.

Error Analysis
Watch for students who use the diameter rather than the radius when finding the area of a circle.
Prevent by stressing that the area formula $A = \pi r^2$ uses only the radius.

24. **Sports** American Matt Ghaffari captured a silver medal in 286-pound Greco-Roman wrestling at the 1996 Olympics. Wrestling uses a 12-by-12 meter mat with a circular ring inside. The ring has an inside radius of 4.5 meters and a width of 10 centimeters. **b. about 63.62 m²**

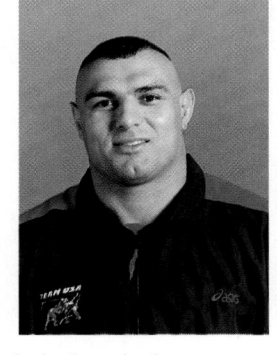

   a. Draw and label a diagram of a wrestling mat and ring. **a, c. See margin.**

   b. What is the area inside a wrestling ring?

   c. What is the area of the ring itself? (*Hint:* Subtract the area of the inside circle from that of the outside circle.)

25. **Critical Thinking** If you double the radius of a circle, how is the area affected? **The area is multiplied by 4.**

Mixed Review

26. What is the area of a triangle with a base 8 meters long and a height of 14 meters? (*Lesson 10-2*) **56 m²**

27. **Test Practice** A recipe calls for $1\frac{2}{3}$ cups water, $\frac{1}{3}$ cup oil, and $2\frac{1}{3}$ cups milk. How much liquid is used? (*Lesson 6–3*) **A**

   A $4\frac{1}{3}$ cups   **B** $3\frac{1}{3}$ cups   **C** 3 cups   **D** $2\frac{3}{3}$ cups   **E** Not Here

28. Change 3,400 centimeters to meters. (*Lesson 4-9*) **34 m**

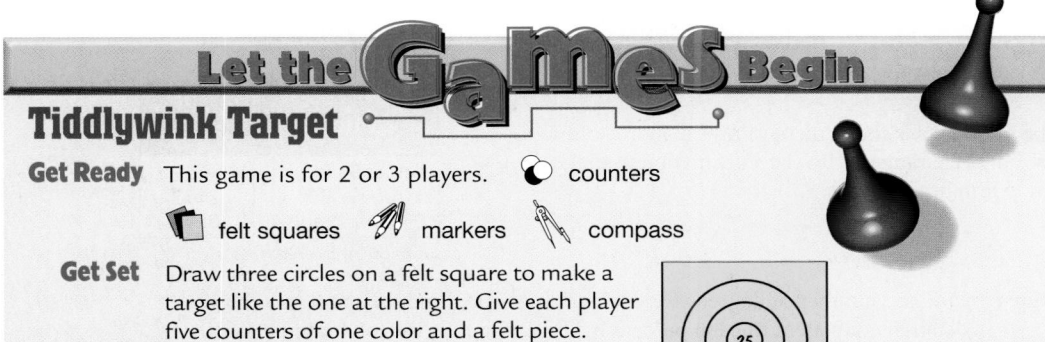

Tiddlywink Target

Get Ready This game is for 2 or 3 players.   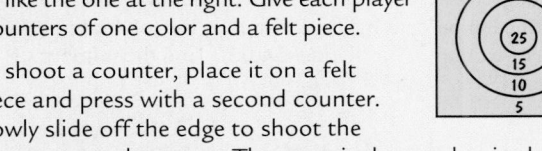 counters

   felt squares   markers   compass

Get Set Draw three circles on a felt square to make a target like the one at the right. Give each player five counters of one color and a felt piece.

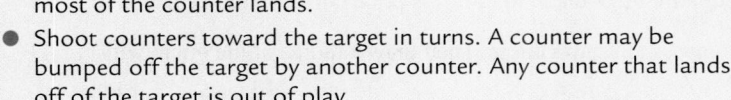

Go • To shoot a counter, place it on a felt piece and press with a second counter. Slowly slide off the edge to shoot the counter onto the target. The score is the number in the area where most of the counter lands.

   • Shoot counters toward the target in turns. A counter may be bumped off the target by another counter. Any counter that lands off of the target is out of play.

   • After each player shoots four counters, the winner is the player with the highest total score.

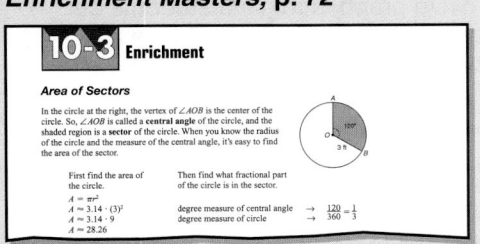 Visit www.glencoe.com/sec/math/mac/mathnet for more games.

Lesson 10-3 Area of Circles   **409**

■ Extending the Lesson ■

Enrichment Masters, p. 72

Let the Games Begin

If a player's counter lands partially in one area and partially in another, have students see how much of the counter is resting in each area. The player is awarded the points allotted to the area where most of the counter lies.

*Additional resources for this game can be found on page 44 of the **Classroom Games**.*

Closing Activity

Modeling Have students gather objects with circular faces, such as clocks, coins, cans, and campaign buttons. Have them work in pairs to determine the area of each object.

Chapter 10, Quiz B (Lesson 10-3) is available in the *Assessment and Evaluation Masters*, p. 267.

Mid-Chapter Test (Lessons 10-1 through 10-3) is available in the *Assessment and Evaluation Masters*, p. 266.

Additional Answers

24a.

12 m

10 cm

4.5 m

12 m

24c. about 2.86 m²

Practice Masters, p. 72

 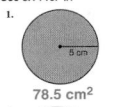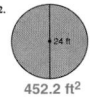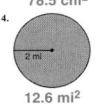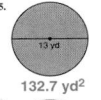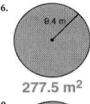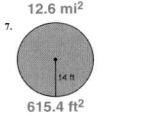
Lesson 10-3   **409**

GET READY

Objective Students construct circle graphs.

Optional Resources
Hands-On Lab Masters
- circle graph template, p. 29
- protractor, p. 22
- worksheet, p. 62

Overhead Manipulative Resources
- compass
- protractor

Manipulative Kit
- ruler
- compass
- protractor

MANAGEMENT TIPS

Recommended Time
45 minutes

Getting Started Review reading circle graphs from Chapter 2, pages 60–63. Ask students how circle graphs show that one part of the whole is less than, greater than, or about equal to another. Then ask them how they think circle graphs are constructed.

The **Activity** demonstrates how to construct a circle graph. Remind students that percents in a circle graph must add up to 100% and that the degrees must add up to 360°. Discuss the possibility that this may not always happen due to rounding. Students may then need to round up or down to get 360°.

HANDS-ON LAB

COOPERATIVE LEARNING

10-3B Making Circle Graphs

A Follow-Up of Lesson 10-3

 colored pencils

 ruler

 compass

 protractor

calculator

Sun tops the list of things Americans look for in a vacation spot. The graph shows how people answered the question "How important is sunny weather in a vacation location?"

Circle graphs are used to compare parts of a whole. Usually the information is expressed in percents as in the circle graph at the right.

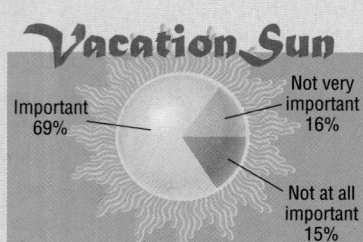

Vacation Sun

Important 69%
Not very important 16%
Not at all important 15%

Source: *Opinion Research Corp.*

TRY THIS

Work with a partner.

The chart shows the percent of toys that are sold in several price ranges. Use the information to make a circle graph.

Price	Percent of Toys
Under $4.00	47%
$4.00–$7.99	21%
$8.00 and above	32%

Source: The TMI Report

Step 1 Find the number of degrees for each price range. To do this, multiply the decimal equivalent of each percent by 360°, which is the total number of degrees in a circle. For example, for "Under $4.00," find the number of degrees as follows.

$$0.47 \; \boxed{\times} \; 360 \; \boxed{=} \; 169.2$$

The section of the circle graph for this price range should be about 169°.

Step 2 Use a compass to draw a circle. Then draw a radius of the circle with the ruler.

Step 3 Draw the angle for "Under $4.00" using your protractor. Repeat this step for each price range.

Step 4 Use the colored pencils to color each section of the graph. Label the sections. Then give the graph a title.

410 Chapter 10 Geometry: Understanding Area and Volume

Classroom Vignette

"Another way to create a circle graph is to have students cut a strip of paper 100 cm long. Have them mark the strip into sections so that each section represents a given percent of the graph (1 cm = 1%). Join the ends of the strip to form a circle and use the marks to define each section."

Gail L. Burrill

Gail Burrill, Teacher
Whitnall High School
Hales Corners, WI

ON YOUR OWN

1. Compare the graph you made to the chart. Which do you think displays the data more clearly? Explain your reasoning.

2. Why were you able to display the data on toy prices in a circle graph? **The data show the relative proportions.**

3. What type of information *cannot* be displayed in a circle graph? How could you display this type of data?

4. *Hobbies* Say cheese! Fuji Photo Film surveyed customers about how often they take pictures. Make a circle graph of the data. **See Answer Appendix.**

How Long Does Film Spend in Your Camera Before Being Developed?					
2 weeks or less	2–4 weeks	1–3 months	3–12 months	more than a year	don't know
40%	19%	19%	12%	3%	7%

Source: Fuji Photo Film U.S.A.

5. *Population* The table below shows the approximate percent of the population of Earth living in each region. Make a circle graph of the data. **See Answer Appendix.**

Region	Percent of Population
Asia	59.39%
Africa	12.68%
Europe	8.78%
Latin America and the Caribbean	8.47%
Former USSR	5.07%
North America	5.11%
Oceania and Australia	0.50%

Source: Bureau of the Census

6. *Entertainment* Do you like to play video games? According to a *Sports Illustrated* survey, 90% of kids do! The graph shows the amount of time kids who play video games spend playing each day.

   Time for Video Games

   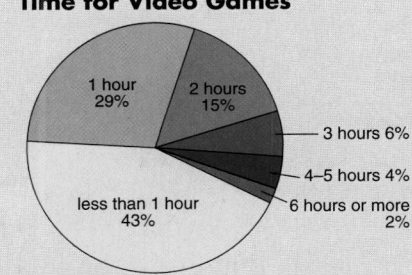

   Source: Sports Illustrated

   a. How much time do most of the kids surveyed spend playing video games each day? **less than 1 hour**

   b. How much time do 15% of the video players spend each day? **2 hours**

   c. What percent of those surveyed spend one hour or less on video games each day? **72%**

7. *Reflect Back* How is the area of a circle related to making a circle graph? **The area of the circle is divided into parts to represent the portion of the whole for each category.**

Lesson 10-3B HANDS-ON **LAB** 411

---

1. Sample answer: The circle graph, because it allows a visual comparison of the proportions.

3. Sample answers: data that doesn't equal 100%; a bar or line graph

---

Have students complete Exercises 1–4. Discuss with students how they could make a circle graph with information that is not given in percents. Make sure they understand that only information that can be shown as a part of a whole should be shown in a circle graph.

Use Exercises 5–7 to determine whether students understand how to construct and analyze a circle graph.

Hands-On Lab Masters, p. 62

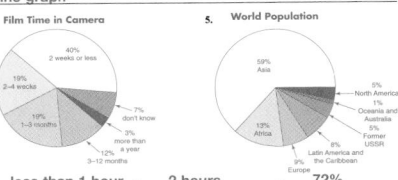

Math Journal

Have students write a paragraph about the uses of circle graphs. Encourage them to include a description of the limitations.

Instructional Resources

* *Study Guide Masters*, p. 73
* *Practice Masters*, p. 73
* *Enrichment Masters*, p. 73
* Transparencies 10-4, A and B
* *Diversity Masters*, p. 10

 CD-ROM Program
  * Resource Lesson 10-4
  * Extended Activity 10-4

Recommended Pacing

Standard	Days 7 & 8 of 13
Honors	Days 6 & 7 of 13
Block	Day 4 of 7

1 FOCUS

5-Minute Check
(Lesson 10-3)

Find the area of each circle to the nearest tenth. Use 3.14 for π.

1. 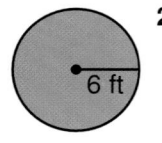 6 ft
2. 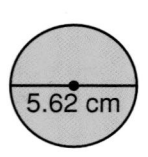 5.62 cm

   113.0 ft²   24.8 cm²

3. radius, 32 m 3,215.4 m²
4. diameter, 4 ft 12.6 ft²
5. Find the area of a circle if the diameter is 24 cm. 452.2 cm²

 The 5-Minute Check is also available on **Transparency 10-4A** for this lesson.

Motivating the Lesson

Hands-On Activity Use a variety of objects such as cans, boxes, marbles, and party hats or use wooden geometric solids. Pass them around and have students identify the flat and curved surfaces. Discuss how the shapes could be named.

10-4 Three-Dimensional Figures

What you'll learn
You'll learn to identify three-dimensional figures.

When am I ever going to use this?
Knowing how to identify three-dimensional figures will help you to solve problems in social studies and earth science.

Word Wise
three-dimensional figure
face
edge
vertex (vertices)
prism
base
rectangular prism
pyramid
square pyramid
cone
cylinder
sphere
center

You can find Captain Eddie's Flying Circus Kite Team in competitions all year. A stunt kite like the team uses is a two-dimensional object. Other kites, like box kites, are three-dimensional objects.

A **three-dimensional figure** encloses a part of space. The flat surfaces of a three-dimensional figure are called **faces**. The **edges** are the segments formed by intersecting faces. The edges intersect at the **vertices**.

Boxes are examples of **prisms**. A prism has at least three lateral faces that are shaped like rectangles. The faces on the top and bottom are the **bases** and are parallel. The shape of the bases tells the name of the prism. Because its bases are rectangles, a cereal box is a **rectangular prism**.

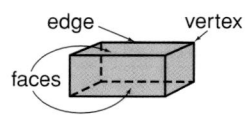

A rectangular prism has 12 edges and 8 vertices.

A **pyramid**, like the Egyptian pyramids, has only one base. The base can be shaped like any polygon. The shape of the base gives the pyramid its name. For example, a pyramid with a square base is called a **square pyramid**. The other faces of the pyramid are triangles.

This pyramid has 8 edges and 5 vertices.

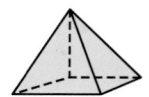
Example CONNECTION

Geography How many faces, edges, and vertices are there in the Great Pyramid of Khufu?

The Great Pyramid of Khufu is a square pyramid. It has one square face and four triangular faces. So there are five faces in all.

There are four edges where the side faces meet the base. Four more edges are formed where the side faces meet each other. So, there are eight edges in the pyramid.

There are four vertices of the square base and the vertex where the other faces meet. So there are five vertices in the pyramid.

Classroom Vignette

"In the spring, students construct three-dimensional tetrahedron kites using straws, string, and tissue paper. Then we go outside and attempt to fly the kites."

Kathy Bessette, Teacher
Dover Middle School
Dover, NH

Kathy Bessette

There are three-dimensional figures that have curved surfaces. One of these figures is a **cone**. Like a pyramid, a cone has one base. Its base is a circle.

A cone has one vertex, but no edges.

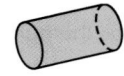

A cylinder has no vertices and no edges.

Another figure with a curved surface is a **cylinder**. A soup can is a model of a cylinder. A cylinder has two circular bases.

A basketball is a model of a sphere. A **sphere** is a three-dimensional figure with no faces, bases, edges, or vertices. All of the points on a sphere are the same distance from a given point called the **center**.

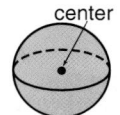

center

A sphere has no vertices or edges.

CHECK FOR UNDERSTANDING

Communicating Mathematics

Read and study the lesson to answer each question.

1. **Tell** what type of pyramid is shown at the right. **square pyramid**

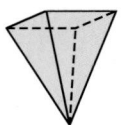

2. **Explain** the difference between a two-dimensional and a three-dimensional figure. **See Answer Appendix.**

3. **List** each type of three-dimensional figure in this lesson. Then give an example of a real-world object with that shape. **See students' work.**

Math Journal

Guided Practice

Name each figure.

4. triangular pyramid

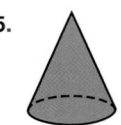

5. cone

State the number of faces, edges, and vertices in each figure.

6. rectangular prism 6; 12; 8

7. triangular prism 5; 9; 6

8. **Hobbies** What type of three-dimensional figure is the base for the box kite shown on page 412? **rectangular prism**

EXERCISES

Practice

Name each figure.

9.

cylinder

10.

sphere

11.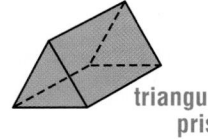

triangular prism

Lesson 10-4 Three-Dimensional Figures **413**

Reteaching the Lesson

Activity Draw the figures shown in this lesson on an overhead transparency and have students identify them. Trace the edges and vertices of each figure as students count them. Then display an object or block and have students identify and count the faces.

Error Analysis
Watch for students who have difficulty interpreting three-dimensional drawings. **Prevent by** suggesting they focus on one face of the object at a time.

---

2 TEACH

 Transparency 10-4B contains a teaching aid for this lesson.

Reading Mathematics Have students use a dictionary to look up the words in the Word Wise list. Suggest that they read all of the definitions for each word and pick out the mathematical one to copy. Then discuss their lists of definitions and make sure they understand what each term means. Emphasize that "vertices" is plural for "vertex."

In-Class Example

For the Example
a. How many faces, edges, and vertices are there in a triangular pyramid? 4, 6, 4
b. How many faces, edges, and vertices are there in a shoebox? 6, 12, 8

3 PRACTICE/APPLY

Check for Understanding
If students need additional practice or instruction after completing Exercises 1–8, one of these options may be helpful.
• Extra Practice, see p. 584
• Reteaching Activity
• *Study Guide Masters,* p. 73
• *Practice Masters,* p. 73

***Study Guide Masters,* p. 73**

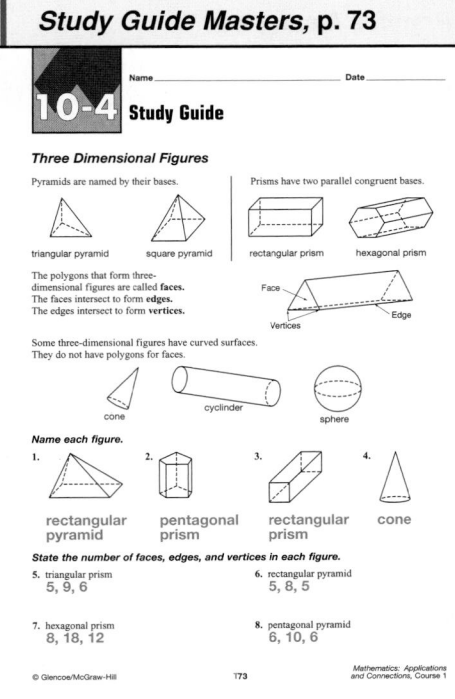

Lesson 10-4 413

Family Activity

Have students give examples of objects that were found. You may want to keep track of these on the chalkboard or overhead.

4 ASSESS

Closing Activity

Speaking Separate students into pairs. Ask one partner to verbally describe an object in the room including the number of faces, edges, and vertices. Have the other partner guess the identity of the object. Then have them reverse roles.

Mid-Chapter Self Test

The Mid-Chapter Self Test reviews the concepts in Lessons 10-1 through 10-4. Lesson references are given so students can review concepts not yet mastered.

Practice Masters, p. 73

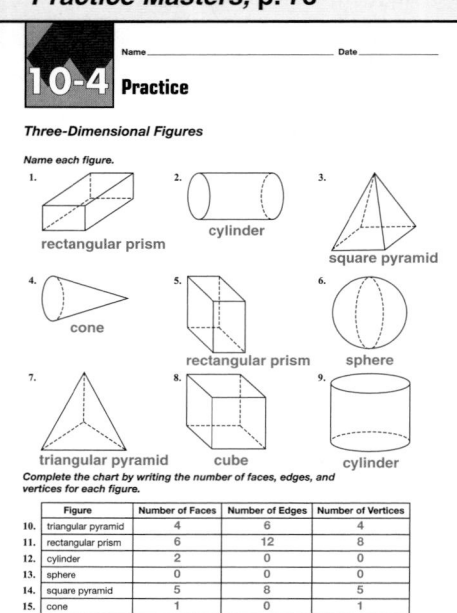

Family Activity

Write a list of three-dimensional figures. Then challenge family members to a scavenger hunt to see who can find the most items that match the figures on the list.

Name each figure.   12. rectangular prism

12.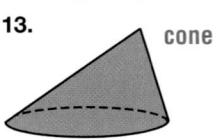
13.   cone
14.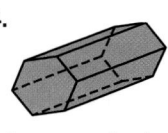
hexagonal prism

Copy and complete the chart for the numbers of faces, edges, and vertices in each figure.

	Figure	Faces	Edges	Vertices
15.	triangular pyramid	4	6	4
16.	cylinder	none	none	none
17.	sphere	none	none	none
18.	hexagonal pyramid	7	12	7
19.	hexagonal prism	8	18	12

Applications and Problem Solving

Crystal of alum

20. **Number Theory** Swiss mathematician Leonard Euler (pronounced OY ler) found that in three-dimensional figures that have no curved surfaces $E = F + V - 2$ for the number of faces F, edges E, and vertices, V. Does this formula agree with what you found in Exercises 15, 18, and 19? **Sample answer: yes**

21. **Earth Science** A crystal of alum is in the form of two square pyramids sharing the same base. **a.8; 12; 6   b. yes**
   a. How many faces, edges, and vertices are in the crystal?
   b. Does your answer to part a agree with Euler's formula above?

22. **Critical Thinking** A pyramid has four faces. What type of pyramid must it be? **triangular**

Mixed Review
23. 1,519.76 in²

23. Find the area of a circle with a radius of 22 inches. *(Lesson 10-3)*

24. How many lines of symmetry are there in a square? *(Lesson 9-5)* **4 lines**

25. **Test Practice**   If you work 22 hours a week and earn $139.70, how much money do you earn per hour? *(Lesson 8-2)* **B**
   A $6.50   B $6.35   C $6.05   D $5.85   E Not Here

CHAPTER 10   Mid-Chapter Self Test

1. Find the area of a parallelogram with a base 50 feet long and a height of 77 feet. *(Lesson 10-1)*

2. What is the area of a triangle with a 21-inch base and a height of 15 inches? *(Lesson 10-2)*

3. What is the area of a circle with a 2-inch radius? *(Lesson 10-3)* **12.6 in²**

4. The diameter of a circle is 88 meters. Find the area of the circle. *(Lesson 10-3)* **6,079.0 m²**

5. A stack of compact discs is a model of what three-dimensional figure? *(Lesson 10-4)* **cylinder**
1. 3,850 ft²   2. 157.5 in²

414   Chapter 10 Geometry: Understanding Area and Volume

Extending the Lesson

Enrichment Masters, p. 73

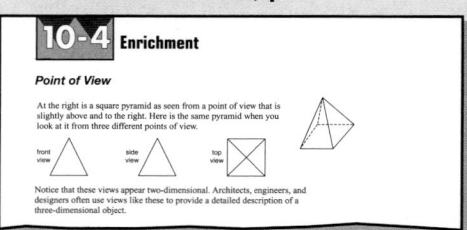

Activity Have groups of students write a brief report on famous examples of various types of three-dimensional buildings such as the Egyptian pyramids. Ask them to look for information such as when and how the buildings were constructed.

HANDS-ON LAB

COOPERATIVE LEARNING

10-4B Three-Dimensional Figures

A Follow-Up of Lesson 10-4

isometric
dot paper

ruler

Sometimes it helps to draw a sketch of a three-dimensional figure when you are trying to solve a problem. You can use isometric dot paper to help you draw a three-dimensional figure.

TRY THIS

Work with a partner.

A rectangular prism has a length of 3 units, a height of 4 units, and a width of 2 units. Use isometric dot paper to sketch the prism.

The rectangular surfaces of the prism are drawn as parallelograms to give a three-dimensional appearance.

Step 1 Draw a parallelogram with sides of 3 units and 2 units. This represents the top of the prism.

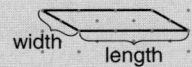
width   length

Step 2 Place your pencil at one of the vertices of the parallelogram. Then draw a line passing through four dots. Repeat for the other three vertices, drawing the hidden edges as dashed lines.

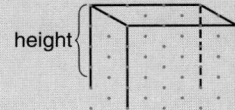
height

Step 3 Finally, connect the ends of the lines to complete the prism.

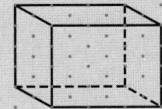

ON YOUR OWN

1. In the prism that you drew, which faces are the bases of the prism?

2. If you turned the drawing on its side, would the figure appear different? **yes**

3. Explain how to draw a prism with a hexagonal base. **See margin.**
1. the sides that are rectangles 3 units long and 2 units wide

Use isometric dot paper to draw a sketch of each figure. 4–5. **See Answer Appendix.**

4. a rectangular prism 4 units long, 2 units high, and 2 units wide

5. cube that is 3 units long, 3 units high, and 3 units wide

6. *Reflect Back* Do you think this method would work well for sketching a sphere? Explain. **Sample answer: No; because the sphere has no faces to draw.**

Lesson 10-4B HANDS-ON **415**

Have students write a description of how to use isometric dot paper to draw a three-dimensional figure.

GET READY

Objective Students draw three-dimensional figures.

Optional Resources
Hands-On Lab Masters
• isometric dot paper, p. 14
• worksheet, p. 63

Overhead Manipulative Resources
• isometric dot paper

Manipulative Kit
• ruler
• rubber stamp (isometric dot paper)

MANAGEMENT TIPS

Recommended Time
30 minutes

Getting Started Discuss with students the differences between two-dimensional and three-dimensional drawings. Have students try to sketch a simple box freehand. Discuss what would make this task easier.

The **Activity** demonstrates drawing a three-dimensional figure using isometric dot paper. Make sure students use the same number of units for parallel sides.

ASSESS

Have students complete Exercises 1–5. Watch for students who have difficulty connecting the points. Remind them that the bases of a prism match.

Use Exercise 6 to determine whether students understand drawing three-dimensional figures.

Additional Answer
3. Draw the hexagonal base first. Then for each vertex, draw a line to represent depth. Then connect the ends of the segments.

Objective
Students solve problems by making models.

Recommended Pacing	
Standard	Day 9 of 13
Honors	Day 8 of 13
Block	Day 5 of 7

1 FOCUS

Getting Started Give students centimeter cubes and have them form various size "squares." Have them record how many cubes it takes to make each square.

2 TEACH

Teaching Tip Have students duplicate Emilio and Terrence's model with small blocks. Have them try the activity with other numbers of cubes.

In-Class Example
Charles has 24 cubes. Which structure can he make?

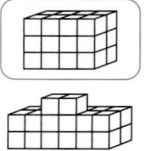

10-5A Make a Model

A Preview of Lesson 10-5

Emilio and Terrence work at Brooks Sporting Goods. The store has just received a shipment of 72 basketballs packaged in 18-inch cubes. The store manager asked Emilio and Terrence to use as many of them as possible to build a pyramid with the boxes in the display window. Let's listen in as they plan.

Emilio

You know, Terrence, there are an awful lot of boxes here. Let's make sure we have a plan that will work before we start to build.

Terrence

You're right. Let's build a model with these smaller baseball boxes over here. The base of the pyramid should be a square. Then each layer above it can be a smaller square.

That sounds good. If the bottom layer is a 4-by-4 square, it will take 16 boxes.

The next layer would be 3 by 3, or 9 boxes. Then 2 by 2 or 4 boxes in the next layer. And 1 box on top.

That's 16 + 9 + 4 + 1 or 30 boxes. Since we have 72 boxes to use, we can probably add another layer.

OK. Then we'd start with a 5-by-5 layer. So we'd use 25 boxes in the first layer and 30 in the ones above it, for a total of 55 boxes.

Great! That leaves 72 − 55 or 17 boxes left over. 17 isn't enough for a 6-by-6 layer, so it will work.

Grab a box and let's go!

■ Reteaching the Lesson ■

Activity Have students work in pairs to create word problems that can be solved by making a model. Have them exchange their problems with others to solve. Be sure they check each other's models.

THINK ABOUT IT

Work with a partner.

1. **Tell** how making a model helped Terrence and Emilio plan their pyramid. **See margin.**

2. **Describe** another way that Terrence and Emilio could have displayed the basketballs. **See margin.**

ON YOUR OWN

3. **See margin.**
3. The last step of the 4-step plan for problem solving is to *examine* the solution. How did Emilio and Terrence examine the solution?

4. **Write** a list of real-life situations where it would be helpful to **make a model** to solve a problem. **Answers will vary.**

5. **Look Ahead** If you built a model of Emilio and Terrence's pyramid with cubes that are 1 cubic inch each, how many cubic inches would there be in the model? **55 cubic inches**

8. Suzy—volleyball; Benito—soccer; Vicky—basketball

MIXED PROBLEM SOLVING

STRATEGIES
Look for a pattern.
Solve a simpler problem.
Act it out.
Guess and check.
Draw a diagram.
Make a chart.
Work backward.

Solve. Use any strategy.

6. **Patterns** A number is doubled and then 9 is subtracted. If the result is 15, what was the original number? **12**

7. **Sales** Karen is making a pyramid-shaped display of laundry detergent. Each box is a rectangular prism. The bottom layer of the pyramid has six boxes. If there is one less box in each layer and there are five layers in the pyramid, how many boxes will Karen need to make the display? **20 boxes**

8. **Sports** Vicky, Benito, and Suzy play volleyball, soccer, and basketball. One of the girls is Benito's next door neighbor. No person's sport begins with the same letter as their first name. Benito's neighbor plays volleyball. Which sport does each person play?

9. **Geometry** A rectangular prism is made of exactly 8 cubes. Find the length, width, and height of the prism. **Sample answer: 4 by 2 by 1**

10. **Geometry** Find the area of the shaded region. Round to the nearest tenth. **7.7 m²**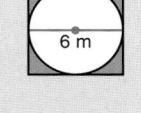

11. **Number Theory** What is the least positive number that you can divide by 7 and get a remainder of 4, divide by 8 and get a remainder of 5, and divide by 9 and get a remainder of 6? **501**

12. **Test Practice** An equilateral triangle has a base of 10 inches and a height of 8.7 inches. If you arranged six triangles like this into a hexagon as shown, what is the area of the hexagon? **D**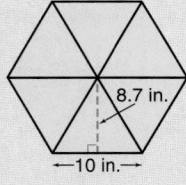

   A 600 in²
   B 522 in²
   C 300 in²
   D 261 in²

Lesson 10-5A THINKING **LAB** **417**

■ Extending the Lesson ■

Activity Ask students what other shapes besides cubes could be useful for making models. Have them look at the lists they made in Exercise 4 of On Your Own and determine for which situations other shapes may be needed to build accurate models.

3 PRACTICE/APPLY

Check for Understanding
Use the results from Exercises 1–2 to determine whether students understand how to solve a problem by modeling.

Extra Practice If students need additional practice in problem solving, extra practice is available on the following pages.
• Make a Model, see p. 584
• Mixed Problem Solving, see pp. 593–594

Assignment Guide

All: 3–12

4 ASSESS

Closing Activity
Writing Have students write an explanation of how a model could be used to decide how to build storage units for a closet.

Additional Answers
1. Sample answer: It allowed them to first build a small pyramid with easy-to-move boxes.
2. Sample answer: They could have built a prism that was 3 boxes long, 3 boxes wide, and 8 boxes tall.
3. Emilio made sure that the number of boxes they planned to use was the most possible for the type of pyramid they were building.

Instructional Resources

- *Study Guide Masters*, p. 74
- *Practice Masters*, p. 74
- *Enrichment Masters*, p. 74
- Transparencies 10-5, A and B
- *Assessment and Evaluation Masters*, p. 268

CD-ROM Program
- Resource Lesson 10-5
- Interactive Lesson 10-5

Recommended Pacing	
Standard	Day 10 of 13
Honors	Day 9 of 13
Block	Day 5 of 7

1 FOCUS

5-Minute Check
(Lesson 10-4)

Name each figure.

1.   2.

pyramid   cylinder

State the number of faces, edges, and vertices for each figure.

3. cone **no faces, no edges, 1 vertex**
4. triangular pyramid **4 faces, 6 edges, 4 vertices**
5. rectangular prism **6 faces, 12 edges, 8 vertices**

The 5-Minute Check is also available on **Transparency 10-5A** for this lesson.

2 TEACH

Transparency 10-5B contains a teaching aid for this lesson.

Using the Mini-Lab You may suggest that students make up dimensions for a fourth prism and predict the number of cubes that would be needed to build the prism.

10-5   Volume of Rectangular Prisms

What you'll learn
You'll learn to find the volume of rectangular prisms.

When am I ever going to use this?
Knowing how to find the volume of a rectangular prism will help you find the amount of water needed to fill an aquarium.

Word Wise
volume

7. The volume of a rectangular prism is the product of its length, width, and height.

The amount of space inside a three-dimensional figure is called its **volume**. You can investigate the volume of a rectangular prism by building models.

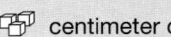

HANDS-ON MINI-LAB

Work with a partner.   centimeter cubes

Try This

1. Build rectangular prism A so that it is 3 units long, 2 units wide, and 4 units high. **See student's work.**
2. Count the number of cubes that you used to build the prism. **24**
3. Find the product of the length, width, and height of the prism. **24**
4. Copy the table below. Complete the first row. **See below.**

Prism	Length	Width	Height	Number of Cubes	Length × Width × Height
A	3	2	4	24	24
B	3	4	4	48	48
C	4	1	3	12	12

Talk About It

5. In prism A, how does the number of cubes compare to the product of its length, width, and height? **They are the same.**
6. Build prisms B and C with the dimensions given. Record the results in your table. **See above.**
7. Write a sentence about how the volume of a prism is related to the length, width, and height.

From the Mini-Lab, we can conclude that the volume of a rectangular prism is directly related to its dimensions.

Volume of a Rectangular Prism	**Words:** The volume (V) of a rectangular prism equals the product of its length (ℓ), its width (w), and its height (h).
	Symbols: $V = \ell w h$ **Model:**

Volume is expressed in cubic units. For example, if the length, width, and height are measured in feet, the volume will be written in cubic feet, or ft^3.

Motivating the Lesson

Communication Review Lessons 4-8 and 7-7. Have students compare capacity and volume. Discuss the different units that are used for volume.

Multiple Learning Styles

Naturalist Have students look around at natural shapes when they are outside. Have them make sketches of the three-dimensional shapes that they see. They should break down complex shapes into the simpler shapes they are learning in this chapter.

1 Find the volume of the rectangular prism.

$\ell = 4$ cm

$w = 6$ cm

$h = 10$ cm

$V = \ell wh$

$V = 4 \times 6 \times 10$

$V = 240$

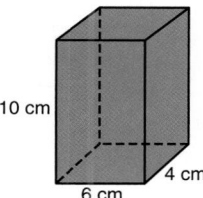

10 cm

6 cm

4 cm

The volume is 240 cm³.

APPLICATION 2 **Money Matters** Firewood is sold in units called *cords*. A cord of wood is 8 feet long, 4 feet wide, and 4 feet high. If you buy a stack of wood that is 8 feet long, 2 feet wide, and 8 feet high, is it equivalent to a cord?

Explore   You know the dimensions of a cord. You need to know if a stack that is 8 feet long, 2 feet wide, and 8 feet high is equivalent to a cord.

Plan   If the volume of your stack is equal to the volume of a cord, then your stack is equivalent to a cord. Find the volume of your stack and of a cord. Then compare.

Solve

Volume of your stack	Volume of a cord
$V = \ell wh$	$V = \ell wh$
$V = 8 \times 2 \times 8$	$V = 8 \times 4 \times 4$
$V = 128$ ft³	$V = 128$ ft³

The volumes are equal. So the 8-by-2-by-8 foot stack is equivalent to a cord.

Examine   Use cubes to build a rectangular prism that is 8 units long, 4 units wide, and 4 units high. Rearrange the cubes to verify that they can make a prism 8 units long, 2 units wide, and 8 units high.

CHECK FOR UNDERSTANDING

Communicating Mathematics

Read and study the lesson to answer each question. 1–2. See margin.

1. *State* why volume is expressed in cubic units.

2. *You Decide* Maise says that when you find the volume of a prism, the order in which you multiply the length, width, and height does not matter. Craig disagrees. Who is correct and why?

HANDS-ON 3. *Build* a rectangular prism that has a volume of 18 cm³ using centimeter cubes.
MATH   Sample answer: a prism 1 cm high, 2 cm wide, and 9 cm long

Lesson 10-5 Volume of Rectangular Prisms   **419**

■ Reteaching the Lesson ■

Activity Have students work in pairs using centimeter cubes to construct models. Ask them to build a model, sketch it, find its volume with a calculator, and then take it apart and count the cubes to check their answer.

Additional Answers

1. Each of the three dimensions being multiplied is expressed in a unit of measure.

2. Maise is correct. Because multiplication is commutative, the order in which the numbers are multiplied is not important.

In-Class Examples

For Example 1
Find the volume of the rectangular prism. **64 ft³**

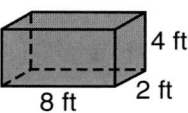

4 ft

8 ft   2 ft

For Example 2
An average cereal box is 19 centimeters long, 6 centimeters wide, and 29 centimeters high. If the cereal box is full, will all the cereal fit in a plastic storage container 17 centimeters long, 8 centimeters wide, and 30 centimeters high? **yes**

Teaching Tip Remind students that it doesn't matter in which order they multiply the length, width, and height when finding volume.

3 PRACTICE/APPLY

Check for Understanding
If students need additional practice or instruction after completing Exercises 1–6, one of these options may be helpful.
• Extra Practice, see p. 584
• Reteaching Activity
• *Study Guide Masters*, p. 74
• *Practice Masters*, p. 74

Study Guide Masters, p. 74

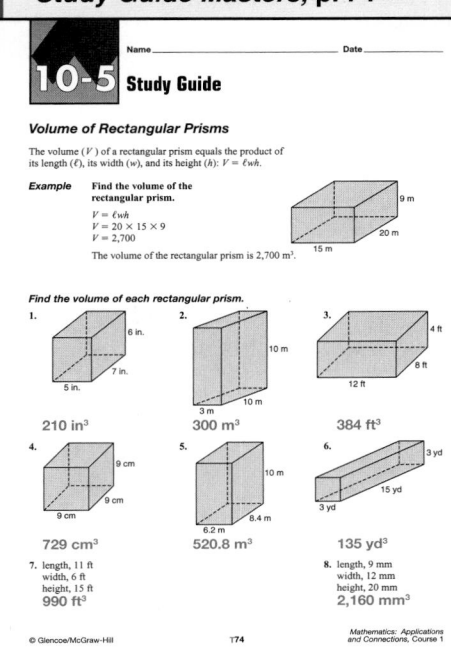

Lesson 10-5   419

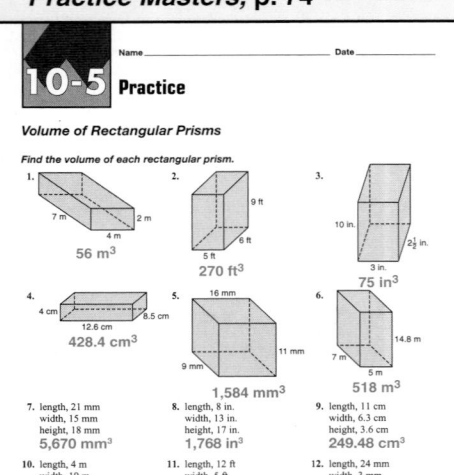
Guided Practice   **Find the volume of each rectangular prism.**

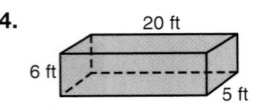 **600 ft³**   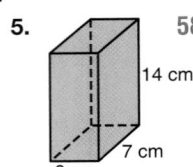 **588 cm³**

6. Find the volume of a rectangular prism 15 meters long, 22 meters wide, and 26 meters tall. **8,580 m³**

EXERCISES

Practice   **Find the volume of each rectangular prism to the nearest tenth.** **9. 1,536 ft³**

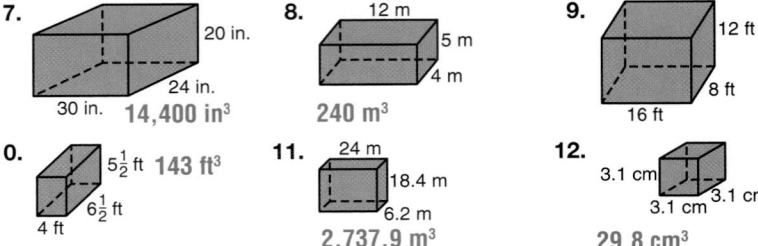

7. **14,400 in³**   8. **240 m³**   9.

10. 5½ ft **143 ft³**   **11. 2,737.9 m³**   **12. 29.8 cm³**

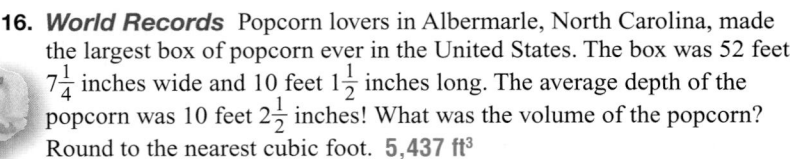
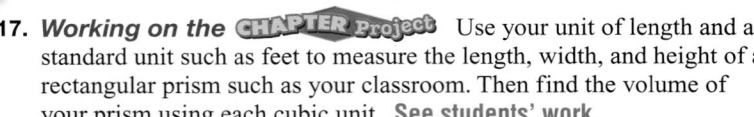

13. Find the volume of a rectangular prism that is 5 mm wide, 3 mm high, and 6 mm long. **90 mm³**

14. What is the volume of a rectangular prism 15 by 12 by 3 feet? **540 ft³**

Applications and Problem Solving

15. *Sports* An Olympic-sized pool is 25 meters wide, 50 meters long, and 3 meters deep.
  a. What is the pool's volume? **3,750 m³**
  b. A liter of water occupies 0.001 cubic meter. How many liters of water are needed to fill an Olympic-sized pool? **3,750,000 liters**

16. *World Records* Popcorn lovers in Albermarle, North Carolina, made the largest box of popcorn ever in the United States. The box was 52 feet 7¼ inches wide and 10 feet 1½ inches long. The average depth of the popcorn was 10 feet 2½ inches! What was the volume of the popcorn? Round to the nearest cubic foot. **5,437 ft³**

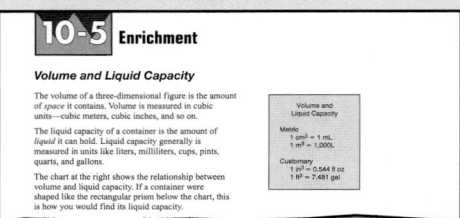

17. *Working on the* CHAPTER Project Use your unit of length and a standard unit such as feet to measure the length, width, and height of a rectangular prism such as your classroom. Then find the volume of your prism using each cubic unit. **See students' work.**

18. Sample answer: 2 in. × 12 in. × 12 in

18. *Critical Thinking* A mailing box in the shape of a rectangular prism has a volume of 288 in³. What are the possible dimensions of the box?

Mixed Review

19. Test Practice The base of a cone is a — *(Lesson 10-4)* **B**
  A triangle.   **B** circle.   **C** radius.   **D** rectangle.

20. *Patterns* Find the next two numbers in the sequence 160, 80, 40, 20, ... *(Lesson 7-8)* **10, 5**

420   Chapter 10 Geometry: Understanding Area and Volume

Extending the Lesson

Enrichment Masters, p. 74

10-5 **Enrichment**

Volume and Liquid Capacity

Activity In energy-efficient classrooms, experts recommend 15 ft³ of fresh air per minute for every student. Ask students how much fresh air should be pumped into their classroom every minute. Have them find the percent this is of all the air in the room.

What you'll learn

You'll learn to find the surface area of rectangular prisms.

When am I ever going to use this?

Knowing how to find the surface area of a rectangular prism will help you find the amount of paint needed to complete a room.

Word Wise

surface area

Just imagine! The NBA changes the rules so that a basketball is a cube! Dribbling, passing, shooting . . . everything would be different. The current NBA ball uses about 284 square inches of material. Would they use more or less material to make a basketball that is shaped like a $9\frac{1}{4}$-inch cube? *This problem will be solved in Example 3.*

Lumpy Gravy by John Long

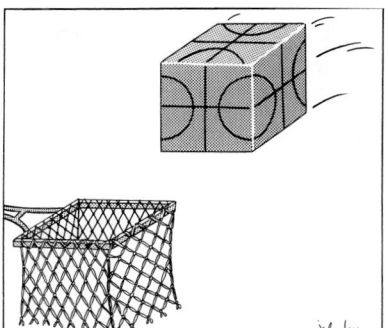

IT'S TIME TO MAKE BASKETBALL
MORE OF A CHALLENGE

The **surface area** of a three-dimensional object is the total area of its faces and curved surfaces. In the Mini-Lab, you will construct a rectangular prism and find its surface area.

HANDS-ON

MINI-LAB

Work with a partner. ruler · centimeter grid paper · scissors · calculator · tape

Try This

- Use the measurements shown to draw a pattern like this on a sheet of grid paper.
- Cut out the pattern along the dark lines.
- Fold the pattern along the dashed lines to form a rectangular prism. Tape the edges.

Talk About It

1. Find the area of each face of the prism.
2. What is the sum of the areas, or the surface area, of the prism?
3. What do you observe about the opposite sides of the prism? How could this simplify finding the surface area? **See margin.**

(Pattern labeled: 8 cm across, 3 cm, 3 cm; Back, Side, Bottom, Side, Front, 3 cm, Top, 5 cm)

1. top and bottom: 40 cm²; front and back: 24 cm²; sides: 15 cm²
2. 158 cm²

Lesson 10-6 Surface Area of Rectangular Prisms **421**

10-6 Lesson Notes

Instructional Resources

- *Study Guide Masters*, p. 75
- *Practice Masters*, p. 75
- *Enrichment Masters*, p. 75
- Transparencies 10-6, A and B
- *Assessment and Evaluation Masters*, p. 268
- *School to Career Masters*, p. 10
- CD-ROM Program
  - Resource Lesson 10-6
  - Interactive Lesson 10-6

Recommended Pacing	
Standard	Day 11 of 13
Honors	Days 10 & 11 of 13
Block	Day 6 of 7

1 FOCUS

5-Minute Check
(Lesson 10-5)

1. A filing cabinet is 42 inches high, 30 inches wide, and 19 inches deep. What is the volume of the cabinet? **23,940 in³**

Find the volume of each rectangular prism to the nearest tenth.

2.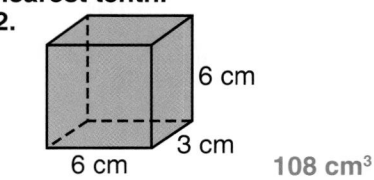
6 cm, 6 cm, 3 cm, 6 cm **108 cm³**

3.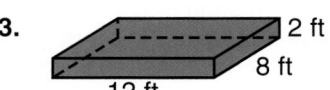
2 ft, 8 ft, 12 ft **192 ft³**

4. length = 16 mm
   width = 7 mm
   height = 4 mm **448 mm³**

5. Find the volume of a rectangular prism whose length is 9 inches, width is 6 inches, and height is 14 inches. **756 in³**

 The 5-Minute Check is also available on **Transparency 10-6A** for this lesson.

Additional Answer for the Mini-Lab

3. They have the same dimensions. You can find the area of each unique shape once and multiply by 2 to find the total.

Motivating the Lesson

Problem Solving Ask students how they would determine which package will require more wrapping paper.
- a rectangular box that is 2 inches high, 12 inches long, and 8 inches wide
- a square box with 8-inch sides

2 TEACH

Transparency 10-6B contains a teaching aid for this lesson.

Using the Mini-Lab Since the pattern will cover almost a full sheet of paper, have students start their drawing at the top of the paper. Designate one student as the model maker and the other as the recorder.

Teaching Tip Have students review the formula for finding the area of a rectangle.

In-Class Examples

Find the surface area of each rectangular prism to the nearest tenth.

For Example 1

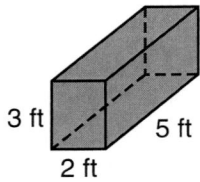

3 ft
5 ft
2 ft

62 ft²

For Example 2

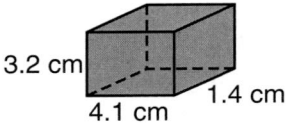

3.2 cm
4.1 cm
1.4 cm

46.7 cm²

For Example 3
Jesse is making enclosed storage cubes for his room. Each cube has faces that are $1\frac{3}{4}$ feet on each side. He has three 32 square-foot sheets of plywood. How many storage containers can he make?
5 containers

Examples

Did you know Physical Education Instructor James Naismith dreamed up the game of basketball in 1891. Volleyball and basketball are the only popular sports invented by Americans.

Find the surface area of each rectangular prism to the nearest tenth.

1

7 ft
6 ft
8 ft

In a rectangular prism, opposite sides have the same dimensions.

top and bottom
   $8 \times 6 = 48$ ft²

front and back
   $8 \times 7 = 56$ ft²

right and left sides
   $7 \times 6 = 42$ ft²

Add the areas.
$2(48) + 2(56) + 2(42) = 292$

The surface area is 292 ft².

2

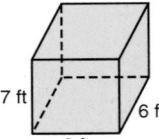

2.1 cm
4.4 cm
3.5 cm

Opposite sides have the same dimensions.

top and bottom
   $3.5 \times 4.4 = 15.4$ cm²

front and back
   $2.1 \times 3.5 = 7.35$ cm²

right and left sides
   $2.1 \times 4.4 = 9.24$ cm²

Find the total of the areas.

2 ⊠ 15.4 ⊞ 2 ⊠ 7.35 ⊞ 2
⊠ 9.24 ⊟ *63.98*

The surface area is about 64.0 cm².

APPLICATION

3 **Sports** Refer to the beginning of the lesson. Would the cube-shaped basketball require more or less material than the current NBA ball?

Explore You know the surface area of the spherical ball. You need to determine whether the surface area of the cube is larger or smaller.

Plan First, find the surface area of the cube. Then compare it with the surface area of the sphere.

Solve All six faces of the cube are squares that are $9\frac{1}{4}$ inches on each side. So the surface area of the cube is 6 times the area of one face.

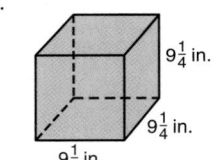

$9\frac{1}{4}$ in.
$9\frac{1}{4}$ in.
$9\frac{1}{4}$ in.

Area of a face: $9\frac{1}{4} \times 9\frac{1}{4} = \frac{37}{4} \times \frac{37}{4} = \frac{1,369}{16}$ in²

Surface area: $6 \times \frac{1,369}{16} = \frac{8,214}{16}$ or $513\frac{3}{8}$ in²

The surface area of the spherical ball is about 284 in². The cube's surface area is greater. It would use more material to make a cube-shaped basketball.

Examine Find the surface areas of several different cubes with edges between 9 and 10 inches long to verify that the solution is reasonable.

Reteaching the Lesson

Activity Give each group of four students a box such as a shoe box. Ask three of the students to each write their initials on two opposite sides and find the area of each. Have the fourth student use a calculator to find the surface area.

Error Analysis
Watch for students who overlook a face when finding the surface area of a rectangular prism.
Prevent by having students check to make sure they have six numbers before adding.

CHECK FOR UNDERSTANDING

Communicating Mathematics

Read and study the lesson to answer each question. 1–3. See margin.

1. *Demonstrate* how to find the surface area of a rectangular prism that is 6 inches long, 4 inches wide, and 10 inches high.

2. *Tell* why you can find the surface area of a rectangular prism after finding the area of just three faces.

3. *Describe* a situation when you would have to find the surface area of a rectangular solid.

Guided Practice

Find the surface area of each rectangular prism.

4. 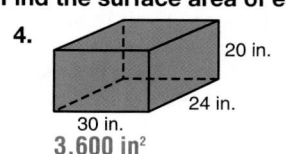 20 in. 24 in. 30 in.
3,600 in²

5. 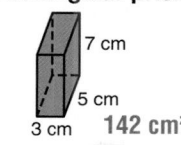 7 cm 5 cm 3 cm **142 cm²**

6. length = 2 mm
width = 7 mm
height = 5 mm
118 mm²

7. *Manufacturing* Kellogg's introduced Cocoa Frosted Flakes in 1997. The package is a rectangular prism 12 inches high, 8 inches wide, and 3 inches deep. How many square inches did the label designer have to cover on the package? **312 in²**

EXERCISES

Practice

Find the surface area of each rectangular prism to the nearest tenth.

8. 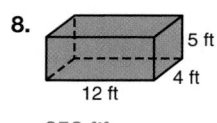 5 ft 4 ft 12 ft
256 ft²

9. 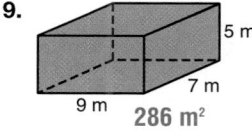 5 m 7 m 9 m
286 m²

10. 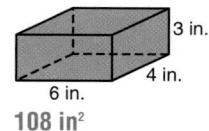 3 in. 4 in. 6 in.
108 in²

11. 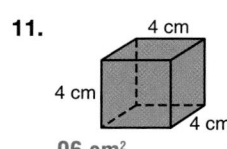 4 cm 4 cm 4 cm
96 cm²

12. 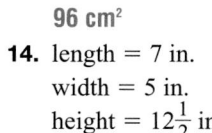 10.2 mm 8.7 mm 16.1 mm
786.1 mm²

13. 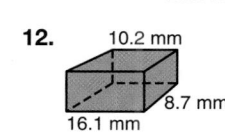 6 ft $16\frac{1}{2}$ ft 10 ft **648 ft²**

14. length = 7 in.
width = 5 in.
height = $12\frac{1}{2}$ in.
370 in²

15. length = 15.5 m
width = 10 m
height = 8 m
718 m²

16. length = 2.4 cm
width = 0.8 cm
height = 6.6 cm
46.1 cm²

Lesson 10-6 Surface Area of Rectangular Prisms **423**

---

3 PRACTICE/APPLY

Check for Understanding
If students need additional practice or instruction after completing Exercises 1–7, one of these options may be helpful.
- Extra Practice, see p. 585
- Reteaching Activity, see p. 422
- *Study Guide Masters*, p. 75
- *Practice Masters*, p. 75

Assignment Guide
Core: 9–21 odd, 23–26
Enriched: 8–18 even, 20, 21, 23–26

Study Guide Masters, p. 75

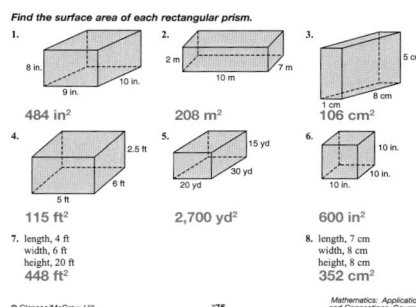

Name_____ Date_____

10-6 Study Guide

Surface Area of Rectangular Prisms

The surface area of a rectangular prism is equal to the sum of the areas of its faces.

Example Find the surface area of the rectangular prism.

Find the area of each face.
front: 6 × 10 = 60 cm²
back: 6 × 10 = 60 cm²
top: 8 × 10 = 80 cm²
bottom: 8 × 10 = 80 cm²
right side: 6 × 8 = 48 cm²
left side: 6 × 8 = 48 cm²

Add the areas: 60 + 60 + 80 + 80 + 48 + 48 = 376.
The surface area of the rectangular prism is 376 cm².

Find the surface area of each rectangular prism.

1. 8 in. 9 in. 10 in.
484 in²

2. 2 m 10 m 7 m
208 m²

3. 1 cm 8 cm 5 cm
106 cm²

4. 2.5 ft 6 ft 5 ft
115 ft²

5. 15 yd 20 yd 30 yd
2,700 yd²

6. 10 in. 10 in. 10 in.
600 in²

7. length, 4 ft
width, 6 ft
height, 20 ft
448 ft²

8. length, 7 cm
width, 8 cm
height, 8 cm
352 cm²

© Glencoe/McGraw-Hill
T75
Mathematics: Applications and Connections, Course 1

Lesson 10-6 423

Exercise 22 asks students to advance to the next stage of work on the Chapter Project. You may wish to have students make a paper model with the dimensions on it of the object they chose.

4 ASSESS

Closing Activity

Writing Have students work in pairs to create word problems about containers that can be solved by finding the surface area.

Chapter 10, Quiz D (Lesson 10-6) is available in the *Assessment and Evaluation Masters*, p. 268.

Practice Masters, p. 75

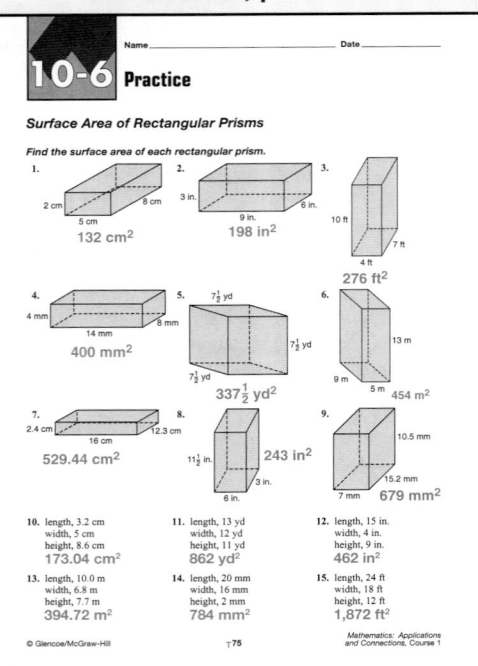

---

17. What is the surface area of a rectangular prism that is 10 inches long, 14 inches wide, and 20 inches high? **1,240 in²**

18. Find the surface area of a rectangular prism that is 9 meters by 5 meters by 11 meters. **398 m²**

19. *Write a Problem* involving a rectangular prism with a surface area of 720 in². **See students' work.**

Applications and Problem Solving

20. *Sports* Wallyball is a variation of volleyball invented in the 1970s. A wallyball court is 20 feet wide, 40 feet long, and 20 feet high. If a gallon of paint will cover 400 square feet, how many gallons will be needed to cover the walls, ceiling, and floor of a wallyball court? **10 gallons**

21. *Life Science* The shark petting tank at Nauticus, in Norfolk, Virginia, is 20 feet long, 8 feet wide, and 3 feet deep. Sometimes the tanks must be resurfaced to prevent leakage. What is the area to be resurfaced if the top of the shark petting tank is open? **328 ft²**

22. *Working on the* **CHAPTER** *Project* Choose an object shaped like a rectangular prism such as a cabinet or your classroom. Find the surface area of the object using your unit of length and a standard unit such as square feet or square meters. **See students' work.**

23. *Critical Thinking* In a certain cube, the measure of the surface area is the same as the measure of its volume. What are its dimensions? **6 units on each side**

Mixed Review

24. **Test Practice** A rectangular fish tank measures 34 inches by 22 inches by 18 inches. If the tank is filled to a height of 15 inches, what is the volume of water in the tank? *(Lesson 10-5)* **B**
A 13,464 in³
B 11,220 in³
C 9,180 in³
D 5,940 in³

25. On December 19, the sunrise was at 7:19 A.M., and the sunset was at 4:28 P.M. How many hours of daylight were there? *(Lesson 6-7)* **9 hours 9 minutes**

26. *Statistics* What scale would you use in making a frequency table for the following set of data? 21, 79, 11, 9, 55, 38, 111, 92 *(Lesson 2-2)* **0 to 120**

424 Chapter 10 Geometry: Understanding Area and Volume

Extending the Lesson

Enrichment Masters, p. 75

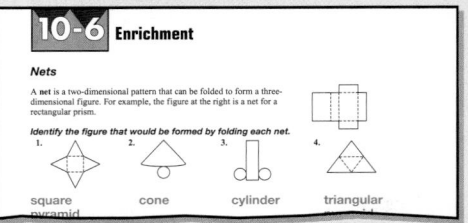

Activity Have several students write to manufacturers for information on how they design their packaging. Encourage students to ask for information on how marketing information, shipping requirements, and materials affect the design.

TECHNOLOGY LAB

SPREADSHEETS

10-6B Surface Area and Volume

A Follow-Up of Lesson 10-6

📟 computer

💽 spreadsheet software

The spreadsheet below can be used to find the volume and surface area of a rectangular prism. To use the spreadsheet, you enter the length, width, and height of the prism. Then the computer finds the volume and the surface area.

The formula in cell D2 tells the computer to multiply the values in cells A2, B2, and C2 together.

	A	B	C	D	E
1	LENGTH	WIDTH	HEIGHT	VOLUME	SURFACE AREA
2	A2	B2	C2	= A2*B2*C2	= 2*A2*B2 + 2*A2*C2 + 2*B2*C2
3	A3	B3	C3	= A3*B3*C3	= 2*A3*B3 + 2*A3*C3 + 2*B3*C3
4	A4	B4	C4	= A4*B4*C4	= 2*A4*B4 + 2*A4*C4 + 2*B4*C4
5	A5	B5	C5	= A5*B5*C5	= 2*A5*B5 + 2*A5*C5 + 2*B5*C5
6	A6	B6	C6	= A6*B6*C6	= 2*A6*B6 + 2*A6*C6 + 2*B6*C6
7	A7	B7	C7	= A7*B7*C7	= 2*A7*B7 + 2*A7*C7 + 2*B7*C7

TRY THIS

Work with a partner.

The result of using the spreadsheet to find the surface area and volume of the rectangular prism in row 2 is shown below.

	A	B	C	D	E
1	LENGTH	WIDTH	HEIGHT	VOLUME	SURFACE AREA
2	3	3	7	63	102
3	10.2	4.1	1.6	66.912	129.4
4	4	4	4	64	96
5	8	4	4	128	160
6	8	8	4	256	256
7	8	8	8	512	384

ON YOUR OWN

1. Use the spreadsheet to determine the volume and surface area for the rectangular prisms in rows 3, 4, 5, 6, and 7. Print or record your results. **See above.**

2. What does the formula in cell E2 tell the computer to do?

3. Show how to modify the spreadsheet to find the area and perimeter of a rectangle. **See margin.**

4. How do the dimensions of the prisms in rows 5, 6, and 7 compare to those of the prism in row 4? **See margin.**

5. Describe the pattern of volumes for prisms in rows 4, 5, 6, and 7. **5–6. See Answer Appendix.**

6. Describe the pattern of surface areas of prisms in rows 4, 5, 6, and 7.

> **2.** Find the surface area of the prism using the quantities in cells A2, B2, and C2.

Lesson 10-6B TECHNOLOGY LAB   **425**

GET READY

Objective Students use spreadsheets to find the surface area and volume of rectangular prisms.

Technology Resources
Suggested spreadsheet software:
- *Lotus 1·2·3*
- *ClarisWorks*
- *Microsoft Excel*

MANAGEMENT TIPS

Recommended Time
30 minutes

Getting Started Remind students how each cell of a spreadsheet is named. Also review how formulas using cell names are entered into the spreadsheet.

Using a Calculator If spreadsheet software is not available, have groups of students create the table on paper and use a calculator to find the values.

ASSESS

Lead a class discussion on the answers for Exercises 4–6. Have students give examples of other series of prisms that would show a similar pattern. **any series where the dimensions increased by a factor of two**

 Have students write a description of how the spreadsheet was used to compare surface area and volume.

Additional Answers

3. Delete the height column. Change "volume" to "area" and change the formulas in column D to A2*B2, A3*B3, and so on. Change "surface area" to "perimeter." Change the formulas in column E to 2*A2 + 2*B2, 2*A3 + 2*B3, and so on.

4. The dimensions of prism 5 are the same as those for prism 4 with the length doubled. The dimensions of prism 6 are the same as those for prism 4 with the length and width doubled. The dimensions of prism 7 are the same as those for prism 4 with the length, width, and height doubled.

Study Guide and Assessment

Vocabulary

This section provides a listing of the new terms, properties, and phrases that were introduced in this chapter. Have students define each term and provide an example or two of it, if appropriate.

Understanding and Using the Vocabulary

These exercises check students' understanding of the terms by using a variety of verbal formats including matching, completion, and true/false.

Glossaries A complete glossary of terms appears on pages 642–648. The glossary also appears in Spanish on pages 649–656.

Additional Answer

10. **Find the area of the two rectangular bases, the rectangular front and back, and the rectangular right and left sides. Add these areas to find the surface area.**

Vocabulary

After completing this chapter, you should be able to define each term, concept, or phrase and give an example or two of each.

Measurement
surface area (p. 421)
volume (p. 418)

Geometry
base (pp. 398, 412)
center (p. 413)
cone (p. 413)
cylinder (p. 413)
edge (p. 412)
face (p. 412)

height (p. 398)
prism (p. 412)
pyramid (p. 412)
rectangular prism (p. 412)
sphere (p. 413)
square pyramid (p. 412)
three-dimensional figure (p. 412)
vertex (p. 412)

Problem Solving
make a model (p. 416)

Understanding and Using the Vocabulary

Choose the correct term or number to complete each sentence.

1. The (<u>height</u>, edge) of a parallelogram is the distance from the base to the opposite side.
2. The flat surfaces of a three-dimensional figure are called (<u>faces</u>, vertices).
3. The faces of three-dimensional figures intersect in (bases, <u>edges</u>).
4. A (<u>pyramid</u>, cylinder) is a three-dimensional figure with one base where all other faces are triangles that meet at one point.
5. A (<u>sphere</u>, cone) is a three-dimensional figure with no faces, bases, edges, or vertices.
6. A three-dimensional figure with two circular bases is a (cone, <u>cylinder</u>).
7. The amount of space that a three-dimensional figure contains is called its (area, <u>volume</u>).
8. A rectangular prism with length 3 meters, width 4 meters, and height 2 meters has a volume of (14, <u>24</u>) cubic meters.
9. The total area of a three-dimensional object's faces and curved surfaces is called its (<u>surface area</u>, volume).

In Your Own Words

10. *Explain* how to find the surface area of the rectangular prism. **See margin.**

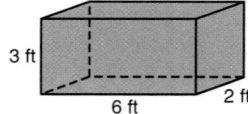
3 ft, 6 ft, 2 ft

 MindJogger Videoquizzes

MindJogger Videoquizzes provide an alternative review of concepts presented in this chapter. Students work in teams to answer questions, gaining points for correct answers. The questions are presented in three rounds.
Round 1 Concepts–5 questions
Round 2 Skills–4 questions
Round 3 Problem Solving–4 questions

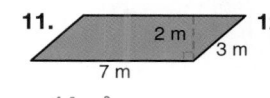

Objectives & Examples

Upon completing this chapter, you should be able to:

● find the area of parallelograms *(Lesson 10-1)*

Find the area.

$A = bh$

$A = 6 \times 5$

$A = 30$

5 in.

6 in.

The area of the parallelogram is 30 in².

Review Exercises

Use these exercises to review and prepare for the chapter test. **12. $48\frac{1}{8}$ in²**

Find the area of each parallelogram.

11. 2 m / 3 m / 7 m
14 m²

12. $5\frac{1}{2}$ in. / $7\frac{5}{8}$ in. / $8\frac{3}{4}$ in.

13. Find the area of a parallelogram that is 12.5 centimeters wide and 7 centimeters high. **87.5 cm²**

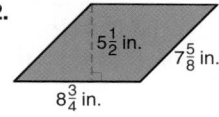

Objectives & Examples

This section reviews the skills and concepts of the chapter and shows completely worked examples.

Review Exercises

These exercises provide practice for the corresponding objectives.

Assessment and Evaluation Masters, pp. 255–256

● find the area of triangles *(Lesson 10-2)*

Find the area.

$A = \frac{1}{2} bh$

$A = \frac{1}{2}(150 \times 50)$

$A = 3,750$

The area of the triangle is 3,750 m².

150 m

50 m

Find the area of each triangle. Round decimals to the nearest tenth.

14. **99 in²**
11 in.
18 in.

15. **72 cm²**
12 cm
12 cm

16. **5.3 m²**
3.4 m
3.1 m

17. $18\frac{1}{2}$ ft / 17 ft **$62\frac{7}{16}$ ft²**
$6\frac{3}{4}$ ft

● find the area of circles *(Lesson 10-3)*

What is the area of the circle?

Use 3.14 for π.

$A = \pi r^2$

$A = \pi \cdot 7^2$

$A = \pi \cdot 49$

$A \approx 153.9 \text{ cm}^2$

7 cm

The area is about 153.9 cm².

Find the area of each circle to the nearest tenth. Use 3.14 for π. 19. 153.9 in²

18. 11 m **379.9 m²**

19. 14 in.

20. radius, 1.5 km **7.1 km²**

21. diameter, $7\frac{1}{6}$ in. **40.3 in²**

Assessment and Evaluation

Six forms of Chapter 10 Test are available in the *Assessment and Evaluation Masters* as shown in the chart.

Chapter 10 Test, Form 1B, is shown at the right. Chapter 10 Test, Form 2B, is shown on the next page.

1A	Multiple Choice	Honors
1B	Multiple Choice	Average
1C	Multiple Choice	Basic
2A	Free Response	Honors
2B	Free Response	Average
2C	Free Response	Basic

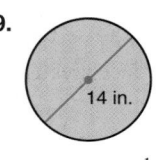

10 Chapter 10 Test, Form 1B

Name_____ Date_____

1. Find the area of the parallelogram shown at the right.
   A. 18 square units
   B. 24 square units
   C. 30 square units
   D. 27 square units
   1. __B__

2. Find the area of a parallelogram if the base is 18 centimeters and the height is 25 centimeters.
   A. 430 cm² B. 86 cm² C. 450 cm² D. none of these
   2. __C__

3. Find the area of the parallelogram shown at the right.
   A. 245 ft² B. 196 ft²
   C. 69 ft² D. none of these
   3. __B__

4. Find the area of the triangle shown at the right.
   A. 49 cm² B. 12.25 cm²
   C. 24.5 cm² D. none of these
   4. __C__

5. What is the area of a triangle with a base of 64 feet and a height of 36 feet?
   A. 1,152 ft² B. 100 ft² C. 2,304 ft² D. none of these
   5. __A__

6. Find the area of the triangle shown at the right.
   A. 20.6 in² B. 50.84 in²
   C. 101.68 in² D. none of these
   6. __B__

7. Find the area of the circle shown at the right. Use 3.14 for π.
   A. 28.26 in² B. 254.34 in²
   C. 2,289.06 in² D. none of these
   7. __B__

8. Find the area of a circle with a diameter of 4.2 centimeters. Use 3.14 for π.
   A. 13.188 cm² B. 13.8474 cm² C. 6.594 cm² D. none of these
   8. __B__

9. State the number of edges in a square pyramid.
   A. 4 B. 3 C. 5 D. none of these
   9. __D__

10. Name the figure shown at the right.
    A. cone B. rectangular prism
    C. square pyramid D. sphere
    10. __B__

---

10 Chapter 10 Test, Form 1B (continued)

11. State the number of faces in a rectangular prism.
    A. 6 B. 12 C. 8 D. none of these
    11. __A__

12. Which pattern can be folded to form a box shaped like a rectangular prism?
    A. B. C. D.
    12. __B__

13. How many different rectangular prisms can be formed by using exactly 15 cubes?
    A. 1 B. 2 C. 3 D. none of these
    13. __B__

14. Find the volume of the rectangular prism shown at the right.
    A. 48 mm³ B. 1,536 mm³
    C. 4,096 mm³ D. none of these
    16 mm
    14. __C__

15. A box is 26 inches long by 15 inches wide by 12 inches high. What is the volume of the box?
    A. 53 in³ B. 4,680 in³ C. 1,764 in³ D. none of these
    15. __B__

16. Find the volume of the rectangular prism described at the right.
    length, 10.5 ft / width, 8.5 ft / height, 5 ft
    A. 223.125 ft³ B. 446 ft³ C. 89.25 ft³ D. none of these
    16. __D__

17. Find the surface area of the rectangular prism shown at the right.
    A. 26 m² B. 576 m²
    C. 432 m² D. none of these
    17. __C__

18. Find the surface area of the rectangular prism described at the right.
    length, 17 in. / width, 13 in. / height, 8 in
    A. 922 in² B. 461 in² C. 1,768 in² D. none of these
    18. __A__

19. A circular tabletop has a radius of 30 inches. What is the area of the tabletop? Use 3.14 for π.
    A. 94.2 in² B. 2,826 in² C. 188.4 in² D. none of these
    19. __B__

20. How much paper would it take to cover a box that is 32 centimeters by 20 centimeters by 10 centimeters if there is no overlap?
    A. 62 cm² B. 6,400 cm² C. 1,160 cm² D. none of these
    20. __D__

Study Guide and Assessment

Additional Answers
22. triangular pyramid
23. rectangular prism

Assessment and Evaluation Masters, pp. 261–262

Chapter 10 Test, Form 2B

Find the area of each figure. Use 3.14 for π.

1.   1. __4,116 mm²__
2.   2. __1,256 m²__
3.   3. __168 cm²__
4.    4. __45 cm²__
5.   5. __2,289.06 ft²__
6.   6. __34.8 cm²__
7. A circular tablecloth has a diameter of 50 inches. What is the area of the tablecloth? Use 3.14 for π.   7. __1,962.5 in²__

Name each figure.
8.   8. __sphere__
9.   9. __square pyramid__
10. State the number of edges for the figure in Question 9.   10. __8__

© Glencoe/McGraw-Hill   261   Mathematics: Applications and Connections, Course 1

Chapter 10 Test, Form 2B (continued)

Find the surface area of each rectangular prism.
11.   11. __3,188 mm²__
12.    12. __369 ft²__
13. length, 42 in.; width, 30 in.; height, 12 in.   13. __4,248 in²__
14. length, 33 cm; width, 24 cm; height, 8 cm   14. __2,496 cm²__

Find the volume of each rectangular prism.
15. 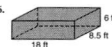   15. __5,304 mm³__
16. 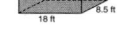   16. __918 ft³__
17. length, 36 in.; width, 22 in.; height, 17 in.   17. __13,464 in³__
18. length, 29 cm; width, 20 cm; height, 11 cm   18. __6,380 cm³__
19. A rectangular aquarium is 20 inches long by 12 inches wide by 10 inches deep. What is the volume of the aquarium?   19. __2,400 in³__
20. How many different rectangular prisms can be formed by using exactly 12 cubes?   20. __4__

© Glencoe/McGraw-Hill   262   Mathematics: Applications and Connections, Course 1

Objectives & Examples

● identify three-dimensional figures
(Lesson 10-4)

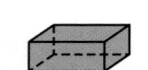
rectangular prism   cylinder   pyramid

cone   sphere

● find the volume of rectangular prisms
(Lesson 10-5)

Find the volume.

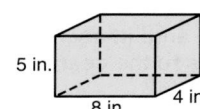

$V = \ell wh$
$V = 8 \times 4 \times 5$
$V = 160 \text{ in}^3$

The volume is 160 cubic inches.

● find the surface area of rectangular prisms
(Lesson 10-6)

What is the surface area of the prism above?

top and bottom: $8 \times 4 = 32 \text{ in}^2$
front and back: $8 \times 5 = 40 \text{ in}^2$
sides: $4 \times 5 = 20 \text{ in}^2$

Surface area $= 2(32) + 2(40) + 2(20)$
$= 64 + 80 + 40$ or 184

The surface area is 184 square inches.

Review Exercises

Name each figure. 22–23. See margin.

22. 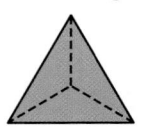   23.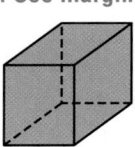

State the number of faces, edges, and vertices in each figure.
24. triangular pyramid **4; 6; 4**
25. square prism **6; 12; 8**

26. **96 m³**
Find the volume of each rectangular prism.

26. 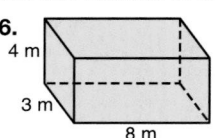   27.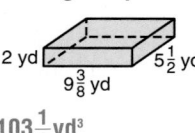
   $103\frac{1}{8}$ yd³

28. A rectangular prism has a length of 3.6 meters, a height of 4.1 meters, and a width of 8.2 meters. What is the volume of the prism to the nearest tenth? **121.0 m²**

Find the surface area of each rectangular prism.

29. 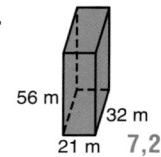   30.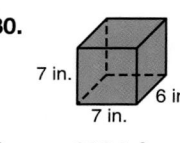
7,280 m²   **266 in²**

31. length = 2 mm
   width = 2.5 mm
   height = 1.7 mm **25.3 mm²**

32. length = $20\frac{1}{3}$ yd
   width = 4 yd
   height = $5\frac{2}{3}$ yd **$438\frac{4}{9}$ yd²**

---

428 Chapter 10 Geometry: Understanding Area and Volume

Test and Review Software

You may use this software, a combination of an item generator and item bank, to create your own tests or worksheets. Types of items include free response, multiple choice, short answer, and open ended.

CD-ROM Program

The CD-ROM Program contains an Assessment Game whose questions review the concepts in this chapter.

Applications & Problem Solving

33. *Architecture* Each year, the Corn Palace in Mitchell, South Dakota, is covered in grain murals. Suppose one mural is as shown below. *(Lessons 10-1 and 10-2)*

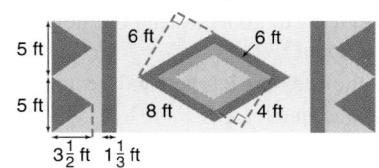

a. What is the area to be covered in dark grain? **See margin.**

b. If a bushel of grain covers 25 square feet, how many bushels of dark grain are needed? **4 bushels**

34. *Gift Wrapping* Mauna Loa of Honolulu, Hawaii, sells gift boxes of macadamia nuts. The 15-pound box is 18 inches long, 8 inches wide, and 5 inches high. Not counting overlap, how much wrapping paper is needed for each 15-pound box? *(Lesson 10-6)* **548 in²**

35. *Plumbing* A plumber digs a rectangular hole in the ground to install water pipes. The hole is dug straight down and has parallel sides. If the hole is 3 meters deep, 1.5 meters wide, and 2 meters long, what volume of dirt must be removed? *(Lesson 10-5)* **9 m³**

36. *Make a Model* The outside of a cube made of 27 small cubes is painted. Find the number of small cubes that are unpainted. *(Lesson 10-5A)* **1**

Alternative Assessment

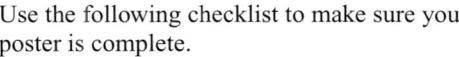

● *Performance Task*

Suppose you are planning to remodel your family room. The rectangular room is 12 feet wide, 10 feet long, and 8 feet high. How can you determine the number of square feet of carpet, wall paint, and ceiling paint needed? Find these amounts. **See margin.**

If carpet costs $5 per square foot and paint costs $15 per gallon (which is enough to cover 400 square feet), how much will these supplies cost? **carpet, $600; paint, $30; remodeling, $630**

A practice test for Chapter 10 is provided on page 604.

● *Completing the* CHAPTER Project

Use the following checklist to make sure your poster is complete.

☑ Describe your unit of measure. Include it or an equivalent length of string.

☑ Show how you found the volume and surface area of your prisms.

☑ Explain why you think measurement systems are now standardized.

Draw examples of the three-dimensional figures you studied in this chapter. Label the faces, edges, and vertices, or center. Place these diagrams in your portfolio.

Applications & Problem Solving

This section provides additional practice in solving real-world problems that involve the skills of this chapter.

Alternative Assessment

The *Performance Task* provides students with a performance assessment opportunity to evaluate their work and understanding.

CHAPTER Project

Students should complete the final stages of their project and prepare a class demonstration of their results. A scoring guide for the project is available in the *Investigations and Projects Masters,* p. 55.

PORTFOLIO Students should add to their portfolios at this time.

Assessment and Evaluation Masters, p. 265

10 Name_____ Date_____

Chapter 10 Performance Assessment

Instructions: Demonstrate your knowledge by giving a clear, concise solution to each problem. Be sure to include all relevant drawings and to justify your answers. You may show your solutions in more than one way or investigate beyond the requirements of the problems.

1. Diana and her mother are making some flags of the states and territories for a report in social studies class. Diana needs to find the amount of material to buy.

   a. How many square inches of material will be needed to make the Alaska flag?    10 in.

   b. How many square inches of blue material will be needed to make the rectangular Virginia flag? How many square inches of white material will be needed to make the circle in the middle? Assume no material will be wasted. Show your work. Use 3.14 for π.    10 in. / diameter of circle: 7 in. / 15 in.

   c. Tell in your own words how to find the area of a triangle.

   d. Find the amount of white material needed to make the triangular part of the American Samoa flag. Show your work.   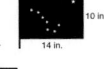 9 in. / 10 in. / 20 in. / triangle height: 18 in.

   e. Tell in your own words how to find the area of a parallelogram.

   f. Find the area of the white parallelogram in the Arkansas flag. Show your work.    ARKANSAS / 10 in. / 15 in. / parallelogram: side 6 in. / height 5 in.

   g. Diana's mother made a Wyoming flag. How would you estimate the area of the buffalo in the center?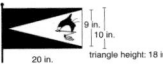

2. Write a word problem that involves finding the surface area and volume of a rectangular prism. Draw an illustration for your word problem.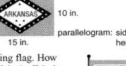

© Glencoe/McGraw-Hill   265   *Mathematics: Applications and Connections, Course 1*

Additional Answer

33a. $85\frac{2}{3}$ ft²

Additional Answer for the Performance Task

Find the area of the floor, ceiling, and walls: floor, 120 ft²; walls, 352 ft²; ceiling, 120 ft²

Performance Assessment

Additional performance assessment tasks for this chapter are included in the *Assessment and Evaluation Masters* on page 265. A scoring guide is also provided on page 277.

Standardized Test Practice

The Standardized Test Practice may be used to help students prepare for standardized tests. The test items are written in the same style as those in state proficiency tests and standardized tests like CAT, CTBS, ITBS, MAT, SAT, and Terra Nova. The test items cover skills and concepts covered up to this point in the text.

The pages can be used as an overnight assessment. After students have completed the pages, discuss how each problem can be solved, or provide copies of the solutions from the *Solutions Manual*.

Assessment and Evaluation Masters, p. 271

10 Name_____ Date_____
Cumulative Review, Chapters 1–10

Solve each equation. Write in simplest form. *(Lessons 3-6, 4-3, and 6-3)*
1. $m = 90 - 9.2$ 2. $3.5 \times 0.6 = y$ 3. $y + \frac{1}{16} = \frac{15}{16}$

4. The sum of a number and its double is 60. Find the number. (Lesson 1-7)

5. In the ordered pair (3, 7), the x-coordinate is __?__. (Lesson 2-9)

6. Find the prime factorization of 115. (Lesson 5-2)

Perform the indicated operation. Write in simplest form. *(Lessons 6-4, 6-6, 7-2, 7-3, 7-5, and 7-6)*
7. $\frac{7}{18} + \frac{1}{12}$ 8. $6\frac{1}{2} - 3\frac{7}{10}$ 9. $\frac{7}{16} \times \frac{8}{9}$
10. $4\frac{2}{3} \times 1\frac{1}{4}$ 11. $\frac{5}{32} + 10$ 12. $6\frac{2}{3} \div 1\frac{2}{3}$

13. Name the figure at the right. (Lesson 9-4)

14. How many lines of symmetry does the figure at the right have? (Lesson 9-5)

15. Express $\frac{19}{25}$ as a percent. (Lesson 8-4)

16. Express 3.8% as a decimal. (Lesson 8-5)

17. Find the area of a circle with a diameter of 26 inches. Round to the nearest tenth. (Lesson 10-3)

Refer to the figure at the right. *(Lessons 10-4, 10-5, and 10-6)*
18. Find the number of edges in the rectangular prism.

19. Find the surface area of the rectangular prism.

20. Find the volume of the rectangular prism.

1.	80.8
2.	2.1
3.	$\frac{7}{8}$
4.	20
5.	3
6.	5 · 23
7.	$\frac{13}{36}$
8.	$2\frac{1}{2}$
9.	$\frac{7}{18}$
10.	$5\frac{3}{4}$
11.	$\frac{1}{64}$
12.	4
13.	pentagon
14.	1
15.	76%
16.	0.038
17.	530.9 in²
18.	12
19.	1,324 in²
20.	3,168 in³

© Glencoe/McGraw-Hill 271 Mathematics: Applications and Connections, Course 1

430 Chapter 10

CHAPTERS 1-10

Standardized Test Practice
Assessing Knowledge & Skills

Section One: Multiple Choice

There are ten multiple choice questions in this section. Choose the best answer. If a correct answer is *not here,* choose the letter for Not Here.

1. Name the percent shaded on the base-ten model. **A**

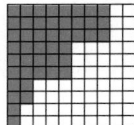

A 45%
B 55%
C 68%
D 72%

2. Find the area of the parallelogram. **H**
F 22.826 cm²
G 62.0 cm²
H 91.53 cm²
J 163.62 cm²

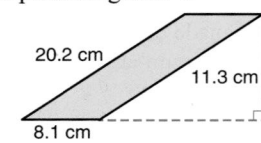
20.2 cm   11.3 cm   8.1 cm

3. Two figures that are the same size and shape are — **B**
A similar.
B congruent.
C neither similar nor congruent.
D bisectors.

4. Find the area of the triangle. **G**
F $\frac{15}{8}$ in²
G $\frac{15}{16}$ in²
H $\frac{30}{8}$ in²
J 2 in²

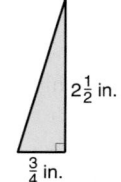
$2\frac{1}{2}$ in. $\frac{3}{4}$ in.

Please note that Questions 5–10 have five answer choices.

5. Lenora bought pencils that cost 29¢ each and pens that cost 89¢ each. What do you need to know to find out how much she spent? **D**
A the cost of each pen and pencil
B how much money she has
C how much change she received
D how many pens and pencils she bought
E how much tax is in her area

6. Which is a reasonable remainder when a number is divided by 8? **H**
F 8
G 9
H 7
J 12
K 10

7. Which is the best estimate for the weight of a 1-year old child? **B**
A 8 g
B 8 kg
C 60 kg
D 60 g
E 60 mg

8. A map has a scale of 1 cm = 20 km. The map distance from Rawson to Delta is 7.2 cm. How far is it from Rawson to Delta? **H**
F 720 km
G 27.2 km
H 144 km
J 20 km
K Not Here

430 Chapters 1–10 Standardized Test Practice

◄◄◄ Instructional Resources
Another cumulative review is shown at the left and is available in the *Assessment and Evaluation Masters,* p. 271.

9. The graph shows the number of boys and girls who signed up for Moosehead Camp. **A**

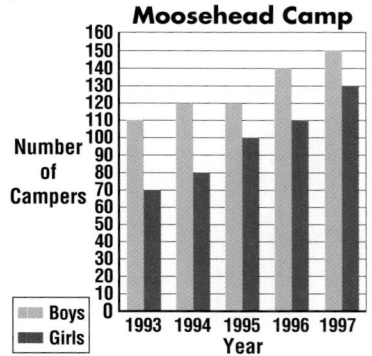

Moosehead Camp

Which is a reasonable conclusion that can be drawn from the information in the graph?

A The percent of campers who are girls increased every year after 1993.

B Girls are less interested in camp than boys.

C The number of boys who signed up for camp went down each year.

D More girls than boys signed up for camp in 1995.

E More people signed up for camp in 1996 than any other year.

10. The figure shows a triangle in the interior of a rectangle. **H**

Which method would you use to find the area of the shaded region?

F perimeter of the rectangle minus perimeter of the triangle

G perimeter of the rectangle plus perimeter of the triangle

H area of the rectangle minus area of the triangle

J area of the rectangle plus area of the triangle

K area of the triangle minus area of the rectangle

Test-Taking Tip

Make sure you know what the rules are for the type of test you are taking. If you are familiar with the method of scoring and the different types of instructions, you may increase your score.

Section Two: Free Response

This section contains six questions for which you will provide short answers. Write your answers on your paper.

11. Dave, Judy, and Marco bought a large submarine sandwich for $19.05. How much did each pay if they shared the price of the submarine equally? **$6.35**

12. Which difference is greater, $28\frac{1}{7} - 6\frac{3}{4}$ $28\frac{1}{7} - 6\frac{3}{4}$ or $30\frac{1}{8} - 8\frac{3}{4}$?   **is greater.**

13. Name the number of faces, edges, and vertices of a triangular pyramid. **4; 6; 4**

14. Find the volume of the rectangular prism. **2,737.92 m³**

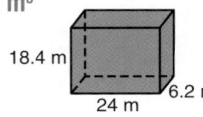

18.4 m   24 m   6.2 m

15. Find the surface area of a rectangular solid with a length of 8 ft, a width of 10 ft, and a height of $15\frac{1}{2}$ ft. **718 ft²**

16. Each serving of pizza is $\frac{1}{16}$ of a pizza. If $\frac{3}{4}$ of the pizza is left, how many servings are left? **12 servings**

Test-Taking Tip

Before students begin the test, suggest they skim through it to check the length and the types of problems it includes. Make sure they know ahead of time whether this particular type of test penalizes them for wrong answers.

Assessment and Evaluation Masters, pp. 269–270

10 Name_____ Date_____
Chapter 10 Standardized Test Practice

1. Kisha's dog is tied to a pole in the backyard. The length of the rope is 14 feet. If the dog can run in a full circle around the pole, how much of the area of the yard can Kisha's dog run on?   **1.   C**
   A. 43.96 ft²   B. 153.86 ft²
   C. 615.44 ft²   D. none of these

2. One lap around the track at the Indianapolis Motor Speedway is $2\frac{1}{2}$ miles. How far would you travel in 18 laps?   **2.   D**
   A. $50\frac{1}{2}$ miles   B. 36 miles   C. $64\frac{1}{2}$ miles   D. 45 miles

3. A rectangular prism has a length of 30 centimeters, a width of 22 centimeters, and a height of 10 centimeters. Find the surface area.   **3.   B**
   A. 1,180 cm²   B. 2,360 cm²   C. 6,600 cm²   D. 6,600 cm³

4. The flag for Trinidad and Tobago has a black strip in the shape of a parallelogram running from top to bottom. If the base of the black parallelogram is 13 inches and the flag is 30 inches high, find the area of the black parallelogram.   **4.   C**
   A. 195 in²   B. 360 in²   C. 390 in²   D. none of these

5. A stone mason is tiling a floor with parallelogram-shaped tiles. If the smaller corner of the tile measures 37°, what is the measure of its supplementary angle?   **5.   C**
   A. 90°   B. 53°   C. 143°   D. 133°

6. If the sales tax is 5.5%, what is the sales tax on a car that costs $11,000?   **6.   C**
   A. $6,050   B. $60.50   C. $605.00   D. $550.00

7. On a square, the diagonals bisect the angles. What is the measure of an angle of a square after it has been bisected?   **7.   D**
   A. 180°   B. 60°   C. 90°   D. 45°

8. It cost Mr. Quinn $16.25 to buy lunch for his family. If each lunch cost $3.25, how many lunches did he buy?   **8.   A**
   A. 5   B. 4   C. 6   D. none of these

9. A person expends 5.2 calories per minute walking, w, and 19.4 calories per minute running, r. Evaluate 5.2w + 19.4r to find the number of calories burned on a 40-minute hike with 35 minutes of walking and 5 minutes of running.   **9.   B**
   A. 705 calories   B. 279 calories
   C. 305 calories   D. none of these

10. A circle graph shows people's preferences for different types of pets. Dog owners account for 40% of pet owners and snake owners for 10%. What part of the circle graph would represent these two groups?   **10.   B**
   A. Insufficient information   B. 50%
   C. 30%   D. none of these

© Glencoe/McGraw-Hill   269   *Mathematics: Applications and Connections, Course 1*

10
Chapter 10 Standardized Test Practice (continued)

11. If a public library collection expanded from 1.91 million volumes to 3.27 million volumes, about how many volumes were added to the collection?   **11.   D**
    A. 2 million   B. 4 million   C. 5 million   D. 1 million

12. If it cost $25.13 for 19.6 gallons of gasoline, how much was the gasoline per gallon rounded to the nearest cent?   **12.   A**
    A. $1.28   B. $0.78   C. $1.30   D. $1.32

13. Express 204 as a product of prime numbers.   **13.   C**
    A. $2^2 \times 51$ B. $4 \times 3 \times 17$ C. $2^2 \times 3 \times 17$ D. $2 \times 3 \times 17$

14. Carol is making picture frames in woodworking class. Each frame uses 3.8 feet of wood. If Carol wants to make 6 frames, estimate how much wood she needs.   **14.   C**
    A. 18 feet   B. 20 feet
    C. 24 feet   D. none of these

15. What is the elapsed time from 7:55 A.M. to 10:10 A.M?   **15.   A**
    A. 2 h 15 min   B. 1 h 15 min
    C. 3 h 45 min   D. none of these

16. Racetown holds a 15-km race every year. Last year, $\frac{1}{3}$ of the 300 runners were men. About how many runners were women?   **16.   B**
    A. 100   B. 200   C. 300   D. 50

17. If the radius of a button is $\frac{3}{8}$ inch, what is the circumference of the button?   **17.   B**
    A. $1\frac{5}{28}$ in.   B. $2\frac{5}{14}$ in.   C. $4\frac{5}{8}$ in.   D. 1 in.

18. If 7 out of 10 students like pizza, how many students in a school of 832 like pizza?   **18.   C**
    A. 522   B. 564   C. 582   D. 5,824

19. Sales tax is 5.75%. Would the tax on a $20 purchase be less than $2?   **19.   A**
    A. yes   B. no
    C. maybe   D. cannot calculate

20. Carla scored 9 out of 10 correct on a test. Singi got 7 out of 8 correct on her test. How do their grades compare?   **20.   B**
    A. the same   B. Carla scored higher.
    C. Singi scored higher.   D. cannot compare

© Glencoe/McGraw-Hill   270   *Mathematics: Applications and Connections, Course 1*

Instructional Resources ▶▶▶

Additional standardized test practice is shown at the right and is available in the *Assessment and Evaluation Masters*, pp. 269–270.

Algebra: Investigating Integers

Previewing the Chapter

Overview

In this chapter, students use models and patterns to compare, order, add, subtract, multiply, and divide integers. Through graphing ordered pairs and transformations, students become familiar with the coordinate system. Students learn to solve problems by working backward.

Lesson (pages)	Lesson Objectives	NCTM Standards	Standardized Tests	State/Local Objectives
11-1 (434–436)	Identify, name, and graph integers.	1–7, 9–13	MAT	
11-2 (437–439)	Compare and order integers.	1–7, 9, 10	CAT, CTBS, ITBS, MAT, SAT, TN	
11-3A (440)	Model integers using counters.	1–3, 5–9	CTBS, MAT, TN	
11-3 (441–444)	Add integers using models.	1–7, 9	MAT	
11-4 (445–448)	Subtract integers using models.	1–7, 9, 10, 12	MAT	
11-5 (449–452)	Multiply integers using models.	1–9, 13	MAT	
11-6A (454–455)	Solve problems by working backward.	1–9	MAT	
11-6 (456–458)	Divide integers using models and patterns.	1–10, 12	MAT	
11-7 (459–461)	Graph ordered pairs of numbers on a coordinate grid.	1–10, 12, 13	MAT	
11-8A (462–463)	Find transformations by using patty paper.	1–3, 8, 12		
11-8 (464–467)	Graph transformations on a coordinate grid.	1–9, 12, 13	MAT	

CAT = California Achievement Tests, CTBS = Comprehensive Tests of Basic Skills, ITBS = Iowa Tests of Basic Skills, MAT = Metropolitan Achievement Tests, SAT = Stanford Achievement Tests, TN = Terra Nova

Organizing the Chapter

CD-ROM

All of the blackline masters in the Teacher's Classroom Resources are available on the **Electronic Teacher's Classroom Resources** CD-ROM.

LESSON PLANNING GUIDE

Lesson	Extra Practice (Student Edition)	Blackline Masters (page numbers)										
		Study Guide	Practice	Enrichment	Assessment & Evaluation	Classroom Games	Diversity	Hands-On Lab	School to Career	Science and Math Lab Manual	Technology	Transparencies A and B
11-1	p. 585	76	76	76								11-1
11-2	p. 585	77	77	77	295				11			11-2
11-3A								64				
11-3	p. 586	78	78	78		29–32						11-3
11-4	p. 586	79	79	79	294, 295							11-4
11-5	p. 586	80	80	80						89–92		11-5
11-6A	p. 587											
11-6	p. 587	81	81	81	296		11				21	11-6
11-7	p. 587	82	82	82				79			22	11-7
11-8A								65				
11-8	p. 588	83	83	83	296							11-8
Study Guide/ Assessment					281–293, 297–299							

Other Chapter Resources

Student Edition

Chapter Project, pp. 433, 436, 444, 461, 471
School to Career, p. 453
Let the Games Begin, p. 444

Technology

 CD-ROM Program

 Interactive Mathematics Tools Software

Teacher's Classroom Resources

Applications
Family Letters and Activities, pp. 21–22
Investigations and Projects Masters, pp. 57–60

Teaching Aids
Answer Key Masters
Block Scheduling Booklet
Lesson Planning Guide
Solutions Manual

Professional Publications
Glencoe Mathematics Professional Series

Planning the Chapter

MindJogger Videoquizzes provide a unique format for reviewing concepts presented in the chapter.

ASSESSMENT RESOURCES

Student Edition
Mixed Review, pp. 436, 439, 444, 448, 452, 458, 461, 467
Mid-Chapter Self Test, p. 452
Math Journal, pp. 435, 458
Study Guide and Assessment, pp. 468–471
Performance Task, p. 471
Portfolio Suggestion, p. 471
Standardized Test Practice, pp. 472–473
Chapter Test, p. 605

Assessment and Evaluation Masters
Multiple-Choice Tests (Forms 1A, 1B, 1C), pp. 281–286
Free-Response Tests (Forms 2A, 2B, 2C), pp. 287–292
Performance Assessment, p. 293
Mid-Chapter Test, p. 294
Quizzes A–D, pp. 295–296
Standardized Test Practice, pp. 297–298
Cumulative Review, p. 299

Teacher's Wraparound Edition
5-Minute Check, pp. 434, 437, 441, 445, 449, 456, 459, 464
Building Portfolios, p. 432
Math Journal, pp. 440, 463
Closing Activity, pp. 436, 439, 444, 448, 452, 455, 458, 461, 467

Technology
Test and Review Software
MindJogger Videoquizzes
CD-ROM Program

MATERIALS AND MANIPULATIVES

Lesson 11-3A
counters*†
integer mat*†

Lesson 11-3
counters*†
integer mat*†
number cubes*

Lesson 11-4
counters*†
integer mat*†

Lesson 11-5
counters*†
integer mat*†

Lesson 11-6
counters*†
integer mat*†

Lesson 11-8A
patty paper
scissors*
notebook paper†

Lesson 11-8
grid paper†
scissors*
geomirror*

*Glencoe Manipulative Kit   †Glencoe Overhead Manipulative Resources

PACING CHART

See pages T25–T27 for the Course Planning Calendar.

COURSE	DAY 1	DAY 2	DAY 3	DAY 4	DAY 5	DAY 6	DAY 7
Standard	Chapter Project	Lesson 11-1	Lesson 11-2		Lessons 11-3A & 11-3		Lesson 11-4
Honors	Chapter Project	Lesson 11-1	Lesson 11-2	Lesson 11-3	Lesson 11-4	Lesson 11-5	Lesson 11-6A
Block	Chapter Project & Lesson 11-1	Lessons 11-2 & 11-3A	Lessons 11-3 & 11-4	Lesson 11-5	Lessons 11-6A & 11-6	Lesson 11-7	Lessons 11-8A & 11-8

Interactive Mathematics:
Activities and Investigations

is an activity-based program that may be used as an enhancement for chapters in *Mathematics: Applications and Connections.*

INTER·ACTIVE
Mathematics
Activities and Investigations

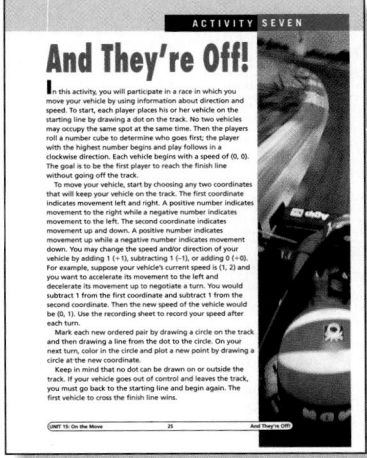

Unit 15, Activity Seven
Use with Lesson 11-7.

Summary Students work in groups to explore velocity and acceleration. Each group selects a race track drawn on graph paper. Then they plot the location of their vehicles as they race around the track. They accelerate and decelerate as they negotiate turns and straightaways. Students write about their strategies and techniques.

Math Connection Students use coordinate graphing and integers in this activity. They record the vehicle's speed as an ordered pair (horizontal movement, vertical movement).

Unit 3, Activity Seven
Use with Lesson 11-8.

Summary Students choose their favorite cartoon or team mascot and trace the figure on grid paper. They then use that tracing to create the same image on another piece of grid paper, but three times as large.

Math Connection Students use proportional relationships to enlarge a given figure to the desired size. This is a geometric transformation known as a dilation. A dilation preserves all of the characteristics and properties of the original figure, except size.

DAY 8	DAY 9	DAY 10	DAY 11	DAY 12	DAY 13	DAY 14	DAY 15
Lesson 11-5	Lesson 11-6A	Lesson 11-6	Lesson 11-7	Lessons 11-8A & 11-8		Study Guide and Assessment	Chapter Test
Lesson 11-6	Lesson 11-7	Lesson 11-8		Study Guide and Assessment	Chapter Test		
Study Guide and Assessment, Chapter Test							

Enhancing the Chapter

APPLICATIONS

Classroom Games,
pp. 29–32

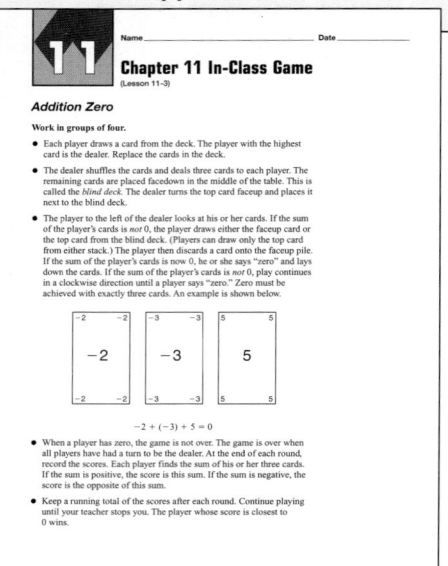

11 Name_____ Date_____

Chapter 11 In-Class Game
(Lesson 11-3)

Addition Zero

Work in groups of four.

● Each player draws a card from the deck. The player with the highest card is the dealer. Replace the cards in the deck.

● The dealer shuffles the cards and deals three cards to each player. The remaining cards are placed facedown in the middle of the table. This is called the *blind deck*. The dealer turns the top card faceup and places it next to the blind deck.

● The player to the left of the dealer looks at his or her cards. If the sum of the player's cards is *not* 0, the player draws either the faceup card or the top card from the blind deck. (Players can draw only the top card from either stack.) The player then discards a card onto the faceup pile. If the sum of the player's cards is now 0, he or she says "zero" and lays down the cards. If the sum does not equal *not* 0, play continues in a clockwise direction until a player says "zero." Zero must be achieved with exactly three cards. An example is shown below.

−2		−3		5
−2		**−3**		**5**
−2		−3		5

$$-2 + (-3) + 5 = 0$$

● When a player has zero, the game is not over. The game is over when all players have had a turn to be the dealer. At the end of each round, record the scores. Each player finds the sum of his or her three cards. If the sum is positive, the score is this sum. If the sum is negative, the score is the opposite of this sum.

● Keep a running total of the scores after each round. Continue playing until your teacher stops you. The player whose score is closest to 0 wins.

© Glencoe/McGraw-Hill   30   *Mathematics: Applications and Connections, Course 1*

Diversity Masters,
p. 11

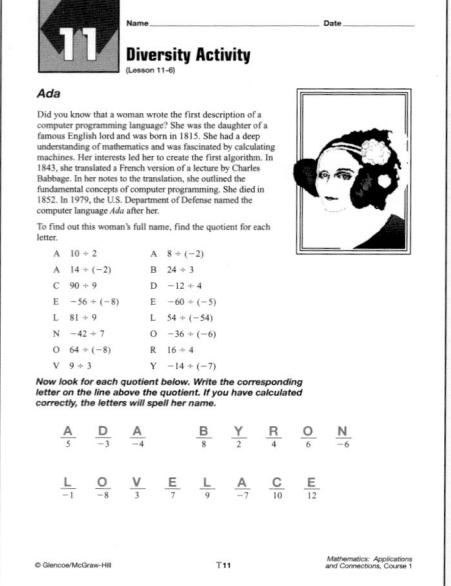

11 Name_____ Date_____

Diversity Activity
(Lesson 11-6)

Ada

Did you know that a woman wrote the first description of a computer programming language? She was the daughter of a famous English lord and was born in 1815. She had a deep understanding of mathematics and was fascinated by calculating machines. Her interests led her to create the first algorithm. In 1843, she translated a French version of a lecture by Charles Babbage. In her notes to the translation, she outlined the fundamental concepts of computer programming. She died in 1852. In 1979, the U.S. Department of Defense named the computer language *Ada* after her.

To find out this woman's full name, find the quotient for each letter.

A	$10 \div 2$	A	$8 \div (-2)$
A	$14 \div (-2)$	B	$24 \div 3$
C	$90 \div 9$	D	$-12 \div 4$
E	$-56 \div (-8)$	E	$-60 \div (-5)$
L	$81 \div 9$	L	$54 \div (-54)$
N	$-42 \div 7$	O	$-36 \div (-6)$
O	$64 \div (-8)$	R	$16 \div 4$
V	$9 \div 3$	Y	$-14 \div (-7)$

Now look for each quotient below. Write the corresponding letter on the line above the quotient. If you have calculated correctly, the letters will spell her name.

$$\underset{5}{\text{A}}\ \underset{-3}{\text{D}}\ \underset{2}{\text{A}} \qquad \underset{8}{\text{B}}\ \underset{2}{\text{Y}}\ \underset{4}{\text{R}}\ \underset{6}{\text{O}}\ \underset{-6}{\text{N}}$$

$$\underset{-1}{\text{L}}\ \underset{-8}{\text{O}}\ \underset{3}{\text{V}}\ \underset{7}{\text{E}}\ \underset{9}{\text{L}}\ \underset{-7}{\text{A}}\ \underset{10}{\text{C}}\ \underset{12}{\text{E}}$$

© Glencoe/McGraw-Hill   T11   *Mathematics: Applications and Connections, Course 1*

School to Career Masters,
p. 11

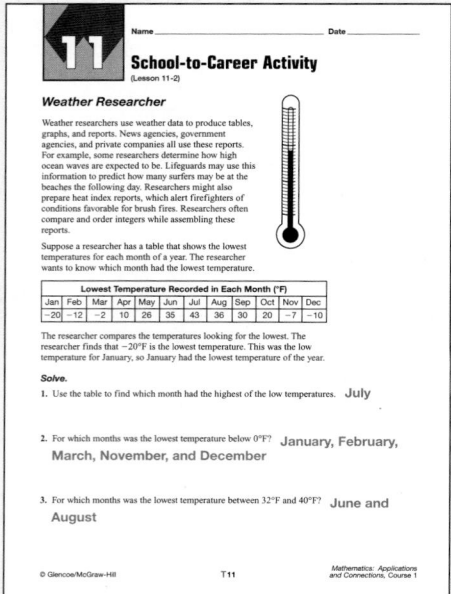

11 Name_____ Date_____

School-to-Career Activity
(Lesson 11-2)

Weather Researcher

Weather researchers use weather data to produce tables, graphs, and reports. News agencies, government agencies, and private companies all use these reports. For example, some researchers determine how high ocean waves are expected to be. Lifeguards may use this information to predict how many surfers may be at the beaches the following day. Researchers might also prepare heat index reports, which alert firefighters of conditions favorable for brush fires. Researchers often compare and order integers while assembling these reports.

Suppose a researcher has a table that shows the lowest temperatures for each month of a year. The researcher wants to know which month had the lowest temperature.

Lowest Temperature Recorded in Each Month (°F)

Jan	Feb	Mar	Apr	May	Jun	Jul	Aug	Sep	Oct	Nov	Dec
−20	−12	2	10	26	35	43	36	20	7	−7	−10

The researcher compares the temperatures looking for the lowest. The researcher finds that −20°F is the lowest temperature. This was the low temperature for January, so January had the lowest temperature of the year.

Solve.

1. Use the table to find which month had the highest of the low temperatures.   July

2. For which months was the lowest temperature below 0°F?   January, February, March, November, and December

3. For which months was the lowest temperature between 32°F and 40°F?   June and August

© Glencoe/McGraw-Hill   T11   *Mathematics: Applications and Connections, Course 1*

Family Letters and Activities,
pp. 21–22

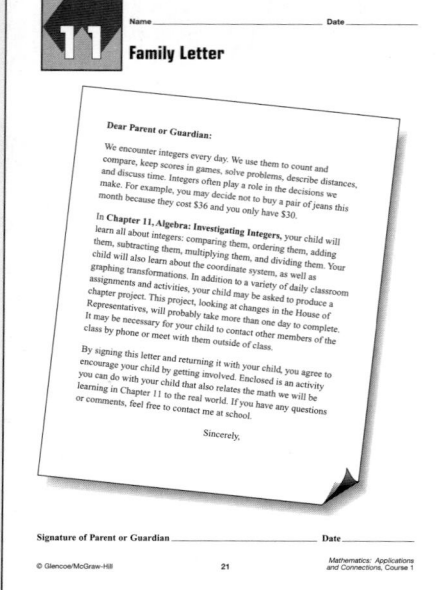

11 Name_____ Date_____

Family Letter

Dear Parent or Guardian:

We encounter integers every day. We use them to count and compare, keep scores in games, solve problems, describe distances, and discuss time. Integers often play a role in the decisions we make. For example, you may decide not to buy a pair of jeans this month because they cost $36 and you only have $30.

In Chapter 11, **Algebra: Investigating Integers**, your child will learn all about integers: comparing them, ordering them, adding them, subtracting them, multiplying them, and dividing them. Your child will also learn about the coordinate system, as well as graphing transformations. In addition to a variety of daily classroom assignments and activities, your child may be asked to produce a chapter project. This project, looking at changes in the House of Representatives, will probably take more than one day to complete. It may be necessary for your child to contact other members of the class by phone or meet with them outside of class.

By signing this letter and returning it with your child, you agree to encourage your child by getting involved. Enclosed is an activity you can do with your child that also relates the math we will be learning in Chapter 11 to the real world. If you have any questions or comments, feel free to contact me at school.

Sincerely,

Signature of Parent or Guardian _____   Date _____

© Glencoe/McGraw-Hill   21   *Mathematics: Applications and Connections, Course 1*

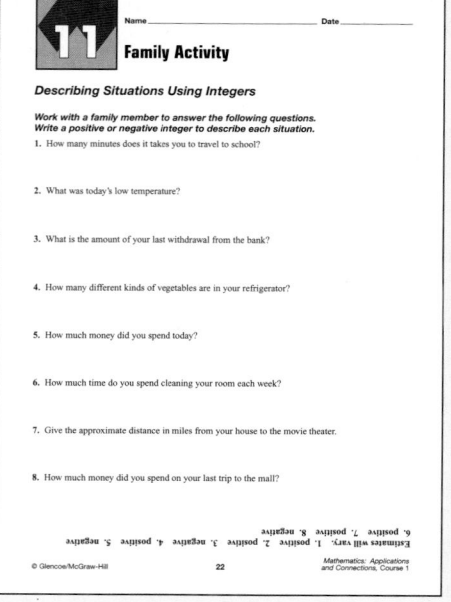

11 Name_____ Date_____

Family Activity

Describing Situations Using Integers

Work with a family member to answer the following questions. Write a positive or negative integer to describe each situation.

1. How many minutes does it takes you to travel to school?

2. What was today's low temperature?

3. What is the amount of your last withdrawal from the bank?

4. How many different kinds of vegetables are in your refrigerator?

5. How much money did you spend today?

6. How much time do you spend cleaning your room each week?

7. Give the approximate distance in miles from your house to the movie theater.

8. How much money did you spend on your last trip to the mall?

Estimates will vary: 1. positive 2. positive 3. negative 4. positive 5. negative 6. positive 7. positive 8. negative

© Glencoe/McGraw-Hill   22   *Mathematics: Applications and Connections, Course 1*

Science and Math Lab Manual,
pp. 89–92

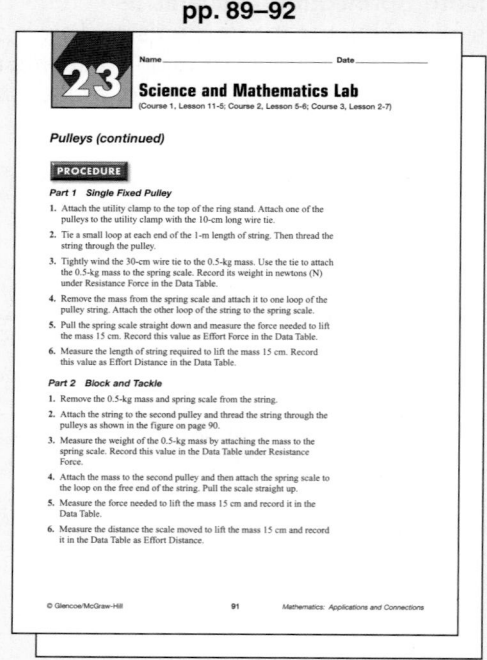

23 Name_____ Date_____

Science and Mathematics Lab
(Course 1, Lesson 11-5; Course 2, Lesson 5-6; Course 3, Lesson 2-7)

Pulleys (continued)

PROCEDURE

Part 1   Single Fixed Pulley

1. Attach the utility clamp to the top of the ring stand. Attach one of the pulleys to the utility clamp with the 10-cm long wire tie.

2. Tie a small loop at each end of the 1-m length of string. Then thread the string through the pulley.

3. Tightly wind the 30-cm wire tie to the 0.5-kg mass. Use the tie to attach the 0.5-kg mass to the spring scale. Record its weight in newtons (N) under Resistance Force in the Data Table.

4. Remove the mass from the spring scale and attach it to one loop of the pulley string. Attach the other loop of the string to the spring scale.

5. Pull the spring scale straight down and measure the force needed to lift the mass 15 cm. Record this value as Effort Force in the Data Table.

6. Measure the length of string required to lift the mass 15 cm. Record this value as Effort Distance in the Data Table.

Part 2   Block and Tackle

1. Remove the 0.5-kg mass and spring scale from the string.

2. Attach the string to the second pulley and thread the string through the pulleys as shown in the figure on page 90.

3. Measure the weight of the 0.5-kg mass by attaching the mass to the spring scale. Record this value in the Data Table under Resistance Force.

4. Attach the mass to the second pulley and then attach the spring scale to the loop on the free end of the string. Pull the scale straight up.

5. Measure the force needed to lift the mass 15 cm and record it in the Data Table.

6. Measure the distance the scale moved to lift the mass 15 cm and record it in the Data Table as Effort Distance.

© Glencoe/McGraw-Hill   91   *Mathematics: Applications and Connections*

Hands-On Lab Masters, p. 79

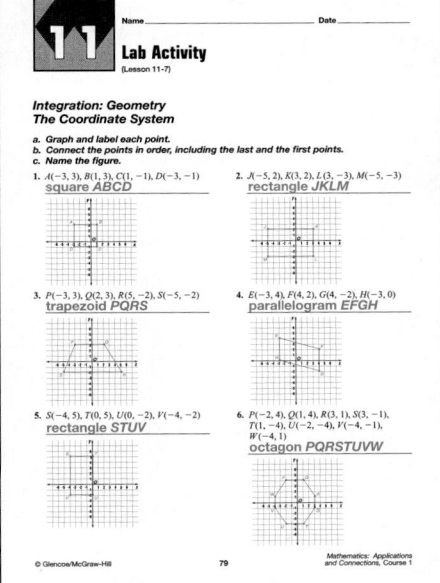

Lab Activity (Lesson 11-7)

Integration: Geometry
The Coordinate System

a. Graph and label each point.
b. Connect the points in order, including the last and the first points.
c. Name the figure.

1. $A(-3, 3)$, $B(1, 3)$, $C(1, -1)$, $D(-3, -1)$
   square ABCD

2. $J(-5, 2)$, $K(3, 2)$, $L(3, -3)$, $M(-5, -3)$
   rectangle JKLM

3. $P(-3, 3)$, $Q(2, 3)$, $R(5, -2)$, $S(-5, -2)$
   trapezoid PQRS

4. $E(-3, 4)$, $F(4, 2)$, $G(4, -2)$, $H(-3, 0)$
   parallelogram EFGH

5. $S(-4, 5)$, $T(0, 5)$, $U(0, -2)$, $V(-4, -2)$
   rectangle STUV

6. $P(-2, 4)$, $Q(1, 4)$, $R(3, 1)$, $S(3, -1)$, $T(1, -4)$, $U(-2, -4)$, $V(-4, -1)$, $W(-4, 1)$
   octagon PQRSTUVW

© Glencoe/McGraw-Hill   79   Mathematics: Applications and Connections, Course 1

Assessment and Evaluation Masters, pp. 294–296

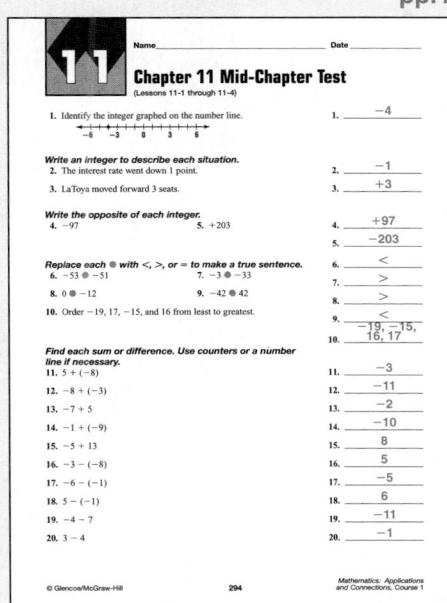

Chapter 11 Mid-Chapter Test (Lessons 11-1 through 11-4)

1. Identify the integer graphed on the number line.
   1. -4

Write an integer to describe each situation.
2. The interest rate went down 1 point.
   2. -1
3. LaToya moved forward 3 seats.
   3. $+3$

Write the opposite of each integer.
4. -97 5. $+203$
   4. $+97$
   5. -203

Replace each ● with <, >, or = to make a true sentence.
6. -53 ● -51 7. -3 ● -33
8. 0 ● -12 9. -42 ● 42
   6. $<$
   7. $>$
   8. $>$
10. Order -19, 17, -15, and 16 from least to greatest.
   9. $<$
   10. $-19, -15, 16, 17$

Find each sum or difference. Use counters or a number line if necessary.
11. $5 + (-8)$
12. $-8 + (-3)$
13. $-7 + 5$
14. $-1 + (-9)$
15. $-5 + 13$
16. $-3 - (-8)$
17. $-6 - (-1)$
18. $5 - (-1)$
19. $-4 - 7$
20. $3 - 4$
   11. -3
   12. -11
   13. -2
   14. -10
   15. 8
   16. 5
   17. -5
   18. 6
   19. -11
   20. -1

© Glencoe/McGraw-Hill   294   Mathematics: Applications and Connections, Course 1

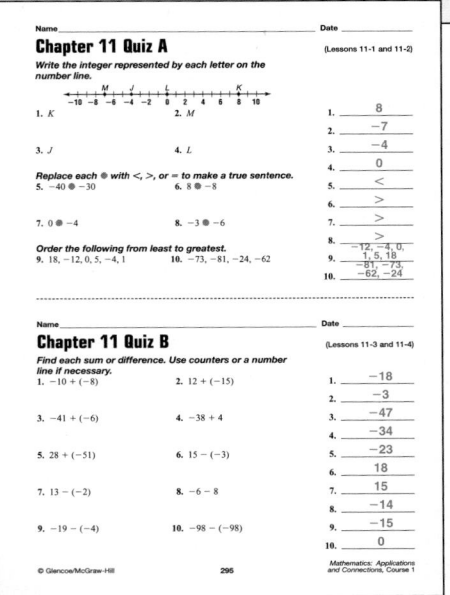

Chapter 11 Quiz A (Lessons 11-1 and 11-2)

Write the integer represented by each letter on the number line.

1. K 2. M
3. J 4. L

Replace each ● with <, >, or = to make a true sentence.
5. -40 ● -30 6. 8 ● -8
7. 0 ● -4 8. -3 ● -6

Order the following from least to greatest.
9. 18, -12, 0, 5, -4, 1 10. -73, -81, -24, -62

   1. 8
   2. -7
   3. -4
   4. 0
   5. $<$
   6. $>$
   7. $>$
   8. $>$
   9. $-12, -4, 0, 1, 5, 18$
   10. $-81, -73, -62, -24$

Chapter 11 Quiz B (Lessons 11-3 and 11-4)

Find each sum or difference. Use counters or a number line if necessary.
1. $-10 + (-8)$ 2. $12 + (-15)$
3. $-41 + (-6)$ 4. $-38 + 4$
5. $28 + (-51)$ 6. $15 - (-3)$
7. $13 - (-2)$ 8. $-6 - 8$
9. $-19 - (-4)$ 10. $-98 - (-98)$

   1. -18
   2. -3
   3. -47
   4. -34
   5. -23
   6. 18
   7. 15
   8. -14
   9. -15
   10. 0

© Glencoe/McGraw-Hill   295   Mathematics: Applications and Connections, Course 1

Technology Masters, pp. 21–22

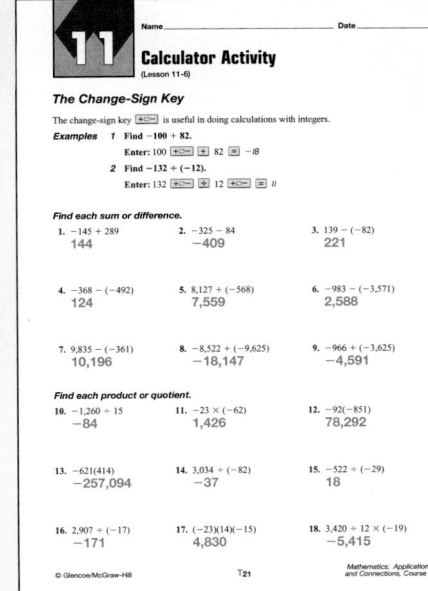

Calculator Activity (Lesson 11-6)

The Change-Sign Key

The change-sign key [+/−] is useful in doing calculations with integers.

Examples   1   Find $-100 + 82$.
   Enter: 100 [+/−] [+] 82 [=] -18
2   Find $-132 \div (-12)$.
   Enter: 132 [+/−] [÷] 12 [+/−] [=] 11

Find each sum or difference.
1. $-145 + 289$ 2. $-325 - 84$ 3. $139 - (-82)$
   144   -409 221

4. $-368 - (-492)$ 5. $8,127 + (-568)$ 6. $-983 - (-3,571)$
   124   7,559   2,588

7. $9,835 - (-361)$ 8. $-8,522 + (-9,625)$ 9. $-966 + (-3,625)$
   10,196 $-18,147$ $-4,591$

Find each product or quotient.
10. $-1,260 \div 15$ 11. $-23 \times (-62)$ 12. $-92(-851)$
   -84 1,426   78,292

13. $-621(414)$ 14. $3,034 \div (-82)$ 15. $-522 \div (-29)$
   $-257,094$ -37 18

16. $2,907 \div (-17)$ 17. $(-23)(14)(-15)$ 18. $3,420 \div 12 \times (-19)$
   -171 4,830 $-5,415$

© Glencoe/McGraw-Hill   T21   Mathematics: Applications and Connections, Course 1

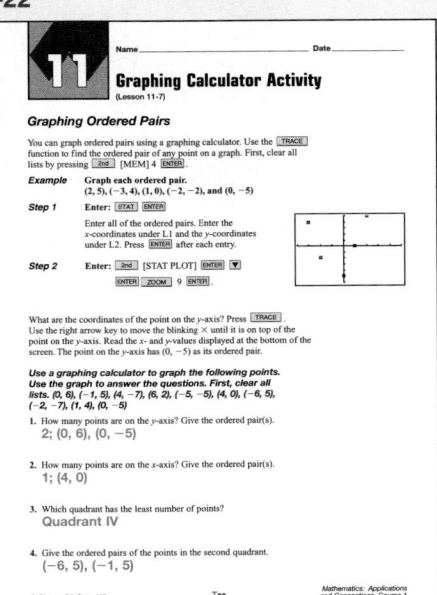

Graphing Calculator Activity (Lesson 11-7)

Graphing Ordered Pairs

You can graph ordered pairs using a graphing calculator. Use the [TRACE] function to find the ordered pair of any point on a graph. First, clear all lists by pressing [2nd] [MEM] 4 [ENTER].

Example   Graph each ordered pair.
   $(2, 5)$, $(-3, 4)$, $(1, 0)$, $(-2, -2)$, and $(0, -5)$.
Step 1   Enter: [STAT] [ENTER]
   Enter all of the ordered pairs. Enter the x-coordinates under L1 and the y-coordinates under L2. Press [ENTER] after each entry.
Step 2   Enter: [2nd] [STAT PLOT] [ENTER] [▼]
   [ENTER] [ZOOM] 9 [ENTER]

What are the coordinates of the point on the y-axis? Press [TRACE]. Use the right arrow key to move the blinking × until it is on top of the point on the y-axis. Read the x- and y-values displayed at the bottom of the screen. The point on the y-axis has $(0, -5)$ as its ordered pair.

Use a graphing calculator to graph the following points. Use the graph to answer the questions. First, clear all lists. $(0, 6)$, $(-1, 5)$, $(4, -7)$, $(6, 2)$, $(-5, -5)$, $(4, 0)$, $(-6, 5)$, $(-2, -7)$, $(1, 4)$, $(0, -5)$

1. How many points are on the y-axis? Give the ordered pair(s).
   2; $(0, 6)$, $(0, -5)$

2. How many points are on the x-axis? Give the ordered pair(s).
   1; $(4, 0)$

3. Which quadrant has the least number of points?
   Quadrant IV

4. Give the ordered pairs of the points in the second quadrant.
   $(-6, 5)$, $(-1, 5)$

© Glencoe/McGraw-Hill   T22   Mathematics: Applications and Connections, Course 1

Algebra: Investigating Integers

Theme: Civics

The United States House of Representatives is the lower house of Congress. Members of the House write ideas for new laws called *bills.* If enough members vote for a bill, it is sent to the Senate (the other house of Congress) and then to the President. When the President signs the bill, it becomes a law.

Question of the Day Based on the changes per state, how would you find the change in the number of members in the House of Representatives for the entire United States? **Add all numbers, making losses negative.**

Assess Prerequisite Skills

Ask students to read through the list of objectives presented in "What you'll learn in Chapter 11." You may wish to ask them what each of the objectives means or if they have experienced or used any of these math concepts before.

 Building Portfolios

Encourage students to revise their portfolios as they study this chapter. Have them include examples of their work that illustrate their understanding of integers.

 **Math and the Family**

In the *Family Letters and Activities* booklet (pp. 21–22), you will find a letter to the parents explaining what students will study in Chapter 11. An activity appropriate for the whole family is also available.

What you'll learn in Chapter 11

- to identify, name, graph, and compare integers,
- to add, subtract, multiply, and divide integers,
- to solve problems by working backward,
- to graph ordered pairs of numbers on a coordinate grid, and
- to graph transformations on a coordinate grid.

Congresswoman Marcy Kaptur, Ohio

432 Chapter 11 Algebra: Investigating Integers

 CD-ROM Program

Activities for Chapter 11

- Chapter 11 Introduction
- Interactive Lessons 11-2, 11-3, 11-4, 11-5
- Extended Activity 11-7
- Assessment Game
- Resource Lessons 11-1 through 11-8

CHAPTER Project

YOU WIN SOME, YOU LOSE SOME

In this project, you will draw a map of the United States and use the map to visually show how the number of members in the House of Representatives has changed for each state. You will also make a line plot and write a paragraph about these changes. You will display your map, line plot, and paragraph on a poster or in a brochure.

Getting Started

- The number of members in the House of Representatives for each state varies according to its population. As a result of the 1990 census, Florida gained 4 members to the House of Representatives, and Pennsylvania lost 2 members. Research how the number of representatives for each state changed after the last census.

- Draw or trace a map of the United States showing each state.

- Color all of the states that lost members in the House of Representatives red. Color all of the states that gained members in the House of Representatives blue. Leave the states that did not change the number of representatives white.

Technology Tips

- Use the **Internet** to find out which states gained and lost members in the House of Representatives.

- Use a **spreadsheet** to keep track of the data you collect.

- Use a **word processor** to write your paragraph about the changes.

inter NET
CONNECTION For up-to-date information on the House of Representatives, visit:
www.glencoe.com/sec/math/mac/mathnet

Working on the Project

You can use what you'll learn in Chapter 11 to help you keep track of the changes in the House of Representatives.

Page	Exercise
436	41
444	34
461	34
471	Alternative Assessment

CHAPTER Project
NOTES

Objectives Students should
- show changes in numbers of members in the House of Representatives for each state.
- make a line plot of the changes.

Project Pointer You may suggest that students begin a *Project Folder* to keep their work as they complete each stage of the Chapter Project. The completed project may also be added to their portfolios.

Explain to students that each citizen of the United States is represented equally in the House of Representatives. To achieve this, census numbers are used to determine the population of each state. The state is then divided into districts, each containing approximately the same number of people. One member of the House of Representatives is elected from each district. When shifts in population occur, the states that lost people are divided into fewer districts, and the states that gained people are divided into more districts.

***Investigations and Projects Masters*, p. 60**

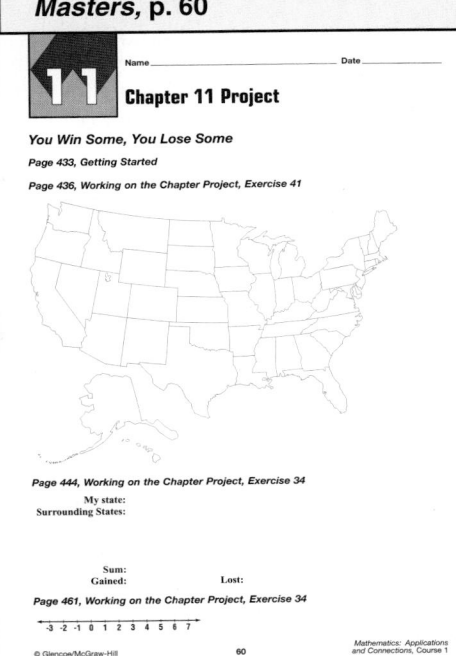

Instructional Resources ▶▶▶
A recording sheet to help students organize their data for the Chapter Project is shown at the right and is available in the *Investigations and Projects Masters,* p. 60.

Instructional Resources

- *Study Guide Masters*, p. 76
- *Practice Masters*, p. 76
- *Enrichment Masters*, p. 76
- Transparencies 11-1, A and B
- CD-ROM Program
  - Resource Lesson 11-1

Recommended Pacing	
Standard	Day 2 of 15
Honors	Day 2 of 13
Block	Day 1 of 8

1 FOCUS

5-Minute Check
(Chapter 10)

Find the area of each figure to the nearest tenth.

1. 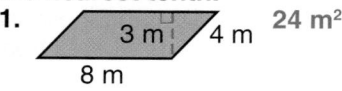 24 m²

2. 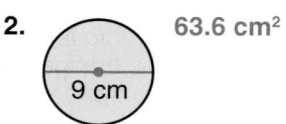 63.6 cm²

3. 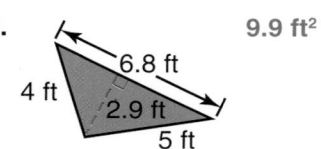 9.9 ft²

4. Find the surface area of a rectangular prism with a length of 16 inches, a width of 8 inches, and a height of 12 inches. **832 in²**

5. Find the volume of the rectangular prism in Exercise 4. **1,536 in³**

 The 5-Minute Check is also available on **Transparency 11-1A** for this lesson.

Motivating the Lesson

Hands-On Activity Show students that a thermometer measures temperature in evenly separated increments. Have them use a thermometer to read the temperature inside and outside the classroom. Tell students that learning to read a thermometer is just one skill in everyday life that involves integers.

What **you'll learn**

You'll learn to identify, name, and graph integers.

When **am I ever going to use this?**

Knowing about integers can help you express temperatures.

Word Wise
integer
positive integer
negative integer
opposite

Each March, a dog-sled race, called the Iditarod, is held between Anchorage and Nome, Alaska. During this time of the year, the average daytime high temperature for Anchorage is 34°F. However, the racers and their dogs can face temperatures as low as 30°F below zero. You can write 30 below zero as -30.

The numbers 34 and -30 are integers. An **integer** is any number from the set $\{...-3, -2, -1, 0, 1, 2, 3,...\}$ where ... means *continues without end.*

Integers that are greater than zero are called **positive integers**. Integers that are less than zero are called **negative integers**. Zero itself is neither positive nor negative. You can show positive and negative numbers on a number line.

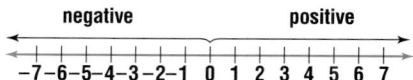

Negative integers are written with a − sign. *Positive integers can be written with or without a + sign.*

You can graph integers on a number line by drawing a dot.

Examples

Study Hint

Reading Math You read −5 as *negative five*. You read +2 as *positive two* or *two*.

CONNECTION

① **Graph −5 on the number line.**

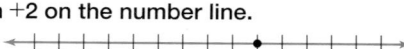

② **Graph +2 on the number line.**

Geography Write an integer to describe each situation.

③ **The Dead Sea is 1,312 feet below sea level.**

You write $-1,312$.

④ **Mt. Everest is 29,028 feet above sea level.**

You write $+29,028$ or $29,028$.

 ### Cross-Curriculum Cue

Inform the teachers on your team that your students are studying integers. Suggestions for curriculum integration are:

Health: body temperature
Geography: longitude and latitude, elevation
Physical Science: electrical charges

Each integer has an opposite. **Opposite** integers are the same distance from zero in opposite directions on the number line. Zero is considered to be the starting point.

Examples

Write the opposite of each integer.

5 +4

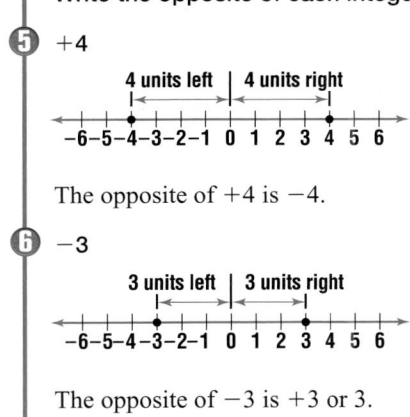

4 units left | 4 units right

−6 −5 −4 −3 −2 −1 0 1 2 3 4 5 6

The opposite of +4 is −4.

6 −3

3 units left | 3 units right

−6 −5 −4 −3 −2 −1 0 1 2 3 4 5 6

The opposite of −3 is +3 or 3.

CHECK FOR UNDERSTANDING

Communicating Mathematics

Read and study the lesson to answer each question.

1. *Identify* the integers graphed on the number line. **−1 and +2**

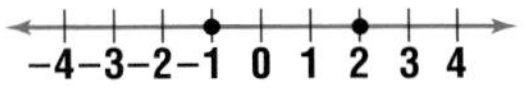

−4 −3 −2 −1 0 1 2 3 4

2. *Show* the number 6 and its opposite by graphing them on a number line. Explain why the two numbers are opposites. **See Answer Appendix.**

Math Journal

3. *Write* about a situation that you could describe using positive and negative integers. Give examples using a positive integer and a negative integer. Explain what each integer means. **See margin.**

Guided Practice

Draw a number line from −10 to 10. Graph each integer on the number line. 4–6. See Answer Appendix.

4. +9   5. −7   6. 3

Write an integer to describe each situation.

7. The quarterback gained 6 yards on the play. **+6**

8. Cecilia lost 5 pounds. **−5**

Write the opposite of each integer.

9. 4 **−4**   10. −9 **+9**   11. +345 **−345**

12. Graph the opposite of −10.

12. See Answer Appendix.

13. *Physical Science* Physicist Paul Ching-Wu Chu discovered materials that conduct electricity at 178°C below zero. Write this number as an integer. **−178**

Lesson 11-1 Integers **435**

■ Reteaching the Lesson ■

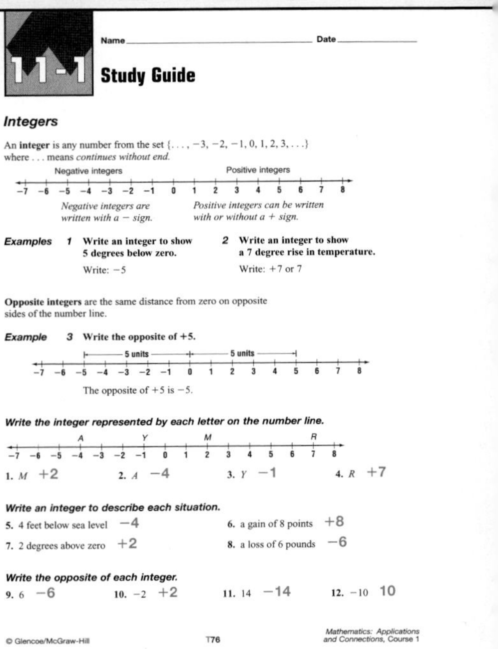

Activity Identify one number cube as a negative cube and another cube as positive. Starting at 0 on a number line, have students take turns rolling each cube and moving their marker left or right. After five turns, ask them where they are in relation to 0.

Additional Answer

3. **Sample answer: The yards lost or gained in one down of a football game. For example, −3 would represent a loss of 3 yards, and +4 would represent a gain of 4 yards.**

Check for Understanding

If students need additional practice or instruction after completing Exercises 1–13, one of these options may be helpful.
- Extra Practice, see p. 585
- Reteaching Activity, see p. 435
- *Study Guide Masters,* p. 76
- *Practice Masters,* p. 76

Assignment Guide

Core: 15–39 odd, 42–47
Enriched: 14–38 even, 39, 40, 42–47

CHAPTER Project

Exercise 41 asks students to advance to the next stage of work on the Chapter Project. Have students place their work in their Project Folders.

4 ASSESS

Closing Activity

Modeling Have pairs of students use a thermometer to measure the temperature of hot, warm, and cold water. Then have them show the temperatures and their opposites on a number line.

Practice Masters, p. 76

11-1 Practice

Integers

Write the integer represented by each letter on the number line.
1. *I* −5   2. *L* +3   3. *M* −2   4. *H* −13
5. *J* +12   6. *K* −16   7. *G* +6   8. *N* −9

Write an integer to describe each situation.
9. a gain of 5 pounds +5   10. 4 degrees below normal −4
11. a loss of 8 yards −8   12. positive 16 +16
13. an increase of 2 inches +2   14. scored 10 fewer points −10
15. negative eighteen −18   16. 15 feet above sea level +15
17. earned 7 dollars interest +7   18. neither positive nor negative 0
19. bowled 9 pins above average +9   20. a decrease of 6 members −6

Write the opposite of each integer.
21. −25 +25   22. 36 −36   23. 54 −54   24. −11 +11
25. 98 −98   26. −47 +47   27. −62 +62   28. 80 −80
29. −73 +73   30. 14 −14   31. 105 −105   32. −29 +29

© Glencoe/McGraw-Hill   T76   *Mathematics: Applications and Connections, Course 1*

Practice   **Draw a number line from −10 to 10. Graph each integer on the number line. 14–21. See Answer Appendix.**

14. +4   15. −1   16. 0   17. −6
18. −8   19. 3   20. 7   21. −2

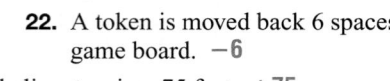

Write an integer to describe each situation.

22. A token is moved back 6 spaces on a game board. −6

23. A helicopter rises 75 feet. +75

24. A withdrawal of $45 is made from a bank account. −45

25. A submarine is 100 meters below the surface of the water. −100

26. The value of a stock decreases by $1. −1

27. An employee receives a $100 bonus. +100

Write the opposite of each integer.

28. −35 +35   29. +23 −23   30. −1 +1   31. 45 −45
32. −250 +250   33. 77 −77   34. −52 +52   35. −110 +110

36. Graph the opposite of 12. **36–38. See Answer Appendix.**
37. Graph −6, −8, 0, 3, and 5 on the same number line.
38. Graph 7, −3, 4, and 0 on the same number line.

Applications and Problem Solving

39. *Earth Science* The temperature of the center of Earth is estimated to be 7,000°C. Express this number as an integer. +7,000

40. *Geography* Jacksonville, Florida, is at sea level. Express the elevation of Jacksonville as an integer. 0

41. *Working on the* **CHAPTER Project** Refer to the map of the United States you made on page 433. In each state, write an integer that represents how the number of members to the House of Representatives has changed for that state. **See students' work.**

42. *Critical Thinking* Compare a number line to a thermometer. How are they alike? How are they different? **See Answer Appendix.**

Mixed Review

43. **Test Practice** Juana is going to wrap a present shaped like a rectangular prism with dimensions 27 inches by 14 inches by 5 inches. What is the minimum amount of wrapping paper Juana will need? *(Lesson 10-6)* C
 A 46 in²   B 583 in²   C 1,166 in²   D 1,890 in²

44. *Geometry* Name the polygon by the number of sides. *(Lesson 9-4)* octagon

45. Express 1.35 as a percent. *(Lesson 8-5)* 135%

46. Find $6\frac{2}{3} - 4\frac{3}{5}$ in simplest form. *(Lesson 6-5)* $2\frac{1}{15}$

47. Round 673.018 to the nearest tenth. *(Lesson 3-4)* 673.0

Extending the Lesson

Enrichment Masters, p. 76

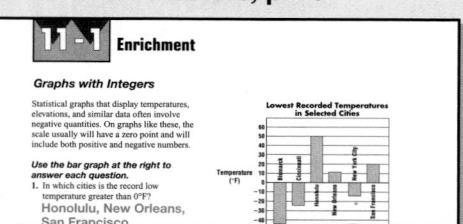

11-1 Enrichment

Graphs with Integers

Statistical graphs that display temperatures, elevations, and similar data often involve negative quantities. On graphs like these, the scale usually will have a zero point and will include both positive and negative numbers.

Use the bar graph at the right to answer each question.
1. In which cities is the record low temperature greater than 0°F?
Honolulu, New Orleans, San Francisco

Lowest Recorded Temperatures in Selected Cities

Activity Ask students the following questions. *Can each integer be written as a fraction?* yes *Can the set of integers be considered a subset of the set of fractions?* yes

11-2 Comparing and Ordering Integers

What you'll learn
You'll learn to compare and order integers.

When am I ever going to use this?
Knowing how to compare integers can help you compare temperatures.

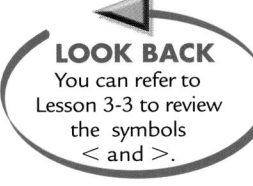

LOOK BACK
You can refer to Lesson 3-3 to review the symbols < and >.

History books use timelines to show the order of events. This timeline shows the establishment of various cities.

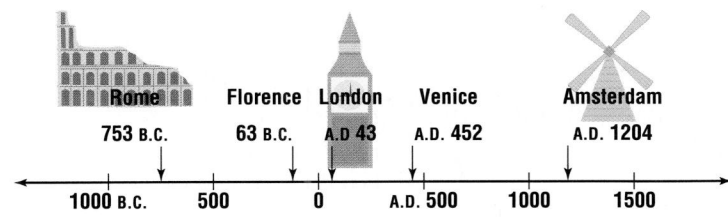

Rome	Florence	London	Venice	Amsterdam
753 B.C.	63 B.C.	A.D 43	A.D. 452	A.D. 1204

1000 B.C.   500   0   A.D. 500   1000   1500

On a timeline, an event depicted to the left always occurred before an event to the right. So, Florence was established after Rome, but before London.

You can use a number line to compare numbers. On a number line, the number to the left is always less than the number to the right.

$$-7\ -6\ -5\ -4\ -3\ -2\ -1\ 0\ 1\ 2\ 3\ 4\ 5\ 6\ 7$$

Notice that -7 is to the left of -3. Therefore, $-7 < -3$. Also, $-3 > -7$ since -3 is to the right of -7.

Examples

Replace each ● with $<$, $>$, or $=$ to make a true sentence.

1 $-5 ● -2$

Graph -5 and -2 on a number line.

$$-7\ -6\ -5\ -4\ -3\ -2\ -1\ 0\ 1\ 2\ 3\ 4\ 5\ 6\ 7$$

-5 is to the left of -2, so $-5 < -2$.

2 $0 ● -5$

Graph 0 and -5 on a number line.

$$-7\ -6\ -5\ -4\ -3\ -2\ -1\ 0\ 1\ 2\ 3\ 4\ 5\ 6\ 7$$

0 is to the right of -5, so $0 > -5$.

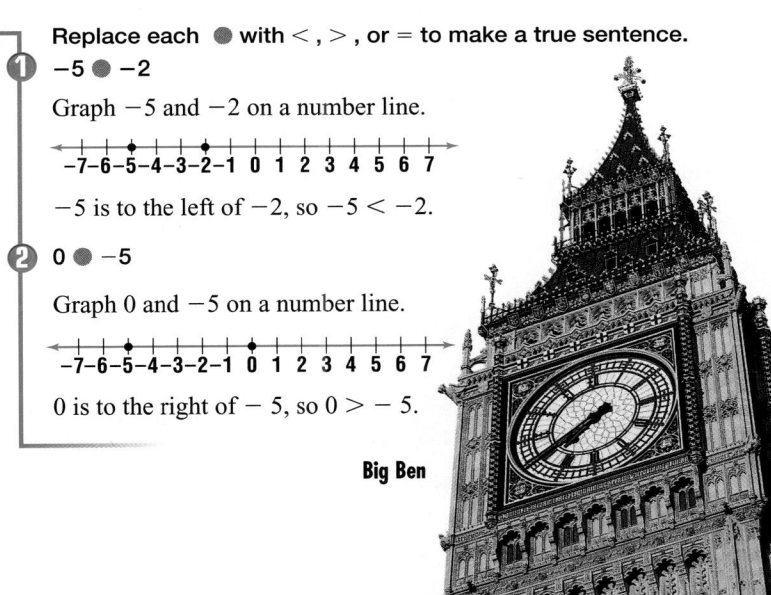

Big Ben

Lesson 11-2 Comparing and Ordering Integers **437**

Multiple Learning Styles

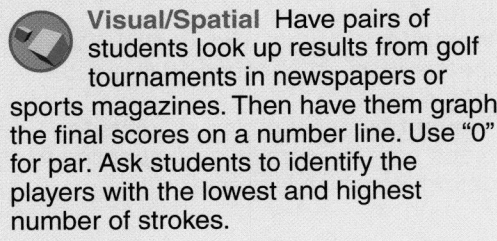

Visual/Spatial Have pairs of students look up results from golf tournaments in newspapers or sports magazines. Then have them graph the final scores on a number line. Use "0" for par. Ask students to identify the players with the lowest and highest number of strokes.

Motivating the Lesson

Communication Draw a number line on the chalkboard, placing a 0 in the middle. Have volunteers label $+1$, -1, $+4$, and -3. Ask: *Which is the greatest number?* $+4$ *the least?* -3 *Where might you see a number line like this?* thermometer

---

11-2 Lesson Notes

Instructional Resources
- *Study Guide Masters*, p. 77
- *Practice Masters*, p. 77
- *Enrichment Masters*, p. 77
- Transparencies 11-2, A and B
- *Assessment and Evaluation Masters*, p. 295
- *School to Career Masters*, p. 11
- CD-ROM Program
  - Resource Lesson 11-2
  - Interactive Lesson 11-2

Recommended Pacing	
Standard	Days 3 & 4 of 15
Honors	Day 3 of 13
Block	Day 2 of 8

1 FOCUS

5-Minute Check (Lesson 11-1)

Draw a number line from -10 to 10. Graph each integer on the number line.

1. -4 **2.** 7

-10 -5 0 5 10

3. Write an integer to describe 13 degrees below zero. -13

Write the opposite of each integer.
4. 26 -26 **5.** -11 11

The 5-Minute Check is also available on **Transparency 11-2A** for this lesson.

2 TEACH

Transparency 11-2B contains a teaching aid for this lesson.

Reading Mathematics Have students locate a timeline in a newspaper, magazine, or history book. Ask them to read the events listed to the class, indicating whether each event occurred before or after the last event mentioned. Ask students how they can tell which event occurred first.

Lesson 11-2 **437**

In-Class Examples

Replace each ● with <, >, or = to make a true sentence.

For Example 1
−6 ● −3 <

For Example 2
2 ● 0 >

For Example 3
Order the integers −5, 4, 1, −4, and −1 from greatest to least.
4, 1, −1, −4, −5

For Example 4
Find the median of the record low temperatures (°F) given for the states in the table. −61°F

AK	CO	ID	MN	MT	HI	WY
−80	−61	−60	−59	−70	12	−66

3 PRACTICE/APPLY

Check for Understanding

If students need additional practice or instruction after completing Exercises 1–8, one of these options may be helpful.
- Extra Practice, see p. 585
- Reteaching Activity
- *Study Guide Masters*, p. 77
- *Practice Masters*, p. 77

Study Guide Masters, p. 77

11-2 Study Guide

Name_____ Date_____

Comparing and Ordering Integers

You can use a number line to compare integers. On a number line, the number on the left is always less than the number on the right.

Examples   **1**   Replace the ○ in −3 ○ −7 with <, >, or =.

−7 is to the left of −3, so −3 > −7.

2 Order the integers −5, 2, 0, −1 from least to greatest.
Write the integers as they appear on the number line from left to right.
−5, −1, 0, 2

Replace each ○ with <, >, or = to make a true sentence.
1. −2 ○ 0 <   2. 5 ○ −1 >   3. −4 ○ 4 <   4. −9 ○ −9 =
5. −44 ○ −4 <   6. −19 ○ 9 <   7. −13 ○ −23 >   8. 0 ○ −54 >

Order each set of integers from least to greatest.
9. −2, 4, 0, −1, 1   −2, −1, 0, 1, 4
10. 0, −4, 4, 7, −6, −5   −6, −5, −4, 0, 4, 7
11. −10, 12, −13, 9, −8, 4   −13, −10, −8, 4, 9, 12
12. 23, 0, 15, −26, −34, −30   −34, −30, −26, 0, 15, 23
13. −7, −19, 19, 0, −25, 30   −25, −19, −7, 0, 19, 30
14. 55, −15, −5, −55, 5, 15   −55, −15, −5, 5, 15, 55

© Glencoe/McGraw-Hill   T77   *Mathematics: Applications and Connections, Course 1*

 **3** Order the integers −4, 3, 0, and −5 from least to greatest.

Graph each number on a number line first.

−7−6−5−4−3−2−1 0 1 2 3 4 5 6 7

Then, write the integers as they appear on the number line from left to right. −5, −4, 0, and 3 are in order from least to greatest.

APPLICATION

LOOK BACK
You can refer to Lesson 2-7 to review median.

4 **Weather** The record cold temperatures for five states are recorded on the map. Find the median of these temperatures.

List the temperatures in order from least to greatest.

−80, −70, −52, −2, 12

The median is the middle number when the numbers are arranged in order. So the median of these low temperatures is −52.

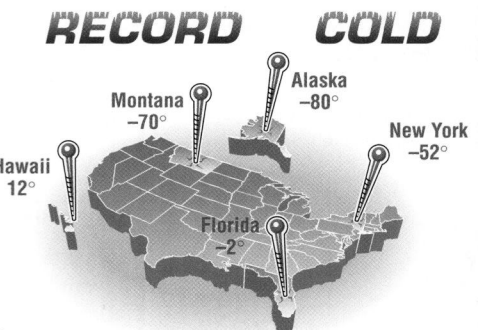

RECORD   COLD

Alaska −80°
Montana −70°
New York −52°
Hawaii 12°
Florida −2°

Source: *The USA Today Weather Almanac*

CHECK FOR UNDERSTANDING

Communicating Mathematics

Read and study the lesson to answer each question. 1–3. See Answer Appendix.

1. *Explain* how a number line can be used to order integers.
2. *Write a Problem* where integers would be compared or ordered.
3. *You Decide* Amy says that 5 is greater than 3, and therefore, −5 > −3. Cordelia disagrees. Who is correct? Explain your reasoning.

Guided Practice

Replace each ● with < , > , or = to make a true sentence.

4. +3 ● −1 >   5. −8 ● −4 <   6. 0 ● −2 >

7. Order −3, 5, 0, and −2 from least to greatest. −3, −2, 0, 5

8. 4 feet below sea level; on a number line −5 is to the left of −4.

8. *Geography* Which is higher, 5 feet below sea level or 4 feet below sea level? Explain.

EXERCISES

Practice

Replace each ● with < , > , or = to make a true sentence.

9. +3 ● −4 >   10. −10 ● −100 >   11. −3 ● 0 <
12. +32 ● −32 >   13. 0 ● +5 <   14. −6 ● −7 >
15. −82 ● −85 >   16. −44 ● −33 <   17. −200 ● +123 <

Reteaching the Lesson

Activity Have pairs of students play a game like "War." Ask them to number a set of cards from −25 to 26, mix them up, and pass them out facedown, 26 cards to each player. Each player reveals the top card. The player with the greater integer takes both cards, placing them to one side. Play continues until both players run out of cards. The player who wins the most rounds wins.

Error Analysis

Watch for students who identify −10 as greater than −4 because 10 is greater than 4.

Prevent by encouraging students to graph both integers on a number line to determine which is greater.

18. Order 4, −5, 6, and −7 from least to greatest. **−7, −5, 4, 6**

19. 41, 10, 3, 0, −10, −20

19. Order 0, 41, 3, −20, −10, and 10 from greatest to least.

20. Which is greater, −45 or −42? **−42**

21. Which is least, 0, −3, −17, or −8? **−17**

22. Is zero greater than, less than, or equal to negative ten? **greater than**

Applications and Problem Solving

23. *Physical Science* The table shows the melting point of some common elements. As the temperature rises above the melting point, the element changes from a solid to a liquid.

23a. −458, −435, −361, −38, 450, 1,542, 1,763, 1,947, 2,795

   a. List the melting points from least to greatest.

   b. The average annual temperature in the interior of Antarctica is −71°F. Would mercury be a solid or a liquid at this temperature? **solid**

   c. Find the median of the melting points. **450**

Element	Melting Point (°F)
Calcium	1,542
Gold	1,947
Helium	−458
Hydrogen	−435
Iron	2,795
Mercury	−38
Oxygen	−361
Silver	1,763
Tin	450

24. *Life Science* Some sea creatures live near the surface while others live in the depths of the ocean. Make a drawing showing the relative habitats of the following creatures. **See Answer Appendix.**

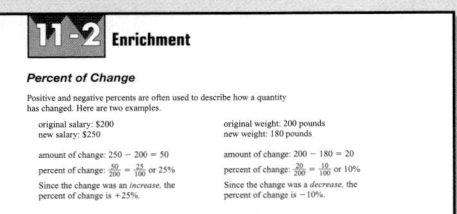

Brittle Star

- ribbon fish: 600 to 3,300 feet below the surface
- blue marlin: 0 to 600 feet below the surface
- brittle stars: 13,200 to 19,800 feet below the surface
- lantern fish: 3,300 to 13,200 feet below the surface

25. *Critical Thinking* Why is any negative integer less than any positive integer? **See margin.**

Mixed Review

26. *Geography* New Orleans is 8 feet below sea level. Express this elevation as an integer. *(Lesson 11-1)* **−8**

27. **Test Practice** Fred had an equilateral triangle with an area of 3 square feet. He traced the triangle several times to make a hexagon. What is the area of the hexagon? *(Lesson 10-2)* **C**

Area = 3 ft²

   A 8 ft²   **B** 12 ft²

   C 18 ft²   **D** 22 ft²

28. Find the GCF of 120 and 150. *(Lesson 5-3)* **30**

Lesson 11-2 Comparing and Ordering Integers   **439**

Extending the Lesson

Enrichment Masters, p. 77

11-2 Enrichment

Percent of Change

Positive and negative percents are often used to describe how a quantity has changed. Here are two examples.

original salary: $200
new salary: $250

original weight: 200 pounds
new weight: 180 pounds

amount of change: 250 − 200 = 50

amount of change: 200 − 180 = 20

percent of change: $\frac{50}{200} = \frac{25}{100}$ or 25%

percent of change: $\frac{20}{200} = \frac{10}{100}$ or 10%

Since the change was an *increase*, the percent of change is +25%.

Since the change was a *decrease*, the percent of change is −10%.

Activity Separate students into groups to research the following questions. *How do temperature extremes on Mercury or Mars compare with those on Earth? How deep are the craters on the moon?* Have groups show their findings on a number line.

---

4 ASSESS

Closing Activity

Modeling Write each integer from −5 to 5 on an index card. Give one card to each of 11 students and ask them to form a human number line, holding the cards in front of them. Ask each student to name one integer greater than and one less than his or her card. Then have students in the line pass their cards to 11 other students and see how quickly they can form a new line. Continue with questions and discussion.

Chapter 11, Quiz A (Lessons 11-1 and 11-2) is available in the *Assessment and Evaluation Masters,* p. 295.

Additional Answer

25. **All negative integers are to the left of 0 on a number line while all positive integers are to the right. The number to the left is always less than the number to the right.**

Practice Masters, p. 77

11-2 Practice

Comparing and Ordering Integers

Fill in each ◯ with <, >, or =.

1. −9 ◯ 8 **<** 2. 0 ◯ −1 **>** 3. −14 ◯ −15 **>** 4. +26 ◯ 26 **=**

5. −32 ◯ 23 **<** 6. −148 ◯ 148 **<** 7. 19 ◯ −91 **>** 8. −67 ◯ −60 **<**

9. 245 ◯ −254 **>** 10. −971 ◯ 791 **<** 11. −830 ◯ −803 **<** 12. −64 ◯ −64 **=**

13. −57 ◯ −75 **>** 14. −33 ◯ 3 **<** 15. 169 ◯ −196 **>** 16. −200 ◯ −201 **>**

Order each set of integers from least to greatest.

17. −6, 16, −26   **−26, −6, 16**

18. −213, −231, 132   **−231, −213, 132**

19. 5, −3, −11, 9, −7   **−11, −7, −3, 5, 9**

20. 8, −6, 4, −10, −2   **−10, −6, −2, 4, 8**

21. −36, 28, −4, −17, −59   **−59, −36, −17, −4, 28**

22. −84, 95, −71, −103, −62   **−103, −84, −71, −62, 95**

23. 21, −34, 65, −12, 43, 0, −56   **−56, −34, −12, 0, 21, 43, 65**

24. −22, 2, 0, 202, 22, −222, −2   **−222, −22, −2, 0, 2, 22, 202**

25. −3, −33, 36, −66, 63, −6, 0   **−66, −33, −6, −3, 0, 36, 63**

26. 0, −172, 1, −127, 7, −171, −117   **−172, −171, −127, −117, 0, 1, 7**

27. Which is greater, negative 21 or negative 22? **−21**

28. Which is greater, 8 degrees above zero or 9 degrees below zero? **8 degrees above zero**

29. Which is less, a checkbook balance of −30 dollars or a balance of −25 dollars? **−30 dollars**

30. Which is greater, negative 98 or positive 89? **+89**

© Glencoe/McGraw-Hill   T77   *Mathematics: Applications and Connections, Course 1*

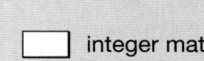

GET READY

Objective Students model integers using counters.

Optional Resources
Hands-On Lab Masters
• counters, p. 6
• integer mat, p. 9
• worksheet, p. 64

Overhead Manipulative Resources
• counters
• integer mat

Manipulative Kit
• counters

MANAGEMENT TIPS

Recommended Time
30 minutes

Getting Started The concept of *zero pair* is very important in modeling not only addition of integers, but other operations as well. It is also used in modeling solving equations and modeling polynomials. Have students practice the zero pair concept by pairing positive and negative counters from a large pile of counters.

The **Activity** has students model $+3 + (-3)$. You may want to model the activity on an overhead transparency or on the chalkboard with colored chalk. If counters are not available, have students cut counters from construction paper.

ASSESS

Have students complete Exercises 1–6. Watch for students who don't understand the meaning of *zero pair*.

Use Exercise 7 to determine whether students understand how to use counters to model integers.

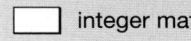

COOPERATIVE LEARNING

11-3A Zero Pairs

A Preview of Lesson 11-3

○ counters of two colors

□ integer mat

You can use counters to help you understand integers. Yellow counters represent positive integers, and red counters represent negative integers. When you pair one counter of each color, the result is zero. We call this pair of counters a *zero pair*.

TRY THIS

Work with a partner.

Step 1 Place three yellow counters on the mat to represent the integer +3.

 $+3$

Step 2 Place three red counters on the same mat to represent the integer −3.

 $+3 + (-3)$

Step 3 Pair the positive and negative counters. Remove as many zero pairs as possible.

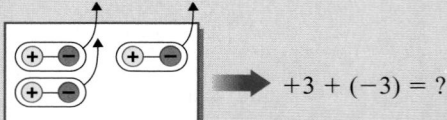 $+3 + (-3) = ?$

ON YOUR OWN

1. How many counters do you have on the mat? **none**
2. What is the sum of $+3 + (-3)$? **0**
3. How are +3 and −3 related? **They are opposites.**
4. What is the sum of any pair of opposites? **0**
5. What is the value of a zero pair? **0**
6. If you removed a zero pair from the mat, what effect does it have on the value of the counters left on the mat? **none**
7. *Look Ahead* How do you think you could use counters to find the sum of +4 and −3? **See margin.**

440 Chapter 11 Algebra: Investigating Integers

Additional Answer
7. Place 4 yellow counters on the mat to represent +4. Place 3 red counters on the mat to represent −3. Remove as many zero pairs as possible. The 1 yellow counter left represents the sum (+1).

 Have students write a paragraph on how to use zero pairs to find the sum of +6 and −6.

Adding Integers

What you'll learn
You'll learn to add integers using models.

When am I ever going to use this?
Knowing how to add integers can help you keep score in games.

Word Wise
zero pair

Remember that yellow counters represent positive integers and red counters represent negative integers.

Monsa and Victor are playing a board game. In the game, each player rolls a number cube and moves a token. Some squares on the game board have further instructions.

- Monsa starts at 0 and rolls a 5.
- The fifth square tells him to roll again. He rolls a 6.
- His token lands on a square that tells him to move back 3 spaces.

How many spaces from the start is Monsa's token?

To find the location of Monsa's token, you will need to add integers. You can add integers using models.

Use yellow counters to represent positive integers and red counters to represent negative integers.

Step 1 Use 5 positive counters to represent Monsa's first roll (+5). Use 6 positive counters to represent Monsa's second roll (+6). Place all of the counters on a mat.

$+5 + (+6) = +11$

Step 2 Use 3 negative counters to represent the 3 spaces backward. Place the 3 negative counters on the mat with the 11 positive counters.

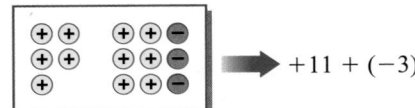

$+11 + (-3)$

Step 3 Pair the positive and negative counters. Remove as many **zero pairs** as possible since it does not change the value on the mat.

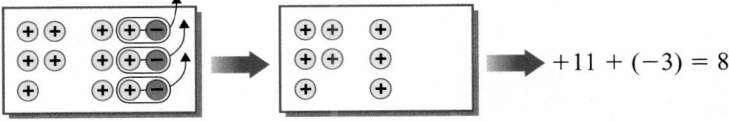

$+11 + (-3) = 8$

There are a total of 8 positive counters left on the mat. Monsa's token is 8 spaces from the starting point.

Lesson 11-3 Adding Integers **441**

Instructional Resources
- *Study Guide Masters*, p. 78
- *Practice Masters*, p. 78
- *Enrichment Masters*, p. 78
- Transparencies 11-3, A and B
- *Classroom Games*, pp. 29–32
- CD-ROM Program
  - Resource Lesson 11-3
  - Interactive Lesson 11-3

Recommended Pacing	
Standard	Days 5 & 6 of 15
Honors	Day 4 of 13
Block	Day 3 of 8

1 FOCUS

5-Minute Check
(Lesson 11-2)

Replace each ● with <, >, or = to make a true sentence.

1. 14 ● −26 **>**
2. −285 ● −42 **<**
3. 0 ● −12 **>**
4. Which is greater, −37 or −9? **−9**
5. Is zero greater than, less than, or equal to negative two? **greater than**

The 5-Minute Check is also available on **Transparency 11-3A** for this lesson.

Motivating the Lesson
Hands-On Activity Give small groups of students a random number of red (negative) and yellow (positive) counters. Ask them to remove as many zero pairs as possible. Then ask them to name the integer that describes the remaining counters.

Classroom Vignette

"To help students picture adding integers on a number line, we call it 'dancing' up and down the number line. We physically move up and down a number line, which is displayed around the perimeter of my classroom."

Karen Shevnock Jamieson, Teacher
Thurman White Middle School
Henderson, NV

Karen Shevnock Jamieson

Transparency 11-3B contains a teaching aid for this lesson.

Using Problem Solving Have students solve the following problem. *In the game "Treadmill," you go backward if you spin an odd number and forward if you spin an even number. What would be your position if you started at 0 and spun 1, 3, 4, and 4?* +4

In-Class Examples

For Example 1
Use counters to find −8 + 5.
−3

For Example 2
Use counters to find −6 + (−6).
−12

For Example 3
A computer consultant earns $2,600.00 in the first month she runs her own company. She spends $3,100.00 to set up her office. How much money does she have at the end of the month? −$500.00

Teaching Tip After Example 3, review the four possible sign combinations of adding integers: negative plus positive, negative plus negative, positive plus positive, and positive plus negative.

Examples

1 Use counters to find −4 + 3.

Step 1 Place 4 negative counters on the mat to represent −4. Place 3 positive counters on the same mat to represent adding 3.

Step 2 Pair the positive and negative counters. Remove as many zero pairs as possible.

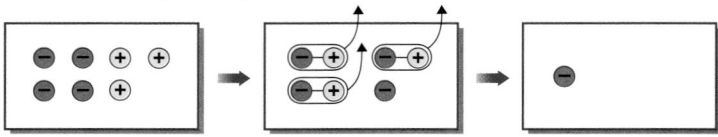

Step 3 Count the counters left on the mat. There is 1 negative counter. This represents −1. So, −4 + 3 = −1.

2 Use counters to find −2 + (−2).

Step 1 Place 2 negative counters on the mat to represent −2. Place 2 more negative counters on the same mat to represent adding −2.

Step 2 Since there are no positive counters, you cannot remove any zero pairs.

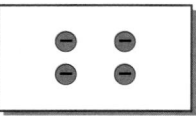

Step 3 Count the counters on the mat. There are 4 negative counters. This represents −4. So −2 + (−2) = −4.

You can also use a number line to add integers.

Example
APPLICATION

3 **Game Shows** On *Jeopardy!*, a contestant has 200 points and then loses 500 points. What is the contestant's score?

You need to find the sum of 200 + (−500). Consider a number line. Start at 0 and go 200 in the positive direction (right).

From that point, go 500 in the negative direction (left).

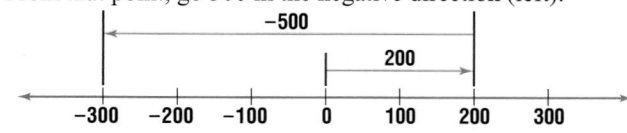

You end at −300. So, 200 + (−500) = −300.

442 Chapter 11 Algebra: Investigating Integers

Communicating Mathematics

Read and study the lesson to answer each question.

1. *Write* an addition sentence represented by the model. $-5 + 2 = -3$

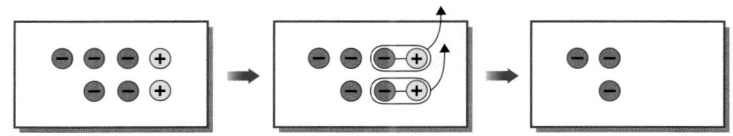

2. *Model* $-3 + 2$ and $2 + (-3)$ using counters. Compare and contrast these two problems. **See margin.**

Guided Practice

State whether each sum is *positive*, *negative*, or *zero*.

3. $-5 + 3$ **negative**

4. $-3 + 7$ **positive**

Find each sum. Use counters or a number line if necessary.

5. $3 + (-1)$ **2**

6. $-4 + (-8)$ **−12**

7. $0 + (-2)$ **−2**

8. $4 + (-6)$ **−2**

9. Find the sum of -4, 8, and -12. **−8**

10. *Football* The Centerville Middle School football team has the ball on the 20-yard line. The team is heading towards the 50-yard line. On the first play, the team gains 6 yards. On the next play, the team loses 8 yards. Where is the ball after the second play? **18 yard line**

EXERCISES

Practice

State whether each sum is *positive*, *negative*, or *zero*.

11. $3 + (-5)$ **negative**
12. $-2 + (-7)$ **negative**
13. $6 + 3$ **positive**
14. $-4 + 6$ **positive**
15. $-3 + (-3)$ **negative**
16. $-10 + 10$ **zero**

Find each sum. Use counters or a number line if necessary.

17. **−9**
25. **−8**

17. $-3 + (-6)$
18. $3 + (-8)$ **−5**
19. $-12 + 12$ **0**
20. $3 + 0$ **3**
21. $-3 + 0$ **−3**
22. $-5 + 4$ **−1**
23. $6 + (-2)$ **4**
24. $-4 + 10$ **6**
25. $-4 + (-4)$
26. $-5 + 5$ **0**
27. $13 + (-3)$ **10**
28. $-11 + (-18)$ **−29**

29. Find the sum of -18, 6, and 20. **8**

30. What is -3 plus -6 plus 4? **−5**

31. *Algebra* Find the value of $a + b$ if $a = -9$ and $b = 6$. **−3**

Applications and Problem Solving

32a. $-21 + 12$

32. *Scuba Diving* A scuba diver dived 21 feet below sea level. Then the diver went up 12 feet.
   a. Write an addition statement representing the dive.
   b. What integer represents the diver's location with respect to sea level? **−9**

33. *Earth Science* At night, the average temperature on the surface of Saturn is $-150°C$. During the day, the temperature rises $27°C$. What is the average temperature on the planet's surface during the day? **−123°C**

Reteaching the Lesson

Activity Use a standard deck of playing cards with the face cards removed to add integers. Groups of students will need 10 cards. Use red cards to represent negative integers and black cards to represent positive integers.

Error Analysis

Watch for students who are having difficulty understanding whether the sum of a positive integer and a negative integer is positive or negative.

Prevent by telling students to use counters to model the problem.

Check for Understanding

If students need additional practice or instruction after completing Exercises 1–10, one of these options may be helpful.
- Extra Practice, see p. 586
- Reteaching Activity
- *Study Guide Masters*, p. 78
- *Practice Masters*, p. 78
- Interactive Mathematics Tools Software

Assignment Guide

Core: 11–33 odd, 35–39
Enriched: 12–30 even, 32, 33, 35–39

Additional Answer

2. See Answer Appendix for art. One has 3 negative counters on the mat and adds 2 positive counters. The other has 2 positive counters and adds 3 negative counters. Both result in 2 zero pairs with just one negative counter left on the mat. The answer to both problems is -1.

Study Guide Masters, p. 78

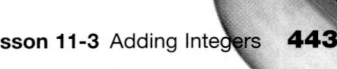

11-3 Study Guide

Adding Integers

You can add integers using models.

Examples 1 Use counters to find $-7 + 2$.

Place 7 negative counters on the mat to represent -7.
Place 2 positive counters on the mat to represent adding 2.
Pair the positive and negative counters.
Remove as many zero pairs as possible.

There are 5 negative counters left on the mat.
So, $-7 + 2 = -5$.

2 Use counters to find $-4 + (-4)$.
Place 4 negative counters on the mat to represent -4.
Place 4 more negative counters on the mat to represent adding -4.
Since there are no positive counters, you cannot remove any zero pairs.

There are 8 negative counters left on the mat.
So, $-4 + (-4) = -8$.

State whether each sum is positive or negative.

1. $-4 + (-2)$ **negative**
2. $5 + (-3)$ **positive**
3. $-10 + 7$ **negative**
4. $9 + (-3)$ **positive**
5. $6 + 0$ **positive**
6. $-8 + (-1)$ **negative**

Find each sum. Use counters or a number line if necessary.

7. $3 + (-6)$ **−3**
8. $-9 + 8$ **−1**
9. $-4 + 7$ **3**
10. $6 + (-6)$ **0**
11. $-8 + (-2)$ **−10**
12. $2 + (-5)$ **−3**

© Glencoe/McGraw-Hill T78 Mathematics: Applications and Connections, Course 1

CHAPTER Project

Exercise 34 asks students to advance to the next stage of work on the Chapter Project. You may have students add integers for another region. Then have students place their work in their Project Folders.

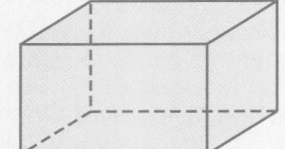

4 ASSESS

Closing Activity

Modeling Have pairs of students use counters and a mat to find sums of examples such as −4 + 6. Then have them add another integer to the sum so that the new sum will be the opposite of the original sum.
2; 2 + (−4) = −2

Additional Answer

37. Sample answer:

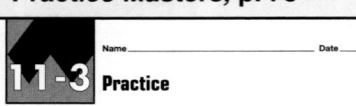

Practice Masters, p. 78

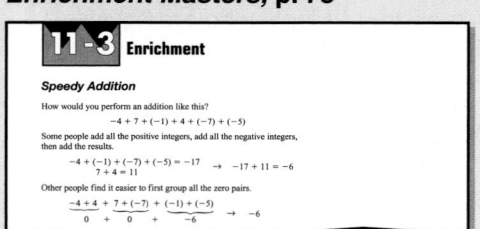

444   Chapter 11

---

34. ***Working on the*** **CHAPTER Project**   Refer to the map of the United States you made on page 433. Write the integers for your state and each state that touches your state. Find the sum of the integers. Has your area of the country gained or lost members in the House of Representatives? **See students' work.**

35. ***Critical Thinking***   Write an addition sentence that satisfies each statement. **a. Sample answer: −2 + (−3) = −5**
   a.   All addends are negative integers, and the sum is −5.
   b.   One addend is zero, and the sum is −5.
   c.   At least one addend is a positive integer, and the sum is −5.

35b. Sample answer:
   0 + (−5) = −5
35c. Sample answer:
   −8 + 3 = −5

Mixed Review

36.   Which is greater, −66 or −75? *(Lesson 11-2)* **−66**

37.   ***Geometry***   Draw a rectangular prism. *(Lesson 10-4)* **See margin.**

38.   ***Algebra***   Evaluate ab if $a = \frac{3}{8}$ and $b = \frac{7}{15}$. *(Lesson 7-2)* $\frac{7}{40}$

39.   **Test Practice**   Before sales tax, what is the total cost of three CDs selling for $13.98 each? *(Lesson 4-1)* **D**
   A $13.98   **B** $20.97   **C** $27.96   **D** $41.94   **E** $48.93

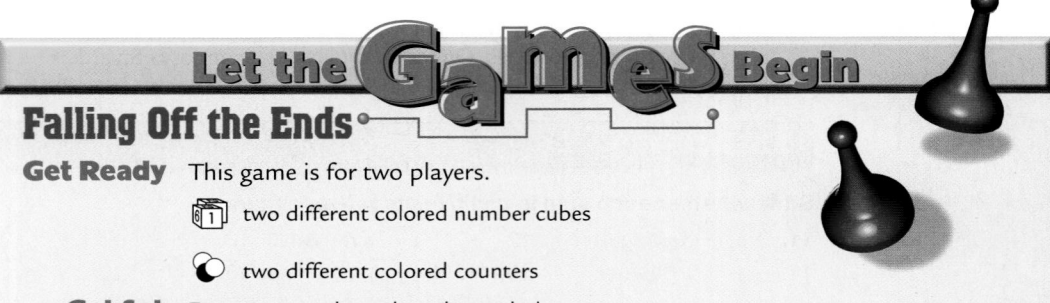

Let the Games Begin

Falling Off the Ends

Get Ready   This game is for two players.

   two different colored number cubes

   two different colored counters

Get Set   Draw a game board as shown below.

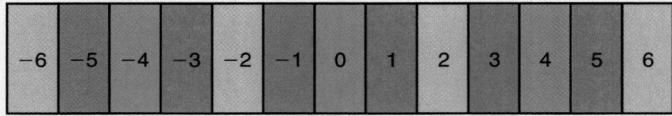

| −6 | −5 | −4 | −3 | −2 | −1 | 0 | 1 | 2 | 3 | 4 | 5 | 6 |

Go
● Let one number cube represent positive integers, and the other represent negative integers.

● Each player places a counter on zero.

● One player rolls the number cubes and adds the numbers shown. The player then moves his or her counter the number of spaces in the direction indicated by the sum.

● Players take turns rolling the number cubes and moving the counters.

● The first player to go off the board in either direction wins the game.

interNET CONNECTION   Visit www.glencoe.com/sec/math/mac/mathnet for more games.

444   **Chapter 11** Algebra: Investigating Integers

■ Extending the Lesson ■

Enrichment Masters, p. 78

Let the Games Begin

Students can use the game to practice the addition of integers. Suggest that students challenge each others' answers if they think a sum is incorrect. The player who supplies the correct answer then moves his or her counter.
*Additional resources for this game can be found on page 45 of the **Classroom Games**.*

Subtracting Integers

What you'll learn

You'll learn to subtract integers using models.

When am I ever going to use this?

Knowing how to subtract integers can help you to compare elevations when studying geography.

Did you know? About 50 people are needed to direct one balloon along the Macy's Thanksgiving Day Parade route.

The 1996 Macy's Thanksgiving Day Parade had 18 helium balloons. Five of the balloons were making their first appearance in the parade. How many of the balloons had been in the parade before?

To answer this question, you must find 18 − 5. You can model this subtraction problem using counters.

Step 1 Place 18 positive counters on a mat to represent the 18 balloons.

 → 18

Step 2 Since subtraction is the opposite of addition, remove 5 of the positive counters from the mat to represent subtracting 5.

→ 18 − 5

Step 3 Count the counters remaining on the mat.

 → 18 − 5 = 13

There are 13 positive counters. This represents 13. So, 18 − 5 = 13. There were 13 balloons that had been in the parade previously.

Lesson 11-4 Subtracting Integers **445**

Instructional Resources

- *Study Guide Masters*, p. 79
- *Practice Masters*, p. 79
- *Enrichment Masters*, p. 79
- Transparencies 11-4, A and B
- *Assessment and Evaluation Masters*, pp. 294, 295
- CD-ROM Program
  - Resource Lesson 11-4
  - Interactive Lesson 11-4

Recommended Pacing	
Standard	Day 7 of 15
Honors	Day 5 of 13
Block	Day 3 of 8

1 FOCUS

5-Minute Check
(Lesson 11-3)

State whether each sum is positive, negative, or zero.
1. −8 + 12 positive
2. −2 + (−6) negative

Find each sum. Use counters or a number line if necessary.
3. −6 + 2 −4
4. 9 + (−4) 5
5. −7 + −8 −15

The 5-Minute Check is also available on **Transparency 11-4A** for this lesson.

Motivating the Lesson

Problem Solving Ask students how they would solve this problem: *A bird is perched on a cliff 12 feet above a lake, watching a fish swim 5 feet underwater. What is the difference between their locations?* 17 feet

Transparency 11-4B contains a teaching aid for this lesson.

Teaching Tip After Example 2, show students that $-5 - 4$ could also be written as $-5 - (+4)$ or $-5 + (-4)$. The answer is the same in all cases.

Using Discussion After working through the examples with students, ask them to discuss the relationship between the subtraction of and the addition of integers.

You can also use counters to model subtraction problems involving negative integers.

Example ➊ Use counters to find $-5 - (-2)$.

Step 1 Place 5 negative counters on the mat to represent -5.

Step 2 Remove 2 negative counters from the mat to represent subtracting -2.

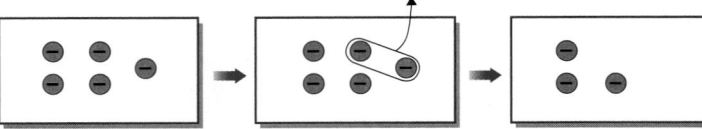

Step 3 Count the counters remaining on the mat. There are 3 negative counters. This represents -3. So, $-5 - (-2) = -3$.

Sometimes, you need to add zero pairs in order to subtract. When you add zero pairs, the value of the integers on the mat does not change.

Examples ➋ Use counters to find $-5 - 4$.

Step 1 Place 5 negative counters on the mat to represent -5.

Step 2 To subtract 4, you must remove 4 positive counters. But you cannot remove 4 positive counters because there are none on the mat. You must add 4 zero pairs to the mat. Then you can remove 4 positive counters.

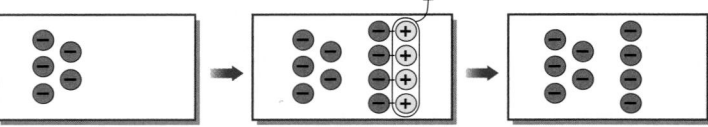

Step 3 Count the counters remaining on the mat. There are 9 negative counters. This represents -9. So, $-5 - 4 = -9$.

APPLICATION ➌ **Weather** One morning when Melissa awoke, the temperature outside was $-5°F$. By noon, the temperature was $10°F$. Find the change in temperature.

Explore You know the starting and ending temperatures. You want to know the change in the temperature.

Plan To find the change in temperature, subtract the starting temperature (-5) from the ending temperature (10). Use counters to find $10 - (-5)$.

Classroom Vignette

"I use students holding a positive or negative sign instead of using counters. Students take turns in front of the room displaying their signs. Those at their seats write the sentence shown by the human counters. Those students remaining after the zero pairs sit down represent the answer."

Sandra J. Hart, Teacher
Emporia Middle School
Emporia, KS

Sandra J. Hart

Solve Place 10 positive counters on the mat to represent +10. To subtract −5, you must remove 5 negative counters. But you cannot remove 5 negative counters because there are none on the mat. You must first add 5 zero pairs to the mat. Then you can remove 5 negative counters.

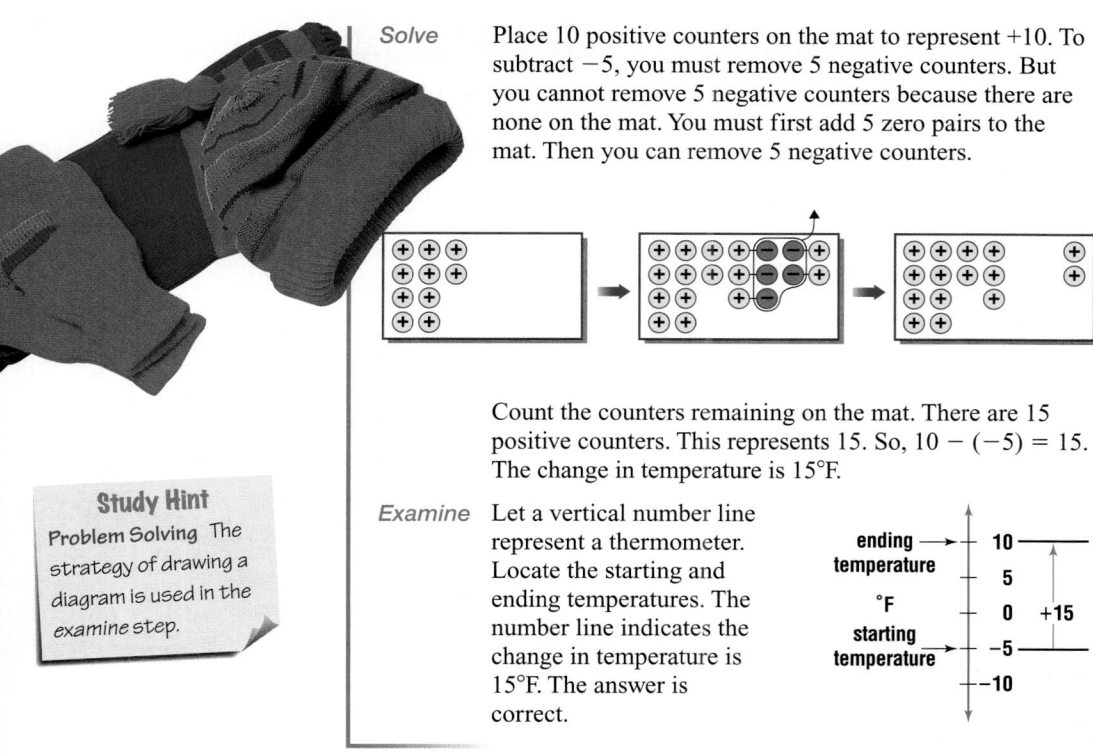

Count the counters remaining on the mat. There are 15 positive counters. This represents 15. So, 10 − (−5) = 15. The change in temperature is 15°F.

Study Hint
Problem Solving The strategy of drawing a diagram is used in the examine step.

Examine Let a vertical number line represent a thermometer. Locate the starting and ending temperatures. The number line indicates the change in temperature is 15°F. The answer is correct.

ending temperature →
°F
starting temperature →

10
5
0   +15
−5
−10

+15

CHECK FOR UNDERSTANDING

Communicating Mathematics

Read and study the lesson to answer each question.

1. *Write* a subtraction sentence represented by the model. 4 − (−2) = 6

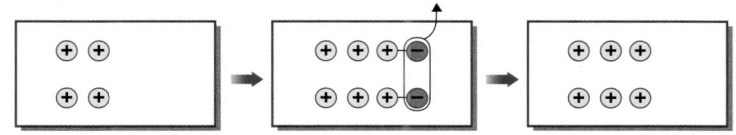

2. You need zero pairs when there are not enough positive or negative counters to remove from the mat.

2. *Explain* when it is necessary to use zero pairs to model subtraction.

3. *You Decide* Andy says that you cannot subtract 8 from 5. Donnell disagrees. Who is correct? Explain. **Donnell; you can subtract 8 from 5 by using negative integers. 5 − 8 = −3**

Guided Practice

Find each difference. Use counters or a number line if necessary.

4. 3 − (−1) **4**   5. −4 − (−4) **0**   6. −2 − 5 **−7**   7. 6 − 8 **−2**

8. *Games* Mai-Lin and Jamal are playing *Mother, May I*. They start by standing next to each other. Mai-Lin moves 4 steps forward and Jamal moves 2 steps backward. How many steps separate the two children? **6 steps**

Lesson 11-4 Subtracting Integers   **447**

3 PRACTICE/APPLY

Check for Understanding
If students need additional practice or instruction after completing Exercises 1–8, one of these options may be helpful.
• Extra Practice, see p. 586
• Reteaching Activity
• *Study Guide Masters*, p. 79
• *Practice Masters*, p. 79

***Study Guide Masters*, p. 79**

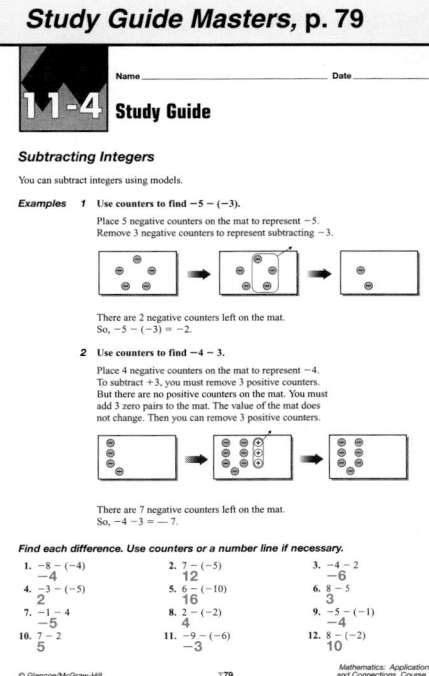

Lesson 11-4   **447**

4 ASSESS

Closing Activity

Writing Have pairs of students use the chart to create word problems that can be solved by subtracting integers.

Average January Temperatures in Four Canadian Cities	
Winnipeg	−2°F
Dawson	−18°F
Quebec	9°F
Toronto	25°F

Chapter 11, Quiz B (Lessons 11-3 and 11-4) is available in the *Assessment and Evaluation Masters,* p. 295.

Mid-Chapter Test (Lessons 11-1 through 11-4) is available in the *Assessment and Evaluation Masters,* p. 294.

Practice Masters, p. 79

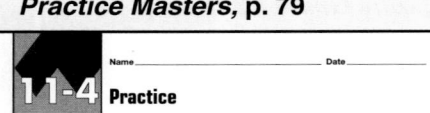

11-4 Practice

Subtracting Integers

Find each difference. Use counters or a number line if necessary.

1. $4 - 9$ −5   2. $-10 - (-7)$ −3   3. $8 - (-5)$ 13

4. $-6 - 12$ −18   5. $-3 - (-11)$ 8   6. $5 - (-9)$ 14

7. $-8 - 7$ −15   8. $2 - 6$ −4   9. $-16 - (-9)$ −7

10. $4 - (-15)$ 19   11. $-18 - 5$ −23   12. $-6 - 6$ −12

13. $-8 - (-14)$ 6   14. $13 - (-7)$ 20   15. $19 - (-6)$ 25

16. $8 - 17$ −9   17. $-12 - (-4)$ −8   18. $-2 - 9$ −11

19. $3 - (-7)$ 10   20. $-13 - 8$ −21   21. $-7 - (-12)$ 5

22. $3 - 5$ −2   23. $-8 - 6$ −14   24. $-2 - (-2)$ 0

25. $7 - (-4)$ 11   26. $-16 - (-8)$ −8   27. $12 - (-12)$ 24

28. $-3 - 10$ −13   29. $-1 - (-4)$ 3   30. $9 - (-6)$ 15

© Glencoe/McGraw-Hill   T79   *Mathematics: Applications and Connections, Course 1*

448 Chapter 11

EXERCISES

Practice

Find each difference. Use counters or a number line if necessary.

9. $2 - 4$ −2   10. $4 - 2$ 2   11. $0 - 5$ −5   12. $0 - (-5)$ 5

13. $-3 - (-6)$ 3   14. $3 - (-8)$ 11   15. $-12 - 12$ −24   16. $-7 - (-5)$ −2

17. $6 - (-2)$ 8   18. $-4 - 10$ −14   19. $-7 - (-7)$ 0   20. $5 - (-5)$ 10

21. Use counters to find $3 + (-6) - (-3)$. **See Answer Appendix; 0.**

22. *Algebra* Find the value of $x - y$ if $x = 9$ and $y = 16$. −7

Applications and Problem Solving

23. *Geography* About one third of the Netherlands is below sea level. The map shows approximate elevations for parts of the country.

   a. What is the difference in elevation between Prins Alexander Polder and Leiden? **18 ft**

   b. What is the difference in elevation between Leiden and a location in the Dunes that is 15 feet above sea level? **19 ft**

24. *Communications* Javier lives in Santa Barbara, California. At 10:00 A.M., he called his grandmother in Orlando, Florida. She told him that it was 1:00 P.M. in Orlando. If we use the number 0 to represent time in Orlando, then what integer would represent the time in Santa Barbara? −3

25. *Critical Thinking* When you subtract a lesser number from a greater number, will the answer always be positive? Give examples to support your answer. **Yes; sample examples: $8 - 3 = 5$, $0 - (-2) = 2$, and $-3 - (-5) = 2$.**

Mixed Review

26. *Algebra* Find the value of $x + y$ if $x = 7$ and $y = -12$. *(Lesson 11-3)* −5

27. Express the ratio, 2 student council representatives out of 28 students in the class, as a fraction in simplest form. *(Lesson 8-1)* $\frac{1}{14}$

28. **Test Practice** There are $18\frac{2}{3}$ cups of jellybeans to be divided among a group of children. If each child gets $\frac{2}{3}$ cup of jellybeans, how many children are there? *(Lesson 7-6)* **D**

   A 25
   B 26
   C 27
   D 28
   E Not Here

Extending the Lesson

Enrichment Masters, p. 79

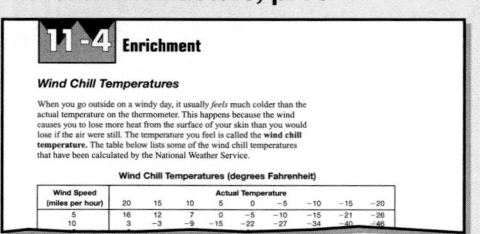

11-4 Enrichment

Wind Chill Temperatures

When you go outside on a windy day, it usually *feels* much colder than the actual temperature on the thermometer. This happens because the wind causes you to lose more heat from the surface of your skin than you would lose if the air were still. The temperature you feel is called the **wind chill temperature.** The table below lists some of the wind chill temperatures that have been calculated by the National Weather Service.

Activity Have students find the value of each pair of expressions.

a. $3 - (-9)$ +12 $3 + 9$ +12

b. $-14 - (-5)$ −9 $-14 + 5$ −9

c. $-2 - (-10)$ +8 $-2 + 10$ +8

d. $+6 - (-6)$ +12 $6 + 6$ +12

Then ask students to write rules for finding sums and differences of integers.

Cultural Kaleidoscope

A dragon head can weigh over 26 pounds. A dragon can be as long as 300 feet and require about 60 dancers to manipulate.

Making dragon heads for parades and celebrations is an ancient Chinese art form. It often takes 2 months to make one dragon head. At this rate, how long does it take to make 5 dragon heads?

To answer this question, you must multiply 5×2. Remember that multiplication is repeated addition. Therefore, 5×2 means $2 + 2 + 2 + 2 + 2$. You can model this multiplication problem using counters.

Step 1 5×2 means to *put in* 5 sets of 2 positive counters. Place these counters on the mat.

Step 2   Count the counters on the mat.

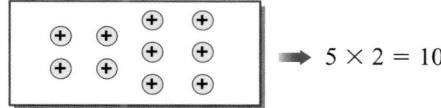

There are 10 positive counters. This represents 10. So, $5 \times 2 = 10$. It will take 10 months to make the dragon heads.

Example 1️⃣   **Use counters to find $4 \times (-2)$.**

Step 1 $4 \times (-2)$ means to *put in* 4 sets of 2 negative counters. Place these counters on the mat.

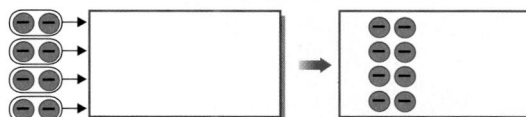

Step 2   Count the counters on the mat. There are 8 negative counters. This represents -8. So, $4 \times (-2) = -8$.

Lesson 11-5 Multiplying Integers   **449**

11-5 Lesson Notes

1 FOCUS

5-Minute Check
(Lesson 11-4)

Find each difference. Use counters or a number line if necessary.
1. $5 - (-7)$ **12**
2. $-6 - 4$ **-10**
3. $-8 - (-3)$ **-5**
4. $9 - 3$ **6**
5. When Jasmine went to bed, the temperature was 3°F. When she woke up the next morning, the temperature was −6°F. How many degrees did the temperature drop during the night?   **9°F**

The 5-Minute Check is also available on **Transparency 11-5A** for this lesson.

Motivating the Lesson
Problem Solving Use a geoboard on the overhead to find 4×3 and 2×5. Ask students how they could model multiplication of negative numbers.

Multiple Learning Styles

Interpersonal Have students work in small groups to create a board game that is played by adding, subtracting, and multiplying integers. The title and illustrations should have a theme that relates to integers, such as weather, exercise, or elevation. Then have groups play each others' games.

2 TEACH

Thinking Algebraically Point out that 6 × 2 is equivalent to 2 × 6. Tell students that this is an example of the commutative property for the multiplication of real numbers. This property is one of the basic rules of algebra.

Teaching Tip Have students compare the models of 4 × (−2), −2 × 4, −4 × 2, and 2 × (−4). What patterns do they notice?

In-Class Examples

For Example 1
Use counters to find 5 × (−3).
−15

For Example 2
Use counters to find −4 × 2.
−8

For Example 3
Use counters to find −3(−2). 6

For Example 4
If the average daily temperature decreased by 2° F each day for a week, how much did the temperature change? −14°F

Teaching Tip After going over the examples, ask students to explain to each other what each model shows. Encourage questions and written demonstrations on the chalkboard.

To multiply a positive integer times another number, you *put in* that many sets. To multiply a negative integer times another number, you do the opposite or *remove* that many sets.

Examples

2 Use counters to find −2 × 3.

Step 1 Since −2 is the opposite of 2, −2 × 3 means to *remove* 2 sets of 3 positive counters. But you cannot remove counters because there are none to remove. You must first add 2 sets of 3 zero pairs. Then you can remove 2 sets of 3 positive counters.

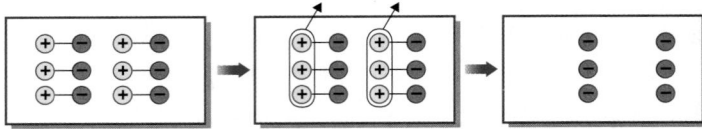

Step 2 Count the counters remaining on the mat. There are 6 negative counters. This represents −6. So, −2 × 3 = −6.

3 Use counters to find −4(−3).

Step 1 Since −4 is opposite of 4, −4(−3) means to *remove* 4 sets of 3 negative counters. But you cannot remove counters because there are none to remove. You must first add 4 sets of 3 zero pairs. Then you can remove 4 sets of 3 negative counters.

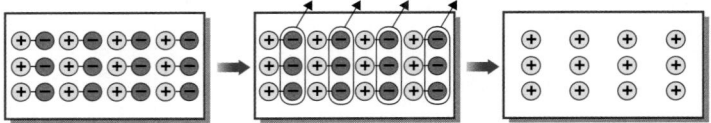

Step 2 Count the counters remaining on the mat. There are 12 positive counters. This represents 12. So, −4(−3) = 12.

CONNECTION **4** **Earth Science** The temperature drops about 7°C for each kilometer above Earth. If the temperature at ground level is 0°C, find the temperature 3 kilometers above Earth.

Explore You know the temperature drop per kilometer above Earth and the temperature at ground level. You want to know the temperature 3 kilometers above Earth.

Plan To find the temperature, multiply 3 times the amount of change per kilometer (−7).

450 Chapter 11 Algebra: Investigating Integers

Solve $3(-7)$ means to *put in* 3 sets of 7 negative counters. Place these counters on the mat.

You have 21 negative counters on the mat. This represents -21. So, $3(-7) = -21$. The temperature is $-21°C$.

Examine   Let a vertical number line represent a thermometer. Start at zero. Mark 3 drops of 7°. The number line indicates that the temperature will be $-21°C$. The answer is correct.

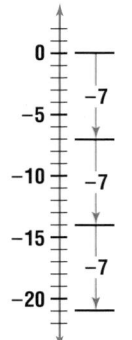

CHECK FOR UNDERSTANDING

Communicating Mathematics

Read and study the lesson to answer each question.

1. ***Write*** a multiplication sentence represented by the model. $-2 \times (-2) = 4$

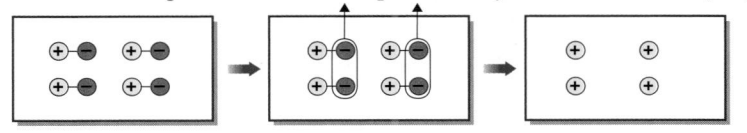

2. ***Model*** -2×6 and $6 \times (-2)$. Compare and contrast these two problems. **See margin.**

Guided Practice

Find each product. Use counters or a number line if necessary.

3. $3 \times (-1)$ **-3**   4. $-4 \times (-8)$ **32**   5. $4(-4)$ **-16**   6. $-2(5)$ **-10**

7. Find the product of -3 and -7. **21**

8. ***Oceanography*** A submarine is at the surface of the water. It starts to descend at a rate of 4 meters per second. How far below the surface will the submarine be after 5 seconds? **20 m below the surface**

EXERCISES

Practice

Find each product. Use counters or a number line if necessary.

9. $-3 \times (-6)$ **18**   10. $3 \times (-8)$ **-24**   11. 7×4 **28**   12. 3×0 **0**

13. $-3(0)$ **0**   14. $5(4)$ **20**   15. $6(-3)$ **-18**   16. $9(-3)$ **-27**

17. $8(-3)$ **-24**   18. $3(-3)$ **-9**   19. $5(-5)$ **-25**   20. $7(-5)$ **-35**

Lesson 11-5 Multiplying Integers   **451**

■ **Reteaching the Lesson** ■

Activity Ask students to use a number line to solve the following problem.
Alyssa has decided to have her long hair cut in stages. On each of her next 4 visits to the salon, she will have 3 inches cut. What is the total length that will be cut? **12 in.**

3 PRACTICE/APPLY

Check for Understanding
If students need additional practice or instruction after completing Exercises 1–8, one of these options may be helpful.
- Extra Practice, see p. 586
- Reteaching Activity
- *Study Guide Masters*, p. 80
- *Practice Masters*, p. 80

Additional Answer
2. See Answer Appendix for models. In -2×6, you place 2 groups of 6 zero pairs and remove 2 groups of 6 positive counters. In $6 \times (-2)$, you add 6 groups of 2 negative counters. In both problems, there are 12 negative counters on the mat at the end. The answer for both problems is -12.

Study Guide Masters, p. 80

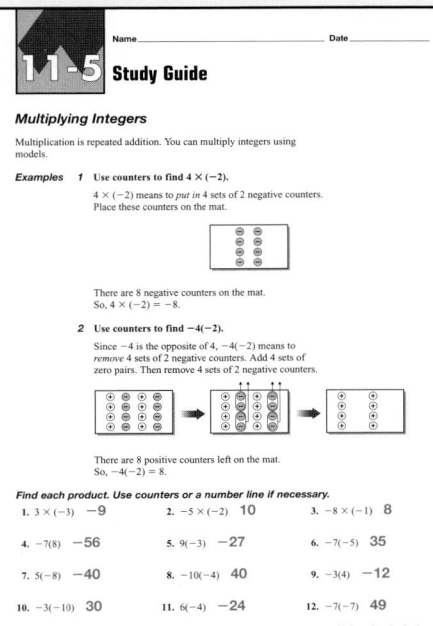

Lesson 11-5   **451**

Assignment Guide

Core: 9–23 odd, 25–28
Enriched: 10–22 even, 23–28
All: Self Test, 1–10

4 ASSESS

Closing Activity

Writing Have pairs of students write an explanation of how they would solve the following problem. Encourage them to include drawings of counters. *A leaky faucet wastes 2 pints of water each day. How much water is wasted in 2 weeks?*

Mid-Chapter Self Test

The Mid-Chapter Self Test reviews the concepts in Lessons 11-1 through 11-5. Lesson references are given so students can review concepts not yet mastered.

Practice Masters, p. 80

11-5 Practice

Multiplying Integers

Find each product. Use counters or a number line if necessary.

1. $6 \times (-4)$ −24
2. -8×7 −56
3. $-2 \times (-9)$ 18
4. $5(-12)$ −60
5. $-15(-3)$ 45
6. $-4(8)$ −32
7. $9(-7)$ −63
8. $-5(-6)$ 30
9. $3(-16)$ −48
10. $-14(2)$ −28
11. $-4(-4)$ 16
12. $-9(6)$ −54
13. $7(-3)$ −21
14. $-12(-8)$ 96
15. $-5(-15)$ −75
16. $2(-18)$ −36
17. $-3(6)$ −18
18. $4(-5)$ −20
19. $-7(14)$ −98
20. $-2(-17)$ 34
21. $-6(-8)$ 48
22. $4(-13)$ −52
23. $-16(-5)$ 80
24. $-9(12)$ −108
25. $-3(-18)$ 54
26. $7(-15)$ −105
27. $-2(19)$ −38
28. $8(-8)$ −64
29. $-9(11)$ −99
30. $-17(-4)$ 68

© Glencoe/McGraw-Hill   T 80   Mathematics: Applications and Connections, Course 1

452   Chapter 11

---

21. Find the product of 7 and −6. −42
22. Solve $2(-13) = b$. −26

Applications and Problem Solving

23. **Pet Care** Sam is a black Labrador retriever who weighs 80 pounds. Her owner puts her on a diet.
    a. If Sam loses 3 pounds each month, how much will she lose in 4 months? 12 lb
    b. What will Sam weigh at the end of the 4 months? 68 lb

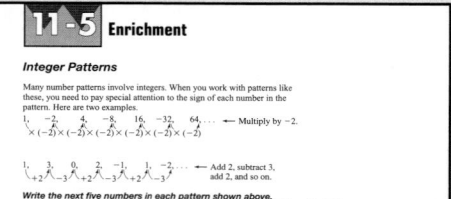

24. **Time** Suppose you had a watch that loses 2 minutes each day.
    a. How many minutes will it lose in a week? 14 min
    b. How many minutes will it lose in the month of April? 60 min

25. **Critical Thinking** The product of 1 times any number is the number itself. What is the product of −1 times any number? the number's opposite

Mixed Review

26. **Algebra** Find the value of $s - t$ if $s = 6$ and $t = -5$. *(Lesson 11-4)* 11

27. **Geometry** Draw a line segment that is $2\frac{1}{2}$ inches long, and then bisect it using a straightedge and compass. *(Lesson 9-3)* See Answer Appendix.

28. **Test Practice** If a can of green beans weighs 13 ounces, how many pounds will a case of 24 cans weigh? *(Lesson 7-7)* C
    A 1.5 lb
    B 15 lb
    C 19.5 lb
    D 312 lb
    E Not Here

CHAPTER 11

Mid-Chapter Self Test

1. **Geography** Death Valley, California, has the lowest altitude in the United States. Its elevation is 282 feet below sea level. Express the elevation of Death Valley as an integer. *(Lesson 11-1)* −282

Replace each ● with $<$, $>$, or $=$ to make a true sentence. *(Lesson 11-2)*

2. $+3 ● -2$ >
3. $-7 ● -3$ <
4. $0 ● -2$ >

Find each sum, difference, or product. Use counters or a number line if necessary. *(Lessons 11-3, 11-4, and 11-5)*

5. $-4 + (-6)$ −10
6. $-7 + 12$ 5
7. $-9 - (-5)$ −4
8. $6 - (-5)$ 11
9. -9×2 −18
10. $-8(-5)$ 40

452 Chapter 11 Algebra: Investigating Integers

Extending the Lesson

Enrichment Masters, p. 80

11-5 Enrichment

Integer Patterns

Many number patterns involve integers. When you work with patterns like these, you need to pay special attention to the sign of each number in the pattern. Here are two examples.

1,   −2,   4,   −8,   16,   −32,   64, ... ← Multiply by −2.
$\times (-2) \times (-2) \times (-2) \times (-2) \times (-2) \times (-2)$

1,   −3,   0,   −2,   1,   −1, ... ← Add 2, subtract 3, add 2, and so on.
$+2 \quad -3 \quad +2 \quad -3 \quad +2 \quad -3$

Write the next five numbers in each pattern shown above.
1, −2, 4, −8, 16, −32, 64 −128 256 −512 1,024 −2,048

Activity Place small stacks of pennies, nickels, and dimes on your desk in the front of the classroom. Tell students the number of coins in any given stack. Ask them to calculate how many cents each stack of coins is worth. Point out that multiplying integers is a skill used in everyday life, and encourage students to name situations that use the skill.

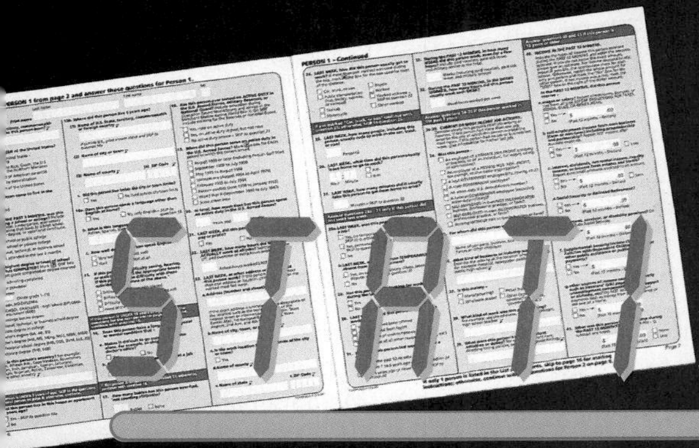

STATISTICS

Robert W. Cleveland
SURVEY STATISTICIAN

Robert Cleveland is a survey statistician for the Bureau of the Census. In the Bureau of the Census, census takers interview people and that information is given to the statisticians. Robert Cleveland works for the Income Statistics Branch of the Bureau of the Census. Each year, he works on the March Current Population Survey. This report includes information about people's jobs, income, and health.

To become a statistician, you will need at least a bachelor's degree. If you would like to become a statistician, you should take as many mathematics and computer courses as possible.

For more information:
American Statistical Association
1429 Duke Street
Alexandria, VA 22314

interNET
CONNECTION
www.glencoe.com/sec/
math/mac/mathnet

Someday, I'd like to study and interpret data like Robert Cleveland.

Your Turn
Find the results of a survey in a newspaper. Write a paragraph describing the data and explaining why it is important.

Motivating Students

Statisticians work in a wide variety of industries, including consumer marketing, politics, medical research, and education. Ask students the following questions.

- Where have you seen statistics used? **sports section of a newspaper, weather data on local television newscasts, public opinion polls**
- Why would a statistician need to understand integers? **to add and subtract numerical data**

Making the Math Connection

Statisticians use math skills on a daily basis. They interpret numerical data using addition, subtraction, multiplication, and division. Often they express a given number as a percent to show how it relates to the other numbers in a set. In addition, statisticians make conclusions based on the mathematical operations they perform.

Working on *Your Turn*

Remind students that statisticians usually work with a large collection of data. Be sure the survey they select to describe contains enough information to analyze. Have students add and subtract survey results to explain the significance of the data. Encourage them to look for trends.

More About Robert W. Cleveland

- Robert Cleveland has been a survey statistician for the Bureau of the Census since 1975. In addition to writing reports, he writes computer programs for evaluating data.
- Cleveland became a statistician because he liked math and found college courses in statistics "quite interesting."

*An additional School to Career activity is available on page 11 of the **School to Career Masters**.*

Objective Students solve problems by working backward.

Recommended Pacing	
Standard	Day 9 of 15
Honors	Day 7 of 13
Block	Day 5 of 8

1 FOCUS

Getting Started Have students act out the situation presented at the beginning of the lesson. Have them use a calendar to verify the dates. Then have them work in pairs to answer Exercises 1 and 2. Have the class discuss the results. Then have them solve Exercise 3.

2 TEACH

Teaching Tip Encourage students to make a diagram of their steps as they work backward through a problem. It will help keep them organized.

In-Class Example

Kyoto is 2 inches shorter than Mike. Mike is 3 inches taller than Amie and 2 inches shorter than Maren. Maren is the tallest. Amie is 53 inches tall. How tall is Maren? **58 in.**

Additional Answers

1. **Callie and Nora used a calendar to locate April 15. Then they counted backward from that date the number of weeks they wanted to spend on each activity.**

2. **Sample answer: They could add the minimum number of weeks for each activity and then count back that number of weeks to find the latest they could start preparations. They could add the maximum number of weeks for each activity and then count back that number of weeks to find the earliest they could start preparations.**

THINKING LAB

PROBLEM SOLVING

11-6A Work Backward

A Preview of Lesson 11-6

Callie and Nora want to get ready for the summer softball season that starts June 15. Before the season starts, they plan to spend 3 to 4 weeks hitting and fielding balls. Before hitting and fielding balls, they plan to spend 6 to 8 weeks running and exercising. The girls are trying to determine the latest and earliest dates to start getting ready for the season. Let's listen in!

Look at a calendar. Three weeks before June 15 is May 25 and 4 weeks before June 15 is May 18.

That means the latest we can begin hitting and fielding is May 25. The earliest we should begin is May 18.

Callie

Six weeks before May 25 is April 13. The latest we can start getting ready for the season is April 13.

Nora

Eight weeks before May 18 is March 23. We could start getting ready as early as March 23.

THINK ABOUT IT

Work with a partner. 1–2. See margin.

1. **Discuss** how using a calendar and **working backward** helped Callie and Nora plan their preparation for the softball season.

2. **Explain** how Callie and Nora could solve their problem differently.

3. **Apply** the work backward strategy to solve this problem.

   Chris and Dani volunteer at the food bank at 9:00 A.M. on Saturday mornings. It takes 30 minutes to get from Dani's house to the food bank. Chris picks up Dani, but it takes him 15 minutes to get to Dani's house. If it takes Chris 45 minutes to get ready in the morning, what is the latest Chris should get out of bed? **7:30 A.M.**

454 Chapter 11 Algebra: Investigating Integers

■ Reteaching the Lesson ■

Activity Give students index cards with a beginning number, an ending number, and several operations (add 5, multiply by 2). Have them start with the ending number and arrange the operation cards in order to get the desired beginning number.

ON YOUR OWN

4. The fourth step of the 4-step plan for problem solving tells you to *examine* your answer. **Tell** how you can check an answer to a problem solved by working backward.
4–6. See margin.

5. *Write a Problem* in which an effective strategy to solve it would be to work backward.

6. *Look Ahead* You know $-4 \times (-3) = 12$. Explain how the work-backward strategy could be used to find $12 \div (-3)$.

MIXED PROBLEM SOLVING

STRATEGIES
Look for a pattern.
Solve a simpler problem.
Act it out.
Guess and check.
Draw a diagram.
Make a chart.
Work backward.

Solve. Use any strategy.

7. *Sales* Books Galore bookstore arranges its best sellers in the front window. In how many different orders can they arrange 4 best sellers?
24 ways

8. *Education* A multiple choice test has 10 questions. A student receives $+3$ for each question answered correctly, -1 for each question answered incorrectly, and 0 for each question not answered. **a. 19 points**

  a. Joe answered 7 questions correctly and 2 questions incorrectly. He did not answer 1 question. What is his score on the test?

  b. Mary scored 23 points on the test. How many did she answer correctly? How many did she answer incorrectly? **See margin.**

9. *Geometry* The area of a square is 49 square feet. **a. 7 ft**

  a. Find the length of each side of the square.

  b. Find the perimeter of the square. **28 ft**

10. *Life Science* A certain bacteria doubles its population every 12 hours. After 3 full days, there are 1,600 bacteria. How many bacteria were there at the beginning of the first day?
25 bacteria

11. *Puzzles* In a magic square, each row, column, and diagonal have the same sum. Copy and complete the magic square.

-2	? 3	? -4
-3	-1	1
? 2	? -5	? 0

12. about 488 times
12. *Geography* The area of Rhode Island is 1,212 square miles. The area of Alaska is 591,004 square miles. About how many times larger is Alaska than Rhode Island?

13. *Money Matters* Gloria bought some watercolor paints and brushes. She spent $3.75 on brushes and 5 times that on paint. After paying for the items, she had $6.89 left. How much money did she have before she went shopping? **$29.39**

14. **Test Practice** Which problem does *not* have -8 as its answer? **C**

  A $-5 + (-3)$
  B $2 - 10$
  C $-4(-2)$
  D $-6 - 2$

Lesson 11-6A THINKING **455**

■ Extending the Lesson ■

Activity Have small groups of students use the results of Olympic games to create word problems that can be solved by working backward. Ask them to display their problems with illustrations or photographs on the bulletin board. Suggested sources for statistics are magazines and almanacs.

Check for Understanding
Use the results from Exercise 3 to determine whether students understand how to solve problems by working backward.

Extra Practice If students need additional practice in problem solving, extra practice is available on the following pages.
• Work Backward, see p. 587
• Mixed Problem Solving, see pp. 593–594

Assignment Guide

All: 4–14

4 ASSESS

Closing Activity
Writing Have students write a brief explanation of how to solve a problem by working backward for a student who missed the lesson.

Additional Answers

4. Start with the answer and then work forward. If the final result checks with the problem, the answer is correct.

5. Sample answer: Sue has a dental appointment at 3:30 P.M. She wants to get there 10 minutes early. If it takes 40 minutes to drive to the dentist's office, what time should she leave her house for the appointment?

6. Division is the opposite of multiplication. Therefore, if $-4 \times (-3) = 12$, $12 \div (-3) = -4$.

8b. 8 questions; 1 question

- *Study Guide Masters*, p. 81
- *Practice Masters*, p. 81
- *Enrichment Masters*, p. 81
- Transparencies 11-6, A and B
- *Assessment and Evaluation Masters*, p. 296
- *Diversity Masters*, p. 11
- *Technology Masters*, p. 21
- CD-ROM Program
  - Resource Lesson 11-6

Recommended Pacing

Standard	Day 10 of 15
Honors	Day 8 of 13
Block	Day 5 of 8

1 FOCUS

5-Minute Check
(Lesson 11-5)

Find each product. Use counters or a number line if necessary.

1. $4 \times (-6)$ **−24**
2. $-7 \times (-4)$ **28**
3. $-5(3)$ **−15**
4. $-2(-9)$ **18**
5. $-5(5)$ **−25**

The 5-Minute Check is also available on **Transparency 11-6A** for this lesson.

Motivating the Lesson

Problem Solving Remind students that $6 \div 3 = 2$ because $2 \times 3 = 6$. Have them write related sentences for these problems:
$15 \div 3 \qquad 24 \div 6 \qquad 36 \div 2$
$5 \times 3 = 15 \quad 4 \times 6 = 24 \quad 18 \times 2 = 36$

2 TEACH

Transparency 11-6B contains a teaching aid for this lesson.

Using Connections
Review the inverse operations of multiplication and division. Then ask students what multiplication expression can help them find the quotient $12 \div (-6)$. **$-2 \times (-6) = 12$**

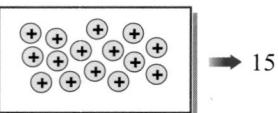

11-6 Dividing Integers

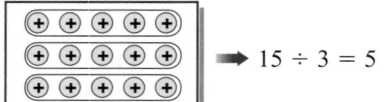

What you'll learn
You'll learn to divide integers using models and patterns.

When am I ever going to use this?
Knowing how to divide integers can help you to find the mean of data with negative integers.

A radio station plans to play 15 minutes of music with no interruptions. How many 3-minute recordings can be played during this time?

To answer this question, you must divide 15 by 3. You can model this division problem using counters.

Step 1 Place 15 positive counters on the mat to represent 15.

➡ 15

Step 2 Separate the 15 counters into 3 equal-sized groups.

➡ $15 \div 3 = 5$

There are 3 groups of 5 positive counters each. So, $15 \div 3 = 5$. The station can play 5 recordings during the 15 minutes.

You can use counters to divide a negative integer by a positive integer.

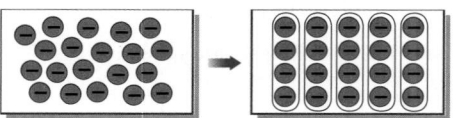

Use counters to find $-20 \div 5$.

Step 1 Place 20 negative counters on the mat to represent -20.

Step 2 Separate the 20 counters into 5 equal-sized groups.

There are 5 groups of 4 negative counters each. So, $-20 \div 5 = -4$.

You can also divide integers by *working backward*. For example, to find $56 \div 7$, think "what number times 7 equals 56?"

$$8 \times 7 = 56, \text{ so } 56 \div 7 = 8. \quad \textit{The quotient is positive.}$$

Examples

2 Find $-10 \div (-2)$.

$5 \times (-2) = -10$, so $-10 \div (-2) = 5$. *The quotient is positive.*

3 Find $18 \div (-6)$.

$-3 \times (-6) = 18$, so $18 \div (-6) = -3$. *The quotient is negative.*

4 Find $-28 \div 7$.

$-4 \times 7 = -28$, so $-28 \div 7 = -4$. *The quotient is negative.*

Do you notice any patterns in Examples 1–4? Notice that when you divide two positive integers or two negative integers, the quotient is positive. When you divide a negative integer and a positive integer, the quotient is negative.

Example 5 — APPLICATION

Sports For each lap of a race, Angeleila was farther behind Paloma. Angeleila finished the race 12 meters behind Paloma. Both runners ran 4 laps. On average, how many meters did Angeleila fall behind each lap?

Represent Angeleila's final position with respect to Paloma as -12. You need to find $-12 \div 4$.

$-3 \times 4 = -12$, so $-12 \div 4 = -3$.

Angeleila lost 3 meters per lap.

Did you know? A male sprinter can run about 22 miles per hour. An African elephant can run 25 miles per hour, and a cheetah can run 62 miles per hour.

CHECK FOR UNDERSTANDING

Communicating Mathematics
1. $-9 \div 3 = -3$
2. a positive number

Read and study the lesson to answer each question.

1. *Write* a division sentence represented by the model.

2. *Describe* the quotient of two negative integers.

Lesson 11-6 Dividing Integers **457**

■ Reteaching the Lesson ■

Activity Have pairs of students use counters and integer mats to make up word problems dealing with division of integers. Suggest that problems can involve temperature, time, distance, or money to include positive and negative integers. Have pairs exchange and complete each others' problems.

Teaching Tip Before Example 1, review the signs of the products when multiplying two integers.

In-Class Examples

For Example 1
Use counters to find $-18 \div 3$.
-6

For Example 2
Find $-16 \div (-4)$. 4

For Example 3
Find $9 \div (-3)$. -3

For Example 4
Find $-25 \div 5$. -5

For Example 5
The Bobcat football team was given penalties for a total loss of 20 yards. What was the yardage lost on each penalty if 4 equal penalties were called on the Bobcats? -5

3 PRACTICE/APPLY

Check for Understanding

If students need additional practice or instruction after completing Exercises 1–8, one of these options may be helpful.
- Extra Practice, see p. 587
- Reteaching Activity
- *Study Guide Masters*, p. 81
- *Practice Masters*, p. 81

***Study Guide Masters*, p. 81**

Name _____ Date _____

11-6 Study Guide

Dividing Integers

You can use counters to help you divide integers.

Example 1 Use counters to find $-8 \div 4$.
Place 8 negative counters on the mat to represent -8.
Separate the 8 counters into 4 equal-sized groups.

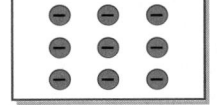

There are 4 groups of 2 negative counters each.
So, $-8 \div 4 = -2$.
You can also use the relationship between multiplication and division to help you divide integers.

Examples 2 Find $-15 \div (-3)$.
$5 \times (-3) = -15$, so $-15 \div (-3) = 5$.

3 Find $18 \div (-2)$.
$-9 \times (-2) = 18$, so $18 \div (-2) = -9$.
When you divide a negative integer and a positive integer, the quotient is negative. When you divide two negative integers, the quotient is positive.

Find each quotient. Use counters or patterns if necessary.

1. $27 \div (-3)$ -9
2. $-40 \div (-8)$ 5
3. $-36 \div 6$ -6
4. $-72 \div 8$ -9
5. $56 \div (-7)$ -8
6. $-81 \div (-9)$ 9
7. $144 \div (-12)$ -12
8. $-100 \div (-5)$ 20
9. $-84 \div 4$ -21
10. $-93 \div (-3)$ 31
11. $77 \div (-7)$ -11
12. $-64 \div (-8)$ 8

© Glencoe/McGraw-Hill   T81   *Mathematics: Applications and Connections, Course 1*

4 ASSESS

Closing Activity

Modeling Have students work in small groups. Review factors. Then ask them to divide 24 negative counters into 3 equal groups and write the division sentence. Continue with other factors of 24.

Chapter 11, Quiz C (Lessons 11-5 and 11-6) is available in the *Assessment and Evaluation Masters,* p. 296.

Additional Answers

3a. Sample answer: $-5 \times (-3) = 15; 15 \div (-5) = -3$ and $15 \div (-3) = -5$

3b. Sample answer: $4 \times (-6) = -24; -24 \div 4 = -6$ and $-24 \div (-6) = 4$

Practice Masters, p. 81

11-6 Practice

Dividing Integers

Find each quotient. Use counters or patterns if necessary.

1. $56 \div (-7)$ −8   2. $-45 \div 9$ −5   3. $-72 \div (-6)$ 12

4. $60 \div (-4)$ −15   5. $-38 \div 2$ −19   6. $-51 \div (-3)$ 17

7. $48 \div (-8)$ −6   8. $-80 \div 5$ −16   9. $-91 \div (-7)$ 13

10. $36 \div (-9)$ −4   11. $-42 \div 3$ −14   12. $-54 \div (-6)$ 9

13. $110 \div (-10)$ −11   14. $-126 \div 7$ −18   15. $-35 \div (-5)$ 7

16. $-27 \div 9$ −3   17. $104 \div (-8)$ −13   18. $-32 \div (-16)$ 2

19. $-68 \div 4$ −17   20. $-120 \div (-8)$ 15   21. $-84 \div (-6)$ 14

22. $132 \div (-11)$ −12   23. $49 \div (-7)$ −7   24. $-18 \div 2$ −9

25. $30 \div (-3)$ −10   26. $-75 \div (-15)$ 5   27. $-28 \div 14$ −2

28. $-116 \div (-29)$ 4   29. $256 \div (-32)$ −8   30. $-144 \div 24$ −6

© Glencoe/McGraw-Hill   T81   *Mathematics: Applications and Connections, Course 1*

458   Chapter 11

Math Journal

3a. *Write* a multiplication sentence with two negative factors. Then, write two related division sentences. **See margin.**

b. *Write* a multiplication sentence with one positive factor and one negative factor. Then, write two related division sentences. **See margin.**

Guided Practice

Find each quotient. Use counters or patterns if necessary.

4. $16 \div 2$ 8   5. $18 \div (-9)$ −2   6. $-27 \div 3$ −9   7. $-81 \div (-9)$ 9

8. *Oceanography* In an undersea exhibition, a self-contained module started at sea level. It descended to a depth of 24 meters. If this descent took 6 seconds, what was the rate of descent?
descent of 4 meters per second

EXERCISES

Practice

Find each quotient. Use counters or patterns if necessary.

9. $12 \div 6$ 2   10. $24 \div (-8)$ −3   11. $-36 \div 9$ −4   12. $-8 \div (-2)$ 4
13. $-35 \div 7$ −5   14. $32 \div (-4)$ −8   15. $-15 \div (-5)$ 3   16. $-12 \div (-2)$ 6
17. $45 \div (-5)$ −9 18. $-40 \div 8$ −5   19. $36 \div (-6)$ −6 20. $-42 \div (-7)$ 6

21. Divide -56 by 8. −7

22. *Algebra* Find the value of $r \div n$ if $r = -12$ and $n = -6$. 2

Applications and Problem Solving

23. *Football* The Madison Middle School football team lost 12 yards in 3 plays. If the team lost an equal amount on each play, how many yards were lost on each play? **loss of 4 yd per play**

24. *Environment* The graph shows the change of the concentration of pollutants and small particles in the air. What is the mean percent change of these pollutants and particles? −3%

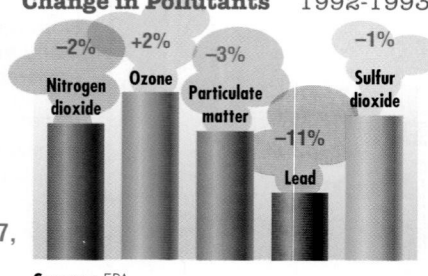
Change in Pollutants 1992-1993
−2% +2% −3% −1%
Nitrogen dioxide   Ozone   Particulate matter   Sulfur dioxide
−11%
Lead
Source: EPA

interNET CONNECTION
For the latest air pollution statistics, visit: www.glencoe.com/sec/math/mac/mathnet

25. Sample answers: $14 \div (-2) = -7$, $-21 \div 3 = -7$, $28 \div (-4) = -7$, $-35 \div 5 = -7$

25. *Critical Thinking* Write four different division problems with a quotient of -7.

Mixed Review

26. **Test Practice** A diver is descending at the rate of 4 meters per minute. How far below the surface will the diver be in 8 minutes? *(Lesson 11-5)* D
A 2 m
B 4 m
C 12 m
D 32 m
E Not Here

27. Estimate 31% of 15. *(Lesson 8-6)* **Sample answer: 5**

28. Find the value of ten plus five divided by five. *(Lesson 1-4)* **11**

Extending the Lesson

Enrichment Masters, p. 81

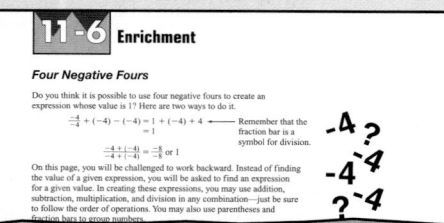

11-6 Enrichment

Four Negative Fours

Do you think it is possible to use four negative fours to create an expression whose value is 1? Here are two ways to do it.

$\frac{-4}{-4} + (-4) - (-4) \div 1 + (-4) + 4$ ← Remember that the fraction bar is a symbol for division.
$= 1$

$\frac{-4 + (-4)}{-4 + (-4)} = \frac{-8}{-8}$ or 1

On this page, you will be challenged to work backward. Instead of finding the value of a given expression, you will be asked to find an expression for a given value. In creating these expressions, you may use addition, subtraction, multiplication, and division in any combination—just be sure to follow the order of operations. You may also use parentheses and fraction bars to group numbers.

Activity Have students solve the following problem. *A steel drill bit was attached to the end of 30-ft sections of steel to drill for oil. About how many sections were needed to drill 23,077 feet below the surface of the Pacific Ocean?*
770 sections

11-7

Integration: Geometry
The Coordinate System

What you'll learn
You'll learn to graph ordered pairs of numbers on a coordinate grid.

When am I ever going to use this?
Knowing how to locate points on a coordinate grid can help you locate places on maps.

Word Wise
coordinate system
coordinate grid
origin
x-axis
y-axis
quadrants
ordered pairs
x-coordinate
y-coordinate

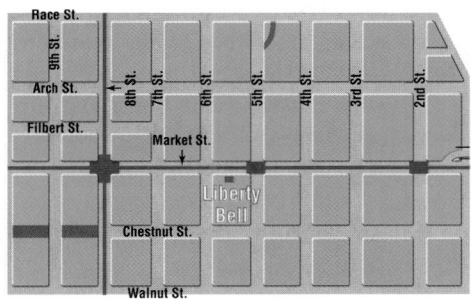

Patrick is studying the map of Philadelphia, Pennsylvania. He wants to take the subway to visit the Liberty Bell. He notices that there is a subway station located at the intersection of Market Street and 5th Street.

Similarly, points can be located on a coordinate system.

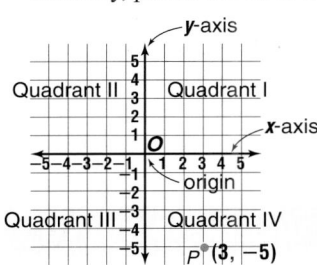

A **coordinate system**, or **coordinate grid**, consists of a horizontal number line and a vertical number line that intersect at their zero points. The point of intersection is called the **origin**. The horizontal line is called the *x*-**axis**, and the vertical line is called the *y*-**axis**.

The *x*-axis and *y*-axis divide the coordinate system into four **quadrants**. Point P is located in the fourth quadrant. It can be named by the **ordered pair** $(3, -5)$. The first number in the ordered pair is the *x*-**coordinate**, and the second number is the *y*-**coordinate**.

Example 1

> **LOOK BACK**
> You can refer to Lesson 2-9 to review graphing ordered pairs in the first quadrant.

Name the ordered pair for point A and identify its quadrant.

- Start at 0. Move left along the *x*-axis until you are directly under point A. Since you moved four units to the left, the first coordinate of the ordered pair is -4.

- Now, move up parallel to the *y*-axis until you reach point A. Since you moved up 3 units, the second coordinate of the ordered pair is 3.

- The ordered pair for point A is $(-4, 3)$. Point A is in the second quadrant.

Lesson 11-7 Integration: Geometry   The Coordinate System   **459**

Multiple Learning Styles

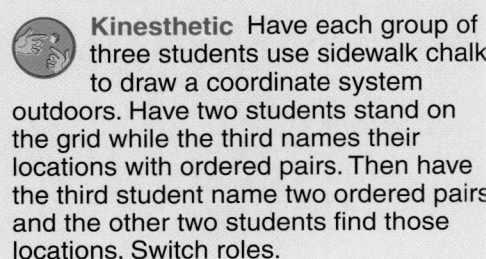

Kinesthetic Have each group of three students use sidewalk chalk to draw a coordinate system outdoors. Have two students stand on the grid while the third names their locations with ordered pairs. Then have the third student name two ordered pairs and the other two students find those locations. Switch roles.

Instructional Resources
- *Study Guide Masters*, p. 82
- *Practice Masters*, p. 82
- *Enrichment Masters*, p. 82
- Transparencies 11-7, A and B
- *Hands-On Lab Masters*, p. 79
- *Technology Masters*, p. 22
- CD-ROM Program
  - Resource Lesson 11-7
  - Extended Activity 11-7

Recommended Pacing	
Standard	Day 11 of 15
Honors	Day 9 of 13
Block	Day 6 of 8

1 FOCUS

5-Minute Check
(Lesson 11-6)

Find each quotient. Use counters or patterns if necessary.
1. $39 \div (-3)$ -13
2. $-42 \div 6$ -7
3. $-22 \div (-2)$ 11
4. $27 \div (-9)$ -3
5. $-32 \div 8$ -4

The 5-Minute Check is also available on **Transparency 11-7A** for this lesson.

Motivating the Lesson
Communication Ask students to determine whether each set of numbers is positive or negative.
- *y*-coordinates above the *x*-axis? positive
  below the *x*-axis? negative
- *x*-coordinates right of the *y*-axis? positive
  left of the *y*-axis? negative

2 TEACH

Transparency 11-7B contains a teaching aid for this lesson.

Reading Mathematics Have students translate ordered pairs into verbal directions to locate a point. For example, $(-2, 3)$ means "from the origin, go 2 units left and 3 units up."

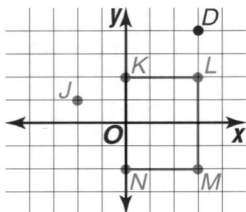

You can also graph a point on a coordinate grid. To graph a point means to place a dot at the point named by an ordered pair.

2 Graph *B*(−3, −4).
• Start at 0. Move 3 units to the left on the *x*-axis.
• Then move 4 units down parallel to the *y*-axis to locate the point.
• Place a dot and label the dot *B*.

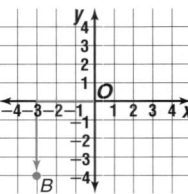

3 Graph *C*(0, 1), *D*(5, 1), *E*(3, −2), and *F*(−2, −2).
a. Draw $\overline{CD}$, $\overline{DE}$, $\overline{EF}$, and $\overline{FC}$.
b. Describe the figure formed.

a. Locate the points and draw the segments.
b. The figure formed looks like a parallelogram.

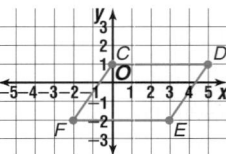

CHECK FOR UNDERSTANDING

Communicating Mathematics

Read and study the lesson to answer each question.

1. ***Tell*** how to graph the ordered pair (−3, 6). **See margin.**
2. ***Draw*** a coordinate grid. **a–b. See Answer Appendix.**
   a. ***Identify*** the portion(s) of the grid where the points are named by coordinates that are both negative. Color the portion(s) blue.
   b. ***Identify*** the portion(s) of the grid where the points are named by coordinates with one negative number and one positive number. Color the portion(s) red.

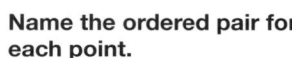

Name the ordered pair for each point.

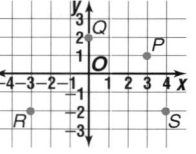

3. *P* (3, 1)   4. *Q* (0, 2)
5. *R* (−3, −2)   6. *S* (4, −2)

Graph and label each point. 7–9. See Answer Appendix.

7. *T*(1, −4)   8. *U*(−3, −2)   9. *V*(−4, 5)

10. ***Maps*** Refer to the map at the beginning of the lesson. If the location of the station at Market Street and 8th Street is at (0, 0), what are the coordinates of the station nearest the Liberty Bell? **(3, 0)**

460 Chapter 11 Algebra: Investigating Integers

EXERCISES

Practice

Name the ordered pair for each point.

11. A $(-3, 1)$ **12.** B $(1, 1)$ **13.** C $(4, -3)$

14. D $(1, -4)$ **15.** E $(2, 3)$ **16.** F $(0, 3)$

17. G $(0, -2)$ **18.** H $(-4, -2)$ **19.** I $(-2, -3)$

20. J $(-2, 0)$ **21.** K $(-4, 4)$ **22.** M $(-2, -1)$

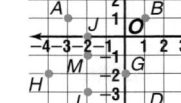

Graph and label each point. 23–30. See Answer Appendix.

23. $N(2, -2)$ **24.** $P(5, 2)$ **25.** $Q(-5, -5)$ **26.** $R(0, 4)$

27. $S(-2, 3)$ **28.** $T(-1, -5)$ **29.** $V(-5, 0)$ **30.** $W(1, 3)$

31a. See Answer Appendix.

31. **a.** Graph $X(-1, -5)$, $Y(0, -2)$, and $Z(1, 1)$ on the same coordinate grid.
b. Are points X, Y, and Z in the same line? **yes**

Applications and Problem Solving

32a. See students' work.

32b. yes; on the Equator just south and west of Nigeria in the Atlantic Ocean

32. *Geography* Longitude and latitude are used to locate places on a map.
a. Find the longitude and latitude of the place where you live.
b. Is there a place with a longitude of 0 and a latitude of 0? If so, where is it?

33. *Weather* Use the chart to form ordered pairs where the day is the x-coordinate and the temperature is the y-coordinate. Graph the point named by each ordered pair. **(1, −5), (2, −3), (3, 0), (4, 5), (5, 8); See Answer Appendix.**

Day	1	2	3	4	5
Temperature (°F)	−5	−3	0	5	8

34. See students' work; 0.

34. *Working on the* **CHAPTER Project** Refer to the map of the United States you made on page 433. For each state that had a change in the number of members to the House of Representatives, represent the data on a line plot showing positive and negative integers. What is the total change in the number of members in the House of Representatives?

35. *Critical Thinking* On a coordinate grid, graph and label each point.

35a–d. See Answer Appendix.

a. $A\left(2\frac{1}{2}, 3\frac{3}{4}\right)$ **b.** $B\left(4\frac{1}{4}, \frac{1}{2}\right)$ **c.** $C\left(0.5, 1.8\right)$ **d.** $D\left(-2.3, -1.5\right)$

Mixed Review

36. Solve $-24 \div 8 = t$. *(Lesson 11-6)* **−3**

37. **Test Practice** Is an angle that measures 92° *acute*, *right*, *obtuse*, or *straight*? *(Lesson 9-1)* **C**
A acute **B** right
C obtuse **D** straight

38. Sample answer: 22

38. Estimate $18\frac{3}{16} + 4\frac{1}{9}$. *(Lesson 6-2)*

39. *Statistics* Find Rita's mean golf score for the season if she played 9 times and her scores were 84, 88, 78, 79, 84, 84, 86, 83, and 81. *(Lesson 2-7)* **83**

Extending the Lesson

Enrichment Masters, p. 82

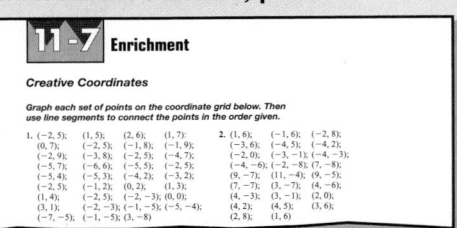

Activity Have students use a United States atlas to find three cities that are on or very near the intersection of grid lines. Then, for each city, have them write the latitude and longitude as an ordered pair. For example, New Orleans, Louisiana, is at (40°N, 75°W). You may wish to have students find cities around the world using a world atlas.

HANDS-ON
LAB

COOPERATIVE LEARNING

 patty paper

 scissors

notebook paper

11-8A Patty Paper Transformations

A Preview of Lesson 11-8

Coordinate grids can be used to graph *transformations*, or movements, of figures. This lab will help you understand two types of transformations, *translations* and *reflections*.

TRY THIS

Work with a partner.

1 To model one type of transformation, follow these steps.

• Draw triangle *A* and point *B* on a sheet of notebook paper. Line up the bottom of triangle *A* with a line on your notebook paper.

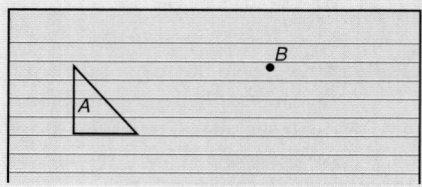

• Place a piece of patty paper on top of triangle *A*.
• Trace triangle *A* onto the patty paper.

• Slowly slide the traced triangle to point *B*. Be sure the bottom of the traced triangle slides along the line on your notebook paper. Stop when the top vertex of triangle *A* is on point *B*.

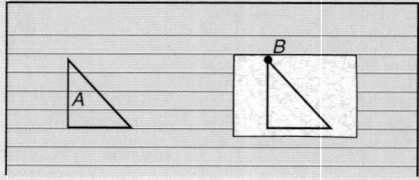

• Draw over the traced triangle. (You will need to press down fairly hard with your pen so that you can see the imprint on your notebook paper.) Then remove the patty paper.
• Trace over this imprint on your notebook paper with your pen. This movement is called a *translation*, or slide, of triangle *A*.

ON YOUR OWN

1. Describe what happened when you made the translation. **The position of the triangle is changed.**
2. How does triangle *A* compare with the new triangle? Is it congruent or similar? **They have the same size and shape; congruent and similar.**

TRY THIS

Work with a partner.

2 To model another type of transformation, follow these steps.

- Draw triangle *C* and line *m* on a sheet of notebook paper.

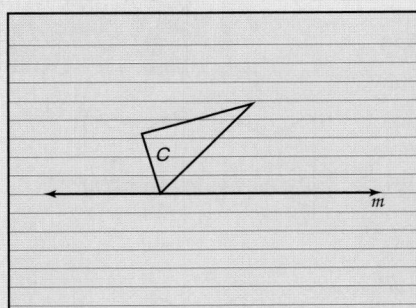

- Place a piece of patty paper on top of your drawing.
- Trace triangle *C* and line *m* onto the patty paper.

- Lift the patty paper up. Turn the patty paper upside down, toward your body. Lay the patty paper on your notebook paper so that line *m* on your patty paper is lined up with the line *m* on your notebook paper.

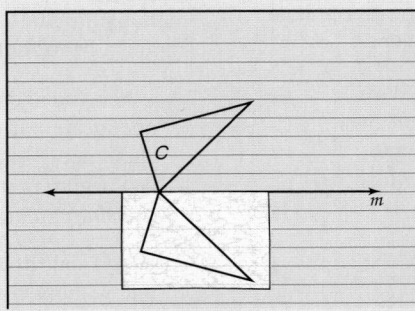

- Draw over the traced triangle, then remove the patty paper.
- Trace over this imprint on your notebook paper with your pen. This movement is called a *reflection*, or flip, of triangle *C*.

3. The new figure is the mirror image of triangle *C*.
4. They are the same size and shape, but are in opposite position.

ON YOUR OWN

3. Describe what happened when you made a reflection.
4. How does triangle *C* compare with the new triangle?
5. **Look Ahead** Look at the clothing you and the other students are wearing. Can you find any patterns in the fabric that have used translations or reflections? Sketch these patterns. **See students' work.**

Lesson 11-8A HANDS-ON **LAB** 463

Hands-On Lab Masters, p. 65

Have students write a paragraph explaining what they have learned about translations and reflections.

Hands-On Lab 11-8A 463

Instructional Resources

- *Study Guide Masters,* p. 83
- *Practice Masters,* p. 83
- *Enrichment Masters,* p. 83
- Transparencies 11-8, A and B
- *Assessment and Evaluation Masters,* p. 296

 CD-ROM Program
  - Resource Lesson 11-8

Recommended Pacing	
Standard	Days 12 & 13 of 15
Honors	Days 10 & 11 of 13
Block	Day 7 of 8

1 FOCUS

5-Minute Check
(Lesson 11-7)

Name the ordered pair for each point.

1. *A* (−3, 2)
2. *B* (1, −2)
3. *C* (0, 1)

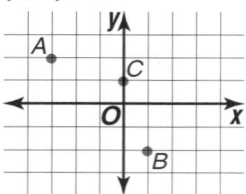

Graph each point.

4. *D*(−2, −4)   5. *E*(1, −3)

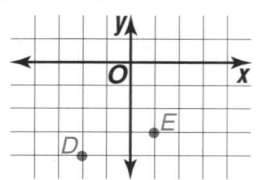

 The 5-Minute Check is also available on **Transparency 11-8A** for this lesson.

Motivating the Lesson

Hands-On Activity Various patterns of cloth, wallpaper, art, and so on will be needed. Give each small group a pattern and ask them how translations and reflections could be used to make the item.

Integration: Geometry
Graphing Transformations

What you'll learn

You'll learn to graph transformations on a coordinate grid.

When am I ever going to use this?

Knowing about transformations can help you do computer animations.

Word Wise

transformation
translation
reflection

The Zapotec Indians in Mexico weave colorful tepetes. Tepetes (tah-PAY-tays) are used as rugs and wall hangings.

A **transformation** is a movement of a figure. What transformations do you see in the design of the tepete?

In Hands-On Lab 11-8A, you learned about two kinds of transformations, a **translation** and a **reflection**.

When you slide a figure from one location to another without changing its size or shape, the new figure is called a *translation image.*

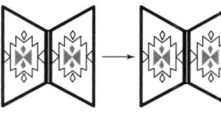

When you flip a figure over a line without changing its size or shape, the new figure is called a *reflection image.*

HANDS-ON MINI-LAB

Work with a partner. grid paper   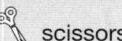 scissors

Study Hint

Reading Math The notation *A′* is read as *A* prime. *A′* is a point related to point *A*.

Try This

- On grid paper, draw a coordinate grid.
- Graph points *A*(−2, 3), *B*(−1, 1), and *C*(−3, 1).
- Connect the points to make △*ABC* .
- On a separate sheet of paper, trace △*ABC* and cut it out.
- Place the cut-out triangle on △*ABC*.
- Slide the cut-out triangle 5 units right. Label the new vertices *A′*, *B′*, and *C′*. Then draw △*A′B′C′*.

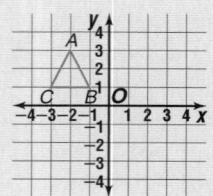

Talk About It 1. *A′*(3, 3), *B′*(4, 1), *C′*(2, 1)

1. What are the coordinates of *A′*, *B′*, and *C′*?
2. What can you say about △*ABC* and △*A′B′C′*? **See margin.**

Additional Answer for the Mini-Lab

2. △*A′B′C′* is a translation of △*ABC*. They are the same size and shape.

Example 1

The vertices of rectangle *DEFG* are *D*(−2, −3), *E*(−2, 1), *F*(−4, 1) and *G*(−4, −3). On a coordinate grid, draw rectangle *DEFG* and its translation image that is 5 units right and 2 units up.

- On grid paper, draw a coordinate grid.

- Graph and label the vertices of rectangle *DEFG*.

- Draw rectangle *DEFG*.

- Translate each vertex 5 units to the right and 2 units up. The coordinates of the new vertices are *D*′(3, −1), *E*′(3, 3), *F*′(1, 3), and *G*′(1, −1).

- Label the new vertices *D*′, *E*′, *F*′, and *G*′.

- Then draw rectangle *D*′*E*′*F*′*G*′.

You can also graph reflections on a coordinate grid.

 HANDS-ON

MINI-LAB

Work with a partner. grid paper geomirror

Try This

- On grid paper, draw a coordinate grid.
- Graph points *R*(−3, 2), *S*(0, 3), and *T*(0, 1).
- Connect the points to make △*RST*.
- Place a geomirror on the *y*-axis. Draw the reflection of △*RST* you see on the other side of the *y*-axis.
- Label the new vertices *R*′, *S*′, and *T*′.

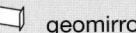

Talk About It 1–3. See margin.

1. What are the coordinates of *R*′, *S*′, and *T*′?

2. What can you say about △*RST* and △*R*′*S*′*T*′?

3. How is using a geomirror like flipping a piece of patty paper?

Lesson 11-8 Integration: Geometry Graphing Transformations **465**

2 TEACH

 Transparency 11-8B contains a teaching aid for this lesson.

Using the Mini-Labs Point out that in the Mini-Lab on page 464, students make a translation image and in the Mini-Lab on page 465, students make a reflection image. In the second Mini-Lab, if geomirrors or other reflective surfaces are unavailable, have students count the squares on the grid to draw the reflection.

In-Class Example

For Example 1
The vertices of square *RSTU* are *R*(−3, 3), *S*(1, 3), *T*(1, −1), and *U*(−3, −1). On a coordinate grid, draw the square *RSTU* and its translation image that is 2 units right and 2 units down.

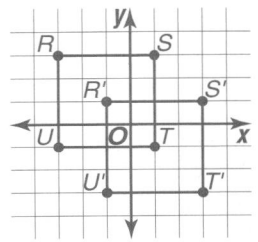

Additional Answers for the Mini-Lab

1. *R*′(3, 2), *S*′(0, 3), *T*′(0, 1)
2. △*R*′*S*′*T*′ is a reflection of △*RST*. They are the same size and shape but are mirror images of each other.
3. Both show the reflection of a shape.

Teaching Tip Before Example 2, review the characteristics of a parallelogram.

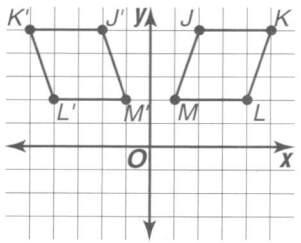
3 PRACTICE/APPLY

Check for Understanding

If students need additional practice or instruction after completing Exercises 1–7, one of these options may be helpful.
• Extra Practice, see p. 588
• Reteaching Activity
• *Study Guide Masters,* p. 83
• *Practice Masters,* p. 83

Study Guide Masters, p. 83

11-8 **Study Guide**

Name_____ Date_____

Integration: Geometry
Graphing Transformations

A **transformation** is the movement of a figure. There are several kinds of transformations. A **translation** is a slide. A **reflection** is a flip.

Examples 1 The vertices of triangle *LMN* are *L*(1, 1), *M*(3, 4), and *N*(5, 2). On a coordinate grid, draw triangle *LMN* and its translation image that is 1 unit to the left and 3 units down.

• Graph and label the vertices of triangle *LMN*.
• Draw triangle *LMN*.
• Translate each vertex 1 unit to the left and 3 units down. The coordinates of the new vertices are *L*′(0, −2), *M*′(2, 1), and *N*′(4, −1).
• Label the new vertices *L*′, *M*′, and *N*′.
• Then draw triangle *L*′*M*′*N*′.

2 The vertices of triangle *QRS* are *Q*(−2, 4), *R*(0, 1), and *S*(−3, −1). On a coordinate grid, draw triangle *QRS* and its reflection image over the *y*-axis.

• Graph and label the vertices of triangle *QRS*.
• Reflect the triangle by flipping it onto the other side of the *y*-axis. To be a mirror image, each new vertex must be the same distance from the *y*-axis as its corresponding vertex. The coordinates of the new vertices are *Q*′(2, 4), *R*′(0, 1), and *R*′(3, −1).
• Label the new vertices *Q*′, *R*′, and *S*′.
• Then draw triangle *Q*′*R*′*S*′.

The vertices of parallelogram *ABCD* are *A*(−1, 1), *B*(1, 4), *C*(4, 4), *D*(2, 1). On the coordinate grid, draw parallelogram *ABCD* and each transformation image.

1. translation image that is 5 units to the left and 1 unit down

2. reflection image over the *x*-axis

© Glencoe/McGraw-Hill   T 83   *Mathematics: Applications and Connections, Course 1*

---

466   **Chapter 11**

The vertices of parallelogram *WXYZ* are *W*(−2, 1), *X*(−1, 3), *Y*(4, 3), and *Z*(3, 1). On a coordinate grid, draw parallelogram *WXYZ* and its reflection image over the *x*-axis.

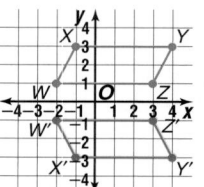

• On grid paper, draw a coordinate grid.
• Graph and label the vertices of parallelogram *WXYZ*.
• Draw parallelogram *WXYZ*.
• Reflect the parallelogram by flipping it onto the other side of the *x*-axis. To be a mirror image, each new vertex must be the same distance from the *x*-axis as its corresponding vertex. The coordinates of the new vertices are *W*′(−2,−1), *X*′(−1,−3), *Y*′(4,−3), and *Z*′(3,−1).
• Label the vertices *W*′, *X*′, *Y*′, and *Z*′.
• Draw parallelogram *W*′*X*′*Y*′*Z*′.

2. Translate each vertex of the polygon 6 units left. Then draw the translation image using the new vertices.

CHECK FOR UNDERSTANDING

Communicating Mathematics

Read and study the lesson to answer each question.

1. ***Compare and contrast*** a translation and a reflection. **See Answer Appendix.**

2. ***Explain*** how to translate a polygon on a coordinate grid 6 units to the left.

HANDS-ON MATH
3. The vertices of rectangle *ABCD* are *A*(1,−2), *B*(1,−4), C(4,−4), and *D*(4,−2). Draw rectangle *ABCD* on a coordinate grid. **a–b. See Answer Appendix.**

  a. On a separate sheet of paper, trace rectangle *ABCD* and cut it out. Place the cut-out rectangle on rectangle *ABCD*. Slide the cut-out rectangle 5 units up. Draw the translation image.

  b. Use a geomirror to reflect rectangle *ABCD* over the *y*-axis.

Guided Practice
4. Tell whether the transformation of △*ABC* to △*A*′*B*′*C*′ is a *translation* or a *reflection*. **translation**

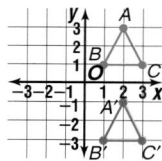

The vertices of △*EFG* are *E*(−3, 2), *F*(−3, 0), and *G*(0, 0). On a coordinate grid, draw △*EFG* and each transformation image.

5. translation image that is 4 units left and 5 units up **See Answer Appendix.**

6. reflection image over the *y*-axis **See Answer Appendix.**

Reteaching the Lesson

Activity Use color transparencies of geometric figures on an overhead projector to model transformations on a coordinate grid. For example, show a colored triangle on a grid and ask students to identify and label its vertices. Flip the triangle on an axis to show its reflection. Ask students to identify and label the reflection's vertices.

Error Analysis
Watch for students who reflect figures instead of translating them and vice versa.

Prevent by discussing the meanings of the terms and by reviewing the procedures for each.

7. **Nature** What type of transformation(s) do you see in the picture of the mountains and the lake? Explain your answer. **Reflection; the mountains are reflected in the lake.**

Assignment Guide

Core: 9–17 odd, 19–22
Enriched: 8–16 even, 17–22

EXERCISES

Have students share their transformations with the class. You may wish to total the number of translations and reflections to see which transformation is more commonly found.

Practice

Tell whether each transformation is a *translation* or a *reflection*.

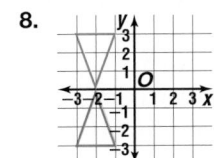

8.

reflection

9.

reflection

10.

translation

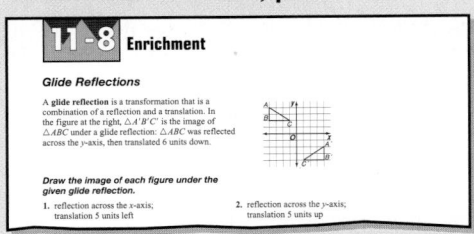

Family Activity

Find some examples of translations and reflections in your home. Make a sketch of each transformation. Label each transformation as a translation, a reflection, or both.

The vertices of △*MNP* are *M*(2, 1), *N*(2, 3), and *P*(4, 4). On a coordinate grid, draw △*MNP* and each transformation image.

11. translation image that is 7 units down **11–16. See Answer Appendix.**

12. translation image that is 6 units left

13. translation image that is 3 units right and 4 units up

14. translation image that is 2 units left and 5 units down

15. reflection image over the *x*-axis

16. reflection image over the *y*-axis

Applications and Problem Solving
17. **Translations; the winged horses are translations of each other.**

17. **Art** What type of transformation(s) do you see in the drawing by M. C. Escher? Explain your answer.

18. **Design** Draw a pattern for wrapping paper that uses both a translation and a reflection. **See Answer Appendix.**

19. **Critical Thinking** The first coordinate of each vertex of a polygon is multiplied by −1. The image of the polygon having these new coordinates is drawn. Describe the relationship between the original polygon and its image. **The image is a reflection of the polygon over the *y*-axis.**

Source: © M.C. Escher/Cordon Art–Baam–Holland Collection Haags Gemeentemuseum–The Hauge

4 ASSESS

Closing Activity

Speaking Ask students to draw a simple pattern of four translations or four reflections using △*PQR* with vertices *P*(2, 3), *Q*(0, 0), and *R*(4, 0). Have several students explain their designs to the class.

Chapter 11, Quiz D (Lessons 11-7 and 11-8) is available in the *Assessment and Evaluation Masters*, p. 296.

Mixed Review

20. **Geometry** Graph *A*(2, −3) and *B*(−3, 2). *(Lesson 11-7)* **See Answer Appendix.**

21. **Test Practice** An architect made an 8-inch by 10-inch scale drawing of a house. If the scale on the drawing is 1 inch = 5 feet, what are the dimensions of the house? *(Lesson 8-3)* **A**

  A 40 ft by 50 ft   **B** 35 ft by 45 ft
  C 20 ft by 25 ft   **D** 4 ft by 5 ft

22. Find the LCM of 15, 20, and 8. *(Lesson 5-7)* **120**

Lesson 11-8 Integration: Geometry   Graphing Transformations   **467**

Practice Masters, p. 83

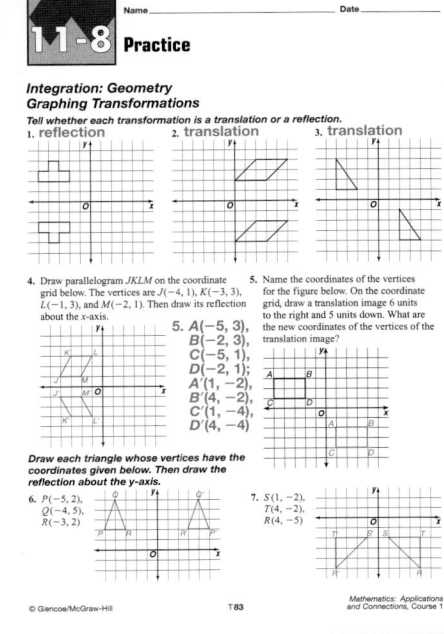

Extending the Lesson

Enrichment Masters, p. 83

11-8 **Enrichment**

Glide Reflections

A glide reflection is a transformation that is a combination of a reflection and a translation. In the figure at the right, △*A'B'C'* is the image of △*ABC* under a glide reflection. △*ABC* was reflected across the *y*-axis, then translated 6 units down.

Draw the image of each figure under the given glide reflection.
1. reflection across the *x*-axis; translation 5 units left
2. reflection across the *y*-axis; translation 5 units up

Activity Have students draw △*TUV* with vertices *T*(−2, 1), *U*(−5, 0), and *V*(−2, −2). Ask them what transformation describes △*T'U'V'* with vertices *T'*(0, 1), *U'*(−3, 0), and *V'*(0, −2). **Translate △*TUV* 2 units to the right.**

Study Guide and Assessment

Vocabulary

This section provides a listing of the new terms, properties, and phrases that were introduced in this chapter. Have students define each term and provide an example or two of each, if appropriate.

Understanding and Using the Vocabulary

These exercises check students' understanding of the terms by using a variety of verbal formats including matching, completion, and true/false.

Glossaries

A complete glossary of terms appears on pages 642–648. The glossary also appears in Spanish on pages 649–656.

Additional Answer

12. In a translation, the position of a figure is changed by sliding it. In a reflection, the position of a figure is changed by flipping it over a line.

Vocabulary

After completing this chapter, you should be able to define each term, concept, or phrase and give an example or two of each.

Geometry
coordinate grid (p. 459)
coordinate system (p. 459)
ordered pair (p. 459)
origin (p. 459)
quadrant (p. 459)
reflection (p. 464)
transformation (p. 464)
translation (p. 464)
x-axis (p. 459)
x-coordinate (p. 459)
y-axis (p. 459)
y-coordinate (p. 459)

Number and Operations
integer (p. 434)
negative integer (p. 434)
opposite (p. 435)
positive integer (p. 434)
zero pair (p. 441)

Problem Solving
work backward (p. 454)

Understanding and Using the Vocabulary

State whether each sentence is *true* or *false*. If false, replace the underlined word or number to make a true sentence.

1. Integers that are greater than zero are called <u>negative</u> integers. **false; positive**
2. The opposite of the number 7 is $\frac{1}{7}$. **false; −7**
3. On the number line -21 is to the left of -8, so <u>$-21 < -8$</u>. **true**
4. When adding two negative integers, the answer will <u>always</u> be negative. **true**
5. When subtracting two negative integers, the answer will <u>never</u> be negative. **false; sometimes**
6. When dividing two negative integers, the answer will <u>always</u> be negative. **false; never**
7. When multiplying two negative integers, the answer will <u>sometimes</u> be negative. **false; never**
8. The <u>horizontal</u> line of a coordinate system is called the x-axis. **true**
9. The first number in an ordered pair is the <u>x-coordinate</u>. **true**
10. The x-axis and the y-axis intersect at the <u>quadrant</u>. **false; origin**
11. When you flip a figure, the transformation is called a <u>translation</u>. **false; reflection**

In Your Own Words

12. *Explain* the difference between a translation and a reflection. **See margin.**

MindJogger Videoquizzes

MindJogger Videoquizzes provide an alternative review of concepts presented in this chapter. Students work in teams to answer questions, gaining points for correct answers. The questions are presented in three rounds.
Round 1 Concepts–5 questions
Round 2 Skills–4 questions
Round 3 Problem Solving–4 questions

Objectives & Examples

Upon completing this chapter, you should be able to:

● identify, name, and graph integers
(Lesson 11-1)

Graph −3 on the number line.

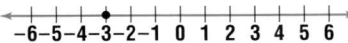

$$-6\ -5\ -4\ -3\ -2\ -1\ \ 0\ \ 1\ \ 2\ \ 3\ \ 4\ \ 5\ \ 6$$

● compare and order integers
(Lesson 11-2)

Compare −7 and −9.

$$-10\ -9\ -8\ -7\ -6\ -5\ -4\ -3\ -2\ -1\ \ 0\ \ 1\ \ 2$$

−7 is to the right of −9, so −7 > −9.

● add integers using models
(Lesson 11-3)

Use counters to find −4 + 6.

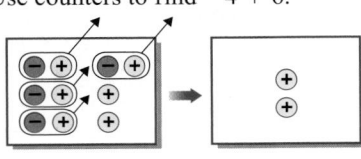

$$-4 + 6 = 2$$

● subtract integers using models
(Lesson 11-4)

Use counters to find −4 − 2.

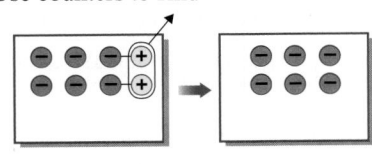

$$-4 − 2 = −6$$

Review Exercises

Use these exercises to review and prepare for the chapter test. **13–15. See Answer Appendix.**

Draw a number line from −10 to 10. Graph each integer on the number line.

13. −9   **14.** 6   **15.** −8

16. Write an integer to describe a temperature drop of 13°. **−13**

Replace each ● with <, >, or = to make a true sentence.

17. +4 ● +7 **<**   **18.** +1 ● −6 **>**

19. −7 ● −12 **>**   **20.** −9 ● +11 **<**

21. Order −15, 0, −1, 6, −7, and 10 from least to greatest. **−15, −7, −1, 0, 6, 10**

Find each sum. Use counters or a number line if necessary.

22. −3 + 5 **2**   **23.** 7 + (−7) **0**

24. −2 + (−5) **−7**   **25.** −7 + 1 **−6**

26. 4 + (−3) **1**   **27.** −6 + 0 **−6**

Find each difference. Use counters or a number line if necessary.

28. −2 − (−5) **3**   **29.** −7 − 1 **−8**

30. 3 − (−4) **7**   **31.** 8 − (−3) **11**

32. 6 − 9 **−3**   **33.** −4 − 7 **−11**

Chapter 11 Study Guide and Assessment   **469**

Objectives & Examples

This section reviews the skills and concepts of the chapter and shows completely worked examples.

Review Exercises

These exercises provide practice for the corresponding objectives.

Assessment and Evaluation Masters, pp. 283–284

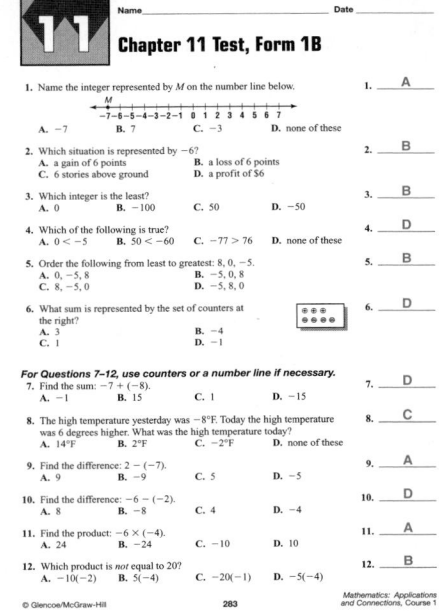

Assessment and Evaluation

Six forms of Chapter 11 Test are available in the *Assessment and Evaluation Masters* as shown in the chart.

Chapter 11 Test, Form 1B, is shown at the right. Chapter 11 Test, Form 2B, is shown on the next page.

1A	Multiple Choice	Honors
1B	Multiple Choice	Average
1C	Multiple Choice	Basic
2A	Free Response	Honors
2B	Free Response	Average
2C	Free Response	Basic

Objectives & Examples

Review Exercises

● multiply integers using models
(Lesson 11-5)

Use counters to find $2(-3)$.

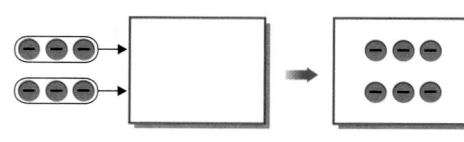

$2(-3) = -6$

Find each product. Use counters or a number line if necessary.

34. $6 \times (-2)$ **−12** **35.** $-3 \times (-5)$ **15**

36. -4×3 **−12** **37.** 5×0 **0**

38. $-5(2)$ **−10** **39.** $7(-1)$ **−7**

● divide integers using models and patterns
(Lesson 11-6)

Find $14 \div (-2)$.

$-7 \times (-2) = 14$, so $14 \div (-2) = -7$.

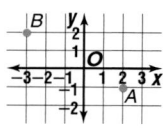

Find each quotient. Use counters or patterns if necessary.

40. $35 \div (-7)$ **−5** **41.** $-36 \div 9$ **−4**

42. $-40 \div 5$ **−8** **43.** $18 \div (-3)$ **−6**

44. $-30 \div (-5)$ **6** **45.** $-24 \div (-6)$ **4**

● graph ordered pairs of numbers on a coordinate grid *(Lesson 11-7)*

Graph $A(2, -1)$.
right 2, down 1

Graph $B(-3, 2)$.
left 3, up 2

Graph and label each point.

46. $M(-2, 2)$ **47.** $N(3, 0)$

48. $P(5, 6)$ **49.** $Q(-2, 4)$

50. $R(-2, -5)$ **51.** $S(7, -1)$

46–51. See Answer Appendix.

● graph transformations on a coordinate grid
(Lesson 11-8)

translation   reflection

The vertices of △ABC are A(−1, 2), B(0, 5), and C(3, 1). On a coordinate grid, draw △ABC and each transformation image.

52. translation image that is 3 units left and 2 units down **52–53. See Answer Appendix.**

53. reflection image over the *x*-axis

Assessment and Evaluation Masters, pp. 289–290

Name_____ Date_____

Chapter 11 Test, Form 2B

1. Write an integer to describe a loss of 20 points. 1. **−20**
2. Write an integer to describe an increase of 5 degrees. 2. **+5**
3. Identify the integer that is the opposite of 15. 3. **−15**
4. Graph the set $\{-6, -3, 0, 3\}$ on the number line. 4. (number line −6 −4 −2 0 2 4 6)

Replace each ● with <, >, or = to make a true sentence.
5. -100 ● 90 5. **<**
6. 0 ● -10 6. **>**
7. -66 ● -666 7. **>**
8. Order 12, −9, 0, −12, and 7 from least to greatest. 8. **−12, −9, 0, 7, 12**

Perform the indicated operation.
9. $-11 + 9$ 9. **−2**
10. $5 + (-4)$ 10. **1**
11. $-8 + (-8)$ 11. **−16**
12. $7 - (-5)$ 12. **12**
13. $-17 - 3$ 13. **−20**
14. $15 - (-2)$ 14. **17**
15. $-14 - 8$ 15. **−22**
16. $9(-2)$ 16. **−18**
17. $(-8)(3)$ 17. **−24**
18. $(-3)(-6)$ 18. **18**
19. $11(-4)$ 19. **−44**
20. $28 \div (-7)$ 20. **−4**

© Glencoe/McGraw-Hill 289 Mathematics: Applications and Connections, Course 1

Chapter 11 Test, Form 2B (continued)

Perform the indicated operation.
21. $-645 \div 43$ 21. **−15**
22. $-48 \div (-8)$ 22. **6**
23. $75 \div (-25)$ 23. **−3**

24. At 1 A.M. the temperature was −2°F. By noon, the temperature had risen 11 degrees. What was the temperature at noon? 24. **9°F**

25. At the end of three days, a stock was selling for $18 per share. On the preceding three days there had been a gain of $2 per share, a loss of $3 per share, and a gain of $1 per share. What was the price per share at the beginning of the three days? 25. **$18**

Graph each point on the grid.
26. $D(5, -2)$
27. $E(-4, -3)$
28. $F(0, -3)$ 26–28. (grid)

Give the coordinates for each point on the grid.
29. D 29. **(5, 3)**
30. E 30. **(−3, 1)**
31. F 31. **(−2, −3)**

The vertices of triangle ABC are A(5, 2), B(5, 5), and C(2, 2). On the coordinate grid, draw each transformation image.
32. a translation of 5 units to the left 32–33. (grid)
33. a reflection of triangle *ABC* over the horizontal axis

© Glencoe/McGraw-Hill 290 Mathematics: Applications and Connections, Course 1

💾 **Test and Review Software**

You may use this software, a combination of an item generator and item bank, to create your own tests or worksheets. Types of items include free response, multiple choice, short answer, and open ended.

💿 **CD-ROM Program**

The CD-ROM Program contains an Assessment Game whose questions review the concepts in this chapter.

Applications & Problem Solving

54. Weather Low temperatures were recorded for one week in January. Order the temperatures from least to greatest. *(Lesson 11-2)* **−7, −5, −3, −2, 0, 1, 2**

S	M	T	W	T	F	S
−5°	0°	1°	−7°	−3°	2°	−2°

55. Football The Branson Middle School football team is 2 yards from its opponent's goal line. On the next two plays, its offense loses 17 yards and then gains 11 yards. How many yards is the team from the goal line? *(Lesson 11-3)* **8 yd**

56. Work Backward Pearl has money in her savings account. She made withdrawals of $100, $43, and $67. Her balance is now $245. How much did she have in her account before the three withdrawals? *(Lesson 11-6B)* **$455**

57. Crafts Diego wants to make a two-sided ornament out of fabric. He uses a pattern to cut one side of the ornament. Should he use a *reflection* or a *translation* of the pattern for the other side of the ornament? *(Lesson 11-8)* **reflection**

Alternative Assessment

Performance Task

Suppose you are playing a game with a friend and you have decided to be the scorekeeper. During the first round, you scored 6 points and your friend scored 4 points. During the second round, you scored 3 points and your friend lost 3 points. During the third round, you lost 2 points and your friend scored 5 points. Finally, during the fourth round, you lost 1 point and your friend scored 5 points. Explain how to organize the information in order to keep score accurately. **See margin.**

Suppose on the fifth round, you scored 2 points and your friend lost 1 point. Find the total scores. Who is winning the game? **See margin.**

A practice test for Chapter 11 is provided on page 605.

Completing the CHAPTER Project

Use the following checklist to make sure your poster or brochure is complete.

☑ The map is clear and easy to read.

☑ The line plot is correct.

☑ The paragraph discussing the changes in the House of Representatives includes whether your part of the country has gained or lost representatives and how the total number of representatives has changed.

Add any finishing touches that you would like to make your poster or brochure attractive.

 Select one exercise that involved graphing from this chapter that you found particularly challenging. Place it in your portfolio.

Applications & Problem Solving

This section provides additional practice in solving real-world problems that involve the skills of this chapter.

Alternative Assessment

The **Performance Task** provides students with a performance assessment opportunity to evaluate their work and understanding.

CHAPTER Project

Students should complete the final stages of their project and prepare a class demonstration of their results. A scoring guide for the project is available in the *Investigations and Projects Masters*, p. 59.

 Students should add to their portfolios at this time.

***Assessment and Evaluation Masters*, p. 293**

Additional Answers for the Performance Task

- **Keep a running tally of each player's score. The first round scores were 6 and 4. After the first round, add the integers represented by each score.**
- **After 5 rounds, your score is 8 and your friend's score is 10. Your friend is winning.**

Performance Assessment

Additional performance assessment tasks for this chapter are included in the *Assessment and Evaluation Masters* on page 293. A scoring guide is also provided on page 305.

The Standardized Test Practice may be used to help students prepare for standardized tests. The test items are written in the same style as those in state proficiency tests and standardized tests like CAT, CTBS, ITBS, MAT, SAT, and Terra Nova. The test items cover skills and concepts covered up to this point in the text.

The pages can be used as an overnight assessment. After students have completed the pages, discuss how each problem can be solved, or provide copies of the solutions from the *Solutions Manual*.

Assessment and Evaluation Masters, p. 299

Section One: Multiple Choice

There are twelve multiple choice questions in this section. Choose the best answer. If a correct answer is *not here*, choose the letter for Not Here.

1. Which numbers are factors of 75? **C**
- **A** 3, 5, and 6
- **B** 3, 5, and 9
- **C** 3 and 5
- **D** 5 and 10

2. If $n = -3$, what is the value of $8 + n$? **H**
- **F** -5
- **G** -3
- **H** 5
- **J** 11

3. How many lines of symmetry does an equilateral triangle have? **D**
- **A** 0
- **B** 1
- **C** 2
- **D** 3

4. What are the coordinates of point G on the graph? **F**

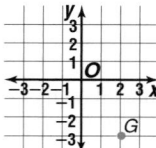

- **F** $(2, -3)$
- **G** $(3, -2)$
- **H** $(-3, 2)$
- **J** $(-2, -3)$

Please note that Questions 5–12 have five answer choices.

5. What integers are graphed? **A**

-5 -4 -3 -2 -1 0 1 2 3 4 5

- **A** $-5, -1, 2$
- **B** $-7, -1, 2$
- **C** $5, 1, 2$
- **D** $-7, 1, 2$
- **E** $-2, 1, 5$

6. Write the sentence represented by the model. **J**

- **F** $4 + (-4) = 0$
- **G** $4 + (-2) = 2$
- **H** $4 + (6) = 10$
- **J** $4 - (-2) = 6$
- **K** $4 - 2 = 2$

7. One boat is 5.05 meters long, and another is 4.8 meters long. How much space is needed to park the boats end to end? **A**
- **A** 9.85 meters
- **B** 9.13 meters
- **C** 10.3 meters
- **D** 5.53 meters
- **E** 0.25 meter

8. A round window in Kevin's house is 40 inches in diameter. Which is a good estimate for the area of the window? **J**
- **F** 60 square inches
- **G** 120 square inches
- **H** 600 square inches
- **J** 1,200 square inches
- **K** 4,800 square inches

◀◀◀ **Instructional Resources**

Another cumulative review is shown at the left and is available in the *Assessment and Evaluation Masters*, p. 299.

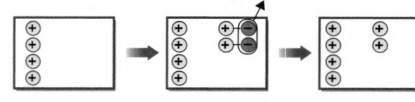

Cumulative Review, Chapters 1–11

Name_____ Date_____

1. Find the next three numbers in the pattern: 5, 7, 9, _?_, _?_, _?_. (Lesson 1-2) **1.** __11, 13, 15__

2. Use the graph at the right to predict Sarah's height at age 12. (Lesson 2-5) **2.** __about 63 in.__

Sarah's Growth Chart

3. Round 8.729 to the tenths place. (Lesson 3-4) **3.** __8.7__

4. Simplify $\frac{48}{88}$. (Lesson 5-4) **4.** __$\frac{6}{11}$__

5. Round $5\frac{13}{14}$ to the nearest half. (Lesson 6-1) **5.** __$5\frac{1}{2}$__

6. Find the next number in the sequence 25, $37\frac{1}{2}$, $56\frac{1}{4}$, $84\frac{3}{8}$, _?_. (Lesson 7-8) **6.** __$126\frac{9}{16}$__

Perform each indicated operation. (Lessons 4-7, 7-5, 11-3, and 11-5)

7. $4.028 \div 3.8$ **7.** __1.06__

8. $\frac{7}{20} + \frac{5}{8}$ **8.** __$\frac{14}{25}$__

9. $-10 + (-30)$ **9.** __-40__

10. $-5(-4)$ **10.** __20__

11. Express 105 miles on 15 gallons as a rate. (Lesson 8-1) **11.** __7 mpg__

12. On a map, the scale reads "1 inch = 25 miles." If the distance between two cities on the map is $2\frac{1}{2}$ inches, what is the actual distance? (Lesson 8-3) **12.** __62.5 miles__

13. How many sides does a hexagon have? (Lesson 9-4) **13.** __six__

14. Tell whether the dashed line is a line of symmetry. Write *yes* or *no*. (Lesson 9-5) **14.** __no__

Find the area of each figure. (Lessons 10-1 and 10-2)

15. **15.** __66.96 cm²__

16. **16.** __588 ft²__

Refer to the rectangular prism at the right.

17. What is the surface area? (Lesson 10-6) **17.** __1,036 cm²__

18. What is the volume? (Lesson 10-5) **18.** __1,960 cm³__

19. Order $-10, 0, 20, -30,$ and 10 from least to greatest. (Lesson 11-2) **19.** __$-30, -10, 0, 10, 20$__

20. Write an integer to describe gaining 2 pounds. (Lesson 11-1) **20.** __$+2$__

© Glencoe/McGraw-Hill   **299**   *Mathematics: Applications and Connections, Course 1*

9. Mr. Orta needs $\frac{7}{8}$ yard of material to upholster a chair seat and $\frac{5}{8}$ yard to upholster the back of the chair. How many yards of material will be needed to cover the chair? **B**

A 1 yard

B $1\frac{1}{2}$ yards

C $1\frac{3}{4}$ yards

D 2 yards

E Not Here

10. Bobbie charges $5.75 per hour for baby-sitting. If she works for 4 hours, how much will she earn? **J**

F $9.75

G $20.75

H $17.25

J $23.00

K Not Here

11. Barbara bought a 3-pound ham and $5\frac{1}{2}$ pounds of steak. How many more pounds of steak did she buy than ham? **B**

A 2 pounds

B $2\frac{1}{2}$ pounds

C $3\frac{1}{2}$ pounds

D $8\frac{1}{2}$ pounds

E Not Here

12. $-3(-13) =$ **H**

F 16

G -16

H 39

J -39

K Not Here

Test-Taking Tip

You could use educated guesses to help improve your scores. A wild guess involves pure chance, but eliminating possible answers as definitely wrong will allow you to increase your chances of getting a right answer.

Section Two: Free Response

This section contains seven questions for which you will provide short answers. Write your answers on your paper.

13. Find the quotient of 32 and -4. **-8**

14. Find the next two numbers in the sequence 64, 16, 4, 1, … . **$\frac{1}{4}, \frac{1}{16}$**

15. Write the ordered pair represented by M. **(3, 3)**

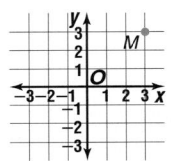

16. Find 98% of 6. **5.88**

17. Draw the reflection of the letter E over a vertical line. **See margin.**

18. How can you find the surface area of a 3-inch by 4-inch by 5-inch rectangular prism? **See margin.**

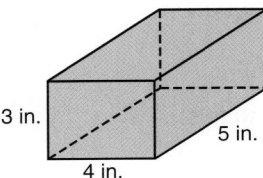

3 in.   5 in.

4 in.

19. Find $3\frac{4}{5} \times 7\frac{1}{2}$. **$28\frac{1}{2}$**

Test-Taking Tip

Tell students to first find out whether they are penalized for wrong answers. If this is the case, suggest they guess only if they are able to narrow down the answer to two choices.

Assessment and Evaluation Masters, pp. 297–298

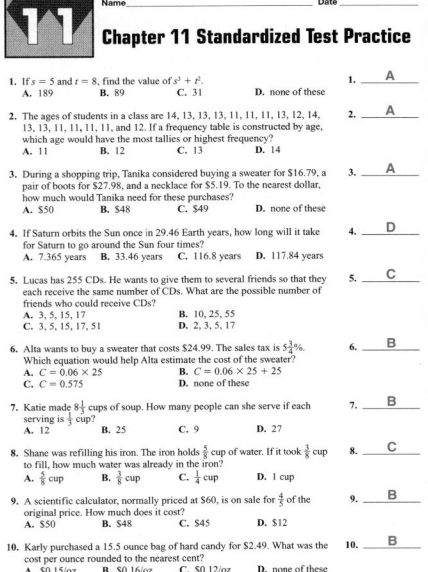

Additional Answers

17. ⊒

18. Add 2(3 × 4) plus 2(4 × 5) plus 2(3 × 5).

Instructional Resources ▶▶▶

Additional standardized test practice is shown at the right and is available in the *Assessment and Evaluation Masters,* pp. 297–298.

Algebra: Exploring Equations

Previewing the Chapter

Overview

This chapter explores algebraic equations and expressions. Students learn to solve equations involving addition, subtraction, multiplication, and division first by using models and then by using properties of equations. Function tables are used to show the output of a given function rule and to determine the function rule when the input and output are given. Coordinate grids are used to graph functions. Students learn to solve problems by using an equation.

Lesson (pages)	Lesson Objectives	NCTM Standards	Standardized Tests	State/Local Objectives
12-1 (476–479)	Solve addition equations by using models.	1–9	MAT	
12-2 (480–483)	Solve subtraction equations by using models.	1–9	MAT	
12-3 (484–487)	Solve multiplication and division equations using models.	1–7, 9, 12, 13	MAT	
12-4 (488–491)	Solve two-step equations using models.	1–3, 5–7, 9	MAT	
12-4B (492–493)	Solve problems by using an equation.	1–7, 9	CAT, MAT	
12-5A (494–495)	Find the input and output for a given function machine.	1–3, 5–9	CTBS, SAT, TN	
12-5 (496–499)	Complete function tables.	1–9	CTBS, SAT, TN	
12-6 (500–503)	Graph functions from function tables.	1–10, 12		

CAT = California Achievement Tests, CTBS = Comprehensive Tests of Basic Skills, ITBS = Iowa Tests of Basic Skills, MAT = Metropolitan Achievement Tests, SAT = Stanford Achievement Tests, TN = Terra Nova

Organizing the Chapter

CD-ROM

All of the blackline masters in the Teacher's Classroom Resources are available on the **Electronic Teacher's Classroom Resources** CD-ROM.

LESSON PLANNING GUIDE

Lesson	Extra Practice (Student Edition)	BLACKLINE MASTERS (PAGE NUMBERS)										
		Study Guide	Practice	Enrichment	Assessment & Evaluation	Classroom Games	Diversity	Hands-On Lab	School to Career	Science and Math Lab Manual	Technology	Transparencies A and B
12-1	p. 588	84	84	84				80				12-1
12-2	p. 588	85	85	85	323	33–35						12-2
12-3	p. 589	86	86	86	322, 323				12			12-3
12-4	p. 589	87	87	87			12					12-4
12-4B	p. 589											
12-5A								66				
12-5	p. 590	88	88	88	324						23	12-5
12-6	p. 590	89	89	89	324					21–24	24	12-6
Study Guide/ Assessment					309–321, 325–327							

OTHER CHAPTER RESOURCES

Student Edition

Chapter Project, pp. 475, 478, 487, 503, 507
Math in the Media, p. 479
Let the Games Begin, p. 491

Technology

 CD-ROM Program

 Interactive Mathematics Tools Software

Teacher's Classroom Resources

Applications
Family Letters and Activities, pp. 23–24
Investigations and Projects Masters, pp. 61–64
Meeting Individual Needs
Investigations for the Special Education Student, pp. 39–42

Teaching Aids
Answer Key Masters
Block Scheduling Booklet
Lesson Planning Guide
Solutions Manual

Professional Publications
Glencoe Mathematics Professional Series

MindJogger Videoquizzes
provide a unique format for reviewing concepts presented in the chapter.

ASSESSMENT RESOURCES

Student Edition

Mixed Review, pp. 479, 483, 487, 491, 499, 503
Mid-Chapter Self Test, p. 487
Math Journal, pp. 482, 501
Study Guide and Assessment, pp. 504–507
Performance Task, p. 507
Portfolio Suggestion, p. 507
Standardized Test Practice, pp. 508–509
Chapter Test, p. 606

Assessment and Evaluation Masters

Multiple-Choice Tests (Forms 1A, 1B, 1C), pp. 309–314
Free-Response Tests (Forms 2A, 2B, 2C), pp. 315–320
Performance Assessment, p. 321
Mid-Chapter Test, p. 322
Quizzes A–D, pp. 323–324
Standardized Test Practice, pp. 325–326
Cumulative Review, p. 327

Teacher's Wraparound Edition

5-Minute Check, pp. 476, 480, 484, 488, 496, 500
Building Portfolios, p. 474
Math Journal, p. 495
Closing Activity, pp. 479, 483, 487, 491, 493, 499, 503

Technology

Test and Review Software
MindJogger Videoquizzes
CD-ROM Program

MATERIALS AND MANIPULATIVES

Lesson 12-1
cups*†
counters*†
equation mat*†

Lesson 12-2
cups*†
counters*†
equation mat*†

Lesson 12-3
cups*†
counters*†
equation mat*†

Lesson 12-4
cups*†
counters*†
equation mat*†
scissors*
beans
index cards
poster board

Lesson 12-5A
scissors*
tape

Lesson 12-6
grid paper†

*Glencoe Manipulative Kit   †Glencoe Overhead Manipulative Resources

PACING CHART

See pages T25–T27 for the Course Planning Calendar.

COURSE	DAY 1	DAY 2	DAY 3	DAY 4	DAY 5	DAY 6	DAY 7
Standard	Chapter Project	Lesson 12-1	Lesson 12-2	Lesson 12-3	Lesson 12-4	Lesson 12-4B	Lessons 12-5A & 12-5
Honors	Chapter Project	Lesson 12-1	Lesson 12-2	Lesson 12-3	Lesson 12-4	Lesson 12-4B	Lesson 12-5
Block	Chapter Project & Lesson 12-1	Lessons 12-2 & 12-3	Lessons 12-4 & 12-4B	Lessons 12-5A, 12-5, & 12-6		Study Guide and Assessment, Chapter Test	

Interactive Mathematics:
Activities and Investigations

is an activity-based program that may be used as an enhancement for chapters in *Mathematics: Applications and Connections.*

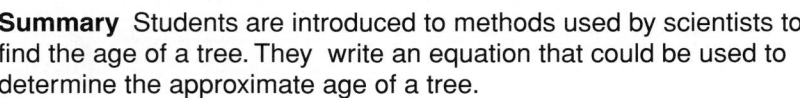

Unit 16,
Pre-Assessment Activity
Use with Lesson 12-4B.

Summary Students are introduced to methods used by scientists to find the age of a tree. They write an equation that could be used to determine the approximate age of a tree.

Math Connection Students use the circumference of a tree to find its age. The circumference can be found by using the formula $C = \pi d$, where d represents the diameter. Students write an equation to determine the approximate age of a tree.

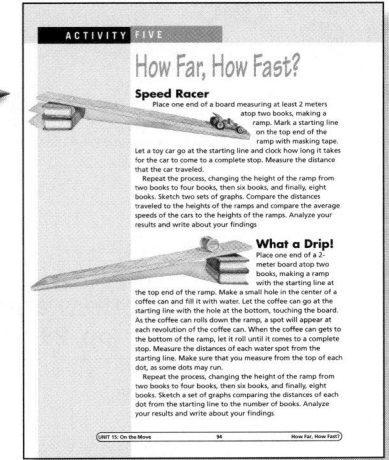

Unit 15, Activity Five
Use with Lesson 12-6.

Summary Students complete tasks that illustrate the relationship between distance and speed, and distance traveled and the number of revolutions a cylindrical object makes. They experiment by rolling objects down ramps and compare the data that they collect.

Math Connection Students conduct experiments rolling objects down ramps of different heights and collecting data to explore momentum, velocity, and acceleration. They sketch graphs of their data and analyze their results.

DAY 8	DAY 9	DAY 10	DAY 11	DAY 12	DAY 13	DAY 14	DAY 15
(continue from Day 7)	Lesson 12-6	Study Guide and Assessment	Chapter Test				
(continue from Day 7)	Lesson 12-6	Study Guide and Assessment	Chapter Test				

Enhancing the Chapter

APPLICATIONS

Classroom Games, pp. 33–35

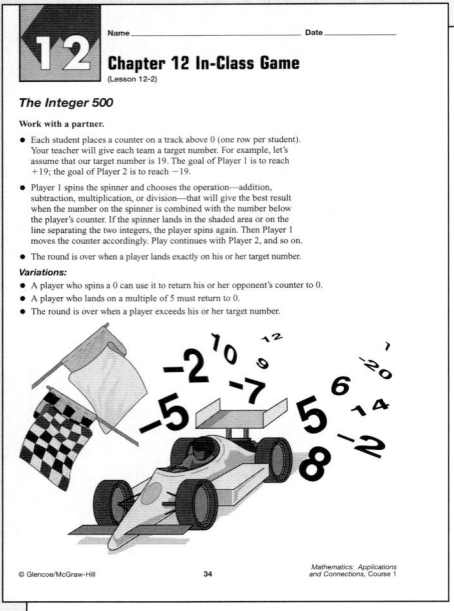

12 Name _____ Date _____

Chapter 12 In-Class Game
(Lesson 12-2)

The Integer 500

Work with a partner.

- Each student places a counter on a track above 0 (one row per student). Your teacher will give each team a target number. For example, let's assume that our target number is 19. The goal of Player 1 is to reach +19; the goal of Player 2 is to reach −19.
- Player 1 spins the spinner and chooses the operation—addition, subtraction, multiplication, or division—that will give the best result when the number on the spinner is combined with the number below the player's counter. If the spinner lands in the shaded area or on the line separating the two integers, the player spins again. Then Player 1 moves the counter accordingly. Play continues with Player 2, and so on.
- The round is over when a player lands exactly on his or her target number.

Variations:

- A player who spins a 0 can use it to return his or her opponent's counter to 0.
- A player who lands on a multiple of 5 must return to 0.
- The round is over when a player exceeds his or her target number.

© Glencoe/McGraw-Hill   34   *Mathematics: Applications and Connections, Course 1*

Diversity Masters, p. 12

12 Name _____ Date _____

Diversity Activity
(Lesson 12-4)

Daniel Hale Williams

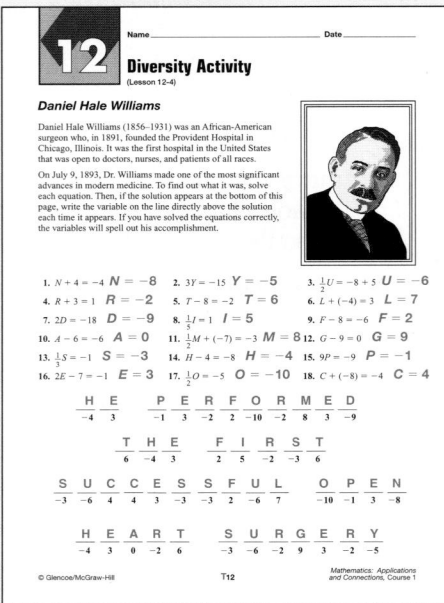

Daniel Hale Williams (1856–1931) was an African-American surgeon who, in 1891, founded the Provident Hospital in Chicago, Illinois. It was the first hospital in the United States that was open to doctors, nurses, and patients of all races.

On July 9, 1893, Dr. Williams made one of the most significant advances in modern medicine. To find out what it was, solve each equation. Then, if the solution appears at the bottom of this page, write the variable on the line directly above the solution each time it appears. If you have solved the equations correctly, the variables will spell out his accomplishment.

1. $N + 4 = -4$ $N = -8$
2. $3Y = -15$ $Y = -5$
3. $\frac{1}{2}U = -8 + 5$ $U = -6$
4. $R + 3 = 1$ $R = -2$
5. $T - 8 = -2$ $T = 6$
6. $L + (-4) = 3$ $L = 7$
7. $2D = -18$ $D = -9$
8. $\frac{1}{3}I = 5$ $I = 15$
9. $F - 8 = -6$ $F = 2$
10. $A - 6 = -6$ $A = 0$
11. $\frac{1}{4}M + (-7) = -3$ $M = 8$
12. $G - 9 = 0$ $G = 9$
13. $\frac{1}{3}S = -1$ $S = -3$
14. $H - 4 = -8$ $H = -4$
15. $9P = -9$ $P = -1$
16. $2E - 7 = -1$ $E = 3$
17. $\frac{1}{2}O = -5$ $O = -10$
18. $C + (-8) = -4$ $C = 4$

H	E		P	E	R	F	O	R	M	E	D
--	--		-	-	-	-	-	-	-	-	-
−4	3		−1	3	−2	−10	−2	8	3	−9

T	H	E		F	I	R	S	T
-	-	-		-	-	-	-	-
6	−4	3		2	5	−2	−3	6

S	U	C	C	E	S	S	F	U	L		O	P	E	N
-	-	-	-	-	-	-	-	-	-		-	-	-	-
−3	−6	4	4	3	−3	−3	2	−6	7		−10	−1	3	−8

H	E	A	R	T		S	U	R	G	E	R	Y
-	-	-	-	-		-	-	-	-	-	-	-
−4	3	0	−2	6		−3	−6	−2	9	3	−2	−5

© Glencoe/McGraw-Hill   T12   *Mathematics: Applications and Connections, Course 1*

School to Career Masters, p. 12

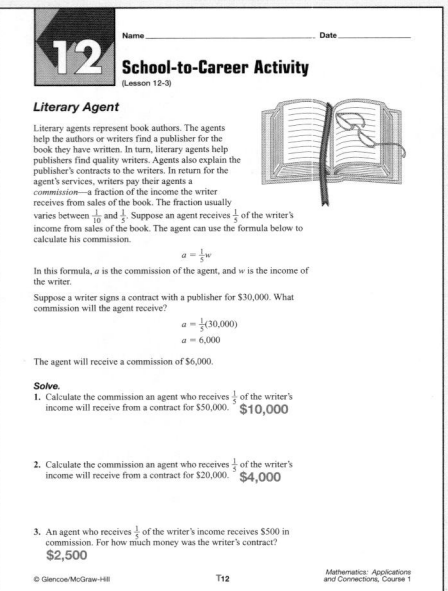

12 Name _____ Date _____

School-to-Career Activity
(Lesson 12-3)

Literary Agent

Literary agents represent book authors. The agents help the authors or writers find a publisher for the book they have written. In turn, literary agents help publishers find quality writers. Agents also explain the publisher's contracts to the writers. In return for the agent's services, writers pay their agents a *commission*—a fraction of the income the writer receives from sales of the book. The fraction usually varies between $\frac{1}{10}$ and $\frac{1}{5}$. Suppose an agent receives $\frac{1}{5}$ of the writer's income from sales of the book. The agent can use the formula below to calculate his commission.

$$a = \frac{1}{5}w$$

In this formula, a is the commission of the agent, and w is the income of the writer.

Suppose a writer signs a contract with a publisher for $30,000. What commission will the agent receive?

$$a = \frac{1}{5}(30,000)$$
$$a = 6,000$$

The agent will receive a commission of $6,000.

Solve.

1. Calculate the commission an agent who receives $\frac{1}{5}$ of the writer's income will receive from a contract for $50,000.   **$10,000**

2. Calculate the commission an agent who receives $\frac{1}{5}$ of the writer's income will receive from a contract for $20,000.   **$4,000**

3. An agent who receives $\frac{1}{5}$ of the writer's income receives $500 in commission. For how much money was the writer's contract?   **$2,500**

© Glencoe/McGraw-Hill   T12   *Mathematics: Applications and Connections, Course 1*

Family Letters and Activities, pp. 23–24

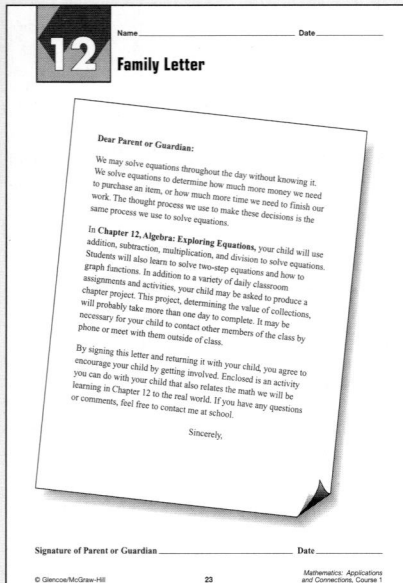

12 Name _____ Date _____

Family Letter

Dear Parent or Guardian:

We may solve equations throughout the day without knowing it. We solve equations to determine how much more money we need to purchase an item, or how much more time we need to finish our work. The thought process we use to make these decisions is the same process we use to solve equations.

In **Chapter 12, Algebra: Exploring Equations**, your child will use addition, subtraction, multiplication, and division to solve equations. Students will also learn to solve two-step equations and how to graph functions. In addition to a variety of daily classroom assignments and activities, your child may be asked to produce a chapter project. This project, determining the value of collections, will probably take more than one day to complete. It may be necessary for your child to contact other members of the class by phone or meet with them outside of class.

By signing this letter and returning it with your child, you agree to encourage your child by getting involved. Enclosed is an activity you can do with your child that also relates the math we will be learning in Chapter 12 to the real world. If you have any questions or comments, feel free to contact me at school.

Sincerely,

Signature of Parent or Guardian _____ **Date** _____

© Glencoe/McGraw-Hill   23   *Mathematics: Applications and Connections, Course 1*

12 Name _____ Date _____

Family Activity

Temperature Differences

Watch the news on television, consult the weather section of a newspaper, or visit a weather-related website. Then work with a family member to collect the high and low temperatures for at least nine cities. Write and solve an equation to find how much warmer the high temperature is than the low temperature.

	City	Low Temperature	High Temperature	Equation	Solution
1.	Austin	72°	91°	$72 + t = 91$	
2.					
3.					
4.					
5.					
6.					
7.					
8.					
9.					
10.					

1. The high temperature was 19° warmer. 2.–10. Answers will vary.

© Glencoe/McGraw-Hill   24   *Mathematics: Applications and Connections, Course 1*

Science and Math Lab Manual, pp. 21–24

6 Name _____ Date _____

Science and Mathematics Lab
(Course 1, Lesson 12-6)

Ivy League Math

INTRODUCTION

Ivies and most other types of vines are members of a group of plants called *angiosperms*, which means "flowering plants." There are many species of ivy, but the one with which most of us are familiar is the English Ivy or *Hedera helix*. Ivies will climb any rigid structure, such as trees or walls, using aerial roots which develop from their stems.

OBJECTIVES

In this lab, you will:

- measure the length of each vine.
- count the number of leaves on each vine.
- enter your data in a function table.
- draw a graph that represents the relationship between the length of the vine and the number of leaves.
- determine the number of leaves per unit length of vine.

MATERIALS

- ivy or similar vine, cut to four different lengths
- large piece of heavy paper such as blotting paper
- pots and potting soil
- ruler
- paper cups

PROCEDURE

1. Work with a partner. Place each vine on the piece of paper in order from shortest to longest.
2. Label each vine by writing a number from 1–4 above each vine.
3. Use your ruler to measure the length of each vine to the nearest inch. One person can hold the vine straight while the other person measures it. Record the measurements in the Data Table as variable x.
4. Count the number of leaves on the vine you have just measured. Record this information in the table as variable y.
5. Repeat Steps 3 and 4 for each vine.

© Glencoe/McGraw-Hill   23   *Mathematics: Applications and Connections*

Hands-On Lab Masters, p. 80

12 Name_____ Date_____

Lab Activity
(Lesson 12-1)

Solving Addition Equations

Write the equation that is represented by each model.

1. $4 + x = 6$

2. $x - 3 = 9$ or $x + (-3) = 9$

3. $x + 2 = 6$

4. $x + (-3) = 4$ or $x - 3 = 4$

Solve each equation using cups and counters. Sketch the arrangement in the boxes.

5. $x + 2 = 6$ $x = 4$

6. $x + (-2) = -7$ $x = -5$

7. $x + 1 = -3$ $x = -4$

8. Solve $x + 4 = -3$ without using models. $x = \underline{-7}$

© Glencoe/McGraw-Hill 80 *Mathematics: Applications and Connections, Course 1*

Assessment and Evaluation Masters, pp. 322–324

12 Name_____ Date_____

Chapter 12 Mid-Chapter Test
(Lessons 12-1 through 12-3)

Solve each equation.

1. $a + 5 = -8$ 1. -13
2. $6 + c = 3$ 2. -3
3. $d - 5 = 13$ 3. 18
4. $r - 9 = -1$ 4. 8
5. $9s = -63$ 5. -7
6. $3x = -15$ 6. -5
7. $b + 22 = 17$ 7. -5
8. $m + 12 = 17$ 8. 5
9. $r - 5 = 4$ 9. 9
10. $s - 7 = -12$ 10. -5
11. $4y = 28$ 11. 7
12. $\frac{1}{3}a = 11$ 12. 33
13. $n + (-4) = 0$ 13. 4
14. $-18 + r = -22$ 14. -4
15. $t - 19 = -22$ 15. -3
16. $w - 52 = -38$ 16. 14
17. $\frac{1}{4}p = -15$ 17. -60
18. $\frac{x}{2} = -21$ 18. -42
19. Lenny rides the elevator up 12 floors and gets off at the 19th floor. Write an equation to represent his movement, and solve it.   19. $x + 12 = 19$; $x = 7$
20. Jill and Amy went out to dinner. They agreed to each pay half the bill. Jill paid $13.24. Write an equation for this situation, and find the amount of the bill.   20. $\frac{1}{2}x = \$13.24$; $x = \$26.48$

© Glencoe/McGraw-Hill 322 *Mathematics: Applications and Connections, Course 1*

Name_____ Date_____

Chapter 12 Quiz A
(Lessons 12-1 and 12-2)

Solve each equation.

1. $m + 7 = 1$ 1. -6
2. $x - 5 = 7$ 2. 12
3. $x + (-6) = -5$ 3. 1
4. $n - 7 = -8$ 4. -1
5. $a + 12 = 7$ 5. -5
6. $b - 11 = -4$ 6. 7
7. $c + (-19) = -24$ 7. -5
8. $t - 18 = 1$ 8. 19
9. $p + 6 = -15$ 9. -21
10. $r - 53 = -38$ 10. 15

Name_____ Date_____

Chapter 12 Quiz B
(Lesson 12-3)

Solve each equation.

1. $12x = -48$ 1. -4
2. $\frac{1}{4}a = -6$ 2. -24
3. $-8d = -32$ 3. 4
4. $51 = -17n$ 4. -3
5. $8s = 5.6$ 5. 0.7
6. $6m = -42$ 6. -7
7. $\frac{1}{3}r = 8$ 7. 24
8. $\frac{1}{2}x = -9$ 8. -18
9. $\frac{1}{3}y = -7$ 9. -21
10. $\frac{1}{5}c = 15$ 10. 75

© Glencoe/McGraw-Hill 323 *Mathematics: Applications and Connections, Course 1*

Technology Masters, pp. 23–24

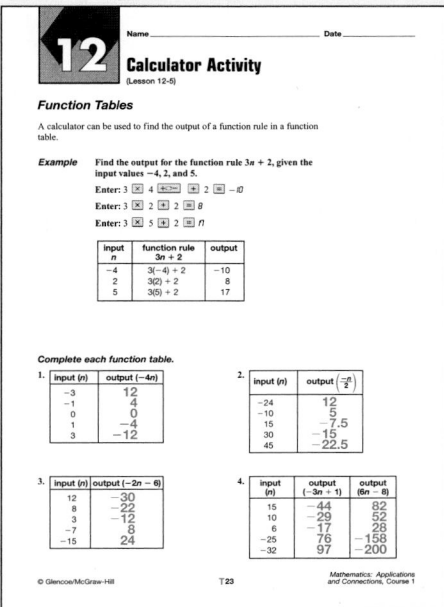

12 Name_____ Date_____

Calculator Activity
(Lesson 12-5)

Function Tables

A calculator can be used to find the output of a function rule in a function table.

Example Find the output for the function rule $3n + 2$, given the input values -4, 2, and 5.

Enter: 3 ✕ 4 +/- + 2 = -10
Enter: 3 ✕ 2 + 2 = 8
Enter: 3 ✕ 5 + 2 = n

input n	function rule $3n + 2$	output
-4	3(-4) + 2	-10
2	3(2) + 2	8
5	3(5) + 2	17

Complete each function table.

1.
input (n)	output ($-4n$)
-3	12
-1	4
0	0
1	-4
3	-12

2.
input (n)	output ($\frac{-n}{2}$)
-24	12
-10	5
15	-7.5
30	-15
45	-22.5

3.
input (n)	output ($-2n - 6$)
12	-30
8	-22
3	-12
-7	8
-15	24

4.
input (n)	output ($-3n + 1$)	output ($6n - 8$)
15	-44	82
10	-29	52
6	-17	28
-25	76	-158
-32	97	-200

© Glencoe/McGraw-Hill T23 *Mathematics: Applications and Connections, Course 1*

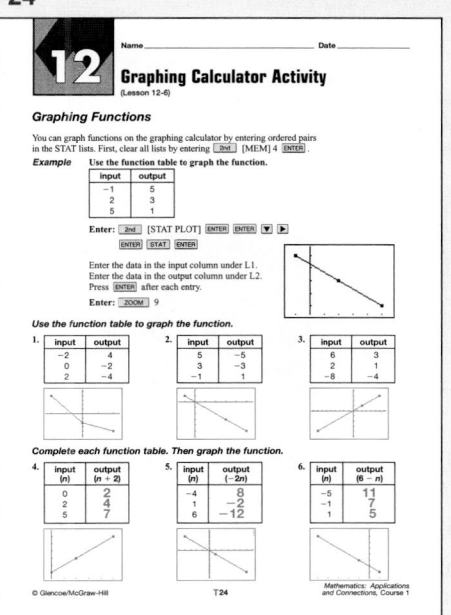

12 Name_____ Date_____

Graphing Calculator Activity
(Lesson 12-6)

Graphing Functions

You can graph functions on the graphing calculator by entering ordered pairs in the STAT lists. First, clear all lists by entering 2nd [MEM] 4 ENTER.

Example Use the function table to graph the function.

input	output
-1	5
2	3
5	1

Enter: 2nd [STAT PLOT] ENTER ENTER ▼ ►
ENTER STAT ENTER

Enter the data in the input column under L1.
Enter the data in the output column under L2.
Press ENTER after each entry.

Enter: ZOOM 9

Use the function table to graph the function.

1.
input	output
-2	4
0	-2
2	-4

2.
input	output
5	-5
3	-3
-1	1

3.
input	output
6	3
2	1
-8	-4

Complete each function table. Then graph the function.

4.
input (n)	output ($n - 2$)
0	-2
2	0
5	3

5.
input (n)	output ($-2n$)
-4	8
1	-2
6	-12

6.
input (n)	output ($6 - n$)
-5	11
1	5
4	2

© Glencoe/McGraw-Hill T24 *Mathematics: Applications and Connections, Course 1*

Investigations for the Special Education Student, pp. 39–42

Use with:
Course 1–Chapter 13
Course 2–Chapter 6
Course 3–Chapter 11

Investigation 13 Teacher's Guide

How Does Your Garden Grow?

Overview
In this investigation, students will experience how mathematics and critical thinking are involved in planting and growing a garden. This goal will be achieved by having each group of students in the class select a plant to put in their designated spot in the class garden. The group will be responsible for tending to their spot each day and charting the plant's growth once it begins. Then, each group will analyze the plant's growth and share their results with the class. Finally, the groups will combine their individual graphs to form a graph of the entire garden using overhead transparencies or a computer.

Activity Goals
Students will:
• plant and care for a garden;
• chart the plants' growth; and
• collect, analyze, and interpret data.

Planning the Instruction
Prerequisite Skills
Students should have a significant amount of practice measuring, collecting, and graphing data.
Materials
• investigation worksheets
• graph paper or graphing aides such as transparencies or computers
• books on plants
• measuring tools
• seeds or seedlings and garden supplies
Time Needed
• two or three 45-minute periods for planting
• five to ten minutes per day during growth period
• two or three 45-minute periods for wrap-up

Procedure
1. Obtain permission to plant a garden somewhere on school grounds.
2. Divide the class into groups of 3-4. Then, divide the garden so that each group will have its own equal-sized spot.
3. Explain to the class that each group is to choose something to plant in their spot. They will tend to it daily and, when it begins to grow, chart its growth.
4. Have each group complete the Planning Page worksheet. Ask groups to consider the location, season and soil type when deciding what to plant.
5. Obtain seeds/seedlings needed for each group. Then, have groups prepare garden and plant.
6. Each day, students will take turns tending to their group's spot. When growth begins, the height will be measured and recorded each time on the Daily Data worksheet.
7. At the end of the investigation, students will interpret the charted results, graph them on their Pattern of Growth worksheet, and answer the questions.
8. Have groups share their growth results with the class. Then, make a "class" graph of the entire garden, giving each group an empty overhead graph with a different color, overlaying the graphs one at a time. Compare and contrast growth of different plants. Discuss what worked, what did not and why.
9. If the growing season in your area is not conducive to planting a garden outside, you might have each group plant seeds in pots, place them in a sunny window, and chart their growth.

Adaptations and Variations
The following are some ideas on how this investigation may be modified depending on student population.

LD • Assist in planning for their spot.
• Monitor data collection to ensure accuracy.
• Simplify the charting process if necessary by using a computer or wall chart.

PH • Assign a supervisory role in the group if student is unable to help plant or measure.
• Provide aid in charting if unable to do by hand.

CD • Take a nonverbal role when groups share.

BD • Monitor closely when planting or measuring.
• If problems arise, prohibit these students from collecting data or tending to the plot.

HI • Give directions in signing or writing.
• Have a student sign when groups share data.

VI • Assist in planting and measuring.
• Enlarge graphs for these students.

© Glencoe/McGraw-Hill 39 *Mathematics: Applications and Connections*

Theme: Collections

Collecting is a popular hobby because a person can collect almost anything of interest. If a person collects objects that are hard to find, the collection can become quite valuable. A single object in a collection can also command a high price. For example, as of 1997, a Michael Jordan retirement basketball card was worth $3,500.00.

Question of the Day

Longaberger® Baskets, which are made in Dresden, Ohio, are collectible, hand-woven, hard-maple baskets. A 1983 Market Basket from the J.W. Collection Series originally sold for $32.95. At an auction in 1997, one of these baskets sold for $1,550. Write an equation to determine how much the value increased. **$32.95 + x = $1,550**

Assess Prerequisite Skills

Ask students to read through the list of objectives presented in "What you'll learn in Chapter 12." You may wish to ask them what each of the objectives means or if they have experienced or used any of these math concepts before.

 Building Portfolios

Encourage students to revise their portfolios as they study this chapter. Have them add to their portfolios examples of algebra concepts they have mastered.

 **Math and the Family**

In the *Family Letters and Activities* booklet (pp. 23–24), you will find a letter to the parents explaining what students will study in Chapter 12. An activity appropriate for the whole family is also available.

What you'll learn in Chapter 12

● to use models to solve equations,
● to solve problems using an equation,
● to complete function tables, and
● to graph functions from function tables.

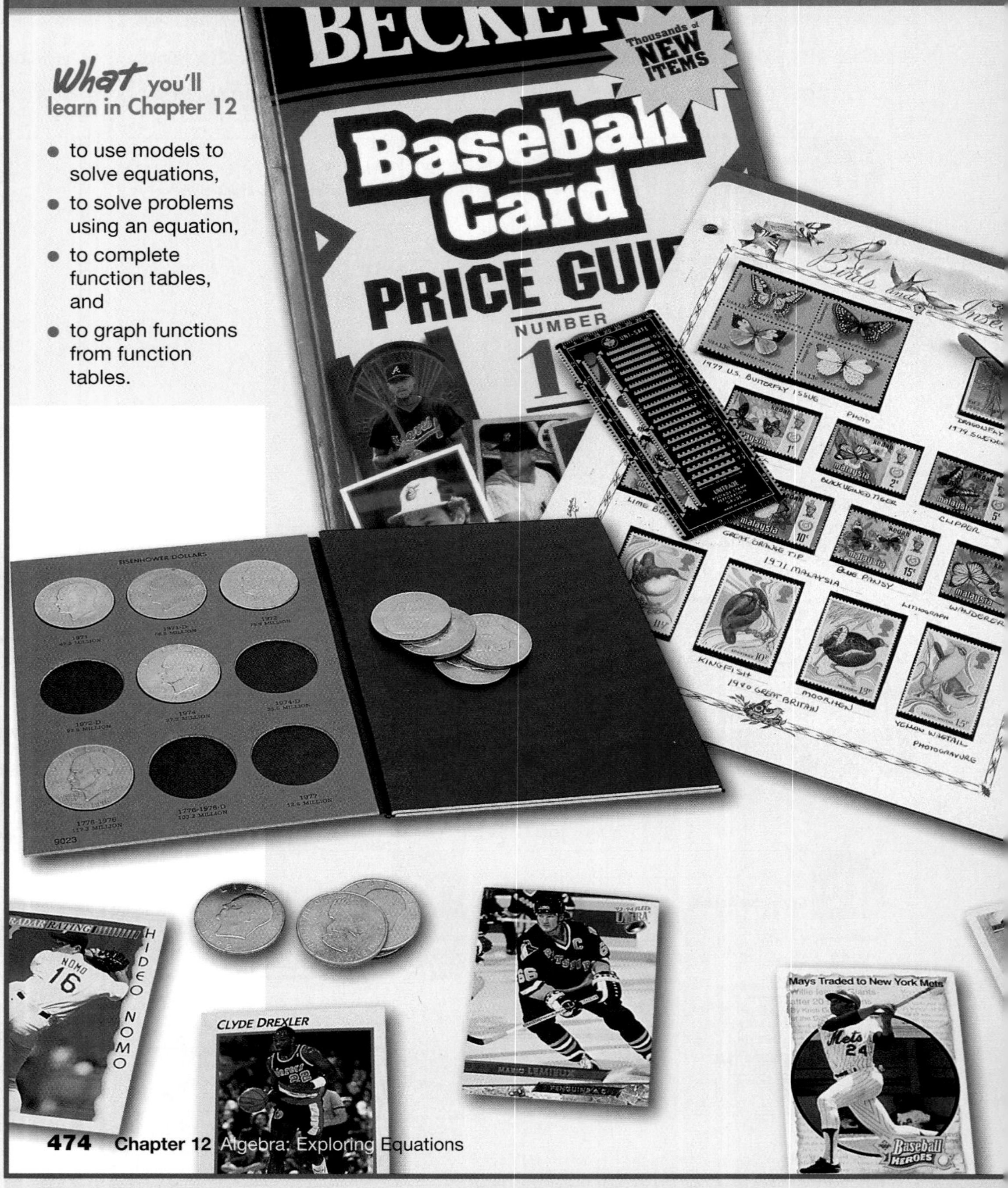

 CD-ROM Program

Activities for Chapter 12

● Chapter 12 Introduction
● Interactive Lessons 12-2, 12-3, 12-4, 12-6
● Extended Activity 12-5
● Assessment Game
● Resource Lessons 12-1 through 12-6

CHAPTER Project

COLLECT A FORTUNE

In this project, you will start an imaginary collection of at least five items such as sport cards, coins, stamps, or dolls. Or you can use a collection you already have. You will need to research the history of the value of these items. You will prepare a table or some type of display showing the value of your collection to present to your class.

Getting Started

- Look at the table. Which card was worth the most in August, 1996? Which card was worth the most in June, 1997?
- How could you determine whether the value of Kristin's six-card collection increased or decreased in the 10 months between August, 1996 and June, 1997?

Kristin's Card Collection

Manufacturer	Player	Value Aug., 1996	Value June, 1997
1995–96 Finest Mystery Borderless Refractors Gold	John Stockton	$10	$7.50
1992–93 Upper Deck All-NBA	Michael Jordan	$25	$20
1985–86 Star Lakers Champs	Larry Bird	$18	$18
1986–87 Fleer	Charles Barkley	$50	$60
1992–93 Hoops Magic's All-Rookies	Shaquille O'Neal	$80	$30
1971–72 Topps	Rudy Tomjan-ovich	$15	$15

Technology Tips

- Use a **spreadsheet** to find the value of a collection.
- Use **computer software** to make graphs.

 interNET
CONNECTION For up-to-date information on card collecting, visit:
www.glencoe.com/sec/math/mac/mathnet

Working on the Project

You can use what you'll learn in Chapter 12 to help you make your presentation.

Page	Exercise
478	32
487	33
503	27
507	Alternative Assessment

Special Edition Barbie

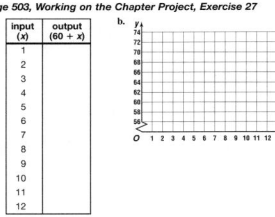

interNET
CONNECTION

Instructional Resources ▶▶▶
A recording sheet to help students organize their data for the Chapter Project is shown at the right and is available in the *Investigations and Projects Masters*, p. 64.

CHAPTER Project
N O T E S

Objectives Students should
- research the history and value of collectibles.
- prepare a display of the values of their collection.

Project Pointer You may suggest that students begin a *Project Folder* to keep their work as they complete each stage of the Chapter Project. The completed project may also be added to their portfolios. Encourage students to bring their collections or pictures of them to class for their presentations.

Using the Table Ask students whether each card in Kristin's card collection increased or decreased in value from August 1996 to June 1997.

Investigations and Projects Masters, p. 64

Name _____ Date _____

12 Chapter 12 Project

Collect a Fortune

Page 478, Working on the Chapter Project, Exercise 32

a.

	A	B	C	D	E
1	Player	Value August, 1996	Value June, 1997	Change in 10 months	Change per month
2	John Stockton	$10	$7.50		
3	Michael Jordan	$25	$20		
4	Larry Bird	$18	$18		
5	Charles Barkley	$50	$60		
6	Shaquille O'Neal	$80	$30		
7	Rudy Tomjanovich	$15	$15		

b. The total change is:

Page 487, Working on the Chapter Project, Exercise 33
Use table above.

Page 503, Working on the Chapter Project, Exercise 27

a.

input (x)	output (60 + x)
1	
2	
3	
4	
5	
6	
7	
8	
9	
10	
11	
12	

b.

c. The ordered pair (5, 65) means:

© Glencoe/McGraw-Hill   64   *Mathematics: Applications and Connections, Course 1*

Instructional Resources

- *Study Guide Masters*, p. 84
- *Practice Masters*, p. 84
- *Enrichment Masters*, p. 84
- Transparencies 12-1, A and B
- *Hands-On Lab Masters*, p. 80

 CD-ROM Program
- Resource Lesson 12-1

Recommended Pacing	
Standard	Day 2 of 11
Honors	Day 2 of 11
Block	Day 1 of 6

1 FOCUS

 5-Minute Check
(Chapter 11)

1. Order −4, 6, 0, and −2 from least to greatest.
   −4, −2, 0, 6

Find each sum, difference, product, or quotient.

2. +5 − (−1) **6**
3. −7 + (−3) **−10**
4. −12(−4) **48**
5. 45 ÷ (−3) **−15**
6. Graph and label point A(−3, 2).

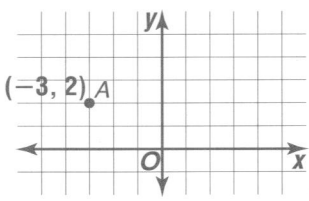

The 5-Minute Check is also available on **Transparency 12-1A** for this lesson.

Motivating the Lesson

Problem Solving Ask students how they would use an addition equation to solve the following problem. *A pie-eating contest is set up for 14 contestants. So far, 8 people have entered the contest. How many openings are left?*

n + 8 = 14; *n* = 6

What **you'll learn**

You'll learn to solve addition equations by using models.

When **am I ever going to use this?**

Knowing how to solve equations can help you determine golf scores.

Justin is playing Monopoly and needs to roll an 8 so he can land on Community Chest. If he rolls a 5 on the first number cube, what must he roll on the second number cube? *This problem will be solved in Example 1.*

You can use cups and counters to solve equations. A cup represents the unknown value, yellow counters represent positive integers, and red counters represent negative integers.

 **Example 1**
APPLICATION

Did you know?
American Charles Darrow invented the game of Monopoly in 1934. Over 85 million sets have been sold worldwide in 19 languages including Braille.

Games Refer to the beginning of the lesson. What must Justin roll on the second number cube?

Explore   The number on the first number cube is 5. Justin wants the total of the two number cubes to be 8.

Plan   Let *d* represent the number tossed on the second number cube. Translate the problem into an equation using the variable *d*. Model the equation and solve.

Solve

number on the first number cube	plus	number on the second number cube	equals	sum of the two numbers
5	+	*d*	=	8

$$5 + d = 8$$

To get the cup by itself, subtract 5 from each side.

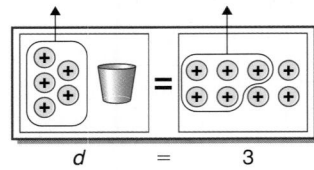

$$d = 3$$

The solution is 3. Justin must roll a 3 on the second number cube.

Examine   Check the solution by replacing the value of the variable in the original equation.

Check: $5 + d = 8$
   $5 + 3 \stackrel{?}{=} 8$ *Replace d with 3.*
   $8 = 8$ ✓

476   **Chapter 12** Algebra: Exploring Equations

Cross-Curriculum Cue

Inform the other teachers on your team that your students are studying algebraic equations. Suggestions for curriculum integration are:

Health: daily dietary needs in grams; weight
Social Studies: shifts in population
Physical Education: score-keeping
Earth Science: barometric pressure

Example —

Use cups and counters to solve $c + 3 = -9$.

Use a cup to represent c. Add 3 positive counters on the left side of the mat to represent +3. Place 9 negative counters on the right side of the mat to represent –9.

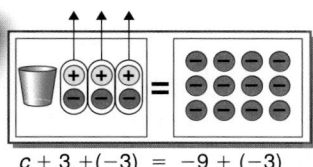

$$c + 3 = -9$$

To get the cup by itself, you need to remove 3 positive counters from each side. Since there are no positive counters on the right side of the mat, add 3 negative counters to each side to make 3 zero pairs on the left side of the mat. Then remove the zero pairs.

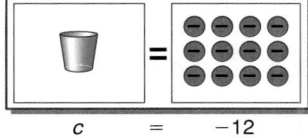

$$c + 3 + (-3) = -9 + (-3)$$

$$c = -12$$

Check: $c + 3 = -9$ *Replace c with –12.*

$$-12 + 3 \stackrel{?}{=} -9$$
$$-9 = -9 \checkmark$$

The solution is -12.

CHECK FOR UNDERSTANDING

Communicating Mathematics

Read and study the lesson to answer each question. 1–2. See margin.

1. *Explain* how the model represents $5 + w = -8$.

2. *Show* how to model the equation $y + 2 = 4$.

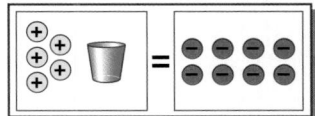

---

2 TEACH

Transparency 12-1B contains a teaching aid for this lesson.

Using Manipulatives Use a pan balance to review equations by placing an equal number of weights on each pan to balance the scale. Then ask students how the balance could be used with integers.

In-Class Examples

For Example 1
Tamara needs to save $16 for a CD she wants to buy. If she has already saved $4, how much more does she need? **$12**

For Example 2
Use cups and counters to solve $6 + x = -5$. **−11**

Teaching Tip After Example 2, remind students that adding or removing zero pairs does not change the value of a side.

3 PRACTICE/APPLY

Check for Understanding
If students need additional practice or instruction after completing Exercises 1–9, one of these options may be helpful.
• Extra Practice, see p. 588
• Reteaching Activity
• *Study Guide Masters,* p. 84
• *Practice Masters,* p. 84

Additional Answers
1. **Sample answer: The 5 positive counters and the cup represent $w + 5$ and the 8 negative counters represent −8.**

2.
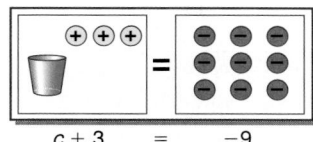

Reteaching the Lesson

Activity Paper strips and a card labeled x will be needed. Without showing students, write a number sentence such as $-1 + 2 = 1$ on a paper strip. Cover one of the numbers with the x card. The student who correctly solves the equation writes the next sentence and so on.

Error Analysis
Watch for students who move the sum to the left side of the equation to add or subtract.
Prevent by reminding them to check their answers by replacing the variable with the solution.

CHAPTER Project

Exercise 32 asks students to advance to the next stage of work on the Chapter Project. You may have students show their work and place it in their Project Folder. Discuss with students what positive and negative changes mean. **an increase or decrease**

Guided Practice | **Solve each equation. Use cups and counters if necessary.**

3. $x + 3 = 9$ **6**   **4.** $t + 25 = 15$ **−10**   **5.** $4 + g = 11$ **7**

6. $r + 2 = -5$ **−7**   **7.** $2 + c = -7$ **−9**   **8.** $-12 + m = 8$ **20**

9. When n is added to 10, the result is 4. What is the value of n? **−6**

EXERCISES

Practice | **Solve each equation. Use cups and counters if necessary.**

10. $y + 7 = 18$ **11**   **11.** $x + 3 = -2$ **−5**   **12.** $4 + g = 6$ **2**

13. $n + 4 = 3$ **−1**   **14.** $-3 + m = -3$ **0**   **15.** $z + 3 = 5$ **2**

16. $2 + s = -1$ **−3**   **17.** $2 + b = -4$ **−6**   **18.** $x + (-3) = -1$ **2**

19. $3 + p = -17$ **−20**   **20.** $d + 1 = -2$ **−3**   **21.** $t + 9 = -34$ **−43**

23. −125

22. $-3 + c = -5$ **−2**   **23.** $a + 25 = -100$ **24.** $-9 + z = -18$ **−9**

25. $x + (-7) = -9$ **−2**   **26.** $f + 6 = -8$ **−14**   **27.** $-2 = m + 6$ **−8**

28. If $8 + c = -12$, what is the value of c? **−20**

29. Find the value of n if $n + (-4) = 6$. **10**

Applications and Problem Solving

30. *Games* In the card game Clubs, it is possible to have a negative score. Suppose your friend had a score of -5 in the second hand. This made her total score after two hands equal -2. What was her score in the first hand? **3**

31. *Sports* On the first day of the 1997 Masters golf tournament, Tiger Woods' score was -2. His score on the second day was added to the first day's score, and the total was -8. What was Tiger's score on the second day? **−6**

32. *Working on the* **CHAPTER** Project   Refer to the table on page 475. Enter the last three columns of the table into a spreadsheet.

   a. Label column D *Change in 10 months*. Write a formula for cells D2, D3, D4, D5, D6, and D7 to find the amount of change in value of each basketball card.

	A	B	C	D
1	Player	Value August, 1996	Value June, 1997	Change in 10 months
2	John Stockton	$10	$7.50	= C2 − B2
3	Michael Jordan	$25	$20	= C3 − B3
4	Larry Bird	$18	$18	= C4 − B4
5	Charles Barkley	$50	$60	= C5 − B5
6	Shaquille O'Neal	$80	$30	= C6 − B6
7	Rudy Tomjanovich	$15	$15	= C7 − B7

   b. Write a formula for cell D8 that will find the total change in value of Kristin's six-card collection over the 10-month period. What was the total change in value? **D2 + D3 + D4 + D5 + D6 + D7; $47.50 decrease**

Study Guide Masters, p. 84

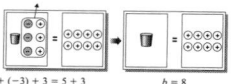

12-1 **Study Guide**

Solving Addition Equations

To **solve an equation** means to find a value for the variable that makes the equation true.

Example   Use cups and counters to solve $b + (-3) = 5$.

Use a cup to represent b. Add 3 negative counters on the left side of the mat to represent -3. Place 5 positive counters on the right side of the mat to represent $+5$.

$b + (-3) = +5$

To get the cup by itself, you need to remove 3 negative counters from each side. Since there are no negative counters on the right side of the mat, add 3 positive counters to each side to make 3 zero pairs on the left side of the mat. Then remove the zero pairs.

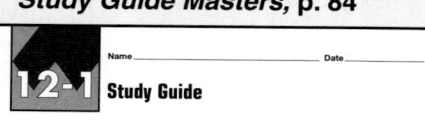

$b + (-3) + 3 = 5 + 3$ $b = 8$

Solve each equation. Use cups and counters if necessary.

1. $k + 5 = 6$ **1**   2. $a + 4 = -5$ **−9**   3. $m + (-8) = -15$ **−7**

4. $v + (-1) = -7$ **−6**   5. $x + 3 = -3$ **−6**   6. $c + 5 = 10$ **5**

7. $4 + d = 7$ **3**   8. $g + (-1) = -3$ **−2**   9. $6 + t = -2$ **−8**

10. If $z + 7 = -12$, what is the value of z? **−19**   11. Find the value of v if $15 + v = -2$. **−17**

© Glencoe/McGraw-Hill   T 84   Mathematics: Applications and Connections, Course 1

33. Critical Thinking Replace the boxes with the numbers 2, 3, 7, 8, and 9 to make a true equation. Use each number exactly once.

$$\blacksquare\,\blacksquare + (-\blacksquare) = \blacksquare\,\blacksquare \quad \text{Sample answer: } 37 + (-9) = 28$$

Mixed Review

34. 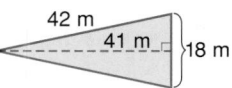 Which of the following illustrates a reflection of the letter L? *(Lesson 11-8)* **A**

A   B   C   D

35. Find the product of −3 and −7. *(Lesson 11-5)* **21**

36. Geometry Find the area of the triangle. *(Lesson 10-2)* **369 m²**

42 m
41 m
18 m

37. Vacations Refer to the circle graph. How much greater is the percent of vacationers that travel and sightsee than the percent that go to a resort? *(Lesson 2-4)* **9%**

How People Spend Vacations

- No vacation/don't know 6%
- Visit family/friends 30%
- Summer/winter resort 17%
- Stay at home 21%
- Travel/sightsee 26%

MATH IN THE MEDIA

$x + 3 = 2x$
$x = 3$

"SAY, WAIT A MINUTE! JUST YESTERDAY SHE SAID X WAS EQUAL TO TWO!"

1. What assumption is the student making about the value of x?

2. How would you check the teacher's solution? **Replace x with 3.**

1. Sample answer: There is a one-to-one correspondence between variables and numbers.

Lesson 12-1 Solving Addition Equations **479**

■ Extending the Lesson ■

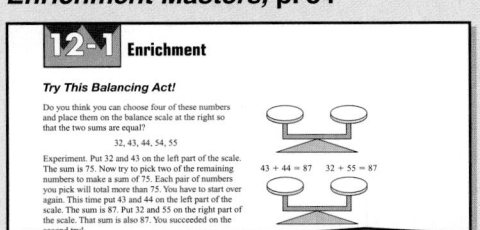
MATH IN THE MEDIA

Discuss with students how the cartoonist found humor in the student's assumption. Ask: *Why do you think the student became confused? How could he have avoided becoming confused?* Have students look through the comics and editorial section of a local Sunday newspaper to find more examples of humor involving mathematics.

Closing Activity

Modeling Have students use cups and counters to model an addition equation whose solution is 8.
Sample answer: $q + (-2) = 6$

Instructional Resources

- *Study Guide Masters*, p. 85
- *Practice Masters*, p. 85
- *Enrichment Masters*, p. 85
- Transparencies 12-2, A and B
- *Assessment and Evaluation Masters*, p. 323
- *Classroom Games*, pp. 33–35

 CD-ROM Program
- Resource Lesson 12-2
- Interactive Lesson 12-2

Recommended Pacing

Standard	Day 3 of 11
Honors	Day 3 of 11
Block	Day 2 of 6

1 FOCUS

 5-Minute Check
(Lesson 12-1)

Solve each equation. Use cups and counters if necessary.
1. $4 + f = 10$ **6**
2. $p + 7 = 15$ **8**
3. $-1 + w = -3$ **−2**
4. $t + 5 = -9$ **−14**
5. $x + (-12) = 1$ **13**

The 5-Minute Check is also available on **Transparency 12-2A** for this lesson.

Motivating the Lesson

Communication Ask students how a subtraction equation can be written as an addition equation. Discuss the function of opposite integers in solving equations. Use the penalty system in American football to illustrate how subtracting an integer is the same as adding its opposite.

12-2 Solving Subtraction Equations

 What you'll learn
You'll learn to solve subtraction equations by using models.

When am I ever going to use this?
Knowing how to solve equations can help you describe changes in temperature.

You probably have heard about the greenhouse effect on Earth. Did you know that it also affects other planets? Scientists have determined that, on Mars, the difference between the air temperature with the greenhouse effect and what it would be if it weren't for the greenhouse effect is 5° F. Suppose the temperature would be −3° F on Mars if it weren't for the greenhouse effect. What is the air temperature with the greenhouse effect?

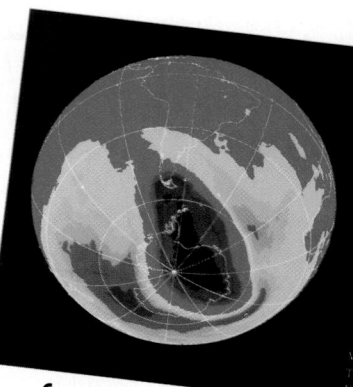
Computer model of Earth's ozone layer

Example 1
CONNECTION

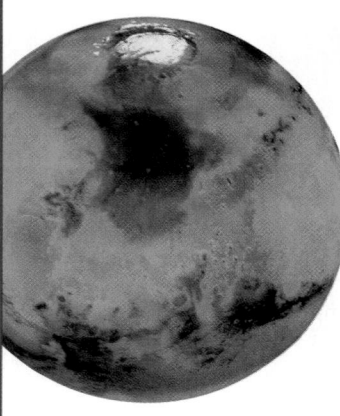

LOOK BACK
Refer to Lesson 11-4 to review subtracting integers.

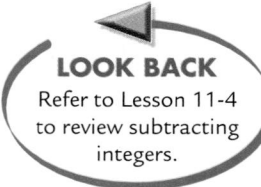

Earth Science Refer to the beginning of the lesson. What would the temperature be with the greenhouse effect?

Let g represent the temperature with the greenhouse effect. Translate the problem into an equation using the variable g. Then model the equation and solve.

temperature with the greenhouse effect	*minus*	*actual temperature*	*equals*	*difference*
g	$-$	-3	$=$	5

Rewrite as an addition equation. Remember that subtracting an integer is the same as adding its opposite.

$$g - (-3) = 5 \rightarrow g + 3 = 5$$

Use a cup to represent g. Add 3 positive counters on the left side of the mat to represent $+3$. Place 5 positive counters on the right side of the mat to represent $+5$.

To get the cup by itself, remove 3 positive counters from each side.

The solution is 2. The temperature with the greenhouse effect is 2° F.

Check the solution by replacing the value of the variable in the original equation.

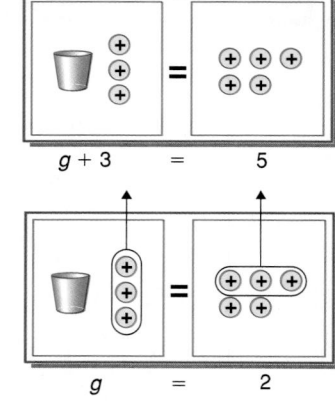

Check: $g - (-3) = 5$
$2 - (-3) \stackrel{?}{=} 5$ *Replace g with 2.*
$2 + 3 \stackrel{?}{=} 5$ *Rewrite as an addition equation.*
$5 = 5$ ✓

480 Chapter 12 Algebra: Exploring Equations

Multiple Learning Styles

Auditory/Musical Have groups of students investigate the speed of sound in various substances. Then have them set up equations to find the difference between speeds in similar substances such as sea water and fresh water.

2 TEACH

Transparency 12-2B contains a teaching aid for this lesson.

2 Use cups and counters to solve $h - (-4) = -3$.

Rewrite as an addition equation.

$$h - (-4) = -3 \rightarrow h + 4 = -3$$

Use a cup to represent h. Add 4 positive counters beside the cup on the left side of the mat to represent $+4$. Place 3 negative counters on the right side of the mat to represent -3.

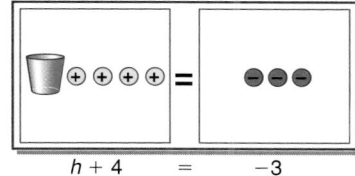

$$h + 4 \quad = \quad -3$$

To get the cup by itself, remove 4 positive counters from each side. Since there are no positive counters on the right side of the mat, add 4 negative counters to each side to make 4 zero pairs on the left side of the mat. Then remove the zero pairs.

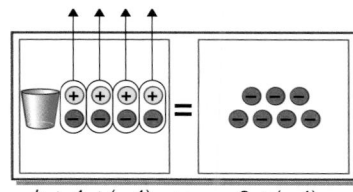

$$h + 4 + (-4) \quad = \quad -3 + (-4)$$

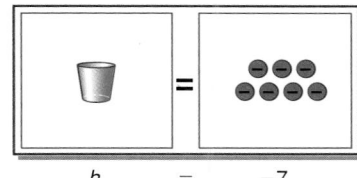

$$h \quad = \quad -7$$

Check: $h - (-4) = -3$

$$-7 - (-4) \overset{?}{=} -3 \quad \textit{Replace h with } -7.$$

$$-7 + 4 \overset{?}{=} -3 \quad \textit{Rewrite as an addition equation.}$$

$$-3 = -3 \quad \checkmark$$

The solution is -7.

3 Use cups and counters to solve $x - 6 = -2$.

Rewrite as an addition equation.

$$x - 6 = -2 \rightarrow x + (-6) = -2$$

Use a cup to represent x. Add 6 negative counters on the left side of the mat to represent -6. Place 2 negative counters on the right side of the mat to represent -2.

(continued on the next page)

Lesson 12-2 Solving Subtraction Equations **481**

Reading Mathematics Have students practice writing an equation from a word problem by using cue words such as difference and total change. Have students write word problems and challenge others to determine the equation needed to solve the problem.

Teaching Tip Before Example 1, have students identify the variable, the number to be subtracted from the variable, and the difference to be sure they are setting up the equation properly.

In-Class Examples

For Example 1
Best Tire Co.'s balance sheet records profits and losses in millions of dollars. Last year's loss was -1. This year's loss was $+2$. What was the change in millions of dollars?
+3 million dollars

For Example 2
Use cups and counters to solve $q - (-8) = 4$. **-4**

For Example 3
Use cups and counters to solve $c - 7 = -1$. **6**

Teaching Tip For students who have difficulty believing that subtracting is the same as adding the opposite, try this activity. To model $h - (-4)$, have students place a cup on one side of the mat. The expression says to subtract -4 from the cup, which is not possible. Add 4 zero pairs to the cup and then have students subtract -4. The result is $h + 4$.

Investigations for the Special Education Student

This blackline master booklet helps you plan for the needs of your special education students by providing long-term projects along with teacher notes. Investigation 13, *How Does Your Garden Grow?*, may be used with this chapter.

Check for Understanding

If students need additional practice or instruction after completing Exercises 1–10, one of these options may be helpful.
- Extra Practice, see p. 588
- Reteaching Activity
- *Study Guide Masters,* p. 85
- *Practice Masters,* p. 85

Additional Answers

1. Sample answer: The 3 negative counters and the cup represent $y - 3$ and the 4 positive counters represent 4.

3. Sample answer: If the equation involves addition, you subtract counters. If the equation involves subtraction, you add counters.

Study Guide Masters, p. 85

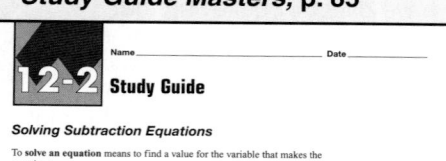

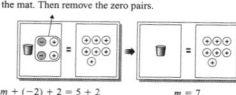

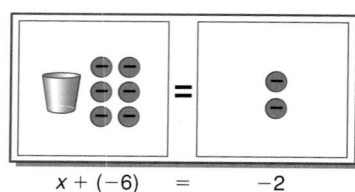

To get the cup by itself, remove 6 negative counters from each side. Since there are not enough negative counters on the right side of the mat, add 6 positive counters to each side to make 6 zero pairs on the left side of the mat. Then remove all zero pairs.

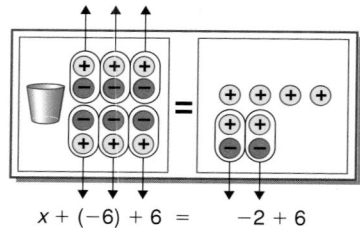

$$x + (-6) + 6 = -2 + 6$$

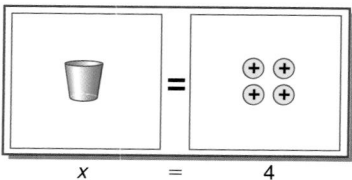

$$x = 4$$

Check: $x - 6 = -2$

$4 - 6 \stackrel{?}{=} -2$ *Replace x with 4.*

$4 + (-6) \stackrel{?}{=} -2$ *Rewrite as an addition equation.*

$-2 = -2$ ✓

The solution is 4.

CHECK FOR UNDERSTANDING

Communicating Mathematics

Read and study the lesson to answer each question.

1. *Explain* how the model represents $y - 3 = 4$. **See margin.**

2. *Show* how to model the equation $x - 2 = 6$. **See students' work.**

3. *Write* a few sentences explaining how you know whether to add or subtract counters from each side of the equation mat. **See margin.**

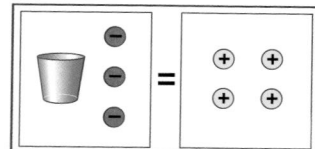

Guided Practice

Use cups and counters to solve each equation.

4. $x - 8 = 2$ **10**

5. $a - 4 = -6$ **-2**

6. $g - 4 = 11$ **15**

7. $y - 2 = 5$ **7**

8. $c - 2 = -7$ **-5**

9. $z - 7 = 9$ **16**

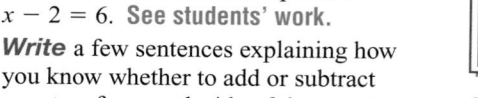

Reteaching the Lesson

Activity Have students create flashcards by writing a subtraction equation on one side of a card and the solution on the other side. Then have pairs of students quiz each other to see who can solve a set of equations first.

Error Analysis

Watch for students who add or subtract counters from only one side of an equation.

Prevent by reminding students that any operation performed on one side of an equation must be performed on the other side as well.

10. Sports After a loss of 5 yards, Ayani had a total of 16 yards. How many yards did he have before the 5-yard loss? **21 yards**

EXERCISES

Practice

Solve each equation. Use cups and counters if necessary.

11. $g - 4 = -6$ **−2**   **12.** $z - (-3) = 5$ **2**   **13.** $x - 1 = -3$ **−2**

14. $c - 5 = -5$ **0**   **15.** $b - 6 = -7$ **−1**   **16.** $r - 7 = -15$ **−8**

17. $z - 10 = -18$ **−8**   **18.** $x - (-2) = -1$ **−3** **19.** $t - 2 = 3$ **5**

20. $h - (-5) = -2$ **−7** **21.** $y - 3 = 4$ **7**   **22.** $x - 9 = 4$ **13**

23. $d - 5 = -2$ **3**   **24.** $s - 8 = -1$ **7**   **25.** $v - 12 = -10$ **2**

26. Find the value of x if $x - (-8) = -14$. **−22**

27. If $t - (-13) = 4$, what is the value of t? **−9**

Applications and Problem Solving

28. Earth Science Refer to the graph.

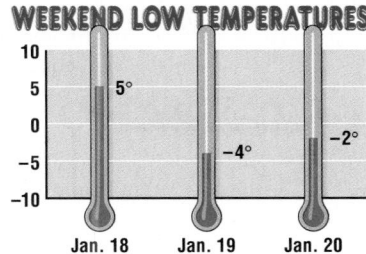

a. How did the temperature change between January 18 and January 19? **It dropped 9°.**

b. The low temperature on January 21 was 5 degrees higher than the low on January 20. What was the low on January 21? **3°**

29. Diving A diver begins to go back to the boat from 140 feet below sea level. A few minutes later, the diver is 35 feet below sea level.

a. Write an addition equation you would use to represent the diver's movement. $-140 + r = -35$

b. How many feet did the diver rise? **105 feet**

30. Critical Thinking Describe how you would solve $5 - x = -4$. **See margin.**

Mixed Review

31. **Test Practice** Sabra collected 6 silver dollars. Her friend Logan gave her some more, and then she had 15. To find out how many silver dollars she was given, Sabra wrote $s + 6 = 15$. What is the value of s? *(Lesson 12-1)* **B**

A 6   **B** 9   **C** 21   **D** 90

32. Measurement Find the area of a circle with a diameter of 6 meters to the nearest tenth. Use 3.14 for pi. *(Lesson 10-3)* **about 28.3 m²**

33. Geometry Use a protractor and draw an angle that measures 70°. *(Lesson 9-2)* **See students' work.**

34. Statistics Refer to the circle graph of students' favorite sports. How much more of the students' votes were for swimming than tennis? *(Lesson 2-4)* **0.02**

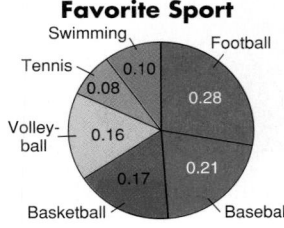

Lesson 12-2 Solving Subtraction Equations **483**

Extending the Lesson

Enrichment Masters, p. 85

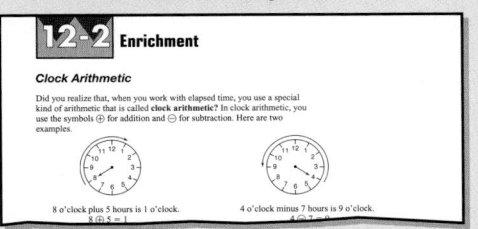

Activity Have groups write subtraction equations with variables on cards placed facedown in a pile. Give each student five marbles. Then have group members take turns selecting a card and solving the equation. Students turn in one marble for each correct answer. The winner is the student who gets rid of his or her marbles first.

Assignment Guide

Core: 11–29 odd, 30–34
Enriched: 12–26 even, 28–34

4 ASSESS

Closing Activity

Speaking Ask students what equation would be used to solve the following question. *How many old pennies did a collector begin with if she traded 36 and has 52 left?* $x - 36 = 52; 88$

Chapter 12, Quiz A (Lessons 12-1 and 12-2) is available in the *Assessment and Evaluation Masters,* p. 323.

Additional Answer

30. Sample answer: Use guess and check. $5 - 3 = 2$; $5 - 6 = -1$; $5 - 9 = -4$; $x = 9$

Practice Masters, p. 85

Name_____ Date_____

12-2 Practice

Solving Subtraction Equations

Solve each equation. Use cups and counters if necessary.

1. $h - (-2) = 6$ **4**   **2.** $v - 7 = -4$ **3**

3. $a - (-6) = -5$ **−11**   **4.** $r - (-3) = -8$ **−11**

5. $j - (-8) = 5$ **−3**   **6.** $x - 8 = -9$ **−1**

7. $c - 26 = 45$ **71**   **8.** $z - (-57) = -39$ **−96**

9. $n - 38 = -19$ **19**   **10.** $w - 23 = 77$ **100**

11. $f - (-26) = 41$ **15** **12.** $p - 47 = 22$ **69**

13. $g - 82 = -63$ **19**   **14.** $t - 14 = 87$ **101**

15. $q - 53 = 27$ **80**   **16.** $b - 48 = 14$ **62**

17. $k - 7 = -2$ **5**   **18.** $y - 47 = -8$ **39**

19. $t - 33 = -51$ **−18**   **20.** $a - 35 = 86$ **121**

21. $n - 84 = 16$ **100**   **22.** $k - 42 = 26$ **68**

23. $x - 33 = -52$ **−19**   **24.** $y - 63 = -19$ **44**

25. $d - 47 = 42$ **89**   **26.** $r - 47 = 84$ **131**

27. $b - 42 = 63$ **105**   **28.** $y - 18 = -47$ **−29**

29. $j - 92 = -20$ **72**   **30.** $s - 26 = -99$ **−73**

Solving Multiplication and Division Equations

Instructional Resources

- *Study Guide Masters*, p. 86
- *Practice Masters*, p. 86
- *Enrichment Masters*, p. 86
- Transparencies 12-3, A and B
- *Assessment and Evaluation Masters*, pp. 322, 323
- *School to Career Masters*, p. 12

CD-ROM Program
- Resource Lesson 12-3
- Interactive Lesson 12-3

Recommended Pacing	
Standard	Day 4 of 11
Honors	Day 4 of 11
Block	Day 2 of 6

1 FOCUS

5-Minute Check
(Lesson 12-2)

Solve each equation. Use cups and counters if necessary.
1. $a - 6 = -8$ **−2**
2. $x - 6 = 2$ **8**
3. $b - 3 = -7$ **−4**
4. $z - 2 = 5$ **7**
5. $y - (-2) = -4$ **−6**

The 5-Minute Check is also available on **Transparency 12-3A** for this lesson.

Motivating the Lesson

Hands-On Activity Bring in several grocery advertisements from the newspaper. Have students calculate cost per unit for several advertised items by using the equation $n \cdot u = t$, where n is the number of units, u is the unit cost, and t is the total cost. Discuss why smart consumers figure the price per unit for multi-item packages.

What you'll learn

You'll learn to solve equations involving multiplication and division using models.

When am I ever going to use this?

Knowing how to solve equations involving multiplication and division can help you find your rate of pay.

If you love your dog, don't let it eat chocolate. An ingredient in chocolate, theobromine, can be poisonous to dogs. Just 2 ounces of chocolate could harm a 10-pound pup. Zina determines that half of her chocolate bar is enough to harm a 10-pound pup. How much does her chocolate bar weigh? *This problem will be solved in Example 3.*

Cups and counters can be used to solve multiplication and division equations.

Example 1

Use cups and counters to solve $3s = 15$.

Use three cups to represent $3s$. Place them on the left side of the mat. Place 15 positive counters on the right side of the mat to represent 15.

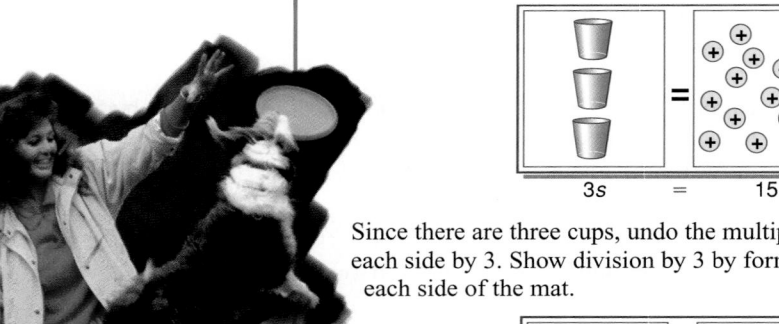

$$3s \quad = \quad 15$$

Since there are three cups, undo the multiplication by dividing each side by 3. Show division by 3 by forming 3 equal groups on each side of the mat.

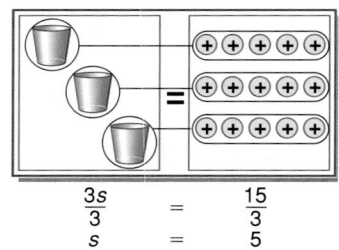

$$\frac{3s}{3} = \frac{15}{3}$$
$$s = 5$$

Check: $3s = 15$
$3(5) \stackrel{?}{=} 15$ *Replace s with 5.*
$15 = 15$ ✓

The solution is 5.

2 Use cups and counters to solve $2x = -4$.

Use two cups to represent $2x$. Place them on the left side of the mat. Place 4 negative counters on the right side of the mat to represent -4.

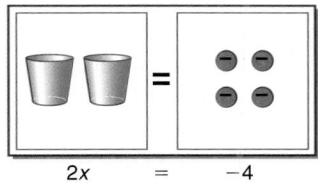

$2x \qquad = \qquad -4$

Undo the multiplication by dividing each side by 2. Show division by 2 by forming 2 equal groups on each side of the mat.

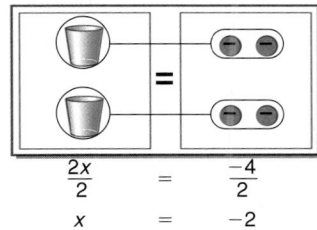

$$\frac{2x}{2} \qquad = \qquad \frac{-4}{2}$$

$$x \qquad = \qquad -2$$

Check: $\qquad 2x = -4$

$\qquad\qquad 2(-2) \stackrel{?}{=} -4 \qquad$ *Replace x with -2.*

$\qquad\qquad -4 = -4 \quad \checkmark$

The solution is -2.

CONNECTION

3 **Life Science** Refer to the beginning of the lesson. How much does Zina's chocolate bar weigh?

First, write an equation to describe the situation. Let b represent the weight of the chocolate bar.

one half of bar	*equals*	*2 ounces*
$\frac{1}{2}b$	$=$	2

Solve the equation $\frac{1}{2}b = 2$.

$$\frac{1}{2}b = 2$$

$$2\left(\frac{1}{2}b\right) = 2(2) \qquad \textit{Undo the division by multiplying each side by 2.}$$

$$b = 4$$

Check: $\frac{1}{2}b = 2$

$\qquad\qquad \frac{1}{2}(4) \stackrel{?}{=} 2 \qquad$ *Replace b with 4.*

$\qquad\qquad 2 = 2 \quad \checkmark$

The solution is 4. Zina's chocolate bar weighs 4 ounces.

Lesson 12-3 Solving Multiplication and Division Equations **485**

 Transparency 12-3B contains a teaching aid for this lesson.

Thinking Algebraically If students ask how to use counters to solve a problem like Example 3, use the following illustration.

$$\frac{1}{2}n = -8$$

Cut a cup in half to represent $\frac{1}{2}n$ and place on the left side of the mat. Place 8 negative markers on the right side of the mat. Double both the half cup and the counters showing $2 \times \frac{1}{2}n = 2(-8)$. Thus, $n = -16$.

Another method is to model $\frac{1}{2}n = -8$ as described above. Add $\frac{1}{2}$ cup to each side. Since $\frac{1}{2}$ cup $= -8$, replace $\frac{1}{2}$ cup on right side with -8 counters. The result is $n = -16$.

In-Class Examples
Use cups and counters to solve each equation.

For Example 1
Use cups and counters to solve $4x = 12$. **3**

For Example 2
Use cups and counters to solve $3m = -15$. **−5**

For Example 3
In a Walk-a-Thon, $\frac{5}{6}$ of the students who participated brought water bottles. If 36 students participated, how many did *not* bring water bottles? **6**

Teaching Tip After Example 3, lead students to see that multiplication and division equations are solved by using the inverse operation.

Check for Understanding

If students need additional practice or instruction after completing Exercises 1–10, one of these options may be helpful.
- Extra Practice, see p. 589
- Reteaching Activity
- *Study Guide Masters*, p. 86
- *Practice Masters*, p. 86

Assignment Guide

Core: 11–31 odd, 34–38
Enriched: 12–30 even, 31, 32, 34–38
All: Self Test, 1–10

Family Activity

Have students share with the class what they used for counters. On the chalkboard, make a list of equations that students used. Which operation was used most often to solve the equations?

Study Guide Masters, p. 86

12-3 Study Guide

Solving Multiplication and Division Equations

When the variable of an equation is multiplied by a number, divide each side of the equation by the number to get the variable by itself.

Example 1 Use cups and counters to solve $3n = -9$.

Use three cups to represent $3n$. Place 9 negative counters on the right side of the mat to represent -9.

Undo the multiplication by dividing each side by 3. Show division by 3 by forming 3 equal groups on each side of the mat.

$3n = -9$ $\frac{3n}{3} = \frac{-9}{3} \longrightarrow n = -3$

When the variable of an equation is divided by a number, multiply each side of the equation by that number to get the variable by itself.

Example 2 Use cups and counters to solve $\frac{1}{3}p = 2$.

Use a cup that is about $\frac{1}{3}$ full to represent $\frac{1}{3}n$. Place 2 positive counters on the right side of the mat to represent $+2$.

Undo the division by multiplying each side by 3. Place 3 sets of 2 positive counters on the right side of the mat.

$\frac{1}{3}p = 2$ $3\left(\frac{1}{3}p\right) = 3(2) \longrightarrow p = 6$

Solve each equation. Use cups and counters if necessary.

1. $5t = 15$ **3**
2. $2k = -14$ **−7**
3. $4p = -16$ **−4**
4. $3m = -15$ **−5**
5. $6n = -24$ **−4**
6. $\frac{1}{2}c = 2$ **4**
7. $\frac{1}{3}r = 4$ **12**
8. $\frac{1}{3}f = -3$ **−6**
9. $\frac{1}{4}y = -1$ **−4**

© Glencoe/McGraw-Hill   T86   Mathematics: Applications and Connections, Course 1

Communicating Mathematics

Read and study the lesson to answer each question.

1. **Tell** what equation is represented by the model. Then solve. $4x = 8$; $x = 2$

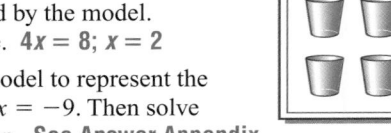

2. **Make** a model to represent the equation $3x = -9$. Then solve the equation. **See Answer Appendix.**

3. Sample answer: They're both correct. Dividing by 2 is the same as multiplying by one-half.

3. **You Decide** Zack says that to solve the equation $2k = 50$ he would divide each side by 2. Aubrey says she would multiply each side by $\frac{1}{2}$. Who is correct? Explain.

Guided Practice

Family Activity

Use items you can find in your home to explain to a family member how you solve an equation involving multiplication or division.

Complete each solution.

4. $3x = 9$ **3, 3; 3**
$\frac{3x}{\blacksquare} = \frac{9}{\blacksquare}$
$x = \blacksquare$

5. $8g = 16$ **8, 8, 2**
$\frac{8g}{\blacksquare} = \frac{16}{\blacksquare}$
$g = \blacksquare$

6. $\frac{1}{4}a = -7$ **4, 4; −28**
$\blacksquare\left(\frac{1}{4}a\right) = \blacksquare(-7)$
$a = \blacksquare$

Solve each equation. Use cups and counters if necessary.

7. $5e = 25$ **5**
8. $2c = 6$ **3**
9. $\frac{1}{2}n = -5$ **−10**

10. **Money Matters** Margie's average weekly earnings in 1997 were three times higher than in 1980. She earned $624 per week in 1997. How much did she earn per week in 1980? **$208**

Practice

Solve each equation. Use cups and counters if necessary.

11. $9a = 18$ **2**
12. $3x = 3$ **1**
13. $2c = -6$ **−3**
14. $4k = -20$ **−5**
15. $4x = -36$ **−9**
16. $5e = -15$ **−3**
17. $16x = 4$ **$\frac{1}{4}$**
18. $5h = 5$ **1**
19. $4m = 20$ **5**
20. $2x = -16$ **−8**
21. $\frac{1}{8}y = 3$ **24**
22. $\frac{1}{3}x = 12$ **36**
23. $\frac{n}{2} = 19$ **38**
24. $53 = \frac{1}{2}y$ **106**
25. $3.5s = 7$ **2**
26. $24.8 = 1.24a$ **20**
27. $\frac{1}{4}w = -9$ **−36**
28. $-12 = \frac{3}{8}g$ **−32**

29. Solve the equation $3.1t = 25.42$. **8.2**

30. What is the value of n if $35 = 3\frac{1}{2}n$? **10**

Applications and Problem Solving

31. **Money Matters** In 1995, Michael Jordan's income from endorsements was 10 times his salary for playing basketball. If his income from endorsements was about $40 million, about how much was his salary for playing basketball? **$4 million**

32. **Money Matters** The track team has 19 members. They stopped at a restaurant on the way home from a meet. If the total bill was $90.25, what was the average cost of each team member's meal? **$4.75**

Reteaching the Lesson

Activity Have students use an even number of positive or negative counters. They should then substitute the number of counters for $\blacksquare$ in $2n = \blacksquare$. Ask: *Can you solve the equation by using the counters? How?* Continue with $\frac{1}{2}n = \blacksquare$.

Error Analysis
Watch for students who omit negative signs in their solutions.
Prevent by having them check their solutions in the equations.

33. Working on the 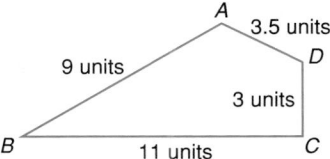 **Project** Refer to the spreadsheet you made for Exercise 32 on page 478. Label column E *Change per month*. Write a formula for cells E2, E3, E4, E5, E6, and E7 to find the change in value per month for each card. **See Answer Appendix.**

34. Critical Thinking Without solving, tell which equation, $\frac{1}{4}x = 13$ or $\frac{1}{8}x = 13$, has the greater solution. Explain. **See Answer Appendix.**

Mixed Review

35. Algebra Solve the equation $y - 11 = -8$. *(Lesson 12-2)* **3**

36. If the figure at the right is a scale drawing with a scale 1 unit = 100 meters, what is the actual length of $\overline{AB}$? *(Lesson 8-3)* **900 meters**

37. Test Practice Melika was making a quilt for her room. She had $9\frac{1}{2}$ yards of material. It took 7 yards for the quilt. How much material was not used for the quilt? *(Lesson 6-6)* **C**

A $16\frac{1}{2}$ yd

B $9\frac{1}{2}$ yd

C $2\frac{1}{2}$ yd

D $1\frac{1}{2}$ yd

E Not Here

Cast of *How to Make an American Quilt*

38. List all the common factors of 20 and 50. *(Lesson 5-3)* **1, 2, 5, 10**

Mid-Chapter Self Test

CHAPTER 12

Solve each equation. Use cups and counters if necessary. *(Lessons 12-1 and 12-2)*

1. $y + 6 = 12$ **6**   **2.** $x + 6 = -3$ **-9**   **3.** $m + (-7) = 14$ **21**   **4.** $t - 8 = 14$ **22**

5. $k - (-10) = -5$ **-15**   **6.** $3g = 18$ **6**   **7.** $2x = -4$ **-2**   **8.** $\frac{1}{3}y = -9$ **-27**

Solve.

9. Transportation Marta's car averages 24 miles per gallon. Her odometer shows that she has driven 72 miles. How many gallons of gasoline has her car used? *(Lesson 12-2)* **3 gallons**

10. Money Matters Lorenzo spends one-third of his monthly allowance on snacks. If he spends $4 on snacks every month, how much money does he get for his allowance? *(Lesson 12-3)* **$12**

Lesson 12-3 Solving Multiplication and Division Equations **487**

Extending the Lesson

Enrichment Masters, p. 86

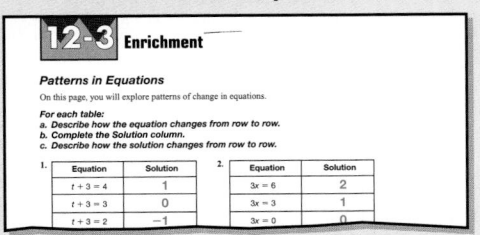

Activity Have students work in pairs to create word problems that can be solved by using multiplication or division equations.

CHAPTER Project

Exercise 33 asks students to advance to the next stage of work on the Chapter Project. You may wish to have students work in pairs to check the formula for column E.

4 ASSESS

Closing Activity

Writing Ask pairs of students to write an equation to solve the following problem. *A $5 fee is charged against a savings account each month that the balance drops below $1,000. If a yearly statement shows charges of $25, how many times was a fee charged during the year?* **5x = 25**

Chapter 12, Quiz B (Lesson 12-3) is available in the *Assessment and Evaluation Masters*, p. 323.

Mid-Chapter Test (Lessons 12-1 through 12-3) is available in the *Assessment and Evaluation Masters*, p. 322.

Mid-Chapter Self Test

The Mid-Chapter Self Test reviews the concepts in Lessons 12-1 through 12-3. Lesson references are given so students can review concepts not yet mastered.

Practice Masters, p. 86

12-3 Practice

Solving Multiplication and Division Equations

Solve each equation. Use cups and counters if necessary.

1. $9k = 54$ $k = 6$	2. $7r = -35$ $r = -5$	3. $\frac{1}{2}y = 6$ $y = 12$
4. $\frac{1}{3}g = -12$ $g = -36$	5. $4a = -28$ $a = -7$	6. $\frac{1}{8}m = -6$ $m = -48$
7. $\frac{1}{5}w = 2$ $w = 10$	8. $6s = 42$ $s = 7$	9. $\frac{1}{4}h = -5$ $h = -20$
10. $\frac{1}{9}x = -8$ $x = -72$	11. $3p = 27$ $p = 9$	12. $\frac{1}{7}t = 9$ $t = 63$
13. $5d = -30$ $d = -6$	14. $\frac{1}{6}j = -12$ $j = -72$	15. $8n = -64$ $n = -8$
16. $2c = 28$ $c = 14$	17. $\frac{1}{5}k = -9$ $k = -45$	18. $7f = -91$ $f = -13$
19. $\frac{1}{3}z = 45$ $z = 135$	20. $4q = -48$ $q = -12$	21. $\frac{1}{9}b = 2$ $b = 18$
22. $\frac{1}{8}e = -11$ $e = -88$	23. $6u = 3$ $u = \frac{1}{2}$	24. $5i = 50$ $i = 10$
25. $\frac{1}{7}y = -7$ $y = -49$	26. $3a = -48$ $a = -16$	27. $\frac{1}{2}r = 20$ $r = 40$
28. $\frac{1}{4}s = -8$ $s = -32$	29. $9p = -108$ $p = -12$	30. $\frac{1}{6}x = 6$ $x = 36$

© Glencoe/McGraw-Hill 86 *Mathematics: Applications and Connections, Course 1*

- *Study Guide Masters*, p. 87
- *Practice Masters*, p. 87
- *Enrichment Masters*, p. 87
- Transparencies 12-4, A and B
- *Diversity Masters*, p. 12
- CD-ROM Program
  - Resource Lesson 12-4
  - Interactive Lesson 12-4

Recommended Pacing	
Standard	Day 5 of 11
Honors	Day 5 of 11
Block	Day 3 of 6

1 FOCUS

5-Minute Check
(Lesson 12-3)

Solve each equation. Use cups and counters if necessary.

1. $4x = -24$ **−6**
2. $\frac{1}{2}y = -7$ **−14**
3. $\frac{1}{5}k = 30$ **150**
4. $3n = 36$ **12**
5. Marta's car averages 24 miles per gallon of gas. If she drove 72 miles on a trip, how many gallons of gas did her car use? **3**

The 5-Minute Check is also available on **Transparency 12-4A** for this lesson.

Motivating the Lesson

Communication Have students read the opening paragraph. Ask students how many operations would be involved in finding the number of games Lupita bowled. Ask for examples of equations that involve two operations.

12-4 Solving Two-Step Equations

What you'll learn

You'll learn to solve two-step equations using models.

When am I ever going to use this?

Knowing how to solve two-step equations can help you determine total costs when extra charges are added.

Lupita went bowling at Great Lanes Sport Center. Shoe rental was $1, and games were $2 each. If she spent $9 on shoe rental and games, how many games did she bowl?

Examples
APPLICATION

LOOK BACK
Refer to Lesson 11-6A to review the work backward strategy.

Sports Refer to the beginning of the lesson. How many games did Lupita bowl?

Let g equal the number of games bowled. Translate the problem into an equation using the variable g. Then model the equation and solve.

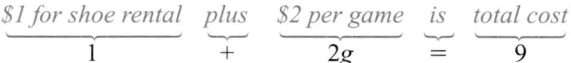

$1 for shoe rental	plus	$2 per game	is	total cost
1	+	2g	=	9

The equation is $1 + 2g = 9$. This is a two-step equation because it involves two different operations, addition and multiplication. To solve this equation, you need to work backward using the reverse of the order of operations.

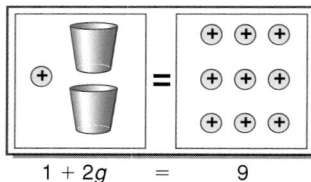

$$1 + 2g = 9$$

To get the cups by themselves, remove 1 positive counter from each side.

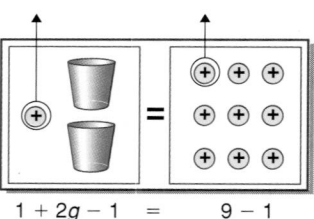

$$1 + 2g - 1 = 9 - 1$$

Multiple Learning Styles

Logical Have students use the work backward strategy to solve equations. For example, use 9 counters to represent $9. Take $1 out for shoe rental; then see how many $2 sets can be formed to find the number of games.

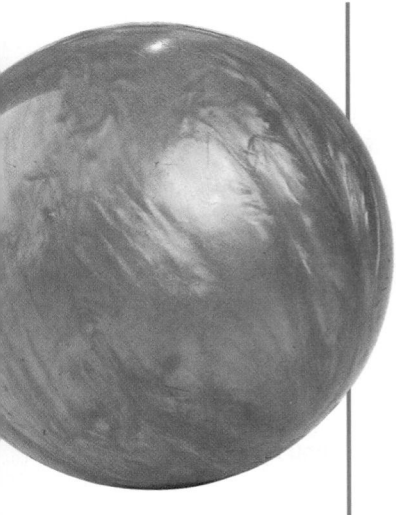

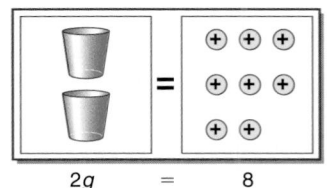

$$2g \quad = \quad 8$$

Since there are 2 cups, undo the multiplication by dividing each side by 2. Form 2 equal groups on each side of the mat.

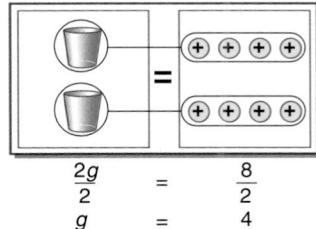

$$\frac{2g}{2} = \frac{8}{2}$$
$$g = 4$$

The solution is 4. Lupita bowled 4 games.

Check by replacing the value of the variable in the original equation.

Check: $1 + 2g = 9$
$1 + 2(4) \stackrel{?}{=} 9$ *Replace g with 4.*
$9 = 9$ ✓

 Use cups and counters to solve $2x + 6 = -4$.

Place two cups and 6 positive counters on the left side of the mat to represent $2x + 6$. Place 4 negative counters on the right side of the mat to represent -4.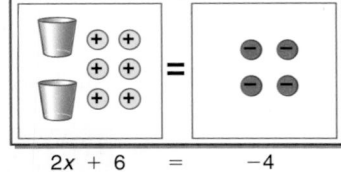

$$2x + 6 \quad = \quad -4$$

To get the cups by themselves, you need to remove 6 positive counters from each side. Since there are no positive counters on the right side of the mat, add 6 negative counters to each side to make 6 zero pairs on the left side of the mat. Then remove the zero pairs.

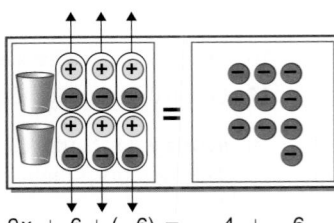

$$2x + 6 + (-6) = \quad -4 + \quad -6$$

(continued on the next page)

Lesson 12-4 Solving Two-Step Equations **489**

 Transparency 12-4B contains a teaching aid for this lesson.

In-Class Examples

For Example 1
A car rental agency charges $10 per day and $2 per mile to rent a car. How many miles did Kohana drive in one day if the car rental bill was $210?
100 miles

For Example 2
Use cups and counters to solve $3x + 7 = -11$. **−6**

Teaching Tip Before Example 2, review the rules for dividing positive and negative integers.

Using Discussion After demonstrating Example 2, ask students why six negative counters were added. Then ask what counters they would add to solve the equation $2x - 3 = -9$.
Add 3 positive counters to each side.

Classroom Vignette

"I find that my students understand solving equations better if we *always* add the opposite kind of counters to make zero pairs on the side with the cup instead of remembering whether they have to add something or take something away."

Cindy J. Boyd
Abilene High School
Abilene, TX

Lesson 12-4   489

Check for Understanding

If students need additional practice or instruction after completing Exercises 1–6, one of these options may be helpful.

- Extra Practice, see p. 589
- Reteaching Activity
- *Study Guide Masters*, p. 87
- *Practice Masters*, p. 87

Assignment Guide
Core: 7–19 odd, 20–23
Enriched: 8–16 even, 18–23

Additional Answers

4. add 3 to each side; 7
5. add 4 to each side; 20

Study Guide Masters, p. 87

Name_____ Date_____

12-4 Study Guide

Solving Two-Step Equations

To solve a two-step equation, undo the addition or subtraction. Then undo the multiplication or division.

Example Use cups and counters to solve $3b - 2 = -5$.

Rewrite as an addition equation.

$$3b - 2 = -5 \longrightarrow 3b + (-2) = -5$$

Place 3 cups and 2 negative counters on the left side of the mat to represent $3b - 2$. Place 5 negative counters on the right side of the mat to represent -5.

$3b - 2 = -5$

To get the cups by themselves, you need to remove 2 negative counters from each side.

$3b + (-2) - (-2) = -5 - (-2)$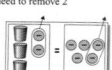

Now the equation is $3b = -3$. Undo the multiplication by dividing each side by 3. Show division by forming 3 equal groups on each side of the mat.

$\frac{3b}{3} = \frac{-3}{3} \longrightarrow b = -1$

Solve each equation.

1. $5x + 3 = 23$ **4**
2. $3a - 14 = 4$ **6**
3. $3y + 5 = -19$ **-8**
4. $5c + 6 = -29$ **-7**
5. $42 = 18 - 4v$ **-6**
6. $8 - 5w = -37$ **9**

7. Five less than 4 times a number is nineteen. What is the number? **6**

© Glencoe/McGraw-Hill   T87   *Mathematics: Applications and Connections, Course 1*

490 Chapter 12

Now the equation is $2x = -10$. Undo the multiplication by dividing each side by 2. Show division by forming 2 equal groups on each side of the mat.

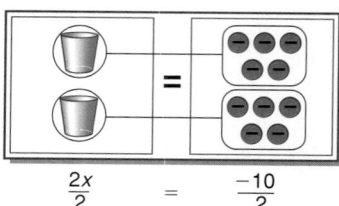

$$\frac{2x}{2} = \frac{-10}{2}$$
$$x = -5$$

Check:
$$2x + 6 = -4$$
$$2(-5) + 6 \overset{?}{=} -4 \quad \textit{Replace x with } -5.$$
$$-10 + 6 \overset{?}{=} -4$$
$$-4 = -4 \quad \checkmark \text{ The solution is } -5.$$

Communicating Mathematics

Read and study the lesson to answer each question.

1. *Write* an equation for the model. Then solve. $3x + 2 = -1; x = -1$
   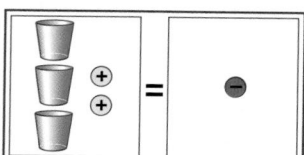

2. *Make* a model to represent the equation $2x + 3 = -9$. Then solve the equation. **See students' work.; -6**

Guided Practice

Name the first step in solving each equation. Then solve. 4–5. See margin.

3. subtract 2 from each side; 4

3. $3x + 2 = 14$ 4. $2a - 3 = 11$ 5. $\frac{1}{2}y - 4 = 6$

6. Twice a number, n, plus 7 is -21. What is the value of n? **-14**

Practice

Solve each equation.

7. $2t + 5 = 13$ **4**   8. $\frac{1}{2}m + 3 = -5$ **-16**   9. $3h - 4 = 5$ **3**

10. $-11 = 4x - 3$ **-2**   11. $2x - (-34) = 16$ **-9**   12. $5r + 1 = 1$ **0**

13. $\frac{1}{4}t - 5 = -13$ **-32**   14. $-1 - 4y = 7$ **-2**   15. $-3y + 15 = 75$ **-20**

16. Ten less than twice a number is sixteen. What is the number? **13**

17. One half a number less five is eleven. Find the number. **32**

Applications and Problem Solving

18. *Money Matters* Claudio ordered three novelty T-shirts from a catalog. The total price including shipping charges was $50. If the total shipping cost was $5, how much did each T-shirt cost? **$15**

19. *Geometry* The perimeter of a rectangle is 40 inches. Find its length if its width is 4 inches. **16 inches**

Reteaching the Lesson

Activity Have students set up the model shown at the right. What equation does the model represent? What counters would you add to both sides to solve the equation? Solve the equation.

$3x + 2 = -1$; 2 negative counters; -1

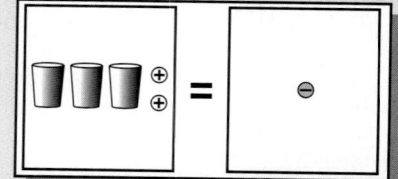

20. Critical Thinking Use what you know about solving two-step equations to solve the equation $\frac{1}{4}(k - 8) = -3$. **−4**

Mixed Review

21. Algebra Solve for q in the equation $-7q = -56$. *(Lesson 12-3)* **8**

22. **Test Practice** Mr. Vega has 232 seedlings to be planted in flowerpots with 8 plants in each. How many flowerpots will he need to plant the seedlings? *(Lesson 11-6)* **B**

    A 32   **B** 29   **C** 27   **D** 19   **E** Not Here

23. Patterns Find the next two numbers in the sequence 34, 36.5, 39, 41.5,... . *(Lesson 7-8)* **44, 46.5**

Closing Activity

Writing Have pairs of students write a general rule for solving two-step equations. Suggest that they include an example.

Let the Games Begin

Four In A Row

Get Ready This game is for two to ten players.

 12 index cards   scissors   ☐ poster board   beans

Get Set
- Cut three index cards in half. Write an equation using the variable *a* on each of the 6 cards. Cut another three index cards in half and write an equation using the variable *b*. Continue until you have six cards with equations for each of the variables, *c*, and *d*. No two equations using the same variable should have the same solution. Make a list of the solutions to the equations for each variable.

- Cut ten 6-inch by 5-inch playing cards from the poster board.

- On each playing card, copy the grid shown. Complete each column of the cards with solutions to the equations containing the indicated variable so that no two cards are identical.

a	b	c	d
	FREE		
		FREE	

Go
- Mix the equation cards and place the deck facedown.

- Each player should choose at least one playing card and cover the free spaces with beans.

- After an equation card is turned up, all players should solve the equation.

- When a solution is found on a card, the player should cover it with a bean.

- The winner is the first player to cover four spaces in a row either vertically, horizontally, or diagonally.

interNET CONNECTION Visit www.glencoe.com/sec/math/mac/mathnet for more games.

Lesson 12-4 Solving Two-Step Equations   **491**

■ Extending the Lesson ■

Enrichment Masters, p. 87

Let the Games Begin

Make variations of Four In A Row by including decks with one-step equations, two-step equations, integer solutions, or non-integer solutions.

*Additional resources for this game can be found on page 46 of the **Classroom Games**.*

Practice Masters, p. 87

Objective Students solve problems by using an equation.

Recommended Pacing

Standard	Day 6 of 11
Honors	Day 6 of 11
Block	Day 3 of 6

1 FOCUS

Getting Started Have students act out the situation presented at the beginning of the lesson. Then have them work in pairs to answer Exercises 1–3. Have the class discuss the results. Then have them solve Exercise 4.

2 TEACH

Teaching Tip Make sure students understand that a variable always represents an unknown number.

In-Class Example

When school pictures were taken, 4 times as many students chose the laser background than the standard background. If 128 students chose the standard background, how many chose the laser background? **512 students**

Teaching Tip Make sure students understand that a variable always represents an unknown number.

Additional Answers

1. **Sample answer: No; Ed's information may be more useful because it contains all the known amounts.**

2. **Sample answer: Maybe they thought CDs were $3 each.**

3. **Sample answer: Subtract 8 from the solution.**

THINKING LAB

PROBLEM SOLVING

12-4B Use an Equation

A Follow-Up of Lesson 12-4

Carol and Ed have just returned from a yard sale where Carol bought a personal CD player and some CDs. Carol bought 8 CDs but thinks she was charged for more. Let's listen in!

Those were good bargains, but I think I was charged for too many CDs.

How many CDs did you get and how much did you pay?

I paid a total of $39 for a $15 CD player and 8 CDs. I can find out if that is correct by adding $15 and $16.

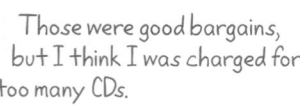
Carol

Ed

Or we could let n represent the number of CDs and translate the problem into an equation. The total is $2 times the number of CDs plus $15. This can be written as $39 = 2n + 15$.

But that's a two-step equation!

SALE	
CD Player	$15
All CDs	$2

It's not difficult. First, subtract 15 from each side. Now the equation is $2n = 24$. Dividing each side by 2 shows that you paid for 12 CDs. We better head back and get this straightened out.

THINK ABOUT IT

1–3. See margin.

1. **Compare and contrast** Carol's method of solving the problem to Ed's method. Do both methods contain the same information? If not, which information is more useful? Explain.

2. **Think** of a way to explain why Carol was charged $24 for the CDs.

3. **Tell** how the solution to Ed's equation could be used to find how many CDs were counted twice.

4. **Apply** the **use an equation** strategy to solve the following problem.

   At the same yard sale, Ed paid $27 for 12 CDs in a carrying case. How much did he pay for the case? **$3**

492 Chapter 12 Algebra: Exploring Equations

■ Reteaching the Lesson ■

Activity Have students play "M&M's Equations." Ask them to choose two M&M colors and assign a point value to each one. Have each player make up problems such as the following. *I have a red one and enough green ones to make 8 points. How many green ones do I have?*

ON YOUR OWN

5. The third step of the 4-step plan for problem solving asks you to *solve*. *Tell* what other problem-solving strategy you could use to solve Carol's problem. **Sample answer: You could use the guess and check strategy.**

6. *Write a Problem* that can be solved by using an equation. **See students' work.**

7. *Reflect Back* Explain how solving a two-step equation is similar to using the work backward strategy. **See margin.**

MIXED PROBLEM SOLVING

STRATEGIES

Look for a pattern.
Solve a simpler problem.
Act it out.
Guess and check.
Draw a diagram.
Make a chart.
Work backward.

Solve. Use any strategy.

8. *Money Matters* Jillisa bought a clock radio for $9 less than the regular price. If she paid $32, what was the regular price? **$41**

9. *School* The sixth grade class is planning a field trip. There are 589 students in the sixth grade. Each bus holds 48 people. About how many buses will they need? **about 12 buses**

10. *Design* A designer wants to arrange 12 glass bricks into a rectangular shape with the least perimeter possible. How many blocks will be in each row? **3 or 4 blocks**

11. *Money Matters* Scott paid $2.50 in sales tax on a Dallas Cowboys sweatshirt. The total cost was $42.49. What was the price of the sweatshirt before taxes? **$39.99**

12. *Geometry* A kite has two pairs of congruent sides. If two sides are 56 centimeters and 34 centimeters, what is the perimeter of the kite? **180 cm**

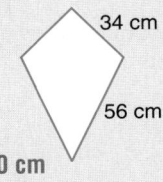

34 cm

56 cm

13. *Fashion* A catalog company offers 3 styles of ski sweaters each in 8 different colors. How many combinations of style and color are possible? **24 combinations**

14. *Language Arts* Science fiction books were the most popular items at the Book Fair. On Monday, 86 science fiction books were sold. This is 8 more than twice the amount that were sold on Thursday. How many science fiction books were sold on Thursday? **39 books**

15. *Travel* Kathy lives in Rockwood and works in Somerset. There is no direct route from Rockwood to Somerset, so Kathy goes through either Boulder Creek or Castleton. There is one road between Boulder Creek and Castleton. How many different ways can Kathy drive to work? **4 ways**

16. The Swann family is going to a play. Ticket prices are shown below. Mr. Swann needs 2 adult tickets, 3 student tickets, and 1 child's ticket. Which number sentence could be used to find T, the cost in dollars of the tickets? **A**

Ticket Prices	
Adult	$7.25
Student	$3.50
Child under 4	$1.75

A $T = (2 \times 7.25) + (3 \times 3.50) + 1.75$

B $T = 6 \times (7.25 + 3.50 + 1.75)$

C $T = (2 + 3) \times (7.25 + 3.50 + 1.75)$

D $T = 7.25 + 3.50 + 1.75 + 6$

E $T = (2 + 7.25) \times (3 + 3.50) \times (1.75)$

Lesson 12-4B **THINKING LAB** **493**

Check for Understanding

Use the results from Exercise 4 to determine whether students comprehend how to solve problems by using an equation.

Extra Practice If students need additional practice in problem solving, extra practice is available on the following pages.
- Use an Equation, see p. 589
- Mixed Problem Solving, see pp. 593–594

Assignment Guide

All: 5–16

4 ASSESS

Closing Activity

Writing Have students work in pairs using department store catalogs to create 2-step word problems that can be solved using equations.

Additional Answer

7. Sample answer: In solving a two-step equation, you undo the operations in reverse order of the order of operations.

■ **Extending the Lesson** ■

Activity Have students solve the following problem. *The Saturday matinee at Cinema East sold 120 student tickets at $4.50 each. Some students also bought popcorn. If the theater took in a total of $600 from students that afternoon, what was the total amount students paid for popcorn?* **$60**

GET READY

Objective Students find the input and output for a given function machine.

Optional Resources
Hands-On Lab Masters
• worksheet, p. 66
Manipulative Kit
• scissors

MANAGEMENT TIPS

Recommended Time
45 minutes

Getting Started Write the pattern 2, 5, 8, 11, 14, . . . on the chalkboard and have students name the next two numbers. Ask them to write a rule for the pattern. *n* + 3 Before students begin the activity, offer the following tips.
• Paper should be cut in two $4\frac{1}{4}$ in. $\times$ 11 in. strips.
• Start numbering the strips about $\frac{1}{3}$ of the way down from the top.

The **Activity** demonstrates using a model to evaluate a function. The activity uses a simple addition function. Students could extend the activity to make function machines for more complex functions and challenge others to make the answer strips.

COOPERATIVE LEARNING

12-5A Function Machines

A Preview of Lesson 12-5

✂ scissors

📼 tape

In this lab, you will use what you have learned about solving equations to help you work with function machines. A *function machine* takes a number called the *input*, performs one or more operations on it, and produces a result called the *output*.

TRY THIS

Work in groups of three.

Make a function machine for the rule $\boxed{+\,5}$.

• Take a sheet of paper and cut it in half lengthwise.

• On one of the halves, cut four slits — two on each side of the paper. Make each slit at least one inch wide.

• From the other half of the paper, cut two narrow strips lengthwise. These strips should be able to pass through the slits you cut in the other half.

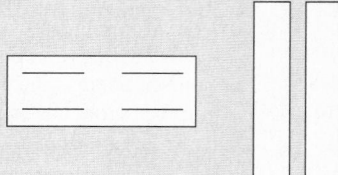

• On one of the narrow strips, write five consecutive numbers starting with 1. On the other narrow strip, write five consecutive numbers starting with 6. The numbers on both strips should align.

1	6
2	7
3	8
4	9
5	10

494 Chapter 12 Algebra: Exploring Equations

- Place the strips into the slits so that the numbers can be seen. Show 1 and 6. Once they appear, tape the ends of the strips together. When you pull the strips, they should move together. Mark the left-hand strip *input,* and the right-hand strip *output.* Write the function rule +5 between the input and output.

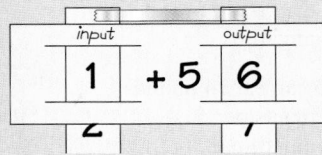

- Make a function table showing the input and the output.

input	output
1	
2	
3	
4	
5	

ON YOUR OWN

1. What is the output when the input is 3? **8**
2. What is the output when the input is 5? **10**
3. Suppose you added more input numbers to the left strip. What would the output be if the input was 8? **13**
4. What is the output if the input is *x*? **x + 5**
5. Make a function machine for the rule ×3.
   a. What is the output when the input is 4? **12**
   b. What is the output when the input is 6? **18**
   c. Suppose you added more input numbers to the left strip. What would the output be if the input was 7? **21**
   d. What was the input if the output was 33? **11**
6. Make up your own function machine. Write pairs of inputs and outputs and have the other members of your group determine the rule. **See students' work.**
7. Temperature is usually measured in Celsius (°C) or Fahrenheit (°F). The formula for changing from Celsius to Fahrenheit is $F = \frac{9}{5} \times C + 32$. This formula is like a function. The input is the Celsius temperature, and the output is the Fahrenheit temperature. Find the output for each input.
   a. 60°C **140°F**   b. 15°C **59°F**   c. 0°C **32°F**   d. −10°C **14°F**

8. *Look Ahead* Use the function machine to find the set of outputs that correspond to the set of inputs 1, 3, 5, and 7. **1, 7, 13, 19**

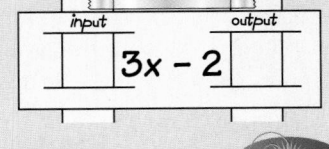

Teaching Tip In Exercise 7, you may suggest that students estimate the Fahrenheit temperatures first by rounding $\frac{9}{5}$ to 2 and 32 to 30. Students can use their estimates to check the reasonableness of their answers.

ASSESS

Have students complete Exercises 1–7. Watch for students who confuse input and output or who don't apply the function rule correctly. Make sure students move the paper strips one number at a time.

Use Exercise 8 to determine whether students understand how to use function machines.

Hands-On Lab Masters, p. 66

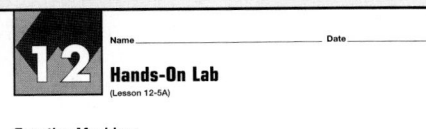

- *Study Guide Masters*, p. 88
- *Practice Masters*, p. 88
- *Enrichment Masters*, p. 88
- Transparencies 12-5, A and B
- *Assessment and Evaluation Masters*, p. 324
- *Technology Masters*, p. 23

 CD-ROM Program
- Resource Lesson 12-5
- Extended Activity 12-5

Recommended Pacing	
Standard	Days 7 & 8 of 11
Honors	Days 7 & 8 of 11
Block	Day 4 of 6

1 FOCUS

5-Minute Check
(Lesson 12-4)

Solve each equation.
1. $3t + 5 = 17$ **4**
2. $-10 = 4x - 2$ **−2**
3. $\frac{1}{4}m + 3 = 8$ **20**
4. $2k + 5 = 5$ **0**
5. Five less than twice a number is seven. Find the number. **6**

The 5-Minute Check is also available on **Transparency 12-5A** for this lesson.

Motivating the Lesson

Hands-On Activity Have pairs of students use centimeter cubes to make a design such as a capital letter *H*. Ask them how many cubes would be needed to make 2 models of the same design, 5 models of the same design, and *n* models of the same design.

12-5 Functions

What you'll learn
You'll learn to complete function tables.

When am I ever going to use this?
Knowing how to complete function tables can help you determine the amount of profit a business will make for selling any number of items.

Word Wise
function
function table

Examples
CONNECTION

Did you know? A newborn joey is the size of a paper clip and weighs about 0.03 of an ounce.

A baby kangaroo is known as a joey. A group of kangaroos is called a mob. The amount of plants that a mob of kangaroos eats depends on how many are in the mob. In other words, the amount of plants eaten is a **function** of the number of kangaroos.

A grown kangaroo can eat 14 pounds of grass and other plants per day. About how many pounds a day would be eaten by a mob of 3 grown kangaroos? 4 kangaroos? 5 kangaroos? This problem can be organized in a **function table**. *This problem will be solved in Example 1.*

1 **Life Science** Refer to the beginning of the lesson. Determine the amount eaten by a mob of 3, 4, and 5 kangaroos.

To find the amount eaten by a mob of kangaroos in a day (output), you need to multiply the number in the mob (input) by 14.

Input	Function Rule	Output
Number in the Mob (*n*)	14*n*	Amount Eaten in a Day (lb)
3	14(3)	42
4	14(4)	56
5	14(5)	70

A mob of 3 eat 42 pounds of grass and plants, 4 eat 56 pounds, and 5 eat 70 pounds.

2 Find the rule for the function table.

Study the relationship between each input and output.

input (*n*)	output (■)
−2	2
0	4
1	5
3	7

input		output
−2	$+4 \rightarrow$	2
0	$+4 \rightarrow$	4
1	$+4 \rightarrow$	5
3	$+4 \rightarrow$	7

The output is 4 more than the input. So, the function rule is $n + 4$.

Life Science Insect-eating bats control insects without chemicals. The brown bat often eats 600 mosquitoes an hour. Make a function table showing the number of mosquitoes eaten in 2, 4, and 6 hours.

You can use the function rule $600h$, where h is the number of hours.

Replace h in the rule $600h$ with the number of hours.

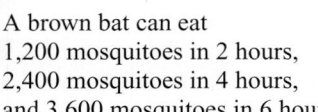

Replace h with 2.   *Replace h with 4.*   *Replace h with 6.*

$600h = 600 \cdot 2$ $600h = 600 \cdot 4$ $600h = 600 \cdot 6$
      $= 1{,}200$ $= 2{,}400$ $= 3{,}600$

A brown bat can eat 1,200 mosquitoes in 2 hours, 2,400 mosquitoes in 4 hours, and 3,600 mosquitoes in 6 hours.

input (h)	output (600h)
2	1,200
4	2,400
6	3,600

CHECK FOR UNDERSTANDING

Communicating Mathematics

Read and study the lesson to answer each question.

1. *Make* a function table for the function rule $y - 2$. Use inputs of -4, -1, 0, and 3. **See margin.**

2. *Write* the function rule for the table. $n \div 3$

input (n)	output (■)
9	3
0	0
-3	-1
-6	-2

3. Sample answer: Juanita is correct. If you use 4, 5, and 6 for inputs, the outputs are 1, 2, and 3. Each output is 3 less than the input.

3. *You Decide* The output of a function table is 3 less than each input. Juanita says that the function rule is $x - 3$. Ruby says the rule is $3 - x$. Who is correct? Explain.

Guided Practice

Copy and complete each function table.

4.

input (n)	output (n+3)
-1	■ 2
0	■ 3
3	■ 6

5.

input (n)	output ($\frac{1}{4}n$)
4	■ 1
8	■ 2
12	■ 3

Find the rule for each function table.

6. $4n$

n	■
1	4
2	8
3	12

7. $n \div 2$

n	■
0	0
2	1
4	2

8. If the input values are -2, 0, and 4 and the corresponding outputs are 3, 5, and 9, what is the function rule? $n + 5$

■ Reteaching the Lesson ■

Activity Have students play "High Card Rules." Cards numbered from -10 to 10 will be needed. Have players start with the rule $n + 2$. Each player takes a card and uses its number as the input. The player with the greatest output designates the next rule.

Additional Answer

1.

input (y)	output (y − 2)
-4	-6
-1	-3
0	-2
3	1

2 TEACH

Transparency 12-5B contains a teaching aid for this lesson.

Reading Mathematics Make sure students understand that a function rule is simply an expression that is going to be evaluated for several numbers. Have students verbally interpret the function rule before beginning each exercise in the Guided Practice.

In-Class Examples

For Example 1
If Ebony sells her handmade earrings for $4 a pair, how much will she make by selling 6 pairs? 8 pairs? 10 pairs? What is the function rule? **24; 32; 40; 4n**

For Example 2
Find the rule for the function table.

input (n)	output (■)
-1	4
0	5
1	6
2	7

$n + 5$

For Example 3
A bicyclist can pedal 15 miles an hour. Make a function table showing the number of miles traveled in 3, 4, and 5 hours.

input (n)	output (15n)
3	45
4	60
5	75

3 PRACTICE/APPLY

Check for Understanding
If students need additional practice or instruction after completing Exercises 1–9, one of these options may be helpful.
- Extra Practice, see p. 590
- Reteaching Activity
- *Study Guide Masters,* p. 88
- *Practice Masters,* p. 88
- Interactive Mathematics Tools Software

9. *Money Matters* Marc Wright of Windsor, Ontario, Canada, started his own greeting card business, called Kiddie Cards. Marc makes a profit of $0.75 for every card he sells.

   a. Write the function rule to represent Marc's profits. **0.75c**

   b. How much profit would Marc earn on a sale of 50 cards? **$37.50**

EXERCISES

Practice   **Copy and complete each function table.**

10.

input (n)	output ($n + 4$)
−3	■ 1
0	■ 4
3	■ 7

11.

input (n)	output ($3n$)
0	■ 0
3	■ 9
4	■ 12

12.

input (n)	output ($5 − n$)
5	■ 0
6	■ −1
7	■ −2

13.

input (n)	output ($\frac{1}{8}n$)
−8	■ −1
0	■ 0
12	■ $1\frac{1}{2}$

Find the rule for each function table.

14. $n - 3$

n	■
0	−3
3	0
6	3

15. $5n$

n	■
0	0
3	15
4	20

16. $-2n$

n	■
−2	4
−1	2
3	−6

17. $n \div (-2)$

n	■
−4	2
−2	1
0	0

18. $6 - n$

n	■
1	5
5	1
9	−3

19. n^2

n	■
5	25
−5	25
0	0
−1	1

20. If a function rule is $2n + 1$, what is the output for an input of 2? **5**

21. If a function rule is $6n - 4$, what is the output for an input of −3? **−22**

22. If the input values are 2 and 7 and the corresponding outputs are −6 and −21, what is the function rule? **−3n**

23. If the input values are 5 and 10 and the corresponding outputs are 1 and 2, what is the function rule? **n ÷ 5**

24. If the output values are −5 and −1 and the function rule is $n - 5$, what are the corresponding input values? **0 and 4**

Study Guide Masters, p. 88

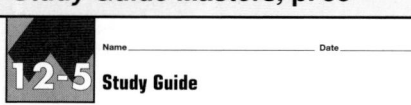

Name _____ Date _____

12-5 Study Guide

Functions

A function connects a number n, the input, to another number, the output, by a rule.

Example   Replace n with −4, −2, 0, 4 in the rule $2n - 1$.

input (n)	output ($2n - 1$)
−4	$2(-4) - 1 = -9$
−2	$2(-2) - 1 = -5$
0	$2(0) - 1 = -1$
4	$2(4) - 1 = 7$

Complete each function table.

1.

input (n)	output ($n + 4$)
−2	2
0	4
1	5
3	7

2.

input (n)	output ($3n$)
−1	−3
2	6
3	9
5	15

3.

input (n)	output [$n + (-1)$]
−5	−6
−1	−2
4	3
8	7

4.

input (n)	output ($\frac{n}{2} + 2$)
−6	−1
0	2
2	3
6	5

Find the rule for each function table.

5.

n	$n + 2$
−4	−2
0	2
2	4
5	7

6.

n	$n - 3$
−3	−6
−1	−4
1	−2
3	0

7.

n	$\frac{n}{2}$
−2	−1
0	0
2	1
4	2

© Glencoe/McGraw-Hill   T88   *Mathematics: Applications and Connections, Course 1*

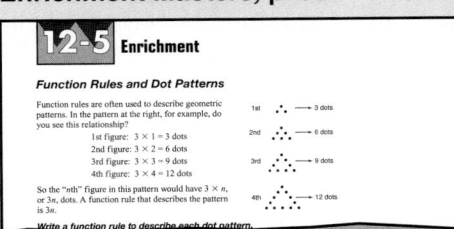

Applications and Problem Solving

25. *Spreadsheets* In the spreadsheet, the number of correct answers on a quiz is entered in column A. The number of questions that were on the quiz are entered in column B. Column C computes the percent correct. The formula in cell C1 is A1/B1*100. The formula acts like the rule of a function. What are the output values for cells C2, C3, C4, and C5? **100, 60, 80, 70**

	A	B	C
1	9	10	90
2	10	10	
3	6	10	
4	8	10	
5	7	10	

26. *Money Matters* Ebony Hood of Washington, D.C., started her own business selling scarves and fashion pins. Suppose she sells a scarf for $2 and a pin for $4.

  a. Write a function rule to represent the total cost of scarves (s) and pins (p). **$2s + 4p$**

  b. How much would 5 scarves and 3 pins cost? **$22**

27. *Critical Thinking* Find the rule for the function table.

$2n + 2$

n	■
-2	-2
-1	0
2	6
3	8

Mixed Review

28. *Geometry* The area A of a trapezoid can be found by multiplying the height h and one-half the sum of the bases b_1 and b_2. The formula is $A = \frac{1}{2}h(b_1 + b_2)$. The area of the trapezoid at the right is 48 square centimeters. It is 6 centimeters high, and the length of one base is 7 centimeters. What is the length of the other base? *(Lesson 12-4)* **9 cm**

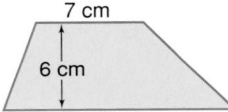
7 cm
6 cm

29. Write an integer to describe a temperature of twelve degrees below zero. *(Lesson 11-1)* **-12**

30. **Test Practice** If $\triangle TUV$ is congruent to $\triangle PQR$, then — *(Lesson 9-6)* **B**

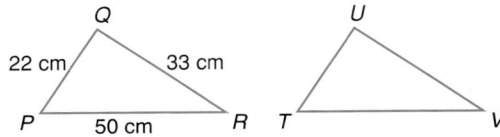
Q U
22 cm 33 cm
P 50 cm R T V

  A the measure of side $\overline{TU}$ is equal to the measure of side $\overline{PR}$.

  B side $\overline{TV}$ measures 50 centimeters.

  C the perimeter of triangle TUV is 100 centimeters.

  D the measure of angle P is equal to the measure of angle V.

  E the measure of side $\overline{TV}$ is twice the measure of side $\overline{TU}$.

Lesson 12-5 Functions **499**

Extending the Lesson

Enrichment Masters, p. 88

Activity Rose's parents offered her an allowance of $5 a week or one that pays 2 cents for Week 1, (2 cents)² for Week 2, (2 cents)³ for Week 3, and so on, through Week 10. The Week 10 rate applies thereafter. Which plan would give Rose the greater allowance in the tenth week? Explain. **second plan; (2 cents)¹⁰ > $5**

Closing Activity

Modeling Have pairs of students make function tables and have other students determine the function rules.

Chapter 12, Quiz C (Lessons 12-4 and 12-5) is available in the *Assessment and Evaluation Masters*, p. 324.

Practice Masters, p. 88

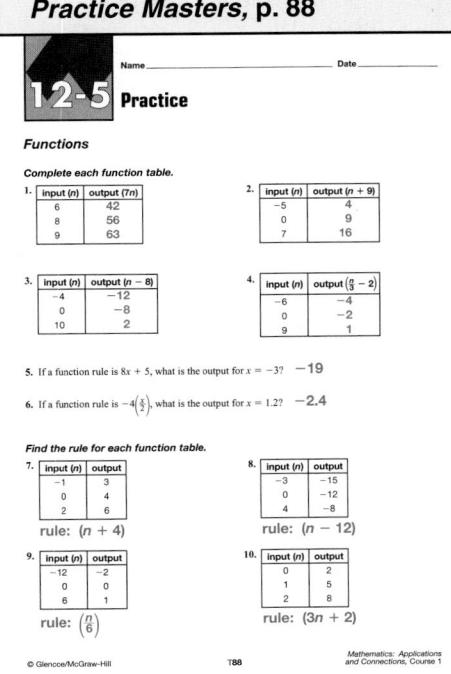
Lesson 12-5 **499**

- *Study Guide Masters*, p. 89
- *Practice Masters*, p. 89
- *Enrichment Masters*, p. 89
- Transparencies 12-6, A and B
- *Assessment and Evaluation Masters*, p. 324
- *Science and Math Lab Manual*, pp. 21–24
- *Technology Masters*, p. 24
- CD-ROM Program
  - Resource Lesson 12-6
  - Interactive Lesson 12-6

Recommended Pacing

Standard	Day 9 of 11
Honors	Day 9 of 11
Block	Day 4 of 6

1 FOCUS

5-Minute Check
(Lesson 12-5)

1. Copy and complete the function table.

input (*n*)	output (2*n* − 3)
−2	■ −7
0	■ −3
2	■ 1
5	■ 7

2. Find the rule for the function table. **3*n***

n	■
3	9
4	12
5	15

The 5-Minute Check is also available on **Transparency 12-6A** for this lesson.

Motivating the Lesson

Problem Solving On a map, locate three cities that lie approximately in a straight line, and find their coordinates. Discuss how a function table can be constructed using the coordinates. Ask: *What would a graph of this function look like?*

12-6 Graphing Functions

What you'll learn
You'll learn to graph functions from function tables.

When am I ever going to use this?
Knowing how to graph functions can help you analyze data at a quick glance.

Getting an allowance of $2 a week is an example of a function. The equation $y = 2x$, where y is the amount received and x is the number of weeks, will give you the total allowance received after the number of weeks you choose. In the equation, $y = 2x$, x is the input, and y is the output. The function rule is $2x$.

We can use a coordinate system to graph an equation or function.

Example — APPLICATION

Money Matters Refer to the beginning of the lesson. Make a function table for the rule $2x$. Then graph the function.

Step 1 Record the input and output in a function table. We chose 0, 2, 4, and 6 for the input. List the input and output as ordered pairs.

input	function rule	output	ordered pairs
x	2*x*	*y*	(*x*, *y*)
0	2(0)	0	(0, 0)
2	2(2)	4	(2, 4)
4	2(4)	8	(4, 8)
6	2(6)	12	(6, 12)

Step 2 Graph the ordered pairs from the table in Step 1 on the coordinate plane.

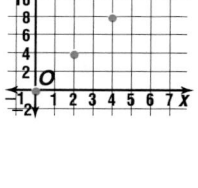

The x-coordinates represent the number of weeks from the start.

The y-coordinates represent the total allowance received after x weeks.

LOOK BACK
You can refer to Lesson 11-7 to review the coordinate system.

Step 3 The points appear to lie on a line. Draw the line that contains these points. The line is the graph of $y = 2x$.

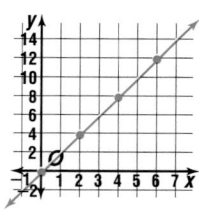

Multiple Learning Styles

Visual/Spatial Have students work in pairs with geoboards to graph functions from given ordered pairs. Set up geoboards with the middle row and middle column forming the *x*- and *y*-axes. Connect (−2, −2) to (2, 2). Connect (−1, −2) to (1, 2).

Connect (−2, −1) to (2, 1). Each band represents a function rule. Have students set up function tables for each band. What is the function rule for each band? $x, 2x, \frac{1}{2}x$

Example **2**

Make a function table for the graph. Then determine the rule.

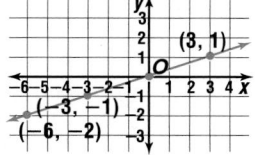

(3, 1)

O

(−3, −1)

(−6, −2)

Use the ordered pairs to make a function table.

input (x)	output (y)	(x, y)
−6	−2	(−6, −2)
−3	−1	(−3, −1)
0	0	(0,0)
3	1	(3,1)

Study the input and output to determine a rule.

input		output
−6	$\times \frac{1}{3} \rightarrow$	−2
−3	$\times \frac{1}{3} \rightarrow$	−1
0	$\times \frac{1}{3} \rightarrow$	0
3	$\times \frac{1}{3} \rightarrow$	1

Each input is multiplied by $\frac{1}{3}$ to get the output.

The function rule is $\frac{1}{3}x$ or $\frac{x}{3}$.

CHECK FOR UNDERSTANDING

Communicating Mathematics

Read and study the lesson to answer each question. 1–2. See Answer Appendix.

1. *Explain* how you can use the function table to graph a function.

2. *Tell* why only those solutions graphed in the first quadrant in Example 1 make sense.

input	output
−1	2
0	0
1	−2
2	−4

3. *Write* a brief paragraph explaining how to graph a function when you know the function rule. **See Answer Appendix.**

Guided Practice

Graph the functions represented by each function table. 4–5. See Answer Appendix.

4.
input	output
0	−3
3	0
6	3

5.
input	output
5	10
6	11
7	12

Lesson 12-6 Graphing Functions **501**

■ Reteaching the Lesson ■

Activity Have students work in small groups. Ask them to draw a line that crosses both the x- and y-axes on a coordinate grid. Have them identify three ordered pairs on the line and then determine the function rule for the line.

---

2 TEACH

Transparency 12-6B contains a teaching aid for this lesson.

Using Discussion Using several ordered pairs, review how to plot ordered pairs on a coordinate system. Ask students for rules that relate to x and y.

In-Class Examples

For Example 1
Make a function table for the rule 4x. Then graph the function.

input	function rule	output	ordered pairs
x	4x	y	(x, y)
0	4(0)	0	(0, 0)
2	4(2)	8	(2, 8)
4	4(4)	16	(4, 16)
6	4(6)	24	(6, 24)

For Example 2
Make a function table for the graph. Then determine the rule.

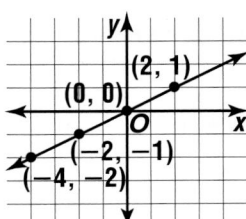

(2, 1)

(0, 0)

O

(−2, −1)

(−4, −2)

input (x)	output (y)	(x, y)
−4	−2	(−4, −2)
−2	−1	(−2, −1)
0	0	(0, 0)
2	1	(2, 1)

$\frac{1}{2}x$ or $\frac{x}{2}$

Teaching Tip Make sure students make the connection that the output value is the y value.

Lesson 12-6 **501**

Check for Understanding

If students need additional practice or instruction after completing Exercises 1–9, one of these options may be helpful.

- Extra Practice, see p. 590
- Reteaching Activity, see p. 501
- *Study Guide Masters*, p. 89
- *Practice Masters*, p. 89
- Interactive Mathematics Tools Software

Assignment Guide

Core: 11–25 odd, 28–32
Enriched: 10–24 even, 25, 26, 28–32

Additional Answer

8.

input (x)	output (x − 6)
1	−5
4	−2
5	−1

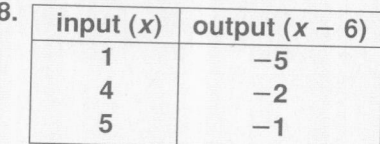

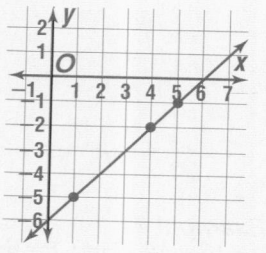

Study Guide Masters, p. 89

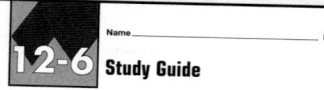

12-6 Study Guide

Graphing Functions

To graph a function, let the input number be the x-coordinate. Let the output number be the y-coordinate.

Example   Make a function table for the rule 2x. Then graph the function.

Choose input values. Find output values. Then graph the ordered pairs and draw a line.

input (x)	output (2x)	ordered pairs
−2	−4	(−2, −4)
0	0	(0, 0)
2	4	(2, 4)
3	6	(3, 6)

Complete each function table. Then graph the function.

1.
input (n)	output (⅓)	ordered pairs
−3	−1	(−3, −1)
0	0	(0, 0)
6	2	(6, 2)

2.
input (n)	output (n + 1)	ordered pairs
−3	−2	(−3, −2)
−1	0	(−1, 0)
2	3	(2, 3)

3.
input (n)	output (n − 2)	ordered pairs
0	−2	(0, −2)
2	0	(2, 0)
5	3	(5, 3)

© Glencoe/McGraw-Hill   T89   Mathematics: Applications and Connections, Course 1

6–7. See Answer Appendix for graphs.

Copy and complete each function table. Then graph the function.

6.
input (n)	output (n − 5)
4	■ −1
0	■ −5
−2	■ −7

7.
input (n)	output ($\frac{n}{2}$)
6.6	■ 3.3
8.4	■ 4.2
10.5	■ 5.25

8. Make a function table for the rule $x − 6$ using 1, 4, and 5 as the input. Then graph the function. **See margin.**

9. Make a function table for the graph. Then determine the rule. **See margin.**

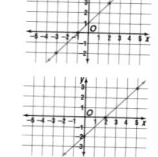

EXERCISES

Practice
10–15. See Answer Appendix.

Graph the functions represented by each function table.

10.
input	output
0	−5
2	−3
4	−1

11.
input	output
5	3
3	1
1	−1

12.
input	output
−2	4
0	0
2	−4

13.
input	output
0	0
4	1
8	2

14.
input	output
2	5
0	−1
−1	−4

15.
input	output
−4	0
0	1
4	2

Copy and complete each function table. Then graph the function.

16–19. See Answer Appendix for graphs.

16.
input (n)	output (2n + 4)
$5\frac{1}{2}$	■ 15
$4\frac{1}{4}$	■ $12\frac{1}{2}$
0	■ 4

17.
input (n)	output (−3n)
3	■ −9
0	■ 0
−2	■ 6

18.
input (n)	output (2n − 3)
−2	■ −7
0	■ −3
2	■ 1
5	■ 7

19.
input (n)	output ($\frac{1}{2}n$ + 1)
−4	■ −1
−2	■ 0
0	■ 1
2	■ 2

20. Make a function table for the rule $a + 5$ using input values of −5, −3, and 0. Then graph the function. **See Answer Appendix.**

21. Make a function table for the rule $3n − 3$ using input values of −2, 1, and 4. Then graph the function. **See Answer Appendix.**

502   Chapter 12 Algebra: Exploring Equations

Additional Answer

9.

input (n)	output (■)
−1	−3
2	0
4	2

The rule is $n − 2$.

Make a function table for each graph. Then determine the rule.

22–24. See Answer
Appendix.

22.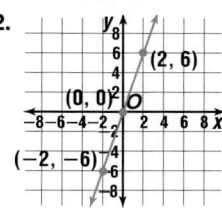

23.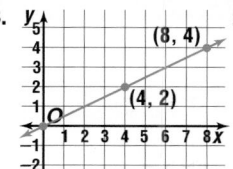

24.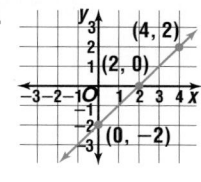

Applications and Problem Solving

25a. Jane: $25w - 20$;
Julie: $20w$

25c. The point of intersection represents when the total earnings are the same for both girls.

27a–c. See Answer Appendix.

25. *Money Matters* During the summer, Jane earned $25 a week and was required to purchase a uniform for $20. No special clothing was required at Julie's summer job. Julie earned $20 a week.

   a. Write the function rule for each girl's wages.

   b. Graph each function on the same coordinate plane. **See Answer Appendix.**

   c. What does the point of intersection of the two graphs represent?

26. *History* The line graph shows the number of cups of chocolate the Aztec emperor Montezuma drank each day. **a–b. See Answer Appendix.**

   a. Make a function table for the graph.

   b. Determine the function rule.

Cups of Chocolate

(3, 150)
(2, 100)
(1, 50)
(0, 0)
Number of Days

27. *Working on the* CHAPTER Project Refer to the table on page 475.

You can graph a function to see how the value of a basketball card might change in the future. In June, 1997, the Charles Barkley card was worth $60. The value had been increasing $1 per month.

   a. Make a function table for the rule $60 + x$. Use inputs of 1 through 12 to predict the value of the card over the next 12 months.

   b. Graph the function using graphing software or a graphing calculator.

   c. What does the ordered pair $(5, 65)$ mean in this situation?

28. *Critical Thinking* Determine the rule for the line that passes through $A(-2, 6)$ and $B(3, -6)$. $-2.4x + 1.2$

Mixed Review

29. *Algebra* Complete the function table.
(Lesson 12-5)

input (n)	output (n + 7)
−5	■ 2
0	■ 7
4	■ 11

30. **Test Practice** If an airplane travels 438 miles per hour, how many miles will it travel in 5 hours? *(Lesson 8-2)* **D**

   A 2,050 mi   **B** 2,090 mi   **C** 2,150 mi   **D** 2,190 mi   **E** Not Here

31. *Geometry* Find the circumference of a circle with a radius of 5 meters. Use 3.14 for pi. *(Lesson 7-4)* **about 31.4 m**

32. Find the prime factorization of 120. *(Lesson 5-2)* $2 \times 2 \times 2 \times 3 \times 5$

▬ Extending the Lesson ▬

Enrichment Masters, p. 89

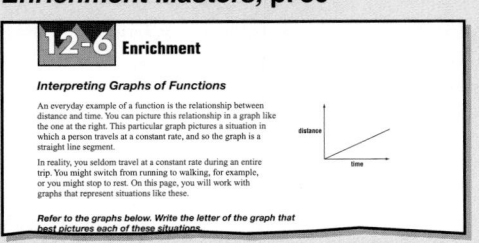

Activity Ask students what role they think math plays in designing a building. Discuss ways in which a coordinate grid could be used to sketch architectural ideas that can be used for scale drawings or blueprints.

Lesson 12-6 503

CHAPTER Project

Exercise 27 asks students to advance to the next stage of work on the Chapter Project. You may have students compare their graphs for part b. How are the graphs similar? How are they different?

4 ASSESS

Closing Activity

Speaking Have students work in pairs to draw a graph. Ask them to exchange their graph with others to make a function table and determine the rules. Have students discuss how they determined the function rule and whether some rules are difficult to find.

Chapter 12, Quiz D (Lesson 12-6) is available in the *Assessment and Evaluation Masters*, p. 324.

Practice Masters, p. 89

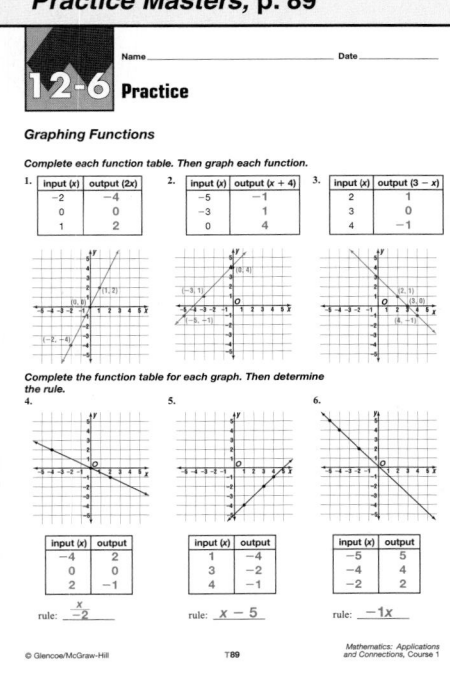

Vocabulary

This section provides a listing of the new terms, properties, and phrases that were introduced in this chapter. Have students define each term and provide an example or two of it, if appropriate.

Understanding and Using the Vocabulary

These exercises check students' understanding of the terms by using a variety of verbal formats including matching, completion, and true/false.

Glossaries A complete glossary of terms appears on pages 642–648. The glossary also appears in Spanish on pages 649–656.

Additional Answer

10. Sample answer: Make a table recording the input and output of the given function. Write the input/output as an ordered pair. Record at least three ordered pairs on the table. Graph these ordered pairs and draw a line connecting the graphed points.

Vocabulary

After completing this chapter, you should be able to define each term, concept, or phrase and give an example or two of each.

Patterns and Functions

function (p. 496)
function machine (p. 494)
function table (p. 496)
input (p. 494)
output (p. 494)

Problem Solving

use an equation (pp. 492–493)

Understanding and Using the Vocabulary

Choose the correct term, number, or equation to complete each sentence.

1. To solve an equation means to find a value for the (coordinate, <u>variable</u>) that makes the equation true.

2. The second step in solving $4b + 3 = 27$ is to (<u>divide</u>, multiply) each side of the equation by 4.

3. In an equation, when a number is (added to, <u>subtracted from</u>) the variable, add the number to each side to solve the equation.

4. A(n) (<u>function</u>, output) describes the relationship between two sets of numbers.

5. A(n) (input, <u>function table</u>) can be helpful in organizing information by using a mathematical rule to assign an output to a given input.

6. The solutions of a function table are found in the (<u>output</u>, input) column.

7. If the function rule is $6n$ and the input value is -3, then the output value is ($\underline{-18}$, -2).

8. On the graph of a function, every point has (<u>coordinates</u>, variables) that satisfy the function.

9. The diagram at the right models the equation ($\underline{2x + 3 = -9}$, $x = 6$).

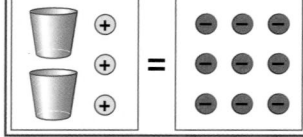

In Your Own Words

10. *Tell* how to graph the function $y = 2x - 1$. **See margin.**

MindJogger Videoquizzes

MindJogger Videoquizzes provide an alternative review of concepts presented in this chapter. Students work in teams to answer questions, gaining points for correct answers. The questions are presented in three rounds.
Round 1 Concepts–5 questions
Round 2 Skills–4 questions
Round 3 Problem Solving–4 questions

Objectives & Examples

Upon completing this chapter, you should be able to:

● solve addition equations using models
(Lesson 12-1)

Solve $m + (-2) = -5$.

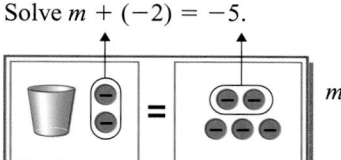

$$m + (-2) = -5$$
$$m = -3$$

● solve subtraction equations by using models
(Lesson 12-2)

Solve $t - 3 = 4$.

Rewrite as an addition equation.

$$t - 3 = 4 \rightarrow t + (-3) = 4$$

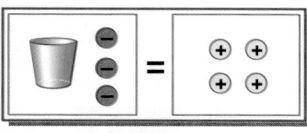

$$t + (-3) = 4$$

Since there are no negative counters on the right side, add 3 positive counters to each side to make 3 zero pairs on the left side. Then remove the zero pairs.

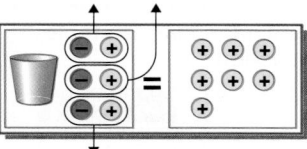

$$t + (-3) + 3 = 4 + 3$$
$$t = 7$$

● solve multiplication and division equations using models *(Lesson 12-3)*

Solve $3x = -9$.

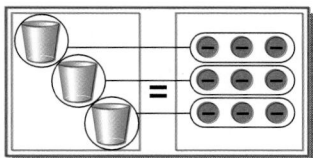

$$3x = -9$$
$$x = -3$$

Review Exercises

Use these exercises to review and prepare for the chapter test.

Solve each equation. Use cups and counters if necessary.

11. $x + 2 = 5$ **3**   **12.** $p + 7 = 4$ **−3**
13. $6 + r = -7$ **−13**   **14.** $-2 + w = -5$ **−3**
15. $a + (-4) = -7$ **−3** **16.** $t + 2 = -6$ **−8**

Solve each equation. Use cups and counters if necessary.

17. $x - 1 = 3$ **4**
18. $y - 3 = 11$ **14**
19. $f - 3 = -1$ **2**
20. $g - 4 = -5$ **−1**
21. $h - 9 = -9$ **0**
22. $c - (-8) = -12$ **−20**

Solve each equation. Use cups and counters if necessary.

23. $7q = 28$ **4**   **24.** $\frac{1}{3}d = 9$ **27**
25. $\frac{1}{5}k = -11$ **−55**   **26.** $8x = 40$ **5**
27. $3m = -15$ **−5**   **28.** $2.5g = 15$ **6**
29. $\frac{1}{4}y = -5$ **−20**   **30.** $-4t = -4$ **1**

Chapter 12 Study Guide and Assessment **505**

Objectives & Examples

This section reviews the skills and concepts of the chapter and shows completely worked examples.

Review Exercises

These exercises provide practice for the corresponding objectives.

***Assessment and Evaluation Masters,* pp. 311–312**

Name_____ Date_____

Chapter 12 Test, Form 1B

1. Solve $x + 5 = 1$.
   A. 4   B. −4   C. 14   D. none of these   1. __B__
2. Solve $y + (-4) = 11$.
   A. 7   B. 15   C. −7   D. none of these   2. __B__
3. Solve $-8 + t = 10$.
   A. 18   B. 2   C. −2   D. none of these   3. __A__
4. Solve $-7 = w + 6$.
   A. 1   B. −1   C. −13   D. none of these   4. __C__
5. Solve $z - 7 = 4$.
   A. 3   B. −3   C. −11   D. none of these   5. __D__
6. Solve $s - (-5) = -12$.
   A. 7   B. −17   C. 17   D. none of these   6. __B__
7. Solve $m - 4 = -2$.
   A. −2   B. 2   C. −6   D. none of these   7. __B__
8. Solve $\frac{1}{2}x = 10$.
   A. 20   B. 5   C. −5   D. none of these   8. __A__
9. Solve $\frac{1}{4}m = -4$.
   A. 1   B. −1   C. 16   D. none of these   9. __D__
10. Solve $4x = 20$.
    A. 5   B. −5   C. 80   D. none of these   10. __A__
11. Solve $3y - 4 = 8$.
    A. −4   B. 4   C. 1   D. none of these   11. __B__
12. Solve $-7 = 2n + 5$.
    A. −2   B. 6   C. −6   D. none of these   12. __C__
13. Solve $-6r + 11 = 13$.
    A. −2   B. 24   C. 2   D. none of these   13. __D__

© Glencoe/McGraw-Hill   311   *Mathematics: Applications and Connections, Course 1*

Chapter 12 Test, Form 1B (continued)

14. Al's father is three times as old as Al. Al's father is 36 years old. If x stands for Al's age, which equation could you solve to find Al's age?
    A. $3 + x = 36$ B. $36x = 3$
    C. $3x = 36$ D. none of these   14. __C__
15. What number should replace ■ in the function table at the right?
    A. 4   B. 10   C. 6   D. none of these

n	$n + 6$
−3	3
0	6
4	■

    15. __B__
16. Find the rule for the function table at the right.
    A. $n + 2$ B. $2n$ C. $n - 2$ D. none of these

n	■
8	6
2	0
−3	−5

    16. __C__
17. Find the rule for the function table at the right.
    A. $3n$ B. $\frac{1}{3}n$ C. $n - 10$ D. none of these

n	■
15	5
6	2
−9	−3

    17. __B__
18. If a function rule is $2n + 4$, what is the output for an input of 5?
    A. 18   B. −18   C. 14   D. none of these   18. __C__
19. Which graph represents the function table at the right?

x	$\frac{x}{3}$
−3	−1
0	0
3	1

    A.   B.   C.   D.   19. __D__
20. Which function table represents the graph at the right?

    A.
input	output
−4	−2
−3	0
0	6

    B.
input	output
−2	−4
0	−3
6	0

    C.
input	output
4	2
3	0
0	6

    D.
input	output
2	4
0	3
6	0

    20. __A__

© Glencoe/McGraw-Hill   312   *Mathematics: Applications and Connections, Course 1*

Assessment and Evaluation

Six forms of Chapter 12 Test are available in the *Assessment and Evaluation Masters* as shown in the chart.

Chapter 12 Test, Form 1B, is shown at the right. Chapter 12 Test, Form 2B, is shown on the next page.

1A	Multiple Choice	Honors
1B	Multiple Choice	Average
1C	Multiple Choice	Basic
2A	Free Response	Honors
2B	Free Response	Average
2C	Free Response	Basic

Assessment and Evaluation Masters, pp. 317–318

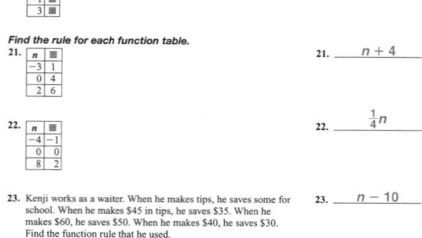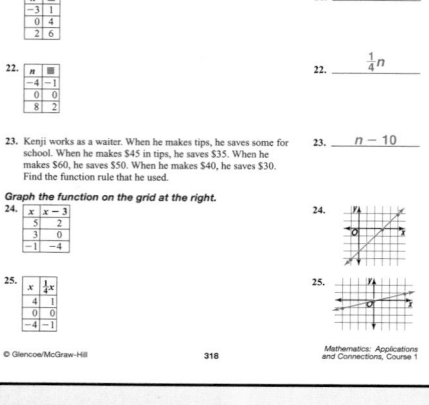
Objectives & Examples

solve two-step equations using models *(Lesson 12-4)*

Solve $2n + 1 = 7$.

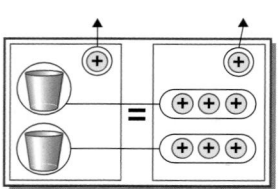

$$2n + 1 = 7$$
$$2n + 1 - 1 = 7 - 1$$
$$2n = 6$$
$$n = 3$$

complete function tables *(Lesson 12-5)*

Find the rule for the function table.

input (n)	output (■)
−1	3
0	4
3	7

$$\left.\begin{array}{l} -1 + 4 = 3 \\ 0 + 4 = 4 \\ 3 + 4 = 7 \end{array}\right\} n + 4$$

graph functions from function tables *(Lesson 12-6)*

Graph the function represented by the function table.

x	x + 4	output	ordered pair
−3	−3 + 4	1	(−3, 1)
0	0 + 4	4	(0, 4)
1	1 + 4	5	(1, 5)

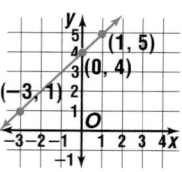

Review Exercises

Solve each equation. Use cups and counters if necessary.

31. $2r - 8 = 4$ **6**

32. $\frac{1}{5}k + 5 = 7$ **10**

33. $2x + 5 = -13$ **−9**

34. $4n + 12.4 = -6.8$ **−4.8**

Copy and complete each function table.

35.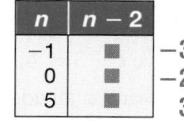
n	3n
−1	■	−3
0	■	0
2	■	6

36.
n	n − 2
−1	■	−3
0	■	−2
5	■	3

37. If a function rule is $2n + 1$, what is the output for $n = -3$?   **−5**

Find the rule for each function table.

38.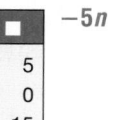
n	■
−1	5
0	0
3	−15
  −5n

39.
n	■
−1	−5
0	−4
3	−1
  n − 4

Copy and complete each function table. Then graph the function.

40.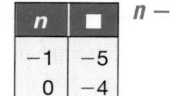
x	4x
−1	■	−4
0	■	0
2	■	8

41.
x	x + 3
−2	■	1
0	■	3
2	■	5

40–41. See Answer Appendix for graphs.

Test and Review Software

You may use this software, a combination of an item generator and item bank, to create your own tests or worksheets. Types of items include free response, multiple choice, short answer, and open ended.

CD-ROM Program

The CD-ROM Program contains an Assessment Game whose questions review the concepts in this chapter.

Applications & Problem Solving

42. Health After dieting for 8 weeks and losing 17 pounds, Jeremy weighed 172 pounds. How much did he weigh before the diet? *(Lesson 12-1)* **189 pounds**

43. Use an Equation Keith swam three times as many laps as Dan in the swim meet. Keith swam 24 laps. How many laps did Dan swim? *(Lesson 12-4B)* **8 laps**

44. Money Matters On her vacation, Mrs. Salgado planned to spend less than $120 each day on her hotel room and meals. She found a hotel room for $75 a night. How much can she spend on meals? *(Lesson 12-4)* **less than $45**

45. Money Matters Mr. Jackson got a new shipment of T-shirts to sell at his souvenir shop. The table shows the amount Mr. Jackson paid for each of the three types of T-shirts, and the selling price he marked on the shirts. Find the function rule he used. *(Lesson 12-5)* $n + 3.50$

Amount Paid	Selling Price
$4.00	$7.50
$5.50	$9.00
$6.00	$9.50

Alternative Assessment

● **Performance Task**

Record the temperature of a cup of hot tap water to the nearest degree. Then record the temperature every minute for the next 10 minutes. **See students' work.**

Graph the ordered pairs (time, temperature) on a coordinate plane. **See students' work.**

Estimate the temperature after 30 minutes. Estimate again after 60 minutes.

Answers will vary.

A practice test for Chapter 12 is provided on page 606.

● **Completing the**

Use the following checklist to make sure your project is complete.

☑ You have included a list of the items in your collection.

☑ You have included your table or other display showing the value of at least five items in your collection.

☑ You may want to include pictures of the items in your collection.

Add any finishing touches that you would like to make your project attractive.

 Select an item from this chapter that you feel shows your best work. Place it in your portfolio. Explain why you selected it.

Applications & Problem Solving

This section provides additional practice in solving real-world problems that involve the skills of this chapter.

Alternative Assessment

The **Performance Task** provides students with a performance assessment opportunity to evaluate their work and understanding.

CHAPTER Project

Students should complete the final stages of their project and prepare a class demonstration of their results. A scoring guide for the project is available in the *Investigations and Projects Masters,* p. 63.

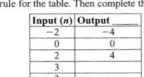 Students should add to their portfolios at this time.

Assessment and Evaluation Masters, p. 321

Performance Assessment

Additional performance assessment tasks for this chapter are included in the *Assessment and Evaluation Masters* on page 321. A scoring guide is also provided on page 333.

12   Name_____   Date_____
Chapter 12 Performance Assessment

Instructions: Demonstrate your knowledge by giving a clear, concise solution to each problem. Be sure to include all relevant drawings and justify your answers. You may show your solutions in more than one way or investigate beyond the requirements of the problems.

1. Draw cups, ▯, positive counters, ⊕, and negative counters, ⊖, to model each of the following.
   a. Model $x - 3 = 4$. Explain your reasoning.
   b. Use modeling to solve the equation in part a. Explain each step.
   c. Model $4x + 3 = -5$. Explain your reasoning.
   d. Use modeling to solve the equation in part c. Show each step.
   e. Write a word problem concerning integers.
   f. Write an equation for the problem in part e. Solve. Explain each step.

2. a. Write the rule for the table. Then complete the table.

Input (n)	Output
−2	−4
0	0
2	4
−3	

   b. Graph the function in part a.

© Glencoe/McGraw-Hill   321   *Mathematics: Applications and Connections, Course 1*

The Standardized Test Practice may be used to help students prepare for standardized tests. The test items are written in the same style as those in state proficiency tests and standardized tests like CAT, CTBS, ITBS, MAT, SAT, and Terra Nova. The test items cover skills and concepts covered up to this point in the text.

The pages can be used as an overnight assessment. After students have completed the pages, discuss how each problem can be solved, or provide copies of the solutions from the *Solutions Manual*.

Assessment and Evaluation Masters, p. 327

Section One: Multiple Choice

There are eleven multiple-choice questions in this section. Choose the best answer. If a correct answer is *not here*, mark the letter for Not Here.

1. Name the number of faces in a square pyramid. **B**
   - A 4
   - B 5
   - C 6
   - D 7

2. What is the solution of the equation $5 + x = 3$? **G**
   - F -8
   - G -2
   - H 2
   - J 8

3. Which is a reflection of the letter H? **D**
   - A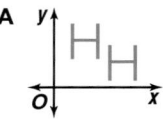
   - B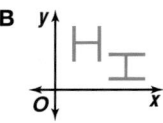
   - C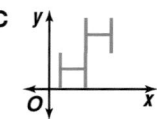
   - D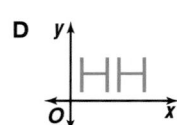

4. What numbers complete the function table? **F**

n	3n
10	▪
0	▪
−5	▪

   - F $30, 0, -15$
   - G $30, 20, -10$
   - H $13, 3, -2$
   - J $13, 3, -10$

5. Which ordered pair is represented by point M? **D**

   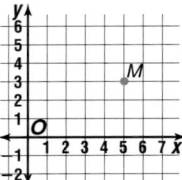

   - A $(5, 4)$
   - B $(3, 5)$
   - C $(2, 5)$
   - D $(5, 3)$

Please note that Questions 6–12 have five answer choices.

6. Diane had 10 coins in her pocket. There were 4 pennies, 1 nickel, 4 dimes, and 1 quarter. If she took out 1 coin at random, what is the probability it would be a dime? **H**
   - F $\frac{1}{10}$
   - G $\frac{1}{4}$
   - H $\frac{2}{5}$
   - J $\frac{1}{2}$
   - K Not Here

7. Sharon is knitting a sweater. She needs 8 skeins of red yarn that cost $1.89 each and 3 skeins of white yarn that cost $1.29 each. Which number sentence could be used to find S, the total cost in dollars? **E**
   - A $S = (1.89 \div 8) + (1.29 \div 3)$
   - B $S = (8 \times 1.89) \times (3 \times 1.29)$
   - C $S = (1.89 + 1.29) \times 11$
   - D $S = (1.89 \div 8) \times (1.29 \div 3)$
   - E $S = (8 \times 1.89) + (3 \times 1.29)$

◄◄◄**Instructional Resources**
Another cumulative review is shown at the left and is available in the *Assessment and Evaluation Masters*, p. 327.

---

12 Name_____ Date_____

Cumulative Review, Chapters 1–12

Replace each ● with <, >, or = to make a true sentence. (Lessons 3-3, 5-8, and 11-2)
1. 0.0004 ● 0.003   2. $\frac{9}{100}$ ● $\frac{4}{25}$ 3. -100 ● -145

4. Find the elapsed time from 7:45 A.M. to 11:35 A.M. (Lesson 6-7)

5. Round 2,387 to the nearest hundred. (Lesson 1-3)

6. What is the mode for the data: 84, 80, 79, 86, 85, 86? (Lesson 2-7)

7. Solve 8 × 152 using the distributive property. (Lesson 4-2)

Estimate. (Lessons 7-1 and 8-6)
8. $\frac{1}{3}$ × 25   9. 21% of 26

10. Use a compass and straightedge to bisect the angle at the right. (Lesson 9-3)

11. Find the area of a triangle with a base of 48 centimeters and a height of 30 centimeters. (Lesson 10-2)

12. Find the area of a circle with a radius of 19 millimeters. Use 3.14 for π. Round to the nearest tenth. (Lesson 10-3)

13. State the number of edges in a rectangular prism. (Lesson 10-4)

Perform the indicated operation. (Lessons 11-3, 11-4, and 11-5)
14. $-14 + 9$

15. $8 - (-4)$

16. $-9(-4)$

17. If a function rule is $2n + 3$, what is the output for $n = -2$? (Lesson 12-5)

Solve each equation. (Lessons 12-1, 12-2, and 12-3)
18. $m + 5 = -9$

19. $x - 6 = -8$

20. $9n = -45$

© Glencoe/McGraw-Hill   327

1. ___<___
2. ___<___
3. ___>___
4. _3 h 50 min_
5. __2,400__
6. ___86___
7. (8 × 100) + (8 × 50) + (8 × 2) = 1,216
8. 3 × 24 = 8
9. 5 × 25 = 5
10. [angle bisector figure]
11. __720 cm²__
12. _1,133.5 mm²_
13. ___12___
14. ___−5___
15. ___12___
16. ___36___
17. ___−1___
18. ___−14___
19. ___−2___
20. ___−5___

Mathematics: Applications and Connections, Course 1

8. Which best describes the two triangles? **H**

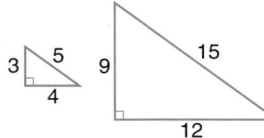

F The triangles are congruent.
G The triangles are reflections.
H The triangles are similar.
J The triangles are equilateral.
K The triangles are symmetrical.

9. Marty bought 5 packages of hamburger. If each package weighed 1.45 pounds, what was the weight of the 5 packages? **D**
   A 3.45 lb
   B 6.45 lb
   C 7.90 lb
   D 7.25 lb
   E Not Here

10. What is the perimeter of a triangle with sides of $3\frac{1}{2}$ cm, $2\frac{1}{2}$ cm, and $1\frac{3}{4}$ cm? **K**
    F 6 cm
    G $6\frac{1}{2}$ cm
    H $6\frac{3}{4}$ cm
    J 7 cm
    K Not Here

11. $0.0125 \div 1.7 =$ **E**
    A 1
    B 0.1
    C 0.01
    D 0.001
    E Not Here

12. $-12 - (-10) =$ **G**
    F -22
    G -2
    H 2
    J 22
    K Not Here

Test-Taking Tip

If you solve a problem and the answer you get is not one of the choices, check to see if your answer can be written in a different form. For example, if you solve a problem and get $\frac{x}{4}$ and it is not listed as a choice, try a different form such as $\frac{1}{4}x$. You may have the right answer, but in a different form.

Section Two: Free Response

This section contains six questions for which you will provide short answers. Write your answers on your paper.

13. Find the volume of the rectangular prism. **240 m³**

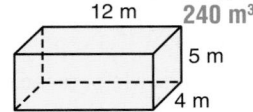

14. Complete the function table.

input (x)	output (x + 7)
−3	■ 4
1	■ 8
5	■ 12

15. In a survey, 17 out of 110 people said they liked boating. How many people would that be out of 100? **about 15 people**

16. The fence around a pool measures 25 feet wide by 60 feet long. If the length is increased by 20 feet, what will the new dimensions be? **25 ft by 80 ft**

17. The base of a cone is a _____. **circle**

18. Make a function table for the rule $x - 2$ using 2, 5, and 8 as the input. Then graph the function. **See Answer Appendix.**

Chapters 1–12 Standardized Test Practice   **509**

Test-Taking Tip

One of the important skills in test-taking is the ability to recognize the correct answer when multiple choices are given. Guide students to consider every possible form their answer could take before deciding that their answer is not one of the choices.

Assessment and Evaluation Masters, pp. 325–326

Instructional Resources ▶▶▶
Additional standardized test practice is shown at the right and is available in the *Assessment and Evaluation Masters, pp. 325–326.*

GET READY

This optional investigation is designed to be completed by a group of 2 or 3 students over several days or several weeks.

Mathematical Overview

This investigation utilizes the concepts from Chapters 1–12.
- measuring circumference
- graphing functions relating the radius, circumference, and area of circles

Time Management	
Gathering Data	30 minutes
Calculations	30 minutes
Creating Graphs	45 minutes
Summarizing Data	15 minutes
Presentation	10 minutes

Instructional Resources

- *Investigations and Projects Masters*, pp. 13–16
- *Manipulative Kit*
  - tape measures

Investigations and Projects Masters, p. 16

Name _____ Date _____

Interdisciplinary Investigation
(Student Edition, Pages 510–511)

Going In Circles

Use this table to record your data.

Name of Item	Radius (inches)	Circumference (inches)	Area (square inches)

3. Using the formula $C = 2\pi r$, the equation to find the radius is:

5. a. (radius, circumference)   b. (radius, area)   c. (circumference, area)

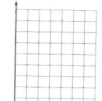 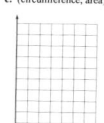

© Glencoe/McGraw-Hill   16   *Mathematics: Applications and Connections, Course 1*

GOING IN CIRCLES

What do these items have in common—a drum, a tambourine, and cymbals? You may have guessed they all have a circular shape. Circles are common in our daily lives.

What You'll Do

In this investigation, you will measure the circumference of common circular items. You will graph functions relating the radius, circumference, and area of circles. You will use your graphs to estimate the area of a circle.

Materials 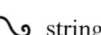 string 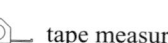 tape measure  calculator

 grid paper   circular objects

Procedure

1. Work in pairs. Measure the circumference of six different circular items to the nearest $\frac{1}{8}$ inch. To do this, wrap a string around the object and then measure the string or use a tape measure.

2. Make a table similar to the one below. Record the name and circumference of each item measured by you and your partner. Arrange the items from least to greatest circumference. If you use a spreadsheet, convert all fractions to decimals.

Name of Item	Radius (inches)	Circumference (inches)	Area (square inches)
pencil			
vegetable can			

3. Recall that the circumference of a circle can be found using the formula $C = 2\pi r$. Write and solve an equation to find the radius of each item. Record each radius in column 2 of your table.

4. Use the formula $A = \pi r^2$ to find the area of each circle. Record the area in column 4 of your table.

5. Draw three coordinate planes.
   a. For each item in the table, write the ordered pair, (radius, circumference). Then graph each ordered pair.

510 Interdisciplinary Investigation Going in Circles

◄◄◄ Instructional Resources

A recording sheet to help students organize their data for this investigation is shown at the left and is available in the *Investigations and Projects Masters*, p. 16.

 ### Cooperative Learning

This investigation offers an excellent opportunity for using cooperative learning groups. For more information on cooperative learning strategies and group management, see *Cooperative Learning in the Mathematics Classroom*.

b. For each item in the table, write the ordered pair (radius, area). Then graph each ordered pair on the second coordinate plane.

c. For each item in the table, write the ordered pair (circumference, area). Then graph each ordered pair on the third coordinate plane.

6. Write a sentence or two describing each graph.

7. Measure the circumference of a new item. The circumference should be between the smallest and largest items you used for your table. Use the graph from Exercise 5c to estimate the area of the circle. How did your estimate compare to the actual area?

Making the Connection

Use the information about your circles as needed to help in these investigations.

Language Arts

Choose two circles from the investigation. Write directions for finding the side length of two squares that have approximately the same area as the circles.

Physical Education

In the sport of archery, athletes often use a target to improve their skill or for competitions. Research this sport and find the diameter of the target. Use your graphs to estimate the circumference and area of the target.

Technology Tips

- Use a **spreadsheet** to calculate radius and area.

- Use a **graphing calculator** or **graphing software** to display your graphs.

Go Further

- Use a graphing calculator to find equations for the three functions you graphed.

- Research the history of π.

- Find the area of each ring on an archery target.

 **inter**NET **CONNECTION** For current information on mathematics, visit:
www.glencoe.com/sec/math/mac/mathnet

 You may want to place your work on this investigation in your portfolio.

Interdisciplinary Investigation Going in Circles **511**

Instructional Resources ▶▶▶

Sample solutions for this investigation are provided in the *Investigations and Projects Masters* on p. 14. The scoring guide for assessing student performance shown at the right is also available on p. 15.

MANAGEMENT TIPS

Working in Teams Have students take turns measuring the circular items. Have one student make the table and then trade off making the graphs.

Making the Connection

You may wish to alert other teachers on your team that your students may need their assistance in this investigation. *Language Arts* Students may trade their directions to see whether other students can follow them easily. *Physical Education* Students may wish to investigate whether there are different-size targets for different levels of competition.

ASSESS

You may wish to have students determine careers that would require knowledge of circumference and area relationships.

Investigations and Projects Masters, p. 15

SCORING GUIDE

Interdisciplinary Investigation
(Student Edition, Pages 510–511)

Going In Circles

Level	Specific Criteria
3 Superior	• Shows a thorough understanding of the concepts of *applying known formulas to find missing information about circles and constructing coordinate planes and graphing coordinate pairs.* • Uses appropriate strategies to solve problems. • Computations are correct. • Written explanations are exemplary. • Charts, model, and any statements included are appropriate and sensible. • Goes beyond the requirements of some or all problems.
2 Satisfactory, with minor flaws	• Shows understanding of the concepts of *applying known formulas to find missing information about circles and constructing coordinate planes and graphing coordinate pairs.* • Uses appropriate strategies to solve problems. • Computations are mostly correct. • Written explanations are effective. • Charts, model, and any statements included are appropriate and sensible. • Satisfies the requirements of problems.
1 Nearly Satisfactory, with obvious flaws	• Shows understanding of most of the concepts of *applying known formulas to find missing information about circles and constructing coordinate planes and graphing coordinate pairs.* • May not use appropriate strategies to solve problems. • Computations are mostly correct. • Written explanations are satisfactory. • Charts, model, and any statements included are appropriate and sensible. • Satisfies the requirements of problems.
0 Unsatisfactory	• Shows little or no understanding of the concepts of *applying known formulas to find missing information about circles and constructing coordinate planes and graphing coordinate pairs.* • Does not use appropriate strategies to solve problems. • Computations are incorrect. • Written explanations are not satisfactory. • Charts, model, and any statements included are not appropriate or sensible. • Does not satisfy the requirements of the problems.

© Glencoe/McGraw-Hill · 15 · Mathematics: Applications and Connections, Course 1

Using Probability

Previewing the Chapter

Overview

This chapter explores probability in a variety of ways. Students learn to find the probability of single events and independent events. They also learn to predict actions of a large group by using a sample. Area models and tree diagrams are used to find outcomes. Students also learn to solve problems by making tables.

Lesson (pages)	Lesson Objectives	NCTM Standards	Standardized Tests	State/Local Objectives
13-1A (514)	Explore fair and unfair games.	1–4, 7, 8, 10, 11		
13-1 (515–518)	Find and interpret the theoretical probability of an event.	1–11	CAT, CTBS, ITBS, MAT, SAT, TN	
13-2A (520–521)	Solve problems by making a table.	1–8, 10	CTBS, MAT, TN	
13-2 (522–525)	Predict the actions of a larger group using a sample.	1–8, 10–13	CTBS, ITBS, MAT, TN	
13-3 (526–529)	Find probability using area models.	1–8, 11–13	MAT	
13-3B (530)	Investigate geometric probability by using a graphing calculator.	1–5, 7–12		
13-4 (531–534)	Find outcomes using lists, tree diagrams, and combinations.	1–8, 11–13	CTBS, ITBS, MAT, SAT, TN	
13-4B (535)	Explore probability by conducting a simulation.	1–4, 10, 11		
13-5 (536–539)	Find the probability of independent events.	1–8, 10–12	CAT, CTBS, ITBS, MAT, SAT, TN	

CAT = California Achievement Tests, CTBS = Comprehensive Tests of Basic Skills, ITBS = Iowa Tests of Basic Skills, MAT = Metropolitan Achievement Tests, SAT = Stanford Achievement Tests, TN = Terra Nova

Organizing the Chapter

CD-ROM

All of the blackline masters in the Teacher's Classroom Resources are available on the **Electronic Teacher's Classroom Resources** CD-ROM.

LESSON PLANNING GUIDE

Lesson	Extra Practice (Student Edition)	BLACKLINE MASTERS (PAGE NUMBERS)										
		Study Guide	Practice	Enrichment	Assessment & Evaluation	Classroom Games	Diversity	Hands-On Lab	School to Career	Science and Math Lab Manual	Technology	Transparencies A and B
13-1A								67				
13-1	p. 590	90	90	90			13					13-1
13-2A	p. 591											
13-2	p. 591	91	91	91	351				13		25	13-2
13-3	p. 591	92	92	92	350, 351			81				13-3
13-3B												
13-4	p. 592	93	93	93	352						26	13-4
13-4B								68				
13-5	p. 592	94	94	94	352	37–40				93–96		13-5
Study Guide/ Assessment					337–349, 353–355							

OTHER CHAPTER RESOURCES

Student Edition
Chapter Project, pp. 513, 518, 529, 543
School to Career, p. 519
Let the Games Begin, p. 539

Technology
 CD-ROM Program

 Interactive Mathematics Tools Software

Teacher's Classroom Resources

Applications
Family Letters and Activities, pp. 25–26
Investigations and Projects Masters, pp. 65–68
Meeting Individual Needs
Investigations for the Special Education Student, pp. 45–46

Teaching Aids
Answer Key Masters
Block Scheduling Booklet
Lesson Planning Guide
Solutions Manual

Professional Publications
Glencoe Mathematics Professional Series

Planning the Chapter

MindJogger Videoquizzes provide a unique format for reviewing concepts presented in the chapter.

ASSESSMENT RESOURCES

Student Edition

Mixed Review, pp. 518, 525, 529, 534, 539
Mid-Chapter Self Test, p. 525
Math Journal, pp. 517, 533
Study Guide and Assessment, pp. 540–543
Performance Task, p. 543
Portfolio Suggestion, p. 543
Standardized Test Practice, pp. 544–545
Chapter Test, p. 607

Assessment and Evaluation Masters

Multiple-Choice Tests (Forms 1A, 1B, 1C), pp. 337–342
Free-Response Tests (Forms 2A, 2B, 2C), pp. 343–348
Performance Assessment, p. 349
Mid-Chapter Test, p. 350
Quizzes A–D, pp. 351–352
Standardized Test Practice, pp. 353–354
Cumulative Review, p. 355

Teacher's Wraparound Edition

5-Minute Check, pp. 515, 522, 526, 531, 536
Building Portfolios, p. 512
Math Journal, pp. 514, 530, 535
Closing Activity, pp. 518, 521, 525, 529, 534, 539

Technology

Test and Review Software
MindJogger Videoquizzes
CD-ROM Program

MATERIALS AND MANIPULATIVES

Lesson 13-1A	**Lesson 13-1**	**Lesson 13-3**	**Lesson 13-4B**
number cubes*	newspaper	dot paper†	two-colored counters*†
	white paper	beans	cups*†
	black paper		
	hole punch	**Lesson 13-3B**	**Lesson 13-5**
		graphing calculator	index cards
	Lesson 13-2	grid paper†	
	newspaper	straightedge*†	

*Glencoe Manipulative Kit   †Glencoe Overhead Manipulative Resources

PACING CHART

See pages T25–T27 for the Course Planning Calendar.

COURSE	DAY 1	DAY 2	DAY 3	DAY 4	DAY 5	DAY 6	DAY 7
Honors	Chapter Project	Lesson 13-1	Lesson 13-2A	Lesson 13-2	Lessons 13-3 & 13-3B		Lessons 13-4 & 13-4B

Interactive Mathematics:
Activities and Investigations

is an activity-based program that may be used as an enhancement for chapters in *Mathematics: Applications and Connections.*

Unit 10, Activity Two
It Happened by Chance
Use with Lesson 13-1.

Summary Students work in groups to complete this activity. Each group selects a fraction card from the deck provided and creates a written situation using that fraction as a probability. The groups write their solutions on posterboard and present them to the class.

Math Connection Students work in groups to explore probability by creating written situations that involve a chosen probability. Probability is the ratio of the number of ways an outcome can occur to the number of possible outcomes.

Unit 10, Activity Three
Use with Lesson 13-5.

Summary Students work in groups to design five random-generating spinners using the clues given on clue cards. Once the group has designed the spinner, students work individually, using compasses and protractors, to make an accurate drawing of the spinner.

Math Connection Students connect probability with geometry and measurement in this activity. They expand their knowledge of circles by designing random-generating spinners. Students use compasses, protractors, and number properties to design and draw their spinners.

DAY 8	DAY 9	DAY 10	DAY 11	DAY 12	DAY 13	DAY 14	DAY 15
(continue from Day 7)	Lesson 13-5	Study Guide and Assessment	Chapter Test				

Enhancing the Chapter

APPLICATIONS

Classroom Games, pp. 37–40

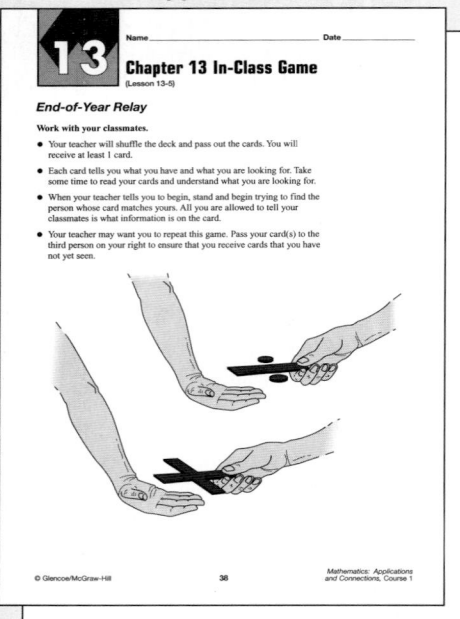

13 Chapter 13 In-Class Game
(Lesson 13-5)

End-of-Year Relay

Work with your classmates.

- Your teacher will shuffle the deck and pass out the cards. You will receive at least 1 card.

- Each card tells you what you have and what you are looking for. Take some time to read your cards and understand what you are looking for.

- When your teacher tells you to begin, stand and begin trying to find the person whose card matches yours. All you are allowed to tell your classmates is what information is on the card.

- Your teacher may want you to repeat this game. Pass your card(s) to the third person on your right to ensure that you receive cards that you have not yet seen.

© Glencoe/McGraw-Hill   38   *Mathematics: Applications and Connections, Course 1*

Diversity Masters, p. 13

13 Diversity Activity
(Lesson 13-1)

Zhū-Shijié

Zhū-Shijié (1280–1303) has been called one of the greatest mathematicians of all time. In 1303, he wrote the classic text *Siyuán yùjiàn (The Precious Mirror of the Four Elements)*, which included the triangular pattern of numbers shown. Zhū-Shijié knew his pattern could help solve advanced equations.

```
          1           ← row 0
        1   1         ← row 1
      1   2   1       ← row 2
    1   3   3   1     ← row 3
  1   4   6   4   1   ← row 4
```

More than 300 years later, French mathematician Blaise Pascal (1623–1662), discovered that the pattern had uses in probability. The pattern became known as **Pascal's Triangle**. In the exercises that follow, you will have a chance to explore some of the special properties of the triangle. Use H for heads and T for tails.

1. If you toss one coin, what are the possible outcomes? two coins? Arrange your answers for two coins into the table below.   **H, T; HH, HT, TH, TT**

Outcomes	All H	1 H, 1 T	All T
Number of Outcomes	1	2	1

2. Compare the numbers in the table to the second row of the triangle. What do you notice?   **The pattern of 1, 2, 1 is the same.**

3. If you toss three coins, what are the possible outcomes? Arrange your answers into the table below.   **HHH, HHT, HTH, THH, HTT, TTH, THT, TTT**

Outcomes	All H	2 H, 1 T	2 T, 1 H	All T
Number of Outcomes	1	3	3	1

4. Compare the numbers in the table to the third row of the triangle. What do you notice?   **The pattern of 1, 3, 3, 1 is the same.**

5. Using Pascal's Triangle, predict the possible outcomes from tossing four coins. Arrange your answer into the table below.

Outcomes	All H	3 H, 1 T	2 H, 2 T	3 T, 1 H	All T
Number of Outcomes	1	4	6	4	1

© Glencoe/McGraw-Hill   T13   *Mathematics: Applications and Connections, Course 1*

School to Career Masters, p. 13

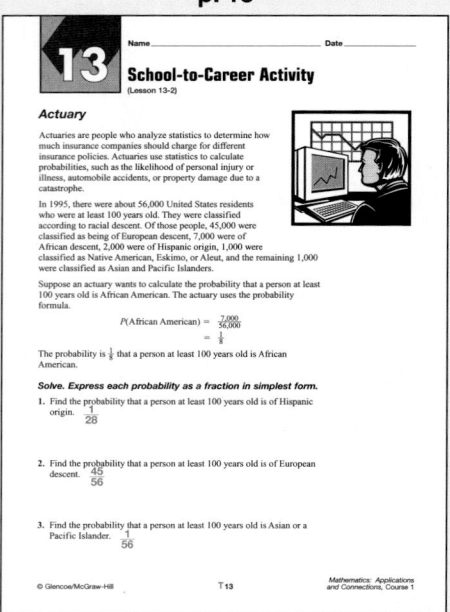

13 School-to-Career Activity
(Lesson 13-2)

Actuary

Actuaries are people who analyze statistics to determine how much insurance companies should charge for different insurance policies. Actuaries use statistics to calculate probabilities, such as the likelihood of personal injury or illness, automobile accidents, or property damage due to a catastrophe.

In 1995, there were about 56,000 United States residents who were at least 100 years old. They were classified according to racial descent. Of those people, 45,000 were classified as being of European descent, 7,000 were of African descent, 2,000 were of Hispanic origin, 1,000 were classified as Native American, Eskimo, or Aleut, and the remaining 1,000 were classified as Asian and Pacific Islanders.

Suppose an actuary wants to calculate the probability that a person at least 100 years old is African American. The actuary uses the probability formula.

$$P(\text{African American}) = \frac{7,000}{56,000}$$
$$= \frac{1}{8}$$

The probability is $\frac{1}{8}$ that a person at least 100 years old is African American.

Solve. Express each probability as a fraction in simplest form.

1. Find the probability that a person at least 100 years old is of Hispanic origin. $\frac{1}{28}$

2. Find the probability that a person at least 100 years old is of European descent. $\frac{45}{56}$

3. Find the probability that a person at least 100 years old is Asian or a Pacific Islander. $\frac{1}{56}$

© Glencoe/McGraw-Hill   T13   *Mathematics: Applications and Connections, Course 1*

Family Letters and Activities, pp. 25–26

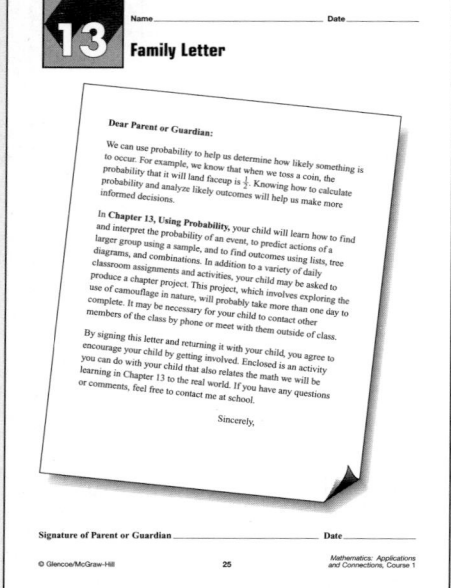

13 Family Letter

Dear Parent or Guardian:

We can use probability to help us determine how likely something is to occur. For example, we know that when we toss a coin, the probability that it will land faceup is $\frac{1}{2}$. Knowing how to calculate probability and analyze likely outcomes will help us make more informed decisions.

In **Chapter 13, Using Probability,** your child will learn how to find and interpret the probability of an event, to predict actions of a larger group using a sample, and to find outcomes using lists, tree diagrams, and combinations. In addition to a variety of daily classroom assignments and activities, your child may be asked to produce a chapter project. This project, which involves exploring the use of camouflage in nature, will probably take more than one day to complete. It may be necessary for your child to contact other members of the class by phone or meet with them outside of class.

By signing this letter and returning it with your child, you agree to encourage your child by getting involved. Enclosed is an activity you can do with your child that also relates the math we will be learning in Chapter 13 to the real world. If you have any questions or comments, feel free to contact me at school.

Sincerely,

Signature of Parent or Guardian _____ Date _____

© Glencoe/McGraw-Hill   25   *Mathematics: Applications and Connections, Course 1*

13 Family Activity

What's Your Preference?

Work with a family member to answer the following questions.

1. Determine the number of parents, siblings, grandparents, aunts, uncles, and close friends you have. Record the number below.

2. Write the name of each family member or friend on a separate piece of paper. Place all of the papers in a hat. Without looking, select five pieces of paper. Record the names of those family members or friends selected in the space below.

3. Ask those five family members or friends what color they prefer. Record the results in the space below.

4. What fraction of your family members or friends selected in Exercise 2 prefers red? blue? yellow? green?

5. Based on your sample, predict how many of your family members and friends listed in Exercise 1 prefer blue.

6. Based on your sample, predict how many of your family members and friends listed in Exercise 1 prefer yellow.

7. Based on your sample, predict how many of your family members and friends listed in Exercise 1 prefer red.

© Glencoe/McGraw-Hill   26   *Mathematics: Applications and Connections, Course 1*

Science and Math Lab Manual, pp. 93–96

24 Science and Mathematics Lab
(Course 1, Lesson 13-5; Course 2, Lesson 13-4; Course 3, Lesson 12-6)

The Gender of Children

INTRODUCTION

There is an equal probability that a child will be born female or male. When a family has more than one child, the gender of each child is an independent event that is not influenced by the gender of previously born children.

OBJECTIVES

In this lab, you will:
- explore a series of independent events in a simulation.
- compare the results of your simulation with those of your classmates.

MATERIALS
- coin

DATA AND OBSERVATIONS

Family	Child 1	Child 2	Child 3
A			
B			
C			
D			
E			
F			
G			
H			
I			
J			

© Glencoe/McGraw-Hill   95   *Mathematics: Applications and Connections*

MANIPULATIVES/MODELING

Hands-On Lab Masters, p. 81

13 Name ____________ Date ____________

Lab Activity
(Lesson 13-3)

Probability and Area

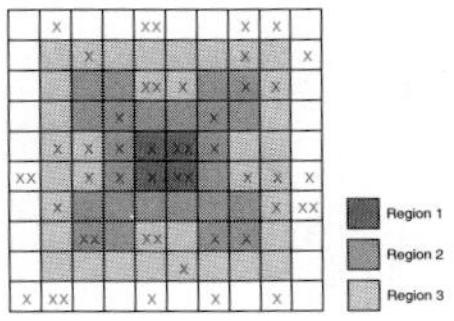

Region 1
Region 2
Region 3

1. Count the number of squares in each region to find the area.
   Region 1: 36 squares
   Region 2: 24 squares
   Region 3: 4 squares
   Total Area: 100 squares
2. Find the ratio $\frac{\text{area of region}}{\text{total area}}$ for each region.
   Region 1: $\frac{9}{25}$
   Region 2: $\frac{6}{25}$
   Region 3: $\frac{1}{25}$

Drop 50 grains of rice onto the square target from a height of 2 or 3 inches. Mark an X wherever a grain lands inside the target. If a grain does not fall completely inside of one square, mark the X on the square that contains most of the grain. See X's on target for sample answers below.

3. Count the number of grains that landed in each region.
   Region 1: 16 grains
   Region 2: 11 grains
   Region 3: 6 grains
   Total dropped on target: 50 grains
4. For each region, find the ratio $\frac{\text{number of grains in region}}{\text{number of grains dropped on target}}$.
   Region 1: $\frac{8}{25}$
   Region 2: $\frac{11}{50}$
   Region 3: $\frac{3}{25}$
5. Compare the ratios in Exercise 2 to the ratios in Exercise 4 and write a summary of your results. Sample answer: The area ratios and grain ratios are close, but not exactly the same.

© Glencoe/McGraw-Hill 81 *Mathematics: Applications and Connections, Course 1*

ASSESSMENT/EVALUATION

Assessment and Evaluation Masters, pp. 350–352

13 Name ____________ Date ____________

Chapter 13 Mid-Chapter Test
(Lessons 13-1 through 13-3)

A bag contains 8 white marbles, 5 black marbles, and 7 red marbles. Marbles are selected randomly. Find the probability of each event.

1. P(white)
2. P(red)
3. P(blue)
4. P(white or black)
5. P(white or black or red)
6. 100 marbles were selected randomly from the bag and replaced. Make a table showing the expected results.

In a survey at a restaurant, it was found that 12 out of 40 people order french fries with their lunch.

7. What is the sample size?
8. What is the probability that someone will order french fries with their lunch?
9. What is the probability that someone will *not* order french fries with their lunch?
10. If 120 people eat at the restaurant during lunchtime, how many will likely order french fries?
11. A survey of favorite radio stations is taken at an arena after a concert sponsored by a local station. Is this a random survey?

The figure shown represents a dartboard. It is equally likely that a dart will land anywhere on the dartboard. Find each probability for randomly thrown darts.

12. P(shaded)
13. P(unshaded)
14. Suppose you threw a dart 180 times at the dartboard. How many times would you expect it to land in the shaded region?
15. How many times would you expect it to land in the unshaded region?
16. A dart is randomly thrown at the dartboard shown at the right. It is equally likely that a dart will land anywhere on the dartboard. Find the probability of a randomly thrown dart landing in the shaded region.

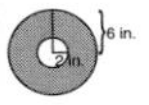

6 in.
2 in.

Answers:
1. $\frac{2}{5}$
2. $\frac{7}{20}$
3. 0
4. $\frac{13}{20}$
5. 1
6.

Color	Frequency
white	40
black	25
red	35

7. 40
8. $\frac{3}{10}$
9. $\frac{7}{10}$
10. 36
11. no
12. $\frac{5}{9}$
13. $\frac{4}{9}$
14. 100
15. 80
16. $\frac{8}{9}$

© Glencoe/McGraw-Hill 350 *Mathematics: Applications and Connections, Course 1*

Name ____________ Date ____________

Chapter 13 Quiz A
(Lessons 13-1 and 13-2)

A number cube is rolled. Find the probability of each event.

1. $P(6)$
2. P(even number)

A sack contains 10 red jelly beans, 8 black jelly beans, and 7 white jelly beans. One jelly bean is chosen at random. Find the probability of each event.

3. P(white)
4. P(red)
5. P(black or red)
6. P(green)

Maria surveyed people buying flats of plants at a garden center. The results are shown in the table.

geraniums	8
petunias	12
impatiens	10
begonias	10

7. What is the size of the sample?
8. What is the probability that a customer will buy petunias?
9. What is the probability that a customer will buy impatiens or begonias?
10. If the garden center manager plans to order 400 flats of plants, how many flats of geraniums should be ordered?

Answers:
1. $\frac{1}{6}$
2. $\frac{1}{2}$
3. $\frac{7}{25}$
4. $\frac{2}{5}$
5. $\frac{18}{25}$
6. 0
7. 40
8. $\frac{3}{10}$
9. $\frac{1}{2}$
10. 80

Name ____________ Date ____________

Chapter 13 Quiz B
(Lesson 13-3)

Suppose you threw a dart 100 times at each dartboard shown. How many times would you expect it to land in the shaded region?

1. 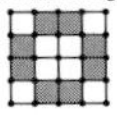
2. 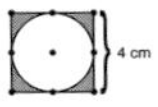

4 cm

3. What is the probability that a randomly thrown dart would land in the shaded region of the dartboard in Question 1?
4. What is the probability that a randomly thrown dart would land in the shaded region of the dartboard in Question 2?
5. What is the probability that a randomly thrown dart would land in the unshaded region of the dartboard in Question 2?

Answers:
1. 50
2. about 21
3. $\frac{1}{2}$
4. about $\frac{1}{4}$
5. about $\frac{3}{4}$

© Glencoe/McGraw-Hill 351 *Mathematics: Applications and Connections, Course 1*

TECHNOLOGY/MULTIMEDIA

Technology Masters, pp. 25–26

13 Name ____________ Date ____________

Calculator Activity
(Lesson 13-2)

Making Predictions Using Samples

Two classes of sixth-grade students were asked to give their favorite brand of athletic shoe. The results are shown at the right.

Favorite Brand of Shoe (Number of Students)

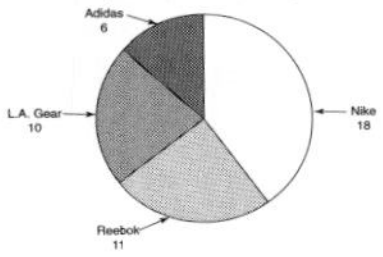

Adidas 6
Nike 18
L.A. Gear 10
Reebok 11

Find the probability that the favorite brand of athletic shoe of a sixth-grade student is Nike.

Find the size of the sample.

Enter: 6 [+] 18 [+] 11 [+] 10 [=] 45

The sample size is 45.

Find the probability that the favorite brand is Nike.

$P(\text{Nike}) = \frac{\text{number who chose Nike}}{\text{size of the sample}}$

Enter: 18 [÷] 45 [=] 0.4

The probability that the favorite brand is Nike is 0.4.

Find the probability of making each of the choices below. Round to the nearest hundredth.

Students' Favorite Colors

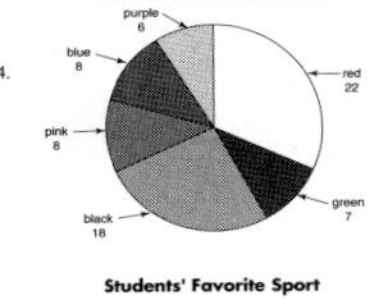

purple 6
blue 8
red 22
pink 8
green 7
black 18

1. survey of sixth-graders' favorite color
   P(red): 0.32 P(blue): 0.12
   P(green): 0.10 P(pink): 0.12
   P(black): 0.26 P(purple): 0.09

Students' Favorite Sport

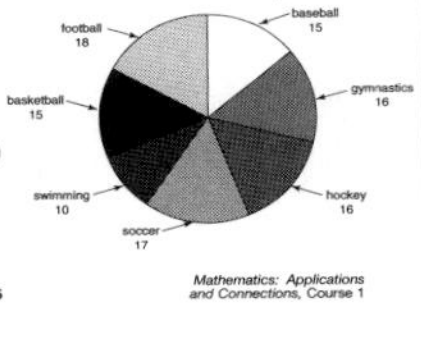

football 18
baseball 15
basketball 15
gymnastics 16
swimming 10
hockey 16
soccer 17

2. survey of the favorite sport of 107 sixth graders
   P(soccer): 0.16 P(baseball): 0.14
   P(basketball): 0.14 P(swimming): 0.09
   P(hockey): 0.15 P(football): 0.17
   P(gymnastics): 0.15

© Glencoe/McGraw-Hill T25 *Mathematics: Applications and Connections, Course 1*

13 Name ____________ Date ____________

Graphing Calculator Activity
(Lesson 13-4)

Expected and Actual Probability

You can use a graphing calculator to conduct simulations.

Example **Suppose you asked 50 people to choose a number from 0 to 9. Find the expected probability of each possible response or outcome.**

Possible Outcome	Expected Frequency	Expected Probability	Possible Outcome	Expected Frequency	Expected Probability
0	5	$\frac{1}{10}$	5	5	$\frac{1}{10}$
1	5	$\frac{1}{10}$	6	5	$\frac{1}{10}$
2	5	$\frac{1}{10}$	7	5	$\frac{1}{10}$
3	5	$\frac{1}{10}$	8	5	$\frac{1}{10}$
4	5	$\frac{1}{10}$	9	5	$\frac{1}{10}$

Instead of asking 50 people, you can simulate the data using a graphing calculator.

Enter: [MATH] [▶] [▶] [▶] [ENTER] [ENTER]

The calculator will display a random decimal. Use the individual digits in the number as if they were responses. Record them in a table like the one below. Repeat this process until you have 50 responses. Find the actual frequency by counting the tallies. Find the actual probability by using the actual frequency.

Outcome	Tally	Actual Frequency	Actual Probability
0			
1			

Solve.

1. Use a graphing calculator to conduct this simulation. Complete the table using your random data. Then find the actual probabilities.
   Check students' tables for the correct actual probabilities.
2. Compare your data with those of a friend. Why are they not exactly the same? Since the calculator gives random numbers, no two sets of data will be exactly the same.

© Glencoe/McGraw-Hill T26 *Mathematics: Applications and Connections, Course 1*

MEETING INDIVIDUAL NEEDS

Investigations for the Special Education Student, pp. 45–46

Use with:
Course 1-Chapter 14
Course 2-Chapter 13
Course 3-Chapter 13

Investigation 15 Teacher's Guide

The Game Show

Overview

The purpose of this investigation is for students to recognize that problem-solving and mathematical strategies are needed to play even the most enjoyable games. The problem-solving and mathematical value of several games will be evaluated by students during a three-day game review period. At the end of this time, the class will compile the results of their reviews into one master list of the "top ten" mathematics-related games. This list will then be forwarded to the appropriate school personnel along with a letter of recommendation to include such games as part of the mathematics program at the school.

Activity Goals

Students will:
- explore the problem-solving and mathematical strategies used in games,
- review and give feedback for certain games, and
- participate in selecting and recommending games for the school to purchase.

Planning the Instruction

Prerequisite Skills
Students should have a significant amount of practice reading and following directions, and computing with whole numbers. A basic understanding of probability may be helpful.

Materials
- investigation worksheet
- assortment of games for the classroom
- manila envelopes (one per game)

Time Needed
six 45-minute periods

Procedure

1. Given an example of a well-known game such as checkers, have students brainstorm the problem-solving and mathematical strategies that are necessary to play the game. Then, explain to the class that they are about to become game consultants for the school by recommending games that will enhance the problem-solving and mathematical skills of its students.
2. Obtain a variety of games for use in the classroom. Depending on the classroom situation, teachers may bring games from home or have students volunteer to bring them.
3. Students will have three days in class to review as many games as possible. To review a game, students must follow the reviewing procedure on the Game Review worksheet. They will play each game twice, the second time answering the review questions. After reviewing each game, students will place the review sheet for that game in the envelope provided. If time permits, students should review another game using a new Game Review worksheet.
4. At the end of the reviewing periods, the results will be tallied, and, as a class, students and teacher will compile a list of the top ten games.
5. Once the list is established, the class will submit it to the appropriate personnel along with a letter explaining and supporting their research. The letter should also address the need for these games to be part of the mathematics program at the school.
6. An extension of this investigation might require students to create their own game involving the problem-solving and mathematical strategies believed to be the most valuable.

Adaptations and Variations

The following are some ideas on how this investigation may be modified depending on student population.

LD
- Simplify rules for easier comprehension.
- Review several games together with the class.
- Allow students to work with a partner.

PH
- Allow students to type or word process their review.
- Be sure to choose games suitable to their needs.

CD
- Select games suitable to their needs.

BD
- Maintain close proximity during the review process.
- If problems arise, have them play alone until they earn their way back to working with others.

HI
- Select games suitable to their needs.

VI
- Enlarge directions to games.
- Assign an aide to help students with any reading.

© Glencoe/McGraw-Hill 45 *Mathematics: Applications and Connections*

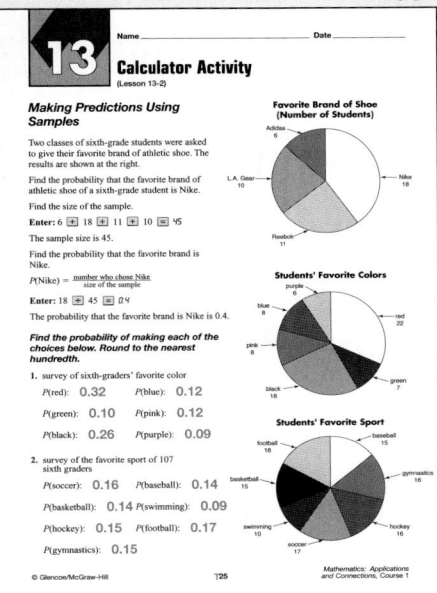

CHAPTER 13
Using Probability

Many animals must protect themselves from predators. Camouflage is an adaptation involving colors, patterns, and sometimes textures that enables animals to blend into their environment and become difficult to see. Some animals such as chameleons, flatfish, and shrimp can change color. Stick insects blend in because they look like twigs.

Question of the Day A sea turtle lays 200 eggs on shore but only 75 survive and make it back to the sea. What is the probability that an egg layed will result in a young turtle making it back to the sea? $\frac{3}{8}$

Assess Prerequisite Skills

Ask students to read through the list of objectives presented in "What you'll learn in Chapter 13." You may wish to ask them what each of the objectives means or if they have experienced or used any of these math concepts before.

 **Building Portfolios**

Encourage students to revise their portfolios as they study this chapter. Have them add work that shows how they determine probability and find outcomes. Encourage students to include predictions for real-world problems along with descriptions of actual outcomes.

 Math and the Family

In the *Family Letters and Activities* booklet (pp. 25–26), you will find a letter to the parents explaining what students will study in Chapter 13. An activity appropriate for the whole family is also available.

***What* you'll learn in Chapter 13**

- to find and interpret the probability of an event or events,
- to solve problems by making a table,
- to predict actions of a larger group using a sample,
- to find probability using area models, and
- to find outcomes using lists, tree diagrams, and combinations.

512 Chapter 13 Using Probability

 CD-ROM Program

Activities for Chapter 13
- Chapter 13 Introduction
- Interactive Lesson 13-3
- Extended Activity 13-1
- Assessment Game
- Resource Lessons 13-1 through 13-5

CHAPTER Project

HERE TODAY, GONE TOMORROW

In order to survive in their environment, many animals are camouflaged, or hidden, from predators. In this project, you will use probability to explore how camouflage works. You will also research how one animal is camouflaged, make a poster about the animal, and write a report on how a probability experiment can be used to illustrate camouflage.

Getting Started

- You will need black paper, white paper, two sections of the classified ads from the newspaper, a hole punch, and poster board.
- Choose one animal that lives in the wild that you would like to research.
- Research the animal.
  - Where does the animal live?
  - What does the animal eat?
  - Is survival of the animal aided by camouflage?
  - Is the animal an endangered species? Find statistics that tell how many of the animals still exist. What is being done to help the animals survive?

Technology Tips

- Use an **electronic encyclopedia** or the **Internet** to research the animal.
- Use a **word processor** to write a report on how the probability experiment illustrates camouflage.
- Use **computer software** to design your animal poster.

interNET CONNECTION For up-to-date information on animal species, visit:
www.glencoe.com/sec/math/mac/mathnet

Working on the Project

You can use what you'll learn in Chapter 13 to help you explore and understand camouflage.

Page	Exercise
518	27
529	23
543	Alternative Assessment

Instructional Resources ►►►

A recording sheet to help students organize their data for the Chapter Project is shown at the right and is available in the *Investigations and Projects Masters*, p. 68.

CHAPTER Project NOTES

Objectives Students should
- use probability to explore how camouflage works.
- research how animals are camouflaged.
- write a report on how a probability experiment can be used to illustrate camouflage.

Project Pointer You may suggest that students begin a *Project Folder* to keep their work as they complete each stage of the Chapter Project. The completed project may also be added to their portfolios.

Have students determine the probability that an animal's camouflage technique will enable it to live for various periods of time. Have them use population estimates for the past several years to predict how many of this species will survive the current year.

Investigations and Projects Masters, p. 68

---

13 Name_____ Date_____

Chapter 13 Project

Here Today, Gone Tomorrow

Page 513, Getting Started

Animal:

Home:

Food:

Behavior:

Type of camouflage:

Endangerment status:

Page 518, Working on the Chapter Project, Exercise 27

Color	Number Gathered	Probability
Classified Ads		
White Paper		
Black Paper		

Page 529, Working on the Chapter Project, Exercise 23

Color	Number Picked from 100	Number Picked from 250
Classified Ads		
White Paper		
Black Paper		

GET READY

Objective Students explore fair and unfair games.

Optional Resources

Hands-On Lab Masters
- number cube patterns, p. 21
- worksheet, p. 67

Manipulative Kit
- number cubes

MANAGEMENT TIPS

Recommended Time
45 minutes

Getting Started Have students discuss what makes a game fair and what makes a game unfair. Students should agree on what to do when a number cube falls on the floor.

In **Activity 1,** students play a fair game with even chances for winning. Have partners take turns recording the results and rolling the number cubes.

In **Activity 2,** students play an unfair game. There are more even products than odd. Again have partners take turns recording the results and rolling the number cubes.

ASSESS

Have students complete Exercises 1–4. Ask students if changing the number of rolls would affect the outcome of the games. Have them explain their answers.

Use Exercise 5 to determine whether students understand the ideas of more likely, equally likely, and less likely outcomes.

COOPERATIVE LEARNING
13-1A Fair and Unfair Games
A Preview of Lesson 13-1

number cubes

In this lab, you will explore fair and unfair games. A *fair game* is a game in which players have an equal chance of winning. In an *unfair game*, players do not have an equal chance of winning.

TRY THIS

Work with a partner.

1 Rules of the game:
- Roll two number cubes.
- Add the two numbers that are face up.
- Player 1 gets one point if the sum is even. Player 2 gets one point if the sum is odd.
- Roll the number cubes 40 times.
- Record each sum in a chart like the one shown.

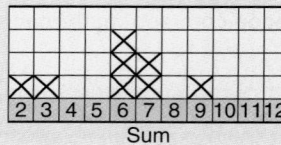
Sum

2 Rules of the game:
- Roll two number cubes.
- Multiply the two numbers that are face up.
- Player 1 gets one point if the product is even. Player 2 gets one point if the product is odd.
- Roll the number cubes 40 times.
- Record each product in a chart like the one shown.

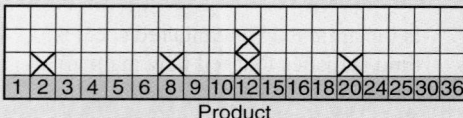
Product

ON YOUR OWN

1–5. See Answer Appendix.
1. Compare the number of even sums tossed to the number of odd sums tossed, and the number of even products tossed to the number of odd products tossed.
2. Which game do you think is fair? Which is unfair? Explain your reasoning.
3. Your charts should resemble a bar graph. Describe the shape of each of your graphs.
4. Which sum occurred most often? least often? Which product occurred most often? least often?
5. *Look Ahead* In Game 1, is getting a sum of 12 impossible, less likely than not getting 12, equally likely with not getting 12, more likely than not getting 12, or certain to occur? Explain your reasoning.

514 Chapter 13 Using Probability

Math Journal Have students write a description of a fair game and an unfair game involving number cubes. Have them include a definition of *fair* and *unfair*.

13-1 Theoretical Probability

What you'll learn
You'll learn to find and interpret the theoretical probability of an event.

When am I ever going to use this?
Knowing about probability can help you determine the chance of winning a game at a festival.

Word Wise
theoretical probability
outcomes
event
sample space
complementary events

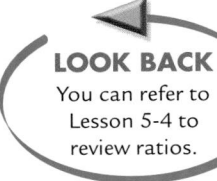

LOOK BACK
You can refer to Lesson 5-4 to review ratios.

Have you ever watched the Showcase Showdown on *The Price is Right*? The contestants spin a wheel that is separated into equal sections marked 5¢, 10¢, 15¢, ... , $1. Each contestant gets two spins. The contestant closest to $1, without going over, wins a spot in the showcase. What is the **theoretical probability**, or chance, that the wheel will land on $1 the first time a contestant spins it?

There are twenty equally likely results, or **outcomes** on the wheel. The specific outcome or type of outcome is called an **event**. In this case, the outcomes are all of the different sections the wheel could stop on, and the event is landing on $1.

The set of all possible outcomes is called the **sample space**. The sample space for the wheel is {5¢, 10¢, 15¢, ..., $1}.

To find the probability of landing on $1, you can use a ratio.

Theoretical Probability	Words:	The theoretical probability of an event is the ratio of the number of ways the event can occur to the number of possible outcomes.
	Symbols:	$P(\text{event}) = \dfrac{\text{number of ways the event can occur}}{\text{number of possible outcomes}}$

In the sample space, there is one way to land on $1 and 20 possible outcomes. So, $P(\$1) = \frac{1}{20}$.

You can express the probability of an event as a fraction, decimal, or percent. The probability that an event will occur is a number from 0 to 1. You can interpret probabilities using a number line.

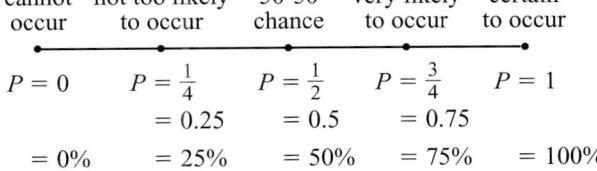

cannot occur	not too likely to occur	50-50 chance	very likely to occur	certain to occur
$P = 0$	$P = \frac{1}{4}$	$P = \frac{1}{2}$	$P = \frac{3}{4}$	$P = 1$
	$= 0.25$	$= 0.5$	$= 0.75$	
$= 0\%$	$= 25\%$	$= 50\%$	$= 75\%$	$= 100\%$

- A probability of 0 means that the event cannot occur.

- A probability of $\frac{1}{2}$ means that there is a 50-50 chance that the event will occur.

- A probability of 1 means that the event is certain to occur.

- The closer a probability is to 1, the more likely the event is to occur.

Lesson 13-1 Theoretical Probability **515**

Instructional Resources
- *Study Guide Masters*, p. 90
- *Practice Masters*, p. 90
- *Enrichment Masters*, p. 90
- Transparencies 13-1, A and B
- *Diversity Masters*, p. 13

CD-ROM Program
- Resource Lesson 13-1
- Extended Activity 13-1

Recommended Pacing	
Honors	Day 2 of 11

1 FOCUS

5-Minute Check
(Chapter 12)

Solve each equation.
1. $r - 7 = -18$ −11
2. $x + (-5) = 14$ 19
3. $4n = -32$ −8
4. $\frac{1}{5}b = 6$ 30
5. Complete the following function table. Then graph the function.

input (x)	output ($\frac{1}{4}x$)
−4	■ −1
0	■ 0
8	■ 2

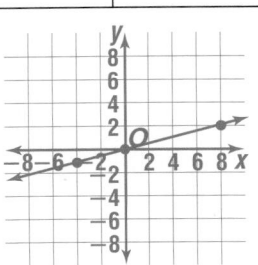

The 5-Minute Check is also available on **Transparency 13-1A** for this lesson.

Cross-Curriculum Cue

Inform the teachers on your team that your students are studying probability and area. Suggestions for curriculum integration are:
Physical Education: archery
Life Science: genetics

Communication Ask these questions about tossing a coin.

- What are the chances that the tossed coin will come up *heads*? **1 out of 2**
- What are the chances that the tossed coin will come up neither *heads* nor *tails*? **0**
- What other situations can you think of that have two equally likely results? **Sample answer: drawing a red or a black card from a deck of playing cards.**

2 TEACH

Transparency 13-1B contains a teaching aid for this lesson.

Reading Mathematics Have students describe each new term in this lesson in their own words. Then have them use magazines to find examples that illustrate each term.

In-Class Examples

For Example 1
A number cube is marked with 1, 2, 3, 4, 5, and 6 on its faces. You roll the cube once. Find the probability of rolling a 1 or 2. Then tell how likely it is to roll a 1 or 2. $\frac{1}{3}$ **The event is not very likely to occur.**

For Example 2
There are four equally likely outcomes on the spinner.

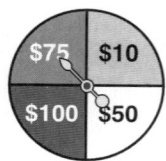

a. Find the probability of spinning $50. $\frac{1}{4}$
b. Find the probability of spinning an amount other than $50. $\frac{3}{4}$

For Example 3
A sportscaster says there is a 60% chance the Bobcats will win the baseball game. What is the probability that the Bobcats will lose the game? **40%**

Examples

1 A set of counters is numbered 1, 2, 3, ... , 10. Suppose you draw one counter without looking. Find the probability of choosing a number less than 3. Then tell how likely it is to choose a number less than 3.

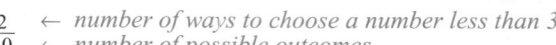

There are two numbers less than 3: 1 and 2.

$\frac{2}{10}$ ← *number of ways to choose a number less than 3*
    ← *number of possible outcomes*

Therefore, $P(\text{less than 3}) = \frac{2}{10}$ or $\frac{1}{5}$.

Since the probability is close to $\frac{1}{4}$, it is not very likely to occur.

> **Study Hint**
> **Reading Math** The notation $P(\text{less than 3})$ is read as *the probability of a number less than 3 occurring.*

2 There are six equally likely outcomes on the spinner.

a. Find the probability of spinning blue.
b. Find the probability of spinning a color other than blue.

a. $\frac{2}{6}$ ← *number of ways to spin blue*
    ← *number of possible outcomes*

Therefore, $P(\text{blue}) = \frac{2}{6}$ or $\frac{1}{3}$.

b. $\frac{4}{6}$ ← *number of ways to spin a color other than blue*
    ← *number of possible outcomes*

Therefore, $P(\text{not blue}) = \frac{4}{6}$ or $\frac{2}{3}$.

In Example 2, either one or the other event must take place, but they cannot both happen at the same time. Notice that the sum of the probabilities of the two events is 1. These events are examples of **complementary events**.

Complementary Events	**Words:**	Complementary events are two events in which either one or the other must take place, but they cannot both happen at the same time. The sum of their probabilities is 1.
	Symbols:	$P(\text{event}_1) + P(\text{event}_2) = 1$

516 Chapter 13 Using Probability

Classroom Vignette

"I have students play several games of Paper-Rock-Scissors and record the outcome of each round. I prepare a worksheet asking questions about the probability of the various outcomes during the game. Then they calculate the statistical measure for this data."

Deana Bobzien

Deana Bobzien, Teacher
Rochester Junior High
Rochester, IL

Example 3
CONNECTION

Earth Science A meteorologist predicts a 30% chance of rain. What is the probability that it will not rain?

The probability of no rain is the complement of the probability of rain. To find the probability of no rain, you can use an equation.

$P(\text{rain}) + P(\text{no rain}) = 1$

$0.3 + P(\text{no rain}) = 1$ *Replace P(rain) with 30% or 0.3.*

$P(\text{no rain}) = 1 - 0.3$ *Subtract 0.3 from each side.*

$P(\text{no rain}) = 1.0 - 0.3$ *Rewrite 1 as 1.0.*

$P(\text{no rain}) = 0.7$

So, $P(\text{no rain}) = 0.7$ or 70%.

CHECK FOR UNDERSTANDING

Communicating Mathematics

Read and study the lesson to answer each question.

1. **Draw** a spinner that shows $P(\text{yellow}) = \frac{2}{5}$. 1–3. See Answer Appendix.

2. **Give two examples** of events in which the probability of each event occurring is 0.

3. **Write** a sentence describing the relationship between complementary events. Then give an example of complementary events.

Guided Practice

A number cube is marked with 1, 2, 3, 4, 5, and 6 on its faces. You roll the cube one time. Find the probability of each event. Then tell how likely the event is to happen. 4–9. For likelihood, see Answer Appendix.

4. $P(3)$ $\frac{1}{6}$ 5. $P(\text{greater than 2})$ $\frac{2}{3}$ 6. $P(\text{odd})$ $\frac{1}{2}$

7. $P(5 \text{ or } 6)$ $\frac{1}{3}$ 8. $P(\text{not } 6)$ $\frac{5}{6}$ 9. $P(7)$ 0

10. **Industry** On a toy assembly line, 3% of the toys are defective. If an inspector selects a product at random, what is the probability the toy will pass inspection? **97%**

EXERCISES

Practice

Suppose you spin the spinner one time. Find the probability of each event. Then tell how likely the event is to happen. 11–17. For likelihood, see Answer Appendix.

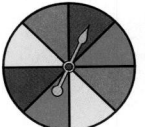

11. $P(\text{red})$ $\frac{1}{4}$ 12. $P(\text{blue or red})$ $\frac{3}{8}$

13. $P(\text{green or red})$ $\frac{1}{2}$ 14. $P(\text{not yellow})$ $\frac{3}{4}$

A set of 26 counters is lettered a, b, c, ... , z. Suppose you choose one counter without looking. Find the probability of each event. Then tell how likely the event is to happen.

15. $P(\text{m, n, or p})$ $\frac{3}{26}$ 16. $P(26)$ 0   17. $P(\text{consonant})$ $\frac{21}{26}$

Lesson 13-1 Theoretical Probability **517**

Check for Understanding

If students need additional practice or instruction after completing Exercises 1–10, one of these options may be helpful.

- Extra Practice, see p. 590
- Reteaching Activity
- *Study Guide Masters*, p. 90
- *Practice Masters*, p. 90

Assignment Guide

Core: 11–25 odd, 28–32
Enriched: 12–24 even, 25, 26, 28–32

Study Guide Masters, p. 90

Name _____ Date _____

13-1 Study Guide

Probability

The **probability** of an event is the ratio of the number of ways the event can occur to the number of possible outcomes.

$P(\text{event}) = \frac{\text{number of ways the event can occur}}{\text{number of possible outcomes}}$

Examples 1 On the spinner below, there are eight equally likely outcomes. Find the probability of spinning a number less than 3.

Numbers less than 3 are 1 and 2. There are 8 possible outcomes.

$P(\text{less than 3}) = \frac{2}{8}$ or $\frac{1}{4}$

2 Find $P(\text{greater than 10})$.

$P(\text{greater than 10}) = \frac{0}{8}$ or 0

3 Find $P(\text{less than 9})$.

$P(\text{less than 9}) = \frac{8}{8}$ or 1

Suppose you choose one of the cards shown without looking. Find the probability of each event.

3	6	9	12
15	18	21	24
27	30	33	36

1. $P(12)$ $\frac{1}{12}$ 2. $P(\text{even})$ $\frac{1}{2}$ 3. $P(2 \text{ digits})$ $\frac{3}{4}$

4. $P(\text{prime})$ $\frac{1}{12}$ 5. $P(\text{odd})$ $\frac{1}{2}$ 6. $P(\text{less than 8})$ $\frac{1}{6}$

7. $P(\text{greater than 40})$ 0   8. $P(\text{divisible by 3})$ 1

John has 15 baseball caps. 4 are red, 6 are blue, 3 are yellow, and 2 are white. If he chooses one without looking, find each probability.

9. $P(\text{yellow})$ $\frac{1}{5}$ 10. $P(\text{red or blue})$ $\frac{2}{3}$ 11. $P(\text{black})$ 0

12. $P(\text{white})$ $\frac{2}{15}$ 13. $P(\text{red or white})$ $\frac{2}{5}$ 14. $P(\text{yellow or white})$ $\frac{1}{3}$

© Glencoe/McGraw-Hill   T90   Mathematics: Applications and Connections, Course 1

Reteaching the Lesson

Activity Place colored centimeter cubes in a basket. Ask students to find the probability of picking a red cube without looking. Then have students take turns picking cubes until someone chooses a red one. Compare experimental results with theoretical probability. Continue with other colors.

Error Analysis
Watch for students who count only different outcomes when determining all possible outcomes.
Prevent by reviewing the rules for counting possible outcomes.

Exercise 27 asks students to advance to the next stage of work on the Chapter Project. In part a, instruct students to pick up the paper circles one at a time. Have students place their paragraph from part c in their Project Folder.

4 ASSESS

Closing Activity

Modeling Have pairs of students design a spinner so that the P(blue) is $\frac{1}{8}$. Then have them find the probability of spinning a color other than blue.

Additional Answers

25b. Sample answer: Sometimes, eliminating the ones that cannot be correct leaves the correct answer.

25c. increase; There would be fewer choices.

28. Sample answer: A set of 10 cards is numbered 1, 2, 3, . . . 10. Suppose you choose one card without looking. Find the probability of choosing a number less than 7.

Practice Masters, p. 90

Name _____ Date _____

13-1 Practice

Probability

Jared keeps his socks in random order in his top dresser drawer. There are two brown socks, eight black socks, four gray socks, and two blue socks in his drawer. Jared reaches into the drawer and, without looking, grabs one sock. Find the probability of each event.

1. P(gray) $\frac{4}{16}$ or $\frac{1}{4}$ 2. P(blue) $\frac{2}{16}$ or $\frac{1}{8}$

3. P(black) $\frac{8}{16}$ or $\frac{1}{2}$ 4. P(white) $\frac{0}{16}$ or 0

5. P(brown or black) $\frac{10}{16}$ or $\frac{5}{8}$ 6. P(gray or blue) $\frac{6}{16}$ or $\frac{3}{8}$

A set of 52 playing cards contains 4 different suits of 13 cards each. Hearts and diamonds are red; spades and clubs are black. Each suit contains cards numbered 2 through 10, a jack, queen, king, and ace. It is equally likely to choose any one card. Find the probability of each event.

7. P(red) $\frac{26}{52}$ or $\frac{1}{2}$ 8. P(clubs) $\frac{13}{52}$ or $\frac{1}{4}$

9. P(ace) $\frac{4}{52}$ or $\frac{1}{13}$ 10. P(jack of diamonds) $\frac{1}{52}$

11. P(black 10) $\frac{2}{52}$ or $\frac{1}{26}$ 12. P(red king or queen) $\frac{4}{52}$ or $\frac{1}{13}$

13. P(black 2, 3, or 4) $\frac{6}{52}$ or $\frac{3}{26}$ 14. P(red 1) $\frac{0}{52}$ or 0

Mrs. Phipps found 10 identical cans without labels in her cupboard. She knew that she originally had two cans of peas, five cans of corn, one can of carrots, and two cans of beets. She opens one can. Find the probability of each event.

15. P(carrots) $\frac{1}{10}$ 16. P(corn) $\frac{5}{10}$ or $\frac{1}{2}$

17. P(beans) $\frac{0}{10}$ or 0   18. P(peas) $\frac{2}{10}$ or $\frac{1}{5}$

19. P(corn or beets) $\frac{7}{10}$ 20. P(carrots or peas) $\frac{3}{10}$

© Glencoe/McGraw-Hill   T90   Mathematics: Applications and Connections, Course 1

A set of 30 cards is numbered 1, 2, 3, ... , 30. Suppose you choose one card without looking. Find the probability of each event. Then describe the complementary event of each event, and find its probability.

18–23. See Answer Appendix.

18. P(12)   19. P(odd)   20. P(1 digit)

21. P(integer)   22. P(less than 1)   23. P(greater than 18)

24. When two number cubes are rolled, the probability of rolling doubles is $\frac{1}{6}$. What is the probability of not rolling doubles? $\frac{5}{6}$

Applications and Problem Solving

25. **School** A multiple-choice test question has five possible answers. Suppose you guess at the answer.
   a. What is the probability of choosing the correct answer? $\frac{1}{5}$
   b. Explain how you can use the problem-solving strategy *eliminating possibilities* to choose the correct answer. **b–c. See margin.**
   c. By eliminating possibilities, would the probability of choosing the correct answer increase or decrease? Explain.

26. **Advertising** A cereal company is having a contest. Each box of cereal contains one of the letters B, A, M, O. In every 20 boxes, there are four Bs, five As, ten Ms, and one O. What is the probability that a box of cereal will have a B? $\frac{1}{5}$

27. **Working on the** 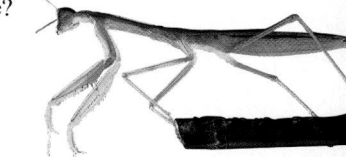 Place one sheet of the classified ads on the floor. Use a hole punch to cut out 100 holes from each sheet of white paper, black paper, and classified ads. Scatter the circles on the spread-out sheet of classifieds.

27a–b. Answers will vary.

   a. Randomly pick up as many paper circles as you can for 10 seconds. How many circles of each kind do you have?
   b. To find the probability of picking up a black circle, divide the number of black circles picked up by the total number of circles picked up. Find the probability of picking up each color.
   c. Write a paragraph that explains how probability relates to camouflage. Tell whether or not this activity is an example of experimental or theoretical probability. **See students' work.**

28. **Critical Thinking** Write a problem in which the probability of an event occurring is 0.6. **See margin.**

Mixed Review
29. See Answer Appendix.

29. **Functions** Graph the function represented by the function table shown. *(Lesson 12-6)*

30. Find $-12 + (-6)$. *(Lesson 11-3)* **−18**

31. What is 32% of 148? *(Lesson 8-7)* **47.36**

input	output
4	4
−2	−2
0	0

32. **Test Practice** Every class that sells 75 tickets to the school play earns an ice cream party. Kate's class has sold 47 tickets. How many more tickets must they sell to earn the ice cream party? *(Lesson 1-1)* **C**

    A 23   **B** 32   **C** 28   **D** 35   **E** Not Here

Extending the Lesson

Enrichment Masters, p. 90

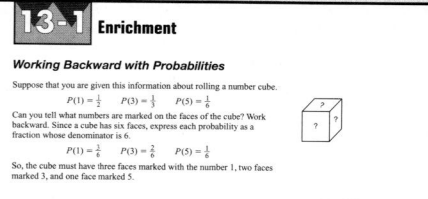

13-1 Enrichment

Working Backward with Probabilities

Suppose that you are given this information about rolling a number cube.

$$P(1) = \frac{1}{2} \quad P(3) = \frac{1}{3} \quad P(5) = \frac{1}{6}$$

Can you tell what numbers are marked on the faces of the cube? Work backward. Since a cube has six faces, express each probability as a fraction whose denominator is 6.

$$P(1) = \frac{3}{6} \quad P(3) = \frac{2}{6} \quad P(5) = \frac{1}{6}$$

So, the cube must have three faces marked with the number 1, two faces marked 3, and one face marked 5.

Activity Ask small groups of students to create a game where the probability of winning can change with each turn. A score of 18 is needed to win. **Sample answer: Roll a number cube 6 times. If the number is even, add it to your score. If the number is odd, subtract it.**

BIOLOGICAL SCIENCE

Lyle Allard
Biological Science Technician

Lyle Allard is a biological science technician for the Fish Technology Center in Bozeman, Montana. At the Center, they are developing healthier fish foods. Mr. Allard takes care of experimental fish and watches their growth and behavior. He is a member of the Turtle Mountain Band of the Chippewa Tribe in North Dakota. One of his goals is to help Native American tribes improve their management of fish and wildlife.

To work in biological science, you will need a bachelor's degree in a biology-related field. Some careers in biological science are zoologist, ecologist, naturalist, park ranger, veterinarian, and zookeeper. A zoologist may use mathematics when recording or interpreting data about animals. If you like animals and being outdoors, then you may want to consider a career in biological science.

For more information:
U.S. Fish and Wildlife Service
1849 C Street
Washington, DC 20240

*inter*NET
CONNECTION
www.glencoe.com/sec/math/mac/mathnet

I love the outdoors and I think that it would be great to work with wildlife.

Your Turn

Interview a person who works with animals. Ask about the advantages and disadvantages of working with animals and how they use mathematics in their job. Then write a report that summarizes your interview.

School to Career: Biological Science **519**

More About Lyle Allard

- Lyle Allard received an associate's degree from the College of Southern Idaho and a bachelor of arts degree from Montana State University. He believes that it is important to focus on a career while still in school because hard work and devotion in youth can lead to a successful career later in life.
- Mr. Allard became a biological scientist because he enjoyed working outdoors, especially with fish. An avid hiker and camper, he has also worked as a professional hunting guide.

Motivating Students

Wildlife management is necessary to prevent specific animal populations from becoming too large or too small. For example, in the past 150 years, more than 60 species or subspecies of mammals have become extinct. To begin a discussion, ask students these questions.

- Why have certain animals become extinct? **their habitat is developed for human use, pollution**
- Why might people who catch fish be concerned about the health of the fish? **If the fish is sick, the person who eats the fish may become ill.**

Making the Math Connection

Biological scientists use sampling and probability to predict population birth, growth, death, and survival rates. They then use these predictions to plan for future wildlife management.

Working on *Your Turn*

Encourage students to interview someone who has a job they might like to have themselves. Have them ask the person what education and experience he or she needed to qualify for the position. Have students include this information in their reports. You may also have students read their reports to the class.

*An additional School to Career activity is available on page 13 of the **School to Career Masters.***

Objective Students solve problems by making a table.

Recommended Pacing	
Honors	Day 3 of 11

1 FOCUS

Getting Started Refer to Lesson 1-2 to review making and interpreting frequency tables. Bring in a newspaper article that contains survey results shown in graphic form. Point out the information about the number of people sampled. Frequency tables or spreadsheets are ways to organize survey information before presenting it in pictorial or graphic form.

2 TEACH

Teaching Tip Have students suggest other questions that could be investigated with a survey.

In-Class Example

A frequency table for the results of a class test is shown.

Test Scores		
Score	**Tally**	**Frequency**
91–100	ЖII	7
81–90	ЖIIII	9
71–80	IIII	4

a. How many tests were graded? **20**

b. How many scores were between 81 and 90? **9**

THINKING LAB

PROBLEM SOLVING

13-2A Make a Table

A Preview of Lesson 13-2

Patsy and Adam are organizing their school's upcoming music festival. They are discussing how to conduct a survey to determine which songs the students would prefer to sing. Let's listen in!

Patsy

I have the list of songs to choose from. I think that we should ask all of the chorus members which of the songs they would like to sing.

There isn't enough time to ask everyone. Let's ask every fourth chorus member who comes into chorus practice tonight to choose six songs.

Adam

Song Survey		
Song	**Tally**	**Frequency**
Yesterday	IIII	4
You've Lost That Lovin' Feeling	ЖI	6
Georgia On My Mind	III	3
Bridge Over Troubled Water		0
A Whole New World	ЖIII	8
Tomorrow	Ж	5
Just Around The River Bend	II	2
Yellow Rose Of Texas	IIII	4

O.K. I'll make a frequency table like this to record the results of the survey. We'll choose the six songs with the most votes.

THINK ABOUT IT

Work with a partner. 1–3. See margin.

1. *Analyze* the way Patsy and Adam conducted the survey. Do you think the survey reflected the opinions of the entire chorus? Why or why not?

2. *Describe* other types of information that can be concluded from the table.

3. *Tell* an advantage of organizing information in a table.

4. *Apply* the **make a table** strategy to solve the following problem.

   A list of test scores is shown.

71	89	65	77	79	98	84
86	70	97	93	80	91	72
100	75	73	86	99	77	68

   a. *Make a frequency table of the test scores.* **See Answer Appendix.**

   b. *How many more students scored 71 to 80 than 91 to 100?* **2**

■ Reteaching the Lesson ■

Activity Have students make a frequency table using the first letter of their first names. Would the table look different if only every third student were surveyed? **possibly** What if they used last names? **If seated in alphabetical order, results could vary if every third student were surveyed.**

Additional Answers

1. **Sample answer: Yes; surveying every fourth member provides an adequate sample.**

2. **Sample answer: The most and least preferred songs.**

3. **Sample answer: The information is easy to locate and it is easier to draw conclusions from organized data.**

ON YOUR OWN

5. The first step of the 4-step plan for problem solving asks you to *explore* the problem. *Tell* the advantages of exploring the set of data before you make a table. **See margin.**

6. *Write a Problem* that can be solved by making a table. Then ask a classmate to solve the problem by making a table. **6. See students' work.**

7. *Explain* how the make a table strategy is used to help conduct the pet survey in Exercise 18 on page 525. **See margin.**

MIXED PROBLEM SOLVING

8. 50 and 36

STRATEGIES
Look for a pattern.
Solve a simpler problem.
Act it out.
Guess and check.
Draw a diagram.
Make a chart.
Work backward.

Solve. Use any strategy.

8. **Number Sense** The difference between two whole numbers is 14. Their product is 1,800. Find the two numbers.

9. **Life Science** A snail at the bottom of a 10-foot hole crawls up 3 feet each day, but slips back 2 feet each night. How many days will it take the snail to reach the top of the hole and escape? **8 days**

10. **School** The list shows the birth month of the students in Miss Miller's geography class.

June	October	May	April
April	May	October	June
July	August	April	May
March	June	September	July
July	April	December	March
June	October	January	June

  a. Make a frequency table of the students' birth months. **See Answer Appendix.**

  b. How many of the students were born in April? **4**

  c. How many more students were born in June than in August? **4**

11. **Money Matters** Harris spent $5.69 on golf tees. About how much change should he expect to receive if he paid with a $20 bill? **Sample answer: $20–$6 or $14**

12. **Sports** Coach Franco is deciding in which order the players will bat. She looks at the number of hits each player made in the previous game. Use the frequency table shown.

Number of Hits		
Player	Tally	Frequency
Parker	II	2
Martinez	III	3
Cruz	IIII	4
Plesich	II	2
Higgins	III	3
Reid	I	1
Hartley		0
Wilson	II	2

  a. Who had the most hits? **Cruz**

  b. Who had the least hits? **Hartley**

  c. Find the total number of hits in the previous game. **17**

13. **Test Practice** There are 4 green, 2 purple, 1 orange, and 5 yellow marbles in a pouch. Forest chooses 1 marble at random, records its color, and replaces it. He repeats this process 25 times. Which color did Forest probably choose the greatest number of times? **D**

  A orange
  B purple
  C green
  D yellow

Lesson 13-2A THINKING **521**

3 PRACTICE/APPLY

Check for Understanding
Have students complete Exercises 1–3 in pairs. Have them ask each other additional questions about the table. Use the results from Exercise 4 to determine whether students understand how to set up a frequency table.

Extra Practice If students need additional practice in problem solving, extra practice is available on the following pages.
• Make a Table, see p. 591
• Mixed Problem Solving, see pp. 593–594

Assignment Guide
All: 5–13

4 ASSESS

Closing Activity
Writing Have students write a paragraph explaining why making a table is a good problem-solving strategy for some problems. Have them include an example of a problem that can be solved by making a table and one that cannot be solved that way.

Additional Answers
5. Sample answer: It enables you to determine a reasonable interval width for the given data.

7. Sample answer: Ashley organized the results of the survey by making a table. The table makes it easy to find information such as the most preferred and the least preferred pet.

Extending the Lesson

Activity Have pairs of students write a survey question that deals with an issue in their school. Discuss how many students would have to be surveyed to answer the question. Have students consider the work involved in a survey and what types of results might be found on the frequency table. Have them conduct the survey and record the data in a frequency table. Sample problem: *What is your favorite cafeteria food? Make a list of food items available in the cafeteria. Survey half the students in the school to find out which items are preferred. Make a frequency table.*

Instructional Resources

- *Study Guide Masters*, p. 91
- *Practice Masters*, p. 91
- *Enrichment Masters*, p. 91
- Transparencies 13-2, A and B
- *Assessment and Evaluation Masters*, p. 351
- *School to Career Masters*, p. 13
- *Technology Masters*, p. 25
- CD-ROM Program
  - Resource Lesson 13-2

Recommended Pacing	
Honors	Day 4 of 11

1 FOCUS

5-Minute Check
(Lesson 13-1)

A set of 20 cards is numbered 86, 87, 88, . . ., 105. Suppose you choose one card without looking. Find the probability of each event. Then tell how likely the event is to happen.

1. $P(88)$ $\frac{1}{20}$; The event is not very likely to occur.

2. $P(\text{divisible by 2})$ $\frac{1}{2}$; The event is equally likely to occur.

3. $P(\text{less than 50})$ 0; The event cannot occur.

4. $P(\text{3 digits})$ $\frac{3}{10}$; The event is not too likely to occur.

5. A soccer team has a 65% chance of winning their next game. What is the probability of the team losing the next game? 35%

The 5-Minute Check is also available on **Transparency 13-2A** for this lesson

Motivating the Lesson

Hands-On Activity Have pairs of students choose 10 paper clips from a package of 100 colored clips. Ask them how their sample could be used to make a prediction about the colors in the package. Multiply each result by 10.

13-2

What you'll learn

You'll learn to predict the actions of a larger group using a sample.

When am I ever going to use this?

You can predict the number of left-handed students in your school by using a sample.

Word Wise

sample
population
random

Examples

Integration: Statistics
Making Predictions Using Samples

The student council at Springfield Middle School is planning a year-end field trip to either Sea World or Six Flags in San Antonio, Texas. Since the committee does not have time to survey every student to find their preference, they can survey a smaller group, or **sample**. The information from the sample will help them predict where the students would prefer to go on the field trip.

All of the students in the school make up the **population**. To predict the population's preference, the survey should represent the preferences of all of the students. Here are some suggestions.

- Survey every fifteenth student named on the school roster.

- Survey every tenth student who exits the school at the end of the day.

- Survey two students from each homeroom class.

Each of these methods is a way of making sure that the sample is **random**, or drawn by chance from the population.

1 Chase wanted to find out which place the students would prefer to go: Sea World or Six Flags. He surveyed every twentieth student who entered the school at the beginning of the day. Is this a random sample?

Chase had no control over the order in which the students entered the school at the beginning of the day. By surveying every twentieth student, he avoided surveying an entire group of friends that might all prefer the same place. This is a random sample.

2 Chase wanted to find out which day students would prefer to go on the trip: Wednesday, Thursday, or Friday. He surveyed a small group of people standing in the hallway before school. Is this a random sample?

By surveying a group of students standing in the hallway, their preferences may be influenced by those in the group. This is *not* a random sample.

Multiple Learning Styles

Intrapersonal Ask students to make a list of times that knowing the probability of an event could have made their decision making easier. For example, many people don't plan vacations to Montana in January because of the likelihood of blizzards.

You can use the results of a survey to make predictions about the actions of the entire population.

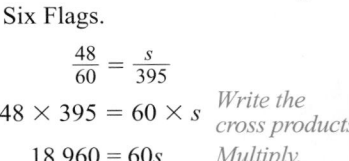

School When Chase conducted the survey to find students' field trip preferences, he learned that 48 out of the 60 students surveyed preferred Six Flags.

a. What is the probability that any given student will want to go to Six Flags?

b. There are 395 students at Springfield Middle School. Predict how many students will want to go to Six Flags.

a. 48 out of 60, or $\frac{4}{5}$, prefer Six Flags. The probability is $\frac{4}{5}$, or 80%.

b. Use a proportion. Let s represent the number of students who will want to go to Six Flags.

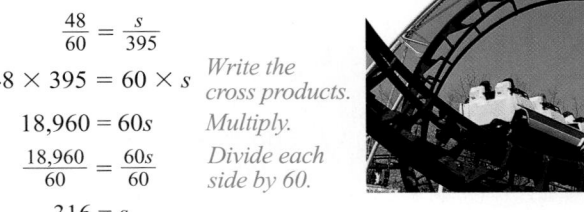

$$\frac{48}{60} = \frac{s}{395}$$

$48 \times 395 = 60 \times s$ *Write the cross products.*

$18,960 = 60s$ *Multiply.*

$\frac{18,960}{60} = \frac{60s}{60}$ *Divide each side by 60.*

$316 = s$

Of the 395 students, about 316 will prefer to go to Six Flags.

> **LOOK BACK**
> You can refer to Lesson 8-2 to review proportions.

In the Mini-Lab, you will conduct a survey to predict the number of left-handed students in your school.

MINI-LAB

Work with a partner.

Try This

- Decide how you will choose a random sample to predict the number of left-handed students in your school.
- With your teacher's permission, conduct your survey.

Talk About It 1–3. See students' work.

1. Find the probability that a student selected at random from your sample is left-handed. Explain why this is an example of experimental probability.

2. Based on your sample, predict the number of left-handed students in your school.

3. Compare your results with the other groups. Explain why different groups may arrive at different conclusions.

Lesson 13-2 Integration: Statistics Making Predictions Using Samples   **523**

 Transparency 13-2B contains a teaching aid for this lesson.

In-Class Examples

For Example 1
Yolanda wanted to find out how many people in her community would support recycling. She surveyed every tenth adult shopper at the local supermarket. Is this a random sample? **yes**

For Example 2
Reiko wanted to know what hours citizens would most like the recycling center to be open. She surveyed the people who arrived at the center Saturday morning. Is this a random sample? **no**

For Example 3
A theater company conducted a random survey of theater-goers to determine their preference in plays. Of the 50 people surveyed, 35 preferred musical comedies.

a. What is the probability that any given theater-goer will buy a ticket to a musical comedy? $\frac{7}{10}$ **or 70%**

b. If ticket information is sent to 200 homes of theater-goers, predict the number that will buy musical comedy tickets. **about 140 people**

Using the Mini-Lab You may wish to set a time limit for completion of the survey. If the total school population is unknown, assign a student to get the information from the school office.

Check for Understanding

If students need additional practice or instruction after completing Exercises 1–7, one of these options may be helpful.
- Extra Practice, see p. 591
- Reteaching Activity
- *Study Guide Masters*, p. 91
- *Practice Masters*, p. 91

Assignment Guide
Core: 9–19 odd, 20–23
Enriched: 8–16 even, 18–23
All: Self Test, 1–5

Additional Answers

1. Sample answer: Population is the entire group while random sample is a subset of the entire group.

2. Sample answer: The conclusions derived from the random sample should reflect the same results derived from the population. Therefore, you should be able to predict the winners of a school board election from the random sample.

Study Guide Masters, p. 91

Name _____ Date _____

13-2 Study Guide

Integration: Statistics
Making Predictions Using Samples

Data gathered by surveying a random sample of the population may be used to make predictions about the entire population.

Example Joyce surveyed every tenth person entering the school to determine whether they would prefer attending a rock concert or a dance. Of the 60 students surveyed, 35 said they preferred a rock concert, and 25 said they preferred a dance. If 800 students attend the school, predict how many would prefer a rock concert.

$\frac{35}{60}$ or 58% of those surveyed said they preferred a concert.

58% of 800 = 0.58 × 800
= 464

About 464 students would prefer a rock concert.

Solve.

1. In a random sample of 600 batteries, 3 were found to be defective. If a factory produces 7,000 batteries each day, predict the number that are defective. **35**

2. Wilson has made 8 out of the last 20 free throws he has attempted. What is the probability that he will make the next free throw? **$\frac{2}{5}$**

3. The graph shows who has bought tickets so far. What is the probability that a senior citizen will buy the next ticket? **$\frac{1}{5}$**

Ticket Sales
Adults 35% Seniors 20% Students 45%

4. In a poll of 200 people, 82 said they would vote for Peterson for mayor, 106 said they would vote for Sanderson, and 12 were undecided. If 5,200 people vote in the election, predict the number that will vote for Peterson. **2,132**

5. In a survey, 35% of the students said they would buy a hot dog at a football game, 45% said they would buy a hamburger, and 85% said they would buy a soft drink. If 350 students are expected to attend a football game, how many hot dogs, hamburgers, and soft drinks should be ordered for the concession stand? **123; 158; 298**

© Glencoe/McGraw-Hill   T91   Mathematics: Applications and Connections, Course 1

Communicating Mathematics

Read and study the lesson to answer each question. 1–2. See margin.

1. *Tell* the difference between population and random sample.

2. *Explain* how you can use a random sample to predict the results of a school board election.

HANDS-ON MATH

3. *Find* a newspaper article or advertisement that refers to a survey. Then write one or two sentences describing how the results of the survey can predict the actions of a larger group. **See students' work.**

Guided Practice

Tell whether each of the following is a random sample. Explain your answer. 4–6. See Answer Appendix for explanations.

Type of Survey	Survey Location
4. favorite entertainment	Broadway show **no**
5. favorite flower	shopping center **yes**
6. favorite holiday	fast-food restaurant **yes**

7. *Sports* In soccer, Karena scored 6 goals in her last 10 attempts.

   a. What is the probability of Karena scoring a goal on her next attempt? **$\frac{3}{5}$ or 60%**

   b. Suppose Karena attempts to score 15 goals. About how many goals will she make? **about 9**

Practice

Tell whether each of the following is a random sample. Explain your answer. 8–16. See Answer Appendix for explanations.

Type of Survey	Survey Location
8. favorite color	grocery store **yes**
9. favorite hobby	model train store **no**
10. favorite musician	rock concert **no**
11. favorite sport	mall **yes**
12. favorite season	public library **yes**
13. favorite car	Chevrolet dealer **no**
14. favorite TV show	skating rink **yes**
15. favorite candy bar	chocolate factory **no**
16. favorite food	Mexican restaurant **no**

17. *Civics* Louisa conducted a survey to find out which political party people in her county preferred. She surveyed every tenth person standing in line at the Democratic headquarters. Is this a random sample? Explain. **No; no Republicans were included in the survey.**

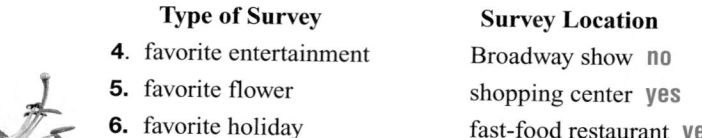

■ Reteaching the Lesson ■

Activity Have groups of students conduct a survey among themselves to find out if they prefer hamburgers or tacos. Ask: *Is your group an example of a random sample? What part of your group chose tacos? What is the probability that another student in your class would choose tacos? Predict how many would prefer tacos.*

Applications and Problem Solving

18. **Life Science** Ashley conducted a survey of students' favorite pets from a random sample of 56 students at Park Middle School. The results are shown in the chart.

a. What is the size of the sample? **56**

b. What is the probability that a student at Park Middle School prefers a gerbil? $\frac{1}{4}$, **or 25%**

c. If there are 252 sixth graders, about how many will prefer gerbils? **about 63**

Pet	Number
Dog	24
Cat	11
Gerbil	14
Bird	7

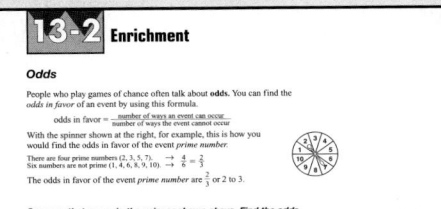

interNET
CONNECTION

For the latest Nielsen ratings, visit: www.glencoe.com/ sec/math/mac/mathnet

19. **Technology** The Nielsen ratings use a sample of about 970 households in the United States to determine the rankings of TV shows. The top rated show in May 1997, was *E.R.* An estimated 38% of all TV sets in use were tuned to *E.R.* How many sets in the sample were tuned to *E.R.*? **about 369**

20. **Critical Thinking** A pre-election poll predicted that a certain candidate for mayor would receive 25% of the vote. The candidate received 15,248 votes. Estimate how many people voted in the mayoral election.
about 60,000 people

Mixed Review

21. **Test Practice** A bag contains 2 yellow counters, 1 red counter, 5 blue counters, and 3 green counters. Alec chooses 1 counter at random 20 times, records its color, and then replaces it in the bag. Which color did Alec probably choose the most number of times? *(Lesson 13-1)* **C**

   A yellow   B red   C blue   D green

22. **Geometry** An angle measures 91.2°. Is it an obtuse or acute angle? *(Lesson 9-1)* **obtuse**

23. **Measurement** How many inches are in 5 yards? *(Lesson 5-6)* **180 inches**

CHAPTER 13

Mid-Chapter Self Test

A bag contains 4 blue marbles, 3 yellow marbles, 6 purple marbles, and 5 green marbles. One marble is chosen at random. Find the probability of each event. *(Lesson 13-1)*

1. $P(\text{yellow})$ $\frac{1}{6}$

2. $P(\text{orange})$ **0**

3. $P(\text{green or blue})$ $\frac{1}{2}$

4. $P(\text{blue or purple})$ $\frac{5}{9}$

5. **Technology** A quality control inspector found that 3 out of 50 computer keyboards were defective. *(Lesson 13-2)*

a. What is the probability that a randomly-chosen computer keyboard will be defective? $\frac{3}{50}$, **or 6%**

b. Suppose the company manufactured 15,000 computer keyboards. How many of the keyboards will be defective? **about 900**

Extending the Lesson

Enrichment Masters, p. 91

13-2 Enrichment

Odds

People who play games of chance often talk about **odds**. You can find the *odds in favor* of an event by using this formula.

odds in favor = $\frac{\text{number of ways an event can occur}}{\text{number of ways the event cannot occur}}$

With the spinner shown at the right, for example, this is how you would find the odds in favor of the event *prime number*.

There are four prime numbers (2, 3, 5, 7). → $\frac{4}{6} = \frac{2}{3}$
Six numbers are not prime (1, 4, 6, 8, 9, 10).

The odds in favor of the event *prime number* are $\frac{2}{3}$ or 2 to 3.

Suppose that you spin the spinner shown above. Find the odds.

Activity Have pairs of students determine how they could use a survey to plan a fund-raising activity. Ask them who they would survey and why. Then have them explain how they would apply the survey results to their fund-raising plan.

---

4 ASSESS

Closing Activity

Speaking Give students sample results that involve their class such as: 2 out of 6 students in this class play on a basketball team. Ask them to use the sample to make a prediction about the whole class.

Chapter 13, Quiz A (Lessons 13-1 and 13-2) is available in the *Assessment and Evaluation Masters,* p. 351.

Mid-Chapter Self Test

The Mid-Chapter Self Test reviews the concepts in Lessons 13-1 through 13-2. Lesson references are given so students can review concepts not yet mastered.

Practice Masters, p. 91

13-2 Practice

Integration: Statistics
Making Predictions

Tell whether each of the following is a random sample. Explain your answer.

Type of Survey	Survey Location
1. favorite brand of sneaker	Reebok outlet store

No; only one brand of sneaker sold.

2. students' favorite pizza topping — one student from each homeroom
Yes; no control over how students are assigned to homerooms.

3. favorite vacation spot — people at the beach in the summer
No; assume people at the beach in summer are on vacation and enjoy the beach.

4. preferred means of traveling to work — people in a weekday traffic jam
No; traffic jam rules out commuting by car as favorite means of transportation.

5. In a random survey of 50 customers in a supermarket, Bonnie found 16 that drive mini vans. If there were 225 customers in the supermarket that day, how many customers are likely to drive a mini van? **72**

Manuel took a survey of students' favorite core subjects from a random sample of 20 sixth graders. The results are shown in the chart below. Use the results to answer each question.

6. What is the size of the sample? **20**

6th Graders' Favorite Subjects	
Math	8
Language Arts	3
Science	5
Social Studies	4

7. What is the probability that a 6th grader prefers
a. math? $\frac{2}{5}$ b. science? $\frac{1}{4}$

8. If there are 100 sixth grade students, what percent prefers
a. social studies? **20%**   b. language arts? **15%**

© Glencoe/McGraw-Hill   T91   *Mathematics: Applications and Connections, Course 1*

1 FOCUS

 5-Minute Check
(Lesson 13-2)

A survey was taken to find out how many local residents supported building a new recreation center.

Support Building a New Recreation Center	
Yes	12
No	8
Undecided	5

1. What is the size of the sample? **25**

2. What is the probability that a resident is undecided? $\frac{1}{5}$, **or 20%**

3. Suppose there are 300 residents. How many could be expected to support building a new recreation center? **144 residents**

The 5-Minute Check is also available on **Transparency13-3A** for this lesson.

Motivating the Lesson
Problem Solving A round birthday cake serves twelve people and will be cut into twelve equal pieces so that only one of the three roses on the cake would be on a piece. Ask students to determine the probability of receiving a piece with a rose on it. **There would be 3 chances out of 12, or a probability of $\frac{1}{4}$.**

 **Integration: Geometry**
Probability and Area

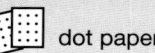 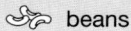

What you'll learn
You'll learn to find probability using area models.

When am I ever going to use this?
Knowing how to relate probability to the area can help you find the probability of a dart landing on a dartboard.

Have you ever played the game Twister? If so, then you know that it is easy to get tied up in knots! The spinner shown is used to play the game. It has 16 equally likely outcomes. What is the probability of spinning left hand yellow?

There are 16 possible outcomes and 1 way the event can occur. Therefore, $P(\text{left hand, yellow}) = \frac{1}{16}$.

In Lesson 13-1, you learned to relate probability to spinners or number cubes. You can also relate probability to the areas of geometric shapes.

 HANDS-ON MINI-LAB

Work with a partner. ⬚ dot paper 🫘 beans

Try This
- Draw two squares on dot paper similar to the ones shown.
- Drop 50 beans onto the paper, from about 8 inches above.
- Record the number of beans that land within the large square. Record the number of beans that land within the small square. Do not count those that land outside the squares.

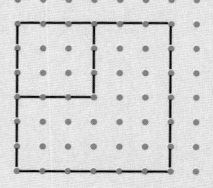

Talk About It
1. Find the ratio $\frac{\text{number of beans in a small square}}{\text{number of beans in a large square}}$. **See students' work.**

2. Find the ratio $\frac{\text{area of small square}}{\text{area of large square}}$. $\frac{1}{4}$

3. Compare the two ratios. **3–4. See students' work.**

4. Combine your results with another group and compare the ratios.

With a very large sample, the experimental probability should be very close to the theoretical probability.

The results of the Mini-Lab suggest the following conclusion.

$$\frac{\text{number landing in small square}}{\text{number landing in large square}} = \frac{\text{area of small square}}{\text{area of large square}}$$

Multiple Learning Styles

Auditory/Musical Have students count and measure the bars on a toy xylophone. Suggest that they strike each bar and listen to the sound it makes. Then have them determine the probability that a toddler playing randomly will strike a particular bar. Assume that each stroke hits a bar, and only one bar.

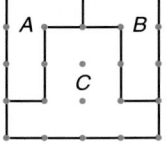

Examples ➊ The figure shown represents a dartboard.

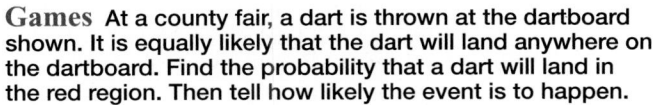

a. Suppose you threw a dart randomly at the board and it hits the board. Find the probability of the dart landing in region C.

b. Suppose you threw the dart 500 times. How many times would you expect it to land in region C?

a. $P(\text{region C}) = \dfrac{\text{area of region C}}{\text{area of target}}$

$= \dfrac{8}{16}$ or $\dfrac{1}{2}$

b. Let c = times the dart lands in region C.

$\dfrac{c}{500} = \dfrac{1}{2}$ ← *area of region C*
                ← *area of dart board*

$c \times 2 = 500 \times 1$ *Find the cross products.*

$2c = 500$

$c = 250$ *Divide each side by 2.*

So, out of 500 times, the dart should land in region C about 250 times.

APPLICATION ➋ **Games** At a county fair, a dart is thrown at the dartboard shown. It is equally likely that the dart will land anywhere on the dartboard. Find the probability that a dart will land in the red region. Then tell how likely the event is to happen.

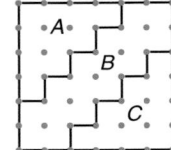

5 in.

1 in.

5 in.

$P(\text{red region}) = \dfrac{\text{area of red region}}{\text{area of target}}$

Area of red region $= \pi r^2$

$\approx 3.14 \times 1 \times 1$

$\approx 3.14 \text{ in}^2$

Area of target $= s^2$

$= 5 \times 5$ or 25 in^2

$P(\text{red region}) \approx \dfrac{3.14}{25}$

$\approx \dfrac{3}{24}$ or $\dfrac{1}{8}$ *3 and 24 are compatible numbers.*

The probability of landing in the red region is about $\dfrac{1}{8}$ or 12.5%.

The event is not very likely to happen.

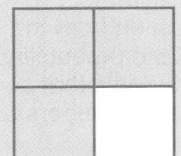
Cultural Kaleidoscope

During the 17th century, mathematicians Blaise Pascal and Pierre de Fermat studied the probability of games of chance.

CHECK FOR UNDERSTANDING

Communicating Mathematics

Read and study the lesson to answer each question.

1. *Tell* the probability of a randomly-thrown dart landing in region B of the figure shown. $\dfrac{4}{9}$

2. *Draw* a dartboard in which the probability of landing in a shaded region is $\dfrac{3}{4}$. **See margin.**

Lesson 13-3 Integration: Geometry   Probability and Area   **527**

Transparency 13-3B contains a teaching aid for this lesson.

Using the Mini-Lab If beans are unavailable, rice, popcorn kernels, or puffed cereal may be substituted. Ask students if they think the number of items that fall within the squares will be affected by the height of the drop. Make sure students discover that the larger the sample, the closer the experimental probability gets to the theoretical probability.

In-Class Examples

For Example 1
The figure represents a dartboard.

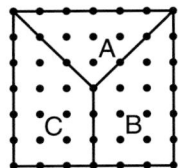

a. Suppose you threw a dart randomly at the board and it hit the board. Find the probability of the dart landing in region A. $\dfrac{1}{4}$

b. Suppose you threw the dart 300 times. How many times would you expect it to land in region A? **75 times**

For Example 2
Irina lost her contact lens while playing basketball. If it is equally likely that she lost the lens on any part of the court, estimate the probability that Irina lost her lens inside the center circle.
about $\dfrac{1}{47}$, or 2%

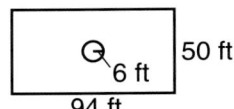

50 ft

6 ft

94 ft

Additional Answer

2. Sample answer:

Lesson 13-3   527

Check for Understanding

If students need additional practice or instruction after completing Exercises 1–8, one of these options may be helpful.
- Extra Practice, see p. 591
- Reteaching Activity
- *Study Guide Masters*, p. 92
- *Practice Masters*, p. 92
- 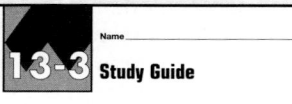 Interactive Mathematics Tools Software

Assignment Guide

Core: 9–21 odd, 24–27
Enriched: 10–20 even, 21, 22, 24–27

HANDS-ON MATH

3. Use your dartboard from Exercise 2. Drop a bean 50 times onto the target. Then find the ratio that compares the number of times the bean lands in the shaded region to the number of times the bean lands in the target. How does this ratio compare to $\frac{3}{4}$, the theoretical probability of landing in the shaded region? **See students' work.**

Guided Practice

Each figure represents a dartboard. It is equally likely that a dart will land anywhere on the dartboard. Find the probability of a randomly-thrown dart landing in the shaded region.

4. 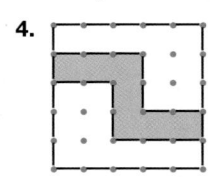 $\frac{7}{25}$ **5.** 12 in. $\frac{1}{8}$

12 in.   6 in.   6 in.

6. about 42
7. about 19

6. Suppose you threw a dart 150 times at the dartboard in Exercise 4. How many times would you expect it to land in the shaded region?

7. Suppose you threw a dart 150 times at the dartboard in Exercise 5. How many times would you expect it to land in the shaded region?

8. *Games* A randomly-thrown dart is thrown at the dartboard shown. It is equally likely that the dart will land anywhere on the dartboard. Find the probability that a dart will land in the shaded region of the figure shown. Then tell how likely the event is to happen.
8 in.   2 in.   2 in.   8 in.
$\frac{15}{16}$; The event is very likely to occur.

EXERCISES

Practice

Each figure represents a dartboard. It is equally likely that a dart will land anywhere on the dartboard. Find the probability of a randomly-thrown dart landing in the shaded region.

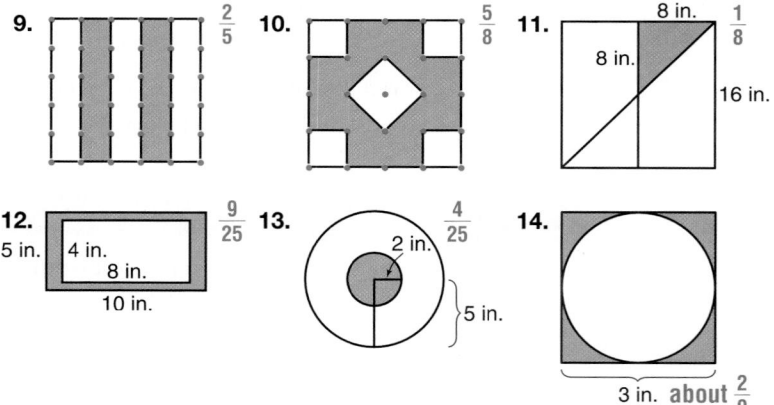

9. $\frac{2}{5}$ **10.** $\frac{5}{8}$ **11.** 8 in. $\frac{1}{8}$
8 in.   16 in.

12. $\frac{9}{25}$ **13.** $\frac{4}{25}$ **14.**
5 in.   4 in.   8 in.   2 in.   5 in.
10 in.
3 in.   about $\frac{2}{9}$

Study Guide Masters, p. 92

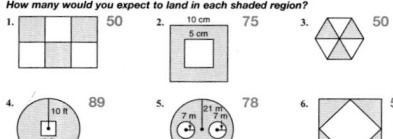

Reteaching the Lesson

Activity Have pairs of students create a square design with red and blue tiles. Have them find the probability of an object landing on a red tile. Then have them use additional tiles to form a new shape and find the probability of an object landing on a red tile in the new shape.

Error Analysis
Watch for students who subtract the area of the region in question from the area of the whole to find the probability.
Prevent by reminding students that a ratio is a comparison of two numbers by division.

Copy and complete.

	Dartboard	Times Dart is Thrown	Times Expected to Land in Shaded Region
15.	Exercise 9	200	about 80
16.	Exercise 10	200	about 125
17.	Exercise 11	250	about 32
18.	Exercise 12	250	about 90
19.	Exercise 13	275	about 44
20.	Exercise 14	275	about 61

Applications and Problem Solving

21. *Entertainment* A sky diver is the featured entertainer for the half-time show of the Johnstown High School homecoming football game. It is equally likely that the sky diver will land on any point of the field. Find the probability that the sky diver will land inside the circle. **about $\frac{1}{40}$, or 2.5%**

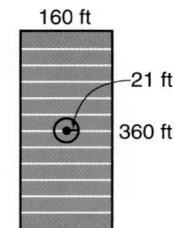
160 ft
21 ft
360 ft

22. *Games* A dart is thrown at the dartboard shown. It is equally likely that the dart will land anywhere on the dartboard. What is the probability of the dart landing on the shaded region? Then tell how likely the event is to happen. **See Answer Appendix.**

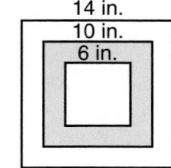
14 in.
10 in.
6 in.

23a–b. See students' work.

23. *Working on the* CHAPTER Project Refer to page 518.
   a. Suppose you had randomly picked up 100 circles. How many circles of each color would you expect to pick up?
   b. Suppose you had randomly picked up 250 circles. How many circles of each color would you expect to pick up?
   c. Write a paragraph that explains how this experiment relates to camouflage. **See margin.**

24. A-50; B-50; C-about 13; D-25; E-about 13; F-25; G-25

24. *Critical Thinking* Suppose you made a dartboard from the tangram shown. If you throw 200 darts at the dartboard, how many would you expect to land in each region? Assume it is equally likely that the dart lands anywhere in the tangram.

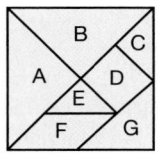
A B C D E F G

Mixed Review
25. No; people in a Levi's store may prefer Levi's jeans.

25. *Statistics* Tell whether a survey on favorite jeans at a Levi's store is a random survey. Explain your answer. *(Lesson 13-2)*

26. Test Practice The area of a rectangle is 153 square feet, and the width is 9 feet. Use the formula $A = \ell w$ to find the length of the rectangle. *(Lesson 12-3)* **A**
   A 17 ft   **B** 144 ft   **C** 162 ft   **D** 1,377 ft

27. *Geometry* What is the area of a triangle whose base is 52 feet and whose height is 38 feet? *(Lesson 10-2)* **988 ft²**

Lesson 13-3 Integration: Geometry   Probability and Area   **529**

Extending the Lesson

Enrichment Masters, p. 92

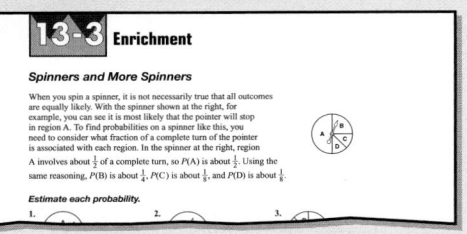

Activity Invite an insurance agent to give a presentation on how probability is used to set rates for accident, life, and hazard insurance. Have students prepare questions and find probabilities with the help of the speaker.

Exercise 23 asks students to advance to the next stage of work on the Chapter Project. You may wish to have students form groups to brainstorm ideas for their paragraphs. Have students place their paragraphs in their Project Folders.

4 ASSESS

Closing Activity
Writing Have students work in pairs to create probability word problems involving any playing field, such as a soccer field or hockey rink.

Chapter 13, Quiz B (Lesson 13-3) is available in the *Assessment and Evaluation Masters,* p. 351.

Mid-Chapter Test (Lessons 13-1 through 13-3) is available in the *Assessment and Evaluation Masters,* p. 350.

Additional Answer
23c. Sample answer: In the experiment, it is easier to see the black and white circles but it is hard to see the newsprint circles. The newsprint circles represent the camouflage trait.

Practice Masters, p. 92

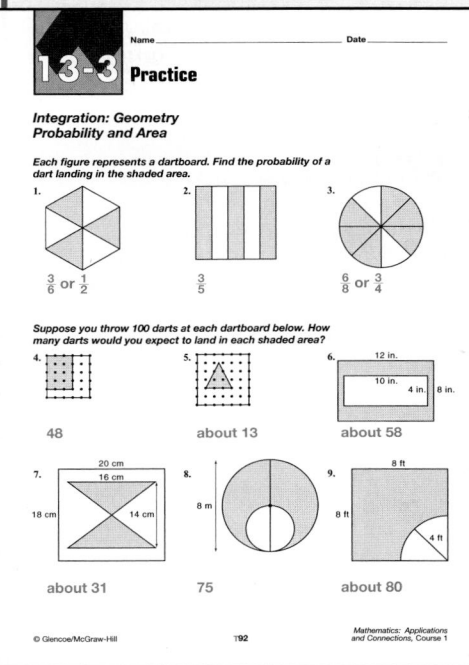

GET READY

Objective Students investigate geometric probability by using a graphing calculator.

Technology Resources
• TI-80, TI-81, TI-82, or TI-83 graphing calculator

MANAGEMENT TIPS

Recommended Time
30 minutes

Getting Started Explain that the calculator program will generate ordered pairs. If students have had little experience with graphing calculators, they will need help entering the program into the memory. Consult the User's Guide for the locations of specific commands.

Explain that the calculator needs an integer seed value to act as a starting point for the random number generator.

Instruct students to count points that fall on the outside edge of the large square or unshaded square as inside the particular square.

Using a Spreadsheet
Spreadsheet software may also be used to generate the needed random numbers.

ASSESS

After students answer Exercises 1–5, ask them to compare their graphs. Have students with similar graphs compare their ratios for Exercises 1 and 4. Are these ratios also close? They should be.

13-3B Probability

A Follow-Up of Lesson 13-3

 graphing calculator

 grid paper

straightedge

In Lesson 13-3, you used area to find a probability. You can experiment with area and probability by using a graphing calculator.

TRY THIS

Work with a partner.

• Use a straightedge to draw the figure shown on grid paper. Then shade the figure as shown.

• Use the program to generate 50 random ordered pairs. When you run the program, it will ask you how many ordered pairs you want: type "50" and push the ENTER key. It will then ask you for a seed number. Type in any number and push the ENTER key. This helps the calculator pick random numbers. The calculator will now start giving you pairs of coordinates. The first number is *x* and the second number is *y*. Each time you push the ENTER key, it will give you another pair of numbers until it has given you 50 of them.

• Graph each pair on the grid paper. Keep a tally of how many points are in the shaded regions and how many points are in the unshaded region.

```
:Fix 1
:Disp "Number of Pairs"
:Input P
:Disp "Seed Number"
:Input N
:N → rand
:For (J,1,P,1)
:10 * rand → X
:Disp X
:10 * rand → Y
:Disp Y
:Disp " "
:Pause
:End
```

ON YOUR OWN

1. Write a ratio that compares the number of points graphed in the shaded regions to the total number of points graphed. **1–4. See Answer Appendix.**
2. Use the formulas for the area of a triangle and the area of a square to find the ratio that compares the area of the shaded regions to the area of the square.
3. How do the ratios compare?
4. Generate and graph 50 more ordered pairs by changing the seed number. Add the results to your previous results. What fraction of the 100 points generated are in the shaded region? How does this fraction compare to the fraction of the square that is shaded?
5. Suppose you wanted to plot points on a grid in which the *x*- and *y*-coordinates went from 0 to 20, instead of 0 to 10. What change would you need to make in the program? **Sample answer: In the program, change 10 to 20.**

530 Chapter 13 Using Probability

 Math Journal Have students write a paragraph about their reaction to the close link between are a and probability. Were they surprised? Why or why not?

What you'll learn

You'll learn to find outcomes using lists, tree diagrams, and combinations.

When am I ever going to use this?

Knowing how to find the number of possible outcomes can help you determine options when ordering a pizza.

Word Wise

tree diagram
combinations

Study Hint

Problem Solving
Use the draw-a-diagram strategy.

Mr. Mason wrote the suffixes -ness, -er, and -ly on the chalkboard. He then asked his students to form words using the given suffixes and the words *quick, slow,* and *happy.* How many different words can be made using the suffixes and the words?

To solve the problem, you can determine the sample space by making a list.

quickness	quicker	quickly
slowness	slower	slowly
happiness	happier	happily

There are 9 different words that can be formed.

You can also use a **tree diagram** to show the sample space.

Word	Suffix	Outcome
quick	-ness	quickness
	-er	quicker
	-ly	quickly
slow	-ness	slowness
	-er	slower
	-ly	slowly
happy	-ness	happiness
	-er	happier
	-ly	happily

Example 1

At Pizza House, you can order a thin crust or deep dish pizza with cheese, pepperoni, or sausage. Draw a tree diagram that shows all of the ways you can order a one-topping pizza.

Pizza	Topping	Outcome
thin crust (T)	cheese (C)	TC
	pepperoni (P)	TP
	sausage (S)	TS
deep dish (D)	cheese (C)	DC
	pepperoni (P)	DP
	sausage (S)	DS

There are six ways to order a one-topping pizza.

Study Hint

Reading Math The outcome TC means Thin crust Cheese pizza.

Lesson 13-4 Finding Outcomes **531**

Multiple Learning Styles

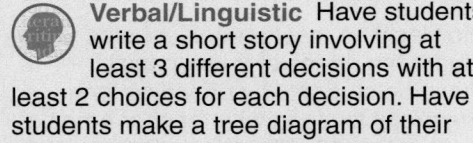 **Verbal/Linguistic** Have students write a short story involving at least 3 different decisions with at least 2 choices for each decision. Have students make a tree diagram of their story and list all possible outcomes.

13-4 Lesson Notes

Instructional Resources

- *Study Guide Masters,* p. 93
- *Practice Masters,* p. 93
- *Enrichment Masters,* p. 93
- Transparencies 13-4, A and B
- *Assessment and Evaluation Masters,* p. 352
- *Technology Masters,* p. 26
- CD-ROM Program
  - Resource Lesson 13-4

Recommended Pacing	
Honors	Day 7 of 11

1 FOCUS

 5-Minute Check
(Lesson 13-3)

Each figure represents a dartboard. It is equally likely that a dart will land anywhere on the dartboard. Find the probability of a randomly thrown dart landing in the shaded region.

1. $\frac{1}{9}$

2. 7 ft   12 ft $\frac{1}{4}$

3. Suppose you threw a dart 100 times at each dartboard shown in Exercise 1. How many times would you expect it to land in each shaded region? 11, 25

 The 5-Minute Check is also available on **Transparency 13-4A** for this lesson.

Example 2 Each spinner is spun once. Find the probability of spinning blue, green, and red in that order.

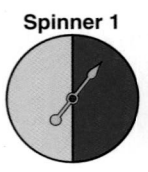 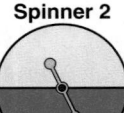 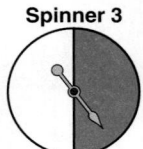

Spinner 1 Spinner 2 Spinner 3

Find all of the possible outcomes.

Spinner 1	Spinner 2	Spinner 3	Outcome

```
                       white (W) ——→ PTW
            tan (T) <
                       red (R)  ——→ PTR
  pink (P) <
                       white (W) ——→ PGW
            green (G) <
                       red (R)  ——→ PGR

                       white (W) ——→ BTW
            tan (T) <
                       red (R)  ——→ BTR
  blue (B) <
                       white (W) ——→ BGW
            green (G) <
                       red (R)  ——→ BGR
```

One outcome has blue, green, red in that order. There are eight possible outcomes. Therefore, $P(\text{BGR}) = \frac{1}{8}$.

Arrangements or listings where order is not important are called **combinations**. Combinations are another way to determine a sample space. To find combinations, you can make a list.

Example 3 **APPLICATION**

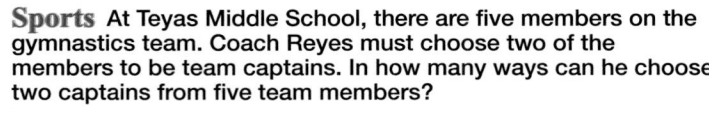

Sports At Teyas Middle School, there are five members on the gymnastics team. Coach Reyes must choose two of the members to be team captains. In how many ways can he choose two captains from five team members?

Let V, W, X, Y, and Z represent the team members. List all of the ways two team captains can be chosen.

VW	VX	VY	VZ	WV
WX	WY	WZ	XV	XW
XY	XZ	YV	YW	YX
YZ	ZV	ZW	ZX	ZY

Then count all of the *different* arrangements. Since order is not important, VW and WV are the same.

VW	VX	VY	VZ	WX
WY	WZ	XY	XZ	YZ

There are 10 ways Coach Reyes can choose two team captains.

Communicating Mathematics

1. See Answer Appendix.

Read and study the lesson to answer each question.

1. *Write a Problem* that can be solved by using the tree diagram shown.

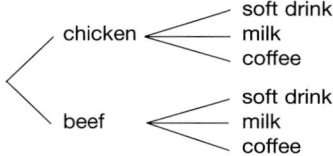

2. *Draw* a tree diagram that shows the results of tossing a nickel, a dime, and a quarter. Then find the probability of getting three tails. **See margin.**

3. *Write* about the advantages of using a tree diagram to count outcomes. Are there any disadvantages of tree diagrams? **See Answer Appendix.**

Guided Practice

For each situation, make a list and draw a tree diagram to show the sample space. 4–5. See Answer Appendix.

4. a choice of a hot dog or hamburger and a choice of iced tea or lemonade

5. a choice of a blue or red shirt with blue, black, or red shorts

6. How many ways can a person choose two magazines from four magazines? **6 ways**

7. *School* A science quiz has one multiple-choice question with answer choices A, B, and C, and two true/false questions.
   a. Draw a tree diagram that shows all of the ways a student can answer the questions. **See Answer Appendix.**
   b. What is the probability of answering all three questions correctly by guessing? $\frac{1}{12}$

Family Activity

Make a diagram of your family tree going back to your great-grandparents. Explain how your diagram is like a tree diagram.

EXERCISES

Practice

For each situation, make a list and draw a tree diagram to show the sample space. 8–13. See Answer Appendix.

8. a choice of lemon, apple, or pecan pie with milk or tea

9. a choice of a leather or nylon backpack in purple, green, black, or brown

10. rolling two number cubes

11. spinning each spinner once

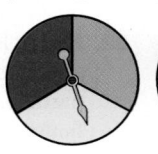 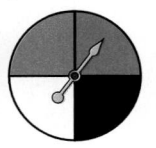

12. a choice of a portable or stationary basketball hoop with a graphite or acrylic backboard

13. a choice of a tower or spinner CD holder made from wood or plastic that holds 60, 100, or 250 CDs

Lesson 13-4 Finding Outcomes **533**

Check for Understanding

If students need additional practice or instruction after completing Exercises 1–7, one of these options may be helpful.
- Extra Practice, see p. 592
- Reteaching Activity
- *Study Guide Masters,* p. 93
- *Practice Masters,* p. 93

Assignment Guide

Core: 9–17 odd, 19–22
Enriched: 8–16 even, 17–22

Family Activity

An alternative activity would be to have students make a tree of how their school staff is organized, beginning with the principal.

Additional Answer

2. See students' diagrams; outcomes are HHH, HHT, HTH, HTT, THH, THT, TTH, TTT. P(TTT) = $\frac{1}{8}$

Study Guide Masters, p. 93

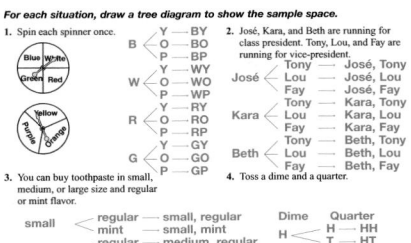

Reteaching the Lesson

Activity Ask students to predict how many different figures they would have if they drew squares and triangles each in red, black, green, and yellow. **8** Have them draw the shapes and make a tree diagram to check their prediction. Ask what the probability would be of randomly choosing a green triangle. $\frac{1}{8}$

Error Analysis
Watch for students who repeat patterns within a tree diagram.
Prevent by having them multiply the number of choices to check the number of outcomes.

Closing Activity

Act It Out In situation one, four students stand in front of the class. How many handshakes need to occur so that each student shakes hands with the other three students? Have a volunteer list the handshakes on the chalkboard.
6 handshakes

In situation two, have four students put their names in a hat for a raffle. The first name drawn wins first prize, and second wins second prize. How many possible first and second place winner arrangements are there? **12 arrangements**

Chapter 13, Quiz C (Lesson 13-4) is available in the *Assessment and Evaluation Masters*, p. 352.

Practice Masters, p. 93

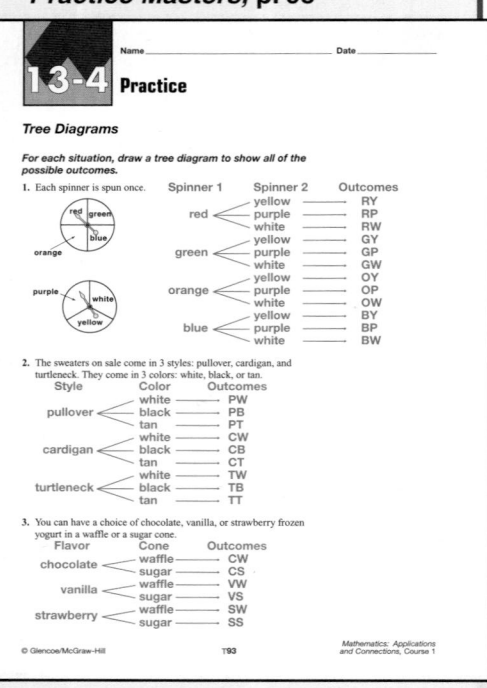

---

14. How many different combinations of 2 recipes can a chef choose from 3 different recipes? **3 ways**

15. How many different ways can a person choose 3 kittens from a litter of 5 kittens? **10 ways**

16. How many different ways can a teacher choose 3 three-dimensional figures to put on a quiz from 4 different three-dimensional figures? **4 ways**

Applications and Problem Solving

17. **Food** At Otani Sushi Restaurant, customers can choose from the following six Nigiri-Sushi, or fish: Maguro (tuna), Ika (squid), Tako (octopus), Ebi (shrimp), Kani (crab), and Saba (mackerel). In how many ways can a customer choose two of the six Nigiri-Sushi? **15 ways**

18. **Fashion** Awenasa is buying a new navy sweater. The sweater will coordinate with her black, white, red, and yellow blouses and her blue, plaid, and black pants.
   a. How many different outfits will she have? **12 outfits**
   b. Suppose Awenasa chooses one blouse and one pair of pants at random. What is the probability that she will choose the yellow blouse and the blue pants or the red blouse and the black pants? $\frac{1}{6}$

19. **Critical Thinking** There are three 2-sided counters in a cup. One is red on one side and blue on the other, one is red and white, and the third is blue and white. Amiri shakes the cup and tosses out the chips. He receives one point if the chips are all different colors. Irene receives one point if two chips match. Use a tree diagram to help you determine whether or not this is a fair game. If this is not a fair game, how could you make it fair? **See Answer Appendix.**

Mixed Review

20. **Probability** The figure shown represents a dartboard. It is equally likely that the dart will land anywhere on the dartboard. Find the probability of a randomly-thrown dart landing in the shaded region. *(Lesson 13-3)* **about $\frac{2}{5}$ or 40%**

21. **Test Practice** Solve $n = -145 \div (-29)$. *(Lesson 11-6)* **C**
   A -5 B 4 C 5 D -3 E Not Here

22. **Food** Refer to the graph. Suppose 2,000 people participated in the peanut butter survey. How many people can be expected to prefer creamy peanut butter? *(Lesson 8-6)* **940 people**

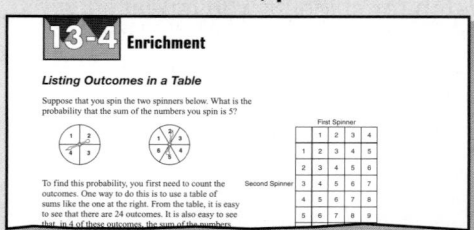

Source: The Adult Peanut Butter Lover's Fan Club, Peanut Advisory Board

---

Extending the Lesson

Enrichment Masters, p. 93

13-4 Enrichment

Listing Outcomes in a Table

Suppose that you spin the two spinners below. What is the probability that the sum of the numbers you spin is 5?

To find this probability, you first need to count the outcomes. One way to do this is to use a table of sums like the one at the right. From the table, it is easy to see that there are 24 outcomes. It is also easy to see that in 4 of these outcomes, the sum of the numbers

Activity On a restaurant menu, a person can order 1 item from column A listing 6 main courses, 2 items from column B listing 10 side dishes, and 1 item from column C listing 5 desserts. How many different meals are possible? Have students discuss their answers.
1,350 meals

COOPERATIVE LEARNING

13-4B Simulations

A Follow-Up of Lesson 13-4

🌓 two-colored
counters

🥤 cups

A simulation is an application of the Acting It Out problem-solving strategy.

A *simulation* is a way of acting out a problem. You can conduct a simulation by using manipulatives such as counters or number cubes. In this lab, you will conduct a simulation to explore the probability that in a family with three children, at least two of them are girls.

It is equally likely that a boy or a girl will be born. Similarly, it is equally likely that a two-colored counter will land on one color or the other. So, you can toss a two-colored counter to simulate outcomes.

TRY THIS

Work with a partner.

To explore the probability that at least two of the three children in a family are girls, follow these steps.

Step 1 Place three counters in a cup and toss them onto your desk.

Step 2 Count the number of counters that land with the red side up. This represents the number of boys. Count the number of counters that land with the yellow side up. This represents the number of girls.

Step 3 Record the results in a table like the one shown.

Step 4 Repeat Steps 1-3 until you have 50 trials.

Outcome			
Trial 1	B	B	G
Trial 2	G	G	G
Trial 3	G	G	B
Trial 4	B	B	G

Suppose 23 of the 50 trials have at least two girls. The experimental probability that at least two of the three children in a family are girls is $\frac{23}{50}$.

You can refer to Lesson 5-4B for information on experimental probability.

ON YOUR OWN

1. Based on your results of the simulation, what is the experimental probability that, in a family with three children, at least two of them are girls? **1–4. See Answer Appendix.**

2. Make a list showing all of the possible outcomes of the experiment. What is the theoretical probability that, in a family with three children, at least two of them are girls?

3. Compare the experimental probability with the theoretical probability.

4. **Reflect Back** Find the probability that in a family with four children, all four of them are girls. Then describe a simulation that you can use to explore this probability. Conduct the simulation. Record and explain your results.

Lesson 13-4B HANDS-ON **535**

 Have students write a paragraph describing another situation where a simulation would be useful. Have them include a sample simulation.

HANDS-ON
13-4B LAB Notes

GET READY

Objective Students explore probability by conducting a simulation.

Optional Resources
Hands-On Lab Masters
• counters, p. 6
• pattern for cup, p. 8
• worksheet, p. 68

Overhead Manipulative Resources
• counters
• cups

Manipulative Kit
• counters
• cups

MANAGEMENT TIPS

Recommended Time
30 minutes

Getting Started Ask students how they think three counters and a cup will be used to investigate the problem of this lab. Then ask: *How many counters would be needed for a family with one child? two children? n children?* **1; 2; n**

The **Activity** requires students to toss 3 counters and record results. The phrase "at least 2 boys" means that students will count all outcomes with 2 or 3 boys for their experimental probability. Have students pool all their experiments to get an experimental probability for the class for no boys, 1 boy, 2 boys, and 3 boys. If there are students who are one of 3 children, use your class data to determine the probability for the siblings in their families.

ASSESS

Have students complete Exercises 1–4. If students are having difficulty with Exercise 4, remind them that a tree diagram is very helpful for finding theoretical probabilities.

1 FOCUS

 5-Minute Check
(Lesson 13-4)

1. For a class project, you can make a display, write a report, or build a model about a person, locale, or object. Draw a tree diagram to show all of the outcomes.

```
Project      Subject   Outcome
             person(P)→ DP
display(D) ┤ locale(L) → DL
             object(O) → DO
             person(P)→ RP
 report(R) ┤ locale(L) → RL
             object(O) → RO
             person(P)→ MP
 model(M)  ┤ locale(L) → ML
             object(O) → MO
```

2. Find the number of games that would be played in the first round of a five-team tournament where each team plays every other team.
10 games

The 5-Minute Check is also available on **Transparency 13-5A** for this lesson.

Motivating the Lesson
Communication Ask students to discuss how they would determine the probability that a set of fraternal twins would be two boys.
Possibilities are BB, BG, GB, GG; $P = \frac{1}{4}$

What you'll learn
You'll learn to find the probability of independent events.

When am I ever going to use this?
Knowing how to find the probability of independent events can help you determine the chance of rolling doubles in Monopoly.

Word Wise
independent event

While on a nature walk with their baby-sitter, Margarita and Billy collected acorns, walnuts, and stones. The table shows the contents of each of their bags. Suppose they each reach into their respective bag and randomly choose an object. What is the probability that Billy chooses an acorn and Margarita chooses a stone?

Name	Acorns	Walnuts	Stones
Billy	6	8	10
Margarita	12	12	16

The object that Billy chooses does not affect the object that Margarita chooses. They are called **independent events**.

To find the probability of two independent events, you multiply their probabilities.

Probability of Two Independent Events		
	Words:	The probability of two independent events, A and B, is the product of the probability of event A and the probability of event B.
	Symbols:	$P(A \text{ and } B) = P(A) \cdot P(B)$

$P(\text{acorn}) = \dfrac{6}{24}$ ← *number of ways Billy can choose an acorn*
← *number of possible outcomes: 6 + 10 + 8 = 24*

$P(\text{stone}) = \dfrac{16}{40}$ ← *number of ways Margarita can choose a stone*
← *number of possible outcomes: 12 + 16 + 12 = 40*

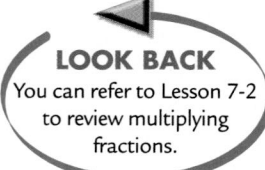 **LOOK BACK**
You can refer to Lesson 7-2 to review multiplying fractions.

Multiply to find the probability.

$$\frac{6}{24} \times \frac{16}{40} = \frac{\overset{1}{6}}{\underset{4}{24}} \times \frac{\overset{2}{16}}{\underset{5}{40}} \quad \text{\textit{The GCF of 6 and 24 is 6.}}$$
$$\text{\textit{The GCF of 16 and 40 is 8.}}$$
$$= \frac{2}{20} \text{ or } \frac{1}{10}$$

The probability that Billy chooses an acorn and Margarita chooses a stone is $\frac{1}{10}$. The event is not too likely to occur.

Multiple Learning Styles

 Logical Have students research sports statistics to determine the probability that a professional athlete will make a free throw in basketball, score a goal in hockey, get a hit in baseball, serve an ace in tennis, and so on.

Examples

1 A number cube is rolled, and the spinner shown is spun. Find the probability of rolling 5 and landing on a consonant.

$P(5) = \frac{1}{6}$ $P(\text{consonant}) = \frac{3}{4}$

$P(5 \text{ and consonant}) = \frac{1}{6} \times \frac{3}{4}$

$= \frac{3}{24} \text{ or } \frac{1}{8}$

The probability of rolling a 5 and landing on a consonant is $\frac{1}{8}$.

CONNECTION **2** **Life Science** Owls lay two to five rounded, white eggs. Suppose an owl lays two eggs. What is the probability that the first egg hatched is a female owl and the second egg hatched is a male owl?

Explore   You know the probability of female and the probability of male. You need to find the probability that the first egg hatched is a female owl and the second egg hatched is a male owl.

Plan   Multiply to find the probability.

Solve $P(\text{female owl}) = \frac{1}{2}$ $P(\text{male owl}) = \frac{1}{2}$

$\frac{1}{2} \times \frac{1}{2} = \frac{1}{4}$

So, $P(\text{female owl, then male owl}) = \frac{1}{4}$.

Examine   Use a tree diagram. One outcome has a female owl and then a male owl. There are four possible outcomes. Therefore, the probability is $\frac{1}{4}$.

First Egg	Second Egg	Outcome
female (F)	female ⟶	FF
	male ⟶	FM
male (M)	female ⟶	MF
	male ⟶	MM

Lesson 13-5 Probability of Independent Events   **537**

Transparency 13-5B contains a teaching aid for this lesson.

Thinking Algebraically Students' knowledge of tree diagrams can help them deduce the formula for the probability of two independent events. For each possibility in A, there are all of the possibilities of B. So if A has 6 branches and B has 3, the total number is 6 · 3 or 18 possibilities.

Teaching Tip After Example 1, have students make a tree diagram to verify their answer.

In-Class Examples

For Example 1
A number cube is rolled and a coin is tossed. Find the probability of rolling a 2 and landing on heads. $\frac{1}{12}$

For Example 2
To win the grand prize, Ted has to choose one of 10 keys to unlock Treasure Chest A, B, or C. The winning combination is Key 3 in Treasure Chest A. What are Ted's chances of winning? $\frac{1}{30}$

Classroom Vignette

"I have students interview adults and write a report about how they use math in their job or in everyday life. They are encouraged to use their creativity in preparing their report. They also use computers to prepare their reports."

Deana Bobzien, Teacher
Rochester Junior High
Rochester, IL

Deana Bobzien

Check for Understanding

If students need additional practice or instruction after completing Exercises 1–7, one of these options may be helpful.
- Extra Practice, see p. 592
- Reteaching Activity
- *Study Guide Masters*, p. 94
- *Practice Masters*, p. 94

Assignment Guide

Core: 9–19 odd, 20–23
Enriched: 8–16 even, 18–23

Additional Answers

1. Multiply the probability of the first event by the probability of the second event.

2. Sample answer: A coin is tossed and a number cube is rolled. Find the probability of the coin landing on heads and the number cube landing on a three.

3. Flora is correct. The probability of a 4 or 5 is $\frac{2}{6}$, and the probability of a 6 is $\frac{1}{6}$. $\frac{1}{3} \times \frac{1}{6} = \frac{1}{18}$

Study Guide Masters, p. 94

13-5 Study Guide

Probability and Independent Events

If the outcome of one event does not affect the outcome of a second event, the two events are **independent**.

The probability of two independent events, A and B, is equal to the probability of event A times the probability of event B.

$$P(A \text{ and } B) = P(A) \times P(B)$$

Example Suppose you spin each of these two spinners. What is the probability of spinning an even number and a vowel?

$P(\text{even}) = \frac{1}{2}$ ← number of ways to spin even / number of possible outcomes

$P(\text{vowel}) = \frac{1}{5}$ ← number of ways to spin a vowel / number of possible outcomes

$P(\text{even, vowel}) = \frac{1}{2} \times \frac{1}{5}$ or $\frac{1}{10}$

The two spinners shown above are spun. Find the probability of each event.

1. $P(6, P)$ $\frac{1}{30}$
2. $P(\text{less than 4, consonant})$ $\frac{2}{5}$
3. $P(\text{odd, S})$ $\frac{1}{10}$
4. $P(5, \text{consonant})$ $\frac{2}{15}$
5. $P(\text{greater than 8, T})$ 0
6. $P(\text{less than 7, vowel})$ $\frac{1}{5}$

A quarter and a dime are tossed. Find the probability of each event.

7. $P(T, H)$ $\frac{1}{4}$
8. $P(\text{both the same})$ $\frac{1}{2}$
9. $P(T, T)$ $\frac{1}{4}$

Suppose you write each letter of your first and last names on a separate index card and select one letter from each name without looking. Find the probability of each event. Answers will vary.

10. $P(\text{vowel, vowel})$
11. $P(\text{consonant, vowel})$
12. $P(M, E)$

© Glencoe/McGraw-Hill   794   Mathematics: Applications and Connections, Course 1

Communicating Mathematics

Read and study the lesson to answer each question. 1–3. See margin.

1. *Explain* how to find the probability of two independent events.

2. *Write* a probability problem involving two events that are independent. Then ask a classmate to solve your problem.

3. *You Decide* Two number cubes are rolled. Madeline says that the probability of the first number cube landing on 4 or 5 and the second number cube landing on 6 is $\frac{1}{9}$. Flora disagrees. She says that the probability is $\frac{1}{18}$. Who is correct? Explain.

Guided Practice

The two spinners shown are spun. Find the probability of each event.

4. $P(\text{green and 5})$ $\frac{1}{15}$
5. $P(\text{orange and odd})$ $\frac{1}{10}$
6. $P(\text{blue and prime})$ $\frac{1}{10}$

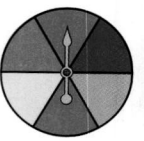 

7. *School* A pencil box contains 4 lead pencils, 2 pens, 3 color pencils, and 6 markers. Another pencil box contains 3 pens, 2 markers, and 5 lead pencils.

   a. What is the probability of choosing a marker from the first pencil box? $\frac{2}{5}$
   b. What is the probability of choosing a pen from the second pencil box? $\frac{3}{10}$
   c. Find the probability of choosing a marker from the first box and then a pen from the second box. $\frac{3}{25}$

EXERCISES

Practice

A coin is tossed and a number cube is rolled. Find the probability of each event.

8. $P(\text{tails and 3})$ $\frac{1}{12}$
9. $P(\text{heads and odd})$ $\frac{1}{4}$
10. $P(\text{heads and 2 or 4})$ $\frac{1}{6}$
11. $P(\text{heads and 7})$ 0
12. $P(\text{tails and prime})$ $\frac{1}{4}$
13. $P(\text{heads or tails and composite})$ $\frac{1}{3}$

The spinner shown is spun, and a card is chosen from the set of cards shown. Find the probability of each event.

14. $P(\text{2-digit and consonant})$ $\frac{5}{16}$
15. $P(\text{even and vowel})$ $\frac{3}{8}$
16. $P(\text{less than 14 and T})$ $\frac{3}{16}$

17. A red number cube and a green one are rolled. Find the probability of tossing a number greater than 4 on the green cube and a number other than 1 on the red one. $\frac{5}{18}$

Applications and Problem Solving

18. *School* A quiz has one true/false question and one multiple-choice question with possible answer choices a, b, c, and d. If you guess each answer, what is the probability of answering both questions correctly? $\frac{1}{8}$

■ Reteaching the Lesson ■

Activity Give one student colored centimeter cubes and another a number cube. Ask one to determine the $P(\text{red})$ and the other to determine the $P(2)$. Then ask them to multiply their answers to determine the $P(\text{red and 2})$. Continue with other examples.

19. **Sports** The probability that the Eagles will win on Saturday is 0.6. The probability that the Lions will win on Sunday is 0.8.
   a. What is the probability that both teams will win? **0.48**
   b. What is the probability that both teams will lose? **0.08**

20. **Critical Thinking** Five students have no absences for the first semester. They are able to draw from a bag that has 7 movie tickets and 3 gift certificates. Each person keeps what is drawn. Paul and Kikuyu are the first two students to draw. What is the probability that they both draw movie tickets? $\frac{7}{15}$

Mixed Review
21. How many ways can a person choose 3 videos from a stack of 6 videos? *(Lesson 13-4)* **20 ways**

22. **Geometry** If a circle has a radius of 27 feet, what is its circumference? *(Lesson 7-4)* **169.6 ft**

23. **Test Practice** Nathan walks $\frac{3}{4}$ of a mile to school. James walks $\frac{5}{8}$ of a mile to school. How much farther does Nathan walk to school than James? *(Lesson 6-4)* **D**
   A $\frac{1}{2}$ mile   **B** $\frac{3}{8}$ mile   **C** $\frac{1}{4}$ mile   **D** $\frac{1}{8}$ mile   **E** Not Here

Let the Games Begin

Mix & Match

Get Ready This game is for two players. 16 index cards

Get Set Make a set of fraction cards containing the numbers $\frac{1}{12}$, $\frac{5}{12}$, $\frac{1}{36}$, $\frac{1}{4}$, $\frac{1}{3}$, 0, $\frac{1}{6}$, $\frac{1}{18}$. Make a set of event cards containing the following independent events: heads and tails, 3 and prime, 4 and 6, heads and less than 5, tails and 8, even and less than 6, odd and greater than 4, 2 and 1 or 4.

Go ● Mix each set of cards and place them facedown as shown.
   ● Suppose that two number cubes are rolled, two coins are tossed, or one of each. Player 1 turns over an event card and a fraction card. If the fraction corresponds to the probability that the event occurs, the player picks up the two cards, a point is scored, and the player turns over two more cards. If the fraction does not correspond to the event, the player turns the cards facedown, no points are scored, and it is the next player's turn.

Event Cards   Fraction Cards

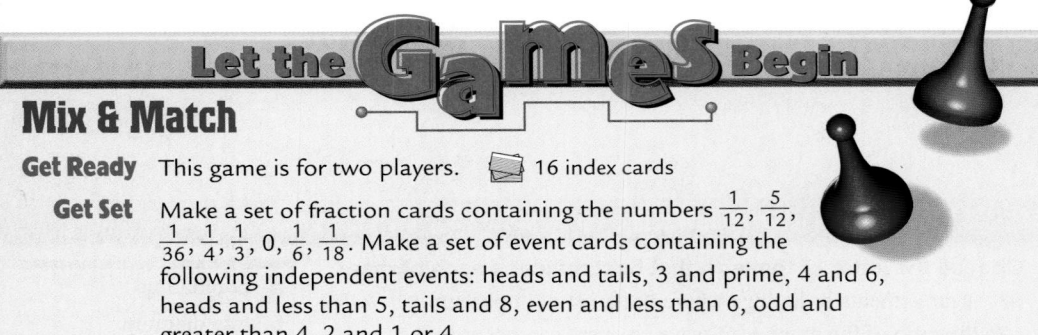

   ● Take turns until all cards are matched. The player with the most points wins.

 Visit www.glencoe.com/sec/math/mac/mathnet for more games.

Lesson 13-5 Probability of Independent Events **539**

■ Extending the Lesson ■

Enrichment Masters, p. 94

13-5 Enrichment

Dependent Events

If the result of one event affects the result of a second event, the events are called **dependent**. For example, suppose you draw one card from the set of cards shown at the right, but do not replace it. Then you draw a second card. What is the probability that you will draw the T, then an L?

First find the probability of drawing the T.
$P(T) = \frac{1}{10}$ ← There is 1 card marked T.
        ← There are 10 cards in all.

Then find the probability of drawing an L after drawing the T.

Let the Games Begin

Students could extend the game by making up more event and fraction cards.
Additional resources for this game can be found on pages 47–48 of the Classroom Games.

4 ASSESS

Closing Activity

Speaking Have small groups discuss how they would find the probability of choosing a red marker from a box of 12 different-colored markers and yellow paper from a stack with six different colors. Have groups compare their solutions. $\frac{1}{12} \times \frac{1}{6} = \frac{1}{72}$

Chapter 13, Quiz D (Lesson 13-5) is available in the *Assessment and Evaluation Masters,* p. 352.

Practice Masters, p. 94

13-5 Practice

Probability of Independent Events

Solve.

1. A bakery shop offers breakfast specials consisting of a muffin and a beverage. There are five kinds of muffins to choose from: corn, blueberry, honey-bran, cranberry-orange, and banana-nut. The beverages available are coffee, tea, orange juice, or milk. Each choice is equally likely.
   a. What is the probability of choosing a blueberry muffin? $\frac{1}{5}$
   b. What is the probability of choosing orange juice? $\frac{1}{4}$
   c. Find P(blueberry muffin and orange juice). $\frac{1}{20}$

The two spinners shown are spun. Find the probability of each event.

2. P(1 and white) $\frac{1}{18}$ 3. P(3 and red) $\frac{1}{6}$
4. P(2 and blue) $\frac{1}{9}$ 5. P(odd and red) $\frac{1}{3}$
6. P(odd and blue) $\frac{2}{9}$ 7. P(4 and white) 0
8. P(even and any color other than white) $\frac{5}{18}$

Suppose you toss a coin and pick a card from a pile of 18 cards, each printed with a letter from the phrase "MATHEMATICS IS FOR ME." Find the probability of each of the following.

9. P(heads and M) $\frac{1}{12}$ 10. P(tails and F) $\frac{1}{36}$
11. P(tails and T) $\frac{1}{18}$ 12. P(heads and vowel) $\frac{7}{36}$
13. P(tails and consonant) $\frac{11}{36}$ 14. P(heads and N) 0

© Glencoe/McGraw-Hill   T94   Mathematics: Applications and Connections, Course 1

Vocabulary

Vocabulary

This section provides a listing of the new terms, properties, and phrases that were introduced in this chapter. Have students define each term and provide an example or two of it, if appropriate.

Understanding and Using the Vocabulary

These exercises check students' understanding of the terms by using a variety of verbal formats including matching, completion, and true/false.

Glossaries A complete glossary of terms appears on pages 642–648. The glossary also appears in Spanish on pages 649–656.

Additional Answer

11. Sample answer: Multiply the probability of the first event by the probability of the second event.

After completing this chapter, you should be able to define each term, concept, or phrase and give an example or two of each.

Problem Solving
make a table (pp. 520-521)

Statistics and Probability
combinations (p. 532)
complementary events (p. 516)
event (p. 515)
fair game (p. 514)
independent events (p. 536)

outcomes (p. 515)
population (p. 522)
random (p. 522)
sample (p. 522)
sample space (p. 515)
simulation (p. 535)
theoretical probability (p. 515)
tree diagram (p. 531)
unfair game (p. 514)

Understanding and Using the Vocabulary

Choose the letter of the term that best matches each phrase.

1. arrangements or listings where order is not important e
2. the ratio of the number of ways an event can occur to the number of possible outcomes i
3. a specific outcome or type of outcome g
4. the entire group from which samples are taken a
5. a diagram used to show all of the possible outcomes b
6. a randomly selected group chosen for the purpose of collecting data h
7. a word that means the same as results c
8. when one event's occurring does not affect another event j
9. two events in which either one or the other can occur, but not both d
10. events that happen by chance f

a. population
b. tree diagram
c. outcomes
d. complementary events
e. combinations
f. random events
g. event
h. sample
i. probability
j. independent events
k. simulation

In Your Own Words

11. *Explain* how to find the probability of two independent events. **See margin.**

MindJogger Videoquizzes

MindJogger Videoquizzes provide an alternative review of concepts presented in this chapter. Students work in teams to answer questions, gaining points for correct answers. The questions are presented in three rounds.
Round 1 Concepts–5 questions
Round 2 Skills–4 questions
Round 3 Problem Solving–4 questions

Objectives & Examples

Upon completing this chapter, you should be able to:

● find and interpret the theoretical probability of an event *(Lesson 13-1)*

There are six equally likely outcomes on the spinner. Find the probability of spinning blue.

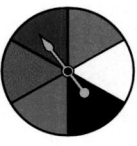

$\frac{1}{6}$ ← *number of ways to spin blue*
  ← *number of possible outcomes*

So, $P(\text{blue}) = \frac{1}{6}$.

● predict the actions of a larger group using a sample *(Lesson 13-2)*

If 12 out of 50 people prefer to watch TV after 11 P.M., how many people out of 1,000 would prefer to watch TV after 11 P.M.?

Let p represent the number of people who would prefer to watch TV after 11 P.M.

$$\frac{12}{50} = \frac{p}{1,000}$$

$$12 \times 1,000 = 50 \times p$$

$$12,000 = 50p$$

$$\frac{12,000}{50} = \frac{50p}{50}$$

$$240 = p$$

Of the 1,000 people, 240 would prefer to watch TV after 11 P.M.

Review Exercises

Use these exercises to review and prepare for the chapter test.

Suppose you spin the spinner shown at the left once. Find the probability of each event. Then tell how likely the event is to happen. 12–15. See Answer Appendix.

12. $P(\text{green})$ **13.** $P(\text{red or white})$
14. $P(\text{pink})$ **15.** $P(\text{not blue})$

A bag contains a nickel, dime, and penny. Suppose you choose one coin without looking. Find the probability of each event. Then tell how likely the event is to happen. 16–17. See Answer Appendix.

16. $P(\text{nickel})$ **17.** $P(\text{dime or penny})$

Tell whether each of the following is a random survey. Explain your answer.

	Type of Survey	Survey Location
18.	favorite fast food	beach
19.	favorite sport	football game
20.	favorite pet	mall

18–20. See Answer Appendix.

21. The results of a survey showed that 14 out of 40 students at West Middle School are interested in publishing a school newspaper. **a. See Answer Appendix.**

  a. What is the probability that a student at this school would be interested in publishing a school newspaper?

  b. If there are 420 students, how many would be interested in publishing a school newspaper? **147 students**

Chapter 13 Study Guide and Assessment **541**

Objectives & Examples

This section reviews the skills and concepts of the chapter and shows completely worked examples.

Review Exercises

These exercises provide practice for the corresponding objectives.

Assessment and Evaluation Masters, pp. 339–340

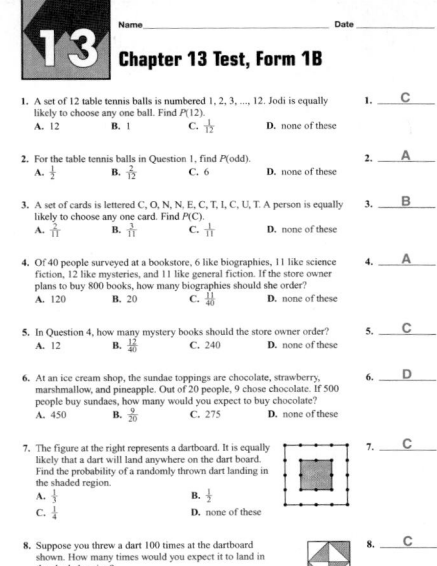

Assessment and Evaluation

Six forms of Chapter 13 Test are available in the *Assessment and Evaluation Masters* as shown in the chart.

Chapter 13 Test, Form 1B, is shown at the right. Chapter 13 Test, Form 2B, is shown on the next page.

1A	Multiple Choice	Honors
1B	Multiple Choice	Average
1C	Multiple Choice	Basic
2A	Free Response	Honors
2B	Free Response	Average
2C	Free Response	Basic

Additional Answers

22. $\frac{5}{6}$

23. about $\frac{5}{16}$ or 31.25%

Assessment and Evaluation Masters, pp. 345–346

Chapter 13 Test, Form 2B

A set of 25 cards is numbered 1, 2, 3, ..., 25. Suppose you draw one card without looking. Find the probability of each event.

1. $P(8)$ — 1. $\frac{1}{25}$

2. $P(25)$ — 2. $\frac{1}{25}$

3. $P(1$ or $9)$ — 3. $\frac{2}{25}$

4. P(an even number) — 4. $\frac{12}{25}$

5. P(a multiple of 4) — 5. $\frac{6}{25}$

In a survey of customers at a video store, 12 customers preferred action films, 5 customers preferred comedies, 3 customers preferred horror films, and 10 customers preferred dramas. The shop owner plans to order 90 new films. How many of each type should she order?

6. action — 6. 36

7. comedy — 7. 15

8. drama — 8. 30

Suppose you threw a dart 100 times at each dartboard below. How many times would you expect it to land in the shaded region of each dartboard?

9. — 9. 25

10. — 10. 25

11. — 11. about 33

© Glencoe/McGraw-Hill   345   Mathematics: Applications and Connections, Course 1

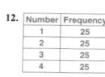

Chapter 13 Test, Form 2B (continued)

12. Make a frequency table that represents the expected results of a spinner numbered 1–4 being spun 100 times.

Number	Frequency
1	25
2	25
3	25
4	25

Use the tree diagram to answer the questions.

13. How many possible outcomes are there? — 13. 9

14. What is the probability of a small vanilla? — 14. $\frac{1}{9}$

15. What is P(medium)? — 15. $\frac{1}{3}$

16. What is P(chocolate)? — 16. $\frac{1}{3}$

Suppose you roll a 6-sided number cube and spin the spinner shown. Find the probability of each of the following.

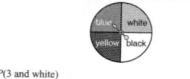

17. $P(3$ and white) — 17. $\frac{1}{24}$

18. P(even number and blue) — 18. $\frac{1}{8}$

19. P(not 1 and blue) — 19. $\frac{5}{24}$

20. P(prime and white) — 20. $\frac{1}{6}$

© Glencoe/McGraw-Hill   346   Mathematics: Applications and Connections, Course 1

● **find probability using area models** *(Lesson 13-3)*

The figure shown represents a dartboard. Find the probability that a randomly-thrown dart lands in the shaded region.

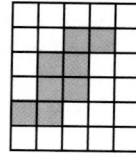

$$P\left(\begin{array}{c}\text{shaded}\\\text{region}\end{array}\right) = \frac{\text{area of shaded region}}{\text{area of target}}$$

$$= \frac{8}{30} \text{ or } \frac{4}{15}$$

● **find outcomes using lists, tree diagrams, and combinations** *(Lesson 13-4)*

In how many ways can a person choose a 1-dip ice cream cone from two types of cones and three flavors of ice cream?

Cone	Flavor	Outcome
sugar (S)	grape (G)	SG
	mint (M)	SM
	peach (P)	SP
waffle (W)	grape (G)	WG
	mint (M)	WM
	peach (P)	WP

There are 6 ways.

● **find the probability of independent events** *(Lesson 13-5)*

Two number cubes are rolled. Find P(odd and 4).

$P(\text{odd}) = \frac{1}{2}$ $P(4) = \frac{1}{6}$

$P(\text{odd and 4}) = \frac{1}{2} \times \frac{1}{6} \text{ or } \frac{1}{12}$

Each figure represents a dartboard. It is equally likely that a dart will land anywhere on the dartboard. Find the probability of a randomly-thrown dart landing in the shaded region.

22. 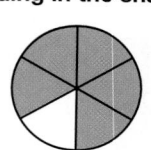   23.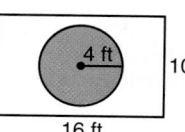

22–23. See margin.

24. Suppose you threw a dart 150 times at the dartboard in Exercise 22. How many times would you expect it to land in the shaded region? **125**

For each situation, make a list and draw a tree diagram to show the sample space.

25. a choice of black or blue jeans in tapered leg, straight leg, or baggy style

26. a choice of soup or salad with beef, chicken, fish, or pasta

27. a choice of going to a basketball game, an amusement park, or a concert on a Friday or a Saturday

28. Tell how many different ways a person can choose four movies from a list of five movies. **5 ways**

25–27. See Answer Appendix.

A coin is tossed and a number cube is rolled. Find the probability of each event.

29. P(heads and even) $\frac{1}{4}$

30. P(heads and 1) $\frac{1}{12}$

31. P(tails and 5 or 6) $\frac{1}{6}$

32. P(tails and 2, 3, or 4) $\frac{1}{4}$

542   Chapter 13 Using Probability

Test and Review Software

You may use this software, a combination of an item generator and item bank, to create your own tests or worksheets. Types of items include free response, multiple choice, short answer, and open ended.

CD-ROM Program

The CD-ROM Program contains an Assessment Game whose questions review the concepts in this chapter.

Applications & Problem Solving

33. *Make a Table* The table shows the results of a survey on the number of hours teenagers use a computer on the weekend. *(Lesson 13-2A)*

Weekend Computer Use		
Hours	**Tally**	**Frequency**
0-2	卌 卌	10
3-4	卌 卌 I	11
5-6	卌	5
7 or more	卌 I	6

a. How many teenagers use a computer 5 or more hours during the weekend? **11**

b. How many teenagers participated in the survey? **32**

34. *Earth Science* The probability of rain on Saturday is 0.6. The probability of rain on Sunday is 0.3. What is the probability that it will rain on both days? *(Lesson 13-5)* **0.18**

35. *School* The cheerleaders at Ross Middle School have three different skirts they can wear as part of their uniforms. One is purple, one is white, and one is yellow. They also have a choice of a yellow vest or a purple sweater to wear. Make a list and a tree diagram to show all of the possible uniform combinations. *(Lesson 13-4)* **See Answer Appendix.**

36. *Party Planning* According to a survey of 30 students, 13 prefer chicken, and 17 prefer hot dogs. If 750 students will be attending the picnic, how many will prefer each selection? *(Lesson 13-2)* **325 will prefer chicken; 425 will prefer hot dogs**

Alternative Assessment

● **Performance Task**

Suppose you are taking pictures to be used in the school yearbook. You can choose either color or black and white film. When the pictures are developed, you can choose a glossy or matte finish, and the pictures can be developed into 3 × 5, 4 × 6, 5 × 5, or 8 × 10 size prints. How can you determine the number of different ways you can develop a photo? **See Answer Appendix.**

Draw a tree diagram to show all of the sample space. If the prints can also be cut with square or rounded corners, how many outcomes are possible? **See Answer Appendix.**

A practice test for Chapter 13 is provided on page 607.

● **Completing the**

Use the following checklist to make sure that your project is complete.

☑ The data about the animal is included on the poster.

☑ The paragraph that explains how probability relates to camouflage is included.

☑ Add any finishing touches that you would like to make your project complete.

 Select one of the assignments from this chapter and place it in your portfolio. Attach a note to it explaining why you selected it.

Applications & Problem Solving

This section provides additional practice in solving real-world problems that involve the skills of this chapter.

Alternative Assessment

The *Performance Task* provides students with a performance assessment opportunity to evaluate their work and understanding.

Students should complete the final stages of their project and prepare a class demonstration of their results. A scoring guide for the project is available in the *Investigations and Projects Masters*, p. 67.

 Students should add to their portfolios at this time.

Assessment and Evaluation Masters, p. 349

Name _____ Date _____

13 Chapter 13 Performance Assessment

Instructions: Demonstrate your knowledge by giving a clear, concise solution to each problem. Be sure to include all relevant drawings and justify your answers. You may show your solutions in more than one way or investigate beyond the requirements of the problems.

1. **a.** Tell what a probability of 0 means. Give an example of an event with probability 0.

   b. Tell what a probability of 1 means. Give an example of an event with probability 1.

   c. Tell what it means for an event to have a probability of $\frac{1}{2}$. Give an example of such an event.

   d. Tell in your own words what is meant by independent events.

   e. Write a word problem that uses the probability of two independent events.

   f. Solve the problem in part e and tell in your own words what the answer means.

2. Wendy is interviewing students on her bus to see for whom they plan to vote for student body president. She found that 18 plan to vote for Kevin and 22 plan to vote for Juanita.

   a. According to Wendy's poll, what is the probability that Kevin will win? that Juanita will win?

   b. Tell why Wendy's sample may not be a random sample of students in the school.

   c. Explain how Wendy could choose a more random sample.

3. A pizza is divided into eight slices. One fourth of the pizza is mushroom, one fourth is pepperoni, one fourth is sausage, and one fourth is green pepper.

   a. If Elaine tosses an anchovy onto the pizza, what is the probability that it will land on a slice with green pepper?

   b. Reggie wants to eat a total of three different pieces of pizza. Draw a tree diagram to show his choices if he has already eaten a slice of mushroom pizza.

© Glencoe/McGraw-Hill   349   *Mathematics: Applications and Connections, Course 1*

 **Performance Assessment**

Additional performance assessment tasks for this chapter are included in the *Assessment and Evaluation Masters* on page 349. A scoring guide is also provided on page 361.

The Standardized Test Practice may be used to help students prepare for standardized tests. The test items are written in the same style as those in state proficiency tests and standardized tests like CAT, CTBS, ITBS, MAT, SAT, and Terra Nova. The test items cover skills and concepts covered up to this point in the text.

The pages can be used as an overnight assessment. After students have completed the pages, discuss how each problem can be solved, or provide copies of the solutions from the *Solutions Manual.*

Assessment and Evaluation Masters, p. 355

Section One: Multiple Choice

There are thirteen multiple-choice questions in this section. Choose the best answer. If a correct answer is *not here,* choose the letter for Not Here.

1. Find $6^3 \div 3^2 + 2$. **D**
   - A 6
   - B 216
   - C 132
   - D 26

2. How many lines of symmetry does the figure have? **G**
   - F 2
   - G 3
   - H 4
   - J 5

3. Which expression represents the volume of a cube with side r? **C**
   - A $6r$
   - B $6r^2$
   - C r^3
   - D $3r^3$

4. What is the probability of a randomly-thrown dart landing in the shaded region of the figure? **G**
   - F 30%
   - G 50%
   - H 60%
   - J 75%

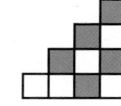

5. Which two-dimensional figure best describes the base of a cone? **B**
   - A square
   - B circle
   - C triangle
   - D rectangle

6. Which number is missing from this sequence? **G**
   $\ldots, 27, 81, \underline{?}, 729, 2{,}187, \ldots$
   - F 357
   - G 243
   - H 261
   - J 539

Please note that Questions 7–13 have five answer choices.

7. Taye bought 3 bags of chips for $0.39 each and 2 cans of soft drink for $0.55 each. Which expression can be used to find the cost in dollars of the items? **D**
   - A $3 \times 0.39 \times 5 \times 0.55$
   - B $3 \div 0.39 + 2 \div 0.55$
   - C $3 \times 0.39 + 5 \times 0.55$
   - D $3 \times 0.39 + 2 \times 0.55$
   - E $3 \div 0.39 \times 2 \div 0.55$

8. In a survey, Akili found that 8 out of 40 students were the oldest child in their family. If 250 students had participated in the survey, how many would be expected to be the oldest child? **J**
   - F 120
   - G 75
   - H 60
   - J 50
   - K 85

9. What percent of the numbers from 1 to 25 contain the digit 3? **A**
   - A about 10%
   - B about 20%
   - C about 30%
   - D about 40%
   - E about 50%

10. In May, about 78 tools were rented each day at The Rent Center. Estimate the number of tools rented for the entire month of May. **G**
    - F 2,000
    - G 2,400
    - H 2,800
    - J 3,200
    - K 3,600

◄◄◄ Instructional Resources

Another cumulative review is shown at the left and is available in the *Assessment and Evaluation Masters,* p. 355.

11. The perimeter of the triangle is $2\frac{3}{8}$ inches. Find the value of y. **C**

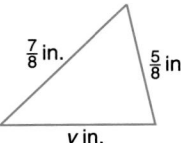
- $\frac{7}{8}$ in.
- $\frac{5}{8}$ in.
- y in.

- **A** $\frac{3}{8}$
- **B** $\frac{1}{2}$
- **C** $\frac{7}{8}$
- **D** $\frac{5}{8}$
- **E** Not Here

12. Ms. Hayashi made a tablecloth for her kitchen. She bought $4\frac{3}{4}$ yards of material. She used $3\frac{1}{8}$ yards of material to make the tablecloth. How much material was not used to make the tablecloth? **K**

- **F** $7\frac{3}{8}$ yd
- **G** $6\frac{1}{2}$ yd
- **H** $7\frac{7}{8}$ yd
- **J** $6\frac{3}{4}$ yd
- **K** Not Here

13. What is the area of the octagon? **C**

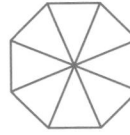
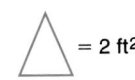 = 2 ft²

- **A** 6 ft²
- **B** 10 ft²
- **C** 16 ft²
- **D** 20 ft²
- **E** Not Here

Section Two: Free Response

This section contains three questions for which you will provide short answers. Write your answers on your paper.

14. A rectangular aquarium has sides measuring 28 inches long, 16 inches wide, and 12 inches high. The aquarium is filled with water to a height of 10 inches. What is the volume of the water? **4,480 in³**

15. A bag contains 3 yellow marbles, 5 red marbles, 6 blue marbles, and 16 green marbles. **a–b. See margin.**

  a. Suppose you choose one marble without looking, record its color, and then replace the marble. What is the probability of choosing a red marble?

  b. Suppose you choose one marble without looking, record its color, replace the marble, and then choose another marble. What is the probability of choosing a yellow marble and then a green marble?

16. How many different words can be made using the words *college, game, season,* and *election* and the prefixes *pre-* and *post-*? **8**

Test-Taking Tip

One of the most important skills in test taking is the ability to follow directions. Have students practice filling in an answer sheet. Remind them that to correct a mistake, they must thoroughly erase the incorrect answer and completely fill in the correct circle.

Assessment and Evaluation Masters, pp. 353–354

Additional Answers

15a. $\frac{1}{6}$

15b. $\frac{4}{75}$

Instructional Resources ▶▶▶

Additional standardized test practice is shown at the right and is available in the *Assessment and Evaluation Masters,* pp. 353–354.

Student Handbook
Table of Contents

Basic Skills

Place Value

Write the place-value position for each digit in 721,056,938,504,550.

1. 4 thousands
2. 2 ten trillions
3. 6 billions
4. 8 millions
5. 3 ten millions
6. 7 hundred trillions
7. 1 trillions
8. 9 hundred millions

Write each number in words.

9. 47,900 forty-seven thousand nine hundred

10. 2,013 two thousand thirteen

11. 540,006,000 five hundred forty million six thousand

12. 7,036,000,000 seven billion thirty-six million

13. 263 two hundred sixty-three

14. 95,000,100,000,000 ninety-five trillion one hundred million

15. 7,261 seven thousand two hundred sixty-one

16. 120,760 one hundred twenty thousand seven hundred sixty

17. 102,000,016 one hundred two million sixteen

18. 582 five hundred eighty-two

19. 8,000,070,000,600 eight trillion seventy million six hundred

20. 67,826 sixty-seven thousand eight hundred twenty-six

Write each number in standard form.

21. seventy-six million 76,000,000
22. nine trillion 9,000,000,000,000
23. six hundred thousand 600,000
24. forty-two 42
25. fifty-five trillion 55,000,000,000,000
26. three billion 3,000,000,000
27. seven hundred seventy-one 771
28. sixteen thousand 16,000
29. one hundred twenty-four million 124,000,000
30. six thousand nine hundred 6,900
31. eighty-eight billion 88,000,000,000
32. three hundred seventy trillion 370,000,000,000,000

Replace each ● with a number to make a true sentence.

33. 67,000 = ● hundreds 670
34. 39,000,000 = ● hundred thousands 390
35. 1,400,000,000,000 = ● millions 1,400,000
36. 760,000 = ● thousands 760
37. 4,200 = ● hundreds 42
38. 86,000,000 = ● thousands 86,000
39. ● = 17 millions 17,000,000
40. ● = 930 ten thousands 9,300,000

EXTRA PRACTICE

Basic Skills

Adding Whole Numbers

1. 40
   + 8
    48

2. 32
   + 5
    37

3. 63
   + 6
    69

4. 41
   + 8
    49

5. 53
   + 4
    57

6. 30
   +60
    90

7. 20
   +50
    70

8. 47
   +20
    67

9. 85
   +10
    95

10. 56
    +33
     89

11. 600
   + 50
    650

12. 506
   + 30
    536

13. 225
   + 40
    265

14. 704
   + 35
    739

15. 628
   + 71
    699

16. 500
   +200
    700

17. 320
   +430
    750

18. 405
   +503
    908

19. 342
   +127
    469

20. 315
    +583
     898

21. 27
   + 4
    31

22. 76
   + 9
    85

23. 59
   + 7
    66

24. 25
   +68
    93

25. 24
    +48
     72

26. 304
   + 57
    361

27. 845
   + 29
    874

28. 637
   + 36
    673

29. 304
   +509
    813

30. 228
    +534
     762

31. 83
   +56
    139

32. 94
   +72
    166

33. 62
   +85
    147

34. 380
   +270
    650

35. 761
    +187
     948

36. 684
   + 67
    751

37. 495
   + 48
    543

38. 347
   + 59
    406

39. 676
   +276
    952

40. 733
    +197
     930

41. 24
    76
   +53
    153

42. 67
    28
   +44
    139

43. 55
    89
   +23
    167

44. 368
    275
   +256
    899

45. 275
    384
    +633
    1,292

46. 4,680
   +3,945
    8,625

47. 5,126
   +2,899
    8,025

48. 2,973
   +1,689
    4,662

49. 52,046
   +41,388
    93,434

50. 96,277
    +27,563
    123,840

Basic Skills

Adding Whole Numbers

1. 50
   + 9

    59

2. 46
   + 3

    49

3. 23
   + 2

    25

4. 62
   + 5

    67

5. 81
   + 4

    85

6. 20
   +60

    80

7. 40
   +30

    70

8. 38
   +20

    58

9. 61
   +10

    71

10. 58
    +11

     69

11. 100
    + 70

     170

12. 207
    + 40

     247

13. 563
    + 20

     583

14. 716
    + 81

     797

15. 334
    + 53

     387

16. 300
    +500

     800

17. 240
    +530

     770

18. 621
    +347

     968

19. 406
    +273

     679

20. 748
    +111

     859

21. 17
    + 5

     22

22. 34
    + 8

     42

23. 78
    + 4

     82

24. 57
    +24

     81

25. 22
    +39

     61

26. 517
    + 64

     581

27. 266
    + 29

     295

28. 742
    + 38

     780

29. 354
    + 37

     391

30. 635
    + 46

     681

31. 44
    +72

     116

32. 35
    +83

     118

33. 56
    +92

     148

34. 580
    +340

     920

35. 174
    +261

     435

36. 674
    + 36

     710

37. 398
    + 43

     441

38. 264
    + 89

     353

39. 593
    +327

     920

40. 786
    +114

     900

41. 62
     37
    +46

     145

42. 28
     54
    +19

     101

43. 71
     96
    +12

     179

44. 265
     612
    +117

     994

45. 422
     536
    +314

     1,272

46. 3,276
    +4,563

     7,839

47. 6,127
    +1,932

     8,059

48. 2,985
    +1,316

     4,301

49. 47,864
    +32,297

     80,161

50. 93,760
    +42,163

     135,923

EXTRA PRACTICE

Basic Skills

Subtracting Whole Numbers

1. $\begin{array}{r} 98 \\ -\ 5 \\ \hline 93 \end{array}$ 2. $\begin{array}{r} 87 \\ -\ 4 \\ \hline 83 \end{array}$ 3. $\begin{array}{r} 56 \\ -\ 3 \\ \hline 53 \end{array}$ 4. $\begin{array}{r} 45 \\ -\ 5 \\ \hline 40 \end{array}$ 5. $\begin{array}{r} 29 \\ -\ 7 \\ \hline 22 \end{array}$

6. $\begin{array}{r} 60 \\ -20 \\ \hline 40 \end{array}$ 7. $\begin{array}{r} 80 \\ -50 \\ \hline 30 \end{array}$ 8. $\begin{array}{r} 56 \\ -40 \\ \hline 16 \end{array}$ 9. $\begin{array}{r} 90 \\ -60 \\ \hline 30 \end{array}$ 10. $\begin{array}{r} 78 \\ -24 \\ \hline 54 \end{array}$

11. $\begin{array}{r} 798 \\ -\ 45 \\ \hline 753 \end{array}$ 12. $\begin{array}{r} 955 \\ -\ 23 \\ \hline 932 \end{array}$ 13. $\begin{array}{r} 354 \\ -\ 34 \\ \hline 320 \end{array}$ 14. $\begin{array}{r} 865 \\ -\ 52 \\ \hline 813 \end{array}$ 15. $\begin{array}{r} 697 \\ -\ 83 \\ \hline 614 \end{array}$

16. $\begin{array}{r} 800 \\ -500 \\ \hline 300 \end{array}$ 17. $\begin{array}{r} 650 \\ -300 \\ \hline 350 \end{array}$ 18. $\begin{array}{r} 854 \\ -630 \\ \hline 224 \end{array}$ 19. $\begin{array}{r} 355 \\ -103 \\ \hline 252 \end{array}$ 20. $\begin{array}{r} 695 \\ -132 \\ \hline 563 \end{array}$

21. $\begin{array}{r} 93 \\ -\ 7 \\ \hline 86 \end{array}$ 22. $\begin{array}{r} 47 \\ -\ 8 \\ \hline 39 \end{array}$ 23. $\begin{array}{r} 54 \\ -\ 5 \\ \hline 49 \end{array}$ 24. $\begin{array}{r} 78 \\ -59 \\ \hline 19 \end{array}$ 25. $\begin{array}{r} 60 \\ -38 \\ \hline 22 \end{array}$

26. $\begin{array}{r} 760 \\ -\ 36 \\ \hline 724 \end{array}$ 27. $\begin{array}{r} 382 \\ -\ 67 \\ \hline 315 \end{array}$ 28. $\begin{array}{r} 630 \\ -\ 23 \\ \hline 607 \end{array}$ 29. $\begin{array}{r} 460 \\ -248 \\ \hline 212 \end{array}$ 30. $\begin{array}{r} 373 \\ -126 \\ \hline 247 \end{array}$

31. $\begin{array}{r} 578 \\ -\ 93 \\ \hline 485 \end{array}$ 32. $\begin{array}{r} 247 \\ -\ 83 \\ \hline 164 \end{array}$ 33. $\begin{array}{r} 623 \\ -\ 93 \\ \hline 530 \end{array}$ 34. $\begin{array}{r} 738 \\ -165 \\ \hline 573 \end{array}$ 35. $\begin{array}{r} 954 \\ -372 \\ \hline 582 \end{array}$

36. $\begin{array}{r} 232 \\ -184 \\ \hline 48 \end{array}$ 37. $\begin{array}{r} 540 \\ -275 \\ \hline 265 \end{array}$ 38. $\begin{array}{r} 727 \\ -538 \\ \hline 189 \end{array}$ 39. $\begin{array}{r} 660 \\ -383 \\ \hline 277 \end{array}$ 40. $\begin{array}{r} 840 \\ -496 \\ \hline 344 \end{array}$

41. $\begin{array}{r} 315 \\ -227 \\ \hline 88 \end{array}$ 42. $\begin{array}{r} 712 \\ -555 \\ \hline 157 \end{array}$ 43. $\begin{array}{r} 408 \\ -209 \\ \hline 199 \end{array}$ 44. $\begin{array}{r} 705 \\ -509 \\ \hline 196 \end{array}$ 45. $\begin{array}{r} 400 \\ -189 \\ \hline 211 \end{array}$

46. $\begin{array}{r} 6{,}791 \\ -\ \ 899 \\ \hline 5{,}892 \end{array}$ 47. $\begin{array}{r} 3{,}406 \\ -\ \ 408 \\ \hline 2{,}998 \end{array}$ 48. $\begin{array}{r} 5{,}690 \\ -\ \ 792 \\ \hline 4{,}898 \end{array}$ 49. $\begin{array}{r} 6{,}243 \\ -4{,}564 \\ \hline 1{,}679 \end{array}$ 50. $\begin{array}{r} 7{,}092 \\ -6{,}895 \\ \hline 197 \end{array}$

51. $\begin{array}{r} 64{,}700 \\ -\ 3{,}792 \\ \hline 60{,}908 \end{array}$ 52. $\begin{array}{r} 41{,}905 \\ -\ 4{,}916 \\ \hline 36{,}989 \end{array}$ 53. $\begin{array}{r} 52{,}009 \\ -\ 7{,}314 \\ \hline 44{,}695 \end{array}$ 54. $\begin{array}{r} 80{,}490 \\ -60{,}495 \\ \hline 19{,}995 \end{array}$ 55. $\begin{array}{r} 68{,}418 \\ -39{,}529 \\ \hline 28{,}889 \end{array}$

Basic Skills

Subtracting Whole Numbers

1. $\begin{array}{r} 67 \\ -\ 4 \\ \hline 63 \end{array}$
2. $\begin{array}{r} 25 \\ -\ 5 \\ \hline 20 \end{array}$
3. $\begin{array}{r} 78 \\ -\ 6 \\ \hline 72 \end{array}$
4. $\begin{array}{r} 93 \\ -\ 2 \\ \hline 91 \end{array}$
5. $\begin{array}{r} 56 \\ -\ 3 \\ \hline 53 \end{array}$

6. $\begin{array}{r} 40 \\ -10 \\ \hline 30 \end{array}$
7. $\begin{array}{r} 70 \\ -30 \\ \hline 40 \end{array}$
8. $\begin{array}{r} 62 \\ -40 \\ \hline 22 \end{array}$
9. $\begin{array}{r} 90 \\ -70 \\ \hline 20 \end{array}$
10. $\begin{array}{r} 86 \\ -32 \\ \hline 54 \end{array}$

11. $\begin{array}{r} 726 \\ -\ 14 \\ \hline 712 \end{array}$
12. $\begin{array}{r} 584 \\ -\ 32 \\ \hline 552 \end{array}$
13. $\begin{array}{r} 963 \\ -\ 51 \\ \hline 912 \end{array}$
14. $\begin{array}{r} 497 \\ -\ 83 \\ \hline 414 \end{array}$
15. $\begin{array}{r} 677 \\ -\ 21 \\ \hline 656 \end{array}$

16. $\begin{array}{r} 600 \\ -400 \\ \hline 200 \end{array}$
17. $\begin{array}{r} 730 \\ -300 \\ \hline 430 \end{array}$
18. $\begin{array}{r} 961 \\ -520 \\ \hline 441 \end{array}$
19. $\begin{array}{r} 874 \\ -352 \\ \hline 522 \end{array}$
20. $\begin{array}{r} 519 \\ -116 \\ \hline 403 \end{array}$

21. $\begin{array}{r} 43 \\ -\ 7 \\ \hline 36 \end{array}$
22. $\begin{array}{r} 72 \\ -\ 5 \\ \hline 67 \end{array}$
23. $\begin{array}{r} 95 \\ -\ 8 \\ \hline 87 \end{array}$
24. $\begin{array}{r} 86 \\ -27 \\ \hline 59 \end{array}$
25. $\begin{array}{r} 40 \\ -13 \\ \hline 27 \end{array}$

26. $\begin{array}{r} 461 \\ -\ 28 \\ \hline 433 \end{array}$
27. $\begin{array}{r} 532 \\ -\ 17 \\ \hline 515 \end{array}$
28. $\begin{array}{r} 670 \\ -\ 52 \\ \hline 618 \end{array}$
29. $\begin{array}{r} 982 \\ -\ 36 \\ \hline 946 \end{array}$
30. $\begin{array}{r} 621 \\ -\ 12 \\ \hline 609 \end{array}$

31. $\begin{array}{r} 726 \\ -\ 43 \\ \hline 683 \end{array}$
32. $\begin{array}{r} 942 \\ -\ 61 \\ \hline 881 \end{array}$
33. $\begin{array}{r} 527 \\ -\ 52 \\ \hline 475 \end{array}$
34. $\begin{array}{r} 438 \\ -229 \\ \hline 209 \end{array}$
35. $\begin{array}{r} 855 \\ -472 \\ \hline 383 \end{array}$

36. $\begin{array}{r} 860 \\ -472 \\ \hline 388 \end{array}$
37. $\begin{array}{r} 215 \\ -167 \\ \hline 48 \end{array}$
38. $\begin{array}{r} 742 \\ -563 \\ \hline 179 \end{array}$
39. $\begin{array}{r} 936 \\ -478 \\ \hline 458 \end{array}$
40. $\begin{array}{r} 487 \\ -199 \\ \hline 288 \end{array}$

41. $\begin{array}{r} 417 \\ -358 \\ \hline 59 \end{array}$
42. $\begin{array}{r} 324 \\ -137 \\ \hline 187 \end{array}$
43. $\begin{array}{r} 707 \\ -309 \\ \hline 398 \end{array}$
44. $\begin{array}{r} 602 \\ -408 \\ \hline 194 \end{array}$
45. $\begin{array}{r} 500 \\ -279 \\ \hline 221 \end{array}$

46. $\begin{array}{r} 7,213 \\ -\ \ 426 \\ \hline 6,787 \end{array}$
47. $\begin{array}{r} 2,146 \\ -\ \ 347 \\ \hline 1,799 \end{array}$
48. $\begin{array}{r} 6,290 \\ -\ \ 794 \\ \hline 5,496 \end{array}$
49. $\begin{array}{r} 7,418 \\ -2,439 \\ \hline 4,979 \end{array}$
50. $\begin{array}{r} 6,052 \\ -5,456 \\ \hline 596 \end{array}$

51. $\begin{array}{r} 64,205 \\ -\ 3,746 \\ \hline 60,459 \end{array}$
52. $\begin{array}{r} 88,644 \\ -\ 9,657 \\ \hline 78,987 \end{array}$
53. $\begin{array}{r} 30,716 \\ -\ 4,755 \\ \hline 25,961 \end{array}$
54. $\begin{array}{r} 64,658 \\ -23,659 \\ \hline 40,999 \end{array}$
55. $\begin{array}{r} 91,273 \\ -86,594 \\ \hline 4,679 \end{array}$

Basic Skills

Multiplying Whole Numbers

1. 40
× 5
200

2. 30
× 6
180

3. 20
× 8
160

4. 60
× 4
240

5. 50
× 7
350

6. 23
× 3
69

7. 44
× 2
88

8. 81
× 6
486

9. 72
× 3
216

10. 61
× 7
427

11. 721
× 4
2,884

12. 513
× 3
1,539

13. 234
× 2
468

14. 634
× 2
1,268

15. 831
× 3
2,493

16. 46
× 5
230

17. 53
× 7
371

18. 82
× 6
492

19. 27
× 4
108

20. 68
× 8
544

21. 704
× 6
4,224

22. 409
× 5
2,045

23. 806
× 8
6,448

24. 307
× 9
2,763

25. 208
× 7
1,456

26. 28
×10
280

27. 86
×10
860

28. 51
×10
510

29. 247
× 10
2,470

30. 4,328
× 10
43,280

31. 52
×20
1,040

32. 37
×50
1,850

33. 26
×40
1,040

34. 175
× 30
5,250

35. 1,469
× 80
117,520

36. 75
×19
1,425

37. 54
×27
1,458

38. 45
×81
3,645

39. 52
×64
3,328

40. 80
×76
6,080

41. 89
×45
4,005

42. 64
×37
2,368

43. 78
×62
4,836

44. 56
×82
4,592

45. 83
×59
4,897

46. 414
× 22
9,108

47. 321
× 43
13,803

48. 522
× 34
17,748

49. 613
× 32
19,616

50. 202
× 24
4,848

Basic Skills

Multiplying Whole Numbers

1. 30
$\times$ 7
210

2. 70
$\times$ 4
280

3. 50
$\times$ 8
400

4. 80
$\times$ 3
240

5. 90
$\times$ 6
540

6. 71
$\times$ 8
568

7. 64
$\times$ 2
128

8. 92
$\times$ 3
276

9. 53
$\times$ 2
106

10. 81
$\times$ 6
486

11. 624
$\times$ 2
1,248

12. 434
$\times$ 2
868

13. 712
$\times$ 3
2,136

14. 221
$\times$ 4
884

15. 511
$\times$ 7
3,577

16. 27
$\times$ 5
135

17. 36
$\times$ 2
72

18. 54
$\times$ 4
216

19. 92
$\times$ 6
552

20. 75
$\times$ 3
225

21. 906
$\times$ 5
4,530

22. 702
$\times$ 7
4,914

23. 503
$\times$ 9
4,527

24. 807
$\times$ 8
6,456

25. 209
$\times$ 3
627

26. 92
$\times$10
920

27. 87
$\times$10
870

28. 43
$\times$10
430

29. 761
$\times$ 10
7,610

30. 5,276
$\times$ 10
52,760

31. 76
$\times$40
3,040

32. 19
$\times$50
950

33. 51
$\times$20
1,020

34. 247
$\times$ 30
7,410

35. 1,236
$\times$ 80
98,880

36. 92
$\times$16
1,472

37. 74
$\times$23
1,702

38. 56
$\times$47
2,632

39. 81
$\times$32
2,592

40. 45
$\times$72
3,240

41. 61
$\times$37
2,257

42. 72
$\times$59
4,248

43. 12
$\times$86
1,032

44. 93
$\times$72
6,696

45. 26
$\times$41
1,066

46. 723
$\times$ 46
33,258

47. 812
$\times$ 51
41,412

48. 245
$\times$ 67
16,415

49. 123
$\times$ 94
11,562

50. 679
$\times$ 77
52,283

Basic Skills

Dividing Whole Numbers

1. $3\overline{)72}$ = 24
2. $4\overline{)96}$ = 24
3. $2\overline{)78}$ = 39
4. $3\overline{)84}$ = 28
5. $3\overline{)57}$ = 19

6. $6\overline{)918}$ = 153
7. $8\overline{)976}$ = 122
8. $5\overline{)965}$ = 193
9. $7\overline{)903}$ = 129
10. $4\overline{)752}$ = 188

11. $12\overline{)60}$ = 5
12. $17\overline{)51}$ = 3
13. $25\overline{)75}$ = 3
14. $15\overline{)90}$ = 6
15. $24\overline{)72}$ = 3

16. $34\overline{)204}$ = 6
17. $18\overline{)126}$ = 7
18. $27\overline{)135}$ = 5
19. $46\overline{)184}$ = 4
20. $53\overline{)424}$ = 8

21. $24\overline{)240}$ = 10
22. $32\overline{)320}$ = 10
23. $25\overline{)500}$ = 20
24. $17\overline{)510}$ = 30
25. $15\overline{)600}$ = 40

26. $6\overline{)384}$ = 64
27. $23\overline{)483}$ = 21
28. $34\overline{)612}$ = 18
29. $14\overline{)546}$ = 39
30. $48\overline{)720}$ = 15

31. $31\overline{)1,953}$ = 63
32. $99\overline{)1,881}$ = 19
33. $47\overline{)1,927}$ = 41
34. $26\overline{)1,742}$ = 67
35. $19\overline{)1,045}$ = 55

36. $18\overline{)3,672}$ = 204
37. $23\overline{)9,223}$ = 401
38. $32\overline{)9,824}$ = 307
39. $15\overline{)7,545}$ = 503
40. $27\overline{)8,154}$ = 302

41. $8\overline{)91}$ = 11 R3
42. $6\overline{)87}$ = 14 R3
43. $5\overline{)99}$ = 19 R4
44. $7\overline{)87}$ = 12 R3
45. $6\overline{)80}$ = 13 R2

46. $8\overline{)685}$ = 85 R5
47. $7\overline{)538}$ = 76 R6
48. $4\overline{)273}$ = 68 R1
49. $6\overline{)580}$ = 96 R4
50. $5\overline{)387}$ = 77 R2

51. $12\overline{)75}$ = 6 R3
52. $23\overline{)97}$ = 4 R5
53. $18\overline{)99}$ = 5 R9
54. $33\overline{)75}$ = 2 R9
55. $27\overline{)56}$ = 2 R2

56. $16\overline{)134}$ = 8 R6
57. $37\overline{)299}$ = 8 R3
58. $53\overline{)483}$ = 9 R6
59. $29\overline{)210}$ = 7 R7
60. $62\overline{)439}$ = 7 R5

61. $49\overline{)29,670}$ = 605 R25
62. $84\overline{)25,880}$ = 308 R8
63. $32\overline{)38,693}$ = 1,209 R5
64. $26\overline{)80,311}$ = 3,088 R23
65. $46\overline{)92,330}$ = 2,007 R8

66. $100\overline{)706}$ = 7 R6
67. $100\overline{)842}$ = 8 R42
68. $200\overline{)900}$ = 4 R100
69. $500\overline{)705}$ = 1 R205
70. $300\overline{)602}$ = 2 R2

71. $400\overline{)1,632}$ = 4 R32
72. $300\overline{)2,205}$ = 7 R105
73. $600\overline{)8,407}$ = 14 R7
74. $200\overline{)9,820}$ = 49 R20
75. $500\overline{)7,513}$ = 15 R13

Basic Skills

Dividing Whole Numbers

1. $5\overline{)95}$ = 19

2. $2\overline{)58}$ = 29

3. $4\overline{)92}$ = 23

4. $3\overline{)78}$ = 26

5. $8\overline{)88}$ = 11

6. $4\overline{)848}$ = 212

7. $6\overline{)984}$ = 164

8. $3\overline{)351}$ = 117

9. $6\overline{)762}$ = 127

10. $2\overline{)276}$ = 138

11. $32\overline{)96}$ = 3

12. $19\overline{)95}$ = 5

13. $13\overline{)78}$ = 6

14. $21\overline{)63}$ = 3

15. $17\overline{)68}$ = 4

16. $32\overline{)256}$ = 8

17. $64\overline{)192}$ = 3

18. $97\overline{)679}$ = 7

19. $72\overline{)360}$ = 5

20. $16\overline{)144}$ = 9

21. $31\overline{)620}$ = 20

22. $14\overline{)980}$ = 70

23. $22\overline{)220}$ = 10

24. $29\overline{)870}$ = 30

25. $12\overline{)600}$ = 50

26. $8\overline{)576}$ = 72

27. $23\overline{)644}$ = 28

28. $56\overline{)952}$ = 17

29. $13\overline{)481}$ = 37

30. $46\overline{)690}$ = 15

31. $27\overline{)2,592}$ = 96

32. $68\overline{)3,060}$ = 45

33. $85\overline{)1,020}$ = 12

34. $53\overline{)4,081}$ = 77

35. $39\overline{)3,198}$ = 82

36. $19\overline{)4,902}$ = 258

37. $59\overline{)9,558}$ = 162

38. $23\overline{)5,175}$ = 225

39. $18\overline{)7,614}$ = 423

40. $12\overline{)6,264}$ = 522

41. $3\overline{)97}$ = 32 R1

42. $4\overline{)65}$ = 16 R1

43. $5\overline{)87}$ = 17 R2

44. $8\overline{)93}$ = 11 R5

45. $7\overline{)74}$ = 10 R4

46. $5\overline{)273}$ = 54 R3

47. $9\overline{)858}$ = 95 R3

48. $6\overline{)412}$ = 68 R4

49. $3\overline{)136}$ = 45 R1

50. $7\overline{)529}$ = 75 R4

51. $13\overline{)68}$ = 5 R3

52. $16\overline{)97}$ = 6 R1

53. $14\overline{)33}$ = 2 R5

54. $24\overline{)54}$ = 2 R6

55. $18\overline{)71}$ = 3 R17

56. $34\overline{)162}$ = 4 R26

57. $19\overline{)135}$ = 7 R2

58. $51\overline{)426}$ = 8 R18

59. $68\overline{)345}$ = 5 R5

60. $46\overline{)234}$ = 5 R4

61. $65\overline{)58,974}$ = 907 R19

62. $47\overline{)32,569}$ = 692 R45

63. $12\overline{)14,586}$ = 1,215 R6

64. $52\overline{)74,853}$ = 1,439 R25

65. $29\overline{)62,185}$ = 2,144 R9

66. $100\overline{)902}$ = 9 R2

67. $100\overline{)764}$ = 7 R64

68. $200\overline{)536}$ = 2 R136

69. $300\overline{)847}$ = 2 R247

70. $500\overline{)604}$ = 1 R104

71. $300\overline{)1,475}$ = 4 R275

72. $400\overline{)3,206}$ = 8 R6

73. $200\overline{)2,568}$ = 12 R168

74. $600\overline{)9,612}$ = 16 R12

75. $500\overline{)8,520}$ = 17 R20

EXTRA PRACTICE

Extra Practice

Lesson 1-1 *(Pages 4–7)*
Use the four-step plan to solve each problem.

1. On a map of Ohio, each inch represents approximately 9 miles. Chad is planning to travel from Cleveland to Cincinnati. If the distance on the map from Cleveland to Cincinnati is about 23 inches, how far will he travel? **207 miles**

2. Sylvia has $102. If she made three purchases of $13, $37, and $29, how much money does she have left? **$23**

3. It took Jason 56 hours to paint a house. It only took Nolan 43 hours and 16 minutes to paint an identical house. How many more hours did it take Jason to paint the house than it took Nolan? **12 h and 44 min**

4. A cassette tape holds sixty minutes of music. If Logan has already taped eight songs that are each five minutes long on the tape, how many more minutes of music can she still tape on the cassette? **20 min**

5. The Cornells want to buy a car that costs $4,260. They plan to make a down payment of $1,500 and pay the rest in twelve equal payments. What will be the amount of each payment? **$230**

Lesson 1-2 *(Pages 8–11)*
Find the next three numbers or shapes for each pattern.

1.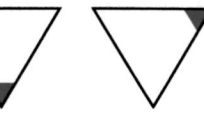
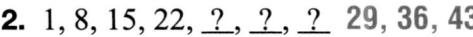

2. 1, 8, 15, 22, _?_, _?_, _?_ **29, 36, 43**   3. 1, 2, 6, 24, _?_, _?_, _?_ **120, 720, 5,040**

Solve by using patterns.

4. Sarah is conditioning for track. On the first day, she ran 2 laps. The second day she ran 3 laps. The third day she ran 5 laps, and on the fourth day she ran 8 laps. If this pattern continues, how many laps will she run on the seventh day? **23 laps**

5. The times that a new movie is showing are 9:30 A.M., 11:12 A.M., 12:54 P.M., and 2:36 P.M. What are the next three times this movie will be shown?
4:18 P.M., 6:00 P.M., 7:42 P.M.

Lesson 1-3 *(Pages 12–15)*
Round each number to the underlined place-value position.

1. 4̲6 **50**
2. 1,2̲49 **1,200**
3. 9̲,499 **9,000**
4. 2,96̲0 **2,960**
5. 6̲,001 **6,000**
6. 16̲3 **160**
7. 6̲58 **700**
8. 6̲,710 **7,000**
9. 12,6̲50 **12,700**
10. 18̲,305 **18,000**
11. 156,9̲99 **157,000**
12. 960,7̲15 **960,700**

Estimate. State whether the answer shown is reasonable.   13–20. See Answer Appendix.

13. $61 \times 5 = 234$
14. $889 - 43 = 846$
15. $2.94 + $6.13 + $9.25 = 18.32
16. $415 \times 4 = 1,660$
17. $385 \div 5 = 65$
18. $2,107 - 182 = 1,625$
19. $108 + 496 + 229 = 833$
20. $5,627 \div 331 = 17$

Lesson 1-4 (Pages 16–19)
Find the value of each expression.

1. $14 - 5 + 7$ **16**
2. $12 + 10 - 5 - 6$ **11**
3. $50 - 6 + 12 + 4$ **60**
4. $12 - 2 \times 3$ **6**
5. $16 + 4 \times 5$ **36**
6. $5 + 3 \times 4 - 7$ **10**
7. $2 \times 3 + 9 \times 2$ **24**
8. $6 \times 8 + 4 \div 2$ **50**
9. $7 \times 6 - 14$ **28**
10. $8 + 12 \times 4 \div 8$ **14**
11. $13 - 6 \times 2 + 1$ **2**
12. $80 \div 10 \times 8$ **64**
13. $1 + 2 + 3 + 4$ **10**
14. $1 \times 2 \times 3 \times 4$ **24**
15. $6 + 6 \times 6$ **42**
16. $14 - 2 \times 7 + 0$ **0**
17. $156 - 6 \times 0$ **156**
18. $30 - 14 \times 2 + 8$ **10**

Lesson 1-5 (Pages 22–25)
Evaluate each expression if $m = 2$ and $n = 4$.

1. $m + m$ **4**
2. $n - m$ **2**
3. mn **8**
4. $2m$ **4**
5. $2n$ **8**
6. $2n + 2m$ **12**
7. $m \times 0$ **0**
8. $64 \div n$ **16**
9. $12 - m$ **10**
10. $2mn$ **16**

Evaluate each expression if $a = 3$, $b = 4$, and $c = 12$.

11. $a + b$ **7**
12. $c - a$ **9**
13. $a + b + c$ **19**
14. $b - a$ **1**
15. $c - a \times b$ **0**
16. $a + 2 \times b$ **11**
17. $b + c \div 2$ **10**
18. ab **12**
19. $a + 3b$ **15**
20. $a + c \div 6$ **5**
21. $25 + c \div b$ **28**
22. abc **144**
23. $144 - abc$ **0**
24. $c \div a + 10$ **14**
25. $2b - a$ **5**
26. $2ab$ **24**

Lesson 1-6 (Pages 28–31)
Write each product using exponents.

1. $2 \cdot 2 \cdot 2 \cdot 2 \cdot 2$ **2^5**
2. $6 \cdot 6 \cdot 6 \cdot 7 \cdot 7$ **$6^3 \cdot 7^2$**
3. $9 \cdot 9 \cdot 9 \cdot 9 \cdot 9 \cdot 9 \cdot 10$ **$9^6 \cdot 10$**
4. $k \cdot k \cdot k \cdot t \cdot t \cdot t$ **$k^3 \cdot t^3$**
5. $14 \cdot 14 \cdot 6$ **$14^2 \cdot 6$**
6. $3 \cdot 3 \cdot 3 \cdot 3 \cdot y \cdot y$ **$3^4 \cdot y^2$**

Write each power as a product.

7. 13^4 **$13 \cdot 13 \cdot 13 \cdot 13$**
8. 9^6 **$9 \cdot 9 \cdot 9 \cdot 9 \cdot 9 \cdot 9$**
9. $2^3 \cdot 3^2$ **$2 \cdot 2 \cdot 2 \cdot 3 \cdot 3$**
10. x^5 **$x \cdot x \cdot x \cdot x \cdot x$**
11. 169^3 **$169 \cdot 169 \cdot 169$**
12. $13,410^2$ **$13,410 \cdot 13,410$**

Evaluate each expression.

13. 5^6 **15,625**
14. 17^3 **4,913**
15. 2^{12} **4,096**
16. $3^5 \cdot 2^3$ **1,944**
17. $6^4 \cdot 3$ **3,888**
18. $2^2 \cdot 3^2 \cdot 4^2$ **576**
19. 176^2 **30,976**
20. $6 \cdot 4^3$ **384**

Lesson 1-7A (Pages 32–33)
Solve.

1. Find an even number between 70 and 80 that is divisible by 2 and 9. **72**

2. Last week the Tri-River Animal Shelter sent a total of 34 cats and dogs to new homes. There were 8 more cats than dogs. How many of each were adopted? **13 dogs, 21 cats**

3. Admission to the Cincinnati Zoo is $8 for adults, $5 for children, and $3 for senior citizens. Eleven members of the Ruiz family paid a total of $55 for admission. If 6 children were in the group, how many adults and senior citizens were in the group? **2 adults, 3 senior citizens**

4. Sylvia's soccer team played a total of 18 matches. Her team won twice as many matches as they lost. How many matches did they win? **12 matches**

5. Jordan makes $5 per hour mowing lawns and $7 per hour painting houses during the summer. This week, Jordan made $102. If he worked twice as many hours mowing lawns as he did painting houses, how many hours did he work at each job? **12 hours mowing, 6 hours painting**

Lesson 1-7 (Pages 34–37)
Tell whether the equation is *true* or *false* by replacing the variable with the given value.

1. $q - 7 = 7; q = 28$ **false**
2. $g - 3 = 10; g = 13$ **true**
3. $r - 3 = 4; r = 7$ **true**
4. $t + 3 = 21; t = 24$ **false**

Identify the solution to each equation from the list given.

5. $7 + a = 10$; 3, 13, 17 **3**
6. $14 + m = 24$; 7, 10, 34 **10**
7. $j \div 3 = 2$; 4, 6, 8 **6**
8. $20 = 24 - n$; 2, 3, 4 **4**

Solve each equation mentally.

9. $b + 7 = 12$ **5**
10. $s + 10 = 23$ **13**
11. $4x = 36$ **9**
12. $6 = t \div 5$ **30**
13. $b - 3 = 12$ **15**
14. $w \div 2 = 8$ **16**

Lesson 2-1 (Pages 46–49)
Make a frequency table for each set of data.

1. To the nearest mile, how many miles did members of the track team run during practice yesterday? **See Answer Appendix.**
   1 4 3 5 2 3 3 2 1 4 3 2 3 1 5 3 4 2 3 1 5 2

2. What were the high temperatures in Indiana cities on March 13? **See Answer Appendix.**
   52 57 48 53 52 49 48 52 51
   47 51 49 57 53 48 52 52 49

Use the frequency table to the right to answer Exercises 3–4.

3. Describe the data shown in the table.

4. Which flavor should the ice cream shop stock the most? **vanilla**

3. **6 flavors of ice cream are being sold, with vanilla being the most purchased and peach the least purchased**

ICE CREAM FLAVORS SOLD IN JULY		
Flavor	Tally	Frequency
Vanilla	JHT JHT JHT JHT IIII	24
Chocolate	JHT JHT JHT III	18
Strawberry	JHT JHT II	12
Chocolate Chip	JHT JHT JHT I	16
Peach	JHT III	8
Butter Pecan	JHT JHT I	11

Lesson 2-2 *(Pages 50–53)*

Choose an appropriate scale and an interval for each set of data.

1. 2, 7, 13, 3, 4, 12, 9 **0 to 13; 1**   **2.** 11, 15, 13, 18, 19, 20 **11 to 20; 1**

3. 56, 85, 23, 78, 42, 63 **21 to 85; 5**   **4.** 10, 25, 88, 64, 99, 37 **0 to 100; 10**

5. 165, 167, 169, 164, 170, 166, 167, 165, 169 **161 to 170; 1**

6. 132, 865, 465, 672, 318, 940, 573, 689 **100 to 1,000; 100**

7. 1,450; 7,896; 5,638; 7,142; 4,287; 8,612 **1,000 to 9,000; 1,000**

Lesson 2-3 *(Pages 54–57)*

Make a bar graph for each set of data.

1. a vertical bar graph

Favorite Subject

Subject	Frequency
Math	4
Science	6
History	2
English	8
Phys. Ed.	12

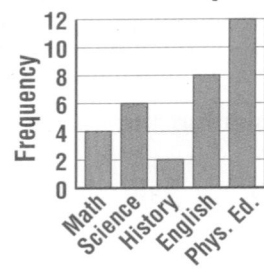

2. a horizontal bar graph

Final Grades

Subject	Score
Math	88
Science	82
History	92
English	94

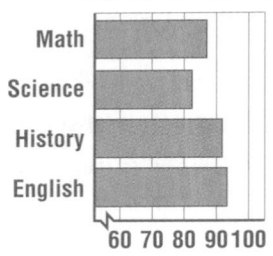

Make a line graph for each set of data.

3.

Test	Score
1	62
2	75
3	81
4	83
5	78
6	92

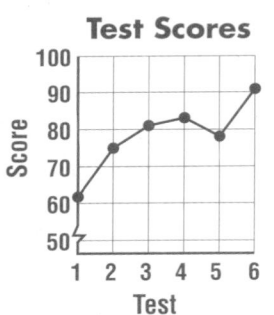

4.

Day	Absences
Mon.	3
Tues.	6
Wed.	2
Thur.	1
Fri.	8

Lesson 2-3B *(Pages 58–59)*

The double bar graph shows the quarterly sales (in millions of dollars) of two companies. Use the graph to answer the following questions.

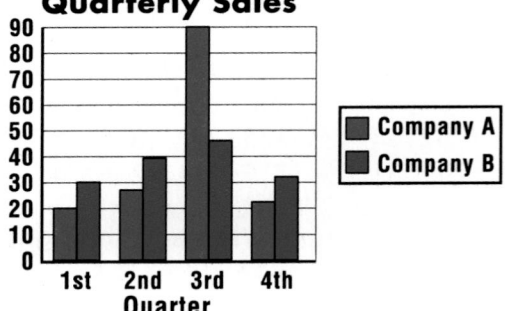

1. In which quarter did Company A have its lowest sales? **1st quarter**

2. About how much higher are the sales for Company A than Company B in the 3rd quarter? **about $45 million**

3. The sales of Company A in the 2nd quarter are about equal to the sales of Company B in which quarter? **1st quarter**

4. In which quarter is there the least difference between the sales of Company A and the sales of Company B? **1st quarter**

Lesson 2-4 *(Pages 60–63)*

The circle graph shows the favorite subject of students at Midland Middle School.

1. The percents are 5%, 12%, 20%, 25%, and 38%. Match each percent with the appropriate section of the graph. **1–2. See Answer Appendix.**

2. Suppose math and history were combined. Would the combination be preferred by $\frac{1}{4}$ of the students?

3. Which two subjects together are preferred by the same percent as English? **history and science**

4. Which two subjects together are preferred by half of the students? **math and phys. ed.**

Students' Favorite Subjects

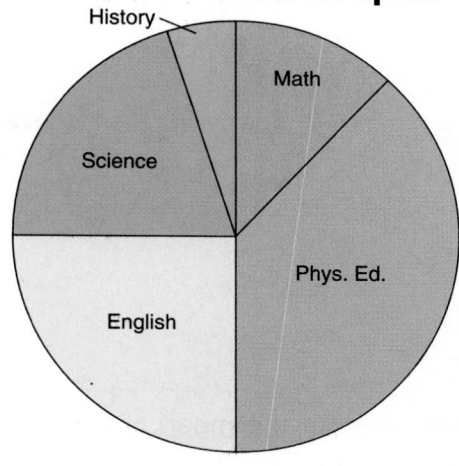

Lesson 2-5 *(Pages 64–67)*

Use the following graph to solve each problem.

Average Starting Salaries

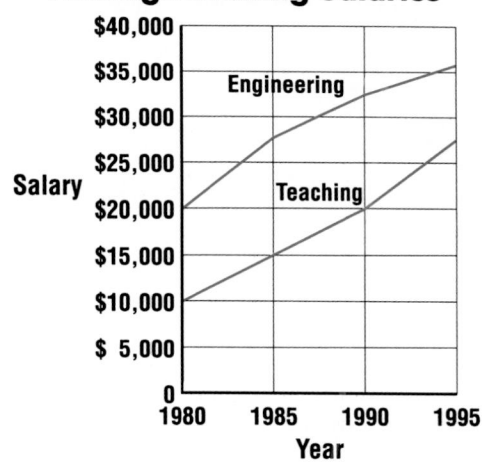

1. Give the expected starting salary in 1988 for:
   a. an engineer **about $31,000**
   b. a teacher **about $18,000**

2. How much more did an engineer make than a teacher in 1980? **$10,000**

3. Which of these two professions do you think will start with a higher salary in 2000? **engineering**

4. How much less was the difference in salaries in 1995 than in 1980? **about $1,000**

Lesson 2-6 *(Pages 68–70)*

Make a stem-and-leaf plot for each set of data.

1. 23, 15, 39, 68, 57, 42, 51, 52, 41, 18, 29

2. 5, 14, 39, 28, 14, 6, 7, 18, 13, 28, 9, 14

3. 189, 182, 196, 184, 197, 183, 196, 194, 184

4. 71, 82, 84, 95, 76, 92, 83, 74, 81, 75, 96

1. | Stem | Leaf |
   |---|---|
   | 1 | 5 8 |
   | 2 | 3 9 |
   | 3 | 9 |
   | 4 | 1 2 |
   | 5 | 1 2 7 |
   | 6 | 8 |

   $1|5 = 15$

2. | Stem | Leaf |
   |---|---|
   | 0 | 5 6 7 9 |
   | 1 | 3 4 4 4 8 |
   | 2 | 8 8 |
   | 3 | 9 |

   $0|5 = 5$

3. | Stem | Leaf |
   |---|---|
   | 18 | 2 3 4 4 9 |
   | 19 | 4 6 6 7 |

   $18|9 = 189$

4. | Stem | Leaf |
   |---|---|
   | 7 | 1 4 5 6 |
   | 8 | 1 2 3 4 |
   | 9 | 2 5 6 |

   $7|1 = 71$

Lesson 2-7 *(Pages 71–75)*

Find the mean, median, mode, and range for each set of data. 6. 710, 721, no mode, 495

1. 1, 5, 9, 1, 2, 4, 8, 2 4; 3; 1, 2, and 8
2. 2, 5, 8, 9, 7, 6, 3, 5 5.6, 5.5, 5, 7
3. 1, 2, 4, 2, 2, 1, 2 2, 2, 2, 3
4. 12, 13, 15, 12, 12, 11 12.5, 12, 12, 4
5. 256, 265, 247, 256 256, 256, 256, 18
6. 957, 562, 462, 848, 721
7. 46, 54, 66, 54, 46, 66
   55.33; 54; 46, 54, and 66; 20
8. 81, 82, 83, 84, 85, 86, 87 84, 84, no mode, 6

Lesson 2-8 *(Pages 78–81)*

Tell whether the mean, median, or mode would be best to describe each set of data. Explain each answer.

1. 627, 452, 573, 602, 498 mean–the numbers are all about the same
2. Favorite bagel flavor mode–want to find the most frequent
3. $42,360; $51,862; $47,650; $23,400; $52,961 median–$23,400 is much lower than the rest

The graphs below display the same information.

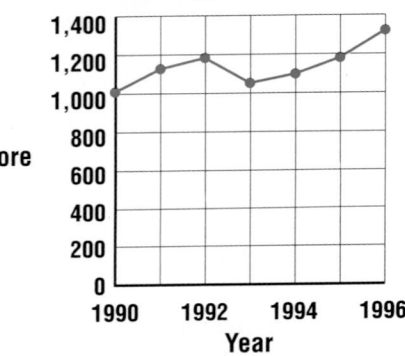

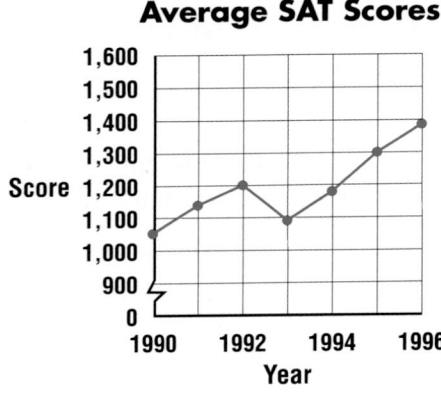

4. Explain why these graphs look different. **They have different vertical scales.**

5. If Mr. Roush wishes to show that SAT scores have improved greatly since he became principal in 1993, which graph should he use?
   Answers may vary. Sample answer: He should use Graph B because the increase looks greater.

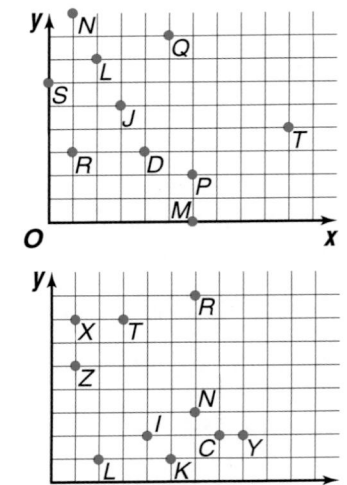

Lesson 2-9 *(Pages 82–85)*

Use the grid at the right to name the point for each ordered pair.

1. (6, 2) *P* 2. (3, 5) *J* 3. (1, 3) *R* 4. (6, 0) *M* 5. (5, 8) *Q*
6. (2, 7) *L* 7. (0, 6) *S* 8. (4, 3) *D* 9. (10, 4) *T* 10. (1, 9) *N*

Use the grid at the right to name the ordered pair for each point.

11. *L* (2, 1) 12. *I* (4, 2) 13. *C* (7, 2) 14. *R* (6, 8) 15. *T* (3, 7)
16. *X* (1, 7) 17. *Z* (1, 5) 18. *K* (5, 1) 19. *Y* (8, 2) 20. *N* (6, 3)

Lesson 3-1 *(Pages 95–98)*

Write each fraction or mixed number as a decimal.

1. $\frac{4}{10}$ **0.4**   2. $\frac{66}{100}$ **0.66**   3. $\frac{73}{100}$ **0.73**   4. $\frac{5}{100}$ **0.05**   5. $\frac{79}{100}$ **0.79**

6. $\frac{94}{100}$ **0.94**   7. $6\frac{85}{1,000}$ **6.085**   8. $\frac{875}{10,000}$ **0.0875**   9. $\frac{1,264}{1,000}$ **1.264**   10. $\frac{527}{10,000}$ **0.0527**

Write each expression as a decimal.

11. two hundredths **0.02**

12. sixteen hundredths **0.16**

13. four tenths **0.4**

14. two and twenty-seven hundredths **2.27**

15. nine and twelve hundredths **9.12**

16. fifty-six and nine tenths **56.9**

17. twenty-seven thousandths **0.027**

18. one hundred ten-thousandths **0.0100**

19. two ten-thousandths **0.0002**

20. twenty and six hundred thousandths **20.600**

21. two thousand, four hundred seventy-five and six tenths **2,475.6**

22. twelve thousand, ninety-seven and sixty-two thousandths **12,097.062**

Lesson 3-2 *(Pages 102–104)*

Use a centimeter ruler to measure each line segment.

1. _____ **3.4 cm**

2. _____ **4.8 cm**

3. _____ **1.9 cm**

4. _____ **3.7 cm**

5. _____ **3.0 cm**

6. _____ **4.1 cm**

7. ___ **1.0 cm**

8. _____ **1.8 cm**

Lesson 3-3 *(Pages 105–108)*

State the greater number in each group.

1. 0.112 or 0.121 **0.121**   2. 0.9985 or 0.998 **0.9985**   3. 0.556 or 0.519 **0.556**

4. 1.19 or 11.9 **11.9**   5. 0.6, 6.0 or 0.06 **6.0**   6. 0.0009 or 0.001 **0.001**

Order each set of decimals from least to greatest.

7. 415.65 **415.56**	8. 0.0256 **0.0009**	9. 1.2356 **1.2335**	10. 50.12 **5.012**
451.65 **415.65**	0.2056 **0.0255**	1.2355 **1.2353**	5.012 **5.901**
451.66 **451.56**	0.0255 **0.0256**	1.25 **1.2355**	50.22 **50.02**
451.56 **451.65**	0.0009 **0.2056**	1.2335 **1.2356**	5.901 **50.12**
415.56 **451.66**	0.2560 **0.2560**	1.2353 **1.25**	50.02 **50.22**

Order each set of decimals from greatest to least.

11. 13.664 **13.664**	12. 26.6987 **26.9688**	13. 1.00065 **1.00165**	14. 2.014 **22.14**
13.446 **13.446**	26.9687 **26.9687**	1.00100 **1.00100**	2.010 **22.00**
1.3666 **1.6333**	26.9666 **26.9666**	1.00165 **1.00065**	22.00 **2.014**
1.6333 **1.3666**	26.9688 **26.6987**	1.00056 **1.00056**	22.14 **2.010**

Lesson 3-4 *(Pages 109–111)*
Round each number to the underlined place-value position.

1. 5.6̲4 **5.6**

2. 12.37̲6 **12.38**

3. 0.053̲62 **0.054**

4. 6̲.17 **6**

5. 15.2̲98 **15.3**

6. 0.0026̲325 **0.0026**

7. 758.99̲9 **759.00**

8. 4̲.25 **4**

9. 32.658̲3 **32.658**

10. 0̲.025 **0**

11. 1.004̲9 **1.005**

12. 9.2̲5 **9.3**

13. 67.49̲2 **67.49**

14. 25.1̲9 **25.2**

15. 26.9̲6 **27.0**

16. 4.00̲098 **4.00**

Lesson 3-5 *(Pages 112–115)*
Estimate using rounding.

1.
$$\begin{array}{r} 0.245 \\ +0.256 \\ \hline \end{array} \quad \begin{array}{r} 0.25 \\ +0.26 \\ \hline 0.51 \end{array}$$

2.
$$\begin{array}{r} 2.45698 \\ -1.26589 \\ \hline \end{array} \quad \begin{array}{r} 2.5 \\ -1.3 \\ \hline 1.2 \end{array}$$

3.
$$\begin{array}{r} 0.5962 \\ +1.2598 \\ \hline \end{array} \quad \begin{array}{r} 0.6 \\ +1.3 \\ \hline 1.9 \end{array}$$

4.
$$\begin{array}{r} 17.985 \\ -9.001 \\ \hline \end{array} \quad \begin{array}{r} 18.0 \\ -9.0 \\ \hline 9 \end{array}$$

5. $0.256 + 0.6589$
0.26 + 0.66; 0.92

6. $1.2568 - 0.1569$
1.3 − 0.2; 1.1

7. $12.999 + 5.048$
13 + 5; 18

Estimate using clustering.

8. $4.5 + 4.95 + 5.2 + 5.49$ **5 × 4 = 20**

9. $2.25 + 1.69 + 2.1 + 2.369$ **2 × 4 = 8**

10. $\$12.15 + \$11.63 + \$12 + \$11.89$
$12 × 4 = $48

11. $0.569 + 1.005 + 1.265 + 0.765$ **1 × 4 = 4**

Lesson 3-6A *(Pages 116–117)*
Solve.

1. Jaleel bought a t-shirt at a concert. He paid for the $17.49 shirt with a $20 bill. Should he expect about $4 or about $2 in change? **about $2**

2. There are 3,261 seats in the Northmore High School stadium. What is a reasonable number of rows in the stadium if each row holds about 55 people? **about 60 rows**

3. Jayson wants to spend his allowance on CD ROM games for his computer. He has saved $80. About how many games can Jayson buy if each CD ROM costs $29.95? **about 2 games**

4. Lucas needs to buy school supplies. He found a backpack for $29.95, a notebook for $8.49, and a package of pens and pencils for $3.70. Is $30 or $50 needed to pay for these supplies? **$50**

5. When Jamilah added 0.00276 and 0.0149 the calculator showed 0.1766. Is this answer reasonable? **no**

Lesson 3-6 (Pages 118–121)

Add or subtract.

1. $0.46 + 0.72$ **1.18**
2. $13.7 + 2.6$ **16.3**
3. $17.9 + 7.41$ **25.31**
4. $19.2 + 7.36$ **26.56**
5. $0.5113 + 0.62148$ **1.13278**
6. $12.56 - 10.21$ **2.35**
7. $0.2154 - 0.1526$ **0.0628**
8. $2.3125 + 1.02$ **3.3325**
9. $1.025 - 0.58697$ **0.43803**
10. $14.526 - 12.654$ **1.872**
11. $2.3568 + 5$ **7.3568**
12. $20 - 5.98671$ **14.01329**
13. $15.256 + 0.236$ **15.492**
14. $3.7 + 1.5 + 0.2$ **5.4**
15. $0.23 + 1.2 + 0.36$ **1.79**
16. $0.896352 - 0.25639$ **0.639962**
17. $25.6 - 2.3$ **23.3**
18. $13.5 - 2.8456$ **10.6544**
19. $1.265 + 1.654$ **2.919**
20. $24.56 - 24.32$ **0.24**
21. $0.256 - 0.255$ **0.001**

Lesson 4-1 (Pages 133–136)

Multiply.

1. $\begin{array}{r} 0.2 \\ \times\ 65 \\ \hline 13 \end{array}$
2. $\begin{array}{r} 0.73 \\ \times\ 12 \\ \hline 8.76 \end{array}$
3. $\begin{array}{r} 0.65 \\ \times\ 27 \\ \hline 17.55 \end{array}$
4. $\begin{array}{r} 9.6 \\ \times\ 13 \\ \hline 124.8 \end{array}$
5. $\begin{array}{r} 12.15 \\ \times\ 6 \\ \hline 72.9 \end{array}$
6. $\begin{array}{r} 0.91 \\ \times\ 16 \\ \hline 14.56 \end{array}$
7. $\begin{array}{r} 0.265 \\ \times\ 7 \\ \hline 1.855 \end{array}$
8. $\begin{array}{r} 2.612 \\ \times\ 14 \\ \hline 36.568 \end{array}$
9. $\begin{array}{r} 0.003 \\ \times\ 55 \\ \hline 0.165 \end{array}$
10. $\begin{array}{r} 0.67 \\ \times\ 21 \\ \hline 14.07 \end{array}$

Solve each equation.

11. $r = 19 \times 0.111$ **2.109**
12. $1.65 \times 72 = a$ **118.8**
13. $9.6 \times 101 = q$ **969.6**
14. $24 \times 1.201 = d$ **28.824**
15. $610 \times 7.5 = j$ **4,575**
16. $z = 0.001 \times 6$ **0.006**
17. $x = 510 \times 0.0135$ **6.885**
18. $b = 9.2 \times 17$ **156.4**
19. $14.1235 \times 4 = m$ **56.494**

Lesson 4-2 (Pages 137–139)

Find each product mentally. Use the distributive property.

1. 5×18 **90**
2. 9×27 **243**
3. 8×83 **664**
4. 7×21 **147**
5. 3×47 **141**
6. 2×106 **212**
7. 6×34 **204**
8. 56×3 **168**
9. 27×8 **216**
10. 5×3.4 **17**
11. 6×40.7 **244.2**
12. 1.5×30 **45**
13. 0.9×71 **63.9**
14. 30×2.08 **62.4**
15. 16×7 **112**
16. 33×4 **132**
17. 0.6×12 **7.2**
18. 80×7.9 **632**

Lesson 4-3 *(Pages 141–143)*
Multiply.

1. $9.6 \cdot 10.5$ **100.8**
2. $3.2 \cdot 0.1$ **0.32**
3. $1.5 \cdot 9.6$ **14.4**
4. $5.42 \cdot 0.21$ **1.1382**
5. $7.42 \cdot 0.2$ **1.484**
6. $0.001 \cdot 0.02$ **0.00002**
7. $0.6 \cdot 542$ **325.2**
8. $6.7 \cdot 5.8$ **38.86**
9. $3.24 \cdot 6.7$ **21.708**

Solve each equation.

10. $9.8 \cdot 4.62 = s$ **45.276**
11. $7.32 \cdot 9.7 = v$ **71.004**
12. $t = 0.008 \cdot 0.007$ **0.000056**
13. $a = 0.001 \cdot 56$ **0.056**
14. $c = 4.5 \cdot 0.2$ **0.9**
15. $9.6 \cdot 2.3 = h$ **22.08**
16. $5.63 \cdot 8.1 = q$ **45.603**
17. $10.35 \cdot 9.1 = u$ **94.185**
18. $t = 28.2 \cdot 3.9$ **109.98**
19. $g = 102.13 \cdot 1.221$ **124.70073**
20. $n = 2.02 \times 1.25$ **2.525**
21. $z = 8.37 \times 89.6$ **749.952**

Lesson 4-4 *(Pages 145–148)*
Find the perimeter of each figure.

1.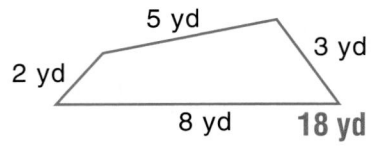
2.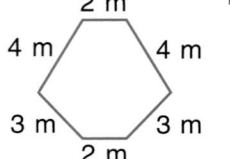
3.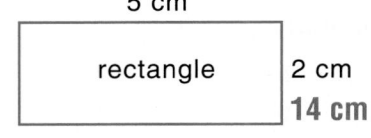

Find the perimeter and area of each figure.

4. $19 \text{ cm} \times 3 \text{ cm}$ **44 cm, 57 cm²**
5. $5 \text{ in.} \times 3 \text{ in.}$ **16 in., 15 in²**
6. 9 ft square **36 ft, 81 ft²**
7. 4.3 m square **17.2 m, 18.49 m²**

8.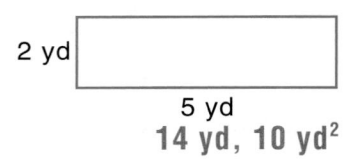
14 yd, 10 yd²
9.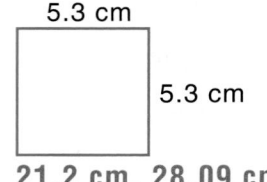
21.2 cm, 28.09 cm²
10. **27.4 in., 39.36 in²**

9.6 in.

4.1 in.

Lesson 4-5A *(Pages 150–151)*
Solve.

1. Genaro wants to carpet a den that measures 18 feet by 24 feet. In the center of the room is a tile hearth for his stove which he does not want to carpet. If the hearth measures 6 feet by 6 feet, how much carpet does he need? **396 ft²**

2. Jamika and Jordan are carving pumpkins for a Halloween party. Jamika can carve 3 pumpkins in 2 hours. Jordan can carve 2 pumpkins in 1 hour. If they work the same amount of time, how long will it take to carve 14 pumpkins? **4 hours**

3. Lonny has 24 feet of fence to make a pen for his rabbit. What dimensions should he make the pen so the area is the greatest possible? **6 ft by 6 ft**

4. How many cuts must be made to divide a 6-foot submarine sandwich equally among 10 people? **9 cuts**

Lesson 4-5 *(Pages 152–155)*
Find each quotient.

1. $6\overline{)1.26}$ **0.21**
2. $8\overline{)23.2}$ **2.9**
3. $6\overline{)89.22}$ **14.87**
4. $15\overline{)54.75}$ **3.65**

5. $13\overline{)128.31}$ **9.87**
6. $9\overline{)2.583}$ **0.287**
7. $47\overline{)11.28}$ **0.24**
8. $26\overline{)32.5}$ **1.25**

9. $37.1 \div 14$ **2.65**
10. $5.88 \div 4$ **1.47**
11. $3.7 \div 5$ **0.74**

12. $41.4 \div 18$ **2.3**
13. $9.87 \div 3$ **3.29**
14. $8.45 \div 25$ **0.338**

Round each quotient to the nearest tenth.

15. $26.5 \div 4$ **6.6**
16. $46.25 \div 8$ **5.8**
17. $19.38 \div 9$ **2.2**

18. $8.5 \div 2$ **4.3**
19. $90.88 \div 14$ **6.5**
20. $23.1 \div 4$ **5.8**

Round each quotient to the nearest hundredth.

21. $19.5 \div 27$ **0.72**
22. $26.5 \div 19$ **1.39**
23. $46.23 \div 25$ **1.85**

24. $46.25 \div 25$ **1.85**
25. $4.26 \div 9$ **0.47**
26. $18.74 \div 19$ **0.99**

Lesson 4-6 *(Pages 157–159)*
Find each quotient.

1. $0.5\overline{)18.45}$ **36.9**
2. $0.08\overline{)5.2}$ **65**
3. $2.6\overline{)0.65}$ **0.25**
4. $1.3\overline{)12.831}$ **9.87**

5. $0.87\overline{)5.133}$ **5.9**
6. $2.54\overline{)24.13}$ **9.5**
7. $3.7\overline{)35.89}$ **9.7**
8. $26\overline{)32.5}$ **1.25**

9. $5.88 \div 0.4$ **14.7**
10. $3.7 \div 0.5$ **7.4**
11. $6.72 \div 2.4$ **2.8**

12. $9.87 \div 0.3$ **32.9**
13. $8.45 \div 2.5$ **3.38**
14. $90.88 \div 14.2$ **6.4**

15. $33.6 \div 8.4$ **4**
16. $25.389 \div 4.03$ **6.3**
17. $85.92 \div 4.8$ **17.9**

18. $63.18 \div 16.2$ **3.9**
19. $18.49 \div 4.3$ **4.3**
20. $9.363 \div 0.003$ **3,121**

21. $1.02 \div 0.3$ **3.4**
22. $6.4 \div 0.8$ **8**
23. $7.2 \div 0.9$ **8**

Lesson 4-7 *(Pages 161–163)*
Find each quotient to the nearest hundredth.

1. $9\overline{)0.36}$ **0.04**
2. $13\overline{)39.39}$ **3.03**
3. $45\overline{)0.585}$ **0.01**
4. $8\overline{)0.24}$ **0.03**

5. $6\overline{)0.312}$ **0.05**
6. $7\overline{)0.161}$ **0.02**
7. $7\overline{)7.21}$ **1.03**
8. $3\overline{)9.18}$ **3.06**

9. $\$0.72 \div 12$ **$0.06**
10. $0.36 \div 9$ **0.04**
11. $0.56 \div 14$ **0.04**

12. $32.2 \div 8$ **4.03**
13. $0.3869 \div 5.3$ **0.07**
14. $0.39 \div 7.8$ **0.05**

Solve each equation.

15. $0.0426 \div 71 = q$ **0.0006**
16. $f = 0.1185 \div 7.9$ **0.015**
17. $j = 0.84 \div 12$ **0.07**

18. $m = 4.544 \div 64$ **0.071**
19. $v = 0.384 \div 9.6$ **0.04**
20. $0.2262 \div 8.7 = d$ **0.026**

Lesson 4-8 (Pages 164–166)

Write the unit that you would use to measure each of the following. Then estimate the mass or capacity.

1. a bag of sugar kilogram; 2 kg
2. a pitcher of fruit punch liter; 2 L
3. the mass of a dime gram; 10 g
4. the amount of water in an ice cube
   milliliter; 20 mL
5. a vitamin gram; 1 g
6. a pencil gram; 20 g
7. the mass of a puppy kilogram; 5 kg
8. a bottle of perfume milliliter; 20 mL
9. a grain of sand milligram; 1 mg
10. the mass of a car kilogram; 1,000 kg

Lesson 4-9 (Pages 167–169)

Complete.

1. 400 mm = ____40____ cm
2. 4 kg = ___4,000___ g
3. 660 cm = ___6.6___ m
4. 0.3 L = ___300___ mL
5. 30 mm = ___3___ cm
6. 84.5 g = ___0.0845___ kg
7. ___0.54___ m = 54 cm
8. ___0.563___ L = 563 mL
9. ___21,000___ mg = 21 g
10. 4 L = ___4,000___ mL
11. 61.2 mg = ___0.0612___ g
12. 4,497 mL = ___4.497___ L
13. ___450___ mm = 45 cm
14. 632 mL = ___0.632___ L
15. 61 g = ___61,000___ mg
16. ___510,000___ mg = 0.51 kg
17. 0.63 L = ___630___ mL
18. 18 km = ___1,800,000___ cm

Lesson 5-1 (Pages 178–180)

Determine whether the first number is divisible by the second number.

1. 89; 3 no
2. 64; 2 yes
3. 125; 5 yes
4. 156; 4 yes
5. 216; 9 yes
6. 330; 10 yes
7. 225; 3 yes
8. 524; 6 no

State whether each number is divisible by 2, 3, 5, 6, 9, and 10.

9. 1,986 2, 3, 6
10. 2,052 2, 3, 6, 9
11. 110 2, 5, 10
12. 315 3, 5, 9
13. 405 3, 5, 9
14. 918 2, 3, 6, 9
15. 243 3, 9
16. 735 3, 5
17. 1,233 3, 9
18. 5,103 3, 9
19. 8,001 3, 9
20. 9,270 2, 3, 5, 6, 9, 10

Lesson 5-2 *(Pages 182–184)*

Tell whether each number is *prime*, *composite*, or *neither*.

1. 20 composite **2.** 65 composite **3.** 37 prime **4.** 26 composite **5.** 54 composite

6. 155 composite **7.** 201 composite **8.** 0 neither **9.** 49 composite **10.** 17 prime

Find the prime factorization of each number.

11. 72 $2^3 \times 3^2$ **12.** 2,648 $2^3 \times 331$ **13.** 32 2^5 **14.** 86 2×43 **15.** 120 $2^3 \times 3 \times 5$

16. 576 $2^6 \times 3^2$ **17.** 68 $2^2 \times 17$ **18.** 240 $2^4 \times 3 \times 5$ **19.** 24 $2^3 \times 3$ **20.** 70 $2 \times 5 \times 7$

21. 102 $2 \times 3 \times 17$ **22.** 121 11^2 **23.** 164 $2^2 \times 41$ **24.** 225 $5^2 \times 3^2$ **25.** 54 2×3^3

Lesson 5-3A *(Pages 186–187)*

Solve.

1. Mychal needs to go to the bank, the post office, and the bicycle shop. In how many different orders can she do this? **6 different orders**

2. How many different four digit numbers can be made from the digits 2, 5, 7, and 9? **24 numbers**

3. Ai-lien usually buys her lunch from the a la carte line in the cafeteria. Today, she has a choice of chicken, fish, or spaghetti; soup or salad; and fresh fruit or yogurt. How many possible choices does she have? **12 choices**

4. A shoe company offers 8 different styles of running shoes in 4 different colors. How many combinations of style and color are possible? **32 combinations**

Lesson 5-3 *(Pages 188–190)*

Find the GCF of each set of numbers by making a list.

1. 8, 18 2 **2.** 6, 9 3 **3.** 4, 12 4 **4.** 18, 24 6

5. 25, 30 5 **6.** 36, 54 9 **7.** 64, 32 32 **8.** 16, 32, 56 8

Find the GCF of each set of numbers by using prime factorization.

9. 6, 15 3 **10.** 8, 24 8 **11.** 14, 22 2 **12.** 12, 27 3

13. 17, 51 17 **14.** 48, 60 12 **15.** 54, 72 18 **16.** 14, 28, 42 14

Find the GCF of each set of numbers using either method.

17. 5, 20 5 **18.** 9, 21 3 **19.** 10, 25 5 **20.** 12, 28 4

21. 16, 24 8 **22.** 42, 48 6 **23.** 60, 75 15 **24.** 27, 45, 63 9

Lesson 5-4 *(Pages 193–196)*

Replace each ▪ with a number so that the fractions are equivalent.

1. $\frac{12}{16} = \frac{▪}{4}$ 3

2. $\frac{7}{8} = \frac{▪}{32}$ 28

3. $\frac{3}{4} = \frac{75}{▪}$ 100

4. $\frac{8}{16} = \frac{▪}{2}$ 1

5. $\frac{6}{18} = \frac{1}{▪}$ 3

6. $\frac{27}{36} = \frac{3}{▪}$ 4

7. $\frac{1}{4} = \frac{16}{▪}$ 64

8. $\frac{9}{18} = \frac{▪}{2}$ 1

State whether each fraction or ratio is in simplest form. If not, write each fraction or ratio in simplest form.

9. $\frac{50}{100}$ no; $\frac{1}{2}$

10. $\frac{24}{40}$ no; $\frac{3}{5}$

11. 2:5 yes

12. 8 out of 24 no; 1 out of 3

13. 20 to 27 yes

14. $\frac{4}{10}$ no; $\frac{2}{5}$

15. $\frac{3}{5}$ yes

16. 14:19 yes

17. 9:12 no; 3:4

18. $\frac{6}{8}$ no; $\frac{3}{4}$

19. $\frac{15}{18}$ no; $\frac{5}{6}$

20. 9 out of 20 yes

Lesson 5-5 *(Pages 198–201)*

Express each mixed number as an improper fraction.

1. $3\frac{1}{16}$ $\frac{49}{16}$

2. $2\frac{3}{4}$ $\frac{11}{4}$

3. $1\frac{3}{8}$ $\frac{11}{8}$

4. $1\frac{5}{12}$ $\frac{17}{12}$

5. $7\frac{3}{5}$ $\frac{38}{5}$

6. $6\frac{5}{8}$ $\frac{53}{8}$

7. $3\frac{1}{3}$ $\frac{10}{3}$

8. $1\frac{7}{9}$ $\frac{16}{9}$

9. $2\frac{3}{16}$ $\frac{35}{16}$

10. $1\frac{2}{3}$ $\frac{5}{3}$

Express each improper fraction as a mixed number.

11. $\frac{33}{10}$ $3\frac{3}{10}$

12. $\frac{103}{25}$ $4\frac{3}{25}$

13. $\frac{22}{5}$ $4\frac{2}{5}$

14. $\frac{13}{2}$ $6\frac{1}{2}$

15. $\frac{29}{6}$ $4\frac{5}{6}$

16. $\frac{101}{100}$ $1\frac{1}{100}$

17. $\frac{21}{8}$ $2\frac{5}{8}$

18. $\frac{19}{6}$ $3\frac{1}{6}$

19. $\frac{23}{5}$ $4\frac{3}{5}$

20. $\frac{99}{50}$ $1\frac{49}{50}$

Lesson 5-6 *(Pages 202–205)*

Draw a line segment of each length. See students' work.

1. $2\frac{1}{4}$ inches

2. $1\frac{3}{8}$ inches

3. $\frac{3}{4}$ inch

4. $1\frac{1}{2}$ inches

5. $3\frac{1}{8}$ inches

6. $2\frac{1}{4}$ inches

Find the length of each line segment to the nearest half, fourth, or eighth inch.

7. ━━━━━━━━ $1\frac{1}{4}$ in.

8. ━━━━━━━━━━ $1\frac{1}{2}$ in.

9. ━━━━━━ 1 in.

10. ━━━━━━━━━ $1\frac{3}{8}$ in.

Lesson 5-7 (Pages 206–209)

Determine whether the first number is a multiple of the second number.

1. 30; 7 no

2. 42; 14 yes

3. 10; 8 no

4. 30; 10 yes

5. 13; 7 no

6. 84; 28 yes

7. 150; 25 yes

8. 21; 14 no

Find the LCM for each set of numbers.

9. 5, 15 15

10. 13, 39 39

11. 16, 24 48

12. 18, 20 180

13. 8, 12 24

14. 12, 15 60

15. 9, 27 27

16. 5, 6 30

17. 12, 18, 3 36

18. 12, 35, 10 420

19. 21, 14, 6 42

20. 3, 6, 9 18

21. 6, 10, 15 30

22. 15, 75, 25 75

Lesson 5-8 (Pages 210–213)

Find the LCD for each pair of fractions.

1. $\frac{2}{5}$ $\frac{4}{15}$ 15

2. $\frac{2}{28}$ $\frac{15}{42}$ 84

3. $\frac{8}{16}$ $\frac{14}{24}$ 48

4. $\frac{9}{25}$ $\frac{21}{30}$ 150

Replace each ● with <, >, or = to make a true sentence.

5. $\frac{1}{2}$ ● $\frac{1}{3}$ >

6. $\frac{2}{3}$ ● $\frac{3}{4}$ <

7. $\frac{5}{9}$ ● $\frac{4}{5}$ <

8. $\frac{3}{6}$ ● $\frac{6}{12}$ =

9. $\frac{12}{23}$ ● $\frac{15}{19}$ <

10. $\frac{9}{27}$ ● $\frac{13}{39}$ =

11. $\frac{7}{8}$ ● $\frac{9}{13}$ >

12. $\frac{5}{9}$ ● $\frac{7}{8}$ <

13. $\frac{25}{100}$ ● $\frac{3}{8}$ <

14. $\frac{6}{7}$ ● $\frac{8}{15}$ >

15. $\frac{5}{9}$ ● $\frac{19}{23}$ <

16. $\frac{120}{567}$ ● $\frac{1}{2}$ <

17. $\frac{5}{7}$ ● $\frac{2}{3}$ >

18. $\frac{9}{36}$ ● $\frac{7}{28}$ =

19. $\frac{2}{5}$ ● $\frac{2}{6}$ >

20. $\frac{5}{9}$ ● $\frac{12}{13}$ <

Lesson 5-9 (Pages 214–216)

Express each decimal as a fraction or mixed number in simplest form.

1. 0.5 $\frac{1}{2}$

2. 0.8 $\frac{4}{5}$

3. 0.32 $\frac{8}{25}$

4. 0.875 $\frac{7}{8}$

5. 0.54 $\frac{27}{50}$

6. 0.38 $\frac{19}{50}$

7. 0.744 $\frac{93}{125}$

8. 0.101 $\frac{101}{1,000}$

9. 0.303 $\frac{303}{1,000}$

10. 0.486 $\frac{243}{500}$

11. 0.626 $\frac{313}{500}$

12. 0.448 $\frac{56}{125}$

13. 0.074 $\frac{37}{500}$

14. 0.008 $\frac{1}{125}$

15. 9.36 $9\frac{9}{25}$

16. 10.18 $10\frac{9}{50}$

17. 0.06 $\frac{3}{50}$

18. 0.75 $\frac{3}{4}$

19. 0.48 $\frac{12}{25}$

20. 0.9 $\frac{9}{10}$

21. 0.005 $\frac{1}{200}$

22. 0.4 $\frac{2}{5}$

23. 1.875 $1\frac{7}{8}$

24. 5.08 $5\frac{2}{25}$

Lesson 5-10 *(Pages 217–219)*

Write each repeating decimal using bar notation.

1. 0.757575... $0.\overline{75}$ **2.** 0.444444... $0.\overline{4}$ **3.** 2.875875... $2.\overline{875}$ **4.** 0.333333... $0.\overline{3}$

5. 6.404040... $6.\overline{40}$ **6.** 5.272727... $5.\overline{27}$ **7.** 0.8521685216... $0.\overline{85216}$ **8.** 0.833333... $0.8\overline{3}$

Express each fraction or mixed number as a decimal. Use bar notation to show a repeating decimal.

9. $\frac{3}{16}$ 0.1875   **10.** $\frac{8}{33}$ $0.\overline{24}$ **11.** $\frac{7}{12}$ $0.58\overline{3}$ **12.** $\frac{14}{25}$ 0.56

13. $\frac{7}{10}$ 0.7   **14.** $\frac{5}{8}$ 0.625   **15.** $\frac{11}{15}$ $0.7\overline{3}$ **16.** $\frac{8}{9}$ $0.\overline{8}$

17. $\frac{15}{16}$ 0.9375   **18.** $\frac{1}{12}$ $0.08\overline{3}$ **19.** $\frac{7}{20}$ 0.35   **20.** $\frac{5}{18}$ $0.2\overline{7}$

Lesson 6-1 *(Pages 228–231)*

Round each number to the nearest half.

1. $\frac{11}{12}$ 1   **2.** $\frac{5}{8}$ $\frac{1}{2}$ **3.** $\frac{2}{5}$ $\frac{1}{2}$ **4.** $\frac{1}{10}$ 0   **5.** $\frac{1}{6}$ 0   **6.** $\frac{2}{3}$ $\frac{1}{2}$

7. $\frac{9}{10}$ 1   **8.** $\frac{1}{8}$ 0   **9.** $\frac{4}{9}$ $\frac{1}{2}$ **10.** $1\frac{1}{8}$ 1   **11.** $\frac{12}{11}$ 1   **12.** $2\frac{4}{5}$ 3

13. $\frac{7}{9}$ 1   **14.** $7\frac{1}{10}$ 7   **15.** $10\frac{2}{3}$ $10\frac{1}{2}$ **16.** $\frac{1}{3}$ $\frac{1}{2}$ **17.** $\frac{7}{16}$ $\frac{1}{2}$ **18.** $\frac{5}{7}$ $\frac{1}{2}$

1. $14 - 7 = 7$	2. $8 + 2 = 10$	3. $5 + 8 = 13$	4. $12 - 5 = 7$
5. $3 - 1 = 2$	6. $4 + 1 = 5$	7. $5 - 2 = 3$	8. $5 + 10 = 15$
9. $7 - 3 = 4$	10. $2 + 1 = 3$	11. $18 - 13 = 5$	12. $13 + 9 = 22$
13. $9 - 6 = 3$	14. $3 - 2\frac{1}{2} = \frac{1}{2}$	15. $9 - 0 = 9$	16. $12 - 1 = 11$
17. $2 + 2 = 4$	18. $9 - 8 = 1$	19. $9 + 5 = 14$	20. $1 + 7 = 8$

Lesson 6-2 *(Pages 232–234)*

Estimate. 1–20. Sample answers are given.

1. $14\frac{1}{10} - 6\frac{4}{5}$ **2.** $8\frac{1}{3} + 2\frac{1}{6}$ **3.** $4\frac{7}{8} + 7\frac{3}{4}$ **4.** $11\frac{11}{12} - 5\frac{1}{4}$

5. $3\frac{2}{5} - 1\frac{1}{4}$ **6.** $4\frac{2}{5} + \frac{5}{6}$ **7.** $4\frac{7}{12} - 1\frac{3}{4}$ **8.** $4\frac{2}{3} + 10\frac{3}{8}$

9. $7\frac{7}{15} - 3\frac{1}{12}$ **10.** $2\frac{1}{20} + 1\frac{1}{3}$ **11.** $18\frac{1}{4} - 12\frac{3}{5}$ **12.** $12\frac{5}{9} + 8\frac{5}{8}$

13. $8\frac{2}{3} - 5\frac{1}{2}$ **14.** $3\frac{1}{8} - 2\frac{3}{5}$ **15.** $9\frac{2}{7} - \frac{1}{3}$ **16.** $11\frac{7}{8} - \frac{5}{6}$

17. $2\frac{2}{5} + 2\frac{1}{4}$ **18.** $8\frac{1}{2} - 7\frac{4}{5}$ **19.** $8\frac{3}{4} + 4\frac{2}{3}$ **20.** $1\frac{1}{8} + 7\frac{1}{10}$

Lesson 6-2B *(Pages 236–237)*
Solve.

1. Ben wants a part-time job. There are three 10 hour per week jobs available. One pays $45 per week. One pays $4.75 per hour. One pays $3.25 per hour plus tips which average $20 per week. If Ben wants to take the job that pays the most, for which job should he apply? **$3.25 per hour plus tips**

2. Micaela gave the clerk $50 to pay for a $21.79 hat and a $17.33 belt. Should she expect $10, $20, or $40 change? **$10**

3. Jackie used her calculator to multiply 678 and 34. Should she expect the product to be about 2,305, 23,050 or 230,500? **23,050**

4. If Carl blinks 15 times per minute, about how many times does he blink in one day? **c**
   a. 216 times
   b. 2,160 times
   c. 21,600 times
   d. 216,000 times

Lesson 6-3 *(Pages 238–241)*
Add or subtract. Write the answer in simplest form.

1. $\frac{2}{5} + \frac{2}{5}$ $\frac{4}{5}$
2. $\frac{5}{8} + \frac{3}{8}$ 1
3. $\frac{9}{11} - \frac{3}{11}$ $\frac{6}{11}$
4. $\frac{3}{14} + \frac{5}{14}$ $\frac{4}{7}$

5. $\frac{7}{8} - \frac{3}{8}$ $\frac{1}{2}$
6. $\frac{3}{4} - \frac{1}{4}$ $\frac{1}{2}$
7. $\frac{15}{27} - \frac{7}{27}$ $\frac{8}{27}$
8. $\frac{1}{36} + \frac{5}{36}$ $\frac{1}{6}$

9. $\frac{2}{9} - \frac{1}{9}$ $\frac{1}{9}$
10. $\frac{7}{8} + \frac{5}{8}$ $1\frac{1}{2}$
11. $\frac{9}{16} - \frac{5}{16}$ $\frac{1}{4}$
12. $\frac{6}{8} + \frac{4}{8}$ $1\frac{1}{4}$

13. $\frac{1}{2} + \frac{1}{2}$ 1
14. $\frac{1}{3} - \frac{1}{3}$ 0
15. $\frac{8}{9} + \frac{7}{9}$ $1\frac{2}{3}$
16. $\frac{5}{6} - \frac{3}{6}$ $\frac{1}{3}$

17. $\frac{3}{9} + \frac{8}{9}$ $1\frac{2}{9}$
18. $\frac{8}{40} + \frac{12}{40}$ $\frac{1}{2}$
19. $\frac{56}{90} - \frac{26}{90}$ $\frac{1}{3}$
20. $\frac{2}{9} + \frac{8}{9}$ $1\frac{1}{9}$

Lesson 6-4 *(Pages 243–245)*
Add or subtract. Write the answer in simplest form.

1. $\frac{1}{3} + \frac{1}{2}$ $\frac{5}{6}$
2. $\frac{2}{9} + \frac{1}{3}$ $\frac{5}{9}$
3. $\frac{1}{2} + \frac{3}{4}$ $1\frac{1}{4}$
4. $\frac{1}{4} + \frac{3}{12}$ $\frac{1}{2}$

5. $\frac{5}{9} - \frac{1}{3}$ $\frac{2}{9}$
6. $\frac{5}{8} - \frac{2}{5}$ $\frac{9}{40}$
7. $\frac{3}{4} - \frac{1}{2}$ $\frac{1}{4}$
8. $\frac{7}{8} - \frac{3}{16}$ $\frac{11}{16}$

9. $\frac{9}{16} + \frac{13}{24}$ $1\frac{5}{48}$
10. $\frac{8}{15} + \frac{2}{3}$ $1\frac{1}{5}$
11. $\frac{5}{14} + \frac{11}{28}$ $\frac{3}{4}$
12. $\frac{11}{12} + \frac{7}{8}$ $1\frac{19}{24}$

13. $\frac{2}{3} - \frac{1}{6}$ $\frac{1}{2}$
14. $\frac{9}{16} - \frac{1}{2}$ $\frac{1}{16}$
15. $\frac{5}{8} - \frac{11}{20}$ $\frac{3}{40}$
16. $\frac{14}{15} - \frac{2}{9}$ $\frac{32}{45}$

17. $\frac{9}{20} + \frac{2}{15}$ $\frac{7}{12}$
18. $\frac{5}{6} + \frac{4}{5}$ $1\frac{19}{30}$
19. $\frac{23}{25} - \frac{27}{50}$ $\frac{19}{50}$
20. $\frac{19}{25} - \frac{1}{2}$ $\frac{13}{50}$

Lesson 6-5 *(Pages 246–249)*

Add or subtract. Write the answer in simplest form.

1. $5\frac{1}{2} + 3\frac{1}{4}$ $8\frac{3}{4}$

2. $2\frac{2}{3} + 4\frac{1}{9}$ $6\frac{7}{9}$

3. $7\frac{4}{5} + 9\frac{3}{10}$ $17\frac{1}{10}$

4. $9\frac{4}{7} - 3\frac{5}{14}$ $6\frac{3}{14}$

5. $13\frac{1}{5} - 10$ $3\frac{1}{5}$

6. $3\frac{3}{4} + 5\frac{5}{8}$ $9\frac{3}{8}$

7. $3\frac{2}{5} + 7\frac{6}{15}$ $10\frac{4}{5}$

8. $10\frac{2}{3} + 5\frac{6}{7}$ $16\frac{11}{21}$

9. $15\frac{6}{9} - 13\frac{5}{12}$ $2\frac{1}{4}$

10. $13\frac{7}{12} - 9\frac{1}{4}$ $4\frac{1}{3}$

11. $5\frac{2}{3} - 3\frac{1}{2}$ $2\frac{1}{6}$

12. $17\frac{2}{9} + 12\frac{1}{3}$ $29\frac{5}{9}$

13. $6\frac{5}{12} + 12\frac{5}{8}$ $19\frac{1}{24}$

14. $8\frac{3}{5} - 2\frac{1}{5}$ $6\frac{2}{5}$

15. $23\frac{2}{3} - 4\frac{1}{2}$ $19\frac{1}{6}$

Lesson 6-6 *(Pages 250–253)*

Subtract. Write the answer in simplest form.

1. $11\frac{2}{3} - 8\frac{11}{12}$ $2\frac{3}{4}$

2. $3\frac{4}{7} - 1\frac{2}{3}$ $1\frac{19}{21}$

3. $7\frac{1}{8} - 4\frac{1}{3}$ $2\frac{19}{24}$

4. $18\frac{1}{9} - 12\frac{2}{5}$ $5\frac{32}{45}$

5. $12\frac{3}{10} - 8\frac{3}{4}$ $3\frac{11}{20}$

6. $43 - 5\frac{1}{5}$ $37\frac{4}{5}$

7. $8\frac{1}{5} - 4\frac{1}{4}$ $3\frac{19}{20}$

8. $14\frac{1}{6} - 3\frac{2}{3}$ $10\frac{1}{2}$

9. $25\frac{4}{7} - 21$ $4\frac{4}{7}$

10. $17\frac{3}{9} - 4\frac{3}{5}$ $12\frac{11}{15}$

11. $18\frac{1}{9} - 1\frac{3}{7}$ $16\frac{43}{63}$

12. $16\frac{1}{4} - 7\frac{1}{5}$ $9\frac{1}{20}$

13. $18\frac{1}{5} - 6\frac{1}{4}$ $11\frac{19}{20}$

14. $4 - 1\frac{2}{3}$ $2\frac{1}{3}$

15. $26 - 4\frac{1}{9}$ $21\frac{8}{9}$

Lesson 6-7 *(Pages 254–257)*

Complete.

1. 2 h 10 min = 1 h __70__ min

2. 3 h 65 min = __4__ h 5 min

3. 3 min 14 s = 2 min __74__ s

4. 1 h 15 min 10 s = __75__ min 10 s

Add or subtract. Rename if necessary.

5. 6 h 14 min
  −2 h 8 min
  ‾‾‾‾‾‾‾‾‾‾‾‾
  4 h 6 min

6. 5 h 35 min 25 s
  +   45 min 35 s
  ‾‾‾‾‾‾‾‾‾‾‾‾‾‾‾‾‾‾
  6 h 21 min 0 s

7. 5 h 4 min 45 s
  −2 h 40 min 5 s
  ‾‾‾‾‾‾‾‾‾‾‾‾‾‾‾‾‾‾
  2 h 24 min 40 s

8. 15 h 16 min
  − 8 h 35 min 16 s
  ‾‾‾‾‾‾‾‾‾‾‾‾‾‾‾‾‾‾
  6 h 40 min 44 s

9. 9 h 20 min 10 s
  +1 h 39 min 55 s
  ‾‾‾‾‾‾‾‾‾‾‾‾‾‾‾‾‾‾
  11 h 0 min 5 s

10. 2 h 40 min 20 s
  +3 h 5 min 50 s
  ‾‾‾‾‾‾‾‾‾‾‾‾‾‾‾‾‾‾
  5 h 46 min 10 s

Find the elapsed time.

11. 10:30 A.M. to 6:00 P.M. $7\frac{1}{2}$ hours

12. 8:45 P.M. to 1:30 A.M. $4\frac{3}{4}$ hours or 4 hours 45 min

Lesson 7-1 *(Pages 268–270)*

Round each fraction to 0, $\frac{1}{2}$, or 1.

1. $\frac{3}{4}$ 1 **2.** $\frac{5}{8}$ $\frac{1}{2}$ **3.** $\frac{3}{25}$ 0 **4.** $\frac{1}{20}$ 0 **5.** $\frac{5}{11}$ $\frac{1}{2}$ **6.** $\frac{11}{18}$ $\frac{1}{2}$ **7.** $\frac{1}{3}$ $\frac{1}{2}$

Round each mixed number to the nearest whole number.

8. $6\frac{4}{9}$ 6 **9.** $19\frac{3}{4}$ 20 **10.** $2\frac{1}{15}$ 2 **11.** $17\frac{2}{7}$ 17

12. $8\frac{9}{16}$ 9 **13.** $1\frac{3}{5}$ 2 **14.** $15\frac{41}{45}$ 16

Estimate each product.

15. $\frac{2}{3} \times \frac{4}{5}$ $\frac{1}{2} \times 1 = \frac{1}{2}$ **16.** $\frac{1}{6} \times \frac{2}{5}$ $0 \times 0 = 0$ **17.** $\frac{4}{9} \times \frac{3}{7}$ $\frac{1}{2} \times \frac{1}{2} = \frac{1}{4}$ **18.** $\frac{5}{12} \times \frac{6}{11}$ $\frac{1}{2} \times \frac{1}{2} = \frac{1}{4}$

19. $\frac{3}{8} \times \frac{8}{9}$ $\frac{1}{2} \times 1 = \frac{1}{2}$ **20.** $\frac{3}{5} \times \frac{5}{12}$ $\frac{1}{2} \times \frac{1}{2} = \frac{1}{4}$ **21.** $\frac{2}{5} \times \frac{5}{8}$ $\frac{1}{2} \times \frac{1}{2} = \frac{1}{4}$ **22.** $5\frac{3}{7} \times \frac{4}{5}$ $5 \times 1 = 5$

Lesson 7-2 *(Pages 273–276)*

Find each product. Write in simplest form.

1. $\frac{5}{6} \times \frac{15}{16}$ $\frac{25}{32}$ **2.** $\frac{6}{14} \times \frac{12}{18}$ $\frac{2}{7}$ **3.** $\frac{2}{3} \times \frac{3}{13}$ $\frac{2}{13}$ **4.** $\frac{4}{9} \times \frac{1}{6}$ $\frac{2}{27}$

5. $\frac{3}{4} \times \frac{5}{6}$ $\frac{5}{8}$ **6.** $\frac{9}{10} \times \frac{3}{4}$ $\frac{27}{40}$ **7.** $\frac{8}{9} \times \frac{2}{3}$ $\frac{16}{27}$ **8.** $\frac{6}{7} \times \frac{4}{5}$ $\frac{24}{35}$

9. $\frac{8}{11} \times \frac{11}{12}$ $\frac{2}{3}$ **10.** $\frac{5}{6} \times \frac{3}{5}$ $\frac{1}{2}$ **11.** $\frac{6}{7} \times \frac{7}{21}$ $\frac{2}{7}$ **12.** $\frac{8}{9} \times \frac{9}{10}$ $\frac{4}{5}$

Solve each equation. Write the solution in simplest form.

13. $q = \frac{7}{11} \times \frac{12}{15}$ $\frac{28}{55}$ **14.** $r = \frac{7}{9} \times \frac{5}{7}$ $\frac{5}{9}$ **15.** $\frac{8}{13} \times \frac{2}{11} = a$ $\frac{16}{143}$

16. $m = \frac{4}{7} \times \frac{2}{9}$ $\frac{8}{63}$ **17.** $d = \frac{4}{9} \times \frac{24}{25}$ $\frac{32}{75}$ **18.** $\frac{1}{9} \times \frac{6}{13} = s$ $\frac{2}{39}$

19. $j = \frac{4}{7} \times 6$ $3\frac{3}{7}$ **20.** $\frac{7}{10} \times 5 = p$ $3\frac{1}{2}$ **21.** $10 \times 3\frac{1}{5} = k$ 32

Lesson 7-3 *(Pages 277–279)*

Express each mixed number as an improper fraction.

1. $2\frac{1}{3}$ $\frac{7}{3}$ **2.** $2\frac{1}{2}$ $\frac{5}{2}$ **3.** $2\frac{2}{3}$ $\frac{8}{3}$ **4.** $3\frac{1}{2}$ $\frac{7}{2}$ **5.** $1\frac{1}{4}$ $\frac{5}{4}$

Find each product. Write in simplest form.

6. $3\frac{5}{8} \times 4\frac{1}{2}$ $16\frac{5}{16}$ **7.** $\frac{4}{5} \times 2\frac{3}{4}$ $2\frac{1}{5}$ **8.** $6\frac{1}{8} \times 5\frac{1}{7}$ $31\frac{1}{2}$ **9.** $2\frac{2}{3} \times 2\frac{1}{4}$ 6

10. $6\frac{2}{3} \times 7\frac{3}{5}$ $50\frac{2}{3}$ **11.** $7\frac{1}{5} \times 2\frac{4}{7}$ $18\frac{18}{35}$ **12.** $8\frac{3}{4} \times 2\frac{2}{5}$ 21 **13.** $4\frac{1}{3} \times 2\frac{1}{7}$ $9\frac{2}{7}$

Solve each equation. Write the solution in simplest form.

14. $4\frac{3}{5} \times 2\frac{1}{2} = c$ $11\frac{1}{2}$ **15.** $z = 5\frac{5}{6} \times 4\frac{2}{7}$ 25

16. $6\frac{8}{9} \times 3\frac{5}{6} = f$ $26\frac{11}{27}$ **17.** $2\frac{1}{9} \times 1\frac{1}{2} = x$ $3\frac{1}{6}$

18. $4\frac{7}{15} \times 3\frac{3}{4} = h$ $16\frac{3}{4}$ **19.** $k = 5\frac{7}{9} \times 6\frac{3}{8}$ $36\frac{5}{6}$

Lesson 7-4 *(Pages 280–283)*

Find the circumference of each circle shown or described to the nearest tenth. Use $\frac{22}{7}$ or 3.14 for π.

1.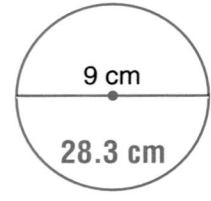
   9 cm
   28.3 cm

2.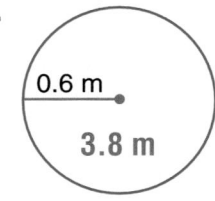
   2.1 yd
   13.2 yd

3.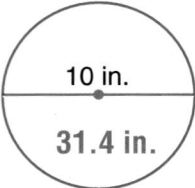
   0.6 m
   3.8 m

4.
   10 in.
   31.4 in.

5. $d = 5.6$ m   **17.6 m**

6. $r = 3.21$ yd   **20.2 yd**

7. $r = 0.5$ in.   **3.1 in.**

8. $d = 4$ m   **12.6 m**

9. $r = 16$ cm   **100.5 cm**

10. $d = 9.1$ m   **28.6 m**

11. $r = 0.1$ yd   **0.6 yd**

12. $d = 65.7$ m   **206.3 m**

13. $r = 1$ cm   **6.3 cm**

Lesson 7-5 *(Pages 285–288)*

Find the reciprocal of each number.

1. $\frac{12}{13}$ $\frac{13}{12}$

2. $\frac{7}{11}$ $\frac{11}{7}$

3. $5\frac{1}{5}$

4. $\frac{1}{4}$ 4

5. $\frac{7}{9}$ $\frac{9}{7}$

6. $\frac{9}{2}$ $\frac{2}{9}$

7. $\frac{1}{5}$ 5

Find each quotient. Write in simplest form.

8. $\frac{2}{3} \div \frac{1}{2}$ $1\frac{1}{3}$

9. $\frac{3}{5} \div \frac{2}{5}$ $1\frac{1}{2}$

10. $\frac{7}{10} \div \frac{3}{8}$ $1\frac{13}{15}$

11. $\frac{5}{9} \div \frac{2}{3}$ $\frac{5}{6}$

12. $4 \div \frac{2}{3}$ 6

13. $8 \div \frac{4}{5}$ 10

14. $9 \div \frac{5}{9}$ $16\frac{1}{5}$

15. $\frac{2}{7} \div 7$ $\frac{2}{49}$

Solve each equation. Write the solution in simplest form.

16. $\frac{1}{14} \div 7 = b$ $\frac{1}{98}$

17. $\frac{2}{13} \div \frac{5}{26} = y$ $\frac{4}{5}$

18. $g = \frac{4}{7} \div \frac{6}{7}$ $\frac{2}{3}$

19. $w = \frac{7}{8} \div \frac{1}{3}$ $2\frac{5}{8}$

20. $15 \div \frac{3}{5} = n$ 25

21. $\frac{9}{14} \div \frac{3}{4} = v$ $\frac{6}{7}$

22. $u = \frac{8}{9} \div \frac{5}{6}$ $1\frac{1}{15}$

23. $j = \frac{4}{9} \div 36$ $\frac{1}{81}$

24. $d = \frac{15}{16} \div \frac{5}{8}$ $1\frac{1}{2}$

Lesson 7-6 *(Pages 289–291)*

Write each mixed number as an improper fraction. Then write its reciprocal.

1. $4\frac{1}{5}$ $\frac{21}{5}; \frac{5}{21}$

2. $3\frac{2}{9}$ $\frac{29}{9}; \frac{9}{29}$

3. $2\frac{3}{8}$ $\frac{19}{8}; \frac{8}{19}$

4. $5\frac{3}{5}$ $\frac{28}{5}; \frac{5}{28}$

Find each quotient. Write in simplest form.

5. $\frac{3}{5} \div 1\frac{2}{3}$ $\frac{9}{25}$

6. $2\frac{1}{2} \div 1\frac{1}{4}$ 2

7. $7 \div 4\frac{9}{10}$ $1\frac{3}{7}$

8. $1\frac{3}{7} \div 10$ $\frac{1}{7}$

9. $3\frac{3}{5} \div \frac{4}{5}$ $4\frac{1}{2}$

10. $8\frac{2}{5} \div 4\frac{1}{2}$ $1\frac{13}{15}$

11. $6\frac{1}{3} \div 2\frac{1}{2}$ $2\frac{8}{15}$

12. $5\frac{1}{4} \div 2\frac{1}{3}$ $2\frac{1}{4}$

Solve each equation. Write the solution in simplest form.

13. $n = 4\frac{1}{8} \div 3\frac{2}{3}$ $1\frac{1}{8}$

14. $p = 2\frac{5}{8} \div \frac{1}{2}$ $5\frac{1}{4}$

15. $1\frac{5}{6} \div 3\frac{2}{3} = z$ $\frac{1}{2}$

16. $k = 21 \div 5\frac{1}{4}$ 4

17. $b = 12 \div 3\frac{3}{5}$ $3\frac{1}{3}$

18. $q = 18 \div 2\frac{1}{4}$ 8

Lesson 7-7 *(Pages 292–294)*

Complete.

1. 3 gal = _?_ pt **24**
2. 24 pt = _?_ gal **3**
3. 20 lb = _?_ oz **320**
4. 2 gal = _?_ fl oz **256**
5. 20 pt = _?_ qt **10**
6. 18 qt = _?_ pt **36**
7. 2,000 lb = _?_ T **1**
8. 3 T = _?_ lb **6,000**
9. 6 lb = _?_ oz **96**
10. 9 lb = _?_ oz **144**
11. 15 qt = _?_ gal $3\frac{3}{4}$
12. 4 pt = _?_ c **8**
13. 4 gal = _?_ qt **16**
14. 4 qt = _?_ fl oz **128**
15. 12 pt = _?_ c **24**
16. 10 pt = _?_ qt **5**
17. 24 fl oz = _?_ c **3**
18. 1.5 pt = _?_ c **3**
19. $\frac{1}{4}$ lb = _?_ oz **4**
20. 5 T = _?_ lb **10,000**
21. 2 lb = _?_ oz **32**

Lesson 7-8A *(Pages 296–297)*

Solve.

1. What are the next three numbers in the pattern? 6, 7, 9, 12, 16, . . . **21, 27, 34**

2. Kip and Celia began working for the same company in 1997. Celia earned $19,000 per year, and Kip earned $16,500 per year. Each year Kip received a $1,500 raise and Celia received a $1,000 raise.
   a. In what year will they earn the same amount of money? **2002**
   b. What will be their annual salary in that year? **$24,000**

3. What is the missing measurement in the pattern?
   . . . $\frac{1}{2}$ in., $\frac{1}{4}$ in., ___ in., $\frac{1}{16}$ in., . . . $\frac{1}{8}$

4. Daria eats half of a pizza at 12:00 P.M. At 2:00 P.M., she eats half of what was left of her pizza from lunch. At 4:00 P.M., she eats half of what was left from the pizza. If Daria continues eating at this rate for six more hours, how much of the pizza has she eaten? $\frac{63}{64}$

5. At Snowfalls Middle School, the bell rings at 8:05, 8:55, 9:00, and 9:50 each morning. If this pattern continues, when would the next three bells ring? **9:55, 10:45, 10:50**

Lesson 7-8 *(Pages 298–300)*

Find the next two numbers in each sequence.

1. 14, 21, 28, 35, . . . **42, 49**
2. 36, 42, 48, 54, . . . **60, 66**
3. 3, 9, 27, 81, . . . **243, 729**
4. 2, 6, 10, 14, . . . **18, 22**
5. 1,600, 800, 400, 200, . . . **100, 50**
6. 192, 96, 48, 24, . . . **12, 6**
7. 15, $14\frac{1}{3}$, $13\frac{2}{3}$, 13, . . . $12\frac{1}{3}, 11\frac{2}{3}$
8. 11, $11\frac{1}{2}$, 12, $12\frac{1}{2}$, . . . $13, 13\frac{1}{2}$
9. $\frac{1}{5}$, 2, 20, 200, . . . **2,000, 20,000**
10. 36, 6, 1, $\frac{1}{6}$, . . . $\frac{1}{36}, \frac{1}{216}$

Find the missing number in each sequence.

11. 5, 10, _?_, 40, . . . **20**
12. _?_, 193, 293, 393, . . . **93**
13. 8, 32, 128, _?_, . . . **512**
14. 11, _?_, 19, 23, . . . **15**
15. 9, $8\frac{3}{4}$, _?_, $8\frac{1}{4}$, . . . $8\frac{1}{2}$
16. 64, _?_, 16, 8, . . . **32**
17. _?_, 19, 26, 33, . . . **12**
18. $\frac{1}{2}$, $1\frac{1}{2}$, $4\frac{1}{2}$, _?_, . . . $13\frac{1}{2}$
19. $\frac{1}{81}$, _?_, 1, 9, . . . $\frac{1}{9}$
20. 12, _?_, 432, 2,592, . . . **72**

Lesson 8-1 *(Pages 312–315)*

Express each ratio as a fraction in simplest form.

1. 21 sugar cookies out of an assortment of 75 cookies. $\frac{7}{25}$

2. 10 girls in a class of 25 students. $\frac{2}{5}$

3. 34 non-smoking tables in a restaurant with 50 tables. $\frac{17}{25}$

4. 7 striped ties out of 21 ties. $\frac{1}{3}$

Express each ratio as a rate.

5. $2.00 for 5 cans of tomato soup **$0.40/can**

6. $200.00 for 40 hours of work **$5/hour**

7. 540 parts produced in 18 hours **30 parts/hour**

Lesson 8-2 *(Pages 317–320)*

Use cross products to determine whether each pair of ratios forms a proportion.

1. $\frac{3}{10}, \frac{7}{25}$ **no**   2. $\frac{5}{12}, \frac{3}{8}$ **no**   3. $\frac{12}{16}, \frac{9}{12}$ **yes**   4. $\frac{5}{4}, \frac{125}{100}$ **yes**   5. $\frac{4}{5}, \frac{80}{100}$ **yes**

Solve each proportion.

6. $\frac{15}{21} = \frac{5}{b}$ **7**   7. $\frac{22}{25} = \frac{n}{100}$ **88**   8. $\frac{24}{48} = \frac{h}{50}$ **25**   9. $\frac{9}{27} = \frac{y}{42}$ **14**   10. $\frac{4}{7} = \frac{16}{x}$ **28**

11. $\frac{4}{6} = \frac{a}{9}$ **6**   12. $\frac{6}{14} = \frac{21}{m}$ **49**   13. $\frac{3}{7} = \frac{21}{d}$ **49**   14. $\frac{4}{10} = \frac{18}{e}$ **45**   15. $\frac{9}{10} = \frac{27}{f}$ **30**

Lesson 8-3A *(Pages 322–323)*   3. Clara, Sam, Toma, Benito, Emma
Solve.

1. Four cars were crossing a bridge. The red car was behind the yellow car. The green car was last. The blue car was not first. In what order did the cars cross the bridge? **yellow, red, blue, green**

2. The streets in Sandra's town are arranged in square blocks. Sandra left school and walked 5 blocks west and 3 blocks south to Marcia's house. She then walked 1 block south and 3 blocks east to the park. Finally she walked 2 blocks east and 2 blocks north to get home. How far is Sandra's home from the school? **2 blocks south**

3. Nida arranged five friends in a line for a photograph. Benito stood between Emma and Toma. Sam stood between Toma and Clara. Clara was on the left. In what order were the five people arranged for the photograph?

4. How many different rectangles can be made with 36 one-inch tiles? **5 rectangles**

5. In how many different ways can Amelia, Bob, Cornelia, and Damien seat themselves in four chairs at a round table? **24 ways**

Lesson 8-3 *(Pages 324–327)*

The map at the right has a scale of $\frac{1}{4}$ in. = 5 km.
Use a ruler to measure each map distance
to the nearest $\frac{1}{4}$ inch. Then find the
actual distances.

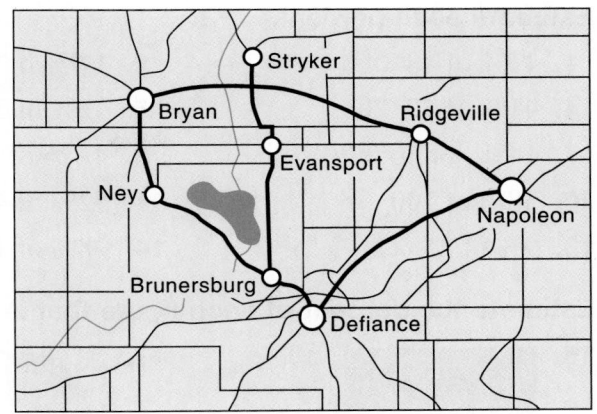

1. Bryan to Napoleon 40 km
2. Stryker to Evansport 10 km
3. Ney to Bryan 10 km
4. Defiance to Napoleon 25 km
5. Ridgeville to Bryan 30 km
6. Brunersburg to Ney 15 km

Lesson 8-4 *(Pages 330–333)*

Express each percent as a fraction in simplest form.

1. 13% $\frac{13}{100}$
2. 25% $\frac{1}{4}$
3. 8% $\frac{2}{25}$
4. 105% $\frac{21}{20}$ or $1\frac{1}{20}$
5. 60% $\frac{3}{5}$

6. 70% $\frac{7}{10}$
7. 80% $\frac{4}{5}$
8. 45% $\frac{9}{20}$
9. 20% $\frac{1}{5}$
10. 14% $\frac{7}{50}$

11. 75% $\frac{3}{4}$
12. 120% $\frac{6}{5}$ or $1\frac{1}{5}$
13. 5% $\frac{1}{20}$
14. 2% $\frac{1}{50}$
15. 450% $\frac{9}{2}$ or $4\frac{1}{2}$

Express each fraction as a percent.

16. $\frac{77}{100}$ 77%
17. $\frac{3}{4}$ 75%
18. $\frac{17}{20}$ 85%
19. $\frac{3}{25}$ 12%
20. $\frac{3}{10}$ 30%

21. $\frac{27}{50}$ 54%
22. $\frac{2}{5}$ 40%
23. $\frac{3}{50}$ 6%
24. $\frac{9}{20}$ 45%
25. $\frac{8}{5}$ 160%

26. $\frac{1}{4}$ 25%
27. $\frac{1}{5}$ 20%
28. $\frac{19}{20}$ 95%
29. $\frac{7}{10}$ 70%
30. $\frac{11}{25}$ 44%

Lesson 8-5 *(Pages 334–336)*

Express each percent as a decimal.

1. 5% 0.05
2. 22% 0.22
3. 50% 0.50
4. 420% 4.2
5. 75% 0.75

6. 1% 0.01
7. 100% 1.00
8. 3.7% 0.037
9. 0.9% 0.009
10. 9% 0.09

11. 90% 0.9
12. 900% 9.00
13. 78% 0.78
14. 62.5% 0.625
15. 15% 0.15

Express each decimal as a percent.

16. 0.02 2%
17. 0.2 20%
18. 0.002 0.2%
19. 1.02 102%
20. 0.66 66%

21. 0.11 11%
22. 0.354 35.4%
23. 0.31 31%
24. 0.09 9%
25. 5.2 520%

26. 2.22 222%
27. 0.008 0.8%
28. 0.275 27.5%
29. 0.3 30%
30. 6.0 600%

Lesson 8-6 *(Pages 337–339)*
Estimate each percent.

1. 11% of 48 **5**
2. 1.9% of 50 **1**
3. 29% of 500 **150**
4. 41% of 50 **20**
5. 32% of 300 **100**
6. 411% of 50 **200**
7. 149% of 60 **90**
8. 4.1% of 50 **2**
9. 62% of 200 **120**
10. 58% of 100 **58**
11. 52% of 400 **200**
12. 68% of 30 **21**
13. 9% of 25 **2.5**
14. 48% of 1000 **500**
15. 98% of 725 **725**

Estimate the percent of each figure that is shaded.

16. 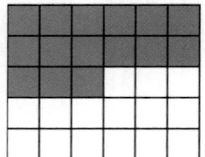 **50%**
17. 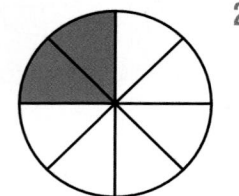 **25%**
18. 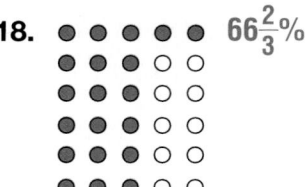 $66\frac{2}{3}\%$

Lesson 8-7 *(Pages 340–343)*
Find the percent of each number.

1. 38% of 150 **57**
2. 20% of 75 **15**
3. 0.2% of 500 **1**
4. 25% of 70 **17.5**
5. 10% of 90 **9**
6. 16% of 30 **4.8**
7. 39% of 40 **15.6**
8. 250% of 100 **250**
9. 6% of 86 **5.16**
10. 12.5% of 160 **20**
11. 9% of 29 **2.61**
12. 3% of 46 **1.38**
13. $66\frac{2}{3}\%$ of 60 **40**
14. 89% of 47 **41.83**
15. 435% of 30 **130.5**
16. 25% of 48 **12**
17. 5% of 420 **21**
18. 55% of 134 **73.7**
19. 28% of 4 **1.12**
20. 14% of 40 **5.6**
21. 14% of 14 **1.96**
22. 90% of 140 **126**
23. 40% of 45 **18**
24. 0.5% of 200 **1**

Lesson 9-1 *(Pages 352–355)*
Use a protractor to find the measure of each angle. Then classify the angle as *acute*, *right*, or *obtuse*.

1. 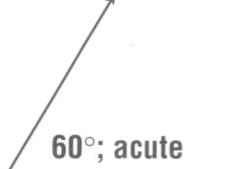 **60°; acute**
2. 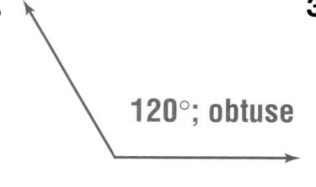 **120°; obtuse**
3. 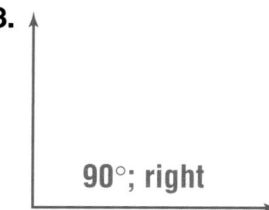 **90°; right**
4. 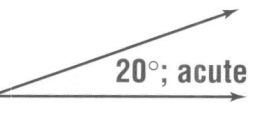 **20°; acute**

Classify each angle measure as *acute*, *right*, or *obtuse*.

5. 86° **acute**
6. 101° **obtuse**
7. 90° **right**
8. 145.6° **obtuse**

Classify each pair of angles as *complementary*, *supplementary*, or *neither*.

9.  **supplementary**
10. 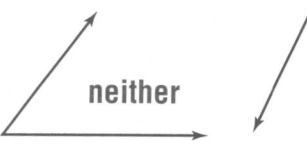 **neither**
11. 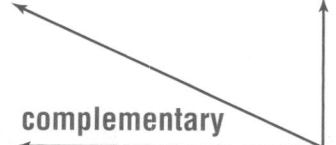 **complementary**

EXTRA PRACTICE

Lesson 9-1B *(Pages 356–357)*

Solve.

1. At Morgon Middle School, 37 of the eighth grade students are involved in a club activity, 63 play sports, and 21 are involved in both sports and clubs.
   a. How many students participate in only clubs? **16**
   b. How many students participate in only sports? **42**
   c. What is the total number of students who participate in extracurricular activities? **79**

2. In a class of 26 students going on a field trip, 8 have backpacks, 13 have lunch bags, and 6 have both backpacks and lunch bags. How many have neither a lunch bag nor a backpack? **11**

3. What are the next two figures in the pattern? ↑ ↖ ← ↙   ↓ ↘

4. Of the 320 students at Lincoln Junior High, 75 are in the orchestra, 120 are in the marching band, and 45 are in both orchestra and marching band. How many students are in neither orchestra nor marching band? **170**

Lesson 9-2 *(Pages 358–361)*   **1–4. See students' work.**

Use a protractor to draw angles having the following measurements.

1. 165°   **2.** 20°   **3.** 90°   **4.** 41°

Estimate the measure of each angle.

5.   **6.**   **7.**   **8.**

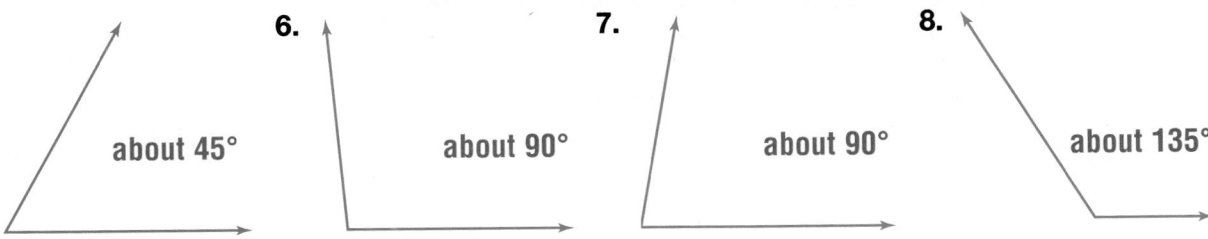

about 45°   about 90°   about 90°   about 135°

Lesson 9-3 *(Pages 364–367)*   **1–8. See students' work.**

Draw the angle or line segment with the given measurement. Then use a straightedge and a compass to bisect each angle or line segment.

1. 3 in.   **2.** 5 cm   **3.** 110°   **4.** 48 mm

5. 70°   **6.** 33 mm   **7.** 25°   **8.** 150°

Lesson 9-4 *(Pages 370–373)*

Name each polygon. Then tell if the polygon is a regular polygon.

1. 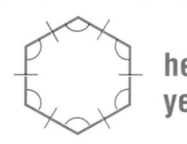 hexagon;
yes

2.  pentagon;
no

3. 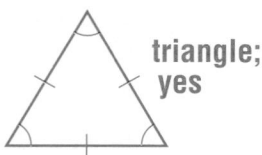 parallelogram; no

4. triangle;
yes

Explain how each pair of figures is alike and how each pair is different. See Answer Appendix.

5.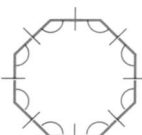

6.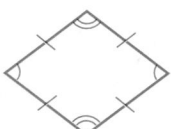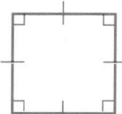

7.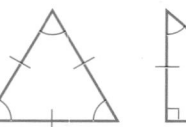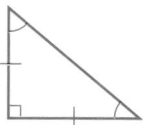

Lesson 9-5 *(Pages 375–378)*

Tell whether the dashed line is a line of symmetry. Write *yes* or *no*.

1. 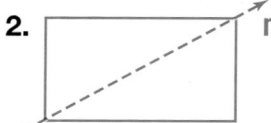 no

2. no

3. 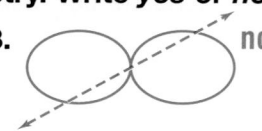 no

4. 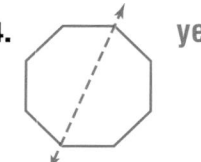 yes

Trace each figure. Draw all lines of symmetry.

5.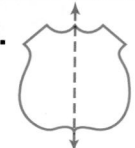

6.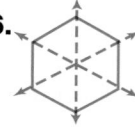

7.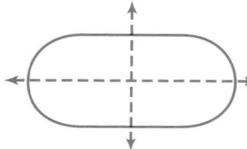

8.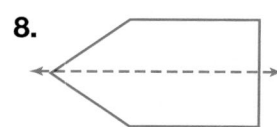

Tell whether the figure shows a reflection. Write *yes* or *no*.

9. 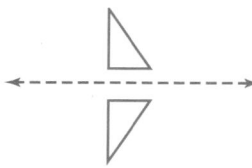 yes

10. 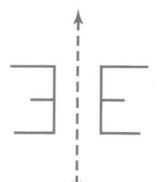 yes

11. 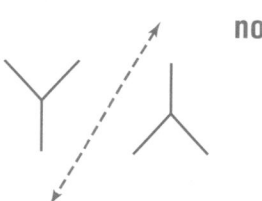 no

Lesson 9-6 *(Pages 379–382)*

Tell whether each pair of polygons is *congruent*, *similar* or *neither*.

1.  similar

2. 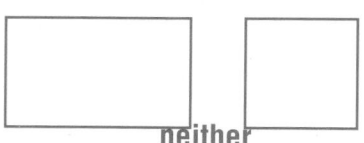 neither

3. 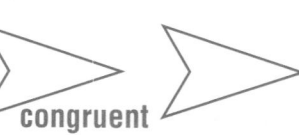 congruent

4.  congruent

5.  similar

6. 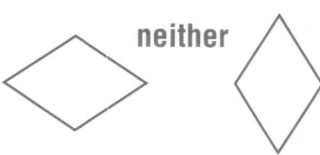 neither

Lesson 10-1 *(Pages 398–401)*   1. 27.75 ft²   2. 1,360 cm²   3. 20 m²   4. 2,920 in²
Find the area of each parallelogram to the nearest tenth.

1.
3.7 ft
7.5 ft

2.
34 cm
40 cm

3.
4 m
5 m

4.
40 in.
73 in.

5.
23 in.
50 in.
1,150 in²

6.
3.5 mm
9 mm
31.5 mm²

7.
5.1 m
2 m
10.2 m²

8.
1.5 ft
11 ft
16.5 ft²

Lesson 10-2 *(Pages 402–405)*
Find the area of each triangle. Round decimal answers to the nearest tenth.

1. base, 6 ft   **9 ft²**
height, 3 ft

2. base, 4.2 in.
height, 6.8 in.
14.3 in²

3. base, 9.1 m
height, 7.2 m
32.8 m²

4. base, 13.2 cm
height, 16.2 cm
106.9 cm²

5.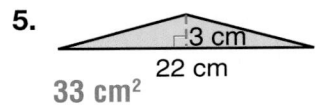
3 cm
22 cm
33 cm²

6.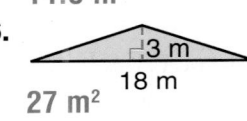
3 m
18 m
27 m²

7.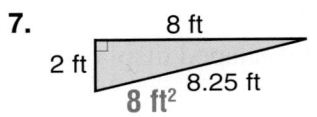
8 ft
2 ft
8 ft²   8.25 ft
8 ft²

8.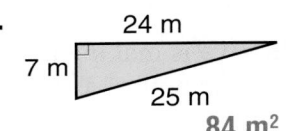
24 m
7 m
25 m
84 m²

9.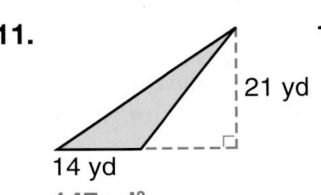
6 yd
5 yd
15 yd²

10.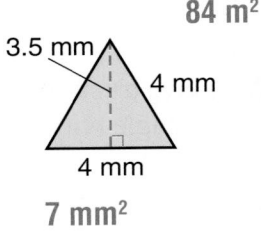
5 in.
4 in.
3 in.
6 in²

11.
21 yd
14 yd
147 yd²

12. 3.5 mm
4 mm
4 mm
7 mm²

Lesson 10-3 *(Pages 406–409)*
Find the area of each circle to the nearest tenth. Use 3.14 for π.

1. radius, 4 m **50.2 m²**

2. diameter, 6 in. **28.3 in²**

3. radius, 16 m **803.8 m²**

4. diameter, 11 in. **95.0 in²**

5. radius, 9 cm **254.3 cm²**

6. diameter, 24 mm **452.2 mm²**

7.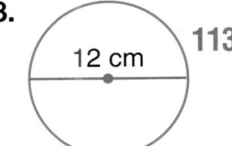
153.9 m²
7 m

8.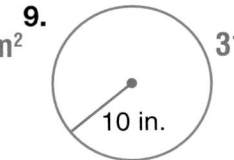
113.0 cm²
12 cm

9.
314 in²
10 in.

10.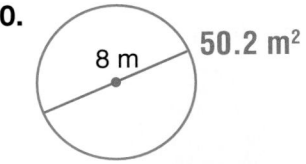
50.2 m²
8 m

11.
7 in. **153.9 in²**

12.
314 in²
20 in.

13.
226.9 mm²
8.5 mm

14.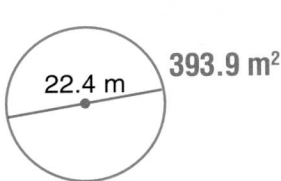
393.9 m²
22.4 m

Lesson 10-4 *(Pages 412–414)*

Name each figure.

1. 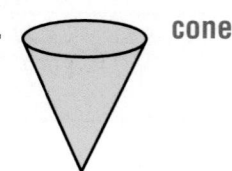 cone

2. 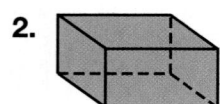 rectangular prism

3. 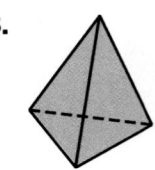 triangular pyramid

4. 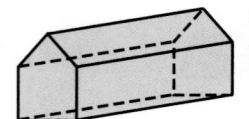 cylinder

5. 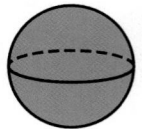 pentagonal prism

6. 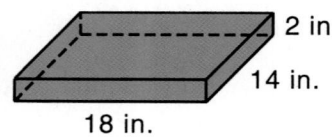 sphere

State the number of faces, edges, and vertices in each figure.

7. square prism 6; 12; 8

8. triangular pyramid 4; 6; 4

9. cone 0; 0; 0

10. pentagonal pyramid 6; 10; 6

Lesson 10-5A *(Pages 416–417)*   **1. Sample answer: 15 in. × 3 in. × 6 in.**
Solve.

1. Ten 3-inch cubes fit in a carton. What are the possible dimensions of the carton?

2. Eve is making a pyramid-shaped display of boxes of soda. Each box is a cube. The bottom layer of the pyramid has 8 boxes. If there is one less box in each layer and there are 6 layers in the pyramid, how many boxes will Eve need to make the display? **33 boxes**

3. A construction worker wants to arrange 20 cement bricks into the shape of a rectangle with the smallest perimeter possible. How many bricks will be in each row? **4 five-rows or 5 four-rows**

4. A rectangular prism can be formed by using exactly 10 cubes. Find the length, width, and height of the prism. **Sample answer: 5 cubes × 2 cubes × 1 cube**

Lesson 10-5 *(Pages 418–420)*
Find the volume of each rectangular prism to the nearest tenth.

1. 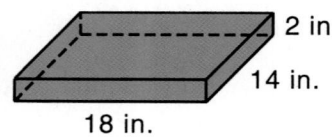 2 in. / 14 in. / 18 in.
504 in³

2. 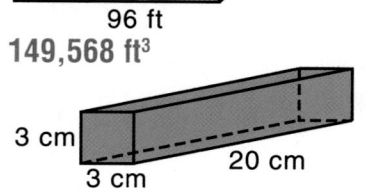 41 ft / 38 ft / 96 ft
149,568 ft³

3. 3 m / 6 m / 5 m
90 m³

4. 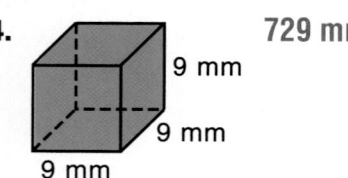 9 mm / 9 mm / 9 mm
729 mm³

5. 3 cm / 3 cm / 20 cm
180 cm³

6. 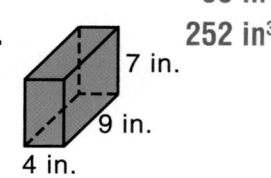 7 in. / 9 in. / 4 in.
252 in³

7. length = 8 in.
width = 5 in.
height = 2 in. **80 in³**

8. length = 10 cm
width = 2 cm
height = 8 cm **160 cm³**

9. length = 20 ft
width = 5 ft
height = 6 ft **600 ft³**

EXTRA PRACTICE

Lesson 10-6 *(Pages 421–424)*

Find the surface area of each rectangular prism to the nearest tenth.

1.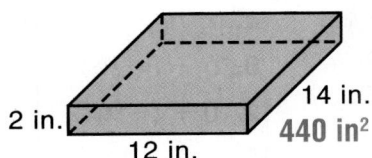
2 in. 12 in. 14 in. **440 in²**

2.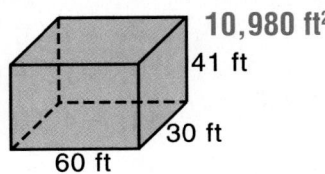
10,980 ft² 41 ft 30 ft 60 ft

3.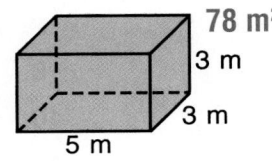
78 m² 3 m 3 m 5 m

4.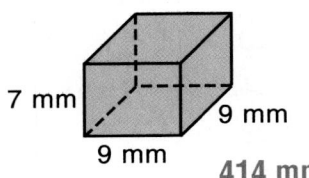
7 mm 9 mm 9 mm **414 mm²**

5.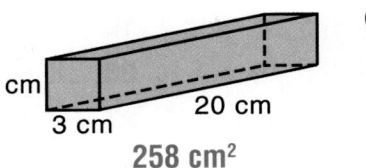
3 cm 3 cm 20 cm **258 cm²**

6.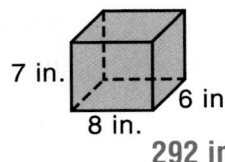
7 in. 6 in. 8 in. **292 in²**

7. length = 4 mm
width = 12 mm
height = 1.5 mm **144 mm²**

8. length = 16 cm
width = 20 cm
height = 20.4 cm **2,108.8 cm²**

9. length = 8.5 m
width = 2.1 m
height = 7.6 m **196.8 m²**

Lesson 11-1 *(Pages 434–436)* 1–7. See Answer Appendix.

Draw a number line from −10 to 10. Graph each integer on the number line.

1. −3   2. 3   3. −1   4. −8   5. 9   6. 10   7. 6

Write an integer to describe each situation.

8. a loss of 15 dollars **−15**

9. 9 degrees below zero **−9**

Write the opposite of each integer.

10. 7 **−7**   11. −3 **3**   12. 11 **−11**   13. −9 **9**   14. −13 **13**   15. 101 **−101**   16. 0 **0**

Lesson 11-2 *(Pages 437–439)*

Replace each ● with <, >, or = to make a true sentence.

1. −5 ● −55 **>**   2. 4 ● −66 **>**   3. −777 ● −77 **<**   4. −75 ● −75 **=**

5. −898 ● −99 **<**   6. 0 ● 44 **<**   7. 56 ● −1 **>**   8. −82 ● −9 **<**

9. −6 ● −7 **>**   10. 90 ● 101 **<**   11. 4 ● −2,000 **>**   12. −3 ● 0 **<**

Order each set of integers from least to greatest.

13. 8, 0, −808, −8, −88, 88, −888 **−888, −808, −88, −8, 0 , 8, 88**

14. 0, 3, −21, 9, −89, 8, −65, −56, **−89, −65, −56, −21, 0, 3, 8, 9**

15. 70, −9, 67, −78, 0, 45, −36, −19 **−78, −36, −19, −9, 0, 45, 67, 70**

16. 0, −90, −56, −29, −92, −87, −35 **−92, −90, −87, −56, −35, −29, 0**

17. −239, −999, 458, −29, −77, 200, −818 **−999, −818, −239, −77, −29, 200, 458**

Lesson 11-3 *(Pages 441–444)*
Find each sum. Use counters or a number line if necessary.

1. $-4 + (-7)$ -11
2. $-1 + 0$ -1
3. $7 + (-13)$ -6
4. $-20 + 2$ -18
5. $4 + (-6)$ -2
6. $-12 + 9$ -3
7. $-12 + (-10)$ -22
8. $5 + (-15)$ -10
9. $17 + 9$ 26
10. $18 + (-18)$ 0
11. $-4 + (-4)$ -8
12. $0 + (-9)$ -9
13. $-12 + (-9)$ -21
14. $-8 + 7$ -1
15. $3 + (-6)$ -3
16. $-9 + 16$ 7
17. $-5 + (-3)$ -8
18. $-5 + 5$ 0
19. $-3 + (-3)$ -6
20. $-11 + 6$ -5
21. $-10 + 6$ -4
22. $-5 + (-9)$ -14
23. $18 + (-20)$ -2
24. $-4 + (-4)$ -8
25. $2 + (-4)$ -2
26. $-3 + (-11)$ -14
27. $-17 + 9$ -8
28. $-11 + 6$ -5
29. $-6 + (-12)$ -18
30. $8 + 8$ 16
31. $-9 + 0$ -9
32. $4 + (-5)$ -1

Lesson 11-4 *(Pages 445–448)*
Find each difference. Use counters or a number line if necessary.

1. $7 - (-4)$ 11
2. $-4 - (-9)$ 5
3. $13 - (-3)$ 16
4. $2 - (-5)$ 7
5. $-9 - 5$ -14
6. $-11 - (-18)$ 7
7. $-4 - (-7)$ 3
8. $-6 - (-6)$ 0
9. $-6 - 6$ -12
10. $17 - 9$ 8
11. $-12 - (-9)$ -3
12. $0 - (-4)$ 4
13. $-7 - 0$ -7
14. $-12 - (-10)$ -2
15. $-2 - (-1)$ -1
16. $3 - (-5)$ 8
17. $5 - (-1)$ 6
18. $-5 - (-6)$ 1
19. $9 - (-1)$ 10
20. $1 - 9$ -8
21. $-5 - 1$ -6
22. $-1 - 4$ -5
23. $0 - (-7)$ 7
24. $8 - 13$ -5
25. $-4 - (-6)$ 2
26. $9 - 9$ 0
27. $-7 - (-7)$ 0
28. $7 - 5$ 2
29. $8 - (-5)$ 13
30. $5 - 8$ -3
31. $1 - 6$ -5
32. $-8 - (-8)$ 0

Lesson 11-5 *(Pages 449–452)*
Find each product. Use counters or a number line if necessary.

1. $3 \times (-5)$ -15
2. -5×1 -5
3. $-8 \times (-4)$ 32
4. $6 \times (-3)$ -18
5. -3×2 -6
6. $-1 \times (-4)$ 4
7. $8 \times (-2)$ -16
8. $-5 \times (-7)$ 35
9. $3 \times (-9)$ -27
10. -9×4 -36
11. $-4 \times (-5)$ 20
12. $5 \times (-2)$ -10
13. $-8(3)$ -24
14. $-9(-1)$ 9
15. $7(-3)$ -21
16. $2(3)$ 6
17. $-6(0)$ 0
18. $-5(-1)$ 5
19. $5(-5)$ -25
20. $-2(-3)$ 6
21. $8(-4)$ -32
22. $-2(4)$ -8
23. $-4(-4)$ 16
24. $2(9)$ 18
25. $-2(-12)$ 24
26. $7 \times (-4)$ -28
27. $-5 \times (-9)$ 45
28. -2×11 -22
29. $4(-2)$ -8
30. $4(-4)$ -16
31. $-3(-11)$ 33
32. $-3(3)$ -9

Lesson 11-6A *(Pages 454–455)*
Solve.

1. Morgan and Stephen want to get ready for the fall football season that starts August 29. Before the season starts, they plan to spend 3 to 4 weeks practicing plays. Before practicing plays, they plan to spend 6 to 8 weeks running and exercising. What are the earliest and latest dates that Morgan and Stephen can begin getting ready for the football season? **June 6, June 27**

2. Daniella and Wayne must arrive at school no later than 7:25 A.M. weekday mornings. It takes 20 minutes to get from Wayne's house to the school. Daniella picks up Wayne, but it takes her 15 minutes to get to Wayne's house. If it takes Daniella 55 minutes to get ready in the morning, what is the latest she should get out of bed? **5:55 A.M.**

3. Melina bought some school supplies. She spent $4.35 on paper, pens, and pencils and 3 times that on notebooks. After paying for the items, she had $7.26 left. How much money did she have before she went shopping? **$24.66**

4. In the trivia bowl, each finalist must answer four questions correctly. Each question is worth twice as much as the question before it. The fourth question is worth $6,000. How much is the first question worth? **$750**

5. Connor, Petra, Daria, and Lyle each handed in term papers of the Industrial Revolution. Petra's paper was 4 pages shorter than Connor's. Daria's paper was 7 pages longer than Petra's. Lyle's paper was half as long as Daria's. Lyle's paper was 14 pages long. How long was Connor's paper? **25 pages long**

Lesson 11-6 *(Pages 456–458)*
Find each quotient. Use counters or patterns if necessary.

1. $12 \div (-6)$ -2	**2.** $-7 \div (-1)$ 7	**3.** $-4 \div 4$ -1	**4.** $6 \div (-6)$ -1
5. $0 \div (-4)$ 0	**6.** $45 \div (-9)$ -5	**7.** $15 \div (-5)$ -3	**8.** $-6 \div 2$ -3
9. $-28 \div (-7)$ 4	**10.** $20 \div (-2)$ -10	**11.** $-40 \div (-8)$ 5	**12.** $12 \div (-4)$ -3
13. $-18 \div 6$ -3	**14.** $9 \div (-1)$ -9	**15.** $-30 \div 6$ -5	**16.** $-54 \div (-9)$ 6
17. $28 \div (-7)$ -4	**18.** $-24 \div 8$ -3	**19.** $24 \div (-4)$ -6	**20.** $-14 \div 7$ -2
21. $9 \div 3$ 3	**22.** $-18 \div (-6)$ 3	**23.** $-9 \div (-1)$ 9	**24.** $18 \div (-9)$ -2
25. $-25 \div (-5)$ 5	**26.** $15 \div (-3)$ -5	**27.** $-36 \div 9$ -4	**28.** $-4 \div 2$ -2
29. $-40 \div 8$ -5	**30.** $-32 \div 4$ -8	**31.** $-27 \div (-9)$ 3	**32.** $-8 \div 8$ -1

Lesson 11-7 *(Pages 459–461)*
Name the ordered pair for each point.

1. M $(-3, 1)$ 2. A $(-4, 4)$ 3. D $(2, -4)$ 4. E $(3, 2)$
5. P $(-4, -3)$ 6. Q $(1, -1)$ 7. B $(-2, 3)$ 8. C $(3, -2)$
9. F $(-2, -4)$ 10. G $(-2, -2)$ 11. N $(-4, -2)$ 12. R $(0, 0)$
13. K $(1, 4)$ 14. H $(1, 3)$

Graph and label each point. 15–22. See Answer Appendix.

15. $S(4, -1)$ 16. $T(-3, -2)$ 17. $W(2, 1)$ 18. $Y(-5, 3)$
19. $Z(-1, -3)$ 20. $U(3, -3)$ 21. $V(1, 2)$ 22. $X(-1, 4)$

Lesson 11-8 *(Pages 464–467)*

The vertices of $\triangle QRS$ are $Q(1, -2)$, $R(3, 4)$, and $S(-2, 2)$. On a coordinate grid, draw $\triangle QRS$ and each transformation image. 1–5. See Answer Appendix.

1. translation image that is 5 units down

2. translation image that is 4 units left

3. translation image that is 2 units right and 3 units up

4. translation image that is 1 unit left and 8 units down

5. reflection image over the *x*-axis

Lesson 12-1 *(Pages 476–479)*

Solve each equation. Use cups and counters if necessary.

1. $x + 4 = 14$ **10**
2. $b + (-10) = 0$ **10**
3. $-2 + w = -5$ **−3**

4. $k + (-3) = -5$ **−2**
5. $-4 + h = 6$ **10**
6. $-7 + d = -3$ **4**

7. $m + 11 = 9$ **−2**
8. $f + (-9) = -19$ **−10**
9. $p + 66 = 22$ **−44**

10. $-34 + t = 41$ **75**
11. $e + 56 = -24$ **−80**
12. $-29 + a = -54$ **−25**

13. $17 + m = -33$ **−50**
14. $b + (-44) = -34$ **10**
15. $w + (-39) = 55$ **94**

Lesson 12-2 *(Pages 480–483)*

Solve each equation. Use cups and counters if necessary.

1. $y - (-7) = 2$ **−5**
2. $a - 10 = -22$ **−12**
3. $g - (-1) = 9$ **8**

4. $c - 8 = 5$ **13**
5. $z - (-2) = 7$ **5**
6. $n - 1 = -87$ **−86**

7. $j - 15 = -22$ **−7**
8. $x - 12 = 45$ **57**
9. $y - 65 = -79$ **−14**

10. $q - 16 = -31$ **−15**
11. $q - (-6) = 12$ **6**
12. $j - 18 = -34$ **−16**

13. $k - (-2) = -8$ **−10**
14. $r - 76 = 41$ **117**
15. $n - 63 = -81$ **−18**

588 Extra Practice

Lesson 12-3 *(Pages 484–487)*
Solve each equation. Use cups and counters if necessary.

1. $5x = 30$ 6
2. $18w = 2$ $\frac{1}{9}$
3. $\frac{1}{2}a = 7$ 14
4. $2d = -28$ -14
5. $\frac{1}{4}c = -3$ -12
6. $11n = 77$ 7
7. $\frac{1}{3}z = 15$ 45
8. $9y = -63$ -7
9. $6m = -54$ -9
10. $5f = -75$ -15
11. $20p = 5$ $\frac{1}{4}$
12. $\frac{1}{4}x = 16$ 64
13. $4t = -24$ -6
14. $7b = 21$ 3
15. $19h = 0$ 0
16. $22d = -66$ -3
17. $\frac{1}{3}y = 11$ 33
18. $3m = -78$ -26
19. $8x = -2$ $-\frac{1}{4}$
20. $9c = -72$ -8
21. $\frac{1}{2}p = 35$ 70
22. $\frac{1}{5}k = 20$ 100
23. $33y = 99$ 3
24. $6z = -5$ $-\frac{5}{6}$

Lesson 12-4 *(Pages 488–491)*
Solve each equation.

1. $3x + 7 = 13$ 2
2. $\frac{1}{2}r + 6 = -3$ -18
3. $2h - 5 = 7$ 6
4. $-10 = 5x + 5$ -3
5. $2x - (-16) = 26$ 5
6. $6r + 2 = 2$ 0
7. $\frac{1}{5}t - 6 = -12$ -30
8. $-2 - 3y = -11$ 3
9. $-4y + 16 = 64$ -12

Lesson 12-4B *(Pages 492–493)*
Solve.

1. Eric bought a personal CD player for $12 less than the regular price. If he paid $54, what was the regular price? $66

2. Lisa paid $2.58 in sales tax on an Ohio State Buckeyes sweatshirt. The total cost was $45.57. What was the price of the sweatshirt before taxes? $42.99

3. School T-shirts are the most popular item sold in the school store. On Monday, 49 T-shirts were sold. This is 3 more than twice the amount that were sold on Tuesday. How many T-shirts were sold on Tuesday? 23

4. Jayson made a long-distance phone call to his grandmother. The first 5 minutes cost $3, and each minute after that cost $0.50. How many minutes did they talk on the phone if the cost of the call was $12? 23 minutes

5. Paige went bowling at Champion Bowling Center. Shoe rental was $1.75 and games were $2.25 each. If she spent $8.50 on shoe rental and games, how many games did she play? 3 games

Lesson 12-5 *(Pages 496–499)*

Copy and complete each function table.

1.

input (n)	output (n − 4)
5	1
2	−2
−1	−5

2.

input (n)	output (3n)
1	3
0	0
−2	−6

Find the rule for each function table.

3. $n + 5$

n	▪
−1	4
0	5
3	8

4. $\frac{n}{2}$

n	▪
−6	−3
0	0
8	4

Lesson 12-6 *(Pages 500–503)* 1–6. See Answer Appendix for graphs.

Copy and complete each function table. Then graph the function.

1.

input (x)	output (x + 1)
2	▪ 3
0	▪ 1
−3	▪ −2

2.

input (x)	output (2x)
2	▪ 4
0	▪ 0
−3	▪ −6

3.

input (x)	output (x − 3)
4	▪ 1
0	▪ −3
−1	▪ −4

4.

input (x)	output $\left(\frac{x}{5}\right)$
10	▪ 2
0	▪ 0
−5	▪ −1

5.

input (x)	output (−3x)
2	▪ −6
−1	▪ 3
−2	▪ 6

6.

input (x)	output (2x − 3)
2	▪ 1
0	▪ −3
−1	▪ −5

Lesson 13-1 *(Pages 515–518)*

A set of 30 tickets are placed in a bag. There are 6 baseball tickets, 4 hockey tickets, 4 basketball tickets, 2 football tickets, 3 symphony tickets, 2 opera tickets, 4 ballet tickets, and 5 theater tickets. One ticket is drawn without looking. Find the probability of each event. Then tell how likely the event is to happen. 1–18. See Answer Appendix.

1. $P(\text{basketball})$ $\frac{2}{15}$

2. $P(\text{sports event})$ $\frac{8}{15}$

3. $P(\text{opera or ballet})$ $\frac{1}{5}$

4. $P(\text{soccer})$ **0**

5. $P(\text{not symphony})$ $\frac{9}{10}$

6. $P(\text{theater})$ $\frac{1}{6}$

7. $P(\text{basketball or hockey})$ $\frac{4}{15}$

8. $P(\text{not a sports event})$ $\frac{7}{15}$

9. $P(\text{not opera})$ $\frac{14}{15}$

10. $P(\text{baseball})$ $\frac{1}{5}$

11. $P(\text{football})$ $\frac{1}{15}$

12. $P(\text{not soccer})$ **1**

13. $P(\text{opera})$ $\frac{1}{15}$

14. $P(\text{not theater})$ $\frac{5}{6}$

15. $P(\text{symphony})$ $\frac{1}{10}$

16. $P(\text{soccer or football})$ $\frac{1}{15}$

17. $P(\text{opera or theater})$ $\frac{7}{30}$

18. $P(\text{hockey})$ $\frac{2}{15}$

EXTRA PRACTICE

Lesson 13-2A *(Pages 520–521)*
Solve.

1. Sean made this frequency table to show the time it took his classmates to run 100 meters.

100 Meter Times (in seconds)		
Time	Tally	Frequency
13.0 – 13.9	III	3
14.0 – 14.9	IIII I	6
15.0 – 15.9	IIII IIII I	11
16.0 – 16.9	IIII IIII III	13
17.0 – 17.9	IIII	4
18.0 – 18.9	II	2

   a. How many students ran 100 meters in less than 16 seconds? **20**
   b. How many students took 17 seconds or longer to run 100 meters? **6**
   c. How many students did Sean survey? **39**
   d. In what interval did the greatest number of times occur? **16.0-16.9**
   e. How many students ran 100 meters in 15.0 - 16.9 seconds? **24**

2. The list shows the birth month of the students in Mrs. Barr's geometry class. a. **See Answer Appendix.**
   a. Make a frequency table of the students birth month.
   b. How many of the students were born in September? **3**
   c. Which months were not represented?
   January, April, May, November

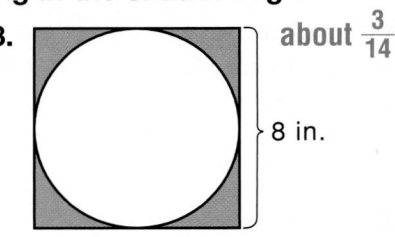

September
June
October
October
February
March
September
October
February
July
October
March
June
August
July
October
July
July
February
December
March
September

Lesson 13-2 *(Pages 522–525)*
Tell whether each of the following is a random sample. Explain your answer. 1–10. **See Answer Appendix.**

	Type of Survey	Survey Location
1.	favorite television show	department store
2.	favorite baseball team	Texas Rangers Ballpark
3.	favorite movie	shopping center
4.	favorite cookie	park
5.	favorite sport	soccer game
6.	favorite ice cream flavor	school
7.	favorite entertainer	Garth Brooks' concert
8.	favorite fruit	apple orchard
9.	favorite food	county fair
10.	favorite vacation	Disney World

Lesson 13-3 *(Pages 526–529)*
Each figure represents a dartboard. It is equally likely that a dart will land anywhere on the dartboard. Find the probability of a randomly-thrown dart landing in the shaded region.

1. 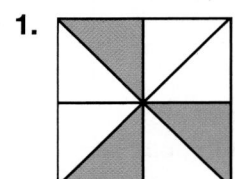 $\frac{3}{8}$

2. $\frac{5}{9}$

   7.5 m
   6 m
   5 m
   4 m
   3 m
   4.5 m

3. about $\frac{3}{14}$

   8 in.

Lesson 13-4 *(Pages 531–534)*

For each situation, make a list and draw a tree diagram to show the sample space.

1. tossing a quarter and rolling a number cube

2. spinning each spinner once

   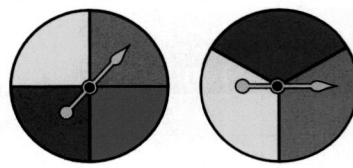

   1–4. See Answer Appendix.

3. a choice of a red, blue, or green sweater with a white, black, or tan skirt.

4. a choice of a chicken, ham, turkey, or bologna sandwich with coffee, milk, juice, or soda

5. How many ways can a person choose to watch two television shows from four television shows? **6 ways**

EXTRA PRACTICE

Lesson 13-5 *(Pages 536–539)*

The spinner shown is spun and a card is chosen from the set of cards shown. Find the probability of each event.

1. P(15 and C) $\frac{1}{36}$

2. P(even and H) $\frac{1}{12}$

3. P(odd and a vowel) $\frac{1}{6}$

4. P(multiple of 5 and O) $\frac{1}{3}$

5. P(multiple of 10 and consonant) $\frac{1}{3}$

6. P(composite and L) $\frac{5}{36}$

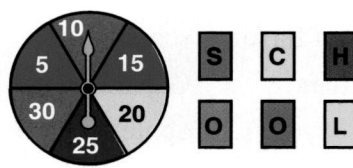

A bag contains 3 quarters and 5 dimes. Another bag contains 12 pennies and 8 nickels. One coin is randomly-chosen from each bag. Find the probability of each event.

7. P(dime and nickel) $\frac{1}{4}$

8. P(quarter) $\frac{3}{8}$

9. P(quarter and penny) $\frac{9}{40}$

10. P(not a nickel) $\frac{3}{5}$

11. P(dime and penny) $\frac{3}{8}$

12. P(not a dime and nickel) $\frac{3}{20}$

Mixed Problem Solving

Solve using any strategy.

1. *Money Matters* The Martin family bought tickets to the Science Museum. Admission is $8 for adults and $5 for children under 12. They spent $49 for admission. How many adult and children tickets did the Martin family purchase? **3 adults, 5 children**

2. How many 3-digit numbers can you make using the even digits from 1 through 9, without repeating the same digit twice in any number? **24**

3. *Life Science* Your heart at rest beats approximately 72 times per minute. At this rate, about how many times per day does your heart beat? **a.**
   a. 103,680
   b. 10,368
   c. 1,368
   d. 1,036,800

4. *Sports* Mrs. Zimmer sent her grandson an autographed souvenir baseball from the Baseball Hall of Fame in Cooperstown, N.Y. The baseball arrived in a cube-shaped box. Each side of the box measured 4 inches. Make a pattern for this box.

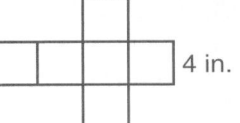
4 in.

5. *Money Matters* Marcus belongs to a teen book club. He purchases a new paperback at $3.95 every other week for a full year. At this rate, how much will he pay over a 3-year period? **$308.10**

6. The bus leaves downtown for the mall at 7:35 A.M., 8:10 A.M., 8:45 A.M., and 9:20 A.M. If the bus continues to run on this schedule, what time does the bus leave between 10:00 A.M. and 11:00 A.M.? **10:30 A.M.**

7. *Civics* The American flag has 50 white stars on a blue background. The stars are arranged in alternating rows of 5 and 6. How many rows are there on the American flag? **9**

8. Cadet Girl Scout Troop 548 held their annual cookie sale. Bethany sold half as many boxes of cookies as Olivia. Julie sold 15 more boxes than Bethany. Kara sold 9 fewer boxes than Alaina. Olivia sold three times as many boxes as Kara. How many boxes of cookies did Troop 548 sell if Julie sold 39 boxes? **152 boxes**

Mixed Problem Solving

EXTRA PRACTICE

Solve using any strategy.

1. **Sports** The Chicago Cubs play 81 games each year. If 2,225,000 people attend the games, what is a reasonable estimate of the number of people that attend each game, 28,000 or 280,000? **28,000**

2. **Money Matters** The Changs had their washing machine repaired. They were charged $45 for the first 15 minutes and $12 for every 15 minutes after that. The Chang's bill, excluding parts and tax, was $93. How many minutes did the repair work take? **75 minutes or 1 hour 15 minutes**

3. **School** Of the 100 6th grade students at the John Glenn Middle School, 30 are taking French, 40 are taking Spanish, and 15 are taking both French and Spanish. How many students are taking either French or Spanish? **55 students**

4. **School** Jared has to read a mystery novel for English class. If the novel contains 168 pages, how many pages must he read per day to finish the book in 7 days? **24 pages**

5. **Geometry** The Parks Commission is preparing a triangular section of Evening Shade Park in order to plant flowers. The base of the triangular region measures 18.4 feet, and the height is 12.5 feet. What is the area of the triangular flower garden? **115 square feet**

6. The graph shows the types of footwear men say they wear in their home.

   a. Which two types of footwear are worn for about the same frequency? **sneakers and barefoot**

   b. Which two types of footwear make up about 60% of the type of footwear worn? **socks and slippers**

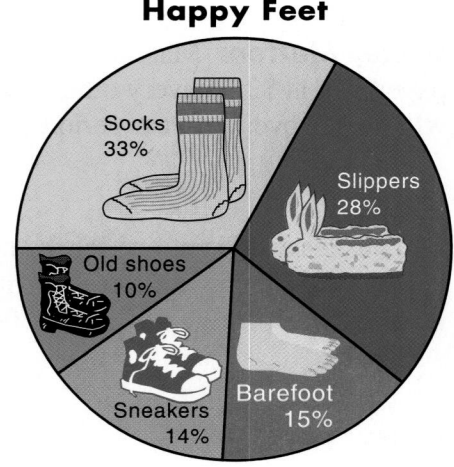

Happy Feet

Socks 33%

Slippers 28%

Old shoes 10%

Sneakers 14%

Barefoot 15%

Source: L.B. Evans

7. **Sports** Vicki wants to try out for the basketball team on Saturday. Tryouts begin at 9:30 A.M. sharp. She wants to arrive 20 minutes early to practice her foul shots. It takes her 35 minutes to eat breakfast and get ready. She plans to walk 10 minutes to her friend Madison's house. Vicki and Madison will then walk the remaining 15 minutes to the tryouts. What time should Vicki plan to get up on Saturday morning? **8:10 A.M.**

CHAPTER 1 **Test**

Use the four-step plan to solve.

1. *Geography* Amad and Marisela went hiking and canoeing down the Danube River, Europe's second largest river. They traveled from its source in the Black Forest 1,760 miles to its end at the Black Sea. If they averaged 11 miles each day, how long was the trip? **160 days**

Find the next three numbers or draw the next three shapes for each pattern.

2. 192, 96, 48, 24, <u>?</u>, <u>?</u>, <u>?</u> **12, 6, 3**

3.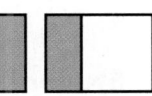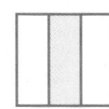

Round each number to the underlined place-value position.

4. 7<u>5</u>4 **750**

5. <u>4</u>,839 **5,000**

Estimate. State whether the answer shown is reasonable.

6. $198 \div 5 = 50$
   $200 \div 5 = 40$; **not reasonable**

7. $1,492 + 1,941 = 3,300$
   $1,000 + 2,000 = 3,000$; **reasonable**

8. *Earth Science* Each year, 16,000,000 thunderstorms occur throughout the world. About how many thunderstorms occur each week? **about 300,000**

Find the value of each expression.

9. $48 - 24 \div 3 \times 6$ **0**

10. $63 \div 7 - 2 \times 3$ **3**

Evaluate each expression if $m = 2$, $n = 7$, and $p = 21$.

11. $p - m$ **19**

12. $m + p \div n$ **5**

13. Write $5 \cdot 5 \cdot 5 \cdot 5$ using exponents. **5^4**

14. Write d^6 as a product. **$d \cdot d \cdot d \cdot d \cdot d \cdot d$**

Evaluate each expression.

15. 10^4 **10,000**

16. 3 cubed **27**

Identify the solution to each equation from the given list.

17. $32 \div x = 2$; 16, 30, 64 **16**

18. $y + 7 = 23$; 5, 16, 7 **16**

Solve each equation mentally.

19. $5z = 55$ **11**

20. $15 - w = 9$ **6**

Test

1. The test scores of the students in Mrs. Wimberly's history class are shown below. Make a frequency table of the data. **See Answer Appendix.**

72	95	87	77	88	79	92	97	100
82	75	93	92	75	77	71	98	99
85	84	77	80	91	91	85	83	89

Geography The lengths, in miles, of the Great Lakes are Lake Superior, 350; Lake Michigan, 307; Lake Huron, 206; Lake Erie, 241; and Lake Ontario, 193.

2. What scale would you use for the vertical axis of a vertical bar graph? **0 to 350**

3. Make a bar graph for this data. **See Answer Appendix.**

4. Find the median length of the lakes. **241 miles**

Students' Favorite Colors

pink 0.17, red 0.33, yellow 0.17, blue 0.25, green 0.08

Refer to the circle graph for Exercises 5–6.

5. Which color is the most popular? **red**

6. Which three colors together are favored by about half the students? **yellow, green, and blue**

Jobs David mows lawns, rakes leaves, shovels snow, and does other outdoor jobs. Last year, his earnings were: January–$55, February–$70, March–$35, April–$23, May–$38, June–$55, July–$74, August–$78, September–$69, October–$55, November–$38, December–$58.

7. Find David's mean monthly earnings. **$54**

8. Find the mode. **$55**

9. Find the median. **$55**

10. David wants to buy a new bike that costs $179. If his earnings this year are the same as last year, when will he have enough to buy the bike? **April**

11. Make a line graph for this data. **See Answer Appendix.**

12. Change your graph to make it misleading. **See Answer Appendix.**

Make a stem-and-leaf plot for each set of data.

13. 86, 85, 92, 73, 75, 96, 84, 92, 74, 87 **See Answer Appendix.**

14. 46¢, 59¢, 42¢, 69¢, 55¢, 48¢, 66¢, 43¢, 85¢ **See Answer Appendix.**

Use the grid to name the point for each ordered pair.

15. (3, 1) *H*   16. (4, 0) *E*   17. (2, 5) *C*

Use the grid to name the ordered pair for each point.

18. *A* **(0, 4)**   19. *B* **(1, 2)**   20. *D* **(3, 4)**

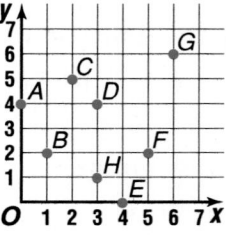

Write each fraction or mixed number as a decimal.

1. $\dfrac{341}{10,000}$ **0.0341**

2. $7\dfrac{13}{1,000}$ **7.013**

3. $\dfrac{9}{100}$ **0.09**

4. Write two hundred and six thousandths as a decimal. **200.006**

Use a centimeter ruler to measure each line segment.

5. _____ **6.8 cm or 68 mm** _____

6. _____ **4.4 cm or 44 mm** _____

Use a centimeter ruler to measure one side of each figure.

7. 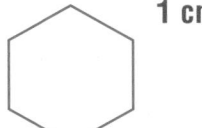 **1 cm**

8.  **2.5 cm**

9. *School* Jennifer's semester math average is 76.7. Her twin sister, Julie, is close with an average of 76.35. Whose average is higher? **Jennifer's**

10. Order 17.4, 1.747, 1.8, 17.36, and 17.09 from least to greatest. **1.747, 1.8, 17.09, 17.36, 17.4**

11. Order 0.89, 0.98, 0.889, 0.982, and 0.88 from greatest to least. **0.982, 0.98, 0.89, 0.889, 0.88**

12. *Money Matters* Sonia and four friends ate at a restaurant. Dividing the bill, each owed $7.146. What should each have paid? **$7.15**

Round each number to the underlined place-value position.

13. 901.2̲63 **901.3**
14. 0.99̲9 **1.00**
15. 2.4̲45 **2.4**
16. 7̲8.1 **78**

Estimate using any strategy. 17. $7 + 7 + 7 + 7 = 28$ 19. $5 + 6 + 2 = 13$

17. $7.14 + 7 + 6.7 + 6.9$
18. $45.9 - 6.12$ **$46 - 6 = 40$**
19. $5.34 + 6.33 + 1.9$

20. *Earth Science* The table shows the five most common elements in Earth's crust along with the percent of the crust each element represents. About how much more is the percent that is silicon than the percent that is aluminum? **about 20%**

Element	Percent
Oxygen	46.60
Silicon	27.72
Aluminum	8.13
Iron	5.00
Calcium	3.63

21. *Money Matters* At breakfast, Phil ordered juice for $0.89, scrambled eggs and toast for $3.69, and a glass of milk for $0.59. Did he spend about $6 or $4? **about $6**

Add or subtract.

22.
$$\begin{array}{r} 9.04 \\ +12.8 \\ \hline \end{array}$$ **21.84**

23. $54.29 - 3.8$ **50.49**

24. $54 + 1.8$ **55.8**

25.
$$\begin{array}{r} 71.34 \\ -43.78 \\ \hline \end{array}$$ **27.56**

Multiply.

1.   0.81 **17.82**
     $\times$ 22

2. 22.38 $\times$ 803 **17,971.14**

3.   30.98 **173.488**
     $\times$ 5.6

Find each product mentally. Use the distributive property.

4. 6 $\times$ 17 **102**

5. 7 $\times$ 9.3 **65.1**

6. 21.8 $\times$ 30 **654**

Solve each equation.

7. $x = 47.3 \times 5.6$ **264.88**

8. $20.86 \cdot 4.11 = m$ **85.7346**

9. *Algebra* Evaluate $b(a + c)$ if $a = 7.8$, $b = 5.05$, and $c = 0.02$. **39.491**

Find the perimeter of each figure.

10.   24 m   **56 m**
      25 m   7 m

11. a square with sides of length 5.4 inches **21.6 in.**

12.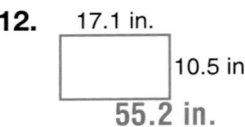
    17.1 in.   10.5 in.
    55.2 in.

Find the area of each figure.

13. rectangle: ℓ, 10.5 mi; w, 17.1 mi **179.55 mi²**

14. square: s, 2.5 cm **6.25 cm²**

Find each quotient.

15. $19.36 \div 44$ **0.44**

16. $21.6 \overline{)49.68}$ **2.3**

Solve each equation.

17. $x = 58.305 \div 11.5$ **5.07**

18. $1,274.7 \div 2.1 = t$ **607**

19. *Algebra* Evaluate $c \div d$ if $c = 108.9$ and $d = 3.3$. **33**

Write the unit that you would use to measure each of the following. Then estimate the mass or capacity.

20. liter, 4 L

20. a bottle of windshield-washer fluid

21. a potato **gram, 200 g**

Complete.

22. 739 mL = _?_ L **0.739**

23. _?_ mg = 2.1 kg **2,100,000**

24. *Transportation* Danielle drove 11.5 kilometers each day for 5 days. How far did she travel in all? **57.5 kilometers**

25. Find the area of the figure. **683 ft²**

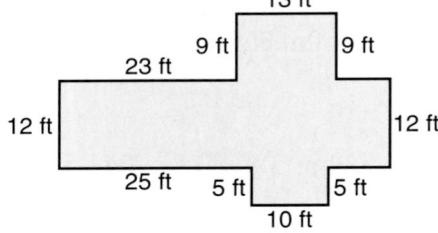
13 ft
9 ft   9 ft
23 ft
12 ft   12 ft
25 ft   5 ft   5 ft
10 ft

State whether each number is divisible by 2, 3, 5, 6, 9, or 10.

1. 75 **3, 5**

2. 864 **2, 3, 6, 9**

3. Games In one card game, all 52 cards are to be dealt out to the players. If there are six players, will all players have the same number of cards? Explain. **Sample answer: No; 52 is not divisible by 3.**

4. Which of the numbers 21, 31, 54, or 75 is prime? **31**

Find the prime factorization of each number.

5. 72 $2^3 \times 3^2$

6. 108 $2^2 \times 3^3$

7. 58 2×29

8. List all of the factors of 20 from least to greatest. **1, 2, 4, 5, 10, 20**

Find the GCF of each set of numbers.

9. 27, 45 **9**

10. 18, 48 **6**

11. 21, 38, 42 **1**

State whether each fraction is in simplest form. If not, write each fraction or ratio in simplest form.

12. $\frac{40}{100}$ **no; $\frac{2}{5}$**

13. $\frac{6}{25}$ **yes**

14. $\frac{45}{50}$ **no; $\frac{9}{10}$**

Express each mixed number as an improper fraction.

15. $2\frac{5}{7}$ $\frac{19}{7}$

16. $4\frac{1}{3}$ $\frac{13}{3}$

17. $12\frac{4}{5}$ $\frac{64}{5}$

Express each improper fraction as a mixed number.

18. $\frac{7}{3}$ $2\frac{1}{3}$

19. $\frac{16}{5}$ $3\frac{1}{5}$

20. $\frac{21}{4}$ $5\frac{1}{4}$

21. Draw a line segment that is $1\frac{7}{8}$ inches long. **See Answer Appendix.**

Find the LCM for each set of numbers.

22. 6, 16 **48**

23. 12, 18 **36**

24. 3, 9, 15 **45**

Replace each ● with < , >, or = to make a true sentence.

25. $\frac{5}{6}$ ● $\frac{7}{9}$ **>**

26. $\frac{6}{11}$ ● $\frac{4}{9}$ **>**

27. $\frac{8}{12}$ ● $\frac{12}{18}$ **=**

Express each decimal as a fraction or mixed number in simplest form.

28. 0.05 $\frac{1}{20}$

29. 5.1 $5\frac{1}{10}$

30. 0.42 $\frac{21}{50}$

Express each fraction or mixed number as a decimal. Use bar notation to show a repeating decimal.

31. $6\frac{5}{6}$ $6.8\overline{3}$

32. $\frac{7}{16}$ **0.4375**

33. *Food* The Pretzel Factory sells salted, unsalted, cinnamon, herb, and chocolate pretzels in two sizes, regular and jumbo. How many different ways can a person buy a pretzel? **10 ways**

Round each number to the nearest half unit.

1. $\frac{7}{15}$ $\frac{1}{2}$

2. $7\frac{2}{11}$ 7

3. $\frac{11}{14}$ 1

4. *Decorating* One wall space in Eric's room is $24\frac{3}{8}$ inches wide. When looking for a poster to fit in that space, should he round up or down? **down**

Estimate.

5. $4\frac{7}{10} - 2\frac{1}{5}$ 5 − 2 = 3

6. $\frac{7}{8} + \frac{1}{6}$ 1 + 0 = 1

7. $\frac{15}{16} - \frac{4}{9}$ $1 - \frac{1}{2} = \frac{1}{2}$

Add or subtract. Write the answer in simplest form.

8. $\frac{3}{17} + \frac{16}{17}$ $1\frac{2}{17}$

9. $\frac{7}{8} + \frac{1}{6}$ $1\frac{1}{24}$

10. $6\frac{3}{10} - 2\frac{9}{10}$ $3\frac{2}{5}$

11. $4\frac{2}{9} + 5\frac{2}{9}$ $9\frac{4}{9}$

12. $12 - 7\frac{3}{8}$ $4\frac{5}{8}$

13. $3\frac{1}{8} - 1\frac{1}{16}$ $2\frac{1}{16}$

14. $\frac{15}{16} - \frac{7}{16}$ $\frac{1}{2}$

15. $\frac{2}{15} + \frac{3}{10} - \frac{1}{5}$ $\frac{7}{30}$

16. $5\frac{7}{8} + 1\frac{5}{6}$ $7\frac{17}{24}$

Solve each equation. Write the solution in simplest form.

17. $m = \frac{15}{16} - \frac{11}{16}$ $\frac{1}{4}$

18. $y = \frac{4}{5} + \frac{3}{5}$ $1\frac{2}{5}$

19. $9\frac{4}{5} = 7\frac{3}{5} + t$ $2\frac{1}{5}$

20. $d - 8\frac{3}{8} = 7\frac{5}{24}$ $15\frac{7}{12}$

21. *Gardening* Mr. Hannon had $\frac{1}{2}$ ton of dirt for his garden delivered on Tuesday. On Thursday, another $\frac{1}{4}$ ton was delivered. How much dirt was delivered in all? $\frac{3}{4}$ **ton**

22. *Tickets* Emeka got to TicketCentral at 4:52 A.M. to wait in line for concert tickets. If TicketCentral opened at 7:30 A.M., how long did he wait? **2 h 38 min**

Add or subtract.   23. **5 h 22 min**

23. 10 h 12 min
  − 4 h 50 min

24. 3 h 4 min 10 s
  +9 h 58 min 28 s
  13 h 2 min 38 s

25. 6 h 55 min
  −5 h 49 min
  1 h 6 min

CHAPTER 7

Test

Estimate each product.

1. $6\frac{7}{8} \times 8\frac{1}{6}$ $7 \times 8 = 56$

2. $37 \times \frac{4}{9}$ $40 \times \frac{1}{2} = 20$

3. $\frac{9}{10} \times \frac{5}{8}$ $1 \times \frac{1}{2} = \frac{1}{2}$

Find the product. Write in simplest form.

4. $1\frac{4}{5} \times 2\frac{2}{3}$ $4\frac{4}{5}$

5. $\frac{9}{10} \times \frac{5}{8}$ $\frac{9}{16}$

6. $3\frac{1}{5} \times 1\frac{1}{4}$ 4

7. *Cooking* Kahlil's chocolate chip pie recipe calls for $1\frac{1}{2}$ cups of sugar. How much sugar will he need to make half of the recipe? $\frac{3}{4}$ cup

8. *Civics* In Homeland, U.S.A., $\frac{3}{4}$ of the adult residents are registered to vote. In the last May primary election, $\frac{2}{3}$ of those registered voted. What portion of the adult residents voted? $\frac{1}{2}$

Solve each equation. Write the solution in simplest form.

9. $4 \times \frac{3}{5} = z$ $2\frac{2}{5}$

10. $r = \frac{1}{8} \div \frac{3}{4}$ $\frac{1}{6}$

11. $k = 3\frac{2}{3} \times 2\frac{1}{3}$ $8\frac{5}{9}$

12. $g = 6 \div 1\frac{4}{5}$ $3\frac{1}{3}$

13. $\frac{3}{8} \times \frac{2}{5} = f$ $\frac{3}{20}$

14. $m = 5\frac{3}{4} \div 1\frac{1}{2}$ $3\frac{5}{6}$

Find the circumference of each circle shown or described. Use $\frac{22}{7}$ or 3.14 for π. Round decimal answers to the nearest tenth.

15.  34.7 m 109.0 m

16. $r = 3\frac{1}{2}$ in. 22 in.

17.  7 cm 44 cm

Find each quotient. Write in simplest form.

18. $\frac{3}{4} \div \frac{4}{7}$ $1\frac{5}{16}$

19. $2\frac{2}{3} \div 1\frac{5}{9}$ $1\frac{5}{7}$

20. $\frac{2}{3} \div \frac{8}{9}$ $\frac{3}{4}$

Complete.

21. $2\frac{1}{2}$ qt = _?_ c 10

22. 9,000 lb = _?_ T $4\frac{1}{2}$

23. 8 c = _?_ fl oz 64

Find the next two numbers in each sequence.

24. 4, $6\frac{1}{2}$, 9, $11\frac{1}{2}$, ... 14, $16\frac{1}{2}$

25. 810, 270, 90, 30,... 10, $3\frac{1}{3}$

Express each ratio as a fraction in simplest form.

1. 2 notebooks out of 6 notebooks are blue. $\frac{1}{3}$

2. 56 calculators out of 100 are graphing calculators. $\frac{14}{25}$

Express each ratio as a rate.

3. 384 kilometers in 4 hours **96 km per h**

4. 5 videos for $79.95 **$15.99 per video**

5. *Civics* In an election, 2,000 registered voters voted for Marilyn Williams, and 500 registered voters voted for Emilio Cardona. What ratio represents the portion of the voters that voted for Emilio Cardona? $\frac{1}{5}$

Solve each proportion.

6. $\frac{4}{6} = \frac{x}{15}$ **10**

7. $\frac{m}{12} = \frac{12}{16}$ **9**

8. $\frac{10}{p} = \frac{4}{14}$ **35**

9. *Mechanics* The machine part shown has a scale of 1 unit = 1.25 centimeters. Find the actual length of the side labeled *b*. **25 cm**

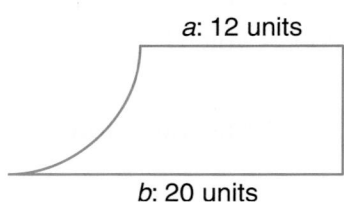

a: 12 units

b: 20 units

10. *Geography* Bartolome is planning to travel from Dallas to Houston. His map has a scale of 1 inch is approximately 30 miles. If the distance on the map from Dallas to Houston is $7\frac{1}{4}$ inches, what is the approximate distance between the two cities? **about 217 miles**

Express each fraction as a percent.

11. $\frac{4}{5}$ **80%**

12. $\frac{9}{100}$ **9%**

13. $\frac{13}{25}$ **52%**

Express each percent as a fraction in simplest form and as a decimal.

14. 80% $\frac{4}{5}$; **0.8**

15. 7% $\frac{7}{100}$; **0.07**

16. 0.1% $\frac{1}{1,000}$; **0.001**

Express each decimal as a percent.

17. 0.012 **1.2%**

18. 0.92 **92%**

19. 3.1 **310%**

Estimate each percent. 20–22. Sample answers are given.

20. 9.5% of 51 $\frac{1}{10} \times 50 = 5$

21. 49% of 26 $\frac{1}{2} \times 26 = 13$

22. 308% of 9 $3 \times 9 = 27$

Find the percent of each number.

23. 60% of 35 **21**

24. 49% of 26 **12.74**

25. 2% of 50 **1**

Use a protractor to find the measure of each angle. Then classify the angle as *acute, right,* or *obtuse.*

1. 130°; obtuse

2. 90°; right

3. 17°; acute

Classify each angle measure as *acute, right,* or *obtuse*.

4. 143° obtuse

5. 68° acute

6. 91.9° obtuse

7. ***Algebra*** Angles R and S are supplementary angles. If $m\angle S = 137°$, find $m\angle R$. **43°**

8. Draw a 105° angle. **See Answer Appendix.**

Estimate the measure of each angle shown.

9. about 60°

10. about 70°

11. about 120°

Draw the angle or line segment with the given measurement. Then use a straightedge and compass to bisect each angle or line segment. 12–15. See Answer Appendix.

12. 4 cm

13. 3 in.

14. 140°

15. 50°

Name each polygon. Then tell if the polygon is a regular polygon.

16. 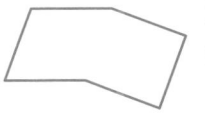 hexagon; no

17. 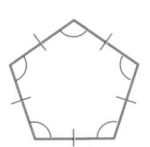 pentagon; yes

18.  quadrilateral; no

19. Draw an example of an octagon. **See students' work.**

Trace each figure. Draw all lines of symmetry.

20.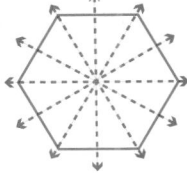

21.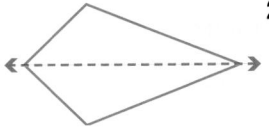

22. ***Safety*** The traffic sign shown warns motorists of a slow-moving vehicle. How many lines of symmetry does the sign have? **3**

Tell whether each pair of polygons is *congruent, similar,* or *neither*.

23. 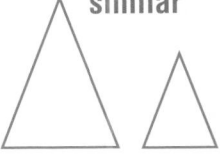 similar

24.  congruent

25.  neither

Find the area of each figure to the nearest tenth.

1. 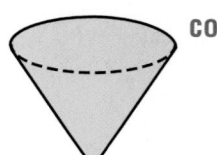 4 m, 4.3 m, 12 m   **48 m²**

2. 5 in.   **78.5 in²**

3. 5 m, 3.5 m, 10 m, 12 m   **21 m²**

4. 1 mi, 1.7 mi, 4 mi   **4 mi²**

5. 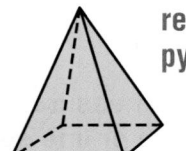 11 cm   **95.0 cm²**

6. 7.3 ft, 2 ft, 7 ft   **7 ft²**

7. *Traffic Signs* A triangular yield sign has a base of 32 inches and a height of 30 inches. Find the area of the sign. **480 in²**

8. *Gardening* Keisha plans to plant bulbs in a circular flower bed that has a radius of 2 meters. If she will plant 40 bulbs per square meter, how many bulbs should she buy? **502 bulbs**

Name each figure.

9. **cone**

10. **rectangular pyramid**

11. 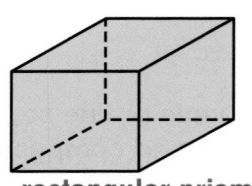 **rectangular prism**

12. State the number of faces, edges, and vertices in a triangular prism. **5; 9; 6**

Find the volume of each rectangular prism in Exercises 13–16 to the nearest tenth.

13. 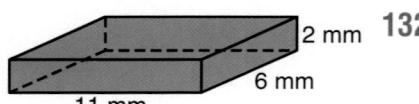 2 mm, 6 mm, 11 mm   **132 mm³**

14. 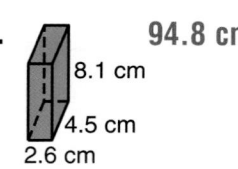 8.1 cm, 4.5 cm, 2.6 cm   **94.8 cm³**

15. length, 11 inches; width, 8 inches; height, $5\frac{1}{2}$ inches **484 in³**

16. length, 4.5 meters; width, 2 meters; height, 5 meters **45 m³**

17. Find the surface area of the rectangular prism in Exercise 13. **200 mm²**

18. Find the surface area of the rectangular prism in Exercise 14. **138.42 cm²**

19. Find the surface area of a rectangular prism whose length is 20 yards, width is 30 yards, and height is 10 yards. **2,200 yd²**

20. *Pools* A rectangular diving pool is 20 feet by 15 feet by 8 feet. How much water is required to fill the pool? **2,400 ft³**

Test

1. Write an integer to describe a loss of 12 dollars. **−12**

2. Write the integer that is the opposite of −5. **+5**

3. Graph −3 on a number line.

$$\longleftarrow \!\!\!\!\!\mid\!\mid\!\mid\!\mid\!\mid\!\mid\!\bullet\!\mid\!\mid\!\mid\!\mid\!\mid\!\mid\!\mid\!\mid\!\mid\!\mid\!\mid\!\mid\!\mid\!\mid \longrightarrow$$
$$-10\ -8\ -6\ -4\ -2\ \ \ 0\ \ \ 2\ \ \ 4\ \ \ 6\ \ \ 8\ \ \ 10$$

Replace each ● with >, <, or = to make a true sentence.

4. 8 ● −12 **>**   5. −77 ● −777 **>**   6. −127 ● −9 **<**

7. Order −10, 0, −12, 3, −1, and 11 from least to greatest. **−12, −10, −1, 0, 3, 11**

Find each sum or difference. Use counters or a number line if necessary.

8. −5 + (−7) **−12**   9. 6 − (−19) **25**   10. −13 + 10 **−3**   11. −4 − (−9) **5**

12. *Weather* The temperature at 6:00 A.M. was −5°F. What was the temperature at 8:00 A.M. if it was 7 degrees warmer? **2°F**

Find each product or quotient. Use counters, a number line, or patterns if necessary.

13. 21 ÷ (−7) **−3**   14. 6 × (−7) **−42**   15. −60 ÷ (−6) **10**   16. −8(3) **−24**

17. *Health* While Jane was recovering from the flu, her temperature dropped 1°F each hour for 3 hours. What was the total drop? **−3°**

Name the ordered pair for each point.

18. *K* **(2, 1)**
19. *R* **(4, −1)**
20. *M* **(−3, −2)**

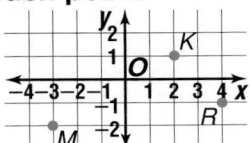

On grid paper, draw a coordinate grid. Then graph and label each point.

21. *A*(5, −1)   22. *B*(−2, −3)   23. *C*(−4, 5)

21–23. See Answer Appendix.

The vertices of △ABC are A(2, 3), B(−4, 1), and C(−1, 0). On a coordinate grid, draw △ABC and each transformation image. **24–25. See Answer Appendix.**

24. a translation 3 units down   25. a reflection over the *x*-axis

Solve each equation. Use cups and counters if necessary.

1. $x + (-3) = -1$ **2**

2. $r - 2 = 5$ **7**

3. $w + 4 = -3$ **−7**

4. $b - 4 = -1$ **3**

5. $\frac{1}{10}p = 2$ **20**

6. $5m = 30$ **6**

7. $\frac{1}{4}w = -8$ **−32**

8. $4x + 1 = -15$ **−4**

9. $10 = \frac{1}{3}n - 2$ **36**

10. What number should replace the cup on the equation mat at the right? **5**

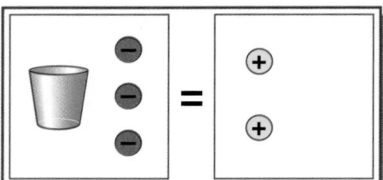

11. *Earth Science* The temperature at 10:00 P.M. was half what it was at noon. If the temperature at 10:00 P.M. was 26°, what was it at noon? **52°**

12. *Sports* Rodney was sacked for a 12-yard loss. On the next play, he had a gain of 5 yards. What was his net yardage on the two plays? **−7**

13. *Money Matters* Jeannette bought a sweater. The sale price was $38, which was $15 less than the original price. Find the original price. **$53**

Copy and complete each function table.

14.

input (*n*)	output (*n* + 5)
−5	■ 0
0	■ 5
1	■ 6

15.

input (*n*)	output (6*n*)
−3	■ −18
−1	■ −6
1	■ 6

Find the rule for each function table.

16. $n - 3$

input (*n*)	output (■)
−2	−5
0	−3
3	0

17. $7n$

input (*n*)	output (■)
−2	−14
0	0
3	21

18. *Age* Bret is 12 years old, and his father is 35 years old. When Bret is 20, his father will be 43 years old. Write a function rule for this relationship. **B + 23**

19. *Patterns* The triangles below were formed using toothpicks.

   a. Write the function rule to find the number of toothpicks for any number of triangles. **2t + 1**

   b. How many toothpicks will be needed to make 8 triangles? **17**

20. Graph the function given by the rule $\frac{1}{2}n$. **See Answer Appendix.**

A set of 20 cards has two cards numbered 10, three cards numbered 15, three cards numbered 20, and one card for each of the numbers 1 through 12. Suppose you choose one card without looking. Find the probability of each event.

1. $P(8)$ $\frac{1}{20}$
2. $P(3 \text{ or } 10)$ $\frac{1}{5}$
3. $P(\text{multiple of } 5)$ $\frac{1}{2}$
4. $P(15)$ $\frac{3}{20}$
5. $P(\text{a multiple of } 3)$ $\frac{7}{20}$
6. $P(\text{prime})$ $\frac{1}{4}$

Tell whether each of the following is a random sample.

Type of Survey	Survey Location	
7. favorite brand of tennis shoes	library	**yes**
8. favorite subject	science fair	**no**
9. favorite TV show	department store	**yes**

10. *Entertainment* According to a survey, 52 out of 100 people prefer listening to country music. Suppose 375 people participated in the survey. How many people can be expected to prefer listening to country music? **about 195 people**

Each figure represents a dartboard. It is equally likely that a dart will land anywhere on the dartboard. Find the probability of a randomly-thrown dart landing in the shaded region.

11.  $\frac{1}{2}$

12. 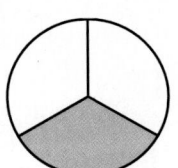 $\frac{1}{3}$

13.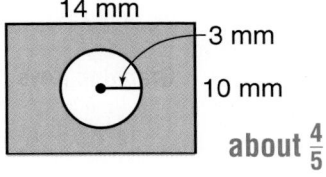
about $\frac{4}{5}$

14. Suppose you throw a dart 135 times at the dartboard in Exercise 12. How many times would you expect it to land in the shaded region? **about 45**

15. Suppose you throw a dart 150 times at the dartboard in Exercise 13. How many times would you expect it to land in the shaded region? **about 120**

16. *Food* At The Snack Shack, a person has a choice of iced tea or soda and nachos, popcorn, or a chocolate bar.
   a. Make a list and draw a tree diagram to show the sample space. **See Answer Appendix.**
   b. How many possible outcomes are there? **6 possible outcomes**
   c. What is the probability the next customer who orders a beverage and a snack will choose an iced tea and nachos? $\frac{1}{6}$

A coin is tossed, and the spinner shown is spun. Find the probability of each event.

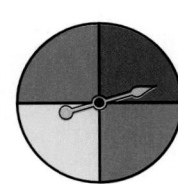

17. $P(\text{tails and blue})$ $\frac{1}{8}$
18. $P(\text{heads and not green})$ $\frac{3}{8}$
19. $P(\text{not heads and red})$ $\frac{1}{8}$
20. $P(\text{heads or tails and yellow})$ $\frac{1}{4}$

Getting Acquainted with the Graphing Calculator

When some students first see a graphing calculator, they think, "Oh, no! Do we *have* to use one?", while others may think, "All right! We get to use these neat calculators!" There are as many thoughts and feelings about graphing calculators as there are students, but one thing is for sure: a graphing calculator *can* help you learn mathematics. Keep reading for answers to some frequently asked questions.

What is it?

So what is a graphing calculator? Very simply, it is a calculator that draws graphs. This means that it will do all of the things that a "regular" calculator will do, *plus* it will draw graphs of equations.

What does it do?

A graphing calculator can do more than just calculate and draw graphs. For example, you can program it and work with data to make statistical graphs and computations. If you need to generate random numbers, you can do that on the graphing calculator. If you need to find the absolute value of a number, you can do that too. It's really a very powerful tool, so powerful that it is often called a pocket computer.

What do all those different keys do?

As you may have noticed, graphing calculators have some keys that other calculators do not. The Texas Instruments TI-83 will be used throughout this text.

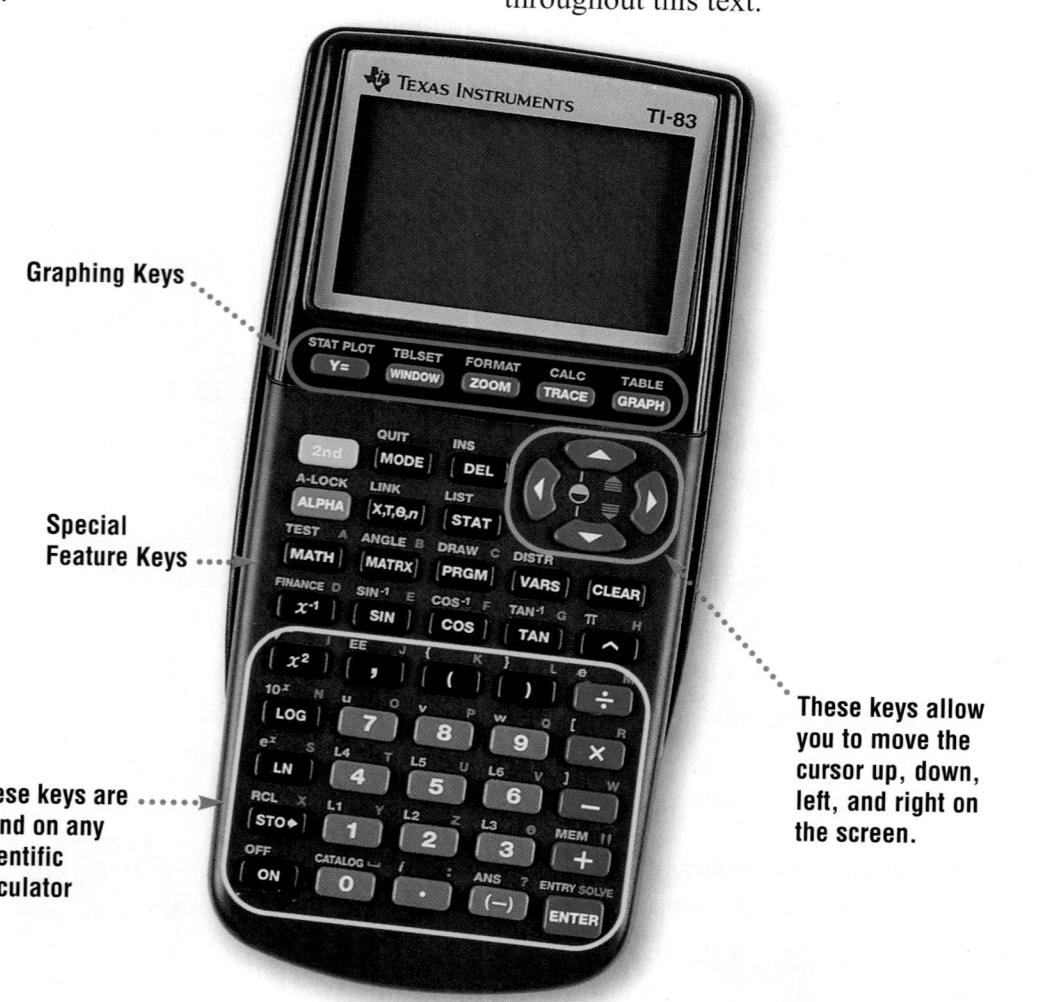

Graphing Keys

Special Feature Keys

These keys are found on any scientific calculator

These keys allow you to move the cursor up, down, left, and right on the screen.

Basic Keystrokes

- The yellow commands written above the calculator keys are accessed with the `2nd` key, which is also yellow. Similarly, the green characters above the keys are accessed with the `ALPHA` key, which is also green. In this text, commands that are accessed by the `2nd` and `ALPHA` keys are shown in brackets. For example, `2nd` [QUIT] means to press the `2nd` key followed by the key below the yellow `QUIT` command.
- `2nd` [ENTRY] copies the previous calculation so you can edit and use it again.
- `2nd` [QUIT] will return you to the home (or text) screen.
- Negative numbers are entered using the `(-)` key, not the minus sign, `-` .
- `2nd` [OFF] turns the calculator off.

Order of Operations

As with any scientific calculator, the graphing calculator observes the order of operations.

Example	Keystrokes	Display
4 + 13	4 `+` 13 `ENTER`	4 + 13 17
5³	5 `^` 3 `ENTER`	5 ^ 3 125
4 (9 + 18)	4 `(` 9 `+` 18 `)` `ENTER`	4(9 + 18) 108
√24	`2nd` [√] 24 `ENTER`	√ (24 4.8989 79486

Programming on the TI-83

The TI-83 has programming features that allow you to write and execute a series of commands for tasks that may be too complex or cumbersome to perform otherwise. Each program is given a name. Commands begin with a colon (:), which the calculator enters automatically, followed by an expression or an instruction.

When you press `PRGM` , you see three menus: EXEC, EDIT, and NEW. EXEC allows you to execute a stored program, EDIT allows you to edit or change a program, and NEW allows you to create a program.

- To begin entering a new program, press `PRGM` `▶` `▶` `ENTER` .
- You do not need to type each letter using the `ALPHA` key. Any command that contains lowercase letters should be entered by choosing it from a menu. Check your user's guide to find any commands that are unfamiliar.
- After a program is entered, press `2nd` [QUIT] to exit the program mode and return to the home screen.
- To execute a program, press `PRGM` . Then use the down arrow key to locate the program name and press `ENTER` twice, or press the number or letter next to the program name followed by `ENTER` .
- If you wish to edit a program, press `PRGM` `▶` and choose the program from the menu.
- To immediately re-execute a program after it is run, press `ENTER` when Done appears on the screen.
- To stop a program during execution, press `ON` or `2nd` [QUIT].

While a graphing calculator cannot do everything, it can make some things easier and help your understanding of math. To prepare for whatever lies ahead, you should try to learn as much as you can. Who knows? Maybe one day you will be designing the next satellite or building the next skyscraper with the help of a graphing calculator!

Getting Acquainted **with** Spreadsheets

hat do you think of when people talk about computers? Maybe you think of computer games or using a word processor to write a school paper. But a computer is a powerful tool that can be used for many things.

One of the most common computer applications is a spreadsheet program. Here are answers to some of the questions you may have if you're new to using spreadsheets.

What is it?
You have probably seen tables of numbers in newspapers and magazines. Similar to those tables, a spreadsheet is a table that you can use to organize information. But a spreadsheet is more than just a table. You can also use a spreadsheet to perform calculations or make graphs.

Why use a spreadsheet?
The advantage a spreadsheet has over a simple calculator is that when a number is changed, the entire spreadsheet is recalculated and the new results are displayed. So with a spreadsheet, you can see patterns in data and investigate what happens if one or more of the numbers is changed.

How do I use a spreadsheet?
A spreadsheet is organized into boxes called *cells*. Throughout this text, we will use the Microsoft Excel spreadsheet program. In Excel, the cells are named by a letter, that identifies the column, and a number, that identifies the row. In the spreadsheet below, cell C4 is highlighted.

	A	B	C
1	Width	Length	Area
2	3	4	12
3	2	10	20
4	5	12	60
5	8	14	112

To enter information in an Excel spreadsheet, simply move the cursor to the cell you want to access and click the mouse. Then type in the information and press Enter.

How do I enter formulas?
If you want to use the spreadsheet as a calculator, begin by choosing the cell where you want the result to appear.

- For a simple calculation, type = followed by the formula. For example, in the spreadsheet above, the formula in cell C2 is entered as "=A2*B2." *Notice that * is the symbol for multiplication in a spreadsheet.*

- Sometimes you will want similar formulas in more than one cell. First type the formula in one cell. Then select the cell and click the copy button. Finally select the cells where you want to copy the formula and click the paste button.

- Often it is useful to find the sum or average of a row or column of numbers. Excel allows you to choose from several functions like this instead of entering the formula manually. To enter a function, click the cell where you want the result to appear. Then click on the = button above the cells. A list of formulas will appear to the left. Click the down arrow button and choose your function. Excel will enter a range, which you may alter. For example, to find the average of row 2 of the spreadsheet below, the function chooses to find the average of cells B2, C2, and D2.

	A	B	C	D	E
1	Student	Test 1	Test 2	Test 3	Average
2	Kathy	88	85	91	88
3	Ben	86	89	92	89
4	Carmen	92	86	92	90
5	Anthony	80	88	87	85

The formula for cell E2 is =(B2+C2+D2)/3.

Spreadsheet software is one of the most common tools used in business today. You should try to learn as much as you can to prepare for your future. Who knows? Maybe you'll use what you're learning today as a company president tomorrow!

Selected Answers

Chapter 1
Problem Solving, Numbers, and Algebra

Pages 6–7   Lesson 1-1
1. Explore–Decide what facts you know and what you need to find out.   Plan–Make a plan for solving the problem and estimate the answer. Solve–Solve the problem. Make a new plan if necessary.   Examine–Check to see if your answer makes sense.   **5.** about 15 million   **7.** 180 miles   **9.** $40   **11.** 65°F   **13.** 84,480 pennies   **15.** Yes; Sample answer: a team can score 7 field goals and 1 safety or 3 touchdowns, 2 two-point conversions, and 1 one-point conversion.

Pages 10–11   Lesson 1-2
3.

Number of Folds	1	2	3	4
Number of Thicknesses	2	4	8	16

Sample answer: The number of thicknesses doubles.
5. 8, 4, 2   **7.** 3, 4, 3   **9.** 512, 2,048, 8,192
11. 17, 22, 28   **13.** 21, 31, 43   **15.** 5:10 P.M.
17. 256   **19.** C

Pages 14–15   Lesson 1-3
1. On the number line, 957 is closer to 1,000 than 900.   **5.** 400   **7.** 300   **9.** 10,000   **11.** 900 − 400 = 500; reasonable   **13.** 500 ÷ 50 = 10; not reasonable   **15.** 100   **17.** 50   **19.** 440
21. 5,500   **23.** 3,700   **25.** 5,600   **27.** 10,000
29. 2,000   **31.** 9,000   **33.** 80 × 3 = 240; reasonable   **35.** 100−20 = 80; not reasonable
37. 5,000 + 5,000 = 10,000; reasonable   **39.** 800 − 200 = 600; reasonable   **41.** $5.00 + $35.00 = $40.00; not reasonable   **43.** Sample answer: 200 + 60 + 600 + 300 = 1,160; about 1,160 lbs
45a. $75 + $36 + $21 + $6 + $2 + $2 + $10 = $152   **47.** 11, 16, 22   **49.** 10:22 A.M., 11:02 A.M., 11:06 A.M.   **51.** C

Pages 18–19   Lesson 1-4
1. Multiply and divide in order from left to right. Then add and subtract in order from left to right.   **3.** +; 14

---

5. ×; 33   **7a.** 12 × $15 + 28 × $6   **7b.** $348
9. 117   **11.** 43   **13.** 15   **15.** 23   **17.** 9   **19.** 55
21a. 2 × 15 + 1 × 14 + 3 × 18   **21b.** 98 g
23a. Sample answer: 3 × $29.95 + 2 × $16.95 + $12.95 = $136.70   **25.** 9,000   **27.** B

Page 19   Mid-Chapter Self Test
1. $7   **3.** 37, 45, 53   **5.** 150   **7.** 18,000 ÷ 300 = 60
9. 9

Pages 20–21   Lesson 1-5A
1. the unknown value   **3.** 13   **5.** the unknown value   **7.** 5 bags

Pages 24–25   Lesson 1-5
1. $5 \times r$ or $5 \cdot r$ **3.** Nicholas; 2 + 14 ÷ 2 = 9
5. 4   **7.** 18   **9.** 16   **11.** 15   **13.** 10   **15.** 22
17. 105   **19.** 60   **21.** 35   **23.** 22   **25.** 9
27. 18   **29.** 54   **31.** 9   **33.** 24   **35.** 30   **37.** 9
39. 12 inches   **41.** yes; 30   **43.** 21   **45.** A

Page 26   Lesson 1-5B
1. 3   **3.** 28   **5.** 14   **7.** 20   **9.** Yes, unless the scientific calculator does not have parentheses keys.

Pages 29–31   Lesson 1-6
1. An exponent is a number that tells how many times a number, called the base, is used as a factor. A power is a number that is expressed using exponents.   **3.** Tamika; $3^2 + 2^4 \cdot 6 = 105$.   **5.** m^5
7. $c \cdot c \cdot c$ **9.** $h \cdot h \cdot h \cdot h \cdot h \cdot h$ **11.** 100,000
13. about 1,000,000,000 people   **15.** 3^6
17. $3^2 \cdot 4^4$ **19.** r^4 **21.** $6^3 \cdot 1^3$ **23.** 14 · 14
25. 7 · 7 · 7 · 7 · 7   **27.** 16 · 16 · 16 · 16 · 16 · 16
29. $x \cdot x \cdot x \cdot y \cdot y \cdot y \cdot y$ **31.** 1,000   **33.** 32
35. 9   **37.** 1,027   **39.** 103   **41.** 2,000
43. 5,832 in³   **45.** 18   **47.** 400 + 700 + 200 = 1,300; reasonable   **49.** 379 students

Pages 32–33   Lesson 1-7A
3. The piece that is beside the top left-hand puzzle piece. It will connect to the corner piece.

5. Sample answer: When making a plan to solve a problem, you can use the guess-and-check strategy to guess an answer and compare the answer to the problem. If your answer does not work, then you know to make another guess. **7.** Sample answer: You could guess values for p to find what number divided by 5 is 45. **9.** 15 dogs, 9 cats **11.** There are 8 bones in the wrist, 14 in the fingers, and 5 in the palm. **13.** 67,740 feet per minute **15.** B

Pages 35–37   Lesson 1-7

1. Sample answer: Both algebraic expressions and equations can contain variables, operations, and numbers. Whereas an equation contains an equals sign, an algebraic expression does not contain an equals sign. **3.** false **5.** 7 **7.** 4 **9.** 9 **11.** 10 **13.** true **15.** true **17.** false **19.** 15 **21.** 7 **23.** 13 **25.** 7 **27.** 44 **29.** 26 **31.** 7 **33.** 3 **35.** 225 **37.** 2 **39.** 16 **41a.** $19 + g = 65$ **41b.** 46 grams **43.** C **45.** 12 **47.** about $70

Pages 38–41   Study Guide and Assessment

1. multiply and divide **3.** 4 **5.** 3 and 4 **7.** 2^5 **9.** 10 **11.** 11 **13.** $39.92 **15.** 531 votes
17.

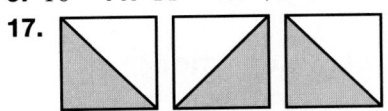

19. 48, 56, 64 **21.** 30 **23.** 650 **25.** 2 **27.** 2 **29.** 7 **31.** 16 **33.** 7 **35.** 21 **37.** 44 **39.** 4,096 **41.** 125 **43.** t^5 **45.** 3 **47.** 63 **49.** 9 **51.** 6 **53.** 13 years 10 months **55.** about 63,000

Pages 42–43   Standardized Test Practice

1. B **3.** B **5.** D **7.** E **9.** B **11.** A **13.** 10 ft **15.** 21 **17.** 27

Chapter 2
Statistics: Graphing Data

Pages 48–49   Lesson 2-1

1. Sample answer: to make it easier to study the data.

5a.

Age	Tally	Frequency
10			1
11			1
12	JHT	5
13					3
14				2

5b. 12

7.

Days Absent	Tally	Frequency
0	JHT	5
1	JHT		6
2				2
3					3
4		0
5				2

9a. Sample answer: All the shoes sold were between sizes 9 and 11. **9b.** 77 pairs of shoes **9c.** 10 **11a.** Sample answer: Based on the data in the table, they should keep potato chips and corn chips well stocked. **11b.** Sample answer: There are too few to tell. **13.** A

Pages 51–53   Lesson 2-2

1. Sample answer: The scale must include all of the data. Since the shortest distance is 12 and the longest distance is 63, a scale from 10 to 70 is appropriate. **5.** 2; 10 is too large.

7a.

Shelby's Video Game Scores		
Score	Tally	Frequency
60,000–69,900			1
50,000–59,900			1
40,000–49,900	||||	4
30,000–39,900	JHT	5
20,000–29,900	||||	4
10,000–19,900	||	2
0–9,900			1

7b. Most of her scores were between 30,000 and 39,900. **9.** a **11.** a **13.** b

15.

Amount	Tally	Frequency
81–100			1
61–80		0
41–60	||||	4
21–40			1
0–20	||	2

17a. No; since Texas is a much larger state, the distances will be larger. If she used the same scale the frequency table would not include all the distances. **17b.** Sample answer: Yes; however, the table will have a large number of rows. **19a.** Sample answer: 5,000–20,999

19b.

Elevation	Tally	Frequency
19,000–20,999	|	1
17,000–18,999		0
15,000–16,999		0
13,000–14,999	ⅢⅡ ||	7
11,000–12,999	||||	4
9,000–10,999		0
7,000–8,999	||	2
5,000–6,999	ⅢⅡ |	6

21.

Number of Hits	Tally	Frequency
0	|	1
1	||	2
2	|||	3
3	||	2
4	|	1

23. D

Pages 55–57   Lesson 2-3

1. Call Them   **3.** Sample answer: If the vertical scale is much higher than the highest value, it makes the graph flatter. Changing the interval does not affect either type of graph.   **5.** 0–70; 10

7.

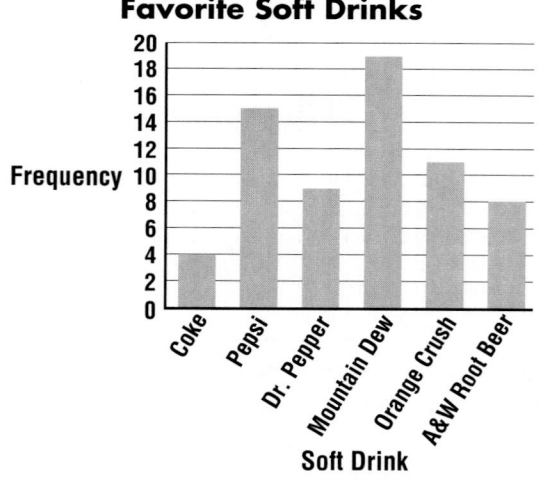

Favorite Soft Drinks

9.

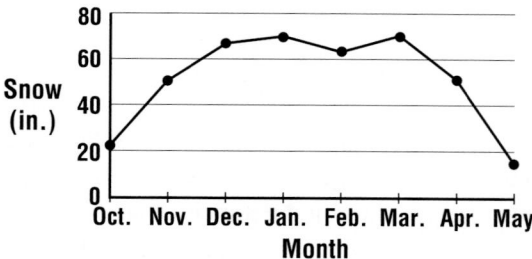

Average Snowfall for Brighton, Utah

11. Sample answer: 0 to 300   **13.** Table A; It shows the change and direction of change over time.

15.

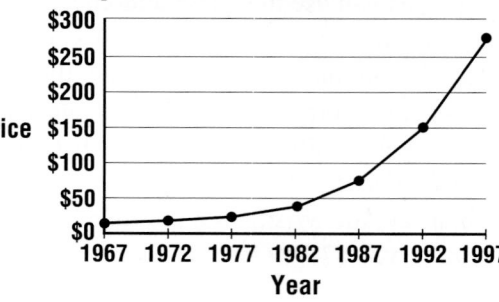

Super Bowl Ticket Prices

19. B   **21.** 4

Pages 58–59   Lesson 2-3B

3. Sample answer: McDonald's; It is the second favorite and the healthiest.   **7.** $30
9. 1,380 miles   **11.** C

Pages 62–63   Lesson 2-4

1. Sample answer: When you want to represent data that looks at part to whole relationships.
5a. 13%–Special; 15%–egg noodles; 31%–short pasta; 41%–long pasta   **5b.** special and egg noodles
5c. 72%   **5d.** special and egg noodles   **5e.** either long pasta and special (54%) or short pasta and egg noodles (46%)   **7a.** LP album, 7-12 in. singles and music videos   **7b.** cassette single–5%; cassette album–25%; CDs–68%

9.

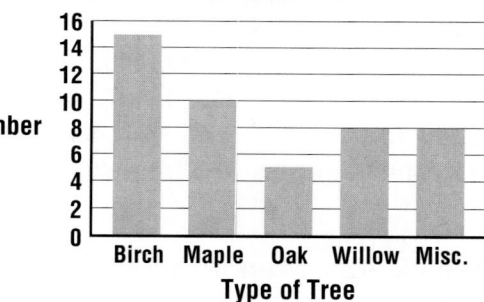

Janet's Leaf Collection

Pages 65–67   Lesson 2-5

1. Sample answer: If the extension aligns with the segment between 1980 to 1990, world population could be about 7 billion.   **5a.** Sample answer: It will continue to bounce higher.   **5b.** Sample answer: It will bounce about as high as it did when dropped from 110 cm.   **5c.** Sample answer: Ball A might be a Super Ball.   **5d.** Sample answer: No; at some point both balls will begin to level off. If you dropped Ball A from 100 feet it will not bounce 100 feet.   **7.** The point where they cross indicates the times were the same.   **9.** D

Page 67 Mid-Chapter Self Test

1.

Height	Tally	Frequency
54	I	1
55	I	1
56	I	1
57	IIII	4
58	I	1
59	I	1
60	II	2
61		0
62		0
63	I	1

3.

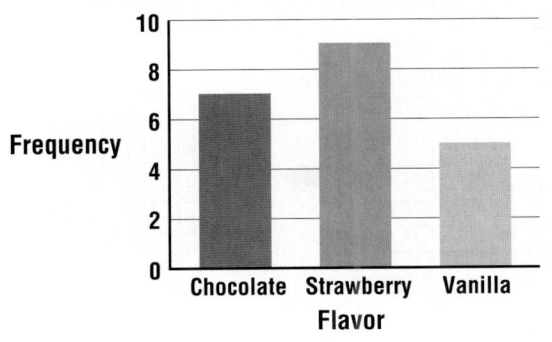

Favorite Ice Cream Flavor

5a. the 5th week **5b.** the 11th week **5c.** Sample answer: No; the horizontal scale doesn't extend far enough.

Pages 69–70 Lesson 2-6

1. when there are many numbers in a set of data

5.

Stem	Leaf
2	0 4 7
3	4 5 6 6 8
4	3 5 7
5	3 4 8 8
6	
7	8 2|7 = 27

7. 1, 2, 3, 4, 5
9. 0, 1, 2, 3, 4

11.

Stem	Leaf
9	2 4 9 9
10	
11	
12	4 4
13	0 2 3
14	0
15	
16	2 7 12|4 = 124

13.

Stem	Leaf
0	9
1	
2	7 8 9
3	4 4 4
4	5 6
5	3 6 6 7 7 7 8

5\|3 = 53°

15a.

Stem	Leaf
0	5 5 6 7 8 9 9 9 9 9
1	0 1 1 3 4 6 7 8 8 9 9
2	2 2 3 6 7 7 9
3	7
4	1 1|0 = 10

15b. 9 days   **17.** Sample answer: 400

Pages 73–74 Lesson 2-7

1. The mean because the numbers do not differ greatly.   **5.** 10, 26, 27, 29, 36, 37, 57, 83, 88; 36; no mode; ≈ 43.7; 78   **7.** 44; 44; no mode; 22
9. 15; 15; 14 and 16; 4   **11.** 31; 31; 31; 26
13. The mean would be lower.   **15.** yes; Sample answer: The total would be higher so the mean would also be higher.   **17.** Sample answer: The mode because she would want to know what size they sold the most shakes.

19.

Stem	Leaf
0	0 2 2 4 5 7 7 9
1	0 2 3 3 8 9
2	0 3 2|0 = 20

Page 77 Lesson 2-7B

1.

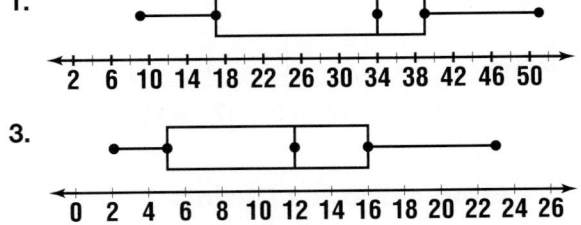

3.

Pages 79–81 Lesson 2-8

1. Sample answer: Using different scales and not having a title.   **3.** Sample answer: If the range is very high the mean could be misleading. If the set of data has a gap between two groups of numbers that are all about the same, the mean could be misleading. For example, 1, 2, 3, 2, and 8, 7, 6, 9, 8.   **5.** mode–want to find the most frequent   **7a.** Graph B is misleading because the distance between 0 and 25 is the same as the distance between 25 and 30.
7b. Graph A because it shows a general rise in price.
9. $1\frac{1}{2}$; median   **11.** $1962; mean   **13.** There may be a few very low priced CDs, thus giving a lower average.   **15a.** mean: 134 ÷ 24 ≈ 5.6 or 6 pets; median: 3 pets; mode: 2 pets

15b. Sample answer: The median best represents the data. The student with 24 pets has a big effect on the mean.   **17.** A   **19.** 25, 36, 49

Pages 84–85   Lesson 2-9

1. From the origin, go 7 units left, then up 9 units.
3. Sample answer: Raul; to graph (8, 3) you move 8 units right and 3 units up and to graph (3, 8) you move 3 units right and 8 units up.   **5.** *F*   **7.** (8, 0)
9. (5, 6)   **11.** *C*   **13.** *K*   **15.** *Q*   **17.** *L*   **19.** *M*
21. (7, 4)   **23.** (3, 7)   **25.** (0, 9)   **27.** (0, 0)
29. (9, 8)   **35.** C   **37.** 7

Pages 86–89   Study Guide and Assessment

1. f   **3.** a   **5.** c   **7.** i   **9.** j   **11.** Sample answer: A bar graph represents data using rectangles to show the frequency of responses. A line graph uses dots at the frequency points connected by segments to show how data changes over time.

13.

Favorite Color	Tally	Frequency
blue	IIII	4
red	IIII	4
pink	II	2
orange	I	1
yellow	I	1
green	II	2
brown	I	1

15. scale = 0 to 50, interval 5   **17.** scale = 0 to 160, interval 20

19.

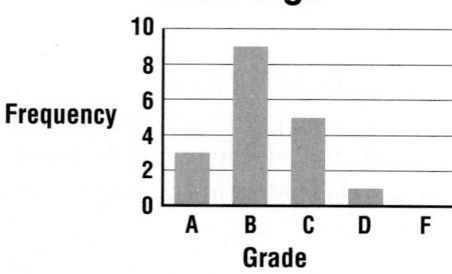

Alma's Grades During Junior High

21. less   **23.** 22   **25.** 25   **27.** median   **29.** mean
31. *A*   **33.** (3, 6)

35.

Stem	Leaf
4	0 5
5	0
6	5 9
7	5 5
8	5
9	0 2 5

9|0 = 90

Pages 90–91   Standardized Test Practice

1. D   **3.** B   **5.** B   **7.** C   **9.** C   **11.** E   **13.** 7 + *r* = 22; 15   **15.** 25

Chapter 3 Adding and Subtracting Decimals

Page 94   Lesson 3-1A

1. thirty hundredths
3. 9 tenths;

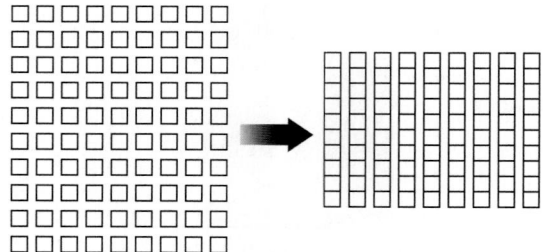

Pages 97–98   Lesson 3-1

1. 0.05, $\frac{5}{100}$, five hundredths   **3.** Mercedes is correct because the last digit, 8, is in the thousandths place.   **5.** 0.43   **7.** 4.0029   **9.** 0.0702   **11.** 0.3
13. 0.07   **15.** 19.53   **17.** 0.009   **19.** 47.047
21. 172.0001   **23.** 20.9   **25.** 3.003   **27.** 0.0801
29. 0.01   **31.** greatest: 0.750; least: 0.057   **33.** B
35. $19

Page 101   Lesson 3-2A

3. Students should notice that the measures are related to powers of 10 and start to notice how decimals are used.

Pages 103–104   Lesson 3-2

5. cm; about 2.5 cm   **7.** 3.2 cm or 32 mm   **9.** cm; about 90 cm   **11.** m; about 5 m   **13.** 1.3 cm or 13 mm   **15.** 3.8 cm or 38 mm   **17.** 3 cm or 30 mm
19. 2 cm or 20 mm   **21.** 300 cm   **25.** C   **27.** 10

Pages 107–108   Lesson 3-3

1.

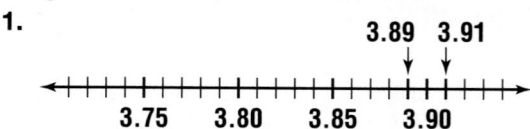

3. 1.63 > 1.54   **5.** 0.06   **7.** 13.05, 13.507, 13.84, 13.9   **9.** 389.225, 388.404, 388.246, 385.867,

385.841   **11.** 0.331   **13.** 1.018   **15.** 47.553
17. 0.7   **19.** 14.95, 15, 15.01, 15.8   **21.** 0.0316,
0.0306, 0.025, 0.0249, 0.0208   **23.** 397.877, 379.9,
379.88, 379.8778, 378.87   **25.** 0.603   **27a.** 2.1,
2.9, 4.2, 5.1, 5.8, 7.0, 9.0, 13.0, 14.0   **27b.** 5.8
29. 5.7 cm or 57 mm   **31.** Sample answer: 95, 91,
93, 89; no number occurs more often than the others

Pages 110–111   Lesson 3-4

1. Sample answer: Since the digit to the right of the
ones place is greater than 5, add one to the ones
place. $10.79 rounded to the nearest dollar is $11.
3. Sharon is correct. Carlos rounded to the nearest
hundred.   **5.** 8.20   **7.** 20   **9.** 18   **11.** 20.5
13. 49.8   **15.** 19.78   **17.** 100.0   **19.** 7.000
21. $2.00   **23.** 102   **25.** Sample answer: Using
the exact data in this case would be most appropriate.
27. No, because no numbers differ greatly.

Page 111   Mid-Chapter Self Test

1. 0.7   **3.** 0.681   **5.** 3 cm   **7.** 638.178   **9.** 3.40

Pages 114–115   Lesson 3-5

1. Sample answer: Round both amounts to the
nearest dollar. $17 − $4 = $13   **5.** 20 + 30 = 50
7. 11 + 11 + 11 + 11 = 44   **9.** 20 + 20 + 20 =
60   **11.** Sample answer: Round 0.38 to 0.4 and
round 0.21 to 0.2. 0.4 + 0.2 = 0.6. Since 0.6 > 0.5
both samples cannot be stored in the 0.5-liter
container.   **13.** 9 − 5 = 4   **15.** 32 + 17 = 49
17. $30 − $20 = $10   **19.** $58 − $27 = $31   **21.** 4
+ 4 + 4 = 12   **23.** 1 + 1 + 1 + 1 = 4   **25.** $54
+ $54 + $54 = $162   **27.** $65 − $40 = $25
29. $200 − $110 = $90   **31.** 102°F − 99°F = 3°F
33. about 1 million square feet   **35a.** Sample
answer: An estimate for Mandy's family is $9 + $9
+ $9 + $4 or $31. Since all costs were rounded up,
her whole family can go on the tour for less than
$35.   **35b.** Sample answer: An estimate for each
meal is $5 for breakfast, $7 for lunch, and $12 for
dinner. The estimate for one person would be $5 +
$7 + $12 or $24. The estimate for a family of 4
would be $24 + $24 + $24 + $24 or $96. This
could be rounded up to $100.   **37.** 0.270   **39.** 21
41. 8

Pages 116–117   Lesson 3-6A

3. Sample answer: If you round to the nearest
hundred, the 1986 and 1987 amounts or the 1987

and 1988 amounts total 200 million. If you round to
the nearest ten the 1986 and 1988 amounts total 200
million.   **9a.** They sold 12 subscriptions during
week three.   **9b.** Since each symbol means 4
subscriptions, a half-magazine means 2
subscriptions.   **9c.** They sold 16 subscriptions
during week two.   **11.** Sample answer: By drawing
a line graph and extending it to match the segment
between 1993 and 1994, over 1.6 million CDs will
be distributed in 2000.   **13.** B

Pages 119–121   Lesson 3-6

1. Sample answer: Annex two zeros. Then line up
the decimal points, rename and subtract.   **5.** 2.24
7. 0.75   **9.** 6.592   **11.** 6.4   **13.** 0.24   **15.** $11.77
17. 1.747   **19.** 2.02   **21.** 13.253   **23.** 151.575
25. $12.10   **27.** 26.42   **29.** 2.1   **31.** 10.72
33a. 2.6 pounds   **33b.** 18 pounds   **35.** 0.2 −
0.1876543 = 0.0123457   **37.** B   **39.** 3

Pages 122–125   Study Guide and Assessment

1. false, less   **3.** false, one hundredth   **5.** false,
centimeters   **7.** false, 600.012   **9.** true   **11.** 0.08
13. 14.017   **15.** 0.2   **17.** 0.053   **19.** eight ten-
thousandths   **21.** 6 cm   **23.** 2 cm   **25.** 11.6
27. 0.0289, 0.0319, 0.032, 0.31   **29.** 6.75, 6.39,
6.32, 6.02   **31.** 7   **33.** 13.6   **35.** 5 − 1 = 4
37. 6 + 6 + 6 + 6 + 6 = 30   **39.** 12.912
41. 77.1   **43.** 7320.4, 7321.5, 7321.539, 7342.98,
7346.24, 7346.4   **45.** about $210

Pages 126–127   Standardized Test Practice

1. A   **3.** C   **5.** A   **7.** D   **9.** B   **11.** D   **13.** divide
120 by 8   **15.** 241 miles   **17.** 68   **19.** mode

Chapter 4
Multiplying and Dividing Decimals

Page 132   Lesson 4-1A

1.

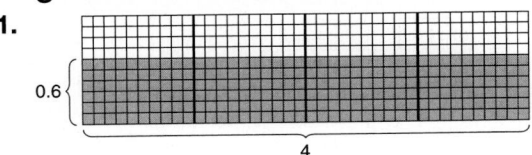

0.6 { ... } 4

3.

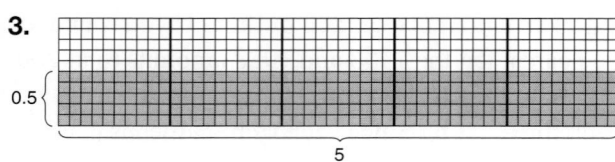

0.5 { }

5

5. 3.2

5.

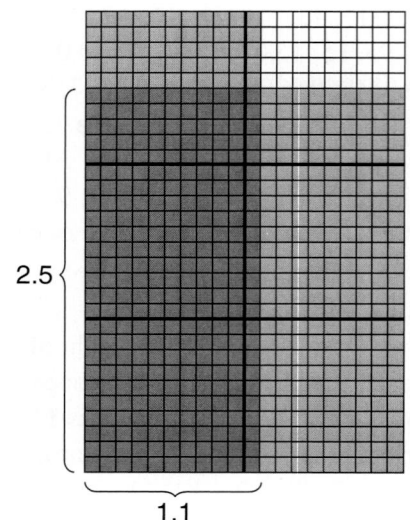

2.5 { }

1.1

Pages 134–136 Lesson 4-1

1. Sample answer: Round 40.32 to 40 and round 251 to 250. You can find 40×250 since 4 and 25 are compatible numbers. **3.** Sample answer: Yes; by rounding 34.78 to 30 and rounding 452 to 500, the product of 30 and 500 is 15,000. **5.** 239.4 **7.** 1.6 **9.** 0.32 **11.** 66,593.12 miles per hour **13.** 1.12 **15.** 30.78 **17.** 64.2 **19.** 2,606.5 **21.** 9.425 **23.** 0.252 **25.** 3,940.2 **27.** 21,286.734 **29.** Sample answer: Since 0.5 is less than 5, 0.5×5 will be less than 25. **31.** $29.25 **33.** Sample answer: 5×9.178 **35.** Sample answer: 80 **37a.** Sample answer: 800 points **37b.** Sample answer: Use clustering to total the hundreds and use rounding to estimate to the nearest ten. **37c.** Sample answer: You would round to the nearest ten because rounding to the nearest hundred they would all be the same.

Pages 142–143 Lesson 4-3

1. Sample answer: The total decimal places of the factors are 5. 0.40563 **3.** 4.08 **5.** 27.9258 **7.** 0.0078 **9.** 69.608 **11.** 0.49 **13.** 13.0968 **15.** 16,688.0428 **17.** 9.49 **19.** 0.0186 **21.** 0.00212 **23.** 5.82788 **25.** 108.8472 **27.** 0.00696 **29.** 5.41 **31.** Sample answer: 0.1×0.6 **33.** D **35.** 216

Page 144 Lesson 4-3B

1. E4 **3.** $28.79, $37.49, $25.41, $12.00, $12.34, $17.95 **5.** No; you would still multiply.

Pages 138–139 Lesson 4-2

1. First find 8^2. Then add 64 and 6. Finally, multiply 70(9). **5.** $80 + 64 = 144$ **7.** $30 + 2 = 32$ **9.** $100 + 25 = 125$ **11.** $3 \times 20 + 3 \times 7$ **13.** 84 **15.** 216 **17.** 48.6 **19.** 124.2 **21.** 92.7 **23.** 630 **25.** $15.75 **27.** $3 \times 5 \rightarrow 3(3 + 2) = (3 \times 3) + (3 \times 2)$ **29.** 35.9 in.

Pages 147–148 Lesson 4-4

5. 7.5 in. **7.** $P = 121.6$ cm; $A = 924.16$ cm^2 **9.** 43 m **11.** 27 m **13.** $P = 6$ in.; $A = 2.25$ in.^{2} **15.** $P = 12.8$ m; $A = 7.68$ m^2 **17.** $P = 46.4$ in.; $A = 135.56$ in.^{2} **19.** 4.5 m **21a.** perimeter **21b.** 880 ft^2

23.

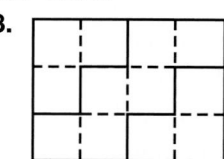

25. C

Page 140 Lesson 4-3A

1. 0.6×1.9

3.

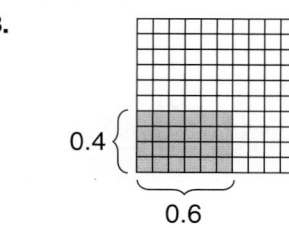

0.4 { }

0.6

Page 149 Lesson 4-4B

1a.

Length	Width	Area
7	1	7
6	2	12
5	3	15
4	4	16

1b.

Length	Width	Area
8	1	8
7	2	14
6	3	18
5	4	20

1c.

Length	Width	Area
9	1	9
8	2	16
7	3	21
6	4	24
5	5	25

3. Sample answer: The longer the length and shorter the width, the lesser the area of the rectangular shape.

Pages 150–151   Lesson 4-5A

1. Sample answer: She thought she was dividing the length into 2-foot sections.   **3.** 17 cuts   **5.** Sample answer: This strategy allows you to break a difficult problem into manageable parts.   **7.** 28 feet
9a. $40 million   **9b.** $50 million   **9c.** $3 million
11. His competition average was 17 pins higher than his previous average.   **13.** D

Pages 153–155   Lesson 4-5

1. Sample answer: The estimate tells you the answer is close to 1.   **3.** Sample answer: Estimate $80 \div 40 = 2$; The estimate shows that the answer is reasonable.   **5.** 0.415   **7.** 7.9   **9.** 2.6   **11.** 0.25   **13.** 21.2   **15.** 9.23   **17.** 2.7   **19.** 3.3   **21.** 22.42   **23.** 9.79   **25.** 5.5   **27.** 4.53   **29.** $14.67   **31.** $0.19, $0.07, $0.06   **33.** Sample answer: $0.9984 \div 8$ **35.** about 27 cm   **37.** B

Page 155   Mid-Chapter Self Test

1. $20,000   **3.** 420   **5.** 0.0204   **7.** 9.8 cm, 0.94 cm^2 **9.** 9.63

Page 156   Lesson 4-6A

1. 2   **3.** 30   **5.** 3

Pages 158–159   Lesson 4-6

1. Sample answer: $1.84 \div 0.8$.   **3.** 19   **5.** 694   **7.** 54.3   **9.** 2.2   **11.** 112   **13.** 0.4   **15.** 0.9   **17.** 2,967   **19.** 0.6   **21.** 8.37   **23.** 107   **25.** 3.9   **27.** 3.41   **29.** 0.84   **31.** 19 tanks   **33.** Sample answer: $0.896 \div 0.35 = 2.56$ **35.** two hundred four and two thousand three hundred ninety-eight ten-thousandths   **37.** 7

Pages 162–163   Lesson 4-7

1. Sample answer: $4.53 \div 15$ **3.** Sample answer: Multiply the divisor and dividend by 10. The problem becomes $1,561.4 \div 148$. Estimate: $1,500 \div 150 = 10$.   **5.** 4.03   **7.** 5.02   **9.** 17.05   **11.** 10.2   **13.** 301   **15.** 5.21   **17.** 10.94   **19.** 20.26   **21.** 10.42   **23.** 2.00   **25.** 101   **27.** 0.807   **29.** 0.0032   **31.** 6.08 m   **33.** never   **35.** D   **37.** Sample answer: 0 to 120

Pages 165–166   Lesson 4-8

1. kilogram   **3.** kilogram; 500 kg   **5.** liter; 2 L   **7.** Sample answer: The fruit is heavier, so the loaf has to be smaller to have the same mass.   **9.** milliliter, 200 mL   **11.** kilogram, 100 kg   **13.** kilogram, 2 kg   **15.** milligram, 1 mg   **17.** gram, 100 g   **19.** milligram, 500 mg   **23.** 1.765 g   **25a.** 4,389 g   **25b.** more than 4 kg   **27.** 201.5   **29.** C

Pages 168–169   Lesson 4-9

1. divide by 1,000   **3.** Sample answer: Parker and Dinh are both incorrect. Parker multiplied by 10 and Dinh divided by 10. They should have multiplied by 1,000.   **5.** multiply; 7   **7.** multiply; 5,020   **9.** divide; 0.15   **11.** 21   **13.** 2,500   **15.** 5.24   **17.** 81.7   **19.** 1.953   **21.** 3,290   **23.** 5,250   **25.** 10   **27.** 0.0067   **29.** 58 cm   **31.** 0.213 L   **33.** about 5.6 cans   **35.** liters   **37.** 77.65

Pages 170–173   Study Guide and Assessment

1. true   **3.** false, perimeter   **5.** true   **7.** true   **9.** true   **11.** false, multiply   **13.** Sample answer: Round 6.9 to 7 and 88 to 90. $7 \times 90 = 630$ **15.** 488.05   **17.** 47.46   **19.** 96,819.2   **21.** 322.8   **23.** 34   **25.** 0.204   **27.** 62.0984   **29.** 539.76   **31.** 27.3 m   **33.** 41 mi; 104.16 mi^2 **35.** 30.4 ft, 57.76 sq ft   **37.** 6.37   **39.** 0.52175   **41.** 19.21   **43.** 141   **45.** 5.9   **47.** 0.01   **49.** 0.08   **51.** 5.436   **53.** kilogram, 5 kg   **55.** milliliter, 15 mL   **57.** 0.001   **59.** 5.2 kilometers   **61.** $319.52

Pages 174–175   Standardized Test Practice

1. D   **3.** C   **5.** C   **7.** C   **9.** B   **11.** C   **13.** 88, 93, 95   **15.** liters   **17.** 0.0235   **19.** Sample answer: The 539-gram can of soup is a better buy. It costs 0.135¢ per gram while the 306-gram can costs 0.137¢ per gram.

Chapter 5
Using Number Patterns, Fractions, and Ratios

Pages 179–180   Lesson 5-1

1. Sample answer:

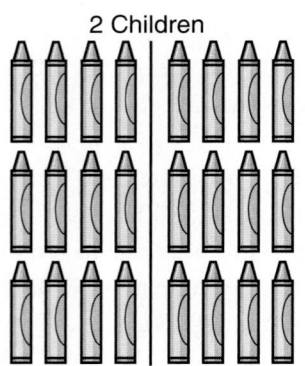
2 Children

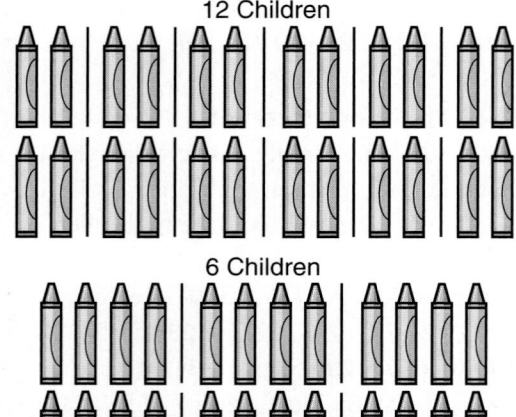
12 Children

6 Children

3. 2: if the ones digit is divisible by 2. 3: if the sum of the digits is divisible by 3. 5: if the ones digit is 0 or 5. 6: if the number is divisible by 2 and 3. 9: if the sum of the digits is divisible by 9. 10: if the ones digit is zero.   **5.** no   **7.** none   **9.** 2, 3, 6   **11.** yes
13. yes   **15.** no   **17.** yes   **19.** yes   **21.** 2, 3, 6, 9
23. none   **25.** 3, 9   **27.** 2, 5, 10   **29.** yes   **31.** no
33. Sample answer: 90   **35.** Yes; each chaperone will have 15 students.   **37.** 50   **39.** 275.856

Page 181   Lesson 5-2A

1.

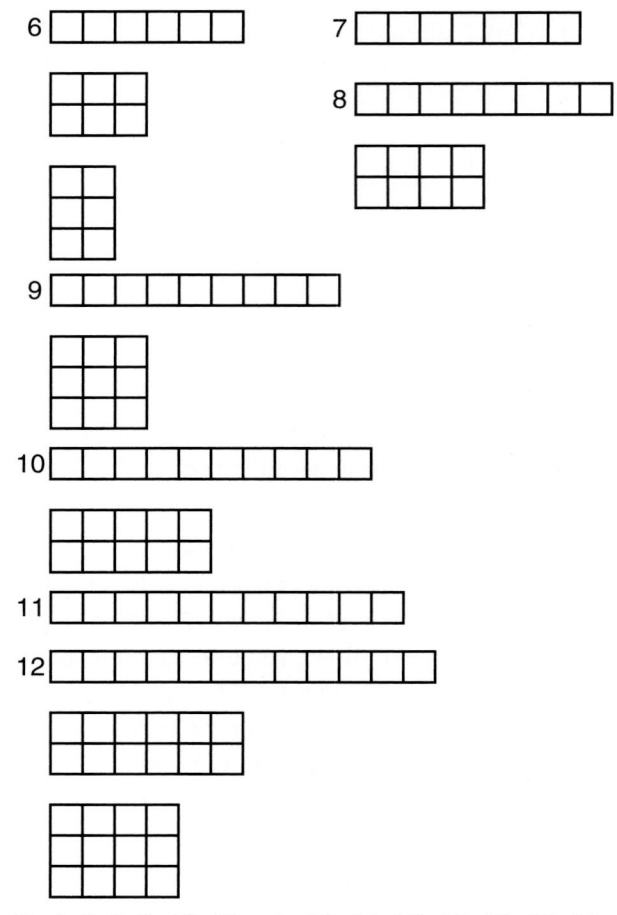

3. 4, 6, 8, 9, 10, 12   **5.** 12, 14, 15, 16, 18, 20, 21, 22, 24; these numbers have several factors.

Pages 183–184   Lesson 5-2

1.

1×10

2×5

So, 10 is a composite number.   **3.** Both are correct. The factors are arranged in a different order.
5. neither   **7.** composite   **9.** 3×5^2 **11.** $2^3 \times 13$
13. neither   **15.** prime   **17.** composite
19. composite   **21.** composite   **23.** composite
25. $2 \times 3 \times 7$ **27.** 5×13 **29.** prime   **31.** 2×3^2
33. $2 \times 3 \times 17$ **35.** $2^3 \times 3 \times 5$ **37.** 83   **39.** 2;
Two is the only even number that has exactly 2 factors, 1 and itself.   **41.** 3 and 5, 5 and 7, 11 and 13, 17 and 19, 29 and 31, 41 and 43, 59 and 61, 71 and 73   **43.** yes   **45.** 4.36   **47.** $17 \times 17 \times 17 \times 17 \times 17 \times 17 \times 17$

Pages 186–187   Lesson 5-3A

3. 12 possible options   **7.** 30   **9.** 6   **11.** $1,680
13. C

Pages 189–190   Lesson 5-3

1. Sample answer: Find the prime factorization of 12 and 18. Multiply all common factors of both numbers to find the greatest common factor.   **3.** 5   **5.** 7   **7.** 8   **9.** 8 ft   **11.** 2   **13.** 15   **15.** 11   **17.** 17   **19.** 12   **21.** 6   **23.** 3   **25.** 17   **27.** 19   **29.** 3   **31a.** 8 in. by 8 in.   **31b.** 18   **33.** Sample answer: 8 and 9   **35.** 24.9 yd^2 **37.** 9.016

Pages 191–192   Lesson 5-4A

1. 5 out of 6, 5 to 6, or 5:6; $\frac{5}{6}$ **3.** 5 out of 9, 5 to 9, or 5:9; $\frac{5}{9}$ **5.** $\frac{3}{4}, \frac{6}{8}$
7.

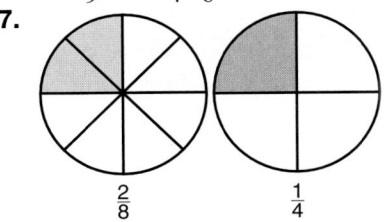

$\frac{2}{8}$ $\frac{1}{4}$

Pages 195–196   Lesson 5-4

1a. $\frac{8}{12}$ **1b.** 4   **1c.** $\frac{2}{3}$ **1d.**
3. 9   **5.** 3   **7.** no; 1 to 9
9. yes   **11.** 4   **13.** 9
15. 24   **17.** 6   **19.** 20   **21.** no; $\frac{5}{19}$ **23.** no; 3 out of 4   **25.** no; $\frac{3}{20}$ **27.** no; 3:5   **29.** no; $\frac{21}{25}$ **31.** no; 1 out of 4   **33.** Sample answer: $\frac{4}{6}$ **35a.** Gwynn, $\frac{2}{5}$; Ripken, $\frac{1}{3}$; Gonzalez, $\frac{11}{36}$; Williams, $\frac{8}{21}$; Bonds, $\frac{1}{3}$; Rodriguez, $\frac{5}{14}$ **35b.** Ripken and Bonds   **37.** $\frac{36}{48}$
39. C   **41.** 26.98, 27.025, 27.13, 27.131, 27.9

Page 197   Lesson 5-4B

1. $\frac{7}{48}$ **3.** $\frac{17}{48}$

Pages 200–201   Lesson 5-5

1. Sample answer: a fraction with a numerator greater than or equal to the denominator.
3.

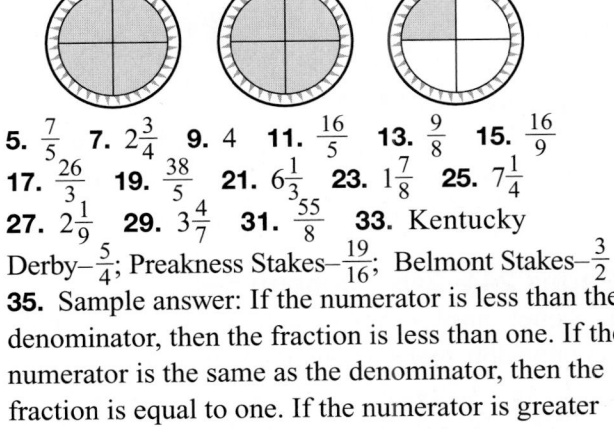

5. $\frac{7}{5}$ **7.** $2\frac{3}{4}$ **9.** 4   **11.** $\frac{16}{5}$ **13.** $\frac{9}{8}$ **15.** $\frac{16}{9}$
17. $\frac{26}{3}$ **19.** $\frac{38}{5}$ **21.** $6\frac{1}{3}$ **23.** $1\frac{7}{8}$ **25.** $7\frac{1}{4}$
27. $2\frac{1}{9}$ **29.** $3\frac{4}{7}$ **31.** $\frac{55}{8}$ **33.** Kentucky Derby–$\frac{5}{4}$; Preakness Stakes–$\frac{19}{16}$; Belmont Stakes–$\frac{3}{2}$
35. Sample answer: If the numerator is less than the denominator, then the fraction is less than one. If the numerator is the same as the denominator, then the fraction is equal to one. If the numerator is greater than the denominator, then the fraction is greater

than one.   **37.** $2^2 \times 3 \times 17$ **39.** 0.13

Page 201   Mid-Chapter Self Test

1. 3, 5   **3.** 2, 5, 10   **5.** $2^3 \times 11$ **7.** 18   **9.** 4:25

Pages 203–205   Lesson 5-6

1. Divide 24 by 12.   **3.** A cubit is about 18 in. A span is about 8 to 9 in.   **5.** 3
7. ▬▬▬▬▬▬▬▬▬
9. $\frac{7}{8}$ in.   **11.** 5   **13.** 108   **15.** 10,560
17. ▬▬▬▬▬▬▬   **19.** ▬
21. ▬▬▬▬▬▬   **23.** $\frac{1}{2}$ in.
25. $1\frac{3}{8}$ in.   **27.** $\frac{1}{4}$ in.   **29.** $1\frac{1}{2}$ ft   **31.** 1 in., $\frac{1}{2}$ in., $\frac{1}{4}$ in., $\frac{1}{8}$ in.   **33a.** 1 in.   **33b.** $1\frac{3}{4}$ in.   **33c.** $2\frac{1}{8}$ in.
33d. $3\frac{7}{8}$ in.   **33e.** 3 in.   **33f.** $\frac{3}{4}$ in.   **35.** $\frac{43}{8}$ **37.** C

Pages 208–209   Lesson 5-7

1. The least common multiple is the least number, other than zero, that is a multiple of two or more numbers.   **3.** 30   **5.** yes   **7.** 9   **9.** 22   **11.** 4 packages of hot dogs, 5 packages of hot dog buns, 2 packages of plates   **13.** yes   **15.** yes   **17.** no   **19.** no   **21.** 12   **23.** 21   **25.** 80   **27.** 208   **29.** 36   **31.** 280   **33.** 18   **35.** 12   **37.** Sample answer: 18 and 24, 36 and 72   **39.** $9\frac{2}{5}$
41. Sample answer: 6 + 234 = 240

Pages 212–213   Lesson 5-8

1. Use the LCM of the denominators to rename fractions so they will have the same denominators. Then compare the numerators.   **3.** The LCD of $\frac{2}{5}$ and $\frac{4}{9}$ is 45. So, rename the fractions with a denominator of 45. $\frac{2}{5} = \frac{18}{45}$ and $\frac{4}{9} = \frac{20}{45}$. Since 18 < 20, $\frac{2}{5} < \frac{4}{9}$.   **5.** 20   **7.** =   **9.** kitchen   **11.** 12
13. 45   **15.** >   **17.** =   **19.** >   **21.** <   **23.** =
25. $\frac{2}{7}$ **27.** $\frac{3}{5}, \frac{3}{7}, \frac{2}{5}, \frac{1}{6}$ **29.** $\frac{16}{90}$ and $\frac{8}{45}$; Since $\frac{16}{90} = \frac{8}{45}$, the cost is the same.   **31.** 300   **33.** $\frac{25}{28}$ **35.** 23

Pages 215–216   Lesson 5-9

1. Sample answer: Write the decimal as a fraction with a denominator of 100. Then simplify by using the GCF.   **3.** Ann; The decimal is written to the hundredths place. So, the denominator will be 100.
5. $\frac{9}{20}$ **7.** $4\frac{3}{8}$ **9.** $3\frac{1}{5}$ **11.** $8\frac{13}{20}$ **13.** $5\frac{16}{25}$
15. $2\frac{2}{5}$ **17.** $13\frac{9}{1,000}$ **19.** $7\frac{89}{100}$ **21.** 0.38; $\frac{19}{50}$
23. McDonald's–$\frac{1}{2}$ mi; Burger King–$\frac{2}{5}$ mi;

Wendy's—$\frac{3}{10}$ mi   **25.** krypton: $83\frac{4}{5}$; selenium: $78\frac{24}{25}$; sulfur: $32\frac{3}{50}$; carbon: $12\frac{11}{1,000}$ **27.** $\frac{3}{7}$ **29.** D

31.

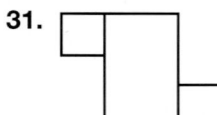

Pages 218–219   Lesson 5-10
1. Divide the numerator by the denominator.   **3.** $0.\overline{4}$
5. 0.375 **7.** $0.58\overline{3}$ **9.** $0.\overline{7}$ **11.** $2.\overline{45}$ **13.** $0.8\overline{31}$
15. 0.75 **17.** $0.\overline{2}$ **19.** $0.\overline{45}$ **21.** $0.91\overline{6}$ **23.** $4.\overline{09}$
25. $0.1\overline{6}$ oz   **29.** $29\frac{3}{4}$ **31.** D

Pages 220–223   Study Guide and Assessment
1. g   **3.** e   **5.** b   **7.** f   **9.** 3   **11.** 3, 9
13. prime   **15.** neither   **17.** 2×3^3 **19.** $2^2 \times 31$
21. 6   **23.** 2   **25.** 20   **27.** no; $\frac{5}{6}$ **29.** no; 7 to 8
31. $\frac{18}{5}$ **33.** $3\frac{4}{5}$ **35.** ▬▬▬▬▬
37. $1\frac{3}{4}$ inches   **39.** 75   **41.** 84   **43.** 20   **45.** >
47. $\frac{4}{5}$ **49.** $3\frac{13}{20}$ **51.** 0.625 **53.** 24 ways
55. $30.40

Pages 224–225   Standardized Test Practice
1. A   **3.** B   **5.** A   **7.** B   **9.** A   **11.** C   **13.** D
15. 129   **17.** $20.49   **19.** $5.63

··

Chapter 6
Adding and Subtracting Fractions

Pages 230–231   Lesson 6-1
1.

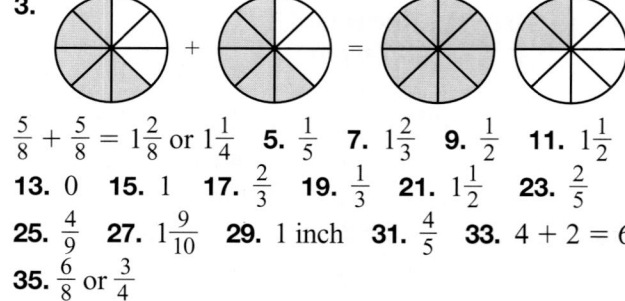

3. Sample answer: If the numerator is close in value to the denominator, round up. If the numerator is about half the denominator, round to $\frac{1}{2}$. If the numerator is much smaller than the denominator, round down.   **5.** 4   **7.** $\frac{1}{2}$ **9.** down
11. $2\frac{1}{2}$ cups   **13.** 1   **15.** $6\frac{1}{2}$ **17.** $\frac{1}{2}$ **19.** 12
21. 4   **23.** $10\frac{1}{2}$ **25.** $5\frac{1}{2}$ **27.** down   **29.** up
31. up   **33a.** $\frac{23}{28}, \frac{2}{7}, \frac{5}{28}, \frac{1}{2}, \frac{11}{14}, \frac{17}{28}, \frac{5}{7}, \frac{3}{4}$
33b. salads, turkey   **33c.** Cracker Jacks, chicken, frozen yogurt   **35.** Sample answer: $5\frac{3}{8}, 5\frac{7}{16}, 5\frac{3}{5}$
37. sixteen ten-thousandths   **39.** D

Pages 233–234   Lesson 6-2
1.

$$2\frac{3}{4}$$

3. Sample answer: Rounding $4\frac{3}{8}$ to 4 is correct, but since the measure was rounded down, there would not be enough material to complete the picture frame. Nina should buy 20 inches of framing material.   **5–9.** Sample answers are given.
5. $7 + 4 = 11$ **7.** $2\frac{1}{2} - \frac{1}{2} = 2$ **9.** $6 + 10 = 16$
11–27. Sample answers are given.   **11.** $1 + \frac{1}{2} = 1\frac{1}{2}$
13. $8 - 1 = 7$ **15.** $1 + \frac{1}{2} = 1\frac{1}{2}$ **17.** $1 + 6 = 7$
19. $9 + 0 = 9$ **21.** $22 + 5 = 27$ **23.** 5 min − 1 min = 4 min   **25.** 20 cups − 11 cups = 9 cups
27. $7 + 7 + 7 + 7 + 7 = 35$ **29.** 64 in. − 62 in. = 2 in.   **31.** Sample answer: $\frac{5}{8}$ and $\frac{7}{16}$ **33.** $\frac{17}{100}$
35. 400,000 cm

Pages 236–237   Lesson 6-2B
1. Sample answer: Franco and Cesar eliminated the routes that they could not take to see which route they could.   **3.** C   **5.** Sample answer: Is $15,840 \div 18$ equal to 80, 88, 880, or 8,800?   **7.** Viho is 8 years old. His father is 32 and his grandfather is 64.
9. 50¢   **11.** about 7 cups   **13.** 12   **15.** D

Pages 240–241   Lesson 6-3
1. Sample answer: Add numerators and keep the denominators the same.
3.

$$\frac{5}{8} + \frac{5}{8} = 1\frac{2}{8} \text{ or } 1\frac{1}{4}$$ **5.** $\frac{1}{5}$ **7.** $1\frac{2}{3}$ **9.** $\frac{1}{2}$ **11.** $1\frac{1}{2}$
13. 0   **15.** 1   **17.** $\frac{2}{3}$ **19.** $\frac{1}{3}$ **21.** $1\frac{1}{2}$ **23.** $\frac{2}{5}$
25. $\frac{4}{9}$ **27.** $1\frac{9}{10}$ **29.** 1 inch   **31.** $\frac{4}{5}$ **33.** $4 + 2 = 6$
35. $\frac{6}{8}$ or $\frac{3}{4}$

Page 242   Lesson 6-4A
1. Sample answer: you rename the units so that the answer makes sense.   **5.** Sample answer: find a common denominator.

Pages 244–245   Lesson 6-4
1. Sample answer: You rename fractions with unlike denominators to get a common unit name.
3. $\frac{7}{16}$ **5.** $\frac{1}{12}$ **7.** $\frac{1}{4}$ **9.** $\frac{7}{8}$ **11.** $\frac{1}{16}$ **13.** $1\frac{1}{10}$
15. $\frac{8}{15}$ **17.** $\frac{7}{24}$ **19.** $\frac{7}{20}$ **21.** $\frac{17}{36}$ **23.** $\frac{3}{8}$

25. $\frac{5}{16}$ inch **27.** $\frac{3}{4}$ **29.** $\frac{1}{12}$ **31.** $1\frac{3}{10}$ **33.** $1\frac{1}{2}$
35. 30 feet **37.** Sample answer: $\frac{2}{8}$ or $\frac{1}{4}$ in. **39.** D

Pages 248–249 Lesson 6-5

1. Sample answer: Find the LCM of the denominators. **3.** Sample answer:

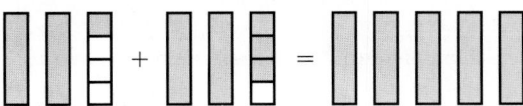

5. $10\frac{3}{4}$ **7.** $28\frac{1}{10}$ **9.** $7\frac{1}{15}$ **11.** 20 inches
13. $2\frac{2}{3}$ **15.** $19\frac{1}{12}$ **17.** $2\frac{1}{4}$ **19.** $8\frac{2}{5}$ **21.** $\frac{5}{24}$
23. $9\frac{13}{80}$ **25.** $14\frac{31}{40}$ **27.** $2\frac{1}{2}$ **29.** $1\frac{1}{4}$ pounds
31. Sample answer: How much more cotton was planted in 1996 than was estimated to be planted in 1997? **33.** D **35.** 0.8 kg

Page 249 Mid-Chapter Self Test

1. Sample answer: To avoid ruining the chili, you should round down to one-half teaspoon of salt. You can always add more. **3.** $22 - 15 = 7$ **5.** $\frac{3}{10}$
7. $\frac{7}{30}$ **9.** $20\frac{1}{6}$

Pages 251–253 Lesson 6-6

1. Sample answer:

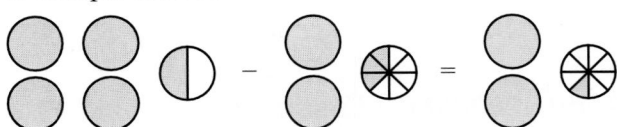

3. Sample answer: When Gail borrowed one from 10 she just carried it next to the 1 in $\frac{1}{5}$ and made it $\frac{11}{5}$. **5.** 30 **7.** $2\frac{5}{6}$ **9.** $\frac{7}{8}$ **11.** 2 **13.** 7 **15.** 29
17. $5\frac{1}{2}$ **19.** $3\frac{7}{10}$ **21.** $4\frac{7}{12}$ **23.** $7\frac{5}{12}$ **25.** $9\frac{3}{4}$
27. $7\frac{34}{35}$ **29.** $\frac{2}{3}$ **31.** $1\frac{23}{30}$ **33.** $1\frac{2}{3}$ **35a.** softball; $2\frac{5}{8}$ to $3\frac{1}{8}$ in. larger **35b.** softball; 1 to 2 ounces heavier **37b.** $1\frac{3}{4}$ in. **39.** $2\frac{17}{24}$ **41.** 1,010.5
43. 11

Pages 255–257 Lesson 6-7

1. 11 minutes **3.** Sample answer: Adding and subtracting measures of time is similar to adding and subtracting mixed numbers because you have to rename. **5.** 10 **7.** 10 min 45 s **9.** 3 h 50 min 40 s
11. 6 h 45 min **13.** $28 **15.** 68 **17.** 6; 27
19. 27; 23 **21.** 10 h 38 min **23.** 16 h 1 min
25. 20 min **27.** 16 min 55 s **29.** 16 h 15 min 20 s
31. 10 h 26 min 20 s **33.** 3 h 50 min 40 s

35. 59 min **37.** 10 h 30 min **39.** 21 h 20 min 10 s
41. 45 min **43.** 10:29 A.M. **45.** $3\frac{3}{4}$ in. **47.** B
49. 100,000,000 miles

Pages 258–261 Study Guide and Assessment

1. 5 **3.** $\frac{4}{5}$ **5.** LCD **7.** $5\frac{1}{2}$ **9.** 80 **11.** $\frac{1}{2}$
13. $11\frac{1}{2}$ **15.** down **17.** up **19.** $8 - 4 = 4$
21. $0 + 5 = 5$ **23.** $1\frac{4}{15}$ **25.** $\frac{3}{4}$ **27.** $\frac{1}{2}$ **29.** $1\frac{5}{18}$
31. $5\frac{2}{7}$ **33.** $1\frac{3}{10}$ **35.** $4\frac{7}{8}$ **37.** $5\frac{1}{3}$ **39.** $5\frac{7}{8}$
41. 68 **43.** 7 h 36 min **45.** 59 min 52 s
47. $8.50 **49.** $1\frac{1}{6}$ miles

Pages 262–263 Standardized Test Practice

1. B **3.** D **5.** C **7.** C **9.** E **11.** A
13. when both factors are less than one **15.** $\frac{5}{9}$
17. $1\frac{1}{16}$ **19.** 14 h 10 min

Chapter 7 Multiplying and Dividing Fractions

Pages 269–270 Lesson 7-1

1. Round $3\frac{1}{2}$ to 4. Round $9\frac{1}{8}$ to 9. Since $4 \times 9 = 36$, $3\frac{1}{2} \times 9\frac{1}{8} \approx 36$. **3.** Both Juan and Odina are correct. If you round $8\frac{1}{2}$ to 9 and $6\frac{1}{4}$ to 6, the product is 9×6, or 54. If you round $8\frac{1}{2}$ to 8 and $6\frac{1}{4}$ to 6, the product is 8×6, or 48. **5.** $\frac{1}{2}$ **7–9.** Sample answers are given. **7.** $\frac{1}{8} \times 16 = 2$ **9.** $3 \times 9 = 27$ **11.** $\frac{1}{2}$
13. $\frac{1}{2}$ **15.** 6 **17.** 3 or 4 **19–29.** Sample answers are given. **19.** $\frac{1}{4} \times 32 = 8$ **21.** $\frac{1}{7} \times 21 = 3$
23. $\frac{1}{6} \times 42 = 7$ **25.** $\frac{4}{9} \times 45 = 20$ **27.** $8 \times 3 = 24$
29. $15 \times 7 = 105$ **31.** about $\frac{2}{3} \times 6$, or about 4 million **33.** 1 h 3 min 51 s **35.** 0.498

Page 271–272 Lesson 7-2A

1. 6 square units **3.** $\frac{1}{6}$ **5.** $\frac{1}{12}$ **7.** To multiply fractions, multiply the numerators and then multiply the denominators.

Pages 275–276 Lesson 7-2

1. The rectangle is separated into thirds. Then the rectangle is separated into halves. The shaded portion shows overlap between $\frac{1}{3}$ of the rectangle and $\frac{1}{2}$ of the rectangle. This area represents the product, $\frac{1}{6}$. **3.** To find the product of fractions, you need to multiply the numerators and then multiply the denominators. If necessary, simplify the answer. **5.** $\frac{3}{20}$ **7.** $\frac{2}{15}$ **9.** $\frac{1}{3}$ **11.** about 105 pounds **13.** $\frac{8}{27}$ **15.** $\frac{1}{9}$ **17.** $\frac{2}{15}$ **19.** $\frac{2}{3}$ **21.** $\frac{10}{27}$ **23.** $\frac{2}{27}$ **25.** $6\frac{2}{3}$ **27.** $\frac{1}{12}$ **29.** $\frac{4}{9}$ **31.** $10\frac{1}{2}$ **33.** $\frac{4}{15}$ **35a.** $\frac{9}{64}$ **35b.** $\frac{16}{25}$ **37.** $\frac{3}{4} \times 36 = 27$ **39.** 2, 3, 6

Pages 278–279 Lesson 7-3

1. To multiply mixed numbers, express each mixed number as an improper fraction and then multiply as with fractions. **3.** $\frac{19}{3}$ **5.** $\frac{23}{4}$ **7.** $3\frac{3}{5}$ **9.** 38 **11.** $1\frac{1}{9}$ **13a.** $\frac{3}{4}$ **13b.** $\frac{3}{8}$ **15.** $\frac{23}{3}$ **17.** $\frac{14}{3}$ **19.** $\frac{33}{5}$ **21.** $\frac{35}{4}$ **23.** $6\frac{2}{3}$ **25.** $10\frac{1}{2}$ **27.** $4\frac{7}{8}$ **29.** 42 **31.** 21 **33.** $8\frac{1}{7}$ **35.** $7\frac{1}{3}$ **37.** 30 **39.** $7\frac{1}{2}$ **41.** $8\frac{1}{4}$ sq ft **43.** more; $2\frac{2}{3} \times 4\frac{1}{2} \approx 3 \times 4$ or 12 **45.** $3\frac{5}{12}$ **47.**

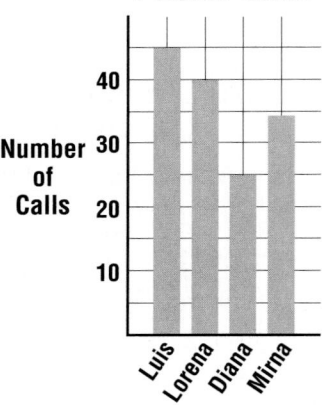

Phone Calls

Pages 282–283 Lesson 7-4

1. Sample answer: Multiply 2 by π by 3. **5.** $\pi = \frac{22}{7}$; $7\frac{6}{7}$ yd **7.** 6.0 m **9.** 12.6 in. **11.** $\pi = \frac{22}{7}$; $7\frac{1}{3}$ in. **13.** 37.7 yd **15.** 25.1 ft **17.** 9.7 yd **19.** 52.8 cm **21.** $\pi = \frac{22}{7}$; $15\frac{5}{7}$ ft **23.** The circumference is twice as long. **25.** $\frac{1}{2}$

Page 283 Mid-Chapter Self Test

1–3. Sample answers are given. **1.** $16 \times \frac{3}{4} = 12$ **3.** $3 \times \frac{1}{2} = 1\frac{1}{2}$ **5.** $\frac{1}{4}$ **7.** $3\frac{3}{8}$ **9.** 42.7 cm

Page 284 Lesson 7-5A

1.

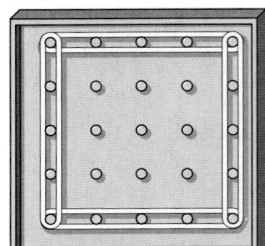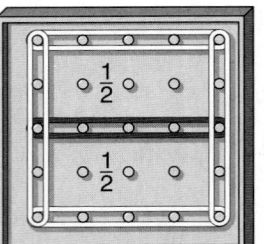

3.

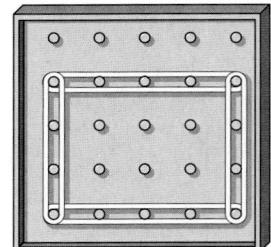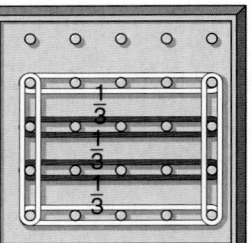

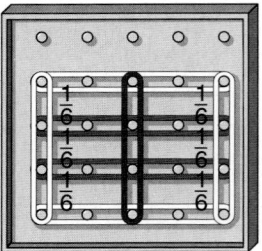

5. 2

Pages 287–288 Lesson 7-5

1. Multiply $\frac{1}{2}$ by $\frac{3}{2}$. **3.** 2 **5.** $\frac{1}{8}$ **7.** $\frac{1}{6}$ **9.** $1\frac{1}{8}$ **11.** 20 **13.** $1\frac{2}{3}$ **15.** $\frac{1}{5}$ **17.** 3 **19.** 7 **21.** 1 **23.** $\frac{5}{9}$ **25.** $\frac{4}{5}$ **27.** $\frac{1}{10}$ **29.** $\frac{1}{3}$ **31.** $\frac{5}{16}$ **33.** $\frac{8}{9}$ **35.** $\frac{2}{3}$ **37.** $\frac{9}{10}$ **39.** $\frac{2}{9}$ **41.** $3\frac{1}{2}$ **43.** 8 quarters **45.** 28 **47.** $0.\overline{7}$ **49.** about 4 inches

Pages 290–291 Lesson 7-6

1. First, write $4\frac{5}{8}$ as the improper fraction $\frac{37}{8}$. Then invert the numerator and the denominator. The reciprocal of $4\frac{5}{8}$ is $\frac{8}{37}$. **3.** Change each mixed number to an improper fraction. Multiply the first fraction by the reciprocal of the second fraction. Divide 8 and 12 by the GCF, 4. Multiply $\frac{3}{1}$ and $\frac{3}{2}$ to find the product $\frac{9}{2}$, or $4\frac{1}{2}$. **5.** $\frac{23}{3}$; $\frac{3}{23}$ **7.** $3\frac{1}{5}$ **9.** $4\frac{1}{2}$ **11.** $1\frac{1}{7}$ **13.** 36 slices **15.** $\frac{32}{5}$; $\frac{5}{32}$ **17.** $\frac{14}{5}$; $\frac{5}{14}$ **19.** $\frac{15}{4}$; $\frac{4}{15}$ **21.** $\frac{37}{4}$; $\frac{4}{37}$ **23.** $7\frac{3}{5}$ **25.** 12 **27.** $\frac{4}{5}$ **29.** $4\frac{3}{8}$ **31.** $\frac{2}{3}$ **33.** 26 **35.** $\frac{2}{3}$

37. $8\frac{1}{3}$ 39. $\frac{4}{15}$ 41. 6 photographs 43. Greater than; $\frac{2}{3}$ is less than $\frac{3}{4}$. 45. $\frac{10}{21}$ 47. B 49. two tenths; seven hundredths; four thousandths

Page 293–294   Lesson 7-7

1. fluid ounce, cup, pint, quart, gallon, ounce, pound, and ton 3. 8 5. 4 7. 2 9. about 13,000 pounds 11. 6 13. 144 15. 2 17. 40 19. 4,500 21. 10 23. $1\frac{1}{2}$ 25. 40 27. 500 pounds 29. 16 31. 3, 5, 9 33. E

Page 295   Lesson 7-7B

1.

Planet/Moon	Amount of water	Weight compared to Earth	Weight compared to Jupiter
Sun	28 cups	more	more
Mercury	$\frac{1}{3}$ cups	less	less
Venus	$\frac{9}{10}$ cups	less	less
Moon	$\frac{1}{6}$ cups	less	less
Mars	$\frac{3}{8}$ cups	less	less
Jupiter	3 cups	more	

3. 66 pounds 5. 6 ounces

Pages 296–297   Lesson 7-8A

3.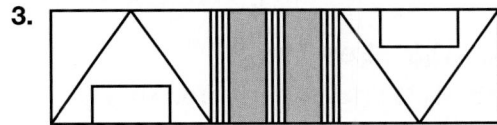

5. Sample answer: Once you identify the pattern, then you can continue the pattern to solve the problem. 9a. $\frac{31}{32}$ 9b. Richard will not finish the sandwich, but the amount will be so small that it will be like he did eat it all. 11. 1,476 sq cm 13. C

Pages 299–300   Lesson 7-8

1. Sample answer: A sequence is a list of numbers in a specific order. 3. Joshua; In the sequence, $2\frac{1}{2}$ is added to each number. The next number is $12\frac{1}{2} + 2\frac{1}{2}$, or 15. 5. 17, 10 7. 125 9. 75, 90 11. $\frac{3}{4}, \frac{3}{8}$ 13. 432, 2,592 15. 33 17. 1 19. 400 21. x^5 23. $\frac{1}{16}$ (sixteenth note), $\frac{1}{32}$ (thirty-second note), and $\frac{1}{64}$ (sixty-fourth note) 25a. $\frac{1}{2}, \frac{1}{4}, \frac{1}{8}, \frac{1}{16}, \frac{1}{32}, \frac{1}{64}, \frac{1}{128}, \frac{1}{256}, \frac{1}{512}, \frac{1}{1,024}$ 25b. 1 27. A

Page 301   Lesson 7-8B

1. Multiply the value in cell B5 by the value in cell B2. 3. 8.000 m, 3.200 m, 1.280 m, 0.512 m, 0.2048 m, 0.08192 m

Pages 302–305   Study Guide and Assessment

1. k 3. i 5. h 7. f 9. d 11–16. Sample answers are given. 11. $10 \times 3 = 30$ 13. $1 \times 13 = 13$ 15. $\frac{2}{3} \times 18 = 12$ 17. $\frac{7}{10}$ 19. $7\frac{1}{2}$ 21. $4\frac{1}{2}$ 23. $13\frac{1}{2}$ 25. 57.1 ft 27. $\pi = \frac{22}{7}$; $5\frac{1}{2}$ yd 29. $1\frac{1}{9}$ 31. $\frac{2}{3}$ 33. $\frac{1}{3}$ 35. $\frac{21}{40}$ 37. 11,000 39. 48 41. 176, 156 43. 2 45. 57 students 47. 144 one-cup servings

Pages 306–307   Standardized Test Practice

1. C 3. B 5. B 7. B 9. D 11. A 13. $5.05 15. $\frac{1}{2}$ yd 17.

Chapter 8
Exploring Ratio, Proportion, and Percent

Page 311   Lesson 8-1A

1a. $\frac{1}{2}$ square unit 1b. Triangle C is $\frac{1}{2}$ the size of triangle B. 3a. $\frac{1}{2}$ 3b. $\frac{1}{4}$ 3c. 1 3d. $\frac{2}{1}$ 3e. $\frac{1}{2}$ 3f. $\frac{1}{2}$

Pages 313–315   Lesson 8-1

1. A ratio is a comparison of two numbers by division. A rate is a ratio of two measurements that have different units. 3. Sample answer: Rates would be useful when shopping in the grocery store. They allow you to determine better buys. 5. Sample answer: 15 to 34, $\frac{15}{34}$, 15 out of 34 7. $\frac{2}{5}$ 9. $0.65 per soft drink 11–15. Sample answers are given. 11. 9 out of 16, 9 to 16, 9:16 13. 18 out of 29, 18 to 29, 18:29 15. 23 out of 25, 23 to 25, 23:25 17. $\frac{1}{3}$ 19. $\frac{7}{9}$ 21. $\frac{4}{9}$ 23. 79 kilometers per hour 25. $8.75 per ticket 27. $0.11 per egg 29. Sample answer: 13 out of 24; 13 to 24; 13:24 31a. $\frac{2}{25}$ 31b. $\frac{6}{25}$ 33. 25 miles 35a. $\frac{1}{144}$ 35b. $\frac{1}{64}$

37. $4\frac{5}{8}$ **39a.** 6.6 billion lb   **39b.** 34.1 billion lb

Page 316   Lesson 8-1B
1. $\frac{1}{4}$; $\frac{3}{4}$ **3.** Sample answer: The results are about the same.   **5.** 0.625

Pages 319–320   Lesson 8-2
1. A proportion is an equation stating that two ratios are equivalent.   **5.** no   **7.** 72   **9.** 120 dentists
11. yes   **13.** no   **15.** no   **17.** 8   **19.** 11
21. 100   **23.** 100   **25.** 6 pairs   **27.** 1,404
29. 15   **31.** $7\frac{1}{6}$ **33.** D

Page 321   Lesson 8-2B
1. Multiply the contents of cell B2 by 2.   **3.** 6 cups of mayonnaise, 24 T of mustard, 24 T of relish, 12 T of vinegar, 3 t of salt, $1\frac{1}{2}$ t of pepper, 60 cups of potatoes, 3 cups of parsley

Pages 322–323   Lesson 8-3A
1.

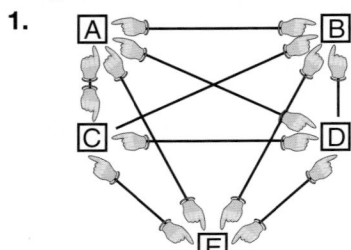

3. 15 handshakes   **5.** A diagram will help you understand and picture the information more clearly.
7.

$\frac{7}{8}$ in. $\left\{\begin{array}{l}\frac{1}{4} = 28 \text{ in.} \\ \frac{1}{4} = 28 \text{ in.} \\ \frac{1}{4} = 28 \text{ in.} \\ \frac{1}{2} \text{ of } \frac{1}{4} = 14 \text{ in.}\end{array}\right.$

9. Reynelda, Juliet, Keela, Pedro, Daniel
11. 417.5 kg   **13.** 6 cuts

Pages 325–327   Lesson 8-3
1. An architect would make a scale drawing or a model in order to show a building exactly as it looks, but smaller.   **3a.** $2\frac{1}{4}$ feet   **3b.** $3\frac{1}{2}$ feet
3c. $3\frac{1}{2}$ feet   **5a.** $1\frac{2}{3}$ yards   **5b.** $1\frac{1}{3}$ yards   **5c.** 2 yards   **7a.** $14\frac{1}{4}$ miles   **7b.** $23\frac{3}{4}$ miles   **7c.** $16\frac{5}{8}$ miles   **7d.** $45\frac{1}{8}$ miles   **7e.** $8\frac{5}{16}$ miles   **9.** 20 feet
11a. length: 5.18 m; width: 2.16 m; height: 1.58 m
11b. length: 4.93 m; width: 1.92 m; height: 1.47 m
13. 5.4

Page 327   Mid-Chapter Self Test
1. Sample answer: 11 to 19, $\frac{11}{19}$, 11:19   **3.** $\frac{3}{7}$
5. 60 mph   **7.** 96   **9.** 2.8

Page 329   Lesson 8-4A

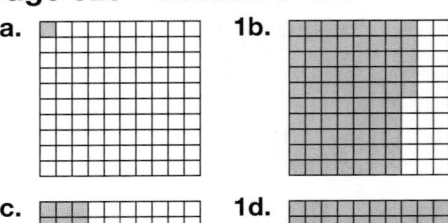

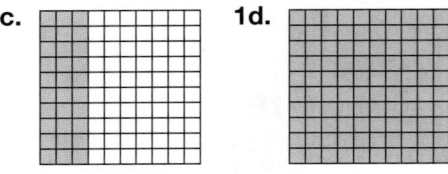

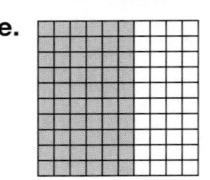

3. Sample answer: Grid C from the first set is the same as grid A from the second set. Grid A from the first set is the same as grid B from the second set. Grid B from the first set is the same as grid C from the second set. Grid D from the first set is the same as grid D from the second set. Grid E from the first set is the same as grid E from the second set.

Pages 332–333   Lesson 8-4
1a. 25%   **1b.** 47%   **1c.** $33\frac{1}{3}$%   **1d.** 80%
1e. 50%   **1f.** $66\frac{2}{3}$%   **5.** $\frac{11}{20}$ **7.** 34%   **9.** 125%
11. $\frac{7}{50}$ **13.** $\frac{1}{100}$ **15.** $1\frac{3}{10}$ **17.** $1\frac{1}{20}$ **19.** $\frac{17}{100}$
21.   **23.**

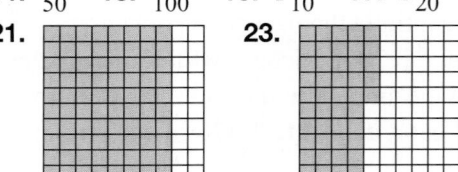

25. 63%   **27.** 150%   **29.** 74%   **31.** 76%
33. 45%   **35a.** 48%   **35b.** 9%   **37a.** $\frac{1}{144}$, 0.01, 1%   **37b.** $\frac{1}{64}$, 0.02, 2%   **39.** 55 miles   **41.** $1\frac{1}{4}$
43. 30

Pages 335–336   Lesson 8-5
1.

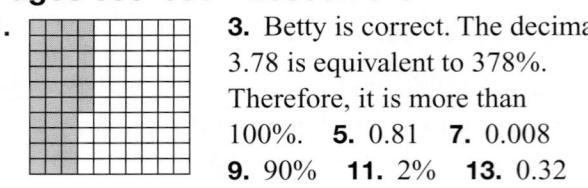

3. Betty is correct. The decimal 3.78 is equivalent to 378%. Therefore, it is more than 100%.   **5.** 0.81   **7.** 0.008
9. 90%   **11.** 2%   **13.** 0.32

15. 0.01 **17.** 0.06 **19.** 0.003 **21.** 0.39 **23.** 0.04
25. 96% **27.** 36.4% **29.** 71.6% **31.** 7%
33. 8% **35.** 20% **37.** 1.8% **41a.** 18%
41b. 82% **43.** $1\frac{14}{25}$ **45.** D **47.** 5.4

Pages 338–339 Lesson 8-6

1. $75\% = \frac{3}{4}$. Round 1,976 to 2,000. $\frac{3}{4} \times 2,000 =$ 1,500. So, 75% of 1,976 is about 1,500. **5.** Sample answer: $\frac{1}{4} \times 200 = 50$ **7–17.** Sample answers are given. **7.** 50% **9.** $\frac{1}{5} \times 35 = \$7$ **11.** $\frac{1}{3} \times 60 = 20$ **13.** $\frac{1}{2} \times 50 = 25$ **15.** $\frac{2}{5} \times 100 = 40$ **17.** $1 \times 300 = 300$ **19.** 75% **21.** $66\frac{2}{3}\%$ **23.** 100% **25–27.** Sample answers are given. **25.** $\frac{3}{8} \times 48 = 18$ **27.** $\frac{1}{100} \times 10 = \frac{1}{10}$ **29.** $50\% \times 5,000 = 2,500$; $20\% \times 5,000 = 1,000$; $2,500 - 1,000 = 1,500$ **31.** c; because a and b equal 11 while c equals 0.11. **33.** 23.04 m² **35.** 38

Pages 342–343 Lesson 8-7

1. Sample answer: **Method 1:** Change the percent to a fraction. First, change the percent to a fraction. Then multiply the number by the fraction. **Method 2:** Change the percent to a decimal. First, change the percent to a decimal. Then multiply the number by the decimal. **Method 4:** Use a calculator. First, enter the percent by entering the number [2nd] [%]. Then press [×] and the number by which you are multiplying. Press [=]. **3.** Kosey is correct. $125\% = 1.25$. $1.25 \times 150 = 187.5$. So, 125% of 150 is 187.5 **5.** 6.72 **7.** 5.88 **9.** 18 **11.** 92 **13.** 6 **15.** 98.98 **17.** 181.25 **19.** 137 **21.** 219 **23.** 6.4 **25.** $34 **27a.** Manuel **27b.** $10.50 **29.** Sample answer: $\frac{3}{8} \times 64 = 24$

Pages 344–347 Study Guide and Assessment

1. false; division **3.** true **5.** false; proportion **7.** true **9.** true **11.** true **13.** false; 34.6% **15.** Sample answer: Express the percent as a fraction with a denominator of 100 and then simplify. **17.** $\frac{11}{20}$ **19.** 15.75 pounds per week **21.** yes **23.** 25 **25.** 30 **27.** 1.5 units **29.** $\frac{3}{100}$ **31.** $1\frac{1}{2}$ **33.** 60% **35.** 55% **37.** 5% **39.** 0.38 **41.** 0.66 **43.** 130% **45.** 59.1% **47.** Sample answer: $\frac{1}{4} \times 80 = 20$ **49.** Sample answer: $\frac{2}{5} \times 240 = 96$ **51.** 16.02 **53.** 17 **55.** 0.162 **57.** 50 mph **59.** 8 cans

Pages 348–349 Standardized Test Practice

1. C **3.** A **5.** D **7.** D **9.** D **11.** C **13.** D **15.** 60% **17.** $1\frac{3}{8}$ inches **19.** 21

Chapter 9
Geometry: Investigating Patterns

Pages 353–355 Lesson 9-1

1. Sample answer: **Step 1:** Place the center of the protractor on the vertex of the angle with the straightedge along one side. **Step 2:** Using the scale that begins with 0° on the side of the angle, read the angle measure where the other side crosses the same scale. **3.** 50°; acute **5.** obtuse **7.** acute **9.** 74° **11.** 65°; acute **13.** 135°; obtuse **15.** 15°; acute **17.** right **19.** obtuse **21.** acute **23.** acute **25.** neither **27.** acute **29.** 18° **31.** Sample answer: A right angle is formed by the two pieces. The angles on the inside corner are obtuse and the angles on the outside corner are acute. **33a.** cut into the hill and flatten the road **33b.** larger **35.** $0.76 = \frac{19}{25}$; $0.15 = \frac{3}{20}$; $0.09 = \frac{9}{100}$ **37.** 52

Pages 356–357 Lesson 9-1B

1. Sample answer: Jesse's and Estella's thinking does make sense because $22 + 33 = 55$, the number of students who play sports. **3.** 82 students **5.** You found that 5 students have only lunch bags, 10 have only backpacks, and 9 have neither a lunch bag nor a backpack. $5 + 10 + 9 = 24$. So, there are 24 students going on the trip. **7.** A right angle measures 90°. The measure of the angle given is 90.4°. Since 90.4° is greater than 90°. It is only logical that the angle is an obtuse angle. **9.** 456
11a. Sample answer:

11b. Each figure has one more side than the previous figure, so the next figure must have six sides. **13.** C

Pages 360–361 Lesson 9-2

3. The corner of a sheet of paper is a 90° angle. A 45° angle is half of a 90° angle. So, to show a 45° angle, you can fold the corner in half.

5.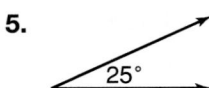

7. Sample answer: about 120°
9. about equal to

11. 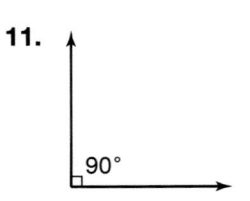 90°

13. 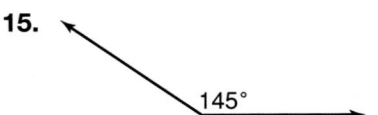 8°

15. 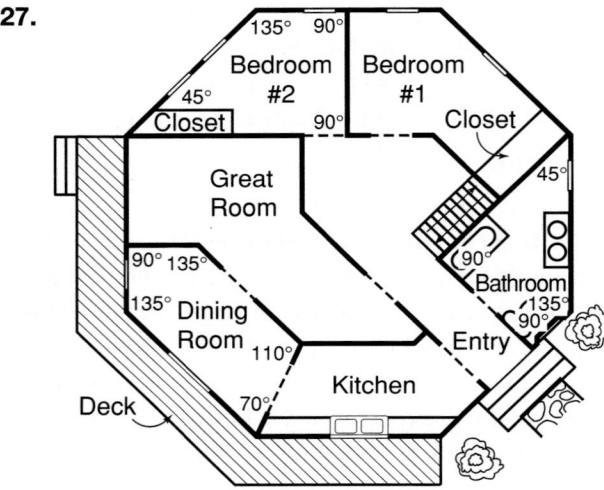 145°

17. 66°

19–23. Sample answers are given. **19.** about 40°
21. about 55° **23.** about 70° **25a.** south **25b.** 180°

27.

29. acute **31.** 120

Pages 362–363 Lesson 9-3A

1.

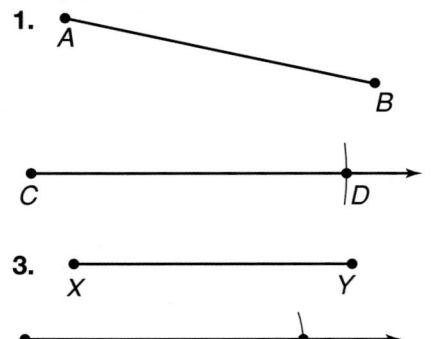

3.

5.

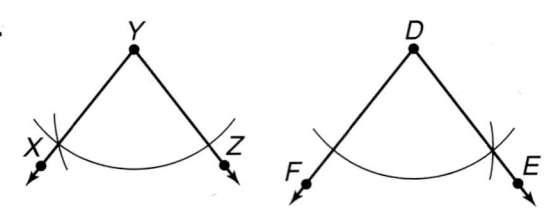

7.

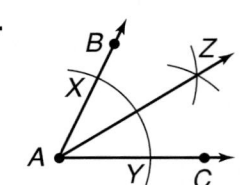

Page 366 Lesson 9-3
1. To separate something into two congruent parts.
3. The crease bisects the angle.

5.

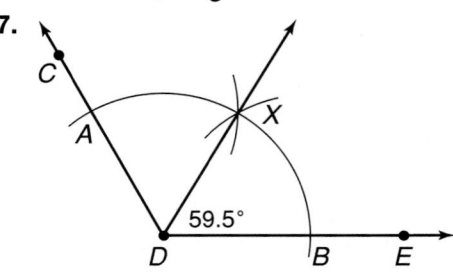

7.

 59.5°

9.

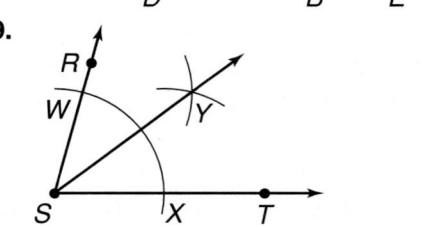

11.

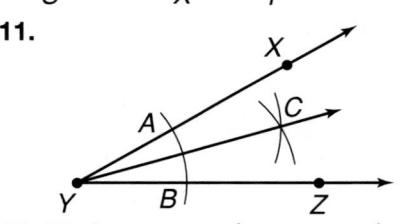

13–15. Answer not drawn to scale.
13.

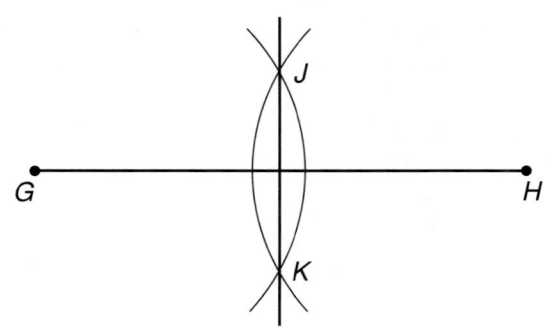

15.

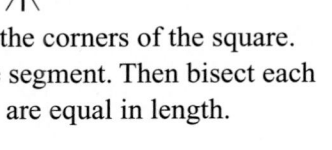

17. The diagonals bisect the corners of the square.

19. Bisect the 7-inch line segment. Then bisect each half of the segment. They are equal in length.

21. $\frac{37}{50}$

Pages 368–369   Lesson 9-4A

1. acute   **3.** right   **5.** Sample answer: The quadrilateral has two sides that are congruent.

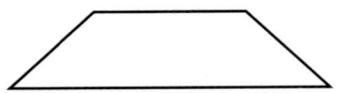

7. Sample answer: The opposite sides of the quadrilateral are parallel.

Pages 372–373   Lesson 9-4

1. Sample answer: A chalkboard is a rectangle and a piece of floor tile is a square.   **3.** Jonathan; All squares are parallelograms since both sets of opposite sides of a square are parallel. Victoria is incorrect. Some parallelograms are not squares since some parallelograms do not have right angles.

5. equilateral triangle; yes   **7.** Alike: All sides are congruent. All angles are congruent.   Different: pentagon has 5 sides and square has 4 sides

9. 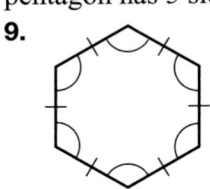   **11.** pentagon; yes   **13.** square; yes   **15.** Alike: all sides congruent; all angles congruent   Different: triangle has three sides and square has four sides

17. Alike: quadrilaterals; at least 1 pair of parallel sides   Different: square has all sides congruent and four right angles

19.    **21.**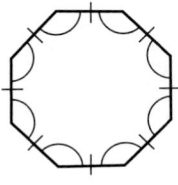

23.    **25.** pentagon, hexagon
27a. quadrilaterals, pentagons, hexagons, octagons   **27b.** 360°; 540°; 360°; 1,080°

29. 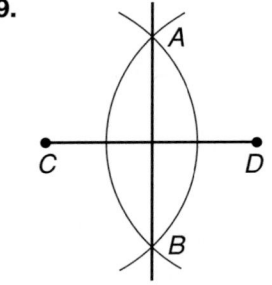   **31.** B

Page 373   Mid-Chapter Self Test

1. 125°; obtuse   **3.** 90°; right   **5.** about 30°

7. 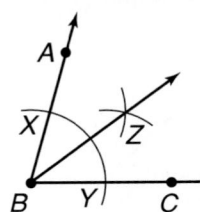   **9.**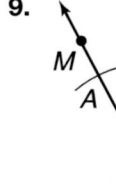

Page 374   Lesson 9-4B

1. The second net did not cover the cube. The "top" and "bottom" covers are on the same side of the cube.

3. Sample answer: Each net contains six squares. Each net has four squares lined up in a row. Each net has two squares positioned on opposite sides of the row of four squares.

Pages 376–378   Lesson 9-5

5. yes

7. 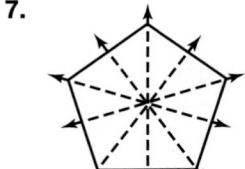   **9.**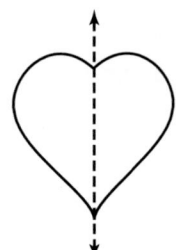

11. yes   **13.** yes   **15.** yes   **17.** yes   **19.** no

21. 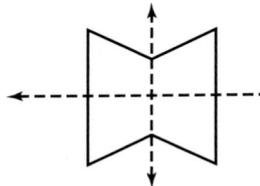   **23.**

25. 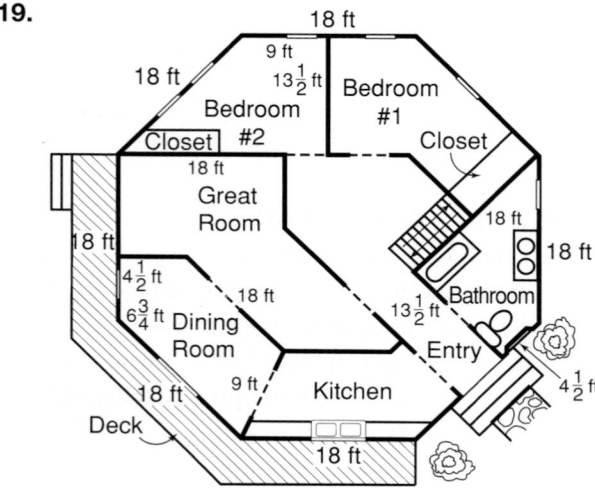 **27.**

29. yes **31.** no **33.** yes **35.** 8 **37.** 1 **41.** $6\frac{1}{3}$
43. 525 people

Pages 380–382   Lesson 9-6

1. The angles are congruent, they have the same shape, and they are different sizes. **5.** congruent **7.** 3 **9.** neither **11.** neither **13.** congruent **15a.** 2.5 m **15b.** $\overline{NM}$ **17.** A and D; A and F; A and G; B and D; B and F; B and G; F and D; A and B; D and G; F and G

19.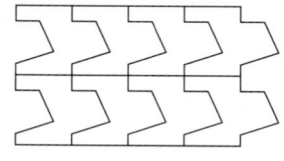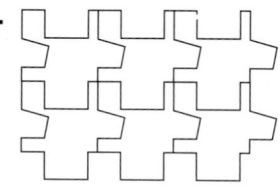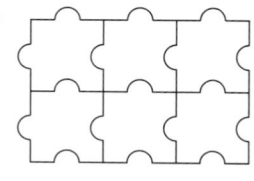

21. B **23.** $5\frac{3}{4}$

Pages 384–385   Lesson 9-6B

1. **3.**

5. **7.**

Pages 386–389   Study Guide and Assessment

1. f **3.** a **5.** e **7.** b **9.** d **11.** If an angle measures less than 90°, it is an acute angle. If an angle measures 90°, it is a right angle. If an angle measures between 90° and 180°, it is an obtuse angle. **13.** 42°; acute **15.** acute **17.** acute **19.** 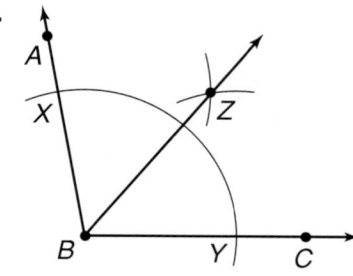 **21.** about 90°

23.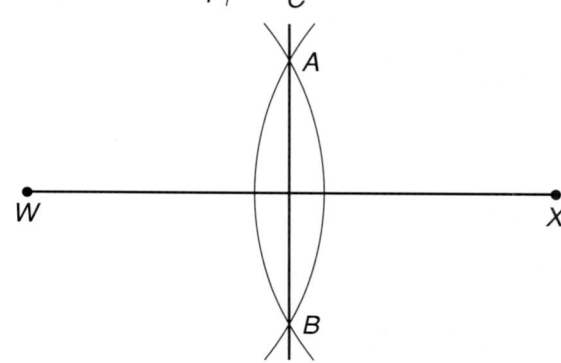

25.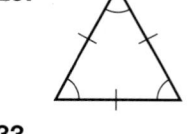

27. pentagon; no

29. 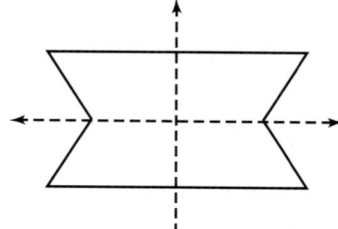 **31.** yes

33. 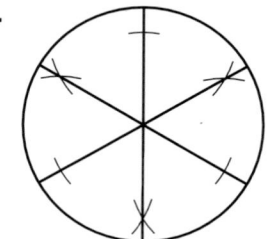 **35.** similar

37.

1. C **3.** D **5.** A **7.** B **9.** A **11.** C
13. 126 feet **15.** $\frac{1}{8}$ **17.** 26 inches by 18 inches

Chapter 10
Geometry: Understanding Area and Volume

Pages 396–397 Lesson 10-1A
7–11. Sample answers are given. **7.** 5 square units
9. $4\frac{1}{2}$ square units **11.** The areas are the same.

Pages 400–401 Lesson 10-1
1. A rectangle is a special parallelogram. **3.** If the lengths of the bases and the heights are the same, the areas will be the same. **5.** 13.6 sq m **7.** 54 in²
9. 38.8 cm² **11.** 70 m² **13.** 42.7 cm²
15. 4,611 cm² **17a.** The area doubles.
17b. The area is increased four times. **19.** 2, 3, 6

Pages 404–405 Lesson 10-2
1. Two congruent triangles form a parallelogram.
3. Ebony is correct. If 5.8 is used as the height, then the base is the length of the side to which the height is drawn, the 7 inch side. **5.** 35 mm² **7.** 1,136 m²
9. 13.5 units² **11.** 51 ft² **13.** 41.3 cm²
15. 1,400 in² **17.** 128 cm² **19.** 32 cm
21. Sample answer: An elephant's ears expands its overall surface area, allowing it more space to release heat and cool itself. **25.** 320 mm²
27. 46.4 ounces

Pages 407–409 Lesson 10-3
1. The circle can be cut and arranged into a figure like a parallelogram to find the area. **5.** 254.3 m²
7. 1.4 ft² **9.** 28.3 in² **11.** 615.4 ft² **13.** 95.0 in²
15. 379.9 in² **17.** 63.6 m² **19.** 379.9 mm²
21. 95.0 m² **23.** about 63.6 m² **25.** The area is multiplied by 4. **27.** A

Page 411 Lesson 10-3B
1. Sample answer: The circle graph, because it allows a visual comparison of the proportions.
3. Sample answers: data that doesn't equal 100%; a bar or line graph

5.

World Population

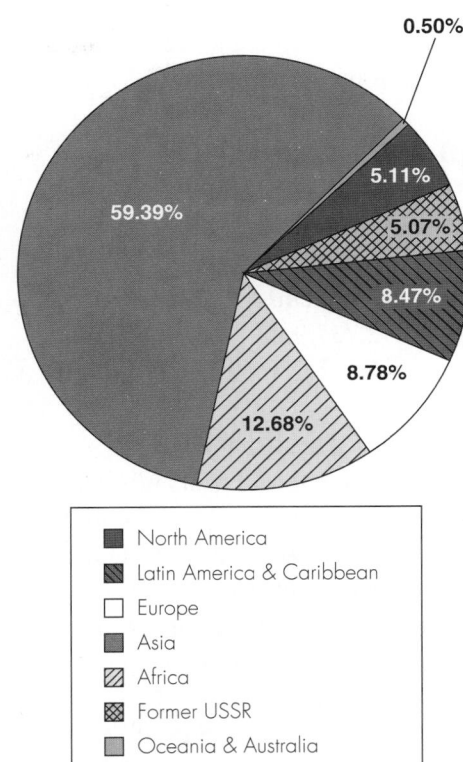

0.50%
5.11%
5.07%
8.47%
8.78%
12.68%
59.39%

- ■ North America
- ▨ Latin America & Caribbean
- ☐ Europe
- ▨ Asia
- ▨ Africa
- ▨ Former USSR
- ▨ Oceania & Australia

7. The area of the circle is divided into parts to represent the portion of the whole for each category.

Pages 413–414 Lesson 10-4
1. square pyramid **5.** cone **7.** 5; 9; 6
9. cylinder **11.** triangular prism **13.** cone
15. 4; 6; 4 **17.** none; none; none **19.** 8; 18; 12
21a. 8; 12; 6 **21b.** yes **23.** 1,519.76 in² **25.** B

Page 414 Mid-Chapter Self Test
1. 3,850 ft² **3.** 12.6 in² **5.** cylinder

Page 415 Lesson 10-4B
1. the sides that are rectangles 3 units long and 2 units wide **2.** yes **3.** Draw the hexagonal base first. Then from each vertex, draw a line to represent the depth. Then connect the ends of the segments.
5.

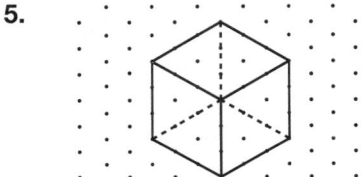

Page 417 Lesson 10-5A
1. Sample answer: It allowed them to build a small pyramid with easy-to-move boxes first. **3.** Emilio

made sure that the number of boxes they planned to use was the most possible for the type of pyramid they were building. **5.** 55 cubic inches **7.** 20 boxes **9.** Sample answer: 4 by 2 by 1 **11.** 501

Pages 419–420 Lesson 10-5

1. Because each of the three dimensions being multiplied is expressed in a unit of measure.
3. Sample answer: a prism 1 cm high, 2 cm wide, and 9 cm long **5.** 588 cm³ **7.** 14,400 in³
9. 1,536 ft³ **11.** 2,737.9 m³ **13.** 90 mm³
15a. 3,750 m³ **15b.** 3,750,000 liters **19.** B

Pages 423–424 Lesson 10-6

1. Find the area of each face. Front and back: 4 × 10 = 40 in² Top and bottom: 4 × 6 = 24 in² Sides: 6 × 10 = 60 in² Then find the total surface area. 2(40) + 2(24) + 2(60) = 248 in² **3.** Sample answers: determining the correct amount of paint for a room; or choosing enough wrapping paper for a package **5.** 142 cm² **7.** 312 in² **9.** 286 m²
11. 96 cm² **13.** 648 ft² **15.** 718 m² **17.** 1,240 in²
21. 328 ft² **23.** 6 units on each side **25.** 9 hours 9 minutes

Page 425 Lesson 10-6B

1.

	A	B	C	D	E
1	LENGTH	WIDTH	HEIGHT	VOLUME	SURFACE AREA
2	3	3	7	63	102
3	10.2	4.1	1.6	66.912	129.4
4	4	4	4	64	96
5	8	4	4	128	160
6	8	8	4	256	256
7	8	8	8	512	384

3. Delete the height column. Change "volume" to "area" and change the formulas in column D to A2*B2, A3*B3, and so on. Change "surface area" to "perimeter". Change the formulas in column E to 2*A2 + 2*B2, 2*A3 + 2*B3, and so on. **5.** When one dimension is doubled, the volume doubles. When two dimensions are doubled, the volume is multiplied by four. When all three dimensions are doubled, the volume is multiplied by eight. The volumes are changed in this way because if any one dimension is multiplied by 2, the volume is multiplied by 2.

Pages 426–429 Study Guide and Assessment

1. height **3.** edges **5.** sphere **7.** volume
9. surface area **11.** 14 m² **13.** 87.5 cm²

15. 72 cm² **17.** $62\frac{7}{16}$ ft² **19.** 153.9 in²
21. 40.3 in² **23.** rectangular prism **25.** 6; 12; 8
27. $103\frac{1}{8}$ yd³ **29.** 7,280 m² **31.** 25.3 mm²
33a. $85\frac{2}{3}$ ft² **33b.** 4 bushels **35.** 9 m³

Pages 430–431 Standardized Test Practice

1. A **3.** B **5.** D **7.** B **9.** A **11.** $6.35
13. 4; 6; 4 **15.** 718 ft²

....

Chapter 11
Algebra: Investigating Integers

Pages 435–436 Lesson 11-1

1. −1 and +2 **3.** Sample answer: The yards lost or gained in one down of a football game. For example, −3 would represent a loss of 3 yards, and +4 would represent a gain of 4 yards.
5.
7. +6 **9.** −4 **11.** −345 **13.** −178
15.
17.
19.
21.
23. +75 **25.** −100 **27.** +100 **29.** −23
31. −45 **33.** −77 **35.** +110
37.
39. +7,000 **43.** C **45.** 135% **47.** 673.0

Pages 438–439 Lesson 11-2

1. Graph the numbers on a number line. The number to the left is less than the number to the right.
3. Cordelia; since −5 is to the left of −3 on the number line, −5 < −3. **5.** < **7.** −3, −2, 0, 5
9. > **11.** < **13.** < **15.** > **17.** < **19.** 41, 10, 3, 0, −10, −20 **21.** −17 **23a.** −458, −435, −361, −38, 450, 1,542, 1,763, 1,947, 2,795
23b. solid **23c.** 450 **25.** All negative integers are to the left of 0 on a number line while all positive integers are to the right. The number to the left is always less than the number to the right. **27.** C

Page 440 Lesson 11-3A

1. none **3.** They are opposites. **5.** 0 **7.** Place 4 yellow counters on the mat to represent +4. Place 3 red counters on the mat to represent −3. Remove as many zero pairs as possible. The 1 yellow counter left represents the sum (+1).

Pages 443–444 Lesson 11-3

1. $-5 + 2 = -3$ **3.** negative **5.** 2 **7.** -2
9. -8 **11.** negative **13.** positive **15.** negative
17. -9 **19.** 0 **21.** -3 **23.** 4 **25.** -8 **27.** 10
29. 8 **31.** -3 **33.** $-123°C$ **35a.** Sample answer: $-2 + (-3) = -5$ **35b.** Sample answer: $0 + (-5) = -5$ **35c.** Sample answer: $-8 + 3 = -5$
37. 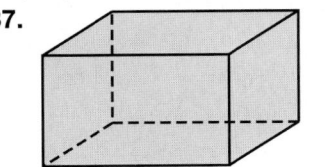 **39.** D

Pages 447–448 Lesson 11-4

1. $4 - (-2) = 6$ **3.** Donnell; you can subtract 8 from 5 by using negative integers. $5 - 8 = -3$
5. 0 **7.** -2 **9.** -2 **11.** -5 **13.** 3 **15.** -24
17. 8 **19.** 0
21. 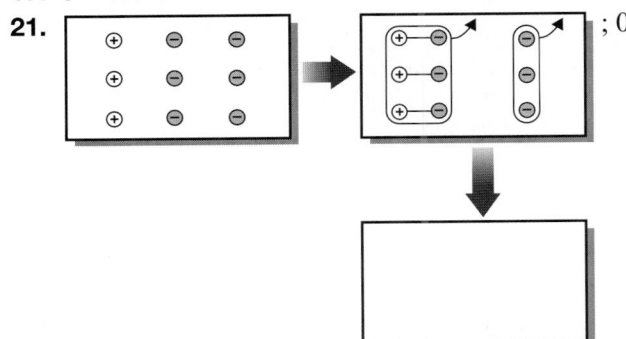 ; 0

23a. 18 ft **23b.** 19 ft **25.** Yes; sample examples: $8 - 3 = 5, 0 - (-2) = 2$, and $-3 - (-5) = 2$.
27. $\frac{1}{14}$

Pages 451–452 Lesson 11-5

1. $-2 \times (-2) = 4$ **3.** -3 **5.** -16 **7.** 21
9. 18 **11.** 28 **13.** 0 **15.** -18 **17.** -24
19. -25 **21.** -42 **23a.** 12 lb **23b.** 68 lb
25. the number's opposite

27.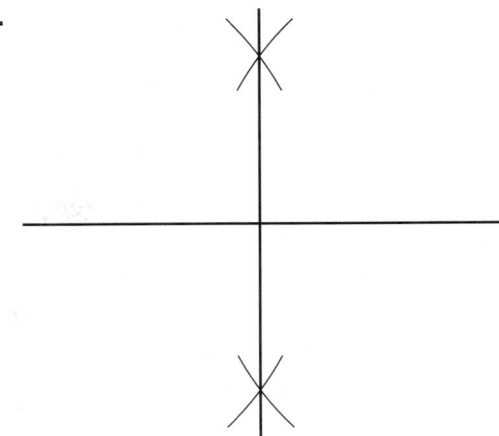

Page 452 Mid-Chapter Self Test

1. -282 **3.** < **5.** -10 **7.** -4 **9.** -18

Pages 454–455 Lesson 11-6A

1. Callie and Nora used a calendar to locate April 15. Then they counted backward from that date the number of weeks they wanted to spend on each activity. **3.** 7:30 A.M. **5.** Sample answer: Sue has a dental appointment at 3:30 P.M. She wants to get there 10 minutes early. If it takes 40 minutes to drive to the dentist's office, what time should she leave her house for the appointment? **7.** 24 ways
9a. 7 ft **9b.** 28 ft
11.

−2	3	−4
−3	−1	1
2	−5	0

13. $29.39

Pages 457–458 Lesson 11-6

1. $-9 \div 3 = -3$ **3a.** Sample answer: $-5 \times (-3) = 15; 15 \div (-5) = -3$ and $15 \div (-3) = -5$
3b. Sample answer: $4 \times (-6) = -24; -24 \div 4 = -6$ and $-24 \div (-6) = 4$ **5.** -2 **7.** 9 **9.** 2
11. -4 **13.** -5 **15.** 3 **17.** -9 **19.** -6
21. -7 **23.** loss of 4 yd per play **25.** Sample answers: $14 \div (-2) = -7, -21 \div 3 = -7, 28 \div (-4) = -7, -35 \div 5 = -7$ **27.** Sample answer: 5

Pages 460–461 Lesson 11-7

1. Start at 0. Move 3 units to the left on the x-axis. Then move 6 units up parallel to the y-axis to locate the point. Place a dot at this location and label the point. **3.** (3, 1) **5.** $(-3, -2)$

7. 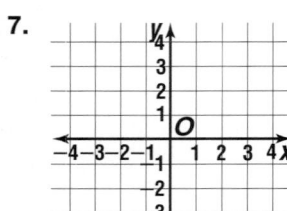 **9.**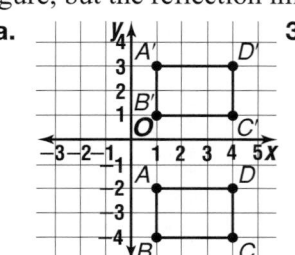

11. $(-3, 1)$ **13.** $(4, -3)$ **15.** $(2, 3)$ **17.** $(0, -2)$
19. $(-2, -3)$ **21.** $(-4, 4)$

23. 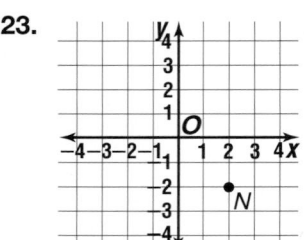   **25.**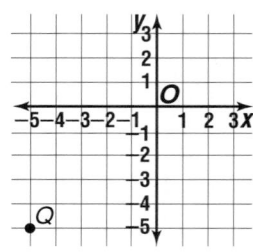

27. 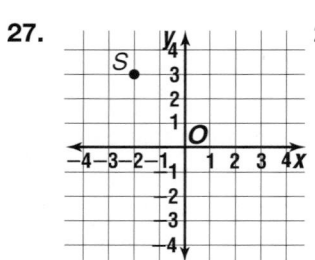   **29.**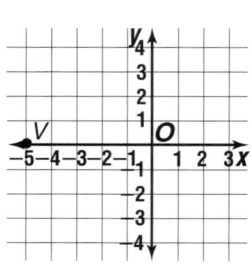

31a. 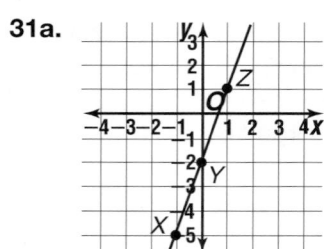   **31b.** yes

33. $(1, -5), (2, -3), (3, 0), (4, 5), (5, 8)$

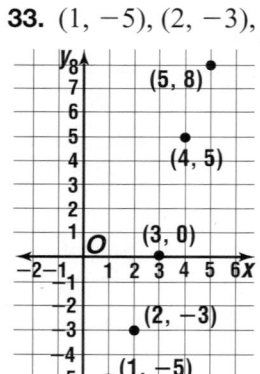   **35.**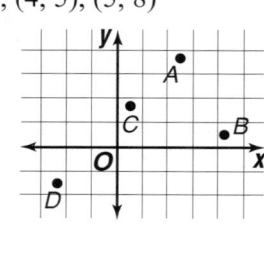

37. C   **39.** 83

Pages 462–463   Lesson 11-8A

1. The position of the triangle is changed.   **3.** The new figure is the mirror image of triangle *C*.

Pages 466–467   Lesson 11-8

1. Both translations and reflections are transformations. In translations and reflections, the image is the same size and shape as the figure. The translation image has the same orientation as the figure, but the reflection image is a flip of the figure.

3a. 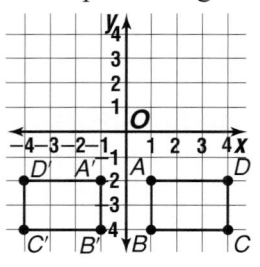   **3b.**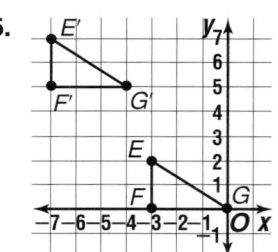

5.

7. Reflection; the mountains are reflected in the lake.   **9.** reflection

11. 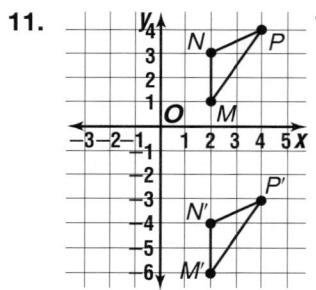   **13.**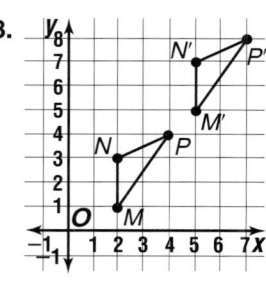

15. 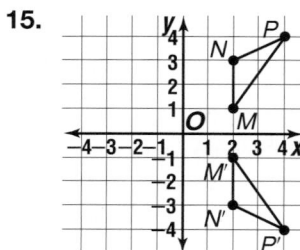   **17.** Translations; the winged horses are translations of each other.   **19.** The image is a reflection of the polygon over the *y*-axis.   **21.** A

Pages 468–471   Study Guide and Assessment

1. false; positive   **3.** true   **5.** false; sometimes
7. false; never   **9.** true   **11.** false; reflection
13.

15.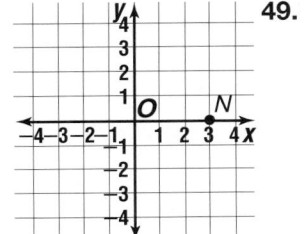

17. < **19.** > **21.** −15, −7, −1, 0, 6, 10 **23.** 0
25. −6 **27.** −6 **29.** −8 **31.** 11 **33.** −11
35. 15 **37.** 0 **39.** −7 **41.** −4 **43.** −6 **45.** 4

47. 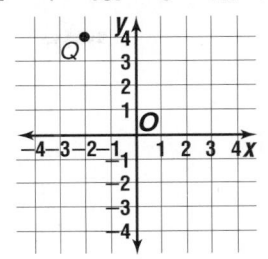 **49.**

51. 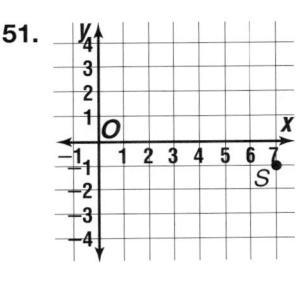 **53.**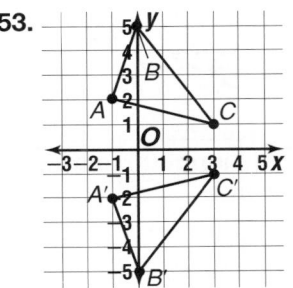

55. 8 yd **57.** reflection

Pages 472–473 Standardized Test Practice

1. C **3.** D **5.** A **7.** A **9.** B **11.** B **13.** −8
15. (3, 3) **17.** ⊐ **19.** $28\frac{1}{2}$

···

Chapter 12
Algebra: Exploring Equations

Pages 477–479 Lesson 12-1

1. Sample answer: The 5 positive counters and the cup represent $x + 5$ and the 8 negative counters represent −8. **3.** 6 **5.** 7 **7.** −9 **9.** −6
11. −5 **13.** −1 **15.** 2 **17.** −6 **19.** −20
21. −43 **23.** −125 **25.** −2 **27.** −8 **29.** 10
31. −6 **33.** Sample answer: 37 + (−9) = 28
35. 21 **37.** 9%

Pages 482–483 Lesson 12-2

1. Sample answer: The 3 negative counters and the cup represent $y − 3$ and the 4 positive counters represent 4. **3.** Sample answer: If the equation involves addition, you subtract counters. If the equation involves subtraction, you add counters.
5. −2 **7.** 7 **9.** 16 **11.** −2 **13.** −2 **15.** −1

17. −8 **19.** 5 **21.** 7 **23.** 3 **25.** 2 **27.** −9
29a. −140 + r = −35 **29b.** 105 feet **31.** B

Pages 486–487 Lesson 12-3

1. 4x = 8; x = 2 **3.** Sample answer: They're both correct. Dividing by 2 is the same as multiplying by one-half. **5.** 8, 8, 2 **7.** 5 **9.** −10 **11.** 2
13. −3 **15.** −9 **17.** $\frac{1}{4}$ **19.** 5 **21.** 24 **23.** 38
25. 2 **27.** −36 **29.** 8.2 **31.** $4 million
33.

E
Change per month
D2/10
D3/10
D4/10
D5/10
D6/1
D7/10

35. 3 **37.** C

Page 487 Mid-Chapter Self Test

1. 6 **3.** 21 **5.** −15 **7.** −2 **9.** 3 gallons

Pages 490–491 Lesson 12-4

1. 3x + 2 = −1; x = −1 **3.** subtract 2 from each side; 4 **5.** add 4 to each side; 20 **7.** 4 **9.** 3
11. −9 **13.** −32 **15.** −20 **17.** 32
19. 16 inches **21.** 8 **23.** 44, 46.5

Pages 492–493 Lesson 12-4B

1. Sample answer: no; Ed's information may be more useful because it contains all the known amounts. **3.** Sample answer: Subtract 8 from the solution. **5.** Sample answer: You could use the guess and check strategy. **7.** Sample answer: In solving a two-step equation, you undo the operations in reverse order of the order of operations. **9.** about 12 buses **11.** $39.99 **13.** 24 combinations
15. 4 ways

Page 495 Lesson 12-5A

1. 8 **3.** 13 **5a.** 12 **5b.** 18 **5c.** 21 **5d.** 11
7. 1, 7, 13, 19

Pages 497–499 Lesson 12-5

1.

input (y)	output (y−2)
−4	−6
−1	−3
0	−2
3	1

3. Sample answer: Juanita is correct. If you use 4, 5, and 6 for inputs, the outputs are 1, 2, and 3. Each output is 3 less than the input.

5.

input (*n*)	output $\left(\frac{1}{4}n\right)$
4	1
8	2
12	3

7. $n \div 2$ **9a.** $0.75c$ **9b.** $37.50

11.

input (*n*)	output (3*n*)
0	0
3	9
4	12

13.

input (*n*)	output $\left(\frac{1}{8}n\right)$
−8	−1
0	0
12	$1\frac{1}{2}$

15. $5n$ **17.** $n \div -2$ **19.** n^2 **21.** -22 **23.** $n \div 5$
25. 100, 60, 80, 70   **27.** $2n + 2$ **29.** -12

Pages 501–503   Lesson 12-6

1. Sample answer: Let the input values represent the *x*-coordinates and let the output values represent the corresponding *y*-coordinates.   **3.** Sample answer: Make a table recording the input and output of the given function. Write the input/output as an ordered pair. Record at least three ordered pairs on the table. Graph these ordered pairs and draw a line connecting the graphed points.

5.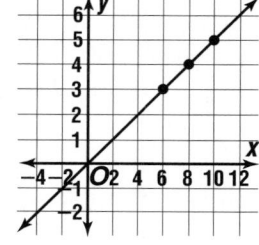

7.

input (*n*)	output $\left(\frac{n}{2}\right)$
6	3
8	4
10	5

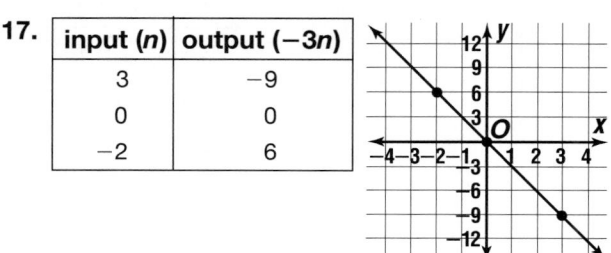

The rule is $n - 2$.

9.

input (*n*)	output (■)
−1	−3
2	0
4	2

11.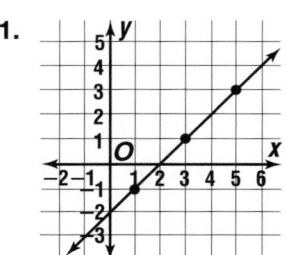

13.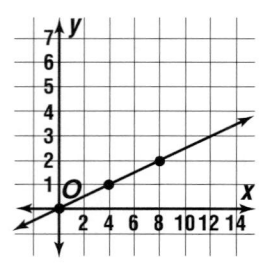

15.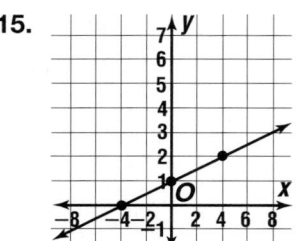

17.

input (*n*)	output (−3*n*)
3	−9
0	0
−2	6

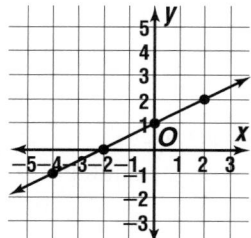

19.

input (*n*)	output $\left(\frac{1}{2}n + 1\right)$
−4	−1
−2	0
0	1
2	2

21.

input (*n*)	output (3*n* − 3)
−2	−9
1	0
4	9

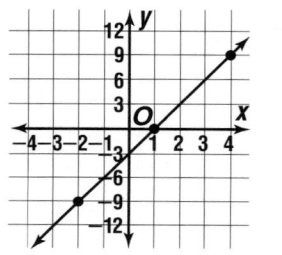

23.

input (*n*)	output (■)
0	0
4	2
8	4

The rule is $\frac{n}{2}$

25a. Jane: $25w - 20$; Julie: $20w$

25b.
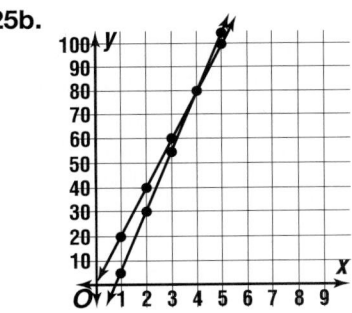

25c. The point of intersection represents when the total earnings are the same for both girls.

27a.

input	output
1	61
2	62
3	63
4	64
5	65
6	66
7	67
8	68
9	69
10	70
11	71
12	72

27b.
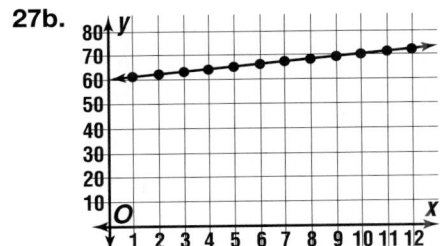

27c. In 5 months, the card will be worth $65.

29.

input (*n*)	output (*n* + 7)
-5	2
0	7
4	11

31. about 31.4 meters

Pages 504–507 Study Guide and Assessment

1. variable **3.** subtracted from **5.** function table
7. -18 **9.** $2x + 3 = -9$ **11.** 3 **13.** -13
15. -3 **17.** 4 **19.** 2 **21.** 0 **23.** 4 **25.** -55
27. -5 **29.** -20 **31.** 6 **33.** -9 **35.** $-3, 0, 6$
37. -5 **39.** $n - 4$
41. 1, 3, 5

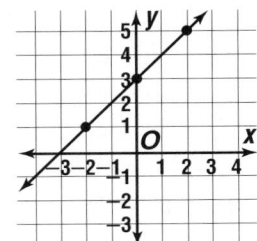

43. 8 laps **45.** $n + 3.50$

Pages 508–509 Standardized Test Practice

1. B **3.** D **5.** D **7.** E **9.** D **11.** E
13. 240 m³ **15.** about 15 people
17. circle

···

Chapter 13
Using Probability

Page 514 Lesson 13-1A

1. Sample answer: The number of even sums tossed and the number of odd sums tossed should be about the same. The number of even products tossed should be greater than the number of odd products tossed. **3.** Sample answer: For the game involving sums, the bar graph resembled an isosceles triangle and is symmetrical about the axis which contains the sum of 7. For the products game, the bar graph reaches a maximum height at products 6 and 12 and the overall graph is not symmetrical. **5.** Sample answer: A sum of 12 is less likely to occur since out of 36 results, a sum of 12 can only occur one way.

Pages 517–518 Lesson 13-1

1. Sample answer:

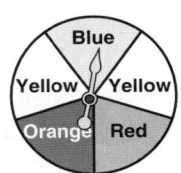

3. Sample answer: Complementary events are two events in which either one or the other can occur, but not both. The sum of their probabilities is 1. An example is the probability of snow and the probability of no snow. **5.** $\frac{2}{3}$; The event is very likely to occur. **7.** $\frac{1}{3}$; The event is not too likely to occur. **9.** 0; The event cannot occur. **11.** $\frac{1}{4}$; The event is not too likely to occur. **13.** $\frac{1}{2}$; The event is equally likely to occur. **15.** $\frac{3}{26}$; The event is not too likely to occur. **17.** $\frac{21}{26}$; The event is very likely to occur. **19.** $\frac{1}{2}$; not odd; $\frac{1}{2}$ **21.** 1; not an integer; 0 **23.** $\frac{2}{5}$; not greater than 18; $\frac{3}{5}$ **25a.** $\frac{1}{5}$ **25b.** Sample answer: Sometimes, eliminating the ones that cannot be correct leaves the correct answer. **25c.** increase; There would be fewer choices.

29.

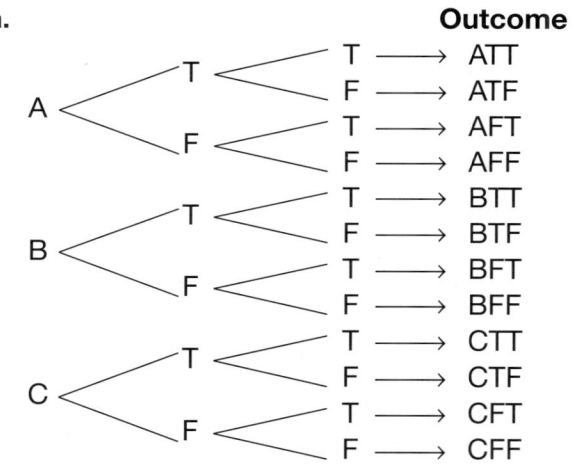

31. 47.36

Pages 520–521 Lesson 13-2A

1. Sample answer: Yes; surveying every fourth member provides an adequate sample. **3.** Sample answer: The information is easy to locate and it is easier to draw conclusions from organized data. **5.** Sample answer: It enables you to determine a reasonable interval width for the given data. **7.** Sample answer: Ashley organized the results of the survey by making a table. The table makes it easy to find information such as the most preferred and the least preferred pet. **9.** 8 days **11.** Sample answer: $20−$6 or $14. **13.** D

Pages 524–525 Lesson 13-2

1. Sample answer: Population is the entire group while random sample is a subset of the entire group. **5.** yes; There is probably a diverse group of people at a shopping center. **7a.** $\frac{3}{5}$, or 60% **7b.** about 9 **9.** no; You would probably survey people whose hobby is model trains. **11.** yes; A mall would have a diverse group of people. **13.** no; You would probably find many people who prefer Chevrolets. **15.** no; A lot of chocolate lovers would be in a chocolate factory. **17.** no; No Republicans were included in the survey. **19.** about 369 **21.** C **23.** 180 inches

1. $\frac{1}{6}$ **3.** $\frac{1}{2}$ **5a.** $\frac{3}{50}$, or 6% **5b.** about 900

Pages 527–529 Lesson 13-3

1. $\frac{4}{9}$ **5.** $\frac{1}{8}$ **7.** about 19 **9.** $\frac{2}{5}$ **11.** $\frac{1}{8}$ **13.** $\frac{4}{25}$ **15.** about 80 **17.** about 32 **19.** about 44 **21.** about $\frac{1}{40}$, or 2.5% **23c.** Sample answer: In the experiment, it is easier to see the black and white circles but it is hard to see the newsprint circles. The newsprint circles represent the camouflage trait. **25.** No; people in a Levi's store may prefer Levi's jeans. **27.** 988 ft^2

Page 530 Lesson 13-3B

1. Sample answer: $\frac{27}{50}$ **3.** Sample answer: The fractions should be about the same. **5.** Sample answer: In the program, change each 10 to 20.

Pages 533–534 Lesson 13-4

1. Sample answer: At a restaurant, a customer can choose either a chicken or beef dinner and one of three beverages. How many possible outcomes are there? **3.** Sample answer: Tree diagrams are useful for organizing the outcomes. One disadvantage is that there could be too many outcomes. **5.** blue shirt, blue shorts; blue shirt, black shorts; blue shirt, red shorts; red shirt, blue shorts; red shirt, black shorts; red shirt, red shorts

7a.

			Outcome
A	T	T →	ATT
		F →	ATF
	F	T →	AFT
		F →	AFF
B	T	T →	BTT
		F →	BTF
	F	T →	BFT
		F →	BFF
C	T	T →	CTT
		F →	CTF
	F	T →	CFT
		F →	CFF

7b. $\frac{1}{2}$ **9.** leather, purple; leather, green; leather, black; leather, brown; nylon, purple; nylon, green; nylon, black; nylon, brown **11.** blue, red; blue, green; blue, white; blue, black; orange, red; orange,

green; orange, white; orange, black; yellow, red; yellow, green; yellow, white; yellow, black
13. tower, wood, 60; tower, wood, 100; tower, wood, 250; tower, plastic, 60; tower, plastic, 100; tower, plastic, 250; spinner, wood, 60; spinner, wood, 100; spinner, wood, 250; spinner, plastic, 60; spinner, plastic, 100; spinner, plastic, 250 **15.** 10 ways **17.** 15 ways **19.** Sample answer: This is not a fair game. Only in two of the eight outcomes are all chips different. It could be made a fair game by giving Amiri 3 points or changing the second counter from red and white to green and white. **21.** C

Page 535 Lesson 13-4B
1. Sample answer: $\frac{22}{50}$ **3.** They are about the same.

Pages 538–539 Lesson 13-5
1. Multiply the probability of the first event by the probability of the second event. **3.** Flora is correct. The probability of a 4 or 5 is $\frac{2}{6}$, and the probability of a 6 is $\frac{1}{6}$. $\frac{1}{3} \times \frac{1}{6} = \frac{1}{18}$. **5.** $\frac{1}{10}$ **7a.** $\frac{2}{5}$ **7b.** $\frac{3}{10}$ **7c.** $\frac{3}{25}$ **9.** $\frac{1}{4}$ **11.** 0 **13.** $\frac{1}{3}$ **15.** $\frac{3}{8}$ **17.** $\frac{5}{18}$ **19a.** 0.48 **19b.** 0.08 **21.** 20 ways **23.** D

Pages 540–543 Study Guide and Assessment
1. e **3.** g **5.** b **7.** c **9.** d **11.** Sample answer: Multiply the probability of the first event by the probability of the second event. **13.** $\frac{1}{3}$; The event is not too likely to occur. **15.** $\frac{5}{6}$; The event is very likely to occur. **17.** $\frac{2}{3}$; The event is very likely to occur. **19.** no; At a football game, you would most likely find people who prefer football. **21a.** $\frac{7}{20}$ or 35% **21b.** 147 students **23.** about $\frac{5}{16}$ or 31.25% **25.** black jeans–tapered; black jeans–straight; black jeans–baggy; blue jeans–tapered; blue jeans–straight; blue jeans–baggy **27.** ball game–Friday; ball game–Saturday; amusement park–Friday; amusement park–Saturday; concert–Friday; concert–Saturday; **29.** $\frac{1}{4}$ **31.** $\frac{1}{6}$ **33a.** 11 **33b.** 32 **35.** purple skirt–yellow vest; purple skirt–purple sweater; white skirt–yellow vest; white skirt–purple sweater; yellow skirt–yellow vest; yellow skirt–purple sweater

Pages 544–545 Standardized Test Practice
1. D **3.** C **5.** B **7.** D **9.** A **11.** C **13.** C **15a.** $\frac{1}{6}$ **15b.** $\frac{4}{75}$

Photo Credits

Cover (bkgd)Index Stock, (hologram)Glencoe photo, (inset)Paul L. Ruben; **iv** Mark Burnett, (bl)courtesy of Arthur Howard; **ix** (l)Hickson & Assoc., (lc)AT&T, (r)Mark Burnett, (rc)Bob Mullenix; **v vii** Mark Burnett; **x** Thomas Kitchin/Tom Stack & Assoc.; **xi** Aaron Haupt; **xii** Don Valentine/Tony Stone Images; **xiii** courtesy Binney & Smith Co.; **xiv** Elisabeth Weiland/Photo Researchers; **xix** Steve Kaufman/Peter Arnold, Inc.; **xv** William Blake/Corbis; **xvi** Doug Martin; **xvii** Bill Bachman/Photo Researchers; **xviii** SuperStock; **xx** Tom McHugh/Photo Researchers; **xxi** GK & Vikki Hart/The Image Bank; **xxii** (l)Frans Lanting/Tony Stone Images, (r) Ann Duncan/Tom Stack & Assoc.; **xxiii** Kenji Kerins; **xxiv** Doug Martin; **xxv** (l)Matt Meadows, (r) David R. Frazier Photo Library; **xxvi** (l)SuperStock, (r) Fred Bavendam/Peter Arnold, Inc.; **xxvii** Van Bucher/Photo Researchers; **xxviii** Mark Steinmetz; **1** (t)KS Studio, (b)Matt Meadows; **2** (t)Matt Meadows, (others)Aaron Haupt **3** (l)Matt Meadows, (r)Aaron Haupt; **2-3** Aaron Haupt; **4** (l)Hickson & Assoc., (r)AT&T; **5** (tl)Bob Mullenix, (tr)Mark Burnett, (b)Aaron Haupt; **7** (l)B & C Alexander/Photo Researchers, (r)Charles Krebs/Tony Stone Images; **8** (tl)Bill Ivy/Tony Stone Images, (tr)Hans Pfletshinger/ Peter Arnold, Inc., (b)Stephen Dalton/Photo Researchers, **9** Mark Steinmetz; **10** Aaron Haupt; **12** Eastcott/Momatuik/Tony Stone Images; **13** Kathi Lamm; **16** (t)Frank Fournier/Contact Press Images/The Stock Market, (b)SuperStock; **17** Tom Brakefield/The Stock Market; **19** SuperStock; **22** Aaron Haupt; **23** Spencer Swanger/Tom Stack & Assoc.; **25 26** Aaron Haupt; **27** (bkgd)Onne Van Der Wal/ Stock Newport, (l)Andy Freeberg, (r)KS Studio; **29** Aaron Haupt; **30** Jerry Wachter/Photo Researchers; **32** KS Studio; **34** (t)Library of Congress/Corbis, (c)Philadelphia Museum of Art/ Corbis, (b)courtesy Chicago Historical Society; **36 37 41** Aaron Haupt; **44** (t)Frans Lanting/Tony Stone Images, (bl)Monika Smith/Cordaiy Photo Library Ltd./Corbis, (br)Ted Horowitz/ The Stock Market; **45** Craig Lovell/Corbis; **46** Aaron Haupt; **47** David Muench/Tony Stone Images; **48** Aaron Haupt; **49** (t)Aaron Haupt, (b)KS Studio; **50** Aaron Haupt, (inset)KS Studio; **51** Aaron Haupt; **52** KS Studio; **53** Thomas Kitchin/ Tom Stack and Assoc.; **54** (l)Glencoe file, (r)StudiOhio; **56** (l)Deborah Davis/PhotoEdit, (r)Ken Frick; **57** Michael Burr/ NFLP; **58** (t)KS Studio, (b)Matt Meadows; **60** StudiOhio; **62** (b)BLT/Brent Turner Productions; **63** Aaron Haupt; **64** David Lees/Corbis; **65** Focus on Sports; **66** Doug Martin; **68** Wolfgang Kaehler/Corbis; **70** (t)Dan Guravich/Corbis, (b)David Madison/ Bruce Coleman, Inc.; **71-72** KS Studio; **76** Aaron Haupt; **78** ChromeSohm Inc./Corbis-Bettmann; **81** Hy Peskin/ FPG; **82** *Close to Home* ©1996 reprinted with permission of Universal Press Syndicate. All rights reserved; **83** (t)StudiOhio, (b)KS Studio; **85** (t)Len Rue Jr./Photo Researchers, (b)Fritz Polking/ Frank Lane Pictures Agency; **92** (bkgd)Terry Donelly/Tony Stone Images, (t)Murry Sill, (c)Aaron Haupt, (b)Gerald & Buff Corsi/Tom Stack and Assoc.; **93** Robert Houser/Index Stock; **95** (t)Barry Rosenthal/FPG, (b)Doug Martin; **96** Robert A. Tyrell; **97** (t)Focus on Sports, (b)R. Scheiber; **98** By permission of Bud Blake, King Features Syndicate; **99** (bkgd)Stuart Westmorland/ Corbis, (t)Tim Wright/Corbis, (bl)Jeffery Salters/SABA, (br)KS Studio; **100** Doug Martin; **102** (t)Focus on Sports, (b)from ©Hammond's Road Atlas; **103** Mike Bacon/Tom Stack & Assoc.; **105** Matt Meadows; **107** Focus on Sports; **109** (t)The Orange County *Register*, (b)Aaron Haupt; **110** courtesy KTXQ; **110-111** Aaron Haupt; **112** Michal Heron/The Stock Market;

113 courtesy Gleim Jewelers; **114** Matt Meadows; **115** David Muench/Corbis, (inset)Matt Meadows; **116** Matt Meadows; **118** (tl)Franz Gorski/Peter Arnold, Inc., (other)Aaron Haupt; **119** Aaron Haupt; **128** Focus on Sports; **128-129** image ©1998 PhotoDisc, Inc.; **129** Focus on Sports; **130** (bkgd)Dick Luria/FPG, (l)Ed Taylor Studio/FPG, (r)Francis Lepine/Earth Scenes; **131** (t)Doug Martin, (b)Dick Luria/FPG; **133** KS Studio; **136** Doug Martin; **137** Art Montes De Oca/FPG; **138** KS Studio; **139** SuperStock; **141** Dominic Oldershaw; **145** KS Studio; **146** Doug Martin; **148** Constance Hansen/The Stock Market; **149** Gary Bumgarner/Tony Stone Images; **150** (t br)KS Studio, (bl)Aaron Haupt; **152** Doug Martin; **153 154** KS Studio; **157** Tom Paiva Photography/FPG; **159** Ken Brate/Photo Researchers; **160** (bkgd)David Barnes/The Stock Market, (t)J. Pickerell/FPG, (bl)John Storey/©1997 *People Weekly*, (br)KS Studio; **161** Matt Meadows; **163** (t)Aaron Haupt, (b)Don Valentine/Tony Stone Images; **164** (t)Rick Stewart/AllSport, (b)Peter Angelo Simon/The Stock Market; **165** Doug Martin; **166** Tom McHugh/Photo Researchers; **167** Chuck Savage/The Stock Market; **169 176** Doug Martin; **178** (t)courtesy Binney & Smith Co., (b)Bob Mullenix; **179** KS Studio; **180** Aaron Haupt; **182** Ron Chapple/FPG; **185** (bkgd)Westlight, (l)courtesy Kathryn Sharar Prusineski, (r)Aaron Haupt; **186** KS Studio; **188** Dominic Oldershaw; **189** M&C Werner/Comstock; **190** (l)Doug Martin, (r)J.T. Miller/The Stock Market; **193** Doug Martin; **194** Aaron Haupt; **195** KS Studio; **196** Doug Martin; **198** (t)KS Studio, (b)Elaine Shay; **199** Gunter Ziesler/Peter Arnold, Inc.; **200** Richard Mackson/Time Inc. Picture Collection; **201** SS Archives/Shooting Star; **202** Aaron Haupt; **205** (t)Matt Meadows, (b)PEANUTS ©United Features Syndicate. Reprinted by Permission; **206 207 208** Doug Martin; **210** Gerard Fritz/Tony Stone Images; **211** Aaron Haupt; **212** SuperStock; **213** (t)Doug Martin, (b)Larry Lefever from Grant Heilman; **214** Aaron Haupt; **215** (t)SuperStock, (b)Doug Martin; **216** (t)KS Studio, (b)Aaron Haupt; **217** Doug Martin; **218** David Woodfall/Tony Stone Images; **219** Morton & White; **226-227** Karen Leeds/The Stock Market; **226** (bkgd)Karen Leeds/The Stock Market, F. Stuart Westmoreland/Photo Researchers; **227** Telegraph Colour Library/FPG; **228** (t)Doug Martin, (b)Aaron Haupt; **230** (t)Doug Martin, (b)Robert Maler/ Animals Animals; **231** (t)Doug Martin, (b)UPI/Corbis-Bettmann; **233** Doug Martin; **234** Erickson Productions/The Stock Market; **235** (bkgd)KS Studio, (l)USGS Photographic Library, Denver CO, (r)Aaron Haupt; **236** (t br)Aaron Haupt, (bl)Stan Obert/ USGS; **238** (t)Matt Meadows, (b)Elaine Shay; **239** Doug Martin; **241** (l)Aaron Haupt, (r)Doug Martin; **242** Doug Martin; **243** Robert Shafer/Tony Stone Images; **245** Elisabeth Welland/Photo Researchers; **246** Aaron Haupt; **248** Bill Meng/Wildlife Conservation Society; **249** Ron Seerford/Mike Agliolo/The Stock Market; **250** Chris Noble/Tony Stone Images; **253** Baron Wolman/Tony Stone Images; **254** NASA/ Science Source/Science Photo Library/Photo Researchers; **256** Tim Courlas; **264 264-265** David R. Frazier Photo Library; **265** (tr)Joseph Pobereskin/Tony Stone Images; **266** (bkgd)William Blake/ Corbis, Doug Martin; **267** Doug Martin; **268** Matt Meadows; **269** (t)Doug Martin, (b)StudiOhio; **270** Aaron Haupt; **273** KS Studio; **275** Bison Archive/Shooting Star; **277** Matt Meadows; **279** (t)Doug Martin, (b)PEANUTS ©United Features Syndicate. Reprinted by Permission; **280** UPI/Corbis-Bettmann; **281** Library of Congress/Corbis; **282** Jack Zehrt/FPG; **283** Corbis-Bettmann; **285** (t)Leonard Lessin/Peter Arnold, Inc., (b)Index Stock; **286** Doug Martin; **288** (l)Latent Image, (r)Glencoe file; **289** Paul Berger/Tony Stone Images;

Glossary

A

acute angle (352) An angle with a measure greater than 0° and less than 90°.

algebra (22) A mathematical language that uses letters along with numbers. The letters stand for numbers that are unknown. $10x - 3 = 17$ is an example of an algebra equation.

algebraic expression (22) A combination of variables, numbers, and at least one operation.

angle (352) Two rays with a common endpoint form an angle.

angle BAC or $\angle BAC$

area (146) The number of square units needed to cover a surface enclosed by a geometric figure.

average (71) The sum of two or more quantities divided by the number of quantities; the mean.

B

bar graph (54) A graph using bars to compare quantities. The height or length of each bar represents a designated number.

bar notation (218) In repeating decimals, the line or bar placed over the digits that repeat. For example, 2.6̄3̄ indicates the digits 63 repeat.

base (28) In a power, the number used as a factor. In 10^3, the base is 10. That is, $10^3 = 10 \times 10 \times 10$.

base (398) Any side of a parallelogram.

base (412) The faces on the top and bottom of a three-dimensional figure.

bisect (364) To separate something into two congruent parts.

box-and-whisker plot (76) A diagram that summarizes data using the median, the upper and lower quartiles, and the extreme values. A box is drawn around the quartile value and whiskers extend from each quartile to the extreme data points.

C

cell (144) The basic unit of a spreadsheet. A cell can contain data, labels, or formulas.

center (280, 413) The given point from which all points on a circle or a sphere are the same distance.

centimeter (100) A metric unit of length. One centimeter equals one-hundredth of a meter.

circle (280) The set of all points in a plane that are the same distance from a given point called the center.

circle graph (60) A graph used to compare parts of a whole. The circle represents the whole and is separated into parts of the whole.

circumference (280) The distance around a circle.

clustering (113) A method used to estimate decimal sums and differences by rounding a group of closely related numbers to the same whole number.

combination (532) An arrangement or listing of objects in which order is not important.

common multiples (206) Multiples that are shared by two or more numbers. For example, some common multiples of 2 and 3 are 6, 12, and 18.

compass (362) An instrument used for drawing circles or parts of circles.

compatible numbers (133, 268) Two numbers that are easy to divide mentally. They are often members of fact families.

complementary (353) Two angles are complementary if the sum of their measures is 90°.

complementary events (516) Two events in which either one or the other must take place, but they cannot both happen at the same time. The sum of their probabilities is 1.

composite number (181, 182) Any whole number greater than 1 that has more than two factors.

cone (413) A three-dimensional figure with curved surfaces, a circular base and one vertex.

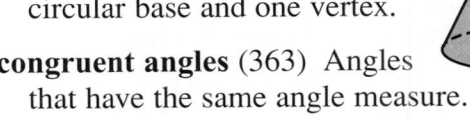
vertex

congruent angles (363) Angles that have the same angle measure.

congruent figures (379) Figures that are the same size and shape. The symbol ≅ means *is congruent to*.

congruent segments (362) Segments having the same length.

coordinate grid (459) Another name for a coordinate system.

coordinate system (82, 459) A plane in which a horizontal number line and a vertical number line intersect at their zero points.

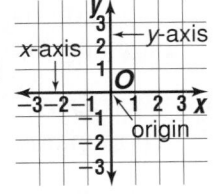

cross products (318) The products of the terms on the diagonals when two ratios are compared. If the cross products are equal, then the ratios form a proportion. In the proportion $\frac{3}{6} = \frac{4}{8}$, the cross products are 3×8 and 6×4.

cubed (28) The product in which a number is a factor three times. Two cubed is 8 because $2 \times 2 \times 2 = 8$.

cup (292) A customary unit of capacity equal to 8 fluid ounces.

cylinder (413) A three-dimensional figure with all curved surfaces, two circular bases and no vertices.

data (46) Numerical information gathered for statistical purposes.

decagon (370) A polygon having ten sides.

degree (352) The most common unit of measure for angles.

diameter (280) The distance across a circle through its center.

diameter

distributive property (137) For any numbers a, b, and c, $a(b + c) = ab + ac$ and $(b + c)a = ba + ca$.

edge (352, 412) The intersection of faces of a three-dimensional figure.

equation (34) A mathematical sentence that contains the equal sign, =.

equilateral triangle (371) A triangle with three congruent sides.

equivalent fractions (193) Fractions that name the same number. $\frac{3}{4}$ and $\frac{6}{8}$ are equivalent.

evaluate (22) To find the value of an expression by replacing variables with numerals.

event (515) A specific outcome or type of outcome.

experimental probability (197) An estimated probability based on the relative frequency of positive outcomes occurring during an experiment.

exponent (28) In a power, the number of times the base is used as a factor. In 5^3, the exponent is 3. That is, $5^3 = 5 \times 5 \times 5$.

extreme value (76) The least value or the greatest value in a set of data.

face (412) The flat surfaces of a three-dimensional figure.

factor (28) A number that divides into a whole number with a remainder of zero. 5 is a factor of 30.

factor tree (183) A diagram showing the prime factorization of a number. The factors branch out from the previous factors until all the factors are prime numbers.

fair game (514) A game in which players have an equal chance of winning.

fluid ounce (292) A customary unit of capacity.

foot (202) A customary unit of length equal to 12 inches.

frequency table (46) A table for organizing a set of data that shows the number of times each item or number appears.

function (496) A relation in which each element of the input is paired with exactly one element of the output according to a specified rule.

function machine (494) A machine that uses a number called the input, performs one or more operations on it, and produces a result called the output.

function table (496) A table organizing the input, rule, and output of a function.

gallon (292) A customary unit of capacity equal to 4 quarts.

gram (164) The basic unit of mass in the metric system.

greatest common factor (GCF) (188) The greatest of the common factors of two or more numbers. The GCF of 24 and 30 is 6.

height (398) The shortest distance from the base of a parallelogram to its opposite side.

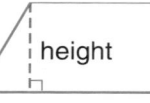

hexagon (370) A polygon having six sides.

improper fraction (198) A fraction that has a numerator that is greater than or equal to the denominator.

inch (202) A customary unit of length. Twelve inches equal one foot.

independent events (536) Two or more events in which the outcome of one event does *not* affect the outcome(s) of the other event(s).

input (494) Information or data given to a function machine to produce output or results.

integer (434) The whole numbers and their opposites. . . . , $-3, -2, -1, 0, 1, 2, 3, . . .$

interval (50) The difference between successive values on a scale.

kilogram (164) A metric unit of mass. One kilogram equals one thousand grams.

leaf (68) The units digit written to the right of the vertical line in a stem-and-leaf plot.

least common denominator (LCD) (210) The least common multiple of the denominators of two or more fractions.

least common multiple (LCM) (206) The least of the common multiples of two or more numbers. The LCM of 2 and 3 is 6.

like fractions (238) Fractions with the same denominator.

line graph (54) A graph used to show change and direction of change over a period of time.

line of symmetry (375) A line that divides a figure into two halves that are reflections of each other.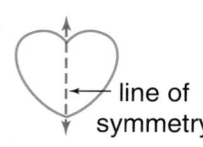

line segment (362) Two endpoints and the straight path between them. A line segment is named by its endpoints. A representation of line segment *ST (ST)* is shown below.

liter (164) The basic unit of capacity in the metric system. A liter is a little more than a quart.

lower quartile (77) The median of the lower half of a set of data.

mean (71) The sum of the numbers in a set of data divided by the number of pieces of data.

measures of central tendency (71) Numbers or pieces of data that can represent the whole set of data. These common measures of central tendency are *mean*, *median*, and *mode*.

median (71) The middle number in a set of data when the data are arranged in numerical order. If the data has an even number, the median is the mean of the two middle numbers.

GLOSSARY

meter (102) The basic unit of length in the metric system.

metric system (102) A base-ten system of weights and measures. The meter is the basic unit of length, the gram is the basic unit of weight, and the liter is the basic unit of capacity.

mile (202) A customary unit of length equal to 5,280 feet or 1,760 yards.

milligram (164) A metric unit of mass. One milligram equals one-thousandth of a gram.

milliliter (164) A metric unit of capacity. One milliliter equals one-thousandth of a liter.

millimeter (100) A metric unit of length. One millimeter equals one-thousandth of a meter.

mixed number (198) The sum of a whole number and a fraction. $1\frac{1}{2}$, $2\frac{3}{4}$, and $4\frac{5}{8}$ are mixed numbers.

mode (71) The number(s) or item(s) that appear most often in a set of data.

multiple (206) The product of the number and any whole number.

negative integer (434) Integer that is less than zero.

net (374) The shape that is formed by "unfolding" a three-dimensional figure. The net shows all the faces that make up the surface area of the figure.

obtuse angle (352) Any angle that measures greater than 90° but less than 180°.

octagon (370) A polygon having eight sides.

opposite (435) Two integers are opposites if they are represented on the number line by points that are the same distance from zero, but in opposite directions from zero. The sum of opposites is zero.

ordered pair (82, 459) A pair of numbers used to locate a point in the coordinate system. The ordered pair is written in this form: (*x*-coordinate, *y*-coordinate).

order of operations (16, 28) The rules to follow when more than one operation is used.
  1. Do all powers before other operations.
  2. Do all multiplications and divisions from left to right.
  3. Then do all additions and subtractions from left to right.

origin (82, 459) The point of intersection of the *x*-axis and *y*-axis in a coordinate system.

ounce (292) A customary unit of weight. 16 ounces equals one pound.

outcomes (515) Possible results of a probability event. For example, 4 is an outcome when a number cube is rolled.

output (494) The result of input that has had one or more operations performed on it in a function machine.

parallel (371) Lines going in the same direction and always being the same distance apart. Parallel lines never meet or cross each other.

parallelogram (371) A quadrilateral that has both pairs of opposite sides equal and parallel.

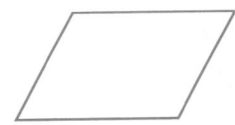

pentagon (370) A polygon with five sides.

percent (330) A ratio with a denominator of 100.

perimeter (145) The distance around any closed figure.

pint (292) A customary unit of capacity equal to two cups.

place value (95) A system for writing numbers in which the position of the digit determines its value.

polygon (368, 370) A simple closed figure in a plane formed by three or more line segments.

population (522) The entire group of items or individuals from which the samples under consideration are taken.

positive integer (434) Integer that is greater than zero.

pound (292) A customary unit of weight equal to 16 ounces.

power (28) A number that can be written using an exponent. The power 3^2 is read *three to the second power*, or *three squared*.

prime factorization (183) Expressing a composite number as the product of prime numbers. For example, the prime factorization of 63 is $3 \times 3 \times 7$.

prime number (181, 182) A whole number greater than 1 that has exactly two factors, 1 and itself.

prism (412) A three-dimensional figure that has two parallel and congruent bases in the shape of polygons and at least three lateral faces shaped like rectangles. The shape of the bases tells the name of the prism.

proportion (317) An equation that shows that two ratios are equivalent, $\frac{a}{b} = \frac{c}{d}$, $b \neq 0$, $d \neq 0$.

protractor (352) An instrument used to measure angles.

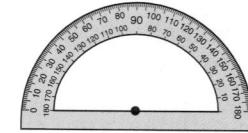

pyramid (412) A solid figure that has a polygon for a base and triangles for sides. A pyramid is named for the shape of its base.

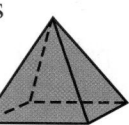

quadrant (459) One of the four regions into which two perpendicular number lines separate the plane.

quadrilateral (368, 370) A polygon with four sides.

quart (292) A customary unit of capacity equal to two pints.

radius (280) The distance from the center of a circle to any point on the circle.

radius

random (522) When an outcome is chosen without any preference the outcome occurs at random.

range (72) The difference between the greatest number and the least number in a set of data.

rate (312) A ratio of two measurements having different units.

ratio (191, 193) A comparison of two numbers by division. The ratio of 2 to 3 can be stated as 2 out of 3, 2 to 3, 2:3, or $\frac{2}{3}$.

ray (363) A part of a line that extends indefinitely from one point in one direction. A representation of ray DE $(\overrightarrow{DE})$ is shown below.

D E

reciprocals (285) Any two numbers whose product is 1. Since $\frac{5}{6} \times \frac{6}{5} = 1$, the reciprocal of $\frac{5}{6}$ is $\frac{6}{5}$.

rectangle (371) A quadrilateral with four congruent angles.

rectangular prism (412) A three-dimensional figure with six rectangular shaped faces. A rectangular prism has a total of six faces, twelve edges, and eight vertices.

reflection (375, 464) A type of transformation where a figure is flipped over a line.

regular polygon (370) A polygon having all sides congruent and all angles congruent.

repeating decimal (218) A decimal whose digits repeat in groups of one or more. 0.181818... can also be written $0.\overline{18}$. The bar above the digits indicates those digits repeat.

right angle (352) An angle that measures 90°.

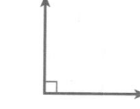

sample (522) A randomly-selected group that is used to represent a whole population.

sample space (515) The set of all possible outcomes.

scale (50) The set of all possible values of a given measurement, including the least and greatest numbers in the set, separated by the intervals used.

scale drawing (324) A drawing that is similar but either larger or smaller than the actual object.

sequence (298) A list of numbers in a certain order, such as, 0, 1, 2, 3, or 2, 4, 6, 8.

sides (145) Line segments that enclose a polygon.

similar figures (379) Figures that have the same shape but different sizes. The symbol ~ means *is similar to*.

simplest form (194) The form of a fraction when the GCF of the numerator and the denominator is 1. The fraction $\frac{3}{4}$ is in simplest form because the GCF of 3 and 4 is 1.

simulation (535) The process of acting out a problem.

solution (34) Any number that makes an equation true. The solution for $12 = x + 7$ is 5.

solve (34) To replace a variable with a number that makes an equation true.

sphere (413) A three-dimensional figure with no faces, bases, edges, or vertices. All of its points are the same distance from a given point called the center. 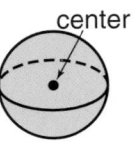 center

spreadsheet (144) A tool used for organizing and analyzing data.

square (371) A parallelogram with all sides congruent and all angles congruent.

squared (28) A number multiplied by itself; 4×4, or 4^2.

square pyramid (412) A pyramid with a square base.

statistics (46) The study of collecting, analyzing, and presenting data.

stem (68) The greatest place value common to all the data that is written to the left of the line in a stem-and-leaf plot.

stem-and-leaf plot (68) A system used to condense a set of data where the greatest place value of the data forms the stem and the next greatest place value forms the leaves.

straightedge (358) Any object with a straight side that can be used to draw a straight line.

supplementary (353) Two angles are supplementary if the sum of their measures is 180°.

surface area (421) The sum of the areas of all the surfaces (faces) of a three-dimensional figure.

 **T**

terminating decimal (214) A decimal whose digits end. Every terminating decimal can be written as a fraction with a denominator of 10, 100, 1000, and so on.

theoretical probability (515) The ratio of the number of ways an event can occur to the number of possible outcomes.

three-dimensional figure (412) A figure that encloses a part of space.

ton (292) A customary unit of weight equal to 2,000 pounds.

transformation (464) Movement of geometric figures to new points in a coordinate system.

translation (384, 464) One type of transformation where a figure is slid horizontally, vertically, or both.

tree diagram (531) A diagram used to show the total number of possible outcomes in a probability experiment.

triangle (368, 370) A polygon with three sides.

 **U**

unfair game (514) A game in which players do not have an equal chance of winning.

upper quartile (77) The median of the upper half of a set of numbers.

variable (22) A symbol, usually a letter, used to represent a number.

vertex (352) A vertex of a polygon is a point where two sides of the polygon intersect.

vertex (412) The point where the edges of a three-dimensional figure intersect.

volume (418) The amount of space that a three-dimensional figure contains. Volume is expressed in cubic units.

x-**axis** (82, 459) The horizontal line of the two perpendicular number lines in a coordinate plane.

x-**coordinate** (82, 459) The first number of an ordered pair.

yard (202) A customary unit of length equal to 3 feet, or 36 inches.

y-**axis** (82, 459) The vertical line of the two perpendicular number lines in a coordinate plane.

y-**coordinate** (82, 459) The second number of an ordered pair.

zero pair (441) The result of pairing one positive counter with one negative counter.

Spanish Glossary/Glosario

acute angle / ángulo agudo (352) Ángulo que mide más de 0° y menos de 90°.

algebra / álgebra (22) Lenguaje matemático que usa letras y números. Las letras representan números desconocidos.

$10x - 3 = 17$ es un ejemplo de ecuación algebraica.

algebraic expression / expresión algebraica (22) Combinación de variables, números y al menos una operación.

angle / ángulo (352) Dos rayos con un extremo común forman un ángulo.

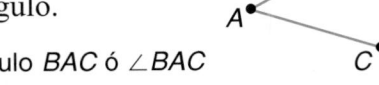

ángulo *BAC* ó ∠*BAC*

area / área (146) Número de unidades cuadradas que se requieren para cubrir la superficie cerrada por una figura geométrica.

average / promedio (71) Suma de dos o más cantidades dividida entre el número de cantidades; la media.

bar graph / gráfica de barras (54) Gráfica que usa barras para comparar cantidades. La altura o longitud de cada barra representa un número específico.

bar notation / notación de barras (218) En los decimales periódicos, la línea o barra que se coloca encima de los dígitos que se repiten. En 2.$\overline{63}$, la barra encima de 63 indica que el bloque de dos dígitos, 63, se repite indefinidamente.

base / base (28) Número que se usa como factor en una potencia. En 10^3, la base es 10, es decir, $10^3 = 10 \times 10 \times 10$.

base / base (398) Cualquier lado de un paralelogramo.

base / base (412) Caras inferior y superior de una figura tridimensional.

bisect / bisecar (364) Separar algo en dos partes congruentes.

box-and-whisker plot / diagrama de caja y patillas (76) Diagrama que resume información usando la mediana, los cuartiles superior e inferior y los valores extremos. Se dibuja una caja alrededor de los cuartiles y se trazan patillas uniendo los cuartiles a los valores extremos respectivos.

cell / celda (144) Unidad básica de una hoja de cálculos. Las celdas pueden contener datos, rótulos o fórmulas.

center / centro (280, 413) Un punto dado del cual equidistan todos los puntos de un círculo o de una esfera.

centimeter / centímetro (100) Unidad métrica de longitud. Un centímetro es igual a la centésima parte de un metro.

circle / círculo (280) Conjunto de todos los puntos en un plano que equidistan de un punto dado llamado centro.

centro ←círculo

circle graph / gráfica circular (60) Gráfica que se usa para comparar las partes de un todo. El círculo representa el todo y aparece dividido en las partes en las que el todo ha sido separado.

circumference / circunferencia (280) La distancia alrededor de un círculo.

clustering / agrupamiento (113) Método que se usa para estimar sumas y restas de decimales, redondeando al mismo número entero un grupo de números estrechamente relacionados.

combination / combinación (532) Arreglo o lista de objetos en la que el orden no es importante.

common multiples / múltiplos comunes (206) Múltiplos compartidos por dos o más números. Por ejemplo, algunos múltiplos comunes de 2 y 3 son 6, 12 y 18.

compass / compás (362) Instrumento que se utiliza para trazar círculos o partes de círculos.

compatible numbers / números compatibles (133, 268) Dos números que son fáciles de dividir mentalmente. A menudo son miembros de la misma familia de factores.

complementary / complementarios (353) Dos ángulos son complementarios si la suma de sus medidas es 90°.

complementary events / eventos complementarios (516) Dos eventos tales, que uno de ellos debe ocurrir, pero ambos no pueden ocurrir simultáneamente. La suma de sus probabilidades es siempre 1.

composite number / número compuesto (181, 182) Cualquier número entero mayor que 1 que posee más de dos factores.

cone / cono (413) Figura tridimensional con superficies curvas, una base circular y un vértice.

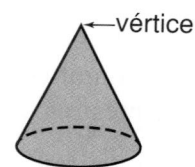
←vértice

congruent angles / ángulos congruentes (363) Ángulos que tienen la misma medida.

congruent figures / figuras congruentes (379) Figuras que tienen la misma forma y tamaño. El símbolo ≅ significa *es congruente a*.

congruent segments / segmentos congruentes (362) Segmentos que tienen la misma longitud.

coordinate grid / plano de coordenadas (459) Otro nombre para el sistema de coordenadas.

coordinate system / sistema de coordenadas (82, 459) Plano en el cual se han trazado dos rectas numéricas, una horizontal y una vertical, las cuales se intersecan en sus puntos cero.

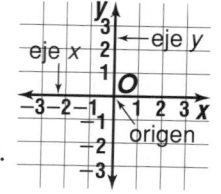
eje *x* — eje *y* — *O* — origen

cross products / productos cruzados (318) Productos que resultan de la comparación de los términos de las diagonales de dos razones. Las razones forman una proporción si y sólo si los productos son iguales. En la proporción $\frac{3}{6} = \frac{4}{8}$, los productos cruzados son 3×8 y 6×4.

cubed / al cubo (28) Producto de un número por sí mismo, tres veces. Dos al cubo es 8 ya que $2 \times 2 \times 2 = 8$.

cup / taza (292) Unidad de capacidad del sistema inglés que equivale a 8 onzas líquidas.

cylinder / cilindro (413) Figura tridimensional que tiene superficies curvas, dos bases circulares y que carece de vértices.

D

data / datos (46) Información numérica que se recoge con fines estadísticos.

decagon / decágono (370) Polígono de diez lados.

degree / grado (352) La unidad de medida angular más común.

diameter / diámetro (280) La distancia a través de un círculo pasando por el centro.

diámetro

distributive property / propiedad distributiva (137) Para números *a*, *b* y *c* cualesquiera, $a(b + c) = ab + ac$ y $(b + c)a = ba + ca$.

E

edge / arista (352, 412) Intersección de las caras de una figura tridimensional.

equation / ecuación (34) Enunciado matemático que contiene el signo de igualdad (=).

equilateral triangle / triángulo equilátero (371) Triángulo cuyos lados son congruentes entre sí.

equivalent fractions / fracciones equivalentes (193) Fracciones que designan el mismo número. $\frac{3}{4}$ y $\frac{6}{8}$ son fracciones equivalentes.

evaluate / evaluar (22) Calcular el valor de una expresión sustituyendo las variables por números.

event / evento (515) Resultado específico o tipo de resultado de un experimento probabilístico.

experimental probability / probabilidad experimental (197) Probabilidad de un evento que se estima basándose en la frecuencia relativa de los resultados favorables al evento en cuestión, que ocurren durante un experimento probabilístico.

exponent / exponente (28) Número de veces que la base de una potencia se usa como factor. En 5^3, el exponente es 3, o sea, $5^3 = 5 \times 5 \times 5$.

extreme value / valor extremo (76) El valor mínimo o máximo de un conjunto de datos.

face / cara (412) Las superficies planas de una figura tridimensional.

factor / factor (28) Número entero que divide otro número entero con un residuo de 0. 5 es un factor de 30.

factor tree / árbol de factores (183) Diagrama que sirve para encontrar la factorización prima de un número. Los factores se ramifican de los factores anteriores hasta que todos los factores son números primos.

fair game / juego justo (514) Juego en el que los jugadores tienen la misma oportunidad de ganar.

fluid ounce / onza líquida (292) Unidad de capacidad del sistema inglés de medidas.

foot / pie (202) Unidad de longitud del sistema inglés de medidas que equivale a 12 pulgadas.

frequency table / tabla de frecuencia (46) Tabla que se utiliza para organizar un conjunto de datos y que muestra cuántas veces aparece cada ítem o dato.

function / función (496) Relación en que cada elemento de entrada es apareado con un único elemento de salida, según una regla específica.

function machine / máquina de funciones (494) Máquina que utiliza un número llamado entrada y que ejecuta una o más operaciones en el número, produciendo un resultado llamado salida.

function table / tabla de funciones (496) Tabla que organiza la entrada, la regla y las salidas de una función.

gallon / galón (292) Unidad de capacidad del sistema inglés de medidas que equivale a 4 cuartos de galón.

gram / gramo (164) Unidad fundamental de masa del sistema métrico.

greatest common factor (GCF) / máximo común divisor (MCD) (188) El mayor factor común de dos o más números. El MCD de 24 y 30 es 6.

height / altura (398) La distancia más corta desde la base de un paralelogramo hasta su lado opuesto.

altura

hexagon / hexágono (370) Polígono de seis lados.

improper fraction / fracción impropia (198) Fracción cuyo numerador es mayor que o igual a su denominador.

inch / pulgada (202) Unidad de longitud del sistema inglés de medidas. Doce pulgadas equivalen a un pie.

independent events / eventos independientes (536) Dos o más eventos en que el resultado de uno de ellos *no* afecta el resultado de los otros eventos.

input / entrada (494) Información o datos que se le proporcionan a una máquina de funciones para que produzca información de salida o resultados.

integer / entero (434) Los números enteros y sus opuestos.

$$\ldots, -3, -2, -1, 0, 1, 2, 3, \ldots$$

interval / intervalo (50) La diferencia entre valores sucesivos de una escala.

kilogram / kilogramo (164) Unidad métrica de masa. Un kilogramo equivale a mil gramos.

leaf / hoja (68) El dígito de las unidades que se escribe a la derecha de la línea divisoria vertical en un diagrama de tallo y hojas.

least common denominator (LCD) / mínimo común denominador (mcd) (210) El menor múltiplo común de los denominadores de dos o más fracciones.

least common multiple (LCM) / mínimo común múltiplo (mcm) (206) El menor múltiplo común de dos o más números. El mcm de 2 y 3 es 6.

like fractions / fracciones con igual denominador (238) Fracciones que tienen el mismo denominador.

line graph / gráfica lineal (54) Gráfica que se usa para mostrar cambio y la dirección del cambio, durante un período de tiempo.

line of symmetry / eje de simetría (375) Recta que divide una figura en dos mitades que son reflexiones una de la otra.

eje de simetría

line segment / segmento de recta (362) Dos puntos y la senda rectilínea entre ellos. Un segmento de recta recibe el nombre de los puntos que la unen. A continuación se muestra una representación del segmento de recta ST $(\overline{ST})$.

S ———— T

liter / litro (164) Unidad fundamental de capacidad del sistema métrico. Un litro es un poco más de un cuarto de galón.

lower quartile / cuartil inferior (77) La mediana de la mitad inferior de un conjunto de datos.

mean / media (71) La suma de los números en un conjunto de datos dividida entre el número total de datos.

measures of central tendency / medidas de tendencia central (71) Números o piezas de datos que pueden representar el conjunto completo de datos. Las medidas comunes de tendencia central son la *media*, la *mediana* y la *modal*.

median / mediana (71) El número central de un conjunto de datos, una vez que los datos se han ordenado numéricamente. Si hay un número par de datos, la mediana es la media de los dos datos centrales.

meter / metro (102) Unidad fundamental de longitud del sistema métrico.

metric system / sistema métrico (102) Sistema de pesos y medidas de base diez. El metro es la unidad fundamental de longitud; el gramo es la unidad fundamental de masa; y el litro es la unidad fundamental de capacidad.

mile / milla (202) Unidad de longitud del sistema inglés que equivale a 5,280 pies ó 1,760 yardas.

milligram / miligramo (164) Unidad métrica de masa. Un miligramo equivale a la milésima parte de un gramo.

milliliter / mililitro (164) Unidad métrica de capacidad. Un mililitro equivale a la milésima parte de un litro.

millimeter / milímetro (100) Unidad métrica de longitud. Un milímetro equivale a la milésima parte de un metro.

mixed number / número mixto (198) La suma de un entero y una fracción. $1\frac{1}{2}$, $2\frac{3}{4}$ y $4\frac{5}{8}$ son números mixtos.

mode / modal (71) Número(s) o ítem(es) de un conjunto de datos que aparece(n) más frecuentemente.

multiple / múltiplo (206) El múltiplo de un número entero es el producto del número por cualquier otro número entero.

negative integer / entero negativo (434) Entero que es menor que cero.

net / red (374) Forma que se obtiene al "desdoblar" una figura tridimensional. La red muestra todas las caras que integran la superficie de una figura.

obtuse angle / ángulo obtuso (352) Cualquier ángulo que mide más de 90° pero menos de 180°.

octagon / octágono (370) Polígono de ocho lados.

opposite / opuestos (435) Dos enteros son opuestos si, en la recta numérica, están representados por puntos que equidistan de cero, pero en direcciones opuestas. La suma de opuestos es cero.

ordered pair / par ordenado (82, 459) Par de números que se utiliza para ubicar un punto en un plano de coordenadas. Se escribe de la siguiente forma: (coordenada x, coordenada y).

order of operations / orden de las operaciones (16, 28) Reglas a seguir cuando hay más de una operación involucrada.
1. Ejecuta todas las potencias antes que cualquier otra operación.
2. Ejecuta todas las multiplicaciones y divisiones de izquierda a derecha, en el orden que aparezcan.
3. Luego ejecuta todas las sumas y restas de izquierda a derecha, en el orden que aparezcan.

origin / origen (82, 459) Punto de intersección axial en un plano de coordenadas.

ounce / onza (292) Unidad de peso del sistema inglés de medidas. 16 onzas equivalen a una libra.

outcomes / resultados (515) Resultados posibles de un experimento probabilístico. Por ejemplo, 4 es un resultado posible cuando se lanza un dado.

output / salida (494) Resultado de una entrada sobre la que se han realizado una o más operaciones, en una máquina de función.

parallel / paralelas (371) Rectas que se extienden en la misma dirección y que se mantienen siempre a la misma distancia una de la otra. Las rectas paralelas nunca se intersecan o cruzan.

parallelogram / paralelogramo (371) Cuadrilátero cuyos pares de lados opuestos son congruentes y paralelos.

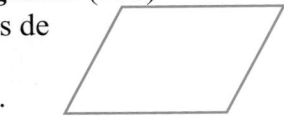

pentagon / pentágono (370) Polígono de cinco lados.

percent / tanto por ciento (330) Razón cuyo denominador es 100. Por ejemplo, 7% y $\frac{7}{100}$ indican el mismo número. 7% se lee 7 *por ciento*.

perimeter / perímetro (145) Medida del contorno de una figura cerrada.

pint / pinta (292) Unidad de capacidad del sistema inglés de medidas que equivale a dos tazas.

place value / valor de posición (95) Sistema de escritura de números en el que la posición de cada dígito de un número determina su valor.

polygon / polígono (368, 370) Figura simple cerrada en un plano, formada por tres o más segmentos de recta.

population / población (522) El grupo total de artículos o individuos del cual se toman las muestras bajo estudio.

positive integer / entero positivo (434) Entero que es mayor que cero.

pound / libra (292) Unidad de peso del sistema inglés de medidas que equivale a 16 onzas.

power / potencia (28) Número que puede escribirse usando un exponente. La potencia

3^2 se lee *tres a la segunda potencia* o *tres al cuadrado*.

prime factorization / factorización prima (183) Forma de expresar un número compuesto como el producto de números primos. La factorización prima de 63, por ejemplo, es $3 \times 3 \times 7$.

prime number / número primo (181, 182) Número entero mayor que 1 que sólo tiene dos factores, 1 y sí mismo.

prism / prisma (412) Figura tridimensional que posee dos bases paralelas y congruentes con forma de polígono, y por lo menos, tres caras laterales con forma de rectángulo. La forma de las bases identifica el prisma.

proportion / proporción (317) Ecuación que demuestra la igualdad de dos razones, $\frac{a}{b} = \frac{c}{d}$, $b \neq 0, d \neq 0$.

protractor / transportador (352) Instrumento que se usa para medir ángulos.

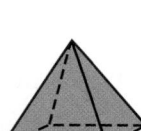

pyramid / pirámide (412) Figura tridimensional que tiene una base poligonal y caras triangulares. Las pirámides se clasifican según la forma de su base.

Q

quadrant / cuadrante (459) Una de las cuatro regiones en que dos rectas perpendiculares dividen un plano.

quadrilateral / cuadrilátero (368, 370) Polígono de cuatro lados.

quart / cuarto de galón (292) Unidad de capacidad del sistema inglés de medidas que equivale a dos pintas.

R

radius / radio (280) Distancia del centro de un círculo a cualquier punto del mismo.

radio

random / aleatorio (522) Un resultado ocurre aleatoriamente o al azar si se escoge sin preferencia alguna.

range / rango (72) Diferencia entre los valores máximo y mínimo de un conjunto de datos.

rate / tasa (312) Razón de dos medidas que tienen distintas unidades de medida.

ratio / razón (191, 193) Comparación de dos números mediante división. La razón de 2 a 3 puede escribirse como 2 de cada 3, 2 a 3, 2:3 ó $\frac{2}{3}$.

ray / rayo (363) Parte de una recta que se extiende indefinidamente en una dirección. A continuación se muestra una representación del rayo DE ($\overrightarrow{DE}$).

reciprocals / recíprocos (285) Dos números cuyo producto es 1. Como $\frac{5}{6} \times \frac{6}{5} = 1$, el recíproco de $\frac{5}{6}$ es $\frac{6}{5}$ y viceversa.

rectangle / rectángulo (371) Cuadrilátero cuyos cuatro ángulos son congruentes entre sí.

rectangular prism / prisma rectangular (412) Figura tridimensional que posee seis caras rectangulares. Un prisma rectangular tiene en total seis caras, doce aristas y ocho vértices.

reflection / reflexión (375, 464) Transformación en que a una figura se le da vuelta de campana por encima de una recta.

regular polygon / polígono regular (370) Polígono cuyos lados, así como sus ángulos, son todos congruentes.

repeating decimal / decimal periódico (218) Decimal en el cual los dígitos, en algún momento, comienzan a repetirse en bloques de uno o más números. 0.181818... puede también escribirse 0.18. La barra encima de los dígitos indica que estos dígitos se repiten.

right angle / ángulo recto (352) Ángulo que mide 90°.

S

sample / muestra (522) Grupo escogido al azar o aleatoriamente que se usa para representar la población entera.

sample space / espacio muestral (515) Conjunto de todos los resultados posibles de un experimento probabilístico.

scale / escala (50) Conjunto de todos los posibles valores de una medida dada, el cual incluye los valores máximo y mínimo del conjunto, separados mediante los intervalos que se han usado.

scale drawing / dibujo a escala (324) Dibujo que es semejante, pero más grande o más pequeño que el objeto real.

sequence / sucesión (298) Lista de números en cierto orden como, por ejemplo, 0, 1, 2, 3 ó 2, 4, 6, 8.

sides / lados (145) Segmentos de recta que encierran un polígono.

similar figures / figuras semejantes (379) Figuras que tienen la misma forma, pero no necesariamente el mismo tamaño. El símbolo ~ se lee *es semejante a.*

simplest form / forma reducida (194) Forma de una fracción en que el MCD de su numerador y denominador es 1. La fracción $\frac{3}{4}$ está escrita en forma reducida pues el MCD de 3 y 4 es 1.

simulation / simulación (535) Proceso de representar un problema.

solution / solución (34) Cualquier número que satisface una ecuación. La solución de $12 = x + 7$ es 5.

solve / resolver (34) Proceso de encontrar el número o números que satisfagan una ecuación.

sphere / esfera (413) Figura tridimensional que carece de caras, bases, aristas o vértices. Conjunto de todos los puntos en el espacio que equidistan de un punto dado llamado centro.

centro

spreadsheet / hoja de cálculos (144) Herramienta que se usa para organizar y analizar datos.

square / cuadrado (371) Paralelogramo cuyos lados, así como sus ángulos, son todos congruentes.

squared / al cuadrado (28) Número multiplicado por sí mismo; 4×4 ó 4^2.

square pyramid / pirámide cuadrada (412) Pirámide cuya base es un cuadrado.

statistics / estadística (46) Estudio de la recolección, análisis y presentación de datos.

stem / tallo (68) El mayor valor de posición común a todos los datos, el cual se escribe a la izquierda de la línea divisoria vertical en un diagrama de tallo y hojas.

stem-and-leaf plot / diagrama de tallo y hojas (68) Sistema que se usa para condensar un conjunto de datos y en el cual el mayor valor de posición de los datos forma el tallo y el segundo mayor valor de posición de los datos forma las hojas.

straightedge / regla (358) Cualquier objeto con un lado recto que se usa para trazar rectas.

supplementary / suplementarios (353) Dos ángulos son suplementarios si la suma de sus medidas es 180°.

surface area / área de superficie (421) Suma de las áreas de todas las superficies de una figura tridimensional.

T

terminating decimal / decimal terminal (214) Decimal cuyos dígitos terminan. Todo decimal terminal puede escribirse como una fracción con un denominador de 10, 100, 1,000, etc.

theoretical probability / probabilidad teórica (515) La razón del número de maneras en que puede ocurrir un evento al número total de resultados posibles.

three-dimensional figure / figura tridimensional (412) Figura que encierra parte del espacio.

ton / tonelada (292) Unidad de peso del sistema inglés que equivale a 2,000 libras.

transformation / transformación (464) Movimiento de figuras geométricas en un sistema de coordenadas a otros puntos del sistema.

translation / traslación (384, 464) Tipo de transformación en que una figura se desliza horizontalmente, verticalmente o de ambas maneras.

tree diagram / diagrama de árbol (531) Diagrama que se utiliza para encontrar y mostrar el número total de resultados posibles de un experimento probabilístico.

triangle / triángulo (368, 370) Polígono de tres lados.

unfair game / juego injusto (514) Juego en que los jugadores no tienen la misma oportunidad de ganar.

upper quartile / cuartil superior (77) La mediana de la mitad superior de un conjunto de números o datos.

variable / variable (22) Un símbolo, por lo general, una letra que se usa para representar números.

vertex / vértice (352) Un vértice de un polígono es el punto de intersección de dos lados del polígono.

vertex / vértice (412) Punto en que se intersecan las aristas de una figura tridimensional.

volume / volumen (418) Cantidad de espacio que encierra una figura tridimensional. Se expresa en unidades cúbicas.

x-**axis / eje** *x* (82, 459) La recta horizontal de las dos rectas numéricas perpendiculares, en un plano de coordenadas.

x-**coordinate / coordenada** *x* (82, 459) Primer número de un par ordenado.

yard / yarda (202) Unidad de longitud del sistema inglés que equivale a 3 pies ó 36 pulgadas.

y-**axis / eje** *y* (82, 459) La recta vertical de las dos rectas numéricas perpendiculares, en un plano de coordenadas.

y-**coordinate / coordenada** *y* (82, 459) Segundo número de un par ordenado.

zero pair / par nulo (441) Resultado de aparear una ficha positiva con una negativa.

SPANISH GLOSSARY/GLOSARIO

Index

INDEX

changing units of capacity,
292–294
cup, 292
fluid ounce, 292
foot, 202–205
gallon, 292
inch, 202–205, 229, 233, 326
length, 202–205
mile, 202–205
to the nearest unit, 203, 229
ounce, 292
pint, 292
pound, 292
quart, 292
ton, 292
yard, 202–205

Cylinder, 413

Data, 46
graphing, 44–91

Data Analysis *See* Applications,
Connections, and Integration
Index on pages xxii–1

Decagon, 370, 372

Decimals
adding, 118–121, 124
bar notation, 218
checking products by adding,
134
comparing, 105–108, 124
dividing by, 156–159, 161–163,
172
dividing by whole numbers,
152–155, 172
estimating differences,
112–115, 124
estimating products, 133–135
estimating quotients, 152–153,
157–158, 161–162
estimating sums, 112–115, 124
fractions as, 95–97, 123
as mixed numbers, 214–215
modeling, 94, 96, 123, 215
models, 132, 140
multiplying, 140–143, 171
multiplying by whole numbers,
132–136, 171
ordering, 106–108, 124
place value, 95, 123
placing the decimal point in
products, 133–134
placing the decimal point in
quotients, 152–153

products using the distributive
property, 137–139, 171
repeating, 218
rounding, 109–111, 124
rounding quotients, 153–155
subtracting, 119–121, 124
terminating, 214
writing in words, 95, 123
writing using numerals, 95, 123
zeros in quotients, 161–162

Decision Making, 24, 29, 65, 84,
93, 97, 110, 131, 135, 153, 168,
183, 215, 233, 251, 269, 299,
372, 404, 419, 438, 447, 486,
497, 538

Degree, 351

Denominator, 193, 238
least common, 210

Diameter, 280

Did You Know? 5, 12, 46, 68, 84,
95, 103, 161, 164, 189, 189,
240, 273, 379, 422, 445, 457,
476, 496, 537

Distributive property, 137–139
mental math using, 138

Diversity, 2e, 44e, 92e, 130e,
176e, 226e, 266e, 308e, 350e,
394e, 432e, 474e, 512e

Divisibility rules, 178–180

Division
by decimals, 156–159, 161–163
of decimals by whole numbers,
152–155
divisibility rules, 178–180
equations, 484–487
estimating, 14
of fractions, 284–288
of integers, 456–458
of mixed numbers, 289–291
in order of operations, 16, 28
of whole numbers, 14

Draw a diagram, 322–323

Edge, 412

Eliminate possibilities, 236–237

Enhancing the Chapter, 2e, 2f,
44e, 44f, 92e, 92f, 130e,
130f, 176e, 176f, 226e, 226f,
266e, 266f, 308e, 308f, 350e,

350f, 394e, 394f, 432e, 432f,
474e, 474f, 512e, 512f

Equations
addition, 476–479
defined, 34
division, 484–487
modeling, 476–477, 480–482,
484–486, 488–490
multiplication, 484–487
solution of, 34
solving, 34–37, 40, 85, 121,
124, 213, 245, 274, 276,
283, 287, 289, 290, 291,
353, 354, 476–493, 487
subtraction, 480–483
two-step, 488–491
using, 492–493
writing, 485, 490

Equilateral triangle, 371

Equivalent fractions, 193–196

Error Analysis, *see Reteaching
the Lesson*

Escher drawings, 384–385

Estimation
of angle measure, 359, 360
area of irregular figures,
396–397
of decimal products, 133–135
of decimal quotients, 152–154,
161–162
of decimals, 93
differences of fractions,
232–234
with percents, 337–339
products of fractions, 268–270
study hints, 35, 119, 153, 239
sums of fractions, 232–234
using clustering, 113–114
using compatible numbers, 133,
268
using graphs, 57
using rounding, 12–15, 39,
112–114, 141, 269

Evaluation, *see Assessment
Resources*

Events, 515
complementary, 516
independent, 536–539

Experimental probability, 197,
526

Exponents, 28, 40

Expressions
algebraic, 22

138–139, 145, 157, 167, 214, 239, 285, 298, 318, 331, 334, 335

number sense, 12, 14, 95, 123, 187, 323, 434, 435, 521

See also problem solving

Mathematical Tools

paper/pencil, 10, 17, 24, 29, 35, 48, 51, 56, 59, 62, 79, 84, 97, 103, 106, 110, 134, 138, 142, 147, 158, 161, 162, 193, 217, 295, 360, 435, 443, 447, 451, 457, 458, 482, 490, 497, 501, 517, 533, 538

real objects, 20, 31, 47, 60, 73, 75, 76, 103, 132, 140, 149, 166, 264, 359, 374, 376, 379, 384, 392–393, 402, 406, 409, 410–411, 421, 444, 462–463, 464, 465, 494–495, 526, 535, 539

base-ten blocks, 94, 95, 96, 123, 156

centimeter cubes, 9, 418

compass, 362, 363, 364, 410–411

counters, 15, 121, 440, 441–443, 445–447, 449–451, 456–457, 476–477, 480–482, 484–486, 488–490, 517, 535

dot paper, 526

geoboard and geobands, 271–272, 284, 368–369, 380

graph paper, 128

grid paper, 106, 132, 140, 149, 238, 239, 383, 384, 396–397, 398–399, 421, 464, 465, 530

integer mat, 440, 441–443, 445–447, 449–451, 456–457, 476–477, 480–482, 484–486, 488–490

isometric dot paper, 415

meterstick, 121

number cubes, 121, 514, 533

protractor, 352, 353, 358, 359, 365, 410–411

ruler, 104, 123, 191, 364, 396–397, 398–399, 401, 410–411, 415, 421

spinner, 75, 515, 517, 532, 533, 538

straightedge, 358, 359, 362, 363, 364, 365, 530

tape measure, 60, 100, 103, 166, 202

technology, 3, 17, 26, 27, 31, 32, 45, 73, 75, 93, 99, 107, 119, 121, 128, 129, 131, 135, 160, 166, 131, 144, 146, 154, 207, 410–411, 421, 425, 530

See also problem solving

Math Journal, 6, 14, 56, 69, 79, 114, 119, 138, 147, 162, 179, 212, 230, 255, 275, 290, 313, 338, 360, 413, 423, 435, 458, 501, 517, 533

Math in the Media, 98, 205, 279, 382, 479

Mean, 71–74, 88, 396

game, 75

Measurement

capacity, 164–166

centimeters, 100–104

cup, 292

customary system, 202–205, 229, 292–296

foot, 202–205

gallon, 292

gram, 164–166

inch, 202–205

kilogram, 164–166

kilometers, 100–104

length, 100–104, 202–205

liter, 164–166

mass, 164–166

meters, 100–104

metric system, 100–104, 164–166, 167–169

mile, 202–205

milligram, 164–166

milliliter, 164–166

millimeters, 100–104

to the nearest unit, 203, 229

nonstandard units, 203

pint, 292

pound, 292

quart, 292

time, 254–257

ton, 292

yard, 202–205

See also Applications, Connections, and Integration Index, pages xxii–1

Measures of central tendency, 71, 79, 80

average, 71

mean, 71–75, 88

median, 71–75, 88

mode, 71–75, 88

range, 72–74, 88

Median, 71–74, 88

game, 75

Meeting Individual Needs, 2f, 44f, 92f, 130f, 176f, 226f, 266f, 308f, 350f, 394f, 432f, 474f, 512f

Mental math, 35, 36, 40, 239

study hints, 145, 157, 167, 214, 285, 298, 318, 331, 334, 335

using the distributive property, 138–139

Metric system, 100–104, 164–166, 167–169

capacity, 164–166

centimeter, 100–104, 122, 167–169

changing units, 167–169

gram, 164–166, 167–169

kilogram, 164–166, 167–169

liter, 164–166, 167–169

mass, 164–166

meter, 100–104, 122, 167–169

milligram, 164–166, 167–169

milliliter, 164–166, 167–169

millimeter, 100–104, 122, 167–169

Mid-Chapter Self Test, 19, 67, 111, 155, 201, 249, 283, 327, 373, 414, 452, 487, 525

Mile, 202–205

Milligram, 164

Milliliter, 164

MindJogger Videoquizzes, *see Technology*

Mini-lab, 9, 17, 47, 48, 51, 60, 73, 106

Mixed numbers, 96, 198–201

adding, 246–249

as decimals, 218

dividing, 289–291

as improper fractions, 198–199, 250

multiplying, 277–279

renaming, 250

subtracting, 246–249, 250–252

Mode, 71–74, 88

game, 75

INDEX

Answer Appendix

CHAPTER 1
Problem Solving, Numbers, and Algebra

Pages 18–19, Lesson 1-4

22a. $24 \times 15 + 24 \times 20 + 24 \times 15$

23a. Sample answer: $3 \times \$29.95 + 2 \times \$16.95 + \$12.95 = \$136.70$

Page 26, Lesson 1-5B

9. Yes; unless the scientific calculator does not have parentheses keys.

Pages 38–41, Chapter 1
Study Guide and Assessment

17.

CHAPTER 2
Statistics: Graphing Data

Pages 48–49, Lesson 2-1

6.

Hours	Tally	Frequency
1	III	3
2	IIII	4
3	II	2
4	I	1

7.

Days Absent	Tally	Frequency
0	HH	5
1	HH I	6
2	II	2
3	III	3
4		0
5	II	2

8.

Temperature	Tally	Frequency
92	I	1
91	HH	5
90	HH IIII	9
89		0
88	HH HH	10
87	IIII	4
86		0
85	I	1

Pages 51–53, Lesson 2-2

7a.

Shelby's Video Game Scores		
Score	Tally	Frequency
60,000–69,000	I	1
50,000–59,900	I	1
40,000–49,900	IIII	4
30,000–39,900	HH	5
20,000–29,900	IIII	4
10,000–19,900	II	2
0–9,900	I	1

14.

Amount	Tally	Frequency
16–20	I	1
11–15	I	1
6–10	II	2
0–5	III	3

15.

Amount	Tally	Frequency
81–100	I	1
61–80		0
41–60	IIII	4
21–40	I	1
0–20	II	2

16.

Amount	Tally	Frequency
401–500	I	1
301–400	I	1
201–300	III	3
101–200	III	3
0–100	II	2

Pages 55–57, Lesson 2-3

6b.

World's Busiest Airports in 1994

7. **Favorite Soft Drinks**

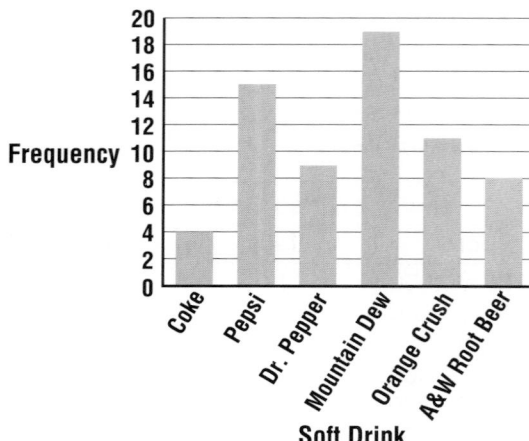

8. **Average Event-Day Attendance**

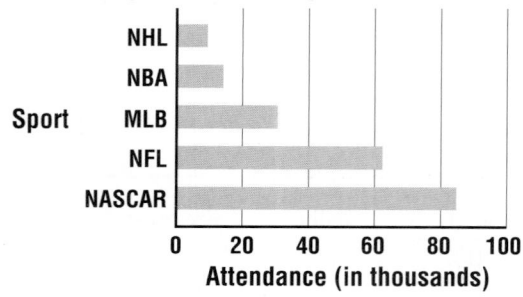

9. **Average Snowfall for Brighton, Utah**

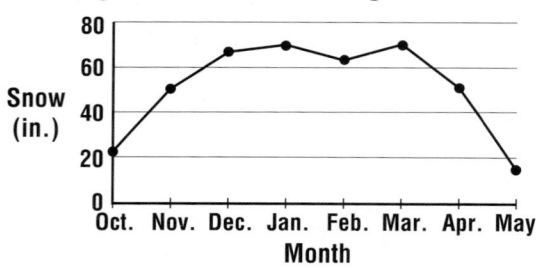

10. **U.S. Sales of Karaoke Players**

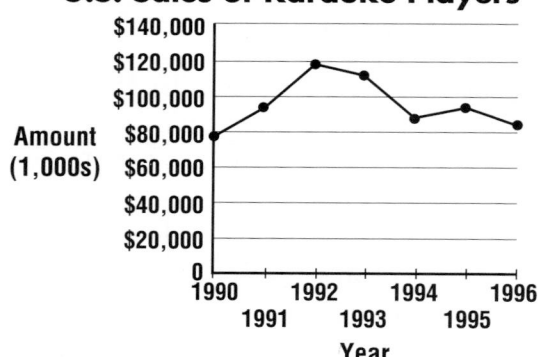

14. **Entertainment Expenses per Person in 1994**

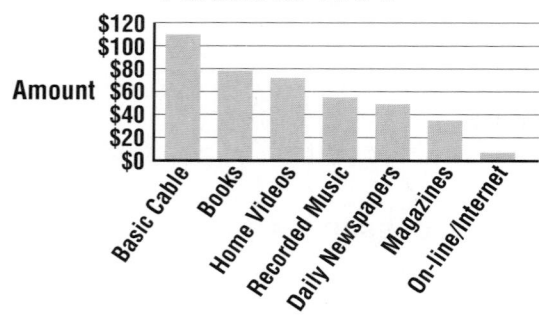

15. **Super Bowl Ticket Prices**

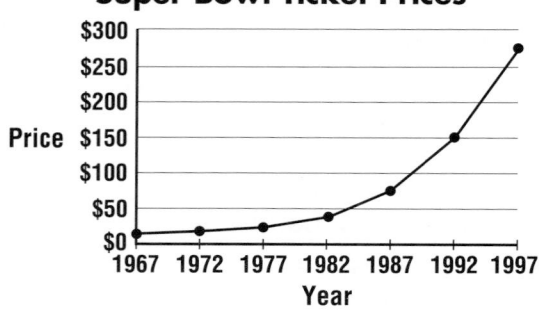

Pages 62–63, Lesson 2-4

9. **Janet's Leaf Collection**

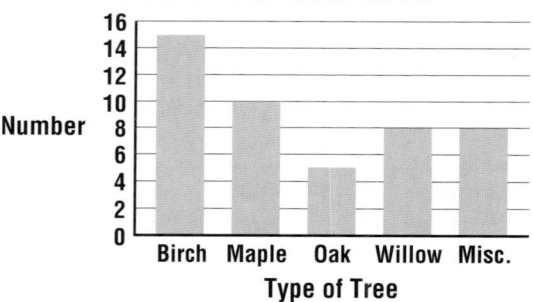

Page 67, Mid-Chapter Self Test

1.

Height	Tally	Frequency
54	I	1
55	I	1
56	I	1
57	IIII	4
58	I	1
59	I	1
60	II	2
61		0
62		0
63	I	1

2c.

Number	Tally	Frequency
81–100	IIII	4
61–80	II	2
41–60	IIII	4
21–40	II	2
0–20	II	2

3. **Favorite Ice Cream Flavor**

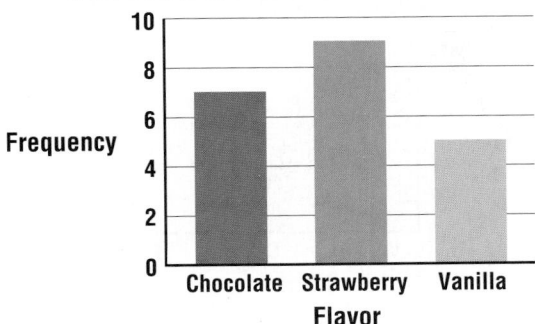

Pages 69–70, Lesson 2-6

5.

Stem	Leaf
2	0 4 7
3	4 5 6 6 8
4	3 5 7
5	3 4 8 8
6	
7	8

$2|7 = 27$

6a.

Stem	Leaf
3	2 3 8 9
4	1 5 6
5	0 1 2 2 3 9
6	1 1 2 6 7 9
7	0 2

$3|2 = 32$

10.

Stem	Leaf
2	1 2 4 7 7 8 8 9
3	0 1 1 2 3 5 8 8
4	4 5

$2|1 = 21$

11.

Stem	Leaf
9	2 4 9 9
10	
11	
12	4 4
13	0 2 3
14	0
15	
16	2 7

$12|4 = 124$

12.

Stem	Leaf
5	4 4
6	
7	5
8	3 5 9
9	2 4 6

$5|4 = 54¢$

13.

Stem	Leaf
0	9
1	
2	7 8 9
3	4 4 4
4	5 6
5	3 6 6 7 7 7 8

$5|3 = 53°$

14a.

Stem	Leaf
0	7 8 9
1	0 1 2 2 2 3 4 5 6 6 8 8 9
2	0 1 2

$1|0 = 10$

15a.

Stem	Leaf
0	5 5 6 7 8 9 9 9 9 9
1	0 1 1 3 4 6 7 8 8 9 9
2	2 2 3 6 7 7 9
3	7
4	1

$1|0 = 10$

Pages 73–74, Lesson 2-7

5. 10, 26, 27, 29, 36, 37, 57, 83, 88; 36; no mode; ≈ 43.7; 78

15. yes; Sample answer: The total would be higher so the mean would also be higher.

17. Sample answer: The mode because she would want to know what size they sold the most shakes.

18. Sample answer: 12, 12, 14, 14, 15, 17

19.

Stem	Leaf
0	0 2 2 4 5 7 7 9
1	0 2 3 3 8 9
2	0 3

$2|0 = 20$

Page 77, Lesson 2-7B

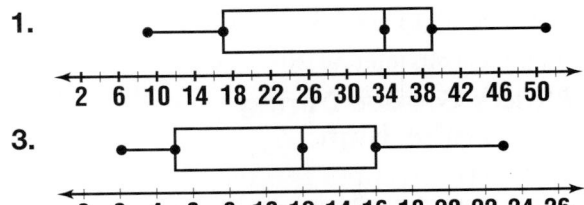

Pages 79–81, Lesson 2-8

3. Sample answer: If the range is very high the mean could be misleading. If the set of data has a gap between two groups of numbers that are all about the same, the mean could be misleading. For example, 1, 2, 3, 2, and 8, 7, 6, 9, 8.

Pages 84–85, Lesson 2-9

36.

Top Animal Speeds

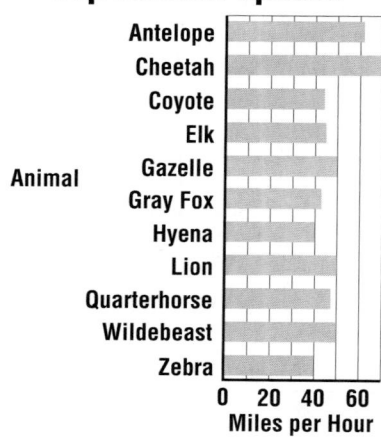

**Pages 86–89, Chapter 2
Study Guide and Assessment**

12.

Number of Brothers and Sisters	Tally	Frequency
0	IIII	4
1	HHII	7
2	III	3
3	III	3
4	I	1

13.

Favorite Color	Tally	Frequency
blue	IIII	4
red	IIII	4
pink	II	2
orange	I	1
yellow	I	1
green	II	2
brown	I	1

15. scale = 0 to 50, interval 5

16. scale = 0 to 20, interval 3

17. scale = 0 to 160, interval 20

18.

Alma's Grades During Junior High

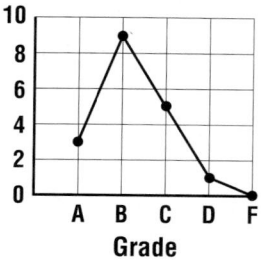

19.

Alma's Grades During Junior High

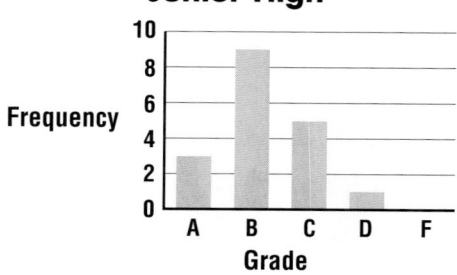

35.

Stem	Leaf
4	0 5
5	0
6	5 9
7	5 5
8	5
9	0 2 5

$9|0 = 90$

CHAPTER 3
Adding and Subtracting Decimals

Page 94, Lesson 3-1A

3.

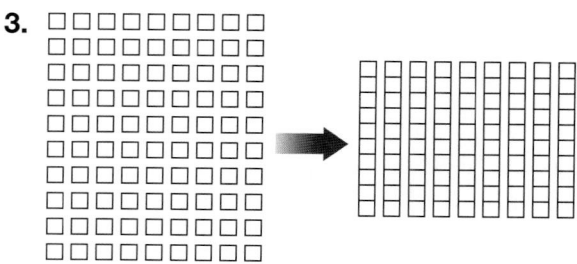

Pages 97-98, Lesson 3-1

2.

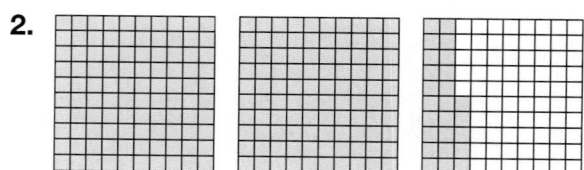

Pages 103-104, Lesson 3-2

2. Sample answer: centimeters; The length of the book is smaller than a meter.

26.

Guinea Pig Weights

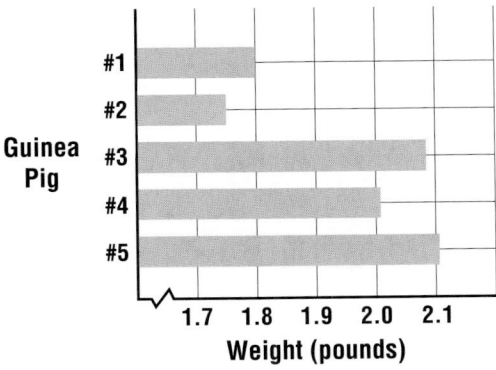

Pages 107–108, Lesson 3-3

1.

Pages 110–111, Lesson 3-4

1. Sample answer: Since the digit to the right of the ones place is greater than 5, add one to the ones place. $10.79 rounded to the nearest dollar is $11.

2.

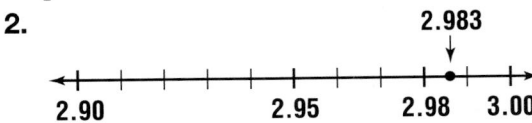

25. Sample answer: Using the exact data would be most appropriate. Since there are only 9 numbers in the set of data, rounding is not necessary.

27. No, because no numbers differ greatly.

Pages 114–115, Lesson 3-5

35a. Answers will vary. Sample answer: An estimate for Mandy's family is $9 + $9 + $9 + $4 or $31. Since all costs were rounded up, her whole family can go on the tour for less than $35.

35b. Answers will vary. Sample answer: An estimate for each meal is $5 for breakfast, $7 for lunch, and $12 for dinner. The estimate for one person would be $5 + $7 + $12 or $24. The estimate for a family of 4 would be $24 + $24 + $24 + $24 or $96. This could be rounded up to $100.

CHAPTER 4
Multiplying and Dividing Decimals

Page 132, Lesson 4-1A

1.

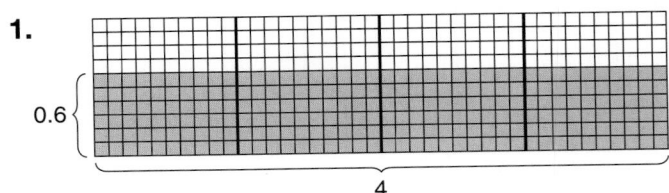

2.

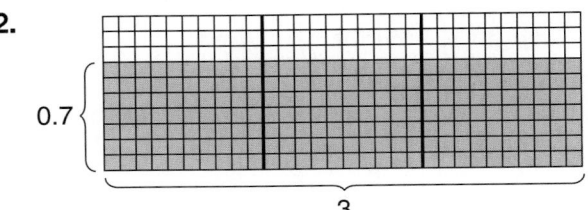

3.

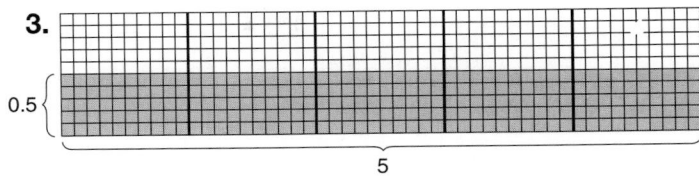

Page 140, Lesson 4-3A

3.   **4.**

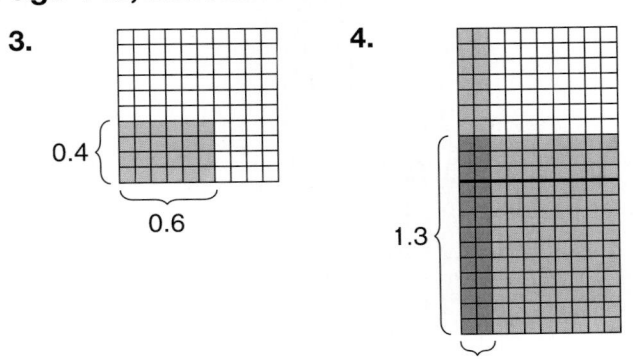

5.

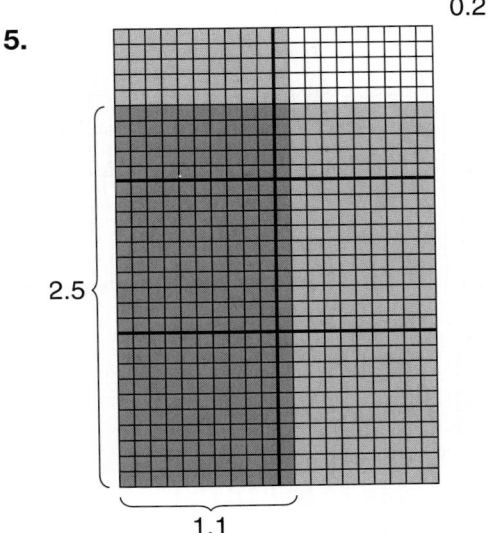

Page 149, Lesson 4-4B

1a.

Length	Width	Area
7	1	7
6	2	12
5	3	15
4	4	16

1b.

Length	Width	Area
8	1	8
7	2	14
6	3	18
5	4	20

1c.

Length	Width	Area
9	1	9
8	2	16
7	3	21
6	4	24

CHAPTER 5
Using Number Patterns, Fractions, and Ratios

Pages 179–180, Lesson 5-1

1. Sample answer:

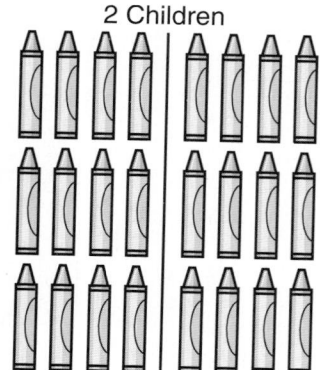

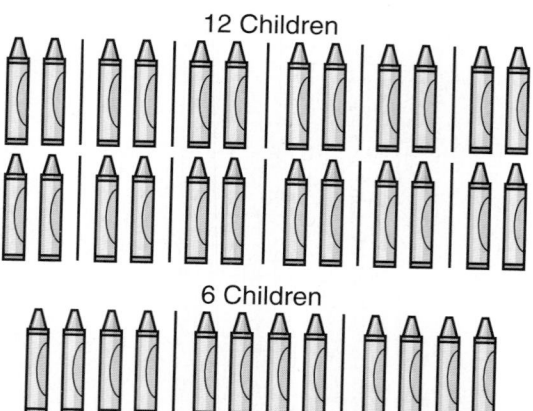

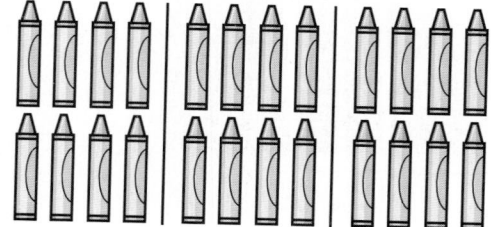

Page 181, Lesson 5-2A

1. 2 ▭
3 ▭
4 ▭
▭
5 ▭

Pages 183–184, Lesson 5-2

1.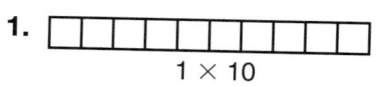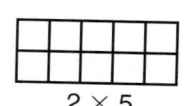

1×10

2×5

So, 10 is a composite number.

2. Sample answer: While a composite number has more than two factors, a prime number has exactly two factors, 1 and the number itself.

3. Both are correct. The factors are arranged in a different order.

Pages 189–190, Lesson 5-3

1. Sample answer: Find the prime factorization of 12 and 18. Multiply all common factors of both numbers to find the greatest common factor.

2. Sample answer: It could be one of the numbers if one number is a factor of the other number. Example #1: The GCF of 3 and 9 is 3. Example #2: The GCF of 4 and 16 is 4.

Pages 191–192, Lesson 5-4A

6.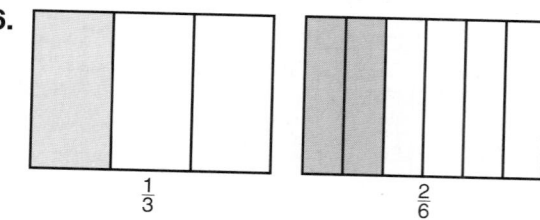

$\frac{1}{3}$ $\frac{2}{6}$

7.

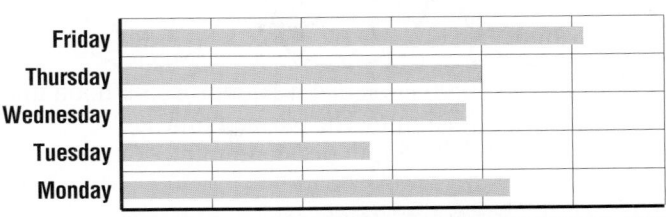

$\frac{2}{8}$ $\frac{1}{4}$

Pages 200–201, Lesson 5-5

34a. *The Lion King:* $\frac{88}{60}$; *Cinderella:* $\frac{76}{60}$; *Snow White and the Seven Dwarfs:* $\frac{84}{60}$; *Bambi:* $\frac{69}{60}$

34b. *The Lion King:* $1\frac{7}{15}$; *Cinderella:* $1\frac{4}{15}$; *Snow White and the Seven Dwarfs:* $1\frac{2}{5}$; *Bambi:* $1\frac{3}{20}$

35. Sample answer: If the numerator is less than the denominator, then the fraction is less than one. If the numerator is the same as the denominator, then the fraction is equal to one. If the numerator is greater than the denominator, then the fraction is greater than one.

Pages 203–205, Lesson 5-6

17. |————————————|
18. |————————————————|
19. |——|
20. |——————————————————————|
21. |——————————————————|
22. |————|

Pages 208–209, Lesson 5-7

38. |———|

Pages 215–216, Lesson 5-9

1. Sample answer: Write the decimal as a fraction with a denominator of 100. Then simplify by using the GCF.

3. Ann; The decimal is written to the hundredths place. So, the denominator will be 100.

Pages 218–219, Lesson 5-10

1. Divide the numerator by the denominator.

2. Sample answer: terminating decimal: 0.8; repeating decimal: 0.890890890…

Pages 220–223, Chapter 5
Study Guide and Assessment

35. |————————————————————|
36. |————————————————————————————|

CHAPTER 6
Adding and Subtracting Fractions

Pages 230–231, Lesson 6-1

38.

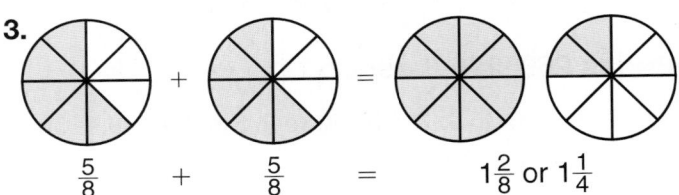

Raul's Earnings

See students' work for the additional graph.

Pages 233–234, Lesson 6-2

1.

$2\frac{3}{4}$

←———|———|———|———→
   0   1   2   3

3. Sample answer: Nina should buy 20 inches of framing material. Rounding $4\frac{3}{8}$ to 4 is correct, but since you rounded down, there would not be enough material to complete the picture frame.

10. Margie has $5\frac{2}{3}$ yards of material. She uses $1\frac{1}{4}$ yards for a scarf. About how much material does she have left?

Pages 240–241, Lesson 6-3

3.

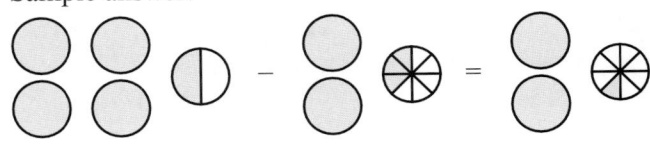

$\frac{5}{8}$ $+$ $\frac{5}{8}$ $=$ $1\frac{2}{8}$ or $1\frac{1}{4}$

Page 248–249, Lesson 6-5

3. Sample answer:

Pages 251–253, Lesson 6-6

1. Sample answer:

3. Sample answer: When Gail borrowed one from 10 she just carried it next to the 1 in $\frac{1}{5}$ and made it $\frac{11}{5}$.

2. Sample answer:

7:50 A.M. to 12:00 noon is 4 h 10 min. 12:00 noon to 2:35 P.M. is 2 h 35 min. The elapsed time is 4 h 10 min + 2 h 35 min or 6 h 45 min.

Page 261, Performance Task

Sample answer: To estimate the sum or difference of mixed numbers round each number to the nearest whole number. Then add or subtract.

Sample answer: Since you need the amounts shown, round all the mixed numbers up.

13 + 11 + 19 = 43 yards

Sample answer: Subtract similar to subtracting decimals.

$$
\begin{array}{ccc}
9{:}18 & \rightarrow & 8{:}78 \\
-7{:}30 & & -7{:}30 \\
\hline
& & 1{:}48 \text{ or } 1 \text{ h } 48 \text{ min}
\end{array}
$$

Sample answer: It is similar to subtracting mixed numbers involving renaming because you rename hours as minutes. For example, 9 h 18 min = 8 h 78 min.

CHAPTER 7
Multiplying and Dividing Fractions

Pages 269–270, Lesson 7-1

1. Round $3\frac{1}{2}$ to 4. Round $9\frac{1}{8}$ to 9. Since $4 \times 9 = 36$, $3\frac{1}{2} \times 9\frac{1}{8} \approx 36$.

2. $\frac{1}{4}$ and 8 are compatible numbers. Since $\frac{1}{4} \times 8 = 2$, it follows that $\frac{3}{4} \times 8 = 2 \times 3$, or 6. So, $\frac{3}{4} \times 9 \approx 6$.

3. Both Juan and Odina are correct. If you round $8\frac{1}{2}$ to 9 and $6\frac{1}{4}$ to 6, the product is 9×6, or 54. If you round $8\frac{1}{2}$ to 8 and $6\frac{1}{4}$ to 6, the product is 8×6, or 48.

19-27. Sample answers are given.

19. $\frac{1}{4} \times 32 = 8$ **20.** $\frac{1}{2} \times \frac{1}{2} = \frac{1}{4}$

21. $\frac{1}{7} \times 21 = 3$ **22.** $2 \times 6 = 12$

23. $\frac{1}{6} \times 42 = 7$ **24.** $1 \times 3 = 3$

25. $\frac{4}{9} \times 45 = 20$ **26.** $\frac{1}{6} \times 1 = \frac{1}{6}$

27. $8 \times 3 = 24$

8.

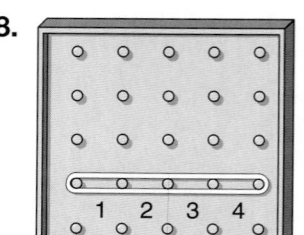
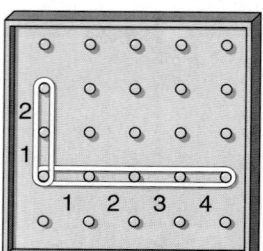
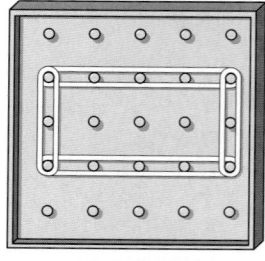
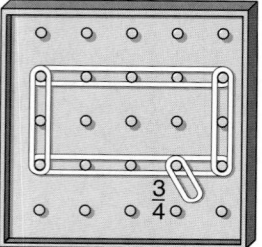
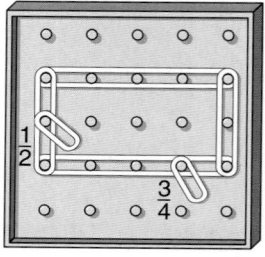
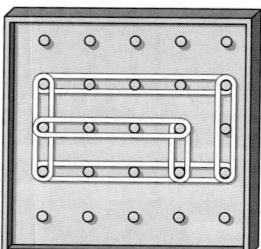

Pages 278–279, Lesson 7-3

1. To multiply mixed numbers, express each mixed number as an improper fraction and then multiply as with fractions.

2. Express $3\frac{1}{2}$ as the improper fraction $\frac{7}{2}$. Express $1\frac{1}{2}$ as the improper fraction $\frac{3}{2}$. Then multiply $\frac{7}{2}$ and $\frac{3}{2}$ $\frac{7}{2} \times \frac{3}{2} = \frac{21}{4}$, or $5\frac{1}{4}$.

Page 284, Lesson 7-5A

1.

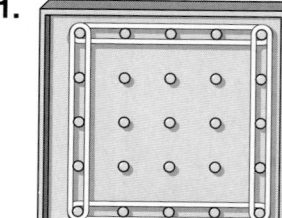
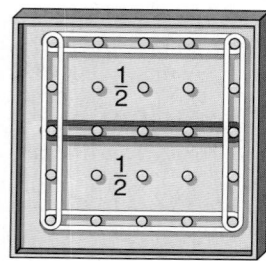

2.

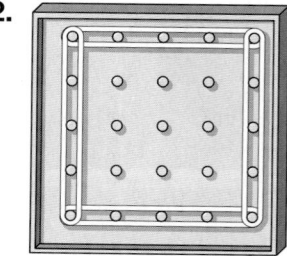
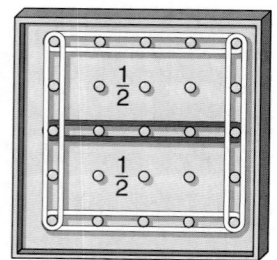

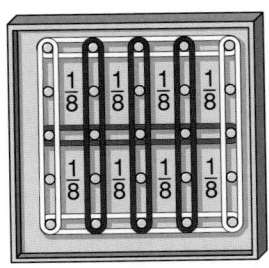

3.

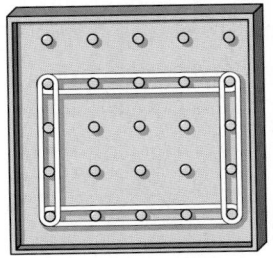 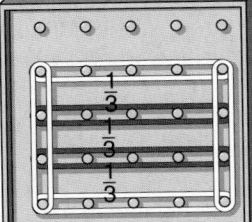

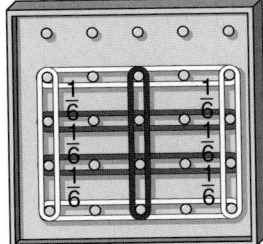

4.

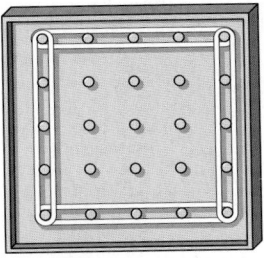 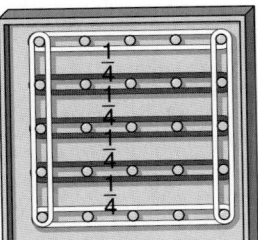

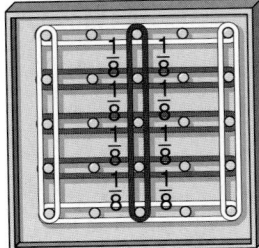

Pages 290–291, Lesson 7-6

1. First, write $4\frac{5}{8}$ as the improper fraction $\frac{37}{8}$. Then invert the numerator and the denominator. The reciprocal of $4\frac{5}{8}$ is $\frac{8}{37}$.

2. More; there are eight $\frac{1}{2}$'s in 4.

3. Change each mixed number to an improper fraction. Multiply the first fraction by the reciprocal of the second fraction. Divide 8 and 12 by the GCF, 4. Multiply $\frac{3}{1}$ and $\frac{3}{2}$ to find the product $\frac{9}{2}$, or $4\frac{1}{2}$.

Page 295, Lesson 7-7B

1.

Planet/Moon	Amount of water	Weight compared to Earth	Weight compared to Jupiter
Sun	28 cups	more	more
Mercury	$\frac{1}{3}$ cups	less	less
Venus	$\frac{9}{10}$ cups	less	less
Moon	$\frac{1}{6}$ cups	less	less
Mars	$\frac{3}{8}$ cups	less	less
Jupiter	3 cups	more	—

2. The container labeled *Jupiter*. It has the most water.

Pages 296–297, Lesson 7-8A

3.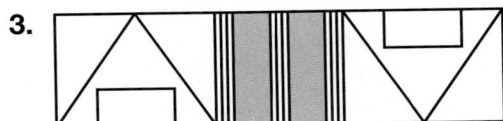

Pages 299–300, Lesson 7-8

3. Joshua is correct. In the sequence, $2\frac{1}{2}$ is added to each number. The next number is $12\frac{1}{2} + 2\frac{1}{2}$, or 15.

Pages 302–305, Chapter 7 Study Guide and Assessment

11-16. Sample answers are given.

11. $10 \times 3 = 30$

12. $4 \times 6 = 24$

13. $1 \times 13 = 13$

14. $1 \times \frac{1}{2} = \frac{1}{2}$

15. $\frac{2}{3} \times 18 = 12$

16. $8 \times \frac{1}{4} = 2$

CHAPTER 8
Exploring Ratio, Proportion, and Percent

Pages 313–315, Lesson 8-1

36.

total number of students	−	number of students at or above 85%	=	number of students below 85%
24	−	18	=	6

Ratio of the number of students scoring below 85% to the total number of students is 6:24.

a. The ratio 18:24 represents the number of students who received 85% or better.

b. The ratio 24:6 is reversed from what Mr. Sarmiento asked. This ratio compares the total students to the number of students below 85%.

c. The ratio 24:18 compares the total number of students to those who received 85% or above.

Page 329, Lesson 8-4A

1a. 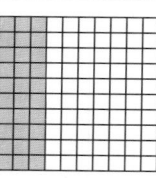 **1b.**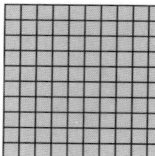

1c. 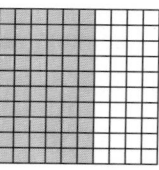 **1d.**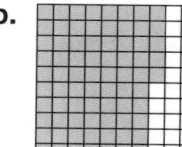

1e.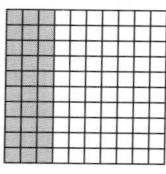

2a. 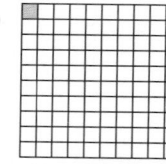 **2b.**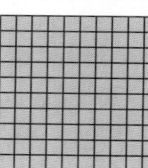

2c. 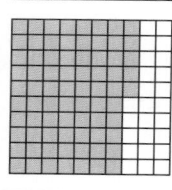 **2d.**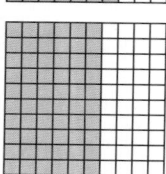

2e.

3. Sample answer: Grid C from the first set is the same as grid A from the second set. Grid A from the first set is the same as grid B from the second set. Grid B from the first set is the same as grid C from the second set. Grid D from the first set is the same as grid D from the second set. Grid E from the first set is the same as grid E from the second set.

Pages 332–333, Lesson 8-4

21. 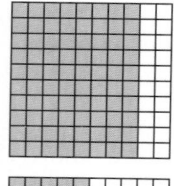 **22.**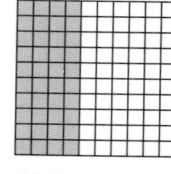

23. 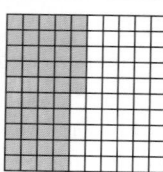 **24.**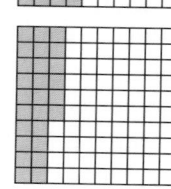

Pages 335–336, Lesson 8-5

1.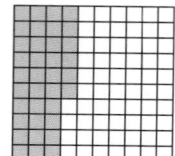

2. First, express 0.008 as the fraction $\frac{8}{1,000}$. Next, divide the numerator and denominator by 10 so that the denominator is 100. $\frac{8}{1,000} = \frac{0.8}{100}$. Then express the fraction as the percent 0.8%.

3. Betty is correct. The decimal 3.78 is equivalent to 378%. Therefore, it is more than 100%.

Pages 338–339, Lesson 8-6

1. $75\% = \frac{3}{4}$. Round 1,976 to 2,000. $\frac{3}{4} \times 2,000 = 1,500$. So, 75% of 1,976 is about 1,500.

16. Sample answer: $\frac{2}{3} \times 90 = 60$

17. Sample answer: $1 \times 300 = 300$

18. Sample answer: $\frac{1}{4} \times 280 = 70$

CHAPTER 9
Geometry: Investigating Patterns

Pages 353–355, Lesson 9-1

1. Sample answer: See students' drawings;

Step 1   Place the center of the protractor on the vertex of the angle with the straightedge along one side.

Step 2   Using the scale that begins with 0° on the side of the angle, read the angle measure where the other side crosses the same scale.

Pages 356–357, Lesson 9-1B

6. Sample answer: In a line of 24 people waiting for a city bus, 12 have umbrellas, 10 have raincoats, and 7 have both umbrellas and raincoats. The rest of the people have neither a raincoat nor an umbrella. How many people only have umbrellas?

Pages 360–361, Lesson 9-2

4.   **5.**

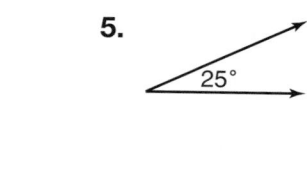

6.   **10.**

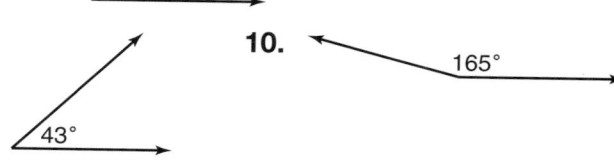

11.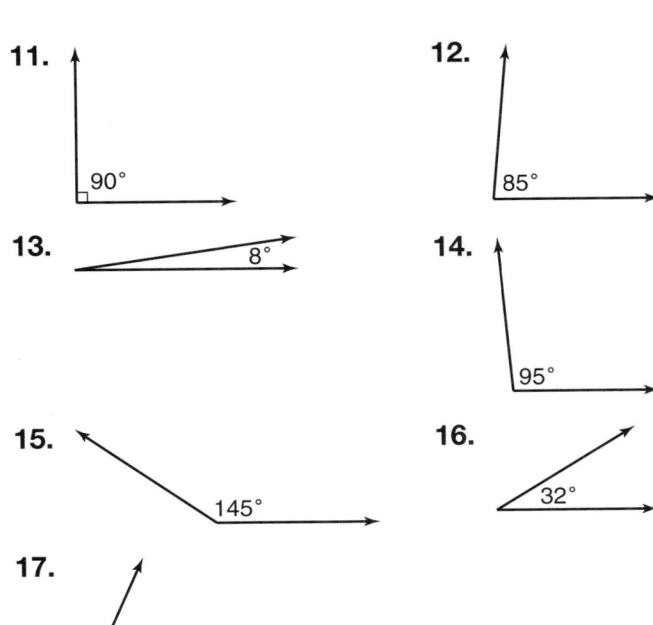
90°

12.
85°

13.
8°

14.
95°

15.
145°

16.
32°

17.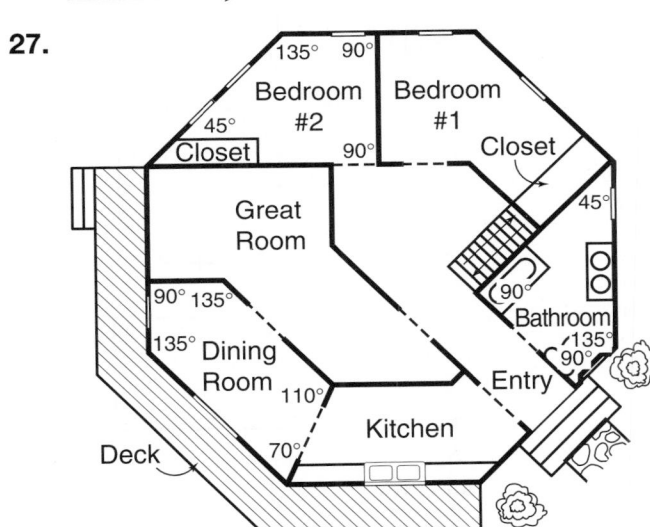
66°

27.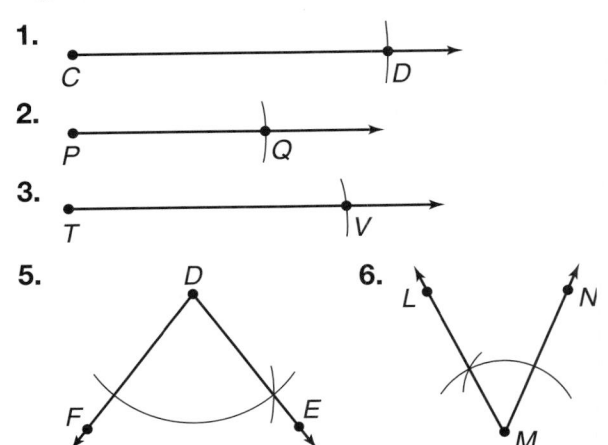
135° 90°
Bedroom #2
Bedroom #1
45°
Closet
90°
Closet
Great Room
45°
90° 135°
135° Dining Room
110°
90° Bathroom
135°
90°
Entry
Kitchen
Deck
70°

Pages 362–363, Lesson 9-3A

1.
C   D

2.
P   Q

3.
T   V

5.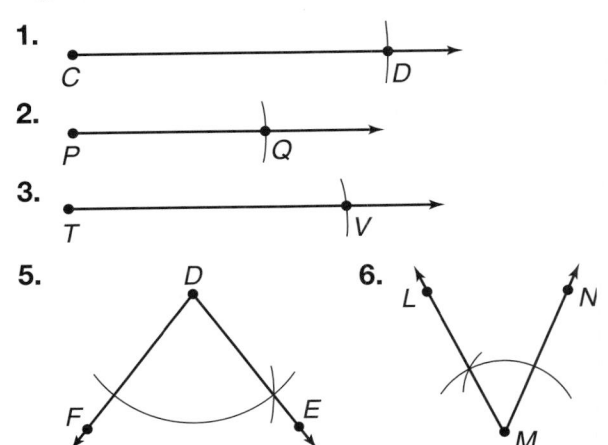
D
F   E

6.
L   N
M

7.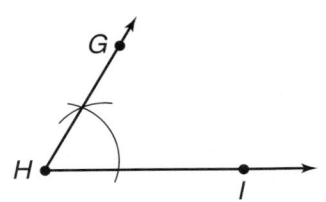
G
H   I

Page 366, Lesson 9-3

1. To separate something into two congruent parts.

2. First, place the compass at one end of the line segment. Then set the compass to more than half the length of the line segment and draw two arcs. Next, using the same compass setting, place the compass at the other end of the line segment and draw two more arcs. Then, using a straightedge, draw a line through the arcs.

3. See students' work. The crease bisects the angle.

4. Note: Art is not drawn to scale.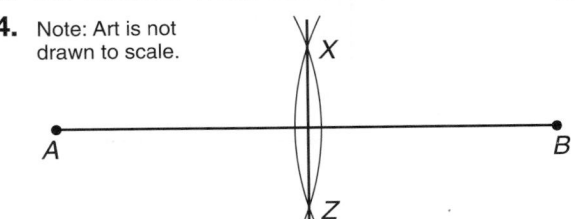
X
A   B
Z

5.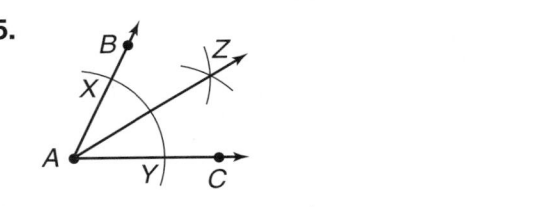
B   Z
X
A   Y   C

6.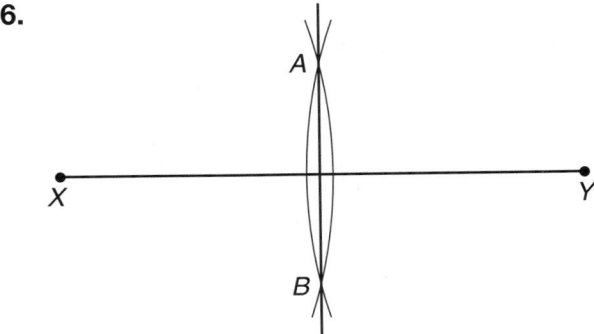
A
X   Y
B

7.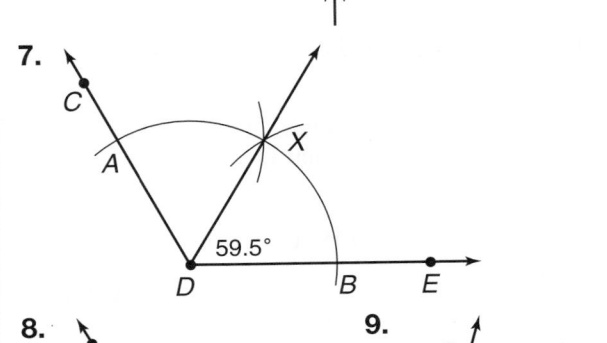
C
A   X
59.5°
D   B   E

8.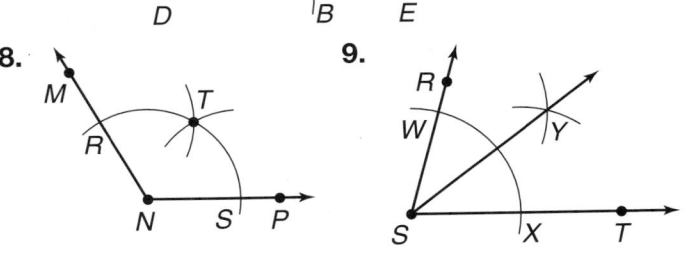
M   T
R
N   S   P

9.
R
W   Y
S   X   T

10.

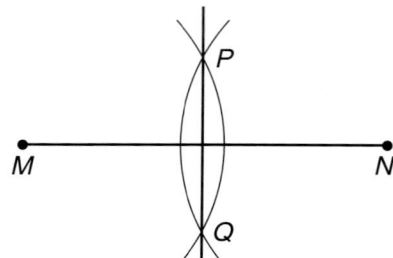

11.

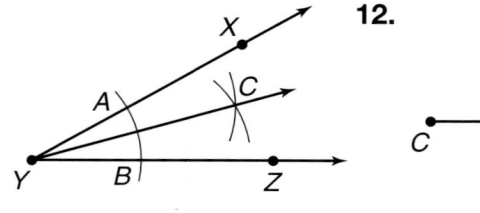

12.

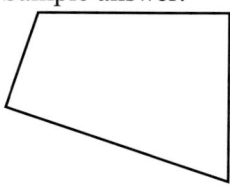

13. Note: Art is not drawn to scale.

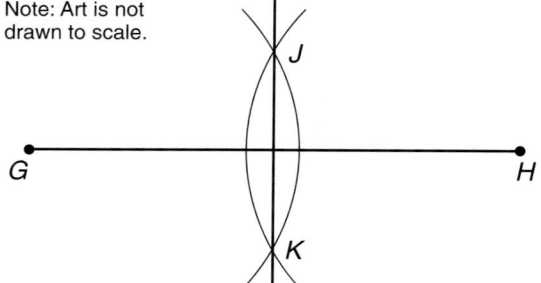

14.

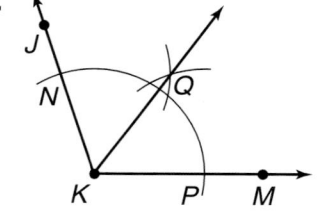

15. Note: Art is not drawn to scale.

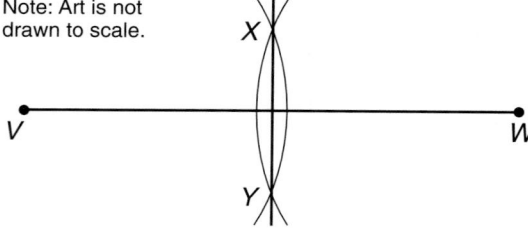

17. The diagonals bisect the corners of the square.

18. See students' work.

19. Bisect the 7-inch line segment. Then bisect each half of the segment. They are equal in length.

Pages 372–373, Lesson 9-4

1. Sample answer: A chalkboard is a rectangle and a piece of floor tile is a square.

2. Sample answer:

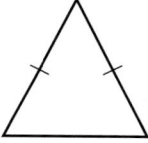

8.

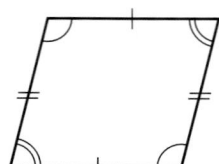

9.

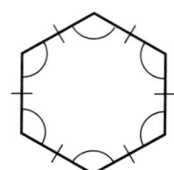

15. Alike: all sides congruent; all angles congruent

Different: triangle has three sides and square has four sides

16. Alike: all sides congruent; all angles congruent

Different: triangle has three sides; octagon has eight sides

17. Alike: quadrilaterals; at least 1 pair of parallel sides

Different: square has all sides congruent and four right angles

18. Sample answer:

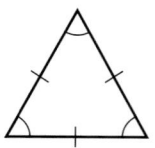

19. Sample answer:

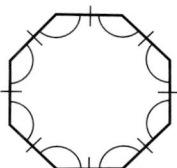

20.

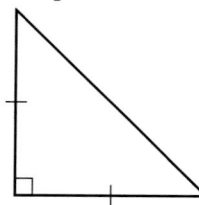

21.

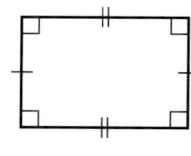

22. Sample answer:

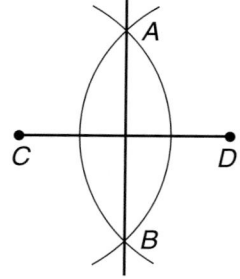

23.

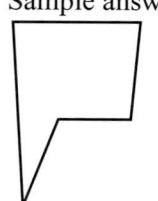

29.

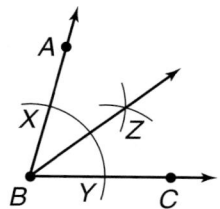

Page 373, Mid-Chapter Self Test

7.

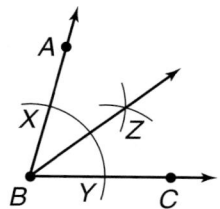

8.

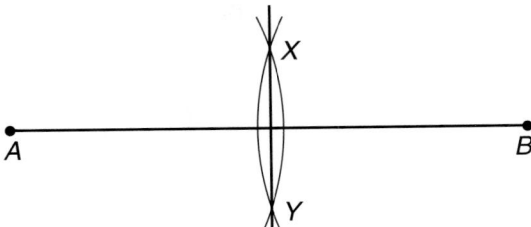

9.

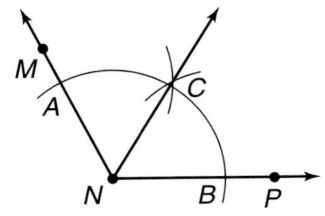

Page 374, Lesson 9-4B

2. Sample answer:

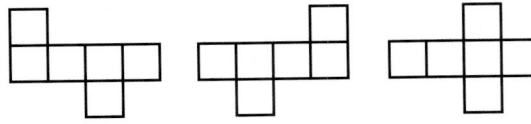

3. Sample answer: Each net contains six squares. Each net has four squares lined up in a row. Each net has two squares positioned on opposite sides of the row of four squares.

Pages 380–382, Lesson 9-6

18a.

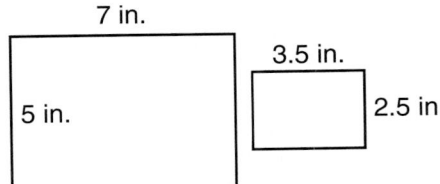

19.

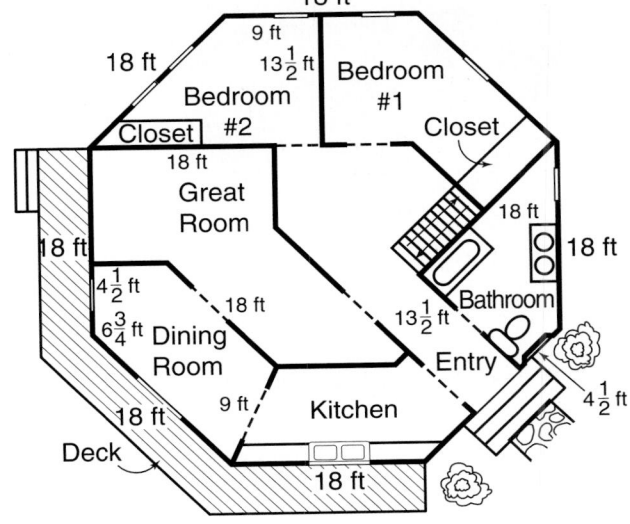

Pages 384–385, Lesson 9-6B

1.

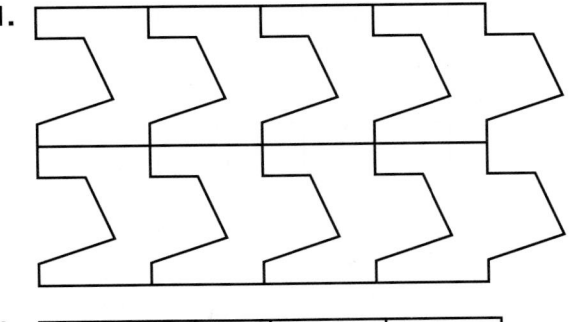

2.

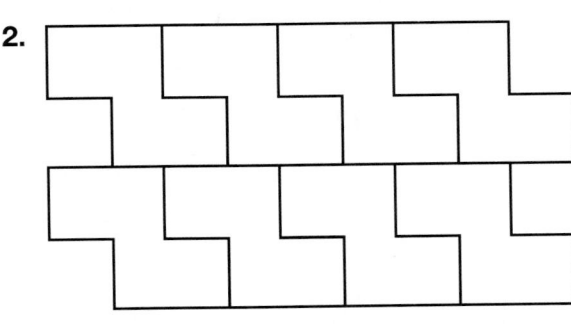

3.

5.

6.

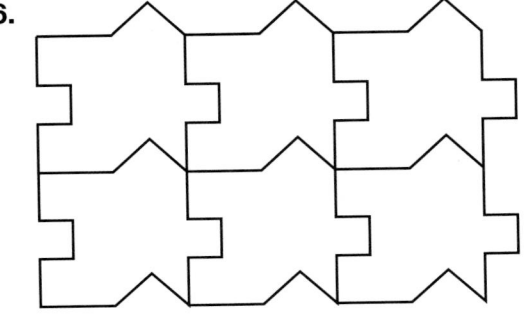

7.

**Pages 386–389, Chapter 9
Study Guide and Assessment**

18.

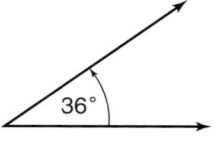

36°

19.

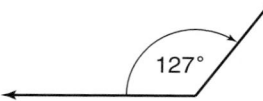

127°

22. Note: Art is not drawn to scale.

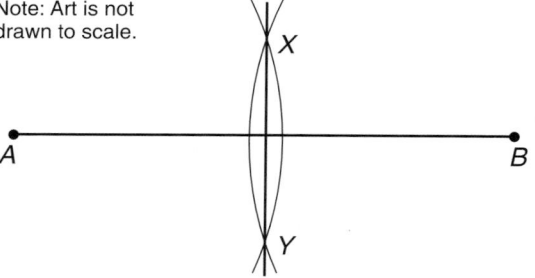

23.

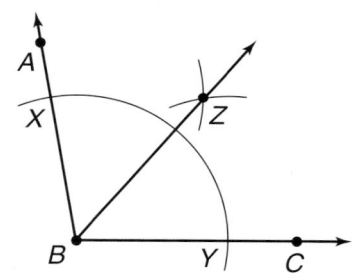

24.

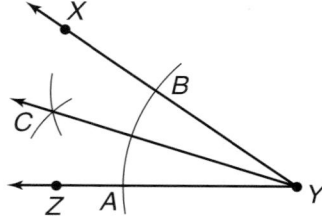

25.

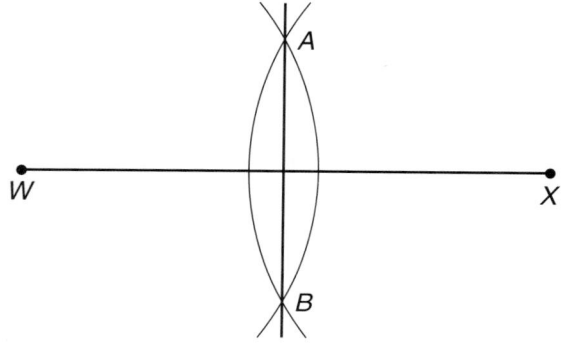

28. Sample answer:

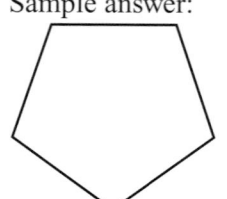

29.

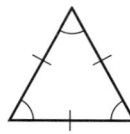

37.

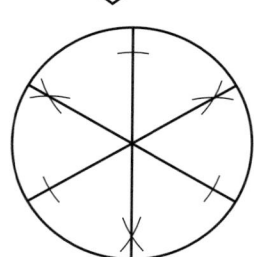

38.

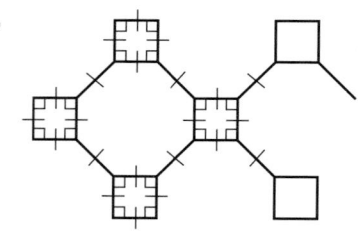

**CHAPTER 10
Geometry: Understanding Area
and Volume**

Page 411, Lesson 10-3B

4.

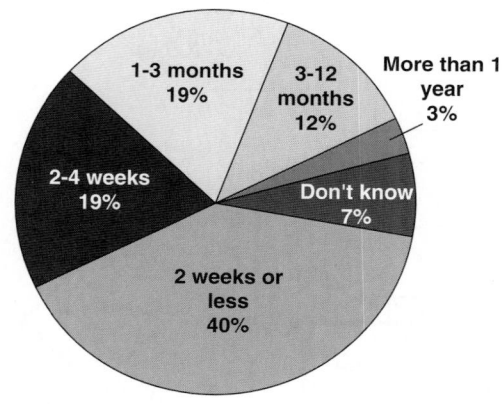

Film Time in Camera

1-3 months 19%
3-12 months 12%
More than 1 year 3%
2-4 weeks 19%
Don't know 7%
2 weeks or less 40%

5. **World Population**

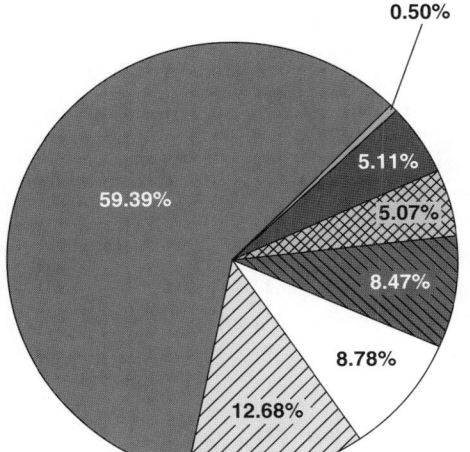

Legend:
- North America
- Latin America & Caribbean
- Europe
- Asia
- Africa
- Former USSR
- Oceania & Australia

Pages 413–414, Lesson 10-4

2. Sample answer: A two-dimensional figure is flat, it has only width and length. A three-dimensional figure is not flat, it has width, length, and depth.

Page 415, Lesson 10-4B

4. 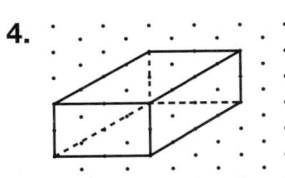 **5.**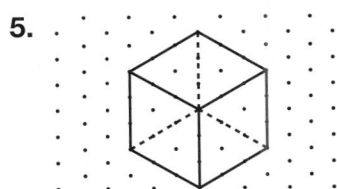

Page 425, Lesson 10-6B

5. When one dimension is doubled, the volume doubles. When two dimensions are doubled, the volume is multiplied by four. When all three dimensions are doubled, the volume is multiplied by eight. The volumes are changed in this way because if any one dimension is multiplied by 2, the volume is multiplied by 2.

6. When one or two dimensions are doubled, the surface area is increased. When all three dimensions are doubled, the surface area is multiplied by four. The surface area is changed because as one dimension is doubled, the areas of four faces are affected, but two are unchanged. When three dimensions are doubled, the areas of all six faces are quadrupled, so the surface area is quadrupled.

CHAPTER 11
Algebra: Investigating Integers

Pages 435–436, Lesson 11-1

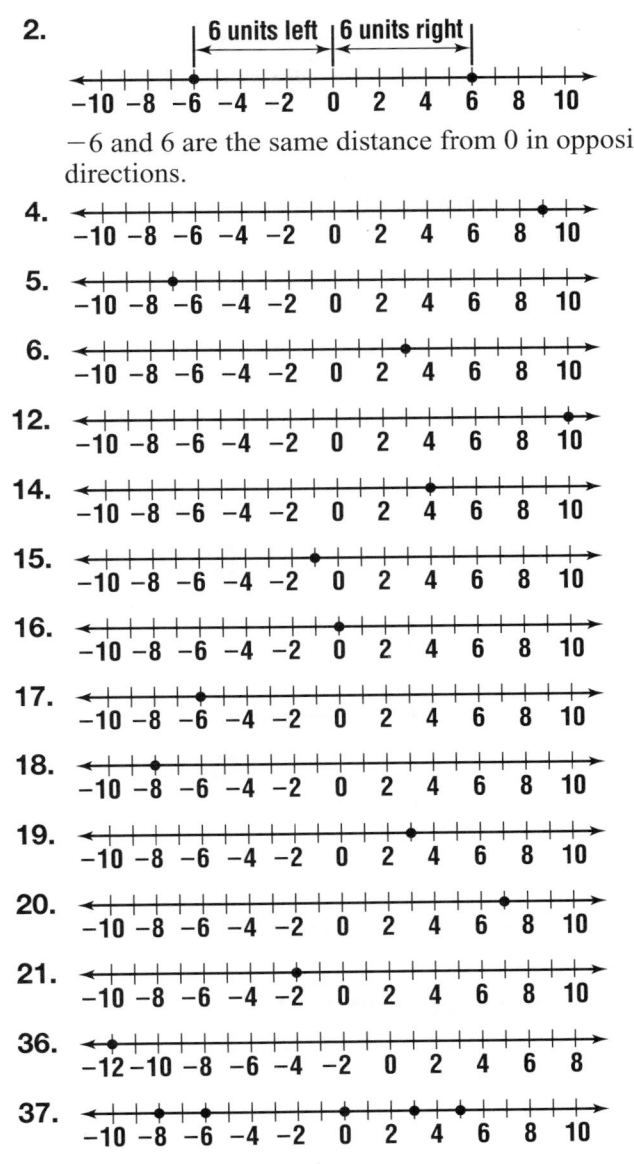

2. −6 and 6 are the same distance from 0 in opposite directions.

42. Both a number line and a thermometer measure positive and negative numbers. A number line is usually horizontal, so the negative values are to the left of 0 and the positive values are to the right of 0. A thermometer is vertical, so the negative values are below 0 and the positive values are above 0.

Pages 438–439, Lesson 11-2

1. Graph the numbers on a number line. The number to the left is less than the number to the right.

2. Sample answer: A scuba diver is 20 ft below the surface of the water. Another diver is 30 ft below the surface. Which diver is farther from the surface of the water?

3. Cordelia; since -5 is to the left of -3 on the number line, $-5 < -3$.

24.

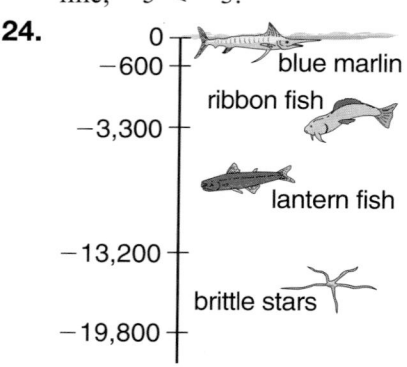

Pages 443–444, Lesson 11-3

2.

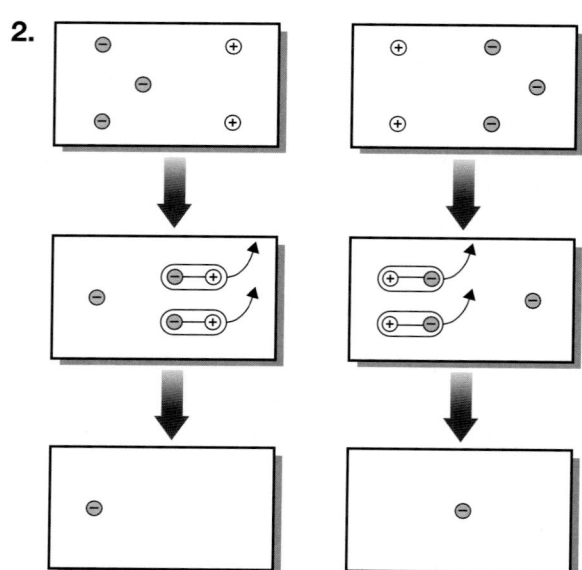

Pages 447–448, Lesson 11-4

21.

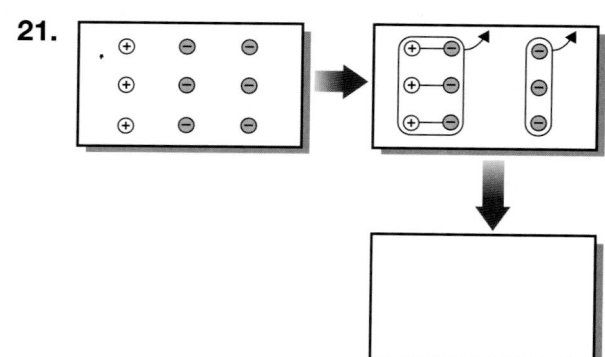

Pages 451–452, Lesson 11-5

2.

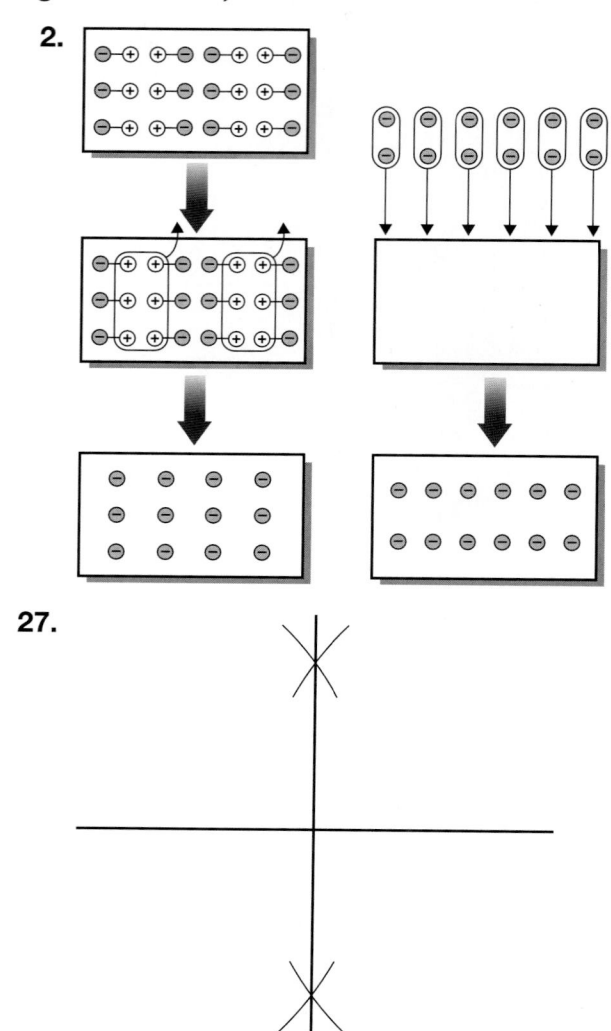

27.

Pages 460–461, Lesson 11-7

2a-b.

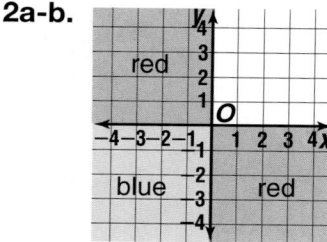

7.

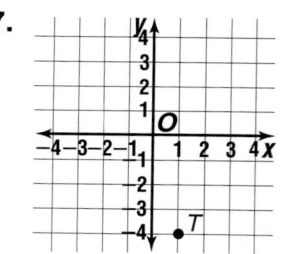

8.

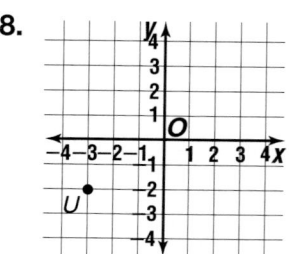

9.

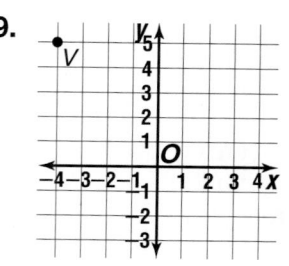

23.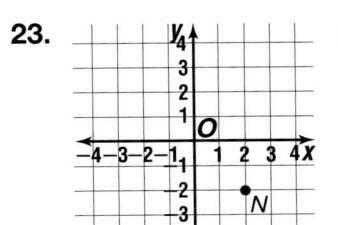

24.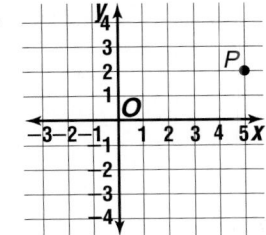

25.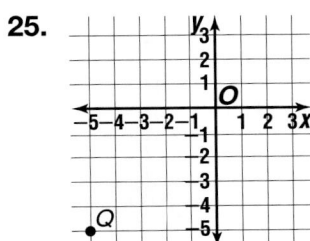

26.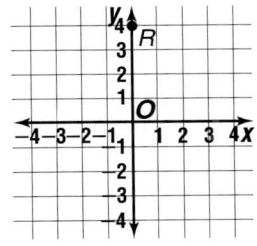

27.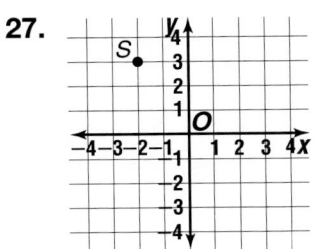

28.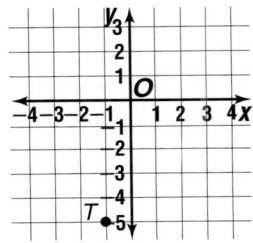

29.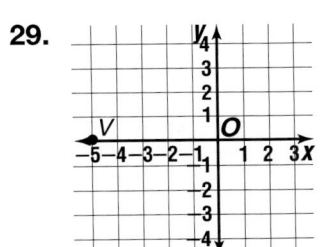

30.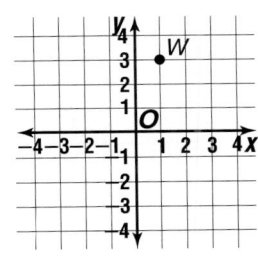

31a.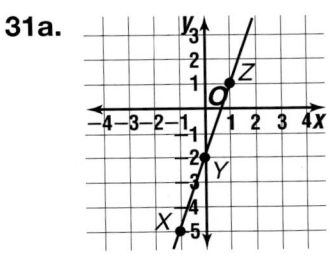

33.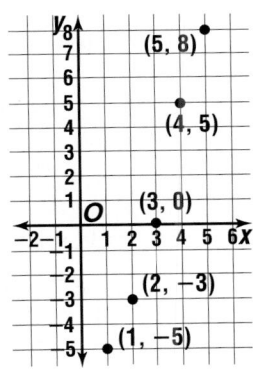

35a-d.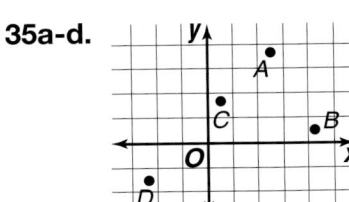

3a.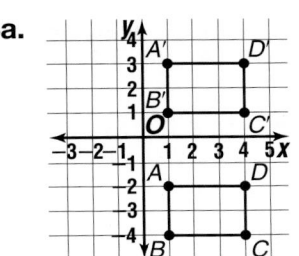

3b.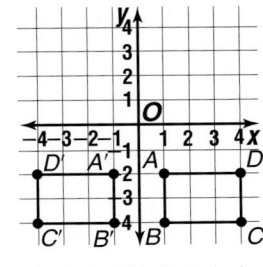

5.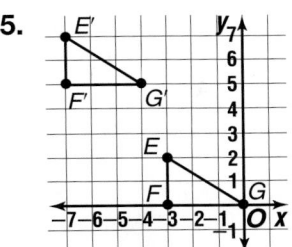

6.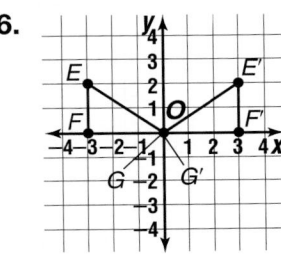

11.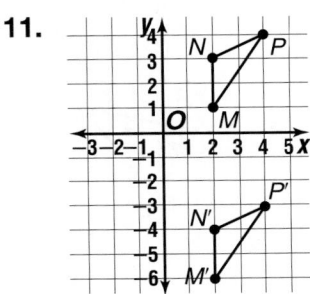

12.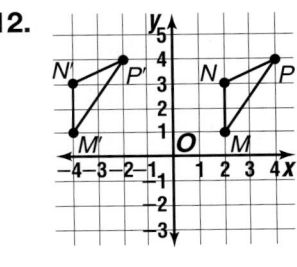

13.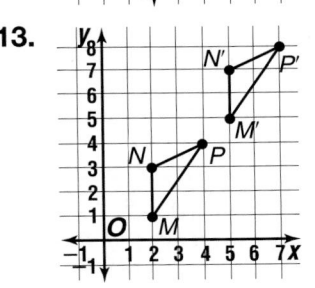

14.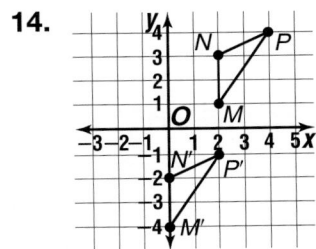

15.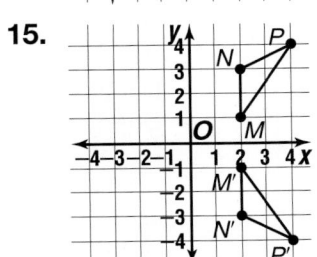

16.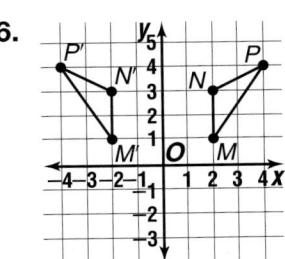

18. Sample answer:

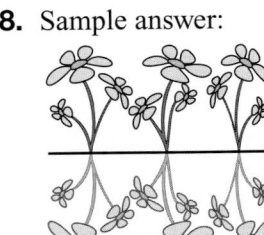

20.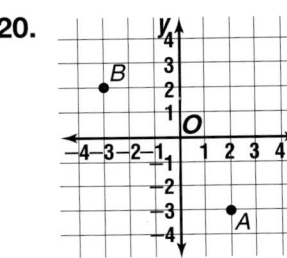

Pages 466–467, Lesson 11-8

1. Both translations and reflections are transformations. In translations and reflections, the image is the same size and shape as the figure. The translation image has the same orientation as the figure, but the reflection image is a flip of the figure.

Pages 468–471, Chapter 11
Study Guide and Assessment

13.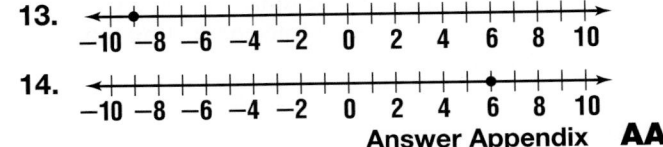

14.

15.

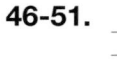

46-51.

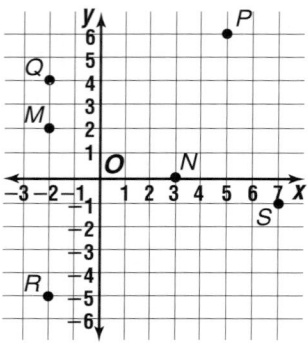

52.

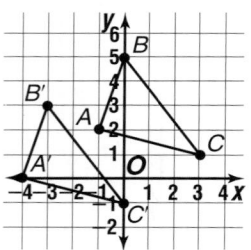

53.

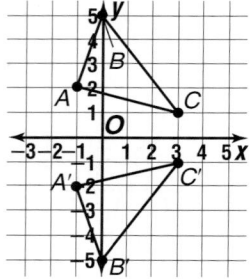

CHAPTER 12
Algebra: Exploring Equations

Pages 486–487, Lesson 12-3

2.

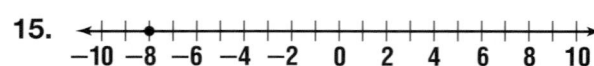

33.

E
Change per month
= D2/10
= D3/10
= D4/10
= D5/10
= D6/10
= D7/10

34. The solution to $\frac{1}{8}x = 13$ will be greater because you have to multiply by a greater number (8) to solve it.

Pages 501–503, Lesson 12-6

1. Sample answer: Let the input values represent the x-coordinates and let the output values represent the corresponding y-coordinates.

2. The graph in the third quadrant represents negative time.

3. Sample answer: Make a table recording the input and output of the given function. Write the input/output as an ordered pair. Record at least three ordered pairs on the table. Graph these ordered pairs and draw a line connecting the graphed points.

4.

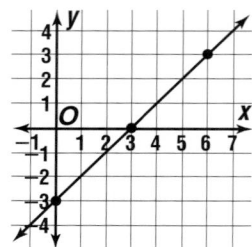

5.

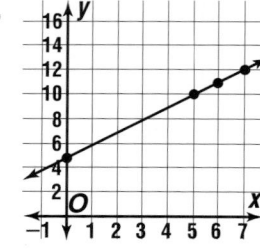

6.

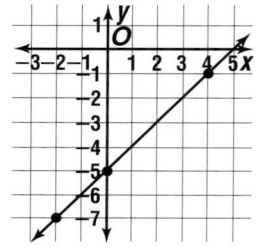

7.

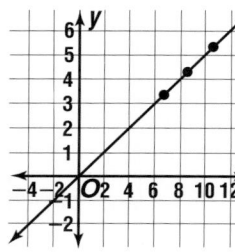

10.

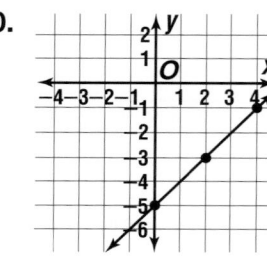

11.

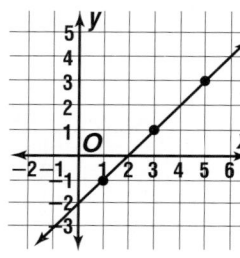

12.

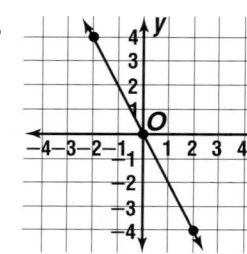

13.

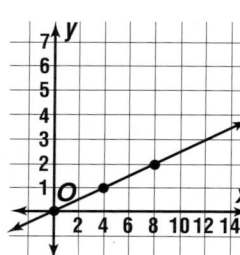

14.

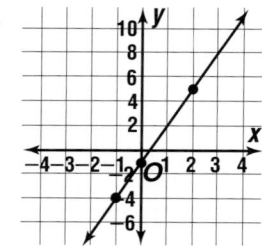

15.

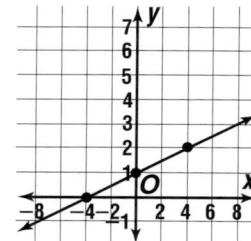

16.

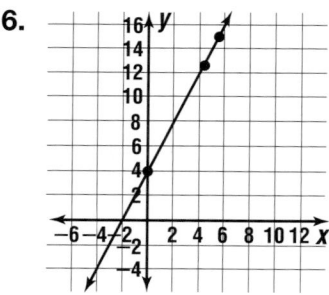

17.

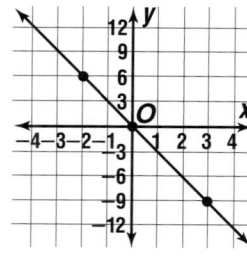

18.

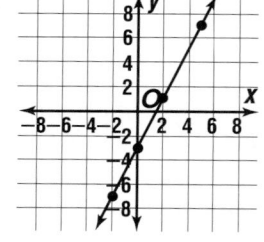

19.

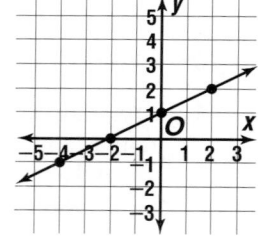

20.

input (*a*)	output (*a* + 5)
−5	0
−3	2
0	5

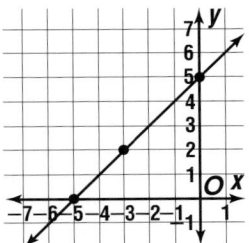

21.

input (*n*)	output (3*n* − 3)
−2	−9
1	0
4	9

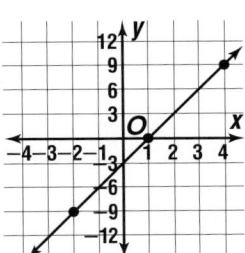

22.

input (*n*)	output (○)
−2	−6
0	0
2	6

The rule is 3*n*.

23.

input (*n*)	output (○)
0	0
4	2
8	4

The rule is $\frac{n}{2}$.

24.

input (*n*)	output (○)
0	−2
2	0
4	2

The rule is *n* − 2.

25b.

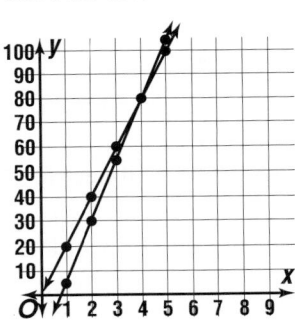

26a.

input	output
0	0
1	50
2	100
3	150

26b. The function rule is 50*n*.

27a.

input	output
1	61
2	62
3	63
4	64
5	65
6	66
7	67
8	68
9	69
10	70
11	71
12	72

27b.

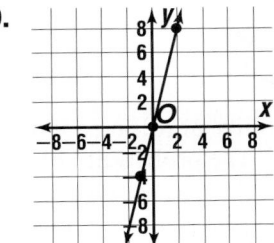

27c. In 5 months, the card will be worth $65.

Pages 504–507, Chapter 12
Study Guide and Assessment

40.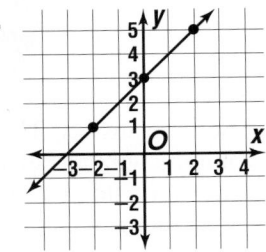

41.

Pages 508–509, Standardized Test Practice

18.

x	*x* − 2
2	0
5	3
8	6

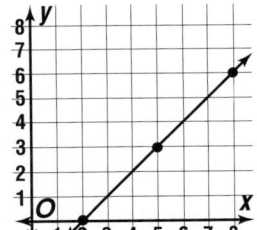

CHAPTER 13
Using Probability

Page 514, Lesson 13-1A

1. Sample answer: The number of even sums tossed and the number of odd sums tossed should be about the same. The number of even products tossed should be greater than the number of odd products tossed.

2. Sample answer: The game involving sums is fair. If you toss two number cubes, there are 36 possible results. Eighteen sums are even and eighteen sums are odd. The game involving products is unfair. If you toss two number cubes, there are 36 possible results. Twenty-seven products are even and nine products are odd. So, you would have a greater chance of tossing an even product than an odd product.

3. Sample answer: For the game involving sums, the bar graph resembled an isosceles triangle and is symmetrical about the axis which contains the sum of 7. For the products game, the bar graph reaches a maximum height at products 6 and 12 and the overall graph is not symmetrical.

4. Sample answer: The sum of 7 occurred most often and the sums of 2 and 12 occurred least often. The products of 6 and 12 occurred most often and the products of 1, 9, 16, 25, and 36 occurred least often.

5. Sample answer: A sum of 12 is less likely to occur since out of 36 results, a sum of 12 can only occur one way.

Pages 517–518, Lesson 13-1

1. Sample answer:

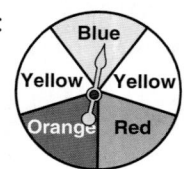

2. Sample answer: The probability of rolling a 9 on a number cube labeled 1 to 6. The probability of the sun not rising.

3. Sample answer: Complementary events are two events in which either one or the other can occur, but not both. The sum of their probabilities is 1. An example is the probability of snow or the probability of no snow.

4. The event is not too likely to occur.
5. The event is very likely to occur.
6. The event is equally likely to occur.
7. The event is not too likely to occur.
8. The event is very likely to occur.
9. The event cannot occur.
11. The event is not too likely to occur.
12. The event is not too likely to occur.
13. The event is equally likely to occur.

14. The event is very likely to occur.
15. The event is not too likely to occur.
16. The event cannot occur.
17. The event is very likely to occur.
18. $\frac{1}{30}$; not 12; $\frac{29}{30}$
19. $\frac{1}{2}$; not odd; $\frac{1}{2}$
20. $\frac{3}{10}$; not a 1 digit number; $\frac{7}{10}$
21. 1; not an integer; 0
22. 0; not less than 1; 1
23. $\frac{2}{5}$; not greater than 18; $\frac{3}{5}$
29.

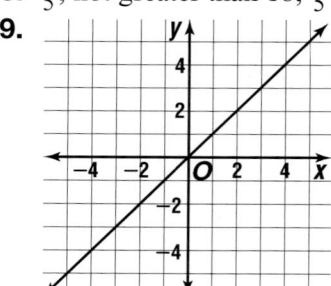

Pages 520–521, Lesson 13-2A

4a.

Test Scores		
Score	Tally	Frequency
91–100	卌I	6
81–90	IIII	4
71–80	卌III	8
61–70	III	3

10a.

Students' Birth Months		
Month	Tally	Frequency
January	I	1
February		0
March	II	2
April	IIII	4
May	III	3
June	卌	5
July	III	3
August	I	1
September	I	1
October	III	3
November		0
December	I	1

Pages 524–525, Lesson 13-2

4. You are probably surveying people whose main interest is a Broadway show.

5. There is probably a diverse group of people at a shopping center.

6. There is probably a fair sampling of people at a fast-food restaurant.

8. There is probably a fair sampling of people at a grocery store.

9. You would probably survey people whose hobby is model trains.

10. You would probably survey people who prefer the musician.

11. A mall would have a diverse group of people.

12. There would be a diverse group of people at the library.

13. You would probably find many people who prefer Chevrolets.

14. There would be a diverse group of people at a skating rink.

15. A lot of chocolate lovers would be in a chocolate factory.

16. The customers may prefer Mexican food.

Pages 527–529, Lesson 13-3

22. $\frac{16}{49}$, or about 32.7%; The event is not too likely to happen.

Page 530, Lesson 13-3B

1. Sample answer: $\frac{27}{50}$
2. $\frac{1}{2}$
3. Sample answer: The fractions should be about the same.
4. Sample answer: The fractions should be closer than the fractions compared in Exercise 3.

Pages 533–534, Lesson 13-4

1. Sample answer: At a restaurant, a customer can choose either a chicken or a beef dinner and one of three beverages. How many possible outcomes are there?

3. Sample answer: Tree diagrams are useful for organizing the outcomes. One disadvantage is that there could be too many outcomes.

4. See students' diagrams; outcomes are hot dog, iced tea; hot dog, lemonade; hamburger, iced tea; hamburger, lemonade.

5. See students' diagrams; outcomes are blue shirt, blue shorts; blue shirt, black shorts; blue shirt, red shorts; red shirt, blue shorts; red shirt, black shorts; red shirt, red shorts.

7a. See students' diagrams; outcomes are ATT, ATF, AFT, AFF, BTT, BTF, BFT, BFF, CTT, CTF, CFT, CFF.

8. See students' diagrams; outcomes are lemon, milk; lemon, tea; apple, milk; apple, tea; pecan, milk; pecan, tea.

9. See students' diagrams; outcomes are leather, purple; leather, green; leather, black; leather, brown; nylon, purple; nylon, green; nylon, black; nylon, brown.

10. See students' diagrams; outcomes are 1, 1; 1, 2; 1, 3; 1, 4; 1, 5; 1, 6; 2, 1; 2, 2; 2, 3; 2, 4; 2, 5; 2, 6; 3, 1; 3, 2; 3, 3; 3, 4; 3, 5; 3, 6; 4, 1; 4, 2; 4, 3; 4, 4; 4, 5; 4, 6; 5, 1; 5, 2; 5, 3; 5, 4; 5, 5; 5, 6; 6, 1; 6, 2; 6, 3; 6, 4; 6, 5; 6, 6.

11. See students' diagrams; outcomes are blue, red; blue, green; blue, white; blue, black; orange, red; orange, green; orange, white; orange, black; yellow, red; yellow, green; yellow, white; yellow, black.

12. See students' diagrams; outcomes are portable, graphite; portable, acrylic; stationary, graphite; stationary, acrylic.

13. See students' diagrams; outcomes are tower, wood, 60; tower, wood, 100; tower, wood, 250; tower, plastic, 60; tower, plastic, 100; tower, plastic, 250; spinner, wood, 60; spinner, wood, 100; spinner, wood, 250; spinner, plastic, 60; spinner, plastic, 100; spinner, plastic, 250.

19. Sample answer: This is not a fair game. Only in two of the eight outcomes are all chips different. It could be made a fair game by giving Amiri 3 points or changing the second counter from red and white to green and white.

Page 535, Lesson 13-4B

1. Sample answer: $\frac{22}{50}$
2. BBB, BBG, BGB, BGG, GBG, GGB, GBB, GGG; $\frac{4}{8}$ or $\frac{1}{2}$
3. They are about the same.
4. $\frac{1}{16}$; Repeat steps 1 to 4, but in step 1, place four counters in a cup. See students' work.

Pages 540–543, Chapter 13 Study Guide and Assessment

12. $\frac{1}{6}$; The event is not too likely to occur.
13. $\frac{1}{3}$; The event is not too likely to occur.
14. 0; The event cannot occur.
15. $\frac{5}{6}$; The event is very likely to occur.
16. $\frac{1}{3}$; The event is not too likely to occur.
17. $\frac{2}{3}$; The event is very likely to occur.
18. yes; There will be many people at the beach with all different interests.
19. no; At a football game, you would most likely find people who prefer football.
20. yes; There would be many people at the mall who would prefer different types of pets.
21a. $\frac{7}{20}$ or 35%
25. See students' diagrams; outcomes are black jeans, tapered; black jeans, straight; black jeans, baggy; blue jeans, tapered; blue jeans, straight; blue jeans, baggy.
26. See students' diagrams; outcomes are soup, beef; soup, chicken; soup, fish; soup, pasta; salad, beef; salad, chicken; salad, fish; salad, pasta.
27. See students' diagrams; outcomes are ball game, Friday; ball game, Saturday; amusement park, Friday; amusement park, Saturday; concert, Friday; concert, Saturday.

35. See students' diagrams; outcomes are purple skirt, yellow vest; purple skirt, purple sweater; white skirt, yellow vest; white skirt, purple sweater; yellow skirt, yellow vest; yellow skirt, purple sweater.

Page 543, Performance Task

- Make a tree diagram to show all of the possibilities; outcomes are color, glossy, 3×5; color, glossy, 4×6; color, glossy, 5×5; color, glossy, 8×10; color, matte, 3×5; color, matte, 4×6; color, matte, 5×5; color, matte, 8×10; black and white, glossy, 3×5; black and white, glossy, 4×6; black and white, glossy, 5×5; black and white, glossy, 8×10; black and white, matte, 3×5; black and white, matte, 4×6; black and white, matte, 5×5; black and white, matte, 8×10.
- There are 16 possible outcomes. If the pictures can also be cut square or rounded, there would be 32 possible outcomes.

Page 557, Extra Practice, Lesson 1-3

13. $60 \times 5 = 300$; not reasonable
14. $900 - 40 = 860$; reasonable
15. $\$3 + \$6 + \$9 = \$18$; reasonable
16. $400 \times 4 = 1,600$; reasonable
17. $400 \div 5 = 80$; not reasonable
18. $2,000 - 200 = 1,800$; not reasonable
19. $100 + 500 + 200 = 800$; reasonable
20. $6,000 \div 300 = 20$; reasonable

Page 559, Extra Practice, Lesson 2-1

1.

# of miles	Tally	Frequency
1	IIII	4
2	IHI	5
3	IHI II	7
4	III	3
5	III	3

2.

Temperature	Tally	Frequency
47	I	1
48	III	3
49	III	3
51	II	2
52	IHI	5
53	II	2
57	II	2

Page 561, Extra Practice, Lesson 2-4

1. 5% – history; 12% – math; 20% – science; 25% – English; 38% – phys. ed.

2. Sample answer: No; the combination of math and history is not $\frac{1}{4}$ of the circle.

Page 582, Extra Practice, Lesson 9-4

Sample answers:

5. same length sides, different number of sides, different angles
6. same length sides, same number of sides, different angles
7. same number of sides, one has 3 equal angles and the other has 2 equal angles and a right angle

Page 585, Extra Practice, Lesson 11-1

1-7.

Page 587, Extra Practice, Lesson 11-7

15-22.
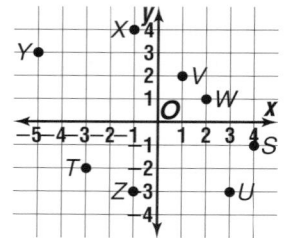

Page 588, Extra Practice, Lesson 11-8

1.
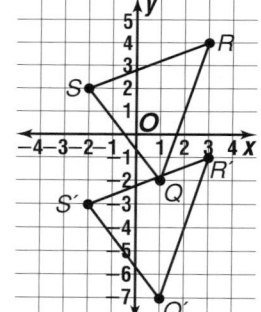

2.
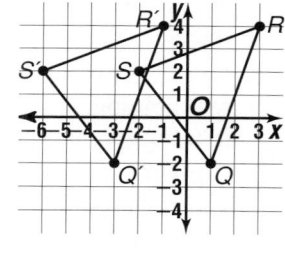

3.
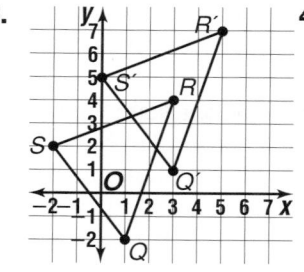

4.
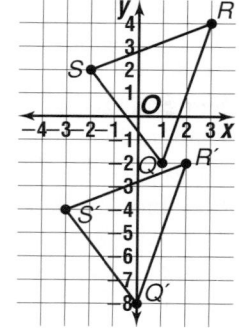

5.

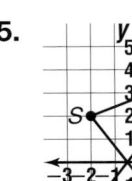

Page 590, Extra Practice, Lesson 12-6

1.

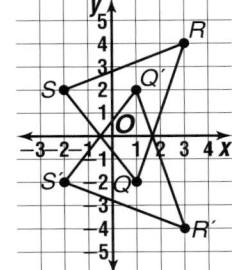

2.

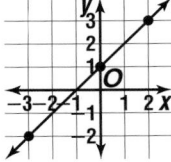

3.

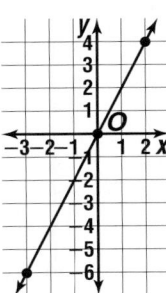

4.

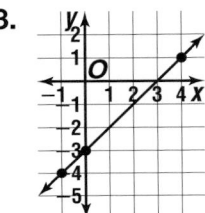

5.

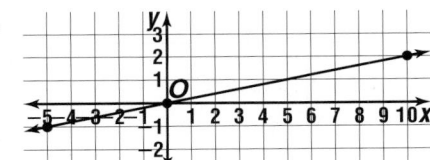

6.

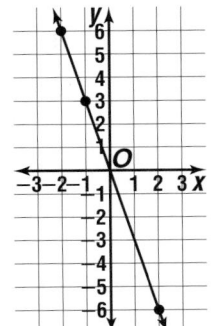

Page 590, Extra Practice, Lesson 13-1

1. not too likely to occur
2. equally likely to occur
3. not too likely to occur
4. cannot occur
5. very likely to occur
6. not too likely to occur
7. not too likely to occur
8. equally likely to occur

9. very likely to occur
10. not too likely to occur
11. not too likely to occur
12. certain to occur
13. not too likely to occur
14. very likely to occur
15. not too likely to occur
16. not too likely to occur
17. not too likely to occur
18. not too likely to occur

Page 591, Extra Practice, Lesson 13-2A

2a.

Month	Tally	Frequency
February	III	3
March	III	3
June	II	2
July	III	3
August	I	1
September	III	3
October	IIII	5
December	I	1

Page 591, Extra Practice, Lesson 13-2

1. yes; A department store would have a diverse group of people.
2. no; You would probably find many fans of the Texas Rangers.
3. yes: There would be a diverse group of people at a shopping center.
4. yes; A park would have a diverse group of people.
5. no; You would probably find many people who prefer soccer.
6. yes; A school would have a diverse group of students.
7. no; You would probably find many people who like Garth Brooks.
8. no; You would probably find many people who like apples.
9. yes; There would be a diverse group of people at a county fair.
10. no; Many of the people would probably prefer to go to Disneyworld for vacation.

Page 592, Extra Practice, Lesson 13-4

1. See students' diagrams; outcomes are heads, 1; heads, 2; heads, 3; heads, 4; heads, 5; heads, 6; tails, 1; tails, 2; tails, 3; tails, 4; tails, 5; tails, 6.
2. See students' diagrams; outcomes are red, red; red, blue; red, yellow; green, red; green, blue; green, yellow; blue, red; blue, blue; blue, yellow; yellow, red; yellow, blue; yellow, yellow.

3. See students' diagrams; outcomes are red, white; red, black; red, tan; blue, white; blue, black; blue, tan; green, white; green, black; green, tan.

4. See students' diagrams; outcomes are chicken, coffee; chicken, milk; chicken, juice; chicken, soda; ham, coffee; ham, milk; ham, juice; ham, soda; turkey, coffee; turkey, milk; turkey, juice; turkey, soda; bologna, coffee; bologna, milk; bologna, juice; bologna, soda.

Page 596, Chapter 2 Test

1.

Score	Tally	Frequency
96–100	IIII	4
91–95	IIIII I	6
86–90	III	3
81–85	IIIII	5
76–80	IIIII	5
71–75	IIII	4

3.
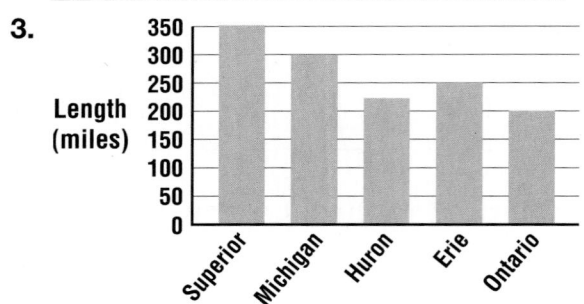

11.
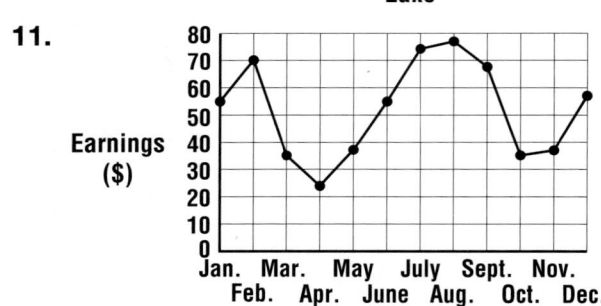

12.
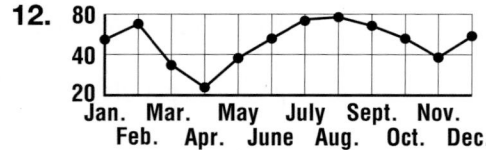

13.
Stem	Leaf
7	3 4 5
8	4 5 6 7
9	2 2 6 $9|2 = 92$

14.
Stem	Leaf
4	2 3 6 8
5	5 9
6	6 9
7	
8	5 $8|5 = 85$

Page 599, Chapter 5 Test

21. ▬▬▬▬▬▬▬▬

Page 603, Chapter 9 Test

8.
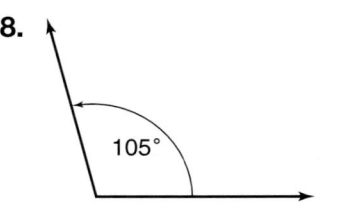
105°

12.
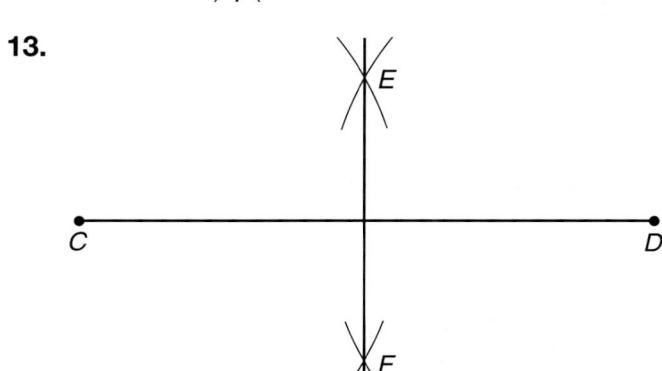

13.
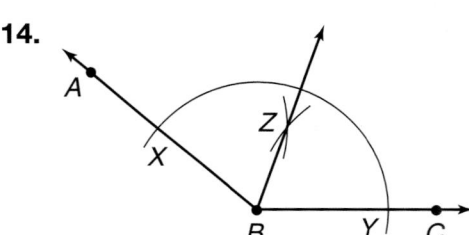

14.
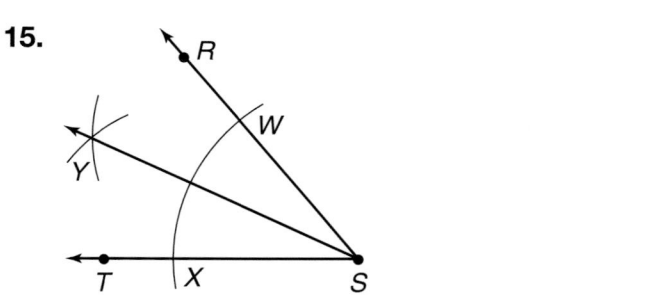

15.

Page 605, Chapter 11 Test

21.
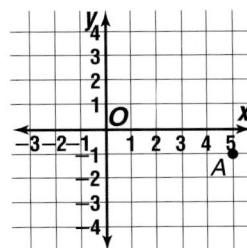

22.
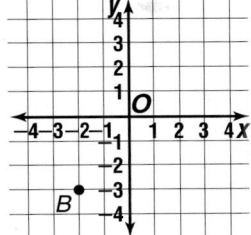

23.
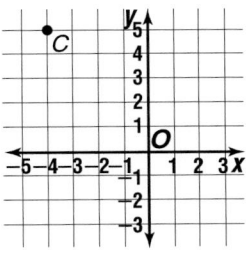

24.
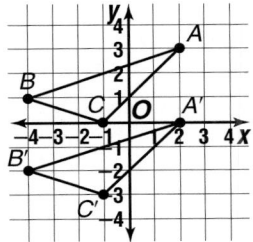

25.
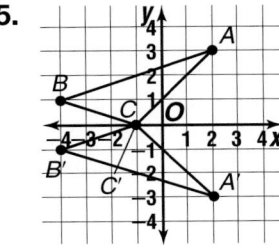

Page 606, Chapter 12 Test

20.
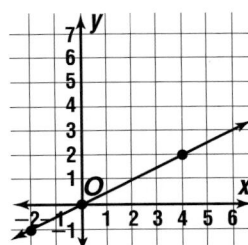

Page 607, Chapter 13 Test

16a. See students' diagrams; outcomes are iced tea, nachos; iced tea, popcorn; iced tea, chocolate; soda, nachos; soda, popcorn; soda, chocolate.

Number and Operations

$+$	plus or positive
$-$	minus or negative
$a \cdot b$	
$a \times b$	a times b
ab or $a(b)$	
$\div$	divided by
$\pm$	positive or negative
$=$	is equal to
$\neq$	is not equal to
$<$	is less than
$>$	is greater than
$\leq$	is less than or equal to
$\geq$	is greater than or equal to
$\approx$	is approximately equal to
$\%$	percent
$a{:}b$	the ratio of a to b, or $\frac{a}{b}$

Geometry and Measurement

$\cong$	is congruent to
$\sim$	is similar to
$^\circ$	degree(s)
$\overleftrightarrow{AB}$	line AB
$\overline{AB}$	segment AB
$\overrightarrow{AB}$	ray AB
$\llcorner$	right angle
$\perp$	is perpendicular to
$\parallel$	is parallel to
AB	length of $\overline{AB}$, distance between A and B
$\triangle ABC$	triangle ABC
$\angle ABC$	angle ABC
$\angle B$	angle B
$m\angle ABC$	measure of angle ABC
$\odot C$	circle C
$\overarc{AB}$	arc AB
π	pi $\left(\text{approximately } 3.14159 \text{ or } \frac{22}{7}\right)$
(a, b)	ordered pair with x-coordinate a and y-coordinate b
$\sin A$	sine of angle A
$\cos A$	cosine of angle A
$\tan A$	tangent of angle A

Algebra and Functions

a'	a prime
a^n	a to the nth power
a^{-n}	$\frac{1}{a^n}$ (one over a to the nth power)
$\lvert x \rvert$	absolute value of x
$\sqrt{x}$	principal (positive) square root of x
$f(n)$	function, f of n

Probability and Statistics

$P(A)$	the probability of event A
$n!$	n factorial
$P(n, r)$	permutation of n things taken r at a time
$C(n, r)$	combination of n things taken r at a time

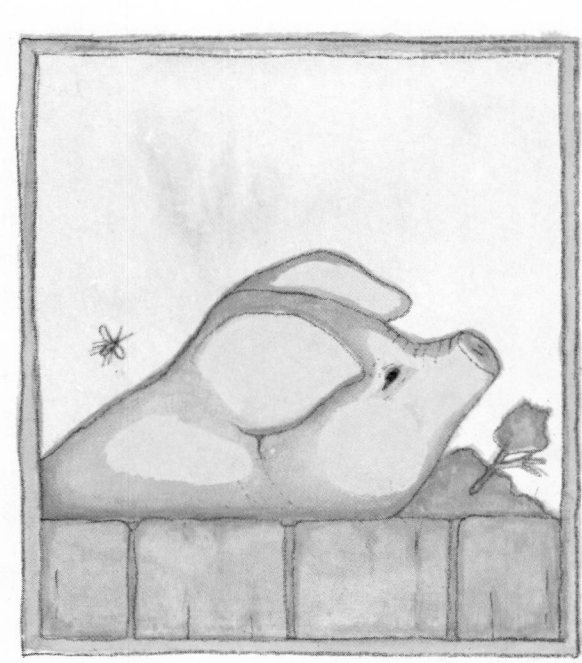

Acknowledgements

Fifteen Bathtubs and *The Tickly Spider* from the book
ANOTHER HERE AND NOW STORY BOOK, edited by
Lucy Sprague Mitchell, copyright 1937 by E. P. Dutton & Co., Inc.,
renewal © 1965 by Lucy Sprague Mitchell, published by
E. P. Dutton & Co., Inc. Used with their permission.

*The Wonderful Kitten, The Shy Little Horse, The Polite Little Polar
Bear, The Steam Roller, The Good Little Bad Little Pig,* and
The Little Girl's Medicine from THE FISH WITH THE DEEP SEA
SMILE by Margaret Wise Brown, copyright 1938 by Margaret Wise
Brown, copyright renewal 1965 by Roberta B. Rauch — *How the
Animals Took a Bath* from JACK AND JILL Magazine, copyright
1939, renewal © 1967 by Curtis Publishing Company, Inc.
Reprinted by permission of Roberta B. Rauch.

Margaret Wise Brown's
WONDERFUL
STORYBOOK

25 Stories & Poems

Illustrated by J. P. Miller

BARNES & NOBLE BOOKS
NEW YORK

A GOLDEN BOOK • NEW YORK
Golden Books Publishing Company, Inc., Racine, Wisconsin 53404

Contents

The Green Eyed Kitten

ONCE THERE WAS one kitten. There were lots of other kittens, there always are, but there was one kitten smaller than all the others. Fatter than all the others and blacker than all the others. He was all black with two green eyes.

The other kittens got fat, their whiskers grew longer, their claws grew longer, their ears grew longer and their eyes got bigger and they crawled and bounced and were carried off to their new homes — all but the little old kitten. "He's too little," everyone said. So there he was. "He's too black," they said. "He's all black except for his two little grass green eyes." And so he was left at home.

But the little kitten didn't care at all. He was alone with his own cat mother and he slept a lot and now that the other kittens were gone he got lots to eat and baths seven times a day.

And more purring
And more blinking
And more warm moments
    than he had ever had before
And a catnip mouse.
That little old kitten,
He was always warm as toast,
    merry as a fiddle,
    bright as electricity,
    and quick as a mouse.

Then one day his mother climbed up in a high and windy tree and forgot to come down.

And there was the little old kitten, more and more and more alone.

He caught seven caterpillars and rolled them over. He jumped for a leaf flying through the air and he caught it.

And then he sat down to think, and these are some of the thoughts he thought. They were all cats' thoughts of course because he was a kitten.

His first thought was *warm fur.*

His next thought was *lions are just big cats so when they are little lions they are kittens.*

Then he thought *I have twenty claws, they pick up sticks and can scratch.*

Then he thought *sweet breezes, a hill has a smell and the moon has not, a tree has a smell and so has a stone. You can hear more and smell more when you're all alone, by yourself.*

That is a dignified way to be, thought the kitten. *I am alone.*

And he looked up into the tree, but all he could see was leaves and leaves and leaves and leaves and leaves and leaves and leaves and they were all green.

Bugs are little,
Beetles are big,
Stars are quiet,
Airplanes buzz
Buzz Buzz Buzz
Bees buzz, flies buzz, dragon flies.

And then his thoughts made him very sleepy, so he went to sleep. He closed his green eyes.

And he dreamed cat dreams.

And his mother climbed down out of the tree.

The Children's Year

SPRING

The wind blows strong across the hill
And slaps the yellow daffodil,
The red feathery trees
And the sharp green leaves
Remind the farmer to plant his seeds
In the brown earth where the old roots stir,
And early in the morning
The small birds sing.

8

SUMMER

Now in July
To lie in the sweet grass
With the earth so near
That you can hear
The murmur of insects,
Summer sound,
And almost hear the earth turn round
While far away in the still blue sky
A sudden storm goes rumbling by.

AUTUMN

Apples heavy and red
Bend the branches down,
Grapes are purple
And nuts are brown,
The apples smell sharp and sweet on the ground
Where the yellow bees go buzzing around.
And way up high
The birds fly southward
Down the sky.

WINTER

December comes;
The white snow comes again;
It falls softly in the night
White
From the dark blue sky.
It covers the earth that the spring kept green
And summer kept warm.
It covers the earth where the autumn leaves fell
Golden on the ground.
All is white.
All is still.
And Christmas trees shine green as emeralds
With rubies and diamonds and sapphires
On Christmas night.

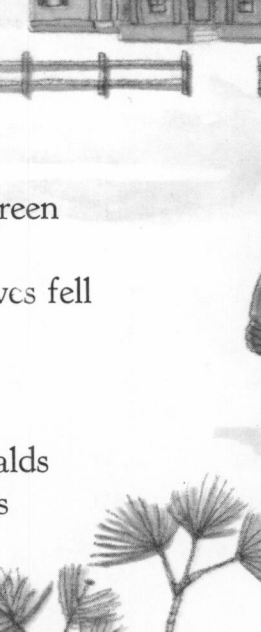

9

The Shy Little Horse

ONCE UPON A TIME in a barnyard there was a shy little horse. Every time he heard anyone coming, he ran away. Not so the donkey, not so the pig.

The old gray donkey rolled his eyes and lowered his head, put forward his ears, and then he trotted over to see who the visitor was. And the old fat pig, if she wasn't eating, wallowed over to the side of her pen and grunted at the visitor. But the little horse was shy. He kicked up his heels and he lowered his head and he galloped across the fields away from the visitor.

He galloped away and his mother galloped with him to the far end of the field where the grass was wet and green from the stream that flowed there.

Then one day a visitor came to the barnyard. The visitor was a tall man with a mustache.

The donkey saw him coming and ran to the fence and stuck forward his long ears and rolled his big brown jackrabbit eyes. But the visitor didn't pay

any attention to the donkey. The old pig blinked at him, and the chickens scratched about as though there were only chickens in the barnyard. But the visitor didn't even see the old pig and the chickens. The visitor was looking at the shy little horse.

And that shy little horse just lowered his head and kicked up his heels and galloped away.

Now this visitor had come just to see the shy little horse. That was why he came to the barnyard. So he climbed the fence and walked across the field after the little horse. But every time the man got near him, the little horse kicked up his heels and tossed his head in the air and away he flew across the field. And his mother galloped with him and stayed by his side.

But the tall man with the mustache knew a lot about shy little horses because he loved them. He had watched horses for a long time. He knew that the shy little horse was also a curious little horse,

10

just chock-full of curiosity. So he just went and leaned against the fence and whistled ♪ away to himself and didn't look at the shy little horse any more.

Now that funny little horse saw the man do this and he heard him whistling; because the shy little horse had brand new eyes and brand new ears, and he heard and saw everything. He lowered his head and nibbled the green grass. But while he was nibbling he peeked at the man and pricked up his ears to hear the man whistling. The man didn't move and kept on whistling. ♪ The shy little horse kicked up his heels and ran farther down the field and nibbled some more grass and peeked at the man. The man didn't move and kept on whistling. Then the shy little horse nibbled some grass nearer to the man. The man didn't move and kept on whistling. ♪

What a funny man, thought the shy little horse. Why doesn't he chase me and try to get me in a cor-

ner and put a halter over my head? The man didn't move and kept on whistling. The little horse nibbled nearer and nearer. The man didn't move and kept on whistling. ♪

The shy little horse's mother put forward her ears and looked hard at the man. Then she snorted and whinnied, kicked up her heels, and galloped far away around the edge of the field. The shy little horse tossed his brand new head in the air and galloped with her. They circled around the field and stopped even nearer to the man than they had been before and nibbled the green grass. The man didn't move and kept on whistling. ♪

By this time the little horse was so curious he was nearly popping inside. He had never seen a man like this before. All the other men had chased him into a corner and caught him and put a halter over his head. The man didn't move and kept on whistling. ♪ The shy little horse stepped nearer and nearer. He was quite near to the man now, and he stood there ready to leap away and gallop to the far ends of the field. The man didn't move and kept on whistling. The shy little horse nosed nearer and nearer. The whistling tickled his ears in a way he liked. And he liked the man to stand so still he could get a good look at him. And he liked the quiet way of the man.

Then the man moved just a tiny little bit. If the shy little horse hadn't been looking so hard, he wouldn't have seen. The man uncurled his fingers, and on the palm of his hand were two white square lumps. The little horse stood there ready to jump away if the man moved any more. The man didn't

11

move and kept on whistling. ♪ The shy little horse's mother stretched her head forward to make sure that she saw what she saw. For on the man's hand were two white square lumps of sugar. Just what the old horse loved, and her mouth began to water as she thought of the sweet prickly taste of sugar. She had been out in the field nibbling green grass for so long with her little horse that no one had given her any sugar. She stepped nearer to the man, almost right up to him. The man didn't move and kept on whistling. This was a wonderful thing. The old mother horse stepped right up to the man and buried her nose in his hand and took one lump of sugar and stepped back and chewed it. Then she stepped up and took the other lump of sugar. The man didn't move and kept on whistling. ♪ Then after a while he walked out of the field the way he had come and went away.

The next day he came back, and he stood there whistling and he gave the mother horse another lump of sugar. The third day when he came, he walked right over to the mother horse and put a halter over her head and gave her a lump of sugar. Then he led her out of the field, and the shy little horse followed after, close to his mother's side. The man led the mother horse and the shy little horse through the barnyard among the chickens, past the old fat pig who was eating potato peelings, past the

old gray donkey who was eating thistle and staring with his big jackrabbit eyes. The man led the mother horse and the shy little horse right down the road where the little horse had never been before. His brand new hooves made a clicking noise on the road as he trotted along beside his mother. And the shy little horse was delighted.

Way down the road they went, until they came to a small dirt road. The man turned up the dirt road, and the shy little horse's brand new hooves didn't make clicking noises any more on the dirt, they just made soft little thuds. They went up the dirt road to a long white house with a big white stable with green doors and windows, all green and white. And there were buckets painted green and white, too, in stripes.

And out from the house came a shy little boy and he looked at the shy little horse.

For the tall man with the mustache was the father of the shy little boy, and he had bought the shy little horse for the little boy's very own. The little boy's mother came out of the house and said what a beautiful young horse it was. And the little boy said, "Some day I will ride him."

It wasn't so long before the shy little boy had taught the shy little horse to eat sugar out of his own hand. And the shy little horse and the shy little boy grew up together, and it wasn't long — maybe a year or two, for there was plenty of time — before the little boy had grown old enough to ride the shy little horse and the shy little horse had grown large enough to carry the little boy on his back.

They rode all over the country, they jumped fences and galloped across the green grassy fields, the boy and his horse. And after a while they weren't even shy any more.

The Seven Weathers

One day a little dog had to stay indoors
because he hated to get his feet wet and

That was the day it rained

and the fog rolled in

and it snowed

and the fog melted

and there was a tornado

and a hurricane

and a breeze

and a shower.

Then the sun came out
and all the flowers came up because it was
spring.

And then the sun shone and the little dog went
out to take a walk before nightfall on his
four soft feet, just in time —

Because then the stars came out and it was
night
And the wind began to blow all sorts of lovely
evening smells to his nose.

15

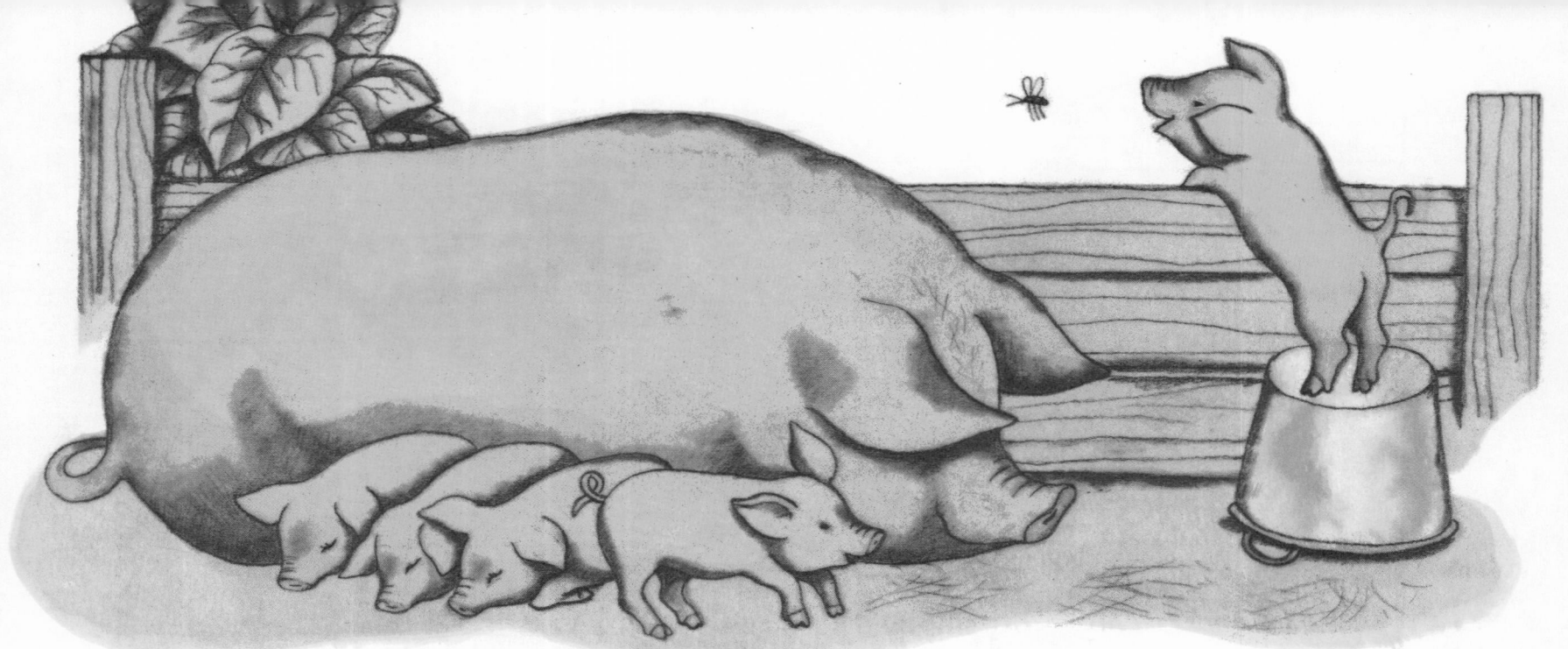

The Good Little Bad Little Pig

POOR LITTLE PIG. He lived in a muddy pigpen, in an old pigpen of garbage and mud, with four other little pigs and an old mother sow. He was a little white-pink pig, but the mud all over him made him look pink and black and gray-pink.

Then one day a little boy named Peter asked his mother if he could have a pig.

"What!" said Peter's mother. "You want a dirty little bad little pig?"

She was very surprised.

"No," said Peter. "I want a clean little pig. And I don't want a bad little pig. And I don't want a good little pig. I want a good little bad little pig."

"I never heard of a clean little pig," said Peter's mother. "But we can always try to find one."

So they sent the farmer who owned the pig a telegram:

TELEGRAM

FARMER, FARMER, I WANT A PIG,
NOT TOO LITTLE AND NOT TOO BIG,
NOT TOO GOOD AND NOT TOO BAD,
THE VERY BEST PIG
THAT THE MOTHER PIG HAD.

The farmer read the telegram, and then he went out to the pigpen and looked at the five little pigs. Three little pigs were fast asleep. "Those," said the farmer, "are good little pigs." And one little pig was jumping all around. "That," said the farmer, "is a bad little pig." And then he heard a little pig squeak, and then he heard a little pig squeal. But when he looked, there was just one little gray-pink pig standing on an old tin pan in the corner of the pen. "That," said the farmer, "is a good little bad little pig." And he reached in and grabbed the little pig by the hind legs and put him in a box and sent him by train to Peter.

When the express man brought the little pig to Peter's front door, his mother said, "What a dirty little pig!"

And the pig said, "Squeak squeeeeeeeeeeeeee ump ump ump."

And Peter said, "Wait till I give this little pig a bath."

But when they let the little pig out, he ran all over the room squealing like a squeaking pig.

"What a bad little pig!" said Peter's father, and he had to catch the little pig by the hind legs to make him hold still while Peter put the red leather dog harness around the little pig's stomach.

"Wait," said Peter, "until the little pig knows us. He is not a bad little pig." And he clipped a red leather leash on the little pig's harness.

The little pig stared at Peter out of his blue squint eyes, and then he shook himself and trotted after Peter on the leash.

"What a good little pig!" said Peter's mother, as she came into the room with a pan of bread and milk for the little pig to eat after his journey.

"Wait," said Peter. "Remember this is a good little bad little pig."

"Galump gump gump gump." The little pig was eating. He seemed to be snuffling and sneezing into his food as he ate.

"What a bad little pig!" said the cook, who had come in to see how the little pig was enjoying bread and milk. "What terrible eating manners he has!"

"But he does enjoy his food," said the little boy.

"Yes," said the cook, "he does enjoy his food." And she beamed with a smile all over, for the cook did dearly love for anyone to enjoy his food. "What a good little pig," she said. "He has eaten up everything in the pan."

"Come on, you good little bad little gray-pink pig," said Peter. "I will give you a bath so you will be a clean little white-pink pig."

So Peter and his mother and his father and the cook all went into the bathroom and put the little pig right into the bathtub and let the warm water run all over him.

Peter's mother held his front legs and his father held the little pig's hind legs, so that the little pig couldn't kick himself or the people who were bathing him. The little boy took a big cake of white soap and rubbed it all along the pig's back until he was all covered with pure white soapsuds. Then he took a scrubbing brush, and he scrubbed and he scrubbed right down through the bristles on the little pig's back to the little pig's skin. He scrubbed and he scrubbed until the pure white soapsuds were all black and gray. Then he poured warm water over the pig's back until there was no soap on it. Then he put some more soapsuds all over the little pig's back. And he scrubbed and he scrubbed and he scrubbed and he scrubbed and he scrubbed until the pure white soapsuds were all gray and black again. Then he rinsed off the pig's back with warm water and put more soapsuds on. But this time the soapsuds stayed almost pure white. So he left it on the little pig's back and washed his stomach and his feet until he was all clean and white and pink from the tip of his tail to the tip of his nose. Then they dried the little pig with a great big bath towel, and Peter took him for a walk in the sunshine.

"Look," said Peter as he showed his little pig to the policeman on the corner. "Did you ever see such a fine little clean little pig?"

"I never did," said the policeman, "see such a good little pig." And he blew his whistle and stopped all the automobiles so that Peter and the little pig could get across.

But the little pig did not want to get across, and he pulled back on the red leather leash and refused to budge. Peter pulled and he pulled, but the little pig would not go across the street.

"What a bad little pig!" said the people in the automobiles, and they began to honk their horns. And the little pig began to squeal and squeak. "Squeak squeeeeeeee ump ump ump." But the policeman held up his hand and wouldn't let the automobiles go. Then he came over to Peter and his pig.

"You pull him, Peter," he said, "and I'll get behind him and push." So they did. And when they got to the middle of the road, the little pig trotted on after Peter just as nicely as you please. "What a good little pig," said the people on the other side of the street.

And so it was that Peter got just what he wanted. A good little bad little pig. Sometimes the little pig was good and sometimes he was bad, but he was the best little pig that a boy ever had.

18

The Boy Who Belonged to the Cat

ONCE THERE WAS a little boy and he belonged to a cat and was brought up by a cat and he was the healthiest, happiest little boy in the world.

Instead of opening his eyes slowly in the morning, he blinked them open. So did the cat.

And then they began to stretch and to yawn. The cat shot out one leg and spread the pads of her paws. So the boy shot one foot forward in bed and spread his toes. He spread his fingers. Then he shot out his other leg and stretched his toes and at the same time on the end of his arms he stretched all his fingers out like cat paws or starfish and yawned again. Then he stretched his little stomach and back and got up.

And while the cat washed her eyes and her whiskers with her tongue and her paw, he washed with a big wet wash rag.

Then they went out and sat and dried themselves in the warm sun and blinked at the world and thought their first thoughts without a word. And the sun shone on them and warmed their bones.

Then they were hungry and they both drank their milk.

And some mornings they would have a little fish.

At noon the cat gave the little boy green vegetables because he didn't like green grass and the cat didn't really care whether she ate hot green string beans or cold green grass. But the boy did.

Then after lunch the cat found a quiet place, curled in a warm ball, purred a little and fell asleep. All day long whenever the cat or the boy had nothing to do or had done too much, they would take these little cat naps — why not?

All the better to grow on, purred the cat.

So the cat taught the little boy to watch quietly and to sit in the sun and to keep out of big smells and big noises and big crowds. And he was the happiest, healthiest little boy in the world and very glad that he belonged to a cat.

19

The Little Toy Train

In the little green house lived a cat and a mouse,
a girl and a boy,
and a man with a little toy train.
One day the train ran away.
It was a very little train.
So it ran down a mouse hole.
And the cat and the mouse
and the girl and the boy
and the man who were all very little too
ran down the mouse hole after the
little toy train.

And the little toy train ran far away following rabbit tracks in the
grass until it popped down a rabbit's hole and woke up seventeen
little rabbits who went right back to sleep again until
the cat
and the mouse
and the girl
and the boy
and the man
came running after the little toy train
and woke all seventeen little rabbits
up again.

20

And so the seventeen little rabbits ran after the little toy train
And the little toy train ran on and away
    through a field of clover and timothy hay,
        through a hole in a tree
            and a hole in a stone
                and a hole in the air
                    near a telephone pole,

        through a log
        and a pond
        near a big green frog,
            through a keyhole into a baker's shop
        and on
            and on
                without a stop
        followed by seventeen rabbits
            a man
                a boy
                    a girl
                        a cat
                            and a mouse
Until it came to the little green house and ran home again,
And that was the trip of the little toy train.

The Little Flat Flounder

ONCE THERE WAS a little flat flounder with two eyes on the top of his head who lived in the depths of the sea.

Around was green water and under the water was sand and seaweed and a few dark rocks where the blue-black lobsters crawled, slowly opening and shutting their claws.

From their long stalked beady black lobster eyes they looked into the cold blue eyes of the flounder and the dark gray-blue eyes of the cod and the pollack and haddock, and into the brown-black eyes of the skate, and into the yellow eyes of the mackerel.

Fish never close their eyes.

Even when they sleep, fish never close their eyes.

So there lay the little flounder flat in the sand with his gray-blue eyes wide open, night and day.

He watched for food and he watched for danger.

And the minute the flounder saw the lobster he hid in the soft wet sand.

For the lobster with the snapping claws was no friend of the flounder.

The little flounder had plenty of enemies waiting to dive through the sea to catch him.

It seemed as though everything in the air and in the water was after the little flounder.

They were all after the little flounder.

The fishhawk was after him ready to plunge down on him, down through the water like a rock from the sky.

If he didn't keep well hidden in the sand at low tide, the blue heron was there waiting to dip his long neck down for him.

Or the gulls were swooping to dive for him.

And always the fishhawk would hover, waiting, his two striped wings spread across the air, his sharp bird eye peering down through the waters for the moving shadow that would be a fish. And, of course, he hoped that the fish would be a flounder — a little flat flounder.

But it was not only the hawks and the herons and the gulls that were after the flounder.

The other fish were after him too — the codfish, the dogfish, the skate. The mackerel, the halibut, and the shad.

And the only way for him to escape from them was to hide.

The little flat flounder could make himself so much like the sand and seaweed he lay on that no one could see he was there. Wherever the little flounder humped himself, his coloring changed with the parts of the sea. He was a chameleon of the ocean deep.

The fishermen were after him, too, with hooks and lines and nets, but he stuck to the clams and sea fleas and sand worms on the bottom of the sea for his food.

He sniffed about in the sand until he knew his food was there. Then he flipped along till he caught it. He kept himself full of his own food. And he never went near a baited hook.

Inside the cage he flipped himself among the crabs and sea urchins and lobsters and took a big bite of the bait. And another bite of the bait. For that was what that good smell was — herring bait, in a lobster pot.

But when he tried to get out of the cage he was trapped. Trapped in a lobster pot!

Poor little flounder, what would he do now? He turned a funny green and black color so that you could scarcely see him there at the bottom of the lobster pot.

But still he couldn't get out.

Night came on and darkened the sea.

The seals were after him when they came diving down looking for food on the bottom of the ocean.

But if the flounder was in the sand he was sand color, if he was in the mud he was mud color, and a seal had a hard time telling where he was. And the little flat flounder always got away.

But one day when he was humping himself along in the sand he smelled something good. He flipped himself along toward that good ripe smell as fast as he could flip himself. For flounders do not swim. They hump and they flip. And so he flipped himself up through the water the way a flounder swims with a flip and a hump and never a wiggle.

He flipped himself closer and he peeked into the cage where the bag of something that smelled good was stuck on a spike. The flounder swam around the cage until he found a hole in it and then he swam in.

Day came on and brightened the sea and the warmth of the sun came down through the green water.

And then there was a pull, a jerk, and there was the lobster pot moving up through the waters toward the sun.

Right out into the air it came. And over the sides of the lobsterman's boat it was hauled. The lobsterman reached in and pulled out the lobsters, pegged their claws so they couldn't bite, and dropped them in the bottom of his boat.

He brushed out the green prickly sea urchins and dropped them back in the sea. He pulled out the sea cucumbers and the conch shells and the sea spiders, and threw them all back into the sea, where they sank slowly down till they disappeared in the water. Last of all, he pulled out the flounder.

"A fine fried flounder I'll make of you," said the fisherman. "Though you are too little for dinner and too small for lunch you will make a fine little tid-bit for breakfast."

And he laid the little flat flounder carefully on the gunnel of his boat.

A tid-bit for breakfast indeed! The fisherman did not realize what a smart, strong little flat flounder he had caught. For with a flip and a hump the little flounder flipped himself into the air and overboard into the sea — just like that.

And night came on again in the sea.

And the lobsterman went home with his lobsters.

And all the sea gulls went home to their sea gull nests.

And the fishhawks to their fishhawk nests.

And the blue heron to where the blue heron goes.

And the fish swam sleepily about in the darkened sea.

And the flounder dozed on the bottom, with his eyes still open.

Fifteen Bathtubs

ONCE THERE WAS a little boy who lived in a house with fifteen bathtubs in it. And you might think that this little boy was the cleanest little boy in the world with fifteen bathtubs in his house. But he wasn't. He was the dirtiest little boy in the world, because he hated to wash and he never used even one of the fifteen bathtubs. Of course, he did take a bath once a month. But when he did take a bath, what do you think he used? He used the garden hose. When the gardener was spraying the flowers in a soft fine spray of water, this dirty little boy would run right through the hose water; and that was the only way he took a bath except when he took a sun bath. He never got into one of the fifteen bathtubs.

Then one morning when all the creatures in the garden were hopping about, the rabbits and squirrels were hopping about and the chipmunks and woodchucks were hopping about and the bees and the butterflies were flying about, the little boy got up from the table where he had been eating toast on jam for breakfast instead of jam on toast; he smeared the jam across his face with the back of his hand and went out in the garden.

Now this was a very lively morning in the garden. It was a clean shining morning, the kind of a

26

morning that made the animals feel as though they owned the garden. They forgot all about the people that usually frightened them away.

When the little boy ran out into the warm buzzing sunlight, the animals didn't notice him. "Just that dirty little boy," they seemed to say and kept on hopping about; and the flowers kept on growing, and the sun kept on shining, clean and shining.

So the little boy lay down to take a sun bath. He turned his sticky jam-smeared face up to the sky and closed his eyes. The rabbits hopped about and the squirrels hopped about and the woodchucks hopped about all around him, and the butterflies flew and the bees went buzzing all around.

They buzzed in the flowers and they droned through the air, looking for something sweet to make into honey. And then they found the little boy's jam-sticky face. The little boy was asleep. Bzzzzzzzzzzzz. A sticky treasure! Then suddenly many bees were flying all around the little boy's face. Zoom Bzzzzzz Szz zzz zz z. They buzzed about his nose and lips, bzzzz. They buzzed nearer to his cheek. Szzzzz buzzz, after the sticky sweet jam on his face.

The little boy jumped awake, and he jumped to his feet and he jumped away. But the bees came buzzing all around right after him, trying to get the jam off his face. He ran and he ran, but still the bees came buzzing after him. He ran all around the

garden, and the bees buzzed all around the garden after him. He ran and he ran into a field, and the bees buzzed through the field after him. He ran and he ran and he ran through a wood, and the bees came buzzing after him. Then he ran and he ran and he ran. Then he ran and he ran down the black tar road, and the bees buzzed down the black tar road after him. He ran and he ran up the gray gravel driveway, and the bees came buzzing up the gray gravel driveway after him. When he got to his house, he ran in the open front door, and the bees came buzzing in the open front door after him; and he ran up the stairs, and the bees came swarming up the stairs after him.

When he got to the fifteen bathrooms, he jumped into one bathtub
   and into another bathtub
      and another bathtub
         and another bathtub
           and another bathtub
         and another bathtub
      and another bathtub
   and another bathtub
into all the fifteen bathtubs fifteen times.

Then the bees flew out the window and back to the flowers in the garden. And they never chased the little boy again, because from that day on he was just as clean as the sunlight.

The Steam Roller

ONCE THERE WAS a little girl and it was Christmas. It was Christmas and her mother and father didn't give her any oranges for Christmas. No dolls. No candy. No new clothes, no books. No sugar plums, no baby carriages. No. They gave her a Steam Roller. A big black steam roller with a silver bell and a brass chimney on it as shiny as gold. Smoke was coming out the chimney and a sort of hissing sound of steam — ssss-wssswwwwwsss.

The little girl climbed up the ladder to the driver's seat, pushed the sticks that made the steam roller go, blew the whistle and rang the bell, and away she went down the road. The big steam roller wheel

crushed all the little pebbles on the road and squashed all the big rocks.

Crunch, Crunch, Crunch, it went rolling down the road making everything flat before it. And sssssss a fine smoke of steam came out of its brass chimney and made the steam roller go. There were a lot of buttons to push and sticks to pull. But the little girl didn't know which to push or which to pull to make the steam roller stop.

An old pig went wallowing across the road and blinked its small red eyes.

"Look out, Old Pig," said the little girl. "Get out of my way because I can't stop. Get out of my way or I'll squash you flat."

But the old pig didn't get out of the way. And the steam roller ran right over it and squashed the old pig flat on the road. Flat as a pig's shadow in the middle of the road.

A chicken came fluttering over the road. The little girl blew the whistle. Screeeeeeeeee — up.

"Get out of my way or I'll squash you flat."

And the old yellow chicken got so excited squawking "Which Way Which Way Which Way," and flapping its yellow feather wings, it flew right under the steam roller. Looking back, the little girl saw it flat on the road like the shadow of a chicken or a feather fan.

But the little girl couldn't stop the big steam roller and it went rolling down the road squashing everything in its way. It squashed her mean old aunt flat on the road. It squashed three people she didn't know, it squashed two automobiles and a garbage truck and a trolley car.

Then a policeman stood up in the middle of the road. He held up his hand and blew his whistle. " S T O P ," he said.

"I can't stop," said the little girl, for by this time she had forgotten how to stop. And she ran right over the policeman and squashed him flat on the road with his hand in the air saying S T O P.

The road ahead was clear. No one was on it. No pigs, No chickens, No aunts, No people she didn't know, No automobiles, No garbage trucks, No trolley cars and No policemen. Then along came the little girl's teacher.

"Merry Christmas, little girl," said the little girl's teacher. And the little girl said, "Merry Christmas," but the steam roller wouldn't stop and it squashed the teacher and the book she was carrying, right flat in the middle of the road with a Merry Christmas smile on her face. Flat in the road like a shadow.

Then the little girl saw her friends coming down the road, the children that were just her age, and they called to her saying, "Give us a ride on your steam roller."

"Look out," called the little girl. "I can't stop this steam roller and it will squash you flat."

But the children didn't seem to hear her and they came running toward her. The little girl blew the whistle and rang the silver bell, but still the children came running toward her. "STOP," she called to them. But still they came running.

This will never do, thought the little girl. I can't squash all the children my own age. So she headed the steam roller across a field and then she jumped out of it while it was still going. The steam roller went rolling over the field squashing the fences until they looked like the shadows of fences around the fields. And away it went. The little girl hurt herself on the road when she jumped from the steam roller going full speed. But when she got up she found that she hadn't hurt herself very much. It didn't hurt for long.

The children her own age came running up to her.

"We wanted a ride," they said.

"Oh, no, you didn't," said the little girl. "That steam roller doesn't stop, whatever stick you pull or button you push. And it squashes everything it comes to flat. It would have squashed you all flat if I hadn't headed it off the road into the fields and jumped out."

"Oh," said the children her own age. And they looked after the steam roller. It was just rolling over the last hill. It rolled over the hill and then disappeared into the distance. It rolled off into the ocean and squashed a few fish flat and stopped.

"Well," said the little girl. "I am glad that old steam roller is gone." And she wished the children a Merry Christmas and ran home to her mother and father.

"Well," said her father, "and how did you like your steam roller ride? And what did you do with your steam roller?"

"At first, it was fun," said the little girl, "but I couldn't stop the steam roller and it squashed everything and everybody it came to flat on the road. And then all the children my own age came running up the road and I couldn't squash them, so I headed the steam roller across the fields and jumped out of it while it was going. And the steam roller rolled away across the fields squashing the fences and went away out of sight. It just rolled away.

"And a pig and a chicken and a policeman all got squashed flat on the road like shadows and I don't know how to get them up again."

"Well," said the little girl's mother and father. "We have another Christmas present for you."

And there at the front door all wrapped up in tissue paper with red ribbons on it was a Giant Steam Shovel.

"Get in that," said her father and mother, "and go scoop up the squashed flat pigs and people and automobiles. Scoop them up out of their shadows and they won't be squashed flat any more."

So the little girl tore off the tissue paper from the steam shovel and away she went down the road. Off she went on the caterpillar steam shovel wheels, and when she came to the pig shadow, *Scoop* she scooped and the old pig grunted up out of its shadow and trotted away. And when she came to the chicken, *Scoop* she scooped it right up out of its shadow; and still cackling Which Way Which Way Which Way, the chicken fluttered away.

And when she came to her mean old aunt and the three people she didn't know, *Scoop* she scooped them right out of their shadows and they all went hurrying down the road.

She scooped up the two automobiles and the garbage truck and the trolley car and off they rattled. And *Scoop* she scooped up the policeman, who raised his other hand and blew his whistle and said Go. Then *Scoop* she scooped up her teacher out of the shadow on the road and her teacher said "Merry Christmas" again and the little girl said "Merry Christmas."

And then all the children her own age came running and she gave them each a ride up in the air in the scoop for the fun of it. *Scoop* she scooped them way high up in the air and then let them down again.

Then she turned the steam shovel around, and off she went home to Christmas dinner.

Count to Ten

(WITH THANKS TO LEAR)

There was once a little owl
  Count to one

1   1
      one little owl

There were two little trolls
  roly
    poly
      Count to two

2   1, 2
         two little trolls

There were three little pigs
  piggy
    wiggy
      dance a jiggy
        Count to three

3   1, 2, 3
           three little pigs

There were four little foxes
  clocksy
    foxy
      doxy
        hide behind the rocks
         Count to four

4   1, 2, 3, 4
           four little foxes

There were five little fish
  fishy
    squishy
      slishy
        in a dishy
         very swishy

5   Count to five
           1, 2, 3, 4, 5
            five little fish

Six little pickles
  tickle
    trickle
      sickle
        fickle
         stickle

6   crickle
           Count to six
           1, 2, 3, 4, 5, 6
            six little pickles

Seven little drums
  bum little
    rum little
      tum little
        dum little
         hum little
          strum little
         some little

7   Count to seven
           1, 2, 3, 4, 5, 6, 7
            seven little drums

Eight little eyes
  blinky
    thinky
      dinky
        linky
         trinky
         minky
        squinky
        winky

8   Count to eight
           1, 2, 3, 4, 5, 6, 7, 8
            eight little eyes

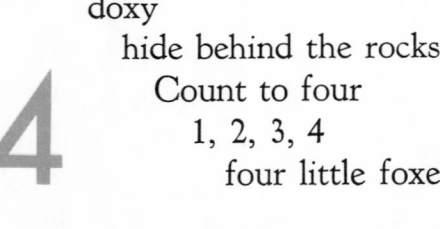

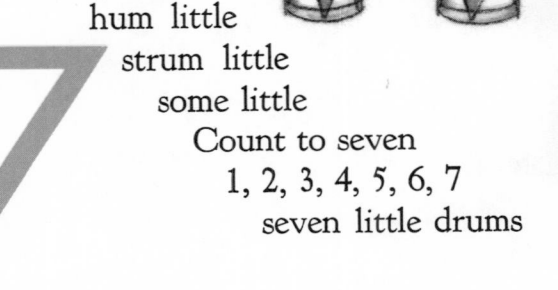
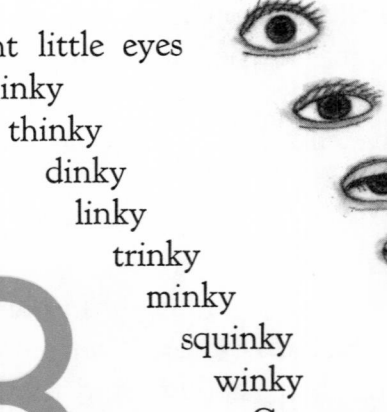

Nine little noses
  sniffy
    whiffy
      miffy
        piffy
          squiffy
           tiffy
            liffy
              diffy
              biffy

9
                Count to nine
                  1, 2, 3, 4, 5, 6, 7, 8, 9
                  nine little noses

Ten little kings
  kingy
    ringy
      singy
        dingy
          lingy
            stringy
              flingy
                mingy
                  quick and clingy
                  wild and wingy
                  Count to ten

10
                  1, 2, 3, 4, 5, 6, 7, 8, 9, 10
                  ten little kings

They Could All Smell It

BUT WHAT WAS IT?

They could all smell it
And it wasn't a lemon.

A daisy? No!

Who remembers the ticklish sneezing smell of
a hot daisy?

It wasn't a cloud
Because a cloud has no smell.

They could all smell it
And it was not a dog's
foot.

Or a kitten's warm paw.

It wasn't a baked potato

Or an onion

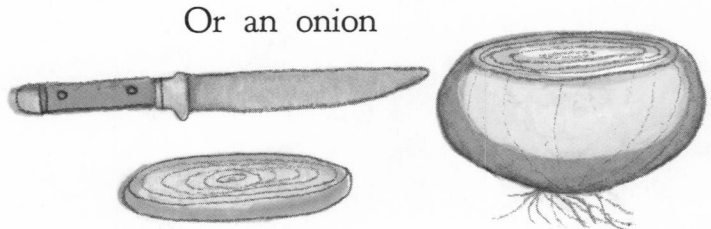

Or a green lima bean

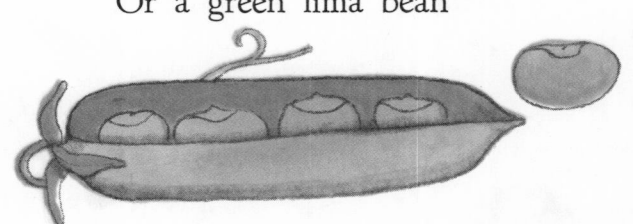

Or a bunch of celery,

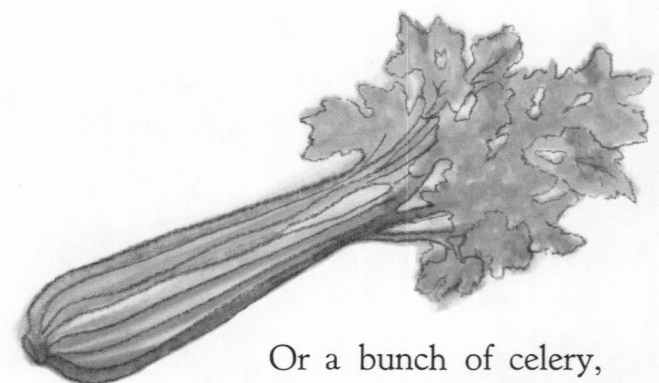

34

A bunny has a smell,

It was not the moon
Because the moon has no smell,
 Nor the moon nor the sun nor the stars.
But they could all smell it coming closer and
 closer and closer.

And a mirror has not.

What could it be?

What do you think it was?

It was a little old skunk far away

Was it a——? No.   Was it a——? No.
Was it a——? No.   Was it a——? No.
Was it a——? No.

 coming closer

 and closer

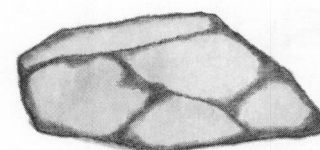

 and closer

 and closer,

And it wasn't a stone.

Carrying a rose in its mouth.

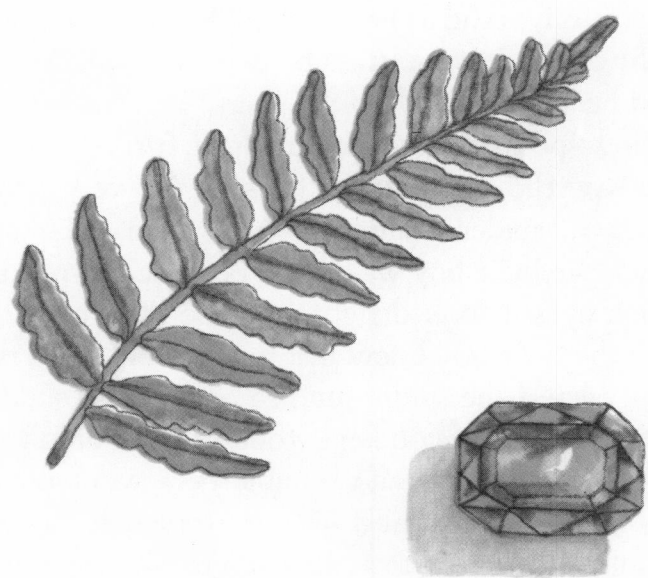

A fern has a smell,

And a ruby has not.

35

The Tickly Spider

A LITTLE BLACK SPIDER lived deep in the grass about three buttercups away from the edge of the field where the little boy was lying on his stomach. The little boy was lying on his stomach peering in between the long green grass blades. Deep, deep, deep in the grassy wilds the little boy looked, even beyond the bright yellow green that the sunlight made as it twisted down the grass blades. And as he watched, he saw strange things. He saw a red lady beetle climbing down a stalk of grass, very slowly taking her time. He saw the shadow of a butterfly that was flying above the grasses. And then——

In and out and around the grass stems came the little black spider. Lugging and puffing along it came, as black spiders come. Over one yellow root and around a brown twig, under a root and over a stick. The black spider was coming straight towards the little boy's nose, just as straight as a spider can come winding in and out of the grasses.

The little boy saw it coming and wondered if it would come all the way up to his nose. He watched it carefully. And as he watched, he saw a ray of sunlight climb down a grass blade to the spider's back, and he got a better look at the spider. He saw little yellow spots on the spider and little furry fuzzy hairs. He saw the dusty gray claws of the very scritchy scrawny spider coming towards him. The spider knew the little boy was there, because with its shiny black eyes it looked right into the little boy's eyes. But the little boy knew just what to do, so he didn't care even if the spider jumped on his nose suddenly. He knew that if you kept stone still and didn't move at all, even wasps and bumble bees wouldn't sting you. So he knew that if he kept stone still, the spider would only walk across him to get into the grass on the other side. He was just like a big hill to the spider, a big hill that had to be climbed to get across to the other side. But if he jumped or jerked, the spider

Then he didn't feel any more tickling. Maybe the spider was gone and was in the grass on the other side of him. But still the little boy lay quiet as a stone. He wasn't sure. After a while, he hadn't felt any more tickly feelings for a long time now, he began to be sure the spider was gone. So he lifted his head slowly; then he turned his head very slowly. It is all right if you move slowly a little bit, just as long as you don't jerk or move suddenly. When he had his head all turned around and his body turned around, too, he looked into the grass on the other side.

At first he didn't see anything. Then he saw the black furry spider crawling off through the grasses, over a root and around a twig, under a root and over a stick, in and out and around the grass stems.

And that was how the little boy stayed very still and saw a lot of things happening around the roots of the grasses.

might think he was not friendly and bite him or sting him. But the little boy knew how to be quiet, so he just lay there stone still. He wasn't brave, he just knew that the spider wouldn't hurt him if he didn't move. He was just a little boy who knew how to be very still.

When the spider came up to the last blade of grass in front of his nose, it stopped. And it looked at the little boy's face. Then it began to climb the tall blade of grass next to the little boy's face. It climbed and it climbed halfway up the grass blade, and it climbed and it climbed until the grass bent towards the little boy's cheek. But it didn't quite touch him. So the spider climbed some more; and when the grass was bending almost all the way over the little boy's cheek, it jumped. Plunk, right in the hollow under the little boy's eye. But the little boy lay very still. Then tickle tickle, the fuzzy old spider started down the little boy's cheek. Tickle tickle, past his nose. Tickle tickle, by the corners of his mouth. The little boy nearly laughed. But he knew better. He lay very still. He didn't even smile. Even though it tickled like a million feathers when the spider climbed down his chin to his neck and went across his neck towards the back, he didn't move or smile.

The Wonderful Kitten

ONCE THERE WERE four little kittens, a little fur pile of kittens. And they lived on a white woolly sheep-skin rug in the clothes closet. When they were born they were as small as a mouse and as big as a bird. They all had whiskers and claws and noses and tiny round ears and tails. But their little eyes were closed tight shut for two weeks, and they didn't have any teeth. All they could do was to kick with their paws in the air and squeak.

Their mother thought they were so beautiful that the minute they were born she began to purr. She purred them a cat song as they lay by her side.

Purr Purr
My cat-whiskered mouse
Come close to my sides
For this is your house
Purr Purr
My Kitten—O—Mouse

Fluff Ball Angel, for that was the mother cat's name, loved her little fur kittens. She loved them with a long steady purr.

Purrrrrrrrrrrrrr. Purrrrrrrrrrrrrr. And her eyes were bright and shiny when she purred. She curled up her paws and waved them in the air.

For they were four beautiful round fur kittens in a little fur pile. One little kitten was white as the snow, and one little kitten was gray. One little kitten was black as the night, and one little kitten had yellow stripes on his black coat like a wee baby tiger. They crawled around their mother's sides for a long time.

Then one day after they had had their little blue eyes open for a week, one of them said, "Peep, Peep. What a wonderful kitten I am. I can Yawn some, Sneeze some, Purr some, Lick some, See some, Hear some, and CRAWL. I'm a wonderful cat the size of a rat and I think I will crawl away."

So he waved his four black paws in the air, because he was on his back when he thought about crawling away. But he didn't get anywhere by

waving his paws in the air. He stayed in the very same spot. So then he wiggled and wriggled until he got his fat little stomach to roll over. And there he was on his legs.

Hic-up.

"Fiddlesticks," said the little kitten. "What's that?" And he opened his eyes, his brand new eyes, and looked all about.

Hic

There it went again!

Hic

He looked at all his brothers and sisters. But there they were all curled up sound asleep. What could that noise be?

39

Hic-up. This time the little kitten shook all over when the noise came.

"Why, it's me!" said the little kitten. "What a wonderful kitten I am. I can Yawn some, Sneeze some, Purr some, Lick some, See some, Hear some, Crawl some, and HIC-UP. I'm a wonderful cat the size of a rat and I think I will crawl away."

And this time, on his little shaky legs with his tail in the air, he began to crawl away. He spread his toes and put one foot forward. Then he spread his toes and put the other foot forward. He crawled all the way across the sheepskin and out the closet door.

There was the little black kitten in the hall. What a big vast hall it was. The little kitten stretched his neck in every direction and looked way out across the gray hall rug.

Zzzzzzz. Bzzzzzz. Along came flying a big black fly.

"Whufff," said the little black kitten. "What is that?" And the hair stood up on him till he looked like an angry black pom-pom.

Zzzzzzz. Zzzzzzzzzzzzzzzzzzzz.

The little kitten arched his back and stood on his shaky legs and——spt! He spit.

Then the zzzzzz got softer, and when the little

kitten finally saw what it was with his brand new eyes, the fly was just a little buzzing spot up in the air.

"Fiddlesticks," said the kitten. "What a tiny little black spot. And what a wonderful kitten I am. I can Yawn some, Sneeze some, Purr some, Lick some, See some, Hear some, Crawl some, Hic-up, and SPIT. And I'm not afraid of anything."

And on he crawled into the middle of the hall, way out across the rug all by himself.

But before long, because he was a very little kitten, and he couldn't think much better than he could crawl, the little black kitten couldn't remember where he was or where he was going. And his brand new eyes couldn't see away back across the rug as far as the white sheepskin inside the closet door.

The little kitten was lost in the middle of the great gray rug. Lost in the great vast hall. And the little kitten was cold without the other warm fur kittens. And his shaky little legs, his brand new legs, shook so that he fell over on his side.

"Oh dear," cried the little kitten. "I don't like this." And then he remembered that he could squeal and squeak.

"Squeeeeeeeeeee squeeeeeeeeeee squeeeee," went the brand new kitten. "I'm a little lost kitten and I don't know where I am."

He squealed so loud that his mother, Fluff Ball Angel, heard him. Fluff Ball Angel had been away all this time on Fluff Ball Angel business in the next room. But when she heard her little black kitten squealing, she came running.

Brrrrrip, mmmrrrrrrrrip.

She ran right up to her little black kitten on the gray rug in the great vast hall and she licked his face with her tongue and she rolled him over with her paw and licked him until he was all warm and happy and didn't squeal any more. He was with his mother, so he didn't care where he was. He just lay on his back and purred to himself. And he thought to himself, What a Wonderful kitten I am. I can Yawn some, Sneeze some, Purr some, Lick some, See some, Hear some, Crawl some, Hic-up, Spit, and SQUEAL. What a WONDERFUL kitten I am. And now I want to go home.

So Fluff Ball Angel picked him up by the loose skin on the back of his neck. She picked him up very gently, and holding her head very high so she wouldn't drag him on the floor, she carried him across the wide gray rug, right through the closet door, and she dropped him right in the pile with his brothers and sisters. Then she lay down by her little pile of kittens. And she scrubbed the little kitten that was white as snow, and she scrubbed her kitten that was gray, and she scrubbed her kitten that looked

like a tiger, and she scrubbed the little black kitten who had run away. She scrubbed them with her tongue. Then she curled herself around them and purred them a cat song.

> *Purr Purr*
> *My cat-whiskered mouse*
> *Come close to my sides*
> *For this is your house*
> *Purr Purr*
> *My kitten—O—mouse*

Then all the little kittens felt full and sleepy. And the little white kitten purred himself to sleep and the little gray kitten purred himself to sleep, and the little tiger cat purred himself to sleep. And the little black kitten just curled in a little black ball and rolled right off to sleep before he could remember all the wonderful things he could do. He just went to sleep.

The Wild Garden

In this garden you will find
No columbine or eglantine.
All the flowers that you will find
Were found by me and they are mine.
You will find white violets from the woods
That I dug up in the Spring,
And you will find daisies from the fields,
And one buttercup.
I planted lots of buttercup seeds
But only one came up.

Here come the bees with a great big buzz,
Yellow and black and covered with fuzz.
Here come the flies in a big black swarm
To tickle a fat little girl's bare arm.

Here come the beetles all shiny and green
And red and yellow and shiny and mean to
    leaves and roots.

Here comes a butterfly first to fly
Out of June and into July.

And here sleeps a sleepy young firefly.

42

Every spring they come again,
Yellow flowers in the rain,
And lambs are born
And birds are born
In the soft green early morn.

The red bird whistles in the tree.
Spring is here
Endlessly
Ceaselessly
For an instant
In a tree.

Wild strawberries
Are wild,
They taste wild
And their roots run wild
Through the ground,
And they hide like
Little wild red strawberries
In the grass,
And you see them
Suddenly
Small and red
And wild.

43

The Wonderful Room

THERE WAS a Wonderful Room with a big red bed
and a vast gray rug with soldiers and drums on it and
a glass fireplace and a big fur chair and a little table
with a cake on it and a book with nothing in it and
a book with something in it and an ocean outside
the window with seven boats sailing on it and a
picture on the wall of a horse with a hat on his
head and a picture of a panda kicking a panda

and a wooden soldier

and a golden hunting horn
and by the fire in the glass fireplace sat a cat and a
dog and a sleepy old owl who lived in this wonderful
room and a hole in the wall where a little mouse lived
and nobody knew he was there.

And there was a great big hump under the
white sheets on the big red bed and who could that
be hiding there with one bare foot wiggling its toes
sticking out from under the covers?

And over the bed was a light like a star and on
the ceiling were starfish that stayed there because
they were glued onto the ceiling. And in four little
vases were wild flowers and wild flowers and wild
flowers, and wild flowers picked from a field at noon
— buttercups, clovers, and daisies. And there were
wild violets from the woods

and wild geraniums

and wild strawberry blossoms

and yellow star flowers

and blue star flowers

and dogtooth violets

and white violets

and one jack-in-the-pulpit

and another buttercup

and three kinds of clover.

44

How the Animals Took a Bath

ONCE THERE WAS a little boy who didn't know how to take a bath. He had never had a bath because his big fat mother was so busy all day scrubbing white clothes that at night she forgot to scrub her little boy.

One day the little boy came to his big fat mother, and he said:

"Mother, I am one dirty little boy. I have jam on my face and chocolate on my knee. I have mud all over my feet and between my toes from walking down the wet, muddy road. I have dust in my hair from the wind blowing yesterday, and I have dirt all over me from all the time. I think I want to take a bath."

"Well, little boy," said his mother, "run along and take a bath."

"Only I never had a bath," said the little boy. "I don't know how to take one."

But his big fat mother was so busy scrubbing the white clothes that she didn't have time to stop and show her little boy how to take a bath. She just said:

"Run along, little boy, and see how the animals take their bath, because I have to wash all these white clothes before dark."

So the little boy started down the road to see how the animals took a bath. He walked on down the road till he saw a little bird sitting on the branch of a tree.

"Little bird," said the boy, "I am dirty and I want to take a bath, only I don't know how. So I have come to watch you take a bath."

The little bird just sat on the branch of the tree. The little boy sat under the tree and waited to see if the bird would take a bath. The sun was shining warm on the road and on the mud puddles in the road from the last night's rain. The road was warm in the sunshine.

"Just the day for a little bird to take a bath," thought the little boy as he waited and waited.

Pretty soon the little bird began to flutter, and began to ruffle up and flutter its feathers. Then it

46

flew off the tree and dove straight into the mud puddle. It hopped around the edge a little, and then, stomach first, it sank into the shallow part of the puddle and shook itself so that the water splashed all over its back and wings. It made a quick winged fluttering noise. *Flitter, flutter, flitter, flutter.* Then it hopped out on the sand in the road and shook some more. *Whirrrrr.* It flapped its wings.

The little boy watched carefully. But the bird wasn't finished. After shaking and shaking and smoothing its feathers down with its bill, it suddenly began to flutter about in the sand with its wings

stretched out, until it was all sandy. Then it shook and shook itself again and bent its head and smoothed out its feathers and flew away.

"So!" thought the little boy, "this is the way to take a bath." And he ran out in the middle of the road and lay on his stomach in the mud puddle. He wiggled about and he splashed the muddy water all over him. Then he got out and rolled in the sand by the side of the road. But when he got up, he was dirtier than ever.

"Oh, dear," said the little boy, "I'm dirtier than I was before. I guess little boys just don't take baths like birds. I guess I had better go on down the road and find some other animals."

So he walked on down the road until he came to a farm. There by the side of the road was the farmer's pigpen with six dirty little pink pigs in it and a big black pool of water.

"Shoo, little pigs, take a bath," shouted the boy. "Shoo, little pigs, take a bath so that this dirty boy can learn how to get clean."

Just then, two of the little pigs went into the pool and wallowed about in its black edges. The boy didn't even wait for them to come out. He jumped the fence and got right into the black, muddy pool

and rolled around squealing with the little pigs. The water and mud was so cool and soft, the boy was sure that this was the only way to take a bath. But when he got out, he was dirtier than ever. He was all covered and black and sticky with mud.

"Oh, dear," said the little boy, "this must not be the way for a boy to take a bath," and he walked on down the road, dirtier than he had ever been before.

After a while he came to a small white house. There on the front porch sat an old yellow cat, licking her paw and then brushing her wet paw against the side of her face. The boy stood and watched her. Then he licked his hand and rubbed it on his cheek, and he licked his other hand and rubbed it on his ear and he licked his hand and rubbed it on his neck. His hands were getting cleaner.

"Oh, shucks," said the boy, "this sure is the way to take a bath."

Through the doorway of the small white house he saw a mirror hanging in the hall and he ran in to look at his face. But in the mirror his face was dirtier than it had ever been before. He had rubbed all the dirt off his hands on to his face.

"Oh, dear," said the boy as he ran down the road, "I washed like a bird and I washed like a pig, and I washed like a cat, and I'm dirtier than I ever was before. How will I ever learn to take a bath and be a clean little boy?"

48

Way down the road he came to a green field with brown shining horses galloping and frisking about in the sunshine. He had never seen such smooth shining horses. And beyond the green field with the shining horses was a big white barn.

"Those horses are certainly clean," thought the little boy, "clean and shining." And he wished that he was clean and shining and running about the green fields.

He went up to the big white barn and stood in the doorway. Two men with brushes were cleaning a horse. The brushes made *scrape, shsh, scrape, shsh* noises in the horse's coat. The little boy watched them. He watched them stir the brush around the horse's coat until all the gray dirt and dust came out of the hair. He watched them take a yellow shiny bristle brush and brush away all the dirt that the iron brush had stirred up. There the horse stood, smooth and brown and clean. And he watched them lead the horse out of the stable and into the field. As soon as they were out of sight, he grabbed the sharp iron brush and he rubbed it over one of his muddy legs. Ouch! The sharp ends scratched him and just made white lines in the dirt on his leg.

"Oh, dear, what shall I do?" said the little boy. "I've washed like a bird, and I've washed like a pig, I've washed like a cat, and I've washed like a horse, and still the dirt will not come off."

So the little boy went back to his big fat mother, dirtier than he had ever been before. His mother was just pulling the last piece of white clothes out of the big soapy tub of water when her little boy came home.

"I declare, little boy," she said, "you are dirtier than I have ever seen you before. Didn't you learn from the animals how to get clean?"

"I washed like a bird, and I washed like a pig, I washed like a cat, and I washed like a horse, Mother, and each time I just got a little bit dirtier than I had ever been before."

"You are no bird, little boy, you are no pig, you ain't no cat, and you ain't no horse. How is it that you didn't figure that out? How is it that you didn't find out how little boys take a bath?"

And she turned on the water and grabbed him by the back of his neck and put him right into a big soapy tub.

"I guess it's your mother who will have to show you how to get clean." And so she scrubbed him and scrubbed him all slippery with soap. And when he came out of his mother's tub, he was cleaner than a bird, he was cleaner than a pig, he was cleaner than a cat. And he was cleaner than a horse. He was clean and shining, like a clean little boy.

Said a Fish to a Fish

Said a fish to a fish,
"Look away up high.
What is that shadow
Under the sky?
That great dark shadow sailing by,
Over the ocean,
Under the sky."

Said a fish to a fish,
"That shadow up high
Is a yellow fish
With a yellow eye
That goes swimming along
Under the sky."

"Come, little fish,
Let us fly fly fly
Up where that shadow is
Under the sky."

So up they swam,
Away up high

Where they could see
With their fishes' eye
The great dark shadow sailing by,
The slow dark shadow
Under the sky.

But the slow dark shadow
Under the sky
Was no yellow fish
With a yellow eye.

For they could see
With their fishes' eye
That the slow dark shadow
Under the sky
Was a fishing boat
Sailing,
Sailing by
Over the ocean
Under the sky.

"Come, little fishes,
Fly fly fly,
Fly from that fishing boat
Under the sky,
Come down in the ocean
Away from the sky."

And the fishing boat sailed by.

50

The Bluefish

A fisherman went fishing,
Went fishing for a fish.
He went fishing for the bluefish
That swims down below.
Jerk Jerk Jerk
Went the fisherman's line,
Pull Pull Pull
His line pulled so.

Ho Ho Ho
Can this be the bluefish
That swims down below?

He pulled in the fish
On the hook on his line
And he looked at the fish
On the hook on his line,
Then the fisherman said,

No
This is just a yellow fish
And I will let him go
So
He let the fish go.

Then the fisherman went fishing
For another fish,
He went fishing for the bluefish
That swims down below.

He caught red fish and green fish
And fast fish and slow,
But he never caught the bluefish
That swims down below.

The Polite Little Polar Bear

ONCE THERE WAS a polite little polar bear who lived in the wide icy regions of the Far North. He lived in a cave of ice in the frozen snow.

This little polar bear had a delightful manner with the fish he ate. "Please, little fish, may I eat you up?" he would say; and the fish were so surprised at such lovely manners in a polar bear that they were frozen with surprise, and the polite little polar bear fished them out from under the ice with one curving swoop of his long fur paw. Then off he loped across the ice on his fur-soled feet, growling happily to himself with his belly full of fish.

One day when the little polar bear was out walking, he met an angry old seal and three rude walruses.

"Bah to you!" said the angry old seal.

"Boo to you!" said the three rude walruses.

But the little polar bear, the polite little polar bear, said, "How very delighted I am to meet three rude walruses and an angry old seal all in one morning!"

"Bah to you, Little Polar Bear," said the angry old seal.

"Boo to you," said the three rude walruses, and shuffled along on their way across the frozen ice to look for fish.

"Unfortunately, Old Seal," growled the polite little polar bear, "I am not very hungry today, so I regret that I will not be able to eat you this minute. But I look forward very much to meeting you again." And the little polar bear went on his way across the frozen ice.

It was such a beautiful icy summer's day, the polite little polar bear felt like a good fifty-mile swim before supper. So when he got to the edge of the great green waters, he slipped from the ice in a walloping dive and swam away off across the icy sea. He rolled through a floating sea meadow of tiny green water plants that grow in the arctic sea, and he batted at some sea butterflies, the kind whales eat. Then he swam on and on through the summer sea, past big ice cakes all white and shining.

He swam and he swam the way polar bears swim miles and miles in the arctic sea for the fun of it; and then suddenly the little polar bear was sleepy, and he wanted to take a nap. But where to

I'm terribly sleepy, and I'm not very old. And as I said before, this water is cold."

But the little baby iceberg just floated right where it was in the arctic sea. Right in the polite little polar bear's way.

"Little Baby Iceberg, if I weren't so polite, I would duck you in the water where you couldn't see the light. Oh, bother!" said the polar bear. "Why am I so polite?"

But the little iceberg stayed right where it was in the arctic sea. All this time the little polar bear was getting sleepier and sleepier, so he just swam right around that silly little iceberg; and there on the other side was the flat ice and a nice fat seal sound asleep for his supper. So he politely ate the seal without even waking it up. He ate sixty pounds of seal before he was full, and then he fell asleep under the midnight sun, looking like a fat little lump of white snow on the rest of the white snow, in the wide icy regions of the Far North.

take it? He was many miles from the flat fields of ice he had come from, and he was in a hurry to get to sleep. So he turned around; and with just his nose out of the water, he swam as fast as he could back towards the ice flats and his own icy cave. He was in an awful hurry to get to sleep.

All of a sudden there was a very small iceberg, just a baby iceberg, in front of him. It was too steep to climb up on; and the little polar bear was in such a hurry, he didn't feel like swimming around it. The little iceberg went too deep down into the water for him to swim under it. So he said:

"Please, Little Iceberg, get out of my way. If you will be so kind, get out of my way."

But the little baby iceberg just floated right where it was in the arctic sea. Right in the polite little polar bear's way.

"Please, Little Iceberg, the water is cold, and

53

Wind in the Corn

I heard the wind in the corn one day,
I knew that it came from far away,
And it rustled the trembling corn to say
That it was going far away
And could not stay,
Could never stay.

The Green Wind

The green wind blew
And the rains came down
And the animals scurried
Right out of the town.

The green wind blew
And it blew the bees
Right out of their flowers
And up in the trees
And it blew the leaves
Right up in the sky
To turn and whirl
And fall and die.

And the bees in the branches
Began to cry
Oh buzzzz for the flowers
And buzzzz for the earth.
And the wind blew backwards in its mirth
And the wind blew backwards
Through the trees
Till the rain came down
And pushed down the bees,
The swift green wind
With rain.

So Many Nights

So many nights,
Blue nights,
Brown nights,
And the sudden lights
In deep black nights
Of stars
And cars
And airplanes
And soft gray nights when it rains
And blue nights with a foggy moon
Smoking in the trees

And pink and red nights
Above great cities
And silver nights all filled with stars
And misty nights when a white mist
Drifts
And lifts over the white-topped fields
And purple nights beyond the lights
Of your own room
And blue snowy nights
And night that is just
Dark bright night.

Eyes in the Night

The moon came up on a summer's night
That was soft and dark except for the light
Of fireflies and the gleaming eyes
Of green-eyed cats and the amber eyes of
    wandering dogs
And the red eyes of frogs and the eye of the
    toad
Picked up by lights coming down the road,
Little emerald eyes, the gleaming eyes,
And the eyes of rabbits and foxes
And more and more flickering fireflies.
Then the car went by, and soon
There was only the light
Of a big warm moon
All over the summer night.

The Little Girl's Medicine

ONCE UPON A TIME there was a little girl who lived way out in the country on a big tobacco farm. She had no brothers and sisters. She had no one to play with, this poor little girl, and she had to play all by herself. She played all by herself year after year and talked to her parents when they ate their meals.

One day, after a while, the little girl became sick. No one knew what was the matter. It was in late August, when they hitched up the two black mules and hauled the tobacco plants away to the big drying barns. But the little girl didn't want to go with them and drive the two big mules. She just sat.

That night there was peach ice cream for dinner, but the little girl didn't want any. She just sat. When it was time to go to bed, she didn't even care.

"Oh dear," said her mother to her father, "our little girl is sick. She loves to drive the two black mules when they haul the tobacco from the fields to the barns, and yet she wouldn't go. She just sat. And she loves peach ice cream, but she wouldn't eat it tonight. She just sat. And she didn't even want to stay up and play when it was time to go to bed. Our little girl must be sick."

So they took the little girl to the doctor in the big city.

The little girl's mother said to the doctor in the big city, "Doctor, my little girl is very sick."

"What is the matter with your little girl?" asked the doctor. "Has she a sore throat?"

And the little girl's mother said, "Doctor, my

little girl doesn't want to drive the mules with the tobacco loads any more, and she doesn't like peach ice cream any more, and she doesn't care whether it is bedtime or whether it isn't bedtime. So I fear that she must be a very sick little girl."

"What!" said the doctor. "Doesn't like peach ice cream! This is serious! Little girl, stick out your tongue."

So the little girl stuck out her tongue, and the doctor looked at it very carefully. "It is a perfectly good tongue that you have in your head, little girl," said the doctor. "Let me see your throat, little girl. Say Ahhhhhhh."

So the little girl leaned her head way back, way back, and opened her mouth so wide that she looked like a baby robin asking for food. The doctor took a flashlight and peered down into the little girl's throat. "Say Ahhhhhhh," he said.

Then he said, "Little girl, let me feel your pulse," So he held the little girl's wrist in his hand, and with his fingers he listened very carefully, count-

ing the beats of the little girl's heart that he could feel in the veins of her wrist. It's a perfectly good heart that you have in you, little girl; but if you don't like to play any more and don't like peach ice cream any more, you are very sick. It would be a pity if your brothers and sisters caught what is wrong with you."

"But I have no brothers and sisters," said the little girl.

"But your cousins and friends might catch it," said the doctor.

"Only I haven't any cousins, and I haven't any friends," said the little girl.

"Then the small animals on the place might catch it," said the doctor.

"There aren't even any small animals," said the little girl, "Not even a little pig. Just two big old black mules that kick every time any one goes near them."

"Well," said the doctor, "this is serious. I will have to prescribe something to make you well."

The doctor sat there for a long time nodding his head, while the little girl and her mother waited. Then he got up and walked around the room three times. Then he opened a book and read three pages of it. Then he coughed three times and he said:

"Little girl, I have just the thing that will make you well." So he picked up a pen and wrote something down on a piece of paper, folded the paper, and handed it to the little girl's mother.

"If you will have this prescription filled," he said to the little girl's mother, "and give it to your little girl right away, I am sure that she will get well."

So the little girl's mother folded up the prescription in her purse without looking at it, thanked the doctor, and went out of his office with the little girl.

"We will go right to the drugstore first thing," said the little girl's mother, "and have this prescription filled before lunch." So they went into the drugstore next door, and the little girl's mother handed the prescription to the druggist, still folded up as the doctor had given it to her.

The druggist was an old man, and he unfolded it slowly.

"Hmmmmp," he said. Then he said it again. "Hmmmmp! Do you expect me to fill this prescription?"

"Why, of course," said the little girl's mother. "Haven't you got that kind of medicine?"

Then the druggist, old as he was, just threw back his head and hollered with laughter. "Do you know what this prescription says?" he asked.

The little girl's mother took the prescription and read it. And this is what the prescription said—

"I have filled prescriptions for thirty years," laughed the druggist, "but never a prescription for a puppy dog. But wait!" he said. "Wait a minute. I think that I can fill this prescription after all. Right across the street. Will you come with me?"

So the little girl and her mother followed the druggist, still chortling and laughing to himself, out the door and across the street to a house that had a back yard.

The little girl and her mother followed him right into the back yard, and there in a box was one furry little puppy dog all by himself.

"This is the last one left," said the druggist. "They belong to my sister, and she is giving them away. So if the doctor says the little girl needs a puppy, this is how we can fill the prescription for her."

"My puppy?" asked the little girl. "All mine?"

"Yes" said the druggist. "That is your puppy if you want him, and you can take him right along home with you this minute."

The little puppy wiggled and jumped around the little girl as if he was just as glad as she was that they would have each other to play with. He had been sitting all by himself in that empty box, just one little puppy all by himself for two long days. He hadn't even eaten the milk that was still in his saucer.

So the little girl took her puppy right home with her. They got back just as the big wagon with the two black mules was going out of the gate. The little girl's father was driving.

"Hey, little girl," he called, "do you want to go out after the last load with me?"

"Indeed I do!" said the little girl. "And look what I am going to bring with me!"

She jumped out of the car with the puppy under her arm and climbed up on the wagon beside her father.

"For goodness sakes!" he said. "What in the world have you got there?"

"This," said the little girl, "is my medicine, and I feel much better already."

"Well, come here, Medicine," said the little girl's father. "Are you going to learn to be a good tobacco farmer like me and the little girl?"

Little fat Medicine (for that was the puppy's name from then on) wiggled right up in her father's arms and licked him on the nose, and they all drove along on the wagon together behind the two black mules. They hauled the last load of tobacco on to the wagon and hauled it away to the big tobacco drying barns and hung it up on the racks to dry. It was

hard hot work. And then they went home to supper.

And what do you think they had for dessert, and the fat little puppy had a spoonful of it, too? Peach ice cream.

And that night when it was time to go to bed, up the steps scampered the little girl, and up the steps scampered the little puppy. And they both scampered right into the little girl's room. And together that night the little girl and her Medicine went right off to sleep, all curled up in their warm little beds in the same room.